D1069018

The Insect Societies

FRONTISPIECE: A worker of *Sphecomyrma freyi,* the most primitive known ant species. Two specimens, including this one, were found in amber in a New Jersey deposit of mid-Cretaceous age. They provide a link between the tiphiid wasps and the most primitive known living ants (photograph courtesy of Frank M. Carpenter).

The Insect Societies

Edward O. Wilson

The Belknap Press of Harvard University Press Cambridge, Massachusetts 1971

Distributed in Great Britain by Oxford University Press, London
Library of Congress Catalog Card Number 74-148941
SBN 674-45490-1
Printed in the United States of America
Book design by David Ford

For my wife Irene, who understands

Contents

Acknowledgments

Anyone who attempts to synthesize insect sociology must draw ideas from many of the most widely separated branches of biology. In undertaking such a task I have been favored by association over many years with friends who conveyed to me a deep respect and some feeling for the kinds of biology in which they have been outstandingly successful: William L. Brown and Frank M. Carpenter in taxonomy and insect evolution; Howard E. Evans, Andrew J. Meyerriecks, and Charles D. Michener in comparative ethology; Thomas Eisner, Martin Lindauer, and Carroll M. Williams in physiology and experimental biology; John H. Law, Barry P. Moore, and Fred Regnier in biochemistry; and Stuart Altmann, William H. Bossert, Robert H. MacArthur, and Lawrence B. Slobodkin in the mathematical theory of population biology. Each will recognize the imprint of his teaching somewhere in these pages, and I am happy to take this particular opportunity to acknowledge my debt.

I am also grateful to colleagues who read parts of the manuscript. They have saved me, and the reader, from more insidious errors than I would like to confess. Their names, with the chapters which each reviewed all or in part, are as follows: R. D. Alexander (7), M. V. Brian (8, 21), W. L. Brown (4, 8, 17, 19), R. H. Crozier (17), H. V. Daly (17), Mary Jane West Eberhard (2, 3), A. E. Emerson (6), H. E. Evans (3, 5, 12, 14, 15), T. H. Goldsmith (11), W. D. Hamilton (3, 17), W. Hangartner (5, 6, 12, 13, 14, 17), B. Hölldobler (7, 19, 20), R. L. Jeanne (3, 21), W. E. Kerr (9, 17), M. Lindauer (5, 11, 13), C. D. Michener (2, 5, 19), C. W. Rettenmeyer (4, 20), O. W. Richards (3, 21), J. W. Truman (8), N. Weaver (9), N. A. Weber (4), P. H. Wells (13), A. M. Wenner (13), R. Zucchi (9, 17). I have also received the loan of specimens, the use of unpublished manuscripts, and other forms of aid from R. D. Alexander, R. H. Arnett, W. L. Brown, J. H. Calaby, F. M. Carpenter, K. A. Christiansen, Mary Jane West Eberhard, A. E. Emerson, T. L. Erwin, W. Faber, R. C. Froeschner, F. J. Gay, B. Hocking, B. Hölldobler, G. A. Horridge, H. F. Howden, D. H. Janzen, B. Krafft, H. Kutter, J. F. Lawrence, H. W. Levi, E. G. MacLeod, M. M. Martin, U. Maschwitz, C. D. Michener, E. L. Mockford, S. A. Peck, C. W. Rettenmeyer, L. M. Roth, S. F. Sakagami, R. Stumper, J. van der Vecht, and R. E. Woodruff. Miss Adaline Wheeler provided several of the original drawings of ants made by her father, W. M. Wheeler, in order to obtain reproduction of comparable quality to that in his book of 1910.

Special credit must be given to Miss Kathleen M. Horton, who typed the manuscript through three difficult drafts while assisting in every phase of the bibliographic research. The illustrations are primarily the work of Mrs. Turid Hölldobler and Mrs. Sarah Landry, who prepared the original drawings, and of Mr. William G. Minty, who prepared the graphs and redrew many of the previously published figures.

Permission to reproduce certain of the illustrations was granted by the following persons: R. D. Alexander, H. -J. Autrum, Suzanne W. T. Batra, M. V. Brian, D. J. Burdick, J. Calaby, F. M. Carpenter, R. Chauvin, W. S. Creighton, J. Delachambre, V. G. Dethier, Mary Jane W. Eberhard, I. Eibl-Eibesfeldt, T. Eisner, H. E. Evans, W. Faber, J. B. Free, K. von Frisch, F. J. Gay, P. -P. Grassé, R. E. Gregg, Dorothy Hodges, B. Hölldobler, J. Ishay, R. Jeanne, W. E. Kerr, T. W. Kirkpatrick, D. H. Kistner, G. Knerer, B. Krafft, H. Kutter, G. Le Masne, M. Lindauer, H. Martin, U. Maschwitz, R. W. Matthews, C. D. Michener, E. M. Miller, H. Oldroyd, A. Raignier, C. W. Rettenmeyer, K. D. Roeder, S. F. Sakagami, S. H. Skaife, J. P. Spradbery, A. M. Stuart, R. Stumper, J. H. Sudd, J. van der Vecht, N. Weber, Adaline Wheeler. The following agencies gave permission to reproduce illustrations for which they hold the copyright: Academic Press, Inc.; Académie des Sciences, Institut de France; American Association for the Advancement of Science (for *Science*); *American Midland Naturalist; Australian Journal of Zoology;* Bee Research Association; British Museum (Natural History) (including *Natural History Magazine*); Cambridge Entomological Club (for *Psyche*); W. Collins Sons and Co., Ltd., London; Columbia University Press; *Ecological Monographs*; Entomological Society of Southern Africa; Field Museum of Natural History; W. H. Freeman Co. (for *Scientific American*); Hawaiian Sugar Planters' Association; International Union for the Study of Social Insects; *Journal of Insect Physiology*; Librairie Plon, Paris; Longmans, Green & Co. Ltd., London; Museum of Zoology, University of Michigan; Naturforschenden Gesellschaft in Zürich; Paul Parey, Berlin and Hamburg; *Quarterly Review of Biology*; Rijksmuseum van Natuurlijke Historie, Leiden; Royal Entomological Society of London; *Science Journal*, London (incorporating *Discovery*); Smithsonian Institution Press; Springer Verlag (for *Zeitschrift für vergleichende Physiologie*); United Nations Educational, Scientific and Cultural Organization (UNESCO); the University of Chicago Press (for *The American Naturalist*); *Wasmann Journal of Biology;* John Wiley and Sons, Inc.; Worth Publishers, Inc.; *Zeitschrift für Tierpsychologie.*

Portions of Chapters 8, 14, and 16 appeared earlier in my articles "The Origin and Evolution of Polymorphism in Ants," in the *Quarterly Review of Biology* (1953), and "The Superorganism Concept and Beyond," in J. Alibert *et al., L'effet de groupe chez les animaux* (1968). They are reproduced here with the permission of the editors of the *Quarterly Review of Biology* and the Centre National de la Recherche Scientifique, Paris.

Most of the personal research reported in this book has been generously supported by an unbroken series of grants from the National Science Foundation that began in 1958. These funds in particular made it possible to undertake a whole new investigation of the role of pheromones in the organization of insect societies.

The Insect Societies

1 Introduction: The Importance of Social Insects

Why do we study these insects? Because, together with man, hummingbirds, and the bristlecone pine, they are among the great achievements of organic evolution. Their social organization—far less than man's because of the feeble intellect and absence of culture, of course, but far greater in respect to cohesion, caste specialization, and individual altruism—is nonpareil. The biologist is invited to consider insect societies because they best exemplify the full sweep of ascending levels of organization, from molecule to society. Among the tens of thousands of species of wasps, ants, bees, and termites, we witness the employment of social design to solve ecological problems ordinarily dealt with by single organisms. The insect colony is often called a superorganism because it displays so many social phenomena that are analogous to the physiological properties of organs and tissues. Yet the holistic properties of the superorganism stem in a straightforward behavioral way from the relatively crude repertories of individual colony members, and they can be dissected and understood much more easily than the molecular basis of physiology.

A second reason for singling out social insects is their ecological dominance on the land. In most parts of the earth ants in particular are among the principal predators of other invertebrates. Their colonies, rooted and perennial like woody plants, send out foragers which comb the terrain day and night. Their biomass and energy consumption exceed those of vertebrates in most terrestrial habitats. Social insects are especially prominent in the tropics. In the seventeenth century Portuguese settlers called ants the "king of Brazil," and later travelers referred to them with such phrases as "the actual owners of the Amazon Valley" and "the real conquerors of Brazil." Brazil, it was claimed, is "one great ants' nest." Similar impressions are invariably gained from other tropical countries. Ants in fact are so abundant that they replace earthworms as the chief earth movers in the tropics (Branner, 1910). Recent research has shown they are nearly as important as earthworms in cold temperate forests as well; in one locality in Massachusetts they bring 50 grams of soil to the surface per square yard each year and add one inch to the topsoil every 250 years (Lyford, 1963). Termites are among the chief decomposers of dead wood and such cellulose detritus as leaf litter and humus in the tropics, and they, too, contribute significantly to the turning of the soil.

When considering ecology, it is useful to think of an insect colony as a diffuse organism, weighing anywhere from less than a gram to as much as a kilogram and possessing from about a hundred to a million or more tiny mouths. It is an animal that forages ameba-like over fixed territories a few square meters in extent. A colony of the common European ant *Tetramorium caespitum,* for example, contains an average of about 10,000 workers who weigh 6.5 g in the aggregate and control 40 m^2 of ground (Brian *et al.,* 1967). The average colony of the American harvester ant *Pogonomyrmex badius* contains 5,000 workers who together weigh 40 g and patrol tens of square meters. The giant of all such "superorganisms" is a colony of the African driver ant *Anomma wilverthi,* which may contain as many as 22 million workers weighing a total of over 20 kg. During the statary phase of their cycle, columns of this species regularly patrol an area of between 40,000 and 50,000 m^2 in extent (Raignier and van Boven, 1955). When all of the resident ant populations are counted, the statistics are even more impressive. In Maryland, a single population of the mound-building ant *Formica exsectoides* comprised 73 nests covering an area of 10 acres and containing approximately 12 million workers (Cory and Haviland, 1938). Since individual workers of this rather large species weigh 11.6 ± 0.13 mg (Dreyer, 1942), the total population weighed about 100 kg, and this was only one of many ant species inhabiting the same area, albeit the most abundant one. Termites have colonies of similar magnitudes, and in tropical habitats their populations approach densities comparable to those of ants. The savannas of Africa are dotted with great mounds of the fungus-growing macrotermitines, some 5 to 20 feet or more in height and containing 2 million workers. The mother of each colony is a grotesquely fattened queen weighing in excess of 10 g.

These superlatives can be made because of an adaptive radiation that took place for the most part between 50 and 100 million years ago in each of the major groups of social insects. In the social wasps, the ants, the social bees, and the termites, evolutionary convergence has resulted in the repeated appearance of the same basic design features: the systems of castes and labor roles changing according to age; the elaborate systems of chemical communication that typically include signals for alarm, recruitment, and recognition; the elaboration of nest structure to enhance temperature and humidity control; and others. One criterion of adaptive radiation that I use half seriously when thinking about evolution is: a group of species sharing common descent can be said to have truly radiated if one or more species is a specialized predator on the others. Ants have achieved this level with some distinction. Many of the army ants (subfamily Dorylinae) feed primarily on ants and other social insects, while all of the Cerapachyinae so far investigated feed entirely on ants. Among the bees, the meliponine *Lestrimelitta limao* specializes in robbing other meliponine species, while the large wasp *Vespa deusta* preys largely on colonies of the wasp genera *Ropalidia* and *Stenogaster*. Social parasitism, in which one species lives inside the nests of another and in some cases receives food and care unilaterally, occurs in all four of the major groups of social insects. It is tempting to speculate (and perhaps impossible to prove) that the social insects as a whole have employed all, or nearly all, of the social strategies permissible within the limits imposed by the arthropod brain and the peculiarities of their colonial system.

This book is an attempt to provide a modern synthesis of insect sociology. The last attempts at a comprehensive treatment, not counting popular reviews, were W. M. Wheeler's classic *The Social Insects* (1928) and Franz Maidl's excellent but nearly forgotten *Die Lebensgewohnheiten und Instinkte der staatenbildenden Insekten* (1934). Not only has the literature on the subject increased enormously (to about 7,000 articles on termites alone, for example, and 12,000 on ants), but a whole new way of studying insect societies has also been created. I refer to the experimental and statistical analysis of insect colonies as populations. We are at last beginning to understand the physiological basis of caste determination, the nature of communication among workers, the principle of a queen's control over the rest of the colony, the means by which social parasites penetrate the heart of the colony communication system, the factors that limit colony growth, and the stimuli that pace the life cycle. Furthermore, we are molding the rudiments of true evolutionary theories of sociality through which explanations can be supplied for the presence of the underlying physiological phenomena in terms of maximal efficiency and fitness at the colony level. The present status of insect sociology can be made clearer by recognizing three stages in its historical development.

The natural history phase. The discovery and description of the social insects and the cataloging and evolutionary interpretation of their behavior and ecology were the unavoidable first steps. This phase, a necessary precursor to succeeding developments, is far from completed. The natural history of the Halictidae, the group containing most of the species of social bees, to take one of many examples, is still in an early period of exploration.

The physiological phase. The experimental analysis of the social systems and their physiological bases constituted the logical next step. This approach to the subject gained its first solid impetus with the work of von Frisch on honeybees in the 1920's. It is currently being applied vigorously to such diverse topics as caste determination, food exchange, communication, circadian activity, nest micrometeorology, and many others.

The population-biology phase. With the first two approaches now yielding much solid information, it has become possible to commence the construction of a new and more rigorous theory of social evolution. This theory does not consist solely of phylogenetic reconstructions in the nineteenth-century manner. It also attempts to account for social phenomena in terms of the first principles of population genetics and population ecology. Relying on the use of mathematical models in the fashion of the hypothetico-deductive method of any mature science, it predicts the existence of still undiscovered phenomena and relations among phenomena, it establishes quantitative laws to describe the underlying processes, and it suggests the best ways to measure and describe complex systems. This approach has only begun to be applied to a few subjects such as the origin of sociality and the determination of caste ratios, and I have tried to exemplify it in Chapters 17, 18, and 21.

In this book an attempt is made not only to review, in an objective and straightforward manner, the substance of insect sociology but also to create a theme in which the relationships of the three historical phases are made clear. Exploration along all three approaches will continue indefinitely to yield exciting discoveries. In time all of this information will be assembled within the framework of population biology and form an important branch of that larger science. A principal theme of this book is, therefore, the expression of insect sociology as population biology.

Now what delight can greater be
Then secrets for to knowe,
Of Sacred Bees, the Muses Birds,
All of which this booke doth shew.
And if commodity thou crave,
Learne here no little gaine
Of their most sweet and sov'raigne fruits,
With no great cost or paine.
If pleasure then, or profit may
To read induce thy minde;
In this smale treatise choice of both,
Good Reader, thou shalt finde.

Charles Butler,
The Feminine Monarchie (1609)

2 The Degrees of Social Behavior

The "truly" social insects, or eusocial insects as they are sometimes more technically labeled, include ants, all termites, and the more highly organized bees and wasps. These insects can be distinguished as a group by their common possession of three traits: individuals of the same species cooperate in caring for the young; there is a reproductive division of labor, with more or less sterile individuals working on behalf of fecund individuals; and there is an overlap of at least two generations in life stages capable of contributing to colony labor, so that offspring assist parents during some period of their life. These are the three qualities by which the majority of entomologists intuitively define eusociality. If we bear in mind that it is possible for the traits to occur independently of one another, we can proceed with a minimum of ambiguity to define *presocial* levels on the basis of combinations of two or less of the three traits. Presocial refers to the expression of any degree of social behavior beyond sexual behavior yet short of eusociality.* Within this broad category there can be recognized a series of lower social stages, for which Michener (1969a) has provided the most recent and sound classification:

Solitary—showing none of the three traits listed immediately above;
Subsocial—the adults care for their own nymphs or larvae for some period of time;
Communal—members of the same generation use the same composite nest without cooperating in brood care;
Quasisocial—members of the same generation use the same composite nest and also cooperate in brood care;
Semisocial—as in quasisocial, but there is also reproductive division of labor, that is, a worker caste cares for the young of the reproductive caste;
Eusocial—as in semisocial, but there is also an overlap in generations so that offspring assist parents.

In this arrangement, presocial applies to all the intermediate stages between solitary and eusocial. Michener (1969a) has introduced yet another term, *parasocial,* to embrace those presocial states in which members of the same generation interact—namely, communal, quasisocial, and semisocial. An agreeable feature of this classification is its explicit recognition that the subsocial and communal states, while undoubtedly serving as early evolutionary steps, contain none of the three intuitive criteria of eusociality. The same can be said of aggregations, including even those of aphids, lepidopteran larvae, and locusts, in which relatively elaborate physiological group effects and coordinated swarming occur (Chauvin and Noirot, 1968). The system also provides a means of making a graphic contrast between the two alternate routes to eusociality, the parasocial and the subsocial, which most students of the subject believe to have been followed in evolution. The logical possibilities framed by the terminology are presented in Tables 2-1 and 2-2. Later in this book virtually all of these states are demonstrated in various groups of insects. In fact, wasps alone exhibit most conceivable degrees of both subsocial and parasocial behavior, as well as both the extreme solitary and eusocial

*Bequaert (1935) invented the term "presocial" for the narrow category of behavior in which one or both of the parents merely guard the offspring temporarily, without providing it with food. This usage has not caught on, and Wilson (1966) suggested that the word, which has the advantage of being mnemonically superior, is best used in the broad sense employed in this book.

TABLE 2-1. Degrees of sociality, showing intermediate parasocial states.

Degrees of sociality	Qualities of sociality		
	Cooperative brood care	Reproductive castes	Overlap between generations
Solitary, subsocial, and communal	–	–	–
Quasisocial	+	–	–
Semisocial	+	+	–
Eusocial	+	+	+

states. In later chapters on the social wasps and presocial insects (Chapters 3 and 7), the successive stages of parental care defined in Table 2-2 will be exemplified in detail.

Before going on to the documentation of social evolution, however, let us quickly review the rather confused history of definitions of the expression "social insect." Most authors have used the phrase interchangeably with "true social insect," or eusocial insect as it has just been defined. There have been semantic problems, stemming almost exclusively from differences in opinion concerning the multiple qualities that go into the definition. This basic aspect of the problem largely escaped the disputants themselves. In essence, one group of writers, starting with W. M. Wheeler, incorporated all three of the traits I have listed; another, smaller group incorporated only two of the traits, omitting the criterion of overlap of generations.

In his last attempt at a major synthesis, Wheeler (1928) defined social insects *sensu stricto* as those in which "progeny are not only protected and fed by the mother, but eventually cooperate with her in rearing additional broods of young, so that parent and offspring live together in an annual or perennial society." Among the many recent authors retaining this usage is Michener (1953b, 1969a), who uses the definition as the terminological cornerstone in his productive writings on the evolution of social bees.

O. W. and Maud J. Richards (1951) evidently tried to apply a more flexible definition to the social wasps when they stated that "Real social life appears when the mother and her offspring (or, less probably, a number of sisters) co-operate in making a nest and feeding their young. The best criterion is that a wasp should feed a larva laid by another individual or help construct its cell." Later O. W. Richards (1965) eliminated the criterion of overlap of generations altogether and made the criterion of reproductive castes optional: "The term 'social' is reserved for communities in which there is more than one female, one or more usually being sterile, unfertilized and nursing the young derived from one or a few of them. The mother that actually lays the eggs usually does no other work in the mature colony." Emerson (1959) also omitted the criterion of overlap of generations: "True social insects are those that live in populations exhibiting division of labor for various functions among mature individuals separated into reproductive and sterile castes." In my opinion nothing is to be gained by broadening the category of true social insects. The stricter Wheeler-Michener usage has the great advantage of permitting the more finely structured classification of presocial states needed to cope with the immense amount of new information on social evolution now finding its way into the literature.

Finally, to complete this circle of nomenclature, what is a "society"? It is a group of individuals that belong to the same species and are organized in a cooperative manner. I believe that the terms society and social must be defined quite broadly in order to prevent the arbitrary exclusion of many interesting phenomena. Such exclusion would cause confusion in all comparative discussions of sociobiology. Not only eusocial insect colonies but also most parasocial and subsocial groups should be designated as societies and their members as social in the most general sense. The same is true of aggregations of locusts and other insects in which organization transcends mere re-

TABLE 2-2. Degrees of sociality, showing intermediate subsocial states.

Degrees of sociality	Qualities of sociality		
	Cooperative brood care	Reproductive castes	Overlap between generations
Primitively subsocial	–	–	–
Intermediate subsocial I	–	–	+
Intermediate subsocial II	+	–	+
Eusocial	+	+	+

productive activity. Reciprocal communication of a cooperative nature is the essential intuitive criterion of a society. Thus it is difficult to think of an egg, a pupa, or even a bee larva sealed into a brood cell as a member of the society that produced it, even though it functions as a true member at other stages of its development. It is also not very satisfying to view the simplest aggregations of organisms, such as swarms of courting males, as true societies. They may be held together initially by mutually attractive stimuli, but if they interact in no other way it seems excessive to refer to them by a term stronger than aggregation. By the same token a pair of animals in courtship can be called a society in the broadest sense but only at the price of diluting the expression to the point of uselessness. Bird flocks, wolf packs, locust swarms, and groups of communally nesting bees are good examples of elementary societies. So are parents and young if they communicate reciprocally. This last, extreme example will seem to approach the trivial and may be questioned by other students of the subject. In my view, however, it is appropriately included and is of special interest because it calls attention to the important topic of the relation of kinship to the evolution of true, organized societies, a relation that can be shown to be of the greatest importance to the study of social insects and will be examined in some depth in later chapters of this book.

3 The Social Wasps

In spite of the relatively small number of wasp species that are truly social, the study of their behavior has repeatedly yielded results of major interest. Four of the basic discoveries of insect sociology—nutritional control of caste (Marchal), the use of behavioral characters in studies of taxonomy and phylogeny (Ducke), trophallaxis (Roubaud), and dominance behavior (Heldmann, Pardi)—either originated in wasp studies or were based primarily on them. Even more importantly, the living species of wasps exhibit in clearest detail the finely divided steps that lead from solitary life to advanced eusocial states.

Eusocial behavior in wasps is limited almost entirely to the family Vespidae. The only known exception is a primitive eusocial organization recently discovered in the sphecid *Microstigmus comes* (Matthews, 1968). In order to put these and other social hymenopterans in perspective, consider the phylogenetic arrangement given in Figure 3-1 of the seven superfamilies of the aculeate Hymenoptera. The aculeates, as they are familiarly labeled by entomologists, include the insects referred to as "wasps" in the strict sense. Also included in this phylogenetic category are ants (Formicoidea), which are considered to have been derived from the scolioid wasp family Tiphiidae, and bees (Apoidea), which are considered to have originated from the wasp superfamily Sphecoidea. The Vespoidea is comprised of three families, the Masaridae, Eumenidae, and Vespidae. These wasps are often called the Diploptera because of the extraordinary ability of the adults to fold their fore wings longitudinally. The trait does not occur in the stenogastrine vespids or in the great majority of Masaridae, but its absence there may be a derived rather than a primitive characteristic. Vespoids are further distinguished from other wasps by the manner in which the combined median vein and radial sector slant obliquely upward and outward from the basal portion of the fore wing. Most can also be recognized by the emarginate (notched) condition of the inner margin of each compound eye.

Most references to the biology of the Vespidae employ the classification of Bequaert (1918, 1928), who recognized 11 subfamilies: Euparagiinae, Gayellinae, Masarinae, Raphiglossinae, Zethinae, Eumeninae, Stenogastrinae, Polybiinae, Ropalidiinae, Polistinae, and Vespinae. Recently Richards (1962), on the basis of new morphological evidence, reduced the number of subfamilies to 9 and split them into 3 families. His arrangement follows the phylogeny shown in Figure 3-2 and the checklist of Table 3-1.

Of approximately 15,000 living species of aculeate wasps believed to occur in the entire world (Hurd, 1955), probably less than 1,000 belong to the Vespidae. The

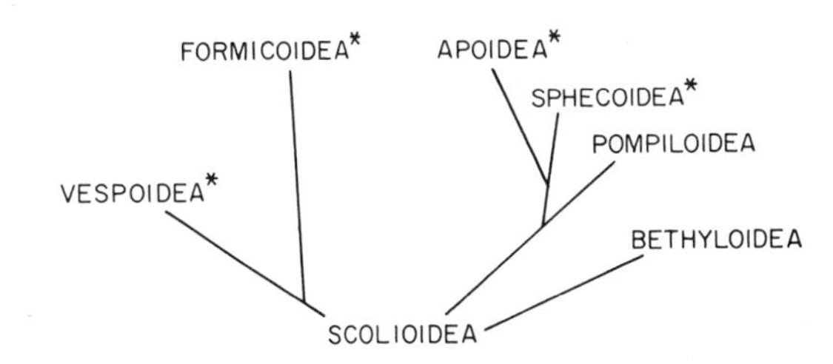

FIGURE 3-1. A simple branching diagram of phylogeny in the aculeate Hymenoptera, or "wasps" in the strict sense. An asterisk indicates the superfamilies in which eusocial behavior has been evolved, probably as at least one independent event in each case (modified from Evans, 1958).

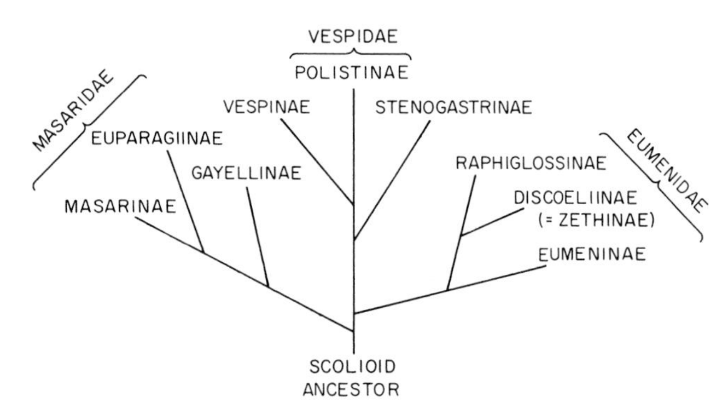

FIGURE 3-2. Phylogeny of the families and subfamilies of the "Diploptera," or wasp superfamily Vespoidea (modified from Richards, 1962).

species of this family are distinguished by the possession of short mandibles that do not cross in the resting position. The ranges of most of the principal tribes and genera are shown in Figure 3-3. The Stenogastrinae, containing about 50 species, range from India to New Guinea. The Vespinae also are centered in southeastern Asia, where about 50 species are known; 12 species also occur in Europe and 16 in North America, including our familiar hornets and yellow jackets. The Polistinae is the most diversified of the subfamilies. The paper wasp genus itself, *Polistes,* occurs over most of the world and, in 1962, was known to contain 189 species and 97 subspecies. Among the social insects it is rivaled in the extent of this range and in the degree of speciation only by several of the largest ant genera such as *Camponotus* and *Pheidole.* The rest of the Polistinae belong to the tribe Polybiini, which is centered in South America and contains no less than 300 species classified in about 20 genera. Approximately 200 additional polybiine species in 4 or 5 genera occur in Africa and tropical Asia.

The species-level taxonomy of the social wasps is only moderately well worked out, and some large and inconvenient gaps remain. Few comprehensive studies of modern vintage exist to which the beginning taxonomist can turn: Bequaert (1918) on the Vespidae of central Africa; Blüthgen (1961) and Kemper and Döhring (1967) on the Vespidae of Europe; van der Vecht (1957) on the Vespinae of southeastern Asia and New Guinea; Ducke (1910) on the South American Polybiini; Willinck (1952, 1953) on the Argentine Polistinae; and Bequaert on the Polybiini of North America and the West Indies (1933) and on the Vespidae of northern South America (1944a). The Poly-

TABLE 3-1. Families, subfamilies, and genera of vespoid wasps, with an indication of the degree of sociality and the distribution of the subfamilies.

Division	Degrees of sociality	Distribution
Family Masaridae		
Subfamily Euparagiinae (*Euparagia*)	Solitary	North America
Subfamily Gayellinae (*Gayella, Paramasaris*)	Solitary	Australia, Mexico to South America
Subfamily Masarinae (*Celonites, Ceramiopsis, Ceramius, Jugurtia, Masaris, Metaparagia, Microtrimeria, Paragia, Pseudomasaris, Quartinia, Quartiniella, Quartinioides, Riekia, Rolandia, Trimeria*)	Solitary	World-wide
Family Eumenidae		
Subfamily Raphiglossinae (*Psiloglossa, Raphiglossa*)	Solitary	Africa
Subfamily Discoeliinae (*Ctenochilus, Discoelius, Elimus, Labus, Zethus*)	Solitary	New World tropics, Africa, tropical Asia
Subfamily Eumeninae (*Ancistrocerus, Cephalodynerus, Dolichodynerus, Eumenes, Euodynerus, Hypalastoroides, Leptochiloides, Leptochilus, Maricopodynerus, Microdynerus, Monobia, Montezumia, Odynerus, Pachodynerus, Pachymenes, Pseudepipona, Pseudodynerus, Pterocheilus, Stenodynerus, Symmorphus*)	Solitary to subsocial	World-wide
Family Vespidae		
Subfamily Stenogastrinae (*Eustenogaster, Liostenogaster, Parischnogaster, Stenogaster*)	Subsocial to communal	Tropical Asia and Australia
Subfamily Polistinae (*Apoica, Belonogaster, Brachygastra, Charterginus, Chartergus, Clypearia, Epipona, Leipomeles, Metapolybia, Mischocyttarus, Parachartergus, Parapolybia, Polistes, Polybia, Polybioides, Protonectarina, Protopolybia, Pseudochartergus, Pseudopolybia, Ropalidia, Stelopolybia, Synoeca, Synoecoides*)	Advanced subsocial to eusocial	Primarily Old and New World tropics; *Polistes* ranges to Canada and Sweden
Subfamily Vespinae (*Provespa, Vespa, Vespula*)	Advanced eusocial	North Temperate Zone, tropical Asia

biini have been partially revised genus by genus: *Ropalidia, Parapolybia,* and *Polybioides* of the Oriental Region (van der Vecht, 1962, 1966); *Belonogaster* in Africa (du Buysson, 1909); and, in the New World, *Brachygastra* (= *Nectarina*) (Naumann, 1968), *Synoeca* (du Buysson, 1906), *Protopolybia* (Bequaert, 1944b), *Mischocyttarus* (Richards, 1945; Zikán, 1949), and *Apoica* (Richards and Richards, 1951: 150–158), with additional notes on *Polybia, Chartergus, Charterginus, Pseudochartergus, Pseudopolybia, Epipona, Tatua,* and the Nearctic species of Polybiini being supplied by Bequaert (1933, 1938). The Nearctic *Vespula* (including *Dolichovespula*) have been revised by Miller (1961). A major remaining task in the taxonomy of the social wasps is the analysis of the very large cosmopolitan genus *Polistes.* At present there exists only the checklist by Yoshikawa (1962c) and a few local faunistic studies such as those of Zimmermann (1930, 1931) and Berland (1942) on the European species. A brief history of aculeate wasp taxonomy has been provided by Hurd (1955).

The culturing and laboratory study of colonies of social wasps have received relatively little attention. Most behavioral work has been concentrated on paper wasps of the genus *Polistes,* which build naked combs in exposed sites and can therefore be observed through the entire colony life cycle without special preparation. A disproportionate amount of the remaining effort has been devoted to a few other groups with similar nesting habits, notably *Belonogaster, Mischocyttarus,* and the Stenogastrinae. The Vespinae and many of the more socially advanced Polistinae, although abundant and relatively accessible, build large carton nests that are difficult to penetrate and study without fatally disrupting the societies inside. Techniques for this purpose have been devised by Gaul (1941), Ishay (1964, 1965), and Montagner (1966). The best form of vespiary appears to be a small outdoor house in which the observer can sit in darkness and view nests constructed by the wasps on the outside of glass partitions.

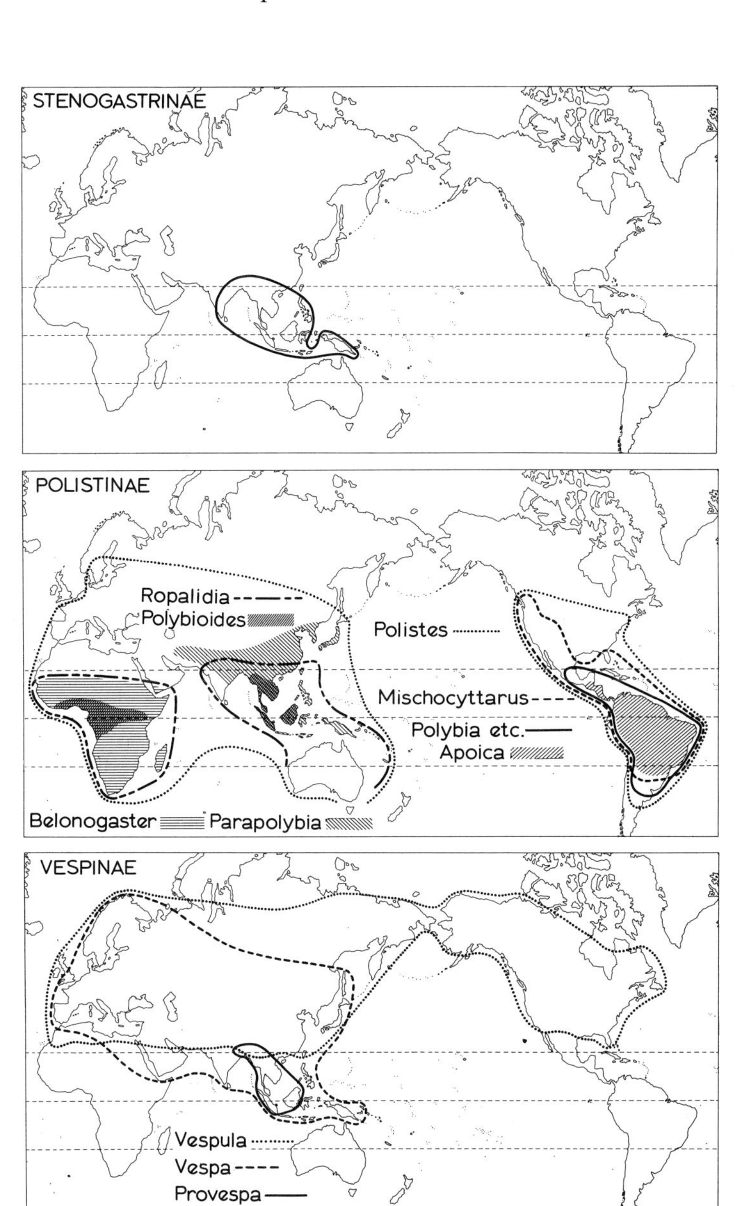

FIGURE 3-3. Geographical distribution of the social wasps. The line for *Polybia* embraces the ranges of the remainder of the more than fifteen highly diverse related genera of Polybiini found in South America, but excludes *Apoica* and *Mischocyttarus* (from van der Vecht, 1967).

The Natural History of Paper Wasps (Genus *Polistes*)

The paper wasp genus *Polistes* is world-wide, and in Europe and North America its colonies outnumber those of all other social wasps combined. The life cycles of four species have been studied in considerable detail: *P. gallicus* in Italy by Pardi (1942, 1948, 1951) and Deleurance (1952a, 1955a,b, 1957), *P. fadwigae* in Japan by Yoshikawa (1962a–d, 1963a–e), and *P. fuscatus* in the United States and *P. canadensis* in South America by Mary Jane West Eberhard (1969). The first three species are quite similar in colony history and individual behavior. Published in-

FIGURE 3-4. A colony of the paper wasp *Polistes fuscatus,* seen from below. The female in the center is heightening one of the cell walls by adding pulp to it. She stands on the central group of cells which were constructed at the beginning of the nest construction and which are now capped and contain pupae. Larvae of gradually decreasing ages occur in progressively newer cells located toward the edge of the comb (photograph courtesy of Mary Jane West Eberhard).

formation on other *Polistes* species, albeit fragmentary, suggests that Eberhard's account of *fuscatus* can be taken as typical for at least the temperate species of the genus.

P. fuscatus is the familiar and abundant brown paper wasp of temperate North America. It is found from Honduras (latitude 15° north), to Chilcotin, British Columbia (52° north). Within this exceptionally broad range there is much geographic variation, some of which may conceal hidden "sibling" species. Bohart (1949), in fact, has elevated seven such species from forms previously considered varieties. Consequently, the most divergent geographic forms, particularly those in Mexico and Central America, may, upon closer examination, prove to be distinct species, and conclusions concerning the geographic variation of social behavior must be drawn with caution.

In the colder parts of the United States, the only individuals to overwinter are the queens. After being fecundated by the short-lived males in late summer and fall, they take refuge in protected places such as the spaces between the inner and outer walls of houses, under shingles, between boards, and beneath the loose bark of trees. In the spring the ovaries begin to develop several weeks before nest initiation, and during this time queens often aggregate in sunny places. Then, presumably when their ovaries reach an advanced stage of development, the queens begin to sit alone on old nests and future nest sites, where they react aggressively to other females who come close.

Eberhard found that nests in Michigan are usually started by a single female. Of 38 nests observed during May when they only contained from one to ten cells, 37 were attended by a single female. A single nest had two foundresses when it was less than 24 hours old. However, by the time the first brood appears in late June, the majority of the foundresses have been joined by from two to six auxiliaries—overwintered queens who for some reason have not managed to start a nest of their own. These individuals are usually subordinate in status and reproductive capacity to the foundresses. Their subordinance is expressed behaviorally in overt ways: the auxiliaries assume submissive postures, undertake food-gathering flights and regurgitate to the dominant foundress, and defer to the foundress in egg laying (see Chapter 15). The foundress not only attempts to prevent her asso-

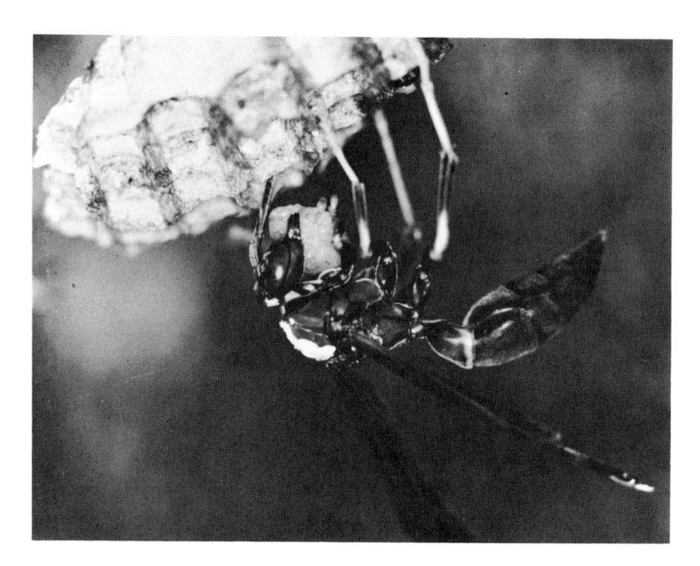

FIGURE 3-5. A female of *Polistes canadensis* on a young nest chews a newly caught arthropod into pulp prior to feeding it to the larvae. She rotates the ball with her forelegs while malaxating it with her mandibles (photograph courtesy of Mary Jane West Eberhard).

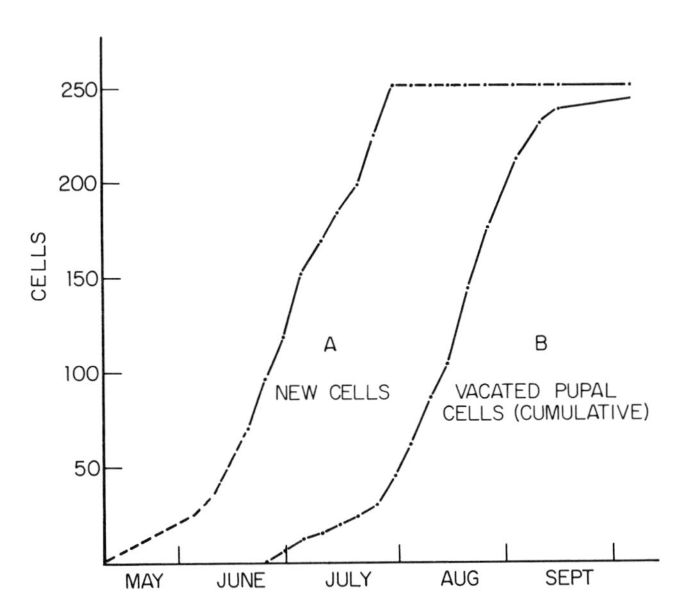

FIGURE 3-6. Parameters of population growth in a typical colony of *Polistes fuscatus:* (*A*) growth of the nest; (*B*) emergence rate of new adults (redrawn from Eberhard, 1969).

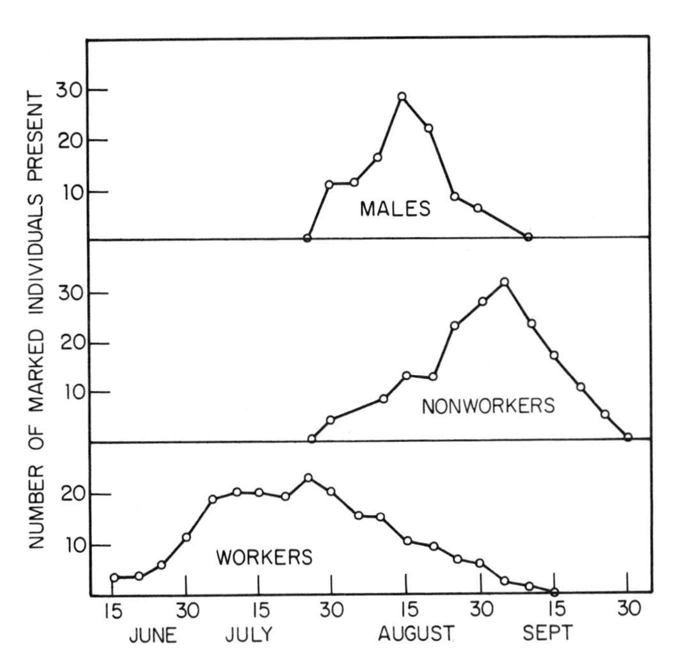

FIGURE 3-7. The progression of castes in the life of a colony of the brown paper wasp *Polistes fuscatus* (redrawn from Eberhard, 1969).

ciates from laying eggs; she also eats their eggs when they occasionally sneak them into unoccupied cells. In time the ovaries of the subordinates regress. Marking experiments by Rau (1940) and Eberhard have revealed that such auxiliaries prefer to associate with foundresses who are sisters. But they move rather readily from nest to nest during the period of colony founding, and a few even attempt to start their own nests while serving as subordinates in established nests.

Through the summer, and on to the onset of the colony's decline and dissolution in early fall, the adult population grows rapidly (Figure 3-6). The complete development from egg to adult takes an average of 48 days, so that roughly three widely overlapping, complete brood sequences can be completed in a season. By the end of summer as many as 200 or more adult individuals may have been reared in a single nest, but their mortality is consistently high, and only a fraction are to be found together at a given time. The first individuals to appear are all workers, that is, females whose wings are generally less than 14 mm in length and whose ovaries are undeveloped. Together with the foundress, and possibly the original auxiliaries, they make up the entire adult population until the end of July. They carry on all of the work of the colony: foraging for insect prey, nectar, and wood pulp for nest construction, building new cells onto the edge of the nest, and caring for the brood and nonworking adults of the colony. In early August males and "queens" (larger females capable of overwintering) begin to emerge; these purely reproductive forms come to replace the workers entirely by fall (Figure 3-7). The reproductives are essentially parasites, and as they grow in number they have an increasingly disruptive influence on the life of the colony. The males are treated aggressively by the workers, and during the peak of male abundance in mid-August the chasing of males is a conspicuous feature of behavior on the nest. According to Eberhard, "The scurrying of males with workers in pursuit, often pulling at a male's hind leg clamped in their mandibles, gives the colony a disrupted appearance, although workers continue to forage and attend the brood." Shortly after the nonworker population appears in late summer, the colony enters its period of fastest growth. But this is destined to be short-lived because the proportion of workers declines rapidly thereafter and their services are spread more and more thinly. In time, the workers returning to the nest with fresh loads of food are mobbed by hungry, soliciting groups of adults. Less food is passed on to the larvae, and colony growth sharply declines. Deleurance (1955b) has described what he calls "abortive" behavior on the part of adults of *Polistes gallicus* during this period of colony waning, during which the brood are not only starved but even pulled out of the cells and killed and eaten. However, Eberhard (1969) found no evidence of such behavior in *P. fuscatus.*

Around the middle of August the *P. fuscatus* males begin to leave the nests and cluster in cracks and on old, abandoned nests. Later, females begin to join these groups. Mating takes place on or close to sunlit structures nearby or within the cavities destined to serve as hibernation sites. With the onset of winter the males die off and the fecundated females hibernate singly to await the coming of spring and the renewal of the colony life cycle.

The *Polistes* life cycle relates to one of the more important generalizations concerning the social biology of wasps. It has been noted by von Ihering (1896), Wheeler (1923), Pardi (1942), Yoshikawa (1962d), and others that the nests of tropical species tend to be founded by multiple foundresses while those of temperate species tend to be founded by single females that overwinter in solitude. The extreme development of the first type is seen in some of the tropical Polybiini, in which new colonies are started by swarms of morphologically similar individuals who leave the old nest about the same time, while the extreme development of the second type is shown by the temperate species of Vespinae, in which new colonies are always begun in the spring by a single fecundated individual belonging to the morphologically very distinct queen caste. The condition illustrated by the polybiines is referred to as primary polygyny (or primary pleometrosis), and that illustrated by the vespines is called primary monogyny (or primary haplometrosis). An extension of the generalization, but not an essential part of it, is that colonies of the tropical swarming species tend to have multiple functional queens but those of temperate species have only one queen. Whether the tropical polygyny is ever secondary, that is, derived in certain cases from original monogynous colony formation, or whether it is always just a holdover from the primary state of polygyny, has never been settled. In fact, it is not even completely certain that true functional polygyny exists in the "polygynous" species. Morphologically similar "queens," with well-developed ovaries, coexist in some polybiine species (Richards and Richards, 1951), but the evidence that all are contributing eggs is for the most part indirect; it is possible that one of these individuals often dominates the others to the point of repressing their oviposition. Recently, however, W. D. Hamilton (personal communication) did observe two queens of *Protopolybia minutissima* ovipositing in the same young nest, and he believes that functional polygyny is a common occurrence in other Polybiini.

Primary monogyny is generally regarded as having evolved from primary polygyny as an adaptation to cold climates. As Wheeler (1923) pointed out, such a transition is easy to visualize: "We might, perhaps, say that our species of *Vespa* and *Polistes* each year produce a swarm of females and workers but that the advent of cold weather destroys the less resistant workers and permits only the dispersed queens to survive and hibernate till the following season." *Polistes* is of special interest because its species display the intermediate steps in this transition. The temperate species *P. fuscatus* is primarily monogynous, to be sure, but the founding queen is usually joined by others within days or even hours after nest construction begins, so that the initial state is nearly polygynous. An even closer approach to swarming is exhibited by *P. canadensis,* a species that ranges (again, possibly as a cluster of sibling species) from the southern United States to northern Patagonia and, in spite of its name, is tropical in origin (Rau, 1933; Eberhard, 1969). In Central and South America a new nest is started by a female which goes directly to the new nest site from the old nest still occupied by her sisters. Often such pioneers are provoked to leave when they fight over the dominant position, a contest which is more overt and evenly matched than in *P. fuscatus.* Just as in *fuscatus,* however, the *canadensis* foundress is quickly joined by other individuals. After quarreling, one female takes precedence, and the colony is functionally monogynous thereafter. Since the primitive species of *Polistes* are tropical (Eberhard, 1969), it seems clear that the cold temperate species have intercalated a hibernation episode in the colony life cycle without having changed social behavior in any important way. In order to find a principal alteration in social organization that can be linked to climatic adaptation, it will be necessary to turn later to the Vespinae.

The Natural History of the Polybiine Wasps

The polybiines encompass every stage of social evolution from the quasisocial to the advanced eusocial. The

African genus *Belonogaster* exemplifies the former stage. In his study of *B. junceus* in Dahomey, Emile Roubaud (1916) showed how the colonies superficially resemble those of *Polistes* in populations and outward appearance but differ from them in several basic details of organization. The nests are composed of naked combs constructed from carton and suspended from aerial supports in forest vegetation. Colonies are usually founded by a single female, but sometimes by 2 or more individuals. The population growth of the colony members is exponential, and its rate is comparable to that of temperate species of *Polistes* during the spring and summer. Within three months, if conditions are ideal, the colony can expand to 20 or more adults which occupy a nest with as many as 200 cells. But only a minority of the colonies reach this stage, and those that do soon enter into a period of decline and dispersal. The reason for the decline appears to be an inevitable buildup of the populations of males and internal parasites. There are no evident caste differences among the females. All individuals are approximately the same size, with well-developed ovaries, and all or nearly all are inseminated within about a week following eclosion. Both before and for some time after insemination the young females serve as workers. According to Roubaud, they are kept sterile by a combination of hard work and lack of sufficient nourishment. However, as they grow older they assume the role of egg layers. Thus there is no permanent caste division, and all the females have essentially the same status when it is considered over a lifetime. One searches in vain through Roubaud's account for evidence of dominance hierarchies, but of course one must bear in mind that Roubaud was not aware of the concept and could easily have neglected to record the pertinent observations. Similar egalitarian societies may exist in the primitively social bees and in the stenogastrine wasp genus *Parischnogaster* (Yoshikawa *et al.,* 1969).

A somewhat higher level of organization, closely approaching that of *Polistes,* occurs in the polybiine genus *Mischocyttarus.* This is a very large group, containing no less than 183 species, and it ranges from northern Argentina northward to British Columbia in the west and through the West Indies to Florida in the east. A South American species, *M. drewseni,* has been the subject of a recent intensive study by Jeanne (1970a,b). His results are of special interest because they elucidate to a much greater extent than the *Belonogaster* data the life cycle of small, short-lived colonies that exist in a relatively uniform tropical climate. *M. drewseni* is a moderately common wasp that ranges from southern Brazil to the northern coast of South America. Along the Amazon River, where Jeanne made his observations, it is limited to open fields and areas of low, shrubby vegetation. The nests consist of single naked combs suspended from very slender pedicels (Figure 3-8). Jeanne found that the wasps regularly coat the pedicels with ant repellent substances that originate from a glandular area near the tip of the abdomen (Jeanne, 1970b). The nests are usually built in low vege-

FIGURE 3-8. A beginning nest of *Mischocyttarus drewseni.* The female is adding ant repellent substance from the sternal glandular area of the abdomen to the pedicel of the nest (photograph courtesy of R. L. Jeanne).

tation or beneath eaves, lintels, and other man-made havens of the kind also favored by some species of *Polistes*. The colonies are small, the largest seldom containing more than thirty adult members. The all-female workers forage during the day for insects, spiders, and nectar. Prey items are malaxated into pulpy masses before being fed to the larvae.

Of the 29 colonies observed from the beginning of their development, 20, or 69 percent, were founded by single females, and the remainder by groups of up to eight females. In the polygynous associations one individual soon came to dominate the others and to take over the exclusive reproductive role. The subordinate foundresses often continued to oviposit for a while, but their eggs were promptly eaten by the dominant female. After a time they gave up the effort altogether and devoted themselves entirely to nest construction, brood care, and foraging. The development of the first brood was rapid; on the average the egg and larval periods were 12 days each, and the pupal period, 13–15 days. The first adults to emerge were all females. They were somewhat smaller in size than the foundresses, about 10 percent shorter in total length, and they all assumed the worker role. Subsequent broods developed more slowly because the larval period stretched out for as much as 32 days. By the end of the next developmental period, approximately 10 to 12 weeks into the colony life cycle, the first males and full-sized females began to emerge. Colonies continued to produce these reproductive forms for another 3 or 4 months, and then they entered a period of rapid decline.

Relatively few colonies reached the terminal, senescent phase. Most were destroyed by external causes, as documented in Table 3-2. The average life span of a colony was therefore quite short. In a sample of 14 colonies followed by Jeanne from near their inception to their natural end, the colony life spans ranged from 6 to 228 days and averaged only 86 days. The causes of natural decline after 6 months or so were not determined with certainty. Unlike the north temperate *Polistes, Mischocyttarus drewseni* is relatively uneffected by climate. Colonies are started at all times of the year and at any given moment a local population contains colonies in every stage of development. The decline appears to be due at least in part to an increasingly heavy investment in males and larger females ("queens") which contribute little to the welfare of the colony and which in any case often disperse after only a short residence on the nest. What would be the advantage of such early programmed colony senility? The answer is very likely that mortality due to external causes is so heavy, and the average life span of a colony correspondingly so short, that it is of selective advantage for colonies to reach maturity quickly and to invest totally in reproductives, rather than to take the chance of producing reproductives at a lower rate over a longer span of time. This is the "explosive" form of colony life cycle, which we will examine later in a more formal analysis of population dynamics (Chapter 21).

The genus *Polybia* can be regarded as representative of the large group of primarily tropical, perennial wasps within the Polybiini that have evolved populous colonies and a more complex social biology than that of *Polistes* or *Mischocyttarus.* It ranges from the tropical portion of Mexico to northern Argentina. Some of its approximately 40 species are exceptionally abundant, forming an ecologically important part of the Neotropical epigaeic fauna (Richards and Richards, 1951).

Polybia occidentalis is the most widespread member of the genus, occurring over almost the entire range of the genus. It is also the best known of the higher polybiines. Richards and Richards, in their treatise on the social wasps of South America, recognize ten geographical subspecies and varieties, plus a complex of additional, closely related species. The colonies of *P. occidentalis* are large; one sample of five collected in Guyana contained from 89 to 596 adults and a closely correlated number of larvae and pupae. As in all polybiine species thus far studied, there are multiple queens (from 4 to 48 in each of the

TABLE 3-2. Causes of mortality of Amazonian colonies of the social wasp *Mischocyttarus drewseni* (from Jeanne, 1970a).

Cause	Number of colonies
Declined normally at end of life cycle	5
Destroyed by ants	2
Destroyed by other predators, probably birds or lizards	11
Queen disappeared, colony declined quickly	1
Abandoned by adults, reason unknown	7
Collected for study purposes	8
Destroyed by unknown agents	13
Total	47

colonies just cited). On the other hand, and this is notably true for the most part of polybiines in general, the only evidence that more than one queen actually lays eggs is the statistical observation that the number of eggs present is correlated with the number of queens (and workers) present. The queens are easily distinguishable from the workers only by their greater ovarian development, and numerous individuals are intermediate even in this character. The queens are also, on the average, smaller in size and possess fewer hamuli (wing hooks), the latter being a curious and as yet unexplained distinction that is the reverse of the usual quantitative queen-worker difference in social insects. In both size and hamulus number the differences are very slight, and there is considerable overlap between the two castes. Some other polybiine species display stronger queen-worker differences, in extreme cases approaching the dimorphism that characterizes the Vespinae (see Chapter 9).

The basic types of nests constructed by polybiines and other social wasps are illustrated in Figures 3-9 to 3-12. All species of *Polybia* build medium-sized to large nests of the "phragmocyttarous" type: rows of horizontal combs are constructed from an arboreal support downward and covered, comb by comb, with a bag-like envelope. The nests of *P. occidentalis* studied by the Richards were hung from large leaves or twigs of trees, at distances of 1 to 8 m from the ground, usually in clearings or open second-growth rain forest. The pyriform envelopes were 6 to 15 cm long and contained 3 to 6 combs. The single entrance holes were located below the level of the lowest comb but on one side of the envelope well above its lowest point. These details are consistent with descriptions of *P. occidentalis* nests studied earlier on Barro Colorado Island, Panama, by Schwarz (1931) and Rau (1933). *P. occidentalis* workers capture a wide variety of insects, especially winged termites abroad during the nuptial flights. The wasps store both malaxated insects and honey in empty brood cells for undetermined periods of time.

In Argentina Bruch (1936) observed both swarming and the formation of new colonies of *P. scutellaris,* a form originally described as a variety of *occidentalis* but now known to be a distinct species (O. W. Richards, personal communication). Although the castes involved were not recorded, it is apparent that *P. scutellaris* conforms to the usual polybiine pattern, in which a group of workers and one or more queens flies from the mother nest, settles on a new site, and constructs a nest. One colony in Argentina was followed through what appeared to be normal development during a period of four years. In the summer month of January 1931 a swarm of adults appeared on a tree and remained as a naked cluster for at least four days, then began constructing a nest—first a convex roof, then a set of six combs. By May, when the colony entered hibernation, the maximum diameter of the envelope had reached 10 cm. The following spring (October 1931) the wasps started a new nest in a less constricted site on another branch, using materials from the old nest. During the ensuing summer the growth of the colony was rapid, and the diameter of the new envelope expanded to 18 centimeters. A year later, in December 1932, several thousand wasps died of an unknown cause and were carried out; but this did not prevent the colony from continuing to add to the nest until, in May 1933, it attained a width of 26 cm and a length of 32 cm. The wasps continued to add combs into the fourth summer, as late as December 1934. But in the following March the nest was found inexplicably abandoned. The pattern through the four-year period, if it can be regarded as typical, indicates that *P. scutellaris* colonies are restless, willing to shift to superior nest sites even during periods of growth.

Most *Polybia* colonies, like the Argentine example, are perennial. Nests of *scutellaris* in Brazil grow to measure as much as a meter in diameter and persist for as long as 25 years. There is very little additional information concerning the pattern of growth of colonies of the *P. occidentalis* species group or the time required to reach "maturity," that is, the stage at which new colonies are generated by swarming. Apparently *P. occidentalis* and its near relatives vary a great deal in these qualities. Richards and Richards (1951), after defining the "developmental cycle" as the period elapsing from the birth of an egg to the eclosion of the adult wasp, made the following generalizations about the life cycles of polybiines in particular and social wasps in general. "While . . . the developmental period is a convenient measure of the age of the nest, the more fundamental unit for the species is the time taken to produce the first males and young queens. It would appear that in this respect there are two main types of behaviour in the POLYBIINAE. In such wasps as *Polybia occidentalis* (Oliv.) (most subspecies), *P. bistriata* (Fab.), *Metapolybia, Protopolybia minutissima* (Spin.), and *Mischocyttarus,* as in *Polistes,* after a few developmental cycles (nests of age 2.0–3.0), males and queens are produced, and the colony breaks up. The

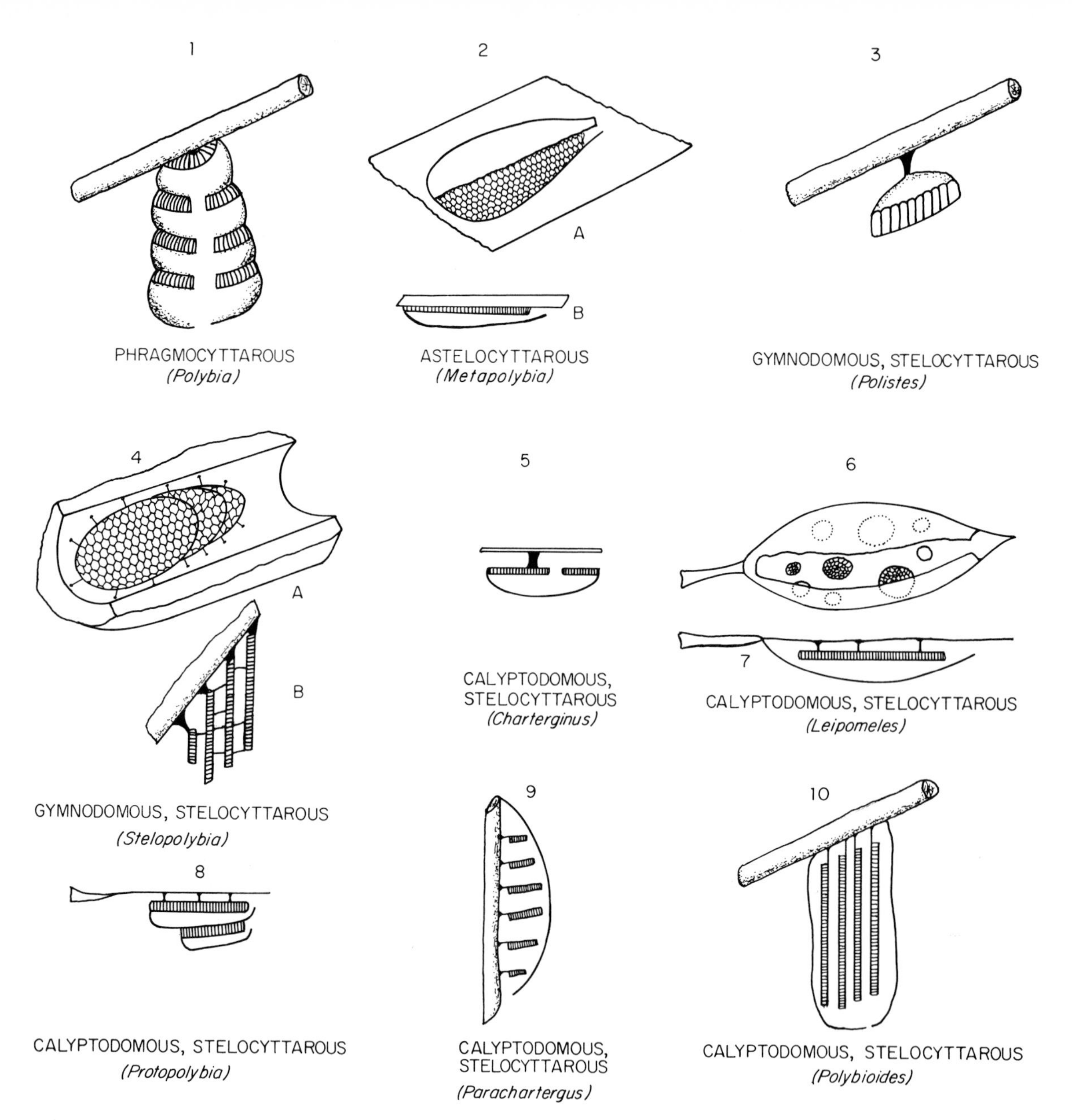

FIGURE 3-9. Diagrams of the basic types of nests of social wasps, according to the modified classification of Henri de Saussure (1853–1858): (1) Cross section of a phragmocyttarous nest, showing the basic plan followed by *Polybia occidentalis* and some other polybiine species. (2*A*) Astelocyttarous nest (attached directly to the support and lacking pillars), surface view with the envelope partly removed; (2*B*) the same in cross section. (3) Cross section of a gymnodomous stelocyttarous nest (that is, lacking an envelope and supported by a pillar); this example contains only one comb and is of the form typically produced by species of *Polistes.* (4*A*) Surface view of another gymnodomous, stelocyttarous nest, this one constructed by a colony of *Stelopolybia* inside a hollow tree trunk; (4*B*) the same in cross section. (5) Cross section of a calyptodomous, stelocyttarous nest (that is, covered by an envelope and supported by a pillar), this one constructed by a colony of *Charterginus.* (6) Calyptodomous, stelocyttarous nest of *Leipomeles* sp., a polybiine that adds new combs laterally; the envelope is partly removed, and the combs are shown at the stage of development prior to fusion. (7) The *Leipomeles* nest after fusion of the combs, in cross section. (8) The nests of *Protopolybia* resemble those of *Leipomeles,* except that the combs are added vertically in the manner shown in this cross section; their individual orientation, however, remains parallel to the substratum. The nests of *Parachartergus* (9) and *Polybioides* (10) also resemble those of *Leipomeles,* except that the combs are hung perpendicular to the surface of the support; that is, they are "laterinidal" instead of "rectinidal" (redrawn from Richards and Richards, 1951, with modifications by W. D. Hamilton and O. W. Richards, personal communications).

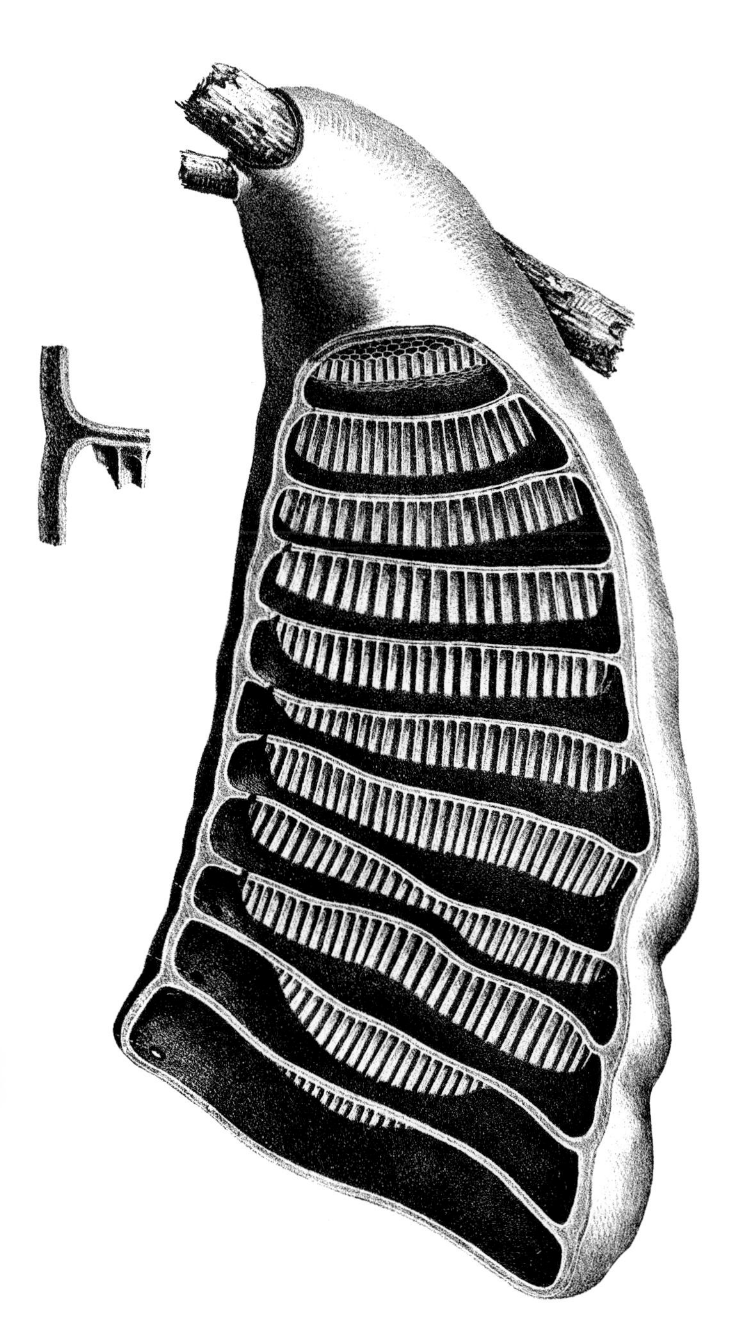

FIGURE 3-10. A nest of the polybiine wasp *Epipona tatua,* with part of the carton envelope removed to show the arrangement of combs inside (from de Saussure, 1853–1858).

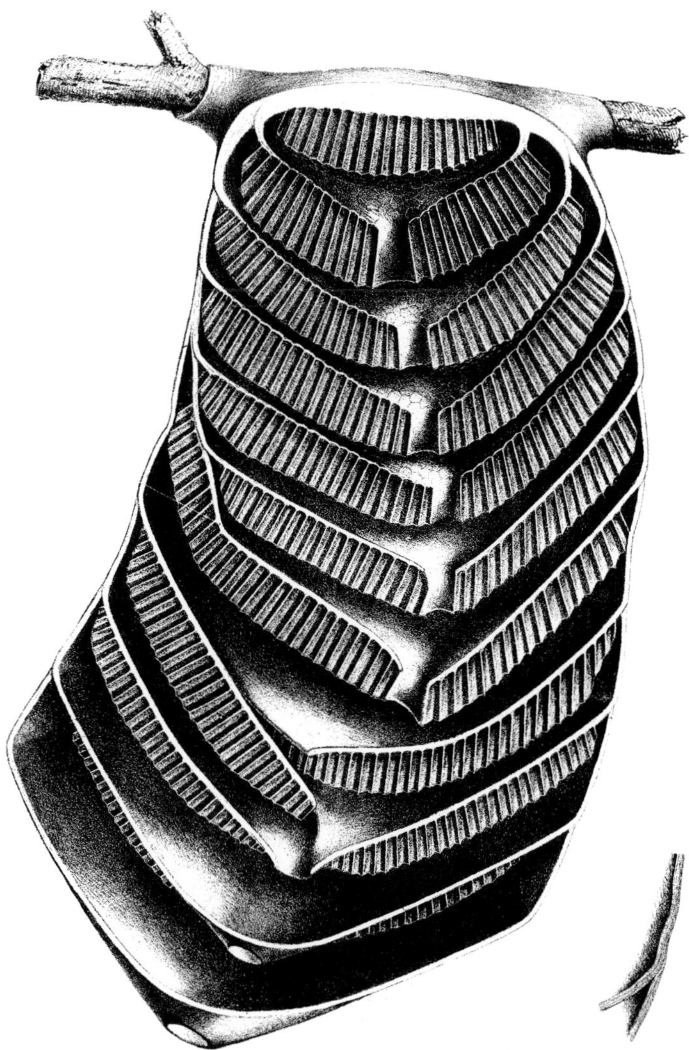

FIGURE 3-11. A nest of the polybiine wasp *Chartergus chartarius,* with part of the carton envelope removed to show the arrangement of combs inside (from de Saussure, 1853–1858).

FIGURE 3-12. A nest of the polybiine *Pseudochartergus fuscatus* from Brazil. The combs are connected by pillars to the leaves of a living tree and surrounded by a semitransparent envelope of hardened glandular secretion (from Jeanne, 1970c).

species of *Vespa s.l.* living in temperate climates essentially follow this plan. There is no evidence that the end of their colonies is due to the weather, much more that it is due to the death of the queen. Colonies of *V. sylvestris* Scop. and *V. rufa* L. [European species] end long before the autumn frosts which, in the other species, only hasten the disintegration which has already started. The ending of the small *Polybia* colonies after a relatively short period is even more curious. Weather can play no part in it and the presence of a number of egg-laying queens makes it less likely that the disintegration is due to their senility . . . The other type of colony, confined to the tropics, is seen in such species as *Polybia rejecta* (Fab.), *Protopolybia pumila* (Sauss.), *Epipona tatua* (Cuv.), and probably *Synoeca surinama* (L.), and at the limits of the tropics in *Polybia occidentalis* var. *scutellaris* (White) and *Brachygastra lecheguana* (Latr.). These wasps make nests which last a long time, probably some years . . . In *P. occidentalis* var. *scutellaris* nests possibly last as long as 25 years (Lucas, 1867 and 1885), and one was observed intermittently for 4 years by Bruch (1936). In *E. tatua,* Fitzgerald observed a nest which lasted 2 years without altering in size and, according to Lacordaire, this species probably sends out swarms at yearly intervals. Cyclical reproduction of this kind is presumably typical of all the species in this group."

The Natural History of the Vespine Wasps

The subfamily Vespinae, comprising what are called the hornets and yellow jackets in English-speaking countries and the "true wasps" in Germany (or *Hornisse* in the case of *Vespa crabro*), is a primarily tropical Asian group that has penetrated deep into the temperate zones of Eurasia and North America. All vespines are eusocial or else social parasites on their eusocial relatives. They are notable for the advanced state of this sociality relative to most of the Polistinae, even though in temperate species the colony life cycle is only annual in nature. The queen is, on the average, much larger in size than the worker caste and is the principal or sole egg layer (Figure 3-13).

The Vespinae are a taxonomically compact group. Workers of the genus *Vespa* are distinguished by their large size and the expanded condition of the rear part of the head. The European *V. crabro,* which has been accidentally introduced by man into the eastern United States (where it is called the European hornet), is a striking species, whose reddish workers average 25 mm in length and whose queens reach 35 mm or more. Its nests are constructed of paper-like carton, ordinarily in hollow tree trunks or between the walls of houses but occasionally in the open air. In the United States most of the species of the subgenus *Dolichovespula* of the genus *Vespula* are referred to as hornets or yellow jackets, the latter in conjunction with species of the subgenus *Vespula,* while *V.* (*D.*) *maculata* is consistently called the bald-faced hornet. The workers are distinguished by their more elongate head shape—in particular, the head capsule is by itself longer than broad—and by the presence of a wide space on the suface of the head between the anterior margin of each eye and the insertion of the nearest mandible. Most species, including *V.* (*D.*) *maculata,* build elaborate carton nests in the open air. In Europe, *Vespula* (*Dolichovespula*) *silvestris* sometimes utilizes cavities in the ground. A third subgenus, *Vespula* (*Paravespula*), can be distinguished, but the differences between it and the subgenus *Vespula* (*Vespula*) are rather weak. Consequently most American writers recognize it at the subgeneric level or not at all, a usage adhered to in the latest revision of North American members of the group by Miller (1961). In contrast, European authorities have chosen to separate *V. austriaca,* a parasitic Holarctic species, and its close relatives as a genus distinct from the majority of the other species, which would then have to be referred to the genus *Paravespula.* Because the grounds for this cleavage seem to me to be weak, I have followed the Europeans in the present work only to the extent of employing the name at the subgeneric rank. Because the name *Paravespula* is used in most of the recent European literature, to drop it altogether would cause unnecessary confusion. *Vespula* and *Paravespula* are both distinguished from members of the subgenus *Dolichovespula* and other vespines by their short head capsule, which contains only a narrow strip of cuticle between the eye and mandible. Most of the species build their carton nests underground, but a few, such as *V.* (*P.*) *germanica* and *V.* (*P.*) *vulgaris,* occasionally nest in cavities in tree trunks.

The life cycle of the vespines is basically similar to that of *Polistes,* except that the queen is not joined by auxiliaries during nest founding in the spring (Duncan, 1939; Kemper and Döhring, 1967; Ishay, Bytinski-Salz, and Shulov, 1967). Little variation in details of the cycle occurs among the species. As a rule only the queens hibernate. A few workers have been found still alive in midwinter in warm climates, but it is doubtful that they play any role in nest founding. In the spring the queen selects the nest site, gathers fragments of dead wood and vegetable fibers, and chews them into a pulp to construct the first cells of the nest. One to three thin paper envelopes are added to enclose the first several cells. At this stage the structure is often called a "queen nest." The queen lays an egg in each cell and, when the first brood of larvae hatches, feeds them with insects caught fresh each day and chewed into a pulp. Soon after the first workers eclose, they begin foraging for insects on their own and adding to the nest. Now the queen only rarely leaves the nest, and, as the season progresses, she gives up all activity except egg laying. Throughout the summer the workers continue to add new cells to the combs as well as new pillars and combs in the stelocyttarous pattern, together with new layers of paper to the cell wall. The entire nest

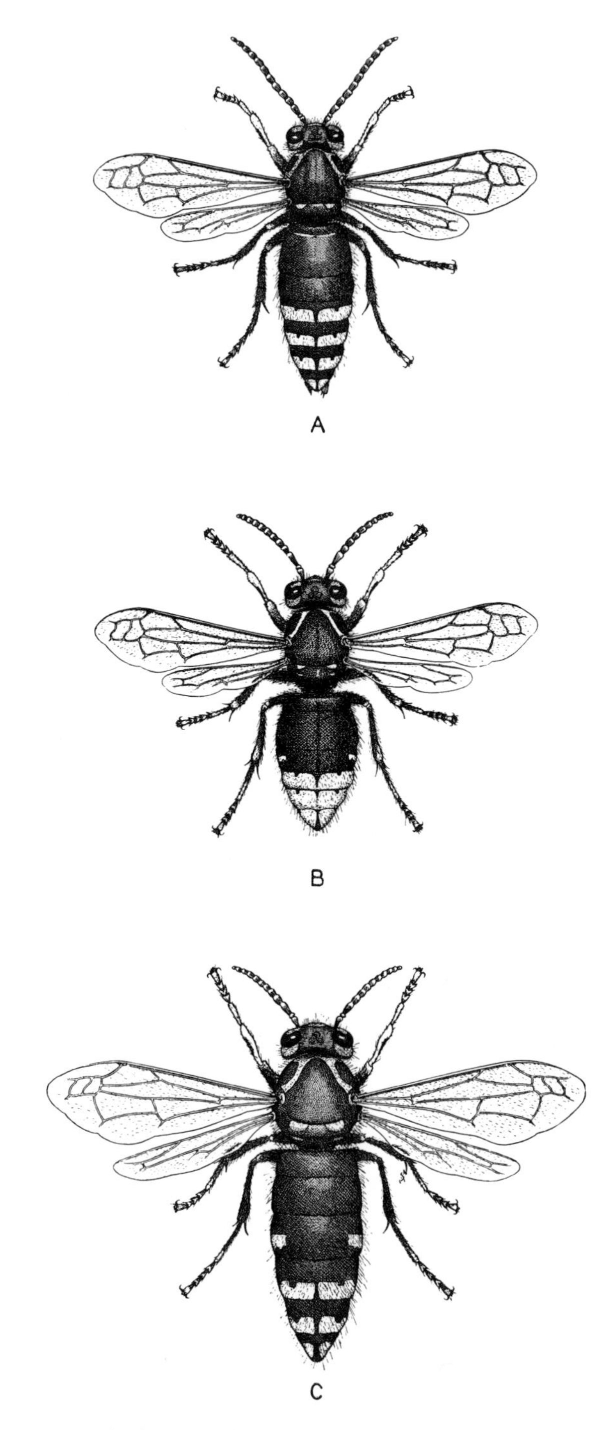

FIGURE 3-13. The three adult castes of the bald-faced hornet, *Vespula* (*Dolichovespula*) *maculata,* a highly social North American vespid wasp: *A*, male; *B*, worker; *C*, queen (from Betz, 1932).

grows outward and downward, assuming an ever larger and more globular shape as the workers tear away old portions and add new material (Figure 3-14).

The workers capture a wide variety of soft-bodied insects to take back to the nest, favoring bees, flies, and both larval and adult lepidopterans. They are rather inept as predators, a characteristic well illustrated by Duncan's following account of the behavior of *Vespula* (*Dolichovespula*) *maculata* workers around a fish cannery in Oregon. "The flies collected on the outside of the walls of the cannery and on the empty fish boxes standing on the docks. Here the wasps came to gather provender. Mostly they flew back and forth making sudden pounces upon the flies from a distance of four to six inches. They were not very skillful, as a dozen or more pounces might be made before a fly was secured . . . The wasps occasionally pounced on one another as they did on the flies. Almost as soon as they came in contact, however, they separated again. Recognition, if this underlay their immediate separation, thus appeared not to occur until actual contact had been made." In Israel the hornet *Vespa orientalis* feeds to a large extent on honeybees, and it has developed a remarkable technique for capturing these difficult prey. In return, the genetic strain of honeybees that is continuously exposed to the attacks of the *Vespa* has apparently evolved an adequate defense. Ishay *et al.* (1967) describe the relationship as follows: "During summer months, the hornets stand or fly at the entrances of beehives, in wait for bees to return from the field. As is known, there are, in Israel, two main races of bees—the local and the Italian ones. The local bees protect themselves against hornets by forming dense clumps at the entrance of the hive. In such instances, a hornet, in order to capture a bee, must first detach it from the rest, a rather difficult feat which is rarely successful. There is the additional danger that the overly venturesome hornet, in coming too close to the apian phalanx, may occasionally become the victim of its prey and be stung to death by the massed bees. In contrast, the bees of the Italian race do not defend themselves against hornets. In most cases, the hornet will approach to within 1.5 cm. of one of the bees standing at the entrance of the bee-hive. It will then entice the bee to chase after it by executing sharp movements of retreat. The bee leaves its hive-mates and follows the retreating hornet, gradually closing the gap between them. To a spectator from the side, the two protagonists seem to be executing a grotesque pas de deux, with the relatively larger hornet simulating retreat and the smaller bee giving chase. When the two have moved some 2–4 cm. from the rest of the bees and are about 3–5 cm. apart, the hornet suddenly pounces on the bee, grabs it with its legs and soars straight up to a nearby branch, where it commences to chew up its prey. In this fashion, hornets deplete bee-hives during the summer and may on occasions totally destroy entire apiaries. The reason entire apiaries are destroyed is that, after the hornets dispose of the sentry bees at the hive entrance, they invade the hive proper and 'gobble up' the adult bees, the young and any food lying about." In addition to hunting for insect prey, the workers of various

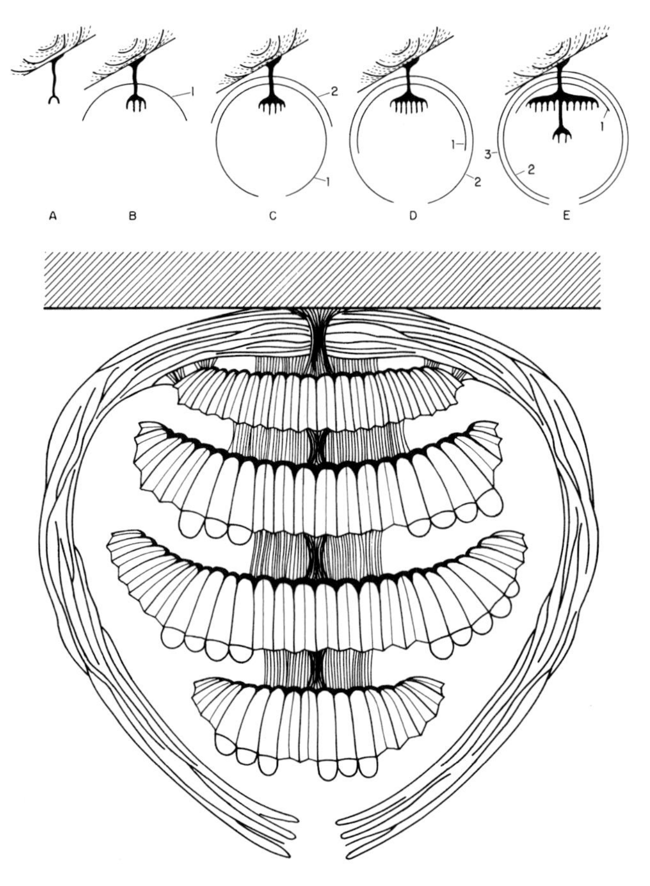

FIGURE 3-14. The nest of a vespine colony: (above) the general form taken in the earliest stages of development, as the queen first founds the nest (*A*) and then as the first brood of workers begins to contribute (*B–E*; successive envelopes are numbered); (*below*) the mature nest of a colony of *Vespula* (*Dolichovespula*) *saxonica* from Europe (from Kemper and Döhring, 1967).

vespine species also carve pieces from fruit and dead vertebrates and gather plant sap and nectar.

Toward the end of summer the workers construct larger cells on the brood combs, and from these they rear a crop of several dozen to several hundred queens and males. About this time the mother queen dies, and brood production ceases. The virgin queens and males leave the nest and mate, and, as cold weather approaches, the last lingering workers in the nest gradually die off. The males, after feeding in solitude on nectar at flowers for a few days or weeks, also perish. But the newly fecundated queens enter hibernacula in the form of spaces under bark of trees, between stacks of cordwood, cracks or abandoned beetle burrows in decaying logs, and so forth, and prepare to wait out the winter.

The Evolution of Social Behavior in Wasps

The natural history of the social wasps has been favored by the attention of a long line of investigators who concentrated on comparative aspects in both tropical and temperate faunas, from de Saussure (1853–1858), Ormerod (1868), Marchal (1896), Janet (1903), von Ihering (1904), Roubaud (1911, 1916), Ducke (1914), F. X. Williams (1919, 1928), Rau (1933), Weyrauch (1935a,b), Duncan (1939), Zikán (1949), and Pardi (1951), to a group of contemporary, more physiologically oriented students of wasp biology, much of whose work has been recently reviewed by Spradbery (1965). By the time of the important publications of Roubaud (1916) and Wheeler (1923), it had been established that the species of aculeate wasps, and in particular the Vespidae, form among themselves a graded series connecting extreme solitary behavior with advanced eusociality. This information provided the foundation for a theory, first conceived by Janet and von Ihering and given complete form by Roubaud and Wheeler, that insect societies arise through subsocial stages. Thus the wasps, in spite of the lack of any significant fossil record, supplied the key information for earlier thinking on the origin of insect societies. The essential idea was crystallized in Wheeler's seven hypothetical, evolutionary steps:

1. The insect mother merely scatters her eggs in the general environment in which the individuals of her species normally live. In some cases the eggs are placed near the larval food.
2. She places her eggs on some portion of the environment (such as leaves) which will serve as food for the hatching larvae.
3. She supplies her eggs with a protective covering. This stage may be combined with step 1 or step 2.
4. She remains with her eggs and young larvae and protects them.
5. She deposits her eggs in a safer or specially prepared situation (nest) with a supply of food put in all at once and made easily accessible to the hatching young (mass provisioning).
6. She remains with the eggs and young and protects and continuously feeds the latter with prepared food (progressive provisioning). By definition, this is "subsocial" behavior.
7. The progeny are not only protected and fed by the mother, but eventually they also cooperate with her in rearing additional broods of young so that parents and offspring live together in an annual or perennial society (eusocial or "truly" social behavior).

This theory departs radically from the simplistic opinion, propounded earlier by Herbert Spencer and Auguste Lameere, that insect societies originated from "voluntary" consociations of individuals of the same generation. Howard E. Evans (1958), drawing on his own extensive knowledge of the solitary Hymenoptera and studies by Richards and Richards (1951) and other contemporary students of the social wasps, proposed a more finely graded ethocline connecting the solitary and advanced eusocial states. His schema, reproduced in Figures 3-15 to 3-19, collates two independent forms of information: first, the morphological similarity and inferred direction of morphological change, expressed in the branching pattern of the phylogenetic trees, and, second, the 13 sequential steps envisioned in behavioral evolution. It is notable that several families of wasps have to be included in order to encompass the whole story. This in no way vitiates the theory. Indeed, if we believe that behavioral evolution is even loosely correlated with morphological evolution, it follows that single taxonomic units such as families and genera should show less behavioral variation than all the wasps taken together.

The group of living Hymenoptera considered on morphological grounds to be closest to the ancestors of the higher Aculeata is the superfamily Scolioidea. It is therefore consistent with the remainder of the comparative data

Steps not taken:
Nest-egg-prey
Nest-sev. prey-egg

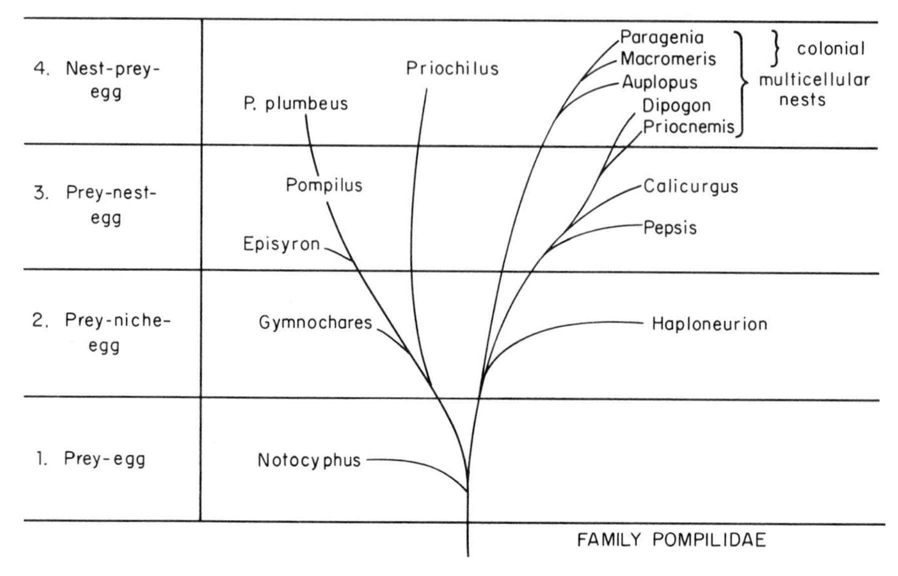

FIGURE 3-15. Inferred evolution of predatory behavior in the spider wasp family Pompilidae. In *Notocyphus* the females sting spiders of certain species within their webs, lay their eggs, and depart; the spider recovers and resumes normal behavior for a time, carrying the *Notocyphus* larva as a parasitoid on its back. In *Haploneurion,* the spider is stung into paralysis, dragged into an abandoned beetle burrow, and left with an egg attached. In more advanced pompilids, the female prepares a nest, either before or after obtaining a paralyzed spider (redrawn from Evans, 1958).

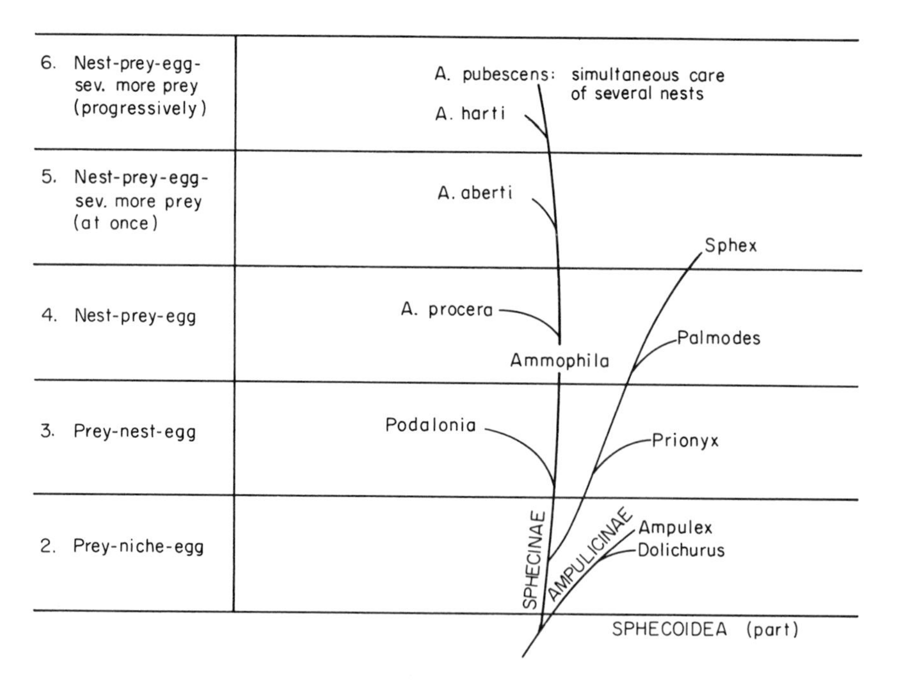

FIGURE 3-16. The transition between primitive mass provisioning and progressive provisioning within two families of the superfamily Sphecoidea. *Ammophila* is notable for exhibiting a variety of degrees of mother-offspring relationships. *A. procera* females place single caterpillars in previously prepared cells and lay one egg on each. Those of *A. aberti* place several caterpillars in one cell and occasionally come into contact with newly hatching larvae before the nest is sealed. In *A. harti* and *A. pubescens,* fresh caterpillars are brought from time to time to the growing larvae (modified and redrawn from Evans, 1958).

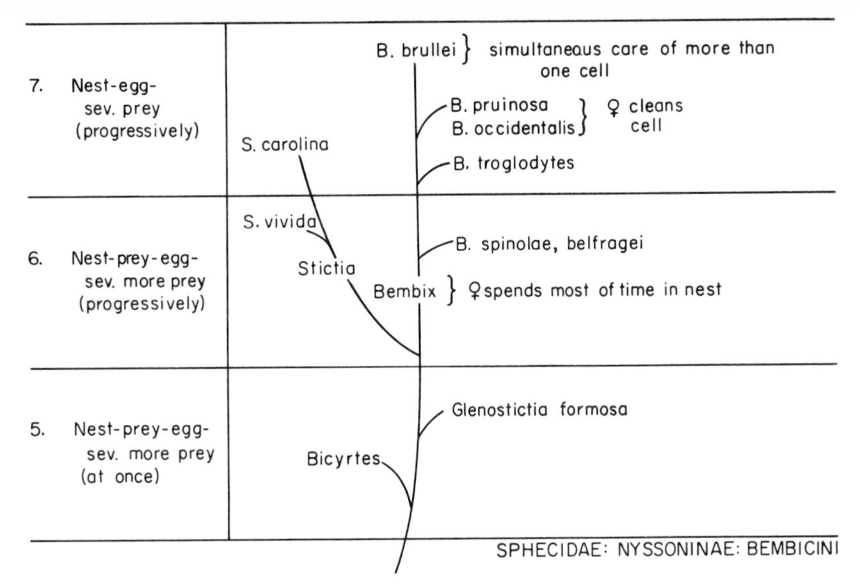

FIGURE 3-17. Evolution of predatory and nesting behavior in the sphecid tribe Bembicini. In some species of *Glenostictia, Bembix,* and *Stictia,* the egg is first laid in the cell, and then a prey (usually a fly) is captured and placed with it. In a second important advance toward sociality, females of the sand wasp genus *Bembix* remain within the nest at all times when not actually feeding or provisioning; consequently they to some extent protect the larvae from predation. Furthermore, females of some species of *Bembix* clean out the cells of the remains of flies which have been consumed, thus reducing infestation by various inquilinous maggots—again presaging an important behavioral component of social organization (redrawn and modified from Evans, 1958).

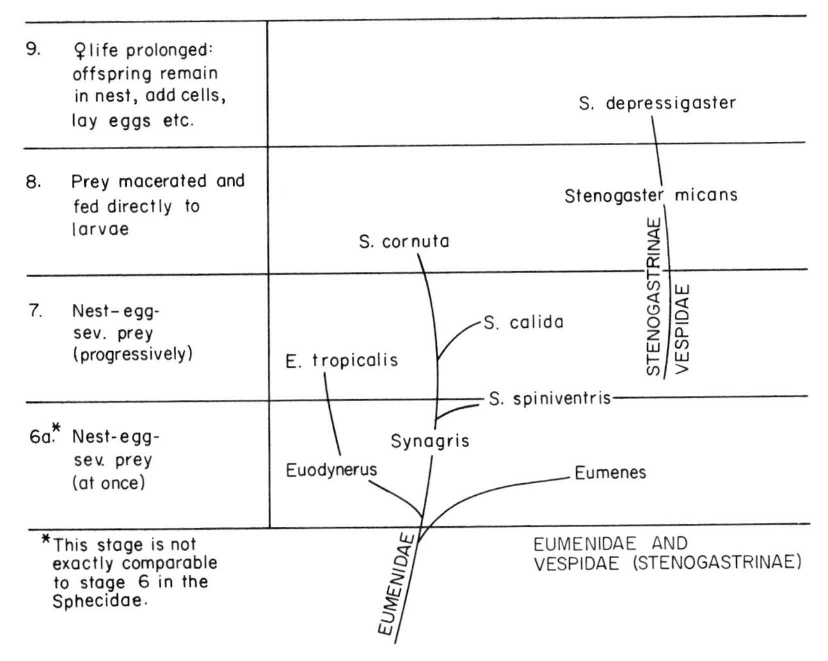

FIGURE 3-18. The vespoid family Eumenidae, which includes the familiar potter wasps, exhibits every conceivable transition from mass to progressive provisioning. *Synagris spiniventris* practices mass provisioning when food is abundant and progressive provisioning when food is scarce. *S. cornuta,* while remaining primitively subsocial, adds an important feature found in all eusocial wasps: the prey are macerated and fed directly to the larvae, rather than being placed in a whole condition next to the larvae. The vespid subfamily Stenogastrinae contains the transition from primitively subsocial to subsocial stage I. Females of *Stenogaster depressigaster* live in small colonies apparently consisting of a mother and daughters. These daughters differ in neither structure nor function from the mother; they merely remain on the nest, adding cells, laying eggs, and caring for their own larvae (redrawn from Evans, 1958; see also Iwata, 1967).

13. Worker caste strongly differentiated, few intermediates	Vespula Vespa	Polybia
12. Differential feeding of larvae; worker caste present but intermediates common	VESPINAE	Protopolybia Stelopolybia
11. Original offspring all♀♀, lay ♂ eggs or none; queen is dominant	Polistes	
10. Trophallaxis; some division of labor but no true workers		Belonogaster

POLISTINAE

VESPIDAE: POLISTINAE AND VESPINAE

FIGURE 3-19. The vespid subfamily Polistinae encompasses the transition from advanced subsocial to advanced eusocial states. *Belonogaster* is in subsocial stage II; mothers and daughters cooperate in nest building and brood rearing, but there is no permanent division into worker and egg-laying castes. This division does take place in other Polistinae, thus completing the three qualities—cooperative brood care, overlap of generations, and division into reproductive castes—needed for the formal designation of eusociality. At least the *Belonogaster* level is also attained in the stenogastrine genus *Parischnogaster,* not shown here. Liquid food exchange (trophallaxis) is added in all of the Polistinae. A slight morphological difference exists between queens and workers of *Polybia*; this difference is strongly marked in the Vespinae (*Vespa, Vespula*) (redrawn from Evans, 1958; see also Yoshikawa *et al.,* 1969).

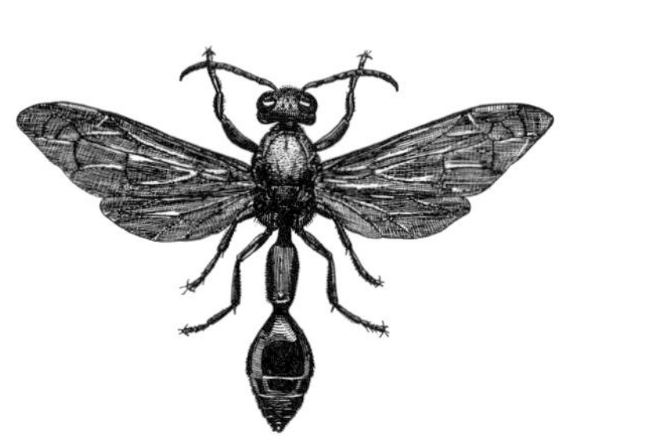

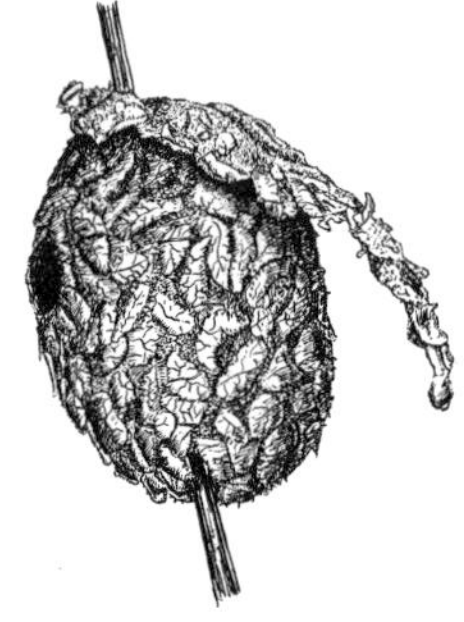

FIGURE 3-20. *Zethus cyanopterus,* a solitary discoeliine eumenid wasp from southeastern Asia: (*left*) adult male; (*right*) nest from a forested area on Luzon. It is constructed from freshly cut leaves and contains four cells, which are progressively provisioned by a single female (from F. X. Williams, 1919).

to find that scolioid females display the most primitive form of predatory behavior—level 1 in the Evans schema. The females of most scolioid species search for beetle grubs, then simply sting them into paralysis, lay an egg on the surface of their victim's body, and leave the grub where it happens to be. This is essentially the behavior of most of the lower parasitoid Hymenoptera from which the aculeates are considered to have been ultimately derived.

Outside the Vespoidea there is one additional group of aculeate wasps, the Sphecoidea, in which higher levels of social behavior have been achieved. In the sphecid tribe Crabronini, communal nesting occurs in at least five species (Evans, 1964). In *Moniaecera asperata,* for example, two or three females of the same generation share the

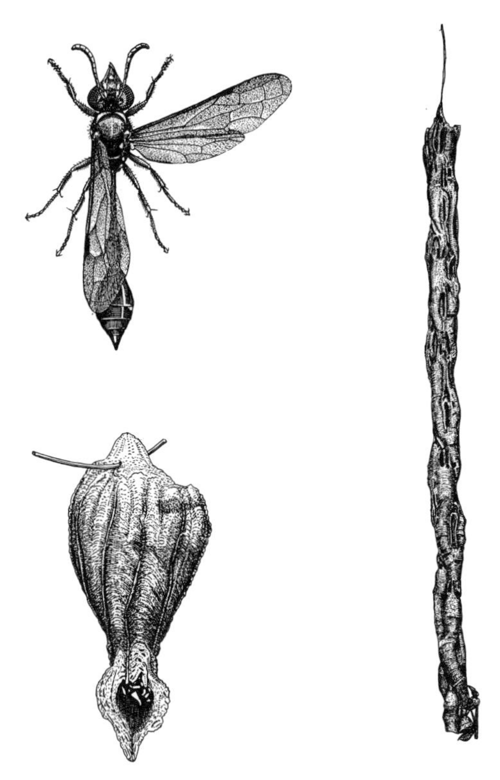

FIGURE 3-21. Wasps of the vespid genus *Stenogaster* and their nests: (*left*) female of *Stenogaster micans luzonensis* and her pear-shaped nest; the nest is constructed of carton (chewed dead wood) and contains as many as 23 cells which are progressively provisioned by a single female; (*right*) nest of the communal species *Stenogaster depressigaster*, with one of the resident females sitting at the bottom. Both species were studied in the Philippine Islands by Williams (1919).

same earthen burrow. They mass-provision their own cells separately (with paralyzed flies and hemipterans) and show no other signs of cooperation. In the Cercerini we find associations of about the same level but this time of a subsocial nature, that is, occurring between mother and daughter, rather than parasocial as in the Crabronini. The best-documented species is *Cerceris rubida* (Grandi, 1961). In Italy this little wasp has two generations during each summer, and as many as four or five females of the second generation remain in the nest with their mother. Although they are smaller in average size than females of the first generation, the daughters carry out their own nesting and provisioning activities. There is a rudimentary form of division of labor, with both mothers and daughters taking turns guarding the communal nest entrance.

True sociality or a state closely approaching it has been achieved by the pemphredonine sphecid *Microstigmus comes* (Matthews, 1968a,b; see Figure 3-22). Females of this Costa Rican species construct diminutive bag-like nests suspended from the underside of leaves of the forest palm *Chrysophila.* Brood cells, which consist of pocket-like cavities in the lower half of the envelope, are mass-provisioned with collembolans. Of 22 nests examined by Matthews, half contained a single female and the other half from 2 to 11 females and a scattering of males. Regardless of the number of females, only one cell was being provisioned at any given moment. The condition of ovarian development varied greatly among the females in a given nest, and a few of the females lacked any trace of oocyte development. From these facts it seems likely that there is a reproductive division of labor, either temporary or permanent with respect to single individuals. The question remains, is there also an overlap of generations? Although many new adults emerged during the brief period of Matthew's study, no nests under construction were found. Also, some of the adult females remaining in the polygynous nests were evidently very young individuals. Thus it appears likely that overlap of generations does occur, and that *Microstigmus comes* is either fully eusocial or at least at an advanced stage of presociality comparable to that already documented in *Belonogaster.*

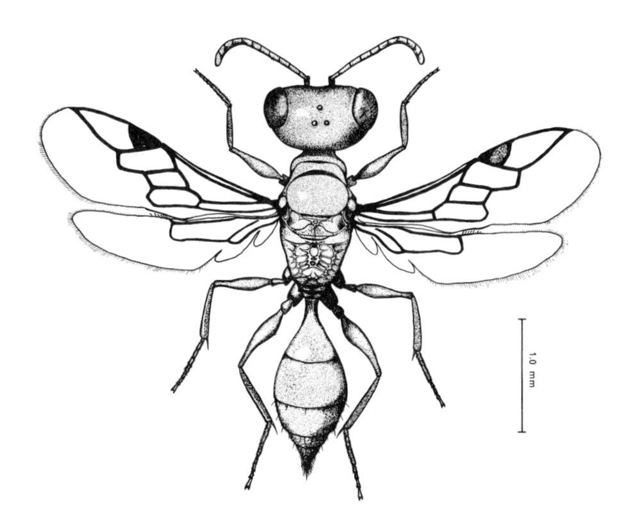

FIGURE 3-22. The female of *Microstigmus comes*, a social sphecid wasp from Costa Rica (from Matthews, 1968a).

Although the traditional reconstruction of social behavior in the vespid and sphecid wasps seems generally persuasive, our information is still relatively weak and uncertain in many crucial aspects. Most needed are thoroughgoing life cycle and population studies, especially in the subsocial and primitively eusocial stages within the span of Evans' stages 7–12. The crucial groups in these stages, most notably the subsocial Eumenidae, the Stenogastrinae, and the primitive Polistinae including *Belonogaster* and *Mischocyttarus,* and *Microstigmus,* are all tropical and still poorly known. It remains an open question whether colonies in most of these groups are founded exclusively by single females or occasionally by groups of cooperating females of the same generation. The degree of reproductive dominance and the relation of division of labor to age and size are still mostly unknown. It seems likely that, as new studies are conducted, the details of the evolutionary scheme will have to be altered frequently, perhaps in some surprising ways.

4 The Ants

On several counts ants can be regarded as the premier social insects. They are the most widely distributed of the major eusocial groups, ranging from the arctic tree line south to Tierra del Fuego, to the tip of South Africa, and to Tasmania, and occurring on virtually every oceanic island between Iceland and the Aleutians in the north and Tristan da Cunha and Campbell in the south. They are numerically the most abundant of social insects. At any given moment there are at least 10^{15} living ants on the earth, assuming that C. B. Williams (1964) is correct in estimating a total of 10^{18} individual insects—and taking 0.1 percent as a conservative estimate of the proportion made up of ants. The ants contain a greater number of known genera and species than all other eusocial groups combined. The diversity of their ecological and social adaptations is truly remarkable. Food specialization is extreme, exemplified by the species of the ponerine genus *Leptogenys* that prey only on isopods; by species of the ponerine genera *Discothyrea* and *Proceratium* that feed only on arthropod eggs (Brown, 1957); by certain members of the myrmicine tribe Dacetini that prey only on entomobryomorph and symphypleonan collembolans (Brown and Wilson, 1959a); and by members of the subfamily Cerapachyinae, all of which, so far as we know, prey exclusively on other ants (Wilson, 1958a; Gotwald and Brown, 1966). The majority of ant groups exhibit a highly variable degree in prey choice, while a few have come to subsist primarily on seeds. Still others rely entirely on the "honeydew" secretions of homopterous insects reared in their nests or on special mutualistic fungi cultured on insect dung or vegetation (Way, 1963; Weber, 1966).

Social parasitism attains its most advanced development in these insects. A finely graded series of stages in the evolution of the phenomenon is displayed by various species up to and including degenerate forms of slavery in which the slave-maker workers are capable only of conducting raids and are totally dependent for minute-to-minute care on their slave workers. Other evolutionary lines lead to total inquilinism, in which the worker caste is lost; in one remarkable species (*Teleutomyrmex schneideri*) the parasite queens have turned into ectoparasites, their bodies having become modified for riding on the backs of the host queens (Chapter 19).

Nesting habits are no less diverse. A few ant species, such as the fungus growers of the genus *Atta* and the extreme desert dwellers *Monomorium salomonis* and *Myrmecocystus melliger,* excavate deep galleries and shafts down into the soil, sometimes to depths of six meters or more (Bernard, 1951; Jacoby, 1952; Creighton and Crandall, 1954). In contrast, there are the arboreal species, some of which are limited to cavities of one or a very few species of plants. A few of the host plants in turn are highly specialized to house and nourish ant colonies and are unable to survive without these insect guests (Janzen, 1967). The tiny myrmicine *Cardiocondyla wroughtoni* sometimes nests in cavities left in dead leaves by leaf-mining caterpillars, while a few formicine species—*Oecophylla longinoda* and *O. smaragdina, Camponotus senex,* and certain species of *Polyrhachis*—have evolved the habit of using silk drawn from their own larvae to construct tent-like arboreal nests (Forel, 1896).

The reason for the numerical preeminence of ants is a matter for conjecture. Surely it has something to do with the innovation, achieved as far back as the mid-Cretaceous period, of a wingless worker caste able to forage deeply into soil and plant crevices. It must also stem partly from the fact that primitive ants began as predators on other arthropods and were not bound, as were the termites, to a cellulose diet and to the restricted nesting sites that place colonies within range of adequate quantities of cellulose. Finally, the success of ants might be explained in part by the ability of all the primitive species and most of their descendants to nest in the soil and leaf mold, a location that gave them an initial advantage in the exploitation of these most energy-rich terrestrial microhabitats. And perhaps this behavioral adaptation was made possible in turn by the origin of the metapleural gland, the acid secretion of which inhibits growth of microorganisms in the nest chambers (Maschwitz *et al.*, 1970).

In 1968, according to Francis Bernard, 7600 living species of ants had been described from the entire world. W. L. Brown (personal communication) believes that the true number, including those remaining to be described, probably falls between 12,000 and 14,000. These comprise about 250 unequivocal genera. There is more variety of species in the tropics, and this diversity declines sharply with increasing latitude (Kusnezov, 1957a; Wilson, 1968a). The fauna of North America north of Mexico, for example, one of the best studied taxonomically, was known to contain only 455 species in 1950 (Creighton). Only 3 of the more widespread of these species make it to the tree line of the American arctic, and they are very rare or absent on the tundra beyond (Weber, 1950; W. Briggs, personal communication).

The ants constitute all of the single superfamily Formicoidea and, within that, the single family Formicidae. Brown (1954) divided the Formicidae into two major branches, the myrmecioid and poneroid "complexes" of subfamilies. The names of the complexes were derived from the most primitive subfamilies known to belong to them at that time, namely the Myrmeciinae and Ponerinae. This taxonomic decision has since come to be supported by a growing amount of solid evidence, and it has some potentially profound evolutionary implications to be discussed later in this chapter. Within the two complexes can be recognized the eleven subfamilies listed in Table 4-1. Some authorities place the Aneuretinae as a tribe within the Dolichoderinae and the Cerapachyinae as a tribe within the Ponerinae; both are clearly no more than borderline candidates for subfamilial rank. Representatives of some of the most diverse ant genera are illustrated in Figures 4-1 to 4-6.

TABLE 4-1. Subfamilies of ants (family Formicidae), together with the principal tribes and genera and their world distribution. (Some authors place the Aneuretinae within the Dolichoderinae and the Cerapachyinae within the Ponerinae.)

Division	Distribution
THE MYRMECIOID COMPLEX	
Subfamily Sphecomyrminae	
Sphecomyrma	Fossil only; Cretaceous of New Jersey
Subfamily Myrmeciinae	
Nothomyrmecia	Australia
Myrmecia	Australia, New Caledonia
Prionomyrmex	Fossil only; Oligocene of Europe
Ameghinoia	Fossil only; early Tertiary of South America
Subfamily Aneuretinae	
Aneuretus	Ceylon
Mianeuretus, Paraneuretus, Protaneuretus	Fossil only; Oligocene of United States and Europe
Subfamily Pseudomyrmecinae	
Pseudomyrmex	New World tropics
Tetraponera	Africa, Asia, Australia
Pachysima, Viticicola	Africa
Subfamily Dolichoderinae	
Tribe Dolichoderini	
Dolichoderus, Monacis	New World tropics
[a] *Hypoclinea*	World-wide
Acanthoclinea, Diceratoclinea	Australia
Tribe Leptomyrmecini	
[a] *Leptomyrmex*	Australia
Tribe Tapinomini	
Tapinoma	World-wide
[a] *Iridomyrmex*	New World, Asia, Australia
[a] *Technomyrmex*	Asia, Australia
Liometopum	Eurasia, North America
Bothriomyrmex	Widespread in Old World
Conomyrma, Dorymyrmex, Forelius	New World
Subfamily Formicinae	
Tribe Myrmecorhynchini	
Myrmecorhyncus	Australia
Tribe Dimorphomyrmecini	
Dimorphomyrmex	Asia
[a] *Gesomyrmex*	Asia
Tribe Myrmoteratini	
Myrmoteras	Asia
Tribe Melophorini	
Notoncus, Prolasius, Melophorus	Australia

Division	Distribution
Subfamily Formicinae (*continued*)	
Tribe Plagiolepidini	
Acantholepis, Anoplolepis, Plagiolepis	Widespread in Old World
Acropyga	Asia, Australia, New World tropics
Tribe Oecophyllini	
[a] *Oecophylla*	Australia, Asia, Africa
Tribe Formicini	
[a] *Formica,* [a] *Lasius, Polyergus*	Eurasia, North America
Acanthomyops	North America
Brachymyrmex	North America
[a] *Prenolepis*	Eurasia, North America
Euprenolepis, Pseudolasius	Asia, Australia
Paratrechina	World-wide
Tribe Camponotini	
[a] *Camponotus*	World-wide
Calomyrmex, Opisthopsis	Australia
Polyrhachis	Asia, Africa, Australia
THE PONEROID COMPLEX	
Subfamily Ponerinae	
Tribe Amblyoponini	
Amblyopone	World-wide
Prionopelta	Asia, Australia, New World tropics
Myopopone	Asia, New Guinea
Mystrium	Asia, Africa, Australia
Onychomyrmex	Australia
Tribe Ectatommini	
Ectatomma	New World tropics
Heteroponera	Pantropical
Rhytidoponera	Asia, Australia
Discothyrea, Proceratium	World-wide
Gnamptogenys	New World, Asia, Melanesia
Tribe Platythyreini	
Platythyrea	Pantropical
Tribe Ponerini	
Hypoponera, [a] *Ponera*	World-wide
Diacamma, Myopias, Ectomomyrmex	Asia, Australia
Bothroponera, Brachyponera, Cryptopone	Old World tropics
Leptogenys, Trachymesopus	Pantropical
Tribe Odontomachini	
Anochetus, Odontomachus	Pantropical
Subfamily Cerapachyinae	
Tribe Sphinctomyrmecini	
Sphinctomyrmex	Asia, Australia, New World tropics
Tribe Cerapachyini	
Cerapachys	Pantropical
Lioponera, Phyracaces, Syscia	Old World tropics
Acanthostichus	New World tropics
Subfamily Leptanillinae	
Leptanilla	Pantropical

Division	Distribution
Subfamily Dorylinae	
Tribe Dorylini	
Aenictus	Old World tropics
Dorylus	Asia, Africa
Anomma	Africa
Tribe Cheliomyrmecini	
Cheliomyrmex	New World tropics
Tribe Ecitonini	
Eciton, Labidus	New World tropics
Neivamyrmex	New World
Subfamily Myrmicinae	
Tribe Myrmicini (broad sense)	
Manica, [a]*Myrmica, Stenamma*	Eurasia, North America
Aphaenogaster, Leptothorax, Monomorium, Myrmecina, Pheidole, Solenopsis	World-wide
Cardiocondyla, Oligomyrmex, Pheidologeton, Pristomyrmex, Rhoptromyrmex, Triglyphothrix	Old World tropics
Lordomyrma, Podomyrma, Vollenhovia	Asia, Australia
Chelaner, Machomyrma	Australia
Xiphomyrmex	Asia, Australia, New World tropics
Rogeria	Melanesia, New World tropics
Messor, Strongylognathus, Tetramorium	Old World
Hylomyrma, Pogonomyrmex	New World
Tribe Ochetomyrmecini	
Blepharidatta, Ochetomyrmex, Wasmannia	New World tropics
Tribe Attini	
Acromyrmex, Apterostigma, Atta, Cyphomyrmex, Trachymyrmex	New World, mostly tropical
Tribe Meranoplini	
Calyptomyrmex, Meranoplus	Old World tropics
Mayriella	Australia
Tribe Crematogastrini	
Crematogaster	World-wide
Tribe Metaponini	
Metapone	Pantropical
Tribe Dacetini	
Acanthognathus, Daceton	New World tropics
Strumigenys	World-wide
Smithistruma	Asia, Australia, New World tropics
Colobostruma, Epopostruma, Mesostruma, Orectognathus	Australia
Tribe Basicerotini	
Basiceros	New World tropics
Eurhopalothrix, Rhopalothrix	Pantropical

[a] Genus also known from early Tertiary fossils.

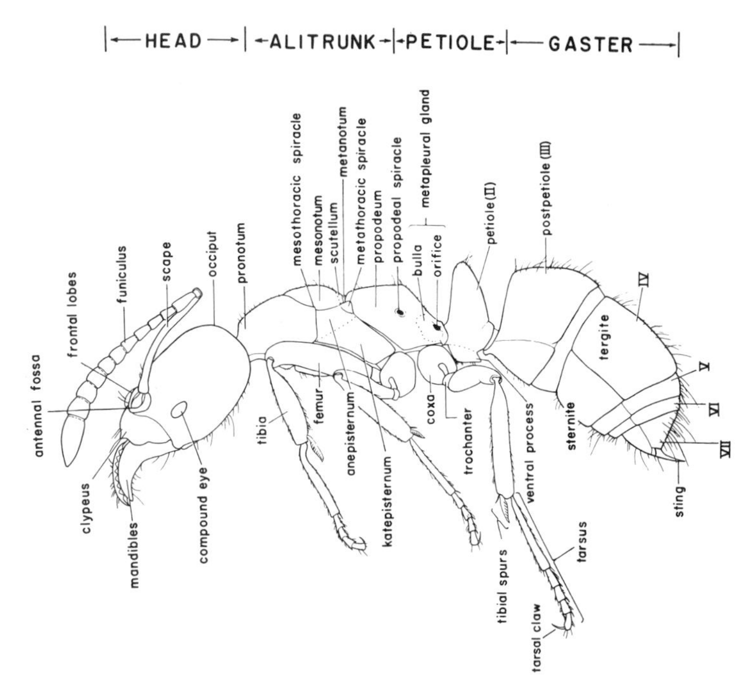

FIGURE 4-1. Worker of the New Zealand ponerine ant *Mesoponera castanea* showing some of the principal morphological features used in taxonomy (from Brown, 1958).

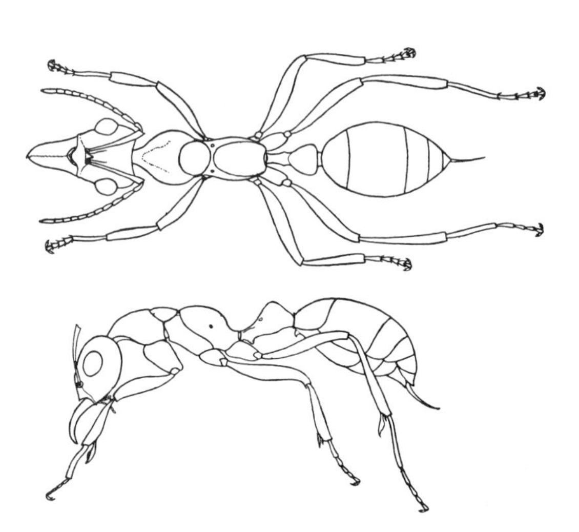

FIGURE 4-2. The worker caste of *Nothomyrmecia macrops*, considered to be the most primitive living member of the Australian subfamily Myrmeciinae and hence of the entire myrmecioid complex (modified from Clark, 1934).

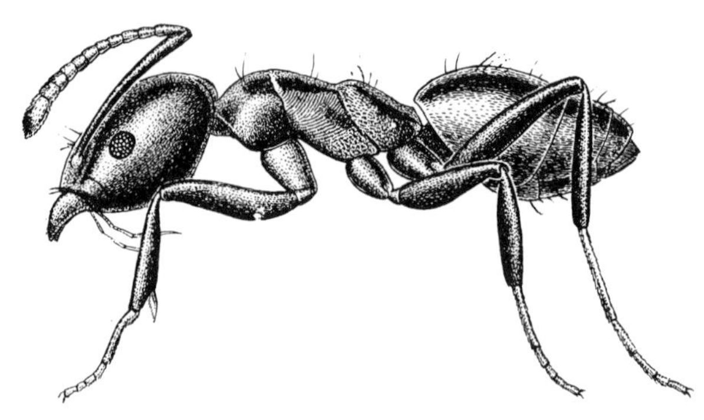

FIGURE 4-3. The worker of *Technomyrmex albipes*, a member of the Dolichoderinae (from Wilson and Taylor, 1967).

FIGURE 4-4. The worker of *Paratrechina longicornis*, a member of the Formicinae (from M. R. Smith, 1947).

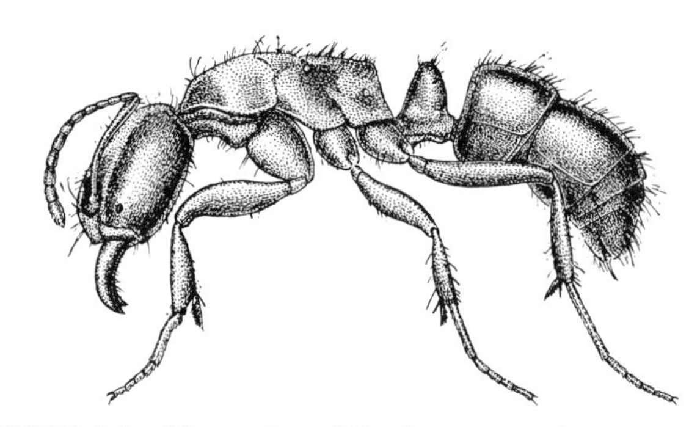

FIGURE 4-5. The worker of *Trachymesopus stigmus*, a member of the Ponerinae (from Wilson and Taylor, 1967).

FIGURE 4-6. The worker of *Triglyphothrix striatidens*, a member of the Myrmicinae (from M. R. Smith, 1947).

The Origin of the Ants

Until very recently the search for the ancestry of ants has resulted in frustration. To be sure, a vast number of fossils were available for study. In examining 9,527 Baltic amber specimens, of Oligocene age, Wheeler (1914) recognized no less than 92 species belonging to 43 genera. Comparably rich Oligocene deposits were described from rock fossils in North America (Carpenter, 1930), and additional finds were made in Miocene amber from Sicily (Emery, 1891), Miocene rocks from East Africa (Wilson and Taylor, 1964), early Tertiary rocks from South America (Viana and Haedo Rossi, 1957), and amber of middle to late Tertiary age from the Dominican Republic and Chiapas, Mexico (Brown and Wilson, unpublished). These Tertiary faunas are for the most part modern in aspect. In the Oligocene deposits, for example, the most abundant genera are *Iridomyrmex* and *Lasius,* in that order. *Iridomyrmex* is still a prominent element of the tropical and warm temperate faunas of the New World and both numerically dominant and species-rich in Australia and New Guinea. *Lasius* remains one of the several most abundant and species-rich genera in the cooler portions of Europe, Asia, and North America. The Oligocene species of *Lasius* fit well within the range of variation exhibited by the living species but are morphologically as distinct from them as the living species are from each other (Wilson, 1955a). This is also true for the other ant genera that have persisted since Oligocene times, including *Camponotus, Formica, Iridomyrmex, Myrmica, Oecophylla, Ponera,* and *Technomyrmex.* There are also quite a few extinct genera and even some primitive forms among the Tertiary fossils. But the most primitive of all of these, *Prionomyrmex longiceps* of the Baltic amber, is a member of the Myrmeciinae, a group still represented by two living genera and many species in Australia and New Caledonia. Furthermore, the most generalized known member of the Myrmeciinae is a living Australian species—*Nothomyrmecia macrops* (see Figure 4-2). This remarkable ant was described by John Clark in 1934 on the basis of two workers collected somewhere within a 10,000-square-mile area of dry, uninhabited heath and eucalyptus forest between Esperance and Balladonia, in southwestern Australia. Several expeditions, including one by C. P. Haskins and the present author, have attempted to recover additional, living specimens, but they have not been successful. The story of the discovery and subsequent unrewarded search for *Nothomyrmecia,* together with a more detailed description of the type specimens, is given by Brown and Wilson (1959b).

In sum, the known Tertiary faunas are essentially modern in character, and they offer no strong clues to the origin of the ants. In order to find fossils that link the primitive ants to the nonsocial wasps, myrmecologists were forced to look to the Mesozoic era. Their hopes were slow in being realized, because insect fossils from Mesozoic times are scarce, and this is particularly true for the Cretaceous period, the most recent and promising section of the Mesozoic era. In 1967 Wilson, Carpenter, and Brown (1967a,b) were fortunate enough to acquire the first ant remains of Cretaceous age. The species, *Sphecomyrma freyi,* and the new subfamily founded on it (Sphecomyrminae), were described from two well-preserved workers in New Jersey (United States) amber dating to the lowermost portion of the upper half of the Cretaceous age (see frontispiece). The age of the specimens was estimated to be about 100 million years, or nearly twice that of the Baltic amber ants. *Sphecomyrma* displays most of the characteristics earlier projected for the hypothetical ancestor of the ant from morphological studies of the rest of the Formicidae, as illustrated in Figure 4-7. It departs in only one important regard: its petiole is ant-like, while its mandibles are wasp-like—that is, very short and bearing only two teeth. We had expected just the reverse. We had guessed that the mandibles, which are the principal working tools of the worker ant and which vary to an extreme degree within the living Formicidae in ways correlated with food habits, would have been altered from the primitive wasp condition early in the evolution of the group. Also, there is a tribe of living ponerine ants, the Amblyoponini, in which the mandibles are modified but the petiole is not (see Figure 4-8).

This second piece of contrary evidence created a seemingly insoluble problem in the reconstruction of early ant phylogeny. The problem can be most clearly phrased in terms of the following abstract question. How is it possible for an ancestral form, in our case the proformicid wasp, to possess two primitive character states a and b, then to give rise to an annectant form (*Sphecomyrma*) with one derived state a', and one original state b, finally for the annectant form to give rise to a more advanced group (the Amblyoponini) with the *reversed* set of character states, a and b'? The only way out seems to be to assume that the Amblyoponini, which are the most primitive known

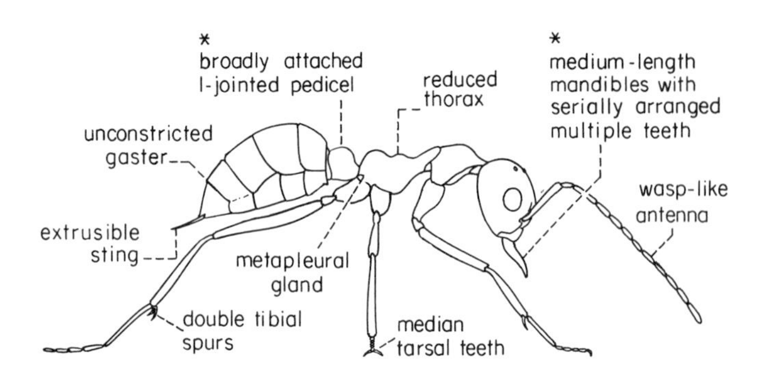

PREVIOUSLY HYPOTHESIZED ANCESTOR

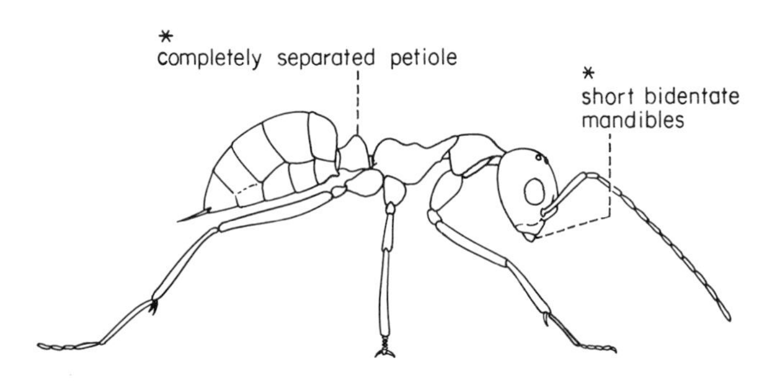

SPHECOMYRMA

FIGURE 4-7. Comparison of the worker of *Sphecomyrma freyi,* the first ant of Mesozoic age discovered, with the hypothetical ancestor projected from earlier morphological studies. The principal character states are indicated, and the trivial details in the hypothetical ancestor, which could not be guessed in advance, are made the same as in *Sphecomyrma* solely for convenience (from Wilson, Carpenter, and Brown, 1967a).

members of the poneroid complex, and *Sphecomyrma,* which is the probable antecedent of the myrmecioid complex, represent independent derivations from the protoformicid Tiphiidae. In other words, the ants may have originated twice in evolution. The possibilities are indicated in Figure 4-9. Of course, we are in no position to know about the degree of sociality of *Sphecomyrma,* so that the moment of origin of eusociality in the Mesozoic era and the number of times it originated independently remain matters for conjecture.

Whatever the source of the Amblyoponini as opposed to the remainder of the poneroid complex, *Sphecomyrma freyi* forms an excellent connecting step between the Myrmeciinae and the nonsocial wasps of the family Tiphiidae. The similarities between *Sphecomyrma* and the tiphiid genus *Methocha* (Figure 4-10) are particularly close. The conditions under which eusociality might have arisen in the early ants will be discussed at greater length later in this chapter.

The Taxonomy of Ants

The taxonomy of the world ant fauna is still relatively inadequate. There are, to begin with, few useful regional monographs. Creighton's (1950) review of the ants of North America north of Mexico is one of the best, and it has the added distinction of being the first major work to dispense with the clumsy and meaningless polynomials that plagued ant taxonomy for a hundred years. Creighton substituted a much simpler and more efficient system of binomials and trinomials based on modern population concepts (for example, *"Camponotus herculeanus pennsylvanicus* var. *whymperi"* was placed as a synonym under *Camponotus herculeanus*). Bernard (1968) has made a similar valuable revision of the ants of Europe. Other useful regional works include compendiums by Gallardo (1916–1932) and Arnold (1915–1926) on the ants of Argentina and South Africa, respectively; the pioneering but fragmentary work by Wheeler (1922) on the fauna of Africa south of the Sahara and the Malagasy Region*; the old but still useful review of the ants of India by Bingham (1903); and monographs on the ants of Puerto Rico by M. R. Smith (1936a), of New Zealand by Brown (1958), and of Polynesia as a whole by Wilson and Taylor (1967). The fauna of New Guinea and surrounding islands has been reviewed in part by Wilson (1959a and included references). Individual taxa revised in recent years include the primitive Australian genus *Myrmecia* (Clark, 1951; Brown, 1953a); the ponerine tribes Amblyoponini, Platythyreini, Ectatommini, and Typhlomyrmecini (Brown, 1960a, 1965, and included references); the genus *Ponera* (Taylor, 1967); the New World doryline army ants (Borgmeier, 1955); the doryline army ants of Asia and Australia (Wilson, 1964); the myrmicine harvesting ants of the genus *Pogonomyrmex* in North America (Cole,

*The massive work in which Wheeler's articles appear (*Bull. Amer. Mus. Nat. Hist.,* 45: 1–1139; 1922), along with contributions on the biology of African ants by I. W. Bailey, J. C. Bequaert, W. M. Mann, and F. Santschi, is lightly referred to by ant taxonomists as the Congo Bible.

1968); most sections of the large myrmicine tribe Dacetini (Brown, 1948, 1952, 1953c,d, 1959, 1964a, and contained references); the myrmicine tribes Solenopsidini (Ettershank, 1966) and Basicerotini (Brown and Kempf, 1960); the fungus-growing myrmicine genera *Atta* (Borgmeier, 1959), *Mycocepurus* (Kempf, 1963), and *Cyphomyrmex* (Kempf, 1965); the myrmicine tribe Cephalotini (Kempf, 1951, 1958); the Aneuretinae (Wilson *et al.,* 1956); the myrmicine genus *Crematogaster* of North America (Buren, 1958, 1968b); the dolichoderine genus *Monacis* (Kempf, 1959); and the formicine genera *Notoncus* (Brown, 1955a), *Lasius* (Wilson, 1955a), and *Acanthomyops* (Wing, 1968). Ant larvae have been systematically described by G. C. and Jeanette Wheeler (1951–1965 and included references). The generic characteristics of male ants in the North American fauna have been treated by M. R. Smith (1943). These studies, along with other, smaller revisions too numerous to list here, cover only a small part of the

FIGURE 4-8. A worker of the primitive ant *Amblyopone australis* stands over a pupa (enclosed in its cocoon) and an assortment of eggs and larvae. The broad posterior attachment of the petiole (a primitive trait) and heavy, elongate mandibles (an advanced trait) can be seen clearly in this individual. In the upper left and lower right corners of the photograph, larvae feed by thrusting their mandibles directly into the bodies of the insect prey (photograph courtesy of R. W. Taylor).

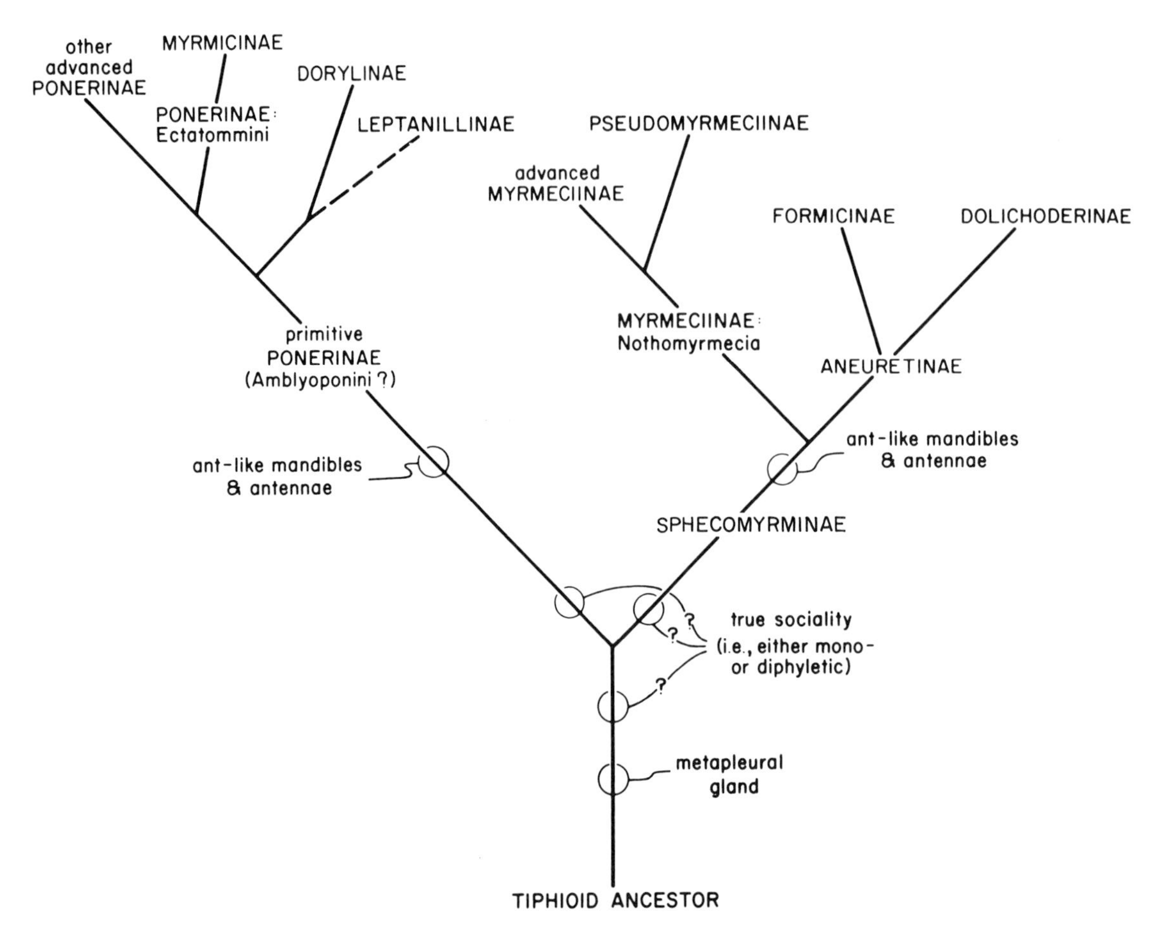

FIGURE 4-9. Phylogenetic diagram of the simple branching form, showing the inferred relationships of the subfamilies of ants. The possible dual origin of eusociality during the early evolution of the ants is indicated (from Wilson, Carpenter, and Brown, 1967a).

world fauna. The tropical groups are especially poorly known. A key to the genera of the world has been presented by Wheeler (1922), who based it in large part on Carlo Emery's earlier key (1902) and genus-level review in the *Genera Insectorum* (1910–1925), but it is cumbersome in use and has grown increasingly out of date. The cytotaxonomy of ants, which is a young but potentially very important subject, has been reviewed by Elisabeth Hauschteck (1963) and Crozier (1968). A perceptive and entertaining history of ant taxonomy has been written by Brown (1955b).

Culturing and Observational Techniques

The culture and study of ants in the laboratory is a simple operation. Colonies of many species can be maintained in nothing more than a bottle containing the natural nesting material (for example, soil or decaying wood) or even a thick pad of moist paper or cotton. Ants are generally very adaptable and will conduct most of their normal activities under conditions more favorable to the observer than to them. Ideally, however, the artificial nest should be constructed so that the dimensions and microclimate of the nest chambers closely simulate those in the wild, the brood chambers should be kept dark or in red light, and the colonies should have a constant food source in a foraging arena apart from the nest. Some species require special food. Many dacetine species, for example, accept only collembolans, while fungus-growing ants need to be provided with substratum suitable for culturing their special kinds of fungi. A variety of culturing methods adaptable to most kinds of ants are described in the popular books by Wheeler (1910) and Skaife (1961). Freeland (1958) invented an excellent vertical observation nest for *Myrmecia* and other very large ants. Wilson (1962a) designed a plastic nest which serves for the simultaneous observation of large colonies inside and outside the nest during foraging activity. A completely defined synthetic diet for ants has been invented by Ettershank (1967), while diets and several mass culturing techniques for various ant species have been reviewed by Carney (1970).

The Natural History of the Primitive Ants of the Genus *Myrmecia*

The "bull-dog ants" of the genus *Myrmecia* have proven to be exceptionally rewarding subjects for the study of social biology in several respects. They are among the largest ants, workers ranging in various species from 10 to 36 mm in length, and yet are easy to culture in the laboratory. They are also, next to *Nothomyrmecia* and perhaps *Amblyopone,* the most primitive of the living ants. The first encounter with foraging *Myrmecia* workers in the field in Australia is always a memorable experience for an entomologist. One gains the strange impression of a wingless wasp just on its way to becoming an ant: "In their incessant restless activity, in their extreme agility and rapidity of motion, in their keen vision and predominant dependence on that sense, in their aggressiveness and proneness to use the powerful sting upon slight provocation, the workers of many species of *Myrmecia* and *Promyrmecia* show more striking superficial resemblances to certain of the Myrmosidae or Mutillidae than they do to higher ants" (Haskins and Haskins, 1950a). According to Brown (1953a), *Myrmecia* and *Promyrmecia* should be joined on morphological grounds into a single genus. Thus enlarged, *Myrmecia* contains approximately 120 species, all of which are limited to Australia except for a single representative on the island of New Caledonia (*M. apicalis*).

Through the efforts of Wheeler (1933a), Clark (1934 and

FIGURE 4-10. Adults of the tiphiid wasp *Methocha fimbricornis* (*left,* female; *right,* male), a representative of the living genus of nonsocial wasps that bear the closest resemblance to *Sphecomyrma freyi* and through it to the more advanced ants of the myrmecioid complex. This species occurs in the Philippine Islands (from F. X. Williams, 1919).

contained references), Haskins and Haskins (1950a), and Freeland (1958), we have begun to assemble much information concerning the natural history of the genus. Unfortunately, no single species has been studied in any great depth, making it necessary to piece together the following account from scattered reports on several species.

The colonies of *Myrmecia* are moderate in size, containing from a few hundred to somewhat over a thousand workers. The nests are typically excavated in the soil and consist of a primarily vertical array of well-formed, large galleries and chambers that extend to a depth of 1-2 m. There is a single large entrance hole, often surrounded by a shallow, crater-like ring of excavated soil or a large mound. At least one species, *M. mjobergi,* nests in epiphytes in the Queensland rain forests.

The workers usually hunt singly over the ground and on low vegetation. They capture a wide variety of living insect prey, which they cut up and feed directly to the larvae. These ants are formidable predators, able, for example, to haul down and paralyze honeybee workers. They also collect nectar from flowers and extrafloral nectaries, which appears to be the main article in their diets when the nest is without larvae. In most species the queens are winged when they emerge from the pupae, whereas the workers are smaller and wingless. Intermediates between the two castes normally occur in some species, and occasionally the usual queen caste has been replaced either by ergatogynes with reduced thoraxes (*froggatti, midas, nobilis, rogeri,* and others) or by mixtures of ergatogynes and brachypterous queens (*rectidens, tarsata*). In *M. gulosa,* a species typical of the larger *Myrmecia* in many respects, the worker caste is differentiated into two subcastes. In a single large colony collected by the Haskins, the head width of the workers varied between 2.1 and 4.5 mm, this size variation being accompanied by slight allometric changes in the proportions of the body. The size-frequency distribution was strongly bimodal and overlapping. (Chapter 8 can be consulted for a fuller discussion of allometry and frequency distributions as qualities of ant polymorphism.) The larger workers do most or all of the foraging, while the smaller ones devote themselves principally to brood care. In some other species of *Myrmecia,* particularly the smaller species formerly grouped under *Promyrmecia,* the worker caste is monomorphic.

In most of the larger species of *Myrmecia* (for example, *pyriformis, nigriscapa, gulosa, forficata,* and *sanguinea*), the virgin queens and males leave their nests and emerge during a spectacular mass nuptial flight. The following account of this behavior in *M. sanguinea* was quoted by Wheeler (1916a) from the notes of W. W. Froggatt:

> On January 30th, after some very hot, stormy weather, while I was at Chevy Chase, near Armidale, N.S.W., I crossed the paddock and climbed to the top of Mt. Roul, an isolated, flat-topped, basaltic hill, which rises about 300 feet above the surrounding open, cleared country. The summit, about half an acre in extent, is covered with low "black-thorn" bushes (*Bursaria spinifera*). I saw no signs of bull-dog ant nests till I reached the summit. Then I was enveloped in a regular cloud of the great winged ants. They were out in thousands and thousands, resting on the rocks and grass. The air was full of them, but they were chiefly flying in great numbers about the bushes where the males were copulating with the females. As soon as a male (and there were hundreds of males to every female) captured a female on a bush, other males surrounded the couple till there was a struggling mass of ants forming a ball as large as one's fist. Then something seemed to give way, the ball would fall to the ground and the ants would scatter. As many as half a dozen of these balls would keep forming on every little bush and this went on throughout the morning. I was a bit frightened at first but the ants took no notice of me, as the males were all so eager in their endeavors to seize the females.

Similar mass flights have been reported in several other subfamilies of ants. Ground swarms and balling behavior in males occur in the formicine genus *Notoncus* (Brown, 1955a), and ground swarms without balling have been reported in some species of *Prenolepis* and *Formica* (Kannowski, 1963). Deviations from the basic pattern occur within *Myrmecia* also, but they evidently represent derived states associated with the loss of the power of flight in the virgin queens. In *M. tarsata,* for example, most of the female reproductives are much like workers and in fact seem to differ from true workers almost solely in their larger size. The males and ergatoid queens leave the nests together, with the males flying off into the air and the queens merely crawling away over the ground. During a subsequent period of wandering that may last for several weeks, the queens are found and fertilized by the males (Haskins and Haskins, 1950a).

After being inseminated, the queen excavates a well-formed cell in the soil under a log or stone and commences rearing the first brood of workers. In 1925 Clark made the important discovery, later confirmed and extended by Wheeler (1933a) and Haskins and Haskins (1950a), that the queens do not follow the typical "claustral" pattern of colony founding seen in higher ants. That is, they do not remain in the initial cell and nourish the young entirely from their own metabolized fat bodies and alary muscle tissues. Instead, they periodically emerge from the cells through an easily opened exit shaft and forage in the open for insect prey (see Figure 4-11). This "partially claustral" mode of colony foundation, which is now known to be shared with most of the Ponerinae, is regarded as a holdover from a more primitive form of progressive provisioning employed by the nonsocial tiphiid ancestors. The following chronicle of colony foundation by a queen of *M. forficata,* made in 1947 by Haskins and Haskins (1950a), is typical for the genus.

August 15.	The newly mated queen is placed in an observation nest. She excavates a typical chamber shortly thereafter and retires into it, emerging regularly at night to forage.
September 3.	Eight eggs are seen in the nest.
September 29.	A group of small, newly hatched larvae are observed feeding on a fragment of mealworm, which has been carried into the nest by the female.
October 4.	Four larvae, apparently about one-quarter grown, are now present.
October 28.	One small but perfect cocoon and two apparently half-grown larvae are present.
October 29.	The first brood now consists of a pupa in a cocoon, one larva, and one larva banked with earth and in the process of spinning.
October 31.	Three pupae in cocoons are present in the nest.
November 8.	Three pupae in cocoons are still present and have been removed to the driest portion of the chamber excavated by the queen, who guards them closely.
December 12.	A small but perfect worker has eclosed from the first cocoon.
December 16.	A second small worker has eclosed from a cocoon. (The third pupa failed to eclose).

As Haskins and Haskins stress, *Myrmecia* displays a mosaic of primitive and advanced traits in its social biology. In Table 4-2 I have classified much of the recorded behavioral traits according to this simple dichotomy. It must be added at once that this effort at a synthesis is no more than a set of phylogenetic hypotheses. The "higher ants" with which *Myrmecia* is compared are all the living subfamilies except the Myrmeciinae and Ponerinae. The last two taxa, which are the most primitive living subfamilies of the myrmecioid and poneroid complexes respectively, share some (but not all) of the primitive traits listed for *Myrmecia.*

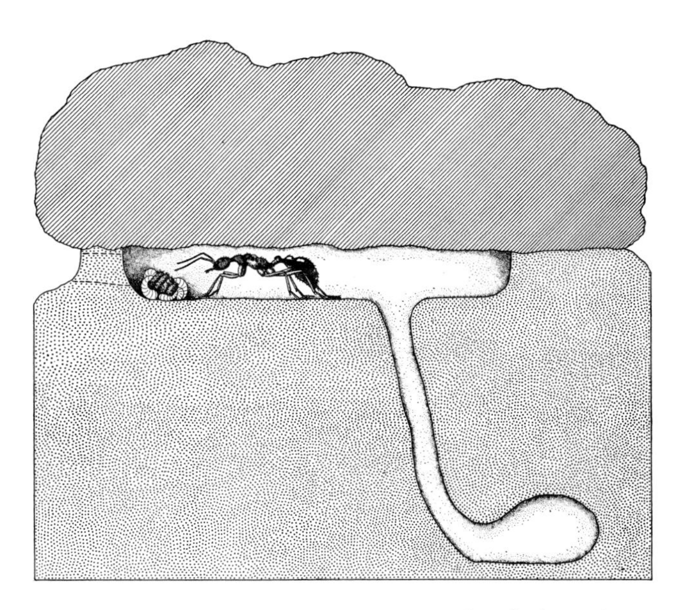

FIGURE 4-11. A queen of *Myrmecia regularis* is shown in the act of founding a colony. From time to time the queen removes the plug in the entrance gallery (outlined in dashed lines in this semidiagrammatic view) and leaves the nest to forage for insect food. This behavior is referred to as the "partially claustral" mode of colony foundation. The nest is also equipped with a lower escape chamber (from Wheeler, 1933a).

Myrmecia behavior is well advanced into the eusocial level in most essential features, yet marked by a residue of primitive traits which gives us an indistinct and tantalizing view of what the behavior of the sphecomyrmine and presphecomyrmine ancestors must have been like. We will return to this aspect of the subject later in the chapter.

The Natural History of a Typical "Higher" Ant, *Myrmica rubra*

I have selected *Myrmica rubra** to follow *Myrmecia* because it displays a life cycle typical in its essential aspects for the majority of the higher ant species or, to be more

* *M. rubra* was often erroneously referred to as *M. laevinodis* prior to Yarrow's (1955a) and Collingwood's (1958) revision of the British species of *Myrmica.* A related species, *M. ruginodis* (= *M. rubra* of some authors), has also been the subject of numerous studies in recent years. It is unfortunate that this important pair of species should have been involved in such a confusing nomenclatural tangle.

TABLE 4-2. Behavioral and other traits of *Myrmecia.*

Primitive traits

1. Multiple queens occur in many nests
2. The eggs are spherical and lie apart from one another on the nest floor
3. The larvae are fed directly with fresh insect fragments
4. The larvae are able to crawl short distances unaided
5. The adults are highly nectarivorous and collect insects mainly as food for the larvae
6. Transport of one adult by another is rare, awkward in execution, and not accompanied by tonic immobility on the part of the transportee
7. There is neither recruitment among workers to food sources nor any other apparent form of cooperation during foraging
8. Alarm communication is slow and inefficient; the nature of the signal is still unknown
9. Colony founding is only partially claustral
10. When deprived of workers, nest queens can revert to colony-founding behavior, including foraging above ground

Advanced traits also found in higher ants:

1. The queen and sterile worker castes are very distinct from each other, and intermediates are rare
2. Worker polymorphism occurs in many species, manifested as the coexistence of two well-defined worker subcastes
3. The colonies are moderately large and the nests regular and fairly elaborate in construction
4. Regurgitation occurs both among adults and between adults and larvae
5. Adults groom each other as well as the brood
6. Trophic eggs are laid by the workers and fed to other workers and the queen
7. The workers cover the larvae with soil just prior to pupation, thus aiding them in spinning cocoons; and they assist the newly eclosed adults in emerging from the cocoons
8. Nest odors exist and territorial behavior among colonies is well developed

specific, for most of the representatives of the Myrmicinae, Pseudomyrmecinae, Dolichoderinae, and Formicinae so far studied. Also, thanks to the large population sizes attained by many species of *Myrmica* in the North Temperate Zone, this genus has been one of the most intensively studied of all ant groups.

On morphological grounds *Myrmica* is a relatively primitive genus within the Myrmicinae. It is similar in many respects to *Hylomyrma,* an obscure genus comprised of several species found in the rain forests of the American tropics. *Hylomyrma* in turn has certain similarities to *Agroecomyrmex duisburgi* of the Baltic amber, the apparent connecting link between the ponerine tribe Ectatommini and the Myrmicinae.* *Myrmica* consists of about 30 species limited to the North Temperate Zone of Europe, Asia, and North America (Weber, 1947, 1948). It reaches its southern limits in the foothills of the Himalayas and the mountains of Mexico. A single species (*M. longispinosa*) is known from the Baltic amber. The great majority of the living species occur in Europe and Asia.

According to Donisthorpe (1915), Elisabeth Skwarra (1929), Gösswald (1932), and other authorities who have studied *Myrmica* closely in the field in Europe, *M. rubra* occurs primarily along woodland trails and borders and less frequently along hedgerows, in fields, and in gardens. Its nests are usually built under flat stones, but a few, perhaps ten or twenty percent of the total, are constructed in dead stumps or in the open soil in wet, grassy places. The workers attend many kinds of aphids for their honeydew, and at least some of the species are protected within the ants' nests. The ants also capture small insects, although this aspect of their behavior has not been carefully studied in the field. Winged forms occur in the nests from June to September, with the largest numbers accumulating just before the nuptial flights in late August or September. The flights are spectacular mass events. The males collect in swarms over trees, tall buildings, and other prominent landmarks, and the queens fly in among them to be mated. Donisthorpe recounts one such flight as seen from a uniquely British viewpoint by an earlier observer: "Farren-White in 1876 observed a swarm of ants near Stonehouse rising and falling over a small beech tree. The effect of those in the air—gyrating and meeting each other in their course, as seen against the deep blue sky—reminded him of the little dodder, with its tiny clustered blossoms and its network of ramifying scarlet threads, over the gorse or heather at Bournemouth. He noticed the swarm about thirty paces off, and it began to assume the appearance of curling smoke; at forty paces he could quite imagine the tree to be on fire. At fifty paces the smoke had nearly vanished into thin air."

After the nuptial flight, the queen sheds her wings and excavates a cell in the soil or rotting log which she seals off completely from the inside. During the following spring she lays a small batch of eggs, which may or may not give rise to workers that summer. If they do not reach

*Brown and Kempf (1967) have discovered another living genus from Costa Rica, *Tatuidris,* which is even closer to *Agroecomyrmex* in the structure of its pedicel. But this form is also distinguished by a bizarre—and obviously derived—mandibular structure.

the adult worker stage by October, they hibernate through the winter as larvae in the third, or final, larval instar (Brian, 1951a). Thus a queen must sometimes wait for nearly two years to obtain worker assistance. Yet, during all this time, she remains sealed in her initial nest chamber. She apparently obtains sustenance for herself and the first brood solely by the dissolution and metabolism of her own fat bodies and flight muscles in the manner of the queen of *Lasius niger,* so well documented by Janet (1907).

With the appearance of the first workers, the colony enters what Brian (1957a) has termed the "juvenile" period of its growth. This is a period of extremely slow increase. In seven years only about 300 workers are added. Then the colony growth spurts toward full maturity, adding about 600 workers in two more years. The latter period (the "adolescent" period) sees the first production of males but no virgin females. Finally, in the later, "mature" period, virgin queens are also produced. Since they and the males now leave the nest on nuptial flights, they add

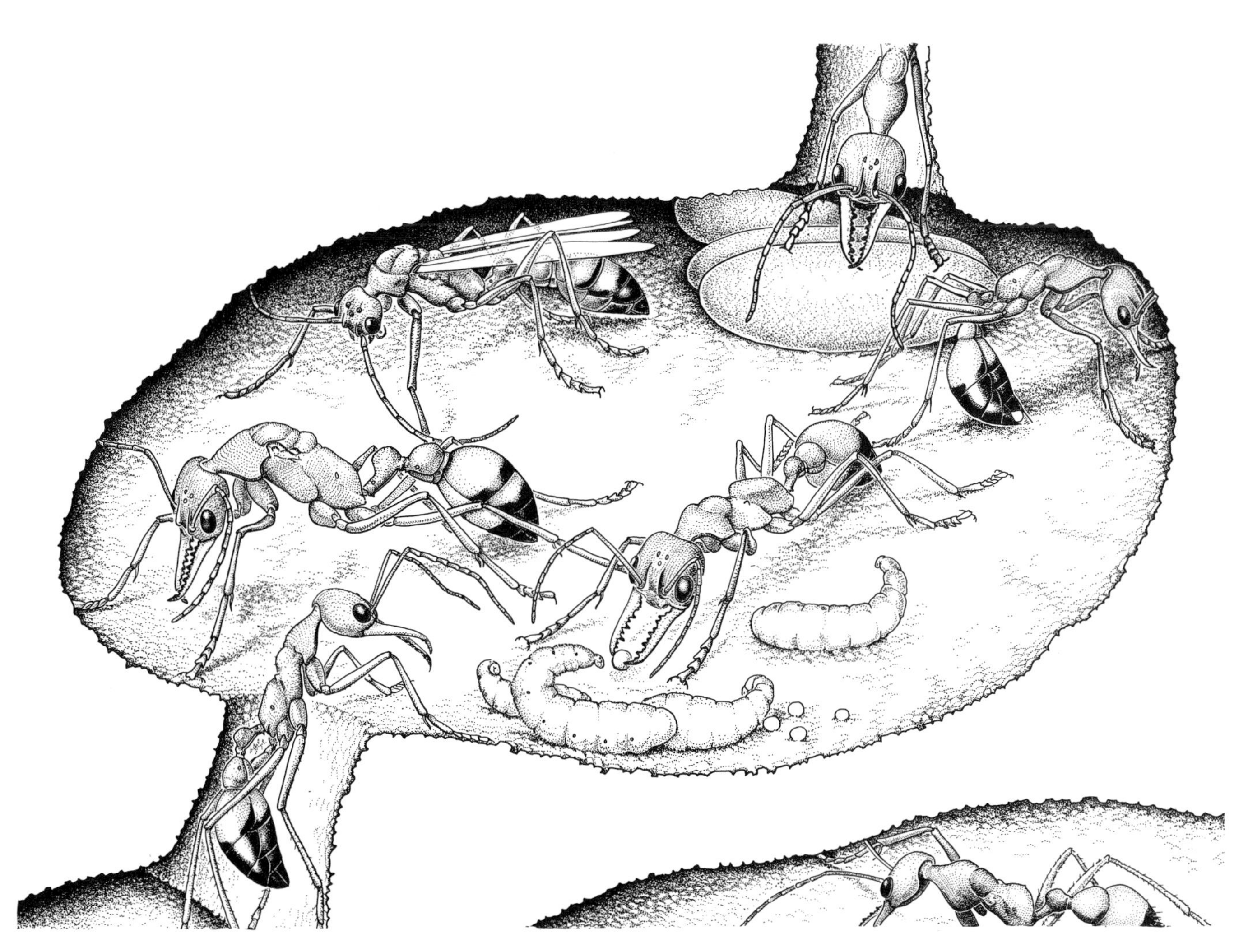

FIGURE 4-12. A view inside the earthen nest of a colony of the primitive ant *Myrmecia gulosa* (to the left is the mother queen and a male; to the right is a worker laying a trophic egg while another worker offers one of its own trophic eggs to a larva). Queen-laid eggs, which will be permitted to hatch, are scattered singly over the nest floor. To the rear of the chamber are three cocoons containing pupae of the ants (original drawing by Sarah Landry, based on Wheeler, 1933a; Haskins and Haskins, 1950a; and Freeland, 1958).

nothing to the growth of the colony. Consequently their production, together with limitations on the egg-laying capacity of the queen herself, puts a brake on further colony growth. Because maturity of the colony comes so late in the life of the founding queen, eight to ten years after the nuptial flight, the colony can produce queens and males for only a few more years before the queen dies. If she is not replaced by another queen through adoption, the colony will then die out altogether.

The colonies of most ant species, in all subfamilies except the Dorylinae and Cerapachyinae, appear to pass through life cycles basically similar to that of *Myrmica*. There is a great deal of variation in the duration of the various growth periods, but the scarcity of exact time measurements in the literature prevents any further generalization concerning the matter.

The seasons of the cold temperate zone impose a strict regime on the *Myrmica* colonies which is reflected in annual cycles in every aspect of their biology (Brian, 1957b). For example, two brood cycles (from egg to adult) run concurrently through the growing season. They are so phased that eggs and pupae occur together and larvae hatch from the eggs while the new workers are eclosing from the pupae. Dephasing experiments performed by Brian on laboratory colonies, in which this scheduling was altered by creating new mixtures of growth stages, failed to produce any evidence of intrinsic controls. Instead, the brood rhythm seems to be paced simply by the independent effects of temperature on the queen and on the brood. In the spring, the warming of the nest causes the queen to lay a burst of eggs at about the same time the hibernated larvae are preparing to pupate. As fall approaches, egg laying ceases and the larvae enter diapause in the third instar. Consequently, by the time cold weather finally inactivates the colony, all of the brood has halted development in the larval stage. The following spring the diapause is broken, the larvae proceed to complete their growth and pupate, the queen starts to lay a new batch of eggs, and the new annual cycle begins. The same cycle occurs in *M. ruginodis* (see Chapter 8), where its role in caste determination has been documented by Brian and his associates.

The Natural History of a Higher Ant with a Modified Life Cycle, *Monomorium pharaonis*

Pharaoh's ant is a prime example of a species that has abandoned the nuptial flight and territorial boundaries among colonies and adopted budding as the mode of colony multiplication. Similar modifications involving partial or total substitution of budding for the nuptial flights occur in a few other higher ants, for example, *Monomorium floricola,* the dolichoderines *Iridomyrmex humilis* and *Tapinoma melanocephalum,* and members of the *Formica exsecta* group (M. R. Smith, 1936b; Creighton, 1950).

Monomorium is a large, relatively specialized myrmicine genus that occurs throughout the tropical and warm temperate parts of the world. *M. pharaonis,* a native of Africa, has been inadvertently spread by human commerce to nearly every part of the world. In tropical climates it nests in plant cavities out of doors. In cold climates, however, it also thrives in greenhouses and in the walls of houses, where it is often a pest. In fact, the abandonment of the nuptial flight and of territorial boundaries has clearly preadapted this species for just such a "tramp" existence in close association with man.

Because of its status as a household pest in Great Britain, *M. pharaonis* was made the object of a special study by a team of entomologists at the University of St. Andrews under the direction of A. D. Peacock (Peacock, Hall, Smith, and Goodfellow, 1950; Peacock, Waterhouse, and Baxter, 1955). These investigators found that the Pharaoh's ants exist in Scotland as little groups scattered through the crevices and cavities of walls in houses. In one sample of seven groups collected at random from a single building, the worker population ranged from 150 to 2,000 and that of the wingless queens from 2 to 110. But since there are no distinct colony boundaries, all the individuals infesting such a building can be regarded as one large colony if the groups are at least joined by odor trails, in which case the "colony" can contain up to millions of workers and thousands of queens.

The workers are nearly omnivorous, feeding on small insects, greasy food particles of many kinds, and sugar. By means of odor trails, single scout workers are able to recruit large numbers of their nestmates to food finds in a matter of minutes.

"Colony" multiplication occurs when groups of workers carrying brood migrate to new nest sites. Although the proximate stimuli have not been identified, the swarming behavior is evidently triggered in some way by overcrowding of the workers in the nest chambers. As Peacock *et al.* (1950) express it, "Large and flourishing artificial colonies have been seen suddenly to develop a state of excitement in which the workers forsake the nest cells, feverishly swarm all over the inside of the container and remove the pre-adult stages to other sites . . . This behaviour to all appearances is the result of overcrowding, and sub-culturing restores quiet."

When a new site is discovered, the ants move to it en masse along odor trails laid by scout workers from the tips of their abdomens. All that is needed to start a new colony is a group of about 50 or more workers and about the same number of immature stages. If the immature stages are not taken along, the colonization attempt fails. Even more surprising, queens are not necessary for colonization. If the founding group is queenless, the workers rear queens and males from the available brood, and these individuals then mate within the nest.

Although winged queens and males are produced in small numbers the year round, nuptial flights apparently never occur. When a virgin queen is isolated with a group of workers, her offspring are all males until she is able to be fertilized—usually by one of her own sons.

The reproductive strategy adopted by *Monomorium pharaonis* has an obviously important consequence at the population level: the species has greatly restricted its degree of outbreeding and its power of active dispersal. Associated with this restriction, and possibly stemming as a direct consequence of it, territorial behavior has been reduced or lost. In exchange, the species has acquired the power to saturate and dominate the special, local habitats to which it is best suited.

The Natural History of the Fungus-growing Ants (Tribe Attini)

Members of the myrmicine tribe Attini share with macrotermitine termites and certain wood-boring beetles the sophisticated habit of culturing and eating fungi. The Attini are a morphologically distinctive group limited to the New World, and most of the 11 genera and approximately 200 species occur in the tropical portions of Mexico and Central and South America. It is conceivable that fungus growing orginated only once in a single ancestral attine living in South America during that continent's long period of geological isolation from late Mesozoic times to approximately four million years ago. The virtual lack of ant fossils from South America has so far prevented us from testing this conjecture or making any confident inferences whatever concerning the early biogeography of the Attini. In Africa, southern Asia, and other parts of the Old World tropics, the Attini are replaced by fungus-growing termites (Macrotermitinae), that do not occur in the New World. We cannot be sure whether this complementary global pattern is due to a mutual preemption involving competitive exclusion of one group by another or whether it is simply one more accidental outcome reflecting the extreme rarity of the evolutionary origin of fungus gardening. I am inclined to believe that the latter is the case, meaning that if attines were to be introduced today into the range of the macrotermitines, or vice versa, the two kinds of insects could coexist with little interference. This is possible because attines utilize insect excrement and fresh plant material, for the most part, while the macrotermitines use dead plant material. Also, ants forage above ground, often even in trees, while termites are primarily subterranean.

The Attini are an enormously successful group where they exist. One species, *Trachymyrmex septentrionalis,* ranges north to the Pine Barrens of New Jersey, while in the opposite direction several species of *Acromyrmex* penetrate to the cold temperate deserts of central Argentina. In the vast subtropical and tropical zones that lie between, attines are among the dominant ants. Many of the species gather pieces of fresh leaves and flowers to nourish the fungus gardens, and *Atta* and *Acromyrmex* rely on this source exclusively. Since they attack most kinds of vegetation, including crop plants, they are serious economic pests. The species of *Atta* in particular are among the scourges of tropical agriculture. They are familiar to local inhabitants as the *wiwi* in Nicaragua and British Honduras, the *bibijagua* in Cuba, the *hormiga arriera* in Mexico, the *bachac* in Trinidad, the *bachaco* in Venezuela, the *saúva* in Brazil, the *cushi* in Guyana, the *coqui* in Peru, and the leaf-cutting or parasol ant in most English-speaking countries, the last name alluding to the fact that an *Atta* worker holding a leaf fragment over its head gives the impression of carrying a parasol. The problems of agriculture in *Atta* country have been humorously epitomized in the following anecdote by V. Wolgang von Hagen, in connection with his attempt to grow a vegetable garden in British Honduras:

> My Indian servants, dusky, kinky-haired Miskito men, lamented all this work. It was useless, quoth a toothless elder, to plant anything but bananas or manioc, as the *Wiwis* were sure to cut off all the leaves. Without the slightest encouragement the Miskito Indians would launch forth on the tales of the ravages of the *Wiwi Laca,* but unswayed by the illustrations, like Pangloss I could only remark that all this was very well but let us cultivate our garden. In two weeks the carrots, the cabbages, the turnips were doing well. The carrots had unfurled their

fernlike tops, the cabbage grew as if by magic. From our small palm-thatched house my wife and I cast admiring eyes over our jungle garden. Our mind called forth dishes of steaming vegetables to replace dehydrated greens and the inevitable beans and yucca. Even the toothless Miskito elder came by and admitted that white man's energy had overcome the lethargy of the Indian. Then the catastrophe fell upon us. We arose one morning and found our garden defoliated: every cabbage leaf was stripped, the naked stem was the only thing above the ground. Of the carrots nothing was seen. In the center of the garden, rising a foot in height, was a conical peak of earth, and about it were dry bits of earth, freshly excavated. Into a hole in the mound, ants, moving in quickened step, were carrying bits of our cabbage, tops of the carrots, the beans—in fact our entire garden was going down that hole. I could see the grinning face of the toothless Miskito Indian. *The Wiwis had come.*

What happens to the vegetation after the wiwis have carried it down their holes is a fascinating story that has been worked out through many decades of research. Bates, in his book *The Naturalist on the River Amazons* (1863), suggested that the ants use the leaves "to thatch the domes which cover the entrances to their subterranean dwellings,

FIGURE 4-13. A young colony of *Atta sexdens* on its fungus garden. Most of the workers belong to the minor and small media subcastes and are dwarfed by the huge queen. A single large media can be seen just above the head of the queen (from Weber, 1966).

thereby protecting from the deluging rains the young broods in the nests beneath." Other early observers believed that the leaves are eaten or used to maintain a constant nest temperature by heat of fermentation. Thomas Belt was the first to surmise the far stranger truth. In *The Naturalist in Nicaragua* (1874) he described the garden chambers deep within the *Atta* nests as being "always about three parts filled with a speckled brown, flocculent, spongy-looking mass of a light and loosely connected substance. Throughout these masses were numerous ants belonging to the smallest division of the workers, and which do not engage in leaf-carrying. Along with them were pupae and larvae, not gathered together, but dispersed, apparently irregularly, throughout the flocculent mass. This mass, which I have called the ant-food, proved, on examination, to be composed of minutely subdivided pieces of leaves, withered to a brown colour, and overgrown and lightly connected together by a minute white fungus that ramified in every direction throughout it . . . That they do not eat the leaves themselves I convinced myself; for I found near the tenanted chambers deserted ones filled with the refuse particles of leaves that had been exhausted as manure for the fungus, and were now left, and served as food for larvae of *Staphylinidae* and other beetles."

It was left to Alfred Moeller (1893) to observe for the first time the actual eating of the fungi. He found that the tips of the hyphae produce peculiar spherical or ellipsoidal swellings (Figure 4-14) which are plucked and eaten. Moeller called these objects "heads of Kohlrabi" because of their fancied resemblance to the vegetable. Later Wheeler relabeled them gongylidia (singular: gongylidium), and this name has stuck.* The gongylidial clusters of *Atta*, averaging about half a millimeter in diameter, are also fed to the larvae. As fresh leaves and other plant cuttings are brought into the nest, they are subjected to a process of degradation before being inserted into the garden substratum. First the ants lick and cut them into pieces 1–2 mm in diameter. Then they chew the fragments along the edges until the pieces become wet and pulpy, sometimes adding a droplet of clear anal liquid to the surface. Then, using side to side movements of the fore tarsi, they carefully insert the fragments into the sub-

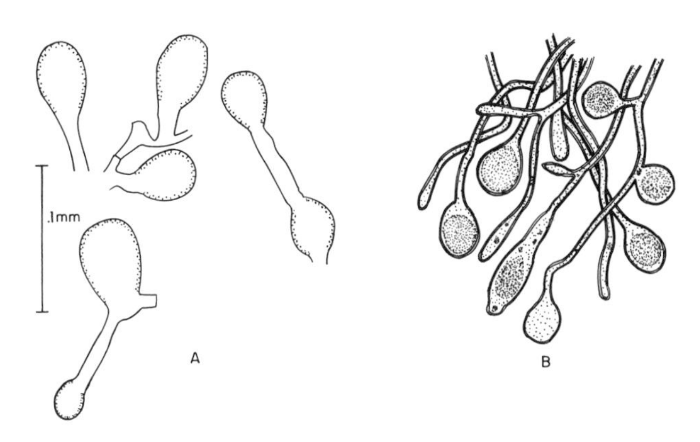

FIGURE 4-14. Hyphae with inflated tips ("gongylidia") of the fungus eaten by *A, Atta colombica,* and *B, Trachymyrmex jamaicensis.* The gongylidia are each 30–50 μ in diameter (redrawn from Weber, 1966).

stratum. Finally, as Weber (1956) has more recently discovered, the ants pluck tufts of mycelia from other parts of the garden and plant them on newly formed portions of the substratum. A newly inserted single leaf section 1 mm in diameter receives up to ten such tufts in five minutes. The transplanted mycelia grow rapidly, as much as 13 μ in length per hour. Within 24 hours they cover most of the substratal surface.

After Moeller's pioneering observations, many entomologists set out to trace the life cycle and biology of *Atta* in detail. Forel, Goeldi, Huber, von Ihering, and Wheeler, all of whose publications are exhaustively reviewed in the classic 1907 study of the North American Attini by Wheeler, and, in more recent years, Autuori, Bitancourt, Bonetto, Borgmeier, Eidmann, Geijskes, Gonçalves, Jacoby, Kerr, Moser, Stahel, and Weber, whose work has been reviewed by Weber (1966), each made significant contributions. Yet, curiously, the single outstanding problem of attine biology, the identity and biological qualities of the symbiotic fungus, has received relatively little attention and is still far from being solved. The principal difficulty has been the reluctance—indeed, the near inability—of the fungus to form sporophores, the elaborate fruiting structures required for taxonomic diagnosis. Evidently the ants do not permit the fungi to form the mushrooms or other spore-bearing bodies under natural conditions. Instead the ants feed exclusively on the special gongylidial tips of the elementary mycelial clusters, a preference that has evidently resulted in the loss of the ability on the part of the fungus to produce sporophores.

*A group of gongylidia, to complete the modern terminology, is sometimes referred to as a staphyla (plural: staphylae), while a piece of the peculiar morel-like fungus of *Cyphomyrmex rimosus* is called a bromatium.

Reciprocally, the fungi utilize the ants for transport and do not have to depend on windborne spores to transfer themselves from nest to nest. Although Moeller did not clarify this problem in *Atta,* he was lucky enough to discover sporophores growing from abandoned *Acromyrmex* nests on four separate occasions. These proved to be agaracine mushrooms, wine-red in color, which Moeller formally named *Rozites gongylophora.* Mycologists have since confirmed their placement in the Agaracaceae, but transferred the species *gongylophora* to the genus *Leucocoprinus.* Subsequent attempts by entomologists to locate sporophores in abandoned attine nests and to culture them in the laboratory from the gardens of various attine genera have only rarely succeeded. Our meager accumulated information for all the Attini is summarized in Table 4-3. The most notable single advance has been Weber's (1957a) use of a medium of sterile oats to rear *Lepiota* sporophores from mycelia originating from a *Cyphomyrmex costatus* garden. If future mycologists ever succeed in isolating a plant hormone that enhances sporophore formation in fungi, dramatic further progress can be expected in this field.

TABLE 4-3. Fungus-growing ants, the material used as substrates in their gardens, and the symbiotic fungi they culture (based principally on Wheeler, 1907, and Weber, 1957a, 1966).

Species of ant	Substratum	Fungus
Cyphomyrmex costatus	Corpses of other arthropods, chiefly ants; insect feces; pieces of fruit; probably other materials	*Lepiota* sp. (Basidiomycetes: Agaricaceae)
Cyphomyrmex rimosus	Caterpillar feces	*Tyridomyces formicarum,* possibly in the genus *Daldinia* or *Xylaria* (Fungi Imperfecti)
Myrmicocrypta buenzlii	Dead vegetable matter; insect corpses	*Lepiota* sp. (Basidiomycetes: Agaricaceae)
Apterostigma mayri	Insect feces; fragments of dead wood	*Auricularia* sp. (Basidiomycetes: Auriculariaceae)
Acromyrmex disciger	Fresh leaves, stems, and flowers	*Leucocoprinus gongylophora,* possibly in the genus *Leucoagaricus* or *Agaricus* (Basidiomycetes: Agaricaceae)
Atta spp.	Fresh leaves, stems, and flowers	Possibly *Leucocoprinus gongylophora* (Basidiomycetes: Agaricaceae)

A second major, largely unsolved problem is the method by which the attines are able to maintain pure cultures. When the ants are removed from their nests in the field, the deserted gardens are quickly overrun by alien fungi belonging to many species. Moeller speculated that ants mechanically weed out alien spores and hyphae. Weber (1957c) confirmed this hypothesis by direct observation. He found that when sporulating alien fungi develop in or near the garden the ants retract all of the lower mouthparts and pull out the alien hyphae with their mandibles. During this operation they pause frequently to clean their fore tarsi and antennae. Weber (1955, 1966) has also postulated that ants utilize fungistatic and bacteristatic substances originating in the anal and salivary secretions. As attractive as this hypothesis seems, it was not confirmed in the careful chemical studies of *Atta colombica* by Martin *et al.* (1969; and personal communication). Not only did these investigators fail to find any trace of fungistatic activities in their fractions, they discovered that the reverse is true: amino acids present in the anal fluid actually promote fungal growth. The fungus cultured by *A. colombica* lacks the full complement of enzymes necessary for it to grow well on substrates in which the nitrogen is present predominantly as protein. This metabolic limitation means that the fungus will be at a particularly strong competitive disadvantage when required to live on such substrates as fresh leaves. The fecal material of the ant compensates for the metabolic deficiences of the fungus, because it contains proteolytic enzymes. Furthermore, the nitrogenous excretory products in the fecal fluid, allantoin, allantoic acid, ammonia, and twenty-one amino acids, provide the fungus with a nutrient supplement that accelerates its initial growth rate. The application of fecal material, together with the weeding activities of the ants, seems to enhance the competitive ability of the symbiotic fungus to the extent that it can exclude other species. The mutualistic association of *A. colombica* and its fungus is thus based on metabolic integration. The ants contribute their protein-degrading ability to the fungus, and the fungus indirectly contributes its cellulose-degrading ability to the ants. Martin and his associates have discovered proteinases in the feces of other species of attines examined, namely in two other species of *Atta,* two of *Acromyrmex,* one of *Trachymyrmex,* and one of *Sericomyrmex.* But fecal proteinases are absent or present only in trace amounts in the nonattine genera examined, including *Myrmica, Aphaenogaster, Formica, Lasius,* and *Polyergus.*

Other evidence concerning the existence of fungistatic secretions in ants has been conflicting. Nancy Lind (personal communication), using a sensitive technique that measures the effect of the contents of single glands on hyphal growth, detected slight fungistatic activity in the venom of the fire ant *Solenopsis saevissima* but none in other glandular products of this species and no trace at all in the secretions of the harvesting ant *Pogonomyrmex badius.* On the other hand, Maschwitz, Koob, and Schildknecht (1970) reported strong fungistatic and antibacterial activity in the metapleural gland secretions of *Atta sexdens* and *Manica rubida,* a nonattine species. The secretion is strongly acid, and the single component identified so far (in *A. sexdens*) is phenylacetic acid, of which 1.4 μg is present on the average per ant. Maschwitz and his co-workers drew the secretory material directly from the metapleural gland bullae of many ants, which is clearly a more effective collecting technique than Lind's method of crushing single glands. This difference could explain why Lind's results were negative, but no explanation is available to explain the contrary negative results reported by Martin and his co-workers.

Fourteen valid species of *Atta* have been described to the present time (Borgmeier, 1959). Insofar as they have been studied, their life histories all appear basically similar. The nuptial flights of some species, such as the infamous *sexdens* of South America, take place in the afternoon, while *texana* of the southern United States and a few others hold their flights at night (Autuori, 1956; Moser, 1967). Because the ponderous females work their way high into the air before the males approach them, actual matings have not been observed. Nevertheless, Kerr (1962a), by counting sperm from the spermathecae of four newly mated *sexdens* queens with the aid of a hemocytometer, was able to show that each individual is inseminated by at least three to eight males. The actual estimated numbers of sperm varied among the queens he examined from 206 to 320 millions, seemingly more than enough to last an individual the ten or more years speculated to be the normal life span of an *Atta* queen.

In 1898 von Ihering made the important discovery of how the fungus is transferred from nest to nest. Prior to departing on the nuptial flight the *Atta sexdens* queen packs a small wad of mycelia into her infrabuccal chamber, a cavity located (in all ants including *Atta*) beneath the opening of the esophagus just to the rear of the base of the labium. Following the nuptial flight, which in Brazil may occur anytime from the end of October to the middle of December, the queen casts off her wings and quickly excavates a little nest in the soil. When finished, the nest consists of a narrow entrance gallery, 12–15 mm in diameter, which descends 20–30 cm to a single room 6 cm long and somewhat less in height. Onto the floor of this room, according to Jakob Huber (1905) and Autuori (1956), the queen now spits out the mycelial wad. By the third day fresh mycelia have begun to grow rapidly in all directions, and the queen has laid the first 3 to 6 eggs. In the beginning the eggs and little fungus garden are kept apart, but by the end of the second week, when more than 20 eggs are present and the fungal mass is ten times its original size, the two are brought together. At the end of the first month the brood, now consisting of eggs, larvae, and possibly pupae as well, is embedded in the center of a mat of proliferating fungi. The first adult workers emerge sometime between 40 and 60 days. During all this time the queen cultivates the little fungus garden herself. At intervals of an hour or so she tears out a small fragment of the garden, bends her abdomen forward between her legs, touches the fragment to the tip of the abdomen, and deposits a clear yellowish or brownish droplet of fecal liquid onto it (Figure 4-15). Then she carefully places the mycelial fragment back into the garden. Although the *A. sexdens* queen does not sacrifice her own eggs as a culture medium, she does consume 90 percent of the eggs herself, and, when the larvae first hatch, they are fed with eggs thrust directly into their mouths. The queen apparently never consumes any of the growing fungi during the rearing of the first brood. Instead, she subsists entirely on her own catabolizing fat body and wing muscles. Soon after the first workers appear, they begin to feed themselves on the gongylidia. They also manure the fungal garden with their fecal emissions and feed their sister larvae with eggs laid by the mother queen. The eggs given to the larvae are larger than those permitted to hatch; a histological study by Bazire-Benazet (1957) has shown that they are in fact "omelets" formed in the oviducts by the fusion of two or more distinct but ill-formed eggs. After about a week the new workers dig their way up

FIGURE 4-15. Colony founding in *Atta:* (*A*) a queen in her first chamber with the beginning fungus garden; (*B*) the queen manures the garden by pulling a hyphal clump free and applying an anal droplet to it; (*C*) three stages in the concurrent development of the fungus garden and first brood (original drawing by Turid Hölldobler; based on Jakob Huber, 1905, and Autuori, 1956).

through the clogged entrance canal and start foraging on the ground in the immediate vicinity of the nest. Bits of leaves are brought in, chewed into pulp, and kneaded into the fungus garden. About this time the queen ceases attending both brood and garden. She turns into a virtual egg-laying machine, in which state she remains for the rest of her life. Now for the first time the workers begin to collect gongylidia from the fungal mass and to feed them directly to the larvae.

The growth of the colony is very slow during the first year. During the second and third years it accelerates quickly and then tapers off as the colony starts producing winged males and queens. Using data provided by Autuori, Bitancourt (1941) demonstrated that the growth of an *A. sexdens* colony, if measured as the increasing number of nest entrances, closely fits the classic formula of logistic growth. This means simply that the rate of growth can be expressed as an elementary function of the population size times the difference between the population size at the given moment and the size finally reached by the colony (see Chapter 21). The ultimate size reached by the *Atta* nests is enormous. Autuori's nest contained slightly over 1,000 entrance holes at the end of the third year. Another three-year-old nest excavated by Autuori (cited in Weber, 1966) contained 1,027 chambers, of which 390 were occupied by fungus gardens and ants. Although in only its first year of production of sexual forms, the colony had generated no less than 38,481 males and 5,339 virgin queens. Still another *A. sexdens* nest, 77 months old, contained 1,920 chambers of which 248 were occupied by fungus gardens and ants. The loose soil that had been brought out and piled on the ground by the ants during the excavation of their nest was shoveled off and measured. It occupied 22.72 m^3 and weighed approximately 40,000 kg. Autuori also estimated that during the short life of the colony the workers had gathered no less than 5,892 kg of leaves to cultivate their fungus gardens!

As these data suggest, the populations of old colonies of *Atta* reach enormous size. Martin calculated that one colony of *A. colombica* which he collected in Central America for chemical analysis contained between 1 and 2.5 million workers. The nests of mature colonies are also structures of extraordinary expanse and complexity, as documented in the studies of Eidmann (1935), Stahel and Geijskes (1939), Jacoby (1937, 1944), and Moser (1963). In well-drained soil the deepest galleries usually penetrate to more than 3 m below the surface, and in some cases they descend to more than 6 m. Their excavation for scientific study requires teams of laborers or, as in Moser's work on *A. texana* in Louisiana, the use of a bulldozer. Stahel and Geijskes systematically observed the movement of small puffs of smoke released over various of the nest entrances; they were thereby able to demonstrate the existence of a primitive ventilation system in the intact nests. Air, it was found, tends to pour into those nest openings located near the nest perimeter and to pass up out of the openings located closer to the nest center. The intake and exhaust openings are about equally numerous. A third kind of opening, through which no movement of air can be detected, is even more common. There is a simple enough explanation for this pattern. Air is heated by metabolism more rapidly in the central zone of the nest, where the fungus gardens and ants are concentrated, and it therefore tends to rise through the central galleries. The movement in turn draws air from the remaining galleries, which are located in the peripheral zones. The "neutral" nest openings probably lead to blocked galleries or gallery systems with relatively few ants and fungus gardens. Thus the construction of large numbers of nest entrances in the *Atta* nests—a feature shared with only a few other kinds of ants—appears to be an adaptation to facilitate ventilation through the exceptionally large biomasses of the leaf-cutter colonies.

The phylogenetic origin of the Attini remains a source of bafflement in spite of decades of speculation on the subject. One authorative opinion is still that of Emery (1895), who, on morphological evidence, placed the Attini near *Ochetomyrmex* and *Wasmannia.* These taxa, together with the aberrant genus *Blepharidatta,* comprise the tribe Ochetomyrmecini (Brown, 1953b). The ochetomyrmecines are exclusively Neotropical, which is at least consistent with the hypothesis of some kind of evolutionary link to the Attini. The overall morphological resemblance between the two tribes is not at all close, however, and in fact the Attini stand well apart from all other ants in their morphology. Forel (1902) offered the contrary opinion that the Attini stemmed from the Dacetini. But recent work on both the phylogeny and natural history of this tribe has contradicted Forel's basic assumptions. Brown and Wilson (1959a) found that all major groups of dacetines are strictly predaceous, and all but the most primitive specialize on collembolans and other soft-bodied arthropods. Furthermore, the most primitive living genus (*Daceton*) is morphologically very distinct from attines and is completely arboreal in its nesting habits, unlike all known attines.

So long as the evolutionary origin of the Attini remains a mystery, we cannot hope to gain any convincing insight into the evolutionary beginnings of fungus gardening. At best we can contemplate two competing hypotheses. The first, due to von Ihering (1894), proposes that attines originated from harvesting ants with slovenly habits: "We know quite a number of ants, like the species of *Pheidole, Pogonomyrmex* and furthermore species of *Aphaenogaster* and even of *Lasius,* which carry in grains and seeds to be stored as food. Such grain carried in while still unripe, would necessarily mould and the ants feeding upon it would eat portions of the fungus. In doing this they might easily come to prefer the fungi to the seeds. If *Atta lundi* still garners grass seeds and in even greater than the natural proportion to the grass blades, this can only be regarded as a custom which has survived from a previous cultural stage." In opposition, Forel (1902) has suggested that the ancestral attines lived in rotting wood and gradually acquired the habit of eating the fungi they chanced to find growing on insect excrement left behind by wood-boring insects. A slight variant of this idea has been offered by Weber (1956), who believes that the ants might have begun feeding on fungi which grew from their own feces.

The notion that insect excrement served as the original, fortuitous substrate of the symbiotic fungi receives some support from our present understanding of evolution *within* the Attini. In Table 4-4 I have listed the known genera of attines, together with those morphological and behavioral characteristics that show the greatest variation among the genera. The evolutionary trends reflected by this variation seem to be consistent with the idea, held by most students of the Attini since the time of Emery and Forel, that *Cyphomyrmex* is primitive, *Atta* is advanced, and the remaining genera occupy positions of varying degrees of intermediacy. Of course such a vertical array is bound to be an oversimplification since the evolution of the Attini, like that of almost all other large animal groups whose histories are better known from the fossil record, almost certainly unfolded in a more complex, dendritic pattern. But the principal evolutionary trends do seem clear enough when considered separately, and they are at least loosely interconsistent. There is a gradual increase in body size and, in a few of the largest species, the appearance of well-marked worker polymorphism. The body develops certain unusual anatomical features such as tuberculation of the body surface, unusual hair structure, and cordate head shape. The mature colony size increases from small (that is, a few tens or hundreds of individuals) through medium (hundreds or thousands) to large (tens of thousands or more), with a corresponding growth in the size and complexity of the nest structure.

Now if these trends do reflect a true evolutionary history, it is reasonable to suppose that feeding behavior also evolved in roughly the same direction, namely from *Cyphomyrmex* to *Atta* and the other, "higher" attine genera. And if that much is granted, we can regard the use of insect feces as the culturing medium to be the primitive trait and the use of fresh vegetation to be the derived trait. It is therefore likely that a closer examination of the biology of *Cyphomyrmex,* along with that of the other presumably primitive attines and perhaps also of the little-known tribe Ochetomyrmecini, will offer some chance of shedding light on the origin of the Attini and the fungus-culturing habit.

The Natural History of Ant Plants and Their Ants

Direct relations between ants and higher plants are diverse and often intricate. *Atta* and other advanced genera of fungus-growing ants depend, as we have just seen, entirely on fresh vegetation to provide the substrate for their fungus gardens. Other ant species utilize seeds for food and are, in turn, important agents in the dispersal of plants (Sernander, 1906; Bequaert, 1922; Tevis, 1958). Some of these, in particular the species of *Goniomma, Messor, Oxyopomyrmex, Pogonomyrmex,* and *Veromessor,* as well as a few of the species of *Melophorus, Meranoplus, Monomorium, Pheidole, Pheidologeton,* and *Tetramorium,* depend principally or entirely on seeds, which they store in their earthen nests in special granary cells. For this reason they are referred to generically as "harvesting ants." These specialists are found mostly in the warmer deserts and semideserts around the world, where they are frequently among the most numerous elements of the insect faunas. The provident ants referred to by Solomon and by classical writers such as Hesiod, Aesop, Aelian, Plutarch, Orus Apollo, Plautus, Horace, Virgil, Ovid, and Pliny, and celebrated in later, derivative writings up to the time of Jeremia Wilde (1615), almost certainly belonged to various *Messor* (*arenarius, barbarus, capitata,*

TABLE 4-4. Characteristics of attine genera believed to exhibit consistent evolutionary trends within the group. The genera are listed in order of their presumed approximate phylogenetic position, with the first genus, *Cyphomyrmex*, being the most primitive (based mostly on Wheeler, 1907a; Weber, 1941, 1946, 1966, and personal communication).

Genus	Morphology	Nest structure	Mature colony size	Garden substrate	Fungus group cultivated
Cyphomyrmex rimosus	Monomorphic; squamiform, appressed hairs; large, widely spaced frontal lobes; smooth body surface; small size	Irregular cavity in soil or rotting wood	Small to medium	Insect feces	Fungi imperfecti
Cyphomyrmex, other species	Monomorphic; hairs simple and sparse; large, widely spaced frontal lobes; smooth to tuberculate body; small size	One symmetrical cell; usually in soil	Small	Insect feces, insect corpses, pieces of fruit	Basidiomycetes
Mycetophylax	Monomorphic; hairs sparse; smooth body surface; frontal lobes of medium size and spacing; small size	One or two symmetrical cells in soil	Small	Dead grass	?
Mycocepurus	Monomorphic; hairs sparse; spinose; frontal lobes approximated and small; small size	One symmetrical cell in soil	Small	Insect feces	?
Myrmicocrypta	Monomorphic; tuberculate thorax bearing squamiform hairs; frontal lobes approximated and small; small size	One large, symmetrical cell in the soil or rotting wood	Medium	Vegetable matter, insect corpses	Basidiomycetes
Apterostigma	Monomorphic; abundant flexuous hairs; smooth body surface; small to medium size	One to several gardens surrounded (in some species) by a very thin, mycelial shroud, built in the open under logs or in cavities under logs, loose bark, or stones	Small	Insect feces and dead, woody matter	Basidiomycetes
Sericomyrmex	Monomorphic; abundant flexuous hairs; body surface tuberculate; head cordate; medium size	One to several symmetrical cells in the soil	Medium	Fruit and possibly dead vegetable matter	Probably Basidiomycetes
Mycetosoritis	Monomorphic; body surface tuberculate with moderately abundant, curved hairs; small size	Several symmetrical cells arranged vertically in the soil	Small	Dead vegetable matter	?
Trachymyrmex	Monomorphic to slightly polymorphic; body surface tuberculate with stiff, hooked hairs; small to medium size	Several symmetrical cells usually arranged vertically in the soil	Small to medium	Insect feces, flower parts, dead vegetable matter	Probably Basidiomycetes

TABLE 4-4 (*continued*).

Genus	Morphology	Nest structure	Mature colony size	Garden substrate	Fungus group cultivated
Acromyrmex	Polymorphic; body surface tuberculate with stiff hairs; occipital lobes developed; large size; strong queen-worker size difference	Complex earthen nests with one very large or many chambers	Large	Fresh leaves, stems, and flowers	Basidiomycetes
Atta	Strongly polymorphic; body surface partly tuberculate; large size; strong queen-worker size difference	Complex earthen nests with many chambers	Large	Fresh leaves, stems, and flowers	Basidiomycetes
Pseudoatta	(A parasitic workerless genus possibly derived from *Acromyrmex*)	—	—	—	—

minor, sancta, semirufus, sordidus, structor), which are the dominant harvesting ants of the Mediterranean region.

The subsequent history of this subject includes a curiously perverse episode. In his book, the first in the modern era devoted to ants, Wilde uncritically accepted the stories of the ancient authors. But the Reverend William Gould (1747), who published the first monograph based on his own observations, could find no evidence in England that "Ants have Magazines or Granaries of Corn, and lay up a Stock of Provisions for the Winter," and he questioned the old stories. For the next hundred years subsequent researchers, most of whom also lived in northern Europe outside the range of harvesting ants, shared Gould's doubts. It was not until studies were extended to the dry tropics and warm temperate zones, particularly by Sykes (1835) and Jerdon (1854) in India, that the existence of harvesting ants was confirmed. Thus for once the tales of the ancients proved right and modern European science proved wrong.

Ant workers of the genus *Formica* appear to be the primary pollinators of *Orthocarpus pusillus,* a moss-like North American member of the Scrophulariaceae (Kincaid, 1963). In tropical forests other ants are believed to be important in the pollination of cauliflorous flowers of great numbers of tree species, including cacao (Bequaert, 1922). Ants are also among the regular visitors to floral and extrafloral nectaries of many other kinds of higher plants; yet, curiously, no known species depends completely on this source for its carbohydrates. Together with earthworms, ants are the principal movers of the soil, in both tropical and temperate regions. And by moving seeds, enriching the earth, and attacking growing plants in a selective and patchy manner, ant colonies play a key role in determining the fine structure of plant communities.

But by far the most intimate and elaborate relationship known is the one that exists between the plants called *myrmecophytes* and the peculiar ants that live in them. Almost any plant with hollow or pithy stems, twigs, thorns, or other preformed cavities can serve as a nest site for colonies of ants. Some species occupy plant cavities only occasionally; some, invariably. On the other side, most plant species only occasionally harbor ants, many frequently do, and a very few almost invariably contain colonies. Myrmecophytes by definition belong in the last category; they are plants that live in constant, mutualistic relationship with ant colonies (Warburg, 1892; Blatter, 1928). In a recent and still unpublished review of the proven and presumptive myrmecophytes, D. H. Janzen has cited the following characteristics as predisposing plant and ant species to enter the mutualistic relations:*

(1) Hollow structures regularly present in or on the living plant.
(2) Ants of the same species or colony tenanting these structures throughout the plant.

*I am grateful to Dr. Janzen for permitting me to quote this list.

(3) Foliar nectaries regularly present.
(4) Twig-inhabiting ants present (implied foraging on foliage).
(5) Ants medium to large in size, possessing a strong chemical or sting defense.
(6) Plant neither a mature forest tree nor a herb.
(7) Food bodies present on stems or leaves.
(8) Ants predators or active scavengers.
(9) Ants and plants occurring between the Tropics of Cancer and Capricorn, not in deserts and not above 2500 meters elevation.

The extreme myrmecophytes, all of which occur in the tropics, have evolved an amazing array of special structures seemingly adapted for harboring ant colonies. These "myrmecodomatia" include the large pseudobulbs of the Indomalayan rubiaceous epiphytes in *Myrmecodia* and *Hydnophytum,* which are riddled with preformed cavities (Figure 4-17). In South America several genera (*Tococa, Maieta, Microphysca, Calophysca, Myrmidone, Hirtella*) produce strange bladder-like swellings on the petiole or at the base of the leaf which are regularly inhabited by colonies of small ants such as *Allomerus* and *Strumigenys.* Some species of *Acacia,* which are native to the drier parts of Africa and the New World tropics, develop large numbers of inflated thorns filled with spongy tissue. Ants chew their way in near the sharp tips of the thorn and occupy the spacious interiors. Other species of trees and shrubs belonging to diverse dicotylodenous families have conspicuously swollen and cavernous trunks or stems that are almost invariably occupied by ant colonies. Examples include species belonging to *Barteria, Cuviera, Endospermum, Macaranga,* and *Vitex* in tropical Africa and Asia, and *Cecropia, Cordia,* and *Triplaris* in the New World tropics.

Besides offering ideal homes for the ants that invade them, the myrmecophytes also produce food from special organs that seemingly serve no other purpose than to nourish the ants. *Pourouma guianensis* and the species of *Cecropia,* common "scrub" trees of the tropical American

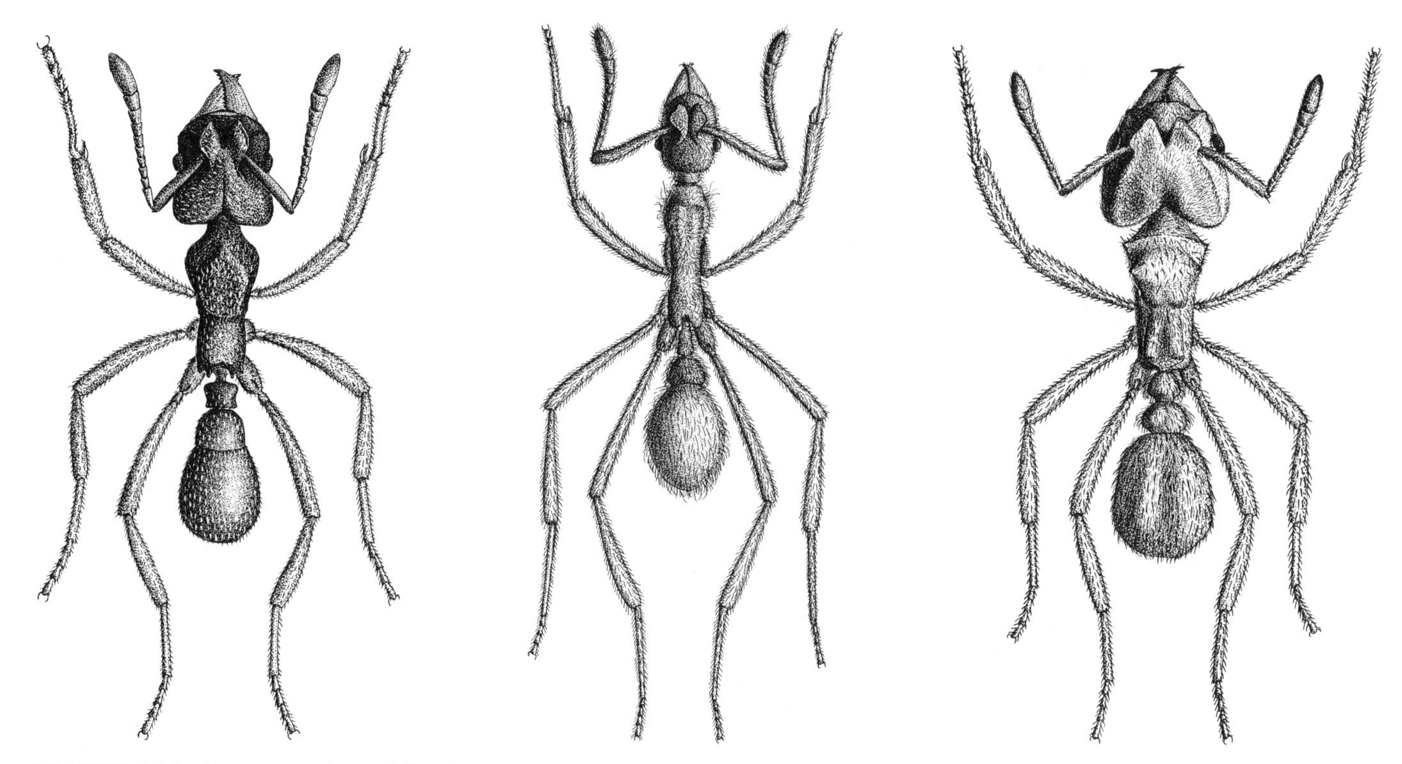

FIGURE 4-16. Representatives of the "lesser" genera of fungus-growing ants: (*left*), *Myrmicocrypta brittoni;* (*middle*), *Apterostigma pilosa;* (*right*), *Sericomyrmex opacus* (from Wheeler, 1910).

FIGURE 4-17. *Myrmecodia pentasperma* of New Britain, Bismarck Archipelago, with its pseudobulb sliced open to reveal a colony of ants (*Iridomyrmex myrmecodiae*) inhabiting the preformed cavities (from Wheeler, 1910; after F. Dahl).

forests, produce "Müllerian bodies," small red or yellow, elliptical corpuscles embedded in dense mats of hair located at the base of each leaf petiole. These objects are rich in oil and protein and are easily detached and carried away by the *Azteca* that ordinarily inhabit the trees. The bull's-horn acacias sprout similar detachable corpuscles, called "Beltian bodies," from the tips of the leaflets (Figure 4-18). Almost all of the Beltian bodies are harvested by the resident colonies of *Pseudomyrmex* ants. The pith of *Cecropia,* which is consumed by the ants, has been assumed by botanists to have nutritive value. The same is true of the "nutritive layer" of the internode walls of *Barteria, Cuviera, Plectronia, Sarcocephalus, Tachigalia,* and *Vitex.* Special structures called "beadglands" (*Perldrüsen*) are characteristic of certain members of the families Bignoniaceae, Melastomaceae, Moraceae, Piperaceae, Sterculiaceae, Urticaceae, and Vitaceae. They are tiny, transparent corpuscles, scattered in large numbers over the surface of the plant and rich in oily and sugary substances attractive to ants. Reviews of most aspects of myrmecophytism, largely still up to date, are given by Wheeler (1910, 1942) and Bequaert (1922).

There remains, however, one basic aspect of the subject concerning which very little has been learned, and even this only in the past decade: the degree of intensity of the symbiosis. In spite of the obvious intimacy of the associations between many of the tropical plant species and their ants, and the extraordinary anatomical features of the plants that seem to have no other function than to service their guests, biologists for many years disagreed about the significance of the association. On one side stood the "protectionist school," to use W. L. Brown's (1960b) expression. It was founded by Thomas Belt, whose observations on the ant acacias in *A Naturalist in Nicaragua* were the beginning of serious studies of myrmecophytism. Belt, and a majority of subsequent writers, including particularly Schimper, Wasmann, and the leading authority on the bull's-horn acacias, W. E. Safford, agreed that the ants provide the plants protection against their natural enemies. They also postulated that in the course of their evolution the acacias developed the hollow thorns, Beltian bodies, and foliar nectaries as devices to promote the welfare of the ants. In short, the protectionist authors believed the symbiosis to be mutualistic. The opposing "exploitationist school," represented chiefly by Elisabeth Skwarra (1934) and W. M. Wheeler (1942), argued that only the ants benefited and that the various "myrmecophilous" structures of the acacias serve some other, still unknown function. This opposition of viewpoints, which is rather oversimplified as I have expressed it here, extended to discussions of other genera of ant plants as well.

Brown, in crystallizing the issue in 1960, developed new evidence favoring the protectionist hypothesis. He pointed out that *Acacia* is a very old and widespread genus containing no less than 700 species. Australia contains a majority of the species as well as the greatest phyletic diversity of any continent. It also contains one of the richest ant faunas of any comparable area in the world. Yet not a single potential myrmecophyte has been found among the acacias of Australia. Moreover, the Australian species have mostly lost their spiniform stipules, in striking contrast to their congeners in other parts of the world. This geographic distribution of characteristics agrees with the

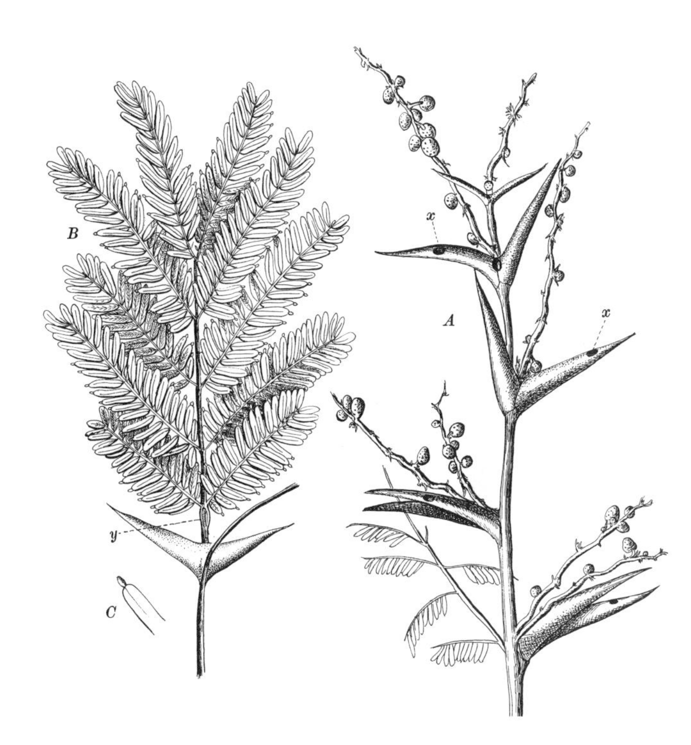

FIGURE 4-18. *Acacia sphaerocephala,* one of the bull's-horn acacias of the American tropics: (*A*) end of a branch showing pairs of the hollow thorns that are normally occupied by ants of the genus *Pseudomyrmex;* (*x*) holes chewed by the ants to form entrances in the thorns; (*B*) a leaf of the same plant; (*y*) extrafloral nectary; (*C*) tip of a leaflet enlarged to show a Beltian body, the organ which is picked and eaten by the ants (from Wheeler, 1910; after A. F. W. Schimper).

known occurrence, in the recent geological past and at present, of large and effective faunas of browsing mammals. Brown inferred, in accordance with Belt's hypothesis, that the development of myrmecophytism and spininess in the African and New World *Acacia* species represents an adaptive response to the presence of such mammals. In other words, they must provide an effective deterrant to browsing. In Australia, where advanced browsing faunas have been unknown (at least in the recent geological past), the species have either failed to develop myrmecodomatia and spines or else lost them secondarily.

It remained for Janzen (1966, 1967, 1969a,b), in a brilliant field experimental study in Mexico, to prove directly that the ants do indeed provide vital protection to the bull's-horn acacias. While conducting a pilot survey Janzen noted, as Bequaert (1922) had found earlier for *Barteria fistulosa* in the Congo, that *Acacia cornigera* shrubs and trees devoid of ants suffer greater damage from attacks by phytophagous insects than do their neighbors harboring ant colonies. They also tend to be overgrown by competing plant species. When Janzen removed the ants (*Pseudomyrmex ferruginea*) with any one of three treatments, namely spraying with parathion, clipping the thorns, or extirpating entire occupied branches, he found that the acacias became decidedly more vulnerable to attack by their insect herbivores. Coreid bugs and membracids sucked on the shoot tips and new leaves; scarabs, chrysomelid beetles, and assorted caterpillars browsed on the leaves; and buprestid beetle larvae girdled the shoots. Moreover, other plants grew in more closely and shaded the stunted shoots. In nearby control trees, still occupied by *Pseudomyrmex* colonies, Janzen observed that the ants attacked the invading insects, in the great majority of cases successfully killing them or driving them off. Alien plants that sprouted within a radius of 40 cm of the occupied acacia trunks were chewed and mauled by the ants until they died. Other plants whose leaves or branches touched the canopy of the acacia were also attacked. Up to one-fourth of the entire ant population were active on the surfaces of the control plants, day and night, constantly patrolling and cleaning them. For the full year during which the experiment was continued, the biomass and growth of the unoccupied acacias steadily fell below that of the occupied ones. In the end it seemed unlikely that they could survive much longer, let alone bear seeds. Thus, Belt's view, that the *Pseudomyrmex* "are really kept by the acacia as a standing army," was substantially confirmed.

Although experimental evidence is still lacking, it seems probable that the ants are also effective against browsing mammals. The *Pseudomyrmex* workers are extremely aggressive toward intruders of all sizes. They become alert at the mere smell of a cow or a man, and, when their tree is brushed or shaken, they swarm out and attack at once. Their stings are very painful, causing a lasting burning and throbbing affect. To brush against an occupied acacia, and thus to acquire a group of vicious, stinging ants on an arm or a leg, is a sensation very much like walking into a large nettle plant. Truculence and fighting prowess are, in fact, general qualities of myrmecophyte-dwelling ants. *Viticicola tessmanni,* the obligate pseudomyrmecine tenant of the verbenaceous creeper *Vitex staudtii* in West Africa, was described by Bequaert (1922) as "exceedingly vicious and alert. When its host plant is ever so slightly disturbed, the workers rush out of the hollow stalks in large numbers and actively explore the plant. Their sting is extremely painful and sometimes produces vesicles on the skin." Even more redoubtable is the African pseudomyrmecine *Pachysima aethiops,* the obligate tenant of the small flacourtiaceous tree *Barteria fistulosa:* "As soon as any portion of their host plant is disturbed, they rush out in numbers and hastily explore the trunk, branches, and leaves. Some of the workers usually also run over the ground about the base of the tree and attack any nearby intruder, be it animal or man. All observers agree that the sting of the *Pachysima* is exceedingly painful and is felt for several hours. Its effects can best be compared with those produced by female velvet ants." The ant is feared by the natives of the Congo, who try to avoid the unpleasant task of cutting the small *Barteria* trees scattered through the forest. As a consequence individuals of *Barteria fistulosa* are often found standing by themselves in the center of clearings or near the sides of forest paths. The species is also abundant in secondary forest growth. It has been my own experience, in both the Old and New World tropics, that the obligate tenants of myrmecophytes are, on the average, much more aggressive than other species belonging to the same genus and occurring at the same localities.

According to Janzen, *Pseudomyrmex ferruginea* is an obligate plant ant that occupies at least five species of *Acacia* (*chiapensis, cornigera, sphaerocephala, hindsii,* and *collinsii*). Its life cycle conforms to the basic claustral pattern of ants generally. After the nuptial flight, which can occur in warm weather in any month of the year, the queen alights, sheds her wings, and searches for a nest site. For a *P. ferruginea* there can be only one such place: an unoccupied acacia thorn. If the thorn has not already been opened by a previous occupant, the queen gnaws a circular hole near the tip of the spine and enters. Then she lays 15–20 eggs and rears her first brood while remaining secluded in the thorn cavity. Although the exact duration of brood development is not known, it is evidently relatively short for an ant species, and the worker population increases at a rapid rate. Within seven months there are about 150 workers, and, three months later, twice this number. The worker population increases to about 1,100 in two years and to over 4,000 in three years. The largest colony collected by Janzen contained 12,269 workers and a single queen. In old colonies the queen is physogastric, heavily attended by workers, and accompanied by masses of hundreds of eggs and young larvae. The production of males and virgin queens begins during the second year and proceeds continuously thereafter. Workers belonging to the youngest colonies leave the protection of the thorn home only long enough to gather nectar and Beltian bodies and, at rare intervals, to take possession of nearby thorns. When their numbers reach 50 to 100, they begin patrolling the open plant surface in the vicinity of the nest thorns. When the population size reaches 200 to 400, the workers become more aggressive and start attacking and destroying other, smaller colonies in nearby thorns. They also become increasingly effective in warding off phytophagous insects that attempt to land in the vicinity. Finally, the dominant colony takes possession of the entire tree, wiping out all competitors in the process. A few colonies are also able to extend their territories to other acacias nearby.

The *Pseudomyrmex ferruginea* colonies appear to subsist primarily on the Beltian bodies and foliar nectar obtained from the host trees. The larvae are fed in part on unaltered fragments of Beltian bodies in the following peculiar manner. The nurse worker first pushes the fragment deep within the trophothylax, the special food pouch located on the lower surface of the thorax just behind the head (and found only in pseudomyrmecine larvae). The larva then starts to rotate its head in and out of the trophothylax, chewing and swallowing the contents. Simultaneously it ejects a droplet of clear fluid, possibly containing a digestive enzyme, into the trophothylax. If the Beltian fragment protrudes from the opening of the pouch, a worker may remove it, cut it up, and redistribute it. From time to time workers also force open the trophothylax and regurgitate droplets of fluid into it. Whether this material consists of elementary crop fluid or some more specialized form of nutritive secretion is unknown. Occasionally the *Pseudomyrmex* workers succeed in capturing insect prey on the nest tree. It is possible that these too are fed to larvae, but they can form a source of protein only secondary in magnitude to the Beltian bodies.

The Natural History of the Army Ants

No spectacle of the tropical world is more exciting than that of a colony of army ants on the march. In his book, *Ants, Their Structure, Development and Behavior,* Wheeler expressed its poetry in the following way: "The driver and legionary ants are the Huns and Tartars of the insect world. Their vast armies of blind but exquisitely coöperating and highly polymorphic workers, filled with an insatiable carnivorous appetite and a longing for perennial migrations, accompanied by a motley host of weird myrmecophilous camp-followers and concealing the nuptials of their strange, fertile castes, and the rearing of their young, in the inaccessible penetralia of the soil—all suggest to the observer who first comes upon these insects in some tropical thicket, the existence of a subtle, relentless and uncanny agency, directing and permeating all their activities."

Eciton burchelli is one of the best understood of the army ants. This big, conspicuous species is abundant in the humid lowland forests from Brazil and Peru north to southern Mexico (Borgmeier, 1955). Its marauding workers, together with those of other species of *Eciton,* are well known to native peoples by such local names as *padicours, tuocas, tepeguas,* and *soldados.* In English they are called army ants, as well as foraging ants, legionary ants, soldier ants, and visiting ants. These insects have understandably been a prime target for study by naturalists for a long time, from Lund (1831) through Bates, Belt, von Ihering, Müller, and Sumichrast in the last century to Beebe, Wheeler, and many others in more recent times. But it was T. C. Schneirla (1933–1965) who, by conducting patient studies over virtually his entire career, first unraveled the complex behavior and life cycle of this and other species of *Eciton.* His results have since been confirmed and greatly extended by C. W. Rettenmeyer (1963a).

A day in the life of an *Eciton burchelli* colony seen through the eyes of Schneirla and Rettenmeyer begins at dawn, as the first light suffuses the heavily shaded forest floor. At this moment the colony is in "bivouac," meaning that it is temporarily camped in a more or less exposed position. The sites most favored for bivouacs are the spaces between the buttresses of forest trees and beneath fallen tree trunks (see Figure 4-26) or any sheltered spot along the trunks and main branches of standing trees to a height of twenty meters or more above the ground. Most of the shelter for the queen and immature forms is provided by the bodies of the workers themselves. As they gather to form the bivouac, they link their legs and bodies together with their strong tarsal claws, forming chains and nets of their own bodies that accumulate layer upon interlocking layer until finally the entire worker force comprises a solid cylindrical or ellipsoidal mass up to a meter across. For this reason Schneirla and others have spoken of the ant swarm itself as the "bivouac." Between 150,000 and 700,000 workers are present. Toward the center of the mass are found thousands of immature forms, a single mother queen, and, for a brief interval in the dry season, a thousand or so males and several virgin queens. The entire dark-brown conglomerate exudes a musky, somewhat fetid odor.

When the light level around the ants exceeds 0.05 foot candle, the bivouac begins to dissolve. The chains and clusters break up and tumble down into a churning mass on the ground. As the pressure builds, the mass flows outward in all directions. Then a raiding column emerges along the path of least resistance and grows away from the bivouac at a rate of up to 20 m an hour. No leaders take command of the raiding column. Instead, workers finding themselves in the van press forward for a few centimeters and then wheel back into the throng behind them, to be supplanted immediately by others who extend the march a little farther. As the workers run on to new ground, they lay down small quantities of chemical trail substance from the tips of their abdomens, guiding others forward. A loose organization emerges in the columns, based on behavioral differences among the castes. The smaller and medium-sized workers race along the chemical trails and extend it at the point, while the larger, clumsier soldiers, unable to keep a secure footing among their nestmates, travel for the most part on either side. The location of the *Eciton* soldiers misled early observers into concluding that they are the leaders. As Thomas Belt put it, "Here and there one of the light-colored officers moves backwards and forwards directing the columns." Actually the soldiers, with their large heads and exceptionally long, sickle-shaped mandibles, have relatively little control over their nestmates and serve instead almost exclusively as a defense force. The minimas and medias, bearing shorter, clamp-shaped mandibles, are the generalists. They capture and transport the prey, choose the bivouac sites, and care for the brood and queen (see Figure 4-19).

E. burchelli has an unusual mode of hunting even for an army ant. It is a "swarm raider," which means that the foraging workers spread out into a fan-shaped swarm with a broad front. Most other army ant species are "column raiders," pressing outward along narrow dendritic odor trails in the pattern exemplified in Figure 4-20. Schneirla (1956a) has described a typical raid as follows:

For an *Eciton burchelli* raid nearing the height of its development in swarming, picture a rectangular body of 15 meters or more in width and 1 to 2 meters in depth, made up of many tens of thousands of scurrying reddish-black individuals, which as a mass manages to move broadside ahead in a fairly direct path. When it starts to develop at dawn, the foray at first has no particular direction, but in the course of time one section acquires a direction through a more rapid advance of its members and soon drains in the other radial expansions. Thereafter this growing mass holds its initial direction in an approximate manner through the pressure of ants arriving in rear columns from the direction of the bivouac. The steady advance in a principal direction, usually with not more than 15° deviation to either side, indicates a considerable degree of internal organization, notwithstanding the chaos and confusion that seem to prevail within the advancing mass. But organization does exist, indicated not only by the maintenance of a general direction but also by the occurrence of flanking movements of limited scope, alternately to right and left, at intervals of 5 to 20 minutes depending on the size of the swarm.

The huge sorties of *burchelli* in particular bring disaster to practically all animal life that lies in their path and fails to escape. Their normal bag includes tarantulas, scorpions, beetles, roaches, grasshoppers, and the adults and broods of other ants and many forest insects; few evade the dragnet. I have seen snakes, lizards, and nestling birds killed on various occasions; undoubtedly a larger vertebrate which, because of injury or for some other reason, could not run off, would be killed by stinging or asphyxiation. But lacking a cutting or shearing edge on their mandibles, unlike their African relatives the "driver ants," these tropical American swarmers cannot tear down their occasional vertebrate victims. Arthropods, such as ticks, escape through their excitatory secretions, stick insects through repellent chemicals, as tests show, as well as through tonic immobility. The swarmers react to movement in particular as well as to the scent of their booty, and a motionless insect has some chance of escaping them. Common exceptions, which may enjoy almost a community invulnerability in many cases, include termites and Azteca ants in their bulb nests in trees, army ants of their own and other species both on raiding parties and in their bivouacs, and leaf-cutter ants in the larger mound communities; in various ways these often manage to fight off or somehow repel the swarmers.

The approach of the massive *burchelli* attack is heralded by three types of sound effect from very different sources. There is a kind of foundation noise from the rattling and rustling of leaves and vegetation as the ants seethe along and a screen of agitated small life is flushed out. This fuses with related sounds such as an irregular staccato produced in the random movements of jumping insects knocking against leaves and wood. This noise, more or less continuous, beats on the ears of an observer until it acquires a distinctive meaning almost as the collective death rattle of the countless victims. When this composite sound is muffled after a rain, as the swarm moves through soaked and heavily dripping vegetation, there is an uncanny effect of inappropriate silence.

Another characteristic accompaniment of the swarm raid is the loud and variable buzzing of the scattered crowd of flies of various species, some types hovering, circling, or darting just ahead of the advancing fringe of the swarm, others over the swarm itself or over the fan of columns behind. To the general hum are added irregular short notes of higher pitch as individuals or small groups of flies swoop down suddenly here or there

FIGURE 4-19. A group of large media workers of the army ant *Eciton burchelli* are shown returning to the bivouac with the tail of a freshly caught scorpion (photograph courtesy of C. W. Rettenmeyer).

upon some probable victim of the ants which has suddenly burst into view . . . No part of the prosaic clatter, but impressive solo effects, are the occasional calls of antbirds. One first catches from a distance the beautiful crescendo of the bicolored antbird, then closer to the scene of action the characteristic low twittering notes of the antwren and other common frequenters of the raid.

On Barro Colorado Island, Panama, where Schneirla conducted most of his studies, the antbirds normally follow only the raids of *Eciton burchelli* and those of another common swarm-raider, *Labidus praedator.* They pay no attention to the less conspicuous forays of *E. hamatum, E. dulcius, E. vagans,* and other column-raiding army ants. There are at least ten species of antbirds on Barro Colorado Island, all members of the family Formicariidae. They feed principally on the insects and other arthropods flushed by the approaching *burchelli* swarms (Johnson, 1954; Willis, 1967). Although a specimen of *Neomorphus geoffroyi* has been recorded with its stomach stuffed with

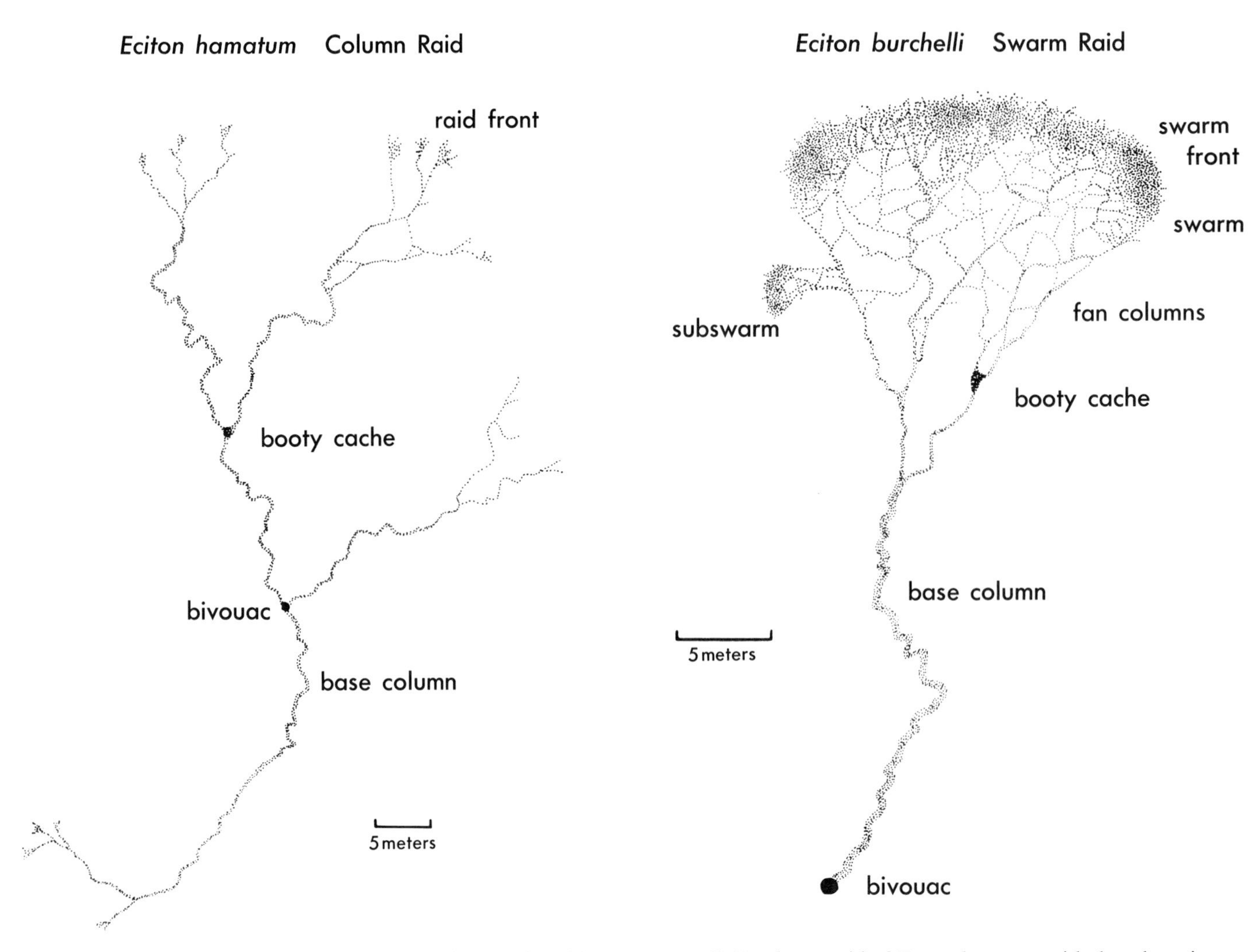

FIGURE 4-20. The two basic patterns of raiding employed by army ants: (*left*) column raid of *Eciton hamatum* with the advancing front made up of narrow bands of workers, the pattern of which is displayed by the majority of army ant species; (*right*) swarm raid of *E. burchelli* with the advancing front made up of a large mass or swarm of workers, followed by the anastomosing columns in the fan area which converge toward the base column and bivouac site (from Rettenmeyer, 1963a).

burchelli workers, most species appear to avoid the ants completely or at most consume them by accident while swallowing other food. The ornithologist Alexander Wetmore (cited by Rettenmeyer, 1963a) and many other biologists have learned to locate *burchelli* swarms by the calls of the antbirds. However, Charles H. Curran, the famous dipterist, always located the swarms by the buzzing of the accompanying flies!

As one might anticipate from these accounts, the *burchelli* colonies and their efficient camp followers have a profound effect on the faunas of those particular parts of the forest over which the swarms pass. E. C. Williams (1941), for example, noted a sharp depletion of the arthropods at those spots on the forest floor where a swarm had struck the previous day. But the total effect on the forest at large may not be very significant. On Barro Colorado Island, which has an area of approximately 16 km^2, there exist only about 50 *burchelli* colonies at any one time. Since each colony travels at most 100 to 200 m every day (and not at all on about half the days), the collective population of *burchelli* colonies raids only a minute fraction of the island's surface in the course of one day, or even in the course of one week.

Even so, it is a fact that the food supply is quickly and drastically reduced in the immediate vicinity of each colony. Early writers, especially Müller (1886) and Vosseler (1905), jumped to the seemingly reasonable conclusion that army ant colonies change their bivouac sites whenever the surrounding food supply is exhausted. At an early stage of his work, however, Schneirla (1933a, 1938) discovered that the emigrations are subject to an endogenous, precisely rhythmic control unconnected to the immediate food supply. He proceeded to demonstrate that each *Eciton* colony alternates between a *statary phase,* in which it remains at the same bivouac site for as long as two to three weeks, and a *nomadic phase,* in which it moves to a new bivouac site at the close of each day, also for a period of two to three weeks. The basic *Eciton* cycle is summarized in Figure 4-21. Its key feature is the correlation between the *reproductive cycle,* in which broods of workers are reared in periodic batches, and the *behavior cycle,* consisting of the alternation of the statary and nomadic phases. The single most important feature of *Eciton* biology to bear in mind in trying to grasp this rather complex relation is the remarkable degree to which development is synchronized within each successive brood. The ovaries of the queen begin developing rapidly

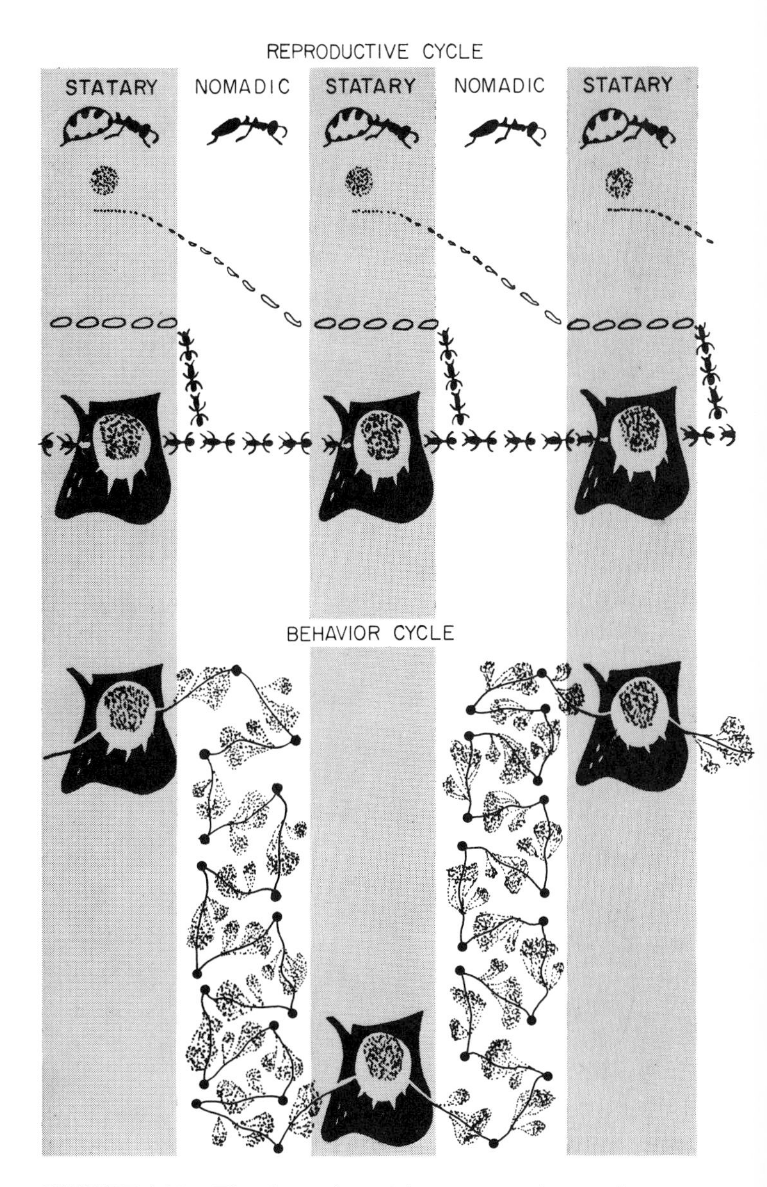

FIGURE 4-21. The alternation of the statary and nomadic phases of the *Eciton burchelli* colony which consists of distinct but tightly synchronized reproductive and behavior cycles. During the statary phase the queen, shown at the top, generates and lays a large batch of eggs, all in a brief span of time; the eggs hatch into larvae; the pupae derived from the previous batch of eggs develop into adults; and, as indicated in the lower diagram, the colony remains in one bivouac site. During the nomadic phase, the larvae complete their development; the new workers emerge from their cocoons; and, as indicated in the lower diagram, the ants change their nest sites after the completion of each day's swarm raid (redrawn from Schneirla and Piel, 1948).

FIGURE 4-22. A queen of *Eciton burchelli* is shown in the egg-laying phase. At this period one individual lays from 100,000 to 300,000 eggs within two weeks (photograph courtesy of C. W. Rettenmeyer).

when the colony enters the statary phase, and within a week her abdomen is greatly swollen by 55,000 to 66,000 eggs. Then, in a burst of prodigious labor lasting for several days in the middle of the statary period, the queen lays from 100,000 to 300,000 eggs (see Figure 4-22). By the end of the third and final week of the statary period, larvae hatch, again all within a few days of each other. A few days later the "callow" workers (so called because they are at first weak and lightly pigmented) emerge from the cocoons. The sudden appearance of tens of thousands of new adult workers has a galvanic effect on their older sisters. The general level of activity increases, the size and intensity of the swarm raids grow, and the colony starts emigrating at the end of each day's marauding. In short, the colony enters the nomadic phase. The nomadic phase itself continues as long as the brood initiated during the previous statary period remains in the larval stage. As soon as the larvae pupate, however, the intensity of the raids diminishes, the emigrations cease, and the colony (by definition) passes into the next statary phase.

The emigration is a dramatic event requiring sudden complex behavioral changes on the part of all adult members of the *Eciton* colony. At dusk or slightly before workers stop carrying food into the old bivouac and start carrying it, along with their own larvae, in an outward direction to some new bivouac site along the pheromone-impregnated trails (Figures 4-23,24). Eventually, usually after most of the larvae have been transported to the site, the queen herself makes the journey. This event usually transpires between 8:00 and 10:00 P.M., well after nightfall. Just before the queen emerges from the bivouac, the workers on the trail nearby become distinctly more excited, and the column of running workers thickens beyond its usual width of 2 to 3 cm, soon widening to as much as 15 cm. Suddenly the queen appears in the thickest part. As she runs along she is crowded in by the "retinue," a shifting mob consisting of an unusual number of soldiers and darkly colored, unladen smaller workers. The members of the retinue jostle her, press in underfoot, climb up on her back, and at times literally envelop her body in a solid mass. But, even with this encumbrance, the queen moves along easily to the new bivouac site. She is guided by the odor trail and can follow it all by herself even if the surrounding workers are taken away. After her passage the emigration tapers off, and it is usually finished by midnight.

The activity cycle of *Eciton* colonies is truly endogenous. It is not linked to any known astronomical rhythm or weather event. It continues at an even tempo month after month, in both wet and dry seasons throughout the entire year. Propelled by the daily emigrations of the nomadic phase, the colony drifts perpetually back and forth over the forest floor (Figure 4-25). The results of experiments performed by Schneirla indicate that the phases of the activity cycle are determined by the stages of development of the brood and their effect on worker behavior. When he deprived *Eciton* colonies in the early nomadic phase of their callow workers, they lapsed into the relatively lethargic state characteristic of the statary phase, and emigrations ceased. Nomadic behavior was not resumed until the larvae present at the start of the experiments had grown much larger and more active. In order to test further the role of larvae in the activation of the workers, Schneirla divided colony fragments into two parts of equal size, one part with larvae and the other without. Those workers left with larvae showed much greater continuous activity.

These results, while provocative, are not decisive and at best solve only half the problem. For if the activity cycle is controlled by the reproductive cycle, what controls the reproductive cycle? The logical place to look would seem

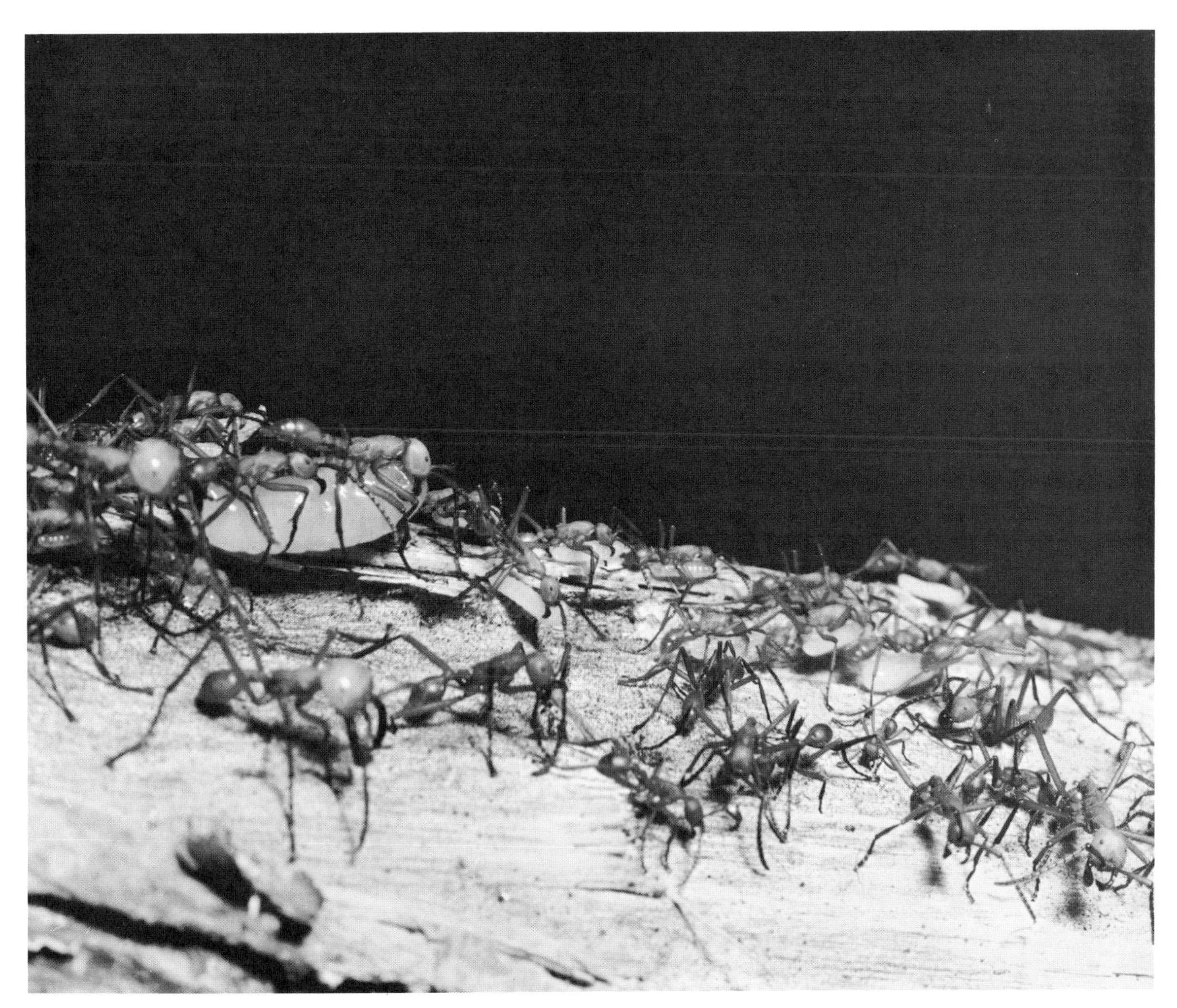

FIGURE 4-23. When army ants move from one bivouac to another during their nomadic phase, the workers transport the larvae by carrying them slung beneath the body. In this photograph of an emigrating *Eciton hamatum* colony, workers are also seen (upper left) carrying a polybiine wasp larva, which had been captured as prey (photograph courtesy of C. W. Rettenmeyer).

FIGURE 4-24. During emigrations army ants sometimes create living bridges of their own bodies. In this photograph the workers of an *Eciton burchelli* colony can be seen linking their legs and, along the top of the bridge, hooking their tarsal claws together to form irregular systems of chains. A symbiotic silverfish, *Trichatelura manni,* is seen crossing the bridge in the center (photograph courtesy of C. W. Rettenmeyer).

to be the queen. By her astonishing capacity to lay all of her eggs in one brief burst, she creates the synchronization of brood development, which is the essential feature for the colonial control of the activity cycle. At first Schneirla (1944) concluded that this reproductive effort by the queen is the "pace-maker," thus implying that the queen herself is the seat of an endogenous rhythm. Later, however, Schneirla (1949b, 1956a) modified his hypothesis by viewing the queen and her colony as reciprocally donating elements in an oscillating system: "When each successive brood approaches larval maturity, the social-stimulative effect upon workers nears its peak. The workers thus energize and carry out some of the greatest raids in the nomadic phase, with their byproduct larger and larger quantities of booty in the bivouac. But our histological studies show that, at the same time, more and more of the larvae (the largest first of all) soon reduce their feeding to zero as they begin to spin their cocoons. Thus in the last few days of each nomadic phase a food surplus inevitably arises. At this time the queen apparently begins to feed voraciously. It is probable that the queen does not overfeed automatically in the presence of plenty, but that she is started and maintained in the process by an augmented stimulation from the greatly enlivened worker population. Within the last few days of each nomadic phase, the queen's gaster begins to swell increasingly, first of all from a recrudescence of the fat bodies, then from an accelerating maturation of eggs. The overfeeding evidently continues into the statary phase, when, with colony food consumption greatly reduced after enclosure of the brood, smaller raids evidently bring in sufficient food to support the processes until the queen becomes maximally physogastric. These occurrences, which are regular and precise events in every *Eciton* colony, are adequate to prepare the queen for the massive egg-laying operation which begins about one week after the nomadic phase has ended." While this interpretation makes a pretty story, it is constructed on nothing more than fragments of very circumstantial evidence. The crucial question is unanswered as to whether the queen really is stimulated to feed in excess by the greater abundance of food or at least by the higher intensity of worker activity associated with the food, as Schneirla posited, or whether her increased feeding is timed by some other, undetermined physiological event. Since work on *Eciton* physiology is still virtually nonexistent, and experimental evidence of any kind

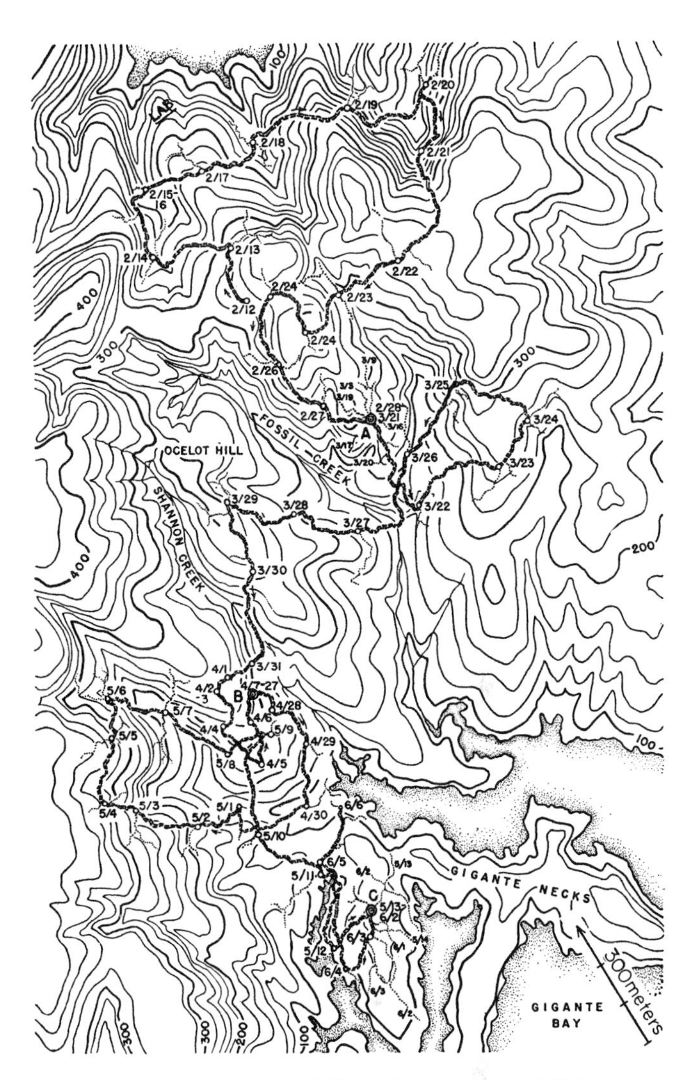

FIGURE 4-25. The route followed by a colony of *Eciton hamatum* for 114 days as it emigrated from one site to another on the eastern side of Barro Colorado Island. Double circles indicate the bivouac sites of the statary phases; small circles, the bivouac sites of the nomadic phases; double broken lines, the successive routes of emigration; and dotted lines, some of the routes followed in daily raiding. The contour intervals are 6.1 m. (from Schneirla, 1949a).

very sparse, one can do no more than reflect on these possibilities as competing hypotheses.

The same hesitant evaluation must be made with respect to our current understanding of the actual communicative stimuli that mediate the activity cycles. In his voluminous theoretical writings on the subject, Schneirla often spoke of "trophallaxis" as the driving force of the cycles, but it is clear that he meant this term to be virtually synonymous with "communication" in the broadest sense. Apparently he had no clear ideas about the nature of the signals utilized. In earlier articles he attributed much of the stimulative effect of the larvae to their "squirming"; later he stressed the probable existence of pheromones as well. But these speculations were based almost entirely on observations of the more obvious outward signs of communication, a level of study usually inadequate to distinguish even the sensory modalities employed in communication within insect colonies and unlikely to identify the signals employed (see Chapter 12). Lappano (1958) discovered that the labial glands of *burchelli* larvae become fully functional on the eighth or ninth nomadic day, about the time raiding activity reaches its peak. She concluded that the labial glands are "probably" producing a pheromone that excites the worker. But again, the only evidence available is the stated coincidence in time of the two events. Our lack of knowledge of the semiotic basis of the *Eciton* cycle is due simply to the lack of any serious attempt to obtain it. This interesting subject does not seem to me to be likely to resist sustained experimental study; any such effort in the future is likely to yield exceptionally interesting results.

Another point on which Schneirla was inadvertantly misleading was his failure to distinguish between ultimate and proximate causation. After he had demonstrated the endogenous nature of the cycle, and its control by synchronous brood development, he dismissed the role of food depletion. The emigrations, he repeatedly asserted, are caused by the appearance of callow workers and the older larvae; they are not caused by food shortage. He overlooked the fuller evolutionary explanation combining the two causations: that the adaptive significance of the emigrations is to take the huge colonies to new food supplies at regular intervals, and that in the course of evolution the emergence of callows and larvae have come to be employed as the timing signals. Stated another way, if there is a selective advantage for colonies to move frequently to new feeding sites (and all of the evidence from the *Eciton* studies suggests that this is so), then worker behavior would tend to evolve in such a way as to synchronize the emigrations precisely with the presence of the life stages that cause the greatest food shortage. The shift in the proximate cause of emigration from exogenous to endogenous controls does not alter the nature of the ultimate cause of emigration, which in the case of army ants seems most probably to be chronic depletion of food sources.

Colony multiplication in *Eciton,* first elucidated by Schneirla and R. Z. Brown (1950, 1952), is a highly specialized and ponderous operation. Through most of the year the mother queen is the paramount center of attraction for the workers, even when she is in competition with the mature worker larvae toward the close of each nomadic phase. By serving as the focal point of the aggregating workers, she literally holds the colony together. The situation changes drastically, however, when the annual sexual brood is produced early in the dry season. This kind of brood contains no workers, but, in *E. hamatum* at least, it consists of about 1,500 males and 6 new queens. Even when the sexual larvae are still very young, a large fraction of the worker force becomes affiliated with the brood as opposed to the mother queen. By the time the larvae are nearly mature, the bivouac can be found to consist of two approximately equal zones: a brood-free zone containing the queen and her affiliated workers, and a zone in which the rest of the workers hold the sexual brood. The colony has not yet split in any overt manner, but important behavioral differences between the two sections do exist. For example, if the queen is removed for a few hours at a time, she is readily accepted back into the brood-free zone from which she originated, but she is rejected by workers belonging to the other zone. Also, there is evidence that workers from the queen zone cannibalize brood from the other zone when they contact them.

The young queens are the first members of the sexual brood to emerge from the cocoons. The workers cluster excitedly over them, paying closest attention to the first one or two to appear (see Figure 4-26). Several days later the new adult males emerge from their cocoons. This event energizes the colony, sets off a maximum raid followed by emigration, and at last splits the bivouac. Raids are conducted along two radial trails from the old bivouac site. As they intensify during the day, the young queens and their nuclei of workers move out along one of the trails, while the old queen with her nucleus moves out

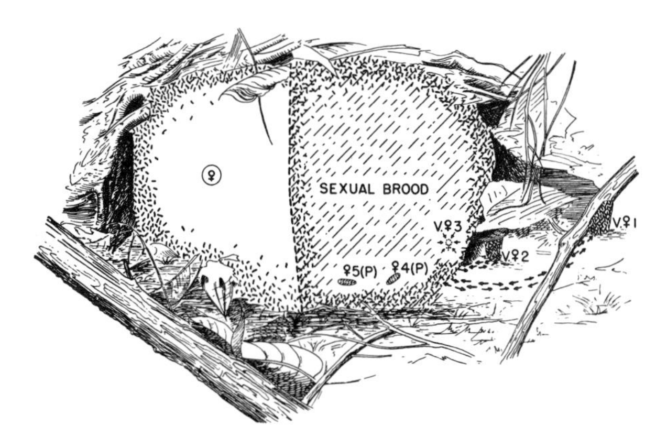

FIGURE 4-26. An *Eciton hamatum* bivouac just prior to colony division. The left portion of the bivouac contains the mother queen but no brood, while the right portion contains the sexual brood. Two of the virgin queens (v.♀1 and v.♀2) have emerged from their cocoons and moved to one side of the bivouac, to be attended by clusters of workers who still run back and forth to the bivouac along odor trails. A third virgin queen (v.♀3) has emerged more recently and is still confined within a knot of workers in the bivouac wall, while two others remain in the pupal stage within cocoons. The males are also still in the pupal stage (from Schneirla, 1956b).

along the other. When the derivative swarm begins to cluster at the new bivouac site, only one of the virgin queens is able to make the journey to it. The others are held back by the clinging and clustering of small groups of workers. They are, to use Schneirla's expression, "sealed off" from the rest of the daughter colony. Eventually they are abandoned and left to die. Now there exist two colonies: one containing the old queen; the other, the successful virgin, daughter queen. In a minority of cases the old queen is also superseded. That is, the old queen herself falls victim to the sealing-off operation, leaving both of the two daughter colonies with new virgin queens. This presumably happens most often when the health and attractive power of the old queen begin to fail prior to colony fission. The maximum age of the *Eciton* queen is not known, but is believed to be relatively great for an insect; a marked queen of *E. burchelli,* for example, was recovered by Rettenmeyer after a period of four and a half years. The males, in contrast, enjoy only one to three weeks of adult existence. Within a few days of their emergence, at least some of them depart on flights away from the home bivouac in search of other colonies. It is also possible that a few remain behind to mate with their sisters; the matter simply has not been documented either way. In any case the new queens are fecundated within a few days of their emergence, and almost all of the males disappear within three weeks after that. Rettenmeyer (1963a) has described the actual mating, and he has presented evidence that a queen sometimes mates more than once in her lifetime and may even mate annually.

The genus *Eciton* represents one of the furthest extensions of an evolutionary trend that has begun independently in many different groups of ants. It is seldom recognized that the behavior patterns characteristic of the doryline "army ants" also occur, at least to a limited extent, in some groups of the primitive subfamily Ponerinae. In the past, entomologists have tended to speak of true army ants as occurring only in the Dorylinae and Leptanillinae, or in the Dorylinae exclusively, but this simple taxonomic qualification is not wholly adequate, particularly in view of the fact that there is now evidence to suggest that the tribe Dorylini and the genus *Aenictus* may each have been derived from stocks separate from the New World members of the "subfamily Dorylinae" (W. L. Brown, 1954; personal communication). The best definition of the expression "army ant" may well be a functional one, of the sort offered in the second edition of *Webster's New International Dictionary:* Any species of ant that goes out in search of food in companies, particularly the driver and legionary ants. It should be added here, in order to clarify the vernacular nomenclature, that there has been a tendency by recent authors to use the terms "army ant" and "legionary ant" interchangeably. According to the same source, the term "legionary ant" refers first to the New World army ants of the genus *Eciton,* and second to the "driver ants" of the genus *Anomma.* Most English-speaking authors of the past fifty years have used this expression to mean specifically *Eciton* and its relatives, while employing "army ant" to cover all of the dorylines. I prefer to use the two terms synonymously in the broader functional sense.

Actually, the definition just quoted is incomplete. Upon closer examination of the subject one finds that there are really *two* discrete features that can be considered fundamental in army ant (legionary) behavior. These diagnostic features have been distinguished under the concepts of *nomadism* and *group-predation* (Wilson, 1958b). Nomadism is defined as relatively frequent colony emigration. Most, if not all, ant species shift their nest site if the environment of the nest area becomes unfavorable, and

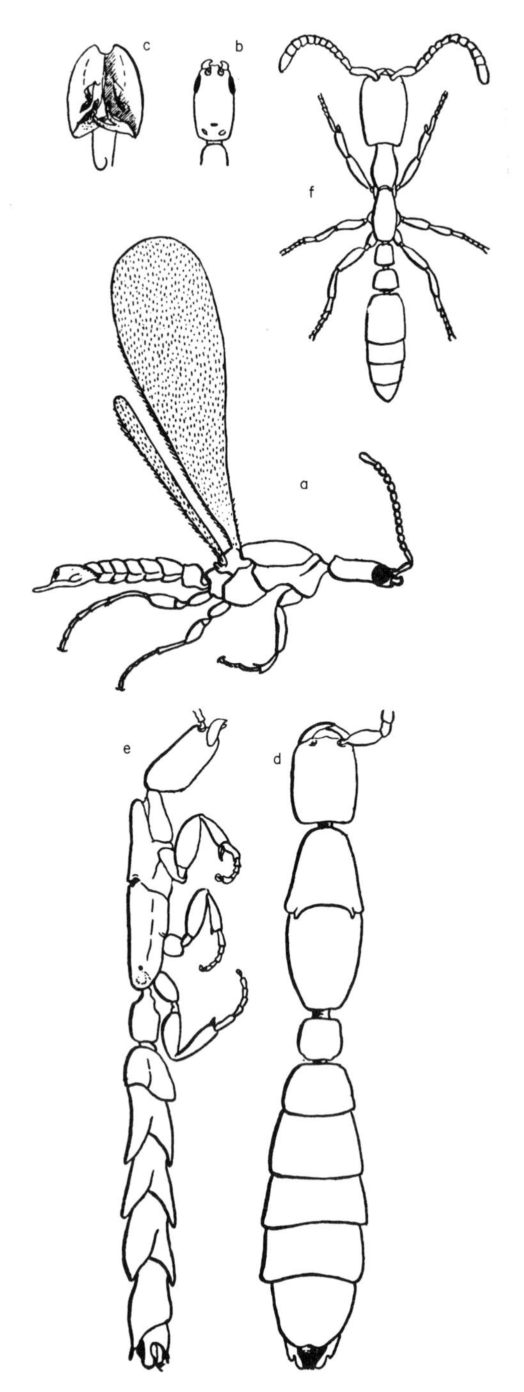

FIGURE 4-27. Representatives of the Leptanillinae, the little-known group of subterranean ants believed to have legionary habits. *Leptanilla minuscula:* (*a*) male; (*b, c*) head and genitalia of male. *L. revelieri:* (*d, e*) dichthadiiform queen; (*f*) worker (from Wheeler, 1910; after Emery and Santschi).

some, for example the famous Argentine ant, *Iridomyrmex humilis,* are exceptionally restless and normally emigrate one or more times during the course of a single season. But hitherto none has been found which undertakes emigration so frequently or accomplishes it in such an orderly fashion to cover so much new territory as do the species of *Eciton* and the other, better-known dorylines. Group-predation includes both group raiding and group retrieving in the process of hunting living prey. These two processes must be carefully distinguished from each other since they involve quite different innate behavior patterns and are not invariably linked. Many ant species, particularly those in the higher subfamilies, engage in group retrieving of prey, but relatively few nondorylines also group raid. Various ponerines that specialize on termites, such as *Leptogenys* of the *processionalis* group and species of *Termitopone, Megaponera, Paltothyreus,* and *Ophthalmopone* (see Wheeler, 1936a), do group raid. The species of the subfamily Cerapachyinae, which feed primarily or exclusively on other ants (Wilson, 1958b), also group raid. Predatory behavior of this kind is distinct from the peculiar form of raiding conducted by the slave-making ants, including *Polyergus, Rossomyrmex, Strongylognathus,* and members of the *Formica sanguinea* group, which are specialized parasites that hunt live captives rather than prey.

How much are nomadism and group-predation associated in ants? Only a very incomplete answer can be supplied to this question since we are still handicapped by a scarcity of information on the extent of nomadism, or, for that matter, emigration of any sort, in most ant groups. Yet it can at least be established that the association of nomadism and group raiding, constituting the most general characteristics of "army ant" behavior, does exist in groups other than the Dorylinae. I have listed the known cases in Table 4-5. An additional characteristic of most of these species is the independent evolutionary development of the peculiar queen form known as the dichthadiigyne. As exemplified in *Eciton* and *Dorylus* (Figures 4-22 and 30), the dichthadiigyne is a permanently wingless form with greatly reduced eyes, massive pedicel and abdomen, and strong legs. Its aberrant morphology clearly contributes to two of its adaptations to nomadic life: its capacity to deliver large quantities of eggs during a short span of time, and its ability to run under its own power from one bivouac site to another.

A great deal more research is needed on the nondoryline army ants. The biology of the enigmatic little

TABLE 4-5. Genera and higher taxa of ants whose species show legionary behavior.

Genus	Distribution	Approximate number of described species	Authority
Subfamily Dorylinae			
Tribe Dorylini			
Dorylus (Subgenera: *Dorylus, Alaopone, Dichthadia, Rhogmus, Typhlopone*)	Africa to tropical Asia	20	Emery (1910), Wheeler (1922), Wilson (1964), Raignier and van Boven (1955)
Anomma	Africa	8	Wheeler (1922), Raignier and van Boven (1955)
Aenictus	Africa to tropical Asia and Queensland	50	Wilson (1964)
Tribe Cheliomyrmecini			
Cheliomyrmex	South America to southern Mexico	5	Wheeler (1921a)
Tribe Ecitonini			
Eciton	South America to southern Mexico	12	Borgmeier (1955)
Labidus	South America to Texas	8	Creighton (1950), Borgmeier (1955)
Nomamyrmex	South America to Texas	2	Creighton (1950), Borgmeier (1955)
Neivamyrmex	South America to Iowa and Virginia	100	Smith (1942), Creighton (1950), Borgmeier (1955)
Subfamily Leptanillinae			
Leptanilla	Africa, tropical Asia, Australia, South America	10	G. C. and J. Wheeler (1965), Petersen (1968)
Leptomesites, Noonilla, Phaulomyrma, Scyphodon	Tropical Asia	4	G. C. and J. Wheeler (1965), Petersen (1968)
Subfamily Ponerinae			
Tribe Ponerini			
Leptogenys (kitteli and *processionalis* groups)	Tropical Asia to Queensland	10	Wilson (1958b, c)
Megaponera	Africa	1	Wheeler (1936a)
Simopelta	Central and South America	8	Wilson (1958b), Gotwald and Brown (1966)
Tribe Amblyoponini			
Onychomyrmex	Australia	3	Wheeler (1916b), Wilson (1958b)

subfamily Leptanillinae, for example, has received almost no attention (see Figure 4-27). Enough is known of the ponerine representatives, on the other hand, to cast some light on the origin and early evolution of the legionary pattern. Elsewhere (Wilson, 1958b), I have pointed out that the key to understanding lies in the adaptive significance of group raiding. It had been stated repeatedly by previous myrmecologists that compact armies of ants are more efficient at flushing and capturing prey than are assemblages of foragers acting independently. This observation is certainly correct, but it is not the whole story. In my opinion there is another, primary function of group raiding that becomes clear only when the prey preferences of the group-raiding ponerines, cerapachyines, and dorylines are compared with those of the ponerines that forage in solitary fashion. Most nonlegionary ponerine species for which the food habits are known take living prey of approximately the same size as their worker caste or smaller. As a rule they must depend on proportionately small animals that can be captured and retrieved by lone

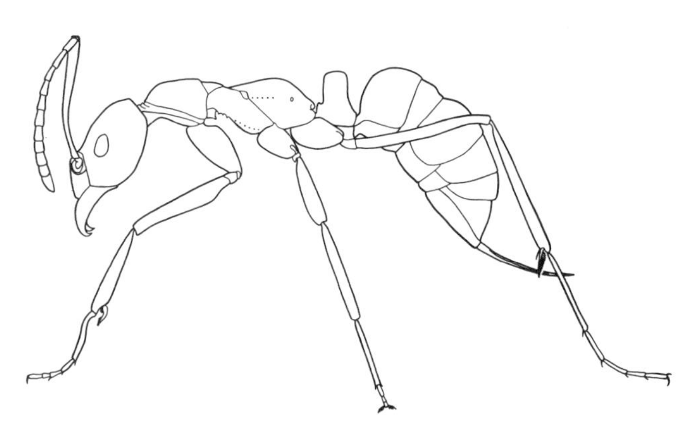

FIGURE 4-28. The worker of *Leptogenys purpurea,* a legionary ponerine ant that occurs on New Guinea (from Wilson, 1958c).

foraging workers. Group-raiding ants, on the other hand, feed on large arthropods or the brood of other social insects, prey not normally accessible to ants foraging solitarily. Thus, the species of *Onychomyrmex* and the *Leptogenys diminuta* group specialize on large arthropods; those of *Eciton* and *Anomma* prey on a wide variety of arthropods that include social wasps and other ants; species of *Simopelta* and the Cerapachyinae specialize on other ants; and *Megaponera foetans* specializes on termites.

With this generalization in mind, it is relatively easy to reconstruct the steps in evolution leading to the full-blown legionary behavior of the Dorylinae.

1. Group raiding is developed to allow specialized feeding on large arthropods and other social insects. Group raiding without nomadism may occur in some of the cerapachyine ants, for example, *Cerapachys* and *Phyracaces,* but this is probably a short-lived stage, soon giving way to the next step.

2. Nomadism is either developed concurrently with group-raiding behavior, or it is added shortly afterward. The reason for this new adaptation is that large arthropods and social insects are more widely dispersed than other types of prey, and the group-predatory colony must constantly shift its trophophoric field to tap new food sources. With the acquisition of both group-raiding and nomadic behavior, the species is now truly "legionary," that is, an army ant in the functional sense. Most of the group-raiding ponerines have evidently reached this adaptive level. Colony size in these species averages larger than in related, nonlegionary species, but it does not approach that attained by *Eciton* and *Anomma.*

3. As group raiding becomes more efficient, large colony size becomes possible. This stage has been attained by many of the Dorylinae, including the species of *Aenictus* and *Neivamyrmex* and at least some of the column-raiding *Eciton* (*dulcius, hamatum, mexicanum, vagans*).

4. The diet may be expanded secondarily to include other smaller and nonsocial arthropods and even small vertebrates and vegetable matter; concurrently, the colony size becomes extremely large. This is the stage reached by the driver ants of Africa and tropical Asia (*Anomma* and *Dorylus*), the species of *Labidus,* and *Eciton burchelli,* most or all of which also utilize the technique of swarm raiding as opposed to column raiding.

The Dorylinae, then, constitute either a phyletic group of species or a conglomerate of two or more convergent phyletic groups that have triumphed as legionary ants over all their competitors. They not only outnumber other kinds of legionary ants in both species and colonies, but they tend to exclude them altogether. Cerapachyines, for example, are relatively scarce throughout the continental tropics wherever dorylines abound, but they are much more common in remote places not yet reached by dorylines—for example, Madagascar, Fiji, New Caledonia, and most of Australia. An idea of the diversity of the Dorylinae can be gained from the synopsis presented in Table 4-6. Among the more interesting and problematical members are the species of *Aenictus.* Most authorities in the past have regarded them as primitive members of the tribe Dorylini, but W. L. Brown (personal communication), who is currently reclassifying ants at the level of the genus, considers them to represent a third phyletic line of advanced army ants, possibly independent of both the Dorylini and Ecitonini. Schneirla (personal communication), while studying *A. gracilis* and *A. laeviceps* in the Philippine Islands in the years just before his death in 1968, detected what he considered to be a relatively primitive behavioral trait. The colonies undergo a continuous alternation between rigidly timed statary and nomadic phases, as in *Eciton.* But whereas *Eciton* conducts one precisely programmed emigration a day during the nomadic phase, *Aenictus* can conduct several, the number being a function of the degree of hunger in the brood. Thus *Aenictus* is somewhat more subject to exogenous influence in its activity cycle than *Eciton,* a characteristic which seems prima facie to be more primitive. *Cheliomyr-*

TABLE 4-6. Comparison of some biological characteristics of the best-known species of legionary ants of the "subfamily Dorylinae."

Species	Number of workers per colony	Nesting habits	Diet	Raiding pattern	Cycle	Authority
Tribe Dorylini						
Anomma wilverthi	2 million (Vosseler) 15 million–22 million (Raignier and van Boven)	Subterranean clustering with much excavation of soil	Many kinds of small animals, mostly arthropods; some vegetable matter	Swarm about 12 m wide in front, tapers to trunk columns in rear; raids begin any time, usually at night	Emigrations irregular, average one about every 20–25 days and continue day and night for 2–3 days for average of 223 m; occur when brood consists mostly of worker pupae	Raignier and van Boven (1955)
Aenictus laeviceps	60,000–110,000	Cluster in sheltered places on the ground surface during nomadic phase, moving underground during statary phase	Mostly other ants, social wasps, and termites; a few other arthropods	Weak dendritic columns, mostly over the surface of the ground; raids occur any time, day or night.	The *Eciton* pattern is followed except that as many as 5 or more emigrations occur daily when larvae are most active during nomadic phase; considered a primitive condition; nomadic phase lasts about 14 days, statary phase 28 days	Schneirla (1965 and personal communication)
Tribe Ecitonini						
Eciton burchelli	150,000–700,000	Exposed clusters above surface, often arboreal; no modification of bivouac site	Arthropods of a wide variety, including immature stages of social wasps and ants	Swarm 10–15 m in front, tapers to trunk column in rear; raid begins at daybreak, ends at dusk	Nomadic phase (11–17 days) alternates with statary phase (19–22 days); emigrations daily during nomadic phase, which begins when adults eclose from pupae	Schneirla (1957a)
Eciton hamatum	100,000–500,000	Exposed clusters above surface, seldom arboreal; no modification of bivouac site	Immature stages of social wasps and other ants	Several dendritic columns lead away from bivouac site; raid begins at daybreak, ends at dusk	As in *E. burchelli*, except that nomadic phase lasts 16–18 days and statary phase 17–22 days	Schneirla (1957a) Rettenmeyer (1963a)
Eciton vagans	?	Mostly in preformed subterranean cavities such as abandoned ant nests, also in and under rotting logs; some excavation is practiced	Primarily immature stages of other ants, plus a few other arthropods	As in *E. hamatum*, except that columns are partly subterranean and partly nocturnal	Emigrate with larvae and callow adults	Fiebrig (1907), Schneirla *et al.* (1954) Rettenmeyer (1963a)
Labidus praedator	"Probably" over 1 million	As in *E. vagans;* some excavation is practiced	Arthropods of a wide variety, including immature forms of other ants; some nuts and other vegetable matter	Swarm variable in size and shape; front usually less than 4 m across, tapering to trunk columns; much of the raiding is underground and occurs day and night	Emigrations occur, but pattern is apparently irregular and not closely correlated with brood development	Rettenmeyer (1963a)
Nomamyrmex esenbecki	?	Subterranean bivouac sites of unknown nature	Mostly brood of other ants; some other arthropods also	Weak dendritic columns, mostly under objects on ground or underground	Probably an irregular *Eciton* pattern, but evidence is incomplete	Rettenmeyer (1963a)
Neivamyrmex nigrescens	80,000–140,000	Mostly subterranean bivouac sites, often under rocks and logs; also occasionally in rotting logs	Primarily other ants; also a few beetles and other small arthropods	Weak dendritic columns, mostly over the surface of the ground but partly subterranean; raid at night or on overcast days	The *Eciton* pattern is followed but less regularly, with emigrations less than daily during the nomadic phase; nomadic phase lasts about 14 days, statary phase 19–21 days	Schneirla (1958, 1961, 1963) Rettenmeyer (1963a)

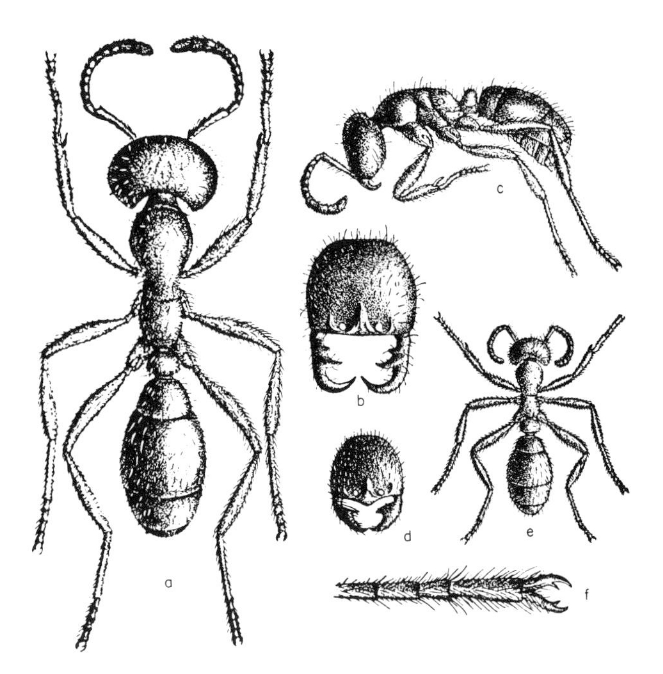

FIGURE 4-29. The army ant *Cheliomyrmex nortoni* of Central America: (*a*) soldier; (*b*) head of soldier seen from above; (*c*) media; (*d*) head of media; (*e*) minor worker; (*f*) tarsus showing toothed claws (from Wheeler, 1910).

mex, a rare genus found in the rain forests of the New World tropics, also has an archaic appearance. Although it is polymorphic and has a specialized mandibular structure placing it close to the Ecitonini, it possesses a single-jointed pedicel and toothed tarsal claws, both regarded as primitive traits (see Figure 4-29). Unfortunately, aside from a single raid and an emigration (with larvae) by a *C. megalonyx* colony reported by Wheeler in 1921, nothing is known of its biology.

The African driver ants of the genus *Anomma* and *Dorylus* differ from the ecitonines in several important details of the activity cycle, which are apparently caused by peculiarities in the queen caste (Raignier and van Boven, 1955). The queens are the largest of all ants (Figure 4-30). Those of *Anomma* vary from 39 to 50 mm in total length and possess as many as 14,000 ovarioles capable of delivering 3 to 4 million eggs in a month. The abdomen is in a permanent state of moderate physogastry, and the queen lays eggs more or less continuously. Most of the eggs, however, are produced in bursts that come in approximately three-week intervals and last five or six days. The ensuing development of the larval brood appears to have little effect on the inducement of emigration. In fact, larvae are usually outnumbered by pupae in the brood of emigrating colonies. No clock-like alternation between statary and nomadic phases of the *Eciton* type is displayed by the colonies of *Anomma.* The emigrations, which take several days to complete, are separated by statary periods that vary in duration from six days to two

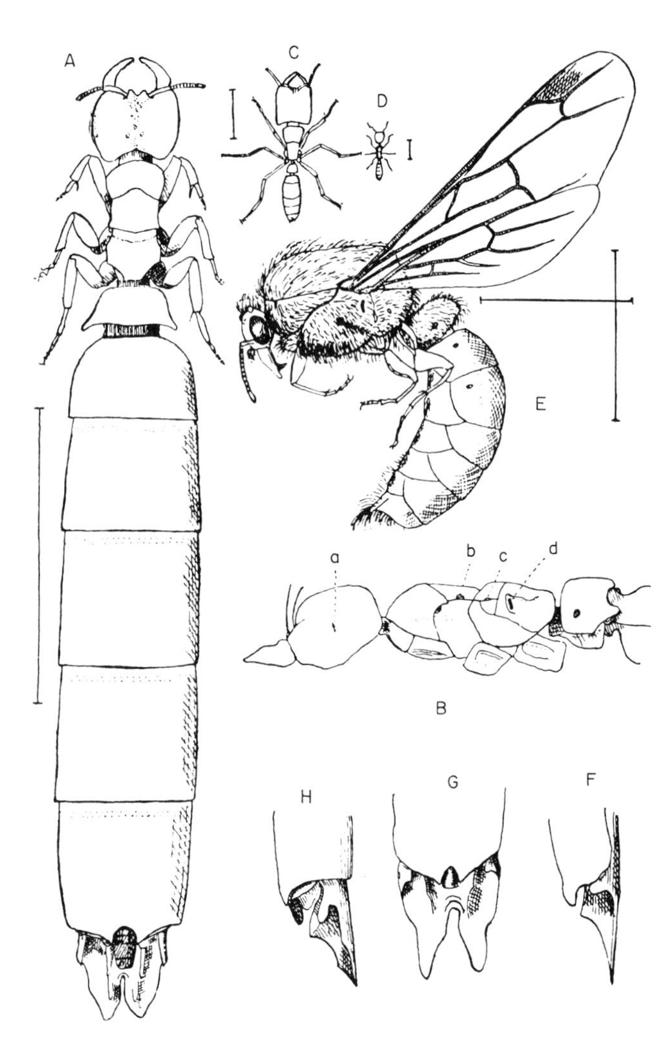

FIGURE 4-30. The castes of the driver ant *Dorylus helveolus* drawn at the same magnification: (*A*) queen (dichthadiigyne) from above; (*B*) queen from the side (*a*, vestige of eye; *b, c,* vestiges of wings; *d,* propodeal spiracle); (*C*) major worker; (*D*) minor worker; (*E*) male; (*F*) tip of queen's abdomen seen from the side. (*G, H*) views of the abdominal tip of the related species *D. furcatus* (from Emery, 1895).

or three months. The bivouacs of *Anomma* are also much more stable affairs than those of *Eciton* and most other army ants. The colonies settle deeply into the soil at the end of the emigration, excavating labyrinthine systems of galleries and chambers to a depth of 1 to 4 m.

From these secure nests the *Anomma* send forth almost daily raids. The swarm pattern, illustrated in Figure 4-31, unfolds like a great pseudopodium. It engulfs all of the ground and low vegetation in its path, and then, after a few hours, drains back to the bivouac site. The advance is leaderless. The excited workers rush back and forth at an average speed of 4 cm per second. Those in the van press forward for a short distance and then retreat back into the mass to give way to new advance runners. The columns resemble thick black ropes lying along the ground. A close examination shows them to be dozens or hundreds of workers wide. The ants are so dense that they pile on top of one another and run along on one another's backs, while some spill away from the column and form scattered crowds to either side, their antennae and mandibles pointed upward in threatening postures. The frontal swarm, which contains up to several millions of workers, advances at a rate of about 20 m per hour, gathering and killing almost all insects and larger animal life too sluggish to get out of the way. Savage's famous account of 1847 expresses the drama of the *Anomma* foray:

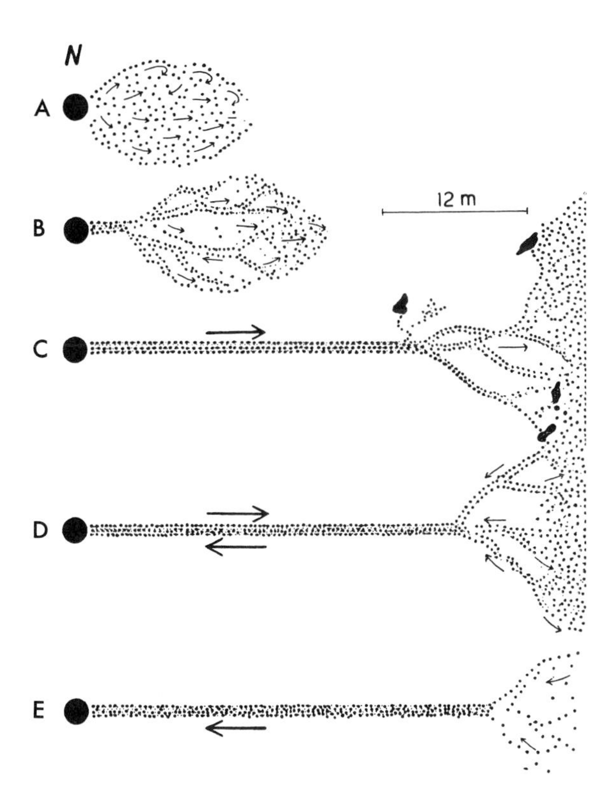

FIGURE 4-31. The general pattern of advance (*A–C*) and retreat (*D, E*) of a swarm raid by an *Anomma* colony from its bivouac site (from Raignier and van Boven, 1955).

> They will soon kill the largest animal if confined. They attack lizards, guanas, snakes, etc. with complete success. We have lost several animals by them,—monkeys, pigs, fowls, etc. The severity of their bite, increased to great intensity by vast numbers, it is impossible to conceive. We may easily believe that it would prove fatal to almost any animal in confinement. They have been known to destroy the *Python natalensis,* our largest serpent. When gorged with prey it lies powerless for days; then, monster as it is, it easily becomes their victim . . . Their entrance into a house is soon known by the simultaneous and universal movement of rats, mice, lizards, *Blapsidae, Blattidae* and of the numerous vermin that infest our dwellings. Not being agreed, they cannot dwell together, which modifies in a good measure the severity of the Driver's habits, and renders their visits sometimes (though very seldom in my view) desirable. Their ascent into our beds we sometimes prevent by placing the feet of the bedsteads into a basin of vinegar, or some other uncongenial fluid; this will generally be successful if the rooms are ceiled, or the floors overhead tight, otherwise they will drop down upon us, bringing along with them their noxious prey in the very act of contending for victory. They move over the house with a good degree of order unless disturbed, occasionally spreading abroad, ransacking one point after another, till, either having found something desirable, they collect upon it, when they can be destroyed "en masse" by hot water; or, disappointed, they abandon the premises as a barren spot, and seek some other more promising for exploration. When they are fairly in we give up the house, and try to await with patience their pleasure, thankful, indeed, if permitted to remain within the narrow limits of our beds or chairs. They are decidedly carnivorous in their propensities. Fresh meat of all kinds is their favourite food; fresh oils they also love, especially that of the *Elais guiniensis,* either in the fruit or expressed. Under my observation they pass by milk, sugar, and pastry of all kinds, also salt meat; the latter, when boiled, they have eaten, but not with the zest of fresh. It is an incorrect statement, often made, that "they devour everything eatable" by us in our houses; there are many articles which form an exception. If a heap of rubbish comes within their route, they invariably explore it when larvae and insects of all orders may be seen borne off in triumph,—especially the former.

In answer to the single question I am asked most frequently about ants, I can give the following answer: No, driver ants are not really the terror of the jungle. Although the driver ant colony is an "animal" weighing in excess of 20 kg and possessing on the order of 20 million mouths and stings and is surely the most formidable creation of the insect world, it still does not match up to the lurid stories told about it. After all, the swarm can only cover about a meter of ground every three minutes. Any competent bush mouse, not to mention man or elephant, can step aside and contemplate the whole grass-roots frenzy at leisure, an object less of menace than of strangeness and wonder, the culmination of an evolutionary story as different from that of mammals as it is possible to conceive in this world.

The Origin of Social Behavior in Ants

Except for some specialized modes of colony reproduction, such as the budding exemplified in *Monomorium pharaonis* and the army ants, the details of the colony life cycle do not vary greatly within the Formicidae. The partially claustral method of colony foundation that characterizes *Myrmecia* is shared with many Ponerinae, in particular the species of *Amblyopone, Odontomachus, Paraponera, Pachycondyla, Bothroponera, Trachymesopus,* and *Proceratium* (Haskins and Haskins, 1951). It occurs, in addition, in at least two myrmicine species: *Acromyrmex octospinosa* (Cordero, 1963) and *Manica* (= *Neomyrma*) *rubida* (Le Masne and Bonavita, 1967). Within the Ponerinae we can further observe the finely graded steps leading from the partial to the fully claustral mode of colony foundation in which the queen never leaves the nest. The queens of *Bothroponera soror,* for example, belong to an exactly intermediate stage. They forage outside the nest, but their wing muscles are still reduced and metabolized as in the higher ants (Haskins, 1941). The queens of *Odontomachus haematodus* are capable of rearing their first brood at least partially on their own oral secretions, but, in one experiment performed by Haskins and Haskins (1950b), the larvae did not reach maturity. Finally, these authors found that the unusually bulky queens of *Brachyponera lutea* are capable of rearing the first brood all the way through on their own secretions, even though they also continue to forage outside the nest for insect prey when given the opportunity.

From *Brachyponera* it is but a short step to the higher formicid stage exemplified by *Myrmica.* The condition described earlier for *Monomorium pharaonis* represents a loss of some of the basic behavioral elements of queen behavior with the addition of some other minor ones. This is generally true of the other specialized life cycles within the ants, including those associated with social parasitism.

Even though evolution of the life cycle within the contemporary Formicidae can be clearly delineated by comparisons of species, the gap between the most primitive living ants (*Myrmecia*) and their presumptive ancestors among the tiphiid wasps remains very great. After the discovery of *Sphecomyrma* pointed to *Methocha* as the closest living wasp genus to at least the myrmecioid complex of ant subfamilies, my co-workers and I consulted the literature to learn whether elements in the behavior of *Methocha* had been detected that might provide hints concerning the origin of social life in ants. It was no help to learn that all of the several species of *Methocha* whose habits are known are specialized parasitoids on tiger beetle larvae. According to Burdick and Wasbauer (1959), for example, the female of *Methocha californica* enters the vertical burrow of its prey, stings it into paralysis, lays an egg, closes the burrow entrance with soil, and leaves to search for new victims (see Figure 4-32.) These observations, along with similar ones on Philippine species by F. X. Williams, indicate that *Methocha* has advanced no further than the lowly first stage of Evans' scheme of the evolution of social wasps (Chapter 3). Nevertheless, in more recent experiments with *M. stygia,* Donald Farish, Barry P. Moore, and I produced evidence that the females are capable of something more. When presented with tiger beetle larvae on the ground away from the burrows, the females stung them into immobility, excavated simple cavities in the soil, and buried the larvae. This sequence is Evans' third stage. Ferton (in Wheeler, 1928) has witnessed similar behavior on the part of *Myzus andrei* toward tenebrionid larvae. The option of handling prey in a somewhat more complicated manner may prove, on closer examination, to be widespread in the Tiphiidae. Even so, the gap between Evans' third stage and the well-established eusociality of *Myrmecia* (stage 13) is vast. Perhaps the biology of *Nothomyrmecia macrops* will narrow the gap somewhat when that enigmatic species is finally rediscovered in Australia, but I am not very hopeful.

The picture is further obscured by the possible independent origin of eusociality, or at least of the thirteenth stage of eusociality, in the poneroid complex of ants. The most primitive living genus of the complex on morphological grounds is *Amblyopone.* Haskins and Haskins (1951), after completing behavioral studies of this genus

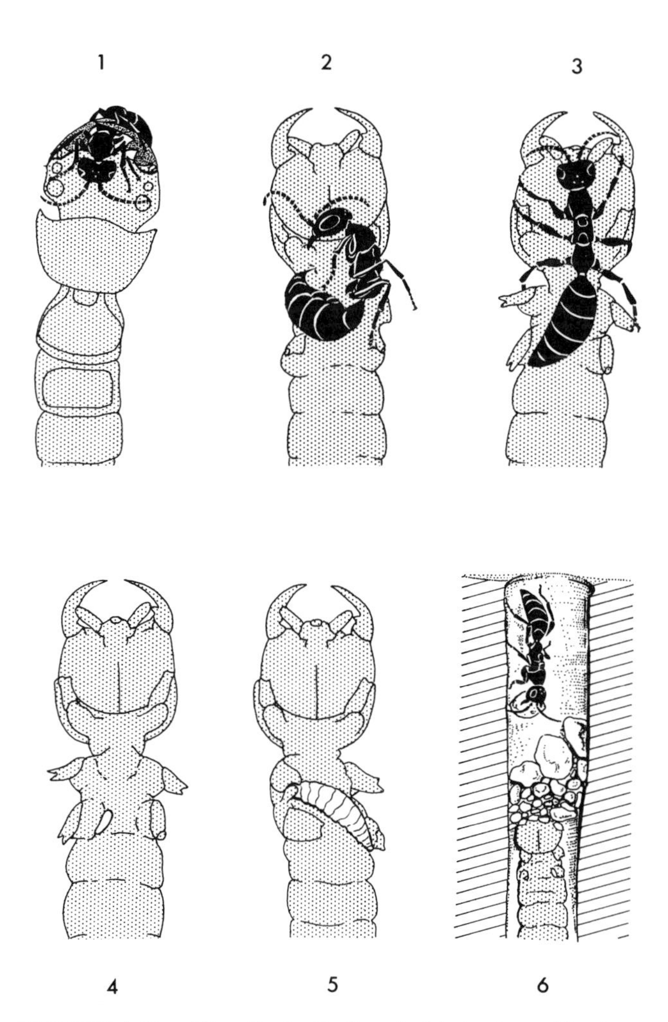

FIGURE 4-32. The attack of a female *Methocha californica* on a tiger beetle larva. In this sequence of drawings, adapted from photographs, the wasp slips through the mandibles of its giant prey (1), stings it into paralysis (2), and prepares to deposit an egg (3). After about 100 hours the egg, as seen in (4), hatches into a rapidly growing larva that feeds on the living but immobile tiger beetle larva (5). Another female is shown closing the burrow of a second prey larva following oviposition (6) (redrawn from Burdick and Wasbauer, 1959).

and *Myrmecia,* drew the following conclusion concerning differences between the two genera: "*Myrmecia* and *Amblyopone* represent widely divergent groups of ants . . . They have, in fact, relatively little immediately in common save the obviously immensely archaic nature of both groups, as unmistakably evidenced in many features of individual morphology, of distribution, and of colonial organization. The marked parallelism in the main features of the pattern of colony foundation to be observed in both groups, therefore, is strongly suggestive that this general pattern may represent an approximate recapitulation of the manner in which Formicid societies arose. It seems likely, in the present state of our knowledge, that the existent wealth and variety of Formicid social structures, with their tremendous range of variation from group to group, took their evolutionary beginnings in the activities of solitary, winged, ground-dwelling wasplike types in which the female, having dealated herself after fertilization, constructed a shelter in the ground and reared a small family to maturity. The larvae were provided with freshly killed prey captured and supplied through a behavior pattern intermediate in its complexity between the simple provisioning of paralyzed insects characteristic of the modern solitary Sphecoid wasps and the malaxated pellets of insects prepared for the larvae by their nurses in such primitive social Vespids as *Polistes.*"

It is still not certain that *Amblyopone* and its close relatives really found colonies in just the same way as *Myrmecia,* as Haskins and Haskins assume. Admittedly, queens of *Amblyopone australis* kept in artificial nests by these investigators behaved essentially like *Myrmecia* queens, and the same is true of other ponerines whose colony founding has proven to be partially claustral. On the other hand, *Amblyopone* and its relatives differ from other primitive ants in a striking way. They favor large arthropods as prey, and they have a strong tendency to move their brood to newly secured prey rather than the other way around. Small, isolated groups of workers and larvae of the North American species *A. pallipes* are often found clustered around recently killed centipedes, the usual prey of the species (W. L. Brown, personal communication). Since the larvae feed directly on the whole centipedes, without previous dismemberment, and since these arthropods are too large and clumsy to be transported for any distance intact, it is clear that the workers transport the larvae to the prey rather than the reverse. It is even conceivable that colonies emigrate in part or in whole in this fashion to follow their food source. I have observed such behavior in the big amblyoponine *Myopopone castanea* in New Guinea (Wilson, 1958b). On several occasions small groups of workers and larvae were found clustered around large wood-boring beetle larvae that could not possibly have been carried any significant distance from the spot where they had been found and presumably killed by the workers. In one rotting log that I thoroughly dissected, workers and larvae of *Myopopone* were found around two such prey at widely separated spots. The conclusion was inescapable in this case that the colony had distributed itself at least in part to be located near the prey. What constitutes "emigration" in *Myopopone* is nevertheless open to question since it can be argued that the entire subcortical surface of the log constitutes the "nest" of the ants.

The late Soviet entomologist S. I. Malyshev (1960, 1968) went so far as to suggest that such behavior was the key step in the socialization of the ants. He postulated that specialization on big prey resulted in the mother wasp staying in the vicinity of her young long enough for them to get to know her and to cooperate with her. Malyshev further postulated that the precursors of the ants must have fed on fungi growing in the nest walls. Otherwise, he contends, the young colonies would have no way of tiding over the period of scarcity after the initial large prey was consumed. But this suggestion is gratuitous with reference to the remainder of the theory and is wholly unsupported by any evidence of fungus eating in the lower ants. The model for the large prey theory is provided by members of the bethylid genus *Scleroderma,* particularly *S. immigrans* and *S. macrogaster,* which were studied in detail by Bridwell (1920) and Wheeler (1928). The female *S. macrogaster,* for example, is only 2.5 to 3 mm long, and she attacks beetle larvae that are hundreds or thousands of times greater in bulk. In a typical sequence, the female first crawls over the surface of her prey, pausing from time to time to grip little folds of the cuticle. As long as the muscles beneath in the cuticle show any sign of contraction, the female stings at that point. Finally, after one to four days, the larva becomes completely paralyzed. The *Scleroderma* now feeds for several days by making little punctures in the cuticle and drinking the hemolymph. Her abdomen then begins to swell as the ovaries develop, and after a time she lays eggs on the surface of the prey. The remainder of the life cycle has been described by Wheeler in the following striking passage:

> The eggs laid on a larva or young pupa produce minute larvae which at first lie on the surface but later become spindle-shaped and erect, so that the host bristles with them like a porcupine. The older larvae acquire the colour of the juices of the prey; those feeding on the pink larvae or pupae of *Liopus* becoming red. They are always spotted with white, owing to the large masses of urate crystals in their fat bodies. The mother *Scleroderma* remains with the larvae, often stands over them and may sometimes lick them, holding them meanwhile in her fore feet. She also continues occasionally to drink the host's blood, which exudes about the deeply inserted heads of her larval offspring. Although she will sometimes eat her eggs I have never seen her attack one of her larvae. The devouring of some of the eggs seems to be due to a tendency to regulate their number according to the volume of the prey. When the larvae are mature they fall away from its shrivelled and exhausted remains and spin snow-white cocoons in a cluster. Pupation covers a period of fourteen to thirty days. The males emerge first from their cocoons, at once eat their way into the female cocoons and fecundate the pupae. They also mate readily with the same individual five to eight times after brief intervals. The same females may also mate with several males in succession. So great is the ardour of the latter that they often attempt to mate with one another. The mother being a long-lived insect may mate with one of her sons and will readily paralyze another beetle larva, rear another brood and mate again with one of her grandsons. (Wheeler, 1928:63).

Although Malyshev, inspired by Wheeler's observations, speaks of the "sclerodermoid ancestors" of the ants, morphological evidence rules against any of the ant groups having been derived from *Scleroderma*-like progenitors or any of the other known Bethylidae. Nevertheless, the possibility exists that eusociality in the Amblyoponini or their tiphiid ancestors originated through a form of subsociality similar to that in *Scleroderma.* The myrmecioids, in contrast, evidently approached eusociality through the *Myrmecia* form of partially claustral nest formation. The behavior patterns of ponerines other than the Amblyoponini are, furthermore, fully consistent with this latter conception. The possibility must therefore be considered that the Amblyoponini represent an independent line of ants which differs from other ant stocks in having passed through a *Scleroderma*-like form of subsociality. This matter might be clarified by comparative studies of the genera of the Amblyoponini (*Amblyopone, Myopopone, Mystrium, Prionopelta, Onychomyrmex*), with close attention to the details of their colony founding. The biology of the great majority of the amblyoponine species is still wholly unknown.

5 The Social Bees

Eusociality has arisen at least eight times within the bee superfamily Apoidea by both the parasocial and subsocial routes, and presociality of nearly every conceivable degree has emerged on an uncounted number of other occasions. This prevalence and great variability of social behavior in bees provides an opportunity to study the evolution of social behavior paralleled only in wasps, an opportunity that has only begun to be exploited. Even more than in wasps, steps in the evolutionary progression of bee sociality can be delineated within small groups of related species and faunal units. In the subgenus *Seladonia* of *Halictus,* for example, are found species ranging from completely solitary to moderately eusocial in behavior. By coincidence, the two extremes are represented in turn by the only two species of *Seladonia* known to occur in Japan (Sakagami and Fukushima, 1961). Such evolutionary spans are commonplace in the Halictinae. When small sets of halictine species are randomly chosen for analysis out of local faunas, the chances are always high that they will embrace much of the full range of sociality. A case in point is Suzanne Batra's (1966c) eleven-month study of halictine bees in India. Among twelve species selected for their accessibility, three were found to be solitary, two communal, two quasisocial, and five eusocial to varying degrees. The replicability of the progression seen in samples of various genera taken from different parts of the world, particularly within the Halictinae, offers the promise that eventually the more important evolutionary hypotheses can be subjected to statistical testing. A further advantage stems from the fact that the honeybee (*Apis mellifera*) has been by far the most intensively studied of all social insects and is available as a base line for comparative studies of most aspects of behavior and physiology.

All the bees together comprise the superfamily Apoidea. On morphological grounds they fall closest to the sphecoid wasps, although the lack of an adequate fossil record has made it impossible to pinpoint the exact ancestral phyletic line. In a word, the Apoidea can be loosely characterized as sphecoid wasps that have specialized on collecting pollen instead of insect prey as larval food. The adults are still wasp-like in that they eat nectar (and sometimes store it, in the form of honey), but, unlike the vast majority of true wasps, including all of the sphecoids, they feed their larvae on pollen or pollen-honey mixtures. Some of the eusocial species feed their larvae on specialized glandular products derived ultimately from pollen and nectar. The adult morphology is modified in peculiar ways to facilitate this dietary specialization. Plumose (branched) hairs occur on the body and evidently serve to collect and to carry pollen during visits to flowers. Most kinds of bees collect the bulk of the pollen with specialized, often curved or hooked hairs on the front tarsi and then transfer it to scopae. In the halictids and many colletids and andrenids, pollen is carried on scopal hairs arrayed along the hind legs all the way from the coxae to the basitarsi, but much of it is also borne by similar hairs on the sides of the propodeum and undersurface of the abdomen. The megachilids are distinguished by the fact that pollen is carried exclusively by an enlarged, dense scopa on the undersurface of the abdomen. In the Apidae the outer surfaces of the hind tibiae and basitarsi have assumed the exclusive carrying role. In the Apinae in particular, which includes the honeybee genus *Apis,* the hind tibiae have been expanded; their inner surfaces are slightly concave and lined on the outside with "pollen combs," rows of stiff hairs in which the pollen is compacted in the form of balls.

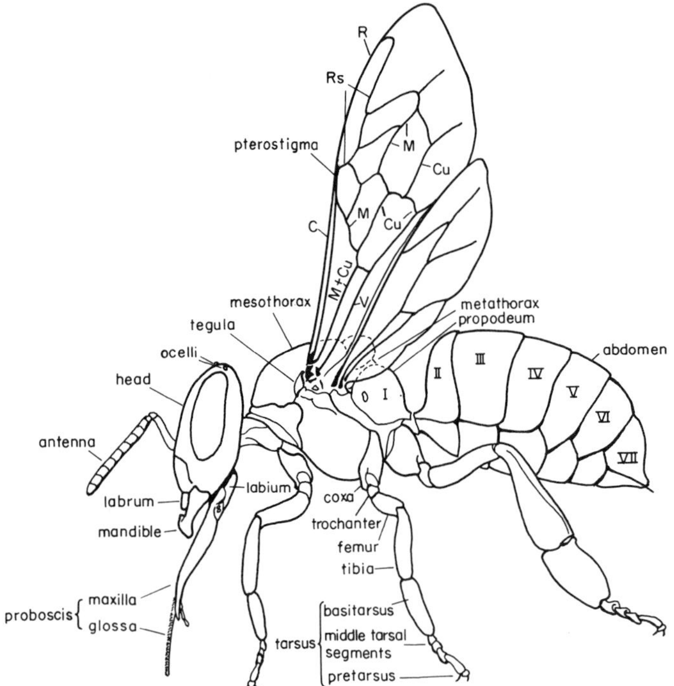

FIGURE 5-1. The principal features of the external anatomy of the honeybee worker (*Apis mellifera*): The principal veins of the fore wing shown are the costa (*C*), the radius (*R*), the radial sector (*Rs*), the media (*M*), the cubitus (*Cu*), the fused portions of the media and cubitus (*M*+ *Cu*), and the vannal veins (*V*). The abdominal segments are labeled with roman numerals. The first of these segments, the propodeum, is fused anteriorly to the thorax (based on Michener, 1944; Snodgrass, 1956).

The entire basket-like structure is called the corbicula. The pollen combs and brushes tend to be lost in parasitic groups such as the halictine *Sphecodes.* Scopal hairs are also lacking in the robber genus *Lestrimelitta* (Apidae: Meliponini) and in some of the Colletidae, the females of which carry their pollen mixed with nectar or honey in the crop.

The present bee fauna probably originated as far back as the Cretaceous era, more than 70 million years ago. H. E. Evans (1969) has recently described two sphecoid wasps from the Cretaceous, one of them generalized enough to be a possible ancestor of the Apoidea. Many specialized bee genera, as well as genera of sphecid wasps, are known from the Baltic amber, Florissant shales, and Oeningen shales, all of Oligocene age. The present bee fauna has evolved for over 50 million years in close concert with the angiosperms on whose flowers they depend. Apparently many groups are still very actively speciating, with the result that bees as a whole are taxonomically difficult.

In 1953 a total of 3,285 bee species had been described from North America, and it seems reasonable that the figure will eventually reach 4,000. Michener (1969a), on the basis of the latter estimate and his personal experience with the faunas of Central and South America, Africa, and Australia, has guessed that the world contains about 20,000 living species of bees. The estimate is the same as that inferred by Friese (1923) on the basis of earlier taxonomic studies. Only a small but as yet undetermined minority of the total are presocial and eusocial.

A partial classification of the bees is given in Table 5-1 and a simple phylogeny of families in Figure 5-2. Modern taxonomic studies of the apoid families containing social species have been only partially completed. Michener's

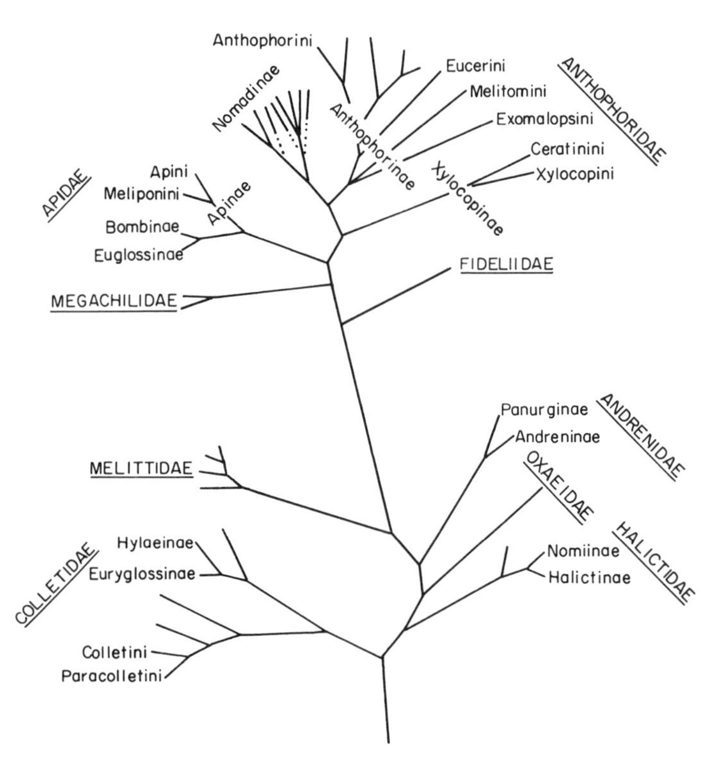

FIGURE 5-2. Simple branching diagram of the phylogeny (cladogram) of the families of the bees (Apoidea) (modified from Michener, 1964a, and personal communication).

TABLE 5-1. Families, subfamilies, and principal tribes of bees (superfamily Apoidea), the degrees of sociality exhibited by various of their species, and their world distribution (based on Michener, 1944, 1965a, 1969a). Selected genera, including those mentioned in the present book, are also listed. The terms designating the degrees of sociality are defined in Chapter 2.

Division	Degrees of sociality	Distribution
Family Colletidae (*Colletes, Hylaeus*)	Solitary	World-wide
Family Halictidae		
Subfamily Nomiinae (*Nomia*)	Solitary, communal, quasisocial	World-wide
Subfamily Halictinae		
Tribe Augochlorini (*Augochlora, Augochlorella, Augochloropsis, Neocorynura, Pseudaugochloropsis*)	Solitary, communal, semisocial, eusocial	New World
Tribe Halictini (*Dialictus, Evylaeus, Halictus, Lasioglossum, Pseudagapostemon, Ruizantheda, Sphecodogastra, Sphecodes*)	Solitary, communal, eusocial	World-wide
Subfamily Dufoureinae (*Dufourea*)	Solitary	Asia, Africa, Europe, North America
Family Andrenidae		
Subfamily Andreninae (*Andrena*)	Solitary, communal (?)	World-wide, except Australia
Subfamily Panurginae (*Panurginus, Perdita*)	Solitary, communal	Africa, Eurasia, New World
Family Oxaeidae (*Oxaea*)	Solitary	New World
Family Melittidae (*Hesperaspis, Melitta*)	Solitary	World-wide
Family Fideliidae (*Fidelia*)	Solitary (?)	Africa, Chile
Family Megachilidae		
Subfamily Lithurginae (*Lithurge*)	Solitary	World-wide
Subfamily Megachilinae		
Tribe Megachilini (*Chalicodoma, Megachile, Osmia*)	Solitary, communal, quasisocial (?)	World-wide
Family Megachilidae (continued)		
Tribe Anthidiini (*Anthidium, Dianthidium, Heteranthidium*)	Solitary, communal	World-wide
Family Anthophoridae		
Subfamily Nomadinae (*Nomada*)	Solitary (parasitic)	World-wide
Subfamily Exomalopsinae (*Exomalopsis*)	Solitary, communal, quasisocial (?)	Africa, Eurasia, New World
Subfamily Anthophorinae		
Tribe Anthophorini (*Amegilla, Anthophora*)	Solitary	World-wide
Tribe Eucerini (*Eucera, Melissodes, Svastra*)	Solitary, communal	World-wide, except Australia
Tribes Melitomini, Centridini, Melectini, Ericrocini, etc.	Solitary (Some are parasitic)	World-wide
Subfamily Xylocopinae		
Tribe Xylocopini (*Xylocopa*)	Solitary	World-wide
Tribe Ceratinini (*Allodape, Allodapula, Allodapulodes, Braunsapis, Ceratina, Eucondylops, Exoneura, Exoneurella, Exoneurida, Exoneurula, Halterapis, Inquilina, Macrogalea, Nasutapis*)	Solitary, subsocial, eusocial, socially parasitic	World-wide
Family Apidae		
Subfamily Euglossinae (*Euglossa, Eulaema, Euplusia*)	Solitary, communal (?), quasisocial	New World tropics
Subfamily Bombinae (*Bombus, Psithyrus*)	Eusocial	Nearly world-wide; introduced into Australia
Subfamily Apinae		
Tribe Meliponini (*Dactylurina, Lestrimelitta, Melipona, Meliponula, Trigona*)	Eusocial	Nearly world-wide; tropical and southern subtropical zones
Tribe Apini (*Apis*)	Eusocial	Africa, Eurasia; introduced into rest of world

1944 generic revision of the bees of the world still remains the key work of its kind, and it has been supplemented by an analysis of larval characters and reassessments of some generic and familial limits (Michener, 1953a, 1964a). Regional studies of entire bee faunas include those of Friese (1923) on Europe, Friese (1909) and Arnold (1947) on Africa, Bingham (1897) on India, Michener (1965a) on Australia and the South Pacific, Mitchell (1960) on the eastern United States, and Michener (1954) on Panama. A list of the North America fauna prepared by Michener, Krombein, and others and complete to 1963 is included in the catalogs of Hymenoptera by Muesebeck *et al.* (1951) and Krombein *et al.* (1958, 1967). Among studies of the social taxa are Schwarz' monograph (1948) on the Meliponini of the New World, Moure's generic revision of the Old World Meliponini (1961) and checklist of the world Euglossinae (1967), Milliron's (1961) generic reclassification of the Bombinae, Michener's (1969c) review of the African allodapine genera, and species-level revisions of the British *Bombus* by O. W. Richards (1927a), the Japanese *Bombus* by Sakagami and Ishikawa (1969), the New World Bombinae by Franklin (1912–1913), the Indomalayan *Halictus* by Blüthgen (1926, 1931), the Japanese Nomiinae by Hirashima (1961), and the honeybees (Apini) by Maa (1953) and Dupraw (1965).

A high level of sophistication has been attained in the techniques of culturing bees and observing their behavior inside the nests. Observation hives for honeybees, accompanied by live colonies and sets of instruction for their care, can be purchased from commercial companies. Special techniques for analysis of bee behavior are given in various recent books, most notably that of von Frisch (1967a). The culturing and study of bumblebees, an old and well-developed art, has been reviewed in detail by Sladen (1912), Plath (1934), Free and Butler (1959), Pouvreau (1965), and Plowright and Jay (1966). The principal trick in studying the social behavior of bumblebees is to induce overwintered females to start their nests in specially constructed observation boxes placed in the field. Sakagami (1966) has described an observation nest which has been used with conspicuous success in the study of stingless bees. Even the halictine bees have been cultured in vertical, glass-walled nests by Batra (1964), a method that permitted the first close studies of behavioral interactions among colony members. Assorted techniques for the study of social halictines in the field have been reviewed by Michener *et al.* (1955).

Let us now examine in some detail current knowledge of the biology of each of the five major groups of social bees: the social Halictinae, the social Ceratini, the Bombinae, the Meliponini, and the Apini. Then we will be in a position to consider the very interesting, albeit tangled, reconstructions of the origin of sociality in bees as a whole.

The Natural History of the Eusocial Sweat Bees in the Subfamily Halictinae

The little sweat bees belonging to *Halictus* and the closely related genera *Dialictus, Evylaeus, Lasioglossum,* and *Sphecodogastra* are notable for containing both solitary species and species in the earliest evolutionary stages of eusociality. Recent authors have differed considerably in their interpretations of generic limits within this complex. What was originally *Halictus* has been divided into two, three, or more genera by some. *Dialictus* is either placed by itself or made a subgenus of *Lasioglossum,* which itself was formerly recognized as no more than a subgenus of *Halictus.* Rather arbitrarily, and in hopes of avoiding nomenclatural confusion, I am following the usage of the "splitters," in particular Mitchell (1960) and Knerer and Atwood (1964). The expression "sweat bee," incidentally, is loosely applied to any small bee of this general appearance, solitary or social, in a number of families including the Halictidae, Andrenidae, and Colletidae. The entire group is taxonomically difficult because of the many sets of sibling species whose members are not only morphologically almost identical to each other but also frequently visit the same flowers and share common nesting areas. In many cases the species are most easily separated by physiological and behavioral characteristics, including the time of appearance, the shape and size of the nest, and the number of annual generations. Knerer and Atwood (1966b) recently distinguished two sibling species of *Evylaeus* for the first time on the basis of differences in nest structure. In France alone there are nearly 150 species in the *Halictus* group of genera, and in North America, north of Mexico, over 250. The majority are solitary. Those that have attained eusociality are still only at a relatively primitive level in this behavioral category. Their colonies contain, depending on the species, from two or three to several hundred individuals which occupy a single burrow in the soil. In most of the eusocial species the queens and workers can be distinguished from each other only by more physiological traits, such as ovarian development, and by behavior. In a minority of species there is also a marked morphological difference (see Chapter 9). Most of the species nest in the soil, a primitive characteristic for bees generally and an apparent holdover from similar behavior in the sphecoid ancestors. A few species nest in rotting wood.

It is a peculiar fact of social life in the Halictinae that little if any positive correlation exists among species between the complexity of nest structure and the degree of sociality (Sakagami and Michener, 1962). Knerer and Atwood (1966b) have even suggested a negative correlation may exist. For example, *Halictus ligatus* and *Evylaeus cinctipes* display a relatively advanced degree of eusociality but a notably simple nest structure, whereas the reverse is the case in *Augochlorella striata.*

Knerer and Cécile Plateaux-Quénu (1966a–c, 1967a,b) have documented the following multiple changes by which advances in eusociality within the Halictinae can be conveniently measured.

1. There is an increased tendency for the workers to keep the brood cells open, to inspect them, and perhaps even to add nectar and pollen (see Figure 5-3). Knerer and Plateaux-Quénu (1966a) and Suzanne Batra (1964) have independently concluded that such direct contact between adults and young is widespread in the eusocial Halictinae and that its intensity is variable and can be taken as a convenient measure of the degree of eusociality.

2. Variation in ovarian development increases among the females that found individual colonies (Figure 5-4). In the more advanced eusocial species, consequently, some of the females of the same generation, if they cooperate in founding colonies, tend to assume the role of workers from the outset, while others assume the role of queens. In other words, during early colony development the females are organized in a semisocial state. As the first generation of workers appears, the colony becomes eusocial.

3. There is an increase in size difference between the queens and the workers. (See Figure 5-5).

4. There is an increased tendency toward the production of two or three annual broods instead of only one and a concomitant reduction in the proportion of males produced. Furthermore, the males tend to be produced with the later broods. There is some question as to how generally this criterion can be applied. At the present time the separation of broods is known only in *Evylaeus,* and, as a consequence, at least one student of the subject (Michener, personal communication) has expressed doubts that the separation of broods is a good measure of social progress.

These four criteria are for the most part both convenient and intuitively satisfying. Even better, they are correlated. Another colony characteristic that appears to be correlated with them is colony size. Still another that may or may not be related in an interesting fashion is the presence or absence of colony odor. Michener (1966b) found that when workers of *Dialictus versatus,* a eusocial species whose colonies nest in small, dense aggregations, occasionally become disoriented during flight and enter the wrong nest, they are amicably received. Thus *D. versatus* appears to depend on visual recognition of the nest site or some other external cue rather than colony odor to segregate its colonies. Information on colony odor in other species is lacking, so it still remains to be seen whether this characteristic occurs within the socially more advanced Halictinae.

The history of studies in halictine sociology is surprisingly recent. The existence of eusociality was not even recognized until the appearance of Stöckhert's classical work (1923) on five species, placed at that time in the old genus *Halictus.* The early history of this and other studies of the behavior of the Halictinae was thoroughly reviewed by Wheeler (1928). His statement that "there may be many surprises in store for us when the life-histories of these seemingly monotonous and uninteresting bees have been subjected to more careful scrutiny and experimentation on other continents, especially in South America and Australia," has proven prophetic. In the past ten years the difficult task of working out life cycles of the multitudinous, look-alike species, typified by what Michener once described as "the painstaking work of making observations and digging deep nests in the hot summer sun," has been achieved with distinction. Due to the efforts of Batra, Knerer, Michener, Plateaux-Quénu, Sakagami, and their associates, there now exists a small but informative body of life history monographs on a diversity of halictine species.

As a paradigm of a primitively eusocial halictine I have chosen *Dialictus zephyrus* (= *Lasioglossum zephyrum*), which was thoroughly studied in Kansas by Batra (1964, 1966a). This small greenish-black bee is widely distributed in North America. It flies from early spring to late November, collecting nectar and pollen from no less than 30 families of flowering plants—a catholicity of diet

FIGURE 5-3. Queen of the eusocial halictine *Evylaeus cinctipes* inspecting her nest. The pupae are workers. Notice the open brood cells, which have been cleaned of feces and molted skins. This closer contact between adults and brood is associated with a higher degree of eusociality in the Halictinae (from Knerer and Plateaux-Quénu, 1966a).

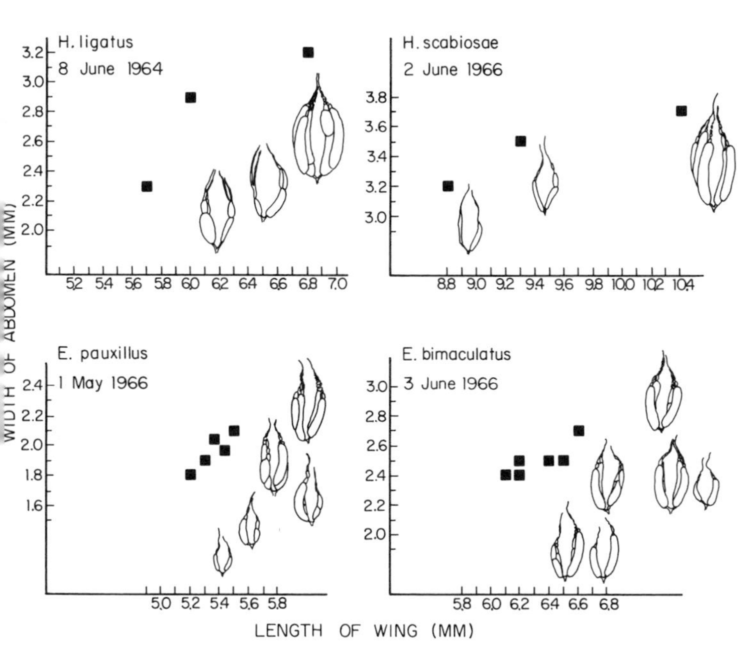

FIGURE 5-4. Variation among halictine species in the amount of ovary polymorphism in groups of females that cooperate in nest founding. A greater amount of polymorphism is considered an advanced social trait. Each quadrant displays the following characteristics for a single colony of cooperating females belonging to a given species: a square represents the position of an individual with reference to the two axes, which give the width of its abdomen and length of its wing; and the drawing shows the relative size and appearance of its ovary. In the *Evylaeus bimaculatus* colony, the six founding females had similar ovaries, and all took part in provisioning the nest. In the other colonies, belonging to "more advanced" species of *Evylaeus* and *Halictus*, the largest female was the principal egg layer, and the others took care of the nest provisioning (modified from Knerer and Plateaux-Quénu, 1966c).

shared with many other social bees. The life cycle commences in the early spring, when the relatively unworn but inseminated females emerge from hibernation in chambers of the old nest where they were born the previous year. These individuals will be the solitary foundresses of new colonies. At first their behavior is identical to that of females belonging to nonsocial species. Some of the females remain in the old nest, but most begin to excavate new nests in the open banks of streams facing south or on bare horizontal ground. The female forages for food, but after three or four cells are provisioned and sealed, she ceases activity and waits. By late May the first new generation of daughters and sons emerges. Soon the young females enlarge the nest by excavating new galleries and chambers and begin to forage for food and to provision cells. One pollen ball is placed in each cell, and the foundress bee lays an egg on it.

At this juncture the colony is evidently in a primitively eusocial state. But a series of events soon alters the situa-

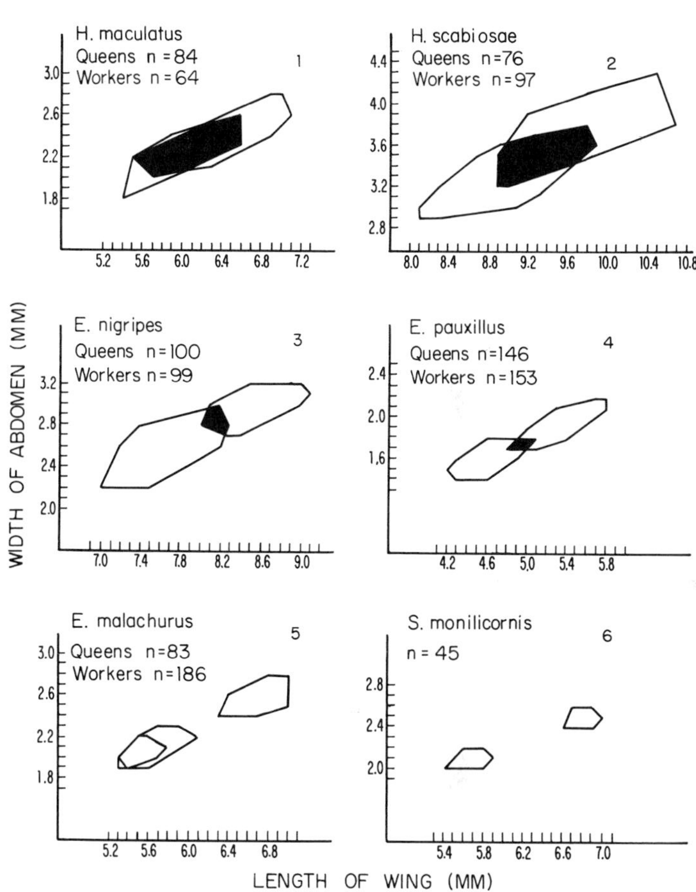

FIGURE 5-5. Variation in the females of six halictine species in two size parameters. The degree of difference between the queens (*right polygon*) and workers (*left polygon*) is an intuitively satisfying measure of the degree of eusociality in these bees. *Evylaeus malachurus* produces two sequential worker broods in a year; the first brood (*smaller polygon on left*) is smaller than the second brood (*larger polygon on left*), while the queens (*right polygon*) are distinct from both. The parasitic species *Sphecodes monilicornis* displays a female polymorphism that has nothing to do with sociality. The small individuals are eclosed in spring nests of *Evylaeus malachurus* and the large ones from summer nests of the same host species (modified from Knerer and Plateaux-Quénu, 1966b).

tion. Some of the new females mate and begin to lay eggs. Within a short time, evidently by the middle of June, the foundress dies and is replaced in her egg-laying role by her fecundated daughters. The colony now consists of an uneasy oligarchy among sisters. The egg layers compete for the available pollen balls, often ovipositing before the ball has been finished by the workers. Batra even observed cannibalism of eggs by competing unfertilized workers in a queenless nest.

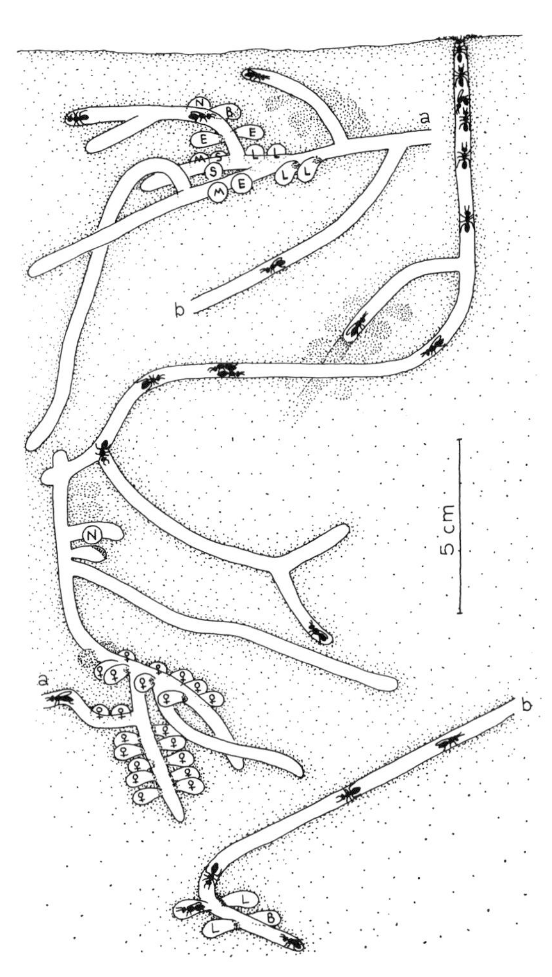

FIGURE 5-6. A nest of *Dialictus zephyrus*, a primitively eusocial halictine bee, dug open in Kansas on July 17. In the chambers of this "mature" summer nest are to be found pollen balls (*B*); eggs (*E*); small (*S*), medium (*M*), and large (*L*) larvae; and female pupae (♀). Empty chambers are designated as *N*. The exit tunnel ascends vertically through the soil and opens through a single hole in the bare, flat ground (from Batra, 1966a).

By midsummer the nest has reached respectable proportions (Figure 5-6). The largest one excavated by Batra reached a depth of 57 cm and contained no less than 38 branch burrows, 85 occupied cells, and 45 adult females. By the criteria of Knerer and Plateaux-Quénu, *D. zephyrus* is nevertheless a primitively eusocial species. The morphological difference between egg layers and workers is very slight, and intermediates are common. Intergradation also occurs in the degree of ovarian development. Male production is programmed in the primitive manner, beginning with the first brood and only gradually increasing over the summer. Soon after eclosion, the males leave the nest and swarm into the air over the nesting areas, where they pursue and mate with females who emerge individually from the nests. The colony cycle is purely annual. In late fall, only the younger inseminated females enter hibernation, and the life cycle reenters its solitary phase.

Behavior within the *D. zephyrus* nest is marked by a lack of contact among the adults other than the inevitable jostling that occurs in the passageways. There is no sign of either alarm communication or recruitment to new food sources. Grooming between bees is evidently absent. As Michener (1964a) believes to be the case in halictine bees generally, communication within *D. zephyrus* is chiefly by means of the signs of work accomplished: a cell constructed by one bee induces another to construct a pollen ball, which in turn induces still another to deposit an egg, and so forth. On the other hand, the females do communicate with the larvae to a limited extent. They periodically open the cells to inspect and clean the interior, and, in the process, they probably touch and antennate the larvae.

In its annual life cycle, tenuous social organization, and moderate colony size, *Dialictus zephyrus* typifies the eusocial halictines found in the North Temperate Zone. A significant departure from this relatively primitive pattern is shown by *Evylaeus marginatus.* This species, studied in France by Plateaux-Quénu (1962), is perennial. The colony is founded by a single queen, which persists until the natural demise of the colony five or six years later. For the first four years colony growth is slow but still exponential in form. By the end of the fourth summer the nest contains the foundress queen plus 50 to 130 workers. Then, in the fifth and what is normally the final year, the adult population increases to between 200 and 400. Many of these individuals are males. They soon leave the nest, engage in solitary nuptial flights, and die. The young females remain together in the old nest until the next spring, when they disperse to start new colonies on their own. The longevity of the queen of *E. marginatus,* the perennial character of the nests, the larger size of the mature colonies, and the periodic release of males are all regarded as evolutionarily advanced traits. On the other hand, there are no external morphological differences between queens and workers. These castes are distinguishable only by their ovarian development and behavior, a situation that Plateaux-Quénu refers to as "imaginal caste determination." Furthermore, the degree of communication between workers and between workers and brood appears to be no more advanced than in other halictine species, including *Dialictus zephyrus.* Even so, the total social structure of *Evylaeus marginatus* is clearly the most advanced of all the halictines studied to date.

The Natural History of the Eusocial Allodapine Bees

The bees belonging to *Allodape* and closely related genera (*Allodapula, Allodapulodes, Braunsapis, Eucondylops, Exoneura, Exoneurella, Exoneurida, Exoneurula, Halterapis, Inquilina, Macrogalea, Nasutapis*) are referred to informally as "allodapine bees" to distinguish them from members of the genus *Ceratina,* the only other major living representatives of the tribe Ceratinini. The allodapines are of special interest to students of insect sociology on two counts. First, in contrast to the larvae of other kinds of bees, those of most allodapines are kept together in open chambers and fed progressively with small meals (Figure 5-7). Second, as a concomitant of this peculiar habit, allodapine species display among themselves the evolutionary transition from solitary to eusocial behavior by way of subsocial stages. These facts were discovered by Brauns (1926) in his work on the South African *Allodape* and have been greatly extended in recent years by field studies in Asia, Australia, and Africa conducted by Iwata, Michener, Rayment, Sakagami, and Skaife.

Allodape angulata is a good example of a eusocial allodapine (Skaife, 1953). It is one of the common species on the Cape Peninsula of South Africa and is easily recognized as a moderate-sized, black bee with yellow markings on its head (Figure 5-8). The colonies nest in dead flower stalks of *Watsonia, Gladiolus, Aristea, Aloe* and a variety of other kinds of plants whose stems have pithy centers. In making the nests the bees simply bite out and remove the pith to a depth of from three to six inches.

The colony life cycle begins when the adults of the new generation emerge in the middle of summer, a period extending from the end of December to early February. They remain together in a largely quiescent state through the remainder of the summer and the following fall, then disperse to new nest sites. Breeding takes place shortly afterward, in July and August. Now the solitary, mated females begin new colonies. In the typical sequence the female digs a short cavity in the pith of a stem and lays a large (3 mm long), white, and slightly curved egg at the bottom. A few days later a second egg is laid and so on at regular intervals until six eggs (on the average) have accumulated. Occasionally, during periods of the cold, wet weather that afflicts the Cape region in August, eggs are eaten and replaced with new ones.

During the four to six weeks required for the eggs to hatch, the mother remains on guard at the nest entrance, and she extrudes the hind portion of her abdomen outward whenever she is disturbed. Skaife reports that such females are quite aggressive and that their stings are "quite painful, more so than that of the much larger carpenter bees, but not so bad as that of the honeybee." While the young are developing, the female arranges them in order of size, with the pupae nearest the entrance, followed by the large larvae and so on down to the eggs, which are always grouped at the bottom of the tube, much as shown in Figure 5-7.

The larvae lie on their sides and are spaced at irregular intervals along the pith cavity. The manner in which they are directly fed is unique to the allodapines.

> The newly hatched young are fed by the mother on a colourless liquid food that she regurgitates on to the abdominal surface of the larva, where it clings as a clear drop just below its mouth. Thus the larva has only to bend its head slightly to reach the food and suck it up. After the larvae are a few days old she feeds them on a mixture of nectar and pollen. Returning to the nest with her hind legs laden with pollen and her crop full of nectar, she first of all scrapes off the pollen into a small heap on the floor of the tube. Then, standing over this heap, she regurgitates a drop of nectar and mixes it with the pollen to form a stiff paste, using her mandibles for the purpose. Picking up the sticky mass between the base of her tongue and the base of her mandibles, she walks along the tube to the larva that has to be fed and thrusts it between the pseudopodia. The larva grasps the food between its pseudopodia and at once begins to feed. The younger larvae, fed on liquid food, do not have these pseudopodia—they only develop when the larva is about

one third grown. The mother does not attempt to store any food in the nest; she brings it from day to day as it is required, and during spells of inclement weather the larvae have to go hungry. I find that they can starve for four or five weeks without coming to any harm (Skaife, 1953).

After seven or eight weeks, with the coming of early summer in November, the first *A. angulata* larvae pupate. There is no cocoon, and the pupae lie loose in the pith cavity. By the middle of December all of the first brood have pupated, and in January they emerge as adults. Just about this time the mother *Allodape,* now a year old, may lay three or four more eggs. Then, after a few more days or weeks have elapsed, she dies. The second brood, tended by their sisters, emerges as adults in late summer or early autumn. During this final episode the males of the first brood occasionally leave the nest to get food for themselves, but they never take part in rearing the later brood.

Most other *Allodape* and *Exoneura* examined thus far exhibit life cycles similar to that of *A. angulata* (Sakagami, 1960; Michener, 1965b). Among the remaining allodapines, *Exoneurella* and an unidentified South African species of *Allodape* studied by Brauns are solitary, while

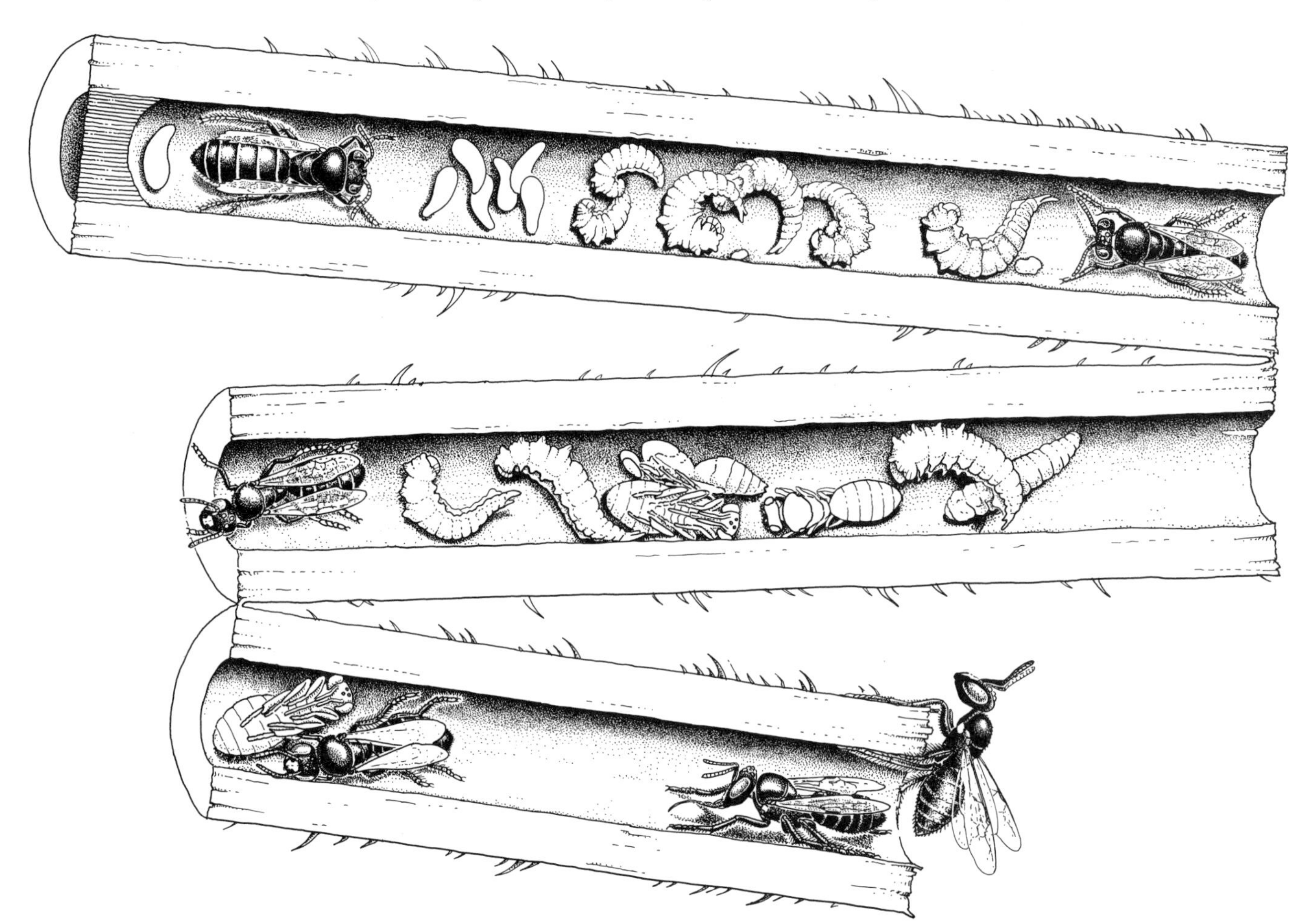

FIGURE 5-7. A populous nest of the Formosan bee *Braunsapis sauteriella* in a hollow stem, showing the free arrangement of the brood in a common chamber and evidences of progressive feeding that characterize the social allodapines. The eggs, whose huge size are a characteristic of this species and some other allodapines, have been placed in a cluster at the bottom of the nest, while the mother queen rests nearby. Pollen is stored briefly in small deposits on the nest wall. The larvae are fed at frequent intervals with little pollen balls (drawing by Sarah Landry; based on Iwata, in Sakagami, 1960). The shrub shown was arbitrarily selected to be *Lantana camara,* a common introduced species in Formosa which the bee is very likely to utilize.

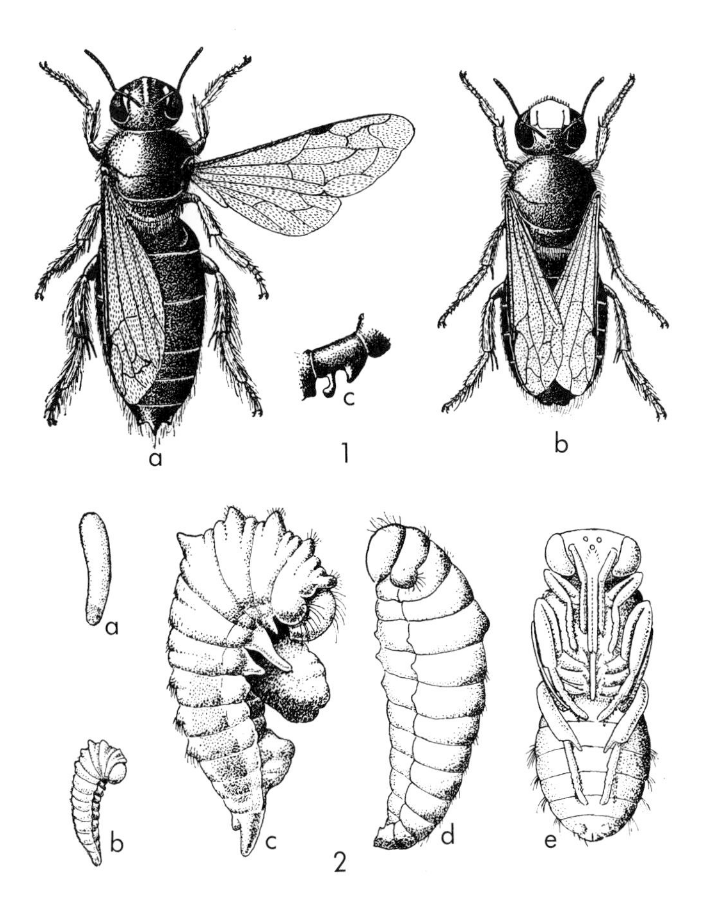

FIGURE 5-8. *Allodape angulata,* a eusocial species from South Africa: (1*a*) female; (1*b*) male; (1*c*) left hind trochanter of the male, a structure useful in species identification; (2*a*) egg; (2*b*) young larva; (2*c*) full-grown larva grasping a pollen ball with its fleshy "arms"; (2*d*) larva in prepupa stage; (2*e*) pupa of male (from Skaife, 1953).

Eucondylops, Inquilina, and some *Braunsapis* are social parasites on other allodapines (Michener, 1964b, 1970). These statements must remain tentative, however, because the life cycles of most of the allodapines are still unknown. The following interesting general observation concerning allodapine evolution can nevertheless be made. In spite of their radical innovations in brood care, which are evidently adaptations to life in hollow stems, the allodapines have progressed only a short distance into eusociality. The most complex societies known in the group, those of *Allodape* and *Exoneura,* are in fact barely eusocial. They have reached just about the level of most of the primitively eusocial halictine bees, as exemplified by *Dialictus zephyrus.* When we consider further that workers of one of the most advanced eusocial bee groups, the Meliponini, seal the brood cells and have no contact with the brood, it is even more apparent that a high degree of worker-larva contact is not a vital factor in social evolution.

The Natural History of the Bumblebees (Genus *Bombus*)

The large, hairy bees of the genus *Bombus* are notable as social insects primarily adapted to colder climates. About 200 species are known, mostly restricted to the temperate zones of North America and Eurasia. Many of these occur only in the northern parts of the continents, and several are found near the Arctic Circle and on the treeless summits of high mountains. Two occur as far north as Ellesmere Island. A few species reach in the other direction as far as Indonesia and Tierra del Fuego in South America, and a single species (*B. atratus*) is even common in the Amazon rain forests, where it retains the basic bumblebee life cycle. All of the species of *Bombus* so far studied are eusocial.

The genus, as such, is not recognizable in Tertiary fossil deposits, but related genera, including the possible ancestral *Protobombus,* together with *Chalcobombus, Electrapis,* and *Sophrobombus,* are represented in the Baltic amber. In view of this greater diversity among the fossil forms, Wheeler (1928) suggested that *Bombus* and its parasitic genus *Psithyrus,* are relics of an adaptive radiation of the Bombinae that took place in the Tertiary.

Because of their primarily temperate distribution and economic importance as crop pollinators, the species of *Bombus* have received close attention from entomologists for many years. F. W. L. Sladen's classic treatise *The Humble-bee* (1912), long out of print and almost unobtainable, has fortunately been replaced by a more recent and very readable review book *Bumblebees* by Free and Butler (1959). The following capsular account of the *Bombus* life cycle is based largely on their much more thorough synthesis and applies principally to the 25 species found in England.

In England as elsewhere in the North Temperate Zone, the life cycle of *Bombus* is annual. Only the fertilized queens hibernate. The history of a colony typically unfolds as follows. In early spring the solitary queen leaves her hibernaculum and searches on wing until she finds an abandoned nest of a field mouse, vole, or shrew, or some

FIGURE 5-9. A colony of the European bumblebee *Bombus lapidarius.* The nest is in an abandoned mouse nest. The large queen sits atop a cluster of cocoons inside which are worker pupae (one pupa has been exposed to show its position). At the upper and lower left are three communal larval cells; the waxen envelopes of the bottom two have been torn open to reveal the larvae inside. Large honeypots occupy the left and center of the ensemble. At the lower right are clusters of abandoned cocoons which are now used to store pollen (original drawing by Sarah Landry; based on photographs by Sladen and by Free and Butler).

other similarly shaped cavity, in an open but relatively undisturbed habitat such as a fallow field or abandoned garden. She pushes her way into the nest and then modifies it for her own use by constructing an entrance tunnel and lining the inner cavity with fine material teased out of the nest walls. While in the nest the queen begins to secrete wax in the form of thin plates from intersegmental glands on the abdomen. From this material she fashions the first egg cell in the form of a shallow cup set onto the floor of the nest cavity. Next she places a pollen ball into the egg cell and then lays 8–14 eggs onto the surface of the ball. Finally, she constructs a dome-shaped roof of wax and other materials over the cell, so that the entire brood cell is now sealed and spherical in shape. About the time the first eggs are laid, the queen also constructs a wax honeypot just inside the entrance of the nest cavity and begins to fill it with some of the nectar gathered in the field. When the first workers emerge, they assist the queen in expanding the nest and caring for additional brood, as shown in Figure 5-10.

Depending on the species of *Bombus* involved, the larvae are fed by one or the other of two very different techniques, which were first recognized by Sladen in 1896. In one group of species, the "pollen-storers," the pollen is placed in abandoned cocoons, which may be extended with wax layers until they form cylinders as high as three inches. From time to time pollen is removed from this modified cocoon and fed into the brood cell in the form of a viscous liquid mixture of pollen and honey. The queen and workers of the pollen-storer species do not feed the larvae directly. Instead, they make a small breach in the larval cell and regurgitate the pollen-honey mixture next to the larvae. In the second group of species, the "pouch-makers" or "pocket-makers," the queens and workers build special wax pouches adjacent to groups of larvae and fill them with pollen. The larvae then feed as a group directly from the pollen mass. Occasionally, the pouch-makers also feed larvae by regurgitation, and groups of larvae destined to become queens are fed exclusively in this manner. The social significance of the difference between the techniques may be that pollen storing and regurgitation permits the bees to exercise day-by-day control over the growth of the larvae, whereas pouch-making makes such control difficult. The life cycle illustrated in Figure 5-10 is that of a typical pollen-storer.

By the end of the summer the colony contains, again according to species, from around 100 to around 400 workers. The largest colony ever recorded anywhere belonged to *Bombus medius* and was found by Michener and LaBerge (1954) in a tropical forest in Mexico. This example contained at least 2,183 workers and possibly had been favored by a growth season longer than that available in temperate zones. As fall approaches the annual colonies of England produce males and queens and begin to break up. The demise of the bumblebee colonies seems to be controlled by endogenous factors. In northern New Zealand, introduced species of *Bombus* fly at all times of the year, and solitary queens can start nests during at least nine months of the year (Cumber, 1954). Colonies sometimes overwinter and attain unusual size. In spite of this opportunity for perennial growth, however, the New Zealand colonies never return to the production of workers after they have reared queens.

Mating behavior varies greatly among the species of *Bombus*. In some, the males hover around the nest entrances and wait for the young queens to emerge. In others, the male selects a prominent object, such as a flower or fence post, and alternately stands on it and hovers over it, ready to dart at any passing object that resembles a queen in flight. In a third group of species,

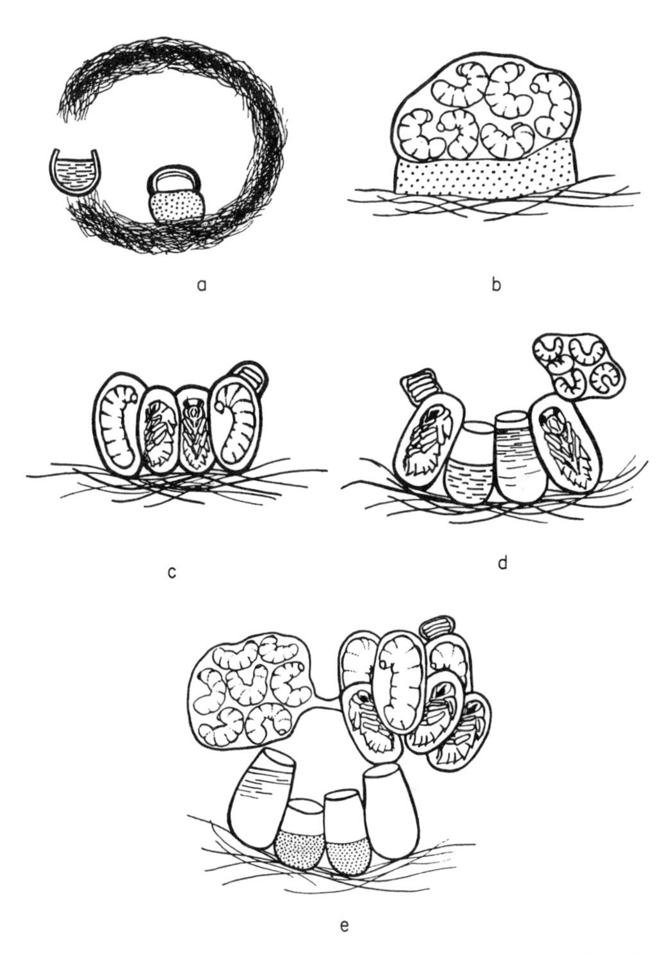

FIGURE 5-10. The growth of a typical bumblebee colony is illustrated in this series of drawings (based on Sladen, by Free and Butler, 1959). (*a*) A cavity is constructed by a single queen in an abandoned rodent nest. The first honeypot is built near the nest entrance and the first brood cell in the center of the nest floor. (*b*) The larvae are shown within the wax brood cell, sitting on a bed of pollen. (*c*) The larvae have spun their cocoons and transformed into pupae in the inside cocoons and into prepupae in the outside cocoons. A second batch of eggs has been laid on top of the right-hand cocoon. (*d*) Adult bees have emerged from the central cocoons of the first brood batch. The vacated cocoons are now used for storing honey and pollen. (*e*) The comb grows upward and outward as new batches of brood are produced.

studied by Frank (1941) and Haas (1946, 1952), the males establish flight paths which they mark at intervals with spots of scent from the mandibular gland dabbed onto objects along the route. The paths shift from day to day and those of different males frequently overlap. The males fly around them hour after hour, day after day, waiting for the approach of the females. After mating, the queens hibernate in specially excavated chambers in the soil, and the following spring they initiate new colonies.

Queens differ from workers only by their larger size and the greater extent of their ovarian development, and intermediates between the two castes are common. There is also great variation in size within the worker caste. The larger workers tend to forage more, and the smaller workers spend more time in nest work. In a few species, the smallest workers do not fly and are thus bound to the nest permanently. Nest guarding occurs in some species and is usually undertaken by workers who possess better-developed ovaries.

Within the Apidae, whose species comprise the *haut*

TABLE 5-2. Primitive (or at least relatively simple) social traits in *Bombus,* compared with the more advanced traits found in the highest social bees, the honeybees of the genus *Apis* and the stingless bees of the tribe Meliponini.

Bombus	*Apis* and Meliponini
Queens and workers differ morphologically to a slight degree, and intermediates are common	Queens and workers are morphologically very different from each other, and intercastes are normally absent
The life cycle is annual, at least among the majority of species; new colonies are founded by single queens; and the mature colony size is small	The life cycle is perennial; new colonies are started by swarming, and colony size is moderate to very large
The queen maintains reproductive dominance by aggressive behavior, and the workers tend to behave toward each other in the same way (see Chapter 15). Workers occasionally steal eggs from each other and the queen	The queen maintains reproductive dominance by pheromones (see Chapter 15), at least in *Apis,* and aggressive behavior is muted or absent. Egg stealing is unknown except as a ritual form of eating by the meliponine queens
The larvae are often reared in groups and must compete with other larvae for food placed indiscriminantly in their vicinity	The larvae are reared in separate cells on the brood comb, greatly increasing the chances for individual attention on the part of the nurses and control of caste determination
The larvae are fed with raw pollen and regurgitated mixtures of pollen and honey	In *Apis,* larvae are fed at least in part with special food manufactured by the mandibular and pharyngeal glands
The adults rarely regurgitate food directly to other adults or try to groom them	Both grooming and direct transfer of food by regurgitation are very frequent and, in the case of *Apis* at least, known to play an important role in communication and regulation (see Chapter 14)
The queen regulates colony growth by building all of the egg cells herself and laying in them, following the same behavior patterns by which she initiates the colony	The queen plays no direct role in colony growth or in the construction of the brood combs. The workers determine these matters and are subject to much more feedback from the environment outside the nest
Temporal division of labor is weakly developed	A temporal division of labor is strongly developed, in which the young adult worker first engages in brood care (or nest work), then nest work (or brood care), and finally in foraging. In *Apis,* at least, this progression is associated with orderly changes in the exocrine glands (see Chapter 9)
Chemical alarm communication is lacking	Chemical alarm communication is well developed and involves pheromones apparently especially evolved for the purpose (see Chapter 12)
Recruitment among workers is lacking	Recruitment is well developed and mediated by special assembling or trail pheromones; in *Apis* there is also a symbolic "waggle dance"

monde of the social bees, *Bombus* occupies a relatively lowly position. Its solutions to the problems of social organization are as a rule crude, and it has not achieved many of the more spectacular control mechanisms that distinguish *Apis* and the Meliponini from the social Halictidae. In Table 5-2 I have indicated the characteristics which, in my opinion, are more primitive, or at least simpler, in the context of the biology of the Apidae as a whole.

Bombus does have a few specialized social traits mixed in with its otherwise elementary behavior. At least one species, *B. agrorum,* is able to maintain its nest temperature at about 30°C (Himmer, 1933). Overheated nests are cooled by fanning. Also, the nest architecture contains one advanced feature already mentioned in passing. "The wax cells of *Bombus,*" as Michener (1964a) has said, "are a most extraordinary development. They are somewhat variable in size and shape, and therefore heteromorphic, but are unique among the bees in that in most species they contain a cluster of eggs or immature stages instead of only one, and in that they grow (*i.e.,* wax is added to them) as their contents grow."

The Natural History of the Stingless Bees (Tribe Meliponini)

The meliponines take their common name from the fact that their stings are vestigial and cannot be used in defense. The workers of most species are nevertheless very effective in defending their nests from human intruders. They swarm over the body, pinching the skin and pulling hair, occasionally locking their mandibles in catatonic spasms so that before the grip can be broken, their heads tear loose from the body. The *Trigona flaveola* group of species in tropical America also eject a burning liquid from their mandibles which in Brazil has earned them the name of *cagafogos,* meaning "fire defecators." An encounter with a colony in Guatemala was one of the worst experiences in the life of William Morton Wheeler. When he got too close to a nest of "these terrible bees," they burned off large patches of skin from his face. Stingless bees are also very effective against other intruders. They are, for example, almost totally immune to the forays of army ants, which are the nemeses of social wasps and other ants.

The meliponines are all eusocial. They form moderately large colonies, comprised of 500 to 4,000 adults in the case of *Melipona,* or 300 to 80,000 in the case of *Trigona* and related genera. The workers range in size among the various species from minute (body length about 2 mm) to somewhat larger than a honeybee. Some are slender in build, others burly; some are nearly hairless and shining, others hairy as honeybees. There are strong morphological and behavioral differences between the queen and worker castes, and intermediates are normally lacking.

Stingless bees are encountered in the tropics and subtropics around the world. In 1964, according to Kerr and Vilma Maule, 183 species were known from the New World tropics, 32 from Africa, 42 from Asia, and 20 from Australia and New Guinea. The most primitive living group appears to be the Neotropical subgenus *Frieseomelitta* of *Trigona,* which has a simple nest structure resembling that of *Bombus.* Waxen tubes are constructed for pollen and simple pots for honey, and the brood cells are assembled singly in clusters rather than as units in combs. *Frieseomelitta* has a haploid chromosome number of 9. *Melipona* also has 9, while the rest of the meliponines, including other subgenera of *Trigona,* have 18 or, rarely, 17. It is assumed that the higher chromosome numbers are a derived condition. The relatively primitive subgenus *Trigona* (*Hypotrigona*) is represented in the Baltic amber, which is of Oligocene age (Kelner-Pillault, 1969). More advanced representatives of *Trigona* have been found in the Miocene amber deposits of Burma, Sicily, and Chiapas (Mexico). The fossils all belong to living, eusocial genera or else are close to them, so that the time and place of the origin of the Meliponini is still a mystery. In the opinion of Kerr and Maule (1964), the much greater diversity of the group in South and Central America, together with the fact that it was widespread by Miocene times, indicates that the Meliponini originated in South America prior to the submergence of the Panamanian land bridge in early Eocene times. In defense of this argument, it should be noted that modern meliponines have great difficulty crossing water gaps. Colonization probably depends on the movement of entire colonies by "rafting" of the nest site on the surface of the water, which must be a rare event at best. It is unlikely that colonies travel far during the act of colony division, which is a prolonged and complicated process and does not ordinarily take place over distances exceeding several hundred meters. Consequently, modern meliponine species are almost strictly limited to continents and islands close to continents. They do not extend east of the Solomon Islands, for example, and only one representative occurs in the West Indies—*Melipona beechei* of Cuba, a species that also occurs in Yucatan and may have been transported to Cuba by early man.

Most of the basic information on the biology of the stingless bees has been reviewed by Wheeler (1923; this work contains a nearly complete bibliography of earlier authors); Schwarz (1948); Nogueira-Neto (1951, 1970); Bassindale (1955); Moure, Nogueira-Neto, and Kerr (1958); Michener (1961a); and Kerr and Maule (1964). There is, unfortunately, no single comprehensive treatise. A great deal of attention has been focused by melittologists on the structure of the meliponine nest, which is complex and unique, and on the variation in nest structure, which lends itself well to evolutionary interpretation. In simplest terms, the basic meliponine nest consists of an inner grouping of brood cells, which may or may not be compacted into combs, and rather large, egg-shaped pots where honey and pollen are stored. The latter structures are quite reminiscent of the pots and cells of the *Bombus* nests. Surrounding the brood cells there may be a soft sheath, or involucrum. A thick, hard outer layer called the batumen surrounds both the pots and the brood cells. Most meliponine species nest in the hollow trunks and branches of trees, although some live only in the deserted nests of ants or termites. A few show no particular preference, accepting many kinds of hollow spaces; these species are often able to thrive in cities in the tropics. The conformations of the nest entrances vary enormously from species to species. Some meliponines construct simple holes, others dome-shaped or trumpet-shaped entrance platforms. Some of the species of *Trigona* cover the nest entrance with fresh, sticky propolis, which is an effective deterrent against ants. The nests are usually constructed of cerumen, which is a brown mixture of wax and propolis. Examples of nests that differ greatly in complexity of structure are illustrated in Figures 5-11 to 5-13. A still more precise conception of the nature of meliponine nests can be gained from the following glossary of special terms that have been assembled and defined by Michener (1961a).

Batumen. A protective layer of propolis or hard cerumen (sometimes vegetable matter, mud, or various mixtures) enclosing the nest cavity. Most commonly it consists of *batumen plates,* sealing off portions of a natural cavity from the nest cavity, and *lining batumen,* which is a thin layer of propolis or brittle cerumen on the walls of the nest cavity. Exposed and partly exposed nests are entirely or partly surrounded by *exposed batumen. Laminate batumen* consists of several layers, with spaces between them in which bees can move about. Laminate batumen is usually exposed. (Batumen is a Brazilian word meaning "wall," used by von Ihering and subsequent authors chiefly for the batumen plates.)

Cells. Brood cells made of soft cerumen, in each of which a single young is reared.

Cerumen. A brown mixture of wax and propolis used for construction. Newly made cerumen is soft, while old cerumen is often brittle.

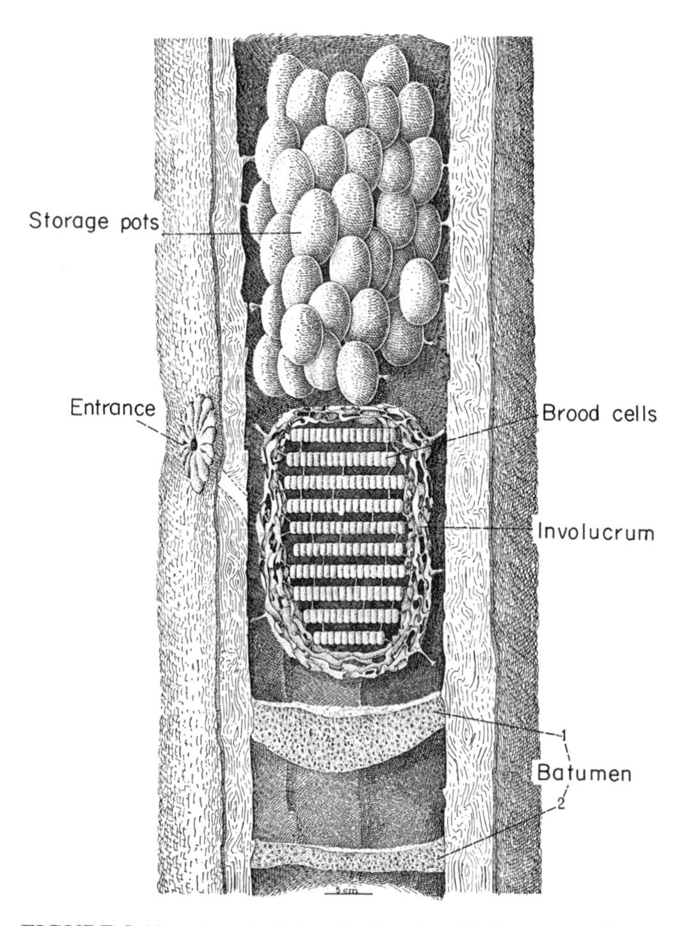

FIGURE 5-11. A nest of the stingless bee *Melipona pseudocentris* constructed in a hollow tree trunk. In this species an involucrum surrounds the very regularly constructed brood combs. The storage pots contain either honey or pollen, but are not differentiated in external structure (drawing by João Maria F. de Camargo, in Kerr *et al.,* 1967).

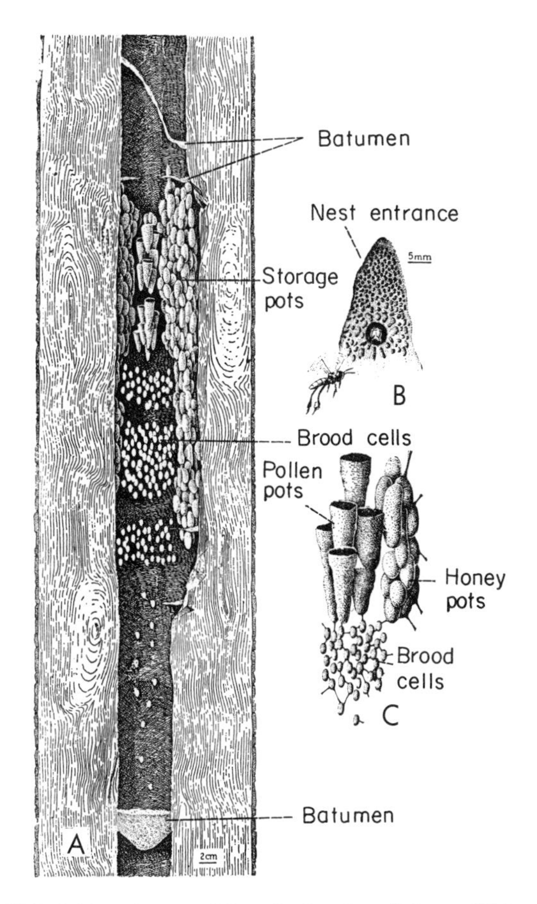

FIGURE 5-12. A nest of the stingless bee *Trigona* (*Frieseomelitta*) *flavicornis* constructed in a hollow tree trunk. In this species the pattern of construction is in a sense the opposite of *Melipona pseudocentris,* illustrated in Figure 5-11. The storage pots are specialized for reception of either honey or pollen, but the brood cells are arranged in irregular clusters rather than in combs and are not enclosed in an involucrum (drawing by João Maria F. de Camargo, in Kerr *et al.,* 1967).

Cluster (*of cells or cocoons*). A group of brood cells or cocoons irregularly arranged, not in combs (see Figure 5-12).

Cocoon. Silk structure spun after defecation by the mature larva around the inner wall of its cell. The worker bees remove and reuse the cerumen of which the cell was constructed, leaving the cocoon largely exposed during the prepupal and pupal periods.

Comb (*of cells or cocoons*). A layer of brood cells or cocoons crowded together in a regular arrangement (see Figure 5-11).

Entrance. The external opening of the nest for coming and going of the bees. It is often contained outside the nest cavity as an *external entrance tube.* It is also often contained inside the nest cavity, usually along the inner wall of the cavity, as an *internal entrance tube.*

Involucrum. A sheath of soft cerumen surrounding the brood chamber. A *laminate involucrum* consists of several layers with spaces between them in which the bees can move about.

Pillars. More or less vertical columns of cerumen (soft or brittle) within the nest. When more or less horizontal, such columns may be called "connectives." There is no real distinction between pillars and connectives.

Propolis. Resins and waxes collected by the bees in the

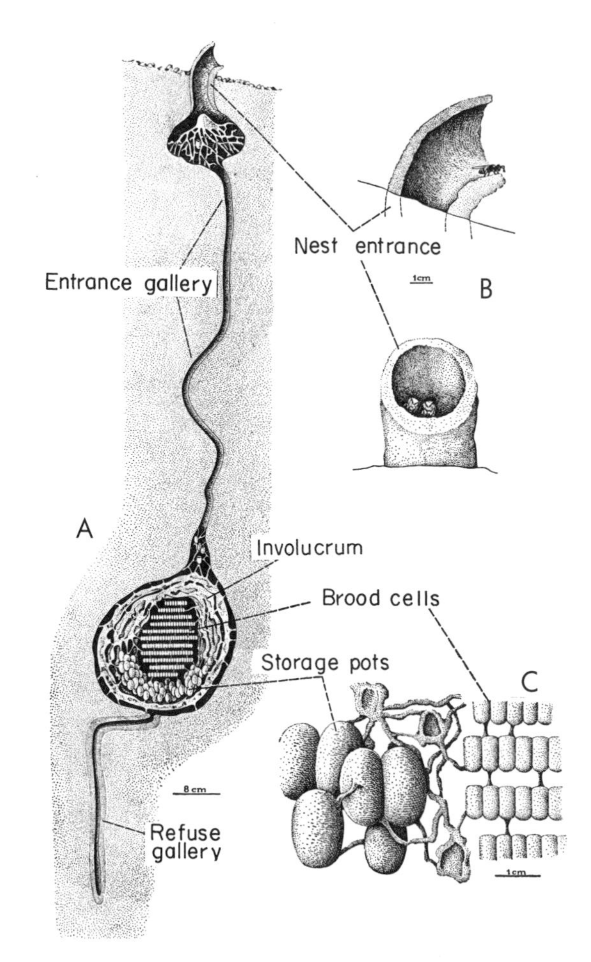

FIGURE 5-13. An underground nest of the stingless bee *Trigona* (*Partamona*) *testacea* (drawing by João Maria F. de Camargo, in Kerr *et al.,* 1967).

field and brought to the nest for construction purposes, especially for sealing fissures in the nest wall.

Storage pots. Containers made of soft cerumen for the storage of honey (honeypots) or pollen (pollen pots).

Wax. White material secreted by the bees and mixed with other substances to make cerumen. Paulo Nogueira-Neto (in letter to Michener) states: "As far as I know, pure wax is never used by meliponines except sometimes in the outermost part of the entrance tube" (as in *Trigona schrottkyi* and others).

The life cycles of only a very few meliponine species have been worked out in full. The colonies are perennial and reproduce by swarming. The cycle begins when worker scouts—perhaps prompted by crowded conditions in the old nest—begin to search for a new nest site. When a site is selected, the workers set about sealing any cracks that may exist around the cavity and preparing the nest entrance. The initial building material is transported from the old nest. As the workers begin to arrive in greater numbers, they construct the involucrum, the pillars, and the pots, and start on the first brood cells. Cerumen continues to be lifted in quantity from the old nest. The workers carry this material in their corbiculae, or "pollen baskets," made up of the long hairs lining the hind tibiae. Honey and pollen are also removed from the pots of the mother nest, transported in liquid suspension in the crops of the workers, and regurgitated by them into the pots of the new nest. Thus a strong, literally physical bond is built between mother and daughter nests.

With intensification of activity, males come in from both the parental nest and other nests in the vicinity and form small groups resting or flying around in front of the new nest. Meanwhile, a young virgin queen has been reared by the workers in the old nest, and for a while she lives side by side with the mother queen. What next occurs has been recorded vividly by Moure, Nogueira-Neto, and Kerr (1958):

> Until this time the young queen is protected in the parental nest by workers, at least in *Plebeia emerina.* In an experimental Nogueira-Neto nest, one of us (Moure) observed that the workers made two globular prisons about one cm. in diameter with two or three holes which permitted passage only to workers. At this time great agitation was noted in the nest. After three days much activity was noted at the entrance and finally a young queen flew, apparently followed by some workers. She was not observed to return during one hour of observation. At that time the nest was opened and found still in great agitation, and two young queens were found being killed by covering with wax. Some workers seemed to be trying to liberate or kill one queen by pulling her antennae. According to Kerr the virgin queen flies to the new nest accompanied or not by a group of bees and she soon makes a mating flight. The mechanism of fecundation has been observed only in an established nest of *Melipona quadrifasciata quadrifasciata* (Kerr and Kraus, 1950). The nuptial flight was 4.5 minutes long. The queen was caught after returning and dissected; it was observed that the whole male genital armature and seminal vesicles were in the vagina. The fact that drones fly outside the new nests is an indication that mating normally occurs in the open, as would be expected to avoid constant inbreeding.

Insofar as these observations are generally applicable, we can see two basic differences in the modes of colony reproduction between the meliponines and the much more familiar honeybee, *Apis mellifera.* In *Apis* the break is swift and clean; the swarm simply leaves the parental nest with the queen in tow, settles at a temporary bivouac, and starts looking for a new site. In the meliponines the queen is not transferred until the new nest has been fully prepared. In *Apis* it is the old queen who migrates. But in the meliponines the old queen has a heavy, swollen abdomen and tattered wings and cannot possibly fly; consequently virgin queens make the flight.

The social organization of the meliponines is quite impressively sophisticated. In fact, now that behavioral studies are being pressed in earnest, new social phenomena are being discovered at such a rate as to suggest that these insects are easily on an evolutionary plane with honeybees. We now know that the meliponines have a well-defined temporal division of labor, correlated at least in part with exocrine gland activity, and also rather advanced systems of chemical alarm and recruitment communication, the latter reinforced by a modulated sound signal similar to that found in the waggle dance of *Apis mellifera.* Further details of these phenomena are furnished in Chapters 9 and 13. Recent research by Sakagami and his co-workers (Sakagami and Oniki, 1963; Beig and Sakagami, 1964; Sakagami, Montenegro, and Kerr, 1965; Sakagami and Zucchi, 1968) has revealed a remarkable form of interaction between the queen and workers of species of *Trigona* whereby the queen acquires her nourishment during the process of provisioning and capping of the brood cells. The whole stereotyped sequence, which is reminiscent of an obeisance rite in some Oriental court, is illustrated in Figure 5-14. This unique behavior pattern, together with the many other peculiarities in behavior and life cycle, indicate that the Meliponini evolved their social organization for the most part after the progenitors of the modern Meliponini and Apini had separated in evolution.

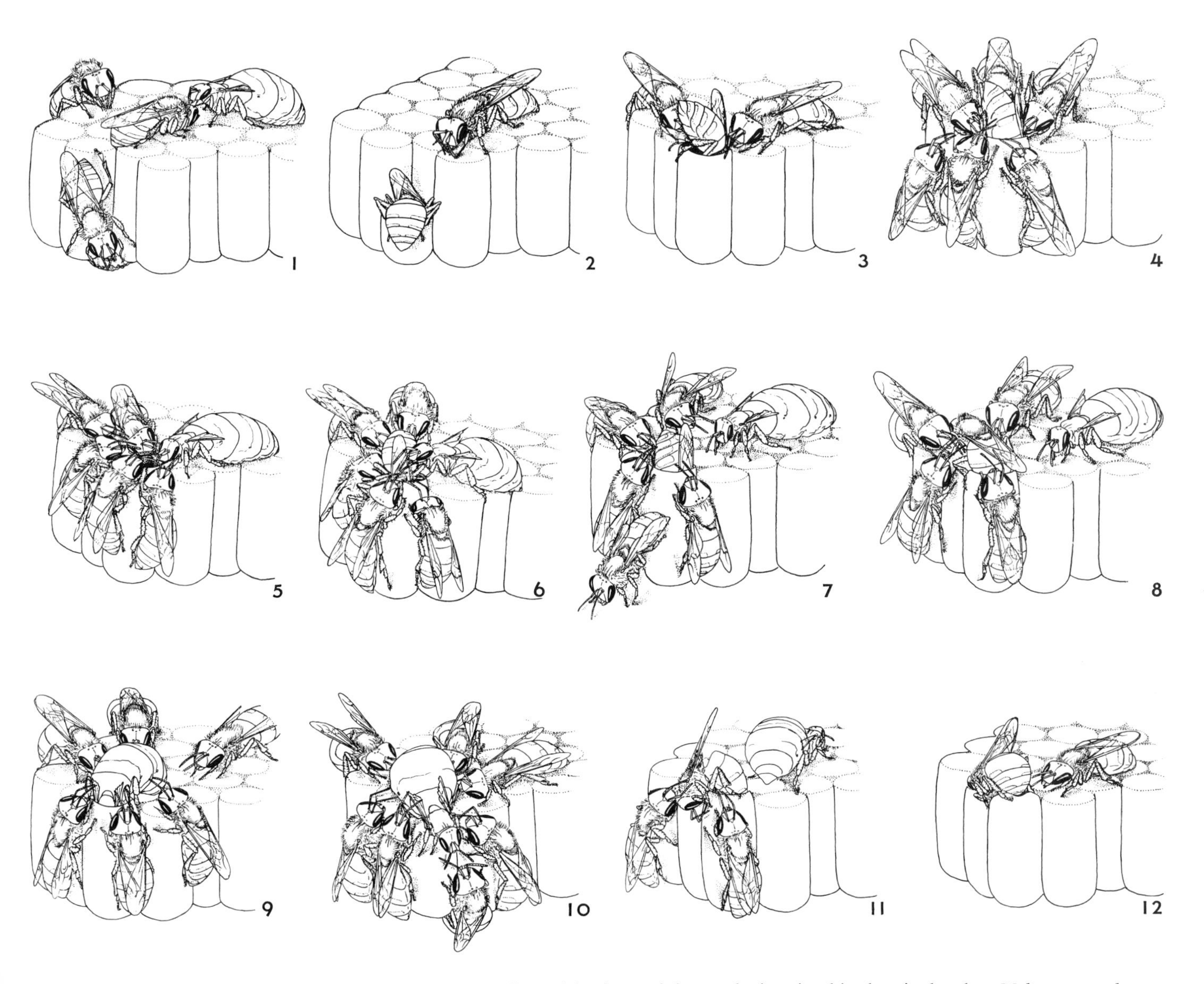

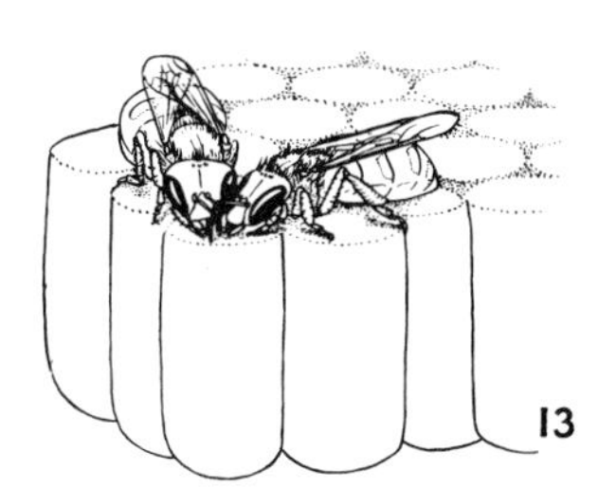

FIGURE 5-14. Cell provisioning and the egg-laying ritual in the stingless bee *Melipona quadrifasciata anthidioides.* In this typical sequence a worker begins construction of a new brood cell on the edge of the comb (1). At this time the queen is seen tapping another worker on the vertex (possibly a dominance gesture), but for the moment she pays no attention to the construction work. Cell building continues quietly (2) until the walls are completed. Then other workers crowd in excitedly (3, 4). Soon the queen is attracted also (5). Within one to several minutes, a succession of workers begin to dip into the cell and to discharge larval food, filling the cell up within a few minutes (6, 7). Now a worker lays a small trophic egg (8). The queen bends into the cell and eats the egg and probably some of the larval food as well (9). Then she lays an egg of her own (10) and finally departs, while a few workers remain behind to seal the cell with a cap (11–13) (modified from Sakagami, Montenegro, and Kerr, 1965).

The Natural History of the Honeybee, *Apis mellifera*

Wheeler (1923) explained the ancient fascination of the honeybee as follows: "Its sustained flight, its powerful sting, its intimacy with flowers and avoidance of all unwholesome things, the attachment of the workers to the queen—regarded throughout antiquity as a king—its singular swarming habits and its astonishing industry in collecting and storing honey and skill in making wax, two unique substances of great value to man, but of mysterious origin, made it a divine being, a prime favorite of the gods, that had somehow survived the golden age or had voluntarily escaped from the garden of Eden with poor fallen man for the purpose of sweetening his bitter lot."

But the insect sociologist must inevitably look at the honeybee as just one more eusocial apoid, compare it objectively with the thousands of other eusocial insects, and ask, "Just how different is *Apis mellifera?*" In a word, it is moderately distinguished as a social insect. By the general intuitive criteria of social complexity—colony size, magnitude of queen-worker difference, altruistic behavior among colony members, periodicity of male production, complexity of chemical communication, thermoregulation of the nest interior and other evidences of homeostatic behavior—the honeybee is at about the level of the other highest eusocial insects, that is to say, the stingless bees, the ants, the higher polistine and vespine wasps, and the higher termites. In one feature alone, the waggle dance, the species comes close to standing truly apart from all other insects. The really remarkable aspect of the waggle dance is that it is a ritualized reenactment of the outward flight to food or new nest sites; it is performed within the nest and somehow understood by other workers in the colony, which are then, and this must be counted the remarkable part, able to translate it back into an actual, unrehearsed flight of their own. A similar ability to interpret modulated symbols is evidently shared by certain meliponine bees, which transmit sound signals correlated in duration and frequency with the distance of food finds. But other cases of symbolical communication have yet to be demonstrated in the social insects. The waggle dance will be discussed in greater detail in Chapter 13.

Apis mellifera has been the subject of scientific observations since ancient times. Aristotle is credited with discovering the principle that individual bees usually stick to one kind of flower during any one foraging trip. He also found evidences of communication of the location of food, but achieved no understanding of the mechanism involved. Pliny used an observation hive with windows of transparent horn to watch the emergence of bees from the brood cells. In the ancient world whole libraries were devoted to the art of beekeeping. In the past hundred years tens of thousands of technical articles on the honeybee have been published, most of them on apiculture, and an uncounted number of journals have been devoted to the subject. The following list includes the principal modern journals, as well as some of the more obscure ones, but it constitutes only a small fraction of the actual number: *L'Abeille, American Bee Journal, Apiculteur, Apicultural Abstracts, Archiv für Bienenkunde, Bee World, Beekeeper's Gazette, Beekeeper's Record, Bee-keepers' Review, Bienenvater, Erlanger Jahrbuch für Bienenkunde, Gleanings in Bee Culture, Indian Bee Journal, Journal of Apicultural Research, Reports of the Bee Research Association, Schweizerische Bienen-zeitung, Zeitschrift für Bienenzucht,* and the *Proceedings* of the International Beekeeping Congresses. Much of the best contemporary basic research on the biology of the honeybee is reported in the following more general periodicals: *Insectes Sociaux, Journal of Economic Entomology,* and *Zeitschrift für Vergleichende Physiologie.* The latter is especially notable since it has been the principal organ utilized by von Frisch and his students. There are also scores of books, the most important of which are Ribbands' *The Behaviour and Social Life of Honeybees* (1953), Butler's *The World of the Honeybee* (1954a), Snodgrass' *Anatomy of the Honey Bee* (1956), *Biene und Bienenzucht* (edited by Büdel and Herold, 1960), Lindauer's *Communication among Social Bees* (1961), Dade's *Anatomy and Dissection of the Honeybee* (1962), von Frisch's *The Dance Language and Orientation of Bees* (1967a), and, most recently, the five-volume *Traité de biologie de l'abeille,* edited by Chauvin (1968). A bibliography of the "other" honeybees, that is, the species of *Apis* other than *mellifera,* has been published in *Bee World,* Volume 48, pages 8–15 (1967).

At the risk of oversimplification, it can be said that the key to understanding the biology of the honeybee lies in its ultimately tropical origin. It seems very probable that the species originated somewhere in the African tropics or subtropics and penetrated colder climates prior to the time it came under human cultivation. In this connection I believe it would be appropriate to quote the authoritative opinion of C. D. Michener (in a letter): "I take the ability of *Apis mellifera* to form a winter cluster as an evolutionary change that adapted it to cold climates. I know of no evidence that it was brought to northern climates by man. The races of Germany, England, etc., differ in obvious genetically controlled features (color, size, proportions of various body parts) from the more southern races and presumably have been selected over a considerable period of time in cold climates. The most distinctive race of all is that of a limited area in southernmost Africa which has a temperate climate. It differs from the other races in the regular production of females from unfertilized eggs, in the large, although presumably not functional, spermatheca of the worker, and in various other respects. Moreover, *Apis mellifera* does not occur in the Asiatic tropics. It is however exceedingly common in the African tropics. Presumably its original distribution, before white man carried it around the world, was the whole of Africa, all but northernmost Europe, and western Asia." Geographic variation in *A. mellifera* has recently been the subject of an intensive review by Rothenbuhler, Kulinčević, and Kerr (1968). These authors stress the complex and apparently long-standing racial variation that occurs through most of the range of this species in Africa, Europe, and the Middle East.

Thus, unlike the vast majority of social bees endemic to the cold temperate zones, the honeybee is perennial, and, being perennial, it is able to grow and sustain large colonies. Having large colonies, it must forage widely and exploit efficiently the flowers within the range of its nests; the waggle dance and Nasanov gland communication system are clearly adaptations to this end. Also, being ultimately tropical in origin, its colonies multiply by swarming; there is no need to have a hibernating solitary phase in the colony life cycle. And finally, since the queen is relieved of the necessity to overwinter and initiate colonies in solitude, she has regressed in evolution toward the role of a simple egg-laying machine, with the result that the queen and worker castes differ strongly from one another in both morphology and physiology. Within the scope of these interlocking effects are to be found just about all of the phenomena that distinguish *Apis mellifera* from the endemic cold temperate bee species. When we turn to the tropical faunas and consider what else has evolved to eusocial levels within the Apoidea, the contrasts are not nearly so sharp. The prevailing group of tropical eusocial bees, the Meliponini, not only resemble *Apis* in their life cycle, but are comparable to it in complexity of social organization. Of course, a great many, perhaps most, of the primitively eusocial bees exist in the tropics, but this does not affect the important generalization that the most advanced bee societies are tropical in origin.

Apis mellifera, the domesticated honeybee, is one of four living species in the genus. The other three are limited to tropical southeastern Asia. One, the Indian honeybee (*Apis indica*) is so close to *A. mellifera* in anatomical features that it ranks as possibly nothing more than a wild southern subspecies. Together with *A. mellifera,* it is distinguished from the other *Apis* by the habit of building several combs, one behind the other, in large cavities. Left to their own devices, colonies of *mellifera* and *indica* normally construct the combs in hollow trees. The abdomen of the *indica* worker is strikingly banded as in the Italian strain of *mellifera,* but its body is only about half as large. The behavior of *indica* colonies, insofar as it is known, is quite similar to that of *mellifera* colonies. They resemble members of the African races of *mellifera* in their strong tendency to abscond from the nest in the face of food shortage or the slightest disturbance of the nest. It appears likely that domestication of *mellifera* in ancient times involved bringing this tendency under genetic control by selecting for less restive strains.

The dwarf honeybee (*Apis florea*) is the smallest member of the genus; its workers are a mere seven to eight millimeters in length. It is also the most strikingly colored; the abdomen of the worker is marked by two brick-red segments and regularly spaced, transverse, silvery bands. The colony constructs a single naked comb, about the size of a dinner plate, which hangs vertically beneath a tree branch. The bodies of the workers are normally massed so thickly over the comb as to constitute a living, shingle-like coating. Lindauer (1961) found that the honey cells are two or three times longer than the brood cells and concentrated at the upper end of the comb, where their combined lateral surfaces form the platform on which horizontal waggle dances are performed. The colonies are even more prone to emigrate than *A. indica* and indeed may be regarded as nomadic since they regularly shift their nest sites to areas of temporarily richest food supply.

The giant honeybee (*Apis dorsata*) is about the size of a hornet (worker length, 16–18 mm). The colonies are large, and the workers are famous for their quick tempers and painful stings. The nests consist of single, naked combs hung vertically from tree branches, cliff overhangs, or the roofs of buildings. The combs are semicircular and often more than a meter in diameter. The colonies are as prone to emigrate as those of *A. florea.* In fact, according to Lindauer (personal communication), those in Ceylon migrate at a fixed season for distances as great as 160 km. *A. dorsata* has recently been the object of an intensive study in the Philippines by Morse and Laigo (1969).

Electrapis meliponoides of the Baltic amber was interpreted by Cockerell (1909) to be intermediate between *Bombus* and *Apis* but nevertheless probably a side branch of early apid evolution, rather than the ancestor of either of the living genera. This form, together with three additional related species of the same geological age, which have been discovered since Cockerell's time, do form a group generically distinct from *Apis.* However, additional bees from the Rott lignite of lower Miocene age have been placed in the genus *Apis* (as the subgenus *Synapis*) and appear to be close to the ancestral line of the living species (Kelner-Pillault, 1969). Thus it appears most likely that the true honeybees arose in the Old World sometime in late Oligocene or very early Miocene times.

The colony life cycle of *Apis mellifera* can be conveniently regarded as beginning with the swarming process inside the parental nest. The causal factors of swarming are not well understood at present. It is at least clear that crowding of the worker bees within the nest plays an important, though not an exclusive, role. At this point the colony has anywhere from about 20,000 to 80,000 workers. The first event is the construction by the workers of a small number of royal cells, which are large, ellipsoidal chambers usually placed along the lower margins of the combs. We know that these cells will not be built so long as the mother queen is producing "queen substance" (*trans*-9-keto-2-decenoic acid) from her mandibular glands in sufficient quantity for each worker to receive on the average of at least 0.1 μg per day. But with the onset of the swarming season in late spring, the queen's production of the 9-ketodecenoic acid falls off, and construction of royal cells ensues. The queen lays one egg in each royal cell, and the hatching larvae are fed special foods by the workers, which insures their development into queens. The growth of a new queen is astonishingly quick, requiring only 16 days from the laying of the egg to the eclosion of the adult bee, as opposed to 21 and 24 days for the worker and drone, respectively. While all of this is going on, the status of the mother queen changes. She still lays a few eggs, but her abdomen is reduced in size, and she begins to behave in an agitated fashion. The workers feed her less and even show mild hostility, pummeling and jumping on top of her. Eventually she is pushed out of the hive and flies off in the company of a large group of workers. Several such swarms may emerge around this time. The "prime" swarm, containing the old queen, usually leaves soon after the first royal cell has been capped, just prior to the pupation of the queen larva inside. The first "afterswarm," containing the first of the new queens, occurs around eight days later, very soon after the new queen emerges from the royal cell and mates. The occurrence of afterswarms depends on the size and health of the colony, and the number of these events varies greatly.

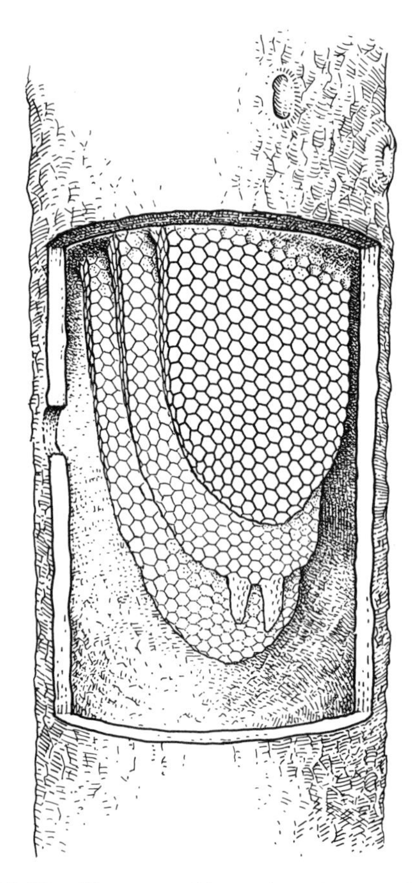

FIGURE 5-15. The nests of honeybees (*Apis mellifera*) in the natural state consist of double-layered waxen combs suspended vertically from a support. In most cases they are constructed in hidden retreats, such as the hollow tree shown in this example.

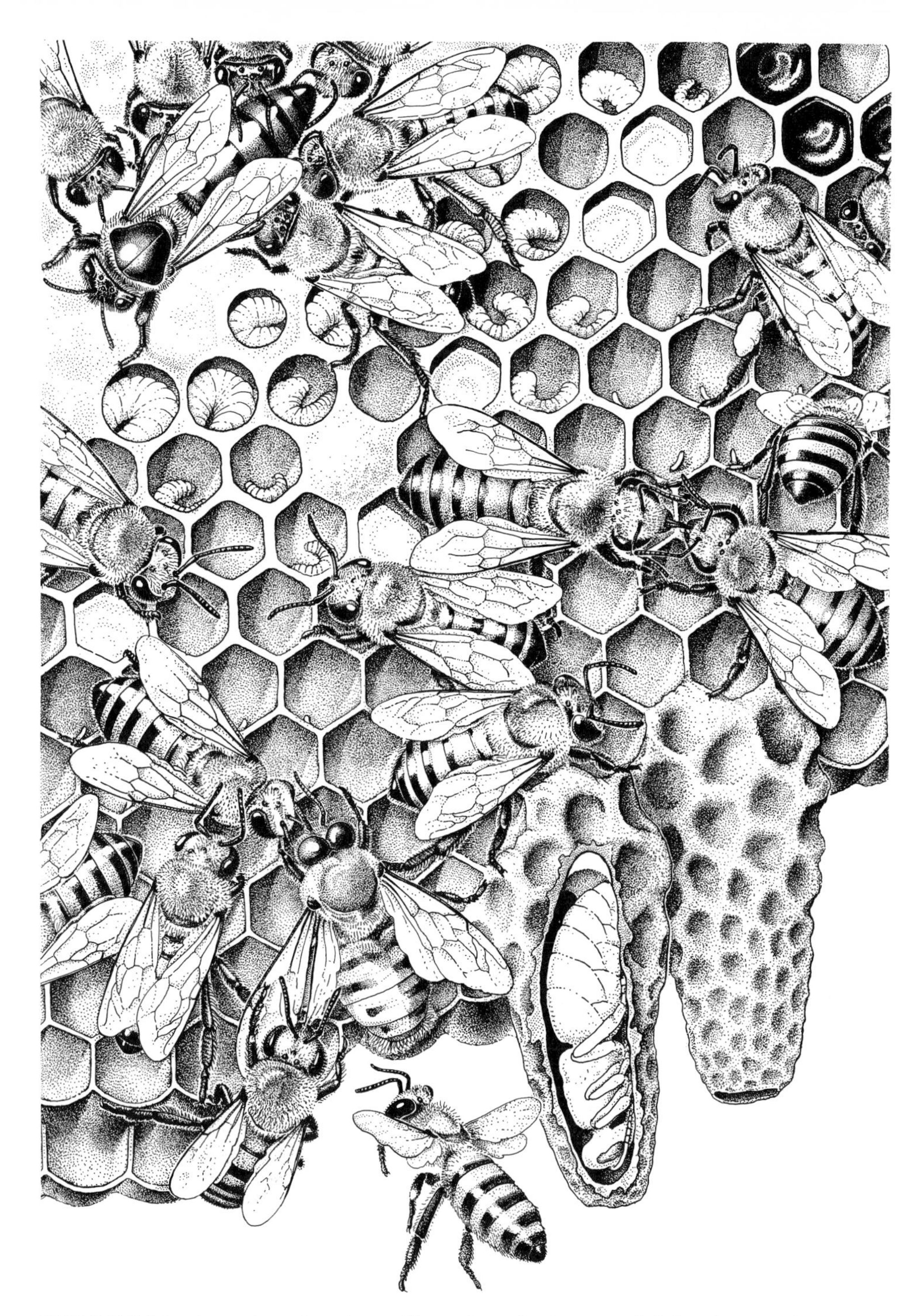

FIGURE 5-16. A portion of a colony of honeybees. In the upper left-hand corner the mother queen is surrounded by a typical retinue of attendants. She rests on a group of capped cells, each of which protects a developing worker pupa. Many of the open cells contain eggs and larvae in various stages of development, while others are partly filled by pollen masses or honey (extreme upper right). Near the center a worker extrudes its tongue to sip regurgitated nectar and pollen from a sister. At the lower left another worker begins to drag a drone away by its wings; the drone will soon be killed or driven from the nest. At the lower margin of the comb are two royal cells, one of which has been cut open to reveal the queen pupa inside (original drawing by Sarah Landry).

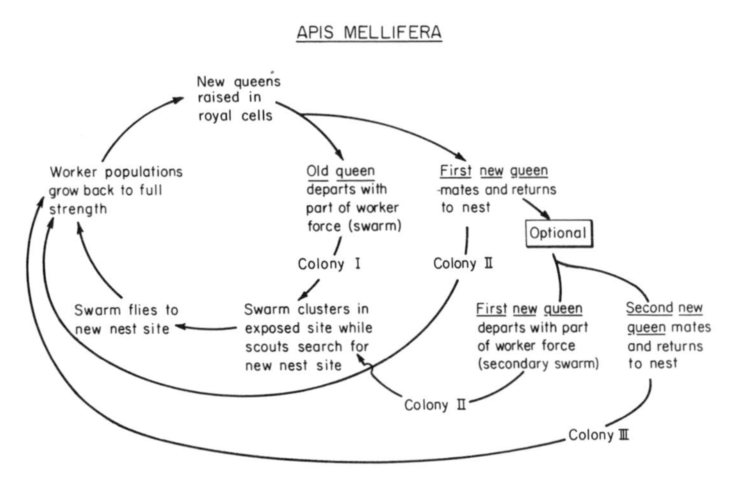

FIGURE 5-17. Principal stages in the life cycle of the honeybee (*Apis mellifera*). Although only a single afterswarm is shown in this particular scheme, two or more occur in extreme cases in nature.

Eventually, however, about two-thirds of all the workers leave the nest.

The swarming bees fly en masse for a short distance from the old hive and settle onto an aerial perch, such as the trunk or branch of a tree, or the side of a building, where they cluster tightly to form a solid mass of bodies. It is known that a second pheromone produced in the queen's mandibular glands, *trans*-9-hydroxy-2-decenoic acid, is necessary for this grouping behavior to be consummated. Scout bees now fly out from the bivouac in all directions in the search for a new permanent nest site. When a suitable site is found—a hollow tree, the enclosed eave of a building, an unoccupied commercial hive—the scouts return and signal the direction and distance of the find. This is accomplished by means of waggle dances performed on the sides of the swarm. Different scouts may announce different sites simultaneously, and a contest ensues. Finally the site being advertised most vigorously by the largest number of workers wins, and the entire swarm flies off to it. Now there are two colonies: the fraction back at the old nest which is about to acquire a newly fecundated daughter queen, and the fraction at the new nest which contains the old mother queen.

For a brief time, the workers at the parental nest are queenless. But the events that ensure requeening have long since been set in motion. Even before the construction of the queen cells that preceded swarming, the workers have built a group of drone cells, which look just like worker cells except that they average slightly larger. Into these the mother queen lays unfertilized eggs, which, true to the haplodiploid mode of sex determination prevalent in most Hymenoptera, develop into males. When they are four days or more into adult life, the males begin leaving on mating flights, traveling short distances from the nests to special areas where they join loose swarms of males from other nests in the vicinity. Here, in sustained flight, they await the approach of the virgin queens.

The first virgin queen to emerge from a royal cell is the only adult member of her caste in the nest. Her mother has already departed, and her sister queens are still in their cells. She now searches through the colony for her rival sisters, exchanging with them special sound signals descriptively labeled as "piping" and "quacking." If her sisters emerge from their own cells while she is present, fighting ensues and is continued until only one queen is left, either by swarming or by sequential killing. The virgin queen is urged out of the nest and on to her nuptial flight by mildly aggressive behavior on the part of the workers. As she approaches the male congregations, she releases small quantities of 9-ketodecenoic acid from her mandibular glands. As this scent disperses downwind, it attracts males from distances of ten meters or more. The mating is quick and violent; the male literally explodes its internal genitalia into the genital chamber of the queen and quickly dies. The queen makes as many as three flights a day for a total of up to 12 flights or more, and on each flight she mates with a different male. Finally, she obtains enough sperm to last her lifetime. Then, she either participates in an afterswarm, making way for the next virgin queen to emerge and mate, or else she destroys the other young queens and takes over the nest. In either case, if conditions are favorable, her own daughter workers will cause the worker population under her control to double within a year, and the colony can divide again.

The Origin of Social Behavior in Bees

In 1923 W. M. Wheeler, enchanted by Emile Roubaud's concept of trophallaxis in wasps, fashioned what was to become the traditional interpretation of the origin of social life in insects. According to this view, which Wheeler documented with new information on primitively social wasps provided by Roubaud, F. X. Williams, and others, eusocial life is reached through a series of subsocial steps in evolution: the mother becomes more intimately associated with her young, then her own longevity increases, and finally she lives long enough to gain the cooperation of her offspring when they reach the adult stage. The theory, which, as we saw in Chapter 3, still fits our knowledge of wasps quite well, was applied indiscriminately to all of the eusocial insects by Wheeler and subsequent authors.

In 1958 C. D. Michener cited new lines of evidence to indicate that a different evolutionary route has been followed by the bees. Except for the allodapines, he suggested, eusociality has been reached by parasocial steps. Adult-larva relationships were thought to play no role because worker bees evidently provision the brood cells with all the food the larva will need prior to the laying of the egg and seal off the brood cell while the larva is developing. In Michener's view, eusociality has arisen in at least some bee groups from a situation in which adult females first cooperate and then, as evolution proceeds, begin to divide the roles of egg laying and foraging among themselves. This bifurcation of evolutionary pathways has been rendered somewhat less neat by the recent discovery that workers of the primitively social halictines often reenter brood cells to clean out waste material and, at least in the case of *Evylaeus malachurus,* to add nectar. In other words there is considerable worker-larva contact after all (Plateaux-Quénu, 1962; Batra, 1964; Knerer, 1969). Yet the matter is not pivotal since the intensity of worker-larva contact cannot contribute as much to social complexity as the amount of reproductive division of labor and the extent of overlap of generations. Even so, Michener's distinction reduces to one of sequence of events in the evolution of the life cycle, the essentials of which can be expressed in the following way.

The parasocial route in evolution (some bees?):

1. Females of the same generation aggregate and cooperate in building the nest (*communal behavior*)
2. Next they cooperate in brood care (*quasisocial behavior*)
3. Then they divide into queen and worker castes (*semisocial behavior*)
4. Finally, some live long enough to coexist in the same colony with their daughters or nieces (*eusocial behavior*)

The subsocial route in evolution (other social insects):

1. A single female builds the nest and stays around to care for her brood (*subsocial behavior*)
2. The female lives long enough to be around when her daughters reach the adult stage, and they help her raise more of her daughters (*eusocial behavior*)

Both routes lead from the solitary state to the accumulation of the same three qualities which I gave earlier (Chapter 2) as tautologically defining the eusocial state, namely cooperative brood care, reproductive castes, and overlap of generations.

Let us now look more closely at the evolutionary steps to eusociality in the bees as they have most recently been envisioned by Michener (1969a). The crux of the whole scheme, as we shall see, is whether or not truly semisocial species exist.

Solitary bees. In the majority of bee species each female makes one or more nests of her own, without reference to the location of nests built by other females of the same species. She mass-provisions the cells in the nest, that is, she places enough pollen and honey in each cell to provide for the larva throughout its development and lays an egg in each cell. Finally she seals the cell and goes on to construct others. In the great majority of solitary species, the mother bee dies before her offspring reach maturity. In a few exceptional cases, however, there is overlap, and direct contact may even occur. In *Halictus quadricinctus,* for example, the mother remains in the nest, which has multiple cells, and is still present when the first of her offspring emerge.

Aggregations of solitary bees. A very common phenomenon in all families of bees except the Apidae is the grouping of nests of solitary bees. The entrances are relatively close together, even though none of the tunnels or brood areas are joined. The numbers of nests in such aggregations vary from only two or three to thousands. Some aggregations clearly come about as a simple effect of limited nesting space. On the other hand, Michener, Lange, Bigarella, and Salamuni (1958) found that aggregations of some solitary species occur even when extensive stretches of seemingly identical but unoccupied habitats occur close by. They concluded that the groupings were due either to a tendency for bees to return to their birthplace or to an innate attractiveness of bees of the same species for each other. The importance of aggregation behavior is that, while not involving cooperation among adults or other obvious forms of sociality, the crowding predisposes species to adopt such behavior.

Communal groups of bees. Michener employs this expression to designate groups of females which utilize a single composite nest but each of which makes and provisions her own cells. The advantages of such a system are obvious: economy of labor results from the necessity of having only one entrance tunnel instead of many, while defense of the tunnel, and hence of all the brood cells, is more persistent and effective. Communal groups are known in many species of the Andrenidae, the Megachilidae, and the halictid subfamilies Halictinae and Nomiinae. Virtually every degree of development of the phenomenon has been documented, from occasional cohabitation of two females belonging to normally solitary species, as, for example, in *Andrena erythronii* and *Nomia punctulata,* to large assemblages that build as many as 1750 cells in one nest, as in *Anthidium repetitum.*

Quasisocial groups of bees. An important step up in sociality has been taken by those species whose females not only build and guard the entrance tunnel in concert but also cooperate in constructing and provisioning brood cells. There is as yet no evidence of division of labor; all of the females work at nest construction and all, presumably, lay eggs from time to time. Michener has postulated the existence of this category in nature, but it has not yet been firmly documented. The strongest indication seems to be in the genus *Euglossa,* where the number of females nesting together (as many as eight) often outnumber the cells being constructed (Roberts and Dodson, 1967; Sakagami, Laroca, and Moure, 1967). The Euglossinae are also significant in that they are the only members of the Apidae that are not eusocial. Their behavior is therefore likely to cast light on the origins of eusociality in the remainder of the family. In addition, Dodson (1966) has reported the existence of a series of species within the single genus *Euglossa* which exhibit stages from wholly solitary to quasisocial nest building. Unfortunately, the euglossines are exclusively tropical, and their nests are notoriously hard to find so that information has been slow in arriving.

Semisocial groups of bees. These societies consist of females of the same generation who cooperate in building and provisioning the nest and who are further divided into two functional castes, inasmuch as some of the females do not lay eggs but serve exclusively as workers. Hamilton (1964) has doubted the existence of such a step in evolution on the a priori grounds that it would be selectively disadvantageous for females to subordinate themselves to other females of the same generation in this fashion. Hamilton's reasoning, which will be explained fully in Chapter 17, is correct as far as it goes. To question the likelihood of the semisocial state is to doubt that eusociality can be attained by the parasocial route as postulated by Michener, that is, communal to quasisocial to semisocial to eusocial. The evidence for the existence of the semisocial stage therefore needs to be closely scrutinized.

Actually, the evidence needed to establish at least the feasibility of semisocial behavior is already firmly in hand, contra Hamilton's inference. As we have seen already, halictine females of the same generation often form associations in the spring to found colonies, and their ovarian development at this time is so variable as virtually to insure some degree of reproductive division of labor. Similar aggregations, with absolute reproductive division of labor, have been documented in the paper wasps of the genus *Polistes* and rationalized in theory by West (1967). Both the halictine and *Polistes* groups go on to produce eusocial colonies as the summer progresses, so that one cannot refer to them as strictly semisocial groups. But the point remains that if initial aggregations can be organized in a semisocial manner, then permanently semisocial species are at least feasible.

Nevertheless, the evidence for a true and persistent semisocial state is still limited to the tribe Augochlorini and is circumstantial in nature. In *Augochloropsis sparsilis* of Brazil, Michener and Lange (1958a, 1959) recorded aspects of the life cycle which indicate that the colonies are semisocial. Each nest consists of a burrow excavated into an earthen bank and ending in one to several cells. In most cases several females, from about two to eight, occupy each nest. The females that overwinter presumably belong to the same generation, although on this crucial point Michener and Lange are less than decisive: "From observations of marked individuals we know that females of various ages go into the winter. The badly worn and tattered bees disappear in the fall, but some bees several months old, although still unworn or nearly so, overwinter with young unworn females." At the beginning of spring all of these females have unworn wings and mandibles, together with weakly developed ovaries. The vast majority (85 of 86 sampled) are fertilized and hence capable of becoming queens. In short time, however, a division of labor becomes manifest. Some of the bees begin to show wing and mandible wear, while their ovaries stay undeveloped; they have evidently become workers. Others undergo marked ovarian development and remain untattered; they are the queens. There are no external morphological differences between the two types. As the season progresses, the differentiation between the castes strengthens. By midsummer 15–20 percent of the females are unmated, and most of these are functioning as workers. The descriptions supplied by Michener and Lange imply a turnover of individuals sufficiently rapid to insure that the adult bees are sisters rather than mothers and daughters, but survivorship data directly supporting such an assumption are lacking. Also, it is not known whether or not the same individual can assume the roles of both egg layer and worker at different times of its life. The *Augochloropsis* case appears to me to be suggestive but not decisive.

The same must be said to be true of the three Costa Rican species of *Pseudaugochloropsis* (*costaricensis, graminea, nigerrima*) studied by Michener and Kerfoot (1967). The evidence for semisociality, as in *Augochloropsis,* is based on composite data from many dissected nests rather than life histories of single colonies. About two-thirds of the nests are inhabited by single females who construct one to six brood cells. The first brood of sisters remain in the nest and sort themselves into primitive castes in the manner of *Augochloropsis,* with the difference that the egg layers are very slightly larger than the workers. The evidence concerning the matter of relationships among the cooperating females is the following. First, Michener and Kerfoot note that the small nests, which make up the majority of all nests, are normally occupied by only one female. The female apparently ceases construction activity after provisioning one to six cells: "Cessation of maternal nesting activity is shown by the numerous small cell clusters containing one to six pupae or prepupae but no younger stages . . . even though the mother is still present. Other nests lack the mother, suggesting that she dies at about the time when her brood reaches maturity." The second piece of evidence indicating a short life for the foundress female is supplied by the circumstance that, in larger colonies, both the "egg layers and workers were comparably worn; they are therefore probably sisters, not mothers and daughters. This implies that the lone egg layers that establish nests do not survive to act as queens of colonies, and that these species are semisocial in the sense of Michener and Lange." These observations are suggestive but not as strong as needed to give definitive support to Michener's theory of parasocial evolution in bees. In any case, Sakagami and Hayashida (1961), as well as Michener himself (1969a), have concluded that evolution by the parasocial route in the strict sense has occurred in only a part, possibly even a small part, of the social bees. The origin of eusociality in bees may, moreover, be even more complex than this degree of uncertainty indicates, as we shall now see.

Consider again the allodapines. Since the discoveries of Brauns, it has always been assumed that the allodapine bees achieved eusociality by the subsocial route. This is because the larvae are progressively fed, requiring the presence of the mother up to at least the time of pupation of the youngest larva. It is an easy step to where the mother lives in the nest long enough to confront her first adult female offspring. If the offspring aid the mother in rearing their own brothers and sisters, even if for only a short interval, eusociality has been attained. This is a precise description of the situation in the eusocial allodapines whose life cycles are known. It is interesting to reflect on why evolution took this particular course in the allodapines. It does not seem to result from any innate advantage accruing to progressive provisioning, and, hence, not from close contacts between adults and larvae. Rather, as Michener has pointed out, progressive provisioning might be simply an adaptation to life in hollow stems. Some species of *Ceratina,* which are solitary relatives of the allodapines and with them the only bees that utilize hollow stems, have the curious habit of provisioning the nest without dividing it entirely into cells. This omission may be a move toward economy in nest construction. In any case, it facilitates the evolution of progressive provisioning and, with this mode of brood care, the attainment of overlap of generations and early eusociality.

Now it is true that most of the allodapines are primitively eusocial, and the condition is intimately associated with progressive provisioning. But it is also the case that, soon after the first adult offspring appear, the mother dies and her adult daughters take over the care of the orphaned larvae. Some begin to lay eggs on their own, while others function as workers, and the association is now semisocial in nature. Exactly the same confused condition occurs in some of the Halictini. This leads to the question of which came first, the overlap of generations or the division of labor among sisters. At this point we can fairly say that the distinction between the parasocial and subsocial routes is no longer a fundamental one. For it is important to remember that two of the defining qualities of eusociality, overlap of generations and reproductive division of labor, can evolve independently, and they can also evolve in concert. It is likely that the two conditions were evolved more or less in concert in the bee groups. Exactly which was achieved first is not a matter of overriding consequence. Put another way, the subsocial and parasocial routes, as originally envisioned by Wheeler and Michener, are two extremes, and they involve the prior, exclusive appearance of either overlap of generations or reproductive division of labor. Many intermediate timetables are possible and seem actually to have been followed in the evolution of the bees.

Finally, it now seems likely that social evolution can be reversed and a eusocial species can revert to a solitary condition. In his study of the biology of Australian allodapines, Michener (1964b, 1965b) determined that the genus *Exoneurella* differs from its close relative *Exoneura* in one detail in the colony life history: when the females mature, they disperse quickly to start new nests instead of tending to remain in the parental nest as is the case in *Exoneura.* If the mother dies, one (but not more than one) of the daughters sometimes remains in the nest to care for the remaining brood. In itself, the difference in behavior is rather subtle, but it results in *Exoneurella* being subsocial, whereas *Exoneura* is by definition eusocial, albeit very primitively so. Furthermore, Michener believes that the solitary condition may be derived from an *Exoneura*-like eusocial state. In evidence of the proposition is the fact that *Exoneurella* males emerge predominantly at the end of the growing season in the fall, a characteristic which Knerer and Plateaux-Quénu have shown to be a concomitant of eusociality. If eusociality is reversible, it is even possible that phyletic lines may fluctuate between the presocial and eusocial states, and back and forth across the various presocial states. Consequently, until substantially more information is available on the comparative life cycle of bees, it will not be possible to do more than set a lower limit on the number of times that sociality has arisen within the bees.

6 The Termites

In an almost literal sense, termites can be called social cockroaches. Comparisons made between the most primitive termite family, the Mastotermitidae, and the primitive blattoid cockroaches, in particular the Cryptocercidae, have revealed detailed resemblances in a multitude of unrelated characters: wing venation (Emerson, 1965); external structure of the terminal abdominal segments (Crampton, 1923; Browman, 1935); internal anatomy of the female genitalia (Ahrens, 1934; McKittrick, 1965); mandibular dentition of the workers and royal imagoes (Ahmad, 1950); tarsal segmentation (Ratcliffe, Gay, and Greaves, 1952); bacteriocytes (Jucci, 1932; Koch, 1938); and the endocrine system (Mosconi, 1958). Of the 25 species of symbiotic hypermastigote and polymastigote flagellates found in the gut of the wood-eating cockroach *Cryptocercus punctulatus* (21 and 22 species, respectively, in the two United States populations, of which 18 are held in common), all belong to families also found in the lower termites. Even one genus, *Trichonympha,* is shared. These intestinal protozoans can be successfully "transfaunated" from cockroach to termite and vice versa (Cleveland, Hall, Sanders, and Collier, 1934). *Mastotermes darwiniensis,* the only living member of its family, also possesses an intestinal bacterial flora similar to that in cockroaches, and it lays its eggs in batches strongly reminiscent of cockroach oothecae. In life the young nymphs of *Cryptocercus* bear a striking superficial resemblance to termite workers. It is of course too much to hope that any of the living cockroaches can be identified as the ancestor of the termites. All known alate blattarians have horny forewings, so that the membranous wings of the termites are more primitive (Ahmad, 1950). *Mastotermes* has an anal lobe resembling that of the winged cockroaches, but it is reduced, including only the second anal vein and its branches. Also, the jugal region and jugal fold found in cockroaches are absent in the termites (Tillyard, 1931; Emerson, 1965). But these are not cardinal differences, and some specialists, McKittrick, for example, have gone so far as to place the termites in the same order (Dictyoptera) as the cockroaches and mantids.

Because the termites have climbed the heights of eusociality from a base extremely remote in evolution from the Hymenoptera, it is of surpassing interest to know whether their social organization differs in any fundamental way. Although value judgments of the degree of convergence of two radically differing stocks are difficult to make, much less to justify quantitatively, I believe the following assessment can reasonably be made. The termites have adopted mechanisms that are mostly but not entirely similar to those in the social Hymenoptera. Also, the level of complexity of termite societies is approximately the same as that in the more advanced hymenopteran societies. In Table 6-1, I have listed the principal known similarities and dissimilarities of the two kinds of societies. This very simplified accounting does not overlook the fact, already stressed in previous chapters, that a great deal of important variation also occurs within the social Hymenoptera. Surely the similarities are remarkable in themselves. They seem to tell us that there are constraints in the machinery of the insect brain that limit not only the options in the evolution of social organization but also the upper limit that the degree of organization can attain. These limits appear to have been reached between 50 and 100 million years ago in both the termites and the social Hymenoptera.

TABLE 6-1. Basic similarities and differences in social biology between termites and higher social Hymenoptera (wasps, ants, bees). Similarities are due to evolutionary convergence.

Similarities	Differences: Termites	Differences: Eusocial Hymenoptera
1. The castes are similar in number and kind, especially between termites and ants	1. Caste determination in the lower termites is based primarily on pheromones; in some of the higher termites it involves sex, but the other factors remain unidentified	1. Caste determination is based primarily on nutrition, although pheromones play a role in some cases
2. Trophallaxis (exchange of liquid food) occurs and is an important mechanism in social regulation	2. The worker castes consist of both females and males	2. The worker castes consist of females only
3. Chemical trails are used in recruitment as in the ants, and the behavior of trail laying and following is closely similar	3. Larvae and nymphs contribute to colony labor, at least in later instars	3. The immature stages (larvae and pupae) are helpless and almost never contribute to colony labor
4. Inhibitory caste pheromones exist, similar in action to those found in honeybees and ants	4. There are no dominance hierarchies among individuals in the same colonies	4. Dominance hierarchies are commonplace, but not universal
5. Grooming between individuals occurs frequently and functions at least partially in the transmission of pheromones	5. Social parasitism between species is almost wholly absent	5. Social parasitism between species is common and widespread
6. Nest odor and territoriality are of general occurrence	6. Exchange of liquid anal food occurs universally in the lower termites, and trophic eggs are unknown	6. Anal trophallaxis is rare, but trophic eggs are exchanged in many species of bees and ants
7. Nest structure is of comparable complexity and, in a few members of the Termitidae (*e.g., Apicotermes, Macrotermes*), of considerably greater complexity. Regulation of temperature and humidity within the nest operates at about the same level of precision	7. The primary reproductive male (the "king") stays with the queen after the nuptial flight, helps her construct the first nest, and fertilizes her intermittently as the colony develops; fertilization does not occur during the nuptial flight	7. The male fertilizes the queen during the nuptial flight and dies soon afterward without helping the queen in nest construction
8. Cannibalism is widespread in both groups (but not universal, at least not in the Hymenoptera)		

A second key to the understanding of termite biology is the principally cellulose diet of these insects. The more primitive species feed directly on the wood in which they nest, while the morphologically advanced species nest in the soil and forage for dead wood, grass, seeds, and other diffuse sources of cellulose. The utilization of their food source has reached impressive levels of efficiency, especially in the family Termitidae. In the rhinotermitid genus *Coptotermes* and termitid genera *Cubitermes, Microcerotermes, Nasutitermes,* and quite likely other termitids as well, cellulose or cellulose products are digested, lignin and siliceous material are passed out in fecal pellets, and the pellets are used in turn to construct the walls of the nest (Snyder and Zetek, 1924; Cohen, 1933; R. M. C. Williams, 1959a). In the fungus-growing termites (subfamily Macrotermitinae) imperfectly digested pellets are built into fungus combs, where symbiotic fungi of the genus *Termitomyces* degrade them further. The combs and fungal pellets are then eaten again by the termites (Grassé and Noirot, 1958a; Sands, 1969). The combs also play a secondary role in the ventilation of the nest by adding heat of fermentation to the central core of the nest which in turn drives used air upward in exit channels through convection (Lüscher, 1961a). The cellulose diet has left its stamp on the social biology of termites in other ways. Perhaps most importantly, it has bound these insects to symbiotic intestinal protozoans and bacteria. In order to transmit the symbionts, the termites engage in a unique form of anal liquid exchange. It is even possible that the symbiosis was the primary cause of social life in termites in the first place, a subject that will be taken up again later in this chapter.

All of the termites together comprise the Order Isoptera. Several versions of higher classification, dividing the order into either five or six families, have been offered in recent years (Grassé, 1949; Snyder, 1949; Roonwal, 1962a). The most authoritative arrangement is that of Emerson (1965), who recognizes six families: Mastotermitidae, Kalotermitidae, Hodotermitidae, Rhinotermitidae, Serritermitidae, and Termitidae. The first four families are referred to as the "lower" termites. They are distinguished as a group by the possession of symbiotic intestinal flagellates. The Termitidae are called the "higher" termites and are exceptionally diverse, being comprised of more genera and species than all the other families put together. The Serritermitidae occupy an uncertain phylogenetic position. The family consists of only one known species, *Serritermes serrifer* of Brazil, whose imagoes and workers possess falcate mandibles of the most specialized type found anywhere in the termites. Holmgren ranked the species as a subfamily of the Rhinotermitidae and Ahmad as a subfamily of the Termitidae, but Emerson does not believe that it fits into any monophyletic grouping of genera comprising either family. As of April 1969 there were a total of about 2,200 living species of termites known, 1,914 described and approximately 300 undescribed in collections available to Emerson (personal communication).

All termites are eusocial. The vast majority are limited to the tropics, and this is especially true of the Termitidae. In Table 6-2 are given the names and distributions of all of the subfamilies and most of the principal genera of the world fauna, while Table 6-3 provides the species counts in the major groups. A good sample of the genera are illustrated in Figures 6-1 to 6-5.

The fossil history of the termites contains curious parallels to that of the ants. The few specimens so far discovered in Eocene rocks belong to the relatively primitive families Mastotermitidae and Kalotermitidae. The Baltic amber fauna, of Oligocene age, is essentially modern (as in the ants), being comprised of representatives from the Hodotermitidae, Kalotermitidae, and Rhinotermitidae. Several of the Baltic amber genera are extinct, but are not of a particularly archaic character. The remainder of these amber genera exist today in warm temperate climates of Europe and Asia. A recently discovered tropical termite fauna in Chiapas (Mexico) amber, of Oligocene-

TABLE 6-2. List of families and principal genera of the world termite fauna, together with distributions of the genera in terms of major zoogeographic regions (based on Emerson, 1952, 1955, 1965, and personal communication).

Division	Distribution
Family Mastotermitidae	
[a] *Mastotermes*	Australia
Miotermes	Fossil only, in Tertiary of Europe
Family Kalotermitidae	
[a] *Kalotermes*	World-wide
[a] *Cryptotermes, Procryptotermes*	World-wide except for temperate Eurasia
[a] *Neotermes*	World-wide
Rugitermes	Polynesia, New World tropics
[a] *Glyptotermes*	World-wide, except for North Temperate Zone
[a] *Calcaritermes*	New World
Electrotermes	Fossil only, in Tertiary of Europe
Family Hodotermitidae	
[a] *Archotermopsis, Hodotermopsis*	Eurasia, Temperate Zone
[a] *Zootermopsis*	North America
Stolotermes	Africa, Australia, New Zealand
Porotermes	Africa, Australia, New World (Chile only)
Hodotermes	Africa
Anacanthotermes	Asia, Africa
Cretatermes	Fossil only, in Cretaceous of Labrador
Termopsis	Fossil only, in Tertiary of Europe
Family Rhinotermitidae	
Psammotermes	Eurasia, Africa
[a] *Coptotermes*	World-wide, mainly tropical
[a] *Heterotermes*	World-wide, mainly tropical
[a] *Reticulitermes*	Eurasia, North America, temperate only
Prorhinotermes	World-wide, tropical islands and shores
Termitogeton	Asia
Parrhinotermes	Asia, Australia
Schedorhinotermes	Asia, Africa, Australia
Rhinotermes	New World tropics
Parastylotermes	Fossil only, in Tertiary of Europe and North America
Family Serritermitidae	
Serritermes	New World tropics
Family Termitidae	
Subfamily Amitermitinae	
Speculitermes	Asia, New World tropics
Anoplotermes	Africa, New World
Euhamitermes, Eurytermes, Indotermes, Protohamitermes	Asia

TABLE 6-2 (*continued*).

Division	Distribution
Family Termitidae (*continued*)	
Ahamitermes, Incolitermes	Australia
[a] *Microcerotermes*	World-wide except North America
Amitermes	World-wide
Drepanotermes	Australia
Gnathamitermes	North America, temperate
Subfamily Termitinae	
Apicotermes, Basidentitermes, Crenetermes, Cubitermes, Euchilotermes, Fastigitermes, Foraminitermes, Hoplognathotermes, Lepidotermes, Megagnathotermes, Noditermes, Ophiotermes, Pericapritermes, Procubitermes, Promirotermes, Thoracotermes, Trichotermes, Unguitermes	Old World tropics
Cavitermes, Dentispicotermes, Neocapritermes, Orthognathotermes, Spicotermes, Spinitermes	New World tropics
Termes	World-wide, in tropics
Angulitermes	Eurasia, Africa
Dicuspiditermes, Homallotermes, Microcapritermes, Procapritermes	Asia
Capritermes	Madagascar
Subfamily Macrotermitinae	
Acanthotermes, Allodontermes, Ancistrotermes, Protermes, Pseudacanthotermes, Sphaerotermes, Synacanthotermes	Africa
[a] *Macrotermes, Microtermes, Odontotermes*	Asia, Africa
Subfamily Nasutitermitinae	
Eutermellus, Mimeutermes, Verrucositermes	Africa
Bulbitermes, Hirtitermes, Hospitalitermes, Lacessititermes	Asia, New Guinea
Grallatotermes, Trinervitermes	Asia, New Guinea, Africa
Armitermes, Constrictotermes, Convexitermes, Cornitermes, Curvitermes, Labiotermes, Obtusitermes, Paracornitermes, Parvitermes, Procornitermes, Rhynchotermes, Subulitermes, Syntermes, Velocitermes	New World tropics
Tenuirostritermes	New World
Nasutitermes	World-wide, in tropics

[a] Also known from Tertiary fossils.

TABLE 6-3. Numbers of known species in the major groups of termites as of April 1969 (A. E. Emerson, personal communication).

Groups	Living species	Pleistocene species	Tertiary species	Cretaceous species
"Lower" termites				
Mastotermitidae	1		13	
Kalotermidae	292	1	10	
Hodotermitidae	30		12	1
Rhinotermitidae	158		13	
Serritermitidae	1			
"Higher" termites				
Termitidae				
Amitermitinae	340	1	1	
Termitinae	333	1		
Macrotermitinae	263			
Nasutitermitinae	476	1	1	
TOTALS	1894	4	50	1
Additional names in the literature needing reclassification	20		6	
GRAND TOTALS	1914	4	56	1

Miocene age, is mostly similar at the generic level to the living Neotropical fauna (Emerson, 1969). An exception is an undescribed mastotermitid represented by an imago, nymphs, and the first known Tertiary soldier. The early Tertiary fossils offer few leads as to the place and time of origin of termites. Termitologists had long looked to the Mesozoic or beyond for traces of a truly archaic termite fauna. Zalessky's *Uralotermes* of the Russian Permian, described in 1937, seemed at first to fill the requirement in a spectacular way, but it has now been taken out of the Isoptera and placed provisionally in the Protothoptera by Emerson (1965). F. M. Carpenter, the principal authority on the Paleozoic insects, believes that the single wing on which *Uralotermes* was based is too incomplete to place it in any order, but concurs that at the very least it cannot be used to extend the Isoptera back to the Paleozoic era (personal communication). In 1967 Emerson reported the discovery of an undoubted termite wing from Labrador rocks that date from the border between Lower and Upper Cretaceous times. This is the first and so far only pre-Tertiary termite fossil to come to light. The species, *Cretatermes carpenteri*, has been placed in a new subfamily of the Hodotermitidae. Plant remains associated with the fossil suggest that *Cretatermes*

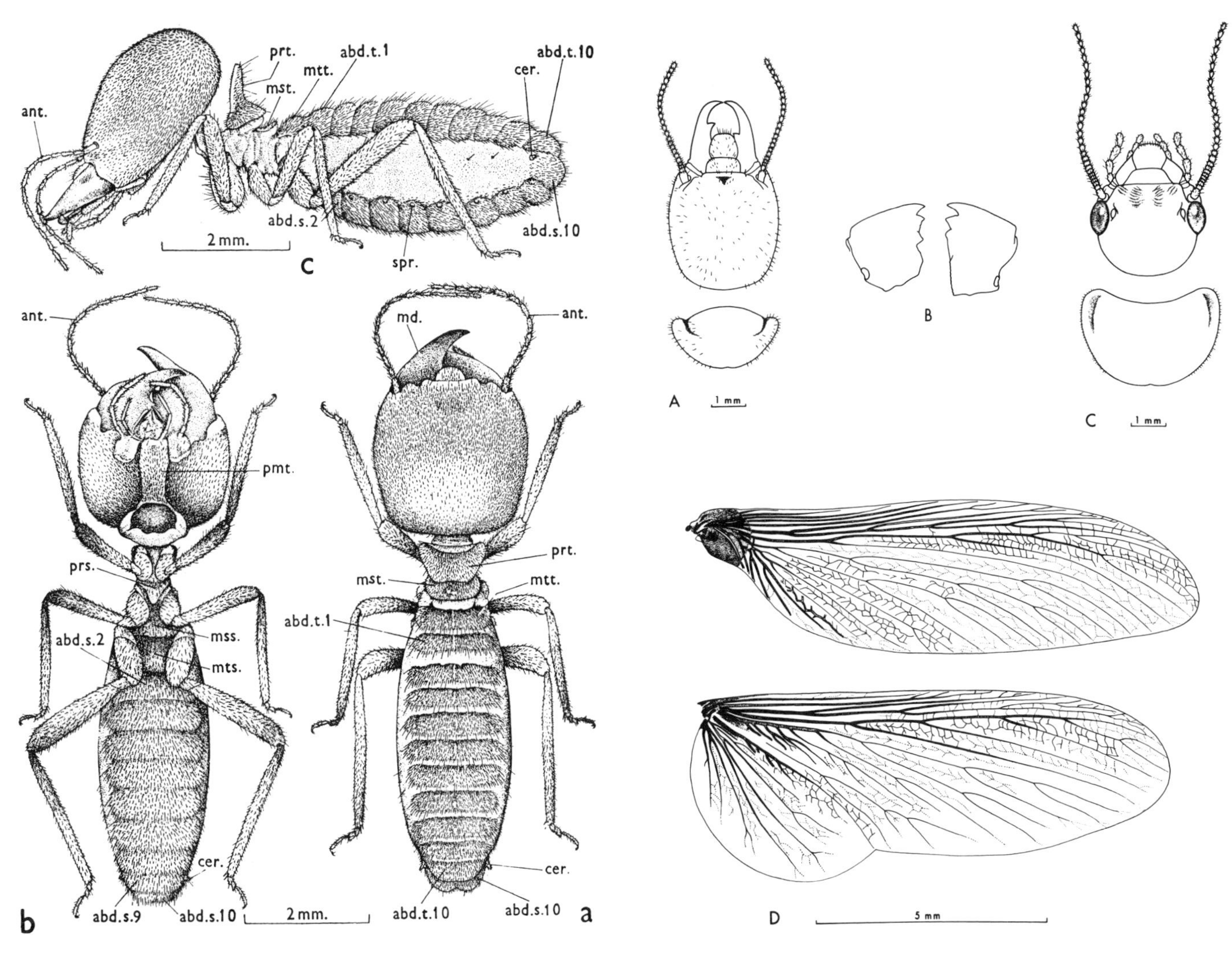

FIGURE 6-1. The soldier of a higher termite, *Indotermes maymensis* (family Termitidae), showing some of the principal external morphological characters used in taxonomy (from Roonwal, 1962a): *abd. s.* 1–*abd. s.* 10, abdominal sterna 1–10, respectively; *abd. t.* 1–*abd. t.* 10, abdominal terga 1–10, respectively; *ant.*, antenna; *cer.*, cercus; *md.*, mandible; *mss.*, mesosternum; *mst.*, mesothorax; *mts.*, metasternum; *mtt.*, metathorax; *prs.*, prosternum; *prt.*, prothorax; *spr.*, spiracle; *pmt.*, postmentum.

FIGURE 6-2. *Mastotermes darwiniensis,* the most primitive of the living termites: (*A*) head and pronotum of soldier; (*B*) imago mandibles; (*C*) head and pronotum of imago; (*D*) wings (from Gay, 1967).

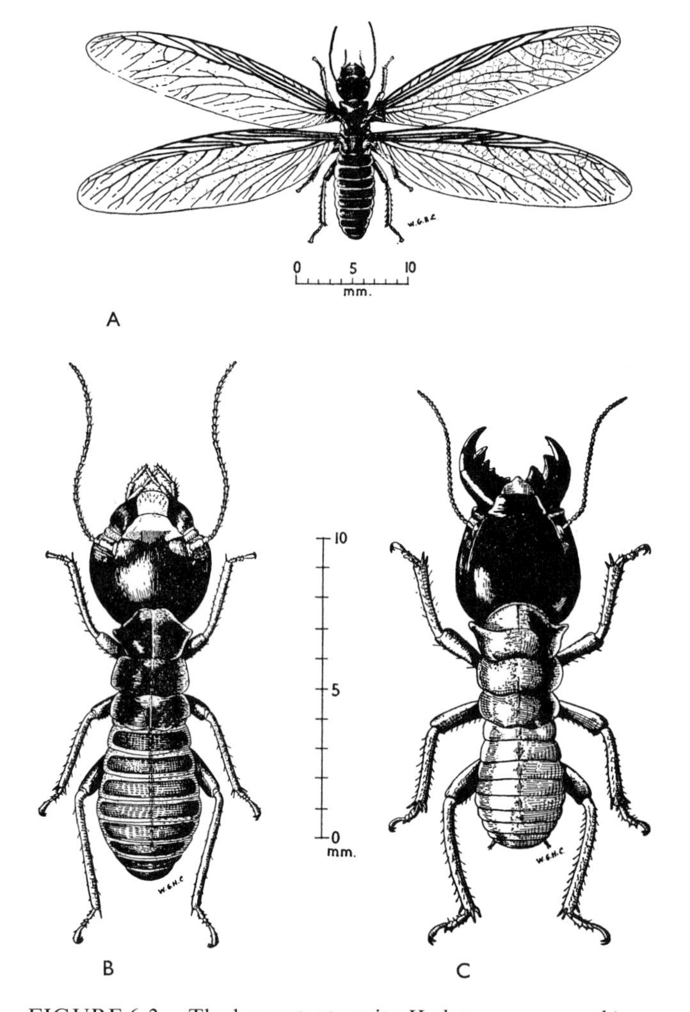

FIGURE 6-3. The harvester termite *Hodotermes mossambicus:* (*A*) alate; (*B*) worker; (*C*) soldier (from Harris, 1961, after W. G. H. Coaton).

lived in a warm temperate climate, just as did the slightly younger Cretaceous ant *Sphecomyrma.* On the basis of its wing venation, Emerson concluded that *Cretatermes* is a fairly advanced member of the otherwise primitive family Hodotermitidae. He inferred from this one important piece of evidence that the Isoptera as a whole may date back to the early Mesozoic or even late Paleozoic times. Although this conclusion is tenuous, it may well be correct. At the very least, the discovery of *Cretatermes* does suggest that the termites came into being before the ants. It will be remembered that although *Cretatermes* is a relatively advanced termite, *Sphecomyrma* is a very primitive ant, representative of the very transition between tiphiid wasps and the earliest myrmecioid ants.

When further Cretaceous discoveries are made, they will probably reveal the Cretaceous termite fauna to be generally primitive in nature. The basis for this prediction is that all the Eocene fossils at hand belong to primitive groups. Moreover, the Mastotermitidae, the most primitive known family, was world-wide in distribution and represented by no less than 4 genera and 13 species during the early and middle Tertiary period (Emerson, 1965). Today it is represented by only one species (*Mastotermes darwiniensis*) restricted to tropical Australia.

A highly simplified phylogeny is presented in Figure 6-6. As pointed out first by Nils Holmgren and amplified by later authors including especially Noirot (1958–1959), evolution in the imago caste of termites has involved relatively minor changes in morphology which tend, moreover, to be regressive in nature, but striking changes in behavior and social organization. The trend has been from small colonies that form irregular nests within the wood on which the termites feed to large colonies that build elaborate nests in the soil or branches of trees with a high degree of temperature and humidity control. The primitive condition is exemplified by the Kalotermitidae and

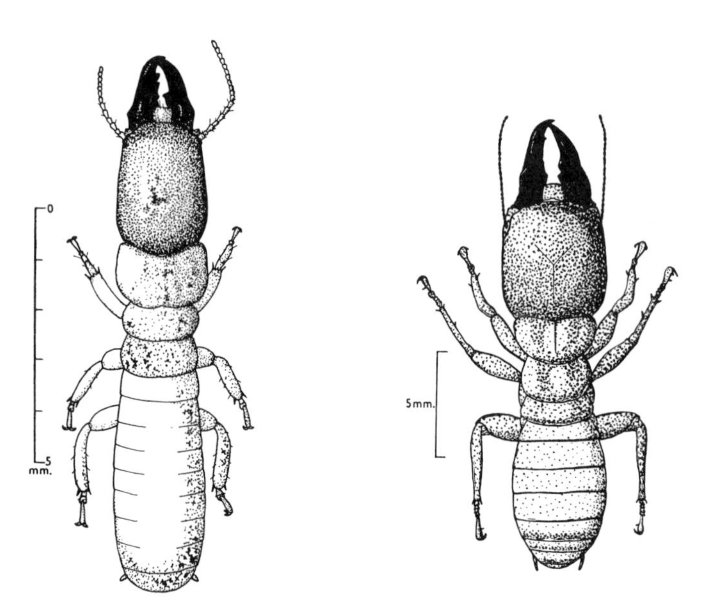

FIGURE 6-4. Soldiers of *Kalotermes flavicollis* (*left*) and *Zootermopsis angusticollis* (*right*) (from Harris, 1961).

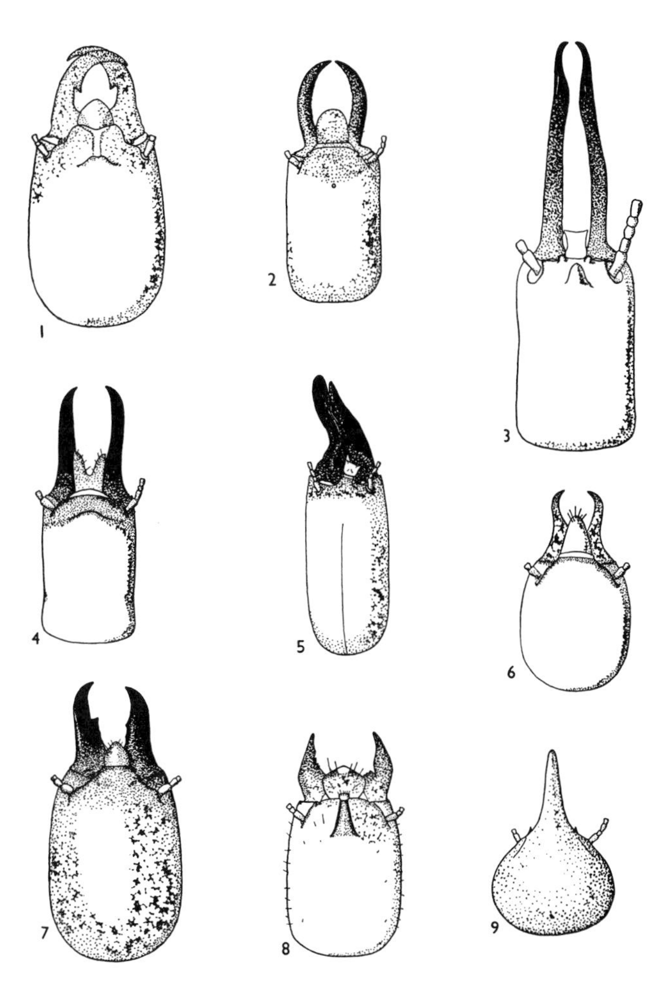

FIGURE 6-5. A gallery of soldier heads from various genera of the higher termites (family Termitidae): (1) *Amitermes messinae;* (2) *Microcerotermes biroi;* (3) *Termes odontomachus;* (4) *Cubitermes minitabundus;* (5) *Pericapritermes dumicola;* (6) *Microtermes luteus;* (7) *Odontotermes montanus;* (8) *Cornitermes silvestri;* (9) *Nasutitermes novarum-hebridarum.* The diversity of head and mandible forms is associated with differences in defensive behavior (from Harris, 1961).

primitive Hodotermitidae and Rhinotermitidae; the most advanced state, by certain elements of the Hodotermitidae and Rhinotermitidae as well as by all of the Termitidae. Once the more specialized termites liberated themselves in this fashion from life within large masses of wood, they were able to penetrate the widest variety of habitats and exploit nearly every conceivable cellulose source, including grass, seeds, leaf litter, and humus as well as live and decaying wood. The adaptive radiation that ensued was mostly accomplished by two families, the Hodotermitidae and Termitidae. A more exact picture of the kinds of changes, and their magnitude, that have occurred in the transition from the most primitive to the most advanced termites can be gained from the list presented in Table 6-4.

Most of the history of termite taxonomy is recent. In 1858 Hagen listed only 60 species, and in 1904 Desneux still listed only 343 in the *Genera Insectorum.* The current classification is in sound condition, evidently better than for any other group of social insects because of the activity of an exceptionally large and able group of taxonomists whose work dates almost entirely from 1900: M. Ahmad, J. H. Calaby, W. G. H. Coaton, J. Desneux, A. E. Emerson, W. W. Froggatt, F. J. Gay, P.-P. Grassé, W. V. Harris, G. D. Haviland, G. F. Hill, N. Holmgren, N. A. Kemner, K. Krishna, M. L. Roonwal, W. A. Sands, F. Silvestri, Y. Sjöstedt, T. E. Snyder, E. Wasmann, H. Weidner, and others. Some indication of the completeness of their work is seen in the fact that the last new termite species to be described from the United States was *Incisitermes milleri* (Emerson), discovered in the Florida Keys in 1941. New species, and even new genera, continue to turn up in the tropics, but the rate has dropped in recent years (Roonwal, 1962). Emerson (personal communication) believes that very few genera remain to be discovered in the world. A superb collection of the world fauna has been gathered at the American Museum of Natural History largely through the efforts of Emerson. By 1968 it contained 1,745 named species, or about 92 percent of the known living world fauna. No less than 80 percent of the named species of the world were represented by primary type specimens. A catalog of the termites of the world was published by Snyder in 1949 and updated by Roonwal in 1962. Snyder has also published a bibliography of termite literature in three parts, complete to 1965. Ahmad (1950) and Emerson (1965, 1969) have reviewed the phylogeny, Emerson (1955) the zoogeography, and Roonwal (1962) the history of the taxonomy of the world fauna. Using these key works as an entrée, it is fairly simple to find one's way through the taxonomic literature and, given a modest amount of time, to identify termites from almost any part of the world. Finally, a comprehensive review of many aspects of termite biology has been presented in the treatise edited by K. Krishna and Frances M. Weesner (1969, 1970) and the popular book by Howse (1970).

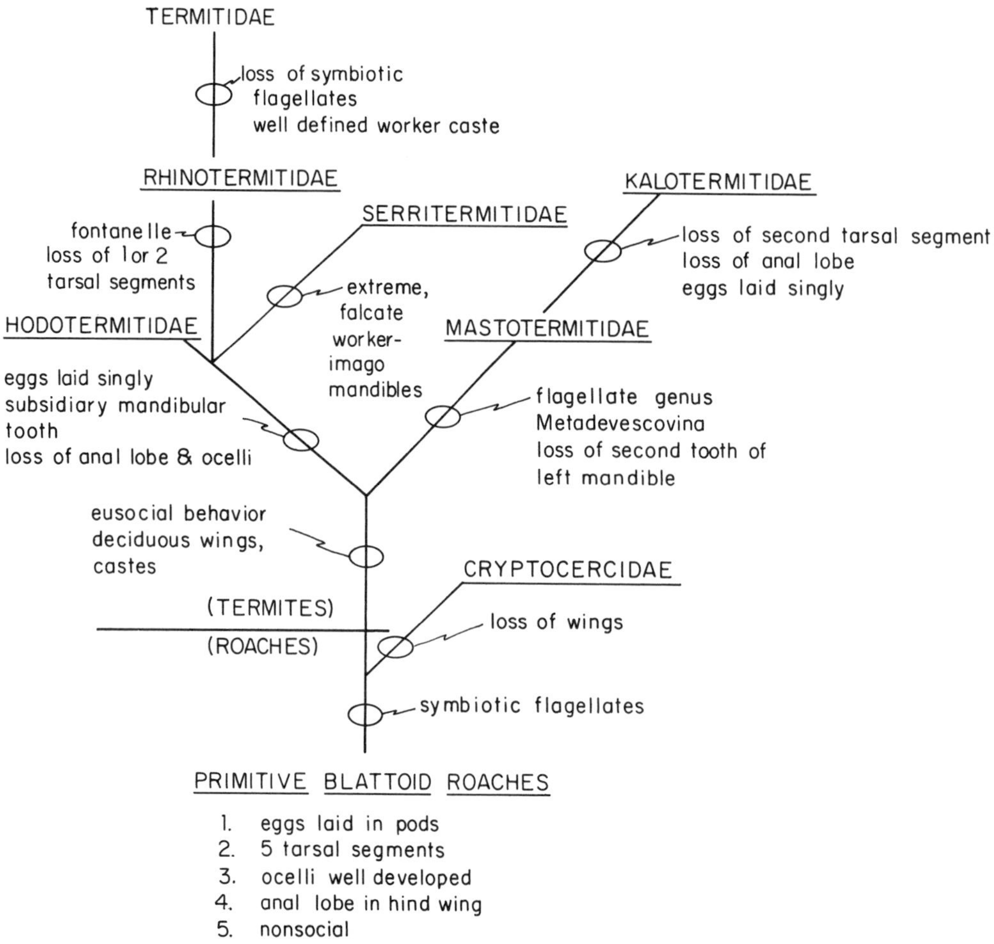

FIGURE 6-6. A simple branching phylogenetic diagram (cladogram) that expresses current opinion on the origin of the termites and the relationships of the termite families. Some of the more important characters involved in the evolutionary divergences are also shown (based chiefly on Imms, 1919; Ahmad, 1950; and Emerson, 1955, 1965, 1967, and personal communication).

The lower termites can be cultured and studied in the laboratory with surprising ease. This is especially true of the wood-eating species that form small or medium-sized colonies. By sandwiching strips of wood or paper between plates of glass, the lateral walls of the nest and the termites' food are made one and the same; then, by making the thickness of the material just exceed that of the largest colony member, the behavior of the colony can be studied in what is effectively two dimensions. Stock cultures require little care. For years the late L. R. Cleveland kept a flourishing colony of *Cryptotermes* in a block of wood in a bottle sitting in his office, with no special attention. Higher termites, especially those with huge colony populations and complex caste systems and nest architecture, are much more difficult to maintain. But it can be done well enough to conduct behavioral studies, as Grassé, Noirot, Skaife, and other students of the African fauna have repeatedly shown. An excellent general account of culturing techniques has been given recently by Becker (1969).

TABLE 6-4. Characters showing evolutionary change within termites. These changes have occurred at various places in termite evolution; the timing of some is indicated in Figure 6-6 (based on Emerson, 1962).

Characters showing advance

Primitive condition	Derivative condition
BEHAVIORAL-PHYSIOLOGICAL	
Small numbers in the mature colony	Large numbers in the mature colony
Excavated nests with little construction	Elaborately constructed carton nests
Relatively little care of eggs, nymphs, and adults	Relatively great care of eggs, nymphs, and adults
Food not stored	Food stored
Nutritive dependence upon symbiotic intestinal flagellates	Nutritive independence from symbiotic intestinal flagellates
Damp-wood eaters	Dry-wood eaters
Wood eaters	Leaf and grass eaters
Wood eaters	Humus eaters
No fungus gardens	Fungus gardens
Few or no termitophiles	Termitophiles, often species-specific and highly modified
Workers forage only within excavated tunnels in wood	Workers forage outside nest, sometimes in covered tunnels and sometimes on exposed odor trails
REPRODUCTIVE CASTES	
Small abdomen of queen with small ovaries and glandular tissue	Large abdomen of queen with large ovaries and glandular tissue
Capacity to produce substitute (neoteinic) kings and queens	Reproduction confined to primary reproductive caste (imago)
Frontal gland absent	Frontal gland present
Front coxae smooth	Front coxae with ridge or projection
WORKERS	
Only ergatoid immatures present	True worker caste present
SOLDIERS	
Head elongate, large, smooth, somewhat flattened, with large, curved, and toothed biting mandibles	Head round, or extremely flat, or phragmotic; surface rough or with ridges and projections; mandibles elongate and thin, or twisted for snapping, or with projection for frontal gland
Mandibles with smooth cutting edges or with reduced teeth	Mandibles with serrated cutting edges
Pronotum wide, flatly convex, and with smooth edges	Pronotum narrow, saddle-shaped and sometimes with serrated edges
Front coxa smooth	Front coxa with ridge or projection

Characters showing regression

Primitive condition	Derivative condition
IMAGOES (PRIMARY REPRODUCTIVES)	
Y-suture of head present	Y-suture of head reduced or absent
Two well-developed ocelli present	Ocelli reduced or absent
Antennae with numerous segments	Antennae with fewer segments
Mandibles with 2 or 3 prominent marginal teeth with sharp basal notches	Mandibles with reduced teeth and notches
Five tarsal joints present	Second tarsal joint reduced or absent
Arolium present between tarsal claws	Arolium absent
Cerci with numerous segments (up to 8)	Cerci reduced, usually with 2 segments
Styli present	Styli absent
Egg mass in cluster	Eggs laid separately
Genitalia similar and clearly homologous to those of cockroaches	Genitalia reduced or absent
Hind wing with anal lobe	Hind wing without anal lobe
Pronotum wide and flat	Pronotum narrow and saddle-shaped
WORKERS	
Compound eyes faceted and pigmented	Compound eyes reduced or absent
Mandibles as in primitive imagoes	Mandibles as in derivative imagoes
SOLDIERS	
Antennae with many segments (up to 29)	Antennae with fewer segments (as few as 10)
Compound eyes pigmented and faceted	Compound eyes unpigmented, or nonfaceted or even absent
Ocellus spot present	Ocellus spot absent
Mandibles with 2 or 3 large marginal teeth	Mandibles with reduced marginal teeth
Soldiers present	Soldiers absent

The Natural History of the Most Primitive Living Termite, *Mastotermes darwiniensis*

The sole surviving species of the once world-wide Mastotermitidae, *Mastotermes darwiniensis,* is found over most of the northern half of Australia. In some ways it behaves most strangely for such an ancient relic. According to Hill (1942) and Ratcliffe, Gay, and Greaves (1952), it is the most destructive termite species in Australia and the most destructive insect of any kind in the northern part of the continent. The colonies, which nest in the soil, are immense, the largest containing over 1 million individuals. The diet of *M. darwiniensis* is the most catholic of any known termite; one might even say it resembles that of the cockroach. Workers have been observed attacking poles, fences, wooden buildings, wharves, bridges, oil-soaked wood, living trees, crop plants, wool, horn, ivory, vegetables, paper, hay, leather, rubber, sugar, flour, bagged salt, human and animal excrement, billiard balls, and the plastic lining of electric cables. Unattended homesteads in the outback have been reduced to dust in only two or three years—house, fences, and all. Colonies of *M. darwiniensis* occupy many kinds of nest sites through a wide range of habitats, and they are able to excavate rapidly in both soil and wood. Their subterranean nests, which are often fragmented and connected by covered passageways constructed on the surface of the ground, are difficult to detect. The galleries run outward for as much as 100 m or more from the nest. Most are shallow, extending no more than 40 cm below the surface. One gallery system, however, was uncovered by quarrying operations at a depth of 4 m.

Considering its phylogenetic position and economic importance, surprisingly little is known concerning the biology of *M. darwiniensis,* including the most basic facts of the life cycle. One curious fact is that the primary reproductives are rare. Multiple supplementary reproductives appear to be the rule, and colony multiplication is often by budding. When groups of nymphs are detached from the main colony, they are able to develop neoteinics from nymphal forms, which then take over the reproductive function. Eggs are laid in packets of about 20 each, in a form reminiscent of the roach ootheca. According to Hill (1942), nuptial flights occur regularly, but their relative contribution to the formation of new colonies is unknown.

Natural History of the Lower Termites of the Family Kalotermitidae

The species of the family Kalotermitidae are referred to as the dry wood termites. (Members of the hodotermitid subfamilies Termopsinae, Stolotermitinae, and Porotermitinae are loosely referred to as the damp wood termites, and the living Hodotermitinae are called harvester termites.) They are limited almost exclusively to dead, dry wood and possess muscular recta that pass hard, dry fecal pellets. The kalotermitids are regarded as comparatively primitive in both their social and morphological features. They appear to have evolved from a mastotermitid base and are not considered by taxonomists to be ancestral to any other known termite family (Emerson, 1969). Their colonies, which rarely contain more than a few hundred individuals, live in ill-defined galleries inside the wood on which they feed. The termites rely on an intestinal flagellate fauna to digest the wood and do not utilize symbiotic fungi or store food. When the primary queens and males are lost, they are quickly replaced by secondary "neoteinics" that transform in one molt from a labile, worker-like caste called pseudergates. When present, the primary reproductives prevent the transformation of pseudergates by means of inhibitory pheromones passed out of their anuses (see Chapter 10). Soldier inhibition also occurs, but the physiological basis is not yet known. The exchange of oral and anal liquids, as well as integumentary exudates, occurs very frequently among all members of the colony. Anal exchange is essential to the transmission of flagellates to young nymphs and newly molted individuals of all ages.

Kalotermes is the best known genus of the Kalotermitidae. It is considered relatively primitive within the family on the basis of, first, the wing venation of its primary reproductives (specifically, the median is midway between the cubitus and the radial sector) and, second, the elongate shape of the heads of its soldier. Along with several other kalotermitid genera, it is represented in European fossil deposits of Eocene to Miocene age. Eighteen living and six fossil species had been described by 1969, taking into account some additions and subtractions subsequent to the revision of the Kalotermitidae by Krishna in 1961. The living species represent every zoogeographic region in the world. Because they are abundant in the North Temperate Zone and easy to culture in artificial nests, the species of

Kalotermes have been favorite subjects for experimental studies on physiology and behavior (Chapter 10).

Incisitermes minor, a member of related genus common in the southern and western United States, has a life cycle typical of the family (Harvey, 1934). In California the alate reproductives swarm during the day, as their dark body color might indicate, from the latter part of September to early May. The optimal weather conditions are brilliant sunlight from a cloudless sky and temperatures over 80°F (27°C). Harvey's description of the emergence behavior can be taken as typical of the genus and, with certain modifications, of termites in general (see the review by Nutting, 1969):

> Let us follow the course of events in an infested pole top. The alates are mature, the temperature is 95°F, it is midday, with brilliant sunshine. When a chip is removed from the outer shell of the pole, exposing the termites, their activity is not unlike that in a hive of bees. This is in marked contrast to the actions of the individuals in the colony at other periods of the year, when they are consistently quiet and plodding. Now, however, the nymphs are highly excited. They run about almost hysterically, with antennae nervously flashing back and forth. The soldiers are as much wrought up as are the nymphs; in fact, they have more the appearance of doing something than at any other time of the year. With legs wide apart and claws attached to the spongy wood of the exposed passageway, they flash their antennae back and forth, and are apparently ready to attack anything that may come near. Agitation appears to be most acute near an emergence hole, an opening about $\frac{1}{16}$ inch in diameter in the shell of infested wood near the top of the experimental pole. . . Apparently, the nymphs made this opening to be used as an exit by the alates, because if a chip of wood is removed near a hole of this nature a group of much disturbed nymphs and soldiers will be exposed and some of the alates that appear to be grouped near the opening escape . . . In fact, the impression is gained that the nymphs and, in particular, the soldiers are covering and protecting, if not actually assisting, the emergence of the alates. The soldiers appear to regulate the exit of the alates, determining the number that emerge at any one time, the recurrent cyclic emergence of alates, and the time of emergence during the day. It should be stated, however, that recent evidence indicates that the alates themselves aid in digging an aperture for swarming.

The alates leave quickly after emergence, and their flight is aimless and wavering. The majority, however, manage to ascend 70 m or more and to fly for distances of at least 100 m and perhaps as much as a kilometer from the parental nest. As soon as it alights the alate breaks off its wings by quickly spreading and lowering the wings until their tips touch the ground, then pivoting back and forth to bring pressure on the wings at the basal sutures. Now the "dealate" runs excitedly in apparently random directions until a member of the opposite sex is encountered. The two individuals stop abruptly, turn face to face, and play their antennae over each other's heads.

> The king makes advances toward the queen, the queen striking at the king with her head. After four or five such overtures, each of which is followed by a pause during which the termites stand facing each other with their antennae fanning slowly, the king is accepted or rejected. If he is rejected the queen turns and runs quickly away and the king goes in the opposite direction. If, however, the king is accepted, the queen turns quickly and speeds away, with the king in close pursuit . . . Although the queen runs rapidly, the king keeps close to her and, when they become separated as occasionally happens, the king rapidly regains contact with her . . . After pairing has been accomplished, separation seldom occurs. It is usually difficult to frighten members of a pair away from each other, and it appears that they seldom, if ever, leave one another for other mates, even though a number of unpaired termites are near (Harvey, 1934).

This sequence of dealation, pairing, and tandem running is universal in the termites. In some groups, the tropical genus *Nasutitermes,* for example, the queens stand still for the most part and "call" males by means of sex pheromones released from intersegmental glands located on the abdominal dorsum. After pairing, the royal couple of *Incisitermes minor* undergo a radical change in behavior. During the nuptial flight and search for mates, the termites are attracted to light. As soon as they pair, however, they are repelled by light and become strongly attracted to wood. When they find a suitable spot, they begin excavating in the wood, alternating shifts, until they complete an entrance tunnel about a centimeter deep. The entrance hole is then sealed off with a putty-like mixture of chewed wood and cement-like secretion. Finally, the pair constructs its first royal cell, a small pear-shaped chamber at the bottom of the entrance tunnel.

When the royal cell has been finished, the queen lays from two to five eggs. Soon after hatching from these eggs, the fragile, chalk white nymphs are fed by regurgitation and, after one or more molts, set about feeding themselves and enlarging the nest. The two activities are, in fact, the same thing! The royal pair stay in the advanced part of the main passage, while the nymphs dig out enlarged feeding chambers and side tunnels. By the end of the second year the young colony has consumed about 3 cm^3 of wood. It consists of the royal pair, a single soldier, and ten or more pseudergates and nymphs. About a year is required for each soldier to develop from an egg to the mature form. The queen's abdomen begins to swell in two years, but the king looks about the same or, if anything, somewhat shrunken in size. After several more years alates are produced, and the colony is now referred to as being in a "mature" condition. The queen's fecundity increases steadily until she is about ten or twelve years old; then it begins to decline, apparently at a rapid rate. About this time one or more secondary queens appear and take over oviposition. The estimated growth of an average colony is given in Table 6-5. The average longevity of colonies has not been measured. The potential longevity, on the other hand, is evidently unlimited due to the capacity of the colony to generate secondary reproductives whenever needed.

TABLE 6-5. Estimated average numbers of individuals of the principal castes in colonies of various kinds in *Incisitermes minor* (after Harvey, 1934).

Age of colony (years)	Reproductives	Soldiers	Pseudergates and nymphs
1	Primaries	0	0
2	Primaries	1	12
3	Primaries	3	50
4	Primaries	10	200
5	Primaries	20	500
6	Primaries	30	700
7	Primaries	40	1000
8	Primaries; possibly one or more supplementaries	50	1200
9	Primaries; possibly one or more supplementaries	60	1400
10	Primaries; possibly one or more supplementaries	70	1600
11	Primaries; possibly one or more supplementaries	80	1800
12	Primaries; possibly one or more supplementaries	90	2000
13	Primaries; possibly one or more supplementaries	100	2200
14	Primaries; possibly one or more supplementaries	100	2400
15	Primaries; possibly one or more supplementaries	120	2600

Natural History of the Higher Termites (Family Termitidae)

Sometime, perhaps in late Mesozoic or early Tertiary times, the Termitidae, or "higher" termites, originated from rhinotermitid ancestors and began a spectacular adaptive radiation. Although fossils of the lower termite families are not uncommon in temperate Eocene and Oligocene deposits, the earliest termitid fossil is an undescribed *Nasutitermes* from the Miocene of Mexico. Emerson (1955) would put the origin of the Termitidae as no later than early Cretaceous times, on the grounds that the family reached South America and radiated there no later than the Tertiary period, during which time the continent was isolated by the Panamanian water gap. I am more inclined to trust the evidence of the fossil record, even if this evidence is largely negative and it means granting to the early termites the ability to cross principal water gaps. It should be kept in mind that a few modern termite genera, such as *Rugitermes* (Kalotermitidae) and *Prorhinotermes* (Rhinotermitidae), display considerable capacity for overseas dispersal. Today the Termitidae make up the bulk of the world termite fauna. Of the 1,894 living species of termites that could be assigned to a genus in early 1969, 1,412 or 75 percent belonged to the Termitidae (Emerson, 1955). This same family also included 153 of the 210 living termite genera. The termitids are heavily concentrated in the tropics: about 85 percent of the species known in 1955 occur in the three principal tropical zoogeographic regions (Oriental, Ethiopian, Neotropical), while only 13 and 17 species were recorded from the Palaearctic and Nearctic regions, respectively. Africa south of the Sahara contained 66 genera, 55 of which are endemic, and 626 species.

The majority of termitids are soil dwellers and responsible for most of the elaborately structured mounds that are such a conspicuous feature of the tropical landscape. Various of their species have specialized on virtually every conceivable cellulose source. To reach this food, workers extend galleries through the soil, or construct covered trailways over the surface of the ground, or even march in columns over exposed odor trails.

The living subfamilies of the Termitidae cannot be arranged in a linear order that exactly reflects phylogeny. After emerging from the ancestral stock, evidently the most primitive Amitermitinae, each has followed a very different course in evolution. The presence of a subsidiary tooth in some of the primitive genera of the Macrotermitinae and in the Indomalayan amitermitine genus *Protohamitermes* supports the notion that the subfamilies had a common origin from the Rhinotermitidae, and it is possible to construct a plausible phylogenetic cladogram connecting the genera within each of the subfamilies (Ahmad, 1950). Otherwise, the most striking single feature that joins all known termitids is the absence of a polymastigote-hypermastigote intestinal flora and the dependence instead on spirochete bacteria for digestion of cellulose (Cleveland, 1926). The inferred evolution of the termite-microorganism symbiosis is summarized in Figure 6-7.

As an example of a relatively unspecialized termitid, we can take *Amitermes hastatus* (= *A. atlanticus*), which has been studied in detail by Skaife (1954a,b, 1955). The species occurs in South Africa, in the mountains of the southwest Cape at elevations from about 100 to 1,000 m above sea level. It nests in the sandy soil of the natural veld, throwing up conspicuous hemispherical or conical mounds constructed of a black soil-excrement mixture. In the late summer months of February and March large numbers of white nymphs with wing pads are to be found in the larger nests. By the end of March, or April at the latest, these individuals transform into alate reproductives. For several weeks the alates wander slowly through the nest. Then, soon after the onset of the autumn rains, the nuptial flight occurs. One day between 11 o'clock in the morning and 4 o'clock in the afternoon, immediately after a ground-soaking rain and with the temperature rising, the exodus begins. The workers first excavate large numbers of tightly grouped exit holes, each about 2 mm in diameter, giving the apex of the mound the appearance of a coarse sieve. True to the pattern of most termite species, this is the only time the workers breach the walls of their nest and expose themselves to the outside air. Workers, soldiers, and alates boil out of the holes in a state of intense excitement, the alates fly off almost immediately, and within three or four minutes the termites retreat back down into the nest, plugging the exit holes after them. Most, but not all, of the alates leave in this first flight. A few remain behind to participate in later departures. The alates are feeble flyers; many do not travel more than 50 or 60 m from the nest before alighting. As soon as they land they break off their wings at the basal fracture line by swiftly pressing the wing tips to the ground. The subsequent pairing and nest-founding behavior follows the sequence basic to termites generally, but differs in certain details from that already described in *Incisitermes:*

In the meantime, other females and males have settled on the ground and herbage in the vicinity and dropped their wings, and they run about, the females with the tip of the abdomen raised and the males obviously seeking eagerly. It is believed that the females give off a scent that attracts the males, but this is not perceptible to our sense of smell. If a seeking male is watched closely he will be seen to run a devious course but in the general direction of a female near him, until he touches her with his antennae. The pair then line up, tandem fashion, with her in front and him close behind, touching the tip of her abdomen with his antennae, and they move off together to seek for their new home. If, while they are running in this way, the male is held down with a matchstick, she will stop after going a short way and wait for him to catch up with her again. If two pairs are guided so that their paths cross, it is easy to manoeuvre a change in partners—the males do not mind what females they follow and the females also are indifferent; any partner will do so long as it is of the opposite sex. Copulation does not take place at this stage.

Having found a suitable stone or half-buried piece of wood, the female begins to burrow and he may or may not help. In any case, by far the greater part of the digging is done by her and she soon sinks a shaft beside the stone or wood and the

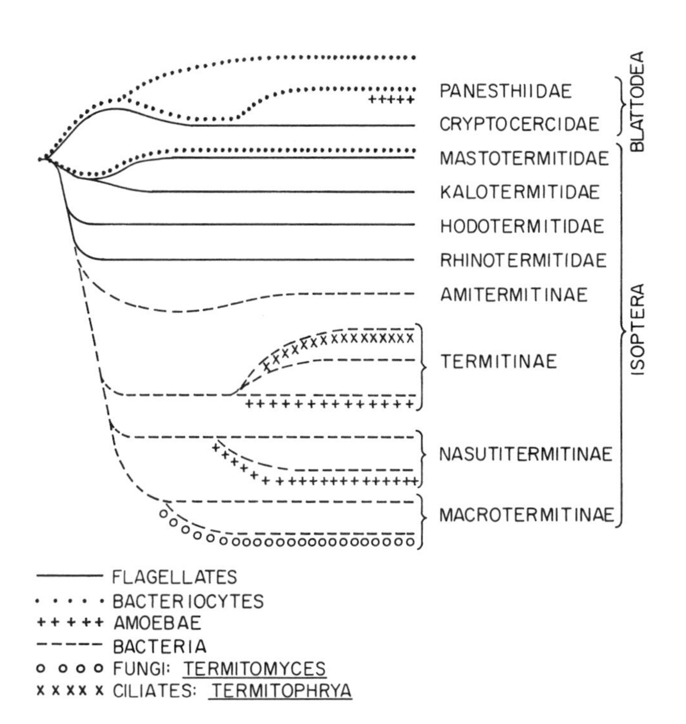

FIGURE 6-7. Simplified branching phylogenetic diagram of the termites and their immediate relatives among the cockroaches, in relation to the evolution of symbiosis with protozoans, bacteria, and fungi (based on Grassé and Noirot, 1959, with some minor alterations in the classification and phylogeny of the termites and cockroaches).

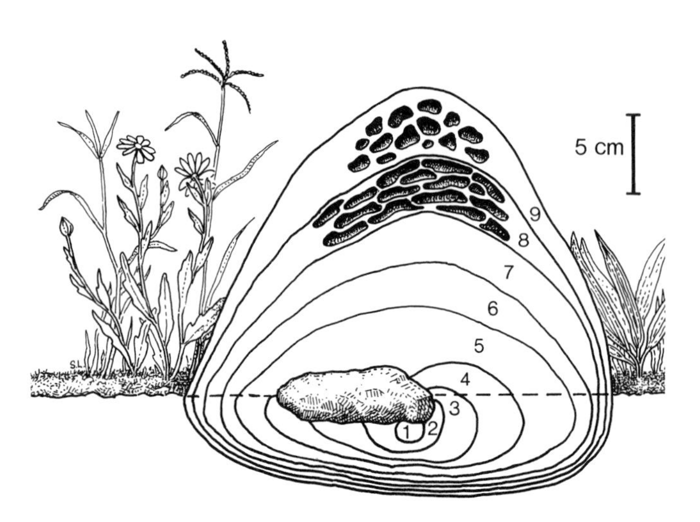

FIGURE 6-8. The growth of a typical mound of the South African termitid *Amitermes hastatus,* over a period of nine years. Each successive year's growth is indicated by a number. Representative outer and inner cells are shown at the top of the mound. There is no royal cell (redrawn from Skaife, 1954a).

two of them disappear into it. An inch or so below the surface a small chamber about a half inch in diameter is hollowed out, and here the pair take up their abode, blocking the tunnel with soil removed in making the chamber. They remain almost motionless in this chamber all through the winter and I do not think that copulation takes place until the arrival of warmer weather in the spring (Skaife, 1952).

In the spring months of October and November the queen lays the first five or six eggs. The individuals of the first brood develop into stunted workers. In later broods, soldiers make their appearance, and, after four years, alate reproductives are finally produced. The growth of a typical nest is shown in Figure 6-8. Skaife has estimated the age of some mounds of *Amitermes hastatus* to be greater than 15 years, but, judging strictly from the size of the mounds, he did not believe any to be more than 25 years old. This mortal state of individual colonies, if true, is an unexpected feature because presumably the colonies are capable of producing secondary reproductives when the queen dies. When the primary queen does fail, the workers put her to death, evidently by licking her abrasively. As Skaife describes it, "She is surrounded by a crowd of workers, all with their mouthparts applied to her skin, and this goes on for three or four days, her body slowly shrinking until no more than the shrivelled skin is left." Secondary and tertiary queens do appear in the absence of the queen—at least sometimes (see Figures 6-9 and 6-10). Skaife, however, was unable to rear them in queenless colonies kept in artificial nests, and he found that only about 20 percent of the natural mounds contained them. Clearly, then, either the supplementary reproductives appear rarely or only under special conditions, or else those colonies that possess them are relatively short lived.

We now turn to a socially more advanced group of the Termitidae. If one had to identify the apex of termite evolution, it would surely have to be the fungus-growing termites of the subfamily Macrotermitinae. The excrement

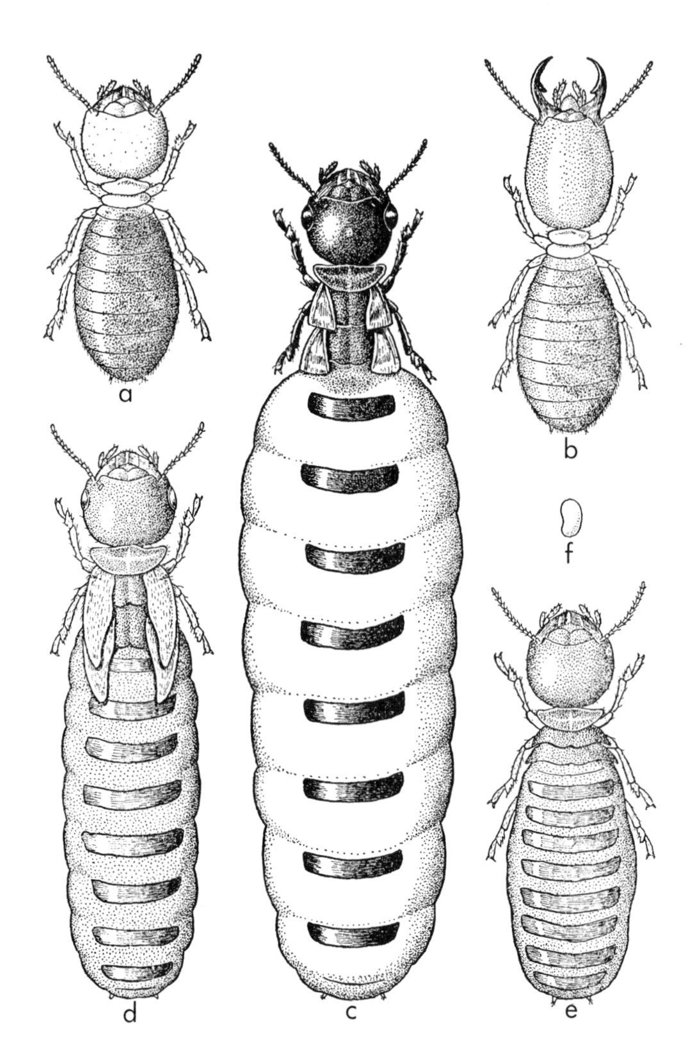

FIGURE 6-9. The five castes that make up the colonies of *Amitermes hastatus,* all drawn to the same scale: (*a*) worker; (*b*) soldier; (*c*) primary queen, about five years old, with greatly swollen abdomen; (*d*) secondary queen; (*e*) tertiary queen; (*f*) egg. The worker is about 5 mm long (after Skaife, 1954a).

of these insects is built into fungus combs, which occupy special gardens in the centers of the nests in large compartments. The combs are often shaped into round forms resembling brains or sponges, with numerous convoluted ridges and tunnels evidently designed to give the maximum surface for growth. The basidiomycete fungus, *Termitomyces,* sprouts round white spherules out from the substratum, but these are not especially favored as food bodies. Instead, the termites consume everything in the comb: substratum, mycelia, spherules, and all. It is a curious fact, possibly coincidental, that the macrotermitines are limited to the Old World tropics, and are concentrated in Africa and southeastern Asia in particular, whereas the attine ants, the only other group of social insects that cultivate special symbiotic fungi, are limited to the New World tropics and warm temperate regions.

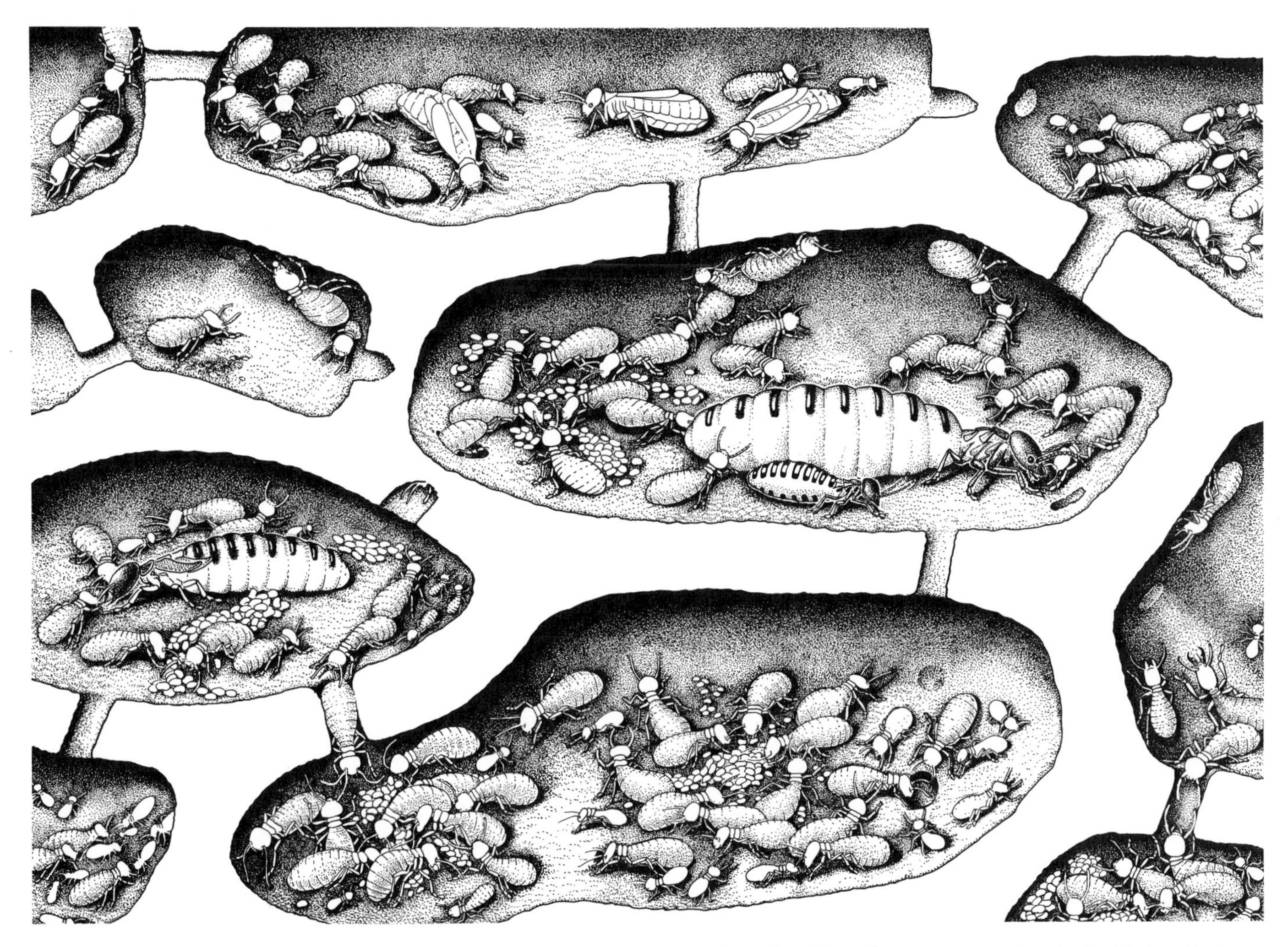

FIGURE 6-10. The interior of a typical nest of *Amitermes hastatus* of South Africa. The primary queen and male sit side by side in the middle cell (there is no specially constructed royal cell). To the lower left can be seen a secondary queen, which is also functional in this case. In the chamber at top center are found reproductive nymphs, characterized by their partially developed wings. Workers attend the queens and are especially attracted to the heads, to which they offer regurgitated food at frequent intervals. Other workers care for the numerous eggs. A soldier and presoldier (nymphal soldier stage) are seen in the lower right chamber, while worker larvae in various stages of development are found scattered through most of the chambers (original drawing by Sarah Landry; based on sketches and photographs by S. H. Skaife).

The exact role of the *Termitomyces* in macrotermitine biology has been the subject of controversy since König and Smeathman discovered the fungus in the eighteenth century. Lüscher (1951) crystallized the problem when he succeeded in rearing young colonies of *Pseudacanthotermes spiniger* with sterile combs. He then questioned whether the fungi are obligate symbionts of this species. Later (1961a), he suggested that the fungus gardens of at least some macrotermitines function not as food but rather as part of the ventilation system of the big nests by providing some of the heat needed for convection. Sands (1960) reared colonies of a second macrotermitine species, *Ancistrotermes guineensis,* with sterile combs. He concluded that the royal couple do not carry basidiospores into the copularium with them during nest founding; rather, workers somehow seem to pick up the spores from the ground around the nest during the first foraging trips and inoculate the beds in the young nest. Further doubt concerning the nutritive role of the *Termitomyces* came from the apparent fact that the fungal bodies themselves are not numerous enough to provide all of the food needed for the colony. Finally, there is even one macrotermitine species, *Sphaerotermes sphaerothorax,* which constructs "fungus gardens," but does not raise fungi on them (Grassé and Noirot, 1948).

Nevertheless, further evidence has simultaneously come to hand which indicates the opposite quite strongly, namely, that *Termitomyces* is an important nutritive symbiont for some of the macrotermitine species. First, there is the circumstance that *Termitomyces* is known only from macrotermitine nests. Also, the only fungi that grow in the nests while termites are present are *Termitomyces* and a common ascomycete species of the genus *Xylaria.* This virtual one-to-one relation parallels that between the attine ants and their peculiar fungi, which are known to be nutritive symbionts. It seems to be an even more exact relation than that found in the Attini, in that each macrotermitine appears to have its own species of *Termitomyces.* Kalshoven (1936) and Grassé and Noirot (1958a) reported that *Macrotermes, Microtermes,* and *Odontotermes* constantly eat the old portions of the combs while adding new material to other parts. Microscopic examination of *Macrotermes* worker gut contents by Grassé and Noirot indicated that the fungi serve to degrade the lignin and to expose the minute fragments of cellulose for quicker digestion by the intestinal bacterial flora. Sands (1956), in the most informative experiment of all, demonstrated that laboratory colonies of *Odontotermes badius* provided only with sterile fungus beds cannot survive. Later (1960, 1969), he analyzed the evolution of fungus growing within the Macrotermitinae and arrived at a new conception which is, in my opinion, the key to understanding the conflicting evidence on the subject. As Sands points out, the structure and size of the fungus gardens vary greatly within the Macrotermitinae. The more primitive genera such as *Pseudacanthotermes* and *Acanthotermes* produce combs of a dense, more homogeneous texture than in the rest of the subfamily; they are organized into simple convoluted lamellae rather than the complex sponge-like structure made by *Macrotermes, Odontotermes,* and other genera. The fungus combs of the lower genera resemble the fecal cartons produced by the Rhinotermitidae, the presumptive ancestors of the macrotermitines and other Termitidae. In particular, the comb of *Pseudacanthotermes* resembles closely the soft carton built by species of the mound-building rhinotermitid genus *Coptotermes.* It is therefore quite plausible that the fungus-growing habit originated by accidental growth of *Termitomyces*-like fungi on soft cartons of primitive macrotermitines not far removed from *Pseudacanthotermes,* that the dependence of *Pseudacanthotermes* on its fungus is not a close one, and that only when the fungus gardens became more complex during the origin of the higher Macrotermitinae did the mutualism become obligatory on the termites. The simplification of the gardens in *Microtermes* and *Sphaerotermes,* and the outright loss of the fungus in the latter genus, can be regarded as secondary evolutionary reductions. This idea can be tested only by a much more intensive study of macrotermitine biology, which is still characterized, as Noirot (1958–1959) has put it, by a "poverty of information and an almost complete absence of precise data and well conducted experiments."

Even more remarkable than the phenomenon of fungus growing is the size and complexity of the nests constructed by the macrotermitines. Figures 6-11 and 6-12 show the growth and ultimate size of nests made by two of the more spectacular species of Africa. We will return to a consideration of the meaning of the various details of this particular nest form in Chapter 16.

A vast number of individuals are required to produce these prodigious structures. A single mature *Macrotermes natalensis* nest contains approximately 2 million living individuals at any given time. The primary queen of this species, according to Fenton (1952), lives for an average of ten years. She lays about 30,000 eggs per day, or 10 million per year, for a total of something like 100 million in her lifetime. Perhaps these estimates are imprecise, as Nutting has recently argued (1969). Paulette Bodot (1964) found that egg production by *Cubitermes* is subject to seasonal variation, being greatest toward the end of the lesser rainy season. But it seems certain that the lifetime production of the primary queen is at least in the tens of millions, and the higher figures do not seem at all fantastic. The activity around the primary queen is intense, with dozens of workers constantly licking her body, feeding her with regurgitated salivary secretions, and taking away her eggs as quickly as they are laid. Nor is *Macrotermes* the most productive of all termites. The record is held by a queen of the macrotermitine *Odontotermes obesus*, observed by Roonwal (1962b) to lay 86,400 eggs in a single day.

FIGURE 6-12. Cross section of a nest of *Macrotermes bellicosus* in Africa. The figure of a man has been added to indicate the great size of the nest, which had a diameter of about 30 m (modified from Grassé and Noirot, 1961).

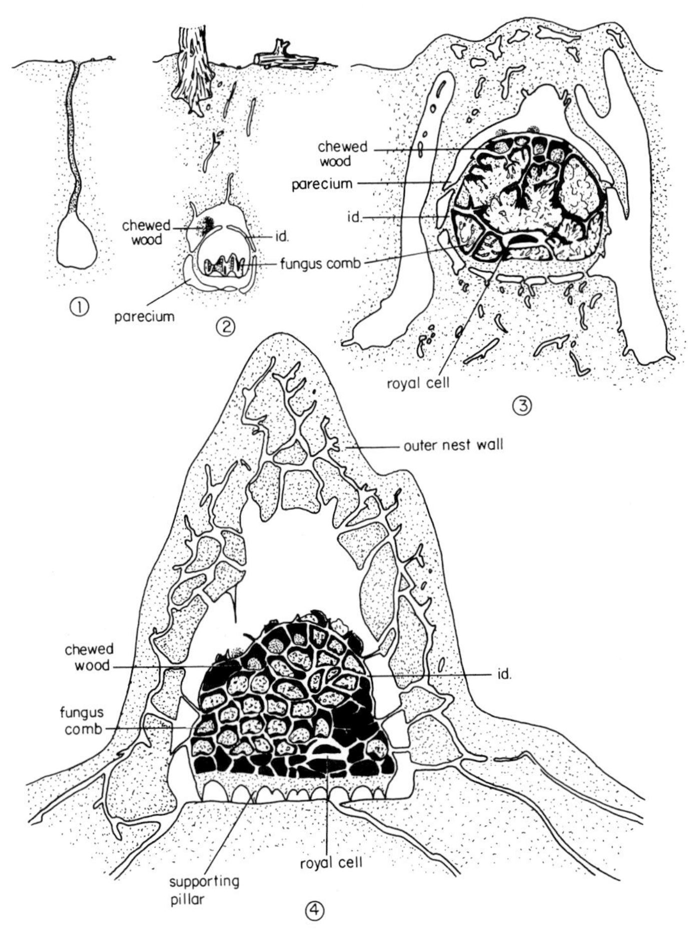

FIGURE 6-11. Development of the nest of the African fungus-growing termite *Macrotermes* (*Bellicositermes*) *natalensis:* (1) "copularium," or first chamber made in the soil by the royal couple; (2, 3) intermediate stages of development; (4) fully developed nest. The wall (*id.*) of the fungus garden (*idiothèque* of Grassé and Noirot) surrounds numerous chambers that contain masses of finely chewed wood that are used as substrate for the fungus; the parecium (*paraécie*) is the air space surrounding the fungus garden (redrawn from Grassé and Noirot, 1958c).

The Origin of Social Behavior in Termites

The termites possess two unique qualities which together provide the clue to their social beginnings: they are the only eusocial insects that do not belong to the order Hymenoptera; and, along with the closely related cryptocercid cockroaches, they are the only wood-eating insects that depend on symbiotic intestinal flagellates. As first pointed out by L. R. Cleveland (in Cleveland *et al.*, 1934), the two phenomena are probably causally related. The flagellates are passed from old to young individuals by anal feeding, an arrangement that necessitates at least a low order of social behavior. Cleveland postulated that termite societies started as feeding communities bound by the necessity of exchanging flagellates and, in a sequence that is the reverse of hymenopteran social evolution, only later evolved social care of the brood. It is not theoretically necessary to the origin of eusociality for sibs to be unusually closely bound by kinship in the hymenopteran manner. Williams and Williams (1957), in an extension of the Wright theory of group selection (1945), demonstrated that eusocial behavior, including the formation of sterile, altruistic castes, can evolve in such insects if competition between groups of sibs is intense enough. The point is that only the termites have gone this far; we should regard their achievement as especially noteworthy and continue to reflect on the special conditions that have made it possible.

7 The Presocial Insects

The evolutionary roads of presociality have been traveled by many kinds of arthropods besides the hymenopterans and termites. Time and again phyletic lines have pressed most of the way to eusociality, in some cases to the very threshold, and then unaccountably stopped. This chapter will describe the most advanced cases of presocial behavior, the evolutionary steps that have led to them, and the special environmental circumstances under which they seem to have arisen.

In a frequently quoted passage in his book *Social Life among the Insects* (1923), Wheeler stated that "social organization" has arisen on at least 24 separate occasions in nearly as many families and subfamilies belonging to 5 different orders of insects. By the time *The Social Insects* was published, in 1928, the count had risen to 30. Today the number is much higher and continues to climb, but the larger it becomes the less it seems a matter of great significance. For Wheeler was including all cases of parental behavior, extending from the slightest protection given to the newly hatched young to the most sophisticated forms of interaction. He defined as an insect society any "family consisting of two parental insects and their offspring or at least of the fecundated mother and her offspring" in which "the two generations live together in more or less intimate, cooperative affiliation." Over the years Wheeler and other authors fell into the habit of referring to the lowest level of social organization as "subsociality," a phenomenon that can now be more rigorously specified as parent-offspring communication that does not include cooperation in the rearing of other offspring. Already a profusion of arthropods are known to be subsocial to some degree: pycnogonids (Kaestner, 1968), crustaceans (Schöne, 1961), spiders (Nørgaard, 1956; Tretzel, 1961), mites (Treat, 1958), scorpions (Vachon, 1952), pseudoscorpions (Weygoldt, 1969), millipedes (Schubart, 1934), centipedes (Weil, 1958; Eason, 1964), symphylans (Jones, 1935), cockroaches (Scott, 1929; Roth and Willis, 1960), crickets (Hahn, 1958; West and Alexander, 1963), earwigs (Fulton, 1924; Weyrauch, 1929; R. Stäger in Maidl, 1934), mantids (Faure, 1940), psocopterans (Mockford, 1957), embiopterans (Ledoux, 1958), hemipterans (Bequaert, 1935; Schaller, 1956; Schorr, 1957; Odhiambo, 1959, 1960), membracids (Beamer, 1930), thrips (Hean, 1943; Mani and Rao, 1950), beetles (von Lengerken, 1939, 1951; Hinton, 1944; Schedl, 1958), bees (Michener, 1969a), and wasps (Evans, 1958; Olberg, 1959). Much more interesting is the fact that behavior and communication in the most advanced subsocial species is as complex as that found in the eusocial insects. In both instances we find such behavior patterns as the building of elaborate nests, nest guarding and cleaning, the transport of food to brood cells, soliciting for food, regurgitation, the laying of trophic eggs, mutual grooming, the transport of young, and alarm signaling.

In line with the system used to designate the evolutionary stages leading to eusociality, the presocial behavior in arthropods to be considered now can be classified into one or the other of three principal categories: the first is subsociality in the strict sense; the second is communal behavior or the aggregation of individuals who cooperate to some extent in foraging for food or in the building of nests; the third is quasisocial behavior, in which individuals cooperate in the rearing of young, whether or not they join in foraging and nest construction (see Chapter 2).

It is too early to systematize our knowledge of presociality by referring classes of phenomena to theory and critical exemplification. What I propose to do instead is to describe a series of case histories that illustrate the diversity of the phenomena concerned and the extremes to which they have been carried. I will also attempt to draw a few empirical generalizations, of the kind that will someday, I hope, lead to the beginnings of a more inclusive theory.

Parental Care in the Hemiptera and Homoptera

In many families of Hemiptera (Acanthosomidae, Aepophilidae, Aradidae, Belostomatidae, Coreidae, Cydnidae, Dysodiidae, Emesidae, Gerridae, Pentatomidae, Reduviidae, Scutelleridae, Tingidae) and in the homopterous family Membracidae, species are known that display some form of parental protection, especially toward young in the earliest developmental stages. In the great majority of cases the female is the guardian. Males of the coreid *Phyllomorpha laciniata* and of certain belostomatid species also perform this service, but it is done more or less involuntarily since their mates fasten the eggs onto their backs, where they remain until the nymphs hatch. According to one observer (J. R. Parker in Hoffman, 1924), males of the belostomatid *Lethocerus americanus,* a giant water bug, actively assist the females in guarding the egg masses. In the African reduviids of the genus *Rhinocoris* the males are the exclusive guardians of the eggs and young nymphs (Bequaert, 1935; Odhiambo, 1959). This is apparently also the case in the dysodiid *Neuroctenus pseudonymus,* observed by McClure (1932) in Texas.

Among a few of the families, notably the Belostomatidae, Coreidae, and Gerridae, parental guardianship is found only in a small minority of the species and is limited to the eggs. Truly subsocial behavior, which involves specialized interaction between the parents and nymphs, occurs sporadically through the other hemipteran families that show parental care, and it is limited in most cases to the first one or two nymphal instars. As a rule, the parent offers shelter with its body, or carries the nymphs around on its back or venter, or simply stands close by. In a few instances a more advanced type of interaction is involved. Nymphs of the scutellerid *Pachycoris fabricii,* for example, orient strongly toward the mother, as described in this account by H. G. Barber (quoted by Bequaert, 1935): "While collecting insects in Porto Rico for the American Museum of Natural History, in the summer of 1914, I noticed a specimen of the brilliantly colored female of this species on the under side of a leaf. Spread over the leaf surface were quite a number of the small dark green nymphs, probably in the second instar. I slightly disturbed the leaf, when suddenly to my great surprise the little bugs scurried to the mother, crowding beneath her robust body in order to gain protection. The mother seemed perfectly conscious of her duty in the matter and remained stationary, covering them over with her body very much as a hen will cover her chicks. No eggs were found on this particular leaf, so that the brood must wander about in the wake of the mother, at least to some extent."

The females of the tingid *Gargaphia solani,* a species observed in Virginia by Fink (1915), stay close to the nymphs throughout most of their development. "When migrating from one leaf to another the female adult usually directs the way and with her long antennae keeps the nymphs together or rebukes any straggler or deserter. It is an interesting sight to observe the migration of a colony of more than a hundred nymphs, with the female adult hurrying from one end of the flock to the other, keeping them together and at the same time urging them in the right direction during the migration." According to Keys (1914), the adults of *Aepophilus bonnairei,* a semimarine bug found along the coast of western Europe, uses alarm signals to get their offspring out of harm's way. The *Aepophilus* families commonly rest under stones. The nymphs form a circle around the parent with their heads pointed toward its body. If the stone is lifted, exposing the hiding place, the adult rushes from one nymph to the other tapping them with her antennae, and the whole group then runs to the other side of the stone. An example of maternal care involving the pentatomid *Mecistorhinus tripterus* is illustrated in Figure 7-1.

Several observers have described incidents in which parent bugs attacked and chased off potential enemies which approached the brood. When Bequaert (1935) placed an ant on the egg mass being protected by a male *Rhinocoris albopunctatus,* "the male at first moved away, evidently frightened by the moving hand; but he soon returned, carefully exploring with the antennae, until the ant was discovered. The bug then proceeded to attack the ant with the beak and finally impaled it and carried it off an inch or so from the egg-mass, where he dropped

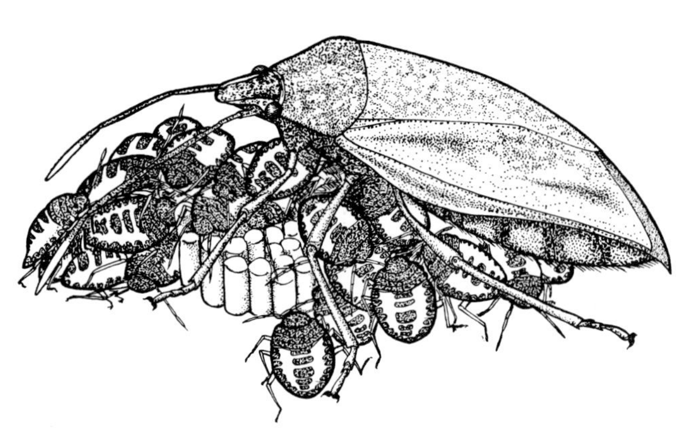

FIGURE 7-1. A female of the cacao stinkbug (*Mecistorhinus tripterus*) of Trinidad is shown guarding her first instar nymphs four days after they hatched (from Kirkpatrick, 1957).

it and then returned to the eggs." Odhiambo (1959) has provided evidence that the chief enemies who cause the evolution of subsocial behavior in the Hemiptera are insect parasites, particularly those that attack the egg stage. In nearly all of the subsocial hemipterans, care is limited to this simple act of protection. However, adults of the Brazilian pentatomid *Phloeophana longirostris* have apparently gone one step further. According to P. S. de Magalhaes and P. Brien, cited by Bequaert, the female is cryptically colored on the bark of *Terminalia catappa.* She uses her body to cover the eight to twelve eggs laid in a batch. After hatching, the nymphs cling to the underside of her abdomen until they reach the final nymphal instar. Since individuals belonging to the first nymphal instars have beaks too short to penetrate the bark of the host tree, they are thought to be fed by the mother, either with some secretory material or by sap delivered through the proboscis. Obviously this case must be studied further before such a singular conclusion can be accepted with any confidence.

Advanced Parental Care in Crickets

Females of some species of mole crickets (Gryllidae: Gryllotalpinae) have the unusual habit, for orthopterans, of sealing themselves in underground burrows with their eggs and young nymphs. Similar behavior is displayed by the "big brown cricket" of India, *Brachytrupes achatinus* (Gryllidae: Brachytrupinae). West and Alexander (1963) discovered that a second brachytrupine, *Anurogryllus muticus* of the southeastern United States, has evolved an even more elaborate form of parental care. After the females mate, they construct a burrow of the characteristic form illustrated in Figure 7-2. At this point they become very aggressive, which is an unusual trait for female crickets at any period of the life cycle. They viciously attack other members of the species who intrude into the nest, including males. West and Alexander, by inducing females to burrow along the glass sides of terraria, were able to observe the details of maternal care. The eggs are laid in batches on the floor of the brood chamber and partly covered with soil. When the nymphs hatch, they cluster together and also attempt to remain near the female. The female's mouthparts and the tip of her abdomen are especially attractive to them. Occasionally the female palpates a nymph or picks it up bodily and gently manipulates it in her mouthparts. Antennal touching is common. The offspring are fed in a unique manner. From time to time the mother lays miniature eggs, which the nymphs immediately carry away and eat. These objects are so attractive to the nymphs that they crowd around them and fight for their possession. They are evidently specialized to serve as baby food, and in fact are closely comparable to the trophic eggs produced by some species of ants. When given the opportunity, the female gathers bits of fresh grass and fruit, carries them into the brood chamber, and shares them with her nymphs. When she defecates, she backs into the lower corner of the brood chamber and deposits a pellet into a lower tunnel. After the fecal chamber has filled up, she begins to pick up fecal pellets and to carry them out through one of the exit tunnels.

Alexander (1961) has pointed out that although the total pattern of maternal behavior of *Anurogryllus muticus* and other brachytrupines and gryllotalpines represents an extraordinary evolutionary advance in sociality, it is constructed directly out of rather simple elements already present in solitary cricket species. Solitary burrowing species in particular are skilled at manipulating soil particles. They are also territorial and stay close to retreats which they excavate themselves. The movements that a subsocial female cricket uses in manipulating her eggs, nymphs, food, and fecal material do not differ significantly from the handling of pieces of soil by solitary, nonsocial crickets during the act of burrowing. Basically, the origin of advanced subsocial behavior of the *Anurogryllus* type re-

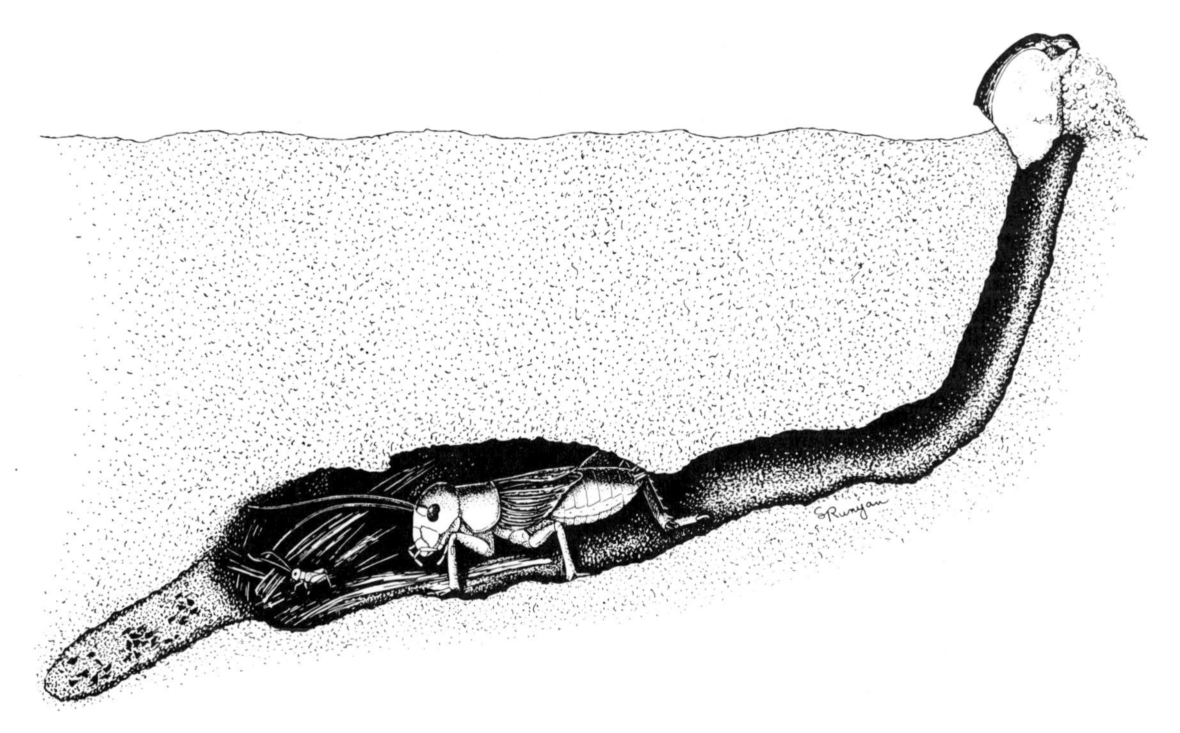

FIGURE 7-2. A female of the burrowing cricket *Anurogryllus muticus* is shown in her sealed brood chamber. A burrow leads (at the right) up to a piece of apple on which the female has been feeding. Pieces of grass have been gathered by the female and piled on the floor of the brood chamber. To the left is a filled defecation chamber, while an egg pile is partly covered by soil beneath the female's abdomen. A single newly hatched nymph stands in front of its mother, and the two individuals touch antennae (from West and Alexander, 1963).

quires only the addition of the capacity to differentiate among classes of objects such as soil, offspring, food, and feces, and a slight specialization in the response to each.

Advanced Parental Care and Division of Labor in Beetles

Subsocial behavior occurs in at least nine families of the Coleoptera, as follows (von Lengerken, 1939; Hinton, 1944; Ellinor Bro Larsen, 1952):

1. Staphylinidae: *Platystethus arenarius, Bledius spectabilis.*
2. Silphidae: *Necrophorus* spp.
3. Hydrophilidae: *Spercheus* spp., *Epimetopus* spp., *Helochares* spp.
4. Scarabaeidae: *Copris hispanus, C. lunaris.*
5. Passalidae: *Popilius* and possibly other genera.
6. Tenebrionidae: *Phrenapates bennetti.*
7. Chrysomelidae: *Omaspides pallidipennis, Pseudomesomphalia thalassina, Selenis spinifex,* and possibly *Phytodecta rufipes.*
8. Scolytidae: the tribes Xyloterini and Xyleborini.
9. Platypodidae: probably all species.

In drawing up the bulk of this list in 1944, Hinton correctly challenged Wheeler's assertion (1921b) that the silvanids *Coccidotrophus socialis* and *Eunausibius wheeleri* of Guyana are subsocial. These beetles have undeniably peculiar habits: the adults live with their larvae in the leaf petioles of the legume *Tachigalia paniculata,* where they first eat out protein-rich pith and exist thereafter by "milking" honeydew from mealybugs (*Pseudococcus breviceps*) who follow them into the cavities. There is really no evidence of cooperation between the adults and offspring. On the contrary, all of the members of the colony compete for the secretions of the *Pseudococcus.* According to Wheeler's own description, when several adults or larvae (or both) are attempting to stroke the same mealybug "they stand around it like so many pigs in a trough, and the larger and stronger individual keeps butting the others away with its head." A third silvanid species, *Coccidotrophus cordiae,* was subsequently discovered living with the same species of mealybug in *Cordia alliodora* (Wheeler, 1928), but it is not known whether its "social" habits differ in any significant way from those of its congener. The case of the neotropical tenebrionid *Phrenapates bennetti* is equally in doubt, but this time only by implication. Ohaus (1909) claimed that the adult beetles excavate brood niches along galleries dug through dead wood, in the manner of the scolytid and platypodid ambrosia beetles, and feed the larvae progressively with fine wood shavings. This much sounds reasonable enough. Unfortunately, Ohaus also claimed that larvae of Brazilian passalids have mouthparts too feebly developed to chew on dead wood directly and therefore have to be fed predigested material by the adults. Both of these claims have been checked and firmly rejected by Heymons (1929). The social behavior of passalids, it should be added, is still largely a mystery despite the fact that some of the species are dominant elements in rotting wood over a large part of the world and can be easily kept in the laboratory. Pearse *et al.* (1936) showed that larvae of the common eastern North American species, *Popilius disjunctus* (still often referred to incorrectly in the literature as "*Passalus cornutus*") can live on decaying wood alone, but they do better with wood that has been chewed and either partly digested or converted into frass by adults. Both adults and larvae stridulate, and Pearse and his co-workers concluded that the sound is used to hold the *P. disjunctus* families together, in line with earlier claims made for Neotropical species by Ohaus and Wheeler. Thus, the passalids appear to be truly subsocial. Very little experimental analysis has ever been made of the stridulation or any other aspect of communal behavior in these insects, however. The single exception of which I am aware is a study by Alexander, Moore, and Woodruff (1963), who found that *P. disjunctus* stridulate when fighting or when suddenly disturbed by another beetle. These observers further noted that passalids stridulate almost continually when together in their normally dense colonies. They agree with the popular view that the sounds serve to keep groups of adults and larvae together, but do not offer any new evidence in support of it.

Some of the verified examples of subsociality in other beetle families involve adaptation at least as extreme as those of the *Anurogryllus* crickets. In addition to the mundane actions of nest construction and guardianship of the young, instances have been reported of nest hygiene, progressive feeding of the young both with raw materials and regurgitated liquid, and even division of labor among the adults. The dung beetles of the family Scarabaeidae offer what seems to be a straightforward evolutionary progression from the elementary to the most advanced forms of parental care. Much of this story was worked out by Jean Henri Fabre at Sérignan. His own account is most conveniently available to English-speaking readers in de Mattos' 1918 translation, *The Sacred Beetle and Others.* More recent research on scarabaeid behavior has been summarized by von Lengerken (1939) and Ritcher (1958). The "Sacred Scarabaeus" (*Scarabaeus sacer*) has the solitary life cycle of a simpler type. When the adult female is ready to oviposit, she fashions a sphere from sheep's dung about her own weight and rolls it to a previously excavated nest in the soil. After settling the pellet into a large elliptical chamber at the bottom of the nest, she makes a crater-shaped depression at one end of the dung, lays an egg in it, and works the material at the edge of the depression up and completely over the egg. Then the nest is abandoned, and the female starts the whole process again elsewhere. A small step toward sociality has been taken by *Sisyphus schaefferi,* a second European species studied by Fabre. Its life cycle is similar to that of the *Scarabaeus,* except that the male accompanies the female throughout the nesting operation. The pair obtains a dung pellet as the first step, and the male guards it while the female excavates the burrow. In *Minotaurus typhaeus* the male is more involved. He forms the fecal pellets and lowers them to the female, who tears the dung into pieces and packs it into the bottom of the burrow. Members of the principally Asiatic genus *Lethrus* have carried division of labor between the sexes still further. The life cycle of these strange insects was worked out principally in Russia by Schreiner (1906). After the pair have formed in the spring and begun to excavate their nest, the male creates a "courtyard" around the entrance. This is a circular area from which humus, stones, and branches are removed. The male is distinguished anatomically by two long tusks that project downward from the lower surfaces of its mandibles. These structures, which appear to be unique among the beetles, are used to rake up the surface of the ground and to shovel away pieces of humus. The female lacks tusks, and she does not join in the clearing of the ground surface. Both partners do cooperate in excavating the nest, however. When the time to provision arrives, labor is again sharply divided between the sexes (see Figure 7-3). The male gathers fresh leaves and buds, which he shears off with his heavy, nipper-like mandibles. These he drags into the nest and turns over to the female, who presses them into elliptical balls fitted to the interior of the brood cells. If the male is eliminated by accident, the female is still able to forage for food on her own and to complete the provisioning of the nest.

Numerous other variations on this scheme exist within the Scarabaeidae. *Bolboceras darlingtoni* of the southeastern United States collects finely divided humus, forms it into a cell, and places it in a cavity above the food mass. Males and females sometimes collaborate, and several pairs may be found in adjoining burrows that appear to interconnect. Larval development is quite rapid, so that generations of adults overlap (H. F. Howden, personal communication). As shown by the studies of Fabre and later confirmed by von Lengerken (1951), at least two species of *Copris* have actually attained a subsocial level of parental care. In both cases the male and female cooperate in the excavation of the nests and the preparation of the larval food. The *C. hispanus* pairs collect cow dung and manufacture four spherical pellets from it in the brood chamber of each nest. When this task is completed, the male leaves, but the female remains behind for the entire four months required for larval development. She stands guard over the pellets, keeping them clean and reshaping them when they begin to deform or crumble. Finally, when the young adults emerge from the pupal stage, she accompanies them out of the nest. Pairs of *C. lunaris* build their nests beneath pats of sheep dung and fashion two pellets in each brood chamber. Both parents remain behind with the young, but only the mother carries on housekeeping.

Subsociality at the *Copris* level or beyond has also been evolved in the staphylinid tribe Oxybelini. Most species in the large family Staphylinidae are predaceous, but certain members of the Oxybelini, together with a few Oxyporinae, are exceptional in feeding on dung or algae. The reliance on such a fixed and locally distributed source of food evidently constitutes a preadaptation for subsocial behavior. *Platystethus arenarius,* which has been studied in England by Hinton (1944), lives in cattle dung. After mating, the little female constructs a broadly oval to nearly spherical chamber 6 to 10 mm in its greatest diameter and lays between 20 and 100 eggs inside. The chamber is kept meticulously clean, if such a phrase can be applied to anything made wholly out of dung. Fungal hyphae that penetrate into the air space are soon chewed off by the female, and arthropods which invade the chamber, including other staphylinid adults and larvae, are repelled or killed. The female, however, does tolerate first instar *P. arenarius* larvae, possibly because of her inability to distinguish them from her own larvae. She continues to protect her brood into the first larval instar. Shortly before the first molt the larvae disperse, and each builds a feeding chamber of its own. Females of the oxybeline genus *Bledius* feed on algae, which they store in nests excavated in the soil. They also protect their eggs. Ellinor Bro Larsen (1952) has shown how one species, *B. spectabilis,* has become highly specialized in related aspects of its ecology and maternal care. This beetle lives in the intertidal mud of the European coast, where it faces extreme hazards both from the high salinity and periodic shortages of oxygen. The female constructs unusually wide tunnels in her brood nest, which are kept ventilated by tidal water movements and by renewed burrowing activity on the part of the female. If she is removed, the brood soon perishes from lack of oxygen. The mother also protects the eggs and larvae from intruders, and from time to time forages outside the nest for a fresh supply of algae.

Approximately the *Bledius spectabilis* level of subsociality has also been attained, independently and in a radically different environment, by certain species of the family Scolytidae commonly referred to as ambrosia beetles. Because the behavior of these insects is adapted to the maintenance of a complex mutualism, and the nomenclature surrounding the mutualism has become confused over a long period of time, a few words concerning the background of the subject are necessary. It was J. Schmidberger who, in 1837, applied the term "ambrosia" to the material lining the burrows of the deep-boring scolytid *Xyleborus dispar.* Without knowing its nature, he stated that this material comes out of the wood and serves as the food of the beetles. In 1844 Theodor Hartig showed that the ambrosia of *X. dispar* is a fungus, and he gave it the formal name *Monilia candida.* Over the next few decades several other kinds of fungus were identified in the galleries of various wood-burrowing species of Scolytidae and Platypodidae. By 1897, when H. G. Hubbard wrote his classic paper on the subject, the

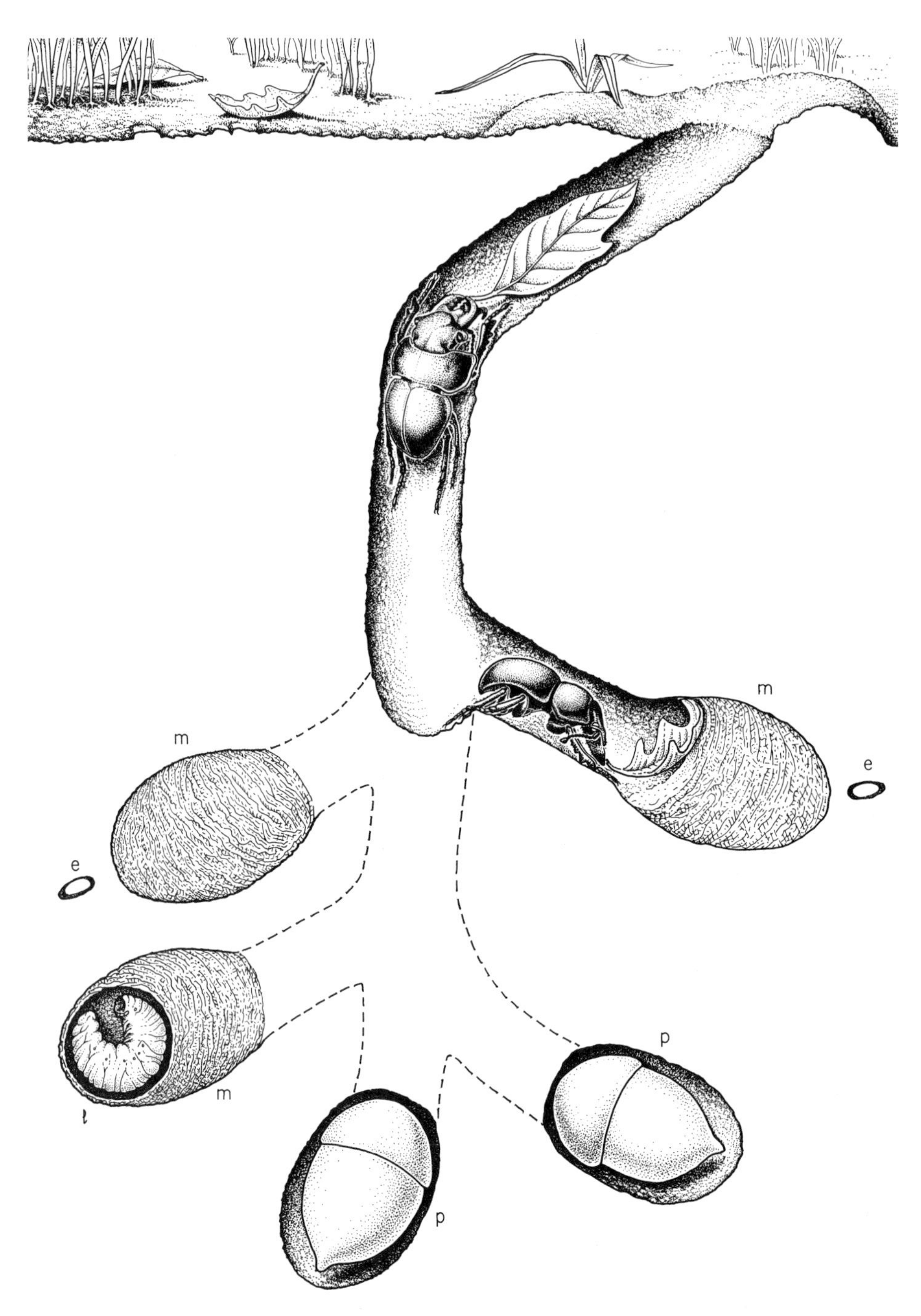

FIGURE 7-3. A pair of *Lethrus apterus* are shown provisioning their nest. The male is pulling a freshly cut leaf backward through the entrance gallery, while the female packs another leaf into the brood cell currently being provisioned: (*m*) leaf masses; (*e*) eggs; (*l*) larva; and (*p*) pupae, which are enclosed in acorn-shaped cocoons. After a cell has been provisioned, the pair pack soil into the main tunnel to that level, as indicated by the limits of the dashed lines (drawing by Sarah Landry, based on von Lengerken, 1939).

species of the Scolytoidea seemed clearly divisible into the following two ecological groups: the "ambrosia beetles," consisting of all of the Platypodidae and some of the Scolytidae, which burrow into the xylem and cultivate the mutualistic ambrosia fungus for food; and the true "bark beetles," comprising most of the Scolytidae, which burrow exclusively within the bark and just beneath it in the phloem. Unfortunately for the initial nomenclature, symbiotic fungi were subsequently discovered to be associated with certain bark-inhabiting scolytids, for example the genus *Ips,* as well as other kinds of beetles, including occasional species of Anobiidae, Brenthidae, Curculionidae (in *Scolytoproctus* and *Gulamentus*), and all of the Lymexylidae so far examined (Schedl, 1958; Graham, 1967). These insects are nevertheless not referred to as "ambrosia beetles." By convention the term is still applied exclusively to the wood-boring scolytoids. Perhaps the term "fungus-growing beetles" can be substituted for the category of species that live in mutualistic association with fungi, parallel with the generally accepted functional category of fungus growers in ants and termites. This minor semantic confusion should not obscure the important fact that the mutualism is an advanced type accompanied by complex adaptations on both sides. The fungi, which belong to the Ascomycetes and several families of the Fungi Imperfecti (Cryptococcaceae, Moniliaceae, Dematiaceae, Stibaceae), are species-specific with respect to the scolytoid beetles they serve. They flourish in the particular conditions of humidity and temperature that arise in the beetle burrows, and they function as the principal source of food for both the adults and the larvae. The "ambrosia" part of the fungus is comprised of special growth forms that are especially capable of proliferating in the open spaces of the burrows. It usually consists of chains of hyphal bodies or cells packed together to form an even palisade lining the gallery walls. Sometimes it is a mat of conidiophores, each bearing a single terminal conidium that is eaten by the beetles and soon replaced by another conidium (L. Batra, 1963). The beetles are typically associated with more than one species of fungus: a primary one brought in by females or (rarely) males, and auxiliary species that make their way into the nests later. Most of the auxiliary forms are either yeasts or filamentous hemiascomycetes (L. Batra, 1966). As discovered independently by Nunberg (1951) and Helene Francke-Grosmann (1956, 1967), the fungus-growing scolytoids have evolved special sac-like receptacles on the body for transporting the spores of the primary symbionts from one tree or log to another. Two of these mycangia, as they are called, are illustrated in Figure 7-4.

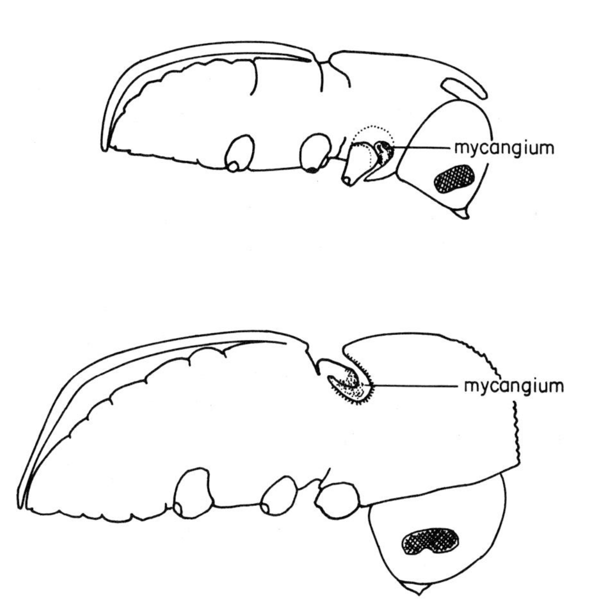

FIGURE 7-4. Location of the mycangia, special structures used to transport spores of symbiotic fungi, are shown in two species of scolytid beetles: *(Above) Monarthrum fasciatum; (below) Anisandrus dispar* (redrawn from L. Batra, 1963).

Some fungus-growing scolytoids are communal in their living habits, but do not display any obvious form of communication between the adults and young. *Xyleborus saxeseni,* for example, feeds on the dead wood of beech, hickory, maple, and oak. It excavates flat, leaf-shaped chambers which stand vertically on end, parallel with the grain of the wood, and which communicate with the bark by one or a few tubular galleries. According to Hubbard, these brood chambers are packed at times with adults of the *Xyleborus* and their eggs, larvae, and pupae in all stages of development. The larvae are able to crawl up and down the vertical walls with the adults. They also assist in enlarging the chambers. Bits of wood are chewed off and swallowed, and, as they pass through the intestine, they turn mustard yellow in color. The beetle colony ejects large quantities of the stained excrement from the nest openings, but a portion is retained and plastered onto the walls of the chambers, where it becomes part of the substratum for the symbiotic moniliaceous fungus. It is not impossible that *Xyleborus* and other communal scolytoids engage in some form of nutritive aid or primitive communication between the adults and larvae, in the manner of *Popilius.* We should bear in mind that it was only recently discovered that adult beetles communicate by stridulation, and, in retrospect, it is hard to see how the phenomenon was overlooked for so long. The 77 species of Scolytidae studied by Barbara Barr (1969) have stridulatory organs of one or the other of three basic types: vertex-pronotal, gula-prosternal, or elytra-abdominal. The presence of the organs among the species is correlated to some extent with sex and ecology. In the large genus *Ips,* fourteen pine-infesting species are known in which only the females possess stridulatory organs, all of the vertex-pronotal type. In two other pine-infesting species, gula-prosternal organs are present in the females and possibly also in the males. Spruce-feeding members of *Ips* appear to lack the organs. Stridulation is used as a discriminating device in the aggregation process. The males of *Ips,* for example, penetrate the host trees first and then call in other members of the same species by means of pheromones (a mixture of terpenoid alcohols produced in the hind gut). In order to gain entrance to the burrows the females must first stridulate. Whether sound is ever used within the brood chambers as a means of communication among members of the growing families remains an open question.

The species of the ipine genera *Gnathotrichus, Monarthrum,* and *Xyloberus* have advanced forms of subsocial organization. The adults rear the young in "cradles," which are short diverticula of the main galleries (Figure 7-5). The adult males and females of *Monarthrum* work together to excavate the nest and to care for the brood. The mother beetle lays eggs singly in circular pits carved into the walls of the main gallery on opposite sides parallel with the grain of the wood. According to Hubbard, the eggs are loosely packed in the pits with chips and material taken from the fungus bed which she has previously prepared in the vicinity and upon which the ambrosia has begun to grow. The fungus referred to in the following account by Hubbard has more recently been identified as a species of *Monilia,* a member of the Fungi Imperfecti. "The young larvae as soon as they hatch out, eat the fungus from these chips and eject the refuse from their cradles. At first they lie curled up in the pit made by the mother, but as they grow larger, with their own jaws they deepen their cradles, until, at full growth, they slightly exceed the length of the larva when fully extended. The larvae swallow the wood which they excavate, but do not digest it. It passes through the intestines unchanged in cellular texture, but cemented by the excrement into pellets and stained a yellowish color. The pellets of excrement are not allowed by the larvae to accumulate in their cradles, but are frequently ejected by them and are removed and cast out of the mouth of the borings by the mother beetle. A portion of the excrement is evidently utilized to form the fungus garden bed. The mother beetle is in constant attendance upon her young during the period of their development, and guards them with jealous care. The mouth of each cradle is closed with a plug of the food fungus, and as fast as this is consumed it is renewed with fresh material. The larvae from time to time perforate this plug and clean out their cells, pushing out the pellets of excrement through the opening. This débris is promptly removed by the mother and the opening again sealed with ambrosia. The young transform to perfect beetles before leaving their cradles and emerging into the galleries." Although the *Monarthrum* life pattern is tantalizingly close to eusociality, there is no evidence that the young adults remain to assist their parents in rearing other young. Schedl (1958) has presented an authoritative account of the evolution of the ambrosia habit, which he calls xylo-mycetophagy, as well as a review of other aspects of scolytoid biology.

The most advanced form of parental care known to have evolved in the Coleoptera is that shown by the

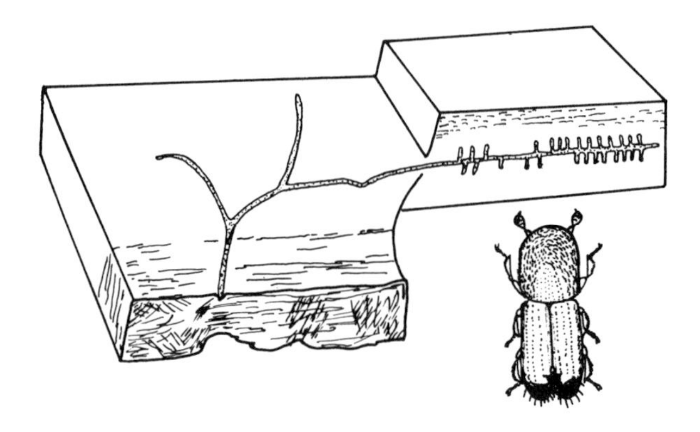

FIGURE 7-5. Adult of *Monarthrum mali* (Scolytidae: Ipinae) and a typical gallery system of the species excavated in the dead wood of maple. Larvae are reared in "cradles," the short chambers leading off from the central gallery (redrawn from Hubbard, 1897).

burying beetles of the genus *Necrophorus*. Erna Pukowski's 1933 report has elucidated many of the details of this very remarkable life cycle. Six species of *Necrophorus* occur in Europe. In May the overwintered adults become active and begin searching for the dead bodies of small vertebrates, preferably birds and such small mammals as mice and shrews. If a male encounters a corpse first, he takes the "sterzelndes" posture, lifting the hind part of the body into the air and releasing a pheromone. Although the substance has not been chemically identified, it appears to be effective only on females of the same species. If more than a single pair assemble on the corpse, and sometimes as many as ten or more do collect at about the same time, fighting ensues, male against male and female against female, until only a single pair is left. The winners now excavate the soil beneath and around the body until their prize is partly buried. At the same time they chew and work the putrefying mass until it is roughly spherical in shape and can be rolled downward into a burrow excavated beneath it. Then the beetles seal the burrow off from below, leaving themselves isolated with the rotting ball in an underground chamber. The female next eats out a crater-shaped depression on the top of the ball and spreads her feces over its surface. When the larvae hatch, they sit in the crater like so many fledgling birds in the nest. As described by Pukowski, they also interact with the parents like the young of altricial birds: "As the female approaches the crater, the larvae lift the fore part of their bodies in unison, so that their legs are grasping air. The beetle stands directly over the larvae and strikes the food ball or the larvae with trembling motions of her fore legs. Now the female opens her mandibles, and one of the larvae swiftly inserts its head between the mandibles and tightly into the mouth opening. If circumstances are right one can see a brown liquid pass from the mother's mouth to the larva. After a few seconds the beetle pulls back and attempts to put her mouth down on the head of another larva. Without doubt the brood is being fed by the female."

This scene is illustrated in Figure 7-6. When the larvae

FIGURE 7-6. A female of the burying beetle *Necrophorus vespillo* regurgitates to a larva, while a second larva attempts to share in the feeding. The larvae, which in this case are in the third instar, rest in a depression dug by the female on top of the carrion ball (original drawing by Sarah Landry; based on photographs by Erna Pukowski).

are only five or six hours old they begin to feed on the ball directly. However, they continue to take meals from the mother at irregular intervals, and for a time after each molt they are wholly dependent on this source of nutrition. If the mother is removed while the larvae are immature, they start to pupate, but are unable to complete adult transformation. In two of the European species (*germanicus* and *vespilloides*) the male assists his mate in feeding the brood, but is still less active in the role.

Cooperation and Division of Labor among Insect Larvae

A very different kind of presocial behavior is shown by the larvae of the jack-pine sawfly (*Neodiprion pratti banksianae*). These caterpillar-like insects feed in tight groups on their coniferous host. Ghent (1960) discovered that the chief advantage to aggregation comes in the first stadium, when the larvae are weak and have great difficulty chewing holes into the tough pine needles on which they depend. In Ghent's experiments, larvae isolated from their fellows suffered an 80 percent mortality, while only 53 percent died among all of those allowed to remain in groups. The effect is a statistical one based on sample size, because even when a larva belongs to a group, it attempts individually to establish its own feeding site. When one does cut through into the succulent inner tissue, whether by luck, superior strength, or greater skill at finding a weak spot, the other larvae are quickly attracted to the spot by the odor of the volatile components of the salivary secretions and plant substances released into the air. Soon the breach is widened, and all of the larvae are able to feed. Comparable group behavior enhances survival of individuals during at least two periods in the life cycle of the Australian sawfly *Perga affinis* (Carne, 1966). The eggs of this species are laid in "pods" within the tissue of the leaf blade. When the larvae hatch, they must rupture the overlying leaf tissue in order to escape and thus to survive. Usually only one or two larvae in a pod succeed in making it to the outside, and they are followed from the exit holes by their brother and sister larvae. It frequently happens that none of the progeny from a small pod succeeds in escaping, in which case all die. In one large sample of infested leaves studied by Carne, the mortality of pods containing fewer than 10 eggs was 66 percent; in those containing more than 30 eggs, only 43 percent. The *Perga* larvae also stay together when they leave the host tree

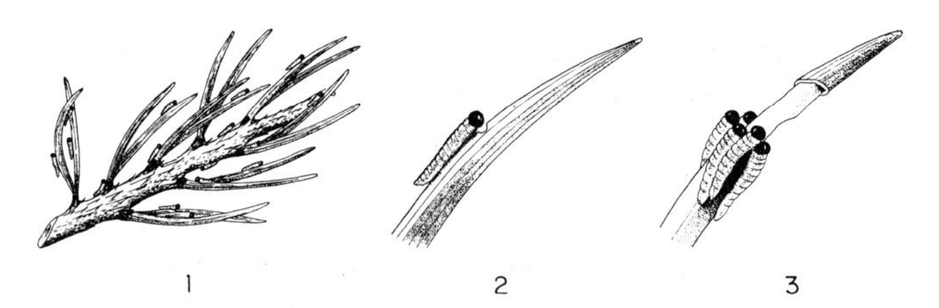

FIGURE 7-7. When larvae of the sawfly *Neodiprion pratti banksianae* first hatch, they have a difficult time cutting through the tough cuticle of the needles of the jack pine in order to feed. At first they wander singly (*left*) until one or more succeeds in cutting through to the inner tissues (*middle*). The smell of damaged tissues attracts other larvae, and they feed thereafter in a group (*right*) (from Sudd, 1963; based on Ghent, 1960; courtesy of *Science Journal*, London, incorporating *Discovery*).

to pupate. In order to cocoon, they must dig into the soil. Since their morphology is poorly adapted for burrowing, most are not able to penetrate the crust, and they face death by dessication unless they can use the entrance burrow of a successful larva. In larger aggregations at least one larva usually succeeds in breaking through, with the result that other individuals are also able to cocoon. But in small groups, complete failure and total mortality are commonplace.

Wellington's study (1957) of the western tent caterpillar, *Malacosoma pluviale* (Lasiocampidae), has revealed a division of labor in feeding aggregations fashioned from strong individual differences in foraging behavior. Some larvae are leaders. They can move from one spot to another on their own without the trails of silk threads that ordinarily guide the groups. Others are wholly followers and seem unable to go anywhere without the aid of the silk trails. Among the followers there exists further variation in the distance over which individuals disperse along the trails away from the silken tents used by the colonies as refuges. Wellington found that the relative frequencies of these behavioral types in a given colony greatly affect the distances covered by the colony when it forages, the number and shape of the tents it manufactures, and its susceptibility to various mortality factors such as predation and disease.

Subsociality and Quasisociality in Spiders

The "social" spiders have been the subject of excellent recent reviews by Tretzel (1961), Kullmann (1968), and Shear (1970). There are no known cases of eusociality in the spiders or even of the formation of reproductive castes among members of the same generation, that is, semisociality. Communal nesting behavior is relatively common, however, and a few examples have been reported in which adult spiders cooperate in the construction of both webs and brood refugia. Also, some truly remarkable cases of communication have been reported between parents and young. Most of the known cases of presociality are found in the phylogenetically more advanced spiders, especially the cribellate groups, which are characterized by the possession of a cribellum or flat spinning plate located anterior to the spinnerets. Among the primitive spiders, parental care is apparently exhibited by some of the tarantuloids. Otherwise, presociality is limited chiefly to the Agelenidae, Eresidae, Oecobiidae, Theridiidae, and Uloboridae.

A hypothetical sequence of events in the evolution of spider presociality is presented in Figure 7-8. This schema is arranged in a way to emphasize that spiders, like the insects, have moved along both of the main phylogenetic pathways open to them. Some have gone the parasocial way, in which adults first aggregate and cooperate without brood care; others have followed the subsocial way, in which parental care is evolved in the absence of adult gregarious behavior. When the two derived behavioral elements are combined, as they have been by some species

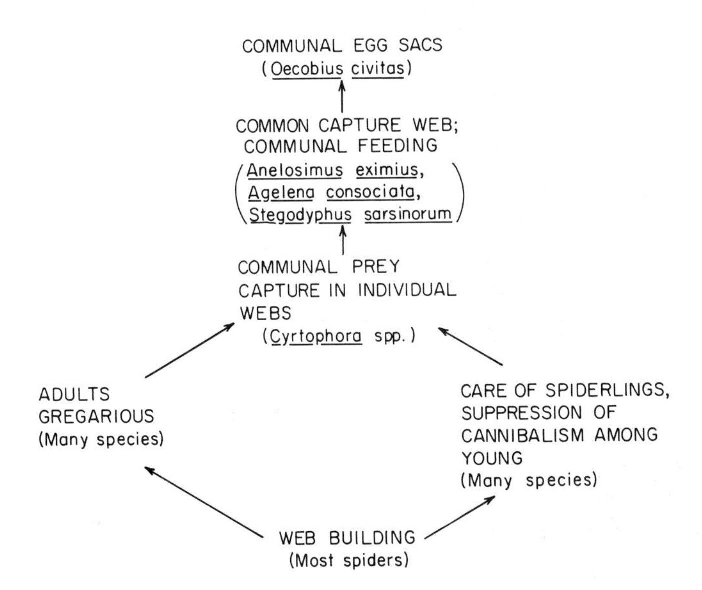

FIGURE 7-8. Steps in the evolution of presocial behavior of spiders (modified from Shear, 1970).

of *Agelena* and *Oecobius*, an advanced state of quasisociality results. It is therefore natural to ask whether there are any eusocial spiders. To rephrase the question somewhat more precisely, do any of the species thought to be quasisocial also exhibit reproductive division of labor? This important possibility seems not to have been investigated explicitly by arachnologists, but I regard it as a basic problem of spider biology, one for which even a firm negative answer will prove significant. It prompts us to examine some of the better-documented examples of the known stages of presociality in the postulated sequence of evolutionary advance.

The Uloboridae contain among themselves two of the key stages of communal behavior. *Uloborus arizonica* and *U. oweni* of the southwestern United States form common web systems, each containing 20 to 200 individuals. No central retreats are constructed, and communal feeding is unknown (Muma and Gertsch, 1964). The Neotropical species, *U. republicanus,* forms larger, more structured communities (Simon, 1891; Schwarz, 1904). Silken lines are spun in a tangle from one piece of vegetation to another, and between the lines the spiders fashion individual webs of the typical uloborid pattern. There is also a central tangle to which the females retreat when they are inactive or threatened. The males remain in the retreat permanently and mate with the females there. The females care for their egg sacs individually. *U. republicanus* nests sometimes are very large. Schwarz observed one in Cuba that occupied nearly the whole crown of a felled tree and was 2 to 3 m wide, about 2 m high, and 1 m deep; it contained approximately 1,000 spiders. "The whole net formed a most perfect trap for all insects that flew through the clearing made by the felled tree, and the individual webs of the spiders were found to be full of insects of all kinds." Other communal webs in the vicinity contained 300 spiders or less.

Anelosimus eximius (family Theridiidae) carries communal life well beyond the stage expressed in the Uloboridae. According to Simon (1891), who studied the species in Venezuela, about a thousand spiders cooperate to build a large tangled web with a central retreat woven around a cluster of dead leaves (Figure 7-9). The spiders touch each other continuously with their forelegs and palps. They attack prey in groups, drag them back into the retreat, and feed on them communally. In spite of this high degree of intimacy, the females still construct and guard their egg sacs individually.

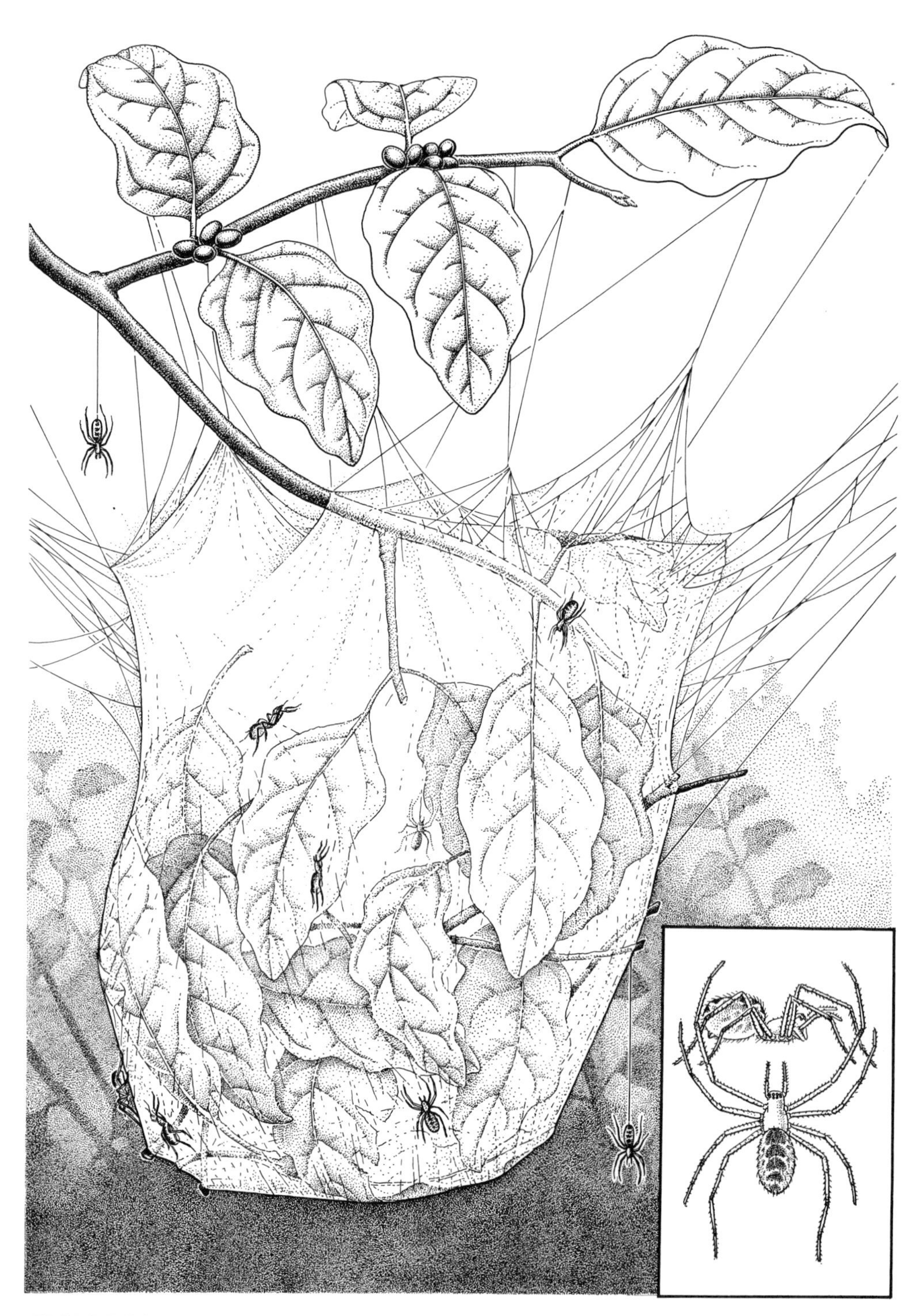

FIGURE 7-9. A communal nest of the South American theridiid spider *Anelosimus eximius* in the vegetation of a coffee plant. The central retreat is constructed of a clump of dead leaves enveloped in silk webbing (redrawn by Sarah Landry; from Simon, 1891).

The Theridiidae are also notable for the exceptional intimacy of their parental care. According to Kaston (1965), some degree of maternal care is manifested by a large percentage of theridiid species. The females of some members of *Theridion,* for example *T. sysiphum,* feed their young by regurgitation and also allow them to share the prey. Nørgaard (1956) has described what appear to be specialized forms of tactile communication between the mother *T. saxatile* and her spiderlings. The young remain in the mother's web for about a month after hatching and feed simultaneously on the prey captured, almost 90 percent of which is ants. While still very young, they remain at the "nest" or retreat, located in the center of the tangled web, and the mother kills the prey. "If they walk too near the struggling ant, the female turns towards them and makes a characteristic movement with its forelegs. They seem to thrum on the threads like a musician plucking the strings of his instrument. The young respond to these 'thrumming'-movements instantly by an escape reaction: they run back to the nest." When the young grow larger, they take part in the capture, and, instead of warning off the spiderlings, the mother appears to summon them with a new, "sweeping" motion of the forelegs. Nørgaard gives the following example as typical. "An ant (*F. pratensis*) is presented to the spider. It sticks to a capture thread, and the spider reacts at once. It makes some tentative movements with its forelegs and climbs to the place where the ant hangs struggling to get loose. The spider touches the ant with its forelegs, then turns it back to the prey, and throws thread upon it. It leaves the prey and goes back again, at the same time making some peculiar 'sweeping'-movements over the web with its forelegs. Now the young swarm down the threads and gather round the ant at some distance. One by one they go to the ant, turn their backs to it, and throw some thread towards it. These threads do not reach the ant, however, on account of the small spiders hanging at too great a distance from the prey. More young arrive at the prey and carry out the 'enswathing'-reaction. One of the young grasps a hindleg of the prey and bites it in the joint between tarsus and tibia. The female comes down to the prey twice, but leaves it again after having made the 'sweeping-'movement over the web with its forelegs." Reciprocal communication occurs in the agelenid *Coelotes terrestris.* As in *Theridion,* the young spiderlings remain in the mother's web for some time after they hatch. They are fed in part by regurgitation, which they actively elicit by stroking the mother's chelicerae with their palps. When the mother dies in the autumn, they feed on her body (Tretzel, 1961).

The West African spider *Agelena consociata* is notable for having both quasisocial and subsocial habits, and is, in fact, as close to the brink of eusociality as any arthropod species known outside the Hymenoptera and Isoptera. It has been the subject of careful field and laboratory studies in Gabon by a group of French entomologists, including Pain (1964), Darchen (1965), Chauvin and Denis (1965), and Krafft (1966a,b; 1967). The webs are up to 3 m in greatest diameter and are woven through shrubby vegetation along the borders of rain forest. They consist of numerous sheets spun in an irregular manner by the hundreds of spiders that inhabit each nest. Small prey are captured and eaten singly by the individuals that happen to be closest at hand, but larger prey are pursued and killed by groups (see Figure 7-10). The *Agelena* are completely friendly toward all their nestmates, frequently touching them with their legs and palps without any show of hostility. They also tolerate the introduction of members of other colonies, and Krafft found that under laboratory conditions they even accept individuals of *Agelena labyrintheca,* a solitary European species. An even more remarkable feature of the behavior of *A. consociata* is the treatment accorded the young. The spiderlings seldom join in hunting. They cluster in groups of 50 or more individuals in the dense silken retreats and wait for food to be brought to them. These groups are apparently true crèches, for the adults show no sign of even being able to distinguish their own offspring, much less of singling them out for favored treatment. As Shear (1970) has suggested, discrimination is made even less likely by the fact that the egg mass laid by each female contains no more than 30 eggs, so that individual spiderling groups must contain the offspring of more than a single female. At least one advantage of social life accruing to *A. consociata* has been made clear by the notes of French observers: when working in groups, the spiders are able to capture prey that would be too large for an individual to handle. This, it will be recalled, is the principal adaptive advantage associated with the origin of group foraging in the early evolution of the army ants (see Chapter 4). A second probable benefit is the added protection given colony members, especially the young, who can remain hidden deep within the silken retreats of the giant nests. Although spiderling mortality data are lacking, indirect evidence of a lower mortality rate is provided by the

FIGURE 7-10. Adults of *Agelena consociata* attack a cricket on their communal web, while the spiderlings wait nearby (original drawing by Sarah Landry; based on photographs by B. Krafft).

existence of a lower natality rate. The sacs made by *A. consociata* each contain 20 to 30 eggs, whereas those of solitary *Agelena* species generally contain 50 to 100 eggs. Moreover, the rate of manufacture of egg sacs is not very high (20 to 100 present per nest at any given time), and the number of eggs that are viable appears to be quite low. Since mortality and natality are approximately equal in stable communities, it seems to follow that mortality in the *A. consociata* colonies is comparatively low.

In a tiny Mexican species of the spider genus *Oecobius civitas* (Oecobiidae), Shear (1970) has discovered a form of quasisocial behavior that represents an advance in one particular aspect beyond *A. consociata*. The webs are only about 25 cm in diameter, but each contains between 110 and 180 spiders in individual retreats. The species is unique among other known social spiders in that the females appear to construct a common egg sac. Each sac found in the nests by Shear contained more than 200 eggs, as opposed to the 2 to 10 eggs per sac that characterize solitary species of *Oecobius,* and surely far more than could be laid by any one female. The newly discovered case of the social *Oecobius* is a reminder that many surprises probably await students of spider biology who pay particular attention to social phenomena. Few species of spiders have been studied intensively anywhere in the world, and this is doubly true for such obscure, primarily tropical groups as the Oecobiidae.

Conditions Favoring the Origin of Presocial Behavior

The evidence shows that presociality is an eclectic and convergent evolutionary phenomenon, generated almost haphazardly in response to any one of a large set of independent environmental forces. Consider the special case of subsocial behavior. Elaborate forms of parental care are seemingly limited to situations where the physical environment is unusually favorable or where the physical environment is unusually harsh. "Ordinary," intermediate environments do not seem especially to promote the evolution of this kind of behavior. At the favorable pole we find a class of highly subsocial species that utilize food sources that are very rich but at the same time scattered and ephemeral: dung (*Platystethus,* Scarabaeidae), dead wood (Passalidae, Platypodidae, Scolytidae), and carrion (*Necrophorus*). These insects have adopted the bonanza strategy. When individuals "strike it rich," they are assured of a more than sufficient food supply to rear their brood. They must, however, exclude others who are seeking to utilize the same bonanza. Territorial behavior is commonplace in all of these groups. Sometimes, as in *Necrophorus,* fighting leads to complete domination of the food site by a single pair. It is surely no coincidence that the males, and to a lesser extent the females, of so many of the species are equipped with horns and heavy mandibles—a generalization that extends to other bonanza strategists which are not subsocial, for example, the Lucanidae, the Ceiidae, and many of the solitary Scarabaeidae. By the same token there is an obvious advantage to remaining in the vicinity of the food site to protect the young. Cleptoparasitism, or cuckooism, the deliberate substitution of an alien brood for the resident brood, accomplished by intruding females, is a constant danger. A great many beetle species, for example, are specialized as cleptoparasites on the nests of the ambrosia scolytoids; these include members of the brenthid tribes Calodromini, Pseudocephalini, and Taphroderini and curculionid genera *Curanigus* and *Scolytoproctus.* At the opposite extreme are subsocial species that are frequently threatened by difficult conditions in the physical environment rather than by the onslaught of competitors in a physically desirable environment. Adults of the intertidal staphylinid *Bledius spectabilis* are forced to remain with their young constantly in order to prevent their suffocation. West and Alexander (1963) have suggested that because the brachytrupine and gryllotalpine crickets nest in deeper soil and feed on moist vegetable material, their attendance is needed to reduce the important threat of invasion by fungi. The same consideration appears to hold for the subsocial earwigs. The clear implication from these examples is that subsocial behavior often permits the utilization of parts of the physical environment that are too harsh to be penetrated by related, solitary species.

The conditions that favor communal and quasisocial associations among adults are similar in certain respects to those that promote specialized parental care. The adult swarms of midges, mosquitoes, braconid wasps, coniopterygid neuropterans, and many other kinds of insects clearly function to bring the sexes together for mating. The huge hibernating aggregations of chysomelid beetles are generally considered to serve the same end (Hagen, 1962). But it is significant that these aggregations are unstructured. There is no division of labor, and communication is, so far as is known, limited to aggregation and reproduction. Nonsexual gregarious behavior, in contrast, appears to have evolved as an adaptation to permit the utilization of parts of the environment that are denied to single individuals. We have already seen how it permits the larvae of certain sawflies to eat an unusually tough kind of foliage and to penetrate hard soil for pupation. It allows the spider *Agelena consociata* to prey on larger arthropods that would otherwise escape. Probably communal and quasisocial associations, which permit the manufacture of large common nests and retreats, give an added edge of protection from predators. This seems clearly to be the case for social spiders, even though the inference is hard to document in any quantitative way.

Finally, what do the presocial arthropods tell us about the origin of eusocial insects? By providing a larger sample of evolutionary experiments they have made possible new and stronger insights into the ecological conditions under which early social evolution is most likely to occur. But the difficult theoretical problem of why some of these groups have not taken the final step to eusociality remains. Independent phyletic lines have repeatedly assembled most of the essential ingredients: long lives for the adults; cooperation and even division of labor among individuals; intimate relationships between adults and young involving specialized, reciprocal communication; the tendency to construct secure nests; and the mental capacity and mechanical ability to manipulate and distinguish between brood and inert objects. Quite a few of the species require only reproductive division of labor among the adults—the existence of reproductive castes, perhaps together with a slight extension of the adult life span—to raise them to the level of the primitively eusocial bees and wasps. That they have not done so is an absorbing mystery that will provide a theme for later discussions in this book.

8 Caste: Ants

The Kinds of Castes

Polymorphism is defined in a special sense in the social insects as the coexistence of two or more functionally different castes within the same sex. The castes must be stable during one or more instars. In the social Hymenoptera, including the ants, they are stable throughout the adult instar. It is also a strong rule, but perhaps not an ineradicable part of the definition of polymorphism, that all the castes make their appearance in the course of development of each normal, mature colony. Slight, continuous variation in color, pilosity, spine shape, sculpturing, and so forth, if it does not meet these requisites, would not ordinarily be classified as polymorphism. Discontinuous genetic morphs, of the kind exemplified by the yellow and black color phases of *Leptothorax schaumi* (Wesson and Wesson, 1940), are not classified as castes. Neither are pathological forms, no matter how deviant and bizarre. Worker ants characteristically change their work preferences with age, but this progression does not represent a change in physical caste. Although the concept of polymorphism is thus sharply defined with reference to social insects, the same term is used in many other ways in biology, often ambiguously (see the discussion of this point by Kennedy *et al.,* 1961). Most commonly it denotes noncontinuous genetic variation within a population, and as such it is especially well entrenched in the literature of genetics. Consequently Mayr (1963) has proposed the alternate term "polyphenism" for nongenetic variation of the sort seen in the caste systems of social insects. But caste variation has been labeled as polymorphism at least as far back as Emery (1896), with little overlap into or confusion with the genetic usage, and a change hardly seems necessary now.

Three basic female castes are found in the ants: the worker, the soldier, and the queen. I refer to them as basic because they exist usually, but not always, as sharply distinctive forms unconnected to other castes by intermediates. The males constitute an additional "caste" only in the loosest sense. No certain case of true caste polymorphism *within* the male sex has yet been discovered. Two forms of the male occur in some species of *Hypoponera,* but even in these cases they are not known to coexist in the same colony. Soldiers are often referred to as major workers, and the smaller coexisting worker forms as minor workers. Where soldiers exist in a species, minor workers are also found. The worker caste has been lost in many socially parasitic species, while in a few free-living species, especially in the primitive subfamily Ponerinae, the queen has been completely supplanted by workers or worker-like forms. In only a minority of species are all three female castes found together. All ant species, on the other hand, produce males in abundance as part of the normal colony life cycle. In other words, no case of obligatory parthenogenesis is known.

In the course of evolution these castes have been elaborated in various, often striking ways. Sometimes the derived form bears little resemblance to the ancestral type, as, for example, the huge, bizarre queens of the army ants. Also, intermediates sometimes connect the basic female castes: ergatogynes between workers and queens; media workers between minor and major workers. Finally, parasites living within the ant nests, or within the bodies of the ants themselves, can alter individuals of various of the castes into extraordinary pathological forms that lack functional significance in the society.

Over the years a complex terminology has been constructed to classify all these caste variants. W. M. Wheel-

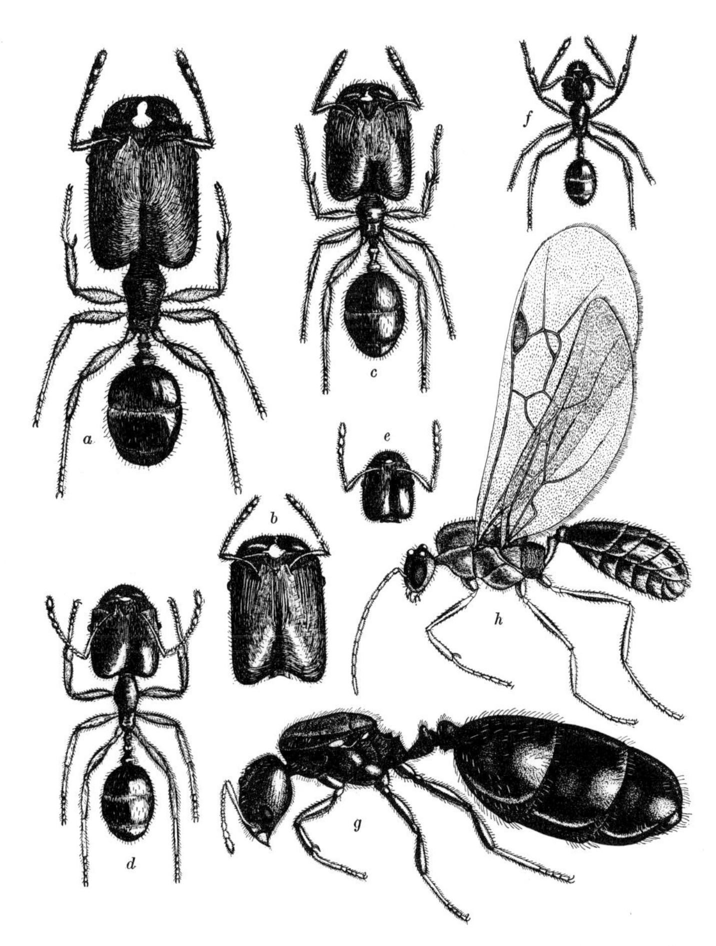

FIGURE 8-1. The female castes and the male of the myrmicine ant *Pheidole kingi instabilis*. The worker caste is comprised of continuously varying subcastes from the major worker (*a*) through media workers (*b–d*) to the minor worker (*e, f*). The queen (*g*) and male (*h*) are also shown (from Wheeler, 1910).

er's system, which was first proposed in 1907 and contains about 30 categories, was the most comprehensive and widely used for many years. Definitions of several of the terms quoted from Wheeler's paper will give a sense of the nature of this classification:

> The *macranēr* is an unusually large form of male which occasionally occurs in populous colonies.
>
> The *micranēr*, or dwarf male, differs from the typical form merely in its smaller stature.
>
> The *dorylanēr* is an unusually large male form peculiar to the driver and legionary ants of the subfamily Dorylinae (*Dorylus* and *Eciton*). It is characterized by its large and peculiarly modified mandibles, long cylindrical gaster and singular genitalia. It may be regarded as an aberrant macranēr that has come to be the typical male of the Dorylinae.
>
> The *phthisanēr* is a pupal male which in its late larval or semipupal state has its juices partially exhausted by an *Orasema* larva . . . The wings are suppressed and the legs, head, thorax, and antennae remain abortive.
>
> The *phthisogyne* arises from a female larva under the same conditions as the phthisanēr, and differs from the typical female in the same characters, namely absence of wings, stenonoty, microcephaly and microphthalmy . . .
>
> The *plerergate*, "replete," or "rotund," is a worker which in its callow stage has acquired the peculiar habit of distending its gaster with stored liquid food ("honey") till this portion of the body is a large spherical sac and locomotion becomes difficult or even impossible . . .
>
> The *pterergate* is a worker or soldier with vestiges of wings on a thorax of the typical ergate or dinergate form, such as I have described in certain species of *Myrmica* and *Cryptocerus*.

Few myrmecologists have had the patience to follow Wheeler in distinguishing plerergates from pterergates or phthisogynes from phthisaners, but some of the more unusual caste terms have been used with varying frequency in the literature. This purely semantic means of analyzing polymorphism has not been successful. During his lifetime and especially in his last work on the subject in 1937, Wheeler believed he had found good reasons for multiplying and naming every qualitatively distinguishable category. On the one hand he considered the parasitogenic forms to be distinctive enough and rare enough to fit easily into the system. More important, however, he believed—erroneously—that most nonparasitogenic castes arise directly by genetic mutations. He saw no difference between normal functional castes and true anomalies. All except the queen and typical males were basically anomalous forms to Wheeler, and he referred to his categories alternately as castes, phases, and anomalies. In a reevaluation of Wheeler's system, Wilson (1953) concluded that some of the names are superfluous, some are for practical purposes synonymous, and some are but stages in an allometric progression. He proposed the much-simplified, new classification given in modified form below.

1. *Male.* Ordinarily possessing a generalized hymenopterous thorax and fully developed, nondeciduous wings. Wingless in the ergatomorphic form in some free-living and mildly parasitic species (see below) and in less worker-like form in many extreme parasitic species such as *Crematogaster* (*Apterocrema*) *atitlanica* and the mem-

bers of *Anergates,* the "*Wheeleriella*" group of *Monomorium, Symmyrmica,* and *Teleutomyrmex.* In *Hypoponera eduardi* and *H. opaciceps,* both alate and ergatomorphic forms are known, but they do not occur in the same colonies. R. W. Taylor (1967) has hypothesized that in the two *Hypoponera* species the same colonies produce ergatomorphs while young and alates when older and larger. It is equally within the realm of possibility that the two male forms represent different genetic morphs or even different sibling species.

2. *Ergatomorphic* (*or ergatoid*) *male.* With normal male genitalia and a worker-like body. The anterior portion of the body converges so close to that of the worker in a few species that it may eventually be shown, as Wheeler has suggested, that the ergatoids are actually persistent gynandromorphs, that is, anterior-posterior mosaics of male and female tissue. Known in species of *Hypoponera, Cardiocondyla,* and *Formicoxenus* (see Figure 8-2).

3. *Queen.* The fully developed reproductive female, possessing a generalized hymenopterous thorax and functional but deciduous wings. The queen is often referred to loosely as "the female" of the colony. The term *gyne* was used synonymously with queen by Wheeler (1907b). Brian (1957c) has employed it to denote more specifically "a sexual female that is not socially a functional reproductive," a confusing secondary application that is unlikely to be widely accepted.

4. *Worker.* The ordinarily sterile female, possessing reduced ovarioles and a greatly simplified thorax, the nota of which are typically represented by no more than a single sclerite each (Tulloch, 1935). Including, in the broadest sense, both minor workers and soldiers in species where these two subcastes occur together.

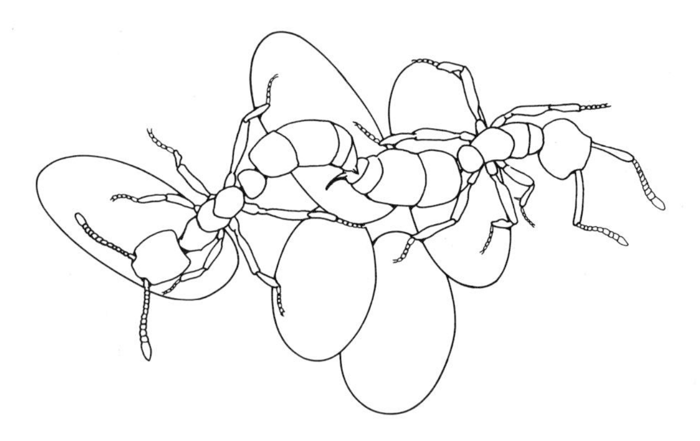

FIGURE 8-2. An ergatomorphic male of *Hypoponera eduardi* mating with a worker in the nest (from Le Masne, 1956a).

5. *Soldier.* The largest worker subcaste, usually possessing a disproportionately large head. Subcastes in a continuous polymorphic worker series are designated approximately according to their size as the *major* (soldier), *media,* and *minor.* In some species the media subcaste drops out, and the soldiers and minor workers are separated by a wide morphological gap. In this case, the soldier is referred to as a full caste.

6. *Ergatogyne.* A form intermediate between the worker and the queen. In a very few species the true queen caste persists, and the ergatogynes form a graded series between it and the typical workers. Such an ergatogynic progression is most often encountered in social parasites that have a degenerate worker caste represented by a small number of individuals, but it also occurs in some free-living species of *Chelaner* and *Monomorium.* In yet other species the ergatogyne replaces the queen entirely and is not connected to the worker caste by a graded series. The second, truly functional type of ergatogyne is especially common in the primitive genus *Myrmecia* and in the subfamily Ponerinae. Haskins and Haskins (1955) have pointed out that this tendency is partly correlated with the tendency of the queens of primitive ants to forage for food outside the nest during colony founding. In higher ants, claustral colony founding is the rule, and queens usually must rear their first brood entirely on the reserves contained in their own fat bodies and degenerating flight muscles. Hence, fully differentiated queens are a necessity when colonies are founded claustrally. Ergatogynes do occur in some free-living higher ants. They are the rule, for example, in the aberrant dolichoderine genus *Leptomyrmex* (Wheeler, 1934) and in the legionary cerapachyines and dorylines (Wilson, 1958b). They are also the sole form of reproductive in a high percentage of the endemic species of New Caledonia, belonging to such phylogenetically advanced genera as *Chelaner, Lordomyrma, Prodicroaspis,* and *Promeranoplus.* New Caledonia is an old, very isolated island, and ergatogyny in its ants corresponds to the flightlessness found so commonly among the endemic species of birds and insects on oceanic islands. Where ergatogynes have replaced true queens in higher ant species, it seems logical to predict that either workers accompany the ergatogynes during colony founding, or else the ergatogynes have reverted to foraging on their own. Field studies of such cases remain to be conducted. Many

ergatogynes of the second, truly functional class show a divergent trend away from the normal queen-worker series in that the development of the gaster and post-petiole outpaces the development of the thorax and head. Such variants possess a gaster that approaches (or surpasses) in size that of the typical queen, while the thorax and head are only slightly ergatogynic. These intercastes either serve as complementals or replace the normal queen.

7. *Dichthadiiform ergatogyne.* This caste is the extreme stage of the phylogenetic trend toward enlargement of the gaster in the ergatogyne. The total size is greatly increased, the gaster is huge, and the postpetiole is expanded to the extent that it has come secondarily to resemble the first gastric segment (see Figures 4-22,30). One gets the impression of a greatly increased allometric growth in the posterior part of the body, with a growth center located in the gaster and a growth gradient declining anteriorly to enlarge also the petiole and propodeum. In addition, the head is broadened and rounded, the mandibles are often falcate, and the petiole is commonly bilaterally cornuate. Dichthadiiform females occur exclusively in groups with a legionary mode of life. The extreme development is seen in the subfamilies Dorylinae and Leptanillinae and the ponerine genus *Onychomyrmex* (Wilson, 1958b). An intermediate stage in the evolution of this form can be seen in the ergatogynes of the ponerine *Acanthostichus quadratus* and species of *Simopelta* (Gotwald and Brown, 1966).

The Evolution of Castes

It was Karl Escherich who, in 1906, first clearly distinguished two very different ways of analyzing the polymorphism of social insects. He pointed out that one can trace the evolution of castes from undifferentiated prototypes by the comparison of species, or one can identify the physiological events that determine caste in individual larvae belonging to a single species at some given evolutionary stage. Considerable progress in both the phylogenetic and physiological approaches has been made since 1950. Let us begin here with a consideration of the evolution of female polymorphism.

Huxley (1932), in the course of his investigations on relative growth in animals, conjectured that some ant castes might be determined simply as a consequence of allometric growth (which he called heterogonic growth). Allometry is defined as the following precise relation between the dimensions of two body parts:

$$\log y = \log b + a \log x$$

or, equivalently,

$$y = bx^a$$

where y and x are measures of the two body parts and a and b are constants. On a double logarithmic plot the relation forms a straight line. Its slope a is determined by the rate of divergence of the two body parts with increasing total size and can be referred to as the allometric constant. If a is equal to unity, no divergence takes place with an overall increase in size; the growth is then referred to as isometric. The greater the departure of a from unity, the more striking the differential growth. Much of relative growth throughout the animal kingdom is allometric, and there is an almost disturbingly simple hypothesis available to explain why. Organs, like organisms and populations, tend to grow exponentially. In growth studies on the dimensions x and y of the same or differing organs, any common factor $\Phi(t)$ affecting transformation rates is easily eliminated, and the number of equations is reduced by one. If the system is a simple one,

$$\frac{1}{y} \cdot \frac{dy}{dt} = \alpha\Phi(t), \frac{1}{x} \cdot \frac{dx}{dt} = \beta\Phi(t)$$

leading to

$$d \log y = \frac{\alpha}{\beta} d \log x$$

which is solved to produce the Huxley allometry equation with

$$a = \alpha/\beta$$

Skellam, Brian, and Proctor (1959) found that in *Myrmica rubra* the imaginal discs of the wings and legs actually do grow in this simple, relative fashion during adult development of queens and males. As a result these organs predictably conform to the Huxley allometry equation in the final adult stage.

Wilson (1953) established that the Huxley equation applies almost universally to continuous variation in worker hard parts throughout the major ant groups. The comparative study of allometry has since proven fruitful in the tracing of the evolution of castes. Most studies have concentrated on the parts of the body that show the greatest degree of allometry: the head in the case of variation within the worker caste, and the alitrunk in the case of ergatogynes. Van Boven (1958) and Hollingsworth (1960) have undertaken comparative analyses of allometric trends in multiple body dimensions in species of *Anomma, Lasius,* and *Formica.*

Polymorphism, as this research has now led us to understand it, can be interpreted as a function of two variable characters of the adult females of any species: allometric growth series and intracolonial size variation. In its most primitive and indistinct stages, relative growth, almost always allometric, is the key quality to look for, and in 1953 I suggested the following definition: *polymorphism is the occurrence of nonisometric relative growth occurring over a sufficient range of size variation within a normal mature colony to produce individuals of distinctly different proportions at the extremes of the size range.*

All degrees of allometry within the worker caste of single colonies can be demonstrated in one ant species or another, grading downward almost insensibly into absolute monomorphism. Each of the most prolific students of ant taxonomy, notably Emery, Forel, and Wheeler, had his own idea of approximately where monomorphism stopped and polymorphism started, with the result that all rarely agreed on the status of feebly polymorphic species, and each tended to judge the matter according to how carefully he had examined the material before him. Within the gradient of increasingly allometric species, it is impossible to draw a precise limit of polymorphism that will at the same time agree with previous usage and remain consistent throughout. It seems necessary to set as a minimum standard the possession of detectable intranidal allometry. This loose criterion is reasonably consistent with past usage; in many species which are generally considered to be polymorphic, such as *Lasius fuliginosus* and many of the species of *Dorylus* and *Anomma,* the allometry is very slight, but it is combined with intranidal size variation great enough to develop it plainly in normal, mature colonies.

Besides nonisometric relative growth, there is another important general feature of caste variation. Some myrmecologists, including Falconer Smith (1942), Cole and Jones (1948), and Haskins and Haskins (1950a) have stressed the multimodality in the size-frequency curves, implying that castes, or at least subcastes, might be defined as modes in the curves. In other words, as workers in certain size groups become more numerous and those in intermediate size groups less numerous, the more frequent size groups, each clustered about a mode in the frequency curve, might be distinguishable as a caste. This is partly true. The measurements of Wilson (1953) show that bimodality in the worker size-frequency curve, emerging in evolution from a skewed unimodal condition, follows closely upon the development of intranidal allometry and is correlated with later changes in the allometric regression curve. Trends toward the development of more than two modes are rare, however, and many strongly polymorphic species are still unimodal. Let us now examine the joint evolutionary changes in allometry and size-frequency curves that have brought about the emergence of subcastes within the worker caste. Five steps can be recognized:

1. *Monomorphism.* The workers of the normal mature colony are either isometric or show limited size variability, or both, and their frequency distribution is unimodal. A typical example is illustrated in Figure 8-3. The worker castes of most ant genera and species are monomorphic, and, in the majority of the cases, monomorphism represents the primitive state. In a few groups—the myrmicine genera *Carebara, Carebarella,* and *Tranopelta*—monomorphism appears to be secondary, however. Phylogenetically, it has succeeded some development of polymorphism and was produced by a dropping out of a large

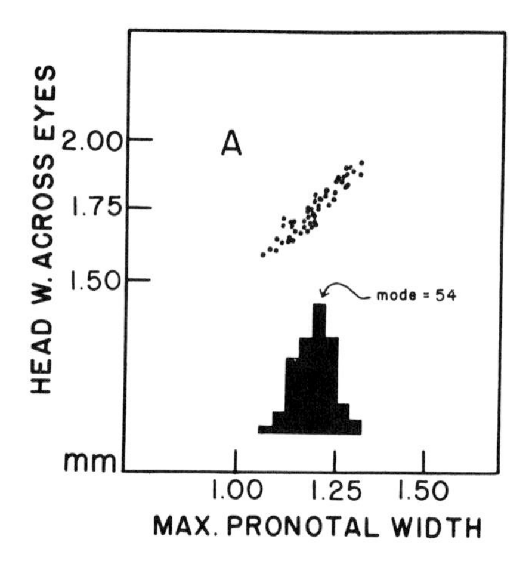

FIGURE 8-3. The worker caste of the formicine ant species *Formica exsectoides* exhibits monomorphism of the primitive type. The body parts are all isometric with reference to one another, or nearly so, as exemplified here by the relation of head width to pronotal width. Also, the size-frequency curve (displayed as the blacked-in histogram) is unimodal (from Wilson, 1953).

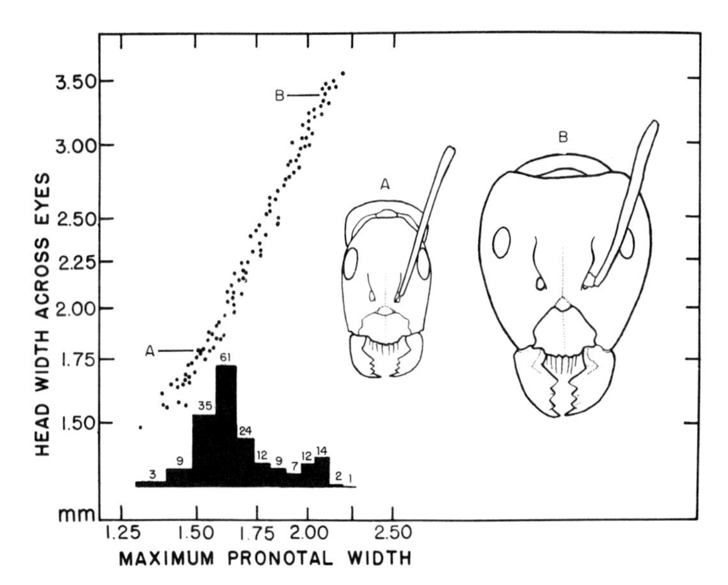

FIGURE 8-4. The workers of the formicine *Camponotus castaneus* show a typical elementary form of polymorphism. Some of the body parts are allometric; in other words, they increase or decrease in relative size with an increase in total body size. Head width, in the example given here, increases faster than pronotal width, while pronotal width is isometric with reference to most of the rest of the body. The allometry is "monophasic," that is, the slope of the relative growth curve remains constant. The size-frequency curve is incipiently bimodal (from Wilson, 1953).

part of the upper segment of the size variation. This results in a caste that is even more uniform than the workers of related, primarily monomorphic species.

2. *Monophasic allometry.* The allometric regression line has a single slope which is greater or less than unity (nonisometric). The example from *Camponotus* illustrated in Figure 8-4 is typical. In the most elementary form of monophasic allometry, and hence of worker polymorphism generally, feeble nonisometric variation is displayed over a short range of size variation, which in turn shows a unimodal frequency. This is a common condition in ants and is exemplified in such diverse forms as the army ant *Neivamyrmex nigrescens* and the formicines *Formica obscuripes* and *Lasius fuliginosus.* An apparently later development in this stage of worker polymorphism involves an increased dispersion of the frequency curve for size and a marked tendency toward bimodality. This is the typical condition in many genera that are often cited in the literature as showing elementary polymorphism, including *Megaponera, Orectognathus, Azteca,* and *Camponotus.* Within *Camponotus,* at least, polymorphism has been demonstrated to be correlated with intracolonial division of labor. Of the various manifestations of nonisometric relative growth in the worker caste, the simple monophasic form is the most common.

3. *Diphasic allometry.* The allometric regression line, when plotted on a double logarithmic scale, "breaks" and consists of two segments of different slopes meeting at an intermediate point. The polymorphism of the fungus-grower genus *Atta* provides a striking example (Figure 8-5). Another well-analyzed case is the driver ant *Anomma nigricans* (Hollingsworth, 1960). In the several species where this condition has been demonstrated, the size-frequency curve is bimodal, and the low point between the two frequency modes falls just above the critical point of the break. Diphasic allometry appears to be a mechanism allowing the stabilization of the body form in a small caste, while at the same time providing for the production of a markedly different soldier caste over a slight size increase. The lower segment of the allometric regression line is nearly isometric, so that individuals falling within a large portion of the size range tend to be uniform in structure, but the upper segment is strongly allometric, with the result that a modest increase in size yields a new morphological type.

4. *Triphasic allometry.* The allometric regression line "breaks" at two points and consists of three segments; the two terminal segments have slight to moderately high slopes, and the middle segment has a very high slope. The effect of triphasic allometry is the stabilization of body proportions in both the major (soldier) and minor castes. It is conceivable that in the course of evolution this condition can succeed diphasic allometry, which first stabilizes the minor caste, or it can emerge directly from monophasic allometry. Triphasic allometry is exemplified in at least three living species, the myrmicine *Pheidole rhea* and the formicines *Camponotus floridanus* and *Oecophylla smaragdina,* the last of which is illustrated in Figure 8-6. All three approach complete dimorphism, with very few intermediates in a given nest series connecting the curves about the two modes. In each case the trough between the frequency modes occupied by medias corresponds almost exactly to the middle segment of the allometric regression line, as shown in Figure 8-6. The medias are not only rare, but are more variable in proportions than the minors and soldiers. Yet one cannot conclude from

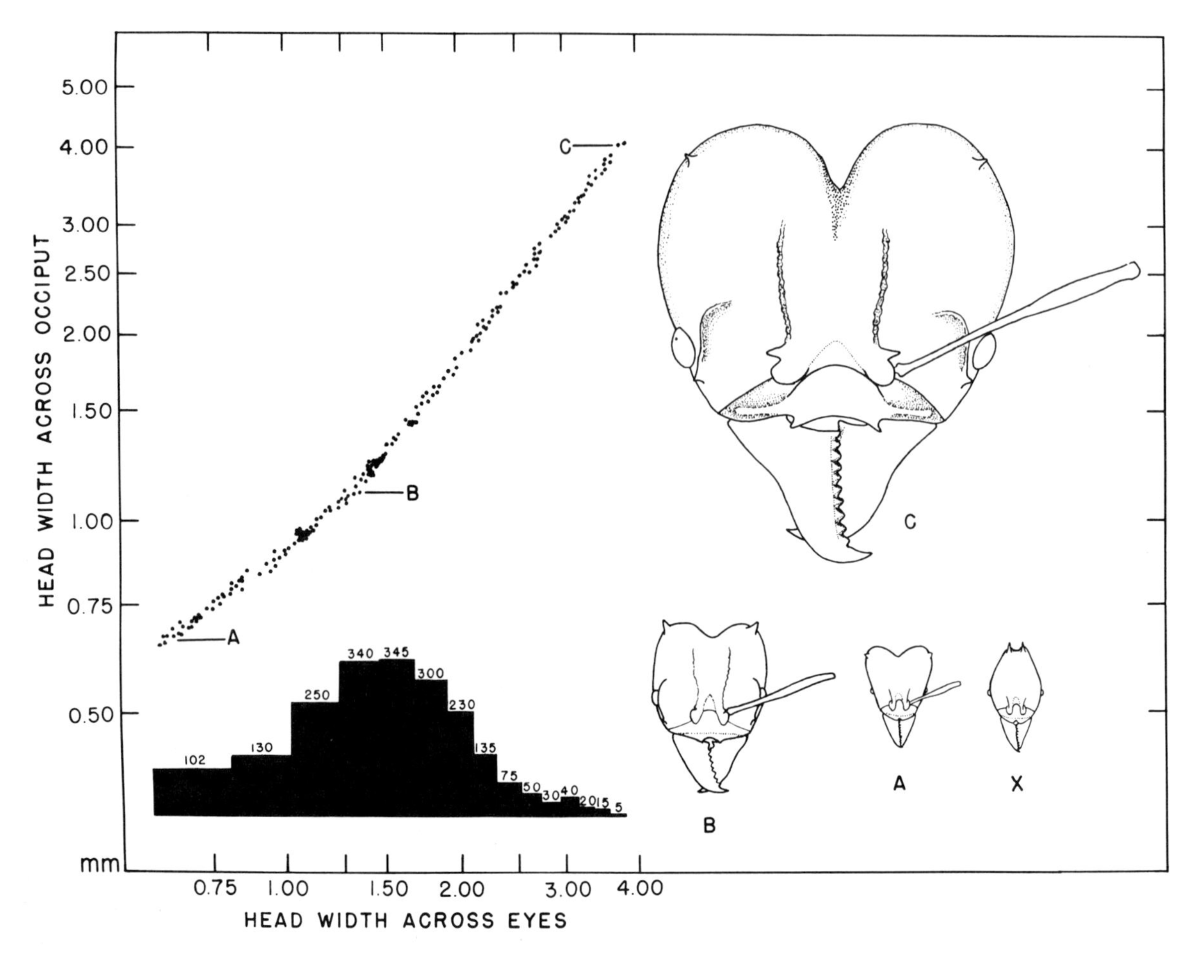

FIGURE 8-5. Workers of the fungus-growing ant *Atta texana* show a polymorphism somewhat advanced over *Camponotus castaneus* (see Figure 8-4). Here the allometry is diphasic, meaning that the relative growth curve changes slope (at point *B*). This change is an adjustment to accommodate the great size variation shown by the species. If the slope of the lower segment of the curve were as high as that of the upper segment, the smallest workers would have the monstrous head shape (*X*) in the lower right hand corner of the figure (from Wilson, 1953).

this fact that triphasic allometry is itself unstable or in any sense a brief stepping stone on the road to complete dimorphism. Wilson and Taylor (1964) have found nearly identical allometry and frequency patterns in a colony of the Miocene ant *Oecophylla leakeyi* (Figure 8-7). This colony, which was preserved as a unit in East Africa and is the only fossil colonial assemblage of any kind in insects that has yet come to light, shows that the distinctive *Oecophylla* morphology and caste pattern have persisted with no major change for a period of over 10 million years.

5. *Complete dimorphism.* Two separate size classes exist, separated by a gap in which no intermediates occur. Each class is nearly isometric, but the allometric regression curves are not aligned, which strongly suggests that this condition can arise directly from triphasic allometry. Examples include most queen-worker divisions and a great many major-minor divisions in a diverse array of genera in the higher subfamilies. Complete dimorphism has clearly evolved independently at least seven times, or, more specifically, at least once in each of the following seven genera: the myrmicines *Pheidole, Oligomyrmex, Acanthomyrmex, Paracryptocerus;* the dolichoderine

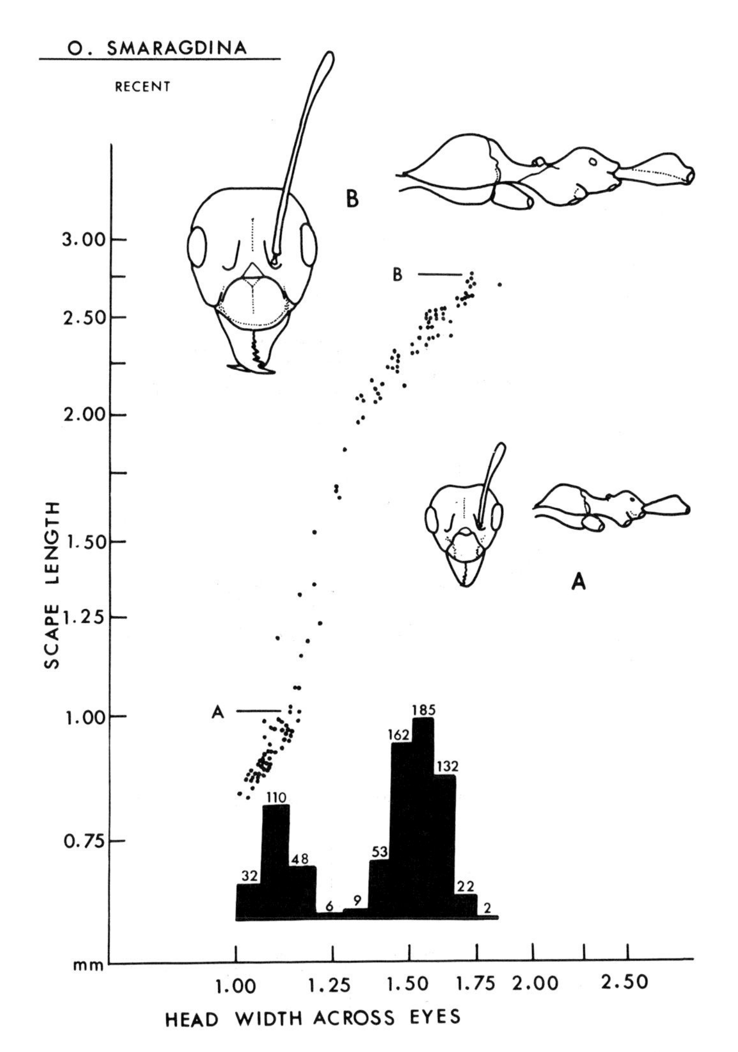

FIGURE 8-6. In *Oecophylla smaragdina* the allometry is triphasic, that is, three different slopes are shown by such allometric characters as scape length taken as a function of body size (in this case represented by head width). The size-frequency curve is bimodal, with majors predominating. The heads, mesosomas, and petioles of selected minor and major workers are also shown (from Wilson, 1953).

Zatapinoma; and the formicines *Camponotus* (subgenera *Tanaemyrmex* and *Colobopsis*) and *Pseudolasius.* Once complete dimorphism is established, the two resultant castes are capable of diverging further in some parts of the body, with no evident intergradient allometry. Profound changes may then ensue in the appearance or size of one or both of the castes, so that their original relationship might never be suspected were earlier phylogenetic stages not still exhibited by other species in the same genus or subfamily. In a few phyletic lines the heads of the soldiers come to surpass in size those of the queens (for example, *Pheidole, Messor,* and *Camponotus*), while in at least one other line the queen has regressed in total size to become smaller than the average worker (*Formica microgyna* group). *Acanthomyrmex* contains species in which some nonallometric parts of the soldier are equal in size to those of the minor worker, or smaller. It seems to be a rule among ants that once the secondary divergence of completely segregated castes becomes profound, the physiological thresholds separating them are no longer capable of being bridged to produce viable intercastes. Thus intergrades between soldiers and queens, between minor workers and queens, and even between strongly divergent soldiers and minor workers, are relatively rare. Soldier-queen intergrades have been reported in the genus *Oligomyrmex* (Kusnezov, 1951a), and soldier-minor

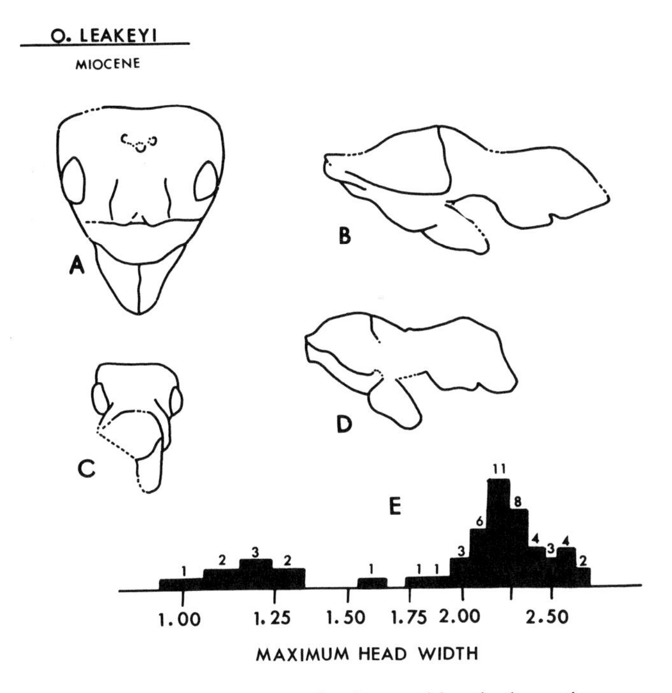

FIGURE 8-7. The pattern of polymorphism is shown in a fossil colony of *Oecophylla leakeyi* from the Miocene of Kenya. A comparison with the pattern of the living *O. smaragdina* presented in Figure 8-6 demonstrates that the same aberrant caste system has persisted in *Oecophylla* for at least 10 million years (from Wilson and Taylor, 1964).

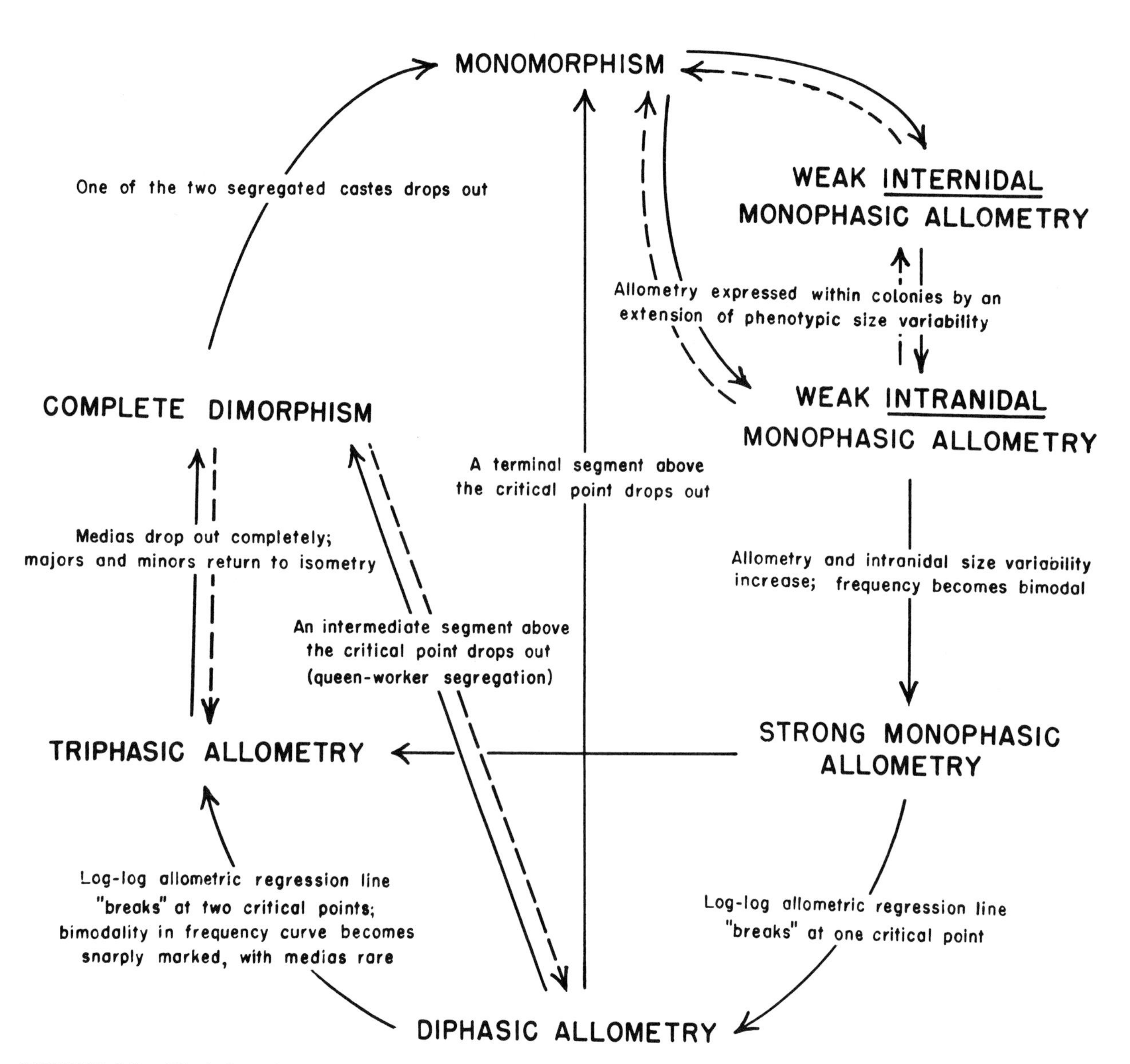

FIGURE 8-8. The inferred pathways in the evolution of polymorphism in ants (from Wilson, 1953).

worker intermediates are found naturally in some species of *Pheidole* (see Figure 8-1). Another general feature of complete dimorphism is that it can be expressed in one character, but not in another. In the queen-worker divergence, for example, the alitrunk becomes dimorphic at a phylogenetically early stage, but recurrent allometric trends in the head may continue to link the two castes by a monophasic regression line.

Other characteristics besides the measurable hard parts are subject to dimorphism. In their study of *Pheidole fallax,* Law *et al.* (1965) have shown just how profound and extensive the divergence can be. The soldier of *P. fallax,* like that of all species in the genus, is radically different in external proportions and sculpturing from the minor worker. Within its abdomen, the soldier has a hypertrophied poison vesicle containing skatole, a fetid liquid employed in chemical defense. Dufour's gland is greatly atrophied or even missing; and, as a result, the soldier cannot lay odor trails. The minor worker, on the other hand, has a normal-sized poison vesicle, which does not contain skatole, and a hypertrophied Dufour's gland, which serves as the source of the odor trail. Outside the nest the soldier is concerned mostly with defense, in which it employs skatole, while only the minor worker lays odor trails.

From species to species through the ants the worker caste displays virtually every conceivable step in a transition from complete monomorphism to complete dimorphism. Certain taxonomic groups embrace large parts of the evolutionary sequence by themselves. For example, various species of the tribe Ponerini display all nuances from monomorphism to strong monophasic allometry. A few genera related to *Pheidole* contain nearly every stage from monomorphism to dimorphism. Comparative studies of groups such as these have led to the schema shown in Figure 8-8.

While the evolutionary pathways of polymorphism within the worker caste can be worked out with some confidence, we can only guess at the manner in which the worker caste originated from a queen-like form in the first place. The reason is that all ant taxa, even the primitive Amblyoponini and Myrmeciinae, display a sharp queen-worker dimorphism as a primitive character. The pre-formicids that traversed the early stages of worker evolution are apparently all extinct and still unknown in the fossil state. Occasionally, however, the queen-worker gap within the ants breaks down secondarily, and we are permitted a glimpse of the innate allometric trends that still separate the two castes. One such example is shown in Figure 8-9. It can be seen that this curve is diphasic. It resembles in important respects the diphasic curve joining the worker subcastes of *Atta texana* (Figure 8-5); namely, the lower segment has a slope of about unity, yielding a uniform worker caste, and the upper segment is steeper, yielding the fully formed queen caste over a small increase in total size.

Although queen-worker dimorphism is a primitive and nearly universal trait in the ants, a great deal of variation

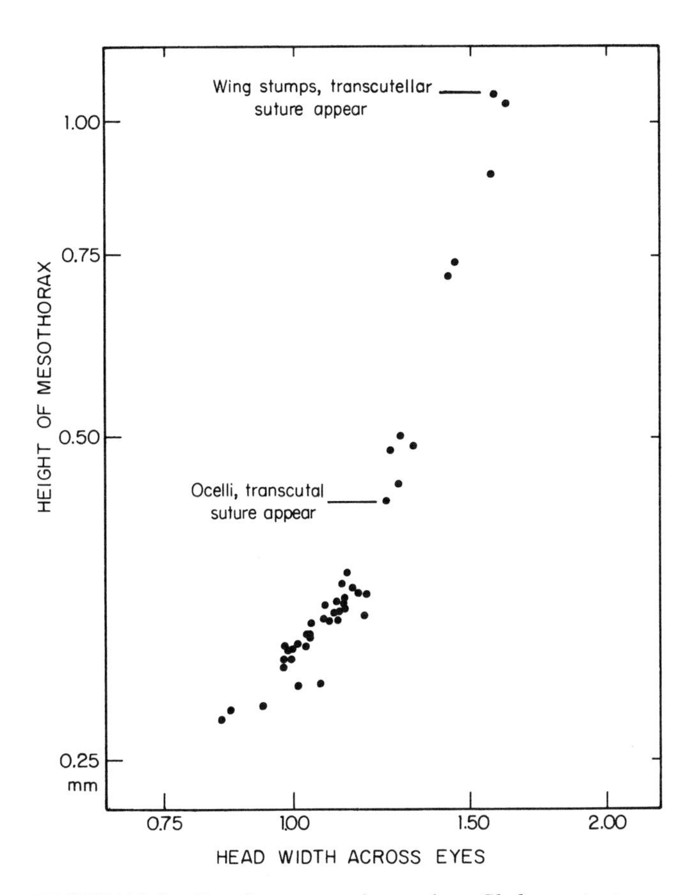

FIGURE 8-9. In a few ant species, such as *Chelaner cinctus* represented here, intergrades between the queen and worker castes have evolved secondarily. The intermediates ("ergatogynes") display diphasic allometry of a kind similar to that differentiating worker subcastes in some ant species. The phenomenon lends support to the idea that the original worker caste in primitive ants could have originated in the same fashion as the later worker subcastes (from Wilson, 1953).

occurs in the degree of separation of the two castes. In general, the most advanced ant taxa manifest the strongest difference. In the primitive genus *Myrmecia,* the queens are not much larger than the largest workers and are morphologically nearly identical to them except for their larger ovaries, possession of wings, and great development of the meso- and metathorax. Intercastes are common in certain species in the genus and, with respect to the alitrunk, fall along a steep incline of the allometric curve connecting the two castes. Queens and workers also closely resemble each other in most species of the primitive subfamily Ponerinae and in certain primitive genera of higher subfamilies, for example, *Hylomyrma* and *Myrmica* in the Myrmicinae, *Dolichoderus s. str.* in the Dolichoderinae, and *Myrmoteras* in the Formicinae. The two castes are most different in groups possessing one of three kinds of adaptations: the queen has transformed into a dichthadiiform ergatogyne as an adaptation to legionary life (see Chapter 4); the queen has become a social parasite (see Chapter 19); or the worker caste has itself undergone an extreme modification, usually in connection with the evolution of a distinct soldier caste. In extreme cases it is difficult or impossible for the taxonomist to place the worker and queen castes in the same species until they are found associated in the same nest. The tiny, yellow, cryptobiotic workers of *Carebara* and related genera, for example, are radically different from their huge, pigmented queens. In fact, the workers of *Carebara* are so much smaller that some catch a ride with each queen on her nuptial flight (Figure 8-10). The queen enters a termite nest after the nuptial flight and is presumably assisted by her little sisters in the rearing of the first brood. To what extent the workers secure food by raiding the host termite colony is not known. The queens of *Rhoptromyrmex,* which are believed to be temporary social parasites of other myrmicines, are so modified in basic external morphological characters that a taxonomist, unaware of their true identity, would probably place them in a different genus from the workers (Brown, 1964b).

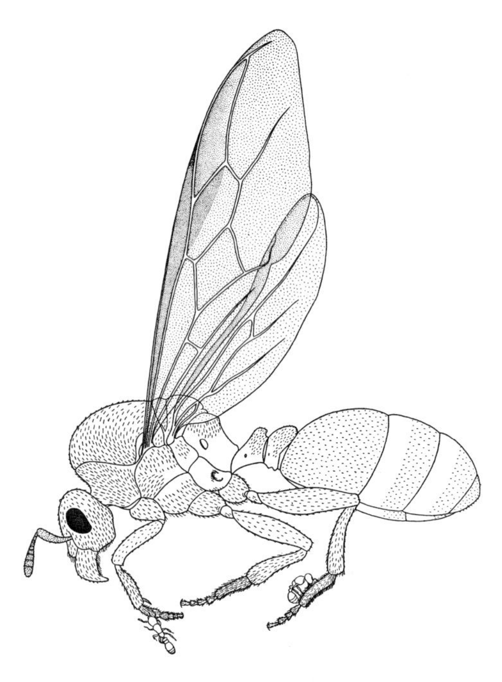

FIGURE 8-10. A winged, recently fecundated queen of the African ant *Carebara vidua,* carrying two minute, blind workers attached to her tarsal hairs (from Wheeler, 1936a).

Determination of the Female Castes

The polymorphism of a given species is a pattern of variation that was shaped by evolution and is displayed by each colony in the course of its life cycle. We may now ask what it is that determines the position of the individual insect within the pattern; in short, what determines caste? This problem is important not only with reference to social insects. It has also figured prominently in the history of biology, and it is best introduced with a brief chronology of ideas.

Prior to about 1950 the problem seemed to resolve itself into a set of two simple alternatives. The question was posed essentially as follows: is caste determined genetically, that is, does each caste arise as a response to a particular set of genes, or is a given caste merely a phase of a very plastic phenotype controlled by a single set of genes? If the latter, phenotypic explanation is true, then determination can occur either in the embryo located within the egg (*blastogenesis* in the modern sense) or else in the larva. If determination occurs in the larva, it seems likely that nutrition must play a role; hence, the expression *trophogenesis* has been used almost synonymously with larval determination. Finally, if control is blastogenetic, it is still possible that nutrition, in the form of yolk or other ovarian nutrient material supplied the embryo, might be crucial in some way.

These alternatives first began to take shape in Darwin's *Origin of Species.* His theory of natural selection led him to propound two novel ideas: nongenetic caste determination and evolution by colony selection. Darwin had reason to be deeply concerned with the problem of castes in ants. He regarded their existence as the greatest challenge to his theory of natural selection. In developing the argument that natural selection does indeed serve as the main evolutionary mechanism in the origin of castes, he first observed that Lamarckian evolution, or evolution by means of the inheritance of acquired characteristics, could not possibly be involved. "For no amount of exercise, or habit, or volition, in the utterly sterile members of a community could possibly have affected the structure or instincts of the fertile members, which alone leave descendants. I am surprised no one has advanced this demonstrative case of neuter insects, against the well-known doctrine of Lamarck." Darwin proposed that selection acts on the fertile parents alone and that the sterile worker castes are to be regarded as no more than highly plastic expressions of the parents' own genetic structure. The quality specifically subject to selection is the parents' ability to produce a variable, sterile, and efficient set of offspring which promote their personal survival and thereby enhance their ability to contribute to the next generation of fertile offspring. In other words, selection occurs at the colony level since the genotype can be transmitted only by fertile individuals. (Colony selection will be treated more fully in Chapters 17 and 18.) It seems to follow that the caste of individual social insects must be environmentally determined.

The genetic implications of ant castes were reexamined during the debate over evolutionary theory conducted by Herbert Spencer and August Weismann in the pages of the *Contemporary Review*, from February 1893 to October 1894. Weismann, in order to make his own theory of the germ line wholly consistent, argued that each caste is controlled by a different set of genetic "determinants." Every member of the colony, he reasoned, has every set of determinants, but only one is activated in a given individual. Thus was born a genetic hypothesis, or a theory of "blastogenesis" as it was often called in the days before the clear distinction was made between true genetic and embryonic determination. Spencer believed, on the other hand, that workers are simply individuals rendered dwarf and sterile by starvation during their larval period. Emery (1894) adopted a similar view, but one refined by the notion that the basic castes are determined by differences in quality of food while subcastes in continuous series are determined by food quantity. The Spencer-Emery thesis was the beginning of the "trophogenesis" theory. This form of the dichotomy between blastogenesis and trophogenesis now seems curiously antiquated and ineffective. Indeed, as Wheeler pointed out in 1907, the Weismann hypothesis is no choice at all but only a "mirror" of the original problem. Nevertheless, the dichotomy dominated thinking on the subject for the next sixty years.

Empirical evidence was slow in coming. The first experiments were few in number and usually lacked controls. Even so, the results obtained are mostly consistent with our present understanding of caste determination. J. H. Hart, quoted by Herbert Spencer in 1894, reported that when queens of fungus-growing ants were removed, new queens developed in the colonies. Similar results were reported by Emery (1921) for *Aphaenogaster testaceopilosa* and *Pheidole pallidula.* In later, better-designed experiments, Gregg (1942) discovered that soldiers of *Pheidole morrisi* tend to inhibit transformation of larvae into their own caste, a result paralleling the well-known inhibition phenomenon in lower termites. Gregg's data do not disclose whether the effect is due to an inhibitory pheromone, as in the termites, or to the simple ineffectiveness of soldiers as nurses permitting them to rear only minor workers. But they do seem to prove that soldier determination is phenotypic rather than genotypic.

Ezhikov (1934) reported that some larvae of *Camponotus* transformed to ergatogynes when starved. Goetsch (1937a) obtained some soldiers from *Pheidole pallidula* larvae fed on flies, but none from a similar group which was fed only sugar or honey. In another, suggestive experiment Goetsch introduced eggs from colony-founding *P. pallidula* queens to larger, older colonies where normal-sized workers were being reared. Despite the richer environment, dwarf-sized minor workers typical of first brood were obtained. This appears to be the first demonstration of some form of ovarian, or truly blastogenic influence.

Wesson (1940a) invented a technique, later employed routinely by Bier and Gösswald in their work on *Formica,* for tracing individual larvae during development in a normal colony. He introduced larvae of the yellow *Leptothorax curvispinosus* into colonies of the black *L. longispinosus* lacking a queen and brood. This permitted a sure distinction between the experimental workers and their hosts. When the *curvispinosus* larvae were fed an excess of protein-rich food through their hosts, they produced a higher percentage of queens than those kept on minimal growth diets. Moreover, larvae kept dormant for five months at 2°C were capable of transforming into queens, whereas those reared straight through in the summer months were not able to do so, no matter how rich their diet. This was the first demonstration of a vernalizing factor in the caste determination of social insects.

At about the same time, Wheeler (1937) became convinced that castes are genetically determined. He was converted to this point of view by the discovery of "gynergates," or individuals containing patches of tissue of both the queen and worker castes. One such anomaly in *Myrmecia* had earlier been described by Tulloch (1932). Then N. A. Weber discovered 46 gynergates in a colony of *Acromyrmex octospinosus,* and an even larger number in a colony of *Cephalotes atratus.* In analyzing the *Acromyrmex* gynergates submitted to him by Weber, Wheeler (1937) found that they had normal soldier bodies with heads seemingly divided into soldier and queen components. He concluded that they were genetic mosaics no different in type from true gynandromorphs, and that female castes must therefore in some way be genetically determined. In the year following, Whiting (1938) offered an alternative explanation which is the one generally accepted today. He pointed out that the *Acromyrmex* mosaics were quite different from true gynandromorphs as recognized in other Hymenoptera. The ant mosaic patterns were all limited to the head and more or less regularly disposed, while in true gynandromorphs the male and female components vary in extent and are distributed at random over the entire body. Whiting proposed that Wheeler's mosaics were likely to be intersexes of the type described in the moth *Porthetria dispar,* and that the head alone was affected because this was the only part not past the initial threshold when the transition from femaleness to maleness began. It follows that the intercaste condition of the head results secondarily from the shift of the entire sex toward maleness; in other words, the sex shift pulls the head away from the worker state and toward queenness.

Ledoux (1950) presented a simple but conclusive demonstration of the nongenetic basis of female castes in the weaver-ant *Oecophylla longinoda.* He showed that parthenogenetic, diploid eggs laid by workers can produce either queens or workers, and in addition the determination appears to occur within the first few days of larval life. Older, queen-determined larvae are white in color, sac-like in shape, and have a mobile head separated from the body by a cuticular fold. Indirect evidence from Ledoux's work suggests that the presence of a functional queen tends to prevent young larvae from taking this course of development.

Thus, by 1950 a modest amount of evidence had accumulated to indicate that the female castes are determined by environmental rather than genetic factors. Further, it was already apparent that although larval nutrition is important, other kinds of influences are involved. The need for detailed experimental studies on the whole life cycle of selected ant species has been met in part by the work of M. V. Brian and his associates in England on *Myrmica,* and that of K. Bier and K. Gösswald in Germany on *Formica.* An evaluation of this work follows.

The Genus Myrmica. Some confusion surrounds the identity of the species studied by Brian because of nomenclatural changes from the time of his first report (1951) to the present. The changes have been required due to taxonomic revisions of the British members of the genus by Yarrow (1955a) and Collingwood (1958).

The various names are related as follows:

Names used by Brian	Taxonomically valid names
M. laevinodis Nylander	*M. rubra* L.
M. rubra var. *macrogyna* Brian	*M. ruginodis* Nylander
M. rubra var. *microgyna* Brian	*M. ruginodis* Nylander

M. rubra and *M. ruginodis* are distinct species, and Brian has employed them both in his various studies on ecology and physiology. The names *macrogyna* and *microgyna* are taxonomically invalid. They probably apply to distinct sibling species, but the published evidence leaves this open to some question (Brian and Brian, 1949, 1955). The principal form used in the caste studies is *"macrogyna,"* the correct species name for which is *M. ruginodis.*

M. ruginodis, like most north temperate myrmicines, overwinters with its brood in the larval stage. In most mature colonies the following spring, some of the female larvae develop into queens and others into workers. The annual cycle is illustrated in simplified form in Figure 8-11. As the "slow" larvae, hatched in early summer, reach a certain point in their development in the late summer or fall, they become dormant. This crucial juncture is in

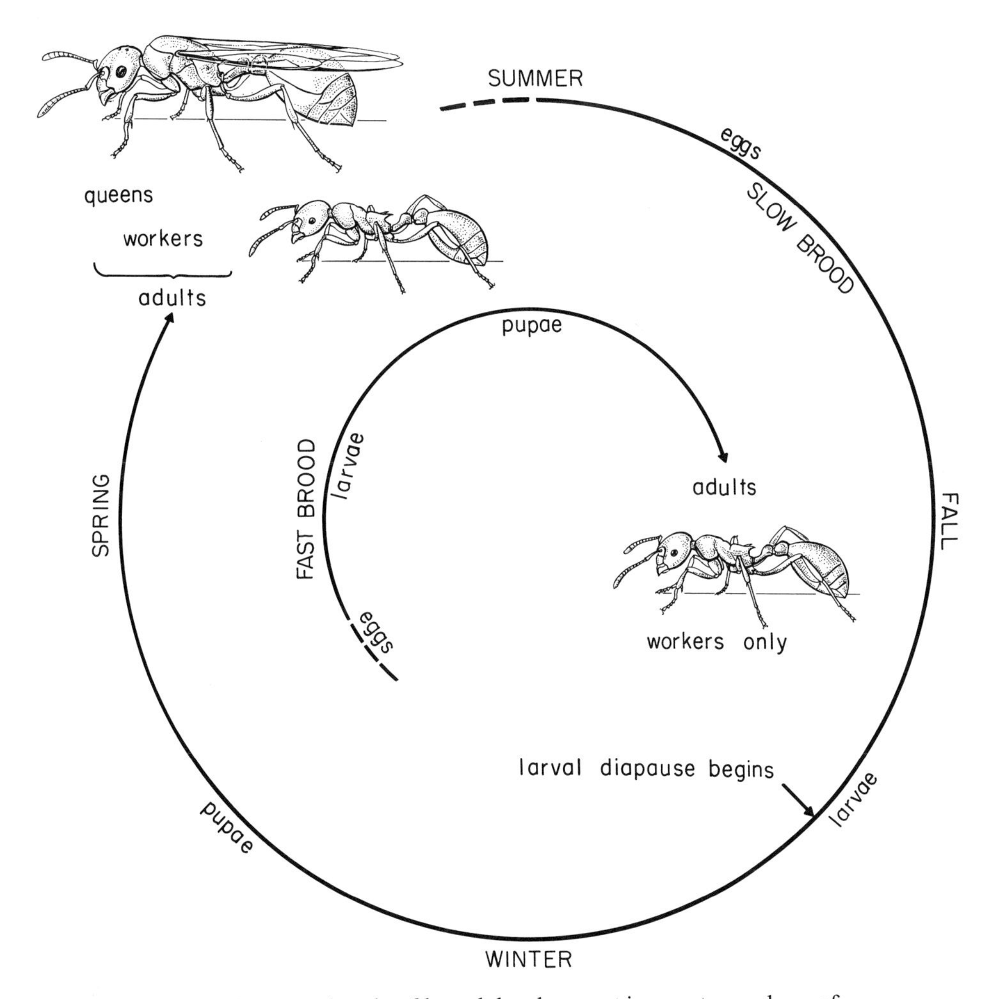

FIGURE 8-11. The annual cycle of brood development in a mature colony of *Myrmica ruginodis.* The mother queen continues to lay eggs intermittently through the spring and summer. Many of the larvae which hatch early in the season are able to complete development by the end of the summer and become workers (fast brood). Others persist as larvae through the winter and can become workers or queens the following spring (slow brood). The full development of fast brood requires about three months: that of slow brood, almost a year.

the terminal (third) instar. More precisely, the larvae are halted at a stage of prepupal development when the brain has migrated halfway back into the larval prothorax. Many larvae do not proceed beyond this 0.5 point until the following spring, a condition referred to as "primary diapause." Others are able to go on until the brain is 0.8 into the prothorax, but they always halt there in a condition of "secondary diapause." Still other, younger larvae do not reach the 0.5 stage at all before cold weather; these pass the winter in a nondiapause state. In any case, all the brood overwinters in the larval stage, and it is from some of these vernalized larvae that the yearly crop of queens is matured in the spring.

Weir (1959b) proved that fall ("serotinal") workers tend to induce dormancy in terminal instar larvae, whereas spring ("vernal") workers cannot. This was achieved by keeping workers in the laboratory at the warm temperature of 25°C for 11 weeks after lengthy chilling to simulate in them the physiological state of wild fall workers, and by keeping other workers at the same temperature for only 5 weeks after chilling to simulate the wild spring condition. The "fall" workers could induce dormancy; the "spring" workers could not. Weir further guessed that diet might be the key, since *Myrmica* workers are known to increase the proportion of protein in their diet as the season progresses, and dormant larvae have a higher nitrogen-to-carbon content in their meconia and fat bodies than do nondormant larvae. When Weir fed spring workers a sufficiently increased amount of protein (by means of a pure diet of *Drosophila*), it turned out that they too were able to induce dormancy. Weir, after considering the matter at length, remained uncertain whether the larval dormancy is true diapause in the sense that it is a shutdown mediated by the endocrine system. But it at least resembles diapause in that, once initiated, it is persistent in its effect, even at raised temperatures.

Dormancy, as pointed out by Brian (1956a), has the effect of holding over until winter all of the late brood in the larval stage. The dormant state can be broken by sustained high temperatures and handling by spring workers. Both of course are normally encountered by larvae in wild colonies in the spring. The important point is that chilling in winter temperatures confers on the larvae the capacity to sustain a high growth rate in critical periods of the final instar and, as a result, to metamorphose into queens. Dormancy itself is not a prerequisite. Some small larvae, as already mentioned, are immobilized by winter cold before they enter stage 0.5 dormancy; yet the following spring they too have the capacity to transform into queens.

Final caste determination occurs in the spring larvae late in their terminal instar. In order to chronicle this event, Brian used several arbitrary morphological changes in pupal development that occur in sequence. These changes, together with the occurrence of plastic periods in caste determination and the approximate time scale, are summarized in Figure 8-12.

The critical periods of caste determination were revealed by experiments in which sets of larvae were starved at different periods in pupal development. Of course, larvae are not normally steered into one path of development or another by any such regimen of sudden starvation or overfeeding. In order to learn about the natural course of determination, it was necessary to follow the development of many larvae being reared individually under relatively undisturbed conditions. When this was done, a clearer picture of the role of the growth rates was obtained. Brian learned first that the time required for spring larvae to develop into worker pupae did not differ from the time required for development into queen pupae. In both cases duration of development ranged from 9 to 21 days and averaged about 13 days. But the final weight attained by larvae transforming into the two castes differed greatly, averaging about 4.5 mg in the worker and 8 mg in the queen. Clearly the larvae destined to be queens either must grow at a faster rate or else start at a higher weight. And either can be the case, as shown in Figure 8-13. In general, queens come from spring larvae

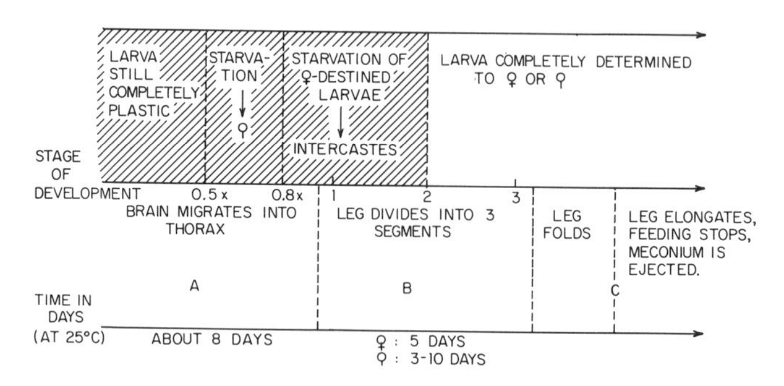

FIGURE 8-12. The plastic periods (crosshatched) of caste determination in the early pupal development in female *Myrmica ruginodis:* ♀ represents the worker caste; ♀, the queen caste (based on data of Brian, 1956a, and contained references).

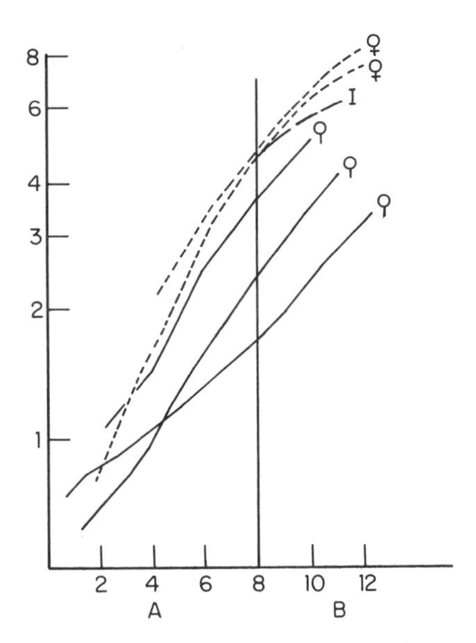

FIGURE 8-13. Typical growth curves of *M. ruginodis* larvae that transformed respectively into queens (♀), intercastes (I), and workers (♀). The time scale was arranged so that the day on which the larva passed from the *A* period to the *B* period (see Figure 8-12) is arbitrarily labeled as the eighth day (modified from Brian, 1955a.)

that either start relatively large and maintain a moderate to high growth rate or else start at a medium weight and maintain a consistently high growth rate. For the most part larvae that have failed to reach a weight of about 3.5 mg by stage A-0.8 are destined to become workers.

An important finding from Brian's studies is that caste determination in *M. ruginodis* is essentially worker determination, or the failure of this event to occur. This means that the worker is an individual diverted from a normal female (that is, queen) course of development by having part of its adult system shut down. For convenience, the imaginal discs of a larva can be divided into a dorsal set, containing wing buds, gonad rudiments, and ocellar buds, and a ventral set, containing leg buds, mouthpart buds, and central nervous system (Brian, 1957c, 1965a). In the case of queen development, leading to a more or less typical hymenopteran female, both sets maintain full growth and development at the onset of pupal development. But in the case of larvae destined to become workers, the dorsal set stops growth and development for the most part, and only the ventral set continues on. The abruptness of the shutdown is quite striking, as shown in Figure 8-14. It is interesting that the dorsal organs do not always shut down together. One of the anomalous "castes" described by Wheeler (1905) were the pterergates, or otherwise normal workers bearing external vestiges of wings. One worker of *Myrmica scabrinodis* figured by Wheeler had wings as long as the thorax itself.

Growth studies, including the starvation experiments, disclosed the important role of chilling in developing the queen potential larvae in larvae, and of nutrition in leading *Myrmica* larvae to growth beyond the thresholds required to sustain development as queens. Subsequent experimentation has revealed the existence of at least four additional factors influencing caste determination in the closely related species *M. rubra* (Brian, 1963; Brian and Hibble, 1964). First, an increase in temperature from 22°C to 24°C, that is, from the optimal temperature for larval survival to slightly above, results in an eightfold increase in the proportion of larvae metamorphosing into workers. Second, the presence of nest queens results in a fourfold increase in the proportion of workers. The latter effect is caused at least in part by a change in the behavior of the workers evoked by the perception of queen pheromones.

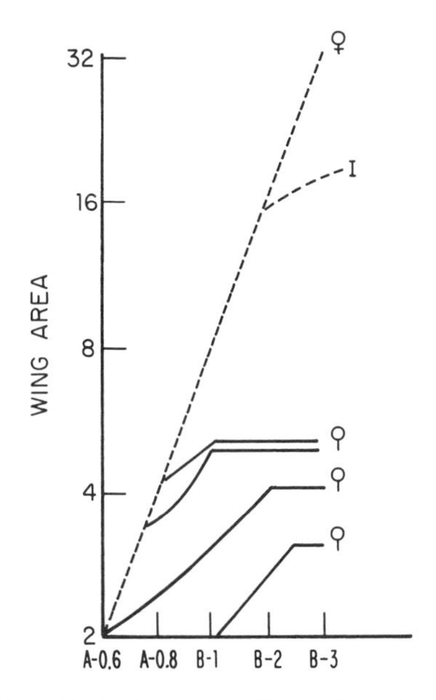

FIGURE 8-14. Typical patterns of growth of the wing bud (imaginal disc) during pupal development of female *Myrmica ruginodis* larvae. Determination to the worker caste (♀) or to an intercaste condition (I) is characterized by a sharp decrease in the growth of this organ. The maturity series is that given in Figure 8-12 (modified from Brian, 1955a).

When two sizes of larvae are presented to adult workers in the presence of a queen, the small larvae are fed more and the larger larvae less than in the case of control groups lacking a queen. Also, the larger larvae are bitten and licked more in the presence of the queen, presumably lessening their chances of survival (Brian and Hibble, 1963). Similar changes are obtained when dead queens are presented daily to queenless colonies (Cecily Carr, 1962). Also, when the sterol fraction of extracts of the heads of queens are fed to larvae or applied topically, larval growth is inhibited (Brian and Blum, 1969).

A third influence is blastogenic. Small eggs laid during periods of most rapid oviposition yield higher numbers of workers. When queens are allowed to emerge from hibernation at 20°C, the rate of oviposition rises during the first three weeks to a maximum that persists through the following three to four weeks. Then the rate gradually declines toward zero, reaching a very low level after about the sixteenth week. Simultaneously the size of the eggs changes, declining rapidly in the first three weeks and then remaining approximately constant. These relations are illustrated in Figure 8-15. Eggs laid during the first three weeks show a greater capacity for transforming to queens than do eggs laid after three weeks when both kinds are cultured under identical conditions.

Finally, a most interesting additional effect discovered by Brian and Hibble (1964) is that eggs laid by different queens differ markedly in their tendency to produce queens or workers. Most of the variation appears to be the result of age; younger queens have a higher tendency to lay worker-biased eggs. This effect, it will be recalled, was foreshadowed by the results of Goetsch's early experiments on *Pheidole pallidula.*

To summarize, there are at least six factors operating in *Myrmica* caste determination: larval nutrition, winter chilling, posthibernation temperature, queen influence, egg size, and queen age. The next question should logically be, what is their relative importance in nature? This is equivalent to asking for the relative contribution of each factor to the variance in caste ratios produced by wild *M. ruginodis* colonies. No answer is likely to be forthcoming without elaborate field studies. The qualitative effect of each factor has so far been revealed in experiments in which the other factors were held constant (more or less) and the critical factor then varied radically in order to produce some kind of effect. The precise contribution of the factors and the degree of their interaction under natural conditions remains unknown.

The Genus Formica. Work on the *Formica rufa* group, paralleling in many respects that on *Myrmica,* was pub-

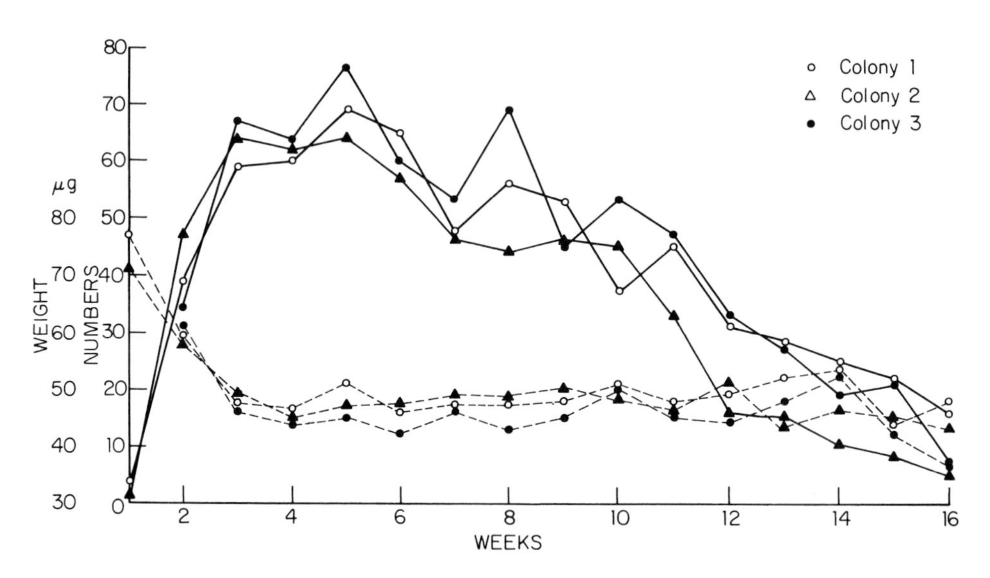

FIGURE 8-15. Oviposition rates (*solid lines*) and size of eggs (*dashed lines*) laid by *M. ruginodis* queens during the first 16 weeks after hibernation (redrawn from Brian and Hibble, 1964).

lished by Bier and Gösswald (1953–1957). As in *Myrmica,* there was originally much confusion in the taxonomy of this difficult species group, but it has been largely straightened out by Yarrow (1955b), Lange (1958), and Betrem (1960). Two species were used in the Bier-Gösswald study: *F. nigricans* (= *rufa pratensis*) and *F. polyctena* (= *rufa rufopratensis minor*). The following account applies primarily to *F. polyctena.*

These formicine ants hibernate without brood. In the spring, when the nest temperature rises to 13°C or above, the queens migrate to the warmest part of the nest near the surface, lay batches of eggs (referred to as the "winter" eggs), and afterward retreat to cooler parts of the nest. Eggs laid at temperatures under 19.5°C remain unfertilized, and as a consequence those first produced in the spring, when the nest temperatures are between 13° and 19.5°C, become males. Also, the smaller the colony, the poorer its thermoregulation, and, hence, the higher the proportion of males produced (Gösswald and Bier, 1955). Later eggs in this first "winter" batch are fertilized and capable of producing either queens or workers. Eggs laid still later—the "summer" eggs—are capable of producing only workers.

The winter and summer eggs differ strikingly from each other in several ways. Viewed in ovarian preparations, the winter eggs contain more RNA around their nuclear membranes and have a much more voluminous polar plasm than is the case in summer eggs; in addition, the nurse cells have larger nuclei (Bier, 1953, 1954a). It has been postulated, quite reasonably, that these cytological differences are in some way intimately connected with the later biassing of larval growth, but the relation has not been proved experimentally.

It is clear in any case that chilling of the eggs in the ovaries makes them bipotent with respect to caste. Final determination, however, occurs during about the first 72 hours of larval life. This was proven in experiments in which Gösswald and Bier (1957 and contained references) introduced eggs of *F. polyctena* into colonies of *F. nigricans* in order to permit the tracing of individual development. When 30 or more host *nigricans* workers, deprived of their own queen, were given small numbers of *polyctena* eggs, they reared queens. Groups of less than 30 workers reared workers. At 27°C the young larvae remained plastic for 72 hours.

Two other factors were discovered that match those in *Myrmica.* The presence of *F. nigricans* queens inhibits development of the *polyctena* eggs into queens. If a queen is present with a large group of workers, the winter eggs transform into either queens or workers, but, if no queen is present, the eggs always transform into queens. Workers just emerged from hibernation, as well as young workers, increase the tendency of winter eggs to yield queens. This ability of workers to influence caste determination directly is not found in *Myrmica,* although it is closely approached by the tendency of fall *Myrmica* workers to induce dormancy in their larvae. Dormancy, as we have seen, holds larvae in the final instar until the onset of winter, and winter chilling gives them the capacity to develop into queens.

In summary, although caste determination in *Formica* differs from that in *Myrmica* in several important details, there is a close resemblance in general pattern. Multiple controls exist; most, perhaps all, of the six factors of *Myrmica* also occur in *Formica.* A close interplay of responses to hibernation and nutrition characterizes both. In *Formica,* as in *Myrmica,* the relative weight and precise degree of interaction of the factors under natural conditions is still unknown.

Caste-determining Factors in Other Ant Genera

Only *Myrmica* and *Formica* have had their caste-determining mechanisms systematically examined. Even so, enough observation and experimentation have been conducted on other genera to suggest that several of the control factors discovered in *Myrmica* and *Formica* are widespread through the ants.

Queen inhibition. Colonies of the African weaver-ant *Oecophylla longinoda* are arboreal. They form multiple, tent-like nests of leaves often separated in space by considerable distances. Queens develop most frequently in nests lacking mated queens (Ledoux, 1950). Large, flourishing colonies of the pantropical myrmicine *Monomorium pharaonis* normally contain numerous mated queens. If a colony that has been producing only workers is deprived of its queens, it will rear both queens and males from the existing brood (Peacock *et al.,* 1954). *M. pharaonis* is the only ant species so far recorded in which the colony can make this immediate response and hence insure its own supply of nest queens. But many other examples probably occur, especially in the almost wholly uninvestigated tropical faunas.

Larval nutrition. It is a well-known generalization that queens are not normally produced until ant colonies reach a relatively flourishing condition. This circumstance is most simply explained as the outcome of larval nutrition. Other hypotheses can be contrived, however. For example, it is conceivable that, as in the honeybee, large colonies produce queens periodically when the queen pheromone becomes sufficiently diluted to make complete inhibition impossible. Too little is known about the vast majority of ant species to make any judgment. One more observation by Ledoux (1950) can be cited that seems to point strongly to a role for larval nutrition in *Oecophylla longinoda.* Under natural conditions, *O. longinoda* queens are produced continuously by large queenless groups of workers but only sporadically by small ones.

Almost nothing is known concerning the quantitative or qualitative differences in diet that affect caste determination. Biochemical studies in larval nutrition are almost wholly lacking. Perhaps the only direct approach to the problem is found in the highly problematical publications of Goetsch (1946, 1947, 1951) on the "Vitamin T complex." Goetsch claimed to have demonstrated a remarkable set of substances in the bodies of insects, fungi, and yeasts that stimulate growth in insects and other animals. Experimental ant colonies were especially strongly affected. In *Acromyrmex, Pheidole,* and *Camponotus,* production of soldiers as well as giant minor workers was stimulated, while in *Camponotus* and *Lasius* egg laying was increased and the rate of larval development advanced. The "T" substances, which were given such labels as "termitin" (from termites), "penicin" (from *Penicillium*), and "torutulin" (from the yeast *Torula*), have never to my knowledge been chemically identified or even purified. Crude extracts containing unknown mixtures were employed on species whose nutritional requirements were equally unknown. As a consequence, it is difficult to see how dosages could have been even approximately regulated or proper controls established. The existence of biochemically distinct growth substances as envisaged by Goetsch remains very much in doubt.

Climatic change. The vernalization (chilling) effect that renders *Myrmica* and *Formica* brood queen-potent can be interpreted as a token stimulus whose main adaptive significance is to permit the colonies to mature their sexual broods simultaneously in the summer. If this is true, it seems safe to predict that the same phenomenon will be found generally in ant species dwelling in cold climates. But we know that the emergence of sexual forms also tends to be seasonal in tropical species. Are these annual cycles also triggered by token stimuli? In species of *Eciton,* about six new queens and thousands of males are produced in a special all-sexual brood at the beginning of the dry season. It is likely, but still unproven, that climatic changes, especially in ground humidity, induce this change (Schneirla and Brown, 1952).

Egg physiology. The biochemical environment of the embryo may well prove to be important in caste determination in most or all ant species. For example, a decline in egg size at the beginning of spring similar to that in *Myrmica ruginodis* (Figure 8-15) has been recorded by Passera (1963a) in the formicine *Plagiolepis pygmaea.* The possibility of caste bias in eggs of various sizes in *Plagiolepis* has not been investigated, but the implication is clear. Similar or opposite trends in egg size may be found in yet other ant genera when deliberate searches are made.

In several articles published between 1945 and 1957, S. E. Flanders argued tenaciously in favor of a dominant role for egg physiology. He predicted, on wholly a priori grounds, that castes are determined primarily by the amount of yolk in the egg and that the amount of yolk is determined by the rate of oviposition. The following sequence was postulated: the faster the rate of oviposition by the queen, the less yolk will be resorbed by the queen's ovaries, the more yolk will be left in the egg, and the greater will be the probability that the embryo will develop into a queen. Although blastogenic influences have been discovered in the interval since 1945, they represent only part of a complex of multiple controls. Ovisorption, the main agent in Flander's model, is not known to play any part at all.

Role of the Endocrine System

In seeking a deeper explanation for the basis of caste determination, attention should be focused on the biochemical events surrounding the shutdown of the dorsal system of imaginal discs. It is a fair guess, not as yet fully established, that the cessation of growth is under the control of the endocrine system. At present we have only a few scattered and doubtful clues concerning the nature of the hormonal control.

It has been shown that the ant larva has a typical insect retrocerebral system (Glöckner, 1957). Ligation experiments suggest indirectly that the sequence of events in the activation of the retrocerebral system and prothoracic gland is similar to that in the better-known insect groups. Brian (1961) has supplied a piece of evidence that is at least consistent with the hypothesis that the development of imaginal discs is under hormonal control. When ovarian discs taken from "worker-presumptive" *Myrmica ruginodis* larvae were transplanted into the hemocoel of larvae about to transform into queens, they grew and differentiated as queen ovaries.

Although the endocrine hypothesis seems wholly reasonable, attempts to distinguish differences in the endocrine systems of larvae of differing caste potential have so far been rather unrewarding. Weir (1959c), employing standard histological techniques and light microscopy, could find "no qualitative differences of any kind" between *Myrmica ruginodis* larvae entering dormancy and those bypassing it at the critical A-0.5 stage (Figure 8-12). Brian (1960) has also examined the endocrine systems of developing larvae. Although his results are difficult to interpret, they seem to imply that in the later B period of *M. ruginodis* development, the only observed caste difference was that the neurosecretory brain cells of the queen become larger, their nucleoli more conspicuous, and their cytoplasm grey, finely granular, and devoid of secretory particles (that is, those particles that stain purple with Gomori's paraldehyde-fuchsin and Groat's iron hemotoxylin). In *Formica polyctena,* G. H. Schmidt (1964) observed that new queen pupae have proportionately larger corpora allata (producing juvenile hormone), while new worker pupae have proportionately larger prothoracic glands (producing ecdysone). However, this particular caste difference is not universal in the social insects. In the honeybee, according to Lukoschus (1955), the relative volume of the worker remains the same as that of the queen through the late larval and pupal instars, but then increases disproportionately in the adult stadium.

Brian (1960) hypothesized that worker determination in *M. ruginodis* is caused by a sharp rise in the ecdysone titer, "so sharp that the relatively small ovaries and wings are incapable of response." Actually, the fragmentary histological evidence just cited can be interpreted in one or the other of several ways. It is equally possible, for example, that the worker larva receives a slightly larger dose of juvenile hormone than the queen larva, enough to inhibit differentiation of the ovarian and wing discs but not of the adult system. By this explanation the larger size of the corpora allata of the queen larva does not result in higher juvenile hormone production, but only signals a preparation for later stimulation of ovarian development that is consonant with the reproductive role of the queen. The difficulty of using the volume of any endocrine gland as an indicator of its activity is that, while in the long run the volume and activity are usually correlated, the relation can fluctuate widely from one day or week to the next.

Determination of the Soldier Caste

I showed earlier that most examples of worker polymorphism are based on some form of adult allometry combined with continuous size variation. Caste determination in such cases offers few mysteries. When Eleanor Rita Lappano (1958) traced the development of labial glands and leg discs through preserved *Eciton burchelli* broods of various ages, she found that the growth was allometric. The precise position a given individual attains in the polymorphic spectrum is therefore simply determined by the size it reaches at the time of the pupal molt, which in turn might depend on how well it had been fed. But a closer examination discloses signs of a more complex control system operating within the colony. For whatever the fate of the individual larva, the final population of larvae reaches a size-frequency distribution that is a relatively invariable characteristic of the species. Moreover, all but the most rudimentary phylogenetic stages of worker polymorphism exhibit some degree of bimodality in the size-frequency distribution. The evolution of polymorphism can be interpreted in part as a progressive separation of the two constituent curves, with the proportion of workers in each remaining an adjustable species characteristic (Wilson, 1954). This fact alone means that the distribution of growth rates among the worker larvae is not haphazard. Lappano has presented some additional indirect evidence of social control in the development of the *E. burchelli* worker brood. The growth rate of the larger, firstborn larvae slows relative to that of the later, smaller larvae, with the result that pupal development is synchronized.

Since bimodality is a nearly universal characteristic of worker polymorphism, it should be profitable to examine the phenomenon for the same kinds of multiple controls that have been discovered in the case of queen-worker determination. As yet very few experiments have been performed. The work of Goetsch (1937a,b) on *Pheidole pallidula* revealed that both larval nutrition and environment of the embryo within the egg can be effective, while the data of Gregg (1942) proved that soldiers of *Pheidole morrisi* in some fashion inhibit development of new members of their own caste. Thus at least three factors have been demonstrated in the course of these quite preliminary studies. Wilson (1954) has called attention to a parallel from the ecology of fish populations to show that bimodality can result solely from food competition in an initially unimodal group, providing it occurs in a simple environment. No doubt competition for food does occur among ant larvae. But in most species the controls are likely to prove more complex.

The Ecology of Caste

One of the more important generalizations that can be drawn from the few patterns of caste determination worked out thus far is the diversity of the patterns themselves. I feel confident in predicting that the striking differences already apparent in a comparison of *Myrmica, Formica, Eciton, Oecophylla,* and *Monomorium* represent only a fraction of the total to be found throughout the ants. Whole tribes and even subfamilies, for example the Myrmeciinae, Ponerinae, and Dolichoderinae, remain largely uninvestigated. Other unexplored aspects of caste determination are the control factors that operate in the tropics, where most of the ant phyla are concentrated. This raises the whole promising subject of the ecological significance of patterns of caste determination. Brian, Weir, Gösswald, and Bier have made an effort to interpret the cyclic features of caste determination in *Myrmica* and *Formica* as adaptations to the peculiar environments of the species concerned. Such ecological analysis should continue to be a part of comparative studies and will eventually result in a deeper understanding of ant evolution in general. Elsewhere (Chapter 18) I will discuss the use of linear programming analysis in the study of the efficiency of various caste ratios and show how it can be related to quantitative ecological studies.

The Division of Labor

A first glimpse into a colony of ants can be compared to the view of a human city from a high building. Order and structure are certainly there, yet the individual inhabitants seem to move about in chaotic patterns. Surprisingly, half or more of the adult ants at any given moment may be idle. The rest divide their labor in a fashion that only a very extended analysis shows to be well suited to the needs of the colony as a whole. The most striking impression obtained in any short period of observation is the great amount of variation in behavior. Again, lengthy study reveals that each worker changes its activities as it grows older, but the progression follows general rules only loosely. This variation in behavior, and hence division of labor, consists of three components. The first two components can be referred to as *caste polyethism* and *age polyethism* (Wilson, 1963a). Polyethism is defined here in the broad sense as synonymous with "division of labor," rather than simply change with age after the earlier usage of Weir (1958a,b). The third component is recognized expediently as all the remaining variation. It is of considerable magnitude and undoubtedly has diverse causes, including the influences of colony size, colony nutrition, and individual learning experience, none of which are well understood at present. In the remainder of this chapter we will concentrate on polyethism, which has been better studied, while illustrating enough of the remaining, third component to indicate its importance.

Caste Polyethism

Male. It is one of the most curious facts of ant biology that the males contribute virtually nothing to the labor of the colony. They do not forage, build, or defend the nest, attend the queen, or nurse the brood. There is only one recorded exception to this rule. Santschi (1907) reported that when he forced a colony of *Cardiocondyla nuda* "var. *mauritanica*" containing an ergatomorphic male to move from one nest to another, the male repeatedly carried pieces of brood. This observation is in need of corroboration. Le Masne (1956a) was unsuccessful in provoking brood-carrying behavior in ergatomorphic males of *Hypoponera eduardi* when he forced nest changes in colonies of that species. However, Le Masne points out that this may not bear directly on Santschi's claim. *H. eduardi* ergatomorphs have small mandibles similar to those of the typical winged males and hence are ill adapted for the transport of solid objects. Similar experiments on the extremely ergatoid males of *H. punctatissima* and *H. gleadowi,* both of which possess heavy, worker-like mandibles, stand a better chance of success.

The possibility of rare instances of brood transport notwithstanding, it appears universally true that males do not join workers in the daily working life of the colony. At the same time it is not true that these individuals are passive toward the female members of the colony. They frequently groom other adults and are groomed in return. More importantly, as Gösswald and Kloft (1960) first showed in *Formica polyctena,* males also regurgitate liquid food to other members of the colony as well as receive it. Bert Hölldobler (1964) has recorded in detail the social activities of males of *Camponotus herculeanus* and *C. ligniperda* through their adult life. These individuals remain in the adult phase for nine to ten months, from eclosion in middle August to the nuptial flight and quickly ensuing death the following early summer. In the first few months before winter the males are in an active "social phase." They receive liquid food from workers and pass it freely to other adult members of the colony, while making little use of solid food lying on the nest floor. With the passage of winter, the males enter their "sexual phase." Even while they are still within the nest prior to the nuptial flight, the amount of food they exchange with other members of the colony by regurgitation sharply declines. The males come to depend increasingly on their own tissue reserves accumulated during the social phase. At the same time they change from a consistently negative to a frequently positive phototactic orientation, while the amount of unoriented locomotion within the nest increases. We know from the observations of Forbes (1954) on *Camponotus americanus* that, sometime prior to the nuptial flight, the males void the gut contents so that large air spaces fill the abdomens. This alteration no doubt improves the males aerodynamically, but hastens their death after the nuptial flight. Hölldobler found that the males of *Formica polyctena* undergo a behavioral sequence like that of *Camponotus,* except that the adult life span is much shorter and contained within one summer season. It would appear that the role of the *Camponotus* and *Formica* males in food distribution is far from vital for the colony as a whole. It is still correct to say that males are in a sense parasitic on the colony, in that they perform no unique or really essential function. Their entire morphology is modified for their primary role as flying sperm dispensers launched by the colony during the nuptial flight.

Queen. The function of the queen as progenitrix of the colony has been described in Chapter 4. The virgin queens lead a mostly parasitic existence in the nest prior to the nuptial flights. Their social behavior, like that of the males, is limited chiefly to mutual grooming and food exchange. In some species they help move the brood when the nest is disturbed. This emergency behavior has so far been observed mainly in genera in which there is little difference between the sizes of the queen and worker castes. Forel (1928) recorded it in *Leptothorax* and *Tapinoma,* for example, and Le Masne (1953), in *Hypoponera, Leptothorax,* and *Myrmica.*

After mating, queens of those species that start new colonies independently show themselves to be behaviorally complete females. They excavate a nest, usually a miniature version of the later, mature nest constructed by workers. They rear the first brood, displaying in proper sequence every act of grooming and cleaning peculiar to the species. In the Attini this includes even the culturing of the first fungus garden. In most species of the primitive subfamilies Myrmeciinae and Ponerinae, and in a very few of the more advanced subfamily Myrmicinae, young mated queens hunt for insects outside the nest. The lifetime behavioral repertories of nonparasitic queens, therefore, far exceed those of the workers. In more primitive groups, at least, the queens exhibit at some stage of their life cycle nearly every fixed-action pattern shown by all the worker subcastes taken together. As the colony commences to flourish, the range of behavior exhibited by the queen progressively narrows. Eventually, as she enters a physogastric condition, her activities become limited almost entirely to locomotion, egg laying, and feeding on the regurgitated liquid or trophic eggs supplied by the workers. It is probable that functional queens of all ant groups reach this passive condition by the time the colony reaches sexual maturity, that is, by the time it begins to produce virgin queens. The process is not wholly irreversible, however. I have observed that, when queens of young *Pheidole fallax* and *Solenopsis saevissima* colonies are suddenly deprived of workers, they resume brood care.

Soldiers in a strongly polymorphic series. We are a long way from understanding why worker polymorphism develops in some species but not in others. The patterns of caste polyethism are likewise diverse and unpredictable. Of the few generalizations that can be made, one of considerable interest concerns the extreme development of the soldier caste. In cases of advanced polymorphism, especially in complete dimorphism where intermediates have dropped out and the soldiers and minor workers have begun to evolve allometric patterns of their own, the greatest modifications in the soldier caste are to be found in the head and mandibles. This rule is connected with another generalization, made perhaps for the first time by Westwood (1838) and since extended repeatedly by other observers, that soldiers of the most strongly polymorphic species are employed principally in nest defense. This caste can serve other functions, of course. We know that the soldiers of a few *Pheidole* species assist in food collecting, while those of some *Oligomyrmex* can become repletes. But it is apparent that the morphological changes that make soldiers such a deviant caste in the first place are directed toward a defensive function. Soldiers are in fact adapted to one of three basic defensive techniques, as follows: (1) Shearing. The mandibles are large but otherwise typical, the head is massive and cordate, and the soldiers are adept at cutting the integument and clipping off the appendages of enemy arthropods. Examples are found in *Pheidole* (Figure 8-1), *Atta* (Figure 8-5), *Camponotus* (Figure 8-4), *Oligomyrmex, Zatapinoma,* and other genera of diverse relationships. The shearing adaptation is the most prevalent in soldier castes generally. Wheeler, in his essay "The physiognomy of insects" (1927), pointed out that the peculiar head shape of this kind of soldier is due simply to an enlargement of the adductor muscles of the mandibles, which imparts to the mandibles greater cutting or crushing power. (2) Piercing. The mandibles are pointed and sickle or hook shaped. The most impressive examples are found in the army ant genus *Eciton* (Figure 4-24). (3) Blocking. The head is shield shaped (many members of the Cephalotini) or plug shaped (*Pheidole lamia,* and several subgenera of *Camponotus,* most notably the polyphyletic assemblage of *Myrmaphaenus, Colobopsis, Pseudocolobopsis, Hypercolobopsis,* and *Paracolobopsis*). Wheeler (1927) coined the word *phragmosis* to label these modifications, as well as similar ones found in the queens of *Crematogaster* (*Colobocrema*) *cylindriceps* and *Colobostruma leae.* Well-developed examples of phragmosis are also known in other groups of insects, spiders, and anurans. All of the earlier reported cases of ant phragmosis involve the head, but W. L. Brown (1967) has recently discovered an Amazonian *Pheidole* (*P. embolopyx*) in which the abdomen of the queen is phragmotic. All of the cephalotines and most of the phragmotic *Camponotus* nest in cavities in dead and living plants, and the soldiers use their heads to block the nest entrances. In the case of the soil-dwelling *Camponotus* (*Myrmaphaenus*) *ulcerosus,* a carton shield is constructed at the ground surface bearing a single aperture that closely approximates the head of the major in size and shape (Creighton, 1953).

The behavior of the shearing-type soldier was first analyzed in detail by Edith Buckingham (1911) in her study of *Pheidole pilifera.* It has been most recently and graphically described by Creighton and Creighton (1959) in their account of *Pheidole militicida.* The soldiers of *P. militicida* have huge heads, nearly as large as all of the rest of the body combined. They play little part in the harvesting of seeds and, in spite of the massive development of their mandibles, do not participate in the breaking open of seeds collected within the nest. Their true métier came to light when the Creightons introduced large workers of *Pogonomyrmex* into a special artificial nest containing a *P. militicida* colony.

> The *Pogonomyrmex* workers moved slowly into the passage but rapidly backed out of it when they became aware of the advancing *militicida* workers. In most cases the *militicida* minors first entered the passage. Some of them would usually be seized and killed by the *Pogonomyrmex* workers but others returned to the nest and alerted the majors. When these entered the passage they . . . advanced very cautiously, with jaws wide open, and made frequent short lunges in the direction of the *Pogonomyrmex* workers. As the *militicida* majors wedged themselves tightly into the passage, three or four ranks deep, the passage was completely blocked and the front face of this block was a highly dangerous area for the *Pogonomyrmex* workers for it consisted of the closely approximated heads and wide open jaws of the *militicida* majors. As to what happened next depended on the *Pogonomyrmex* workers, who would charge up to the barrier and slash at the *militicida* majors with their mandibles. These attacks were usually futile, for the only exposed parts of the *militicida* major which could be damaged were the antennae and these were held so closely against the head that the *Pogonomyrmex* workers were seldom able to grasp them. If these attacks were vigorously pressed the *militicida* major usually stood perfectly still and waited until the mandibles of its opponent were near its own. It then lunged forward, closed its jaws on the mandible of the *Pogonomyrmex* worker and attempted to break off the crushed mandible . . . After a number of *Pogonomyrmex* workers had been put out of action with useless mandibles, or sooner if the *Pogonomyrmex* did not press the attack vigorously, the *militicida* majors emerged from the passage and began a different sort of action. They no longer faced their opponents and struck at their mandibles but approached them from the rear and struck at the thorax or the petiolar nodes. As a result, most of the *Pogonomyrmex* workers were ultimately cut in two, either at the petiole or behind the pronotum . . . In short, there is nothing haphazard about the way in which the *militicida* majors deal with their opponents; they only strike at parts which will put their opponents out of action or kill them.

When W. M. Wheeler discovered *P. militicida* in Arizona in the late fall of 1915, he found the remains of numerous soldiers scattered around the nests, but was unable to locate live soldiers in the nests. As a result, he developed an ingenious and subsequently celebrated theory about caste regulation in *P. militicida.* "It appears," he wrote, "that all the individuals of this caste (the majors) are regularly killed off by the workers on the approach of winter, probably after they have broken open all the hard seeds collected by the workers. Such a slaughter of the members of a large caste during the season when their activities are no longer required, when they would simply be a burden on the colony by consuming stored food and when fresh food cannot be collected, must have great advantages." The Creightons showed that all of Wheeler's premises were wrong, and they reinterpreted the dead soldiers found around the original nests as casualties in some unseen defensive skirmish.

I have observed stereotyped attack behavior similar to that described in *P. militicida* displayed by soldiers of *P. fallax* and *P. caffeicola.* It is likely that the habit is characteristic of the hundreds of species in this cosmopolitan genus. At the same time, my field observations show that the proportion of soldiers and the degree to which they are specialized for defensive behavior vary widely. In many species they also assist the minor workers in foraging.

The behavior of soldiers with piercing-type mandibles is at least equally specialized for colony defense. In the army ant genera *Anomma* and *Eciton,* these individuals sometimes line up along the flanks of the moving columns, their heads facing outward and mandibles gaping. An identical guard posture is assumed around nest entrances by the saber-jawed soldiers of the formicine *Cataglyphis bombycina* in the Sahara Desert (Délye, 1957). These formidable looking individuals rush at any moving object when the nest is disturbed. They seldom engage in other tasks. In particular, Délye found that the soldiers are not well suited, as once suggested by Santschi, for carting sand particles during the excavation of nests.

The responses of soldiers adapted for blocking nest entrances are the most specialized of all. The members of the *truncatus* group of *Camponotus* can be taken as fairly typical. *C. truncatus* of Europe has been studied in detail by Forel (1874) and later authors; Wheeler (1910) described the American *C. etiolatus,* and I have made extensive observations myself on colonies of the American

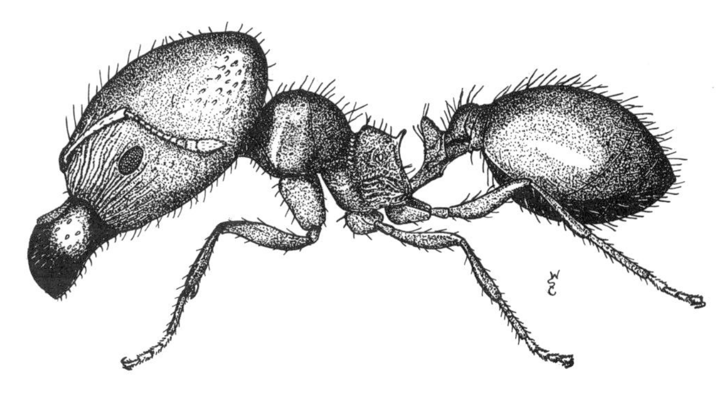

FIGURE 8-16. *Pheidole militicida* soldier in defensive posture (from Creighton and Creighton, 1959).

C. fraxinicola housed in glass tubes. The soldiers seldom leave the nests, which consist of narrow cavities in the dead wood of standing trees and shrubs. One or more of them stand guard at the nest entrances, where they act literally as living doorways. When minor workers approach them from either end and give the right signal (presumably a combination of simple touch and colony-recognition scent, although the matter has never been experimentally analyzed) the soldiers pull back into the nest to allow their nestmates free passage. The nest entrances are cut into wood or plastered with carton so as to just accommodate the head of a soldier. If the entrance is larger, several soldiers may cooperate to plug it with the combined mass of their heads. Both arrangements are shown in the accompanying illustrations of *Camponotus (Colobopsis) truncatus* by Szabó-Patay.

A different form of blocking behavior is exhibited by the North American cephalotine *Paracryptocerus texanus.* The entrance hole to the arboreal nest is somewhat larger than the head of the soldier and is blocked by the combined mass of the head and expanded prothorax, the latter structure being heavily armored and pitted like the head. The head is held obliquely, rather like an animated blade of a miniature bulldozer. This posture, combined with the thrust and pull of the short, powerful legs, allows the soldier to press intruders right out of the nest. When a minor worker returns to the entrance hole, the following sequence unfolds:

> The returning minor may or may not touch the antennae of the guard, although it usually does so. Thereafter the guard crouches down. This brings the anterior rim of the head below the level of the floor of the passage or, if the guard stands

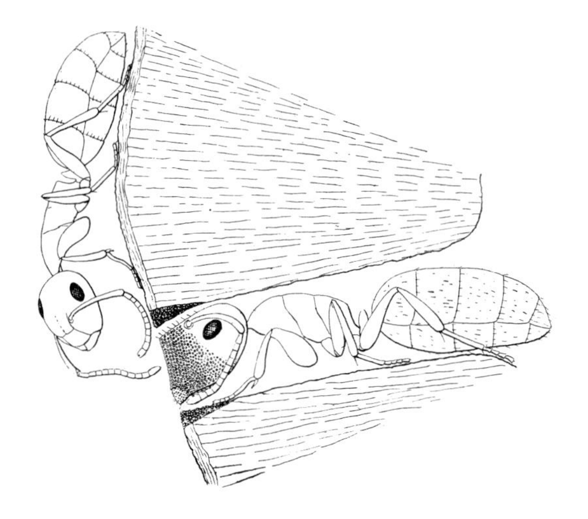

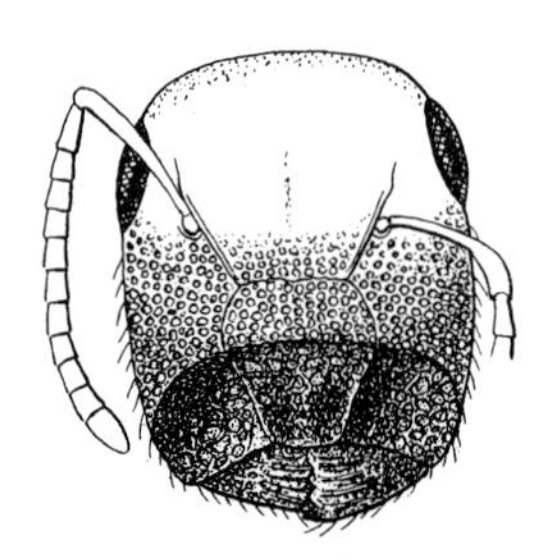

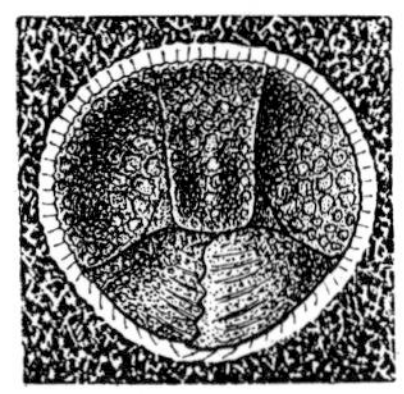

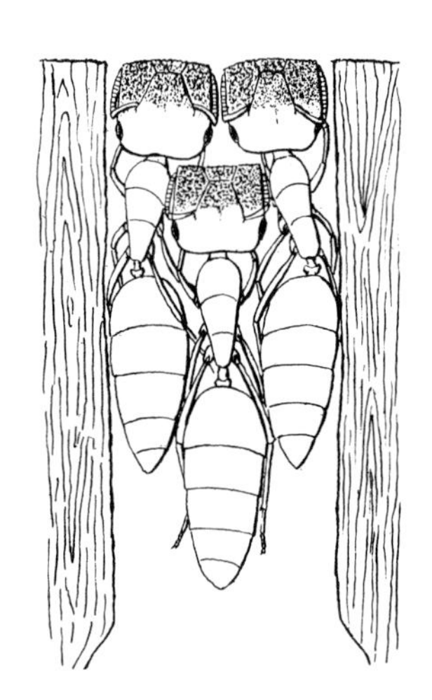

FIGURE 8-17. Nest-guarding behavior by soldiers of the European ant *Camponotus* (*Colobopsis*) *truncatus*: (*top*) a minor worker approaches a soldier that is blocking a nest entrance with its head; (*lower left*) two views of the head of the soldier of *Camponotus* (*Colobopsis*) *truncatus*, the frontal view being framed by a small circular entrance to the nest, demonstrating how the head serves as a living "gate" to the nest; (*lower right*) a group of workers assume the position they use in plugging a large entrance to their nest (from Szabó-Patay, 1928).

completely inside the passage, the front of the head is raised as the guard crouches. The dorsum of the guard's thorax is now no longer close to the roof of the passage and the minor can, if it is sufficiently active, wriggle between the dorsum of the thorax and the roof of the passage . . . If the passageways are made large enough to accommodate two majors simultaneously, they ordinarily assume a position where they are back to back. Under such circumstances the two opposed cephalic discs form a V-shaped area. The bottom of this V is open but the space behind it is closed by the closely approximated thoracic dorsi of the two guards. When minors are admitted to the nest both majors crouch and the entering worker struggles through the narrow space between the thoraces of the guards. It seems scarcely necessary to state that there is no part of this behavior which at all resembles that of the *Colobopsis* major, which must back away from the nest entrance to admit the returning minor (Creighton and Gregg, 1954).

On the occasions when soldiers of *P. texanus* leave the nest, the guard soldiers must come out of the nest entrance altogether in order to readmit them (Creighton, 1963).

Worker subcastes in moderately or weakly polymorphic series. Here we encounter a confusing diversity of polyethic patterns among the small set of species that have been studied. Each genus has so far revealed a pattern more or less peculiar unto itself. Furthermore, although the degree of polyethism is loosely correlated with the degree of polymorphism, the patterns of the two phenomena cannot be said to be linked in any consistent way among the genera. In other words, it is not yet possible to predict the pattern of caste polyethism from a knowledge of the polymorphism alone. The following species show, among themselves, the greatest range of patterns disclosed to date.

Solenopsis saevissima. Workers of this fire ant are moderately polymorphic. Their adult allometry is relatively slight (the allometric exponent for head width versus pronotum width is 1.1 in contrast to 1.5 for *S. geminata*), but they vary greatly in size and have an overlapping bimodal size-frequency distribution. Polyethism is much less developed (Wilson, unpublished observations). In one study, samples of workers employed in various tasks were taken from a single large colony in a field in Alabama. The frequency distributions of workers engaged in nest repair, food retrieval, and nest defense, respectively, did not differ significantly from each other or from the distribution of a sample collected randomly from the interior of the nest. Only workers collected while gathering nest materials outside the nest perimeter differed, as it happened, by containing a slightly higher proportion of larger workers. In this species, therefore, we see a moderate degree of polymorphism associated with a relatively weak degree of polyethism.

Formica obscuripes and *F. polyctena.* The worker caste of *F. obscuripes* (= *F. rufa melanotica*) is weakly poly-

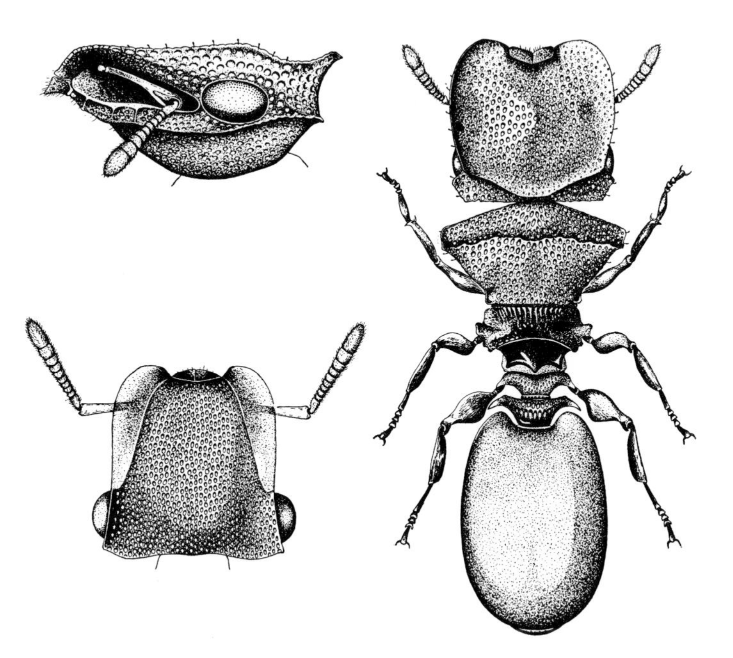

FIGURE 8-18. Soldier of *Paracryptocerus texanus:* (*right and upper left*) the shield-shaped head used to block the entrances to the arboreal nests; (*lower left*) front view of the head of a minor worker included for comparison (from Creighton and Gregg, 1954).

morphic. It shows only slight allometry. Size variation is moderate, and the size-frequency distribution is unimodal. Yet King and Walters (1950) have found that the workers are strongly polyethic. The larger individuals remain mostly in the nest, while the smaller ones do most of the foraging. In this case we have a relation between the degrees of polymorphism and polyethism that is the reverse of that found in *Solenopsis saevissima.* Polyethism of the *F. obscuripes* type has been reported for *F. rufa* by Adlerz (1886) and *F. sanguinea* by Janina Dobrzańska (1959), although neither species has been subjected to equivalent statistical analyses. *F. polyctena* has been analyzed in a slightly different manner by Otto (1958a) but with essentially the same result. Worker size is weakly correlated with division of labor. It accounts for only a small fraction of the total behavioral variation, which consists mainly of age polyethism and individual peculiarities. There is hardly any difference in size between workers engaged in work inside the nest and those building the nest and foraging outside. On the other hand, among those employed outside the nest, the foragers average smaller in size and the nest builders, together with those retrieving prey objects, average larger. If the samples had been taken according to the cruder procedure of King and Walters, the resulting curves would have been similar in kind, if not in degree, to those of *F. obscuripes.*

An extreme development of polyethism of the *Formica* type is found in the formicine honey ants. In weakly polymorphic species of *Myrmecocystus, Camponotus,* and *Proformica,* it is the largest-sized class from which some individuals are drawn to transform into the grotesque "honeypot" repletes. A fuller account of the behavior of this primarily physiological caste will be given in Chapter 14.

Oecophylla longinoda. The African weaver-ant exhibits a polyethic pattern essentially the reverse of that of *Formica obscuripes.* The major workers do most of the foraging, and they pull the leaves together from which the nests are constructed, while the minors remain in and around the nest, caring for the brood and queen and holding the larvae during the weaving operation in nest building (Weber, 1949b; Ledoux, 1950). The worker caste is moderately polymorphic, with allometry affecting mainly the thorax and petiole (Figure 8-6). The frequency distribution is unusual, containing two very distinct modes, the majors outnumbering the minors. In retrospect, these qualities in the allometry and frequency distribution seem well suited for the type of polyethism they serve. In particular, it seems appropriate that the major class, with responsibility for foraging, should be both abundant and "normal" in morphology instead of assuming the grotesque combative form of soldier found in other genera. A roughly similar polyethism exists in the primitive genus *Myrmecia* and in *Daceton armigerum.*

Camponotus americanus. This American species, studied by Buckingham (1911), has a moderately polymorphic worker caste. The allometry and frequency distribution are similar to those of the related *C. castaneus,* which is illustrated in Figure 8-4. Classes of every size participated to some extent in all the work categories examined by Buckingham, but there was a preponderance of soldiers in fighting, of small and medium workers in nest building and foraging, and of small workers in brood care. Patterns similar to this have been described in *C. noveboracensis* (= *C. pictus*) by Buckingham, in *C. japonicus* "var. *aterrimus*" by Lee (1938), and in *C. ligniperda* by Kiil (1934). Two of Kiil's polyethism curves for *C. ligniperda* are shown in Figure 8-19. The pattern described in *Messor*

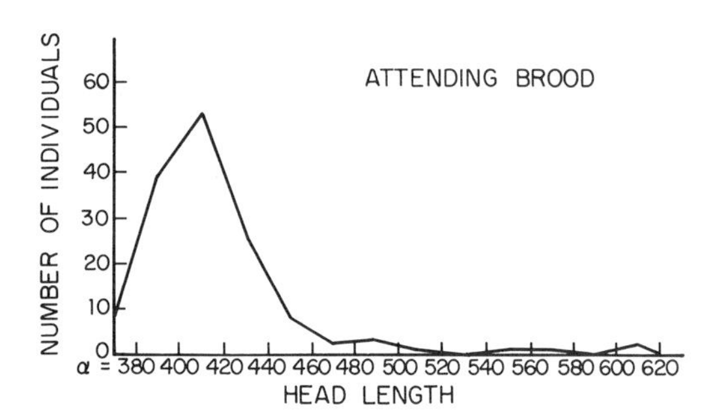

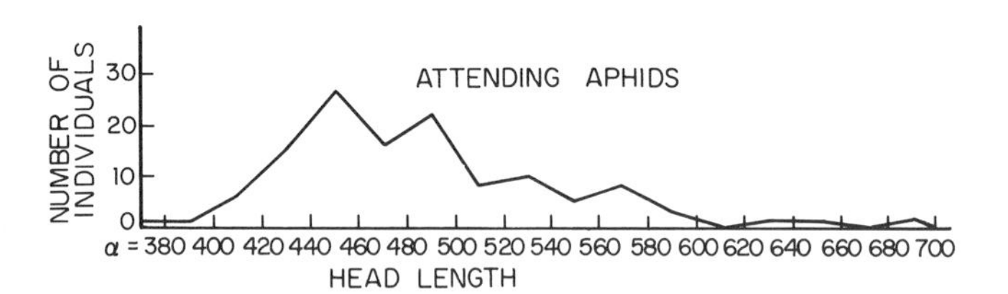

FIGURE 8-19. Polyethism curves showing the frequency distributions of different-sized classes of the ant *Camponotus ligniperda* engaged in two kinds of activities. The head widths are given in arbitrary units (redrawn from Kiil, 1934).

by Goetsch and H. Eisner (1930) resembles that of *Camponotus* except that the soldiers appear to be involved heavily in foraging as well as in nest defense.

Daceton armigerum. This arboreal South American species, one of the most primitive members of the tribe Dacetini, displays what is perhaps the most complex pattern of caste polyethism analyzed so far. The species is moderately polymorphic, combining weak allometry with extensive size variation and an imperfectly bimodal size-frequency curve (Figure 14-3 shows the extreme size variants). The most striking feature of the polyethism is the nearly total restriction of the labor of the smallest-sized class to the care of the egg-microlarva pile. Beyond that, other tasks are divided in differing ways among the size groups, as exemplified by the data presented in Table 8-1. Outside the nest, the major workers behave differently in yet other ways from the medias. They hunt less, spend more time resting in "bivouac areas" on the foraging ground, and are relatively more successful in taking prey objects away from the medias and carrying them back to the nest (Wilson, 1962b).

The Dacetini are one group in which it might be possible to gain some understanding of the adaptive significance of worker polymorphism. As a rule the most primitive members of the tribe, including *Daceton armigerum,* are polymorphic; while the morphologically more advanced members are monomorphic. Brown and Wilson (1959a) have sought the ecological correlates of this loss of polymorphism. The loss is associated, at least roughly, with an increased reliance on collembolans as prey, and—perhaps as a consequence of this specialization in diet—decreased body size, smaller colony size, and more cryptobiotic foraging behavior. The assumption of a more specialized, restricted feeding regime on smaller prey objects could account, at least in part, for the surrendering of the worker differentiation and polyethism displayed to such an extreme degree by *Daceton armigerum.*

Studies of other ant species with polymorphism resembling that of *Daceton armigerum* are likely to reveal similarly complex and unpredictable systems of polyethism. A probable case is the fungus-growing *Atta cephalotes,* where minor workers perform a bizarre function during food-gathering expeditions (Figure 8-20). These miniature

TABLE 8-1. Division of labor among workers of *Daceton armigerum* by head width (from Wilson, 1962b).

Type of labor	Head width (mm) 1	2	3	4	Total number of observations
Surinam colony:					
Total population (in artificial nest, April 5)[a]	13	60	20	9	102
Disposing of corpses and refuse[b]	0	19	12	2	33
Dismembering and feeding on fresh prey in nest[b]	0	14	25	5	44
Feeding larvae by regurgitation[b]	8	15[c]	3[c]	1[c]	27
Attending egg-microlarva pile[b]	24	3	0	0	27
Foraging in the field[a]	0	0	4	10	14
Trinidad colony:					
Foraging in the field[a]	1	91	77	12	181
Resting in way-station[a]	0	8	19	10	37
Carrying prey[a]	0	1	1	10	12

[a] Numbers refer to separate, individual workers.
[b] Numbers refer to separate behavioral acts, without regard to the number of workers engaged.
[c] Consisting mostly of callows.

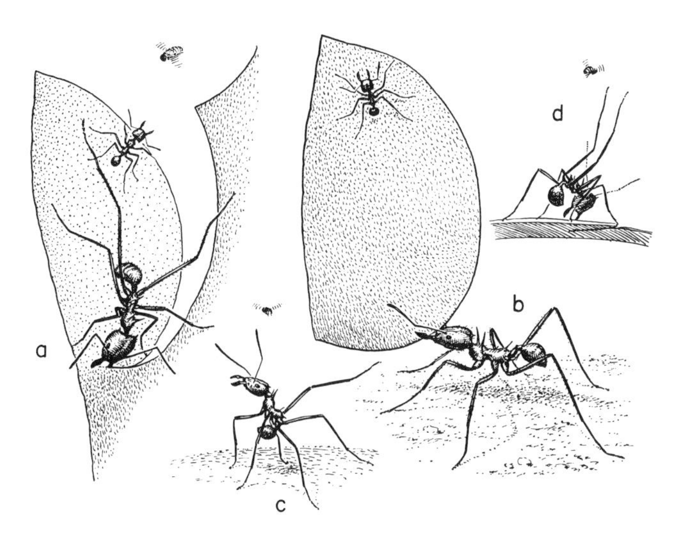

FIGURE 8-20. The minor workers of the fungus-growing ant *Atta cephalotes* accompany the medias during leaf-gathering expeditions (*a*), often hitchhiking on the cut pieces of leaves (*b*). Their chief function at this time is to protect the large workers from parasitic phorid flies (*c, d*) (from Eibl-Eibesfeldt and Eibl-Eibesfeldt, 1967).

insects travel along with the media workers, whose principal activity is the cutting and gathering of leaves. Often they ride on the cut portions of the leaves, but they do not participate in leaf cutting itself. Instead, they protect the large workers from parasitic phorid flies by snapping with their mandibles and "fencing" with their hind legs (Eibl-Eibesfeldt and Eibl-Eibesfeldt, 1967).

Age Polyethism

Adults. It has been well known since the time of Forel (1874) and Lubbock (1894) that young workers tend to remain in the nest and nurse the brood, while older workers spend more time outside the nest. The histories of individual ants have been documented in varying degrees of detail in *Pheidole* and *Camponotus* by Edith Buckingham (1911); in *Myrmica, Lasius, Camponotus,* and *Formica* by Kaethe Heyde (1924); in *Messor* by Goetsch and Eisner (1930); in *Camponotus* and *Formica* by Kiil (1934); in *Formica* by Økland (1930), Otto (1958a), and Dobrzańska (1959); in *Myrmica* and *Messor* by Ehrhardt (1931); in *Myrmica* by Weir (1958a,b); and in *Oecophylla* by Ledoux (1950). Økland first employed the technique of marking ants engaged in different activities in order to trace ontogenies of large samples of workers, while Heyde discovered that the behavioral ontogenies differ among species and made the first, albeit not very conclusive, attempt to relate these differences to interspecific variation in behavioral flexibility. The most detailed protocols are found in the publications of Ehrhardt, Otto, Dobrzańska, and Weir. The best-known species by far are the members of the *Formica rufa* group and particularly *F. polyctena,* whose study by Otto will now be reviewed.

As in all other ant species thus far analyzed in much detail, colonies of *F. polyctena* are not structured into groups of workers who associate in cliques or consistently cooperate in the performance of tasks. Each worker, while completely bound to the colony, addresses its activities indiscriminately to all members of the colony or, at most, to all members of a given caste or stadium. About half the time of the worker is spent at rest and half engaged in some social activity or foraging. There is a tendency for workers to spend at least fifty days after their emergence from the cocoon in what Otto and other German authors call the *Innendienst*—service inside the nest. The activities of the *Innendienst* include care of the brood, care of the queens, care of other adult workers, handling of dead prey in the nest chambers, and nest cleaning. There is a tendency for individual workers to specialize on one or a couple of these tasks, although the majority of workers perform most or all at some time. None of the workers devote themselves to "guard duty" in or around the nest. After about forty days or longer, most workers shift permanently to the *Aussendienst,* during which they forage and work on nest construction. There is a further specialization possible in nest construction, in that some workers concentrate on excavating within the nest while others gather materials for roofing. As in other species of *Formica* and *Lasius,* each forager tends to patrol a certain spot within the colony territory. Individual behavioral ontogenies vary greatly in both content and timing. For example, many workers pass through the *Innendienst* without attending the brood at all.

During the *Innendienst,* the ovaries of the workers contain eggs. Toward the end of this period, resorption of the eggs begins, and, by the onset of the *Aussendienst,* the resorption is total. There is little sign of the extensive glandular alteration that marks the phases of age polyethism in the honeybee. A few suggestive changes do occur, however. The cells of the maxillary gland, which are the homologs of the hypopharyngeal glands in the

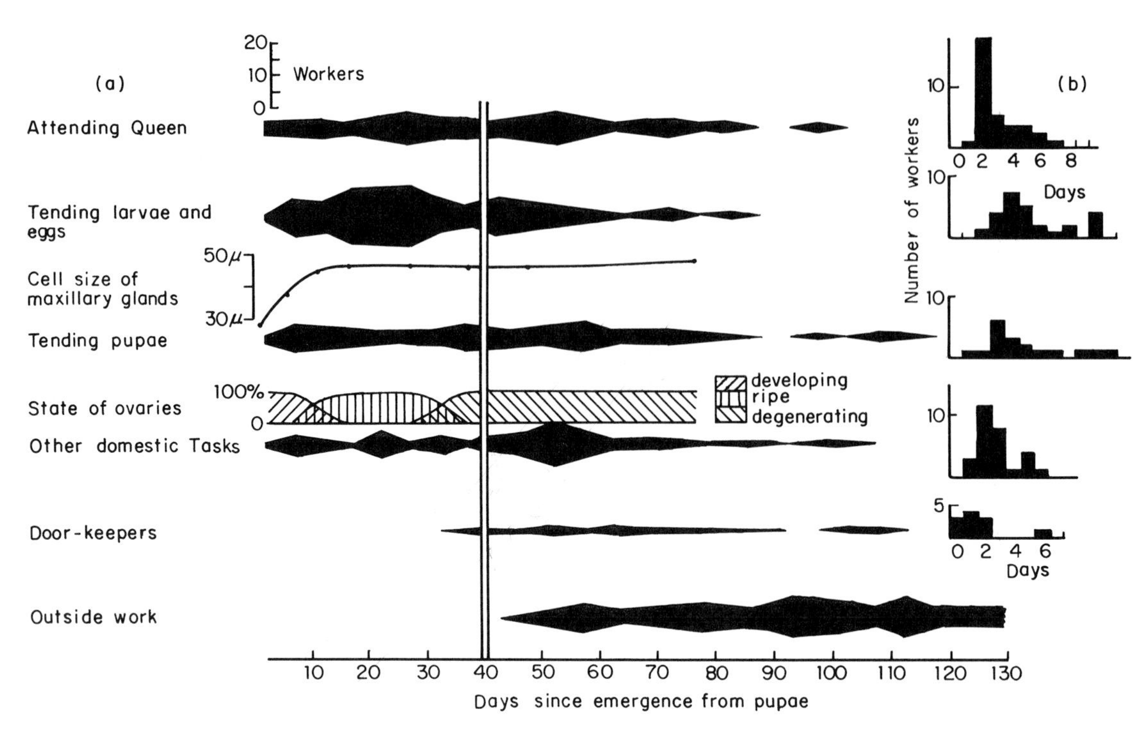

FIGURE 8-21. Workers of *Formica polyctena:* (*a*) change through time in behavior and glandular development; (*b*) histograms showing the number of days that workers persisted in the given task. These data reveal, for example, that few workers specialize in attending the queen, while a comparatively large number specialize in caring for the brood. The double vertical line roughly separates the period of nest work from that of foraging service (redrawn from Sudd, 1967; based on the data of Otto, 1958a).

honeybee, decrease in size somewhat during the *Aussendienst.* The cells of the postpharyngeal glands increase in size with age, while their nuclei become more rounded, but these alterations are not closely associated with the shift to the *Aussendienst* or with any apparent behavioral specialization. Workers concentrating on excavation within the nest, and hence the handling of the finer building materials, have mandibular gland nuclei somewhat larger than in other workers. Other exocrine glands studied by Otto, namely the labial glands, hypostomal glands, and metapleural glands, showed no significant changes.

It would be difficult to overestimate the complexity of age polyethism in ants. Every category of social and individual behavior studied in detail so far has proven to be affected to some degree by age. A case in point is aggressiveness and nest defense behavior in *Formica sanguinea,* as documented in the analysis of Dobrzańska (1959). It is a commonplace observation that, when nests of most ant species are torn open, some of the workers retreat, some appear indifferent, some move toward the outside in a relatively calm exploratory manner, and some attack blindly. Dobrzańska showed that this series of responses, which can be viewed as comprising a gradient of aggressive tendencies, is strongly correlated with the usual location of the workers with reference to their nest, which in turn is correlated with their age (Table 8-2). In *Myrmica ruginodis,* Weir (1958a,b) studied three behavioral classes of workers that probably also represent different age groups: those wandering far away from the brood ("foragers"), those standing near the brood ("domestics"), and those standing on top of the brood ("nurses"). These groups proved to differ in many ways. The nurses were far superior to the others in withstanding starvation, and they laid the most eggs. They were also the least efficient at killing blowfly larvae and, surprisingly, at finding larvae "lost" from their own colony. The foragers were at the opposite extreme of variation in these properties, while the domestics occupied an intermediate position.

TABLE 8-2. Percentages of three major behavioral classes of *Formica sanguinea* workers responding in different ways to disturbance of their nest (modified from Dobrzańska, 1959).

	Response to disturbance					
Workers	Escape	Hiding	Indif-ference	Mild inves-tigation	Attack	Violent attack
Remaining always in the nest	8	64	14	14	0	0
Occasionally leaving the nest	0	25	25	37	13	0
Remaining outside the nest	0	0	4	24	35	37

Much of the variability in behavior not connected to caste and age polyethism must be attributable to individual differences in experience. When young *Innendienst* workers of *Messor* and *Formica* are forced out of the nest prematurely by a disturbance, they often remain outside. (Goetsch and Eisner, 1930; Dobrzańska, 1959). Ants resemble honeybees in that their work preferences can change in response to alterations in their social environment. This was first demonstrated in a simple but conclusive experiment by Ehrhardt (1931). Individual workers of *Myrmica rubra* were initially classified according to their work preferences as members of a normal, undisturbed colony. The group showing an extreme predilection for brood nursing was then transferred to isolated cells containing only earth, while the group that engaged most exclusively in excavation and foraging was isolated in other cells containing brood but no earth. The erstwhile nurses soon devoted themselves to soil excavation, while the erstwhile foragers turned into nurses. When returned to their nestmates in the home nest, however, both groups reverted to their original functions.

Egg. In a real sense the individual ant can perform a service to the colony while still in the egg stage—by serving as a source of food. Egg cannibalism, a common phenomenon in many ant species and of major social significance in some, will be discussed in Chapter 14.

Larvae and pupae. Individuals in various stages of larval and pupal development are occasionally cannibalized, although less regularly than eggs. Injured larvae and pupae are always promptly eaten. When the colony is starving, the entire brood is gradually consumed. It is clear that while the brood comprises the growing point of the colony, it also serves a function analogous to storage tissue in an organism—as energy capital, which, once invested, can still be partly recalled when the colony falls on hard times.

It is highly probable that larvae perform other, more direct services. The glandular secretions, anal effluvia, and cast skins that larvae constantly produce are mostly consumed by the workers. No study has yet been attempted to test the possible nutritive and communicative functions of these elements. Ant larvae also commonly release liquid from their hindguts and mouthparts (the latter evidently salivary in nature) which is rapidly consumed by workers (Le Masne, 1953). At least some of the secretion is likely to have nutritive value. Maschwitz (1966a) found that the oral secretion of *Tetramorium caespitum* is higher in amino acid content than the hemolymph. The phenomenon of liquid exchange and its role in social organization will be examined at greater length in Chapter 14.

9 Caste: Social Bees and Wasps

Caste, like social organization itself, finds a nearly total range of expression among the species of social bees and wasps. In the primitively eusocial Halictinae, it emerges as a mere psychological difference among morphologically similar adults, but goes on to include, in a few species, several forms of striking queen-worker dimorphism. In the honeybee species *Apis mellifera* strong morphological and physiological differences exist between queens and workers, and the caste of individuals is determined by a complex interaction between pheromone-mediated behavior on the part of nurse workers and specialized diets fed to the larvae. Finally, at least one group of species of stingless bees, the genus *Melipona,* has superimposed a genetic control of caste upon the conventional physiological device employed by related groups. Most of these phylogenetic advances, with the most conspicuous exception of the invention of genetic control, have been paralleled in the evolution of the social wasps. Together the social bees and wasps differ from the ants and termites in one major respect: for some reason none of them has fashioned well-defined worker subcastes. It is true that the species with very large colonies display a division of labor comparable to that of the most advanced ants and termites. But where the division in the latter insects is based in part on morphological subcastes and in part on programmed, temporal polyethism, in most bees and wasps it is based almost entirely on temporal polyethism.

Certain other evolutionary rules are recognizable. As colony size has grown in the course of evolution, the differences between the queen and worker castes have been exaggerated, intermediate forms have disappeared, and the behavior of the queen has become increasingly specialized and parasitic. The ultimate stage is attained in the honeybees and meliponines, whose queens never attempt to start colonies on their own and are reduced to the status of little more than egg-laying machines. Correlated with this trend has been a subtle shift in the power structure of the colony. Among the primitively social groups, particularly the halictine bees, the bumblebees, and the primitive polistine wasps, the queen maintains a dominant position primarily by aggressive behavior toward her sisters, daughters, and nieces. In more complexly social species, reproductive control by the queen is exercised through inhibitory pheromones.

Although morphological worker subcastes are generally so weakly developed as to be almost nonexistent when compared with those of ants and termites, size effects do occur. Larger members of a given colony tend to forage more, and smaller members tend to devote themselves to brood care and nest work. In honeybees, the larger an individual bee the more quickly it passes through the normal ontogenetic stages of behavior, terminating in a period devoted principally to foraging. Throughout the course of evolution to higher levels of eusociality, there has been a tendency to produce ever more elaborate patterns of temporal division of labor, the most extreme cases again being those of the honeybees and meliponines. This temporal polyethism, like that of ants and termites, is typically a sequence leading from nest work and brood care to foraging. To my knowledge only one exception has been reported, that of the wasp *Polistes fadwigae,* in which a very weak polyethism follows the opposite sequence. With these several broad principles in mind, let us now review the caste systems of the major groups of social bees and wasps in the seeming approximate order of their phylogenetic advance.

The Halictine Bees: From Behavioral to Morphological Castes

The semisocial colonies of *Augochloropsis* and *Pseudaugochloropsis* contain some of the most elementary caste systems known in all of the social insects (Michener, 1969a). Females of *A. sparsilis,* which work cooperatively to establish nests, are morphologically identical to each other or nearly so. Most have been inseminated and possess normal ovaries. However, some behave as "workers": they do most of the foraging and undergo only limited ovarian development. Whether they remain in this repressed condition all their lives or, at some later time, increase their ovary size and partake in egg laying is not known. In addition, there exist a few unfertilized females. Michener and Lange (1958a) examined 86 females in a colony during the first generation of the year and found only one that was unfertilized, but, in the second generation, between 15 and 20 percent were in this condition. These individuals are true workers. Not only are they unable to contribute to the next female generation, they work harder than their fertilized sisters, with the result that their mandibles wear down and their wings fray more quickly. In the related species *Pseudaugochloropsis costaricensis* as many as one-half of the females are morphologically distinguishable workers. They average smaller in size, in spite of the fact that a majority are fertilized.

Equally primitive caste systems exist within certain eusocial species of Halictinae. According to Cécile Plateaux-Quénu (1960, 1962), caste determination in the perennial colonies of *Evylaeus marginatus* is "imaginal," by which it is meant that the adult females are morphologically indistinguishable and, as in *Augochloropsis sparsilis,* status is merely a matter of behavior and ovarian development. Also, the workers remain unfertilized and live for shorter periods of time. Each *E. marginatus* colony is started by a single female, the "queen," who loses the power of flight after the first year and devotes herself exclusively and permanently to egg laying thereafter. Her presence somehow inhibits ovarian development in her daughters. When she is removed, a large percentage of them undergo quick ovarian development. During the first four years of colony existence, the workers remain unfertilized. In the fifth and final year, some of them mate (even though they are still morphologically identical to their sisters), hibernate, and, in the following spring, commence nests of their own.

Other halictine species show ascending degrees of differentiation between queen and worker castes (Ellen Ordway, 1965; Knerer and Plateaux-Quénu, 1966b). Figure 9-1 gives the percentage difference in average length as a function of the percentage of females that appear in the first brood after nest founding. As I pointed out in Chapter 5, both characteristics can be regarded as independent measures of the degree of social development in halictines. The average size difference ranges from zero—for example, in *E. marginatus*—to as much as 18 percent. In species with a smaller average size difference, the size frequency curves of the two castes overlap. In other words, intercastes occur. The amount of caste

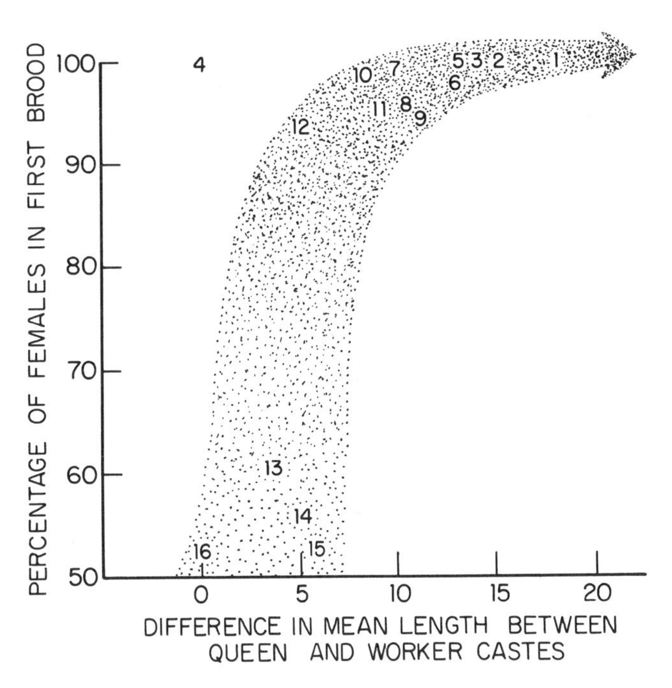

FIGURE 9-1. Among the species of halictine bees, the average size difference between queens and workers belonging to the same species varies from zero to about 18 percent. This caste difference is correlated with a second social characteristic: the percentage of females appearing in the first brood. (1) *Evylaeus malachurus;* (2) *E. linearis;* (3) *E. cinctipes;* (4) *E. marginatus;* (5) *Halictus ligatus* (Indiana); (6) *E. calceatus;* (7) *Lasioglossum rohweri;* (8) *E. nigripes;* (9) *H. ligatus* (Ontario); (10) *L. imitatum* (Kansas); (11) *Halictus scabiosae;* (12) *H. maculatus;* (13) *L. rhytidophorum;* (14) *L. imitatum* (Ontario); (15) *Augochlorella striata;* (16) *H. confusus. Evylaeus marginatus* is also exceptional in having perennial colonies (modified from Plateaux-Quénu, 1967; based on data from Knerer and Plateaux-Quénu, 1966b).

differentiation is also correlated with the size of colony populations, a third intuitive indicator of the level of social evolution. For example, *Evylaeus duplex* and *Lasioglossum rhytidophorum* have slight queen-worker differences and moderate colony populations, while *Evylaeus malachurus* has strong queen-worker differences and large colony populations. Exceptions to the rule include the aberrant *E. marginatus,* which has no morphological caste differentiation but relatively enormous colony populations.

In a few species that display marked average size differences between queens and workers, the size variation is allometric. This phenomenon has been reported in *Halictus scabiosae* in France (Quénu, 1957), in *H. aerarius* in Japan (Sakagami and Fukushima, 1961), and in *H. latisignatus* in India (Sakagami and Wain, 1966). In each of these species the heads of the larger females are proportionately broader and deeper. Such individuals are consequently referred to as "macrocephalic," and the color is generally darker. The extremes of variation are illustrated in Figure 9-2. In *H. aerarius* (and perhaps other allometric halictine species) the size-frequency distribution of the females seems to be continuous; it is skewed to the right and may be incipiently bimodal (Figure 9-3). In short, the queen-worker differentiation of *H. aerarius* closely resembles the primitive level of worker subcaste differentiation exhibited by such ant taxa as the genera *Solenopsis* and *Camponotus.* It is worth emphasizing that, although the queen-worker differentiation in certain lines of halictines has begun to head down this evolutionary path, it has gone only partway. In contrast, all living ant groups have not only completed the transition to complete dimorphism in queen-worker differentiation, but many have repeated the whole process in the evolution of worker subcaste differentiation.

The weak differentiation of the queen and worker castes that characterizes the Halictinae is paralleled by a relatively imprecise division of labor. The females of *Evylaeus malachurus* participate in all tasks, regardless of size, but the larger fertilized individuals do most of the egg laying (Legewie, 1925; Noll, 1931). A broad overlap of functions exists even in the relatively well-marked castes of *Halictus aerarius* and *H. latisignatus,* as Sakagami and his associates have shown. In the Halictinae generally, according to Michener (1969a), the smallest individuals may be incapable of becoming egg layers, and the largest may do nothing else; the great majority are, however, capable of

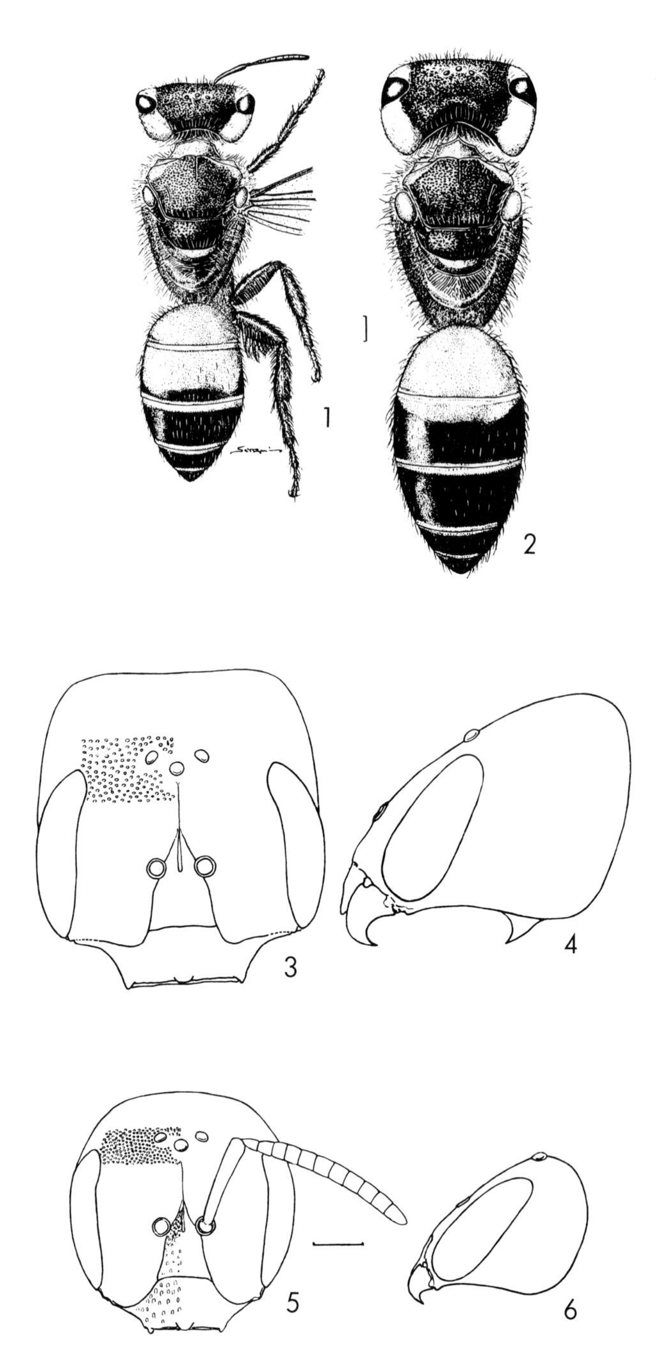

FIGURE 9-2. The social bee *Halictus latisignatus:* (1, 2) small and large ("macrocephalic") females, showing allometric variation in body proportions and color; (3-6) front and side views of the heads of the large and small females, showing strong allometric variation in the proportions. The small individuals specialize as workers; the large ones, as egg layers (from Sakagami and Wain, 1966).

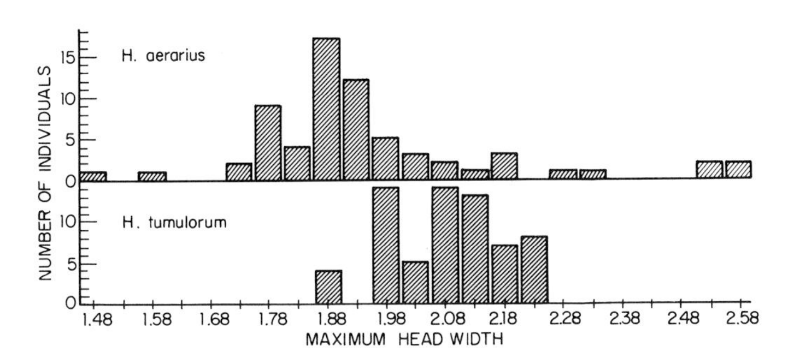

FIGURE 9-3. The size-frequency distribution of the allometric halictine bee species, *Halictus aerarius*, is compared with that of a nonallometric one, *H. tumulorum*. The *H. aerarius* curve is unusual in displaying skewness (or incipient bimodality) of the kind typifying primitive worker subcaste differentiation in ant species. Each sample was drawn from several nests (adapted from Sakagami and Fukushima, 1961).

any role, which does not become completely determined until they reach the adult stage. Mating sometimes shunts females permanently into the queen role, while failure to mate sometimes limits them wholly to the worker role. But exceptions do exist: inseminated females have been observed to continue foraging, and unfertilized females have been seen to lay eggs.

A temporal division of labor is nevertheless pronounced throughout the Halictinae, even in species displaying little or no differentiation of physical castes. Young females of *Augochloropsis sparsilis* guard and excavate the nest and forage for food. As they grow older they switch entirely to egg laying (Michener and Lange, 1959). Those of *Lasioglossum rhytidophorum* first guard the nest entrances, then shift to excavation combined with either foraging or egg laying (Michener and Lange, 1958b). The pattern of *Dialictus zephyrus* is different again: as a rule the younger bees excavate, while the older bees either continue to excavate and forage in addition, or else they shift entirely to egg laying. Guard duty is conducted by working females both in the preforaging and foraging periods of adult life. Suzanne Batra's meticulous study of this species (1964) has shown that the temporal patterns of individual females is extremely variable. In one perverse but illuminating example a bee was seen to oviposit shortly after returning to the nest with a pollen load. In addition, the relative proportion of *D. zephyrus* castes in single nests also fluctuates widely through the season (Batra, 1966a).

A positive correlation exists within individual halictine species between the total size of adult females and the degree of development of their ovaries. Large size also predisposes individuals to mate and to settle into the role of egg laying. These two relations hold whether or not other anatomical characteristics vary allometrically. It follows that size alone is a critical factor in halictine caste determination. In his pioneering study of the social biology of *Evylaeus malachurus*, Legewie (1925) hypothesized that caste might be determined simply by the size of the pollen ball with which the female larva finds herself in the sealed brood cell. He noted that the balls assembled in late summer are larger than those made earlier in the season, and they produce large queens capable of overwintering. Legewie's idea has found substantial support in the observations of Knerer and Atwood (1966a) on *Evylaeus cinctipes.* In this species the queens average 15 percent longer than the workers. Spring larvae were found to receive pollen masses averaging only 33.9 mg in weight. These individuals invariably turned into small workers. Summer larvae on the other hand received masses averaging 61.7 mg in weight, and they turned into either large workers or queens. Parallel observations have been made on *Lasioglossum duplex* by Sakagami and Hayashida (1960) and on *Dialictus zephyrus* by Batra (1964). Other factors, in particular, temperature of the brood cell and quality of the pollen, have not been ruled out by these studies. But the fact that larvae consume all or almost all of their provisions suggests that the food supply at least constrains the maximum adult size to a degree that can be decisive in caste determination.

A most peculiar phenomenon that does not fit in with the remainder of our knowledge of halictid sociobiology is the existence of male dimorphism in at least two Australian species of the subgenus *Chilalictus* of *Lasioglossum* (Rayment, 1955; Houston, 1970). One male form is "normal," that is, typical for halictids generally. The second has reduced wings, is apparently flightless, and has exceptionally large heads and mandibles. Houston has suggested that the second form is a "male soldier," which functions in nest defense. Although he has no direct evidence to support this guess, he points out that normal males of a third *Halictus* species, *H.* (*Chilalictus*) *seductus*, have been seen guarding nest entrances (Rayment, 1935). Houston's hypothesis, while fascinating, must be regarded with great reservation. The Australian *Chilalictus* are not even fully social; at most they are communal in nesting habits. Furthermore, no other cases of male castes have ever been recorded in the social Hymenoptera. In the social bees, not even the females are divided into soldiers and minor workers as supposed for the *Chilalictus* males.

Bumblebees: Caste as Part of a Dominance Hierarchy

In steps paralleling those taken by the Halictinae, the bumblebees have evolved primitive caste systems that are regulated jointly by larval feeding and the behavioral and physiological consequences of queen control. Among the pocket-making species of *Bombus,* there exists little or no difference in average size between workers and egg layers. The queens and workers of *B. hortorum,* for example, are only slightly different in average size and display a broad overlap in size range. The related species, *B. agrorum,* apparently possesses no morphological caste differences. However, strong physiological differences do exist. Cumber (1949a) found that, in addition to the expected distinction in ovarian development, there is also a marked difference in metabolism. When he fed large quantities of food to both queens and workers, the queens gained substantial weight by the addition of fat bodies, but the workers did not change. This result is not too surprising when it is recalled that the queens are the only individuals who overwinter.

Morphological caste differentiation among the pollen-storing species, on the other hand, varies from almost nonexistent to strongly marked. *B. lucorum,* a large European species, is characterized by complete dimorphism. Workers collected by Cumber weighed from 40 to 320 mg; and queens, from 460 to 700 mg. Yet little or no external allometry is associated with the size difference; in outward appearance, the workers are small simulacra of the queens.

Only a single female functions as the mother queen in bumblebee colonies. This individual maintains her position by aggressive behavior, ranging in intensity from head butting with mandibles agape to biting and grappling (Free, 1955a). Other females in the colony are organized into a loose dominance hierarchy. Besides monopolizing egg laying, the queen and her immediate subordinates reveal their status by performing most of the fanning within the nest. The rank of individual bees is closely correlated with the degree of development of their ovaries. The queen directs most of her hostility toward those subordinates who have the largest ovaries. When she is removed, the level of hostility among the surviving females increases until one establishes herself as the new alpha queen, which, predictably and normally, is the individual with the largest ovaries at the outset. During the interregnum some increase in ovarian development also occurs among a few of the workers, and an increased proportion of the larvae transform into morphological queens.

Division of labor among bumblebee workers was discovered in 1890 by Frederick V. Coville, who noticed that the larger members of a *Bombus borealis* colony conduct most of the foraging and rearrangement of nest material, while the smaller individuals largely devote themselves to work within the nest. Since Coville's paper the subject has been investigated in increasing depth by Meidell (1934), Jordan (1936), O. W. Richards (1946), Cumber (1949a), Anne D. Brian (1952), Free (1955b, 1961a), and Sakagami and Zucchi (1965). The *B. borealis* sequence is now recognized to occur generally in the bumblebees. The smallest individuals of some species not only do not leave the nest; they are also unable to fly. For the great majority of individuals, however, the correlation between size and labor remains very loose. It consists chiefly of a tendency for smaller bees to commence foraging at a later age than their larger nestmates, so that the colony members observed foraging on any given day are larger on the average than those remaining in the nest. The raison d'être of the pattern has been conceived by Free and Butler (1959) as follows: "Two advantages can be immediately suggested. Firstly, it is clear that the smaller workers must be able to move more easily through the intricate galleries of the comb than their larger sisters, and thus are better able to perform household duties; the larger bees, on the other hand, are able to collect greater loads of nectar and pollen and, one would suppose, may also be able to fly when conditions are less favourable." The lifetime pattern of behavior of individual bees is variable in the extreme. Most are capable of laying eggs, but usually do so only when the suppressive influence of dominant nestmates is removed. The Amazonian species *Bombus atratus* forms an exception in that the medium-sized workers frequently lay unfertilized eggs and are in fact responsible for practically all of the production of males (Zucchi, personal communication). In a study of seven British species, Free (1955b) discovered that each worker tends to remain either a forager or a house bee for long stretches of time, often for the duration of its adult life. The longer an individual keeps at a given task, the less likely it will change in the future, even in the face of colony need. Nevertheless, about a third of the workers do shift periodically from one task to another. Their behavior, along with that of some of their more specialized nestmates, can be altered according to the needs of the colony. When foragers were deliberately removed by Free, many of the house bees became foragers. The reverse ablation resulted in some foragers becoming house bees, but in a less clear-cut manner. The less consistent a worker had been in its selection of choice of tasks prior to the ablation, the more easily it shifted to the task demanded by the new situation.

The physiological basis of caste determination in bumblebees has not been elucidated fully. As part of a general hypothesis applied to all social bees, Wheeler (1923) proposed that bumblebee workers arise from undernourished larvae. Cumber (1949a) noted that queens do in fact appear only during the later stages of colony development, at which time at least one worker is present for every larva. This finding is consistent with the notion that extra feeding tends to speed the growth of larvae and convert them into queens. Free (1955c) has cited further experimental results consistent with the nutrition hypothesis. He established ten colonies of *Bombus pratorum* in laboratory nests, each comprised of a queen and one or more workers, and supplied them with abundant pollen and honey. Five of the colonies eventually produced adult female bees. In three of these, each queen had only one worker to assist her, and all of the adult offspring turned out to be workers. In the remaining two colonies, one queen had two workers and the second queen five workers to assist her. Under these slightly more favorable circumstances, both queens and workers were produced. The data are too few to be of more than marginal significance, but they suggest that the worker/larva ratio is a limiting factor in queen determination.

Sladen (1912) held the contrary opinion that bumblebee caste is not determined simply by nutrition. He believed that intrinsic changes cause eggs and larvae to become queen-potent only toward the end of the season. Subsequently, Cumber (1949a) and Free (1955c) observed that the presence of the mother queen suppresses the production of new queens until late summer. Colonies which have lost their queens prematurely produce new queens earlier than queenright colonies. This effect might be due to the existence of an inhibitory pheromone secreted by the mother queen, but it is more likely the outcome of the simple reduction of egg production and the consequent rise of the worker/larva ratio resulting from the death of the mother queen. Cumber in fact noted that the mother queen reduces her oviposition rate prior to the production of new queens at the end of the summer. Thus the demonstration that the presence of the queen earlier in the year is associated with the inability of the colony to generate queens is not inconsistent with the simple trophic hypothesis first proposed by Wheeler.

The pocket-maker bumblebees display a peculiarity of behavior that deserves special comment. Plath (1934) noted that, although worker larvae feed from pollen stores placed next to them in the wax pockets, the queen larvae are fed directly by the nurse workers, who tear holes in the tops of the queen cells for this purpose. In other words, pocket-maker species feed their queen larvae in the same fashion that the pollen-storer species feed both their queen and worker larvae. The occurrence of such a preferential treatment provides one more hint that nutrition plays a key role in caste determination. In order to test this possibility, Plowright and Jay (1968) forced workers of pocket-maker colonies to feed larvae directly through the pockets at various times in the season. Their results unfortunately proved inconclusive. Queens still did not appear until the later stages of colony development. The new evidence still does not conflict with the simple trophic hypothesis, since it can be explained as a result of an increased worker/larva ratio later in the season. But neither does it exclude Sladen's view that other, less obvious factors can be decisive in queen determination.

Honeybees: Advanced Trophogenesis and Division of Labor

The queen and worker castes of *Apis mellifera* are so different from each other in appearance as to seem to belong to different species of insects. Lukoschus (1955, 1956) has cited 53 morphological characteristics by which a queen can be recognized, many of them extreme in quality, and these are matched by equally pervasive physiological and behavioral differences. In general these differences can be linked directly to the peculiarities of the honeybee life cycle. It will be recalled that, as a species, the honeybee is distinguished by the fact that its colonies are large, the life span of its individual workers is short, and the rate of oviposition required to sustain the population is consequently very high. It is not, therefore, surprising that the queens are larger than the workers. They possess proportionately even bigger abdomens, which are packed with 300 or more elongate ovarioles. Each queen commonly lays over 1,000 eggs a day, and her basal metabolism is constantly higher than that of the workers surrounding her. Honeybee colonies are also unusual in that they multiply by a process of fission and swarming in which resident queens leave the old hive in the company of a large fraction of her daughter workers (see Chapter 5). At no time does the queen participate in the ordinary duties of the hive. She is highly specialized for reproduction, and her most complicated behavior occurs early in her adult life when she challenges the rival sister queens, which emerge at approximately the same time, and then conducts her nuptial flight. For the rest of her existence she functions as little more than an egg-laying machine. The mark of this regressive existence has fallen everywhere upon the morphology of the queen. Her mouthparts are reduced; her eyes are smaller than those of the workers (an average of 4,920 ommatidia versus 6,300); her antennae are shorter and bear fewer sensilla; she has a conspicuously smaller brain; she lacks pollen-collecting hairs; and her hypopharyngeal and wax glands, which in the worker are the principal sources of larval food and building material, are underdeveloped. The mandibular glands, which are the source of pheromones used to control worker behavior, are among the few non-reproductive organs developed to a greater degree than in the worker caste.

As one might expect in the case of such strong dimorphism, the determination of individual female honeybees to the queen or worker caste occurs at a very early stage in larval development. The nurse workers exercise a tight control over the development of their sister larvae. During most of the year they are inhibited by the presence of the mother queen from making any attempt to produce new queens from available larvae. But at the beginning of the spring reproductive season, or any other time that the mother queen dies or loses her vitality, new queens are produced. The inhibition is due specifically to pheromones, a principal constituent of which is the "queen substance," *trans*-9-keto-2-decenoic acid, a compound manufactured in the mandibular glands of the mother queen (see Chapter 15 for further details). The first step taken by the workers to produce queens is the construction of one or more "royal cells," or "queen cells" as they are frequently called. These are conspicuous waxen structures that are larger than the usual, worker-producing brood cells. They are vertically aligned on the outer surface of the brood comb and are variably oblong in shape (as opposed to the invariable hexagonal form of the brood cells) and strongly pitted on the outside. Any egg of female genetic constitution which is placed in a royal cell will be reared as a queen. Eggs transplanted experimentally from worker cells to royal cells develop into queens, while those transferred in the reverse direction become workers. Taking advantage of this simple effect, Weaver (1957) was able to time the events of differentiation precisely. He found that up to three days into larval development a larva transferred from a worker to a royal cell turns into a queen. At three days some worker characteristics begin to appear in the final adult product: the ovaries are smaller than average for a queen, and a few other anatomical features are intermediate or worker-like. If larvae are left for three and a half to four days before being transferred, some of their adult qualities become irrevocably worker-like, while other features are intermediate between the queen and worker castes. Jay (1963) found that the commitment to queen development occurs later and is much more tenuous. When he removed larvae at a more advanced stage of development from the royal cells, the smallest ones became worker-like, while larger individuals became either queen-worker intercastes or queens, depending on their initial size.

What is put into the royal cells that turns young honeybee larvae into queens? The answer has been known a long time: it is royal jelly, a material supplied chiefly by the hypopharyngeal glands of the nurse workers. A lesser fraction of the royal jelly is supplied by the mandibular glands, and it is not impossible that still further components are added by the postcerebral and thoracic glands. All of these glands together are sometimes referred to loosely as the "salivary gland complex" (Townsend and Shuel, 1962). Von Rhein (1933, 1956) first demonstrated that bee larvae can be reared on royal jelly alone in constant temperature cabinets. He thus succeeded in creating a technique that had been envisioned 150 years earlier by Charles Bonnet (in François Huber, 1792). Bonnet's idea, contained in his letter to Huber of August 18, 1789, is worth quoting in full, because it exemplifies the extraordinarily high quality of the correspondence between these two blind Swiss entomologists: "The royal eggs and those producing drones have not yet been carefully compared with the eggs that give rise to the workers.

But this should be done, so that we can determine whether these two kinds of eggs differ by hidden characteristics. The food supplied to the queen larva is not the same as that given to the common larva. Could we not try, with the point of a pencil, to remove a little of the royal food and give it to a common larva deposited in a cell of the largest dimensions? I have seen common cells hanging almost vertically, in which the queen had laid eggs; and these I would prefer for such an experiment." In 1955 Weaver followed approximately this procedure. He discovered that, if the larvae are given fresh royal jelly every two hours, they will develop into queens. An alternative method was employed by Jay (1964), who found that if larvae are transferred to the deposits of royal jelly three times a day, about half will develop into adults with predominantly queen-like features. Conversely, larvae reared on "worker jelly"—the salivary secretion deposited by workers into regular brood cells—invariably develop into workers (Rembold and Hanser, 1964). Honeybees are unique among the social bees in the amount of care the nurse bees give to the growing larvae. One apparently typical larva observed continuously by Lindauer (1952) and his assistants was visited 2,069 times by nurse workers, who spent a total of 181 minutes and 38 seconds with it. The larva was fed during 143 of the visits, during a summed period of 109 of the 181 minutes. Thus a clear opportunity exists for the nurse workers to appraise the developmental status of larvae at frequent intervals and to adjust the feeding schedule accordingly. Whether they do so remains the key question. Weaver (1957) noted that nurse bees accept larvae transferred to cells of the other caste with increasing reluctance as the larvae age, so it is apparent that they do have at least a crude means of assessing the emerging caste characteristics. Free (1960) interpreted data from studies by Lindauer and himself to indicate that the nurse workers do in fact make regular evaluations of the caste and age of the larva. He regarded it as probable that each nurse deliberately varies the composition of her secretions to conform to the status of each larva visited. Nevertheless, the evidence does not entirely exclude the alternative possibility that each bee specializes in producing a narrow range of secretions and limits her visits to the larvae suited to receive them.

Ever since Weaver found that adult queens can be reared in vitro on a diet consisting solely of royal jelly, the way has been open for the chemical identification of the compounds essential for queen determination. However, two formidable technical difficulties have so far thwarted all efforts to do so. First, it turns out that the critical substance (or combination of substances) is very labile. Royal jelly loses its potency after a few months when stored in ordinary refrigeration and after a few hours when kept at room temperature. The decomposition is of course hastened when the materials are subjected to the ordinary diluting and heating techniques of organic microanalysis. The second difficulty stems from the extraordinary chemical complexity of raw royal jelly. Not even the anatomical sources of the material have been identified with full certainty. The hypopharyngeal and mandibular glands are principal contributors, but still other cephalic glands may be involved. Honey and pollen are known to be added to the worker diets from around the third day of larval life. These are believed to originate in the crop of the adult nurse workers, but this conclusion is no more than a weak inference from indirect evidence (Ribbands, 1953; Irmgard Hoffman, 1960). If these substances are indeed regurgitated, they may bring with them other active compounds from the foregut of the nurse workers. The function (if any) of honey and pollen in caste determination is still unknown. Simpson (1960) has gone so far as to question whether pollen even plays a role in larval nutrition.

The biochemistry of royal jelly, a subject surveyed from differing points of view by Townsend and Shuel (1962), Rembold (1965), Weaver (1966), Weaver *et al.* (1968), Barbier (1968), and Painter (1969), is a discouraging morass of largely unrelated details. The identified components include water, sugars, and proteins. In addition there exists a large quantity of an unusual fatty acid, 10-hydroxy-*trans*-2-decenoic acid, which has been suggestively referred to as "royal jelly acid." In spite of its abundance and close structural similarity to the queen substance (9-keto-*trans*-2-decenoic acid), a proven pheromone of great potency, royal jelly acid has not been implicated in caste determination or any other social role. Recently several other aliphatic hydroxy-acids, dicarboxylic acids, and aromatic acids have been isolated from royal jelly. Still other compounds identified include all of the common amino acids, several B vitamins (pantothenic acid is especially abundant), acetylcholine, 24-methylene cholesterol, adenosine di- and triphosphate, and "biopterin," which is 2-amino-4-hydroxy-6-(L-erythro-1′,2′-dihydroxypropyl)-pteridine. There also exist astonishingly large amounts of nucleic acid bases. In a one-gram sample of lyophilized jelly analyzed by Marko *et al.* (1964) there were 47.0 mg of RNA associated with phosphorus and an equal amount of phosphorus-associated DNA. Judging from the current rate of identification of new components, it seems safe to predict that the chemistry of royal jelly has been no more than marginally explored. The very meaning of its particular composition still eludes us, and there is an excellent chance that the critical caste substance remains among the fraction still to be identified.

In the absence of the crucial biochemical information, no less than six competing hypotheses are extant concerning the physiological basis of caste determination in *Apis mellifera.* The multiplicity of hypotheses is nevertheless useful, not only because they state the problem more clearly but also because they delineate the subtle distinctions that are possible in any set of physiological models of caste determination.

First hypothesis. The larger quantity of food given to the queen larva in the first three days causes both faster growth and a stimulation of hormone production adequate to produce queen characteristics (Haydak, 1943).

Second hypothesis. A specific, labile compound is present in royal jelly that causes early queen determination (Weaver, 1955).

Third hypothesis. Larvae are deficient in a growth hormone when reared on ordinary diets, a condition reflected by low mitochondrial content and reduced respiratory rate, and they thus become workers. Royal jelly contains a substance, perhaps the hormone itself, which makes up for this deficiency and permits the development of complete females, that is, queens (Osanai and Rembold, 1968; Rembold, 1969).

Fourth hypothesis. The early diet of the queen larva contains a factor that inhibits metamorphosis and hence promotes the eventual attainment of greater adult size. The later diet of the queen contains another factor that promotes development of reproductive organs (von Rhein, 1956).

Fifth hypothesis. The nutrient balance in the early larval diets, controlled carefully by the nurse workers, causes variation in hormonal balance which in turn leads to the establishment of caste differences (Shuel and Dixon, 1960).

Sixth hypothesis. Substances in the food of worker larvae promote the development of workers or inhibit the development of queens (Weaver, 1966).

Of course not all of these models are mutually exclusive. If any consensus exists among contemporary students of the subject, it is that some factor in the queen diet triggers an endocrine change in the earliest stage of larval development. This change in turn brings about an important physiological alteration, the most visible manifestation of which is an increased mitochondrial content together with a quickened respiratory rate, both lasting through most or all of the larval period.

Although the royal jelly problem has a perennial fascination, the most significant aspect of honeybee polymorphism is, in my judgment, the extraordinarily elaborate division of labor displayed by the workers. Very little morphological variation occurs within the worker caste. The weight of individuals ranges from 80 to 110 mg, and allometric variation among the body parts is negligible. Size has some influence on behavior, but this factor is small compared to the profound changes that occur in a regular progression through the 30-odd days in the adult life of the average worker bee. Temporal polyethism appears to have been first recorded, and with reasonable accuracy, by Charles Butler in 1609:

> *The young Bees as best able, beare the greatest burdens: for they not only worke abroad but also watch and ward at home both early & late: whē need is, they hazard their lives in defense of the nest, they beat away the drones, & fight with other Bees and waspes, and assaule with their speeres whatsoever else offendeth them, they carry their dead forth to be buried, and performe al other offices. But the labour of the old ones is only in gathering, which they wil never give over, while their wings can bear them: & then when they cease to worke, they wil cease also to eate: such enemies are they to idlenes* (The Feminine Monarchie).

The same basic pattern was rediscovered in 1855 by Dönhoff. He found that he was able to follow the activities of single workers when he requeened a colony of black *Apis mellifera* with a yellow queen and thus made it possible to detect her offspring from the moment they emerged among their black nestmates. Dönhoff noted that during the first part of their life the workers stayed in the brood area. By the age of ten days they were assisting in the repair of broken comb and the construction of new comb from wax. Not until they were 15 days old did they begin to forage outside the hive. Based on limited observations of his own, Gerstung (1891–1926) proposed a detailed scheme for the allocation of tasks which he thought depended on a rigid time schedule of glandular development. Gerstung's theory was partially confirmed and greatly extended by the meticulous studies of Rösch

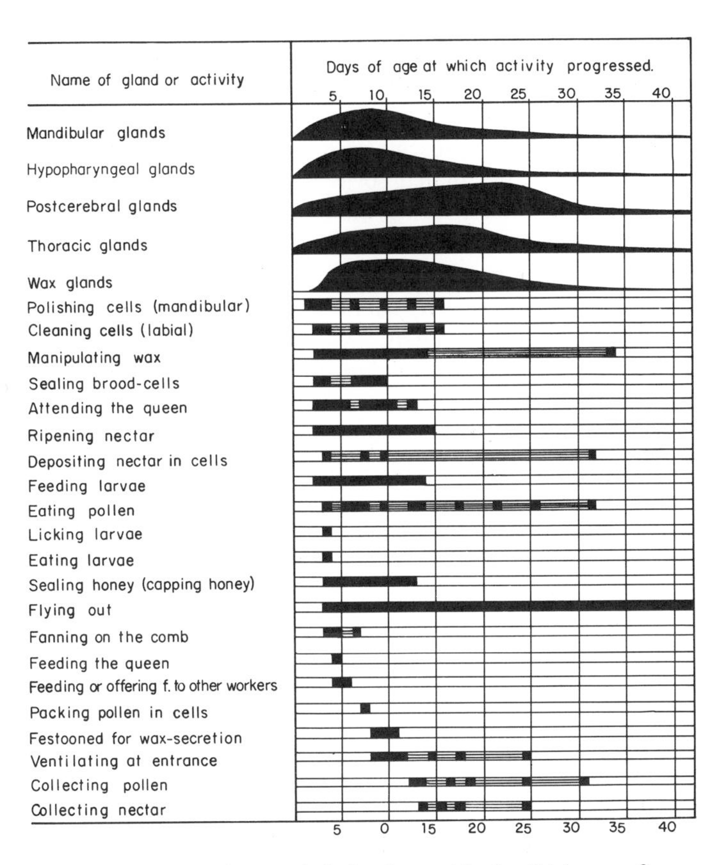

FIGURE 9-4. Changes in behavior and in the thickness of four exocrine glands during the adult life span of worker honeybees (modified from G. E. King, 1933).

(1925, 1927, 1930). Since that time polyethism has been investigated with increasing depth by other investigators, among whom G. E. King (1933), Ribbands (1952), Lindauer (1952), and Sekiguchi and Sakagami (1966) deserve special mention. The amount of effort that has gone into these studies is in fact quite extraordinary—Sekiguchi and Sakagami, for example, spent 720 hours observation time to collect data on 2,700 individually marked bees, while Lindauer watched a single worker for a total of 176 hours and 45 minutes!

The essential story of temporal division of labor is presented in Figures 9-4 to 9-6. These curves and diagrams contain information extracted from the work of King and Lindauer. Figure 9-4 illustrates how in the first days of adult life two of the principal sources of larval food, the hypopharyngeal glands (usually called the pharyngeal glands in the older literature) and the mandibular glands, reach the peak of their development. This phase coincides approximately with the nursing period of the bees, during which time they feed the larvae and queen with the glandular secretions. Shortly after the onset of the first phase, indeed almost coincidentally with it, the wax glands undergo a rapid growth and are maintained in a functioning condition during a three-week period. During this time the bee is active in the construction and sealing of brood and honey cells. The nursing and construction periods are so broadly overlapping as to be almost coincidental. At two to three weeks into adult life, as the three principal glands shrink in size and become less productive, the worker becomes a field bee. Under normal circumstances it remains in this condition for the rest of its life. The exact timing of the principal episodes, and the insertion of lesser activities into the timetable, vary greatly among individual workers. Furthermore, as Kerr and Hebling (1964) have shown, larger bees become field workers about a week before their smaller sisters. Figure 9-5 shows the record of one worker watched by Lindauer for several hours a day for the first 24 days of her life. In this particular case, the overlap of the nursing and construction periods is broader than usual, but the sharp changeover from the status of house bee to field bee is typical. The still more detailed protocol of Figure 9-6 points up a very important feature of the labor schedule of the worker bee, namely that the bee is not propelled from task to task according to any internally guided program. Instead, the individual bee is very labile in its behavior and seems to respond to exigencies as they are encountered. Moreover, the bee spends about two-thirds of its time either resting or wandering through the interior of the nest, an activity that Lindauer has referred to as patrolling. In Lindauer's view these two outwardly unproductive activities enhance the capacity of the colony as a whole to respond to capricious changes in the environment. Patrolling bees assess the needs of the colony from moment to moment and are thus able to respond to local requirements with less delay. Resting bees constitute a reserve force, available for major emergencies, such as overheating of the nest or invasion by a predator, that require the simultaneous employment of many individuals. In the case of the working bees, a superabundance of individuals at any given task—for example, the building of brood cells or the collection of water—forces some into other, less crowded functions, and the division of

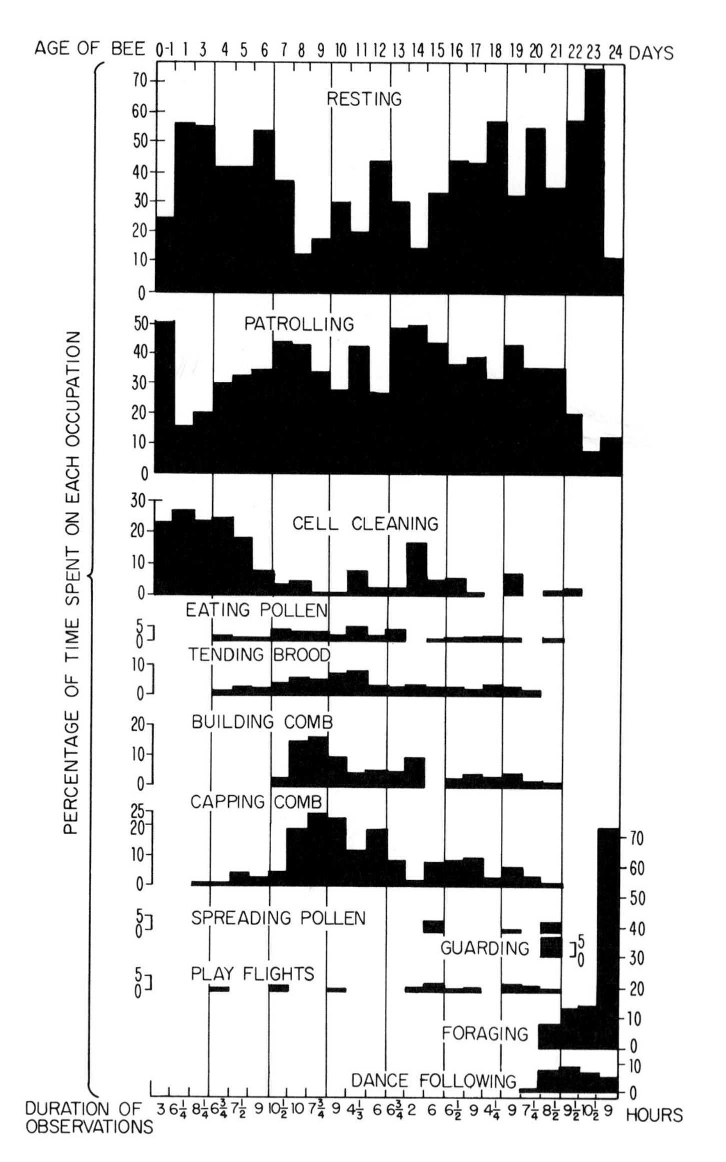

FIGURE 9-5. The activities of a single worker honeybee during the first 24 days of her adult life (redrawn from Ribbands, 1953; based on data of Lindauer, 1952).

labor tends always to attain the proportions appropriate to the needs of the colony as a whole (Free, 1965).

The flexibility of individual labor schedules extends even to functions limited by glandular activity. Nolan (1924) and Rösch (1930) discovered that when a shortage of wax-producing bees arises, some of the older workers redevelop their wax glands and recommence comb building. This result has been corroborated by experiments of Orösi-Pál (1956), who showed that the critical stimulus is the presence of empty space in the hive suitable for comb building. Rösch also found that when honeybee colonies were decimated so as to leave only older bees that had previously flown, the hypopharyngeal glands of many of these individuals regenerated and became functional. According to Moskovljevíc-Filipovíc (1956), some of the older workers regenerate their glands and return to nursing, but others do not, so that the result is a redress of the earlier numerical balance between nurse and field bees. Free (1961b), in conducting more detailed experiments of the Rösch type, found that the presence of larvae is necessary for the redevelopment of the hypopharyngeal glands. Precisely which signals from the larvae are critical is still unknown. Other factors not mediated by colony communication may be involved. For example, the hypopharyngeal glands will not develop if the workers have no pollen in their diet, this being the principal source of

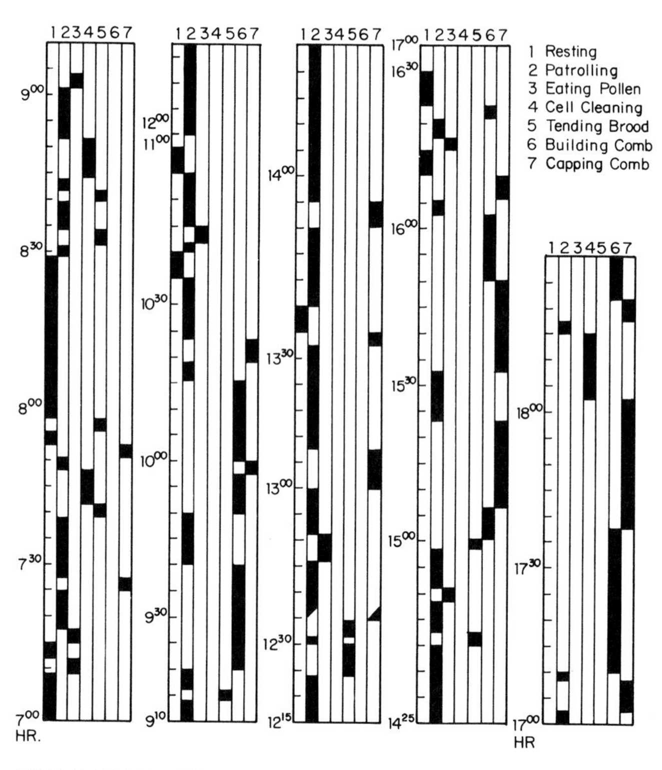

FIGURE 9-6. The activities of a single worker honeybee during the eighth day of her adult life (redrawn from Ribbands, 1953; based on data of Lindauer, 1952).

proteins, vitamins, and fats for bees (Kratky, 1931). Finally, Milojevic (1940) found that the original development of the hypopharyngeal glands is prolonged beyond the normal time for regression if the workers are forced to continue on duty as nurse bees.

Stingless Bees: From Trophogenesis to Genetic Determination

The Meliponini are similar to the honeybees in the degree of queen-worker dimorphism. The queen is approximately the same weight as the worker, but she possesses a smaller head and thorax, which is compensated for by a proportionately larger abdomen. Like the honeybee queen, she exhibits many other peculiarities in the conformation and color of all her major body parts, and she is equally highly specialized in behavior. Since colonies multiply by fission and swarming, rather than by the emission of single founding queens, the meliponine queen has little to do in her life besides feed, mate, and lay eggs. She never aids the workers in the daily chores of nest building, brood care, and foraging. The stingless bees further resemble the honeybees in lacking morphological worker subcastes. The workers vary only moderately in size, and the smallest and largest individuals are closely similar in body proportions and behavior. Division of labor is achieved through an orderly progression of behavioral changes that occur through the life of individual workers, in conjunction with predictable events in development of some of the principal exocrine glands.

The meliponines have nevertheless staged a radical departure from the honeybees and all other social insects by evolving a genetic control of caste determination. In the primitive state, exemplified by *Lestrimelitta* and *Trigona,* the determination of individuals to either the queen or the worker caste is still trophogenic. In this case two kinds of brood cells are constructed by the colonies, small ones for workers and drones and large ones for queens. According to Kerr (1950a), the queen cells are most often located near the margins of the brood combs. Thus the workers are able to determine the caste of the individual female larvae. The physiological control is presumably exercised by means of differential feeding. In some species, for example *Trigona carbonaria* (Rayment, 1932), the cells are kept open for several days, and new food is added after the larvae hatch. Even after its cell is closed, the larva feeds on a sequence of nutriments: first the glandular secretion provided by the workers just before the cell capping, then the honey and pollen mixed with some glandular food. There are abundant opportunities for the nurse workers to influence caste development by fine adjustments in the absolute and relative quantities of these principal constituents.

In the genus *Melipona,* by contrast, the sizes of the cells producing workers and queens are identical, and the queen cells are scattered randomly through the brood comb. Kerr (1950a,b) observed that the ratios of queen pupae to worker pupae during the reproductive season tend to be constant and to conform at least approximately to the phenotypic ratios that would be expected if the queens were heterozygous for two or three independently assorting loci. The Kerr hypothesis can be easily grasped by examining the diagram in Figure 9-7. The key feature is that only individuals who are completely heterozygous for paired allelomorphs at all of the loci are capable of developing into queens. It follows that the inheritance of female caste (queen versus worker) can be represented in the simple fashion displayed in the figure, which is roughly the equivalent of an F_2 hybrid cross of the kind familiar in elementary genetics. The reason for this felicitous correspondence is the following. Since the male is a hemizygous individual derived from an unfertilized egg (this generalization will be examined further in Chapter 17), its genotype will be the same as one of the four equally possible haploid products of female meiosis; in other words, it will be either AB, or Ab, or aB, or ab. It carries one of these four possible combinations of female caste genes, even though its parthenogenetic origin has predetermined it to be a male displaying neither queen nor worker characteristics. It mates with a queen whose complete heterozygosity for the caste genes dictates that her eggs will bear the four combinations (AB, Ab, aB, ab) in equal abundance. At fertilization an average one-fourth of the diploid products will be heterozygous for both loci and hence capable of developing into queens, as shown in Figure 9-7. Kerr's hypothesis assumes that each queen will mate with only a single male. Evidence for single matings in *Melipona quadrifasciata* has been provided by Kerr *et al.* (1962). If three loci are involved in caste determination, one-eighth of the zygotes can be expected to be queen-potent, by a straightforward extension of the reasoning from the two-locus model.

The fit of actual queen-worker ratios to theoretical values has been less than perfect. In Table 9-1 are given

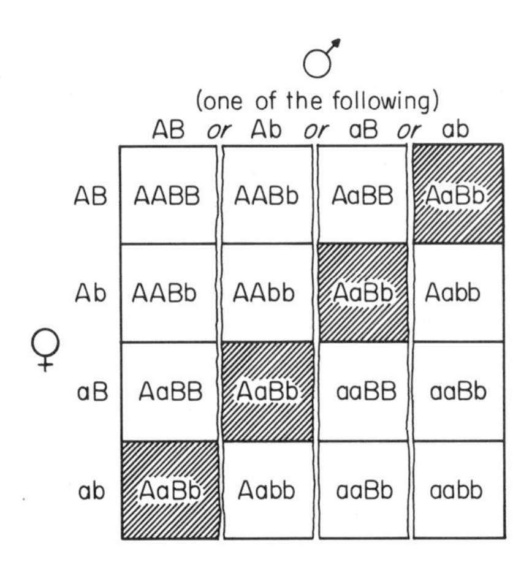

FIGURE 9-7. The Kerr hypothesis of genetic caste determination in stingless bees of the genus *Melipona.* This example is of the two-locus case found in some species of the genus. Only the complete heterozygotes, which are indicated by crosshatching, are capable of developing into queens. Regardless of the genotype of the parental male, the genetic queens will always constitute one-fourth of the offspring.

the maximum ratios in samples accumulated by Kerr and his co-workers to the present time. The convergence of most of these maximum values to the 25 percent level is impressive and supports the idea of a widespread occurrence of two-locus caste control in *Melipona.* Where maximum values fall below 25 percent, they do not conform to the next lower level of 12.5 percent predicted by the three-locus model. The simple genetic hypothesis also does

TABLE 9-1. Maximum percentage of queens found in different species of *Melipona* (from Kerr, 1969).

Species	Segregation: Workers	Segregation: Queens	Percent queens
M. marginata	177	62	25.9
M. quadrifasciata	18	7	28.0
M. nigra	11	4	26.7
M. rufiventris[a]	18	4	18.1
M. interrupta fasciculata	27	7	20.6
M. favosa orbignii[a]	49	8	14.0
M. pseudocentris	142	39	21.5
M. melanoventer	111	15	11.9
M. quinquafasciata	22	9	29.0
M. flavipennis[a]	38	3	7.3

[a]Species in which only this sample was taken; the others were based on many samples with about 25 percent queens.

not explain why queens are not produced at all in winter or in poorly fed colonies. The obvious explanation, which has been tested by Kerr and Nielsen (1966) and Kerr, Stort, and Montenegro (1966), is that both the caste genes and environmental factors (most likely nutrition) are limiting in caste determination in *Melipona.* A breakthrough in methodology came when it was discovered that queens invariably have four abdominal ganglia while workers have four or five (Figure 9-8). Kerr and his co-workers deduced that all individuals who end up with four ganglia are capable of developing into queens but require the right food to do so, while all individuals with five ganglia are destined to become workers regardless of the state of their nutrition. The important test is that ganglion numbers segregate strictly according to the 3:1 ratio (three females with five ganglia to each female with four ganglia) predicted by the two-locus model. But the failure of some of the queen-potent larvae with four abdominal ganglia to receive adequate nutrition causes most or all of the remainder of their adult morphology to become worker-like. These individuals can therefore be regarded as genetic queens who are phenocopies of workers, and they are referred to as "queen-like workers." During the winter, or whenever the colony falls on hard times, all of the 25 percent of the female population with queen-potent genes are turned into "queen-like workers," so that the offspring of the colony are exclusively composed of functional workers. What are the environmental factors that regulate the development of the queen-potent larvae? In experimental studies on *M. quadrifasciata,* Kerr, Stort, and Montenegro (1966) eliminated from consideration most of the conceivable factors including temperature, the nutritional status of the mother queen, the number of eggs laid by the mother queen in the brood cell, and the proportion of protein in the colony diet. Instead, the ultimate caste of the queen-potent larva was shown to be closely related to the quantity of food present in the closed brood cell. Each cell is filled by four to eleven workers, an operation requiring 15 to 64 seconds to complete. Queens were observed to emerge from cells provisioned by four to eight workers, but none were ever seen to be produced in cells provisioned by nine to eleven workers. It turns out, curiously, that the more nurse workers participating in the provisioning, the less the total amount of material contributed, so that queens actually emerge from cells in which larger quantities of food were placed. This fact is reflected in the greater average size of queen pupae. Pupae

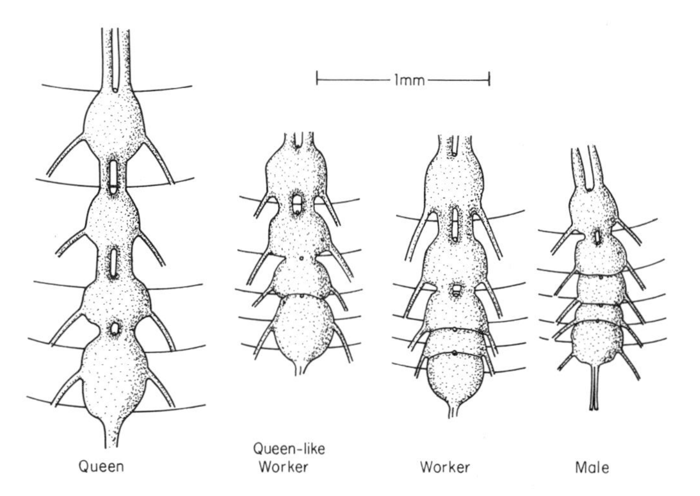

FIGURE 9-8. Abdominal section of the ventral nerve of *Melipona marginata,* showing the sex and caste differences in gross structure (redrawn from Kerr and Nielsen, 1966).

weighing less than 72.0 mg failed to show the expected 3:1 ratio of workers to queens.

Studies on the ontogeny of adult behavior in meliponine workers have revealed patterns basically similar to that of *Apis mellifera* but differing in three principal respects: the period of wax cell construction begins earlier and is more prolonged, the life span is greater, and foraging begins at a later age. Kerr and dos Santos Netos (1956) marked eight workers of *Melipona quadrifasciata* with different combinations of numbers and colors and followed them for four continuous hours every day throughout their lives. The oldest bee lived for 49 days. In the first 11 days, the bees secreted and manipulated wax and participated in the cleaning of the brood area. From the ninth day through the twenty-first day, they constructed and filled brood cells, and fed younger bees and the queen, after which they manipulated wax without constructing brood cells and began to specialize more on nest cleaning and (for the first time) the reception of regurgitated nectar from the older field bees. Sometime between the ages of 34 and 37 days they shifted to field activities. Hebling, Kerr, and Kerr (1964) conducted a parallel study on *Trigona xanthotricha,* this time using the method of introducing worker pupae into nests of *T. postica,* where they could easily be distinguished on the basis of a striking color difference in the adult state (*xanthotricha* are yellow, and *postica* are black). Sixty workers watched in this manner lived an average of 94 days. In the first several days they began working with wax and cerumen. This activity was continued, at a varying pace, for the remainder of their lives. For from 9 to 30 days they provisioned cells and fed the queen. Then, during a brief period lasting from 31 to 36 days into adult life, they secreted wax. Guarding of the nest began at 33 days on the average and was continued for the remainder of their lives. Foraging was initiated at 42 days and also continued to the end of life. After 52 days the first individuals began following odor trails laid down from the mandibular glands of other workers, but it was not until 56 days that they began laying trails of their own. Carminda da Cruz-Landim and Ferreira (1968), in a separate study of *T. postica,* discovered that this curious inability to lay trails until after trail following has been commenced is caused by the temporary failure of the mandibular glands to fill their reservoirs. It thus appears that the behavioral sequence does not result from a lag in learning but rather from a more elementary form of physiological programming.

Caste in Social Wasps

Among the species of social wasps are found almost all conceivable stages in the evolution of caste from the most primitive behavioral differentiation of female nestmates to a well-defined queen-worker dimorphism. Wherever morphological differences exist, they appear to have a relatively simple nutritional basis. Division of labor among workers and programmed behavioral changes during the lives of individual workers are either absent or, at best, very weakly developed. In short, caste systems of the social wasps have not evolved as far as those of the other major groups of social insects.

The origin of caste systems is adumbrated in the life cycle of colonies of the quasisocial species of the polistine genus *Belonogaster* and probably also those of *Ropalidia* and the Stenogastrinae (Roubaud, 1916; Spradbery, 1965). There are no externally visible morphological differences among the collaborating females. Young individuals begin life as foragers and later join the parent female in egg laying. Thus subordinate reproductive castes do not exist. Among the species of *Polistes,* all of which are truly social, the worker caste is kept in a subordinate position by aggressive dominance behavior on the part of the egg-laying females. The degree of the development of the ovaries, and hence the very capacity to lay eggs, is tightly correlated with the position of the female in the dominance hierarchy. Mary Jane West Eberhard (1969) has found that in Colombian populations of *P. canadensis,* no external morphological differences are apparent between the dominant "queens" and the subordinate "workers." Status is maintained by precise aggressive relations among the colony members. In the temperate species *P. fuscatus,* on the other hand, there is a significant difference in mean size in the spring and early summer between the foundress queen and her worker offspring, but with a broad overlap in the size frequency curves of the two classes. Even in species showing no visible external caste differences, there may still exist differentiation at the physiological level. In the autumnal nests of *P. exclamans,* for example, future queens (that is, those destined to overwinter and start colonies the following spring) are indistinguishable from their worker sisters in external morphology, but they show a strikingly greater development of the parietal fat bodies located beneath the abdominal tergites (Kathleen Eickwort, 1969a). Among the remaining species of the Polistinae, comprising the so-called tribe Polybiini, the degree of polymorphism varies greatly (Richards and Richards, 1951). The most advanced case known is that in *Stelopolybia flavipennis* (Figure 9-9). Here we see a large size difference accompanied by an allometric thickening of the body. The entire subfamily Vespinae is also characterized by a complete dimorphism of queen and worker associated with similar allometric thickening (Blackith, 1958a; see Figure 3-13). The Vespinae appear to have evolved their moderately advanced dimorphism along the same lines as the Polistinae, but are no longer represented by the early and intermediate phylogenetic stages in the living fauna.

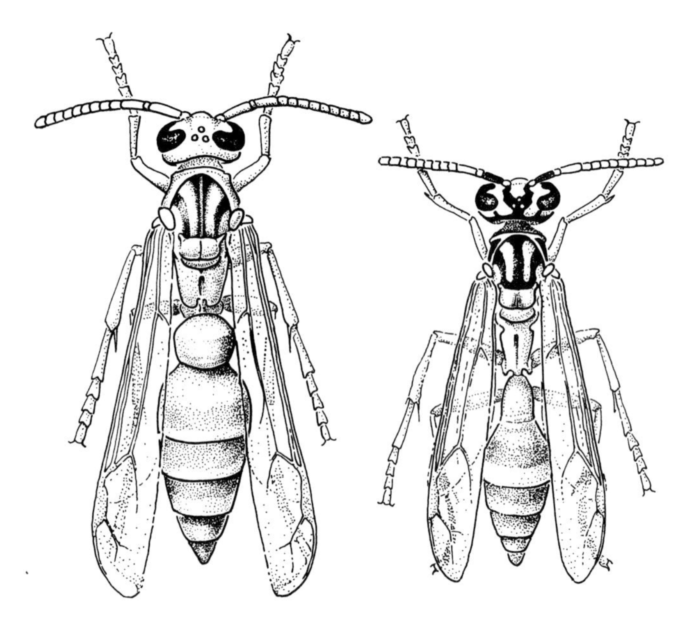

FIGURE 9-9. A queen and a typical worker of the wasp *Stelopolybia flavipennis* from Brazil. This is the most extreme example of queen-worker dimorphism known in the Polistinae (from Evans and Eberhard, 1970).

In two articles published in 1896 and 1897 Paul Marchal postulated that the physical divergence of the queen and worker castes of the Vespinae is based on nutritional discrimination during larval growth. He considered the workers to be victims of "nutritional castration" (*castration nutriciale*). Deprived of adequate nourishment in their larval period, these individuals have ended up as stunted adults with underdeveloped ovaries. In spite of its simplistic sound, Marchal's idea may well be close to the truth in social wasps generally. It can be fitted rather exactly to what is known about the adult caste system of *Polistes.* Workers of this genus are subordinate individuals who must commit relatively large fractions of their time and energy to foraging trips. When they return to the nests, the food they have collected is transferred in disproportionate amounts to the dominant females, who use the energy to make more eggs and thereby entrench themselves still further. Some evidence also exists to suggest that differential larval growth leading to morphological caste differences among the adults has a relatively direct origin in larval nutrition, much as conceived by Marchal. The life cycles of social wasp species generally are programmed to insure a gradually increasing food supply for individual larvae as the colony grows older. Richards and Richards (1951) demonstrated this to be the case in the Polybiini, and they provided a population growth model which predicts a higher ratio of workers to larvae as an automatic by-product of ordinary colony demography (see Chapter 21). A similar effect has been documented in *Vespula germanica* by Spradbery (1965) (Figure 9-10) and in *Polistes fuscatus* by Eberhard (1969). Eberhard noted a second seasonal trend that progressively increases the food supply of individual larvae: from June to August there is a gradual increment in the percentage of loads brought back by *Polistes* workers consisting of food as opposed to building materials. The average size of newly emerging females also increases gradually until finally, in

August and September, many individuals are produced that are as large as the foundress queen. These females are the ones that mate in the fall, go into hibernation, and emerge the following spring to become foundresses of new colonies.

Still other circumstantial evidence can be cited that is consistent with the elementary nutritional hypothesis of caste determination. When Deleurance (1952b) chilled *Polistes gallicus* larvae from late-season colonies at night (to 5°C), they still metamorphosed into queens. But when he chilled the nurse workers, presumably slowing their activity during the ensuing days, some of the larvae they were attending transformed into workers. The queens of *Vespula* are reared in larger cells than those used to produce workers, so that the nurse workers evidently exercise some control over caste. An analysis of *Vespula* by Montagner and Courtois (1963) revealed that the queen larvae start receiving more food than the worker larvae when both are between 0.4 and 0.6 cm in length, and the two castes diverge in size decisively thereafter. Since the food given them by the nurse workers appears to consist primarily or even entirely of nectar, honeydew, fruit pulp, and chewed insect prey, it seems to follow that the quantity of food, rather than its quality, is crucial. The possibility

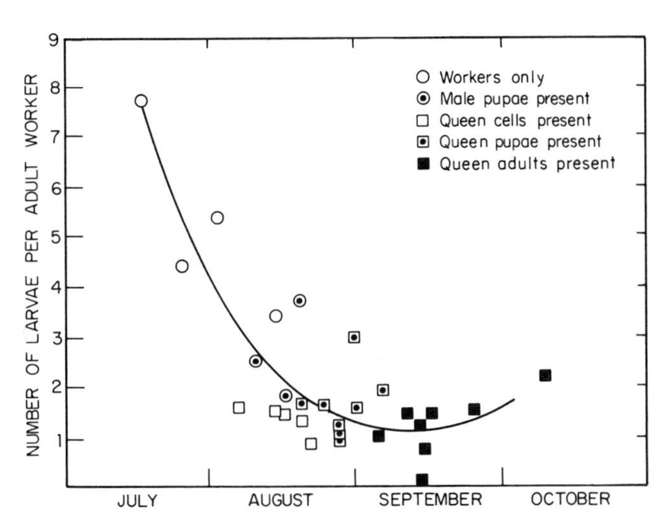

FIGURE 9-10. The larva/worker ratio in colonies of *Vespula germanica* collected at different times of the active season in England. As the colony population grows, the larva/worker ratio drops, providing more food for individual larvae and presumably improving the opportunity for them to mature as queens rather than as workers (redrawn from Spradbery, 1965).

that special additives are secreted into this raw material by the nurses cannot be ruled out. Deleurance (1955a) reported that the salivary glands of *Polistes gallicus* workers change progressively with age, and he has gone so far as to hypothesize that salivary secretions play a role in caste determination.

Division of labor appears to be quite elementary throughout the social wasps. A slight morphological bias has been detected in *Vespula germanica* by Spradbery (1965). Workers weighing 25 mg or less, and with wings 10 mm in length or less, spend most of their time in the nests; larger workers from the same colonies usually forage. Gaul (1947, 1948) reported that in North American species of *Vespula* a distinct class of nurse wasps develops about the time virgin queens are produced, and occasional wingless individuals appear (in *V. squamosa*) which are, of course, wholly confined to the nest. Polyethism according to age is also weakly defined in the wasps. *Polistes fadwigae* workers do not forage until about a week following eclosion, and the older workers have a greater tendency to participate in nest work, but there is otherwise no clear scheduling of tasks (Yoshikawa, 1963a). According to Montagner (1967), *Vespula* workers assist in nest building during the first few days after emergence, and then they also begin to feed larvae; finally, some individuals add foraging to their repertory. A very similar pattern has been recorded in *Vespa orientalis* by Ishay *et al.* (1967).

Caste Evolution as a Function of Colony Size

Evolutionary trends in caste in both the social bees and wasps display an unmistakable correlation with evolution in colony size. At the risk of oversimplification I would like to suggest that this relation involves the following four steps, in each of which the mature colony size is given followed by the caste system generally associated with it.

1. *From 2 to approximately 50 adults.* In species having this marginally low range of mature colony size, the females are either considered to be semisocial, such as the halictine bees in *Augochloropsis* and *Pseudaugochloropsis,* or else they are ambimorphic—starting life as workers but later becoming egg layers—as in *Belonogaster* and possibly also some of the Stenogastrinae.

2. *From approximately 10 to 400 adults.* The queen caste is still identical to the worker caste in external morphology or is at most weakly differentiated and connected by

intercastes. But there is a distinct functional worker caste, usually held in a subordinate position by aggressive dominance behavior on the part of the egg-laying females, which behavior includes the stealing and eating of eggs laid by rivals. Temporal polyethism is weakly developed or absent in the worker caste. Examples include the species of *Bombus* and *Polistes*. Among the Halictinae the dominance is less overt, but still involves egg stealing.

3. *From approximately 100 to 5,000 adults.* A modest degree of morphological queen-worker dimorphism is at least partly under the control of the nurse workers, which practice preferential feeding during the rearing of the larvae. Overt dominance behavior is usually no longer displayed by the queen, although it may still exist within the worker caste, as in the genus *Vespula*. Temporal polyethism is only weakly developed in the worker caste. Most or all of the Vespinae exemplify this evolutionary stage.

4. *From approximately 300 to 80,000 adults.* The queen-worker dimorphism is very strong in ways that appear to be correlated with swarming as the mode of colony multiplication. Dominance behavior has nearly disappeared, and, in *Apis* at least, control by the queen is achieved through inhibitory pheromones. In the meliponines there are still traces of what might be interpreted as dominance: the workers generally avoid direct approaches to the queen and often crouch when face to face with her. Temporal polyethism is strongly developed in the worker caste. Examples include the higher Apinae, both the honeybees and stingless bees (Meliponini).

These trends should provide a rich source for future investigation and theoretical analysis. At present it is possible to summarize them in part by the simplifying generalization that, as mature colony size increases, the degree of caste differentiation increases. But there is much more to the story than that, and deeper analysis is needed. In the early stages of social evolution the relations among the females are very flexible and permissive. As colony size increases, a primitive form of dominance contest makes its appearance as the arbiter of caste. Both the queens and the reproductively competent workers treat each other with overt aggression whenever the reproductive function is subject to dispute. Finally, in the very large colonies of the Apinae, hostility among colony members is all but eradicated, even at times when the nest queen departs or loses vitality to the point that she surrenders her reproductive role. Why should this be? Sakagami (1954) argued correctly that in the honeybee colony the population is too large, and the lives of the workers too short, to permit the existence of individual recognition leading to the establishment of hierarchies. But Montagner's analysis of *Vespula* has demonstrated that hierarchies need not be based on such recognition. Rather, rank works itself out in a statistical manner from the innumerable and impersonal aggressive interactions of the individual colony members. We can go further and conclude that the concept of dominance in the traditional sense of vertebrate ethology has to be abandoned altogether when considering the honeybee. The queen of *Apis mellifera* "controls" the worker caste only in a very narrow sense. Her attractive scents make her a rallying point for colonies, and during colony fission swarming can be consummated only if she is present. The queen also suppresses the rearing of new queens and the development of worker ovaries by means of a pheromone. Otherwise, the workers are very much in charge of colony reproduction. They build the royal cells, create new queens, and initiate and guide the swarms. They even prepare the virgin queens for the nuptial flights by reducing their food supply and encouraging them on their way by aggressive shaking movements (M. Delia Allen, 1965). For several days before a swarm emerges from the hive, the workers give the queen less food than usual and frequently behave aggressively toward her; the queen loses weight and consequently is able to fly the longer distances required in colony fission. Workers initiate the swarm by the buzzing run, and they often actively pursue the queen and force her to leave the hive (von Frisch, 1967a). In short, female reproductive behavior itself has undergone a division of labor in which new supporting roles have been secondarily assigned to the worker caste. The new properties of specialization and integration coupled together transcend the simple dominance hierarchies of the bumblebees and social wasps and provide new evidence that it is the entire colony, and not just the kinship group within the colony, that constitutes the unit of natural selection.

10 Caste: Termites

Although termites and ants are phylogenetically remote from each other, they have evolved a caste system that is remarkably similar in several major respects. Both have produced a soldier caste that is highly specialized in both head structure and behavior for colony defense, and both are characterized by a worker caste that is numerically dominant in the colonies, morphologically similar from species to species, and behaviorally versatile. The number of physical castes in the phylogenetically most advanced termite species is somewhat greater than in the most highly evolved ant species, but the average degree of specialization of individual castes is about the same. Finally, the higher termites have developed temporal polyethism that resembles that of the ants in broad outline.

Some differences also exist between the caste systems of termites and those of ants and their fellow social hymenopterans. The neuter castes of termites are constituted of both sexes, and there are no termite "drones," which live solely for the act of mating and are programmed for an early postreproductive death. The social Hymenoptera, it will be recalled, are holometabolous, and their grub-like larvae are incapable of contributing to the labor of the colony. The termites have the more primitive trait: they are hemimetabolous, which means that the immature stages are not radically different in form and behavior from the mature stages. In the lower termites the nymphs contribute to the work of the colony; in other words, there is an employment of "child labor." This is not the case in the higher termites (the family Termitidae), where the immature forms are wholly dependent on a well-differentiated worker caste. Caste determination in termites has several unique features. The reproductive castes of lower termites secrete sex-specific pheromones that directly inhibit the metamorphosis of immature individuals into reproductive forms. "Regressive molting" occurs, in which individuals already in the process of developing into a particular caste actually go back to a less differentiated stage. Finally, the termites generally have a wide array of "supplementary reproductives," fertile but wingless individuals of both sexes which develop in colonies whenever the primary reproductives are removed. The nearly universal occurrence of these substitute castes provides termite colonies with a degree of resiliency, leading in extreme cases to potential immortality, that is seldom encountered in the social Hymenoptera.

The Kinds of Castes

The classification of termite castes has reached relative stability in the literature of the last three decades. The following glossary has been collated from the recent excellent reviews by E. M. Miller (1969) and Charles Noirot (1969b). Familiarity with it is needed for a full understanding of the detailed accounts of caste systems, to follow later in this chapter.

Larva (apterous nymph). An immature individual lacking any external trace of wing buds or soldier characteristics. The number of larval stages varies among the species and (in the lower termites) in differing environments.

Nymph (brachypterous nymph). Individuals derived from the larval stages who possess external wing buds and enlarged gonads and who are capable of developing into functional reproductives by further molting. In the course of subsequent molts, the wing pads develop in a regular manner, and the eyes become differentiated.

Worker. In the higher termites (Termitidae), individuals characterized by a complete absence of wings and a consequent reduction of the pterothorax. The compound eyes and ocelli are absent or greatly reduced, the genital apparatus is rudimentary, the head is rounded and proportionately more voluminous than in other nonsoldier castes, the mandibular muscles are powerfully developed, and the digestive tract is exceptionally large, occupying a major portion of the abdominal cavity. Molting glands are always present, and in some cases the workers are capable of transforming into other castes. In lower termites there is generally no true worker caste, the worker function being filled by nymphs and pseudergates. An apparent exception has been reported by Buchli (1958), who states that *Reticulitermes lucifugus* contains a worker caste which is identical to the pseudergate except that its mesonotum is narrower than the pronotum. Actually, the distinction between a "true" worker caste and a worker-like caste seems to be arbitrary. A worker caste in general can be defined as a set of more or less distinctive individuals, belonging to one or more instars, who contribute to such ordinary chores of the colony as nest construction, cleaning, nursing, and foraging.

Pseudergate. This term, coined by Grassé and Noirot (1947) in the course of their studies on *Kalotermes flavicollis,* means literally "false worker." The caste occurs only in the lower termites and is comprised of individuals who have either regressed from nymphal stages by molts that reduced or eliminated the wing buds, or else were derived from larvae by undergoing "stationary," nondifferentiating molts.

Soldier. A form with morphological features specialized for defense, such as enlarged mandibles, stopper-like heads, or hypertrophied glands capable of discharging large quantities of defensive secretions. Generally the head is heavily pigmented and sclerotized. A presoldier (also known as a "white soldier," a pseudosoldier, or a soldier nymph) is a developmental stage intermediate between the larva, pseudergate, or nymph, on the one hand, and the definitive soldier form, on the other. The presoldier is evidently incapable of functioning in defense.

Primary reproductive (first-form reproductive, imago). The colony-founding type of queen or male derived from the winged adult.

The following three castes sometimes appear in the nests when the primary reproductives are removed, and they are referred to collectively as *supplementary reproductives, replacement reproductives,* or *neoteinic reproductives.*

Adultoid reproductive. Found in the higher termites (Termitidae) only, this is an imago, closely similar or identical to the primary reproductive in morphology. It arises in the nest after the disappearance of the primary reproductive belonging to the same sex. It may be an imago already present, which then merely assumes the behavioral role, or a nymph reared to the imago stage in apparent response to the absence of the primary individual. The formation of imaginal replacements may not be universal in the Termitidae, having been reported so far only in *Anoplotermes, Macrotermes,* and *Nasutitermes.*

Nymphoid reproductive (second-form reproductive, secondary reproductive, brachypterous neoteinic). A supplementary male or female derived from a nymph and retaining wing buds.

Ergatoid reproductive (third-form reproductive, tertiary reproductive, apterous neoteinic). A supplementary male or female without evidence of wing buds, usually larval in external form but with a more or less pigmented exoskeleton.

Division of Labor

After they meet during the nuptial flight, the primary reproductives construct an initial nest cell (the "copularium") and set about rearing the first brood. In the case of the lower termites, the young couple nourish themselves by feeding on wood or soil surrounding the nest. Under these circumstances they are capable of surviving for long periods of time in the absence of workers. It used to be thought that the royal pair of higher termites (species of the family Termitidae) take no food until the first brood are nearly mature. It was assumed that the pair provide for both themselves and their first brood entirely with nutritional reserves invested in the fat body, flight muscles, and whatever materials are present in the gut at the time of flight. However, recent studies on *Cubitermes* and *Tenuirostritermes* indicate that the reproductives supplement their reserves by feeding on soil, cast skins, and even some of the brood (Nutting, 1969). There is no firm evidence that any of the higher termites have adopted the totally claustral mode of colony founding used by all the phylogenetically advanced subfamilies of ants.

The termites are nevertheless similar to the ants in the care they lavish on their first brood. Among the lower termites the larvae are groomed and provided with both stomodeal and proctodeal food, the latter containing the protozoans vital to their later capacity to feed independently on cellulose. During this time, according to Wilkinson (1962), the royal couple of *Cryptotermes havilandi* exchange proctodeal material reciprocally, and they occasionally engage in autoproctodeal feeding as well. By the fourth instar the larvae have acquired protozoan faunas of their own and are self-sustaining. In *Kalotermes flavicollis* and *Reticulitermes lucifugus* independence is achieved in the third instar (Grassé and Noirot, 1958b; Buchli, 1950). The royal pair of higher termites nurse their first larvae in the same manner, except that the food supplied is exclusively stomodeal in origin. It is a curious fact that in both *K. flavicollis* and *Cubitermes ugandensis,* members of lower and higher termite families, respectively, the males do most of the nursing (Buchli, 1950; Williams, 1959a). Soon after the workers and nymphs of the first brood acquire their own digestive capability, they take over all of the nest construction, nursing, and foraging. They also begin to feed the royal pair with salivary secretions of their own, and the queen and male function thereafter as little more than passive reproductive machines. As the colony grows larger, the abdomen of the queen swells with the expansion of her ovarioles and fat bodies. In extreme cases the abdomen finally comes to resemble a small sausage, its abdominal tergites pulled far away from each other by the stretching of the white, balloon-like, intersegmental membranes, and the thorax and head are perched on one end like some accessory appendage. The queens of *Macrotermes* develop abdomens with capacities of as much as ten milliliters, making them the largest of all individual social insects. Such physogastric queens, each the mother of all the million or more neuter termites around her, are confined to a special royal cell constructed near the center of the fungus comb. Surprisingly, these individuals are still able to move very slowly on their own power by a combination of leg movement and waves of peristaltic contraction of the abdomen.

In the case of the lower termites, that is, the Mastotermitidae, Kalotermitidae, Hodotermitidae, and Rhinotermitidae, labor is divided among nymphs and pseudergates, both of which are superficially worker-like forms. Among the higher termites, comprising the family Termitidae, work is performed solely by a true worker caste, and the nymphs and other immature forms are nonfunctional. Studies of the division of labor have so far been unaccountably sparse and fragmentary, and only a few worthwhile generalizations can be made now. In most termite species there appears to be little or no division of labor based on the size of workers. An exception is provided by certain genera of the phylogenetically advanced Macrotermitinae, in which all of the large workers are males. Among termitid species that regularly forage in groups outside the nest, such as *Macrotermes ivorensis, M. mülleri,* and *Odontotermes magdalenae,* the columns are formed almost entirely of the large male workers. The smaller, female workers confine themselves to work within the nest. Existing records, compiled chiefly by Grassé and Noirot (1951a) and Noirot (1955, 1969b), indicate that such patterns are highly diversified within the Termitidae. Male workers are larger than female workers in the Macrotermitinae and in the nasutitermitine genus *Syntermes,* but smaller in the amitermitine *Microcerotermes parvus* and nasutitermitine genera *Nasutitermes* and *Trinervitermes,* and about equal in size in *Amitermes evuncifer* and the Termitinae. The small workers of *Macrotermes natalensis* participate in nest construction, but this is not the case in *Odontotermes magdalenae.* According to Noirot, a clear division of labor between large and small workers is missing in macrotermitine species that confine their foraging to subterranean passages. But the evidence still seems much too incomplete to accept even such a rough generalization as this.

Studies on the temporal division of labor in termite colonies, as compared with caste polyethism, have scarcely begun. It was not until the recent study by Pasteels (1965) on *Nasutitermes lujae* that temporal patterns of behavior were clearly demonstrated in the termites. *N. lujae* workers in stage I, both the small males and their somewhat larger sisters, do not join older nestmates in laying odor trails or in conducting exploratory foraging trips. Their sternal glands (the source of the trail pheromone in termites generally) are underdeveloped. However, their salivary glands are better developed than those of older workers. Since the salivary glands have been identified in other species as the probable source of larval food, Pasteels postulated that stage I workers are specialized as nurses. If this guess proves true, the basic pattern of temporal division of labor in the *Nasutitermes* will have been shown to be the same as that adopted by the great majority of the social Hymenoptera. A roughly similar pattern has been described in the primitive termite

Zootermopsis nevadensis by Howse (1968). Most of the excavation and nest building is carried out by the nymphs and the older larvae. As they progress in development, the larvae spend less and less of their time in food exchange. The colony members do not forage outside the nest, so that further comparisons with *Nasutitermes* cannot be made.

It is a general characteristic of termite caste systems that the worker caste is morphologically uniform but behaviorally very diverse when different species are compared; the soldier caste is morphologically diverse and behaviorally uniform. The workers construct the nests that vary so drastically among the termite species, and they are also responsible for the greatest part of the phyletic diversity in food habits, foraging styles, and other behavioral properties of the colony. The soldiers, on the other hand, are wholly specialized for the single function of defense. This is true all the way from the very primitive Mastotermitidae to the phylogenetically most advanced elements among the Termitidae. It has often been stressed by Emerson and others that the soldier was the first distinct neuter physical caste to evolve in termite evolution, antedating the "true" worker caste of the Termitidae by what appears to have been a very long stretch of time. Within the limits of their strictly defined role, however, the soldiers vary enormously in their morphology and in the particular defensive techniques they employ. In most species the soldiers are "mandibulate" forms, characterized by large, heavily sclerotized heads, powerful adductor muscles, and sharp, elongate mandibles that seem clearly designed for biting and cutting rather than for such nondefensive functions as digging or handling of the brood. The following description of the behavior of *Zootermopsis* soldiers by Stuart (1969) is typical for many mandibulate species. "In small colonies they are normally present near the reproductives and quite often assume a characteristic position of 'rest.' At other times (for example, periods of flight activity) they can be found in numbers at the periphery of the log they inhabit. When confronted with an intruder the soldier points its head in the direction of the intruder (presumably by a combination of kinesthetic and chemical sense) and the mandibles open. The animal then closes its mandibles quickly at the same time moving its body, but not its legs, forward then backward in the horizontal plane. It is this action which can disembowel another termite. This response can also occur as a low intensity reaction when the mandibles partly open but the lunge does not take place and the mandibles then close slowly." There seems to be little more than this to the repertory of the *Zootermopsis* soldier insofar as its contributions to the colony are concerned. Even the role it plays in defense is specialized. Stuart points out that it is relatively ineffectual against insects smaller than itself. When ant workers about three millimeters in length are introduced into *Zootermopsis* nests, they either manage to escape quickly or else get trapped in some cul-de-sac in the nest interior. In the latter case they are usually then eliminated by the termite nymphs, who wall them off with deposits of fecal cement and pellets.

The specialization of defensive function directed at larger animals has been carried still further in some of the Termitidae (Deligne, 1965). In *Termes, Acanthotermes, Macrotermes,* and *Syntermes* (see Figure 10-1), the soldier mandibles are shaped like scimitars, and their points are as sharp as needles. When snapped together, they can plunge deeply into the flesh. I learned this to my sorrow once while examining *Syntermes* colonies in Surinam. A soldier snapped its mandibles through a fleshy fold of one of my thumbs, rather like passing needles through a piece of cloth, and then folded them tightly in place. Its death grip was released only when the head was carefully severed and each mandible separately extracted.

The soldiers of the kalotermitid genus *Cryptotermes* possess cyclindrical heads that serve as living plugs for

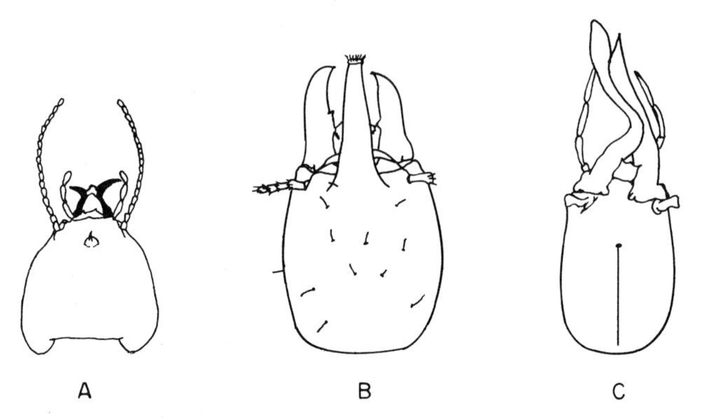

FIGURE 10-1. Heads of soldiers from three termite species specialized for different defensive techniques: (*A*) *Acanthotermes acanthothorax,* a large "mandibulate" soldier; (*B*) *Armitermes parvidens,* a "nasute" soldier still possessing defensive mandibles; (*C*) *Pericapritermes urgens,* a soldier with snapping mandibles (redrawn from Emerson and Banks, 1957; Noirot, 1969b).

the galleries of the nests. Their mandibles are short and not particularly suited for defense. The head capsule, on the other hand, is thick walled, heavily pitted, and truncate in front, so that it forms a barrier in the narrow galleries of the nest which any invading arthropod finds very difficult to push around. This "phragmosis" of the *Cryptotermes* soldiers (to use Wheeler's apt term) is closely paralleled in structure and function to that developed by certain ant species in the genus *Camponotus* and tribe Cephalotini.

Even more bizarre are the "snapping" soldiers of *Capritermes, Neocapritermes,* and *Pericapritermes* (Kaiser, 1954; Deligne, 1965). As illustrated in Figure 10-1, the mandibles are asymmetrical and so arranged that their flat inner surfaces press against one another as the adductor muscles contract. When the muscles pull strongly enough, the mandibles slip past one another with a convulsive snap, in the same way that we snap our fingers by pulling the middle finger past the thumb with just enough pressure to make it slide off with sudden force. If the mandibles strike a hard surface, the force is enough to throw the soldier backward through the air. If they strike another insect, which seems to be the primary purpose of the adaptation, a stunning blow is delivered. Even vertebrates receive a painful flick. The mandibles of *Pericapritermes* in particular are modified in such a way that the left mandible alone strikes out, so that the target can be hit only if it is located on the right side of the soldier's head.

Finally, the premier combat specialists are the soldiers that employ chemical defense. The mandibulate soldiers of the very primitive Australian termite *Mastotermes darwiniensis* produce almost pure *p*-benzoquinone from glands that open into the buccal cavity (Moore, 1968). When a soldier bites an adversary, the quinone is mixed with amino acids and protein in the saliva, soon producing a dark, rubber-like material that entangles the victim. Excess quinone probably acts as an irritant. The mandibulate soldiers of some of the Termitidae have independently modified their salivary glands to the same end. When *Protermes* soldiers attack, they emit a drop of pure white saliva which spreads between the opened mandibles. When they bite, the liquid spreads over the opponent. The large soldiers of *Macrotermes* release a brown, corrosive salivary liquid, while those of *Odontotermes* produce a substance that is either creamy in color or hyaline. None of these secretions have been chemically identified as yet, but it is commonly stated that they are all either toxic or else undergo coagulation in the air, which renders them glue-like. In general, the salivary glands of the soldiers are better developed than those of their worker nestmates, and they sometimes reach a huge size in proportion to the remainder of the body. The salivary reservoirs of *Odontotermes magdalenae* swell out posteriorly to fill most of the anterior segments of the abdomen. Those of *Pseudacanthotermes spiniger* fill nine-tenths of the abdomen. The soldiers of *Globitermes sulphureus* are quite literally walking chemical bombs. Their reservoirs fill the anterior half of the abdomen. When attacking, they eject a large amount of yellow liquid through their mouths, which congeals in the air and often fatally entangles both the termites and their victims. The spray is evidently powered by contractions of the abdominal wall. Occasionally these contractions become so violent that the wall bursts, shedding defensive fluid in all directions.

In certain Rhinotermitidae and other Termitidae it is the frontal gland, or cephalic gland as this organ is often called, that has been evolutionarily modified in the soldier caste for defensive purposes. The organ opens through a small circular or slit-shaped orifice near the center of the head, called the frontal pore or fontanelle. Tentorial-fontanellar muscles, which are elements peculiar to the termites, attach to the anterior portion of the gland inside the head capsule. The frontal gland, like the salivary gland, is greatly enlarged in those soldiers that rely on it for defense. In the soldiers of the genus *Rhinotermes,* for example, it occupies most of the abdominal cavity and forces the bulk of the digestive apparatus to the rear. Soldiers of the rhinotermitid genus *Coptotermes* have been observed to spray a white liquid through the frontal pore; it quickly thickens to a glue-like consistency and entangles both the termite and its adversary. According to Moore (1968, 1969), this material consists of a suspension of lipid material in an aqueous solution of mucopolysaccharide. The lipid fraction "has virtually the same composition as the cuticular lipids of the species, and its purpose appears to be to lend a more resilient texture to the dried latex, which forms a tenacious coating on any adversary coming into contact with it." The mixture appears to be nontoxic to arthropods. In a wholly independent evolutionary development, the members of the termitid subfamily Nasutitermitinae have carried chemical defense to its extreme. In the advanced species the frontal gland has not only been enlarged but the fontanellar region of the head has been drawn out into a conical organ which roughly resembles a great nose on the front of the soldier's head—hence the expressions "nasus" to describe the organ and "nasute soldier" to describe the caste (Figure 10-2). The most primitive nasutitermitine genera, namely *Syntermes, Cornitermes, Procornitermes, Paracornitermes,* and *Labiotermes,* have typically mandibulate soldiers. Certain phylogenetically intermediate genera, such as *Rhynchotermes* and *Armitermes,* are characterized by soldiers that possess both large hooked mandibles and nasute head capsules. These individuals are therefore "double threats" in their defensive roles. According to Sands (1957), the nasus has evolved twice through such an intermediate step. The mandibles have been subsequently reduced in size within several independent phyletic lines. The extreme form of the nasute soldier, in which the mandibles have become small, nonfunctionable lobes, originated independently in at least nine instances within *Convexitermes, Eutermellus, Mimeutermes, Nasutitermes, Obtusitermes, Subulitermes, Trinervitermes,* and *Tumulitermes.* This remarkable flurry of convergent evolution, together with the outstanding diversity and abundance of the higher Nasutitermitinae in the tropics, is evidence that the nasute technique of chemical defense is highly successful. With the aid of their fontanellar "gun," fired by a contraction of powerful mandibular muscles, nasute soldiers are able to eject the frontal gland material over a distance of many centimeters. Their aim is quite accurate in spite of the fact that they are wholly blind. The nature of their orientation device has yet to be studied although, by process of elimination, it seems almost certainly to be olfactory or auditory in nature. After firing the soldier wipes its nasus on the ground and retreats into the nest, apparently lacking enough material to make a rapid series of shots. Because nasute soldiers are able to strike and disable an adversary at a considerable distance, they seldom become fatally entangled in their own secretions. They therefore have an advantage over the soldiers of many other termitid species who are forced to apply mandibular gland secretions at short range. According to Ernst (1959), the frontal gland secretion of *Nasutitermes* is nontoxic and functions solely as a mechanical entrapment device. Moore (1964a, 1969), who has studied the chemistry of the Australian species, reports that the defensive secretion consists primarily or wholly of terpenoids. The volatile fraction contains α-pinene as the principal component and β-pinene, limonene, and other monocyclic isomers as minor components. The "resinous" fraction consists of a number of closely related polyacetoxy diterpenoids, which become increasingly viscous and sticky when exposed to the air. As the volatile components evaporate, they also serve as alarm substances, so that when one soldier fires at a given target others are likely to follow close suit.

Paulette Bodot (1969) has recently discovered that the production of presoldiers in nests of *Cubitermes severus* reaches a maximum in April; the proportion of mature soldiers consequently peaks in May and June when the nuptial flights are held. It is tempting to attribute a causal relation to these two events. Termite colonies are more vulnerable to attacks from predators during the brief period when they dig exit holes to release the reproductives, and at this time the soldiers typically stand guard in the holes and on the ground outside. Noirot (1969b) states that soldier production is also maximal just prior to the nuptial flights in *Cephalotermes rectangularis.* We need to learn if the same correlation holds in other termite species. A search should also be made for independent evidence to test the more general hypothesis that on certain occasions termite colonies do alter the proportions of their neuter physical castes to meet urgent contingencies in the environment.

Caste Determination in the Lower Termites

The modern study of termite castes began when Battista Grassi (in Grassi and Andrea Sandias, 1893–1894) made the following elementary but far-reaching discovery concerning *Kalotermes flavicollis:* "If from three to twenty *Calotermes* of different ages are placed in a glass tube three to eight centimeters long, closed with a cork and kept warm—for example, in the waistcoat pocket, except in the summertime—they continue to live and constitute a family, or better, an independent colony; if orphaned, they rear a fresh king and queen, or if lacking soldiers, they rear soldiers, and so on. In short, after a certain time the tube will contain a complete little colony, if this had not been the case from the beginning." In the course of their subsequent experiments on this primitive European species, Grassi and Sandias (1893–1894, 1896) found that colonies deprived of primary reproductives for no more than 24 hours produced supplementary reproductives from four to seven days later. Supplementary forms never appeared in the laboratory colonies if the original royal

FIGURE 10-2. Nasute soldiers of *Rhynchotermes perarmatus* stand guard along the flank of a foraging column of workers in a Panamanian rain forest. This particular species still retains well-developed mandibles that are used in defense in conjunction with the chemical sprays (photograph courtesy of Thomas Eisner).

pair had been continuously present. Thus was born the "extrinsic school" of theory concerning caste determination in termites. Grassi and Sandias, as well as such later authorities as Escherich and Holmgren, reasoned that all termites have equal potential upon hatching and that their caste is fixed by environmental influences acting on them later. These extrinsic factors were believed to include differential nutrition or physiologically active exudates that originate from nestmates. The opposing, "intrinsic" explanation was first hypothesized by Imms (1919) as an outgrowth of his study of *Archotermopsis wroughtoni.* Stimulated by the then popular new science of genetics, he suggested that castes are Mendelian segregants. Caroline B. Thompson (1919, 1922) presented what appeared to be supporting evidence of a genetic or at least blastogenic origin of castes in *Zootermopsis.* She reported that there are two egg sizes and, possibly correlated with these two classes, two kinds of newly hatched larvae: those possessing small heads, large brains, and large gonads, destined to develop into reproductive forms; and those with proportionately larger heads but smaller brains and gonads, destined to become workers or soldiers. However, Heath (1927, 1928), who repeated Thompson's work with much larger samples (17,000 individuals in several hundred colonies), failed to detect such a dichotomy, and nothing like it has been found in the many later studies of various lower termite species.

With the evidence growing strong in favor of the hypothesis of equipotency of newly hatched larvae, Pickens suggested in 1932 that the queen produces a specific chemical substance that inhibits the development of female nymphs into mature reproductives. This was the first really interesting hypothesis from the extrinsic school that could be tested by experiment. Shortly afterward Castle (1934) did bring forward some evidence favoring the existence of inhibitory "ectohormones," or pheromones as they are now called. First he demonstrated that when individual nymphs of *Zootermopsis* are kept in isolation for 45 days or longer they develop into supplementary reproductives. Castle's result neatly eliminated Holmgren's "Exsudattheorie," the hypothesis that special feeding, which is stimulated by certain exudates or other unknown factors, sets nymphs on the road to development as reproductives. In a second experiment Castle fed isolated groups of nymphs on filter paper treated with ether and alcohol extracts of the whole bodies of supplementary reproductives. Compared with control groups, the experimental nymphs developed more slowly toward the reproductive state. Light (1944a,b) repeated the experiment with larger samples and a greater range of extraction techniques, and he arrived at essentially the same conclusion—that developmental rate can be retarded with extracts. Although the artificial inhibition falls short of the total repression imposed by living reproductives, the *Zootermopsis* data still offer strong evidence for the existence of inhibitory pheromones (Light and Weesner, 1951; Miller, 1969).

The stage was now set for the important research of Martin Lüscher on caste determination in *Kalotermes flavicollis.* Building on the earlier work of Grassi and Sandias, Grassé (1949), and Grassé and Noirot (1946, 1947), Lüscher first traced the developmental caste history of *K. flavicollis.* Then, in a series of experimental studies that have extended over a period of nearly twenty years (1952–1969), he has gradually elucidated the complex inhibition system that prevails in this primitive species. Much of the basic information is summarized in Figures 10-3 and 10-4. Several of Lüscher's experiments are especially significant, and are also worth citing on the basis of their ingenuity alone. First, individuals were marked with paint spots and followed on a daily basis throughout their development. By this means Lüscher was able to show that the ability of a pseudergate to transform into a supplementary reproductive, providing the functional reproductive has been removed from the nest, is highest immediately following its molt to the pseudergate stage. Its degree of "competence" falls off exponentially thereafter, as expressed by the following data: at 5 days into the pseudergate stadium, 80 percent can transform into reproductives at the next molt; at 20 days, 50 percent; at 60 days, 20 percent; and at 90 days, almost none at all. The exponential form of the decline led Lüscher to hypothesize a random decay model since this is the simplest process which can generate such a curve. He inferred that the mechanism involved might be as simple as the deactivation of a single species of molecule at a constant rate.

Lüscher located the source and mode of transmission of the inhibitory pheromones by the following procedures. He divided the colony by means of a wire gauze barrier, leaving one portion of the colony with the functional reproductives and the other portion "orphaned," upon which one or more pseudergates in the orphaned group transformed into supplementary reproductives. But these

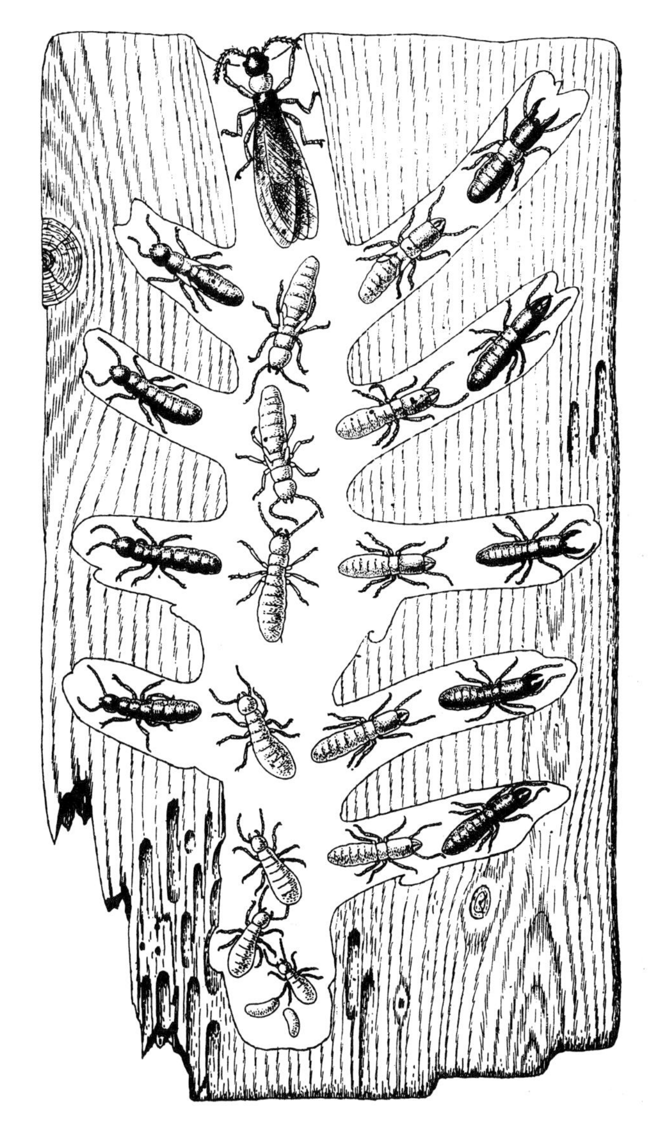

FIGURE 10-3. The development of castes in the dry-wood termite *Kalotermes flavicollis.* At the bottom of the drawing are shown the eggs and the first instar nymphs that hatch from them. The nymphs molt five to seven times to reach the pseudergate stage, represented by the termite in the middle of the drawing. At this point the termite can molt repeatedly without further growth or differentiation, but at any molt it can also transform either into a supplementary reproductive (to the left side of the drawing) or, through a presoldier instar, into a soldier (right side of drawing). From the pseudergate stage the termite can also change, by way of an intermediate stage with external wing pads, into a primary reproductive (at the top of the drawing). Supplementary reproductives and soldiers can also originate from this intermediate stage (from Lüscher, 1953).

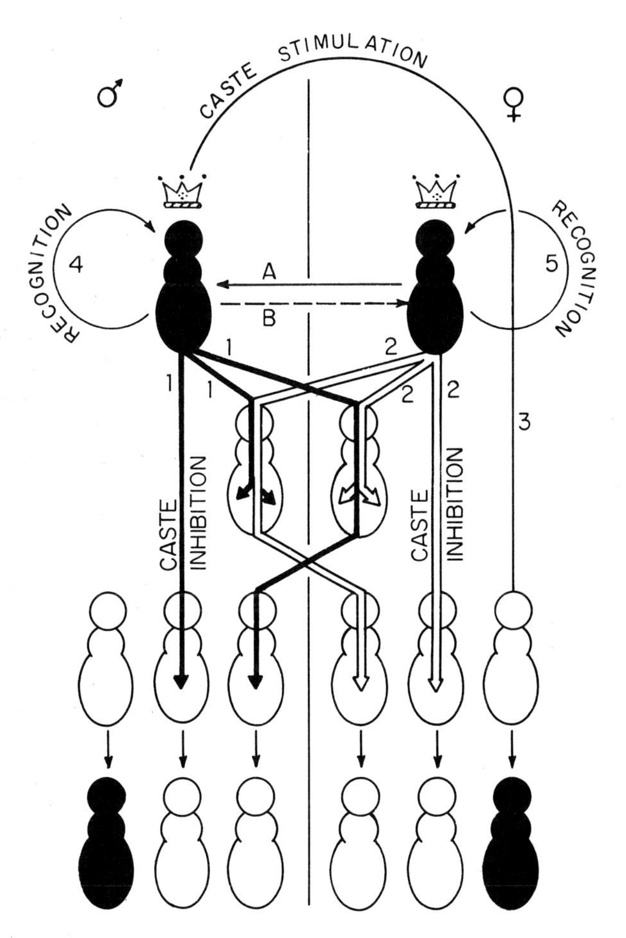

FIGURE 10-4. The known pathways of pheromone action in the control of reproductive caste formation in the termite *Kalotermes flavicollis.* In the top row the pair of "crowned" figures represents the reproductives: the functional male (king) to the left and the functional female (queen) to the right. The remaining figures all represent pseudergates. The king and queen produce substances (labeled 1 and 2, respectively) that inhibit transformation of pseudergates into their own royal castes. These inhibitory pheromones are passed directly from the reproductives to the pseudergates and are also circulated indirectly through the digestive tracts of the pseudergates. Another male substance (pheromone 3) stimulates the female pseudergates to change into the reproductive caste, but the reverse relation does not hold. When supernumerary royal males are present, they recognize each other (through pheromone 4) and fight; similarly, supernumerary royal females recognize each other (through pheromone 5) and fight. Finally, royal males stimulate production of pheromone 2 in royal females, and royal females stimulate production of pheromone 1 in royal males; the nature of the stimuli labeled *A* and *B* is unknown (modified from Lüscher, 1961b).

individuals were soon attacked and eaten by their nestmates. Lüscher then divided other colonies with *double* wire gauze barriers. When the supplementary reproductives appeared this time in the orphaned groups, they were not molested. The implication is clear: a single barrier prevents regurgitation and grooming between the two groups but not antennal contact and hence close-range olfaction. The double barrier prevents both. The inhibitory pheromones, the results seem to suggest, are exchanged by regurgitation or grooming or both. The single barrier therefore enables competent pseudergates to become supplementary reproductives, but their groupmates can still detect the presence of the original reproductives on the other side of the barrier, and they therefore attack the "usurpers." A double barrier, on the other hand, prevents detection as well, and the supplementary reproductives are allowed to live in peace. A fascinating aspect of the experiment is that if the single barrier were to be kept in place indefinitely, the orphaned group might gradually consume itself! The autocannibalism would be hastened if two or more reproductives of the same sex were produced because they fight among themselves. As soon as one is injured, pseudergates move in to kill and eat it (Ruppli and Lüscher, 1964).

Lüscher also fastened reproductives in his wire fences in such a way that their front ends faced one group and their rear ends the second group. Inhibition continued in the groups exposed to the rear ends, but not in the other groups, from which Lüscher concluded that the pheromones are dispensed from the hind gut. In still another experiment, Lüscher fixed pseudergates in the fences so that their heads faced the group containing the reproductives and their abdomens faced the orphaned group. Under these conditions the orphaned groups failed to produce supplementary reproductives, showing that inhibitory pheromones can be transmitted through the intestines of the pseudergates. In fact this oral-anal trophallaxis is probably the principal means by which pheromones are transmitted through large colonies under natural conditions.

None of the inhibitory or recognition pheromones have been chemically identified, so that the concluding chapter of the absorbing *Kalotermes* story cannot be written. Even so, Lüscher and his co-workers have made some progress on the involvement of the endocrine system in caste differentiation. The general conceptual scheme by which this research has been oriented is easily grasped. We assume that growth and development of termites is regulated by the same fundamental endocrine relations found in other insects, particularly the nonsocial cockroaches, which gave rise to the termites. Caste differentiation evolved as a diversification of developmental pathways. The caste determination of an individual termite is the shunting of its development along one pathway or another as the development reaches each crucial juncture. The caste pheromones, together with selective nutrition and whatever additional factors are programmed by a given species into its caste-determining system, serve as the token stimuli by which individual development is guided at the junctures. Since these stimuli are functions of the current caste composition of the colony, as well as of its overall physiological condition, they insure an apportionment of the developing individuals that is statistically appropriate to the colony's current needs. But the endocrine system is the machinery that receives and relays this information to the various organ systems. Therefore the physiology of caste determination is very much a problem within the domain of endocrinology.

The essential facts provided by the work of Lüscher (1960, 1963, 1969) and Lüscher and Springhetti (1960) can be summarized as follows:

1. When ecdysone, the general molting and differentiation hormone of insects, is injected into second-stage nymphs, a whole series of intermediate types are produced, ranging from primarily nymph-like forms to nearly fully adultoid reproductives. Thus ecdysone promotes transformation into the reproductive caste.

2. It was already well known that the corpora allata of insects produce juvenile hormone. This substance usually has the effect of inhibiting differentiation into the next development stage, even though it permits growth to proceed and, providing the ecdysone titer is also sufficiently high, allows static molting to occur. When the corpora allata of pseudergates are implanted into other pseudergates, no caste transformation occurs. But when the corpora allata of primary or secondary reproductives are implanted, or a relatively large dose of synthetic juvenile hormone is fed or injected, most of the pseudergates transform into presoldiers and later into soldiers.

3. As shown in Figure 10-5, the volume of the corpora allata, which is assumed to be an index of their hormone production, varies drastically among castes and in single individuals through time in conjunction with the molting cycle. The reproductives have much larger corpora allata

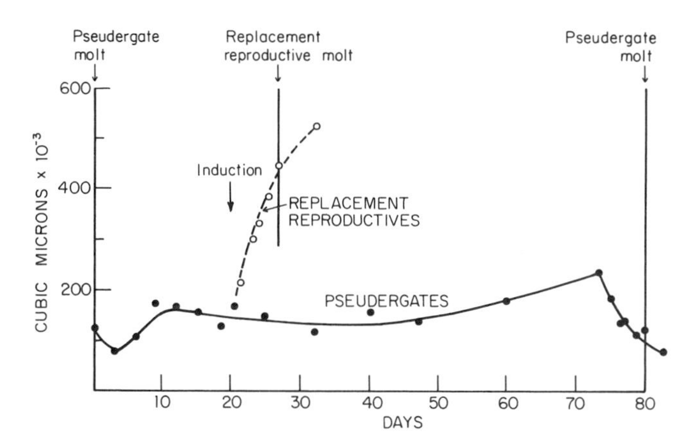

FIGURE 10-5. The volume of the corpora allata of pseudergates between two pseudergate molts and that of replacement reproductives before and after their molt. Each point represents the mean of the volume of the paired glands taken together from three to six individuals (redrawn from Lüscher, 1963).

than the pseudergates, whose corpora allata are smallest at the time of molting and gradually increase in size throughout the duration of the stadium.

These tantalizing fragments of information make it seem very probable that a juvenile hormone inhibits transformation into the final reproductive (hence, adult) form. The conclusion is consistent with both the data from the *Kalotermes* experiments and evidence concerning the action of a juvenile hormone in insects generally. Thus large injections of ecdysone can overcome its effects in the second-stage nymphs, as expected. And, judging from the corpora allata volume alone, juvenile hormone is at its lowest ebb in pseudergates immediately following each molt, at the same time that the pseudergates are most competent to transform into reproductive forms. Although the results of the experiments on reproductives therefore tie together neatly, the inducement of soldier development by implantation of large corpora allata remains something of an enigma. Lüscher (1963) originally postulated the existence of a second corpora allata pheromone, in addition to the juvenile hormone, to account for this special effect. However, it now appears that the juvenile hormone in large doses causes transformation into soldiers, while in lower doses it has either a juvenile or gonadotropic effect, depending on whether the individual is a pseudergate or reproductive form (Lüscher, 1969). It has thus become possible to account for all the known facts of caste determination in *Kalotermes* by a single model involving the competing action of the two principal developmental hormones. Finally, it is generally assumed that the inhibitory pheromones act on the endocrine system via the brain and its neurosecretory cells. Until precise biochemical information is adduced, however, we can only guess at the nature of the coupling mechanisms that enable the two regulatory systems to work together.

Miller (1969) has synthesized comparative information on the developmental histories of the better-known species of lower termites. A study of his schema in Figure 10-6

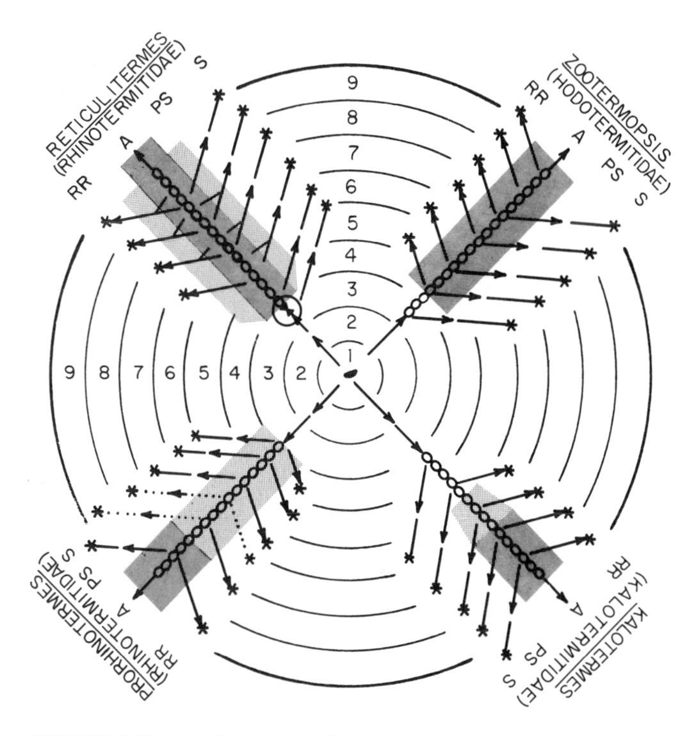

FIGURE 10-6. The potential developmental pathways open to each newly hatched larva in four lower termite species: (*A*) alate or primary reproductive; (*PS*) presoldier; (*RR*) replacement or supplementary reproductive; (*S*) mature soldier. After emerging from the egg, the termite must proceed through nine instars to reach the primary reproductive stage. At any instar from the third or fourth on it can be diverted to the soldier or replacement reproductive pathways. The central chain line represents the instars from which either forward or regressive molts are possible. The fine screen covers the pseudergate period; the coarse screen, the nymphal period. A small circle in the central *Reticulitermes* line marks the position of Buchli's worker-like form (redrawn from Miller, 1969; based on data from Buchli, Castle, Grassé, Light, Lüscher, Miller, Noirot, and Weesner).

shows that across three families there exists a basic pattern, evidently the primitive one for the termites as a whole, comprised of the following features: in going from the egg to the final primary reproductive, the individual must pass through approximately nine instars; from the third or fourth instar onward, it is capable of either forward or regressive molts, depending on its momentary physiological state, and it is also capable at most points of being shunted to either the soldier or replacement reproductive pathways, again depending on its current physiological condition. The developing termite belonging to one of these lower families can be thought of abstractly as an object that is guided (by the physiological state of the colony) back and forth for most of the length of a central track. If it reaches the terminus, it halts. If it leaves and starts down a sidetrack, it must continue moving in that direction, as though caught on a ratchet, until it reaches the end.

The four species differ in the exact positioning of the sidetracks and in the timing of the appearance of wing pads, hence, the stadia in which some individuals qualify technically as "nymphs" rather than as just larvae or pseudergates. Still more significant variation appears to exist in the extrinsic controls that guide the individual termites along these pathways. In particular, *Reticulitermes* seems to differ from *Kalotermes* and *Zootermopsis* in the relative degree of reliance placed on nutrition and pheromones in the determination of the sexual forms. Buchli (1958), who has made an exceptionally careful study of caste in *Reticulitermes lucifugus,* was unable to find clear evidence for the existence of inhibitory pheromones, although he did not exclude the possibility of their operating as a minor factor. He discovered that poor nutritive conditions in the colony inhibit the appearance of both supplementary reproductives and soldiers, and that, under optimal conditions, both castes are generated even when the primary reproductives are still present.

Caste Determination in the Higher Termites

The caste systems of the Termitidae differ from those of the lower termites in several major respects. In general, they display an increased degree of complexity and of developmental rigidity which together constitute a phylogenetic advance beyond the condition of the lower termites. We see this trend exemplified in the case of *Amitermes hastatus* (= *A. atlanticus*), a species primitive by termitid standards but advanced with respect to the termites as a whole (Figure 10-7). Each individual, regardless of sex, passes through two larval instars during which it cannot be assigned to caste on the basis of external morphological criteria. At the succeeding molt it enters one or the other of five definitive caste forms: the worker, the soldier, and the three kinds of reproductives. According to Skaife (1954b), new primary reproductives are reared only in strong, flourishing colonies whose age is greater than ten years. Nymphoid (secondary) reproductives occur in about 20 percent of the nests, even in the presence of the original primary reproductives. They assist the workers

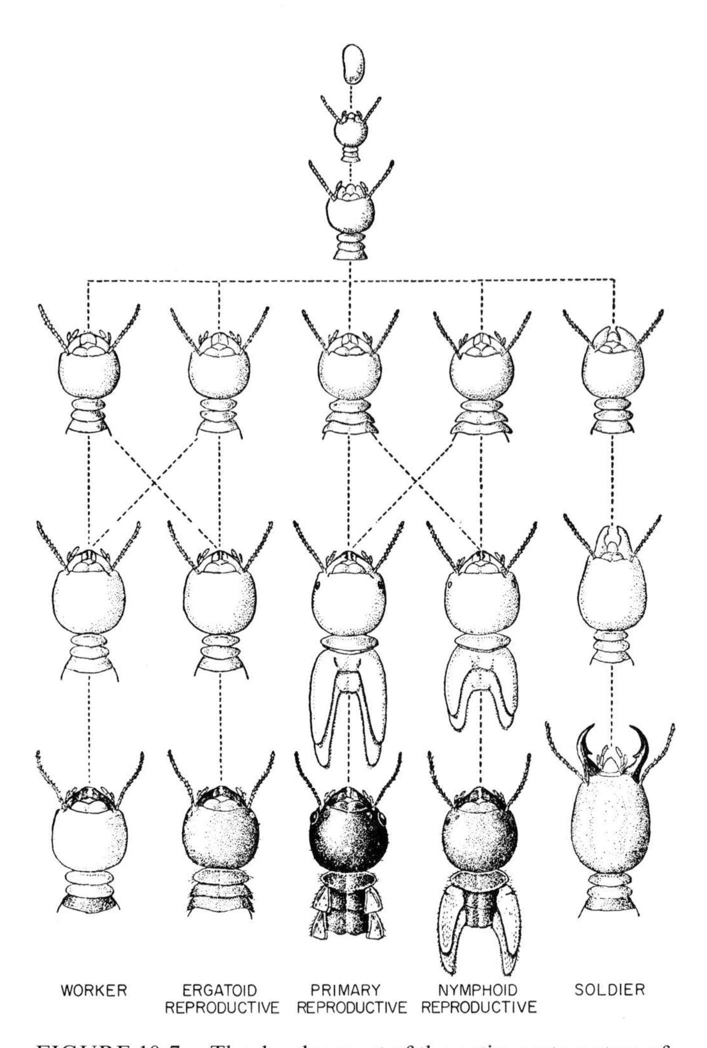

FIGURE 10-7. The development of the entire caste system of the primitive termitid *Amitermes hastatus.* After passing through the egg stage and the first two larval instars (*top*), each male and female enters one or the other of the five developmental lines indicated (modified from Skaife, 1954b).

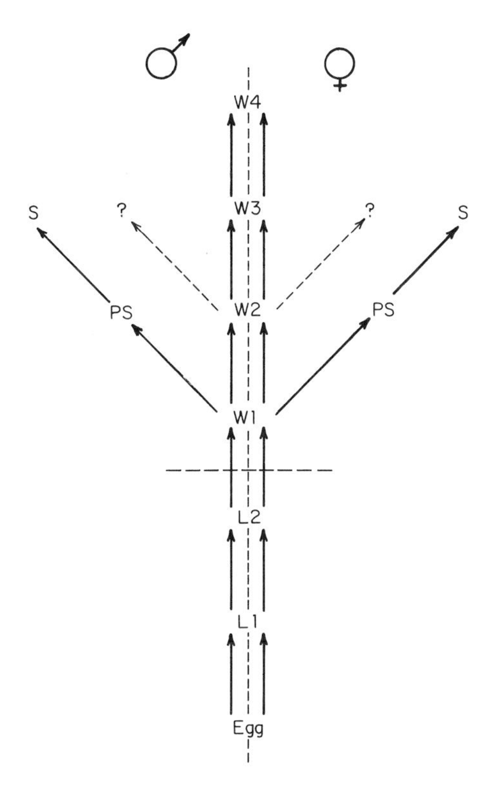

FIGURE 10-8. Development of the neuter castes of *Amitermes evuncifer* (Termitidae: Amitermitinae). In this relatively primitive "higher" termite, soldiers are formed from both sexes, and the development of males and females is closely parallel in both the soldier and worker castes: (*L*) larva; (*PS*) presoldier; (*S*) soldier; (*W*) worker. The larval and worker instars are numbered sequentially (redrawn from Noirot, 1969b).

in foraging, subsist on the same coarse food, and serve as an instant reserve-reproductive caste capable of taking over from the primary reproductives. Ergatoid (tertiary) reproductives are rare in *Amitermes hastatus,* having been found in only 3 out of 150 nests dissected by Skaife. Their significance in relation to the other two reproductive castes is unknown. The Amitermitinae are considered primitive with reference to the remainder of the Termitidae principally on morphological grounds, but the caste system of *A. hastatus* is also primitive in the following two respects: it has two larval instars instead of one, and there is no sexual dimorphism in the worker and soldier castes.

The subsequent evolution of caste systems within the Termitidae has recently been reviewed by Noirot (1969a); to a large extent, our knowledge of the subject is based on his own earlier research (1954, 1955, 1956). The diagrams in Figures 10-8 to 10-10 illustrate several of the key phylogenetic steps. First, note that there is an increasing tendency to produce sexual dimorphism in both worker and soldier castes. In the nasutitermitine genus *Syntermes* and the Macrotermitinae the large definitive workers are males, and the small definitive workers are females; in the amitermitine *Microcerotermes* and in the higher Nasutitermitinae the reverse is true. An association with sex is even more pervasive in the soldier caste. The species of the relatively primitive genus *Amitermes* have about equal proportions of male and female soldiers, and this

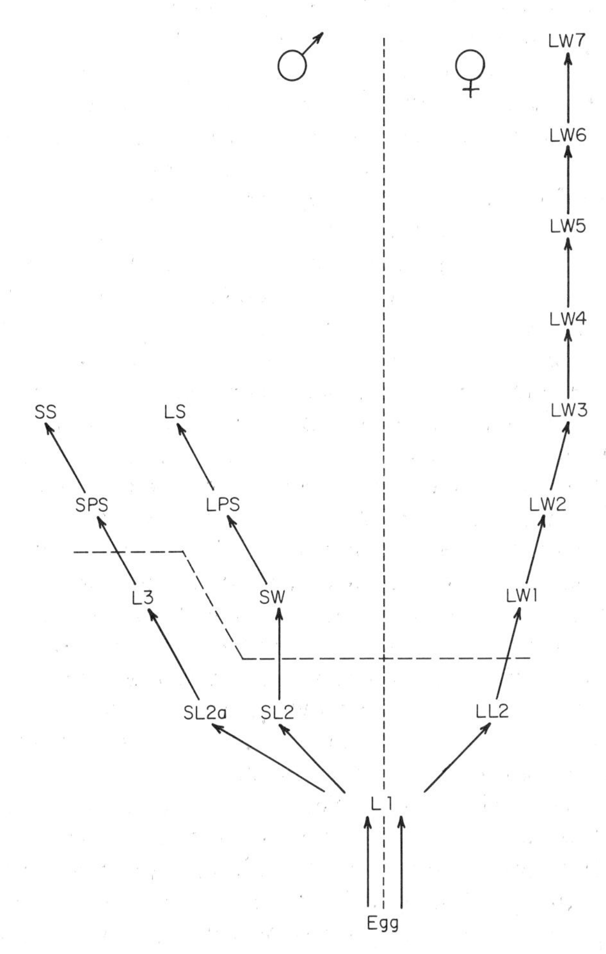

FIGURE 10-9. Development of the neuter castes of a species of *Trinervitermes* (Termitidae: Nasutitermitinae). This species is typical of its subfamily in that only males become soldiers. Two lines of soldier development separate early, one leading to small soldiers (*SS*) through a small presoldier instar (*SPS*) and the other to large soldiers through a small worker instar (*SW*) and a large presoldier instar (*LPS*) in that order. The females produce a large worker line (*LW*), exclusively (redrawn from Noirot, 1969b).

is also the case in the nasutitermitine *Leptomyxotermes doriae.* But in the remainder of the Nasutitermitinae the soldiers are normally all males, while in the Macrotermitinae and Termitinae they are normally all females. A second strong evolutionary trend within the Termitidae as a whole is the increase in the morphological specialization of the worker caste, which comes to monopolize the labor functions carried on in lower termites by the pseudergates and nymphs. This is accompanied by a decrease in the number of larval molts leading to the worker caste, a trend that can be seen by comparing *Amitermes* (Figure 10-8) and *Trinervitermes* (Figure 10-9). A decrease in the number of instars contained within the functional worker caste has also occurred; this is illustrated by the contrast between the eight worker instars (one male, seven female) of *Trinervitermes* and the single worker instar of *Acanthotermes* (Figure 10-10). These trends are countervailing with reference to caste diversity and in one sense produce an ambiguous final result in evolution. On the one hand we see an increase in the number of castes through the appearance of sexual dimorphism coupled with the multiplication of soldier lines within single sexes. Opposite this is the trend toward homogenization of the worker caste. As a consequence, the highly advanced *Acanthotermes acanthothorax* has ended up with five distinct castes, exactly the same number as *Amitermes enuncifer* with its obviously much more primitive system. Nevertheless, the *Acanthotermes* has achieved an important advance as a consequence of all of its developmental evolution—each of its castes is morphologically more specialized. In the absence of any other obvious explanation, therefore, it seems reasonable to conclude that efficiency has been enhanced during caste evolution in the termitids by means of caste specialization rather than through an increase in caste numbers.

In all species of Termitidae so far studied, the segregation of individuals into the neuter and imago lines becomes morphologically visible following the first molt. According to Kaiser (1956), histological evidence of the separation exists even prior to the first molt. Caste determination therefore must occur during the first instar at the latest. Unfortunately, the physiological basis of the determination has not yet been investigated. It is conceivable that the crucial event occurs as far back as during development of the embryo in the egg, or even beyond, by which is meant that in addition to the documented sexual factors there may be some degree of genetic bias influencing whether a given individual will join one caste

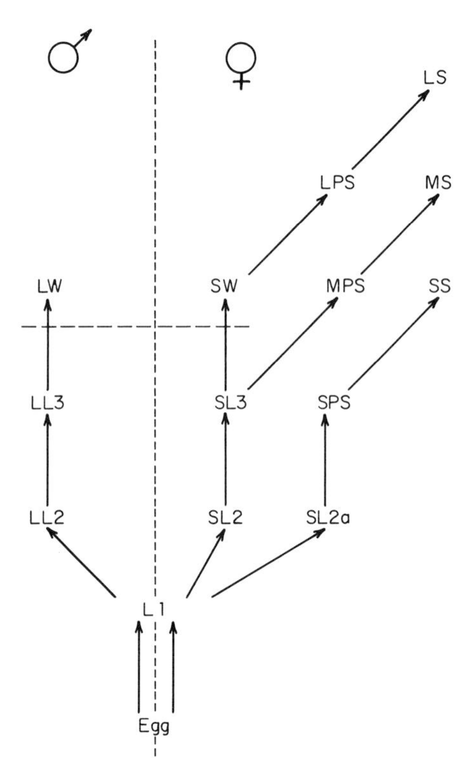

FIGURE 10-10. Development of the neuter castes in *Acanthotermes acanthothorax* (Termitidae: Macrotermitinae). The caste system of this species is one of the most advanced in all the termites. The pathway leading to the large worker caste (*LW*) has been shortened to four instars, and three distinct soldier castes, all female, have been evolved: (*LS*) large soldier; (*MS*) medium soldier; (*SS*) small soldier (redrawn from Noirot, 1969b).

or another. The existence of inhibitory pheromones has been neither demonstrated nor disproven. In many species of Termitidae, the removal of the primary reproductives is usually followed by the appearance of replacement reproductives, but no evidence is yet available to demonstrate that the inhibition is pheromonal in nature. Moreover, according to Noirot (1969b) colonies of at least one species, *Cubitermes fungifaber,* never develop replacement reproductives. The whole subject is in need of experimental investigation of the kind that has proven so successful in studies of the lower termites. Such research has been slow to start not because of the quality of the work devoted to it but because of the fact that the vast majority of species of Termitidae are limited to the tropics and form large, complex colonies that are relatively difficult to culture.

11 The Elements of Behavior

The mind makes the world in which it lives. This is not just a metaphor. Of the near infinitude of stimuli that impinge on the body moment by moment, the sense organs select only a minute fraction to relay to the central nervous system; these messages are transmitted by the afferent nerves and by the associative relays of the central nervous system, which codify the information by reissuing impulses according to rules built into the structure and arrangement of their respective neurons. Each organism therefore creates an *Umwelt* within the brain, a highly imperfect monitoring device by means of which it picks its way through the real world.

It follows that a first step in understanding the behavior of a given species is a thorough examination of the sensory physiology of the species. It is also axiomatic that as complete a repertory of behavioral acts as possible should be cataloged—"the ethogram"—and the functions of the acts as adaptations to the natural environment analyzed. Few behavioral biologists ask any longer whether the responses are instinctive or learned, this having proved to be an inefficient classification. Instead, a series of questions with a more operational basis are posed. For example, exactly how stereotyped is a given behavioral act? If it varies among individuals belonging to the same species, we next inquire how much of the variation is genetic. Under ideal conditions the genetic component of the variance can be precisely measured, with the fraction it constitutes being labeled the heritability. The remainder of the variance by definition is acquired by interaction with the environment, the effects of which are manifested as deviations in organogenesis, as physiological change in established organs, or as learning in the strict sense. Intelligence is variously measured as the diversity, precision, and persistence of the separate acts of learning, most particularly the ability to perform "rational operations," that is, to generalize learned information by transferring it from one set of circumstances to another. Finally, each of these levels—sensory input, behavioral output, and genetic, physiological, and experiential constraints on the relation between the two—is the subject of a different major branch of behavioral biology.

In line with this brief statement of the scope of behavioral analysis, the object of the present chapter is threefold. First, to review our present knowledge of the sensory world of individual social insects. Second, to evaluate their mental capacities, particularly in comparison with those of the vertebrates and nonsocial insects, in a search for peculiarities that might relate to their social achievements. And, finally, to take up what I believe to be the most provocative challenge that the social insects offer to behavioral biologists: the problem of accounting for the complex mass effects of social behavior as the product of the behaviors of the individual colony members.

The Sensory Capacities of Social Insects

The Honeybee. The best-studied of all insect species in the world is *Apis mellifera,* and this qualification applies with special strength to the subject of sensory physiology. Following the lead of von Frisch, whose contributions date from 1914 to the present time, two generations of entomologists have carefully measured the powers of discrimination of the honeybee in every known sensory modality. Their chief technique has been "Pavlovian," meaning that training is employed according to the following set procedure. First the bees are allowed to feed at a sugar source while being exposed simultaneously to the stimulus to be studied. Next tests are conducted to see if the bees are attracted by the stimulus in the absence of a reward. If they are found to have been successfully trained in this simplest possible conditioned response, they are then permitted to choose between the stimulus and a similar but unfamiliar stimulus to see if they can discriminate between the two. Finally, the unfamiliar stimulus is changed until the minimal difference between it and the conditioned stimulus detectable to the insect is determined. The honeybee worker is an exceptionally favorable subject for such behavioral measurements because it can be induced to fly patiently back and forth between the nest and the experimental bait. The performances of numbers of individuals can therefore be repeatedly checked over short periods of time. The conditioned stimulus technique perfected by von Frisch and his associates has been strikingly successful in every sensory modality. However, it now appears to have been pushed to the limits of its power. An increasing number of younger researchers are turning to electrophysiological and biochemical techniques combined with electron microscopy in order to make new discoveries and refine old ones.

The composite picture assembled by these behavioral and physiological studies can be roughly grasped by comparing the honeybee's sensory capacities with those of man. The honeybee can see in almost all directions around its body simultaneously, but, compared with man, it is very myopic and receives fuzzy images, even of large, nearby objects. It is not aware of shapes as we appraise them, but it is very sensitive to broken patterns, the flickering of light, and sudden movement. It requires approximately the same amount of light we need to see any image at all. It has color vision, but, instead of the familiar spectrum ranging from blue-violet to red, its sensitivity starts in the ultraviolet and ends in the yellow or the near red. Its ability to sense ultraviolet light allows it to see the sun through an overcast sky on some days when we are unable to accomplish this feat. Moreover, the colors of many flowers and butterfly wings look radically different to it because they bear ultraviolet markings invisible to us, and in a few cases red markings obvious to us but invisible to the bee. The bee can also evaluate the plane of polarization of sunlight, a quality totally alien to our own vision. The bee is virtually deaf to airborne sound but moderately sensitive to groundborne sound, which it detects through its feet. It has a sense of smell almost identical to ours. Its sense of taste is similar, but appears to be generally less sensitive and to have a coarser discriminating power. The bee is very sensitive to touch all over its body, but it apparently has far less capacity for judging the texture of surfaces. It has a comparatively excellent sense of balance and can perform at least one feat beyond our capacity, namely, orientation to gravity at a constant angle while walking up and down a vertical surface. It also possesses at least a limited responsiveness to the earth's magnetic field.

Let us now review the evidence behind these generalizations, starting with that on vision. The principal, paired eyes of the honeybee worker are compound, each consisting of approximately 6,300 separate visual units called ommatidia (Figure 11-1). The ommatidia are packed together somewhat like straws in a box. On the outside each terminates in a tiny, convex, transparent lens; all the lenses together form the glassy, ellipsoidal outer shell of the eye. Light is focused by each lens and its associated crystalline cone onto the rhabdom, the central sensitive element of the inner ommatidium that lies beneath it. The rhabdom itself is highly modified nervous tissue, comprised of six or eight juxtaposed rhabdomeres, each of which in turn consists of the ending of a single monopolar neuron that leads back into the nearest optic lobe of the brain. Electron micrographs of the rhabdomeres (Fernández-Morán, 1958; Goldsmith, 1962) show that each is packed with microvilli, finger-like invaginations of the cell membrane approximately 400 Å wide. The rhabdomeres appear to be fused into pairs so that each rhabdom actually consists of only four distinct morphological units. Within each unit the microvilli are parallel to each other and aligned in such a way that the central ones pass radially out from the longitudinal axis of the rhabdom. As a result the microvilli of each pair of rhabdomeres are parallel to the microvilli of the pair opposite them and at right angles to the microvilli of the two adjacent pairs. Von Frisch (1967a), operating from the assumption that the microvilli contain the visual pigments, has compared the entire rhabdom to a four-branched polarizer, postulating that the arrangement of the microvilli alone is sufficient to account for the honeybee's observed ability to measure the plane of polarized light. However, the exact nature of the visual transducer is unknown, and the role of the microvilli can be no more than a matter of speculation. The biochemistry of insect photoreception is also in too early a stage of exploration to permit morphological inference. Goldsmith (1958) reported the existence of retinal (= retinene_1) in the

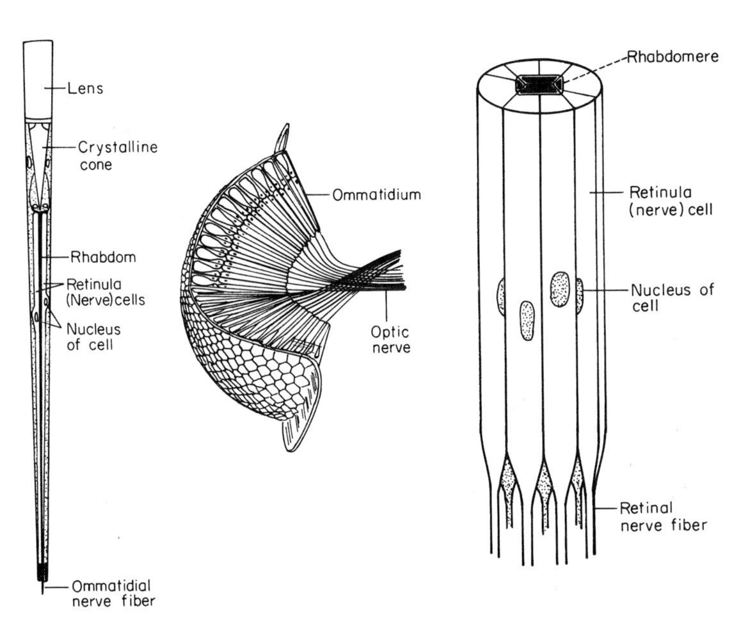

FIGURE 11-1. The structure of the compound eye of the honeybee: (*center*) entire eye of a worker, from which an angular section has been removed to reveal the packing of the ommatidia and the way their neural elements lead inward to form the optic nerve; (*left*) entire rhabdom in longitudinal section; (*right*) lower part of the eye of a drone diagramed with a cross section of the middle parts of the eight nerve cells to show that two are smaller than the remaining six and that the central portions (the rhabdomeres) are arranged into a parallelogram-shaped package of four units (based on Lindauer, 1960a; Neese, 1968, from H. Autrum; Perrelet and Baumann, 1969).

honeybee eye, but its exact location has not been ascertained (Goldsmith and Fernández, 1966).

Light enters each compound eye as a sprinkling of beams transmitted separately by each of the 6,300 ommatidial lens and crystalline cones onto their respective rhabdoms. Contemporary opinion has it that each of these multiple images is not transmitted as such by the ommatidia. In spite of the seeming complexity of the microvilli, nothing comparable to the hundreds of thousands of rods and cones of the vertebrate retina exists in the eight retinula cells of the ommatidial unit. Instead, each unit is thought to transmit information about one point of light—its intensity, its flicker pattern, its color, and its plane of polarization. Although 6,000 such points can make a picture of fair resolution, it would still be far short of that attained in human vision. The human eye and brain can distinguish two points subtending an angle of approximately 0.01°. The honeybee, judged by its responses in training experiments, can distinguish points that subtend an angle of approximately 1°. Lindauer (1957), while recording dances by worker bees on the tops of hives in Ceylon, observed that they became disoriented when the sun, their central reference point, was closer than 2.5° to the zenith. These results are consistent with the "Müller theory" of insect vision (J. Müller, 1826), which holds that an image in the brain is based on some kind of mosaic assembled from the simple pinpoint messages of the ommatidia. The lens systems of honeybee ommatidia, in fact, are directed in such a way as to form angles of from 1° to 4° with reference to an adjacent lens (Figure 11-2). In terms of the more conventional measure of visual acuity, that is, the reciprocal of the minimal angle of visual resolution, the honeybee worker scores approximately 0.02, which is low compared to that of man (2 to 2.5) but still much better than many other kinds of insects, including *Drosophila* (.002).

By means of his training technique, von Frisch proved in 1914 that honeybee workers can see colors. Their vision, unlike that of man, reaches into the ultraviolet, but, in the opposite direction, is cut short in the orange (Figure 11-3). When monochromatic lights are adjusted to equal intensity, the bee shows the following descending order of awareness: 5.6 (ultraviolet 360 mμ), 1.5 (blue-violet 440 mμ), 1.0 (green 530 mμ), 0.8 (yellow 588 mμ), 0.5 (blue-green 490 mμ), 0.3 (orange 616 mμ). More exactly, it is the "stimulus effect" (*Reizwirksamkeit*) that is measured. This is the amount of intensity of a new color component that needs to be added to a mixture; the lower its value, according to the standard interpretation, the greater the bee's awareness of the color. The spectral sensitivity curve (preference plotted as a function of wavelength) is thus bimodal and of approximately the same shape reported in other insects (Dethier, 1963; von Frisch, 1967a). Since the worker bee is essentially blind to reds of greater wavelength than 650 mμ but sensitive to ultraviolet, it should be able to distinguish certain other color combinations alien to us. Ultraviolet, for example, must be added to white to complete the mixture of all colors perceptible to the bee and, in this way, to create what the bee considers to be white, or what insect physiologists refer to as "bee white." Also, in a manner analogous to human vision, a wholly new color is created for the bee when the colors from the two ends of the visible

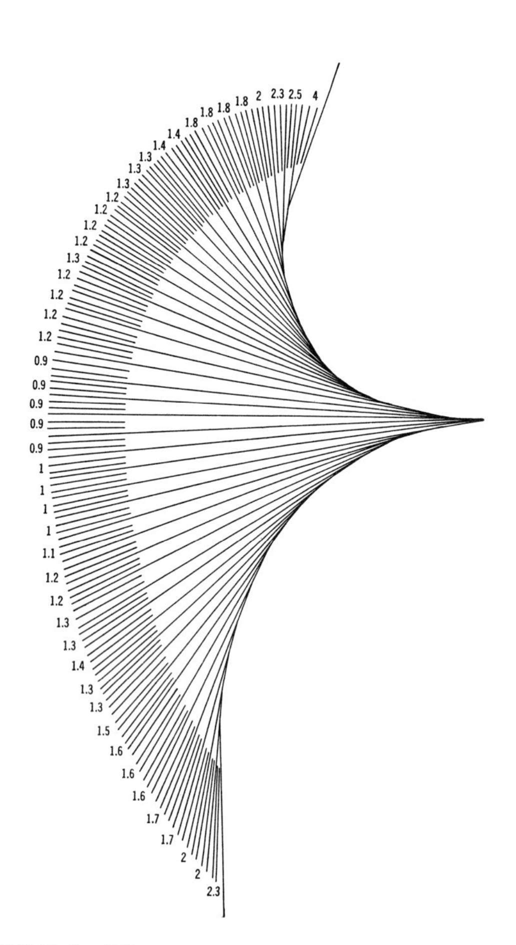

FIGURE 11-2. Diagram of the variation in the angles formed by differences in orientation of adjacent pairs of ommatidia in different positions of the honeybee compound eye. This array represents an approximately vertical section of the entire eye. The angles are consistent with the visual acuity estimated from behavioral tests (from Dethier, 1953; based on Baumgärtner, 1928).

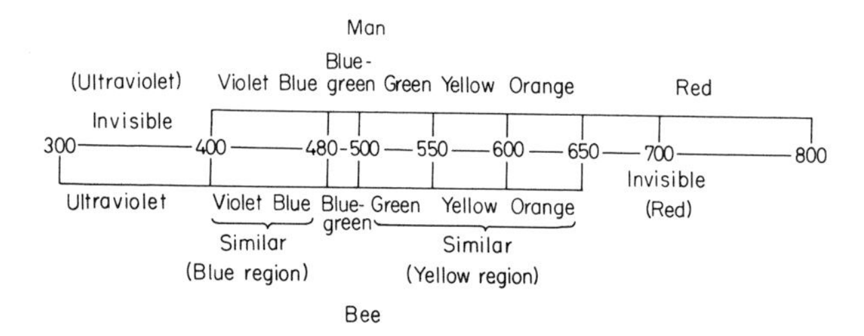

FIGURE 11-3. The colors of the spectrum: (*above*) seen through the human eye; (*below*) seen through the compound eye of the bee (from von Frisch, 1967a).

spectrum are mixed. In our case, purple is created when red and violet are combined. In the case of the bee, "bee purple" is created when its extreme colors, orange and ultraviolet, are mixed. Training experiments by Kühn (1927), Kuwabara (1957), and Daumer (1956, 1958) revealed that bees can easily distinguish four regions of the spectrum: near ultraviolet (300–400 mμ), blue and violet (400–480 mμ), blue-green (480–500 mμ), and yellow-green and orange (510–650 mμ). The insects also can discriminate between wavelengths within each of these bands, but with less precision. Daumer was further able to establish that for the trained bee any color can be matched by appropriate combinations of three monochromatic lights (ultraviolet 360 mμ, blue-violet 440 mμ, and yellow 588 mμ); from this he concluded that three kinds of corresponding receptors exist, each maximally (but not necessarily exclusively) sensitive to one of the given wavelengths. This brilliant inference was subsequently confirmed by Autrum and Vera von Zwehl (1964), who used electrophysiological recordings from individual receptor cells in ommatidia to demonstrate the existence of three types of cells whose response maxima correspond approximately to the three primary colors of Daumer (Figure 11-4). In addition, it was found that the ultraviolet receptors are concentrated on the top of the compound eye, where they are most likely to be stimulated under natural conditions.

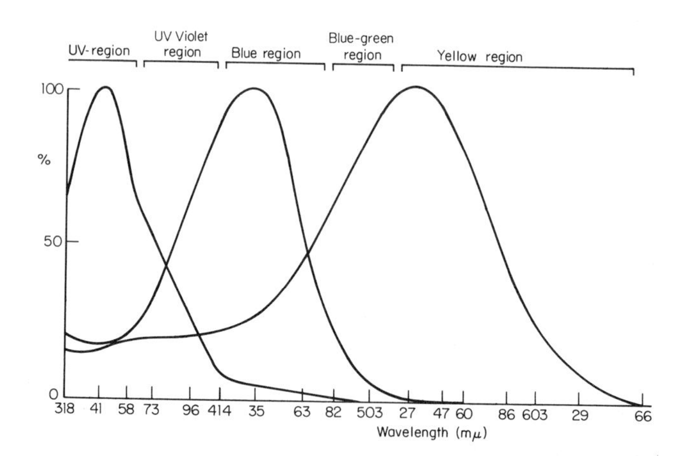

FIGURE 11-4. Sensitivity curves for three different receptors in the compound eye of the drone honeybee. The ordinate represents the magnitude of response in terms of percentage of the maximum response. The drone differs from the worker in lacking yellow receptors (redrawn from Autrum and von Zwehl, 1964).

It was also von Frisch (1949) who discovered that honeybees can utilize the plane of vibration of polarization in orientation. Other investigators have gone on to prove the existence of the same capacity in a wide variety of other insects, crustaceans, spiders, and mites, and even in squids and octopi. The original discovery was made in the following way. It had already been noted that workers of the domestic honeybee (*Apis mellifera*) engaged in waggle dances will continue dancing if the surface of the comb is tilted from the usual vertical position to a horizontal position, provided they are permitted to see the sun. In this case the bees use the sum as the reference point rather than as a straight-up direction indicated by their gravity receptors. If the sky is heavily overcast, their dances become disoriented. But von Frisch also found that if, somewhere in view, there is a blue patch of sky large enough to provide a visual angle of at least 10°–15°, the dances remain correctly oriented, even though the bees cannot see the sun directly. Subsequent experiments revealed that the bees somehow utilize information about the polarization pattern of the sky, which is linked with the position of the sun relative to the observer's position (Figure 11-5). The bees learn this relationship while orienting on clear days, so that on partly cloudy days they are able to deduce the position of the sun from partial information of the polarization pattern seen through a small portion of clear but sunless sky. When they succeed in finding food and return to the hive to report it by means of the waggle dance, the dance is performed with reference to the deduced position of the sun. The follower bees orient by flying at the correct angle to the sun, the position of which they also deduce from the plane of polarization seen through a patch of blue sky.

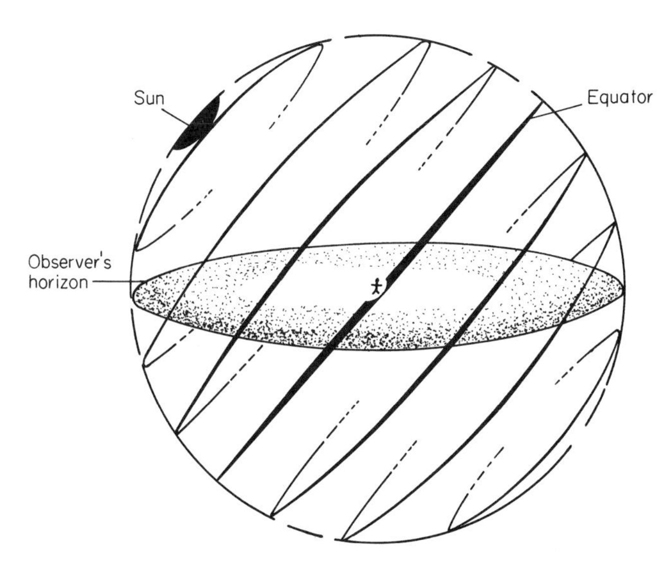

FIGURE 11-5. Perception of polarized light. The relation between the sun and the plane of polarization is shown as it would seem if this property could be directly perceived. The earth is represented by a flat, shaded disk positioned at the center of a transparent globe representing the sky (from Ribbands, 1953).

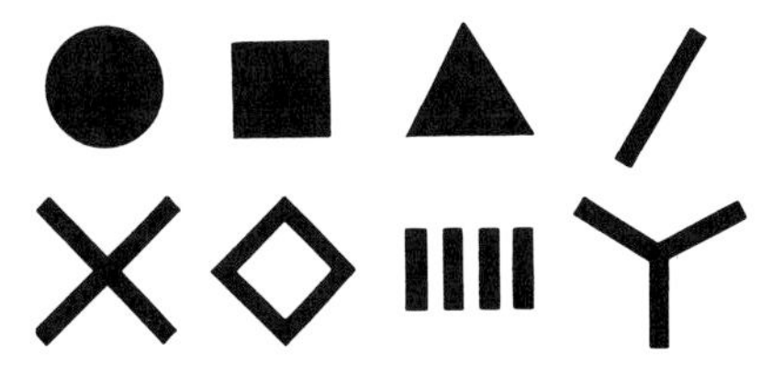

FIGURE 11-6. The honeybee worker, with its emphasis on degree of dissection rather than on geometric shape, can distinguish any figure in the upper row from any one in the lower row and vice versa; it cannot, however, distinguish between figures in the same row (from Hertz, 1930).

The honeybee worker's sense of visual pattern is quite bizarre to our own way of thinking. Using the von Frisch training technique, Mathilde Hertz (1930, 1935) discovered that the degree of dissection of a figure, rather than its outline, is the quality perceived. For example, we easily perceive that each of the figures in the upper row of Figure 11-6 is radically different from the others, but the bee cannot tell them apart. We also see the figures in the lower row as differing greatly, but these, too, the bee is unable to distinguish. The bee is, however, able to tell any one of the figures in the upper row from all of those in the lower row, and vice versa. Evidently what matters is the number of borders, rather than their alignment. Gertrud Zerrahn (1934) found that more precisely it is the length of the contours surrounding a given area that is distinguished. In further experiments Hertz discovered that bees have a spontaneous preference for the most dissected figures while they are searching for food. At first this response was considered to be a fundamental attraction to complex patterns, but later experiments (Una Jacobs-Jessen, 1959) revealed that, once the bees have fed and are attempting to return home, they prefer the opposite stimulus, namely, the simplest possible figures. All of these experiments were performed on the horizontal surface of a table. More recently Wehner and Lindauer (1966) have found that the bees also can distinguish the angle at which simple figures are tilted.

This peculiar way of looking at the world makes some sense if we also consider how the bees behave in their natural environment. When flying low over the ground in search of flowers, they experience the passing of a radial cluster of petals as a burst of flickering over their ommatidial surfaces. The regularity and high frequency of the pattern will provide the needed information as readily as the radial shape provides it to us. In support of this explanation, Wolf (1934) was able to prove that searching bees spontaneously prefer rapidly flickering lights to those flickering at low frequencies. The capacity to recognize flickering is highly advanced in honeybees, as it is in other insects with large compound eyes. Man can distinguish separate flashes of light up to a frequency of about 20 per second; at higher flicker frequencies the light appears to fuse into one steady beam. But the honeybee has a much higher flicker fusion rate, up to about 265 flashes per second (Autrum and Stoecker, 1950). Even the bees' temporary reversal to a preference for less complexity (= lower flickering rate) on its homeward journey seems easily explained. The object it seeks at close range under these circumstances is the nest entrance, which normally appears as a simple dark hole in a vertical surface.

In addition to the two compound eyes, all honeybees possess three ocelli, which are small, simple eyes located centrally on the posterior dorsal surface of the head. The function of ocelli in insects as a whole has always been a matter of conflicting evidence and opinion—most insects seem to be able to orient quite well without them—but recent experiments by Schricker (1965) indicate that the honeybee uses them to monitor light intensity. Worker bees whose ocelli had been blinded left the hive to collect food later in the morning and ceased flying earlier in the evening than unoperated workers. The light intensity required for the initiation and prolongation of flight increased as the number of ocelli blinded was increased. On the other hand, the flying activity of ocellus-blinded bees did not differ in any other perceptible way, and they danced normally on the comb surfaces after successful flights. These results suggest that the ocelli play a major role in the monitoring of light intensity but not in the formation of images.

Honeybees have no ears. That is, they have no organs known to be specialized for the reception of airborne sound. They lack the tympana or hairs designed, as in male mosquitoes and midges, for transmitting vibrations to Johnston's organ in the antenna. Bees are also deaf to airborne sound, or nearly so; they seem to be indifferent to loud noise. Attempts, such as those by Kröning (1925), to train them to respond to particular airborne sounds have failed, but they have proved very sensitive to sound carried through solids. The vibrations are picked up by the feet, transmitted by the lower part of the leg, and sensed by special mechanoreceptors called subgenual organs (Figure 11-7). The subgenual organs, so called because of their location in the proximal portion of the tibia just below the insect's equivalent of the "knee," is comprised of typical chordotonal sensilla—or scolopoid sensilla, or scolopidia as they are often alternatively labeled. The sense cell is a bipolar neuron, one terminus of which is a peg-like organ that penetrates an adjacent cell. The sense cells and associated cells stretch out together like a taut sail in the body fluid of the legs. They respond preferentially to vibrations between 200 and 6,000 cycles per second, with a maximum sensitivity in most insect species between 1,000 and 2,000 cycles per second (Figure 11-8). Without doubt this hearing capacity is

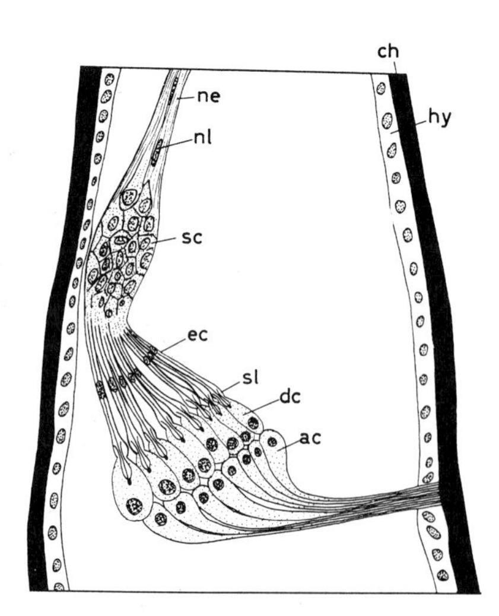

FIGURE 11-7. A subgenual organ in the leg of the ant *Formica sanguinea*. This structure, the principal receptor of groundborne vibrations in ants and honeybees, is associated with a portion of the leg cuticle (*ch*) and hypodermis (*hy*) and contains accessory cells (*ac*); cap cells (*dc*); enveloping cells (*ec*); subgenual nerve (*ne*); nucleus of a neurilemma cell (*nl*); sense cell, or scolopidium (*sc*), with its apical body, or scolops (*sl*), intruding into the adjacent cap cell (from Autrum, 1959; based on Schön, 1911).

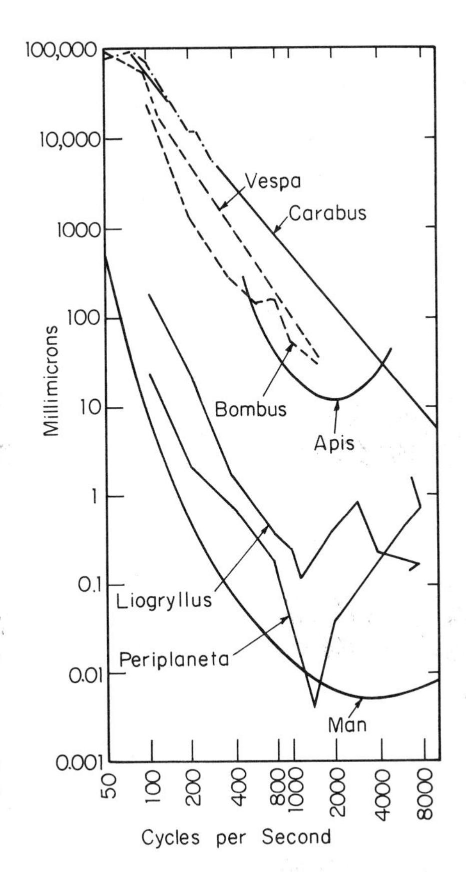

FIGURE 11-8. Vibration thresholds at different frequencies for a few species of insects, including the honeybee (*Apis*), a bumblebee (*Bombus*), and a social wasp (*Vespa*) (modified from Autrum and Schneider, 1948; from Ribbands, 1953, after A. Wilska's human data).

functional. Honeybees react strongly to tapping, scraping, and other substratal sounds of just the kind that would be associated with invasion of the hive by a larger animal. Newly emerged queens communicate back and forth with "toots" and "quacks" that are perceived through the hive floors and walls (see page 278). Worker bees emit a characteristic sound during the straight run of the waggle dance that is almost certainly detected through the substratum and may play a role in recruitment communication.

Gravity reception is vital to the honeybee worker not only for orientation during walking and flight and the maintenance of a normal standing posture, but also for the capacity to climb up and down vertical walls at a constant angle. The latter behavior, called "geomenotactic orientation," is utilized during the communicative waggle dances. While bees are dancing on the combs, a straight-up direction in the dance symbolically represents a flight toward the sun; 10° clockwise in the dance represents a flight 10° to the right of the sun; 45° clockwise in the dance represents 45° to the right of the sun; straight down (180° from straight up) means directly away (180°) from the sun, and so on around the compass. (A fuller account will be given in Chapter 13). The bees are thus able to fix in their minds the position of a flower bed or some other desirable target with respect to the position of the sun and to translate this information to geomenotactic movement on the vertical comb surface. They are able, moreover, to accomplish the transposition with an error of less than 10 percent. Markl (1966a) has been able to train worker bees to run up and down on combs at the following angles measured clockwise from the vertical: 0°, 30°, 60°, 80°, 90°, 100°, 120°, 150°, and 180°. In 1959 Lindauer and Nedel discovered the gravity receptor organs of the worker honeybees. These did not prove to be complex statocysts responding to the movement of free particles or air bubbles, as one might have anticipated from previous research on arthropods. Instead they consist of six rather simple-looking clusters of external sensory bristles, one pair on the neck (or, more accurately, lobes of the episternum that protrude into the neck region) and two pairs on the petiole (Figure 11-9). Markl (1962) subsequently detected similar organs on the legs and anten-

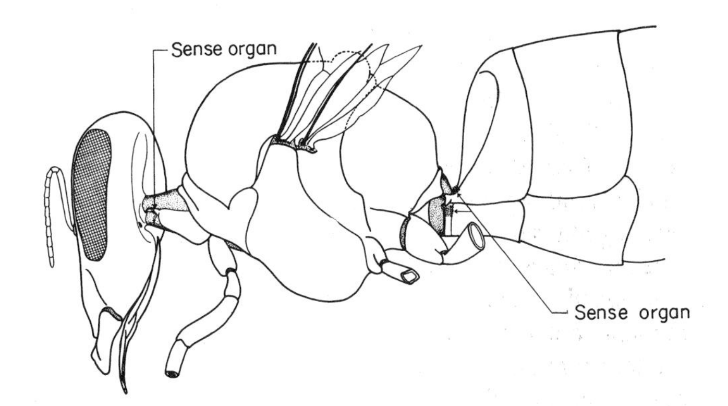

FIGURE 11-9. Location of the hair plates in the honeybee worker, one pair on the episternal projections of the neck region and two pairs around the petiole; other hair plates occur on the coxae and trochanters of the legs and on the antennae. These organs are used to detect gravity and possibly also to measure acceleration during flight (from Lindauer, 1961).

nae. As the bee shifts position, the head and abdomen swing on their articulations with the thorax, the segments of the antennae flex, and the body bends slightly on the upper leg articulations. These slight movements press the sensilla of the hair plates to one side or another, and the degree varies according to position. This causes the underlying neuron clusters to fire differentially. The composite information is then relayed directly to the two large thoracic ganglia and thence to the brain. Lindauer and Nedel (1959) and later Markl (1966b) were able to destroy the gravity sense by interfering with the bristles in various ways and by severing the afferent nerves leading to the thoracic ganglia. They also succeeded in breaking down the precision of gravity orientation in a stepwise fashion by carefully removing groups of bristles one by one. Although experimental proof is elusive, it is probable that the same receptors are used by the bees to measure acceleration in the flying bee, in the fashion postulated by Mittelstaedt (1950) for similar bristle fields in the neck region of dragonflies. If this is true, they supplement the antennae in this function (Heran, 1959).

A bristle of the honeybee gravity receptors represents but one form of sensillum trichodeum, or sensitive hair, which in turn is but one of a variety of classes of sensilla located on the surface of the bee's body (Figures 11-10 to 11-12).

Many other sensilla trichodea, devoted to the reception of simple touch, occur over the entire body (Ribbands, 1953; Thurm, 1964). It is difficult to touch the body surface of a bee with an object the size of a needle without bending at least one of these hairs and arousing the bee. The sensory hairs are especially thick on the mandibles and antennae. One antenna of a representative bee studied by Brigitte Dostal (1958) contained a total of 8,408 sensilla trichodea; the greatest number (1,113) were on the terminal segment, and the lowest numbers, on the first two flagellar segments (334 on the first, 548 on the second). The antenna also carried a total of 2,888 sensilla placodea, 114 sensilla basiconica, and 236 sensilla ampullacea and sensilla coeloconica combined. The sensilla trichodea on the terminal antennal segments are used to sense surfaces and to aid in adjusting the thickness of the honeycomb cell walls and the degree of wall smoothness (Martin and Lindauer, 1966). While adding wax to a cell, the worker repeatedly pushes one side of the cell wall with its mandibles, setting up an aperiodic vibration. By detecting the movement against the mandibles, the bee is evidently able to judge the elasticity and hence the thickness of the walls. As a result the walls have a consistent thickness of 73 μ, with an error not exceeding 4 percent. Martin succeeded in training worker bees to distinguish several basically different artificial surface textures, including perforation,

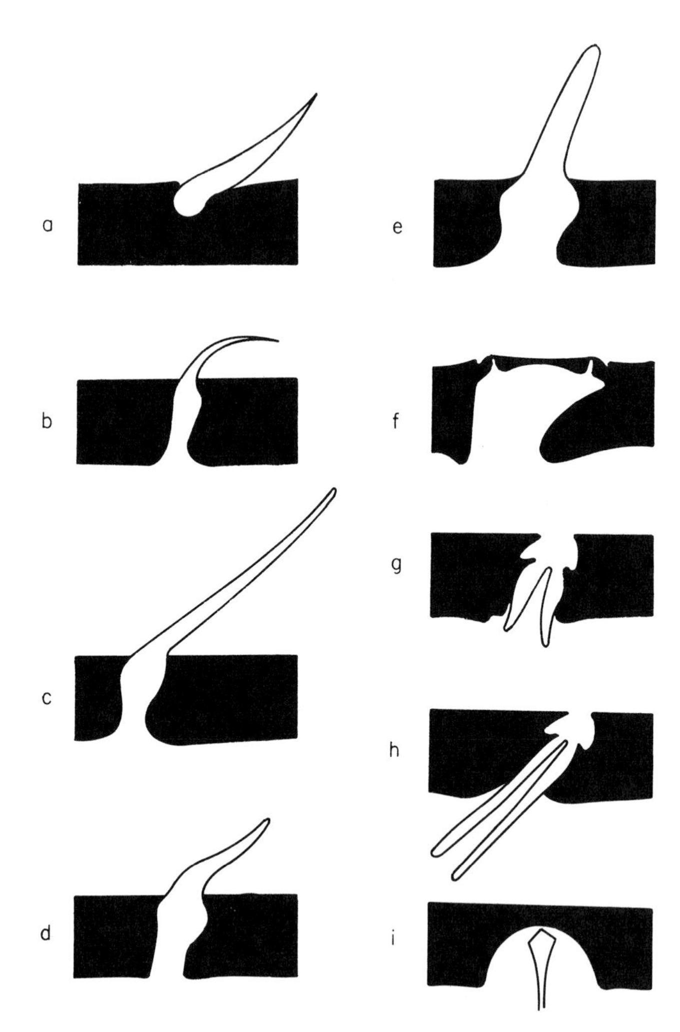

FIGURE 11-10. The types of sensilla found on the surface of the antenna of the worker honeybee. The sensilla are modifications of the cuticular surface, which is shown here in black: (*a–c*) sensilla trichodea; (*d*) sensillum trichodeum used in olfaction; (*e*) sensillum basiconicum; (*f*) sensillum placodeum; (*g*) sensillum coeloconicum; (*h*) sensillum ampullaceum; (*i*) sensillum campaniformium. The unmodified parts of the sensillar walls are 10-15 μ thick. One or more neurons lead away from each of these structures and are discharged more or less directly by the stimuli impinging on the surface (redrawn from Lacher, 1964).

ribbing, and smoothness. In addition to the sensory hairs, the bee possesses an undetermined number of stretch receptors, in the form of campaniform sensilla (sensilla campaniformia) and internally located chordotonal sensilla. These specialized organs provide additional information to the bee concerning its posture and the external pressures exerted against its body.

The ability of masses of honeybees to finely regulate the nest temperature, by fanning their wings and regurgitating water droplets on hot days and clustering in swarms on cold days (Chapter 16), is proof that the individual worker bee possesses a well-developed temperature sense. Heran (1952) allowed single bees of differing age and physiological condition to move freely in a long, glass-covered box heated at one end in such a way as to provide

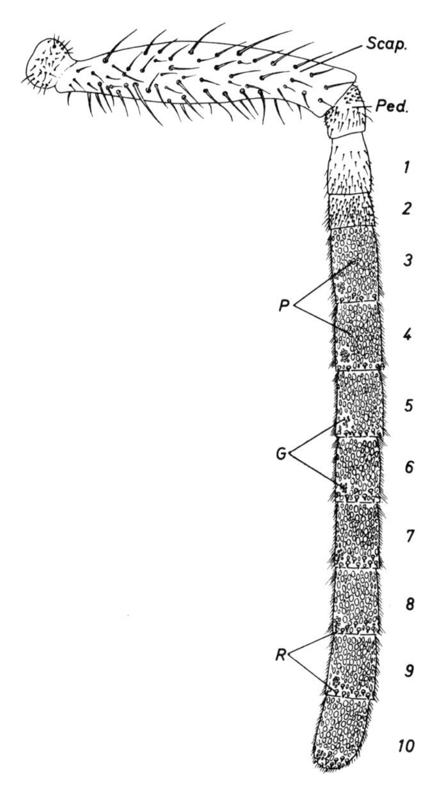

FIGURE 11-11. The left antenna of a worker honeybee, showing the twelve segments, from the scape (first member) and pedicel (second member) to the ten succeeding segments referred to collectively as the flagellum. The location of a few representative sensilla are shown: (*P*) plate organs (sensilla placodea), (*G*) pit organs (sensilla ampullacea and s. coeloconica); and (*R*) sensilla basiconica (from von Frisch, 1967a).

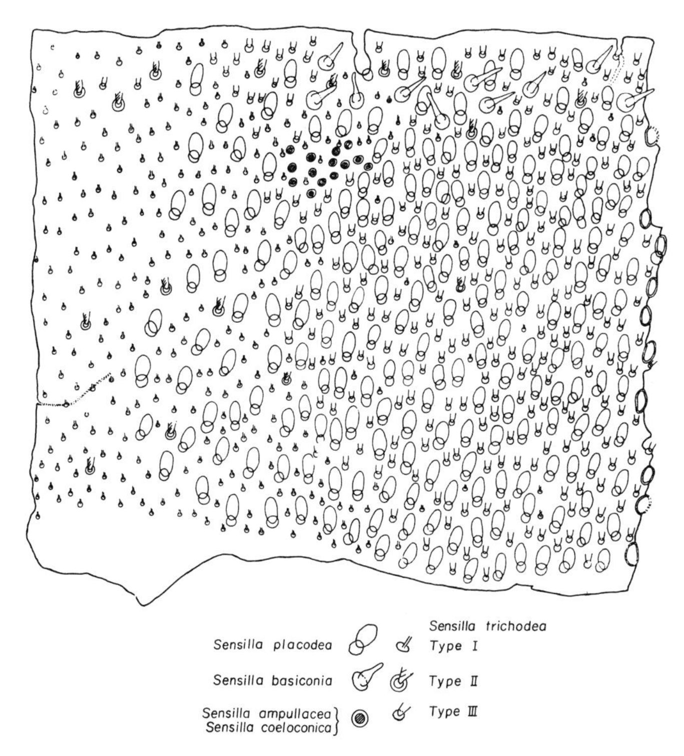

FIGURE 11-12. A map of the sensilla occupying half of the third flagellar segment. On this segment, as on most of the flagellar segments, there is a preponderance of sensilla trichodea (touch receptors) and sensilla placodea (smell organs). Also present are a few sensilla basiconica, possibly used in taste, and sensilla ampullacea and s. coeloconica, which perceive carbon dioxide, humidity, and temperature (modified from Dostal, 1958).

a temperature gradient down the length of the box. Young workers, up to seven days old, came to rest at 35–37.5°C, which is the temperature normally maintained around the brood combs by the colony. Older workers proved more variable in their choice, settling between 31.5° and 36.5°C. Bees confined at lower temperatures, in a simulation of cooling winter conditions, came to prefer lower temperatures when later given a choice in the gradient box. After two hours at 30°C, they chose 32.8°C; after five days at 13.7°C, their preference dropped all the way to 27.8°C. By cooling the box slowly and observing the shift of the bees in response, Heran estimated that they can detect changes of as little as one-fourth of a degree. Finally, in the best von Frisch tradition, Heran proceeded to train bees to take sugar solution from dishes heated (or cooled) to different temperatures. He found that they could distinguish target dishes at 20°, 25°, 32°, and 36°C from control dishes only 2°C warmer or cooler. Individuals deprived of five or more of their terminal antennal segments lose most of their sensitivity to temperature. Subsequently Lacher (1964), in the course of making tungsten-microelectrode recordings from single sensilla, discovered that the thermoreceptors are a small number of sensilla ampullacea and sensilla coeloconica located on the antennae.

Training experiments by von Frisch (1919, 1921), Ribbands (1953), Schwarz (1955), Fischer (1957), and Martin (1964, 1965) have established that the sense of smell of the honeybee worker is closely comparable to that of man. This is true in the sense that both species can detect approximately the same set of compounds in the gaseous phase. But even more significantly, both manifest about the same threshold concentrations, as illustrated in Table 11-1. There are a few noteworthy exceptions. The odors of bee's wax, of the Nasanov gland secretion, and of the queen substance (9-ketodecenoic acid), all of whose recognition is vital to the bee, are perceived by the insect at lower concentrations than by man. The bee is also sensitive to carbon dioxide and water vapor, which are "smelled" in a straightforward manner by sensilla ampullacea and sensilla coeloconica on the antenna (Lacher, 1964, 1967). Other odors are sensed mostly or perhaps even exclusively by the sensilla placodea on the antennae. Certain of these pore plates are specialized for responding to the Nasanov gland secretion and others to the queen substance (Kaissling and Renner, 1968), but the majority can be activated by at least several substances. Thus the basic dichotomy of insect sensilla into "odor specialists" (for specific compounds, such as pheromones) and "odor generalists," first proposed by Schneider (1965) and Boeckh, Kaissling, and Schneider (1965), appears to apply well to bees. Individual bees can learn odors more quickly and retain the memory longer than any other kind of stimulus, including visual (Opfinger, 1949).

TABLE 11-1. Olfactory thresholds (minimum concentrations in molecules per cubic centimeter of air) of selected compounds. The factor in the last column indicates how much worse (−) or better (+) the acuity of the honeybee worker is compared with that of man (from Schwarz, 1955).

Odorant	Olfactory threshold: Man	Olfactory threshold: Bee	Relative acuity factor
Propionic acid	4.2×10^{11}	4.3×10^{11}	1.02−
Butyric acid	7.0×10^{9}	1.1×10^{11}	15.7−
iso-Valeric acid	4.5×10^{10}	1.6×10^{11}	3.5−
Caproic acid	2.0×10^{11}	2.2×10^{11}	1.1−
Ethyl caproate	1.3×10^{11}	3.8×10^{11}	2.9−
Ethyl caprylate	3.7×10^{10}	5.4×10^{10}	1.5−
Ethyl nonanoate	3.1×10^{10}	3.7×10^{10}	1.2−
Ethyl decanoate	4.2×10^{9}	5.6×10^{9}	1.2−
Ethyl hendecanoate	1.4×10^{10}	1.8×10^{10}	1.3−
Methyl anthranilate	2.6×10^{10}	1.9×10^{9}	13.6+
Phenyl propyl alcohol	6.5×10^{9}	2.2×10^{9}	3.0+
Nerol	5.7×10^{9}	3.2×10^{9}	1.8+
α-Ionone	3.1×10^{8}	1.5×10^{10}	48.5−
Eugenol	8.5×10^{11}	2.0×10^{10}	42.5+
Citral	4.0×10^{11}	6.0×10^{10}	6.6+

When a worker bee is presented with a diffusing attractant, it swings its body around to face the direction of the source and moves toward it. This form of response suggests that bees in some fashion measure the odor gradient and move along it. In a series of incisive experiments, Lindauer and Martin (1963) and Martin (1964) proved that the orientation is indeed true osmotropotaxis and that it is achieved by an estimation of differential stimulation of the two antennae. Bees were trained to walk up the lower arm of a Y-tube and, upon reaching the fork, to choose the upper arm out of which the appropriate odor emanated. When Martin crossed the antennae of a bee and glued them in place so that the left antenna projected to the right and the right to the left, the bee took the wrong turn. This indicates that the bee weighs information received from the two antennae concerning the relative concentration of the odorant on either side of the head. Confirmation was obtained by means of the following experiment. After a bee had been trained to approach a certain odor, she was narcotized with carbon dioxide and glass capillary tubes were slipped over her antennae and sealed tight at the base (Figure 11-13). Droplets containing mixtures of the odorant in liquid phase and paraffin oil were then inserted into the ends of the tubes, so that the odorant diffused down over the antennae at a constant rate. By varying the proportions of the liquid compound and the oil in the droplet, the desired concentration of odorant in the gaseous phase was obtained. The bee was then allowed to grasp a freely turning cork hemisphere, and the direction of her walking movements were recorded. The results were conclusive: bees treated in this way attempted to turn toward the antennae receiving the stronger stimulation. Martin found that so long as relatively low initial concentrations are employed, bees can distinguish concentrations of the same substance differing by as little as 1:2.5. This power of resolution is enough to enable the bee to find its way around the surface of a flower, for example, by orienting to the odorous nectar guides alone.

The sense of taste has been investigated intensively by von Frisch (1934) and certain aspects studied further by Wykes (1952) and Jamieson and Austin (1958). Although the absolute dependence of honeybee colonies on nectar in their diet would seem to make it logical for worker bees to be unusually sensitive to sugars, the opposite turns out to be true. The response threshold for sucrose solution, for instance, is about $\frac{1}{8}$ to $\frac{1}{4}$ molar, or ten times higher than for man. Many substances that taste sweet to us are neutral or even slightly repellent to bees; examples include such natural sugars as erythritol, quercitol, mannitol, sorbitol, α-methyl mannoside, trimethyl glucose, xylose, rhamnose, sorbose, cellobiose, and raffinose, as well as artificial sweeteners with very different chemical structures such as saccharin. Of 34 sugars and closely related substances tested by von Frisch, only 7 were attractive to the bees. But herein lies an important point in comprehending honeybee biology: 5 of these substances (sucrose, glucose, fructose, melezitose, and trehalose) are known to be important constituents of nectar or honeydew. The bees have evidently applied economy in the evolution of their passband width. They also display efficiency, or at least no marked inefficiency, in their relatively high response thresholds. This is because the nectar of flowers usually contains a very high concentration of sugars (20 to 70 percent of saturation), which do not require much sensitivity in the insects that utilize them.

Von Frisch found that bees are able to distinguish what our own taste sense regards as bitter, sour, salty, and sweet substances, although the insects are not particularly sensitive to any of them. The precise organs that mediate taste have not been identified, but they probably include the same kind of sensilla basiconica on the mandibles that have been implicated in flies, beetles, and lepidopterans. In addition, Minnich (1932) demonstrated that sugar receptors are present on the forelegs and antennae of worker bees.

Most recently, and very surprisingly, bees have been

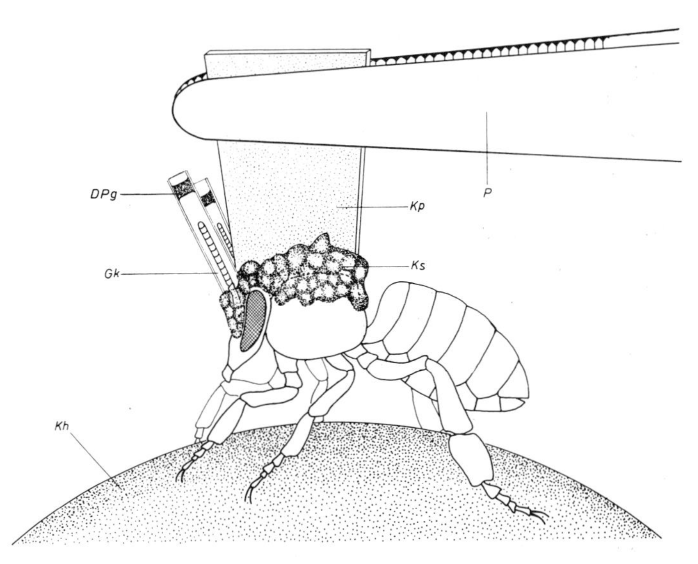

FIGURE 11-13. H. Martin's technique for measuring osmotropotaxis in the honeybee worker. The purpose of the experiment is to determine the minimal differences in concentration of an attractive odorant around the two antennae that causes the bee to turn to the antenna with the higher concentration. Mixtures of the odorant in liquid phase and paraffin oil are placed as droplets (*DPg*) in the upper ends of capillary tubes (*Gk*) surrounding each antenna. The proportions of the two components are varied to obtain the desired concentration of the odorant. The bee is held in place by a piece of cardboard (*Kp*) thrust into an adhesive mass (*Ks*) and grasped by forceps (*P*). The bee then turns the rotating hemisphere of cork as it attempts to walk in one direction or another (from Martin, 1964).

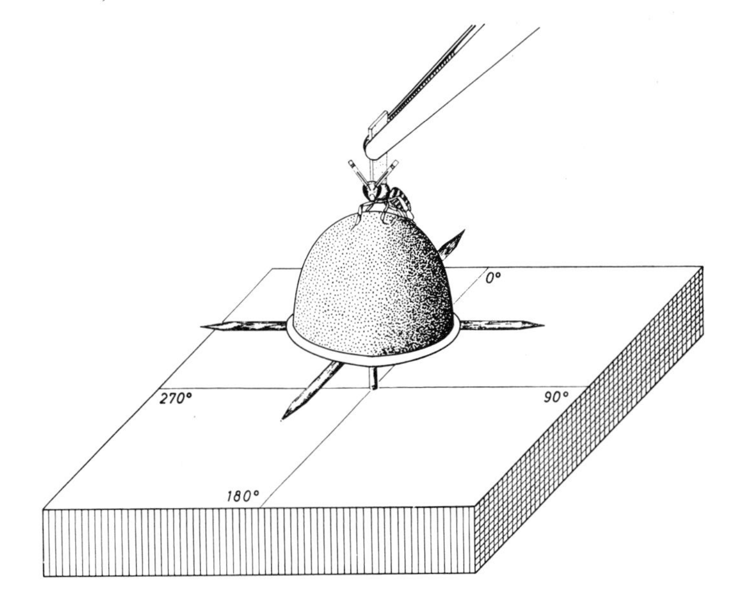

found to respond to the earth's magnetic field. When successful foragers dance on the vertical comb surface, the direction of their straight run displays systematic errors (*"Restmissweisung"*) as follows: if the run is directed obliquely upward or obliquely downward, it tends to be deflected to the vertical; if it approaches within 30° of the horizontal, the run tends to be deflected toward this latter direction (von Frisch, 1967a). When Lindauer and Martin (1968) placed dancing bees in an artificial magnetic field of about 1,000 γ intensity and made appropriate compensations of 4 percent, the errors disappeared. And when they reversed the situation by amplifying the magnetic field to 13 times and countercompensating, the errors increased above the normal level. Although this information is very suggestive, it is not yet known whether the bees utilize their magnetic sense in orientation.

We now come to the amazing ability of honeybees to keep precise time. The phenomenon was first noticed by Auguste Forel (1910b) at his country house in Switzerland. Candied fruits regularly set outdoors at midmorning breakfast and afternoon tea were quickly discovered by worker bees. Once they had become accustomed to the daily routine, the bees showed up at the correct times each day whether or not fruit was present. They gave up only when their visits went unrewarded with candied fruit for several days in a row. Stimulated by Forel's observation, von Frisch proceeded to train bees to visit feeding places under controlled conditions, and later investigators have studied the subject thoroughly in most of its behavioral aspects. Now we understand that it is highly adaptive for bees to remember at what time of day they have been rewarded because each species of plant opens its flowers only during certain hours of the day and visits at other times would be in vain. Von Frisch (1967a) has summarized the situation in the following way:

If at an artificial feeding station one offers sugar water at a set time of day, within a day or two the visitors adjust themselves to the schedule. Thenceforth they come at the designated time, whereas before and after the hour of feeding even informational flights are almost entirely omitted. The foragers remain sitting at home, saving their strength and risking no unnecessary flights. The other situation too, which most often obtains in the actual production of nectar, is readily set up: at the feeding station there is sugar water from morning till night, but at certain hours in greater amounts or—in other experiments—at higher concentrations during the rest of the day. The artificial flower has an 'optimal time.' The bees quickly note this. If the visits are recorded on a 'day of observation,' during which the feeding dish offers nothing from morning till night, lively coming and going prevails at the optimal time. All members of the foraging group come repeatedly and search obstinately. During the other hours, when feeding was more meager or less sweet, searching is much less zealous and only part of the foragers appear (Wahl 1933). That under natural conditions bees behave in the same way relative to flowers was shown by Kleber . . .

As a rule such pauses in foraging activity are spent at home by the bees. But they may give up the recess and visit a second food-yielding plant whose optimal time falls in the hours when the first one is unproductive. Their time sense and their adaptability are competent for this task. The foragers can be trained to visit a feeding place two or three times a day and to remain home meanwhile. Even training to five different times of day has been successful (Beling 1929; Wahl 1932). But they also learn very rapidly to seek out one feeding station at certain hours in the morning and another feeding place during afternoon hours, when at the time fixed upon food is offered now here and then there (Wahl 1932; Lindauer 1957:2). The bees even learned to appear at four different feeding places at four different times of day. But this difficult task is no longer mastered perfectly by all (Finke 1958).

Kleber was conducting experiments at an artificial bed of poppies, whose yield of pollen became exhausted every day between 9:30 and 10:00 A.M. Hence the foragers became trained to a morning time. One day the productive period was lengthened by offering at the end of the usual harvesting time poppy flowers bearing copious pollen that had been cut earlier and protected against visiting bees. With astonishment the observer noticed that individual bees, which kept on coming and carrying in the new harvest, were not mobilizing their group comrades but nevertheless recruited newcomers. It seemed as if the foragers would not respond to dances outside of the regular training time. Anyone who had seen often enough the arousing effect of the dances would hardly believe such a thing. And in fact a different, astounding solution of the riddle was found: bees trained to a definite time, once their good hours are past, withdraw to the margin or into the upper portion of the comb, or, in the observation hive—whose entrance opens on one side—very frequently to the opposite, quieter side of the comb . . . Thus they go away from the dance floor and seek out resting places where they may doze undisturbed and where they will not be reached by a comrade dancing at the wrong hour, there to remain until the accustomed harvesting time approaches again. But then—a charming performance when the bees are marked with colors—they wander slowly from all sides toward the dancing place, until they are assembled there in a dense throng and can be alerted in a moment by the first successful informant (Ilse Körner 1939; v. Frisch 1940).

The chronometric powers of the bees become even more impressive when witnessed as part of the sun orientation mechanism. Highly stimulated bees occasionally become "marathon dancers"; they continue performing the waggle dance for hours at a time without attempting new flights to the target in the interim. Under such conditions, the angle of the sun is shifting in the sky outside the hive (Lindauer, 1954a; von Frisch, 1967a). To take a concrete example, one bee was observed dancing inside a hive in a closed room from 7:05 until 10:46 A.M. in the course of a single morning. During this time the azimuth of the sun shifted 54.5° clockwise, while the bee's direction of dancing shifted 53.5° in the opposite sense. Thus, without any clue whatsoever except its memory of the movement of the sun and its own internal clock, the bee adjusted almost perfectly to the sun's movement and still performed straight runs that pointed symbolically to the target. Lindauer (1961) investigated the matter further by raising colonies in a cellar and allowing bees to observe the transit of the sun for varying periods during the day. Bees never allowed to see the sun were completely disoriented. Bees allowed to make about 60 collecting trips all on one day still could not use the sun as a compass, but this ability was eventually acquired after about 300 training flights in the course of three days. Finally, after a total of 500 flights spaced over five days, the bees acquired knowledge of both sun-compass orientation and the path and rate of passage of the sun in the sky. The truly amazing thing about Lindauer's results is that all of the training was conducted in the afternoon; yet the bees could also calculate the position of the sun in the morning and correct their dances accordingly. In other words, they were able to extrapolate the path of the sun all the way around the clock! Lindauer's marathon dancers were even able to calculate the position of the sun at night, as it might have been seen if they could view it directly through the earth. This discovery led to an intriguing question: if bees were trained to go to a food plate at one place very late in the afternoon and to another food plate at a second place very early in the morning, to which food plate would the dances point at night? This experiment was performed by Lindauer (1957) with the following outcome. Two dances in the evening (8:36 and 9:31 P.M.) pointed to the afternoon site; and three dances in the early morning (3:54, 4:11, 4:18 A.M.) to the morning site. Of seven dances occurring around midnight (11:10 P.M.–1:14 A.M.), five were incorrect or disoriented, one indicated the approximate resultant of the directions of the two feeding places, and one indicated the morning site with a strong error. Thus the bees chose the target that was closer in time to the dance, but became confused during a 2-hour transitional period around midnight.

The temporal memory of the honeybee worker is exclusively circadian. That is, the bee can be readily trained to arouse itself at 24-hour intervals, but attempts to train it to fly at other intervals, for example, every 19 or 48 hours, have always failed. Once entrained, the bee's internal clock keeps it informed of the arrival of the appointed time each day for days or weeks on end. But this experience must be reinforced: if the bee makes one or two flights to the right place at the right time but goes unrewarded, it ceases making the attempt. Also, the memory can be erased under deep carbon dioxide narcosis (Medugorac and Lindauer, 1967).

For the reader who wishes to explore more deeply into the subject of honeybee sensory physiology, the following reviews are recommended: Dethier (1963) and Markl and Lindauer (1965) on the general aspects of insect sensory physiology; Jander (1963) on the principles of orientation; von Frisch (1967a) on most aspects of honeybee sensory physiology and orientation; and Neese (1968), Chauvin (1968c), Markl (1968a), and Heran (1968) on various of the separate senses. The anatomy of the brain and remainder of the nervous system are treated by Hüsing (1968) and Chauvin (1968b). The neurophysiology of the honeybee is still a very lopsided subject. Although the behavior of individual receptor neurons has been studied with a high degree of sophistication, as, for example, in the work of Autrum, Lacher, and Thurm, extremely little has been attempted with the central nervous system. The meager information available, which has been well covered by Franz Huber (1965) and Hoyle (1965), is consistent with the more extensive data on ants to be reviewed shortly. It is a curious fact that what we know of integrative neurophysiology and the control of motor patterns is based mostly on such insects as roaches, mantes, crickets, grasshoppers, and ants, and almost totally neglects *Apis mellifera,* which is otherwise the most thoroughly understood species of arthropod.

Ants. Far less is known about the collective sensory physiology of the thousands of species of ants than about that of the single honeybee species *Apis mellifera.* No more than one-tenth as many articles have been published on the subject, and this difference truly reflects the disparity in information. Enough is known, nevertheless, to indicate that the various species of ants are generally similar to the honeybee in their *Umwelten.* Vision differs drastically among the species, from a total blindness in the workers of some subterranean species to an *Apis*-like acuity in certain large-eyed, epigaeic forms. Hearing, smell, and the sense of gravity, including skills in geomenotactic orientation, are closely comparable. The three species of ants in which taste thresholds have been measured are more sensitive to sugars than *Apis* and, in fact, approach man in this respect. The possible existence of a magnetic sense, so recently demonstrated in honeybees, has not been investigated in ants.

To examine the evidence behind these statements, let us begin again with vision. The number of ommatidia in the compound eye of the worker caste varies among ant species in a manner correlated with their life habit. Most phylogenetically advanced army ants, including the species of *Anomma, Dorylus, Eciton,* and *Leptanilla,* lack eyes altogether, and their optical lobes degenerate in the pupal stage (Werringloer, 1932). Other eyeless workers are found in certain subterranean genera such as *Carebara, Erebomyra, Liomyrmex, Paedalgus,* and *Wadeura.* (The males of all these groups, who must leave the colonies at some time to conduct nuptial flights in the open air, have very large eyes.) The workers of most ant species have from a few tens to a few hundreds of ommatidia per compound eye; the number is roughly correlated with the amount of time spent foraging above ground. For example, the workers of species in the *Formica rufa* group, which forage almost exclusively above ground, have approximately 600 ommatidia per eye. Although the angle of divergence between the optical axes of adjacent ommatidia is a relatively gross nine to ten degrees, these ants can distinguish black and white strips that present visual angles of as little as one-half a degree (Christiane Voss, 1967). Thus in visual acuity *Formica rufa* is equal or slightly superior to the honeybee but still decidedly inferior to man. A few arboreal tropical ants, including those belonging to the formicine genera *Gesomyrmex, Gigantiops, Opisthopsis,* and *Santschiella,* have much larger eyes with thousands of ommatidia each. My own experience with *Gigantiops* and *Opisthopsis* has impressed me with their keen vision and alertness, which makes them exceptionally difficult ants to catch in the field. No physiological studies have been made of the vision of these large-eyed species, and we can only speculate as to the level of their visual acuity.

Nor has color vision in ants been investigated sufficiently to provide any satisfactory conclusion. Marak and Wolken (1965) derived an action spectrum of the fire ant *Solenopsis saevissima* by means of the following procedure. Workers returning to the nest after finding a sugar solution or other food orient at a constant angle to the principal light source. In nature the light source is the sun (or at night possibly the moon). In the laboratory an artificial light can be placed to one side of the foraging platform. When the light is switched off and another one on the opposite side immediately switched on, the ant reverses its direction by 180°. An investigator, by turning to first one light and then the other, can march the ant back and forth like a robot, leaving no doubt that it is reacting to the light. In this fashion Marak and Wolken measured the threshold intensities of a series of monochromatic lights, the combined data on which are referred to as the action spectrum. They found that the ant has a spectral sensitivity similar to that of the honeybee, ranging from below 350 mμ in the near ultraviolet to 650 mμ in the orange, with peak sensitivity at three points near 350, 505, and 620 mμ respectively. But the mere ability to detect monochromatic lights at varying intensities does not necessarily imply that the fire ant can distinguish one from the other, or, in other words, that it has color vision. In order to establish this capability unequivocally, one needs training experiments in which the ant is given a choice between lights of different colors but equal intensities. Tsuneki (1953) trained workers of *Leptothorax congruus* and *Camponotus herculeanus* to orient to monochromatic lights while returning pupae to laboratory nests. He then required them to choose between the training color and some other monochromatic light. Tsuneki found no evidence that his ants could select one color over another under these conditions. He did concede, however, that his negative result might be due to technical reasons, such as the difficulty of orientation under bilateral stimulation and the heavy reliance placed by the ants on relative intensity. Therefore, the question of color vision in ants remains unanswered.

Recent experiments by Christiane Voss (1967), paralleling the honeybee studies of Wehner and Lindauer (1966), have revealed that the large-eyed members of the *Formica rufa* group distinguish shapes in essentially the same way as honeybees. Workers can be trained to select between figures on the basis of the degree of disruption, independent of the manner in which the figure is broken up. The ants also have a strong tendency to run spontaneously toward certain visual patterns, especially toward white areas on black, black areas on white, and vertical black-white borders. Furthermore, they prefer closed to open figures of the same configuration, regardless of the nature of the configuration.

Sun-compass orientation in ants, and in animals generally, was discovered by the ant taxonomist Felix Santschi in Tunisia in 1911. Santschi's attention had been caught by the problem of how workers of desert ants manage to leave their nests, forage at a distance, and then find their way back home over the featureless desert sands, even when strong winds make odor trails impracticable. He found the answer by means of his now famous "mirror experiment." When workers of *Cataglyphis bicolor* and *Messor barbarus* returning home with booty were shaded from the sun on one side and presented with the image of the sun by means of a mirror held on the opposite side, they reversed their direction 180° and headed confidently away from home. When the shade and mirror were removed, they again turned about by 180° and ran homeward. In other words, it was apparent that the ants were reckoning the angle subtended by the sun and the nest and holding it constant as they returned home. Later von Buddenbrock (1917) established that this "light-compass orientation" occurs widely among insects, and, it has since been discovered, in many other invertebrate groups as well. For nocturnally foraging ants a moon-compass response is equally feasible, and has, in fact, been demonstrated in *Monomorium* by Santschi (1923) and *Formica* by Jander (1957).

Since the sun moves through an arc of 15° every hour, what will happen if an ant is trapped on its way home and not permitted to see the sun for a substantial period of time? If it always keeps a constant angle to the sun regardless of the passage of time—what German investigators call *winkeltreue* orientation—then the error the trapped ant makes on being released again should equal the arc through which the sun has passed in the interim. Brun (1914) performed just such an experiment with workers of *Lasius* and found that they did behave as though complying with the *winkeltreue* rule. For example, one worker was confined in darkness from 4:00 to 5:30 one afternoon, during which time the sun moved through an arc of 22.5°; when released, it set off in a direction that attempted to follow the original angle to the sun and consequently deviated from its original, true path by 23.5°, approximately the amount the sun had traveled. This would seem to be a relatively poor way for an ant to get around, especially if it spends hours at a time away from its nest, and Brun's result never seemed to provide a full explanation of visual orientation. In 1957 Jander showed that experienced *Formica rufa* workers are really able to do much better. They duplicate the feat of the honeybee whereby they keep track of the sun's movement and constantly adjust the angle of their return journeys. Newly eclosed *rufa* workers and those which have just emerged from overwintering must learn the sun's movement; until they accomplish this, they orient to the sun in the *winkeltreue* manner. It is not known whether Brun's *Lasius* were similarly naïve, but at least his results are not inconsistent with the time-compensated orientation demonstrated in *Formica* by Jander.

In his early experiments Santschi found that ants can continue to orient correctly if shaded from the sun, provided they are allowed to view a patch of blue sky. This peculiar result strongly indicated that ants were able to utilize the pattern of polarized light to calculate the position of the sun, as von Frisch later proved in his experiments on honeybees. When appropriate equipment became available, Vowles (1950, 1954b) was able to demonstrate orientation to polarized light in the ant *Myrmica ruginodis*. Other investigators established its existence in *Tetramorium, Tapinoma, Lasius,* and *Camponotus* (Gertraud Schifferer in von Frisch, 1950; Carthy, 1951a; Jander, 1957; Jacobs-Jessen, 1959).

Hearing in ants is basically similar to that in honeybees, consisting almost entirely of the reception of groundborne vibrations evidently perceived by the subgenual organs of the legs (see Figure 11-7). Although response threshold curves have not been drawn for ants, the extreme sensitivity of these insects to groundborne sound has been noted by numerous observers (see Haskins and Enzmann, 1938). A slight tap on the edge of an artificial nest containing ants is usually enough to send the colony into a state of alarm. Furthermore, Markl (1967, 1968b) proved experimentally that the stridulation of worker leaf-cutting ants (*Atta cephalotes*) is heard through the soil by nestmates which respond to it as an underground alarm call.

Ants have a gravity receptor system nearly identical to that of the honeybee (Markl, 1962). Its hair plates are located in the same positions and show about the same degree of sensitivity. Markl was able to train workers of *Formica polyctena* to walk up slopes of as little as 2° from the horizontal. He also succeeded in training them to maintain a constant angle with reference to gravity while running up and down a vertical surface. Another feat of the honeybees duplicated by ant workers is the ability to substitute gravity for light signals in maintaining a constant angle. In order to test for the presence of this phenomenon, Vowles (1954a) first permitted workers of *Lasius niger* and *Myrmica ruginodis* to run over the surface of a horizontal board while keeping a constant angle to an artificial light. This proved easy to do because escaping ants tend to run in a straight line at an arbitrary angle to the light source. When Vowles then turned off the light and simultaneously tilted the board to a vertical position, the ants altered their direction to maintain the same angle—but this time with reference to gravity. Unlike honeybees, the ants made no distinction as to the left or right. For example, individuals that had originally oriented toward the light now tended to run straight up; those that had originally run 20° to the right of the light now ran about 20° to the right or to the left of the vertical direction, and so on. Large errors were common in the transposition, but the existence of a correlation was unmistakable. Vowles also induced his ants to perform the reverse substitution, namely the light-compass response for an original gravity orientation. This capacity to interchange such radically different sensory inputs has also been discovered in water striders and beetles, but differences exist in the details of execution (Birukow, 1954). The dung beetle *Geotrupes silvaticus,* for instance, can transpose light-compass into gravity orientation, but, unlike bees and ants, it reads the light source as the downward direction on a vertical surface. The phenomenon in ants and beetles may not have an overt function; instead it may merely represent an outcome of the necessarily economical organization of the small insect brain when two sensory inputs have been connected with one steering device. In the honeybee, on the other hand, the substitution has acquired a wholly new and overt function—becoming the vital component in the waggle dance that permits information about the location of loci outside the hive to be communicated on the vertical combs inside the darkened hives.

Ants have about as much sense of smell as bees and humans. This generalization is based on olfactory threshold measurements made on workers of the formicine species *Acanthomyops claviger* and *Lasius alienus* by Regnier and Wilson (1968, 1969). Employing the diffusion-model technique (Wilson, Bossert, and Regnier, 1969), we estimated the threshold concentrations of a wide variety of alkanes, aldehydes, alcohols, ketones, esters, and ethers required to elicit the beginnings of an alarm response in resting workers. Those with more than about seven carbon members (and molecular weights exceeding 100), regardless of molecular structure, caused responses in concentrations somewhere in the range of 10^9 to 10^{12} molecules per cm^3. This range of magnitudes is similar to that obtained for the same classes of substances in experiments on honeybees and humans, but different techniques were used (see Table 11-1). As illustrated in Figure 11-14, ants are progressively less sensitive to substances of lower molecular weights within the same homologous series, in accordance with a principle that has already been established widely in other insect groups

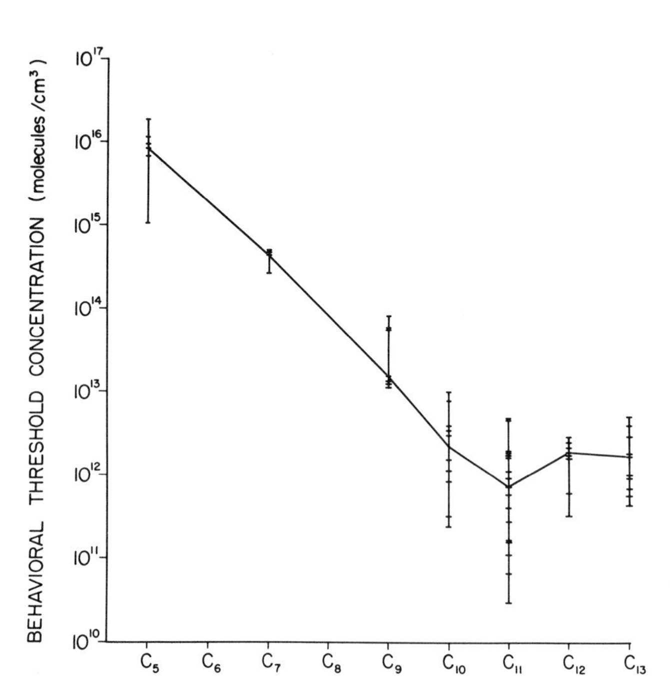

FIGURE 11-14. The minimal concentrations of various members of the alkane series, from pentane (C_5H_{12}) to tridecane ($C_{13}H_{28}$), that elicit the beginnings of an alarm response from workers of the ant *Acanthomyops claviger.* Each point represents a separate measurement. The resulting curve illustrates the positive relation between molecular weight and olfactory efficiency that exists widely in the insects (from Regnier and Wilson, 1968).

(Dethier, 1953, 1963).* Ants, for example, are insensitive to the odor of formic acid, the organic acid of lowest molecular weight. However, there are striking exceptions to these rules. Fire ants (*Solenopsis*), and possibly other ants as well, can detect differences in CO_2 concentrations, an ability shared with bees but not with man (Wilson, 1962a; Hangartner, 1969a). Also, our experiments with the trail substance of *Solenopsis saevissima* indicate that the threshold concentration must be extremely low. Although the chemical structure of the pheromone has not yet been identified (and its olfactory threshold is therefore unobtainable) we have identified it as a single peak in gas chromatograms and made rough estimates of the amount present in various preparations. Each worker appears to carry less than 10^{-9} g at any given moment. Also, less than 10^{-9} g was recovered from an odor trail half a meter in length along which an entire colony migrated in the laboratory. We have concluded that the response threshold must be exceedingly low. A comparable result has been obtained in rough estimates of the olfactory efficiency of dodecatrienol, the trail substance of *Reticulitermes flavipes* recently identified by Matsumura *et al.* (1968).

Our experiments utilizing the diffusion-model method show that ants of various species are able to detect and move up odor gradients (osmotropotaxis), apparently by lateral movements of both antennae. In following an odor trail the worker of *Lasius fuliginosus* walks through a cloud of pheromone vapor, continuing forward so long as it remains inside the space within which the molecular concentration is at or above the threshold value. Experiments by Hangartner (1967), designed along approximately the lines of those by Martin on honeybees, have proved that *fuliginosus* workers keep to the active space by testing the odor gradient with both antennae simultaneously. Analyses of the olfactory sensilla of ants, or of other antennal sensilla in these insects, have not been conducted. There is a rich field for investigation here, particularly in the comparison of different species of ants belonging to radically different phyletic groups and adaptive types. For example, important differences might be found between large-eyed ants that hunt arboreally and blind army ants that rely exclusively on odor trails and the sense of touch during their forays.

*The relationship is not necessarily always so simple. Boeckh (1967) found that an optimal middle range exists in the attractiveness of homologous fatty acids for migratory locusts, with both short-chain and long-chain molecules having lower olfactory efficiency.

TABLE 11-2. Threshold molar concentrations for sweet substances in the worker caste of three species of ants (from A. Schmidt, 1938).

Substance	*Lasius niger*	*Myrmica rubra*	*Manica rubida*
Alcohols			
Erythritol	0	0	0
Mannitol	0	1	0
Sorbitol	1/4	1/16	0
Dulcitol	0	0	0
Glucoside			
α-Methyl glucoside	1/8	1/50	1/8
Pentoses			
Arabinose	0	0	0
Xylose	—	—	0
Methylpentose			
Rhamnose	0	1	0
Hexoses			
Glucose	1/8	1/50	1/32
Fructose	1/64	1/50	1/32
Galactose	0	1/2	0
Mannose	0	1/4	0
Disaccharides			
Saccharose	1/200	1/150	1/400-1/800
Maltose	1/100	1/100	1/100
Lactose	0	0	0
Cellobiose	0	—	0
Melibiose	—	0	0
Trisaccharides			
Melizitose	1/100	1/200	1/64
Raffinose	1/400	1/400	1/100

The sense of taste, particularly for sweet substances, was the subject of a careful investigation by Anneliese Schmidt (1938). Her results, the essentials of which are summarized in Table 11-2, show that ant species vary somewhat in their sensitivity to sweet substances. As a group they share the honeybee's inability to detect some compounds which taste sweet to us, but they are much more sensitive to the effective compounds than bees, and in this latter respect they equal or surpass man. Schmidt was also able to prove that ants, like bees, can sense substances by contact chemoreception with the antennae.

Ant workers kept for a short time in dry chambers are

attracted to drops of water or moist strips of filter paper placed near them. Many investigators, for example Soulié (1961), have noticed that the search for moisture under such conditions overrides even the ants' natural avoidance of light, so that workers will place brood under strong artificial light if by so doing they reach a favorable zone of humidity. The workers are seemingly able to orient up humidity gradients, which, if true, implies the use of the antennae as the sensing organs. However, I know of no attempt that has been made to locate the specific sensilla that serve as hygroreceptors.

There also exists some indirect evidence, albeit very tenuous, that ants can sense the earth's magnetic field. When Markl (1962) trained workers of *Formica polyctena* to hold a constant angle while running over a vertical surface, he found that they displayed a pattern of systematic "residual errors" closely similar to the *Restmissweisung* of honeybees performing waggle dances on their vertical combs. I have already mentioned that Lindauer and Martin were able to correct the honeybee *Restmissweisung* by placing hives in artificial magnetic fields that compensated for the earth's magnetic field. The same experiment has yet to be performed with ants, but an adjustment like that displayed by bees seems a possibility.

A time sense certainly exists in ants. But it varies strikingly among castes in a manner that is just beginning to be elucidated, and it must be carefully defined for each particular case. McCluskey (1958, 1965) discovered that male ants generally show a clear circadian rhythm, with sharp increases in restlessness at just the hour each day at which the nuptial flights occur. The nuptial hour varies among species, occurring in midmorning for some species, in late afternoon for others, in the hours just before dawn for still others, and so on. The peak of male activity at the species-specific time persists in the laboratory under conditions of constant illumination. These rhythms are notable in that they do not require feeding or any other rewarding stimulus to be initiated or sustained indefinitely at full strength. The virgin winged queens of *Pogonomyrmex californicus* and *Veromessor pergandei* also display circadian activity peaks corresponding to the times of the nuptial flight in nature, but this is wholly lost as soon as the queens mate (McCluskey, 1967; McCluskey and Carter, 1969). It remains to be seen whether further investigation will reveal this to be a general phenomenon in the queen caste of ants. The existence of true, innately circadian activity peaks in workers of *Veromessor, Iridomyrmex,* and *Formica* have been denied by Otto (1958a) and McCluskey (1963). However, it may be that these investigators simply did not monitor the appropriate behavior patterns under the right conditions. Baroni Urbani (1965) has reported circadian rhythms in the foraging activity of *Camponotus nylanderi,* a nocturnal Mediterranean species. Even more significantly, Bert Hölldobler detected strong circadian peaks in the prenuptial activities of workers of *Camponotus herculeanus* and *C. ligniperda,* and he has demonstrated the same phenomenon in the evening homing and trail-following behavior of workers of *Pogonomyrmex badius* (personal communication).

Each ant species has a distinctive daily foraging schedule, and some are active only for a few set hours each day. In the heath of southwestern Australia, for example, I discovered the existence of a remarkable degree of precision in the changeover of ant species at dusk. In midafternoon, the ground and the branches and leaves of the low bushes that dominate the vegetation contained hordes of workers, mostly brown, red, or black in color and with medium-sized compound eyes, belonging to ten or so species of *Myrmecia, Rhytidoponera, Dacryon, Iridomyrmex,* and other typically Australian genera. As dusk fell, first one species, then another, began to pull back into their nests, while the nocturnal species—pale-colored, mostly large-eyed species in *Colobostruma, Iridomyrmex,* and *Camponotus*—made their appearance in a regularly staggered succession. So orderly was the changeover that approximately the same number of foraging workers remained on the bushes throughout. It is nevertheless not known whether the Australian ants are influenced by an internal, circadian time sense. Possibly the foraging times are controlled instead by the external conditions of light, humidity, and temperature. It is also likely that cases exist in which a circadian time sense and environmental conditions interact. A suggestive example is provided by Hodgson's study (1955) of the leaf-cutting ant *Atta cephalotes.* Workers of this American tropical species begin to move up to the entrance holes of their nest in the dark minutes just before dawn. As the first light spreads over the forest floor, between 5:30 and 6:00 A.M., the ants commence their daylong foraging. Their departure can be delayed by shading the nest entrance at this time, but no amount of light cast into the entrance holes before dawn can cause a premature start. Thus the workers appear to assemble in accord with a circadian time sense (or, less likely, in response to a periodic change in unknown environmental factors in the nest), but they require illumination to carry through on the first foraging trip.

A circadian memory was reported in *Lasius fuliginosus* and *Formica rufa* by Autrum (1936), who trained workers to search for food at the same time each day. The process required three to five days of successive rewards, and the results resemble closely the circadian learning previously reported by Beling for the honeybee. A noncircadian time sense was reported by Grabensberger (1933), who claimed to have trained workers of several ant species (*Myrmica rubra, Lasius niger, Formica fusca, Camponotus ligniperda*) to search for food at intervals of 3, 5, 21, 22, 26, and 27 hours as well as 24 hours. This is an extraordinary result, especially in view of the fact that no such noncircadian rhythms have ever been taught to honeybees. It appears that Grabensberger was in error, however. Later experiments by Reichle (1943) and Dobrzański (1956), utilizing the same as well as other species and genera, yielded wholly negative results. Neither of these investigators was able to induce so much as a 24-hour rhythm, but this, too, is an extraordinary result. The whole subject must be regarded as remaining in an uncertain state.

Other Social Hymenoptera. Current information on the sensory physiology of social Hymenoptera other than *Apis mellifera* and ants is too meager to bear serious generalization. The interested specialist will be able to glean the bulk of the facts from several articles devoted primarily to other, better-documented subjects: Dethier (1963) on insect sensory physiology; Blackith (1958a) and Jacobs-Jessen (1959) on vision in social wasps and bumblebees; Spradbery (1965) on general aspects of wasp behavior; Free and Butler (1959) on general aspects of bumblebee behavior; and von Frisch (1967a) on the many comparative aspects of the sensory physiology of various social Hymenoptera with *Apis mellifera* as the central paradigm. Perhaps it will seem gratuitous to offer an opinion on the matter, but I believe it unlikely that further research on the other social Hymenoptera will reveal sensory capabilities of any great novelty. In the first place, the few quantitative data available are commensurate with the measurements on honeybees and ants. Second, honeybees and ants are phylogenetically relatively remote from each other, insofar as that is possible within the aculeate Hymenoptera, the bees having sprung from sphecoid wasps and the ants from scolioid wasps. Yet their sensory capacities have proved remarkably similar in all studies to date, differing principally in quantitative aspects of vision and taste. In view of this fact, it is improbable that other social bees, which belong to the same subfamily (Apoidea) as *Apis,* or the social wasps, which are behaviorally less specialized than the bees, have diverged very far in the evolution of sensory physiology.

Termites. Surprisingly little attention has been paid to the sensory physiology of these insects. Richard (1950) has thoroughly analyzed the phototactic responses of *Kalotermes flavicollis,* while Abushama (1966) has used electroantennograms to study olfactory receptors in the antennae of *Zootermopsis angusticollis.* The peripheral nervous system of *K. flavicollis* has been mapped in detail by Richard (1969). Histological studies of several genera of both lower and higher genera of termites, with special reference to caste differences, have been published by Zuberi (1963) and Hecker (1966). From what we now understand of social behavior in the termites, the most promising field of inquiry would seem to be chemoreception, in all of its aspects.

The Mental Capacities of Social Insects

Training experiments designed to bracket sensory capabilities have had the agreeable side effect of providing measures of the capacity of bees to learn in a wide range of environmental circumstances. This learning capacity is impressive in several respects. To sum it all up briefly, the training experiments have shown that worker bees are able to learn signals in every known sensory modality. They can learn them quickly in most cases, and they can master multiple tasks dependent on several modalities simultaneously. Tasks can be memorized and performed in a sequence, as in the programs of visits to different flowers at specific times of the day. Isolated worker bees can be trained to walk through relatively complex mazes, taking as many as five turns in sequence in response to such clues as the distance between two spots, the color of a marker, and the angle of a turn in the maze (Kalmus, 1937; Weiss, 1953, 1957). After associating a given color once with a reward of 2-molar sucrose solution, they can remember it for as long as six days. If given the experience three times in a row they remember the color for at least two weeks (Menzel, 1968). The location of a food site in the field can be remembered for a period of six to eight days; on one occasion worker bees were observed dancing out the location of a site after two months of winter confinement (Lindauer, 1960b).

Ants can perform comparable feats. Workers of *Formica pallidefulva* learned a six-point maze with relative ease at a rate only two to three times slower than that achieved by laboratory rats (Figure 11-15). Workers of *Formica polyctena* can remember their way through mazes for periods of up to four days (Chauvin, 1964), while those of *F. rufa*, operating under more natural circumstances,

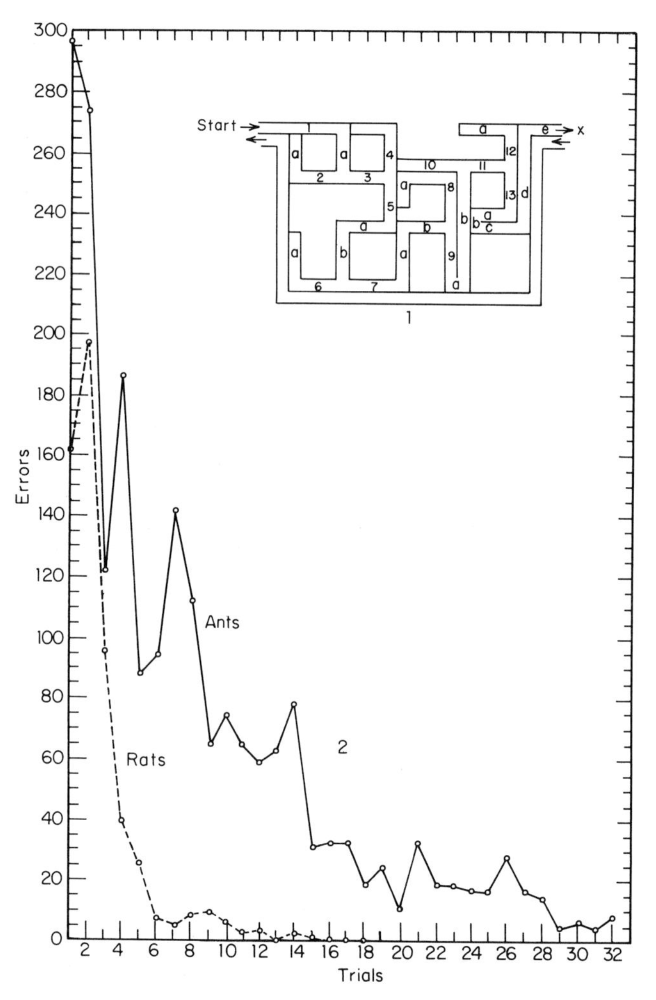

FIGURE 11-15. At the top (1) is shown the plan of a maze with six blind alleys. The curves (2) represent the number of errors made in successive trials for eight hooded rats (*broken line*) and eight ants, *Formica pallidefulva* (*solid line*). Both groups of animals ran from 1 to 13e: the rats to a food place at *x;* the ants to their nest at *x* (redrawn from Schneirla, 1953b).

can simultaneously memorize the position of four separate landmarks and remember them well enough for use in orientation as much as a week later (Jander, 1957). The almost fantastic ability of both ants and bees to memorize the path and angular velocity of the sun has already been described.

Equally impressive is the integrative process that takes place in the brains of bees and ants during foraging trips. The outward bound worker typically winds and loops in tortuous searching patterns until it encounters food. But then it takes a relatively direct route (the "bee-line") in its return trip to the nest. On the basis of his experiments with *Formica* Jander suggested that the insect performs a continuous series of calculations analogous to the simplest possible mathematical operation. As it runs outward, according to Jander's interpretation, the ant perceives the constant light source, the sun, and it is aware of the angles it takes relative to that source during each of its twists and turns. For every new direction taken, the product of the angle to the sun times the duration of the outward leg of the run is calculated, and the sum of all these products is divided by the total running time to produce the average (weighted) movement angle to the light. When the insect is ready to come home, it need only reverse this mean angle by 180°. The neural machinery for accomplishing such a feat—which in our case would require a compass, a stopwatch, and integral vector calculus—is of course quite unknown.

It would be tempting to succumb to a sense of wonder and to conclude from these fragments of information that social insects are mentally comparable to vertebrates—and superior to their relatives among the solitary insect species. But in both their learning capacities and their innate behavior patterns they suffer from severe constraints that hold their total behavioral performances to levels far below those attained by the higher vertebrates. These constraints could not have been anticipated by theoretical considerations; they have been revealed item by item only in the course of empirical research.

Learning is restricted to special conditions and has immediate adaptive value. Each of the outstanding learning feats documented in honeybees and ants is related to a narrow, particular challenge regularly encountered by the insects in the course of their daily activities. To take the ultimate example, individual workers can memorize the angle of their outward journey relative to the sun (even when following a twisting path), while simultaneously accounting for the motion of the sun through the sky. Of course, this is the one feat they must accomplish whenever they are outside the nest in order to complete a simple foraging trip. By the same token, defective worker bees who could not tell time and memorize the hour would lose much of their pollen and nectar crop each day; bees and ants who could not memorize odors with a high degree of precision would soon see their colony boundaries collapse; and so on.

Social insects have achieved most of the basic forms of learning employed by mammals. To use Thorpe's (1956) classification of learning performance, they are capable of habituation—the gradual lessening of responses to stimuli that are first disturbing but found by experience to be harmless or at least unavoidable. Ants can be "tamed" by human observers; if handled frequently enough, they respond calmly to being picked up and moved about. When learning a maze, Schneirla's *Formica* (1941) first went through a stage of habituation ("generalized stage"). In this stage progress occurred through a simultaneous decrease in excited and erratic behavior and an increase in a tendency to continue running when obstacles were encountered. Social insects are also capable of simple associative learning, which means the acquisition of conditioned responses by the Pavlovian association of rewards with previously meaningless stimuli. This is of course the mode of learning exploited by experimenters to delimit the sensory capacities of insects. A third basic mode is latent learning, exemplified in the social insects by the memorization of landmarks by foraging workers during orientation trips. These stimuli are not associated with immediate rewards, and in some cases the memories are not called upon until a later time. Social insects, however, are not believed to be capable of the fourth, most advanced mode, that of insight learning. This means that they apparently cannot duplicate the mammalian feat of reorganizing their memories to construct a new response in the face of a novel problem. Dogs can, if given time, walk purposefully around a transparent barrier instead of trying to push through or over it. A chimpanzee, without coaching, can deduce how to pile boxes in order to reach a banana previously out of reach. No behavior equaling these feats has been observed in social insects. At the same time, there is really no sharp distinction between latent learning and insight learning, so that the gap between the social insects and mammals is probably only a quantitative one.

A closer examination of the learning process reveals other peculiar shortcomings. Honeybees can learn quickly to orient with respect to attractive odors but not at all with respect to repellents (von Frisch, 1919). *Formica pallidefulva* workers tend to "follow their noses" to an extreme degree in maze learning. If the passage in a maze turns to the left, the ant tends to follow it around on the outer, or right-hand, side. If it next comes to a T-choice, it will usually proceed around the right-hand corner and into the right-hand arm of the choice. Consequently ants can learn a maze much more quickly if the arms they tend to follow by momentum have also been made the correct ones by the maze designer. Also, when the *pallidefulva* workers do make a wrong turn in the early stages of learning, they usually follow the blind alley to its very end before turning back. Only in later stages do they develop the ability to turn back shortly after a mistake has been made (Schneirla, 1943).

The insects have only a limited ability to transfer memories to assist in the learning of new situations. By far the most severe restriction in learning by insects, one demonstrated in all social species thus far studied, is the near absence of the process of transfer learning. When *Formica* workers are confronted with the problem of running a mastered maze in reverse, they treat the change as a whole new problem; while rats, in contrast, are able to make a considerable saving in learning time by transfer of the previous information (Schneirla, 1946a). The inability to transpose information to a reversal maze, or its approximate equivalent in field situations, has also been reported in bumblebees (Wagner, 1906) and honeybees (Kalmus, 1937; K. Weiss, 1953). Ants are even more simpleminded than this information suggests. When workers of *Formica schaufussi* are required to learn a maze leading to a food box, they are unable to use the information in finding their way through the same kind of maze when returning to the nest (B. A. Weiss and Schneirla, 1967).

The only exception to this negative rule of which I am aware is the capacity of honeybee workers to generalize very simple visual patterns. Mazokhin-Porshnyakov (1968, 1969) found that bees trained to come to a square divided into quadrants of different colors later responded to the same arrangement of quadrants even when the colors were different. A summary of some of Mazokhin-Porshnyakov's results is presented in Figure 11-16. Parallel experiments have revealed the ability to generalize elementary arrangements of distinctive shapes.

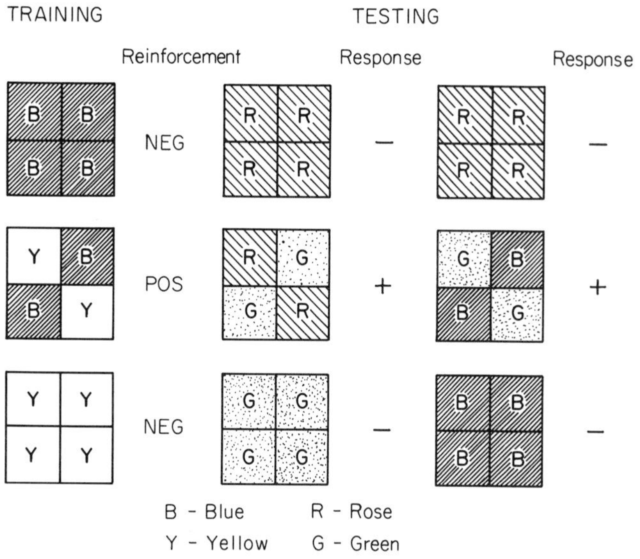

FIGURE 11-16. One of Mazokhin-Porshnyakov's experiments demonstrating the ability of honeybee workers to generalize visual patterns. Workers trained to come to a diagonal arrangement of two colors still responded to the same arrangement in later tests even when the particular colors of the squares were changed.

Social insects do not play. Play, as Hinde (1966) has tried to define it biologically, is "a general term for activities which seem to the observer to make no immediate contribution to survival." In mammals, play is comprised largely of rehearsals performed in a nonfunctional context of the serious activities of searching, fighting, courtship, hunting, and copulation. Among younger individuals, who display it most prominently, play appears to have two functions: first, exploring the environment and social partners; second, perfecting adaptive responses to both.

Several authors have seriously claimed that ants engage in play or something closely resembling it. Pierre Huber (1810) described the following peculiar behavior in a mound-building species of *Formica:*

> One day I approached some of their mounds, which were exposed to the sun and sheltered from the north. The ants had gathered in large numbers and seemed to be enjoying the heat which prevailed on the surfaces of their nests . . . But when I undertook to follow each ant separately, I saw that they approached one another waving their antennae with astonishing rapidity; with the front feet they lightly stroked the sides of the heads of other ants; and after these preliminary gestures, which resembled caresses, they stood up on their hind legs, two by two, and wrestled one another, seizing a mandible, a leg, or antenna, and letting it go immediately, to return to the attack; clinging to one another's waist or abdomen, embracing one another, overturning one another, falling and scrambling up again, and taking revenge for their defeat without appearing to inflict an injury; they did not eject their venom, as they do in their battles, and they did not grip their opponents with the tenacity seen in serious fights; they soon released the ants which they had seized, and tried to catch others; I saw some who were so ardent in their efforts that they pursued several workers in succession, and wrestled with them for a few moments, and the contest was concluded only when the less lively ant, having overthrown her opponent, succeeded in escaping and hiding herself in some gallery.

Huber observed the same behavior repeatedly on the one nest, and he remained especially impressed by the fact that it never seemed to result in death or injury. In 1921 Stumper described a similar form of *combat amical* in the little myrmicine ant *Formicoxenus nitidulus,* which he attributed to play functioning to get rid of excess "muscular energy." But later the same author (1949) observed that the fighting was in deadly earnest and was, in fact, being carried on between members of different colonies! The territorial wars of the pavement ant *Tetramorium caespitum* have much the same appearance as the activity described by Huber: prolonged "wrestling" by pairs of ants in the midst of a struggling mass of contending workers, seldom resulting in injury and death. In short, these activities have a simple explanation having nothing to do with play. I know of no behavior in ants or any other social insects that can be construed as play or social practice behavior approaching the mammalian type.

Social insects exercise a severe economy in communication and response patterns. The queen substance of honeybees, *trans*-9-keto-2-decenoic acid, is used to inhibit ovariole development in workers, to inhibit royal-cell building by workers, to attract workers during swarming; it also serves as a long-distance sex attractant and an aphrodisiac inducing copulation by males who reach the queen in flight. A closely similar substance, *trans*-9-hydroxy-2-decenoic acid, causes clustering by workers during the swarm. The trail substance of fire ants (*Solenopsis saevissima*), possibly a single compound, is used to organize both food retrieval by masses of workers and colony emigration, and it is also employed in alarm communication. Almost the entire social organization of fire ants might be mediated by as few as ten pheromones (Wilson, 1962a).

During orientation, ants and honeybees are able to

make use of cues in almost any sensory modality. But, as we have seen, two of the most different classes of stimuli, gravitational and visual, can be interchanged without difficulty in the course of a single act of orientation. Thus, both appear to be employed by the same steering mechanism in the brain, and this innate limitation has been economically turned to advantage by the honeybees to evolve their waggle dance communication. As a final act of neurophysiological economy, the waggle dance is used as a highly versatile device to recruit workers to every kind of food discovery, including water when that is needed by the colony, as well as to new nest sites during either colony division or absconding.

One intriguing analogy that may exist between social insects and vertebrates is in the category of displacement activities. When workers of social insects are excited by alarm substances or some other, more direct disturbance, they tend to increase their rate of self-grooming. This could be interpreted as displacement behavior in the strict sense of vertebrate ethology, but it is just as reasonably explained as but one consequence of a heightened level of general activity. Recently Pflumm (1969) has presented evidence of a more sophisticated nature favoring the former hypothesis, at least under certain special circumstances. He found that honeybee workers greatly increase their rate of self-grooming when torn between the tendency to drink sugar water at a feeder and the opposing tendency to depart for the nest. Direct behavioral observations have shown that the two impulses are most nearly in balance at the beginning of a feeding session and toward the end, and it is at just these times that the self-grooming increases. Accordingly, Pflumm interprets the behavior to be comparable to displacement activities of vertebrates that arise in the face of conflicting stimuli of nearly equal effectiveness. But there is still no known example, of which I am aware, of displacement activities being ritualized to produce more complex signals, one of the most common and conspicuous events in vertebrate evolution.

To summarize this discussion of the limits of behavioral capacity, experiments on the learning performance of honeybees and ants have already revealed many kinds of constraints which, by themselves, must hold the potential intelligence of these insects far below that of the mammals. Further limitations of the insect brain are indicated by the frequent use of the same communicative signals and responses for two or more very different purposes.

These purely empirical findings seem to be consistent with certain anatomical limitations independently observed in the insect nervous system (Vowles, 1961). For example, the insect brain is vastly smaller and contains only a minute fraction of the number of neurons encountered in the brain of even a primitive vertebrate. The insect neuron furthermore differs in basic structure from most kinds of vertebrate neurons. It has a small cell body almost entirely filled with nucleus. The great majority of the cell bodies lie at the periphery of the central nervous system and, like the dorsal root ganglion cells of vertebrates, they do not appear to play any role in integration. Instead, a short axon leads inward from each cell body, soon dividing into two or three branches that subdivide into dense arborizations resembling the frayed ends of strings. The arborizations from different cells are closely entangled to form the synaptic junctions. Thus the surface of the cell body, which in some parts of the vertebrate nervous system serves as an important receptive and integrative surface for incoming nerve impulses, is not used for that purpose in insect neurons. The insect neuron must receive its input entirely over its dendritic surfaces, a mode of transmission which Vowles considers to be intrinsically less efficient. He has also pointed out that, because the dendritic branches are shorter and fewer in number than in vertebrates, their receptive surface area is relatively much smaller. Also, the compactness of the dendritic arborizations and the intricacy of their entanglement makes it likely that synaptic transmission depends more on how many presynaptic fibers are active than on which particular ones are firing. This coding rule, if it really exists, would automatically reduce the amount of information a neuron can transmit. Finally, insect neurons, because of the thinness of their myelin sheath and their lack of medullation, conduct impulses at only about one-tenth the speed of vertebrate neurons. Vowles draws the following provocative conclusion: "The properties of the insect neuron and the small size of the insect nervous system render necessary a functional organization far simpler than is often supposed. What should arouse our wonder is the success with which simplicity has been crowned."

It is nevertheless clear that the insect brain has made a heroic effort to overcome the limitations of small size in the course of evolution. Many of the shortcomings listed by Vowles are better regarded as compromises in design whose final effect is to increase the level of performance over what would have been obtainable otherwise. By stripping away most of the myelin sheath and cell body, and by displacing the cell body to one side, it has been possible to increase the total number of neurons. And by shortening the axons and increasing the compactness of the intermingled dendritic arborizations, a great, not a lesser, total number of neuronal connections in the central nervous system was made possible. Recent electron microscope studies by Steiger (1967) on the neuropiles of the brain of the ants *Formica rufa* and *Camponotus ligniperda* show how still further evolutionary refinements might have enhanced mental capacity. The neuropiles are dense feltworks of dendritic fibers and glial elements located in the corpora pedunculata of the brain. They are generally considered to be the main integrative areas of the brain in which complex behaviors are generated and organized (Franz Huber, 1965). Within the neuropiles are scattered minute bodies, called glomeruli, which are just visible under ordinary light microscopy. Electron micrographs reveal that each glomerulus contains a large presynaptic end knob with dozens of small postsynaptic end feet attached to its surface. These attachments are packed onto the knob surface at high densities comparable to those of the most elaborate vertebrate glomeruli. Also, there are two types of junctions discernible between the presynaptic and postsynaptic membranes. First are the synapses, or "active regions," which have all the features of chemical junctions seen in the vertebrate nervous system. In addition, there are peculiar "tight junctions" which are located on both the external and internal surfaces of the presynaptic end knobs. The internal tight junctions are formed by the invagination of postsynaptic dendritic branches, whose membranes are fused throughout with the presynaptic membranes.

It is tempting to speculate on the possible correspondence of the synaptic junctions with learned responses and the tight junctions with innate, or "programmed" behaviors. It is also tempting to guess about the degree to which such modifications as the shortening of the axons, compaction of the dendritic arborizations, and differentiation of the glomerular junctions have compensated for the intrinsically small size of the insect brain. The empirical evidence concerning learned behaviors shows that the compensation has been entirely inadequate to bring bees and ants to anywhere near the level of higher vertebrates. The evidence concerning innate behaviors is less clear; the two groups seem in fact to be comparable. Otherwise, the neurophysiology of insects, particularly social insects, is still in too primitive a condition to offer firm correlates between the degree of behavioral complexity and the structure and function of the nervous system. Nevertheless, there is the exciting possibility that the demonstration of such a connection between anatomy and behavior lies close at hand.

We now come to an equally interesting but even more difficult comparison, that between the social insects and their nonsocial relatives. No one doubts that the *colonies* of some social species are capable of more elaborate behavior than the *individuals* of any nonsocial species. But the question now before us is the degree of difference between adult individuals belonging to the two types. It is desirable to know how much of the exceptional complexity of colonial behavior is due to the complexity of individual behavior patterns, and how much to mass effects that come from the meshing of varying, but otherwise simple, individual patterns. Put another way, we need to ask whether new levels of complexity in the behavior of individuals is required before they can form advanced societies, or whether mere specialization and coordination suffice.

This question has never been subjected to careful and systematic scrutiny. Quantitative studies of learning and innate behavior are too scarce, especially among the solitary relatives of social insects, to permit extensive comparisons. Nevertheless, I feel confident enough to offer the following opinion: Within the Hymenoptera, which contain the great majority of social insects, the individuals of many solitary species have mental capacities at least equaling those of social species. One need only think about the following extreme examples of complex behavior in solitary wasps—the orientation flights and prey searching of *Philanthus,* the complicated, species-specific nests constructed by single eumenine females, the extraordinary ability of *Ammophila* females to memorize the location and contents of several nests simultaneously, and many others (see, for example, the review by Markl and Lindauer, 1965)—to realize that this is probably true. It so happens that almost all controlled learning experiments have been confined to social species, particularly in the honeybees and ants. It is therefore little wonder that a quick reading of the literature gives the impression that the behavior of social species is more plastic. I believe that when such experiments are extended to the most closely related solitary groups, and are designed to allow precise comparisons, this apparent difference will disappear. At the same time it is not true that just any kind of insect has the capacity to evolve into a social species. Social life requires at least moderate degrees of complexity and plasticity that may have developed in only a few orders of insects, including the Hymenoptera.

Up to this point I have avoided pursuing the inevitable discussion of innate versus learned behavior. It is interesting for more than just historical purposes to note that extreme opposite philosophies concerning this dichotomy have been expressed in behavioral studies on ants within this century. Albrecht Bethe, an extreme reductionist, believed that ants are "reflex machines," meaning that all of their behavior can eventually be dissected into simple, automatic responses to specific stimuli. Theodore C. Schneirla, a leading American psychologist of the behaviorist school, took a position as close to the opposite as was possible. He stressed always the role of learning, from his early maze training of *Formica* to his later numerous theoretical discussions of comparative psychology. His intent, I believe, was also reductionist, but in this case because he wanted to account for as much behavior as possible in terms of the aggregation of simple responses based on experience. The idea of a programmed nervous system waiting to direct a complex sequence of specific behavioral acts solely upon receipt of a simple, "releaser" stimulus is uncongenial to this philosophy. Schneirla's own excellent work on the army ants, those insects so justly famous for the blind stereotypy of their behavior, must have created a serious quandary in his own mind (so it seemed to me in my own personal conversations with him). However, Schneirla usually attempted to be reductionist in his interpretation of army ant behavior, even when he conceded the existence of innate elements. He searched always for general classes of stimuli, most frequently under the rubric of "trophallaxis," a concept which he believed could explain the full behavior of the ants in terms of such equally simplified categories of response as levels of excitation, attraction, and repulsion (see Chapter 14).

It is now very clear that neither of these opposing simplistic schemes accurately identified the innate and experiential elements of behavior. No conceptual shortcut can be substituted for the acquisition of empirical information on physiology and heritability along the lines roughly outlined at the beginning of this chapter. It is also clear that some of the most elaborate behavior patterns in social insects are inherited in blocks that are relatively insensitive to outside perturbation, and they are thus likely to prove relatively intractable to analysis. The factual basis of this generalization is extensive, but the most persuasive evidence comes from the preservation of species-specific behavior in individual social insects placed in colonies belonging to other species. The following examples are known to me. Pieces of brood comb belonging to one species of meliponine bee can be introduced into the hive of another species. When the alien workers emerge from the pupal stage, they are accepted by their hosts, but they proceed to attempt a reconstruction of the host nest according to the plan characteristic of their own species. The result of the competition between hosts and guests is a hybrid nest of intermediate but very variable form (Nogueira-Neto, 1950). Workers of *Formica* captured as pupae by slave-making species of *Formica* and *Polyergus* still behave much like workers in normal, unmixed colonies of their own species, even to the construction of nests conforming to their species type (Wilson, 1955b). Similar fidelity to species-specific patterns is displayed in the laboratory by ants of differing species that are added to mixed colonies in the pupal stage (Kaethe Heyde, 1924).

Finally there is the important matter of differences in mental capacity between castes belonging to the same colony. Schneirla (1933b) found strong variation in maze-learning ability even within the single monomorphic worker caste of *Formica pallidelfulva.* In ant species with complete dimorphism the behavioral differences between the subcastes are sometimes extreme in degree. Witness the contrast between the brutish soldiers of *Pheidole,* distinguished by an extremely limited repertory of responses, and their versatile, nimble nestmates of the minor worker subcaste—the two might seem to belong to different species.

There are three rules in the evolution of caste differences which I believe hold generally among the social insects. The first is that in the early stages of social evolution the queen is totipotent. That is, the queen can perform not only her reproductive functions but also, at least for a while, the functions of the worker caste. The second rule is that, as evolution progresses, the behaviors of the queen and worker castes tend to diverge, and in the course of the divergence the queen loses her totipotency while the worker comes no closer to acquiring it. The third rule is that, while some new behavioral acts are created in the course of caste evolution, most of the evolution consists of the loss of behavioral acts. The prevalence of loss results in a further drift from totipotency on the part of individual castes, yet without reducing the repertory of all the castes combined. The specialization and differentiation of castes actually tends to result in an increase in the total repertory of the colony as a whole.

A formal theory of caste specialization and its role in the increase of colony fitness will be presented later, in Chapter 18. For the present, let us examine briefly some actual examples of the evolutionary trends just cited. The queens of the primitive ant genus *Myrmecia* can be taken as a case of a seemingly totipotent caste. They go out on the nuptial flights and mate away from the parent nest. Then, in the course of founding their own colony, they perform every known behavioral act in the repertory of the worker caste, including construction of the nest, rearing of an entire brood, and repeated foraging away from the nest for food. The queens of many, perhaps most, of the species of Ponerinae are also totipotent. A condition approaching totipotency is even to be encountered in the fungus-growing ants of the genus *Acromyrmex,* which are phylogenetically among the more advanced myrmicines. The *Acromyrmex* queens go through a normal nuptial flight after which they build a nest, construct a small fungus garden, and rear the first brood of workers, all on their own. In addition to these many complicated procedures, they occasionally leave the nest to forage for garden materials. Whether they can also lay odor trails and cut out leaf fragments, or whether these behaviors are the prerogative of the worker caste, is not known. But it is at least true that the queens alone are capable of most of the repertory of the entire colony, whereas this is decidedly not the case for the individual worker subcastes. In other higher ant groups we find various stages in the progressive loss of behavioral competency on the part of the queen caste. An extreme case of reduction is seen in the dichthadiigynes of the army ants, which do not start new colonies (these are created instead by fission of preexisting colonies) and therefore do not engage in workerlike activities in the early part of their adult life. The dichthadiigynes also fail to conduct a nuptial flight, depending instead upon visits from males in the midst of their home bivouac.

The queens of the bumblebees, the primitively eusocial halictine bees, the vespine wasps, and many polistine wasps also approach behavioral totipotency. On the basis of conventional phylogenetic criteria applied independently to the morphological data, theirs is the primitive condition. At the extreme opposite position are found the queens of the honeybees and stingless bees, which like the army ant dichthadiigynes depend on fission of worker forces for colony multiplication and have lost the capacity to carry out most of the worker tasks.

A persistent idea in insect sociology has been that the absolute size of the brain and the proportions of the corpora pedunculata relative to the rest of the brain can be taken as rough measures of mental capacity. Forel (1901a) was the first to phrase the matter with respect to the castes of ants: "Complicated instincts and the more apparent mental qualities (memory, plasticity, etc.) are possessed above all by the workers and to a much lesser degree by the queens. The males are incredibly stupid . . . These facts are clarified by a comparison of the organs of thought, that is the brains (*corpora pedunculata*) of the three castes. These organs are very large in the workers, much smaller in the queens, and almost wholly atrophied in the males." The same idea was repeated and documented at greater length, at least with reference to morphology, by Wheeler (1910), Pietschker (1911), Brun (1923), Pandazis (1930), Goll (1967), and others in the course of studies of various species of ants, and by Jonescu (1909) with respect to honeybees. In Figure 11-17 are presented diagrams of the ant and honeybee brains to show the location of the principal areas, including the corpora pedunculata; Figure 11-18 illustrates the differences in the relative size of the corpora pedunculata among the castes of a typical ant species. More recent experimental work has yielded evidence supporting the idea that the corpora pedunculata, and particularly the neuropiles, are major centers of organization of both innate and learned behavior. When Vowles (1954c), for example, ablated the corpora pedunculata of *Formica* workers on one side with high-frequency radio waves, the ants lost sensitivity in the antenna on that side and ran in circles to the opposite side. Other experiments, conducted on several very different kinds of insects including honeybees, crickets, and other orthopterans, have prompted the following general conclusions (Huber, 1965): (1) the general inhibitory system of the brain is located at least in part in the corpora pedunculata; (2) the corpora pedunculata act at least in part as antagonists to the central body in locomotory control; (3) the corpora pedunculata play a major integrative role in the regulation of complex (that is, multisegmental) muscle patterns.

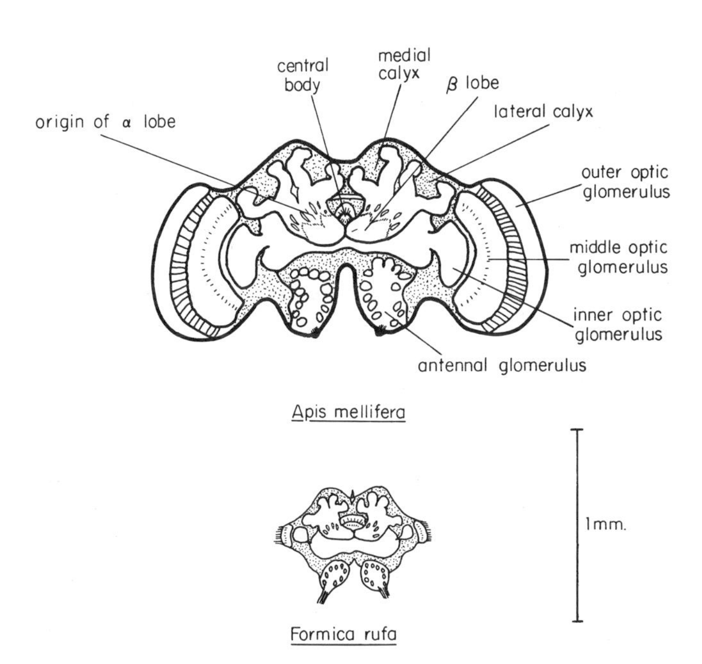

FIGURE 11-17. Diagrams of the brains of a worker honeybee and a worker ant (*Formica*) (redrawn from Vowles, 1955).

Recently Bernstein and Bernstein (1969) were able to put the Forel hypothesis to a somewhat more direct test. They studied variation among workers of *Formica rufa* in the ability to run through a Schneirla maze and then searched for correlates among a broad selection of anatomical characteristics. The "navigational efficiency" of the ants was measured in terms of the time required and the distance covered during six runs between food and nest. The Bernsteins found that this efficiency was strongly and positively correlated with the size of the head, the diameter of the compound eyes and median ocellus, and the dimensions of the calyxes of the corpora pedunculata. It was distinctly less well correlated with the size of certain other anatomical structures, including the tibiae, optic lobes, antennal lobes, and, most significantly, the brain as a whole.

No doubt changes in the corpora pedunculata are implicated in the evolution of complex behavior in insects in ways that we have hardly begun to understand. Even so, it would be an oversimplification and possibly an outright error to suppose, as most previous authors have done, that the size of the brain or any of its parts can be used as an index of mental capacity in comparing insect species or even castes. Notice, for example, the great difference in size between the brain of *Formica* and the brain of the honeybee (Figure 11-17); yet the considerable amount of information now available indicates that the two insects have comparable mental ability. Looking over the insects as a whole, it is notable that the Protura and Thysanura have no corpora pedunculata, while the Ephemeroptera, Megaloptera, and Lepidoptera have only small corpora pedunculata; it is true, as was expected, that these are groups characterized by relatively limited behaviors in comparison with such phylogenetically more advanced orders as the Coleoptera and Hymenoptera. It is also true, however, that roaches (Blattaria) and mantes (Dictyoptera), relatively primitive insects not celebrated for their mental prowess either, have relatively huge corpora pedunculata (Bullock and Horridge, 1965). Finally, Forel was simply in error when he stated that

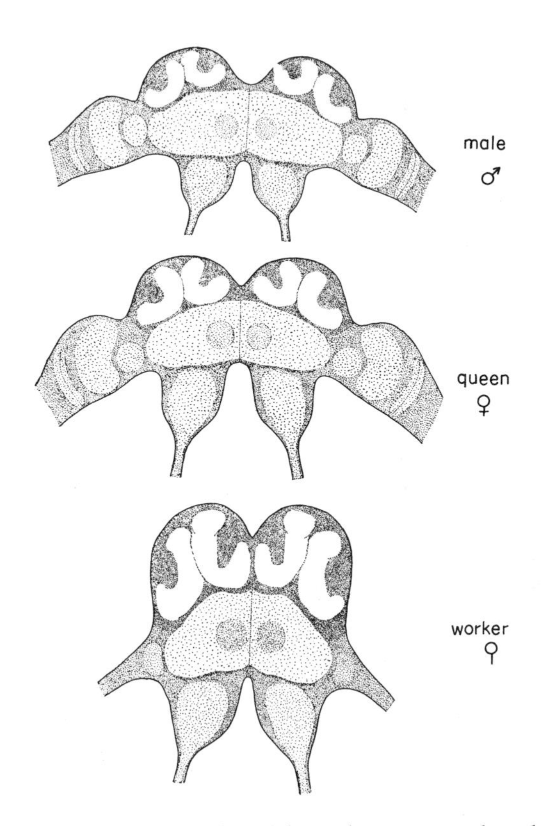

FIGURE 11-18. The brains of the male, queen, and worker of the fungus-growing ant *Acromyrmex multicinodis,* showing the different sizes of the protocerebrum and in particular the corpora pedunculata, which are shown in white (modified from Pandazis, 1930).

worker ants have greater mental capacities than queens of the same species. The reverse is plainly the case in all but a few specialized groups, as I have already pointed out. Yet the corpora pedunculata of worker ants are generally larger than those of queens. Surely better correlates between brain structure and mental capacity will be found, but they await more penetrating studies than those conducted in the past.

The Building of Complexity in Societies

The individual social insect, in comparison with the individual solitary insect, displays behavior patterns that are neither exceptionally ingenious nor exceptionally complex. The remarkable qualities of social life are mass phenomena that emerge from the meshing of these simple individual patterns by means of communication. In this principle lies the greatest challenge and opportunity of insect sociology. Wheeler, in his tract *Emergent Evolution and the Social* (1927), anticipated this property of the subject. He predicted that the integrative phenomena of societies of insects and other organisms will prove relatively easy to analyze because whole organisms are their interacting determining parts and, as such, are easier to manipulate than molecules or cells. "Owing, moreover, to the loose and primitive character of the integration and the size of the components even in the densest societies, it is possible to ascertain the behaviour of the parts and to experiment with them more extensively than with chemical and organismal wholes, since the parts of the latter are either microscopic or ultra microscopic and are always so compactly integrated that analysis becomes very difficult and involves a considerable amount of statistical inference. Experiments in subdividing, castrating and grafting, and in introducing foreign elements with a view to observing their effects on animal and plant societies as emergent wholes, can be carried far beyond the limits of such experiments on the single living organism."

Wheeler was right about this, of course. It was no fault of his that by force majeure molecular biology and cellular biology have momentarily pressed so far ahead of behavioral and social biology. We are only now beginning to acquire the information required to begin comparable systems analyses in insect societies. Some of it has been reviewed in the chapters on caste, and more will be presented in later chapters on communication. At this point I wish to attempt to formulate in a very general way the principles of mass action by which insect colonies translate the numerous individual behavioral acts of its members into higher order effects.

An important first rule concerning mass action is that it usually results from conflicting actions of many workers. The individual workers pay only limited attention to the behavior of nestmates near them, and they are largely unaware of the moment-by-moment condition of the colony as a whole. Anyone who has watched an ant colony emigrating from one nest site to another has seen this principle vividly illustrated. As workers stream outward carrying eggs, larvae, and pupae in their mandibles, other workers are busy carrying them back again. Still other workers run back and forth carrying nothing. Individuals are guided by the odor trail, if one exists, and each inspects the nest site on its own. There is no sign of decision making at a higher level. On the contrary, the choice of nest site is decided by a sort of plebiscite, in which the will of the majority of the workers finally comes to prevail by virtue of their superior combined effort. As Lindauer (1961) first showed, a similar "democratic" process is employed by honeybee swarms in choosing new nest sites. Each scout bee indicates its own choice of a nest site by dancing on the surface of the swarm. Its degree of "commitment" is indicated by the liveliness and duration of its dances. Follower bees inspect the sites recommended by the competing scouts and, if sufficiently stimulated, commence dances of their own. Eventually performances indicating one of the sites come to predominate, and the swarm flies off in that direction. If the competition is too evenly divided, the swarm will either lack direction indefinitely or else be physically split by the opposing factions. The building of nests is also partly the result of antagonistic efforts by individual workers. Individuals in *Polistes* colonies often build up brood cells while others are tearing them down, so that a greater effort in one direction or the other settles how many cells are to be constructed and where they are to be located (Deleurance, 1950). The same process occurs in the construction of comb cells by honeybees. In order to obtain pieces of wax for cells of their own, the workers regularly tear away walls that are in the process of being constructed by other nestmates (Lindauer, 1952). Workers of *Formica* and other kinds of ants constantly work at cross purposes in excavating nests and transporting nest materials (Kloft, 1959a; Chauvin, 1960a). Although these various antagonistic actions seem chaotic when viewed at close range, their final result is almost invariably a well-constructed nest that closely conforms to the plan exhibited throughout the species.

The emergence of statistical order from competing elements is displayed with striking speed and clarity in the marching patterns of army ants. At close range the movements of individual workers of the swarm-raiding species *Eciton burchelli* seem erratic. Schneirla (1940), for example, describes the behavior of a typical worker at the forward edge of the swarm as follows:

> Upon arriving in new terrain the worker slows up and meanders noticeably in her course, now with a jerky movement of the anterior body. Within the limited advance made before she withdraws, the worker's body is held closer to the ground than before in a characteristic sprawly posture, legs extended and moving somewhat stiffly. Together with the wavering of the anterior body there is a rapid wasplike semirotatory vibration of the antennal funiculi. The extended antennae are bent downward and in their rapid beating tap the ground at frequent intervals. After having advanced hesitantly a few centimeters in this manner, the worker leans forward in an abrupt pause which may be repeated very rapidly or followed by another short advance, then she quickly turns and runs back into the swarm.

During the brief advance into new territory the pioneer worker lays a small amount of trail pheromone from the tip of her abdomen, which draws other workers in the same direction. Meanwhile most of the swarm is moving forward in a chaotic manner:

> It is important not to understate the great variability of individual behavior in the swarm, in describing constant trends. When *Eciton* workers cross paths in the swarm there occur all degrees of contact from momentary brushing of antennae or legs to a forcible collision. Ants that collide head-on draw back more or less abruptly and both may turn away or (if running slowly) slip past each other; those running against each other sidewise usually change their courses somewhat according to the force of the contact; or when a worker is overtaken from the rear her pace is accelerated by the bump if she is not actually overrun.

Yet out of all this disorder the characteristic swarm of *Eciton burchelli* emerges: a roughly elliptical mass of workers, 10–15 m or more across and 1–2 m in depth, connected by two or more thick feeder columns of workers leading back to the point of origin at the bivouac site, with the forward edge growing at the speed of 30 cm a minute. How is it created? Schneirla noted that two antagonistic forces are constantly at work on the individual ants in the swarm. The first is pressure: the tendency of ants to move

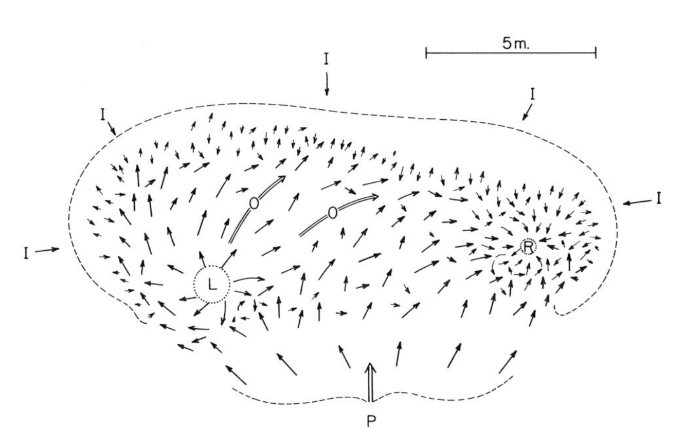

FIGURE 11-19. A schematic representation of movement within a swarm of the army ant *Eciton burchelli.* The limits of the swarm are indicated by a dashed line. The directions of movement of various of the 100,000 or more workers are indicated by arrows. In this instance the swarm is executing a flanking movement to the right: (*P*) the direction of pressure caused by the influx of newly arriving workers along the feeder columns; (*R*) the right flank undergoing a concentration; (*L*) the left flank undergoing an expansion; (*O*) the general direction of the oblique flanking turn; and (*I*) the impedance of mass progress by the slow advance of the pioneer ants along the anterior and lateral borders of the swarm (redrawn from Schneirla, 1940).

away from places where crowding becomes too tight. As newcomers press in, mostly from the direction of the bivouac, they stimulate workers already present to turn and move away from them. This activity in turn induces workers still farther away to move outward, which generates a centrifugal wave of excitation and movement. The second force is drainage: as places are vacated by workers, other workers in adjacent crowded areas tend to fill them again. Drainage is thus the simple opposite of pressure, and it, too, exerts its influence by wave-like propagations through the swarm. The basic pattern of movement within a *burchelli* swarm during a flanking turn is illustrated in Figure 11-19. As pressure builds from the rear by the constant influx of newly arriving workers, the ants already constituting the swarm attempt to move forward and to the side. However, the slow progress of the pioneer ants at the edge of the swarm impedes the movement of other ants at the heads of the columns and causes them to fan out into the terminal swarm formation. For an unknown reason, the impedance is greater at the front than along the sides so that the swarm flattens into an elliptical shape.

When large prey or concentrations of smaller prey are encountered, or the colony runs into difficult terrain, movement also slows down. Ants concentrate in such sites, while at the same time draining away from other sites less able to check their movement. The result is a flanking movement of the kind shown in Figure 11-19. As it inches forward, yielding first to one flanking movement and then to another, the swarm swings repeatedly back and forth.

Elsewhere (Wilson, 1962a) I have referred to such interaction as "mass communication." It is defined as the transfer among groups of information that a single individual could not pass to another. Other examples, which will be described in some detail later in this book, include the regulation of numbers of workers on odor trails (Chapter 13) and thermoregulation in the nests of social insects (Chapter 16).

If group effects are typically the products of friendly competition among large numbers of individuals in the same colony, it is equally true that the action of each individual results from competing stimuli impinging on it, including those produced by other members of the colony. The individual member of a large colony cannot possibly perceive the actions of more than a minute fraction of its nestmates; nor can it monitor the physiological condition of the colony as a whole. Yet somehow everything balances out, a fact that keeps drawing the mind back to Maeterlinck's poetic question about the termite colony: "What is it that governs here, that issues orders, foresees the future . . . ?" The phenomenon of mass communication as I have just described it provides a small part of the answer; the rest can be only dimly visualized at the present time. It is surely true that the behaviors of individual social insects are programmed by their inborn neural organization—not in a way that permits them to communicate simultaneously with all other members of the colony (obviously beyond the capacity of insects), or to execute a series of predestined maneuvers like an elementary servomechanism, but rather in a way that causes them to respond with certain behaviors, in accord with certain probabilities, to the stimuli normally present in the colonial environment. If the single social insect does not comprehend the environment in which it lives, at least it is able to make, on the average, the correct contribution to colony activity.

The important feature of this stochastic theory of mass behavior is that not only the acts themselves are programmed but also the frequency in which they occur. By adjusting the probabilities, and more exactly the transition probabilities leading from one act to another, the mass behavior of the colony can be radically altered. Perhaps the probability matrices are repeatedly altered in the course of evolution to produce patterns of response that are efficient in meeting the particular environmental exigencies in which the species finds itself. Without knowing the exact numbers in the matrices, we can be certain that they differ between castes and change with age. Consequently, due attention will be paid later in this book to the subjects of optimization of caste ratios (Chapter 18) and age distribution (Chapter 21) as they relate to efficiency of performance.

Earlier authors have looked at group behavior in a similar way. In the *Histoire naturelle, générale et particulière,* George Louis Leclerc Buffon proposed that social insects are automata and that societies are organized mindlessly as the outcome of their multitudinous separate activities. This same theme was struck again, in one form or another, by Verlaine, Weyrauch, and other students of "instinct theory" in the 1920's and 1930's. Among contemporary authors, O. W. Richards (1965) suggested that "As a rule the organization cannot depend on the use of language but rather on the automatic effect on the individual of a variety of stimuli which, *on the average,* are likely to act in the right order and with the necessary intensity. Such are effects of age, of physiological state, of size or of structural type, of shared food or pheromones, and a certain amount of active interference by a fertile queen." A. M. Wenner (1961) recommended, correctly I believe, that descriptions of behavior in honeybee colonies be in the form of matrices of transition probabilities. He also predicted that the sequences of behaviors of individual worker bees will prove to conform to a Markov process, as opposed to a "deterministic" or "chain reflex" sequence. Wenner was apparently unaware that a deterministic sequence is but one kind of Markov process, and his definition of a Markov process is not the same one used by mathematicians. Apparently he only meant that more than one kind of response is possible following another given response—an argument that will certainly be accepted by all students of social insects. However, in forecasting the development of a stochastic *theory* of mass behavior, I am proposing a series of more widely ranging evolutionary hypotheses which, because of their necessary involvement with problems of caste determination and population dynamics, are neither self-evident nor susceptible of easy proof: (1) the individual social insect, being unaware of most of what is going on in the colony to which it belongs, responds in an ad hoc manner to the stimuli it encounters moment by moment; (2) the responses and the probabilities of their occurrence are programmed genetically so that mass behavior of the colony is efficient with respect to the particular environmental conditions experienced through evolutionary time by the species; (3) the program evolves as the environment changes, always in the direction of increasing colony efficiency; (4) caste ratios, the age structure of individuals in the colony, and communication also evolve so as to provide the responses and their probability structure with greater efficiency at the colony level.

If these ideas are correct, it follows that the reconstruction of mass behavior from a knowledge of the behavior of single colony members is the central problem of insect sociology. It is easy to go astray in predicting the form that such research will eventually assume. But for purposes of clarifying the problem a couple of imaginary examples will be useful. Table 11-3 gives three matrices of probability values of the kind that could be constructed at the beginning of a systems analysis of an insect society. Each is comprised of the probabilities of response to a particular kind of stimulus, which, in this imaginary example, is the expelling of a droplet of larval salivary secretion. In wasps and ants such material is utilized as food by the workers; in these and other groups it might also serve an excretory function. It is only one of many kinds of social stimuli that members of a colony encounter at high rates around the clock, and for which similar matrices could be written. Matrix A contains the probabilities of workers encountering the stimulus after completing certain other acts; these values are arbitrarily chosen to vary both according to the previous act and to the age of the worker. Matrix B contains the probabilities that the workers will imbibe the salivary droplet after encountering it; these values are selected to vary according to the previous act but not according to age. The probability that a worker of a given age and experience will both encounter the stimulus and display the response is the product of the two values applying to it in matrix A and matrix B. In matrix C are given the probabilities that a response to the stimulus will be given by workers of various given experience and age. Since these values also depend on the numbers of workers belonging to each class in the colony, they cannot be derived entirely from the information given in A and B.

The A and B matrices would tell us how individuals of different castes and age groups organize their time and

TABLE 11-3. Imaginary set of probabilities of response by different colony members to a particular stimulus (larval salivary secretion).

Age (days)	After feeding	After foraging	After excavating	After resting
	A. Probability that a worker of a given age group will encounter larval salivary secretion after completing one of four possible acts			
1–4	0.15	0.04	0.02	0.04
5–9	.10	.03	.01	.03
10–14	.07	.02	.01	.01
	B. Probability of feeding on the secretion after it has been encountered			
1–4	0.22	0.42	0.35	0.65
5–9	.22	.42	.35	.65
10–14	.22	.42	.35	.65
	C. Probability that a given larval secretion will be imbibed by a worker of a given classification of age and immediate past experience			
1–4	0.30	0.10	0.05	0.10
5–9	.20	.05	.01	.05
10–14	.10	.01	.01	.02

divide labor. The C matrices, depending as they do on A and B, would report on which kind of workers handle given tasks and, therefore, how efficiently the tasks are performed. From an accumulation of such information a picture of the functioning of the entire colony could ultimately be drawn up. Among the worthy mathematical problems that will be raised by such a statistical description is the nature of the transition probabilities. These numbers cannot be expected to conform to an independent trials process because what an individual social insect does depends at least to some extent on the physiological and behavioral state it is in and hence on what it has been doing just prior to receiving a new stimulus. At the same time, the transition probabilities will not conform with any great fidelity to elementary Markov states. By this is meant that behavior is not solely prescribed by the action just completed. Social insects are not quite as simpleminded as that. As shown earlier, they are capable of some relatively complicated and long-term learning. There is also some evidence that social insects become fixated on certain objects or tasks from experience alone. Bruns (1954a) discovered that workers of the ant *Formica rufa* come to specialize on the dominant insects in the vicinity of their nests, but they require one or two weeks to develop the preference. Free's experiments

(1955b) on division of labor in bumblebees revealed that the longer an individual bee has been carrying out either household or foraging activities, the more likely it is to continue in the same line of duty when conditions in the colony change. The capacity to learn and particularly the predilection to repeat successful acts mean that probability matrices will vary greatly even among workers belonging to the same caste and age groups. Yet the final effect on the colony will not depend much on individual variability since, in the long run, what matters is the average of the matrices of all the members of a particular caste-age class. Finally, as exemplified in the diagram in Figure 11-20, the behavioral sequences can be expected to be ergodic. This means that, for most or all of its life, a colony member can pass from any stimulus to any response and back again to any stimulus. Included in the scheme is the fact that individual insects often seek the stimulus and are not always just passive recipients. In order for ergodicity to be realized, one category of response would have to be the act of failing to respond. Also, certain transition probabilities will be very low. Suppose that, within each caste and age class, the transition probabilities ($p_{i,j}$ and $q_{j,i}$) are constant, or fluctuate around a constant mean value. Assume also that the number of workers in each caste and age group is constant, or fluctuates around a constant mean, or changes in a regular manner during the life of the colony. Assume finally that the stimuli can be identified, in both kind and frequency, during the life of the colony. All of these conditions are reasonable and likely, albeit beyond our present capacity to measure. When measurements are made, it should then be possible to specify the mass behavior of the colony under various stimulus conditions. Furthermore, because of the ergodic property of the system, a more or less constant environment during a period of, say, a few days or weeks would result in a more or less steady state of behavior in the colony as a whole. The stochastic theory of mass behavior implies that, insofar as the environment is predictable, the caste structure, the age structure, and the behavioral transition probabilities $p_{i,j}$ and $q_{j,i}$ will all evolve to maximize the efficiency of the colony and hence its reproductive fitness. (A more complete theory of the roles of caste and age structure in colony fitness will be provided in Chapter 18.)

The most promising form of group behavior on which a complete analysis can be undertaken is the cooperative construction of nests. First, the outcome of the behavior

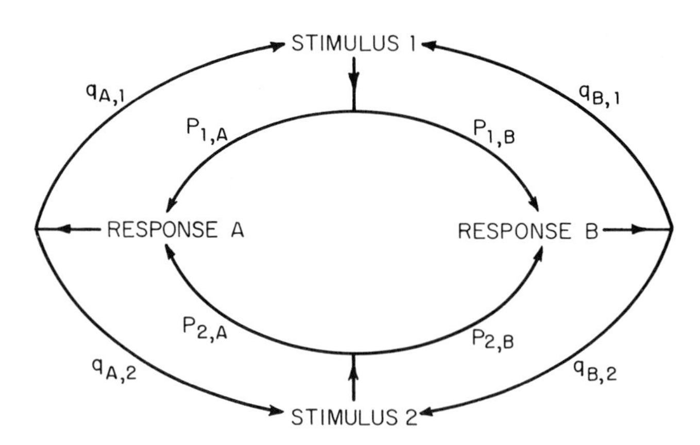

FIGURE 11-20. An imaginary and oversimplified action scheme of behavior in workers of a given caste and age group, comprised of two kinds of stimuli and two kinds of responses. The transition probabilities between the stimuli and responses are given by the eight possible $p_{i,j}$ and $q_{j,i}$. This kind of ergodic structure, if combined with steady caste and age compositions in the colony, leads to a constant division of labor within the colony.

is very predictable. By selectively damaging portions of the nest the experimenter can induce rebuilding activity on the part of the insects that is both rapid and reproducible. An engrossing enigma is presented by the very large, complicated nests of such social insects as the ants *Azteca* and *Formica,* the social wasps *Polybia* and *Vespa,* and the fungus-growing termites *Macrotermes.* It is all but impossible to conceive how one colony member can oversee more than a minute fraction of the construction work or envision in its entirety the plan of such a finished product. Some of these nests require many worker lifetimes to complete, and each new addition must somehow be brought into a proper relationship with the previous parts. The existence of such nests (see, for example, Figure 6-11) leads inevitably to the conclusion that the workers interact in a very orderly and predictable manner. But how can the workers communicate so effectively over such long periods of time? Also, to use another Maeterlinckian phrase, who has the blueprint of the nest? It was with such questions in mind that Pierre-Paul Grassé (1959, 1967) conducted studies of nest building by the termites *Cubitermes* and *Macrotermes.* His explanations do not "perfectly" explain the whole problem, as he has claimed, but they do contain important insight on how construction is initiated and carried along. The key process is "stigmergy"—Grassé's term coined from the Greek phrase meaning "incite to work." In stigmergic labor it is the product of work previously accomplished, rather than direct communication among nestmates, that induces the insects to perform additional labor. Even if the work force is constantly renewed, the nest structure already completed determines, by its location, its height, its shape, and probably also its odor, what further work will be done. Grassé distinguished three successive stages in the stigmergic initiation of a large structure, which are exemplified in the construction of a single foundation arch by workers of *Macrotermes* (*Bellicositermes*) *natalensis.* When workers of this species are placed in a container with some building material consisting of pellets of soil and excrement, they first pass through a state of "uncoordination," during which each explores the container individually. In the next stage, that of "uncoordinated work," the pellets of soil and excrement are carried about and put down in a seemingly haphazard fashion. Although crude passageways may begin to take shape, the termites, for the most part, still appear to act independently of each other. A pellet placed in one spot by one worker is often quickly picked up and placed somewhere else by another worker. Finally, seemingly by chance, two or three pellets get stuck on top of one another. This little structure proves much more attractive to termites than single pellets. They quickly begin to add more pellets on top, and a column starts to grow. If the column is the only one in the immediate vicinity, construction on it will cease after a while. If another column is located nearby, however, the termites continue adding pellets, and, after a certain height is reached, they bend the column at an angle in the direction of the neighboring column (Figure 11-21). When the tilted growing ends of the two columns meet, the arch is finished, and the workers move away. The sense by which the termites detect the proximity of the second column has not been identified. By process of elimination, it seems most likely to be olfaction. The insects cannot see the other column, and it is improbable that, in the midst of all the confused scampering in the vicinity, they can recognize distinct sounds from the column by conduction through substrate. Perhaps they run back and forth and measure distance between columns that way, but, if so, Grassé has not recorded it. This leaves the odor of the columns, which (in view of the skill of termites at detecting and following other odorous substances, including pheromones) might be sufficient to permit location at a distance.

Several authors have challenged the completeness of Grassé's explanation. Harris and Sands (1965) and Chauvin (1968d) have pointed out that stigmergy can work only if the rules change at different times in various parts of the nest. Otherwise the architecture would be homogeneous, and such major elements as temporary launching platforms for nuptial flights and the components devoted to thermoregulation could not be incorporated. Stuart (1967, 1969) has stressed another basic failing of the theory: the inability of a simple stigmergic machine to shut down when the job is finished. It is easy to see that work on an arch would halt when the ends of the two columns meet, but what finally stops the termites from building new arches? In his own studies on the repair of nest walls by workers of *Nasutitermes* and *Zootermopsis,* Stuart discovered that the termites continue building until the disturbing stimuli caused by the breach, namely air currents and lowered humidity, are removed. During the emergency additional workers are recruited to the scene by odor trails, and this communication ceases when the breach is closed. Stuart's finding is especially significant in view of the fact that it reveals, contrary to Grassé's simplifying assumption, that chemical communication is employed by the termites in the coordination of building activity.

Thus it has been easy to show that the basic behavioral patterns employed in nest construction are more complex than envisioned by Grassé. Nevertheless, stigmergic responses are evidently major elements in nest construction by social insects generally, and their explicit identification and study help clarify some baffling phenomena. Consider, for example, the remarkable cooperative labor of the weaver ants (*Oecophylla longinoda* and *O. smaragdina*) illustrated in Figure 11-22. These insects are wholly arboreal, and they construct their nests of green leaves held together by sticky larval silk. In order to make a nest wall, it is necessary for groups of workers to pull leaves together simultaneously while others move the larvae back and forth like animated shuttles. How is this cooperation achieved? The solution discovered by Sudd (1963) involves a simple form of stigmergy. As shown in Figure 11-23, workers work independently in their first attempts to pull down or roll up leaves. When success is achieved by one or more of them at any part of a leaf, other workers in the vicinity abandon their own efforts and join in. Such forms of coordination in social insects are closely analogous to the operations of "majority organs" in general cybernetic systems (Meyer, 1966).

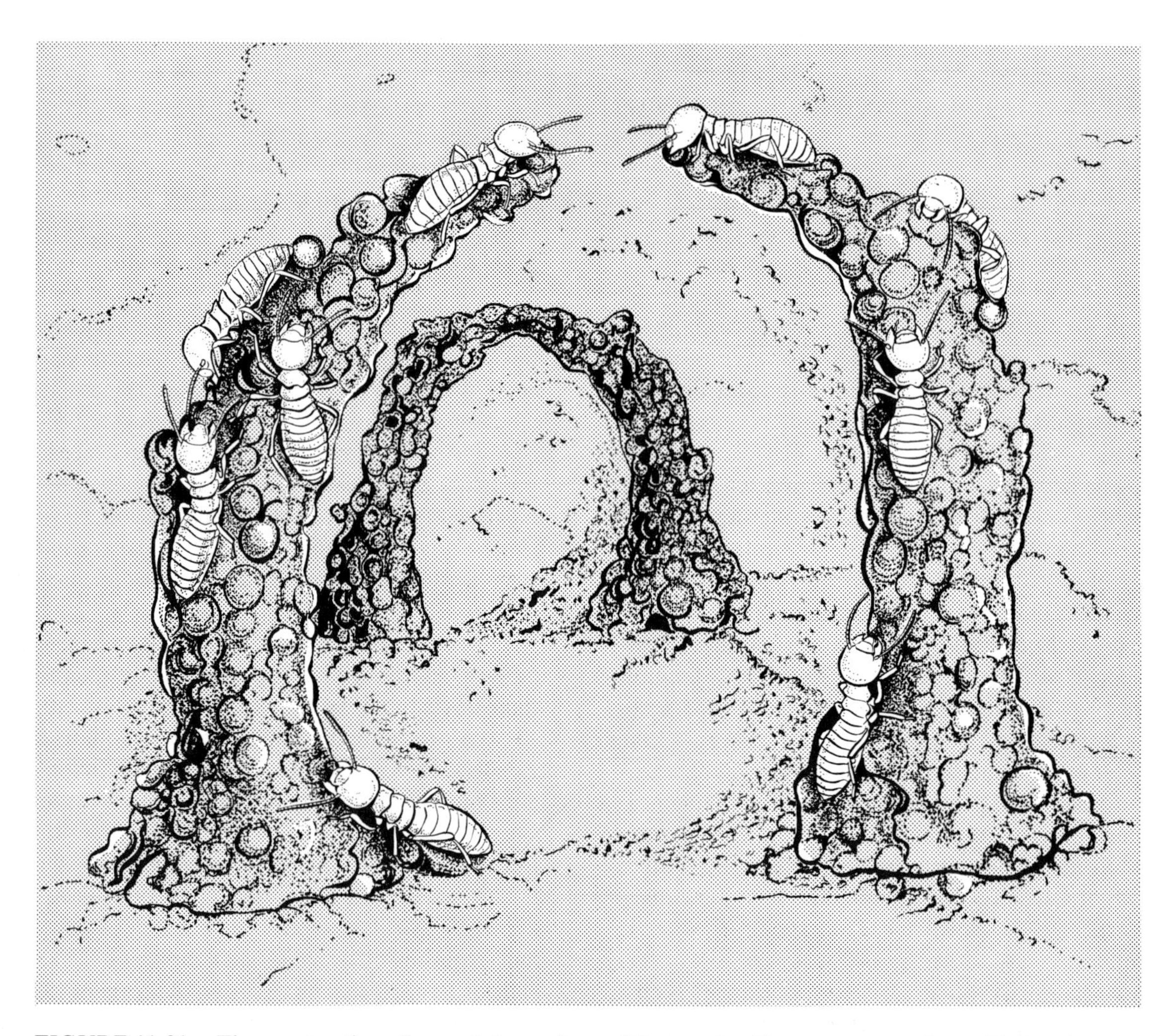

FIGURE 11-21. The construction of an arch by workers of the termite *Macrotermes natalensis.* Each column is built up by the addition of pellets of soil and excrement. On the outer part of the left column a worker is seen depositing a round fecal pellet. Other workers, having carried pellets in their mandibles up the columns, are now placing them at the growing ends of the columns. When a column reaches a certain height the termites, evidently guided by odor, begin to extend it at an angle in the direction of a neighboring column. A completed arch is shown in the background (drawing by Turid Hölldobler; based on Grassé, 1959, and Chauvin, 1968d).

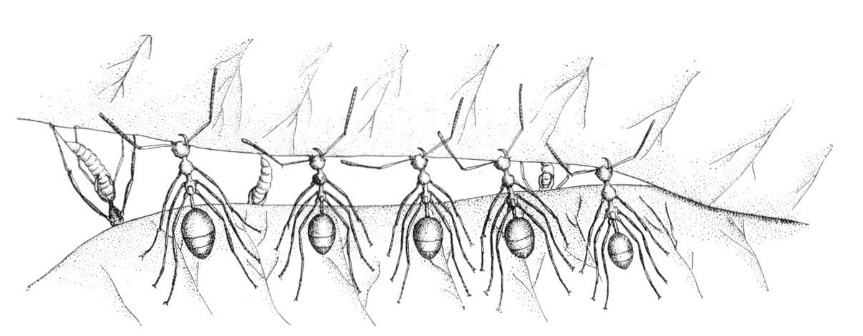

FIGURE 11-22. The repairing of a rent in a nest of the weaver ant *Oecophylla smaragdina.* The nest is constructed of living leaves bound together by larval silk. One group of workers holds the leaves together while another moves the spinning larvae back and forth across the gap (from Wheeler, 1910, after F. Doflein).

Freisling (1938) interpreted nest construction by European social wasps to be governed essentially by what Grassé later called stigmergy. Michener (1964a) has pointed out that stigmergic behavior must be the principal, if not the sole, means by which the primitively eusocial halictid bees build their nests. Very little contact occurs between the bees in or around their nests, and construction often proceeds when only one is present. Even more cogent is the observation that the solitary relatives of the eusocial species build nests of equal or greater complexity. According to Batra (1968) the solitary and social species show no visible differences in the details of nest construction, or for that matter of provisioning, pollen formation, and brood inspection. Finally, it would be unfair to the perspicacious Pierre Huber for me to neglect to point out that it was really he who, in 1810, first conceived the basic idea of stigmergy. Speaking of nest building in *Formica fusca,* he said, "From these observations, and a thousand like them, I am convinced that each ant acts independently of its companions. The first that hits upon an easy plan of execution immediately produces the outline of it; others only have to continue along these same lines, guided by an inspection of the first efforts."

The total simulation of construction of complex nests from a knowledge of the summed behaviors of the individual insects has not been accomplished and stands as a challenge to both biologists and mathematicians. The eventual achievement of such a simulation will be the evidence of a fairly high level of sophistication in our understanding of social behavior. It will also constitute a technical breakthrough of exciting proportions, for it will then be possible, by artificially changing the probability matrices, to estimate the true amount of behavioral evolution required to go from the nest form of one species to that of another. Figure 11-24 presents a phylogenetic dendrogram of nest structure in *Apicotermes,* a genus of termites famous for the elaborateness of its nest architecture and the marked changes that occur from species to species. Many years ago the Brazilian entomologist Adolfo Ducke (1914) actually employed comparisons of nest structure to help work out his now outmoded phylogeny of the social Vespidae. Later, Emerson (1938) argued that such variation in nest structure provides an unparalleled opportunity to study the evolution of instinct because each nest is a frozen product of behavior that can be weighed, measured, and geometrically analyzed. This conception is true to a point, yet it must be stressed that a thorough understanding of behavioral evolution, in nest construction as well as other forms of colonial activity, will come only when the probability matrices can be written and artificially manipulated. One of the effects that might then be analyzed for the first time is amplification, a process conceived a priori to arise as follows: because transition probabilities accumulate two or more steps as the product of the probabilities of the separate steps, it is possible for small alterations in probability values to have large effects on responses several steps removed. Alternatively, effects can be reduced by appropriate

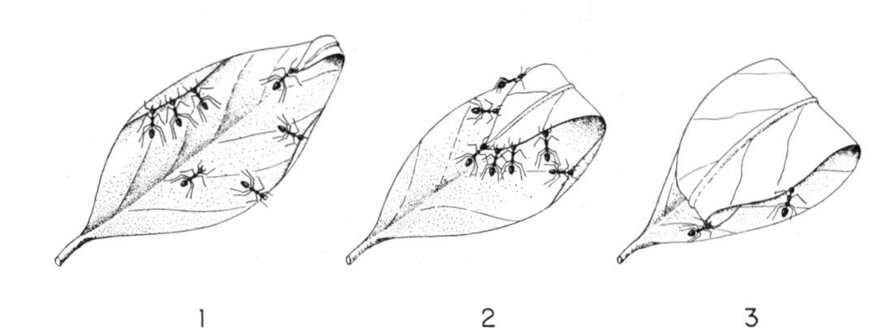

FIGURE 11-23. The initiation of cooperative nest building in the weaver ant. When workers first attempt to fold a leaf (*left*) they spread over its surface and pull up on the edge wherever they can get a grip. One part (in this case, the tip) is turned more easily than the others, and the initial success draws other workers who add their effort, abandoning the rest of the leaf margin (*center*). The result (*right*) is a rolled leaf of a kind frequently encountered in *Oecophylla* nests. (From Sudd, 1963; courtesy of *Science Journal,* London, incorporating *Discovery*).

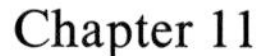

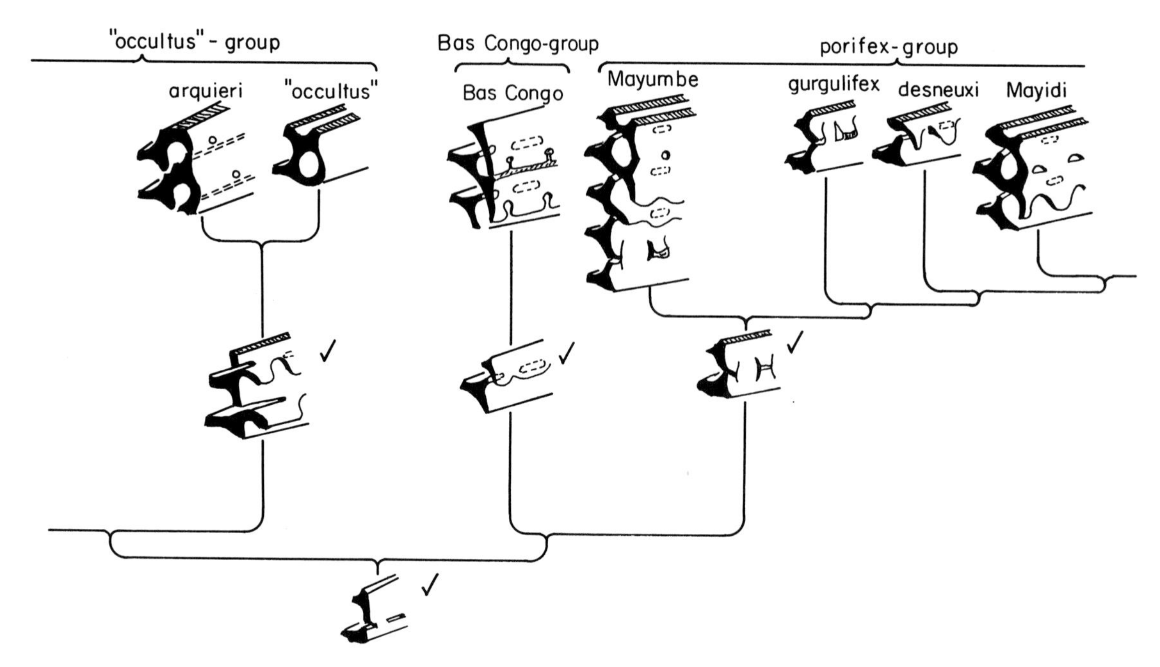

FIGURE 11-24. A phylogenetic tree hypothesizing the evolution of nest structure at the species level within the termite genus *Apicotermes.* The structure shown is a small cross section of the nest wall. The known living species are arrayed in the top row; the hypothetical ancestral species are indicated by a check (√) (redrawn from Markl and Lindauer, 1965; based on Schmidt, 1955).

changes in the length of the chain of events and the probability values separating them.

It seems intuitively likely that important behavioral evolution results from relatively small changes in the probability matrices. In evidence is the fact that behavior evolves at least as rapidly as morphology in social insects. In the termites in particular, behavioral diversity far outstrips morphological diversity at the level of species and higher taxonomic categories (Noirot, 1958–1959). Certain species of the African genus *Apicotermes* can be most easily distinguished from their closest relatives on the basis of nest structure, and in one case (*A. arquieri* versus *A. occultus*) the diagnosis so far has been made solely on the basis of this behavioral character (Emerson, 1956a). Comparable examples have recently been discovered in the halictine bee genus *Dialictus* (Knerer and Atwood, 1966b) and the wasp genus *Stenogaster* (Sakagami and Yoshikawa, 1968). Emerson has gone so far as to label such sibling pairs as "ethospecies." However, there is no reason to believe that the populations constitute anything more or less than full, reproductively isolated species in which behavior has simply outpaced morphology in the early stages of evolutionary divergence.

12 Communication: Alarm and Assembly

General Properties of Communication

Biological communication can be defined as action on the part of one organism (or cell) that alters the probability pattern of behavior in another organism (or cell) in an adaptive fashion. By adaptive I mean that the signaling, or the response, or both, have been genetically programmed to some extent by natural selection. This broad definition of communication still leaves some kinds of interaction in nature ambiguously classified, particularly various predator-prey relations among species; for the moment, however, it is adequate for the social insects.

The modes of communication found in the social insects are awesomely diverse. There exist the expected tappings, stridulations, strokings, graspings, antennations, tastings, and puffings and streakings of chemicals which evoke various responses from simple recognition to recruitment and alarm. We must add to this list other, often subtle and sometimes even bizarre effects: the exchange of pheromones in liquid food that inhibit caste development, the soliciting and exchange of special "trophic" eggs that exist only to be eaten, the acceleration or inhibition of work performance by the presence of other colony members nearby, various forms of dominance and submission relationships, programmed execution and cannibalism, necrophoresis, and still others. Much of the information on communication in social insects is fairly new, and its synthesis and classification present a novel task.

Three generalizations are useful in gaining perspective on this subject. First, most communication systems in the social insects appear to be based on chemical signals. The known visual signals are sparse and simple. In some groups, such as the termites, they play no role in the day-to-day life of the colony. Airborne sound is only weakly perceived by these insects and has not been definitely implicated in any important communication system. Some social insects, on the other hand, are extremely sensitive to sound carried by the substrate, but they evidently employ it only in limited fashion, chiefly during aggressive encounters and alarm signaling. Modulated sound signals evidently play a role in recruitment in the advanced stingless bees of the genus *Melipona* and in the honeybees, which have incorporated them in the waggle dance. Touch is universally employed within insect colonies but, with the possible exception of dominance and trophallaxis control in the vespine wasps, it has not been molded into a Morse-like system capable of transmitting higher loads of information.

In contrast, chemical signals, or pheromones as they are now generally called, have been implicated in almost every category of communication. Thirteen years ago I realized that the separation of these substances by the dissection of their glandular sources could provide the means of analyzing much social behavior that had previously seemed intractable: "The complex social behavior of ants appears to be mediated in large part by chemoreceptors. If it can be assumed that 'instinctive' behavior of these insects is organized in a fashion similar to that demonstrated for the better known invertebrates, a useful hypothesis would seem to be that there exists a series of behavioral 'releasers,' in this case chemical substances voided by individual ants that evoke specific responses in other members of the same species. It is further useful for purposes of investigation to suppose that the releasers are produced at least in part as glandular secretions and tend to be accumulated and stored in glandular reservoirs"

(Wilson, 1958d). With each improvement in organic microanalysis permitting the separation and bioassay of secretory substances, new evidence has been added to support this conjecture. Pheromones, as the chemical releasers were first called by Karlson and Butenandt (1959), may be classified as olfactory or oral according to the site of their reception. Also, their various actions can be distinguished as releaser effects, comprising the classical stimulus-response mediated wholly by the nervous system (the stimulus being thus by definition a chemical "releaser" in the terminology of animal behaviorists), or primer effects, in which endocrine and reproductive systems are altered physiologically. In the latter case, the body is in a sense "primed" for new biological activity, and it responds afterward with an altered behavioral repertory when presented with appropriate stimuli (Wilson and Bossert, 1963). The sum of current evidence, which will be presented in detail in the next four chapters, indicates that pheromones play the central role in the organization of insect societies. The prevalence of chemical signals in the still imperfectly known communication system of a typical ant species, *Solenopsis saevissima,* is documented in Table 12-1. A different view of the diversity and importance of chemical signals used in communication of ants and honeybees can be gained from Figure 12-1 and Table 12-2.

The second generalization is that most of the communication systems have parallels in behavior patterns already present in some form or other in solitary and presocial insects. Nest building is a case in point. The primitive ants, termites, and social wasps build nests which are scarcely more complicated than those of many of their solitary relatives. The nests of primitively social bees are frequently simpler than those of their solitary relatives. Elaboration of nest structure occurred in certain phyletic lines after the eusocial state was attained, and its evolution can be easily traced. The dominance hierarchies that play a key role in bumblebee and wasp societies have a precedent in the territorial behavior of many solitary insect species, including at least a few hymenopterans. Elaborate brood care, a hallmark of higher sociality, has its precursor in progressive larval feeding in a multitude of subsocial species belonging to several insect orders. Alarm substances are in many cases simply modified defensive secretions, and trail substances have a parallel at least in the odor spots used to mark the nuptial flight paths of the males of some solitary Hymenoptera. Even the elements of the honeybee waggle dance, the distant apex of insect social evolution in the eyes of most biologists, have precursors: the modulated rocking behavior of saturniid moths, which varies in duration according to the length of the flight just completed and thus resembles the straight run of the bee dance; the oriented "dances" of hungry *Phormia regina* flies which have been given a small drop of sugar water; and the ability of some solitary insects to shift from light to gravity orientation when placed on dark, vertical surfaces.

This brings us finally to the third generalization about communication in insect societies. The remarkable qualities of social life are mass phenomena that emerge from the integration of much simpler individual patterns by means of communication. If communication itself is first treated as a discrete phenomenon, the entire subject be-

TABLE 12-1. Known categories of communication among workers of the fire ant *Solenopsis saevissima* (modified from Wilson, 1962a).

Stimulus	Transmission	Response
Nest odor	Chemical	Nil, if odor is undisturbed
Casual antennal or bodily contact	Tactile	Turning-toward movement or increased undirected movement
Abdominal vibration	Sound from stridulation (?)	Function, if any, unknown
Body surface attractants	Chemical	Oral grooming
Carbon dioxide	Chemical	Clustering and digging
Ingluvial food solicitation	Probably tactile	Regurgitation
Regurgitation	Chemical, at least in part	Feeding
Emission of Dufour's gland substance as trail	Chemical	Attraction, followed by movement along trail; used in mass foraging and colony emigration
Emission of Dufour's gland substance during attack	Chemical	Attraction to disturbed worker
Emission of cephalic substance	Chemical	Alarm behavior

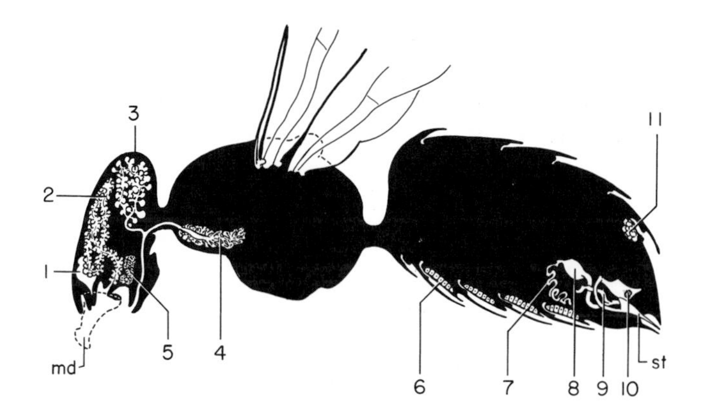

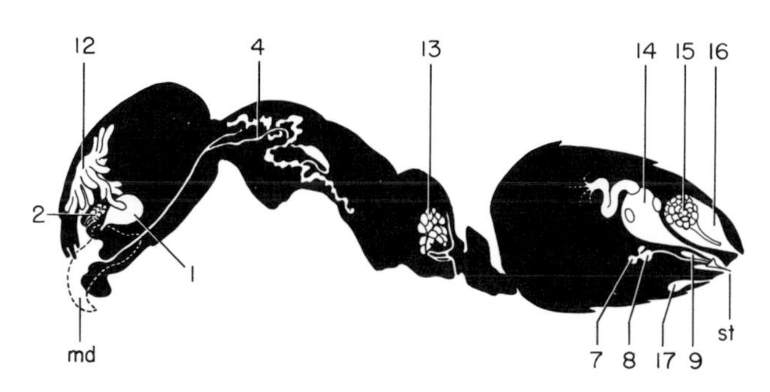

FIGURE 12-1. Exocrine gland systems of (*top*) the honeybee worker (*Apis mellifera*) and (*bottom*) the worker of the ant species *Iridomyrmex humilis*. Glands of the two species are labeled with the same number if they are considered homologous; where different names have been used in the literature, the name used for ants is given in parentheses in the key to follow; several minor glands of unknown function are omitted. (Pavan's gland and the anal gland are found only in the ant subfamily Dolichoderinae.) (1) Mandibular glands; (2) hypopharyngeal (= maxillary) gland; (3) head labial gland; (4) thorax labial gland; (5) hypostomal (postgenal) gland; (6) wax glands; (7) poison gland; (8) vesicle of poison gland; (9) Dufour's gland; (10) Koschevnikov's gland; (11) Nasanov's gland; (12) postpharyngeal gland; (13) metapleural gland; (14) hind gut (glandular nature uncertain); (15) anal gland; (16) reservoir of anal gland; (17) Pavan's gland (diagram of honeybee system, modified from Ribbands, 1953, after R. E. Snodgrass, with Koschevnikov's gland added; diagram of the ant system, modified from Pavan and Ronchetti, 1955; homologies of cephalic glands according to Otto, 1958b and Lind, 1970).

comes much more readily analyzed. In fact, students of animal behavior have begun to treat the analysis of animal communication as a discrete discipline; it has been referred to as zoosemiotics (Sebeok, 1965). To date we have found it convenient to recognize about nine categories of responses, as given in the following list:

1. Alarm
2. Simple attraction (multiple attraction = "assembly")
3. Recruitment, as to a new food source or nest site
4. Grooming, including assistance at molting
5. Trophallaxis (the exchange of oral and anal liquid)
6. Exchange of solid food particles
7. Group effect: either increasing a given activity ("facilitation") or inhibiting it
8. Recognition, of both nestmates and members of particular castes
9. Caste determination, either by inhibition or by stimulation

No doubt this elementary classification is destined for early obsolescence due to the current rapid increase in our information. For the moment, it is the best available framework on which to arrange a highly eclectic set of empirical observations.

Alarm Communication

There exist in the social insects a diversity of responses to threatening stimuli that can be described as strict alarm reactions and a set of signals passed back and forth among members of the same colony that can be unambiguously labeled alarm communication. Yet in other cases the functions are compound and less neatly pigeonholed, as, for example, the combined alarm-defense systems of ants and termites, where the same chemical secretions are used both as signals and defensive weapons. Even more complex are the alarm-recruitment-construction systems of termites, where chemical trails are used to recruit workers to breaches in the nest wall and other irregularities in nest structure, which in turn may or may not be an immediate source of danger.

Chemical Alarm Signals. It is common knowledge among beekeepers that, when one honeybee stings an intruder, her nestmates often move in swiftly to join the attack. That the signal provoking such mass assaults is an odorous chemical released from the sting was realized long ago by Charles Butler (1609). The following passage from

TABLE 12-2. Glandular source, chemical identity, and function of pheromones in the honeybee (*Apis mellifera*) and in several species of ants (Formicidae). A zero indicates that the insect does not have the gland in question; a question mark, that it is present but with unknown function. Unless indicated otherwise, the information applies to the worker caste. From the data of M. S. Blum, R. Boch, D. Otto, M. Pavan, F. E. Regnier, M. Renner, D. A. Shearer, J. Simpson, and E. O. Wilson (cited in Wilson, 1965a).

	Source									
Species	Mandibular glands	Hypopharyngeal glands	Labial glands	Nasanov's gland	Hind gut	Dufour's gland	Poison gland	Glands of sting chamber	Pavan's gland	Anal gland
Apidae	(Worker) 2-Heptanone: alarm (weak)	Royal jelly	Cleaning and dissolving	Geraniol, citral, geranic acid, nerolic acid: attraction leading to assembly	?	?	?	(Worker) Isoamyl acetate: alarm	0	0
Apis mellifera	(Queen) 9-ketodecenoic acid: inhibition of queen rearing and of worker ovary development; sex pheromone 9-hydroxydecenoic acid: settling of worker swarms							(Queen) Unknown scent: attraction		
Formicidae										
Atta spp.	Citral: alarm	?	?	0	?	?	Trail	0	0	0
Iridomyrmex humilis	?	?	?	0	?	?	?	0	Trail	2-Heptanone: alarm
Acanthomyops claviger	Citronellal, citral, etc.: alarm	?	?	0	Trail	Undecane, tridecane, tridecanone: alarm	?	0	0	0

The Feminine Monarchie may, in fact, be the earliest explicit description of an insect pheromone of any kind:

When you are stung, or any in the company, yea though a Bee have strike but your clothes, specially in hot weather, you were best be packing as fast as you can: for the other Bees smelling the ranke favour of the poison cast out with the sting will come about you as thicke as haile: so that fitly and lively did he expresse the multitude & fierceness of his enimies that said They came about me like Bees.

Recent research has identified one of the components of the sting pheromone as isoamyl acetate and pinpointed its glandular source in cells lining the sting pouch (Ghent and Gary, 1962; Shearer and Boch, 1965). The barbed sting of the honeybee worker catches in the skin of the victim, and, when the bee attempts to fly away, it frequently leaves behind the sting along with the attached poison and Dufour's gland and parts of its viscera. The isoamyl acetate is exposed, probably along with other, unidentified alarm substances. It evaporates rapidly and attracts other workers to the source. When pure isoamyl acetate is placed in cotton balls, it is still attractive to worker bees, even in the absence of the sting, accessory pheromones, and other clues that might be left behind by an attacking worker.

Let us now examine in some detail an example of a relatively uncomplicated chemical alarm system, this time in the ants. When a worker of the subterranean formicine ant *Acanthomyops claviger* is strongly disturbed, say placed under attack by a member of a rival colony or an insect predator, it reacts strongly by simultaneously discharging the reservoirs of its mandibular and Dufour's glands. After a brief delay, other workers resting a short distance away display the following response: the antennae are raised, extended, and swept in an exploratory fashion through the air; the mandibles are opened; and the ant begins to walk, then run, in the general direction of the disturbance. Workers sitting a few millimeters away will begin to react within seconds, while those a few centimeters distant may take a minute or longer. In other words, the signal appears to obey the laws of gas diffusion. Experiments have implicated some of the terpenes, hydrocarbons, and ketones displayed in Figure 12-2 as the alarm pheromones. Undecane and the mandibular gland

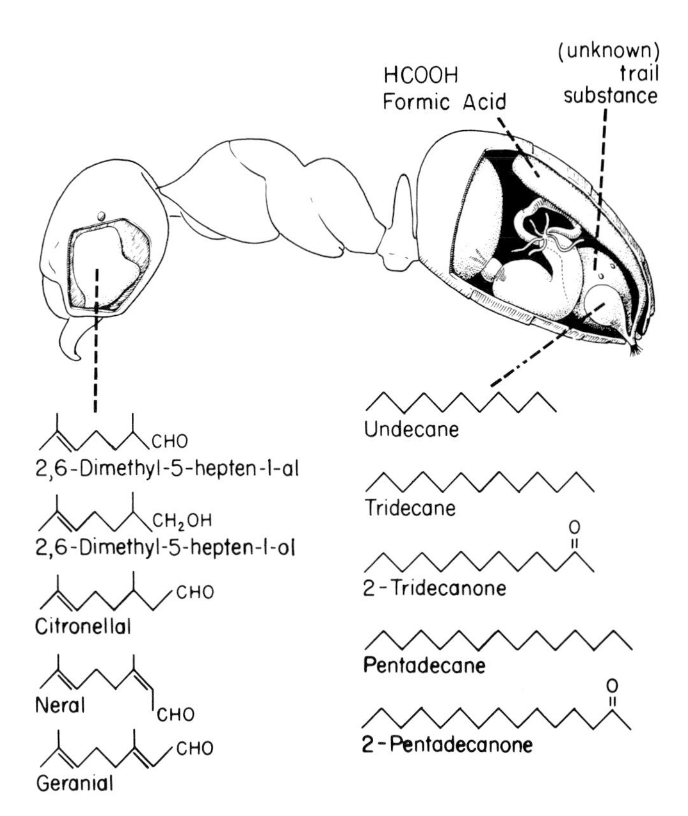

FIGURE 12-2. The structural formulas of volatile substances found in the mandibular gland (*front*) and Dufour's gland (*rear*) of *Acanthomyops claviger.* Formic acid, produced by the poison gland, is used exclusively in defense, while tridecane, tridecanone, pentadecane, and pentadecanone are mostly or entirely defensive in function. Undecane and the mandibular gland substances operate both as defensive substances and as alarm substances over distances of several centimeters (from Regnier and Wilson, 1968).

substances (the latter all terpenes) evoke the alarm response at concentrations of $10^9 - 10^{12}$ molecules per cubic centimeter. These same substances are individually present in amounts ranging from as low as 44 ng to as high as 4.3 μg per ant, and altogether they total about 8 μg. Released in gaseous form during experiments, similar quantities of the synthetic pheromones produce the same responses. Apparently the *A. claviger* workers rely entirely on these pheromones for alarm communication. Their system seems designed to bring workers to the aid of a distressed nestmate over distances of up to 10 cm. Unless the signal is then reinforced by additional emissions, it dies out within a few minutes. The alerted workers approach their target in a truculent manner. This overall defensive strategy is in keeping with the structure of the *Acanthomyops* colonies, which are large in size and often densely concentrated in the constricted subterranean galleries. According to our interpretation, it would not pay for the colonies to try to disperse when their nests are invaded, and, consequently, the workers have evolved so as to meet danger head on (Regnier and Wilson, 1968).

A different strategy is employed in the chemical alarm-defense system of the related ant *Lasius alienus.* Colonies of this species are smaller and normally nest under rocks or in pieces of rotting wood on the ground; such nest sites give the ants ready egress when the colonies are seriously disturbed. *L. alienus* workers produce the same volatile substances as *Acanthomyops claviger,* with the exception of citronellal, neral, and citral. Their principle volatile component is undecane. When they smell the pheromones, the *Lasius* workers scatter and run frantically in a comparatively unoriented fashion. They are more sensitive to undecane than the *Acanthomyops* workers, being activated by only $10^7 - 10^{10}$ molecules per cubic centimeter. It can be concluded, therefore, that, in contrast to *A. claviger, L. alienus* utilizes an "early warning" system and subsequent evacuation in coping with serious intrusion (Regnier and Wilson, 1969).

The harvesting ant *Pogonomyrmex badius* has a somewhat more elaborate response. Its principle alarm pheromone is 4-methyl-3-heptanone, which is stored in quantities of 0.2–34.0 μg (average: about 16 μg) in the mandibular gland reservoir (Vick *et al.,* 1969; Nancy Lind, personal communication). Workers respond to threshold concentrations averaging 10^{10} molecules per cubic centimeter by moving toward the odor source; when a zone of concentration one or more orders of magnitude greater than this amount is reached, the ants switch into an aggressive alarm frenzy (Wilson, 1958d; Lind, personal communication).

The recruitment component of alarm signaling is wholly separated in the fire ant *Solenopsis saevissima.* When a worker of this aggressive species is physically constrained, as it might be in combat with an alien ant worker, it simultaneously discharges an alarm substance from the head and one or more trail substances from the Dufour's gland of the abdomen. The alarm substance excites other workers without orienting them, while the trail substance does the reverse, at least in low concentrations. In combination the two pheromones both attract and excite (Wilson, 1965a).

Chemical alarm systems of one design or another are widespread in the social insects. Maschwitz (1964, 1966b) found evidence of alarm pheromones in all 23 of the more highly social species of Hymenoptera he surveyed in Europe. Several well-formed exocrine glands were implicated: the mandibular gland in the honeybee and many species of ants, the poison gland in *Vespa* and a few ant species, and Dufour's gland and anal gland in still other ant species. On the other hand, the more primitively social Hymenoptera, in particular the bumblebees and wasps of the genus *Polistes,* showed no evidence of utilizing such pheromones. Some of the phylogenetically more advanced termites produce volatile substances that act as straightforward alarm signals: for example, pinenes from the cephalic glands of the nasute soldiers of *Nasutitermes* and limonene from the same glands in the soldiers of *Drepanotermes* (Moore, 1968).

Bossert and Wilson (1963) predicted, on the basis of a priori considerations of potential molecular diversity in the pheromones and of olfactory efficiency in insect chemoreceptors, that most alarm substances in social insects would prove to have between five and ten carbon members and to have molecular weights between 100 and 200. Most alarm substances are not species-specific. This is not at all surprising in view of the consideration that it would seem to be of advantage to social insects to be able to "read" alarm signals coming from other species, especially those that are potentially dangerous to them. What limited specificity does occur in the 100–200 range of molecular weight can be easily obtained with the available diversity of molecular structure. We also concluded that in order for a substance to have a short fading time (a necessary property for an efficient alarm system), an intermediate threshold concentration is needed, that is, one neither very high nor very low in comparison with other pheromone systems. Insect chemoreceptors appear to have been so constructed in the course of evolution as to respond with increasing efficiency to molecules of increasing size in a given homologous series. They can be very efficiently adjusted to volatile odorants with molecular weights in the 100–200 range.

The molecular weight principle is illustrated in the alarm system of *Pogonomyrmex badius.* By directly measuring the effects of 4-methyl-3-heptanone from whole crushed heads, we obtained estimates for the "Q/K ratio"—the ratio of (*i*) pheromone molecules released to (*ii*) response-threshold concentrations (in molecules per cubic centimeter)—ranging from between 939 and 1,800. The Q/K ratio for this substance falls far below the ratios calculated for the sex attractants of moths (10^{11}) and is well above that of the trail substance of the fire ant (about 1). As a consequence of the intermediate Q/K value, the entire contents of the paired mandibular glands of *P. badius* provide a brief signal when discharged in air. A small "active space" (that is, the space within which the concentration is at or above the response threshold) is generated, attaining a maximum radius of only about 6 cm. After approximately 35 seconds, further diffusion reduces the active space to nearly zero, and the signal vanishes. The *Pogonomyrmex* colony is thus able to localize its alarm communication sharply in time and space.

The same design feature occurs in other ant species. In *Acanthomyops claviger* the Q/K ratios are on the order of $10^3 - 10^4$. If the entire contents of the Dufour's gland, containing about 2.5 μg of undecane, is discharged as a puff from the poison funnel, the diffusion model of Bossert and Wilson predicts that the pheromone signal would reach a maximum of about 20 cm in still air. If, on the other hand, only 0.1 percent were discharged, this active space would still reach a maximum of 2 cm. Experiments with *Acanthomyops* colonies disclosed that the signals, generated by all the volatile substances of a single worker combined, actually reach a maximum of 10 cm or slightly more.

Consider what would happen if the Q/K ratios of the alarm substances were very different from the values we estimated. If they had the same value as the fire ant trail substance, the active space would reach a maximum of at most a few millimeters and fade out in seconds. In other words, the signal would not travel beyond the distance within which other ants could perceive the danger directly themselves. If, on the other hand, the Q/K ratio approached the extremely high values of the moth sex pheromones, a single puff of undecane could theoretically generate an active space that would expand outward for a distance of kilometers and persist for years! This is impossible in practice, of course, but at the very least such an overly effective alarm signal would keep ant colonies in perpetual turmoil.

The rule that airborne alarm substances have small, relatively simple molecules has continued to hold since we first formulated it in 1963. Most of the volatile substances identified to date that are known or probable alarm pheromones are listed in Table 12-3.

In some cases the discharge of the pheromones is accompanied by characteristic running patterns or postures on the part of the signaling animal. For example, the alarmed honeybee worker dispenses isoamyl acetate from its sting chamber, and probably other alarm pheromones as well, while assuming the striking pose pictured in Figure 12-3. A disturbed honeybee worker is seen in the act of dispensing isoamyl acetate and possibly other alarm substances. The sting chamber is open, the sting is protruded, and the hairy membrane of the sting chamber, in which the pheromones are stored in liquid form, is everted. The bee is also beating its wings. This behavior draws in other workers, who investigate the vicinity actively and aggressively (Maschwitz, 1964). Some dolichoderine ants, such as the members of *Forelius,* lift their abdomens to a characteristic vertical position while releasing alarm substances (Goetsch, 1953a). Approximately the same behavior, with or without the alarm component, is displayed by species of the distinctive myrmicine genus *Crematogaster,* whose workers dispense drops of poisonous substances from the blunted tips of their stings while discharging alarm pheromones from the head (Buren, 1958; Blum *et al.,* 1969). Species of *Formica* spray mixtures of formic acid and Dufour's gland secretions, the latter serving as both defensive substances and alarm pheromones (F. E. Regnier and E. O. Wilson, unpublished observations).

Other odorant substances from ants have been reported that have not been assayed behaviorally but are likely candidates for alarm pheromones. These include acetic, propionic, and isovaleric acids from the myrmicine *Myrmicaria natalensis* (Quilico *et al.,* 1960), 2-hexenal from the myrmicine *Crematogaster africana* (Bevan *et al.,*

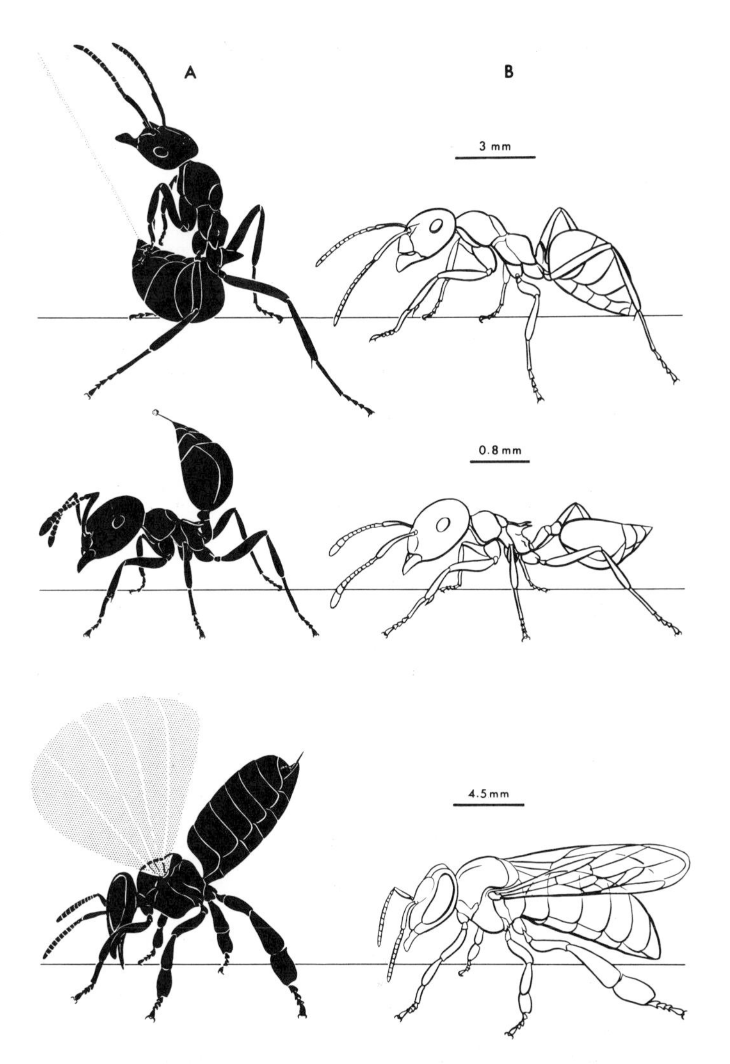

FIGURE 12-3. The alarm-defense posture (*A*) is contrasted with the resting posture (*B*) in workers of two ant species and the honeybee: (*top*) *Formica polyctena;* (*middle*) *Crematogaster ashmeadi;* (*bottom*) *Apis mellifera* (prepared by Bert and Turid Hölldobler).

1961), α-farnesene from the Dufour's gland of the myrmicine *Aphaenogaster longiceps* (Cavill *et al.,* 1967), 4-methyl-2-hexanone from the dolichoderine *Diceratoclinea clarki* (Cavill and Robertson, 1965), and *n*-decyl-, *n*-undecyl-, and *n*-dodecyl acetate from *Formica sanguinea* (Bergström and Löfqvist, 1968).

A general characteristic of the chemical alarm signals is that they fade quickly unless the danger stimulus is applied continuously. In other words, there is no positive

TABLE 12-3. Known and probable alarm substances in worker castes of social insects.

Species	Substance	Glandular source	Effect	Authority
ANTS				
Myrmicinae				
Myrmica brevinodis	3-Octanone, 3-Octanol	Mandibular glands	Excitement, attraction, threat posture	Crewe and Blum (1970)
Pogonomyrmex spp.	4-Methyl-3-heptanone (major component)	Mandibular glands	Low concentrations: attraction High concentrations: excitement, digging	Wilson (1958d); Bossert and Wilson (1963); McGurk *et al.* (1966); Vick *et al.* (1969)
	4-Methyl-3-heptanol (minor component)	Mandibular glands		
Atta sexdens	Citral, 5-Methyl-3-heptanone	Mandibular glands	Excitement, attraction	Butenandt *et al.* (1959); Blum *et al.* (1968)
Atta texana	4-Methyl-3-heptanone (major component)	Mandibular glands	Low concentrations: attract and alarm High concentrations: repel and alarm	Moser *et al.* (1968)
	2-Heptanone (minor component)	Mandibular glands		
Crematogaster (*C.*) *peringueyi*	3-Octanone, 3-Octanol, 3-Nonanone	Mandibular glands	Excitement	Crewe *et al.* (1970)
Crematogaster (*Atopogyne*) sp.	2-Hexenal	Head: probably mandibular glands	Excitement, attraction or repellancy	Blum *et al.* (1969)
Dolichoderinae				
Iridomyrmex pruinosus	2-Heptanone	Anal gland	Excited running, some attraction	Wilson and Pavan (1959); Blum *et al.* (1966)
Conomyrma pyramica	6-Methyl-5-hepten-2-one	Anal gland	Excited running, some attraction	McGurk *et al.* (1968)
C. bicolor	6-Methyl-5-hepten-2-one	Anal gland	Excited running, some attraction	McGurk *et al.* (1968)
Tapinoma nigerrimum	2-Methyl-4-heptanone, 2-Methylhept-2-en-6-one	Anal gland	Probably alarm; pheromone active on *T. sessile*	Wilson and Pavan (1959)
Formicinae				
Acanthomyops claviger	Undecane, etc. (see Fig. 12-2)	Dufour's gland	Attraction, excitement	Regnier and Wilson (1968)
	Citronellal, etc. (see Fig. 12-2)	Mandibular gland	Attraction, excitement	Regnier and Wilson (1968)
Lasius alienus	Undecane, etc.	Dufour's gland	Excitement, some attraction	Regnier and Wilson (1969)
	2,6 Dimethyl-5-hepten-1-al	Mandibular gland	Excitement, some attraction	Regnier and Wilson (1969)
Formica fusca	Undecane	Dufour's gland	Excitement, some attraction	Maschwitz (1964); Regnier and Wilson (unpublished)
BEES				
Apidae				
Apis mellifera	Isoamyl acetate	Lining of sting pouch	Attraction, investigation	Charles Butler (1609); Ghent and Gary (1962); Boch *et al.* (1962); Maschwitz (1964); Morse *et al.* (1967)
	2-Heptanone	Mandibular gland	Alarm; also foraging workers repelled in the field	Shearer and Boch (1965); Simpson (1966); Morse *et al.* (1967); Colin G. Butler (1966a, 1969)

TABLE 12-3 (*continued*).

Species	Substance	Glandular source	Effect	Authority
TERMITES				
Termitidae				
Drepanotermes rubriceps	Limonene (83–93%)	Cephalic gland of soldiers	Aggressive frenzy in soldiers	Moore (1968)
	Terpinolene (1–8%)	Cephalic gland of soldiers	Probably alarm	Moore (1968)
Nasutitermes spp.	α-Pinene (principal component)	Cephalic gland of soldiers	Probably alarm	Moore (1964a, 1968); Ernst (1959)
	β-Pinene (minor component)	Cephalic gland of soldiers		
	Limonene (minor component)	Cephalic gland of soldiers		
Tumulitermes pastinator	α-Pinene (principal component)	Cephalic gland of soldiers	Probably alarm	Moore (1968)
	β-Pinene (minor component)	Cephalic gland of soldiers	Probably alarm	Moore (1968)
	Limonene (minor component)	Cephalic gland of soldiers	Probably alarm	Moore (1968)
Amitermes herbertensis	Terpinolene	Cephalic gland of soldiers	Probably alarm	Moore (1968)
A. laurensis	Limonene (principal component)	Cephalic gland of soldiers	Probably alarm	Moore (1968)
	α-Pinene (minor component)	Cephalic gland of soldiers	Probably alarm	Moore (1968)
	β-Pinene (minor component	Cephalic gland of soldiers	Probably alarm	Moore (1968)
	Terpinolene (minor component)	Cephalic gland of soldiers	Probably alarm	Moore (1968)
A. vitiosus	α-Pinene (major component)	Cephalic gland of soldiers	Probably alarm	Moore (1968)
	Limonene (major component)	Cephalic gland of soldiers	Probably alarm	Moore (1968)
	β-Pinene (minor component)	Cephalic gland of soldiers	Probably alarm	Moore (1968)
	Terpinolene (minor component)	Cephalic gland of soldiers	Probably alarm	Moore (1968)

feedback; the detection of an alarm pheromone does not, according to present information, induce other workers to release more of the same pheromone. Consequently, when an alarmed worker enters a nest it serves as a focus of disturbance as it runs through the nest interior, but the excitement it causes dies down quickly after the worker passes.

Tactile Alarm Signals. The role of tactile stimuli in alarm communication is less well understood. In ants it seems to play a minor role. Wasmann (1899) interpreted the play of antennae on the bodies of sister ants as a codified "antennal language," and some of the tappings were interpreted by him to be alarm signals. Nevertheless we now believe that most antennation serves to receive information, specifically chemical information such as recognition odors, rather than to send it. Antennation is used as a signal to induce regurgitation in the honeybee and some ant, wasp, and termite species, but the behavior associated with this form of exchange is very different from alarm. Several observers have casually commented on the apparent alarming effects of one ant worker shaking or bumping another, yet these purely mechanical stimuli have seldom been isolated from chemical ones. One very interesting exception comes from Mary Jane West Eberhard's 1969 study of *Polistes.* The nests of the paper wasps are relatively small and brittle in composition, and they usually rest on a single thin pedicel. When a resting female is alerted by the sight of a large object moving nearby, she may dart suddenly from one part of the nest to the other. If her movement is violent enough, it causes an audible vibration throughout the entire nest. Either the movement or the sound, or perhaps both, instantly alerts the remainder of the colony, and the nest now bristles with aggressive wasps. Tactile signals, insofar as they can be distinguished in this case from substratal sound, seem also to be involved in the special parasite-alarm behavior, which Eberhard describes as follows. "The most intense alarm signal I have observed in *P. fuscatus* is that stimulated by the presence of the ichneumonid parasite *Pachysomoides fulvus* (Cresson), which lays its eggs in cells containing developing *Polistes* pupae. The *Pachysomoides* female lands near a *Polistes* nest and sits absolutely immobile, initially several inches away. It remains immobile until those present turn away. It then advances suddenly toward the nest, sometimes only a fraction of an inch. Unless chased away it alights on the nest and thrusts its ovipositor through a cell wall. When a *Polistes* female detects a *Pachysomoides* female on or near the nest, she darts at it, causing it to fly away, and begins hyperaggressively darting and wing-flipping about the nest, particularly on the periphery and sometimes even off the nest near its edge. This causes a general rise in activity and other females begin to dart and flip their wings, although with less intensity than the original alarmist. I have seen darting and wing-flipping continue for as long as ten minutes after the parasite has taken flight. In the several times I have observed this behavior, I have never seen wing-flipping by many females in *P. fuscatus* without later finding a female *Pachysomoides* nearby. Sometimes I noticed the small ichneumonid's presence only because the nesting wasps had begun their aggressive display." On one occasion a female was seen to take up the display upon returning to the nest with a load of water after the parasite had been routed. Females of *Polistes canadensis* show the same alarm behavior when their nest is approached by a similar parasite, *Pachysomoides stupidus.*

My impression is that, in communication among ants, ordinary mechanical stimuli play only a minor role. In laboratory colonies of *Formica, Pogonomyrmex,* and *Solenopsis,* I have often witnessed unalarmed but rapidly moving workers bump into other workers hard enough to shift their position, but without causing them to react even to the extent of walking away. When glass rods were inserted into the interior of nests long enough for the workers to become habituated to the alien odors on the glass surface, and the rods were then used to prod, depress and knock the workers about, the responses were very mild, never approaching the alarm behavior elicited by even minute quantities of pheromone. On the other hand, when the rods were first inserted and waved back and forth, they induced attack and alarm behavior, a response I interpreted as being dependent on the presence of unfamiliar odors.

Auditory Alarm Signals. In most of the species belonging to the principal subfamilies Myrmeciinae, Ponerinae, and Myrmicinae, the workers stridulate, that is, they rub specialized parts of the body together to produce a "chirp," a sound which Broughton (1963) has defined as "the shortest unitary rhythm-element that can be readily distinguished as such by the unaided human ear." In all ant species studied to date the chirps are specifically produced by raising and lowering the gaster (the most posterior discrete section of the abdomen) in such a way as to cause a dense row of fine ridges located on the middle portion of the first segment of the gaster (the fourth abdominal tergite) to rub against a scraper situated near the border of the preceding segment (Haskins and Enzmann, 1938; Helen F. Forrest, 1963). Spangler (1967) has given a complete characterization of stridulation in the harvesting ant *Pogonomyrmex occidentalis.* As the gaster is pulled down, a relatively weak sound is produced that lasts about $\frac{1}{10}$ of a second and generates its principal energy between 1 and 4 kc per second (1–4 khz). The gaster is then jerked back up, producing a second and similar sound, also lasting about $\frac{1}{10}$ of a second but overall louder and containing in addition some higher frequencies (7–9 khz). All of the detected frequencies are within the range of human hearing, and the result is that the stridulating *Pogonomyrmex* worker, when held close to the ear, seems to squeak faintly and more or less continuously.

The function of stridulation in ants remained a tantalizing mystery for a long time. Although Wheeler (1910) guessed that the sound serves in communication and the idea was favored thereafter, subsequent attempts to establish correlations between the onset of stridulation and responses of nonstridulating workers were invariably frustrated. No less acute an observer than Autrum (1936) failed to detect a function in the course of detailed experiments on *Myrmica.* In my own studies of the fire ant *Solenopsis saevissima,* I observed workers perform what appeared to be stridulating movements at odd intervals and seemingly quite spontaneously. The action occurred more frequently near the brood or during grooming, but it never seemed to produce any response in workers nearby. A first clue to the function of stridulation in one species was produced when Alan D. Grinnell and I (see Haskell, 1961) discovered that *Pogonomyrmex badius* workers usually stridulate only when they are confined, in particular when they are held by a pair of forceps, pressed under some heavy object, or restricted to a small space in a container. We hypothesized that the sound signal was a call for help sent out by trapped workers, but we could offer no supporting evidence. Finally, careful experiments by Markl (1965, 1967, 1968b) resolved the problem in the cases of *Atta cephalotes* and *Acromyrmex octospinosa.* The chirp produced by these leaf-cutting ants differs in quality from that of *Pogonomyrmex.* They are much louder (as much as 74 db at 0.5 cm from a soldier of *A. cephalotes* as opposed to less than 25 db in *Pogonomyrmex*) and higher pitched, consisting primarily of frequencies between 20 khz and 100 khz, thus ranging from near the limit of human hearing to well into the ultrasonic range. Like *Pogonomyrmex,* the leaf-cutting ants stridulate whenever they are prevented from moving freely. As Adele Fielde and Parker (1904) and Haskins and Enzmann (1938) had shown in other ant species, the leaf-cutting ants are nearly deaf to airborne vibrations, but extremely sensitive to vibrations carried through the substratum—sensitive enough, in fact, to detect stridulation sounds through the soil. Markl found that workers move to nestmates pinned to the ground as much as 8 cm away, and if the stridulating worker is buried under soil in the vicinity of the nest, other workers begin to dig where the sound is loudest. Workers buried to a depth of 3 cm induced digging, while those buried more deeply, to a depth of 5 cm, only attracted workers to the vicinity. It can be reasonably inferred that the stridulation serves as an underground alarm system, employed most frequently when a part of the colony is buried by a cave-in of the nest.

This is not necessarily the whole story in the ants, however. It is possible that in other species stridulation occurs in various situations and serves other functions that are still unsuspected. In any case, I doubt that sound communication will prove to be either elaborate in any particular case or connected with the most basic forms of communication. For one thing, stridulation has been dispensed with entirely in many ant groups. For another, the sounds produced are not known to contain any meaningful modulation. Although the sounds differ greatly from species to species (Forrest, 1963), they do not appear to vary much within species. Spangler (1967) showed that the two phases of the *Pogonomyrmex occidentalis* chirp are broken at regular intervals by exceedingly brief pauses in the abdominal movement, but the pulses thus created do not form any discernible temporal pattern; nor is it even certain that such patterns, if they exist at all, could be distinguished in transmission through the ground. An essentially similar result was obtained in the analysis by Markl (1968b) of sound production in *Atta cephalotes.* So far as we know, stridulation in ants produces a monotonous series of chirps with limited meaning.

Our present understanding of sound communication in termites is less secure. This is true even though sound production is one of the most conspicuous and best-studied aspects of termite behavior. Since the time of König (1779) and Smeathman (1781), it has been frequently recorded that disturbed termites make audible sounds by vibrating their heads or abdomens against the ground. It is an easy guess that the behavior serves to alarm other members of the same colony, and this hypothesis was duly declared by Smeathman, by Grassi and Sandias (1893–1894), and by later authors.

Recently a thorough analysis of sound production and its significance in *Zootermopsis angusticollis* was performed by Howse (1964). The agitated soldier of this species vibrates the forward part of its body by convulsively rocking its head upward and then down to a normal position again, over and over again about 24 times a second. With each upward thrust the forelegs are lifted off the floor and the head is banged against the ceiling of the nest; the overall effect to the human ear is a faint rustling sound. Howse confirmed the earlier observations of Andrews (1911) and Emerson (1929) who found that termites are virtually deaf to airborne sound but extremely sensitive to vibrations carried in the substratum. The latter signals are picked up by the "subgenual organs," scolopophorous sensilla located in the legs. This sensitivity is probably responsible for the curious phenomenon, noted by Thomas Snyder (in Emerson, 1929), that buildings in the southern United States occupied by constantly vibrating cotton-processing machinery, as well as railroad ties on well-used roads, are immune to termite attack. Howse's belief that the vibrations of the *Zootermopsis* soldiers are communicative in function rests on the following indirect argument. Electrophysiological recordings show that the subgenual organs are maximally sensitive to sounds ranging between 1,000 and 1,150 cycles per second, and the head rattling produces sounds at about 1,000 cycles per second. Artificial sounds pulsed into the substratum at the rate of 3–20 per second caused discharges in the afferent subgenual neurons that died out after 10 to 30 pulses, while 30 pulses per second caused activity that died out after only 1 to 8 responses. Evidently the subgenual organs are at least roughly tuned to the pattern of the sound produced by head rattling.

However, the matter is not quite settled by this evidence. No one has yet induced alarm behavior by transmitting a recorded or synthetic version of head rattling to a termite colony. More importantly, the response is usually associated with major disturbances that are apt to disturb the colony as a whole. Stuart (1963b) has distinguished between "Specific Alarm," in which a trail-laying termite actually jostles a nestmate by a stereotyped zigzag movement and "General Alarm," in which most of the colony is disturbed by a jolt to the nest, a sudden increase in illumination, or a sudden air movement. General alarm, the same phenomenon studied by Howse and earlier authors, is marked by excited running, thigmotaxis (a tendency to press against nearby objects), movement downward and away from light, and head banging. Under these circumstances it is possible that head banging is a defensive maneuver designed to frighten off intruders in the same way that many araneid spiders violently shake their webs and buprestid beetles and mutillid wasps stridulate.

Nevertheless, I am personally inclined to believe that head banging must serve as an alarm signal, at least in part. It must happen often in nature that a portion of a colony can enter a state of general alarm before the entire colony is alerted. This is most likely to happen when a small breach is made in the nest wall. Howse (1966) found that *Zootermopsis* workers are extremely sensitive to air movement; they begin to show avoidance when air currents strike their antennae with a velocity of only about a thousandth of the air flow typical for a closed room. When one end of a *Zootermopsis* nest is penetrated by an intruder such as a small mammal or burrowing insect, it is likely that the resulting air movements alone would generate a local state of general alarm which, in turn, would be communicated by head rattling through the remainder of the nest. As in the case of ants, however, this appears to be the upper limit of information content in sound communication. Not only is the original sound signal crude and repetitious in the extreme, but the pattern of subgenual organ response recorded by Howse (1964) allows for very little signal modulation at the sensory level.

Attraction and Assembly

Clustering behavior is universal in the social insects. When workers are removed from their nest and placed in a bare container, most quickly aggregate in one or several little groups. If the mother queen or some larvae are with them, the grouping will occur more quickly and be tighter in the end. In a series of experiments with the primitive termite *Kalotermes flavicollis,* Verron (1963) investigated the phenomenon of simple attraction and clustering in detail. He used the simple device of lining up six cylindrical containers in a row and placing a closed walkway over their tops. The termites whose attractiveness he wished to test were placed in the bottom of one of the tubes, and termites whose responsiveness was to be tested were released singly on the walkway. The latter individuals were allowed to move up and down the walkway testing the air from each of the six containers through a porous partition. Finally, their positions were recorded at set intervals. By definition, termites lingered longest over the most attractive tubes. Verron found that the larvae and nymphs are most responsive to odors of nestmates and that the larvae also produce the most attractive scent. Secondary reproductives react strongly to larvae and nymphs, but are themselves not very attractive. Soldiers are the least attractive and responsive of all. Much, perhaps all, of the volatile attractant is produced in the hind gut. When Verron destroyed the protozoan intestinal faunas of termites so they no longer could digest wood, or fed them pure cellulose, they lost their attractiveness. It seems to follow that some component of natural wood other than cellulose is the source of the attractant, and the component has to be metabolized to produce the attractant. The active pheromone was identified as the cis form of 3-hexen-1-ol. It is still uncertain whether the hexenol is produced directly from digestion of wood or is manufactured by a more circuitous route in exocrine glands associated with the hind gut; the latter alternative seems much more likely.

An even more elementary situation appears to exist in the fire ants of the genus *Solenopsis.* When away from the nest and in close quarters, workers tend to move up CO_2 gradients and thus in the direction of the largest nearby clusters of ants, and they attempt to dig through barriers placed in their way (Wilson, 1962a; Hangartner, 1969a). Carbon dioxide has thereby been implicated as the simplest pheromone used in any known animal communication system. It is likely that the CO_2 gradients are used in the orientation of ants in the immediate vicinity of nests. High concentrations of the order used in our experiments have been reported in ant nests (Portier and Duval, 1929; Raffy, 1929). The ability to detect CO_2 and CO_2 concentration differences has also been reported in the honeybee (Lacher, 1967), but we do not know whether this strange capacity is used in orientation, in monitoring air quality, or in some other, unsuspected function.

It is, of course, probable that other pheromones are involved in the clustering phenomenon in termites and ants. I would like to speculate that there exist substances of moderate or high molecular weight and low volatility that are located in the epicuticle and generate a very shallow active space so that they come into play only when two nestmates are in direct physical contact. At the moment we know almost nothing about these "surface pheromones" or even the volatile attractants of most other social insects.

Somewhat more complex forms of pheromone-mediated attraction and assembly has been demonstrated in the honeybee. In a pioneering study on chemical communication, Sladen (1901, 1902) found that workers near the hive entrance often elevate their abdomens, expose their Nasanov glands (sometimes written as Nassanoff glands), and fan their wings. He noted that a strong scent is produced by the Nasanov glands and postulated that it serves as an attractant to other workers. Later, von Frisch and Rösch (1926) discovered that workers expose their glands at newly discovered food sources and are able to attract other workers over considerable distances. The scent has been subsequently identified as a mixture of geraniol, nerolic acid, geranic acid, and citral (Boch and Shearer, 1962, 1964; Weaver *et al.,* 1964). Citral constitutes only three percent of the total volume of the fresh secretion, but it is by far the most potent in attractive power (Weaver *et al.,* 1964; Butler and Calam, 1969). For many years it was thought that the Nasanov gland secretion is colony-specific, but Renner (1960) provided detailed evidence from behavioral experiments to indicate that it is not. This conclusion has since been backed up by the chemical studies just cited. Renner was also able to clear up the question of why workers sometimes release the pheromones even though they do not have a food site to report. He found that the bees display this behavior whenever they have lost customary contact with their companions for a long period of time. It has also been demonstrated (Velthuis and van Es, 1964) that swarming bees release the pheromone when they first encounter the queen, thus attracting other workers to the vicinity. The Nasanov substances are therefore true assembly pheromones. Evidently the discovery of new food lowers the threshold of the response and thus turns the pheromones secondarily into recruitment signals.

A second, more pervasive assembly scent is found in the hive odor. The existence of an attractant that assists honeybees in finding their hives was suggested by Ribbands and Speirs (1953) and Lecomte (1956) and demonstrated experimentally by Butler, Fletcher, and Watler (1969). The odor appears not to be specific to colonies and may be identical with the "footprint substance" which is laid down continuously around the nest and food sites by the worker bee. The latter pheromone is sometimes paid out in the form of a trail around the hive and serves as a rudimentary guide to honeybees seeking the nest entrance (Butler, 1966b). A similar footprint substance occurs in the wasp *Vespula vulgaris,* but it has not yet been shown to be laid in trails (Butler *et al.,* 1969). Soil from around the nest entrance is highly attractive to workers of the harvesting ant *Pogonomyrmex badius.* Members of each colony are able to recognize their own nest material (Hangartner, Reichson, and Wilson, 1970). Bert Hölldobler and I have also found that the odor from around *P. badius* workers is attractive to other workers of the same species, even when separated from the nest. The attractant is specific at least at the species level. It is possible, but as yet unproven, that this substance comprises part of the nest odor.

By far the most dramatic form of assembly is exercised by the mother queens of colonies. Except in the most primitively social species, any well-nourished, fertilized queen continually attracts a retinue of workers who tend to press close in with their heads facing her. The attendants also lick the surface of the queen's body and offer her food, most often in regurgitated form. The effect is most pronounced—often to a dramatic extent—in "physogastric" individuals, that is, queens with the abdomen distended with hypertrophied ovaries and fat bodies. Nest queens of army ants and fire ants are completely covered and hidden from view by tightly packed masses of workers, which number into the tens or hundreds. It is least marked where the colonies are small or where there are multiple laying queens. Stumper (1956) showed that it is possible to extract scent in ether from single queens of *Lasius niger* and transfer it to dummies constructed from pith. Workers then treat the dummies very much like the original queen. I have repeated Stumper's experiment and obtained the same result in the myrmicine *Aphaenogaster rudis.* Watkins and Cole (1966) found that when queens of army ants (*Labidus, Neivamyrmex*) are allowed to do no more than sit on untreated balsa strips, enough scent is transferred to make the strips attractive to workers. Given a choice, the workers prefer the residue from their own queen over that from other colonies belonging to the same species.

The same phenomenon in honeybees has been analyzed at length by Barbier *et al.* (1960), Gary (1961), Morse (1963), Velthuis and van Es (1964), and most especially by Pain (1961a,b) and Butler and his collaborators (Butler *et al.,* 1961; Simpson, 1963; Butler *et al.,* 1964; Butler and Simpson, 1965, 1967; and contained references). There exist at least two pheromones that attract worker bees to the queen. One is the "queen substance," *trans*-9-keto-2-decenoic acid, a remarkable substance produced in the mandibular glands whose multiple functions in the life of the colony will be dealt with more extensively in Chapter 15. The second is a fat-soluble substance of unknown identity produced by Koschevnikov's gland, a tiny cluster of cells located in the sting chamber and whose principal duct opens between the overlapping spiracle and quadrate plates. These two pheromones are responsible at least in part for the formation of the retinue that surrounds the queen at all times. When the colony divides by the process of swarming, the attractive power of the 9-ketodecenoic acid comes to play a new role. Workers are drawn to the queen in midair by flying upwind when they smell the pheromone. As the queen flies to the swarm site, and later to the new nest (in both cases guided by scout bees), the bulk of the worker force follows along in the wake of her evaporating 9-ketodecenoic acid. Once a new destination is reached, however, this substance is not adequate to settle the flying workers. Now a second mandibular gland pheromone produced by the queen, *trans*-9-hydroxy-2-decenoic acid, comes into play. Workers can smell this substance only over a short distance. Those that do smell it begin to call in other workers by dispersing their own Nasanov gland scent, and, in a short time, the entire colony forms a quiet, stable cluster. There may be still more to the story. The substituted acids are but 2 of at least 32 components present in the mandibular glands of the queen. Other substances identified include methyl-9-ketodecanoate, nononoic acid, and a variety of other esters and acids (Callow *et al.,* 1964). The possibility that these and other as yet unidentified secretions manufactured by other glands in the queen's body (Renner and Margot Baumann, 1964) have some communicative function remains largely unexplored.

13 Communication: Recruitment

Recruitment can be defined as communication that brings nestmates to some point in space where work is required. The social insects have evolved a multitude of ingenious devices to assemble workers for joint efforts in food retrieval, nest construction, nest defense, and migration. The category of recruitment is thus a very loose one, and it cannot be clearly separated from the more elementary systems of alarm and assembly. In the lower termites, nest repair and alarm recruitment by trail laying comprise elements of a complex behavioral pattern under the control of a single pheromone.

Visual Communication

The large-eyed ants have excellent form vision and are especially keen at detecting moving objects (Jander, 1957; Ayre, 1963; Voss, 1967). Workers generally do not respond to prey insects that are standing still, but run toward them as soon as they begin to move. Stäger (1931) noticed that, when lone workers of *Formica lugubris* foraging in a field encounter an insect, they dash in erratic circles around it and thereby attract other workers exploring in the close vicinity. This communication is evidently mediated by vision alone. Stäger labeled the phenomenon "kinopsis." Later Sturdza (1942) proved with laboratory experiments that the sight of a running *Formica nigricans* worker alone is enough to set another worker running. The question needs to be raised whether the excited, broken running pattern that is set off in ants who have just discovered prey is a "ritualized" variant of locomotory running evolved specifically to function as a visual signal. The possibility is intriguing, but no real evidence has yet been adduced in favor of it. During my own studies of *Daceton armigerum* (Wilson, 1962b), a large-eyed predatory ant found in the canopy of South American rain forests, Stäger's effect was observed repeatedly. It often led to the quicker trapping and killing of insect prey, but this result appeared to be incidental to the increased movement necessitated by the chase rather than a signal directed at other workers. Nevertheless, there is little doubt that at least the response of the other workers is adaptive and that the interaction is a true, albeit elementary, form of visual communication.

Simple Forms of Chemical Recruitment

When individual workers of *Pogonomyrmex badius* attack large, active insect prey in the vicinity of the nest, they discharge the alarm pheromone 4-methyl-3-heptenone from their mandibular glands. This substance both attracts and excites other workers within distances of 10 cm or so (just as it does in the presence of dangerous stimuli), with the result that the prey is more quickly subdued. Thus, in *P. badius,* and probably other predaceous ant species that employ alarm pheromones, recruitment is a felicitous by-product of alarm communication. A parallel relation between two quite different behavioral functions exists in the social life of the honeybee, where the Nasanov gland pheromones are used in some instances to assemble workers that have become lost while foraging and participating in colony swarming and in other instances to recruit nestmates to newly discovered pollen and nectar sources (Renner, 1960; Butler and Simpson, 1967).

There is some evidence to suggest that social insects leave chemical "signposts" around food discoveries although few studies have been conducted to characterize the phenomenon. Glass feeding dishes that have been visited by worker honeybees are preferred by newcomer bees over unvisited dishes, even when each container holds identical food (Chauvin, 1960b). The substance can be extracted and is said to come from Arnhardt's glands in the tarsi. The same substance may be responsible for the odor trails sometimes laid by walking honeybees in the vicinity of the hive, as described by Lecomte (1956) and Butler *et al.* (1969).

The odor of food brought to the nests can also influence the behavior of nestmates and therefore serve in a primitive form of recruitment communication. Honeybee workers recognize the odor of food sources both from the smells adhering to the bodies of successful foragers and the scent of the nectar regurgitated to them. If they have had experience in the field with flowers or honeydew bearing the same odor, they will then revisit the site searching for food. This response can be induced in the absence of waggle dancing or other forms of communication. Russian apiarists have used the principle to guide bees to crop plants they wish pollinated. To take a typical example, the colonies are trained to red clover by being fed with sugar water in which clover blossoms have been soaked for several hours. After this exposure, the foraging workers search preferentially for red clover in the vicinity of the hive. The same method has been used to increase pollination rates of vetch, alfalfa, sunflowers, and fruit trees (von Frisch, 1967a). Free (1969) has recently been able to demonstrate that the odor of food stores has a similar effect on bumblebees.

Tandem Running

The next step up the ladder of sophistication in chemical recruitment techniques is tandem running (Wilson, 1959b). When a worker of the little African myrmicine ant *Cardiocondyla venustula* finds a food particle too large to carry, it returns to the nest and contacts another worker, which it leads from the nest. The interaction follows a stereotyped sequence. First the leader remains perfectly still until touched on the abdomen by the follower ant. Then it runs for a distance of approximately 3 to 10 mm, or about one to several times its own body length, coming again to a complete halt. The follower ant, now in an excited state apparently due to a secretion released by the leader, runs swiftly behind, makes fresh contact, and "drives" the leader forward. Other workers approaching the leader become similarly excited, even if the latter is completely immobile at the time. After each contact and subsequent forward drive of the leader, the follower may press immediately behind and move it again. More commonly, it circles widely about in a hurried movement that lasts for several seconds and may take it as far as a centimeter from the path set by the leader. In a short time, however, the circling brings the follower once again into contact with the leader. In the great majority of cases, tandem running in *C. venustula* involves only two workers. Occasionally a third worker crowds in closely behind the leader, but it does not continue following for more than a few centimeters. The leader evidently does not depend on an odor for its initial guidance. Rather, the straight courses it takes over rough terrain suggest that it is orienting visually. Although tandem running is apparently concerned mostly or entirely with the communication of food finds, it is employed by only a fraction of the workers engaged in successful foraging. Less than 10 percent of the *C. venustula* I observed running from the nest to food masses were coupled in tandem pairs. This is in contrast to the high degree of participation shown by a typical trail-laying ant species such as *Solenopsis saevissima,* in which more than 90 percent of the workers returning from rich food sources contribute material to a common odor trail.

Tandem running was first described by Hingston (1929) in the Indian formicine ant *Camponotus sericeus,* and I have recorded it in other species of *Camponotus* in Australia and South America, as well as in a second species of *Cardiocondyla* (*C. emeryi*). Recently Dobrzański (1966) has discovered the same behavioral pattern in *Leptothorax acervorum* in Poland, but serving a wholly different and unexpected function. Foraging workers of this small species, which superficially resemble *Cardiocondyla* and like them form small colonies, apparently never use tandem running in the communication of food finds. The true function was discovered by accident when, in the course of experiments on territorial behavior, Dobrzański picked up sticks containing colonies and placed them in new environments. Under this circumstance, tandem running was observed repeatedly, and it became apparent that this form of communication is used in the exploration of new environments. The pioneer foragers tend to go out in pairs, rather like swimmers venturing forth in what Americans refer to as the "buddy system." Perhaps, as in the buddy system, tandem running serves to increase safety in the face of unknown dangers.

Odor Trails in Ants

The most elaborate of all the known forms of chemical communication is the odor trail system. Trail communication has evidently evolved, at least in some groups of ants, from tandem running. Hingston (1929) found what appear to be excellent intermediate stages between the two forms of communication in the Indian species *Camponotus paria* and *C. compressus.* In *paria,* the leader ant does not halt and wait to be touched, while the follower often drops behind 5 or 10 cm and seems to orient along a short-lived odor trail. The behavior of *compressus* resembles that of *paria* except that as many as 10 or 20 workers follow single file behind the leader. These intermediate stages are not at all rare in ants. I have observed *compressus*-like files in an unidentified South American *Camponotus* as well as in *C. beebei* from Trinidad (Wilson, 1965b). Wesson (1939, 1940b) observed approximately the same kind of behavior in the slave raids of the myrmicine slave-making ants *Harpagoxenus americanus* and *Leptothorax duloticus.* Here is his account of the organization of a raid on the nest of a *Leptothorax curvispinosus* colony by a group of *H. americanus* workers accompanied by *L. curvispinosus* slaves. The action began when a successful *H. americanus* scout returned to her nest:

> Inside the nest she quickly combed her antennae, then ran excitedly about, causing considerable excitement among both the *americanus* and their slaves. [After three minutes] she emerged from the nest followed by a compact file of eight *americanus* and five *curvispinosus.* The gait of the leader was peculiar. She ran with a rather slow, sprawling, stiff-legged movement, running forward a few steps, then waiting until a following worker touched her on the gaster before moving a few steps farther. By gently touching her gaster with a hair it was possible to cause her to move as though touched by another ant. In her progress, she kept her gaster deflected stiffly downward, in all probability laying a scent trail from it. This was demonstrated by the fact that following workers followed precisely the path taken by the leader even when that path was very irregular. On several occasions workers were observed following her exact path when separated from her by several inches. The file moved rather slowly due to the uncertainty and slowness with which the leader moved, arriving at the *curvispinosus* nest about four minutes after starting. The *curvispinosus* slaves tended to drop back and become lost as the file proceeded until there remained only two.

It would seem to be only a short step in evolution from trail-guided processions such as these, where the followers must stay close behind the leader, to typical trail following, where followers are guided by odor alone over long distances in the absence of the trail layer. Trail following in ants has long fascinated biologists. In 1779 Charles Bonnet presented the first crude experimental evidence that the trail is composed of an odorant chemical. When he drew his finger across the trail of a small ant species, the workers were thrown into disorder and were unable to proceed for some time. Of course, the flaw in Bonnet's experiment was the inadvertant addition of an odor of his own from his finger tip. He could not really be sure that the effect observed was due to removal of a trail substance instead of the addition of an alien substance. But subsequent experiments by many authors have shown that disorientation follows the removal of sections of trails by methods free of odor contamination. The best technique is the direct substitution of fresh sections of the substratum after the trails are laid.

In 1959 I showed that it is possible to induce the complete recruitment process in the fire ant *Solenopsis saevissima* with artificial trails made from extracts or smears of single glands containing the trail substance. This technique has since made it possible to pinpoint the glandular source of trail pheromones in various social insect species and also to devise reliable bioassays for use in chemical studies of the trail substances. Subsequently I undertook a detailed analysis of trail communication in *Solenopsis saevissima* (Wilson, 1962a). When fire ant workers leave the nest in search of food, they may follow odor trails for a short while, but they eventually separate from each other and begin to explore singly. When alone they orient visually, a fact that was established in the following way. Single workers were permitted to explore a foraging table in the laboratory for distances up to a meter from the nest. The only source of illumination was a lamp beamed from one side of the table. The workers were first allowed to discover a drop of sugar water. When they had fed and started home, laying an odor trail behind them, the light was switched off, and a second light located on the opposite side of the table was simultaneously switched on. After the direction of the light source had thus been changed by 180°, the ants almost invariably performed a complete about-face. By repeatedly switching the light source from one side to another, individual workers could be marched back and forth like puppets and finally brought home at the discretion of the investigator. When both lights were turned off and the ants observed by red light (invisible to them), they became disoriented. Later, Marak and Wolken (1965) employed monochromatic lights in the same switching technique to measure color vision in fire ants (see Chapter 11).

As a fire ant worker heads home after discovering a food source, it walks at a slower, more deliberate pace. Its entire body is held closer to the ground. At frequent intervals the sting is extruded, and its tip drawn lightly over the ground surface, much as a pen is used to ink a thin line (Figures 13-1,2). As the sting touches the surface, a pheromone flows down from the Dufour's gland. This substance, which has been isolated in pure form by Walsh *et al.* (1965) but not yet chemically identified, is dispensed in exceedingly minute quantities. Each worker contains only a small fraction of a nanogram (billionth of a gram) at any given moment. When B. P. Moore and I used odor traps to collect the pheromone from entire colonies migrating along strong odor trails from one nest to another, we were able to recover no more than a single nanogram per hour. It follows that the pheromone must be a very potent attractant. Artificial trails made from a single Dufour's gland induce following by dozens of individuals over a meter or more. When the concentrated pheromone is allowed to diffuse from a glass rod held in the air near the nest, workers mass beneath it, and they can be led along by the vapor alone if the rod is moved slowly away. When I presented large quantities of the substance at the entrance of artificial nests, I was able to draw out most of the inhabitants, including both workers carrying larvae and pupae and, on a single occasion, the mother queen.

While laying a trail, the worker sometimes loops back in the direction of the food find, but only for short distances, before turning nestward again. If another worker is contacted, the homebound worker turns toward it. It may do no more than rush against the encountered worker before moving on again, but sometimes the reaction is stronger: it climbs partly on top of the worker and, in some instances, shakes its body lightly but vigorously in a vertical plane. The vibrating movement, which is unique to these encounters, has also been described in *Monomorium* and *Tapinoma* by Szlep and Jacobi (1967). These authors believe that the movement by itself conveys a message. In *Solenopsis,* however, the vertical shaking does not appear to impart any essential information about the food find because contacted workers do not seem to exhibit trail-following behavior different from those not contacted. Moreover, the pheromone is by itself sufficient to induce immediate and full trail-following behavior when laid down in artificial trails.

Most workers encountering a freshly laid trail respond at once by following it outward from the nest. They are able to detect it by smelling the vapor over distances as great as a centimeter. The workers do not follow a liquid odor trace on the ground. Instead, they move through the vapor created by diffusion of the pheromone into the air. There is a space, which theoretical calculations show to be semiellipsoidal in shape, within which the pheromone is detected by the ants (Figure 13-2). As follower workers travel through this "vapor tunnel," they sweep their antennae from side to side, evidently testing the air for odorant molecules. In fact, they are able not only to detect these molecules in the gaseous state but also to move up gradients of molecular concentration, a process of orientation referred to as osmotropotaxis (Martin, 1964). When fire ants are presented with trail substance evaporating in still air from the tip of a glass rod, they run directly to the glass rod. The sensory mechanism enabling fire ants to orient in this fashion is still unknown. However, in a set of ingenious experiments, summarized in Figure 13-3, Hangartner (1967) was able to demonstrate the basis of osmotropotaxis in another trail-following ant species, *Lasius fuliginosus.* The method of following odor trails disclosed by these experiments makes it very unlikely that directional signals can be built into the trails. In other words, the odor streaks may or may not be tapered or shaped in some other way so as to point the way home—as discovered for example in *Myrmica ruginodis* trails by Macgregor (1948)—but it would be difficult for the follower ant to "read" this information. Additional experiments performed on *Lasius, Acanthomyops,* and *Solenopsis* have indicated in fact that no such information is transmitted (Carthy, 1951b; Wilson, 1962a). These findings refute the early hypothesis of Bethe (1898) that trails are effectively polarized, as well as the conjecture of Piéron (1904) that ants find their way back and forth by a special kinesthetic sense.

Our present knowledge makes it desirable to reassess the value of Auguste Forel's mysterious theory of the "topochemical sense." Although this notion is frequently mentioned in the literature, especially in connection with odor trails, there is general confusion over what it really means. Forel seems to have been talking about the per-

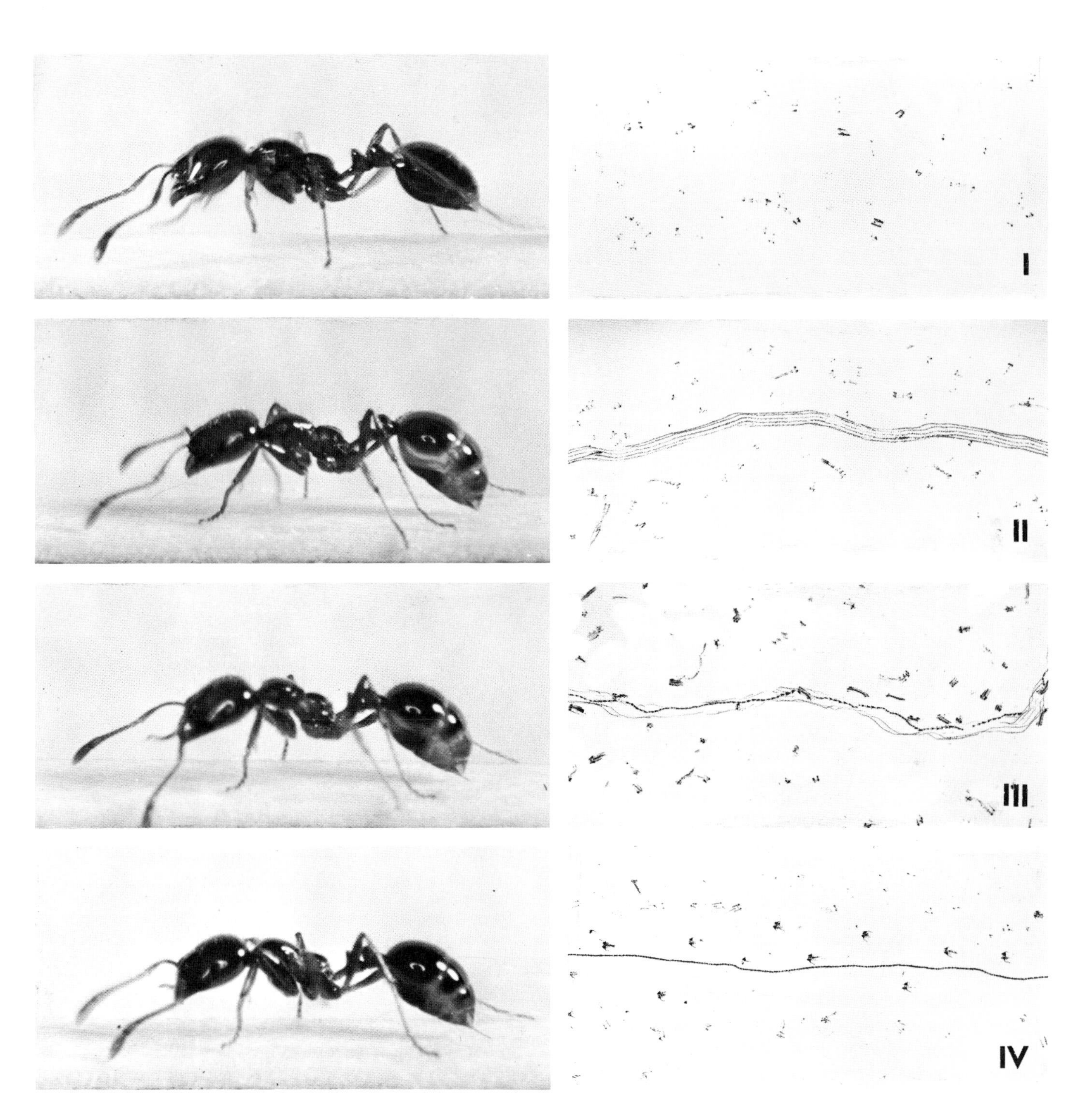

FIGURE 13-1. Workers of the fire ant *Solenopsis geminata:* (*left*) laying odor trails from their extruded stings; (*right*) tracks made by the sting, by the hairs at the tip of the abdomen, and, along the side, by the feet of the trail-laying ants. The latter figures are the negatives of tracks made by ants allowed to walk over pieces of smoked glass (from Hangartner, 1969b).

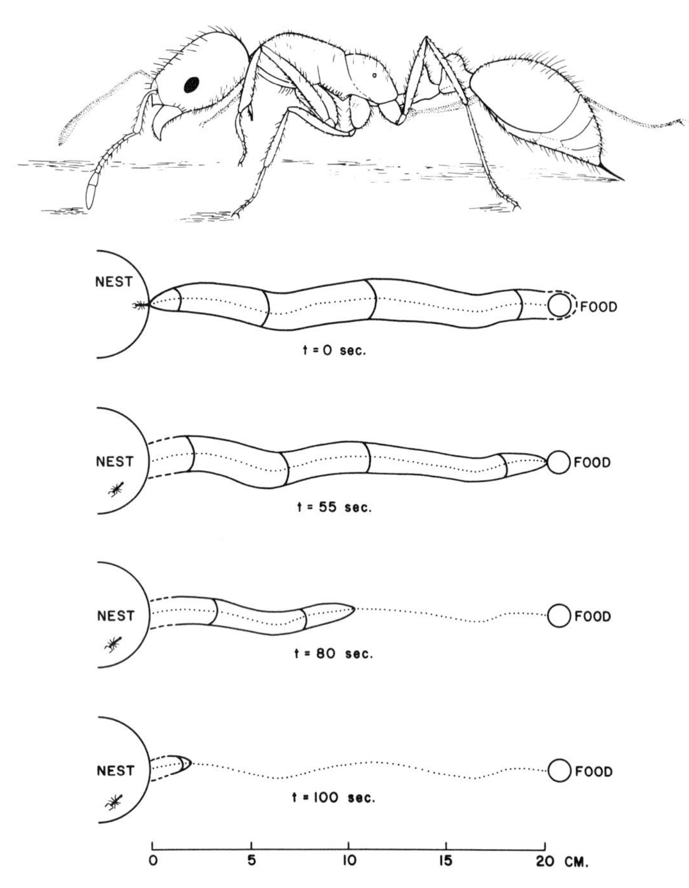

FIGURE 13-2. The form of the odor trail of *Solenopsis saevissima* laid on glass. As the trail substance diffuses from its line of application on the surface, it forms a semiellipsoidal active space within which the pheromone is at or above threshold concentration. This space, and therefore the entire signal, fades after about a hundred seconds. The times shown here are given from the moment a worker reaches the nest after laying a trail in a straight or slightly wavering line from a food source 20 cm away. The trail is shown in this model as continuous. In nature it is irregularly segmental, but the dimensions and fade-out times remain nearly the same. At the top is an enlarged view of a worker laying a trail from right to left (from Wilson, 1962a, and Wilson and Bossert, 1963).

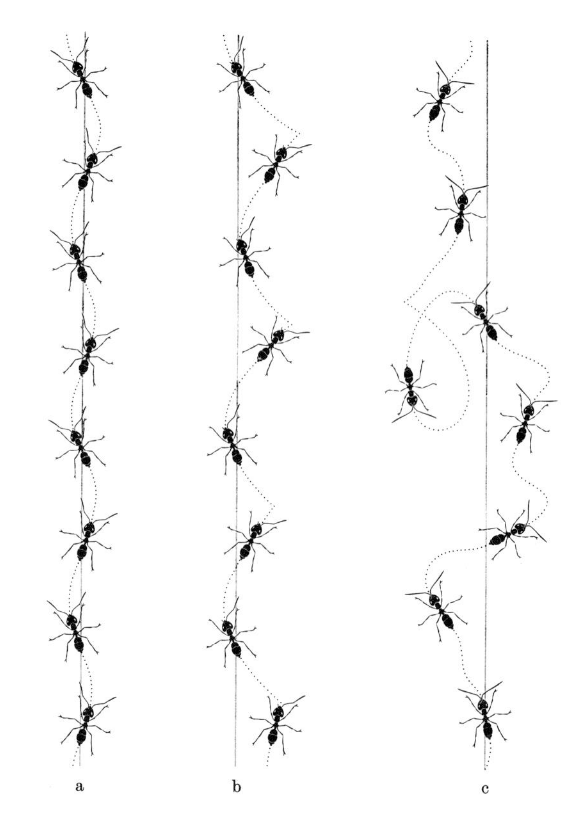

FIGURE 13-3. Hangartner's demonstration that worker ants (*Lasius fuliginosus*) orient along odor trails by detecting varying concentrations of the trail pheromone in vapor form. The straight line represents (in idealized form) the point of application of the pheromone onto the ground. (*a*) As a worker follows the trail, it tends to move out of the active vapor space, first to one side and then to the other. As one antenna, containing the odor receptors, leaves the active space the ant swings back in the opposite direction. Consequently, it moves in a weaving pattern. (*b*) When the left antenna is amputated, the ant repeatedly overcorrects on the right side. (*c*) When the antennae are crossed and glued in this position, the ant is disoriented and recovers the active space only with difficulty. Nevertheless it is still able to move along in one direction, presumably with the aid of light-compass orientation (from Hangartner, 1967).

ception of form and the spatial relation of discrete objects by means of smell, but his pronouncements on the subject were discouragingly obscure. Here is the clearest I have encountered: "By topochemical I mean a sense of smell which informs the ant as to the topography of the places surrounding it by means of chemical emanations, which give an odor to objects" (Forel, 1928:I, 116). The sense was said to be located in the terminal segments of the antennae. The perception of the spatial relation of objects implies an integration of sensory input in the central nervous system considerably more complex even than the following of an odor trail by osmotropotaxis. This has been well documented in the case of visual input (Jander, 1957) but not yet in the case of olfactory input.

A more important finding from the work on *Solenopsis saevissima* involves the way in which mass communication is achieved. By mass communication I mean information that can be transmitted only from one group of individuals to another group of individuals. In the case of *S. saevissima,* the number of workers leaving the nest is controlled by the amount of trail substance being emitted by foragers already in the field. Tests involving the use of enriched trail pheromone showed that the number of individuals drawn outside the nest is a linear function of the amount of the substance presented to the colony as a whole. Under natural conditions this quantitative relation results in the adjustment of the outflow of workers to the level needed at the food source. An equilibration is then achieved in the following manner. The initial buildup of workers at a newly discovered food source is exponential, and it decelerates toward a limit as workers become crowded on the food mass because workers unable to reach the mass turn back without laying trails and because trail deposits made by single workers evaporate within a few minutes. As a result, the number of workers at food masses tends to stabilize at a level which is a linear function of the area of the food mass. Sometimes, for example when the food find is of poor quality or far away or when the colony is already well fed, the workers do not cover it entirely, but equilibrate at a lower density. This mass communication of quality is achieved by means of an "electorate" response, in which individuals choose whether to lay trails after inspecting the food find. If they do lay trails, they adjust the quantity of pheromone according to circumstances (Hangartner, 1969b). The more desirable the food find, the higher the percentage of positive responses, the greater the trail-laying effort by individuals, the more the trail pheromone presented to the colony, and, hence, the more the newcomer ants that emerge from the nest.

Thus, the trail pheromone, through the mass effect, provides a control that is more complex than might have been assumed from knowledge of the relatively elementary individual response alone. This complexity is increased still further by the fact that the pheromone assumes different meanings in at least two different contexts. When colonies move from one nest site to another, a common event in the life of fire ants as well as many other kinds of ants, the new site is chosen by scout workers which then lay odor trails back to the old nest. Other workers are drawn out by the pheromone. They investigate the new site and, if satisfied, add their own pheromone to the trail. In this fashion the number of workers traveling back and forth builds up exponentially. In time the brood is transferred, the queen walks over, and the emigration is completed. The pheromone also functions as an auxiliary signal in alarm communication. When a worker is seriously disturbed, it releases some of the trail substance simultaneously with alarm substance from the head, so that nearby workers are not only alarmed but also attracted to the threatened nestmate.

In the course of these same studies (Wilson, 1962a) I adapted a mathematical technique, first outlined by Haldane and Helen Spurway (1954), to measure the amount of information transmitted by the fire ant odor trail and to compare it with the amount transmitted by the waggle dance of the honeybee. The method is straightforward. Workers were allowed to lay odor trails away from a food find (a drop of sugar water), and the starting point of the trail was marked in order to measure its distance from the food find—hence, the initial error committed by the trail-laying ant. Similarly, the paths taken by the outward-bound follower ants and the distances of their nearest approaches on the trail to the food find were each recorded. A typical set of data of the second category is shown in Figure 13-4, and the procedure used to analyze them appears in the caption. The desired unit in such an information study is of course the *bit,* defined as the amount of information needed to make a choice between two equiprobable alternatives. If n alternatives are present, a choice provides the following quantity of information:

$$H = \log_2 n.$$

Thus, in a most elementary communication system, the sending of one of n equiprobable messages reduces $\log_2 n$ amount of uncertainty. By definition, it provides that much information.

Suppose we were communicating information about direction, and our system permitted us to transmit with perfect accuracy any one of 16 directions. Then the message "go north by northwest," one of the 16 equiprobable messages, conveys $\log_2 16 = 4$ bits of information. Human beings intuitively think of information in such language and numbers, and it would be desirable if we could translate the efficiency of an animal communication system into bits and thence back into easily comprehended analogues such as compasses. This can be accomplished for the odor trail of the fire ant, and perhaps also for the waggle dance

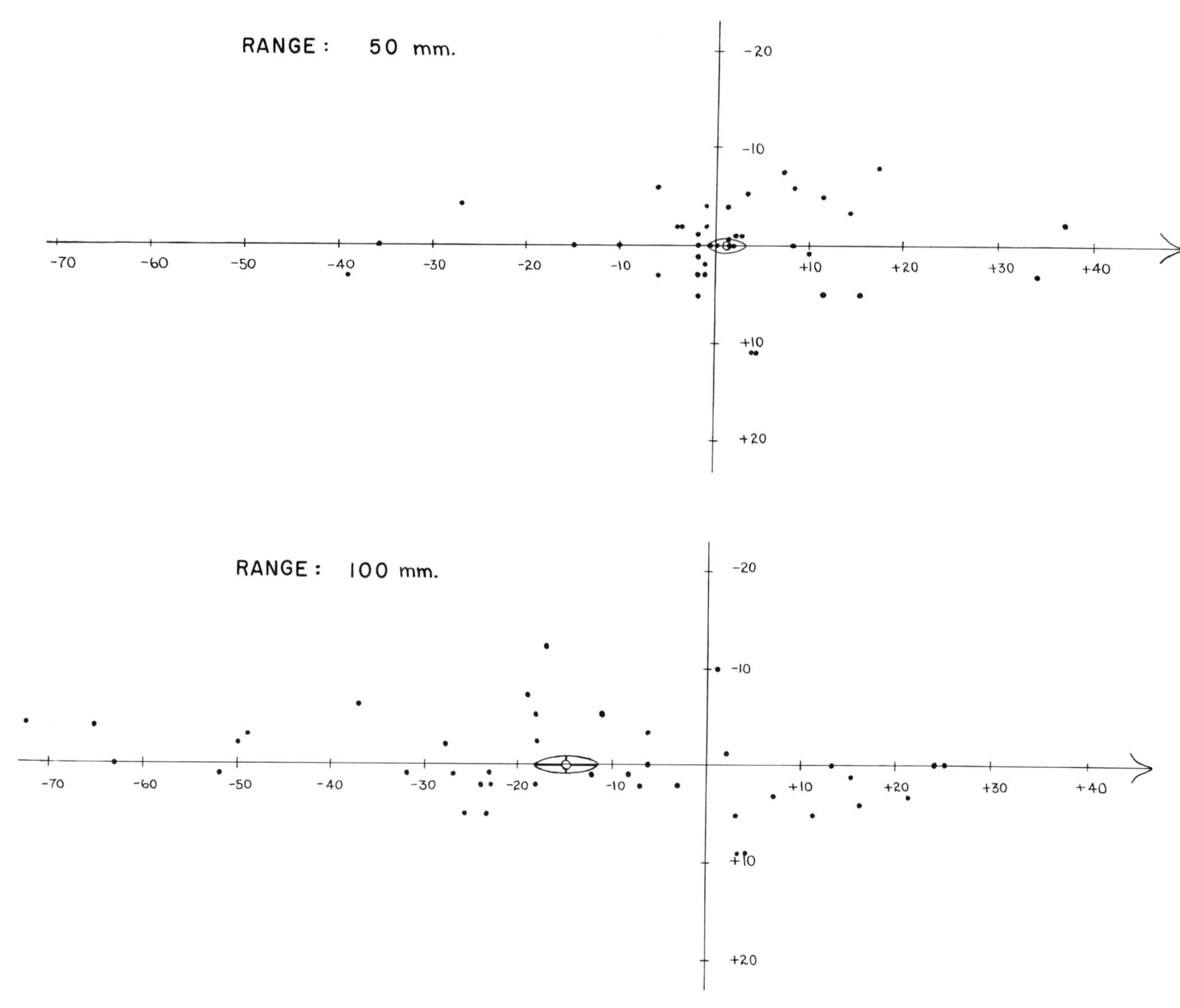

FIGURE 13-4. Points are shown at which different fire ant workers turned back on the odor trail or lost the odor trail in their nearest approach to a food find. The food find was located at the origin of the axes and was removed before trail following began in order to eliminate clues that did not come from the odor trail itself. The error made by each ant was measured in two dimensions, that is, the distance of its turn-back point from each of the two axes. Estimates were then made of the mean position of all the turn-back points, shown here as the small open circle, and the standard deviations of the errors in two dimensions, given here as the long and short axes respectively of the small ellipse. These are the measurements needed to estimate the quantity of information transferred by the odor trail (from Wilson, 1962a).

of the honeybee, by utilizing our measurements of error. Shannon (in Shannon and Weaver, 1949) has shown that in the special case of a one-dimensional Gaussian distribution, where the standard deviation is σ,

$$H = \log_2 \sqrt{2\pi e}\, \sigma$$

Haldane and Spurway applied the latter formula to the angular scattering of newcomer bees around target baits following waggle dances, the dispersion of which was considered to be Gaussian. In the absence of any message at all,

$$H_1 = \log_2 360°$$

where H_1 is the number of bits required to reduce the uncertainty to an interval of one degree. If H_2 is the uncertainty remaining after the message is received by the newcomer bees, then $H_1 - H_2$ is the amount of information transmitted. Hence,

$$H_\theta = H_1 - H_2 = \log_2 360° - H_2$$

If it is assumed that the dispersion has a one-dimensional Gaussian distribution, then

$$H_\theta = \log_2 360° - \log_2 \sqrt{2\pi e}\, \sigma_\theta$$

$$= \frac{\log_{10} \frac{360°}{\sigma_\theta}}{\log_{10} 2} - 2.0471$$

Wilson (1962a) adapted the method to distance communication as well and used it to make estimates from the fire ant data and the data of von Frisch and Jander (1957) on honeybees. The essential results are shown in Figures 13-5 and 13-6. Notice that the two systems transmit roughly comparable amounts of information with reference to both direction and distance. The amount of directional information in the ant odor trail, however, increases with the length of the trail. This is because the width of the active space remains constant, and the follower ants stay about as close to the true path of the trail layer all along the length of the trail. Consequently the angular deviation away from the true path, with reference to the nest, decreases as the trail lengthens away from the nest. Thus the directional errors committed by the followers decrease, and the amount of directional information in the trail itself increases as the trail lengthens.

The estimates of distance information in the waggle dance contain a relatively small error resulting from a mistake in the original von Frisch-Jander statistics. R. Boch

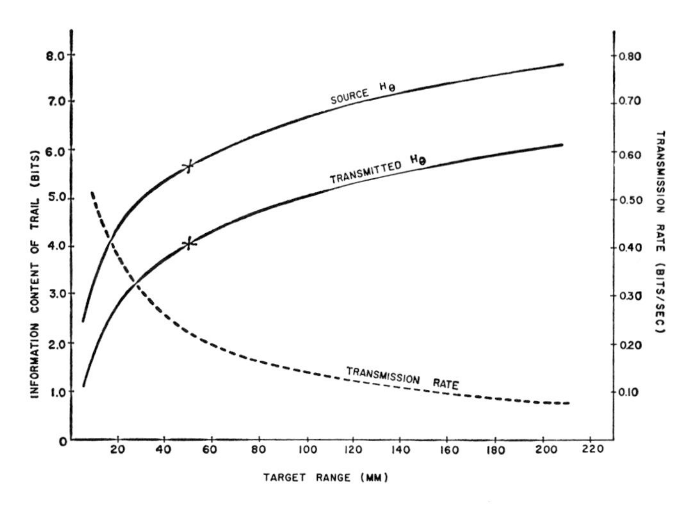

FIGURE 13-5. Source entropy (initial error in the signal), information transmitted, and information transmission rates of fire ant odor trails with reference to direction of the target. The curves are extrapolated from data of the kind presented in Figure 13-4.

(personal communication) has pointed out that, with the exception of one experiment conducted in 1949, von Frisch and his associates always captured the bees landing on the target dishes without counting them. In most cases, therefore, the performances of the "errorless" bees were not entered into the calculations of the standard deviations, and my 1962 estimates of H are too low. A reasonable adjustment can be made. From the 1949 data, together with comparisons made with the proportion of erring versus nonerring bees in direction experiments, it seems very unlikely that the errorless bees were more than three times as numerous as the bees in error. A likely upper limit for transmitted distance information is therefore 2.3–4.3 bits, with 3 bits (instead of 2 bits as shown) being a "typical" intermediate value. Only new data can establish the true value with any confidence.

There is at present no way to assess the contribution of pheromones to the transmitted information in the honeybee. But at least it can be said that the information combined from the waggle dance and odor is comparable to that transmitted by the fire ant odor trail. It is useful to ask how it came about that two such radically different systems have been shaped in evolution to similar levels of accuracy. We may have here yet another instance of the optimal value of a trait differing from its potential maximum. It is quite possible that both the fire ant and

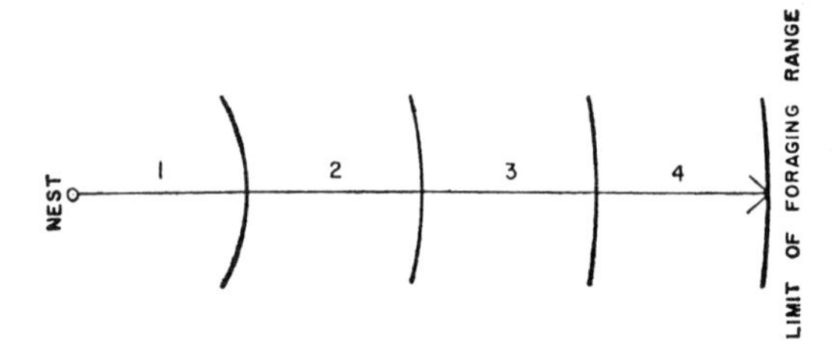

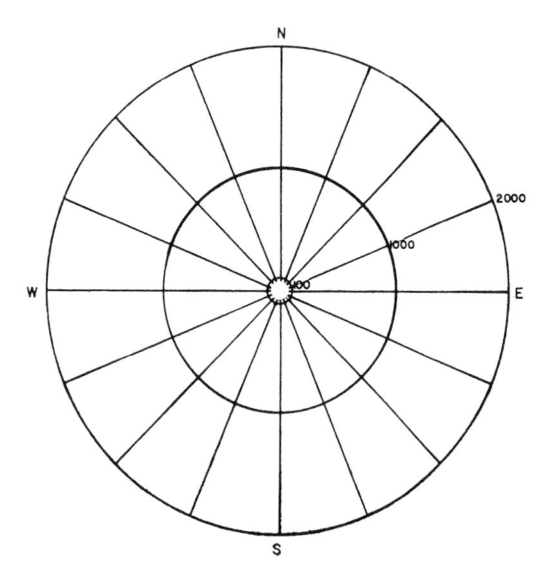

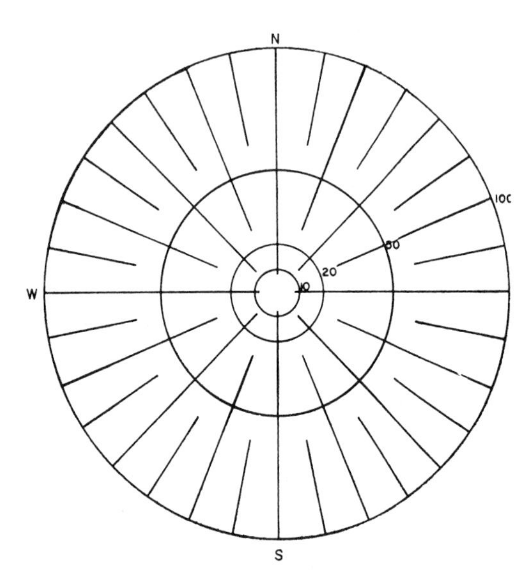

the honeybee are near the optimal level of accuracy of the mass response. As Haldane and Spurway suggested with reference to the honeybee, "natural selection is always acting so as to reduce the error of the mean direction, while acting less intensely, if at all, to reduce individual errors which lead to a scatter round this direction." They cite the analogous problem in naval gunnery, "where a superior force pursuing ships with less fire power should fire salvoes with a considerable scatter, in the hope that at least one shell will hit a hostile ship and slow it down." This same analogy appears at first glance even more relevant in the case of the fire ant. One of the chief problems faced by this species in the course of food gathering is the recruitment of sufficient workers in time to immobilize small prey animals detected while passing through the colony territory. On many occasions I have seen recruited workers succeed in capturing insects on laboratory foraging tables only because they had deviated from odor trails that had been rendered inaccurate by the continued locomotion of the prey.

Even so, there is an even simpler, more plausible explanation. It is possible that the level of accuracy of the fire ant has been arrived at as a compromise between the utmost effort of the ants' chemosensory apparatus to follow trails accurately and, simultaneously, the need to reduce the quantity and to increase the volatility of the trail substance in order to minimize overcompensation in the mass response. The amount of trail substance is indeed

FIGURE 13-6. These abstract figures represent the amounts of information transmitted by the honeybee (race *carnica*) around the time it performs the waggle dance, and by the fire ant when it lays an odor trail. (*center*) The "bee compass" indicates that the worker honeybee receives up to four bits of information with respect to distance, or the equivalent of acquiring information necessary to allow it to pinpoint a target within one of sixteen equiprobable angular sectors. The compass lines are represented arbitrarily as bisecting the sectors. The amount of direction information remains independent of distance, given here in meters. This last estimate is probably subject to revision, as explained in the text. (*top*) The "distance scale" of both bee and fire ant communication shows that approximately two bits are transmitted, providing sufficient information for the worker to pinpoint a target within one of four equal concentric divisions between the nest and the maximum distance over which a single message can apply. (*bottom*) The "fire ant compass" shows approximately how direction information increases with distance, given here in millimeters (from Wilson, 1962a).

microscopic, which seems logical if the mass response is to be governed finely. Which of the two evolutionary hypotheses is closer to the truth cannot be determined until deeper physiological analyses have been made.

The trail pheromones, while they tend to be species-specific in their effectiveness, are not precisely so. A fair generalization from our present knowledge would be that trail pheromones are usually peculiar to genera or to groups of species within genera, and only occasionally to species. For example, the trails of *Solenopsis saevissima* are fully attractive to the closely related *S. xyloni* but not to the more distantly related *S. geminata. S. xyloni* trails are fully attractive to *geminata,* but *geminata* trails are not attractive to either *xyloni* or *saevissima* (Wilson, 1962a). The *Solenopsis* trails are not active on the species of *Crematogaster, Xenomyrmex,* and several dolichoderine genera tested. Similar results have been obtained by Wilson and Pavan (1959) for several dolichoderine genera, by Blum (1966a) for *Cardiocondyla, Huberia,* and *Monomorium,* and by Margaret Dix (personal communication) for the genera of the fungus-growing ants of the tribe Attini. Watkins (1964) found no evidence of pheromone differences within the army ant genus *Neivamyrmex.* Differences among the army ant genera have not, to my knowledge, been tested.

There are a very few cases in which ants utilize the odor trails of other species. The parabiotic ant species of the tropical American forests follow one another's trails (see Chapter 19). In Europe, workers of *Camponotus lateralis* sometimes follow the trails of *Crematogaster scutellaris* in large numbers to the *Crematogaster* feeding grounds and exploit the same food resources. In Trinidad, West Indies, workers of *Camponotus beebei* regularly if not invariably utilize the odor trails of a locally dominant dolichoderine species, *Azteca chartifex.* The *Camponotus* "borrow" the *Azteca* trails during the day, when *Azteca* foraging is at a low ebb. The *Camponotus* are treated as enemies by the *Azteca,* but they are too swift and agile to be caught (Wilson, 1965b).

By means of the artificial trail technique, a respectable amount of information has now been accumulated concerning the anatomical origin of the trail substances in various ant groups. From the data summarized in Table 13-1, it can be seen that there are major differences among the principal groups of ants and that these differences are at least roughly consistent with the current theory of ant phylogeny expressed in Chapter 4. The data also strongly indicate that odor trail communication has originated a number of times independently within the ants. It is further true that many genera and species of ants do not lay odor trails. None of the Myrmeciinae, the most primitive

Table 13-1. Glandular sources of trail substances in ant species.

Species	Glandular source	Authority
Ponerinae		
Termitopone laevigata	Hind gut	Blum (1966a)
Dorylinae		
Neivamyrmex spp.	Hind gut	Watkins (1964); Watkins *et al.* (1967)
Eciton hamatum	Hind gut	Blum and Portocarrero (1964)
Myrmicinae		
Atta texana	Poison gland	Moser and Blum (1963)
A. cephalotes	Poison gland	Blum *et al.* (1964)
Acromyrmex octospinosus	Poison gland	Blum *et al.* (1964)
Crematogaster ashmeadi	Tarsal glands	Leuthold (1968a, b)
C. peringueyi	Tarsal glands	Fletcher and Brand (1968)
Pheidole fallax	Dufour's gland	Law *et al.* (1965)
Solenopsis saevissima	Dufour's gland	Wilson (1959c)
S. geminata	Dufour's gland	Wilson (1962a)
S. xyloni	Dufour's gland	Wilson (1962a)
Monomorium floricola	Poison gland	Blum (1966a)
M. minimum	Poison gland	Blum (1966a)
Huberia striata	Poison gland	Blum (1966a)
Tetramorium caespitum	Poison gland	Blum and Ross (1965)
Dolichoderinae		
Monacis bispinosa	Pavan's gland	Wilson and Pavan (1959)
Iridomyrmex humilis	Pavan's gland	Wilson and Pavan (1959)
I. pruinosum	Pavan's gland	Wilson and Pavan (1959)
Formicinae		
Lasius fuliginosus	Hind gut	Carthy (1950); Hangartner and Bernstein (1964)
Acanthomyops interjectus	Hind gut	Hangartner (1969c)
Myrmelachista ramulorum	Hind gut	Blum and Wilson (1964)
Paratrechina longicornis	Hind gut	Blum and Wilson (1964)

myrmecioid subfamily, do so; among the Ponerinae, the most primitive poneroid subfamily, only a few specialized legionary groups, such as *Termitopone* and *Leptogenys,* do so. All of the Dorylinae and Cerapachyinae lay trails so far as is known. In the higher subfamilies the habit is irregularly distributed. The great majority, perhaps all, of the Dolichoderinae employ trails. An undetermined fraction of species, perhaps a minority, of the Pseudomyrmecinae, Myrmicinae, and Formicinae also employ trails. It appears that the habit is readily acquired or lost during the course of evolution at the level of the genus or species. Species of *Monomorium* generally lay trails, but the related *Chelaner antarcticum* does not (Blum, 1966a). Some species of *Formica* and *Lasius* employ elaborate trail systems, while others do not lay trails of any kind. In general, species with large colony size have a greater tendency to evolve trail communication (Wilson, 1961).

Finally, it is useful to distinguish between "exploratory trails" and "recruitment trails" (Wilson, 1963a). The former are found in legionary (army ant) species among the Ponerinae and Dorylinae and are laid more or less continuously by exploring workers in the van. The contribution of any given worker appears to be relatively slight, but the accumulated contribution of the foraging columns is heavy enough to last for days or even weeks (Schneirla, 1965; Wilson, 1958b). So far, all of the exploratory trail substances studied have proven to originate in the hind gut. In contrast, recruitment trails are laid only by workers that are returning to the nest following the discovery of food, of lost nestmates, or of superior nest sites. They are either ephemeral or built up by accretion into persistent "trunk routes" (Sudd, 1957, 1960). Recruitment trails are characteristic of the trail-laying Myrmicinae, Dolichoderinae, and Formicinae. In most cases analyzed to date (Table 13-1), the recruitment trail substances have turned out to be strong attractants. An exception is the tarsal gland substance of *Crematogaster ashmeadi,* which appears to be a weak attractant functioning more to orient than to impel (R. Leuthold, 1968a,b). To date no trail substances in ants have been chemically identified.

Odor Trails in Termites

Those who live in the North Temperate Zone are accustomed to thinking of termites as white, timorous insects that stay forever sealed in their homes of crumbling wood. In fact most of the termites that live in the tropics—and by species count they constitute the great bulk of the world fauna—live in the soil; of these, a large percentage regularly leave their nests to forage in the open air. The "harvesting termites" in the Hodotermitidae and Termitidae gather humus, leaf litter, and lichen, in other words most of the forms of cellulose-containing debris available on the ground and on tree trunks. Their foraging columns sometimes provide a startling spectacle. Edouard Bugnion (in Forel, 1928) described nocturnal forays of several colonies of *Hospitalitermes monoceros,* the Black Termite of Ceylon. The columns set forth each day as the sky began to darken, as early as 4:30 o'clock when the sky was overcast but otherwise around 6:00 o'clock, and it took about five hours for the entire army to make the march. Bugnion's colonies visited cacao trees 15–20 m from the nest to collect lichen, and soil nearby to collect humus. The return took place during four to five hours around dawn and was completed only around 7:00 o'clock, well after the sun was up. The foraging armies were immense. The columns consisted chiefly of 10 or so workers marching abreast, flanked at regular intervals by nasute soldiers whose heads faced outward, and stretching all the way from the nest to the hunting ground. An average of approximately 1,000 termites were counted passing a given point every minute, and from this figure Bugnion estimated the total army to contain 300,000 individuals.

Bugnion, like most of the other early termitologists, believed that such foraging columns are guided by chemical trails laid down in fecal matter. The workers of *H. monoceros,* he stated, "leave behind them little black, elongated tracks (which are very distinct on a white road, for instance, or a wall), and apparently serve to guide the army over its usual trail. As the intestines are filled with a black substance . . . we may assume that the black marks are produced by a substance expelled from the rectum." Experimental studies performed over the past ten years have shown that these phylogenetically advanced termites are indeed guided by odor trails, but that the pheromone is produced by the sternal gland of the abdomen (Stuart, 1969).

In 1960 Lüscher and Müller in Switzerland and Stuart in the United States independently discovered that the primitive termites also lay odor trails, even though these insects never leave their nest to forage in the open air. Nymphs of *Zootermopsis nevadensis* guide other nymphs through the rotten wood galleries of the nest by means of substances streaked from the sternal gland. Subsequently Stuart (1963a, 1963b, 1964, 1967) elucidated the function of the odor trail throughout the termites. In *Zootermopsis* and probably other primitive forms as well, trails are used to recruit nestmates to breaches in the wall of the nest. Virtually all dangerous situations in the life of the colony can be translated to this single proximate stimulus—a breach in the wall. Termite nymphs are extremely sensitive to the increased light intensities and to microcurrents of air associated with such an event, and when thus disturbed they run back into the interior of the nest laying an odor trail behind them (Figure 13-7). The pheromone is an attractant that operates in much the same fashion as the fire ant pheromone. That is, it "compels" the outward march of nymphs encountering it, and it is adequate in itself to guide them to their destination. When recruited nymphs arrive at the damaged portion of the nest, they set about to repair it. If the breach is too extensive to be repaired at once, the newcomers remain in an alarmed state and lay trails of their own back into the interior of the nest. In this fashion a repair crew is built up in numbers sufficient for the work to be done. Once the repair is completed, alarm ceases, trails are no longer laid, and the activity dies out.

In the higher termites the pheromone has acquired a second function. Now the scent is laid to food sources as well as to damaged portions of the nest, and a recruitment system almost identical to that of the ants prevails. This similarity is the outcome of a remarkable case of evolutionary convergence. In termites, recruitment to food sources has evolved as a secondary function from an elaborate alarm system, which in turn is intimately connected with nest building. In ants, the same behavioral pattern has evidently been evolved from tandem running. In higher termites, moreover, the alarm function has been taken over in part by volatile substances produced in the head.

The sternal gland is found throughout the termites, and its location in the termite abdomen has changed in evolution in a way that is consistent with accepted ideas of phylogeny (Noirot and Noirot-Timothée, 1965a,b). In the very primitive genus *Mastotermes*, there are three of the glands—located under abdominal sternites III, IV, and V, respectively. In the subfamilies Stolotermitinae, Porotermitinae, and Hodotermitinae, there is a single gland beneath sternite IV, and in all other termites so far studied there is a single gland under sternite V. The sternal glands

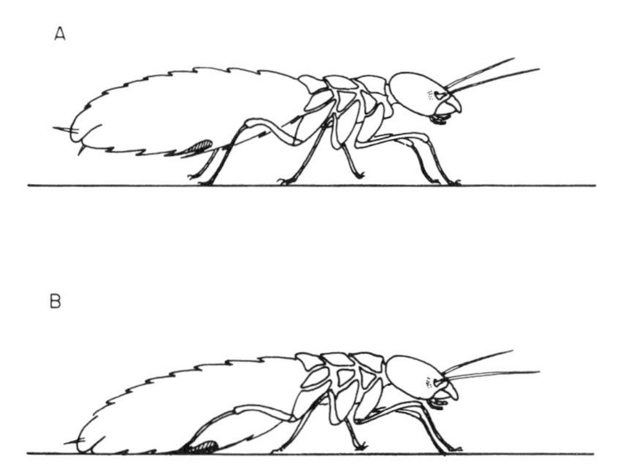

FIGURE 13-7. The nymph of *Zootermopsis nevadensis*, on being alarmed, lays odor trails into the interior of the nest. The location of the gland beneath abdominal sternite V is indicated by cross hatching in this individual, which is shown in both resting (*A*) and trail-laying (*B*) postures (from Stuart, 1969).

are present in every caste, although they regress in functional reproductives. Pasteels (1965) found that in *Nasutitermes lujae* the glands are best developed in older workers which, together with the soldiers, do most of the foraging and trail laying.

Electron microscopic studies of the sternal gland of several species have been conducted by Noirot and Noirot-Timothée (1965a) and Stuart and Satir (1968). The gland is unusual in that it lacks a reservoir and duct, so that the pheromone has to be transported directly from the gland cells out through the cuticle. In 1966 Moore succeeded in purifying the trail substance of the Australian species *Nasutitermes exitiosus*. He characterized it as an unsaturated, relatively involatile diterpenoid hydrocarbon with the empirical formula $C_{20}H_{32}$. The substance is very potent, causing workers to follow it when laid in artificial trails in concentrations of only 10^{-8} to 10^{-5} g/ml. An apparently identical substance was obtained from two other Australian *Nasutitermes*, but *Coptotermes lacteus* was found to produce something different. This fragment of evidence suggests that specificity may be at the genus but not the species level, much as in the ants. Hummel and Karlson (1968) reported a nonterpenoid hydrocarbon, with the formula $C_{11}H_{20}$, to be an active component of the trail substance of *Zootermopsis nevadensis*. They have also reported that hexanoic acid (*n*-caproic acid) is a very active natural component. One mg contains 50,000 "trail units," where a unit is defined as the amount sufficient to induce following by at least three out of ten *Zootermopsis* nymphs along an artificial trail 10 cm in length. Curiously, Karlson, Lüscher, and Hummel (1968) have found that farnesol, which is not a natural trail substance, is active on *Zootermopsis* at just this same level. A large variety of other substances tested, however, were far less potent, in most cases by several orders of magnitude.

Finally, Matsumura, Coppel, and Tai (1968) have reported the complete structure of the trail pheromone in the North American rhinotermitid *Reticulitermes virginicus.* The substance, which occurs in rotting wood also and thus is believed to serve as a beacon for the workers hunting food, is one or more geometric isomers of a dodecatrienol. The most potent isomer thus far synthesized is *cis*-3,*cis*-6,*trans*-8, whose complete structure is as follows:

$$CH_3\text{-}CH_2\text{-}CH_2\text{-}\underset{trans}{CH{=}CH}\text{-}\underset{cis}{CH{=}CH}\text{-}CH_2\text{-}\underset{cis}{CH{=}CH}\text{-}CH_2\text{-}CH_2\text{-}OH$$

This compound is as fully effective as might be expected of a true trail pheromone. When less than 10^{-12} g dissolved in 10 μl of solvent was streaked over a glass surface along a 10 cm path, it was consistently followed by *Reticulitermes* workers.

Odor Trails in Bees

When Lindauer and Kerr (1958, 1960) made a routine check of the ability of eleven species of stingless bees (Meliponini) in South America to communicate the location of food finds, they obtained some striking results. The workers of all of the species were able to alert nestmates to the presence of food. Those of seven of the species, in *Melipona* and *Trigona,* showed no ability to direct their sisters to the target, but those of four of the larger *Trigona* species (*T. capitata, T. mombuca, T. postica, T. ruficrus*) were able somehow to transmit directional information as rapidly and efficiently as honeybees. Further experiments with *T. postica* revealed one form of the communication. When a forager bee finds a feeding table, she makes three or more normal collecting flights straight back and forth between the hive and the table. Then she begins to stop in her homeward flight every two or three meters, settling onto a blade of grass, a pebble, or a clump of earth, opening her mandibles, and depositing a droplet of secretion from her mandibular glands (see Figure 13-8). Other bees now leave the nest and begin to follow the odor trail.

Nedel (1960) subsequently found that the mandibular glands of the trail-laying Trigonas are greatly enlarged in comparison with those of other bee species. Furthermore, after being emptied, the gland reservoirs are refilled in as little time as 20 minutes. It is clear that larger quantities of trail pheromones are used by these insects than by the trail-laying ants and termites. In fact, human observers can smell the typical mandibular gland odor around odor spots deposited by workers in nature. According to Kerr, Ferreira, and Simões de Mattos (1963) the overall *Trigona* trails are polarized; that is, larger quantities of scent are laid down nearer the food source, as shown in the example in Figure 13-8. In three species studied by these investigators in Brazil, the active spaces of the odor spots endured between 9 and 14 minutes. Much as in the ants and possibly the termites as well, closely related species within *Trigona* confuse each others' trails, but distantly related species do not. The alerting stimulus in *Trigona* commu-

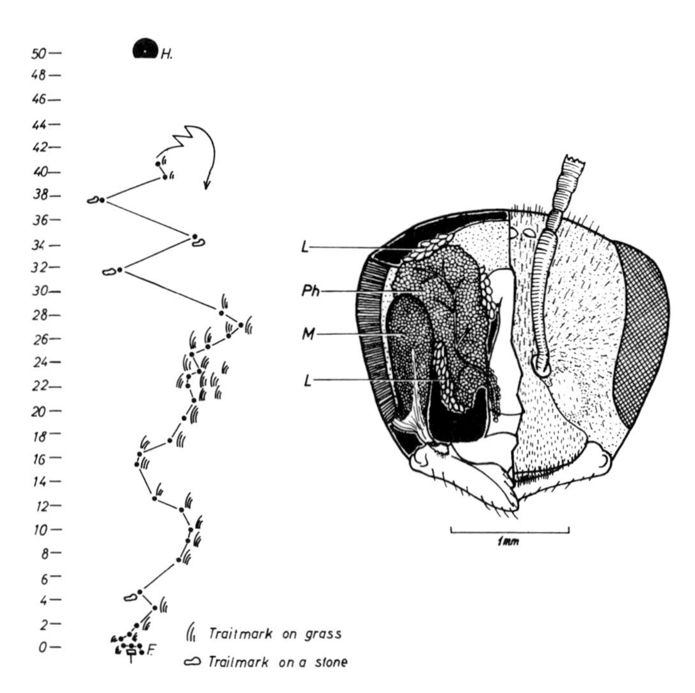

FIGURE 13-8. (*left*) A typical trail-marking flight of a worker of the stingless bee *Trigona postica* between the feeding table (*F*) and hive (*H*). The dots indicate the points where the bee stopped and deposited a droplet of mandibular gland secretion on a blade of grass or pebble. (*right*) The head of a *Trigona postica* with the right half of the anterior capsule cut away to show the principal exocrine glands, including the enlarged mandibular glands: (*L*) labial glands; (*M*) mandibular glands; (*Ph*) hypopharyngeal glands (from Lindauer, 1961).

nication, that is, the stimulus that arouses other workers before they move out along the odor trail, is believed by Kerr, Esch (1965, 1967a,b), and their co-workers to be a buzzing sound made by successful foragers shortly after returning to the nest. According to Esch, the length of a particular pulse increases with the distance of the journey in a highly predictable way. In fact the curve relating pulse length to distance traveled is remarkably similar in shape and slope to the curve relating the duration of the individual waggle dance to distance traveled in the honeybee (*Apis mellifera*). Similar sounds are made by successful foragers in meliponine species that do not lay trails and hence are unable to give directional information.

Most species of stingless bees nest and forage in tropical forests, and odor trail communication seems ideally suited for orientation in this habitat. The individual forager bee can best thread its way through tree trunks and understory vegetation if guided point by point by frequently repeated cues. Odor trails offer an advantage over waggle dances in an important additional aspect. The waggle dance of the honeybee transmits information in only two dimensions, that is, they indicate the location of a point on a plane. The odor trail, which can lead up and down tree trunks as well as over the ground, transmits three-dimensional information, which is more useful in forests.

The honeybee has a rudimentary form of odor trail which is employed for short-distance orientation in special circumstances (Lecomte, 1956; Butler, 1966b; Butler *et al.*, 1969). The most suggestive evidence has been reported by Butler in his initial article on the subject. When he changed the position of the hive entrance of one of his colonies, the first foragers to return managed to find the new entrance only after random searching. When a few had succeeded in walking the whole distance from the former location of the old entrance to the new entrance, however, newcomers were able to follow their trail on foot. The "footprint pheromone" has not yet been identified, but it appears to be the same substance that is routinely deposited on and around the nests and food dishes and makes them both attractive. As mentioned earlier in this chapter, the most probable source, as yet untested, is Arnhardt's glands in the tarsi.

Adult Transport

One of the methods of recruitment used by ants is extremely direct and simple: workers are merely dragged or carried to the target area by their nestmates. The behavior is almost wholly limited to emigrations to new nest sites. It is generally true among ant species that, if a colony is forced to move from one nest site to another, the most active and experienced workers carry the brood and, especially during the later stages of the emigration, pull or bodily carry other adult members to the new nest site. The behavior can be readily induced in the laboratory simply by spilling an entire colony into the foraging chamber next to the artificial nest and allowing it to reoccupy the nest. When the colony occupies multiple nest sites, adult transport sometimes occurs continuously from one site to another. Økland (1934), in the course of his studies of the European *Formica rufa*, was the first to realize that this phenomenon can be an important means of colony integration. In the closely related *F. polyctena*, adult transport among multiple nest sites is seasonal, reaching its maximum in spring and autumn. In one colony of approximately a million workers studied by Kneitz (1964) in Germany, between 200,000 and 300,000 transportations occurred in the course of a year. Most of the workers doing the transporting were older foragers, and most of the workers being carried were younger individuals, of the kind that engage principally in nursing and ingluvial food storage. The adaptive value of such strenuous behavior remains to be demonstrated. It is conceivable that the transporters are engaged in allocating labor resources according to overall colony needs which, by virtue of these insects' more extensive wandering, they are the most qualified to sense.

The value of simple emigration from a bad nest site to a good one is clear enough, and the function can be said to be basic and primitive in the ants. In the higher ants, adult transport has evolved into an elaborate, stereotyped form of communication. In the Formicinae, the transporter approaches the transportee face to face, antennates it rapidly on the surface of the head, and attempts to seize it by the mandibles. If the transportee is receptive it folds its antennae and legs in against the body in the pupal position and allows itself to be lifted from the ground. As it is pulled up, it curls its abdomen forward. The transporter then swiftly carries it to the destination. Arnoldi (1932) was the first to recognize consistent differences among the ant groups in the postures assumed during adult transport. In contrast to the Formicinae, the transporting worker in most Myrmicinae (*e.g., Myrmica, Cardiocondyla*) seizes the transportee just beneath the mandibles or by the neck, and the transportee curls its body over the head of the transporter with its abdomen pointing upward or to the rear (Figure 13-9). In *Crematogaster,* seizure is from above by the postpetiole. In the New World army ants (Dorylinae), adults are carried in the same strange fashion as the larvae and pupae, that is, slung back beneath the body and between the long legs of the workers. According to Rettenmeyer (1963a) the frequency with which adult workers are transported decreases with their size and age and increases near the end of the emigration from one bivouac site to another. In the Australian *Myrmecia,* among the most primitive living members of the myrmecioid phylad, adult transport occurs, but is not stereotyped. It is a rare response, directed only at aged and ailing individuals, callow workers, nest queens, and males. The transporting worker faces the transportee, seizes it by the mandibles or (in the case of males) the antennae, and drags it over the ground while walking backward. The transported individual does not fold its appendages in the pupal position or cooperate actively in any other way (Haskins and Haskins, 1950a). It is evident from these fragments of information that adult transport has evolved considerably within the ants and shows at least a rough correspondence with the principal taxonomic groupings. The behavior pattern is nevertheless not found in every ant species. It is absent, for example, in the arboreal myrmicine *Daceton armigerum* (Wilson, 1962b).

The basic transporting behavior has been adapted to new ends by a few ant species. In *Manica rubida* and *Leptothorax acervorum* it is used to remove alien workers from the colony territory (Le Masne, 1965; Dobrzański, 1966). Interestingly enough, the subdued aliens respond with the same submissive behavior as nestmates. This is the case to a limited extent even when the transportee is a different species, as, in particular, when *Myrmica scabrinodis* workers are carried by *Manica* workers. In the Russian slave-making formicine *Rossomyrmex proformicarum* adult transport is used in recruitment during raids (Arnoldi, 1932). The workers travel to the target nest of *Proformica* in pairs, one ant carrying the other in typical formicine fashion.

In termites the transport of nestmates other than eggs is a rare event. It does occur in at least a few higher termite species on those occasions when colonies or fragments of colonies emigrate from one nest site to another. During colony multiplication by fission ("sociotomy"), workers of African species of *Anoplotermes* and *Trinervitermes* run over the surface of the ground between the home nest and the daughter nest. During this precarious time, they carry the young larvae in their mandibles. Older larvae are sometimes carried, but usually walk, to the new nest site. In *Trinervitermes* the injured are also carried (Grassé and Noirot, 1951b).

I know of no examples of adult or even brood transport in the winged social Hymenoptera. This omission seems to be explainable by the simple fact that older larvae, pupae, and adult nestmates are too heavy to be carried very far in flight. When a colony of honeybees, stingless bees, or polistine wasps emigrates, in the course of either absconding or colony multiplication, the brood is left behind, and the new nest is peopled entirely by queens and workers who travel under their own power.

The Dances of the Honeybee

The waggle dance of the honeybee, called the *Schwänzeltanz* in German and tail-wagging dance in some English-language reports, is the most intensively studied and famous of all the forms of animal behavior. Its fame stems both from the uniqueness of the mode of communication involved and the thoroughness and craftsmanship of the work of Karl von Frisch and his students who have devoted their lives to this and related aspects of honeybee behavior. The waggle dance is unique in an especially notable way. Here we have a signal that is constructed from a ritualized and miniaturized imitation of the journey that the signaling bee has taken in the past and upon which some of its sister bees are about to embark. By following the dance, the receiving bees rehearse the journey in miniature and prepare to translate it into a real flight. When they do make the flight, it can be said that they were sent and not led to the goal. What is different about the waggle dance, then, is that it is a truly symbolical message that guides a complex response after the message has been given. In almost all other known forms of animal communication, individual signals contain much less information than a single waggle dance and, unlike the dance, are effective only while in existence. Even in most acts of chemical communication the behavioral effects cease very soon after the active pheromonal spaces disappear.

Let us examine a typical instance of communication by the waggle dance in *Apis mellifera carnica,* the gray Carniolan race studied by von Frisch in the fields of Germany and Austria. A scout worker has returned to the

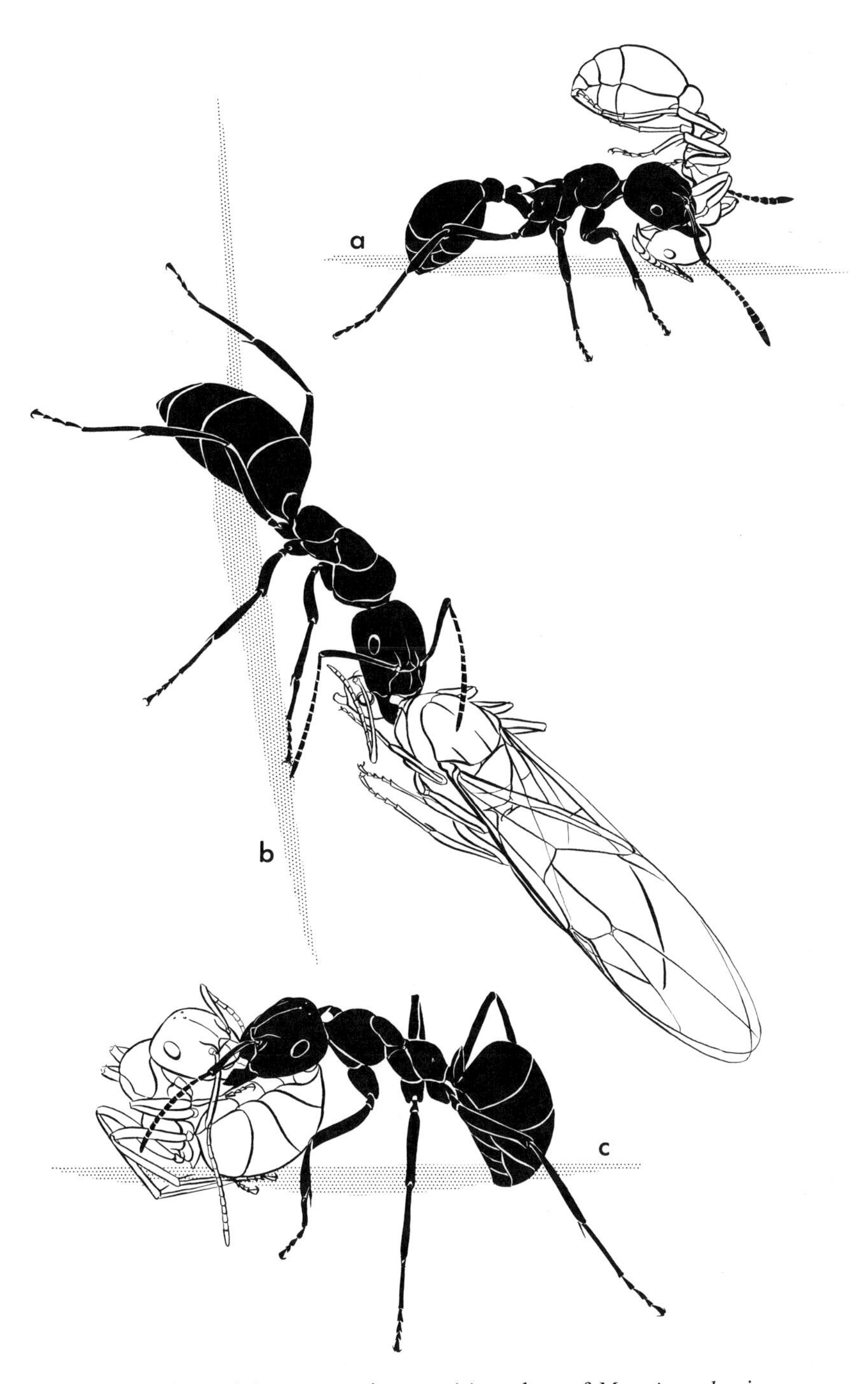

FIGURE 13-9. Adult transport in ants: (*a*) workers of *Myrmica rubra* in one of the carrying postures most frequently seen in the Myrmicinae; (*b*) a worker of *Camponotus ligniperda* carries a male, this form of transport being used only to handle males; (*c*) a worker of *Formica polyctena* carries another in the posture typifying the formicine ants (original drawing by Turid Hölldobler; *Camponotus* based on B. Hölldobler and Maschwitz, 1965).

hive after discovering a rich new food source located a moderate distance, say 300 m, from the hive. The food source will, in most cases, be a cluster of flowers bearing nectar and pollen, the normal food of the honeybee. Suppose, for purposes of illustration, that our food find is located in a direction 20° to the right of the sun with reference to the hive; in other words, if a bee flies straight toward the food on leaving the hive, its path will form a 20° angle with the line running from the hive over the ground in the direction of the sun (Figure 13-10). Now the scout worker enters the hive, mounts one of the vertical combs, and crawls to a position that is determined in part by the distance from the food find to the hive. The farther away the find, the deeper into the nest she penetrates. The bee next regurgitates nectar from her crop to the surrounding nestmates and then begins the dance in the midst of her crowded sisters. She runs through a figure-eight pattern: a straight run, then a turn to the left and circle back to about the original position, another straight run, a turn and circle to the right to the starting point, a straight run, and so on.

The straight run is the most distinctive and informative element of the dance. It is given a particular emphasis, at least to the eye of the human observer, by a rapid lateral vibration of the body—the "waggle"—that is greatest at the tip of the abdomen and least marked at the head. The complete back-and-forth shake of the body is performed 13 to 15 times a second. At the same time the bee emits an audible buzzing sound by vibrating its wings. The whole performance conforms to a pattern that students of vertebrate behavior would probably refer to as a ritualized intention flight movement. Each episode of wing vibration lasts about 15 msec and is separated from the next one by the same interval. Thus 30 vibrations per second ensue, and it is this low-pitched frequency that the human ear interprets as a buzz. The vibration episodes themselves contain oscillations of 250 cycles per second, corresponding to the frequency of the wingbeats. Some evidence exists (for example, Esch, 1967a) to indicate that the sounds are an essential part of the waggle dance message.

The direction of the straight run on the vertical comb and its duration are closely correlated with the direction and distance, respectively, of the food find with reference to the hive. In the case represented in Figure 13-10 the food find is located about 20° to the right of the sun. The straight run of the dance is correspondingly directed at

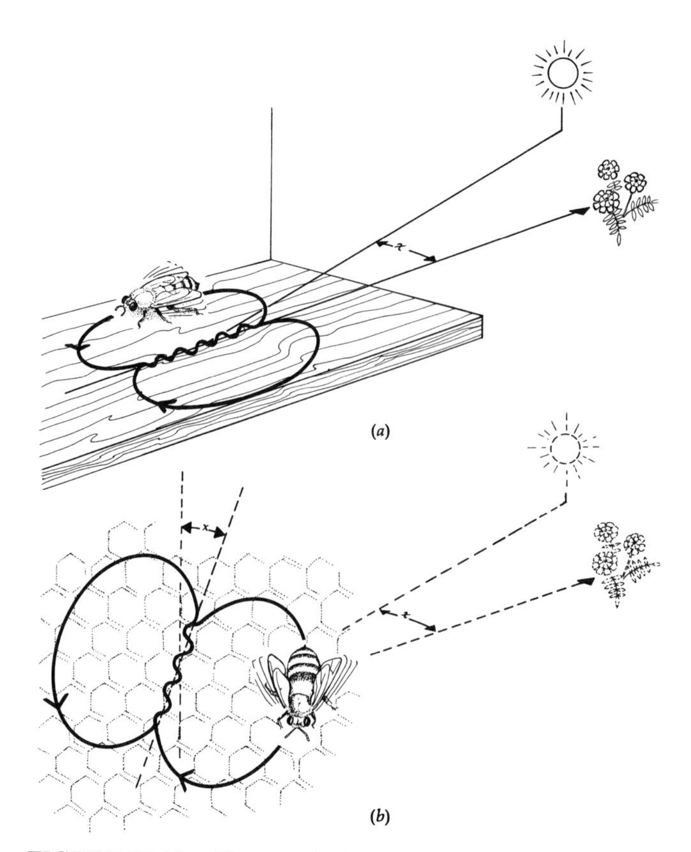

FIGURE 13-10. The waggle dance of the honeybee. As the bee passes through the straight run she vibrates ("waggles") her body laterally, with the greatest movement occurring in the tip of the abdomen and the least in the head. At the conclusion of the straight run, she circles back to about the starting position, as a rule alternately to the left and right. The follower bees acquire the information about the food find during the straight run. In this case the run indicates a food find 20° to the right of the sun as the bee leaves the nest. If the bee performs the dance outside the hive (*a*), the straight run of the dance points directly toward the food source. If she performs the dance inside the hive (*b*), she orients herself by gravity, and the point directly overhead takes the place of gravity. The angle x (= 20°) is the same for both dances (from Curtis, 1968).

an angle of 20° to the right of the vertical. In other words, the bee transposes the angle between the food and the sun (seen in the outward flight) to the angle between the straight run and the vertical (judged by gravitation at the starting point of the straight run). Simultaneously, the duration of the straight run increases with the length of the journey. A precise exemplification of this correlation is given in Figure 13-11. In timing the straight run, the bee actually refers to the outward journey that she has taken. Typically, the scout will complete several round trips by herself before she begins dancing so that she has the opportunity to acquire an accurate impression of the location of the target. Furthermore, the duration of the run is based not on the absolute distance of the target but rather on the energy expended to get there. If the flight outward is favored by a following wind, the subsequent straight run performed in the hive will be shorter.

The follower bees crowd in around the scout, their antennae extended and touching her much of the time. Within minutes some begin to leave the hive and fly to the food find. Their searching is quite accurate; the great majority come to search close to the ground within twenty percent of the correct distance.

In the Carniolan race the waggle dance is consistently performed only when the target is more than 80 m from the hive. If it is less than 50 m away, the bee performs the round dance instead (Figure 13-12). The round dance is similar to the waggle dance except that it lacks the all-important straight run. Workers incited by it who have had no previous experience in the nest vicinity search at random close to the nest. Those who have foraged in the nest vicinity may recognize that the scent of the flowers adhering to the dancer's body and head was from the spot where they had previously encountered the same kind of flowers.

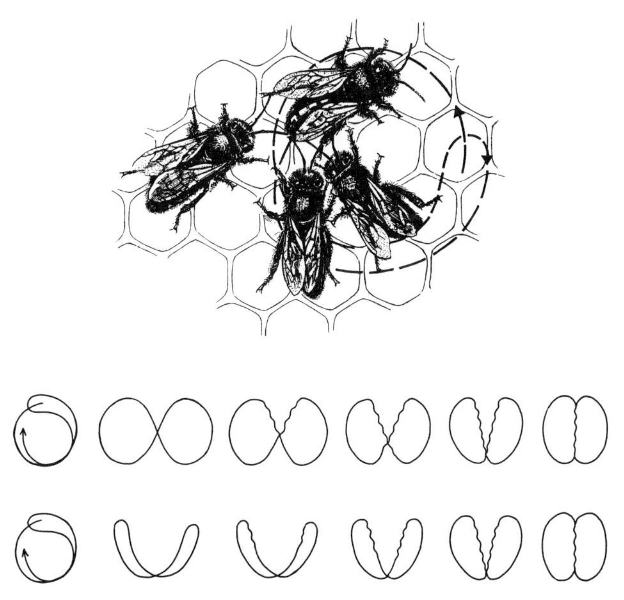

FIGURE 13-12. The round dance performed by the honeybee worker when the target is close to the hive. In the upper figure the dance is followed by three other bees, who are incited to search close to the nest but are given no information concerning direction. As the distance of the target is increased (beyond 50 m in the case of the Carniolan race) the round dance gradually changes to the waggle dance by adding a straight run in the middle. The intermediate form in the lower row is called a sickle dance (from von Frisch, 1967a).

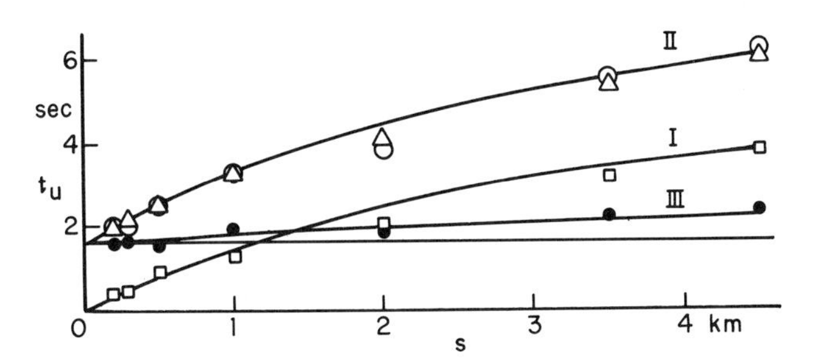

FIGURE 13-11. The durations of elements of the waggle dance of the Carniolan race, and their sum, are given as functions of the distance of the food find from the hive. The duration in seconds is indicated in the ordinate as t_u. The duration of the individual straight run (I) is the component that increases most with the distance of the target. The duration of the return run (III) increases only slightly. Consequently, the duration of the entire dance (II), that is, the sum of I and III, increases only about as fast as that of the straight run. The open squares, triangles, and circles represent different methods by which the data were obtained (redrawn from von Frisch, 1967a).

In trying to keep all of these facts in mind, it is useful to imagine oneself in the position of a worker bee just returning to the hive after making a food discovery. The problem is to perform a charade of the trip out to the food. The straight run has been "chosen" in evolution to represent this activity. It is the symbolic flight, with wings vibrating in nonfunctional manner and the abdomen also vibrating to add emphasis. Gravity must replace the position of the sun since the bee is now in the darkened interior of the hive; straight up is a convenient substitute for moving toward the sun itself. Statistical analyses have shown that the duration of the straight run (or the duration of the wing buzzing, which is virtually the same thing) is the component of the dance most closely correlated with the distance to the target or, more precisely, the effort expended to get there. The circling return run evidently serves the chief and perhaps exclusive purpose of bringing the dancing bee back to the start of the straight run.

If the comb is tilted to a horizontal alignment and the hive is opened enough to admit sunlight, or at least a view of open sky, the dancing bees now direct their straight runs right at the target outside. In *Apis florea,* the socially more primitive dwarf honeybee of southeastern Asia, dancing is performed exclusively in this fashion. The combs are permanently exposed to the outside and are surmounted by a broadened area that serves as a dancing platform (Lindauer, 1961). The domestic honeybee *Apis mellifera* originated from populations that adapted to preformed cavities, such as natural hollows in logs and trees. In thus changing its nest site, which in turn made it amenable to domestication in artificial hives and allowed it to penetrate temperate climates, *A. mellifera* had to switch to gravity cues in order to direct the straight run in total darkness.

The evolutionary origin of the waggle dance and its incorporation of gravity cues are not really so mysterious as they might at first seem. Ritualization of functional motor patterns is commonplace in communication systems of other animal species. In fact, almost all signaling among members of the same species consists of a reenactment of the desired response, a "follow me" invitation often in the form of an incomplete intention movement or ritualized imitation. Esch (1967a,b) has discovered a communication form in the stingless bee *Melipona quadrifasciata* that not only illustrates this principle in graphic manner but also reveals what may be the equivalent of the intermediate stage in evolution of the waggle dance of *Apis mellifera.* Successful foragers of *M. quadrifasciata* lead their hivemates away from the nest by means of a conspicuous zigzag flight in the direction of the feeding place. The hivemates lose contact after 30–50 m and return to the nest. However, after 20 to 30 "guiding flights" the hivemates fly on beyond in the generally correct direction and try to find the feeding place on their own. It is easy to imagine as a next evolutionary step the abbreviation of the guiding flight and its encapsulation into the waggle dance of *Apis.* Esch has also pointed out that the duration of the sound pulses emitted by *M. quadrifasciata* and *M. merillae* are related to the distance to the feeding place in a manner very similar to the relation of duration of the waggle dance and distance to feeding place in the honeybee, *Apis mellifera.* In *Trigona,* sounds are also produced, but they contain no distance information. Esch has argued that sound production is the primitive mode of communication, serving at first (as in *Trigona*) only to alert the colony. Later the duration of the sound, and then the zigzag flights, came to imitate the journey to the feeding place, as in *M. quadrifasciata.* Finally, the zigzag flight, still intimately associated with meaningful sound production, became ritualized further into the waggle dances of *Apis.* Both the capacity for switching light compass and gravity orientation and some of the elements of the waggle dance itself occur even in a few nonsocial insects (Blest, 1960; von Frisch, 1967a).

No one disputes the precision of the waggle dance, which has been repeatedly measured, or the fact that it is associated with a rapid exodus and buildup of worker bees at the food site. But how can we be sure that the relationship is a causal one, that in fact the follower bees have been directed to the goal by information received within the hive from the waggle dance? A. M. Wenner and his associates (Wenner, Wells, and Rohlf, 1967; Wenner and Johnson, 1967; Wenner, Wells, and Johnson, 1969) have taken a contrary position, arguing that the communicative function of the waggle dance has not been proven. They believe that follower bees find their way in large part, if not exclusively, by past experience with the food find, which the bees recognize by odors clinging to the body of the dancer; by odor cues in the field such as secretions of the Nasanov gland; and by visual cues such as the movement of other forager bees and the geometric patterns of the food plates used in the experiments. Their data show that, under at least some circumstances, the foraging bees overwhelmingly favor odor cues to the information in waggle dances—to an extent that the very functioning of the dance can be doubted.

There is a widespread impression among biologists that the communicative function of the waggle dance has been disproved by both the critique and the new experimental results published by Wenner and his co-workers. Wenner, Wells, and Johnson (1969) have, in fact, flatly stated that "Honey bee recruits locate food sources by olfaction and not by use of distance and direction information contained in the recruitment dance." Clearly they desire to abandon the dance language hypothesis and pursue experiments to elucidate their own simpler explanation.

I think it is true to say that, on first hearing the critique, I was, like most other biologists, excited and intrigued. After all, if the classic waggle dance story can be discarded, honeybee communication would be stripped of its unique feature and insect behavior would become that much easier to comprehend. Parsimony alone seemed to favor some other explanation. Furthermore, the waggle dance had become something of a sacred cow and it needed a critical examination by an independent group of investigators. But now it seems to me that the main thrust of the critique was wrong. The communicative function is supported decisively by experimental evidence already in the literature. Furthermore, corroborative experiments have been reported by other investigators. The data from the experiments of Wenner and his associates on the role of learning and additional cues can be explained without affecting the von Frisch interpretation, and it now seems that even those were already anticipated by the much more extensive previous work of the German experimentalists.

In a separate brief article von Frisch (1967b) cited the following three lines of experimental evidence which in his view most firmly establish the communicative function of the waggle dance:

1. When the experimental food plate is 25 m or less from the hive, the foragers perform only the round dance, and the followers swarm out equally in all directions. If the plate is now moved farther away, the dance begins to add the straight run, and the followers fly with increasing accuracy toward the target. At 100 m, the waggle dance is performed exclusively, and, simultaneously, the followers come to orient with maximum accuracy.

2. If forced to dance on a horizontal surface, the scout bee still performs a waggle dance providing she is allowed to see the sun or a portion of blue sky. And the followers still swarm forth with the same accuracy as if the dance were performed in the dark on a vertical surface. If the view of the sky is now cut off, the scout performs disorganized dances that indicate neither distance nor direction. The follower bees are still incited to leave the nest, but they respond as though alerted by a round dance, searching randomly in all directions close to the hive.

3. In "detour experiments," scout bees are made to fly around the edge of a building or a high cliff by gradually moving the food plate on the ground, step by step, until the plate is positioned on the opposite side. The resulting waggle dances cannot indicate the detour. Instead, the bees show a remarkable ability to integrate their outward movements and to translate them into the correct, straight-line direction over the obstacle. Thus the scout flies around the detour, but its subsequent dance instructs the followers to go straight through (or over!) the obstacle. The followers are observed in fact to fly directly over the obstacle on their way to the target rather than to follow the scout around the detour. Thus it is clear that they obey the instructions in the dance and do not rely on visual or odor cues left by the scout along its true flight route.

Lionel Gonçalves (1969) has recently performed a set of experiments in Brazil that test the waggle dance hypothesis in yet another way. He permitted single scout workers to enter long tubes, walk their length, and feed from sugar water supplies at the end. When the scouts returned to the hive, they performed waggle dances that signified not the length of the tube but, rather, the amount of expenditure required to walk the length of the tube. The follower workers then directed their flights primarily to the areas around feeding places indicated by the waggle dances and not to the tubes themselves. Over the short distances used in the experiments (4 to 40 m from nest to feeding places), orientation to odors played the major role, but it was also clear that the waggle dance was contributing an important fraction of the information.

Although the communicative function of the waggle dance is firmly established by these experiments, it is still a remarkable and disturbing fact that the sensory modality in which the communication occurs remains to be disclosed. Do the follower bees perceive the straight run by touching the dancer with their antennae, or do they reckon it by the sound or air currents it makes during the run? Is it possible that it might even be smelled along the track of the straight run? Still another possibility is that the distance is judged by listening to the wing sounds through the substratum and the direction by touch or some other means. The reason for the difficulty is that no one has succeeded in getting follower bees to obey a dummy placed inside the comb and manipulated by the investigator. Only by this means can stimuli be readily screened out in a systematic fashion and the relevant components of the dance precisely identified. Steche (1957) claimed to have directed flights with dummies, but later attempts to duplicate his work have not succeeded. Surely to achieve such a control over the hive would be a tour de force worth any effort.

There is a second gap in current information on bee dancing. It has been well established that bees are capable of learning odors and that they orient part of the time according to odor memory. For example, if a foraging worker has been successful in the past in collecting from a certain kind of flower and is presented with the odor of the same flower in the nest, it will fly out to the place where it did the original collecting. What is missing is exact information on the extent to which odor memory is the guide, as opposed to the abstract information received in the waggle dance. Both function under natural conditions, but in what proportions? Also, there is a scarcity of measurements of the amount of information added to the waggle dance by additional cues, in particular the assembly pheromones of the Nasanov glands released in the vicinity of new food finds and the sight of flying workers.

A surprising aspect of the communication is that the precision of the newcomers in achieving the goal is greater than that of the dances from which they received the information. In other words, the variance of components in the straight run is greater than the variance of the responses of the newcomers in the field. This difference may well be due to the flaws in von Frisch's experimental design, which I pointed out earlier. But it might also be explained in the following way. The newcomers typically follow more than one dance of each dancer and even more than one dancer before leaving the nest in search of the food site. By averaging the information, they are able to obtain a more precise estimate than is available in many individual dances. There is nothing especially remarkable about such a calculation. What occurs may be analogous to the way in which we ourselves quickly deal with such information. When a gun, for example, is fired repeatedly at a fixed target, we are able to glance at the pattern of hits and instantly estimate the point at which the gun was aimed, even when most of the shots are far off the mark.

Most of the research by von Frisch and his students has been concentrated on the Carniolan race of the honeybee, *Apis mellifera carnica.* When other genetic strains were tested, they were found to differ in the details of the waggle dance, in particular the distance intervals at which the round and waggle dances are performed, the presence or absence of the sickle dance as an intermediate form of dancing, and the tempo of the waggle dance (Figures 13-13 and 13-14). These variations have been appropriately referred to as "dialects." The dialect phenomenon has been exploited in an ingenious way by Steche (in von Frisch, 1967a) and Boch (1957) to further test the communicative function of the waggle dance. The central European strain with the fastest tempo is the Carniolan race (*carnica*) and the one with the slowest tempo is the Italian race (*ligustica*). This means that a *ligustica* worker should read the relatively shorter straight run of a *carnica* dancer as indicating a goal closer to the nest than it really is; and a *carnica* worker should conversely overestimate the distance of a goal from the relatively longer straight run of a *ligustica* dancer. Mixed colonies of the two strains were established, and workers of first one strain and then the other were allowed to feed at experimental food tables

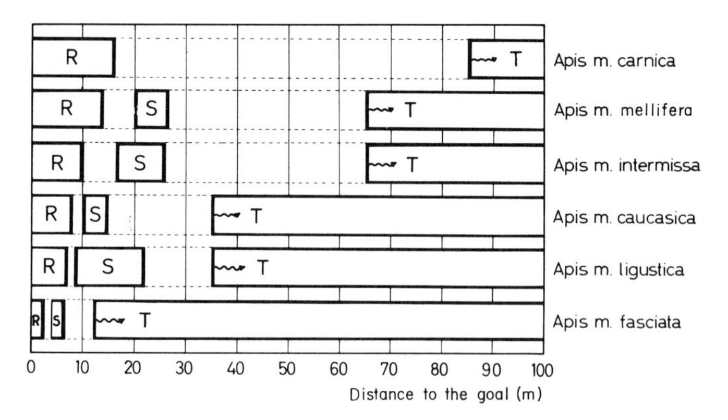

FIGURE 13-13. The differences that exist among several of the principal strains ("races") of the honeybee in the indication of direction provided by their dances: (*R*) round dances; (*S*) sickle dances; (*T*) tail-wagging (waggle) dance. The distance of the target from the hive is given in the abscissa. The spaces between the three forms of dancing indicate gradually changing transitional forms (after von Frisch, 1967a; from Boch, 1957).

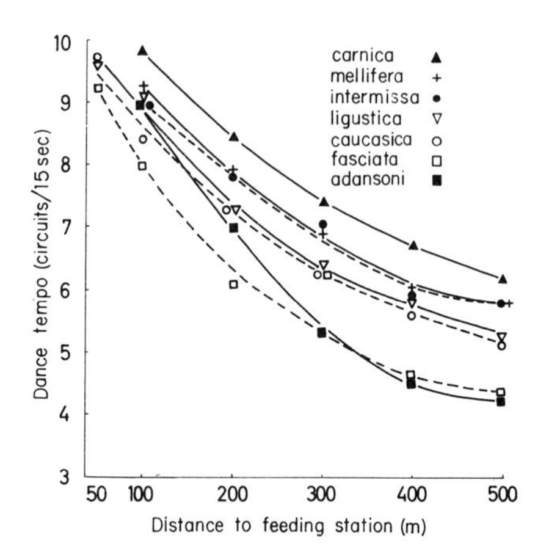

FIGURE 13-14. Racial differences in the tempo of the waggle dance given as a function of the distance of the target. The faster the tempo, the briefer the straight run for a given target distance (from von Frisch, 1967a).

and to return to the hives to dance out their messages. Newcomer bees were then counted at food tables set at varying distances from the hive, following the usual procedure of the "step experiment." It was found that, as predicted by the von Frisch communicative hypothesis, the *ligustica* workers underestimated from the *carnica* dances and the *carnica* workers overestimated from the *ligustica* dances.

There are at least three additional features of the waggle dance that could impart information. Von Frisch and others have spoken of the "liveliness" or "vivacity" of the dance, a quality that is readily perceived by the human observer but so far has resisted measurement. The duration of the dancing—not the duration of a single dance circuit but the entire period of dancing—also varies greatly. Both of these qualities increase when the quality of either the food or the weather improves. The dancing ranges from relatively sluggish initiation of a single circuit to a vigorous, nonstop performance that lasts for minutes. The precise factors known to promote liveliness and endurance are as follows: the sweetness of the sugar solution; the purity of the sweet taste; the ease with which the food is secured, including its nearness to the nest; a flower-like fragrance; the flower-like form of the food container; a uniform, continuous flow from the food source; the starved condition of the colony; favorable weather conditions; and competition from other food sources (von Frisch, 1967a). The third feature in the dance that contains potential information is the frequency of the sound bursts during the straight run. According to Esch (1963) the frequency of the pulses increased in one set of observations from 22 to 30.5 per second as he increased the concentration of sugar solution from 0.5 molar to 2 molar. It needs to be stressed that, although the human observer can, with the aid of instruments, "read" the information in these additional components of the dance, this does not mean that the bees are able to do so. The technique for monitoring responses to the dances of different qualities remains to be devised.

Just as in the odor trails of ants, the waggle dances serve to recruit nestmates both to food and to new nest sites. Largely through the researches of Lindauer (1961) we have a clear picture of how the emigration to new nest sites is organized. Shortly after the swarm containing the old queen has left the hive and clustered elsewhere (see the description of the honeybee life cycle in Chapter 5), scouts fly out in all directions searching for a suitable cavity in which to build the new combs and set up housekeeping. When such a place is discovered, the bee returns to the swarm and begins to dance on its surface, indicating the location of its discovery. It often happens that two or more sites are discovered before the swarm moves from the bivouac. In this case, the scout bees advertise their respective discoveries in competition with each other. Other bees follow their leads, examine the sites, and, if sufficiently stimulated by the quality of the location, commence dances on their own. The most attractive sites evoke the largest number of dances and the most persistent dancing. The scouts reporting inferior sites are gradually won over, and in time a preponderance of the dancing bees come to advertise a single site. Then the swarm leaves for the site chosen in this "democratic" fashion (Figure 13-15). Occasionally the democratic method breaks down. In 2 of the 19 swarms watched by Lindauer, the colonies had great difficulty reaching a decision. The results were startling:

In the first case two groups of messengers had got into competition; one group announced a nesting place to the northwest, the other one to the northeast. Neither of the two wished to yield. The swarm then finally flew off and I could scarcely believe my eyes—it sought to divide itself. The one half wanted to fly to the northwest, the other to the northeast. Apparently each group of scouting bees wanted to abduct the swarm to the nesting place of its choice. But that was naturally not possible, for one group was always without the queen, and there resulted a remarkable tug of war in the air, once 100 meters to the northwest, then again 150 meters to the northeast, until finally after half an hour the swarm gathered at the old location. Immediately both groups began again with their soliciting dances, and it was not until the next day that the northeast group finally yielded; they ended their dances and thus an agreement was reached on the nesting place in the northwest.

The second case ended in a completely unexpected way; for 14 days no agreement had been reached, and then, when a period of rain set in, the scouting bees gave up their search for a dwelling and occupied themselves with the collection of nectar and pollen. The traveling stores of the swarm bees were apparently used up and it was high time for a replacement of provisions. Thus the activity of the hunters of quarters was completely suppressed, and the swarm made its abode at its first landing place, built honeycombs in the bushes, and set up a normal nest for its brood . . . In our climatic region the winter would put an end to this attempt, for the bees cannot survive a heavy frost in the open. (Lindauer, 1961).

On three occasions Lindauer (1955) estimated the position of the favored nest site by watching the dances alone.

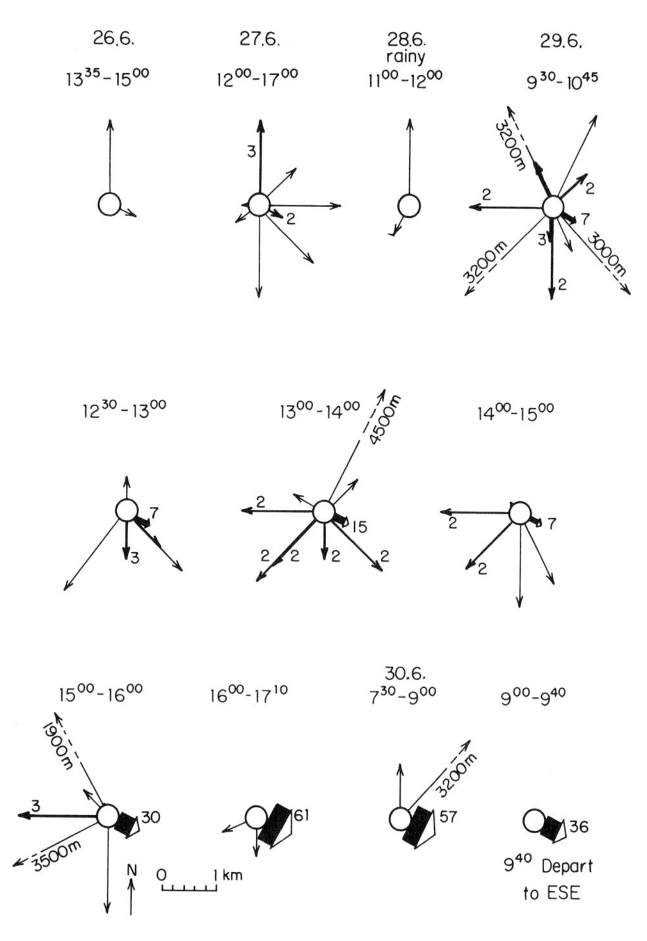

FIGURE 13-15. The history of the search for a new site is shown in diagrammatic form. As the swarm of honeybees hangs in a cluster in the open air, scout bees find suitable nest cavities and advertise their direction and distance by dances on the surface of the swarm. Groups of bees compete with each other in announcing different sites; the thicknesses of the arrows indicate the number favoring a given site at a given time, while their lengths and direction indicate to scale the distances and directions of the actual sites announced. In this typical example, the colony took from 1:35 P.M. on June 26 to 9:40 A.M. on June 30 to arrive at a decision (from Lindauer, 1961).

Then he proceeded to the location, found the site, and was waiting to greet the swarm when it arrived.

The round dance and waggle dance and the forms intermediate between them are not the only stereotyped locomotory forms performed in honeybee hives. At least some of these additional "dance forms" have communicative functions and deserve closer study.

Jostling run (*Rumpeltanz*). As soon as they enter the hive, successful foragers run excitedly through groups of workers, bumping into them as though intentionally. The jostling run occurs after the first successful flight to a food source, whereas the waggle dance more often occurs only after several such flights. The jostling run serves to excite other workers and to draw attention to the forager, and it has its exact parallel in the charging and bumping behavior of trail-laying ants and termites (Schmid, 1964).

Spasmodic dance (*Rucktanz*). Returning bees sometimes run along the comb, distributing food and performing fragments of the straight run. Although the partial straight runs are correctly oriented, there is doubt as to whether they are effective as signals or comprise anything more than incomplete intention movements preliminary to the performance of complete waggle dances (Hein, 1954).

Buzzing run (*Schwirrlauf*). This is the signal that initiates swarming. Just before the swarm occurs, most of the bees are still sitting idly in the hive or outside in front of the entrance. As midday approaches and the air temperature rises, one or several bees begin to force their way through the throngs with great excitement, running in a zigzag pattern, butting into other workers, and vibrating their abdomens and wings in a similar fashion to that observed during the straight run of the waggle dance. The sound produced is very different from that produced during the straight run, however, raising the possibility that it is an important part of the signal to swarm (Esch, 1967c). The *Schwirrlauf* is swiftly contagious, and, within a minute or two, a dozen or more workers are engaged. As Lindauer (1955) describes it, "Like an avalanche the number of buzzing runners grows, many of them rush to the hive entrance, arousing similarly those slothful ones who had gathered together like a tuft before the flight opening, others hover briefly about the hive but return once again to continue their buzzing runs. In about 10 minutes the moment for departure has arrived . . . then the bees nearest the hive entrance rush forth and in a dense stream all follow. The queen too has been aroused, and if she does not follow the swarming bees out at once she is badgered without interruption by bees buzzing and running until she has found the hive entrance and hurls herself into the swarm cloud."* The phenomenon is remarkable in that it is the only example I know of an autocatalytic reaction in an animal communication sys-

*Translation by L. E. Chadwick in von Frisch (1967a).

tem. The signal itself produces the same signal in others, with the result that a chain reaction and a behavioral "explosion" occur. Of course this is just the effect that is needed to ensure a simultaneous action by the ten thousand or more individuals who fly from the hive.

Grooming dance (*shaking dance*). The worker shakes her body rapidly back and forth and from side to side, while attempting to comb her thoracic hairs with her middle legs. Often, but not always, this behavior induces a nearby worker to approach and employ her mandibles to groom the hairy coat on the petiole and base of the wings. These are the parts which a bee is unable to clean herself (Haydak, 1945; Milum, 1955).

Jerking dance (*D-VAV*). Occasionally a worker bee touches a nestmate with her antennae or, more typically, seizes its body with her forelegs or climbs on top of it, then makes seven or eight rapid up-and-down strokes with her abdomen. Milum has referred to the abdominal movement as the "dorso-ventral abdominal vibration" (D-VAV). The function is still unknown (Milum, 1955; Eleonore Hammann, 1957; Allen, 1959; von Frisch, 1967a).

Trembling dance. The legs of the bee twitch, causing the body to quiver and stagger around in random directions. This behavior is uncommon and has no visible effect on nestmates (von Frisch, 1967a). Several lines of evidence suggest that it is pathological, being caused by poisons picked up during foraging trips or even by a neurotic state of the kind induced experimentally by Florey (1954). I have seen a similar phenomenon in laboratory colonies of the ant *Pogonomyrmex badius.* Occasionally a worker "freezes," its appendages straighten, and it trembles lightly for a few minutes. This behavior has a distinctly abnormal appearance and no visible effect on other workers.

In the foregoing account of the dances of the honeybee, I have summarized those aspects of the subject that are most relevant to a comparative treatment of insect social behavior. Most of the vast available documentation has been deliberately omitted in this one particular case. I have also left out a great deal of additional information on lesser aspects of the dances and the underlying chemosensory behavior that does not bear directly on social organization. The interested reader is referred to von Frisch's recently published master work *Tanzsprache und Orientierung der Bienen* (1965) or its English translation by Chadwick (1967).

14 Communication: Recognition, Food Exchange, and Grooming

The Role of Surface Pheromones

The recognition of nestmates and the close bodily interplay dependent on it together form much of the central repertory of social life. The underlying behavior is poorly understood because it is controlled in large part by pheromones that are at once typically complex, present only in extremely small quantities, and for various other reasons difficult to separate and bioassay. Other pheromone classes, in particular those that influence caste determination and those that are transmitted through the air to attract or alarm, have proved more tractable, even yielding in a few cases to complete chemical and biological analysis. In contrast, the substances employed in recognition, food exchange, and perhaps also regurgitation are "surface pheromones," that is, they generate such shallow active spaces in air that they must be perceived by contact chemoreception or a close approach to it. They are, furthermore, usually effective only in the complex and delicate stimulus context in which they function in the normal life of the colony. When removed from the other surface odors or masked by alien substances of the kind employed in conventional extraction techniques, they no longer produce the same behavioral effects. At present we lack the suitable extraction techniques and bioassays to permit the study of the chemistry of surface pheromones, but there is no reason to believe these substances will be permanently intractable. Entomologists have, meanwhile, come to understand the purely behavioral aspects reasonably well and in some cases have begun to characterize indirectly the chemical and auxiliary stimuli that control recognition.

Colony Odors

In all of the social insects recognition of a nestmate seems outwardly a casual matter, usually no more than a pause and sweep of the antennae over the other's body. There is little sign of any notable ability to discriminate odors. The magnitude of this ability is revealed unmistakably, however, whenever an alien insect is encountered within the nest. If the intruder belongs to a different species, it is almost always attacked, violently and without hesitation. If it is a member of the same species but from a different colony, the hostility it evokes falls somewhere within a broad gradient of responses. At one extreme, the insect can be overtly accepted but offered less food until it has time to acquire the colony odor. This subtle distinction, involving only the rate of exchange of liquid food, was first discovered by von Frisch and Rösch (1926) in honeybees and has since been used as the basis of a sensitive bioassay in studies of colony odors in ants by Lange (1960a, 1967) and in vespine wasps by Montagner (1963a,b) and Montagner and Courtois (1963). In the next higher degree of hostility, the alien is intensively investigated by workers it encounters; there is no attack, and acceptance comes with time. In the most frequent, and violent, response, the alien is swiftly inspected, attacked, and killed or driven from the nest. The method of attack varies greatly among species. In the honeybees, the social wasps, and those ants equipped with a well-developed sting, the worker seizes the alien by some protruding part of the body and attempts to insert its sting through an intersegmental membrane. Species equipped with chemical sprays, such as dolichoderine and formicine ants along with a few of the higher termites, often combine these defensive substances with biting and pulling. Species of termites and ants that possess a mandibulate soldier caste rely heavily on the mandibular strikes of these insects (see Chapter 8). Although soldiers are typically the most sluggish caste and least responsive to ordinary minute-by-minute communication within the colony, they are very sensitive to alien colony odors, and their true function is quickly revealed when the nest is invaded.

How can we be sure that recognition is based on odor and not on some deviation in behavior characteristically shown only by aliens? The idea of such a distinctive behavior pattern is not at all farfetched. Workers that invade other honeybee colonies to steal food, sometimes called "robber bees," do have a distinctive flight, and this fact has led such observers as Cale (1946), Adam (1951), and Butler and Free (1952) to conclude that behavior, rather than odor, is the basis of nestmate recognition. However, abundant evidence exists to indicate that recognition is based principally, if not exclusively, on colony odor. Rau (1930) observed that workers of *Polistes* are hostile to both live and dead workers from other colonies but not to the corpses of nestmates. Free himself (1958) later used the same phenomenon in bumblebees to devise a simple but convincing test of the role of odor. He found that workers of *Bombus agrorum* and *B. lucorum* sting anesthetized alien workers introduced into their nest but not anesthetized nestmates. Moreover, when nestmates were left in an alien nest for one to two hours and then returned to their own colony, they were attacked. Kalmus and Ribbands (1952) devised a more elaborate approach to demonstrate the existence of colony odors in the honeybee. They made use of an earlier discovery of von Frisch and Rösch (1926) whereby workers from unrelated colonies were trained to collect sugar water from two different dishes only a few feet apart, and each group attracted recruits predominantly from its own colony. Kalmus and Ribbands refined this *Zwei-Völker-Versuche,* as the German experimentalists call the procedure. A pair of feeding dishes, *a* and *b,* were set up approximately 1 m apart, and two colonies of honeybees, *A* and *B,* were placed about 50 m away. A group of 50 bees from colony *A* were first trained to feed from dish *a* and marked with spots of white paint daubed on their thoraxes, all in the course of a single day. During this time bees from colony *B* were not allowed to fly. On the following day, colony *A* was confined at home, and 50 bees from colony *B* were trained at dish *b* and marked with daubs of blue paint on their thoraxes. As soon as the *B* colony members were marked, both *a* and *b* dishes were filled, and both colonies were allowed to fly. Newcomers to dish *a* were then marked with white paint on their abdomens (to distinguish them from the original scouts), and newcomers to dish *b* were similarly marked with blue paint. When about 100 workers were labeled in this fashion, the experiment was stopped, the hives were opened, and the distribution of white-marked and blue-marked bees was recorded. In repeated trials it was found that *A* workers went predominantly to dish *a*; *B* workers, to dish *b*. The conclusion drawn was that like attracted like at the food dishes, and that the only cue available to the bees was a difference in colony odor (see also Ribbands, 1956). The obvious flaw in the experiment, as outlined so far, is that some other cue, for example a difference in the color, or even in the odor, of the paint marks, could have sufficed. This possibility was eliminated by the results of a second experiment. When Kalmus and Ribbands fed two colonies on the same diet, presumably reducing or abolishing a principal source of colony odor difference, the bees mixed freely at the two dishes.

The fact that such experimentation is still in a relatively primitive state reflects the delicacy of the subject rather than the poverty of ideas about it. In fact, the very existence of colony odors in the vast majority of social insect species remains a supposition only, supported by the observation that hostility among aliens ordinarily follows upon antennal inspection (hence, chemoreception) and does not depend on any other forms of behavior or sensory modalities immediately obvious to the human observer. It is also true that, within a few species, hostility among colonies is negligible or absent. In *Dialictus versatus,* a primitively social halictine bee, alien bees are freely admitted by guards if they make the mistake of approaching the wrong nest entrance. Segregation of colonies in this species depends on the homing ability of the individual foragers (Michener, 1966b). Workers of the halictine *Evylaeus marginatus* are also freely interchangeable among nests (Plateaux-Quénu, 1962), indicating that the absence of a colony odor or other form of nestmate recognition is a primitive trait in social bees generally. Yet specific odors from the nest environs seem to be implicated in the homing. Michener (1960) observed females of the solitary anthophorine bee *Amegilla salteri* enter a laboratory containing earth and cells from an *Amegilla* nest aggregation and fly back and forth over this material, even when it was completely hidden from sight. Similarly, Butler (1965) observed females of solitary *Andrena* being attracted to nest sites previously unseen by the bees.

Lack of colony recognition occurs in a few higher social insects also. The colonies of some ant species, such as members of the *Formica exsecta* group, intermingle freely, so that it is appropriate to speak of any local, cohesive population as one large colony. To be more exact, it is inappropriate to attempt to distinguish among colonies at all, even at the population level. The absence of intraspecific colony odor in ants, unlike its absence in some halictid bees, has clearly been derived as a secondary condition in evolution. A well-marked hostility among colonies, seemingly based on odor, is characteristic of the very primitive ant genus *Myrmecia* (Haskins and Haskins, 1950a), and it is widespread in other primitive ant groups. It is my impression, from many fragmentary accounts in the literature, that colony recognition and hostility is widespread, if not universal, in the social wasps. The same is true for termites, although the very primitive Australian species *Mastotermes darwiniensis* has a sprawling, diffuse nest structure that may indicate the absence of behavioral colony boundaries.

The chemical nature of colony odors is still quite unknown. Several intriguing generalizations about them can be drawn from less direct, behavioral observations, however. First, it has been reasonably well established that at least some of the distinctive elements in colony odor are drawn more or less directly from the environment. Kalmus and Ribbands (1952) and Ribbands, Kalmus, and Nixon (1952), using the *Zwei-Völker* technique just described, were able to demonstrate the pronounced effect of diet on nestmate recognition in honeybees. In a first set of experiments, single colonies were divided into portions, and each group was fed a different kind of honey. In a short time workers were able to discriminate members of their own group from members of other groups. In a second set of experiments, the opposite effect was achieved. All of the food supplies were removed from several distinct colonies, and their hives were taken to a Welsh moor where they had access only to a single species of flowering plant, namely ling (*Calluna*). Every one of the colonies thereafter lost the ability to discern alien workers. The hypothesis offered by Ribbands, Kalmus, and Nixon is that odorants associated with nectar and honey are somehow passed to the epicuticle, either in pure or metabolized form, and that the bees have the ability quickly to learn, and relearn, subtle differences in the epicuticular odor blends. This idea is further supported by experiments that have revealed a powerful sensitivity to differences in artificial odor blends on the part of worker bees (Ribbands, 1955; von Frisch, 1967a). Renner (1960) subsequently repeated the Kalmus-Ribbands experiment and confirmed the principal result. He was, moreover, able to refine the technique in a way to show that odors can be picked up directly by absorption in the cuticle while not requiring ingestion of odor-bearing substances. When black treacle was placed in hives in such a way that the bees could not feed on it, its presence alone was enough to create a distinctive colony odor.

Lecomte (1952) and Chauvin (1968d) performed a related experiment on honeybees with different results. Chauvin summarizes the situation as follows: "If you take a hundred or so bees from *any one hive* and put them into a cage on different sides of a sheet of glass dividing it, you can see that if on the first day you draw out the glass the bees mingle without difficulty; but if you do it on the second day fights start, which become fierce and end in extermination if the mixing of the original sisters is put off to the fourth day." The effect is claimed to be just the same whether the two portions are fed on the same or different foods during the period of separation. Lecomte and Chauvin have offered a genetic hypothesis in opposition to the acquired odor hypothesis of Ribbands and his co-workers. They point out that honeybee queens are typically inseminated by multiple males, and, as a result, the worker offspring are genetically heterogeneous. (In fact, a large proportion are half sisters.) It follows that samples of workers taken at random from the same colony must differ to some extent in gene frequencies from each other. If fragments of the same colony are isolated in addition, these genetic differences might generate average differences in the metabolic production and absorption of epicuticular odorants which, in turn, would be perceived and acted on by the members of the various factions.

Köhler (1955) performed an experiment not very different from that of Lecomte and Chauvin, but obtained results that seem to offer support for the opposite interpretation. When he fed three adjacent colonies on sugar solutions flavored with the same substances, workers began penetrating and robbing other hives without opposition from the guard bees. When the colonies were then fed on sugar solutions containing different scents, however, the robbers came under attack by the guard bees.

To summarize this conflicting information, I believe that the evidence strongly indicates that the colony odors of honeybees are at least in part acquired from the environment and that the contretemps created by the Lecomte-Chauvin experiments can be provisionally explained without rejecting their conclusions outright. One explanation to be considered is that workers are simply more sensitive to odor differences; in other words, their aggressive thresholds are lower when they are suddenly mixed in large numbers in the interior of their own nests—as in the Lecomte-Chauvin procedure. A second possible explanation is that the use of different genetic strains of bees, or of different odorants, could cause variation in response sufficient to include the kind of neutrality reported by Ribbands and his co-workers and Köhler.

Meanwhile, the existence of directly acquired components in the colony odors has been unambiguously documented in other social insects. When Free (1961a) confined anesthetized bumblebee workers next to the combs of strange colonies or suspended them in cages above the combs for periods of one or two hours before returning them to their home nests, they were attacked by their own sisters. It seems to follow, as Renner's experiments also indicated in the case of the honeybee, that the bumblebee nest odors are acquired at least in part by direct absorption of odorants into the cuticle. A similar result was obtained by Lange (1960a) in his studies of the ant *Formica polyctena.* Using the rate of exchange of liquid food by regurgitation among worker groups as a sensitive indicator of the degree of recognition and acceptance among the workers, he showed that differences in either diet or nest material are enough to cause differences in colony odor. His results might be explicable wholly as the outcome of direct absorption of odorants into the cuticle since odorants in food can be picked up as food is exchanged in regurgitation; the food need not have been assimilated by the digestive system first. This simpler explanation is supported by certain experiments performed much earlier by Lubbock (1894). When he placed pupae of *Formica fusca* or *Lasius niger* in an alien colony of the same species and then returned them to their home nest, they were attacked. Since pupae cannot be fed, the implication is that the new odor was absorbed directly by their cuticle.

The available evidence concerning colony odors in termites is much weaker. Andrews (1911) reported that alien *Nasutitermes rippertii* workers are treated with less hostility if washed in plain water or in an aqueous extract made up of whole workers from the colony into which the aliens are introduced. But there is no information on the source of the colony odor in this species or, to the best of my knowledge, in any other termite species.

No evidence has yet been adduced that any component of the colony odor originates as a secretion from an exocrine gland. The hypothesis of Buttel-Reepen (1915) that colony odor in the honeybee originates in the Nasanov gland, an idea favored by many later writers, now seems to have been firmly negated by the experiments of Renner (1960). In a similar vein, Carminda da Cruz-Landim (1963) suggested that the mandibular gland is the source of the colony odors in the meliponine genus *Trigona,* but she provided no experimental evidence. Janet (1898) speculated that the colony odor of ants originates in the metapleural gland. W. L. Brown (1968) has persuasively extended this idea, pointing out that the gland is reduced or lost in just those castes and species that show the most signs of possessing a neutral odor, which in effect is no colony odor at all. But this argument does not eliminate the possibility that the metapleural gland functions in some other social manner that produces the observed correlates, and it overlooks the fact that at least some of the components of colony odor in ants have already been demonstrated to originate more or less directly from the environment. Recent experiments by Maschwitz, Koob, and Schildknecht (1970) also indicate that the metapleural gland secretion is strongly fungistatic and bacteriostatic. In *Atta sexdens* one component has been identified, phenylacetic acid, of which an average of 0.7 μg is present in each gland at any given moment. Still another function has been reported in the African ponerine *Paltothyreus tarsatus* by Sudd (1962). The metapleural gland secretion of this species possesses a strong odor and is said, on the basis of rather unconvincing evidence, to serve as an alarm pheromone among workers.

Colony odors probably possess both innate and acquired components. Innate components are most likely to operate at the level of the species and above. It is a rule among ants that, where the colonies of a given species show any degree of hostility among themselves, they will invariably show a high degree of hostility to all colonies belonging to another species. Furthermore, no one has succeeded in mixing colonies of hostile species merely by feeding them the same diet or keeping them in nests constructed of the same materials. Brian (1956c) has provided an interesting fragment of evidence which may indicate that the odors of alien species are more quickly perceived than colony odors of the same species. He noticed that workers of *Myrmica ruginodis* respond to queens belonging to other *Myrmica* species over a distance, but react to queens of their own species only following contact chemoreception. In some cases genetically-based odor differences probably exist among colonies of the same species as well. When Wallis (1962) isolated groups of *Formica fusca* from the same colony for seven months, and then recombined them, the degree of hostility was only slightly greater among members of groups fed on different diets than among those fed on the same diet, and it was far less than when the workers of either group were placed with workers from a different colony altogether. The crucial experiment permitting the quantitative measurement of innate versus acquired components in colony odors has yet to be designed. Information is also lacking on the total number of distinct colony odors that can exist in a population of social insects. The question must be raised whether every colony of a given territorial species can recognize as an alien every other colony among, say, the millions of colonies belonging to the same species. Or whether there is, instead, a finite number of distinguishable "odor groups" for each species, perhaps in the hundreds or thousands. If the latter is the case, it should be possible to combine a certain number of colonies taken directly from the field without any show of aggressiveness on the part of the workers. Hangartner (personal communication) has in fact found this to be true in the European ants *Lasius fuliginosus* and *L. niger.* It remains, however, for systematic surveys to be undertaken.

Learning apparently plays a key role in the recognition of colony odor, at least among colonies belonging to the same species. Forel (1874) created mixed colonies in the ant genus *Formica* by placing pupae of different species together and allowing them to eclose in the absence of the adult ants. Later Adele M. Fielde (1903) achieved the same remarkable result by combining "callow" workers within twelve hours of eclosion. She went much further than Forel, obtaining mixed colonies of such extremely disparate species as *Amblyopone pallipes* (subfamily Ponerinae), *Aphaenogaster rudis* (subfamily Myrmicinae), and *Formica fusca* (subfamily Formicinae). The implication to be drawn from this demonstration is that newly eclosed workers accept whatever odors they first encounter, even if the odors are very different from those that they would normally experience in undisturbed nests. Very likely this form of learning is imprinting, that is, a fixation on certain stimuli encountered during an early, initial interval in life. Imprinting has been well documented in young vertebrates, but its existence can only be said to be suggested by the imperfect data on colony odor in ants. Soulié (1960b) has provided a second form of evidence for the existence of learning in colony recognition. When he combined fragments of two colonies of the same species of *Crematogaster,* fighting ensued, but gradually the workers calmed down, accepted one another, and finally united as a single colony. When additional workers from the older source colonies were then introduced into the mixed colonies, they were attacked by both the members of other colonies and their own former nestmates. This result, if confirmed, can only be interpreted as a relearning of colony odor by the workers placed into the mixed colonies.

Fielde (1901, 1904a, 1905) discovered that the odor of workers changes with age. Workers of *Aphaenogaster rudis,* segregated at the pupal stage and kept apart as a group until from 16 to 20 days of adult life, accepted additional alien *rudis* workers up to 40 days in age, but attacked workers older than that as well as queens.*

If the conditions are extreme enough, even older ant workers can be habituated to alien species and mixed colonies produced. Kutter (1963, 1964) found that when he placed colonies of *Lasius fuliginosus* and various species of *Formica* (*exsecta, pratensis, rufa, truncorum*) in containers close together and connected them by wooden bridges, intense fighting broke out just as expected. But eventually the surviving workers grew more friendly, and finally they accepted one another, even going so far as to cooperate in nest building and to engage in mutual grooming and feeding. The *Lasius* workers nevertheless remained hostile to the *Formica* queens and hunted them down, so that in time the former "alliance-colony" turned into a pure *Lasius* colony. This situation is not radically

*Fielde's work on colony odor and odor reception has been reviewed by Forel (1928) and, in a less thorough but more critical manner, by Ribbands (1965). Her results, while crude, variable in quality, and often encumbered by elaborate hypotheses, are, nevertheless, the most extensive published to date and contain much suggestive information that should help future investigators.

different from the true life cycle of *L. fuliginosus* and other temporarily parasitic species of *Lasius* and *Formica,* in which the queen penetrates the colony of another species belonging to the same genus, the host queen is killed, and the colony finally comes to be populated wholly by the parasite queen and her offspring.

Finally, an important feature of aggressiveness among alien social insects is that it varies greatly in intensity within the same colony as a function of other social stimuli. The workers of social insect workers generally lose most or all of their hostility when removed from the nest, unless they are defending a food source or are still in the company of large numbers of their nestmates. Honeybee workers become increasingly intolerant of alien workers entering their hive as food supplies become scarcer. Workers of all kinds of social insects also become more receptive if deprived of their queens. To cite an extreme example, colonies of the army ant *Eciton,* normally hostile to each other, will readily fuse if one has been deprived of its queen for a few days.

Recognition of Castes and Life Stages

The capacity of single workers of insect colonies to classify other individuals goes far beyond the mere testing of colony odor. In order for the society to function, workers must be able to distinguish the caste and life stages of nestmates and act accordingly. Nest queens, for example, are treated very differently from other individuals, including virgin members of the same caste. In larger colonies they are as a rule heavily attended by nurse workers, who constantly lick their bodies and offer them regurgitated food or trophic eggs. Three unique pheromones have been implicated in this special treatment in the case of the honeybee: *trans*-9-keto-2-decenoic acid, a mandibular gland substance that causes some attraction; *trans*-9-hydroxy-2-decenoic acid, a second mandibular gland substance that causes clustering; and an unidentified volatile attractant from Koschevnikov's gland (see Chapter 12). In the more elementary societies of bumblebees (*Bombus*) and paper wasps (*Polistes*), certain individuals are recognized as dominant by others and given preference in food exchange and easier access to brood cells. Those with the greatest ovarian development recognize each other as rivals and display open hostility to each other. Workers of stingless bees engaged in brood cell construction give way at the approach of the nest queen and permit her to eat the provender placed in the cell, as well as any of their own eggs previously laid on top. Different castes of the termite genus *Kalotermes* manufacture widely differing quantities of the principal volatile attractant 2-hexenol and hence vary in their capacity to serve as foci of clustering. In the social Hymenoptera, males are generally discriminated against as a group. They are offered less food by the workers (all of whom are female), and in times of starvation they are frequently driven from the nest or killed.

In addition to these instances of caste identification by workers, there exists abundant evidence of an even finer ability to detect differences among life stages. In the relatively primitive myrmicine ant genus *Myrmica,* workers are seemingly not capable of distinguishing the tiny first instar larvae from eggs, so that when eggs hatch the larvae are left for a time in the midst of the egg pile. They feed during the first instar by breaking into a single adjacent egg. As soon as they molt and enter the second instar, however, the larvae are removed by the workers and placed in a separate pile (Weir, 1959a). In the third instar the larvae vary greatly in size; the smaller ones are destined to metamorphose into workers, and the larger ones retain, for a time at least, the ability to develop into queens. If a nest queen is present, the large larvae receive proportionally less food, and they are also licked and bitten more by the workers, an action that may further reduce their growth. The ultimate effect is that in the presence of a nest queen fewer new queens are produced (Brian and Hibble, 1963). The tendency to segregate eggs, larvae, and pupae into separate piles is a nearly universal trait in ants. Workers of many of the species are, furthermore, able to distinguish larvae of two or more size classes (Le Masne, 1953). The same trait is possessed to an extreme degree by the primitively social allodapine bees (Skaife, 1953; Sakagami, 1960). Ant workers, and to some extent ant larvae, are able to distinguish trophic eggs (special eggs produced for consumption) from the normal eggs destined to develop into larvae. In *Monomorium pharaonis* the workers are further able to tell male eggs from female eggs (Peacock *et al.,* 1954).

In most cases identification appears to be by antennal contact. This fact in itself suggests chemoreception, even though Brian (1968) has speculated that several age classes of *Myrmica* larvae might also be distinguished by certain differences in hairiness which are quite apparent to the human observer. There are at least two examples, however, in which the communication appears to be by means of odors transmitted over a distance. When workers of the ant *Pogonomyrmex badius* lay trophic eggs, they search for hungry larvae while sweeping their antennae through the air. When they come within about a centimeter of the head of the larva they move directly to it and unerringly place the egg onto its mouth parts (Wilson, unpublished). Free (1965, 1969) has demonstrated that the smell of the brood alone causes honeybee workers to forage for pollen. The effect is said to be enhanced if the bees are given direct access to the larvae.

Young honeybee queens have a remarkable ability to identify each other by special sounds, a phenomenon discovered by Hansson (1945) and corroborated by Wenner (1962). After the old queen has left with a swarm of workers (the "preswarm"), many workers remain in the hive, together with several queens still developing in the closed royal cells. The first young queen to emerge produces at intervals a "piping" or "tooting" (*Tüten*) as it has been variously called; queens that have eclosed to the adult stage but remain in their cells respond with a seemingly deeper "quacking" (*Quaken*). The duet may continue for days. The sounds are the loudest produced by honeybees under any condition and are easily perceived by the human ear. Both beekeepers and entomologists believe that the communication prevents the premature emergence of the supernumerary queens, which would result in mortal combat with the young queen already on the outside. After the latter individual has mated and departed with a second group of workers (the "afterswarm"), it is safe for another queen to emerge from her royal cell. Whether this interpretation is true or not, the sound communication has at least one other function. Simpson and Cherry (1969) have recently shown that piping invariably precedes swarming and that swarming can even be induced with recorded piping sounds.

Hansson estimated the average frequency of piping to be 435–493 cycles per second, increasing with the age of the bee, and the average frequency of quacking as 323 cycles per second. Simpson (1964) reports that the sounds, like those emitted during the waggle dance of the worker, are produced by contractions of the flight muscles and transmitted to the substrate by pressing the thorax against it. Using a tone generator, Hansson imitated piping and quacking sounds and transmitted them into a hive near both free and enclosed queens. Both of his sounds evoked piping by the free queen and quacking by the enclosed queens, but he was unable to prevent emergence of the enclosed queens, leading to the inference that some other stimulus is essential for this effect, perhaps the odor of the free queen.

Wenner, who apparently worked without knowledge of Hansson's report, obtained somewhat different results. He concluded that the piping tones are about 1,300 cycles per second, while the quacking tones are much higher, as much as 2,500 cycles per second. The deeper sound from quacking that the human ear perceives is apparently due to a strong contribution from lower overtones. Wenner also induced responses by playing artificial sounds into the hive, but, in contradiction to Hansson's finding, he could obtain the result only with artificial piping.

The Recognition and Transport of the Dead

The interiors of the nests of social insects, and in particular the brood chambers, are kept meticulously clean. The workers of most species also tamp down the walls of the nests, and those belonging to some species of bees, wasps, and termites coat the nest walls with secretions that harden to form a crust-like surface. In the halictid bees the coat is waterproof and is believed to originate at least in part from the Dufour's glands of the females (Batra, 1968). Alien objects, including particles of waste material and defeated enemies, are typically dragged out of the nest and dumped onto the ground nearby. When honeybees are unable to remove such objects, they cover them with propolis. Ant workers respond to disagreeable but immovable objects by covering them with pieces of soil and nest material. This same behavior has been modified to serve a new function in some of the species that keep aphids and other honeydew-producing "cattle" outside the nest; they enclose their charges in chambers irregularly constructed of soil or vegetable matter. In a few species, such as some of the members of *Crematogaster*, the behavior pattern has advanced to the point where an elaborate carton shelter is constructed from chewed vegetable fibers (Wheeler, 1919a). Ant workers also occasionally try to cover small pools of water or other liquid in the nest vicinity. The casual observation of this phenomenon has misled some authors to report erroneously that ants construct "bridges" to cross obstacles.

Social insects are especially fastidious when dealing with corpses. The dead of social wasps and bees are dragged out of the nest and abandoned (Lubbock, 1894).

The dead of ants are either eaten (see the following section on cannibalism) or carried out and discarded. Some kinds of ants, for example, army ants of the genus *Eciton* (Rettenmeyer, 1963a), pile the dead among the general refuse in kitchen middens located a short distance from the nest or bivouac, while others, including the fungus growers of the genus *Atta* (Moser, 1963), use deserted nest chambers or galleries. One species, the predatory *Strumigenys lopotyle* of New Guinea, piles fragments of corpses of various kinds of insects in a tight ring around the entrance of its nest in the soil of the rain forest floor (Brown, 1969). On the other hand, despite the claim of some authors in both ancient and early modern times (Wilde, 1615), there is no creditable evidence of the existence of "ant cemeteries," to which only the bodies of fallen nestmates are consigned. Nor is there any documented case of ants burying their dead in anything approaching a ritualistic or organized fashion. The transport of dead nestmates from the nest is nevertheless one of the most conspicuous and stereotyped patterns of behavior exhibited by ants. A full description of the behavior is given, for example, by McCook (1879b) in his monograph on the harvesting ant *Pogonomyrmex barbatus.* Wilson, Durlach, and Roth (1958) analyzed the stimuli that release this "necrophoric" pattern in *Pogonomyrmex* and *Solenopsis.* When a corpse of a *Pogonomyrmex badius* worker is allowed to decompose in the open air for a day or more and is then placed in the nest or outside near the nest entrance, the first sister worker to encounter it ordinarily investigates it briefly by repeated antennal contact, then picks it up and carries it directly away toward the refuse piles. In the laboratory nests we used, the most distant walls of the foraging arenas were less than a meter from the nest entrances, so that the ants had built the refuse piles against them. The distance was evidently inadequate to allow the rapid consummation of the corpse removal response because workers bearing corpses frequently wandered for many minutes back and forth along the back wall before dropping their burdens on the refuse piles. Others were seen to approach the back wall unburdened, pick up the corpses already on the piles, and transport them in similarly restless fashion before redepositing them. We were, therefore, provided with a distinctive and easily repeated bioassay. It was soon established that bits of paper treated with acetone extracts of *Pogonomyrmex* corpses were treated just like intact corpses by both *P. badius* and *Solenopsis saevissima* workers. Separation and behavioral assays of principal components of the extract implicated long-chain fatty acids and their esters. Furthermore, it turned out that oleic acid, a common decomposition product in insect corpses, is fully effective. M. S. Blum (personal communication) has subsequently identified this substance in *Solenopsis* corpses. On the other hand, many other principal products of insect decomposition, including short-chain fatty acids, amines, indoles, and mercaptans, were ineffective. When *Pogonomyrmex* corpses were thoroughly leached in solvents, dried, and presented to colonies, they were seldom transported as corpses, but were more commonly eaten instead. Thus, the worker ants appear to recognize corpses on the basis of a limited array of chemical breakdown products. They are, moreover, very narrow minded on the subject. Almost any object possessing an otherwise inoffensive odor is treated as a corpse when daubed with oleic acid. This classification even extends to living nestmates. When we put a small amount of the substance on live workers, they were picked up and carried, unprotesting, to the refuse pile. After being deposited, they cleaned themselves and returned to the nest. If the cleaning was not thorough enough, they were sometimes mistaken a second or third time as corpses and taken back to the refuse piles.

Perhaps even more remarkable than the simplicity of this control of necrophoric behavior is the tendency of the workers of some ant species to remove themselves from the nest when they are about to die. I have repeatedly observed that injured and dying *P. badius* loiter more in the vicinity of the nest entrance or outside the nest than do normal workers. Injured *Solenopsis saevissima,* particularly those that have lost their abdomens or one or more appendages, tend to leave the nest more readily when it is disturbed. Marikovsky (1962) reports that workers of *Formica rufa* fatally infected with the fungus *Alternaria tenuis* leave the nest and cling fast to blades of grass with their mandibles and legs just before dying.

Cannibalism and Trophic Eggs

In all termite species so far investigated the colonies eat their own dead and injured (Ratcliffe *et al.,* 1952). Cannibalism is in fact so pervasive in termites that it can be said to be a way of life in these insects. The primary reproductives of some of the African termitids perform ritualized cannibalism (Grassé, 1942). In *Cubitermes ugandensis,* for example, the royal couple first build and seal their nuptial cell, then they eat from one to five of the terminal segments of each other's antennae. About three days later they copulate for the first time. In the subsequent rearing of the first brood, the couple consume some of the eggs and young larvae, as well as all of the egg shells and larval exuviae (Williams, 1959a). Buchli (1950) observed that workers of *Reticulitermes lucifugus* sometimes eat their nestmates even when the latter are apparently in a healthy condition. This happens when grooming is carried too far; the cuticle of a leg is broken, then the leg is eaten up, and, finally, the whole unfortunate animal is consumed. When supernumerary reproductives of *Kalotermes flavicollis* are produced in laboratory colonies, they are soon pulled apart and eaten by the workers (Lüscher, 1952b). Winged reproductives of *Coptotermes lacteus* prevented from leaving the nest on a normal nuptial flight are eventually killed and eaten by the workers (Ratcliffe *et al.,* 1952). In general, when alien workers chance into a nest belonging to a colony of the same species, they are first disabled, typically by a mandibular strike from a soldier, and then consumed. Cook and Scott (1933) found that cannibalism became intense in colonies of *Zootermopsis angusticollis* when they were kept on a diet of pure cellulose and hence deprived of protein. When sufficient quantities of casein were added, cannibalism dropped almost to zero. It seems probable that the diets of termites are generally low in proteins even under natural conditions. If so, the degree of cannibalism exhibited by these insects, more intense by far than in any other social insect group, can be interpreted as a protein-conserving device.

There are only a few records of worker ants cannibalizing adult members of their own colonies. Creighton and Gregg (1954) observed workers of *Paracryptocerus texanus* eating an injured queen. It has been my impression that ants will eat other adult members, at least partially, if they are crushed to expose fresh fatty tissue, and queens are more likely to be attacked in this way than workers. In the majority of the many species belonging to all of the principal subfamilies which I have kept in the laboratory, intact corpses are usually discarded, partial eating of crushed corpses is uncommon or rare, and the total consumption of the dead is very rare. Workers of *Solenopsis saevissima* are exceptional in that they appear to consume a large part of their dead, at least under laboratory conditions. Social bees and wasps apparently do not engage in cannibalism of adults at all.

The eating of immature stages, by contrast, is common in the social Hymenoptera. In ant colonies all injured eggs, larvae, and pupae are quickly eaten. When colonies are starved, workers begin attacking healthy brood as well. In fact, there exists a direct relation between colony hunger and the amount of brood cannibalism which is precise enough to warrant the suggestion that the brood functions normally as a last-ditch food supply to keep the queen and workers alive. In the army ants of the genus *Eciton,* cannibalism has been apparently further adapted to the purposes of caste determination. According to Schneirla (1949a, 1952, 1953a), most of the female larvae in the sexual generation (the generation destined to transform into males and queens) are consumed by workers. The protein is converted into hundreds or thousands of males and several of the very large virgin queens. It seems to follow, but is far from proved, that female larvae are determined as queens with the aid of this special protein-rich diet. Brian (1968) has suggested that brood consumption may be widespread in ants as a factor in queen determination, but the necessary physiological data have not been gathered to test the idea in even a preliminary fashion. According to Rettenmeyer (1963a), *Eciton* colonies also consume a fraction of the worker brood, but not nearly as much as in the case of the female sexual broods.

Nest-founding queens of the social wasp *Polistes fadwigae* often invade one another's combs and steal and eat the larvae (Sakagami and Fukushima, 1957a). The cannibalism of larvae and pupae by adults belonging to the same colony commonly occurs in temperate species of the wasp genera *Polistes, Vespa,* and *Vespula,* but usually only in colonies that are in a state of decline in the fall (Janet, 1903; Duncan, 1939). As the proportion of workers to the nonproductive adult queens and males falls off during this time, the food supply of the colony declines, and the brood is increasingly neglected. Finally, the hungry adults begin to attack and eat the larvae and pupae. Predation of healthy larvae or pupae by larvae belonging to the same species is, on the other hand, almost nonexistent in the social Hymenoptera. Eidmann (1929) observed a larva of the ant *Formica fusca* eating another larva in an incipient colony, but it was not ascertained whether the victim was injured or dead prior to the attack.

Egg cannibalism (oophagy) is widespread in the social Hymenoptera, where it serves various purposes. In the more primitively social groups it is crude and exploitative in character. The alpha female of *Polistes,* for example, eats the eggs laid by subordinate females so that only her progeny are reared (Heldmann, 1936a,b; Pardi, 1948; Gervet, 1964). Egg stealing and eating has been reported among the females of bumblebees, with the queen receiving the lion's share (Plath, 1934; Free and Butler, 1959), and it is commonplace in the primitively social halictine bee *Dialictus zephyrus* (Batra, 1964).

In the higher social Hymenoptera oophagy has lost its competitive quality and has been transformed into an important form of food exchange among cooperating members of the same colony. When rearing their first brood, young queens commonly eat some of their own eggs or feed them to the first group of larvae. When colonies in a more advanced stage of growth are starved, the eggs are the first brood stage to be eaten by the workers. First instar larvae of *Myrmecia* and *Myrmica* normally consume one or more eggs adjacent to them in the egg-microlarva pile before they molt to the second instar and are removed by the workers (Freeland, 1958; Weir, 1959a). The workers of many ant species have functional ovaries (Eidmann, 1928; Goetsch, 1937a; Ledoux, 1949; Erhardt, 1962). In some species worker-laid eggs are usually consumed by the queen, larvae, and, less commonly, by other workers shortly after being laid; thus they deserve to be called "trophic" eggs (*oeufs alimentaire*). Trophic eggs have been reported in a wide diversity of both primitive and advanced ant genera: *Myrmecia* (Freeland, 1958); *Myrmica* (Brian, 1953a); *Pogonomyrmex* (Wilson, unpublished); *Leptothorax* (Gösswald, 1933; Le Masne, 1953); *Atta* (Bazire-Benazet, 1957); *Hypoclinea* (Torossian, 1959); *Iridomyrmex* (Torossian, 1961);* *Plagiolepis* (Passera, 1966); *Formica* (Weyer, 1929). The frequency with which trophic eggs are laid varies enormously among ant species. In *Iridomyrmex humilis* it is a rare event, while in *Pogonomyrmex badius* and *Hypoclinea quadripunctata* trophic eggs form the usual diet of the larger larvae and nest queens. As a rule among species, the more frequent the exchange of trophic eggs, the less frequent the exchange of liquid food by regurgitation. The two systems often occur in the same species, but their overall pattern of distribution among species is such as to suggest that they tend to replace each other in evolution. In *Pogonomyrmex badius* the trophic eggs are misshapen and more flaccid than reproductive eggs. Those of *Plagiolepis pygmaea* are smaller in size as well, and they are formed when the trophocytes and follicular epithelium degenerate prematurely so that the oocyte remains small and fails to acquire a chorion (Passera *et al.,* 1968). In *Atta rubropilosa* the oocytes even fuse together in masses, so that what emerges from the oviduct is an extraordinary trophic "omelette" (Madeleine Bazire-Benazet, 1957).

*Torossian at first confused the phenomenon in *Hypoclinea* and *Iridomyrmex* with "proctodeal trophallaxis" of the kind that occurs in termites, but he later (1965) correctly identified the objects he had observed as trophic eggs.

Trophic eggs occur to a limited extent in groups of insects other than ants. Brooding mothers of the subsocial burrowing cricket *Anurogryllus muticus* normally lay and feed such eggs to their nymphs (West and Alexander, 1963). The workers of many, perhaps all, of the meliponine bees lay trophic eggs in a ritualized manner as part of the process of building and provisioning a brood cell. In a typical sequence the nest queen approaches the newly constructed cell, eats the trophic egg just laid in it together with some of the stored provender, and then lays her own egg in the cell (Sakagami and Zucchi, 1967, 1968; Kerr, 1969; see also Chapter 5).

Trophallaxis (Liquid Food Exchange)

Trophallaxis, the exchange of liquids among members of the same colony, plays a key role in the social organization of most species of social insects. Even so, the general significance of the phenomenon has been exaggerated by many earlier writers, and the very meaning of the word trophallaxis has become twisted until it is often a source of confusion rather than understanding. It is therefore necessary to start by disentangling the subject by means of a brief historical review.

When a larva of a social wasp is fed by an adult, it almost invariably secretes a droplet of salivary fluid in return, which is then eaten by the adult. This form of exchange is easy to see; it was described as long ago as 1742 by Réaumur and in more detail by du Buysson and Janet in 1903. Parallel observations were made by Grassi and Sandias (1893–1894) on termites and Wheeler (1918) on ants. It remained for Emile Roubaud (1916), on the basis of his studies of *Belonogaster, Icaria,* and *Polistes* in Africa, to draw the large conclusion from the observed facts: "The retention of the young females in the nest, the associations between isolated females, and the cooperative rearing of a great number of larvae are all rationally explained, in our opinion, by the attachment of the wasps to the larval secretion. The name *oecotrophobiosis* (from *οιχοσ*, family) may be given to this peculiar family symbiosis which is characterized by reciprocal exchanges of nutriment between larvae and parents, and is the raison d'être of the colonies of the social wasps. The associations of the higher vespids has, in our opinion, as its first cause the trophic exploitation of the larvae by the adults. This is, however, merely a particular case of the *trophobiosis* of which the social insects, particularly the ants that cultivate aphids and coccids, furnish so many examples."

With the passage of time W. M. Wheeler has somehow acquired most of the credit for this idea, but his principal contributions to it were, first, to give it a new name, and, second, to confuse the subject totally, almost beyond recall. In his 1918 article on ant larvae, he objected to Roubaud's use of the term "exploitation," and he wrote, "As the relationship is clearly coöperative or mutualistic, I suggest the term *trophallaxis* (from *τροφη*, nourishment and *αλλαττειν*, to exchange) as less awkward and more appropriate than 'oecotrophobiosis.'"

Otherwise, Wheeler agreed enthusiastically in all respects with Roubaud. Later, in his 1928 book, Wheeler wrote a most peculiar but interesting essay on the same subject. He was very disturbed by the fact that Erich Wasmann and his fellow Jesuit, A. Reichensperger, had ridiculed his earlier paper in an attempt to bolster Wasmann's own contradictory theory of symphilic instincts. In response, Wheeler stretched and qualified the trophallaxis concept to the point of virtual uselessness, stating that: "There is no doubt that the glandular secretions of social insects are emitted in greater volume at times of excitement, but since even the persisting individual, caste, colony and nest odours are important means of recognition and communication, there is no reason why the odours should not be included with the gustatory stimuli as trophallactic [page 243]." He went on to say that: "If we compare the distribution of food in the colony regarded as a superorganism with the circulating blood current ('internal medium') in the individual insect or Vertebrate, trophallaxis, as the reciprocal exchange of food between the individuals of the colony, may be compared with the chemical exchanges between the tissue elements and the blood and between the various cells themselves [pages 244–245]."

These two statements, of course, imply very different definitions, and the ambiguity persists through Wheeler's protean writings on the subject. If we select the broadest definition allowed by Wheeler, illustrated in the first of the two statements, trophallaxis must be the equivalent of all of chemical communication in the modern sense. In 1946, T. C. Schneirla, having misunderstood Wheeler (a forgivable mistake), extended trophallaxis to include tactile stimuli also. It was then but a short step for Le Masne (1953) to suggest the *reductio ad absurdum* by defining trophallaxis as synonymous with communication: "By this extension, all the life of the society is encompassed by the trophallaxis concept." Now the term had become what Garrett Hardin has called a panchreston—a word covering a range of different phenomena and loaded with a different meaning to each user, a word that attempts to "explain" everything but explains nothing.

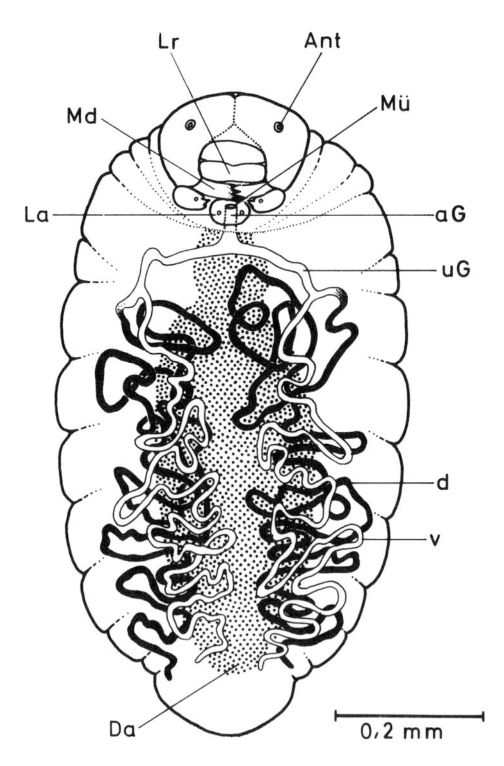

FIGURE 14-1. Larva of *Vespula* (*Paravespula*), showing the large salivary glands used to produce sugars for adult consumption: (*aG*) central duct of the salivary gland; (*Ant*) antenna; (*d*) dorsal ramus of the left salivary gland; (*Da*) intestine; (*La*) labium; (*Lr*) labrum; (*Md*) mandible; (*Mü*) opening of the gland; (*uG*) left confluent duct of the gland; (*v*) ventral ramus of the left gland (from Maschwitz, 1966a).

Perhaps it would have been best if the term trophallaxis had just died at this point, but panchrestons seldom do. Most students of social insects now use the term in its primitive sense, meaning simply the exchange of alimentary liquid, either mutually or unilaterally, and I propose that this is the usage we should keep.

What is the function of larval trophallaxis defined in this way? A number of writers have denied the powerful role assigned it by Roubaud and Wheeler. Weyrauch (1936) postulated that larval regurgitation serves in the regulation of nest temperature and humidity. Brian and Brian (1952), on the basis of what appears to be incorrect experimental evidence (Maschwitz, 1966a), suggested that it has an excretory function. Spradbery (1965) hypothesized that larval fluid might serve secondarily as an aid to the intake and digestion of the relatively dry food provided them. The first solid clue was secured by Montagner (1963b, 1964), who found that males of *Vespula* (subgenus *Paravespula*) obtain nutrition from larval saliva. When queen rearing begins, the males are rebuffed by the workers in their begging attempts, and they then turn to larval secretions as their principal source of food. In chemical studies of the same species of *Vespula* (*Paravespula*), Maschwitz (1966a) confirmed that the larval salivary secretion is both attractive and nutritive. It contains an average of 9 percent trehalose and glucose, approximately four times the concentration in the larval hemolymph. Maschwitz claims that the sugars alone are enough to induce adult feeding, and there is no evidence of additional attractants. Amino acids and proteins are present in the saliva but at only one-fifth the concentration in the hemolymph. Uric acid and ammonia also occur only in small amounts.

Montagner and Maschwitz both consider the larval secretions to represent a colony food reserve, serving exactly the same function as the ingluvial (crop) liquid passed among adults. The salivary glands are thus the functional analog of the crop of adult workers (see Figures 14-1 and 14-2). Maschwitz has calculated that a microliter of larval saliva provides the energy to keep a worker *Vespula* alive up to 1.8 hours, and the sugar released by a large larva in a single "milking" suffices a worker for half a day. More recently, Ikan *et al.* (1968) and Ishay and Ikan (1969) have added a fascinating new twist to this story. They discovered that not only do the larvae of the wasp species they studied in Israel, *Vespa orientalis,* supply the adults with salivary carbohydrates, but also that the larvae are the only colony members capable of converting proteins to carbohydrates in the first place. Only these individuals possess chymotrypsin and carboxypeptidase A and B. There is no evidence that the adults can engage in protein digestion at all. The larvae manufacture glucose, fructose, and sucrose, as well as unidentified trisaccharides and tetrasaccharides and feed them back to the worker nurses. The inability of the adult wasps to engage in gluconeogenesis is unusual among insects, and it was, of course, quite unexpected.

FIGURE 14-2. Worker of *Vespula* (*Paravespula*) *vulgaris* feeding on the salivary secretion of a larva (from Maschwitz, 1966a).

The principal significance of the findings on the vespine wasps is, in my view, that they demonstrate for the first time that larvae can behave in an altruistic manner toward adults and that they thus contribute, by virtue of their behavioral patterns, to the homeostatic machinery of the colony. Their monopoly on gluconeogenesis shows that wasp colonies have gone very far in arranging a biochemical division of labor between adults and young. At the same time, trophallaxis is seen to be not at all the driving force behind social evolution, as envisaged by Roubaud and Wheeler, but only one of a number of forms of communication and nutritive exchange that have been built, with variations, in the course of social evolution. It is curious that J. B. S. Haldane (in Huxley, 1930) very nearly guessed the situation correctly long before any facts were at hand. He pointed out that it would make sense for the larvae to exchange sugar, for which it has little use, for protein, which it absolutely requires. The reverse, of course, is true of the adult wasp. The demonstration of such a function does not, however, mean that it is necessarily the exclusive function. It is still possible that the larval regurgitation also serves in excretion or contributes to humidity and temperature control. These alternatives have not yet been adequately investigated.

The transfer of regurgitated liquid, both among adults and between adults and larvae, occurs generally through the eusocial wasps, including the socially rather primitive *Polistes* (Morimoto, 1960). But behavioral surveys are not nearly complete enough to permit a guess as to whether the habit is universal (Roubaud, 1916; Montagner, 1964; Spradbery, 1965). We have, for example, no information on behavior within the nests of the primitively social sphecid *Microstigmus comes* or of most of the genera and species of the Polybiini.

Trophallaxis is highly variable among species of bees, reflecting both the phylogenetic position of the species concerned and the constraints placed on it by its ecology, especially its nest form. In the bumblebees (Bombini), a primitively social group, the workers simply place pollen on the eggs or larvae, and very little direct contact occurs between adults and larvae. Furthermore, the exchange of liquid food among adults is extremely rare (Jordan, 1936; Anne D. Brian, 1954; Free, 1955b). The halictine bees seal their brood cells, but the females of at least some of the lower social species open the cells to add provisions at frequent intervals, while those of higher social species keep the cells open all the time and attend the larvae regularly (Suzanne Batra, 1964; Knerer and Plateaux-Quénu, 1966a). Even so, there is as yet no evidence that the adults directly regurgitate to the larvae or even to each other, and deliberate efforts to induce such exchange in laboratory colonies of *Evylaeus duplex* have failed (Sakagami, 1960; Sakagami and Hayashida, 1968). Sealed brood cells prevent the adults of stingless bees (Meliponini) from feeding the larvae, but regurgitation among adults is a common event (Sakagami and Oniki, 1963). On the other hand, the queen seldom receives food in this fashion, depending instead on ritualized robbing of the brood cells and consumption of eggs laid by workers prior to oviposition (Sakagami and Zucchi, 1968; Kerr, 1969). Although the brood cells of honeybee colonies are kept open and workers provision them continuously, they do not feed the larvae by direct regurgitation onto the mouthparts (Lindauer, 1952). Adult honeybees, on the other hand, regurgitate to each other at a very high rate (see Figure 5-16). Workers regurgitate water, nectar, and honey to each other out of their crops, but larvae and queens receive most of their protein from "royal jelly" or "brood food" secreted by the hypopharyngeal glands (Rösch, 1930; Free, 1961b). Finally, female allodapine bees regurgitate to their larvae, which are kept exposed in the central nest cavity, but not to each other (Rayment, 1951; Skaife, 1953; Sakagami, 1960).

The species of ants are also highly variable in the rate of trophallaxis, in a way that reflects both phylogenetic position and food habits. The workers of all species of the myrmecioid phyletic complex so far studied engage in liquid food exchange. In the primitive bulldog ants, comprising the subfamily Myrmeciinae, the habit is either rare or else frequent but poorly executed (Haskins and Whelden, 1954; Freeland, 1958). In the higher myrmecioid subfamilies (Aneuretinae, Dolichoderinae, Formicinae) the exchange is frequent, and in the last two subfamilies it is prevalent enough to result in a fairly even distribution of liquid food throughout the worker force of the colony

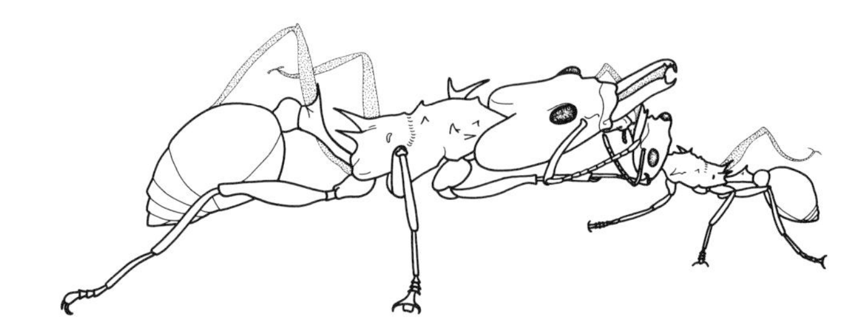

FIGURE 14-3. The exchange of regurgitated crop liquid between workers of the ant *Daceton armigerum*. The larger worker is doing the begging; it holds the head of the smaller worker with its forelegs as it takes the liquid directly from the donor's mouthparts (from Wilson, 1962b).

(Wilson *et al.,* 1956; Wilson and Eisner, 1957; Gösswald and Kloft, 1960). Among the major groups of the poneroid complex, trophallaxis is much more variable and shows less extremes of development than among the myrmecioids. It is apparently absent altogether in *Amblyopone,* one of the most primitive living ponerines; but it has been noted to occur to a limited extent in other ponerines wherever a special search has been made for it (Haskins and Whelden, 1954). In the true army ants (Dorylinae) trophallaxis is evidently weak or absent. The queens of *Eciton* are never seen to be fed by regurgitation, while oral exchange among workers, if it occurs at all, is brief and rare. On the other hand, the workers feed on droplets secreted from the tip of the abdomen by the queen during egg-laying bouts (Schneirla, 1944; Rettenmeyer, 1963a). Most species of Myrmicinae engage in frequent oral exchange, but at least one form, *Pogonomyrmex badius,* apparently never does (Wilson, unpublished). The passing of regurgitated liquid food from workers to larvae seems to follow the same phylogenetic pattern as that for worker-to-worker exchange (Le Masne, 1953; Freeland, 1958). As a rule the frequency of liquid exchange of all kinds in a given species increases with the relative amount of liquid in the natural diet (Gösswald and Kloft, 1960). It is also true that the more trophic eggs are used in the feeding of larvae and queens, the less liquid food is exchanged among workers and between workers and larvae. All adult castes of ants participate in food exchange, including even the males (Hölldobler, 1966). Le Masne and Torossian (1965) have even recorded a unique example of an "altruistic" myrmecophile, the brenthid beetle *Amorphocephalus coronatus,* which regurgitates liquid to its hosts (*Camponotus*) in addition to receiving it.

The crops of most ant species that feed on nectar and homopteran-secreted honeydew are capable of considerable distension. Individual foragers are consequently able to carry home large loads of carbohydrates and to serve as living reservoirs during lean periods. The liquid, digested only to a limited extent while held in the crop, is freely passed from one ant to another. Thus the crops of all the workers taken together serve as a "social stomach" from which the colony as a whole draws nourishment. Eisner (1957), adding extensively to the original discoveries of Forel (1878), showed how the proventriculus has evolved in ants to facilitate this communal function. The proventriculus forms a tight constriction at the anterior end of the crop. Elaborate sepal- and dome-shaped structures have been added to the organ in the Dolichoderinae and are evidently designed to provide automatic, primarily nonmuscular valves to withstand mounting fluid pressure in feeding workers. So distinctive are the structures that they provide useful characters for the study of phylogeny at the genus and tribal level (Eisner and Brown, 1958). The peculiar infrabuccal chamber, a sizable cavity located just beneath the tongue of worker ants, filters out and compacts solid materials that would otherwise clog the narrow, rigid proventricular channels (Eisner and Happ, 1962; see Figure 14-4). From time to time workers of most ant species disgorge the infrabuccal waste material in the form of a pellet. In one arboreal subfamily, the Pseudomyrmecinae, the workers feed them to the larvae. The pseudomyrmecine larvae receive the pellets in a special pouch called a "trophothylax," located on the ventral part of the thorax just beneath the mouthparts, and grind them between two striated plates, called the "trophorhinium," located within the mouth (Wheeler and Bailey, 1920; see Figure 14-5).

The storage of liquid food in the crop has been carried to great heights by the "repletes" of certain ant species, individuals whose abdomens are so distended they have difficulty moving and are forced to remain permanently in the nests as "living honey casks" (Figure 14-6). The extreme examples are ground-dwelling species that live in arid habitats: species of *Myrmecocystus,* a genus confined to the western United States and Mexico; *Camponotus inflatus, Melophorus bagoti,* and *M. cowlei* of the deserts of Australia; some species of *Leptomyrmex* in Australia, New Guinea, and New Caledonia; and *Plagiolepis trimeni* of Natal (McCook, 1882; Wheeler, 1910; Creighton, 1950). *Leptomyrmex* is a dolichoderine, and the remainder belong to the Formicinae. Australian aborigines dig up and eat repletes of *Camponotus* and *Melophorus* as a kind of candy. Intermediate stages of repletion are seen in such diverse genera as *Prenolepis, Proformica,* and *Oligomyrmex,* the latter a dweller in rain forests of the Old World tropics. Repletes are usually drawn from the ranks of the largest workers, who apparently enter their servile role as callows while their abdomens are still soft and elastic (McCook, 1882; Wheeler, 1908, 1910). No less than 1,500 such individuals were recovered from the nest of a colony of *Myrmecocystus melliger* by Creighton and Crandall (1954). Crandall, with the aid of professional gravediggers, followed the nest galleries through 16 ft 3 in of Arizona desert soil until he recovered the nest queen

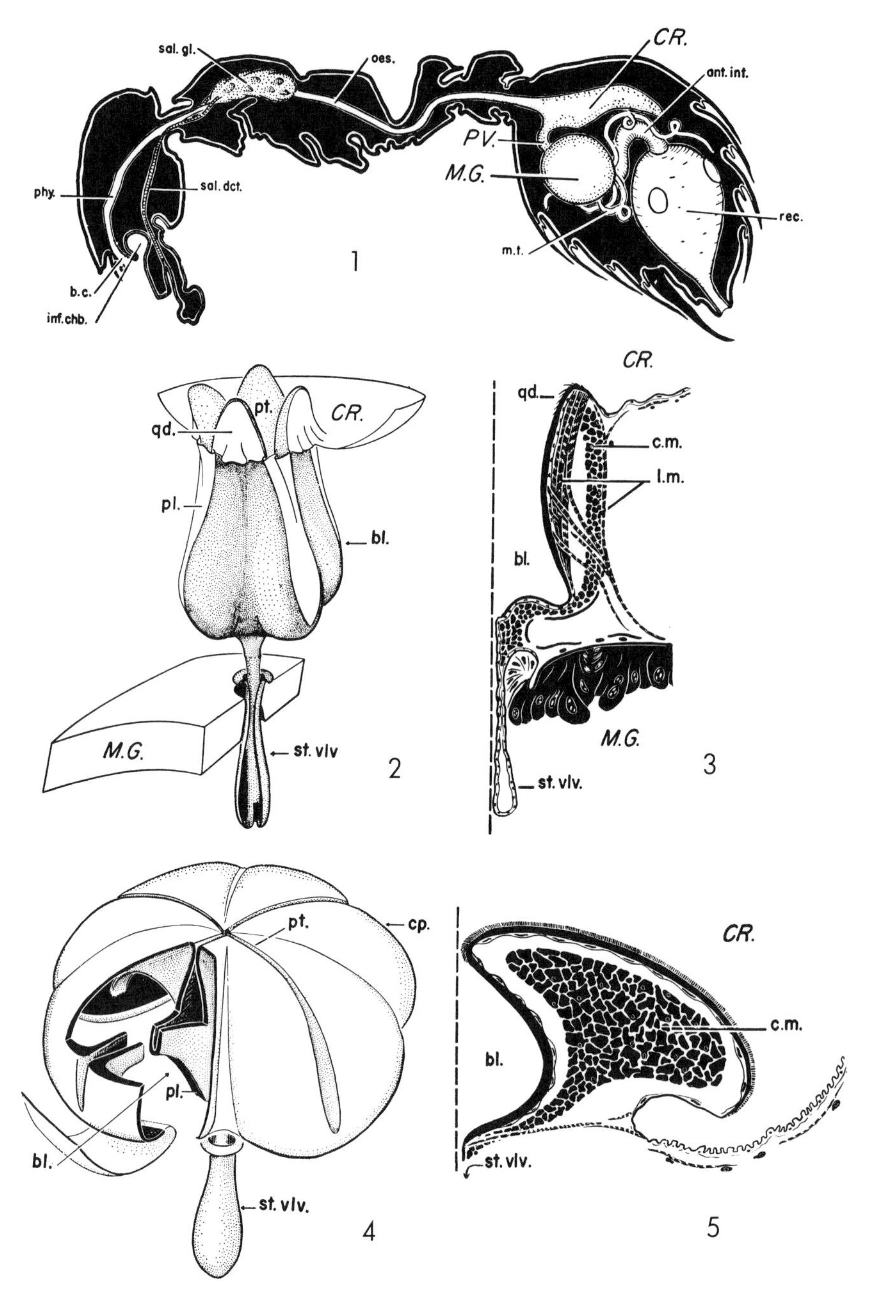

FIGURE 14-4. The intestinal tracts of ants are modified in several ways to store and distribute food communally among nestmates. The upper diagram of a worker of *Myrmica rubra* (1) shows the essential parts: the infrabuccal chamber (*inf. chb.*) filters out most of the solid material from the food; the liquid passes back to the crop (*CR.*), where it is stored; the proventriculus (*PV.*) regulates the flow of the liquid back to the midgut (*M.G.*) where it is digested, and thus serves to segregate the communal supply in the crop from the personal supply in the midgut. The proventriculus of *Myrmecia regularis* (2, 3) is typical of primitive ants, while that of *Iridomyrmex detectus* (4, 5) possesses one of the more advanced forms found in the Dolichoderinae. Other parts shown here are the pharynx (*phy.*); buccal cavity (*b.c.*); salivary gland (*sal. gl.*) and duct (*sal. dct.*); esophagus (*oes.*); anterior intestine (*ant. int.*); Malpighian tubule (*m.t.*); rectum (*rec.*); and parts of the proventriculus, including portal (*pt.*), quadrant (*qd.*), plica (*pl.*), bulb (*bl.*), cupola (*cp.*), stomodeal valve (*st. vlv.*), circular muscles (*c.m.*), and longitudinal muscles (*l.m.*) (from Eisner and Brown, 1958).

from a small chamber at the very bottom. Thus the earlier conjecture by Wheeler (1908) that the *Myrmecocystus* nests are shallow and the population of replete workers is small has been thoroughly refuted. Stumper (1961) used radioactive tracers in honey to study the behavior of the repletes of the European ant *Proformica nasuta.* His results go far to resolve the long-standing controversy about the adaptive significance of repletism, which had been most recently reviewed by Creighton (1950). At moderate temperatures, food passes chiefly from foraging workers to the repletes, but at 30° to 31°C, when the metabolism of the colony sharply increases, the direction of flow is reversed. Stumper infers that in nature the communal supply in the crops of the *Proformica* repletes is built up in relatively cool, moist weather and tapped in hot, dry weather. These results perhaps explain why the species that produce the extreme replete forms are also for the most part desert dwellers. A parallel phenomenon, "adipogastry," involves the exceptional development of the abdominal fat bodies of certain nocturnal, deserticolous species of *Camponotus* (Emery, 1898).

The material regurgitated by worker ants does not consist solely of liquid stored in the crop. Tracer studies have revealed that sugar water held in this fashion is passed chiefly among workers, but larvae and queens receive something else (Wilson and Eisner, 1957). Using autoradiographs Gösswald and Kloft (1960) discovered that P^{32} placed with carbohydrate samples passed from the crop to the midgut and was then concentrated primarily in the abdomen and head over a period of several days. Gösswald and Kloft concluded that the reservoir in the head is the labial glands, and that these organs are therefore to be considered the source of the "salivary" liquid fed as special food to queens and larvae. However, their autoradiographs are not good enough to draw this particular conclusion, and Naarman (1963), using more refined techniques on *Formica,* has, indeed, implicated the postpharyngeal gland instead (see Chapter 12). Even more detailed studies by Markin (1970) on the Argentine ant *Iridomyrmex humilis* have shown that not only is P^{32} concentrated in the postpharyngeal glands but that substantial amounts are subsequently transferred with regurgitated materials to queens and small larvae. Other workers receive only negligible quantities of postpharyngeal secretion and are fed directly from the crop. This new evidence supports an early conjecture by Bugnion (1930) that the postpharyngeal gland is the principal

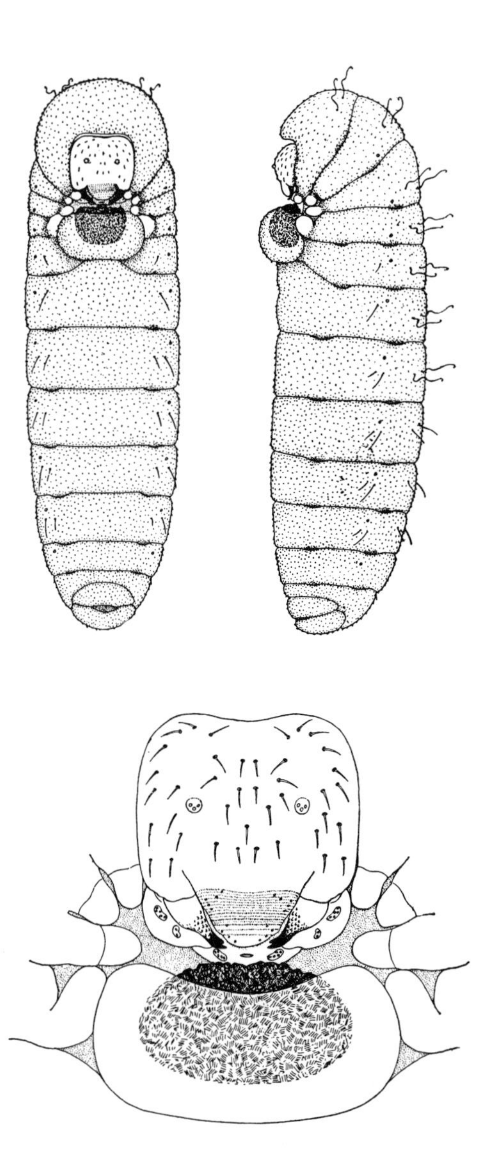

FIGURE 14-5. The larva of the ant *Pseudomyrmex gracilis:* (*top*) front and side views of the larva; (*bottom*) enlarged front view of the head and ventral surface of the thorax. The pouch beneath the head (trophothylax) is filled with a dark infrabuccal pellet given to the larva by a worker, on which the larva feeds. The striated grinding surface of the mouth (trophorhinium) can be seen between the paired mandibles (from Wheeler and Bailey, 1920).

FIGURE 14-6. Repletes of the honey ants belonging to the genus *Myrmecocystus,* photographed in their natural position on the roof of a chamber deep within a newly opened nest (courtesy of T. Eisner).

source of larval food. The gland is a large, elaborate organ which so far has been found only in ants. It seems to be one of those notable organs which, like Pavan's gland in ants and Nasanov's gland in honeybees, has evolved *de novo* to serve in social organization.

In the Ponerinae, Myrmicinae, and Formicinae the larvae periodically discharge small quantities of clear liquid (Le Masne, 1953; Maschwitz, 1966a). The habit appears to be absent in the Dolichoderinae. These offerings are quickly licked up by the first workers to encounter them. The workers also seem to seek the fluid since they lick the head region more frequently than the remainder of the body and occasionally solicit regurgitation by the same kind of antennal stroking used to offer food to larvae (Stäger, 1923; Le Masne, 1953). The source of the liquid is not known, but the most likely source is the paired salivary glands, the only well-developed exocrine glands that open into the mouth. Wheeler (1918) pointed out that these glands are hypertrophied in the myrmicine *Paedalgus termitolestes,* and he speculated that they serve to produce secretions attractive to the workers. Wheeler also suggested that the remarkable thoracic abdominal appendages of larvae of the pseudomyrmecine *Pachysima,* which he called "exudatoria," produce liquids that serve to attract and bind the affection of workers (see Figure 14-7). Similar structures are found in certain species of *Crematogaster,* including *C. rivai* (Menozzi, 1930). As

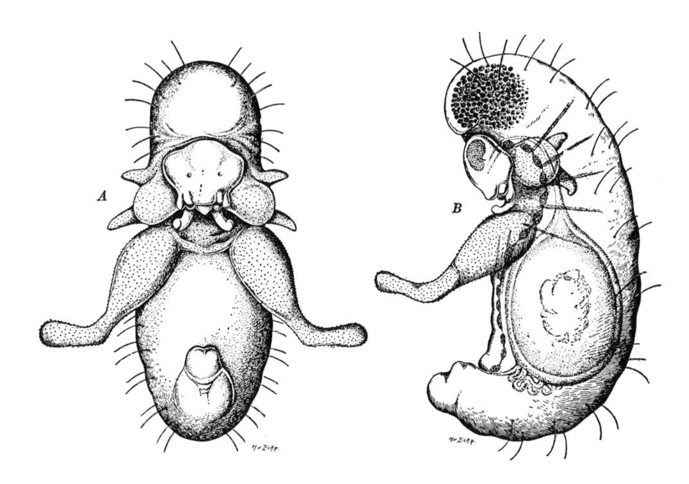

FIGURE 14-7. First instar larvae of the African arboreal ant *Pachysima latifrons,* showing the peculiar thoracic and abdominal organs considered by Wheeler to be the source of worker attractants: (*A*) front view; (*B*) side view (from Wheeler, 1918).

simple and beguiling as Wheeler's mutual attraction hypothesis is, it has not yet been subjected to critical experimental investigation. Maschwitz (1966a) found that the stomodeal contents of *Tetramorium caespitum* larvae have much higher concentrations of amino acids than their hemolymph. He supposed that the stomodeal fraction "very probably originates from the salivary secretion," and thus the secretion must have nutritive value. This relation is by no means proved: it is even possible that the salivary secretions of ant larvae, like those of wasp larvae, are high in carbohydrates instead. No information at all exists on the nature or function of the abdominal "exudatoria" or other possible glandular tissue located elsewhere on the bodies of larvae. Ant larvae also produce a clear liquid from the anal region from time to time, which Le Masne (1953) believes contains waste substances and originates from the Malpighian tubules. The workers often solicit this material by stroking the tip of the larva's abdomen with their antennae.

Trophallaxis in termites, with reference to both its multiple functions and its evolution, has been thoroughly described by Grassé and Noirot (1945), Noirot (1952, 1969a) and Alibert (1968). In all the lower termite species examined to date, belonging to the families Kalotermitidae and Rhinotermitidae, the members of the colony feed each other with both "stomodeal food," which originates from the salivary glands and crop, and "proctodeal food," which originates from the hindgut. The stomodeal material is a principal source of nutriment for the royal pair and the larvae. It is a clear liquid, apparently mostly secretory in origin but with an occasional admixture of woody fragments. The proctodeal material is emitted from the anus. It is quite different from ordinary feces since it contains symbiotic flagellates that feces completely lack, and it has a more watery consistency. Evidently the principal function of proctodeal trophallaxis is the donation of flagellates to nestmates which lost them while molting. Termites have the typically insectan trait of shedding the chitinous linings of both their foreguts and hindguts with each molt (Weyer, 1935). The lining of both parts of the intestine are eliminated through the anus one or two days after the molt, and they carry with them the vital symbiotic protozoans from the hindgut. The newly molted termite, now deprived of its only means of digesting cellulose, must obtain a new protozoan fauna from its nestmates. This it accomplishes by caressing the terminal abdominal segments of another individual with its antennae and mouthparts, causing the extrusion of a proctodeal droplet. The proctodeal fluid almost certainly serves as a secondary source of nutriment as well, but its importance in this regard has not been analyzed. All instars of the Kalotermitidae, a relatively primitive group, possess flagellates. In the somewhat more advanced Rhinotermitidae, on the other hand, the first two instars lack flagellates and depend on regurgitated food from older workers to live. The young rhinotermitids are fed mostly with stomodeal liquid, although Buchli (1950) has observed newborn larvae of *Reticulitermes* feeding on proctodeal liquid.

The higher termites (Termitidae) do not depend on symbiotic flagellates for digestion of cellulose, and they have also lost the habit of proctodeal trophallaxis. At the same time, the immature stages have become completely dependent on stomodeal liquid. Unlike the larvae of the lower termites, termitid larvae are morphologically very distinct from older individuals that undergo radical transformation in either the third molt (subfamily Macrotermitinae) or second molt (other termitid subfamilies). Until this occurs they are entirely white, with soft exoskeletons and nonfunctional mandibles. Noirot (1952) believes that the liquid fed to them consists of pure saliva. The nymphs of the Termitidae also receive stomodeal liquid, but are able to feed on woody material or mycelial debris as well.

An important characteristic of trophallaxis in the majority of social insects is that it is an open system; each worker freely engages in it with an unlimited number of nestmates of all castes. This trait has been documented in detail with the aid of radioactive tracer studies and by prolonged direct observations. When Nixon and Ribbands (1952), for example, allowed six foraging honeybees to collect 20 ml of syrup treated with radioactive phosphorus and carry it back to the colony, they found traces of the material in 27 percent of the nest workers after only 5 hours, and in 55 percent after 24 hours. The rapidity of the exchange can be fully appreciated when it is recalled that honeybee colonies contain tens of thousands of workers. K. P. Istomina-Tsvetkova (in Free, 1959) found that workers of all ages engage in frequent food exchange during the normal life of the colony. She followed two workers from the age of 5 days to the age of 28 days, watching each one for 8 hours daily. These two individuals gave food on the average of 20.0 and 10.0 times, respectively, during each 8-hour period and received it on an average of 16.2 and 8.8 occasions, respectively. However, many of the contacts were brief and did not necessarily involve the actual exchange of liquid. Similar results have been obtained in ants by Gösswald and Kloft (1956, 1960), Wilson and Eisner (1957), Markin (1970), and others, and in termites by Alibert (1968). An actual case involving *Formica fusca,* a species that engages in rapid oral exchange, is given in Figure 14-8. Here it can be seen that, within one day following the feeding of a single worker, every member of the colony had received some of the liquid food and the variation in the amount per ant was rapidly decreasing. It is, therefore, not at all incorrect to refer to the combined crops of the adult population in a species such as *F. fusca* as the communal stomach. Not only do most or all of the workers share the same diet, but each worker is kept informed, on a virtually minute-by-minute basis, of the nutritional status of the colony as a whole. Thus, when a worker reacts to stimuli resulting from its own hunger, or satiation, the chances are great that it is acting for the colony as a whole.

This higher degree of integration in trophallactic species is reflected in a curious ambiguity in the interplay between the "opposite" acts of begging and offering. A begging honeybee attempts to thrust its proboscis between the mouthparts of a nestmate, while a bee that is offering food moves its still-folded proboscis slightly downward and forward from the position of rest, often extruding a drop of regurgitated liquid between the mandibles and base of the proboscis (Free, 1959). Both activities are accompanied by antennal stroking. Yet which role a given worker assumes appears to depend largely on the state of its crop content at the moment. When one well-fed bee is offered food by another, it often responds by offering food in return; when a hungry bee is solicited, it often solicits in return. When two bees are attempting to assume the same role, they actually appear to compete over the matter until one gives way and commences the opposite behavior (Lindauer, 1954b; Free, 1957b). Similarly, in the ants *Formica fusca* and *F. sanguinea* both antennal and foreleg stroking are used by donors and acceptors alike (Wallis, 1961). Foreleg stroking, like vigorous antennal play in the honeybee, indicates strong motivation. It also occurs more frequently in the acceptor, other things being equal. Whether a donor or acceptor *Formica* initiates the exchange depends on the relative strengths of the motivation of the two participants. A set of workers can be shifted from predominantly begging behavior to predominantly donor behavior merely by feeding it to satiation. In sum,

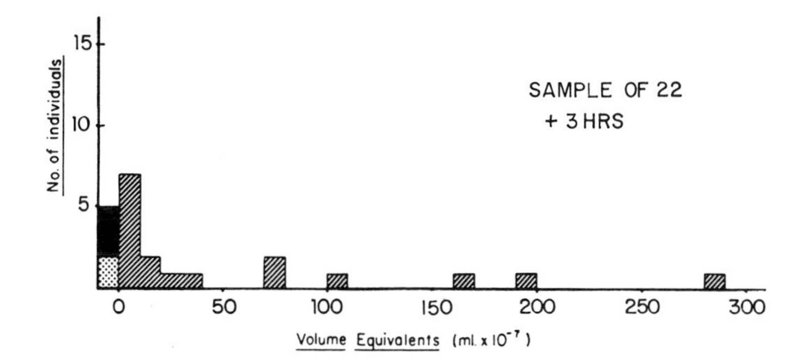

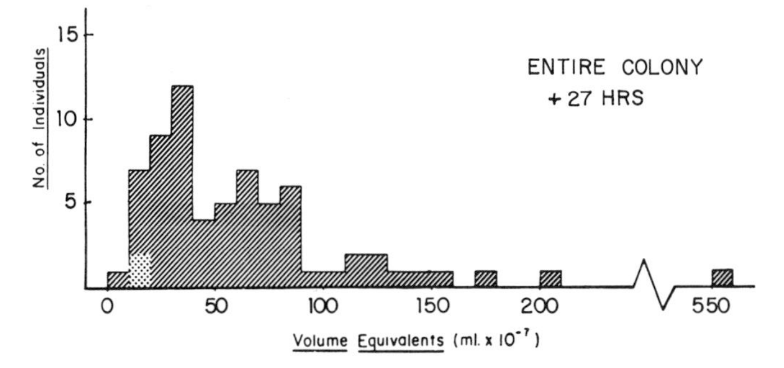

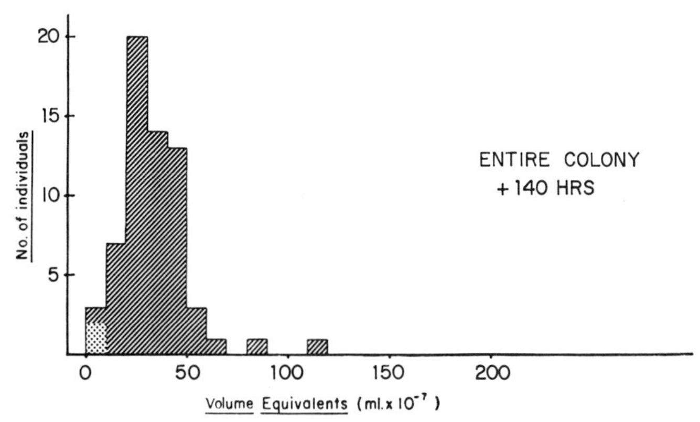

FIGURE 14-8. After a single worker of the ant *Formica fusca* was allowed to imbibe somewhat in excess of 1.9×10^{-4} ml of honey mixed with radioactive iodide (NaI^{131}) and return to its colony, this material was rapidly distributed throughout the colony. Within 27 hours every member of the colony had received some of the liquid, and the frequency distribution of individual shares began to assume a normal shape. The share of the two nest queens is indicated by the stippled portion of the histogram; as in the case of most carbohydrate feedings, this amount is very small (from Wilson and Eisner, 1957).

the lack of selfishness on the part of individual honeybee and ant workers and the ease with which individuals shift from one role to another insures a rapid and relatively even distribution of liquid food through the colony.

Free (1956) used a series of ingenious experiments to analyze the releasers of trophallactic behavior in the honeybee. More soliciting and offering is directed at the head than any other part of the body, and a freshly severed head is sufficient to elicit either reaction. Free noted that heads belonging to nestmates were favored over those belonging to aliens. So important is odor in fact that he even obtained occasional responses with small balls of cotton that had been rubbed against bees' heads. The antennae are also potent stimuli. Heads lacking antennae are less effective than those that possess them, and the loss can be restored by inserting imitation wire antennae of the right length and diameter into heads lacking antennae. Apparently the antennae serve not only as releasers but also as guides for the bees when they touch each other with their lower mouthparts.

The pattern of distribution of food materials in honeybee and ant colonies is nevertheless not wholly equitable. Queens and males receive more than they give. Also favored in the exchanges are younger workers, larger workers, and workers that frequent the same part of the nest. In his studies of *Formica polyctena,* Lange (1967) found that honey is distributed preferentially to workers with reduced ovarioles, while the hemolymph of prey is given more to those with developed ovarioles containing well-formed oocytes. This partition has a straightforward explanation: the former group consists of the older, foraging workers, which need energy-rich food; the latter group consists of the nurses, which require protein to feed the larvae and queens. It is not yet known whether the selection is performed by the donor or the acceptor, but the latter seems more likely.

Montagner's study (1966) of the social wasps *Vespula* (*Paravespula*) *germanica* and *V.* (*P.*) *vulgaris* indicates that trophallaxis in these insects is both more complex and structured than in honeybees and ants. When Montagner repeated Free's experiments, using the wasps instead of honeybees, the results were mostly negative. It is clear that the odor of severed worker heads attracts other workers who seem prepared to engage in food exchange, but the inert head is not sufficient by itself to induce begging or offering. Artificial wire antennae fixed to severed heads and vibrated at 20 to 100 cycles per second induced some

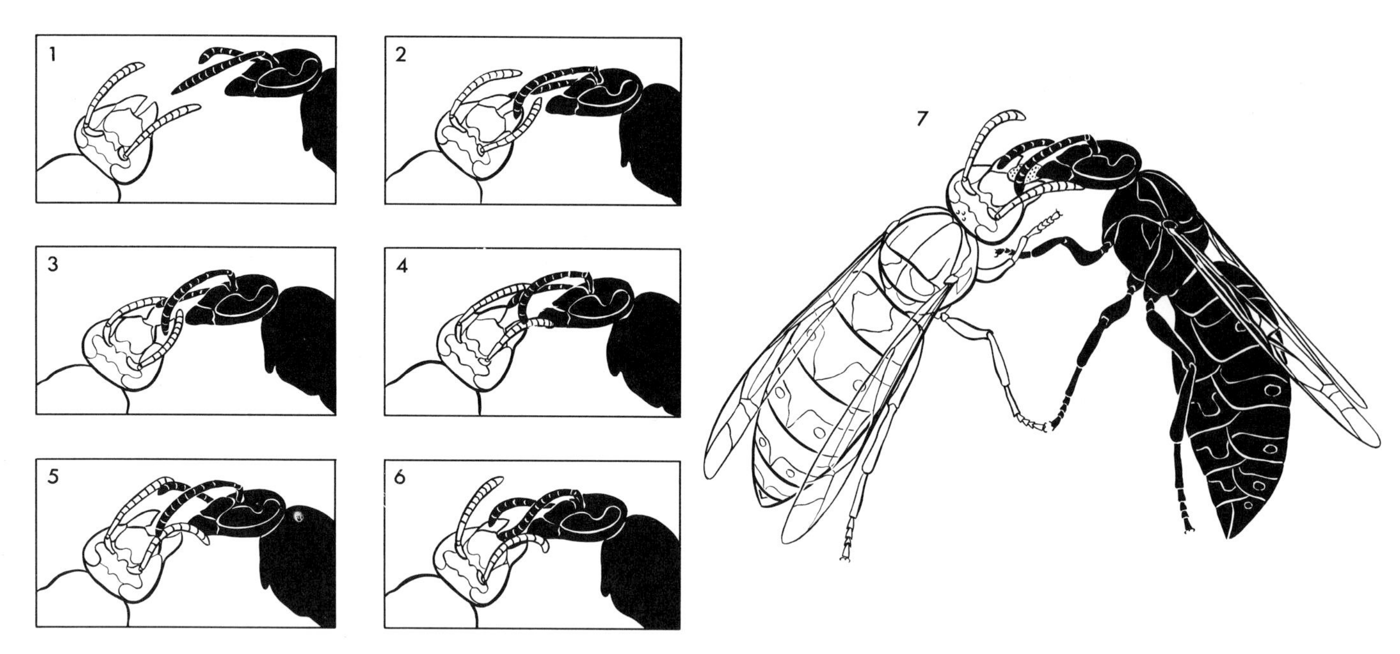

FIGURE 14-9. The initiation of regurgitation between two workers of the wasp *Vespula* (*Paravespula*) *germanica:* The solicitor on the right approaches the donor and places the tips of her flexible antennae on the donor's lower mouthparts (1, 2); the donor responds by closing her antennae onto those of the solicitor (3), who then begins gently to stroke her antennae up and down over the lower mouthparts (4–7). If this interaction continues, the donor will begin to regurgitate, and the solicitor is able to feed (based on Montagner, 1966).

regurgitation, but the live workers broke off contact in seven seconds or less. Montagner has shown that trophallaxis is sustained only when the pair engage in continuous reciprocal antennal signaling according to the specific pattern illustrated in Figure 14-9. The direction of transmission among *Vespula* workers proceeds according to their dominance relationships. Dominant workers receive more food than they give, so that donors are often seen assuming the crouching position taken by subordinate individuals. Strongly dominant workers who are thwarted in their begging attempts even go so far as to threaten the reluctant donors. The mother queen is the alpha individual who always receives and seldom if ever gives, while virgin queens are usually dominant over their worker sisters. Males have a quite different antennal form from females and are consequently inept at initiating trophallaxis. According to Montagner, they feed principally by two techniques: inserting their mouthparts into the regurgitated liquid being passed between two workers, and stimulating larvae to give up salivary secretions. Montagner's account of the state of affairs in *Vespula* is quite surprising since it depicts these insects as having a haphazard, not to say brutish, organization compared to that of ants and honeybees. The constraints placed on food exchange by the dominance relations would seem, at the very least, to slow the distribution of food and to increase the variance in crop contents among workers. However, the effect has not yet been measured by tracer studies of the kind employed on other social insects.

Chain Transport

Some ant species have evolved methods of relaying food from worker to worker so that the transport of the food over long distances from the foraging grounds to the nest is speeded up. The phenomenon, which can be conveniently referred to as "chain transport" (Stäger, 1935; Janina Dobrzańska, 1966), is best described by citing three good cases. The first two examples involve caste. In the arboreal myrmicine ant *Daceton armigerum* of South America, there exists a marked division of labor among the different-sized classes of foraging workers. The bigger foragers tend to loiter at way stations between the nest entrance and the hunting grounds, which are located chiefly in the forest canopy. When insect prey are captured, ordinarily by workers of medium size, they are usually pulled away from their captors by the larger individuals, which complete the homeward journey. The result is that the prey, regardless of which ant originally captures it, is carried home with what must be close to the maximum possible speed (Wilson, 1962b). A closely similar transfer system is used by the polymorphic African ponerine *Megaponera foetens,* which feeds principally on termites. Most of the prey are captured by the smaller workers, which then turn them over to major workers for transport back to the nest (Levieux, 1966). A third case involving castes and a well-marked division of labor has been described in *Formica obscuripes* (= *F. "melanotica"*) by King and Walters (1950). In this species a dramatic polyethism exists: the smaller, dark-colored workers forage for honeydew in low vegetation, and the larger, lighter-colored workers collect the material from them and transport it back to the nest. It seems probable, but cannot be proved from the limited data, that the small workers are more efficient at "milking" the homopterans because of the small size of the latter insects, while the large workers are better at transporting material by virtue of the greater volumes of their crops.

According to Dobrzańska (1966), the monomorphic formicine ant *Lasius fuliginosus* possesses an effective chain transport system which is not related to caste. The foraging workers remain away from the central nests for days on end, patrolling small sections of the trophophoric field with which they become personally familiar and resting from time to time in subterranean way stations. When an insect prey is secured well away from the nest, it is relayed from worker to worker. As the object reaches the border of a new patrolling area, it is transferred over to a worker belonging to that area who carries it across and passes it on to a worker in the next area, and so on, all the way back to the nest. Since workers move more rapidly and surely when on familiar ground, the relay system appears to be swifter than the more primitive method in which one worker carries a burden all the way by itself.

These examples, while suggestive, by no means prove that chain transport is intrinsically a more effective way of moving materials over long distances or that the behavioral interactions underlying it are necessarily adaptive in this context. The interactions, for example, could conceivably be side effects of other forms of communication, such as dominance relations or the strong drive to regurgitate food that would result in a tendency to transfer food whenever nestmates meet. In fact, Stäger (1935, 1939) concluded that chain transport in *Formica rufa* even slows the movement of food. Since workers of this species, unlike those of *Lasius fuliginosus,* are not very familiar with particular sections of the foraging field, it would seemingly be best if single individuals carried loads all the way home. When the transfers observed by Stäger were made, they consumed time and visibly slowed the transporting process.

Grooming

All social insects engage in frequent cleaning of their own bodies. The self-grooming movements are stereotyped and vary from group to group in a way that provides valuable aids to classification and the analysis of phylogeny (Ursula Jander, 1966). The cleaning instruments are the legs—especially (in Hymenoptera) the forelegs, whose curved pretarsus and large pectinate spurs are peculiarly modified for the purpose—and the lower mouthparts, including the tongue. The following thirteen basic movements have been recorded in ants (Wilson, 1962b; Farish, 1969).

1. Cleaning of the antennae, one at a time, by the ipsilateral fore tarsus and spur (Figure 14-10).
2. Cleaning of both antennae simultaneously by the respective ipsilateral fore tarsi and spurs.
3. The head is stroked by the fore tarsi.
4. The thoracic pleura and dorsum are stroked by the femora and tibiae of the forelegs.
5. The abdomen is stroked by the hind legs.
6. The abdomen is bent forward between the legs, and its tip is either stroked by the fore tarsi or washed by the lower mouthparts or both. This movement is unique to the ants (Figure 14-10).
7. The fore tarsi are washed by the lower mouthparts.
8. One fore tarsus is passed through the tibial-tarsal brush of the other fore tarsus, and then the action is reversed.
9. The fore and middle legs on one side are used to clean each other in the fashion described in the eighth movement.

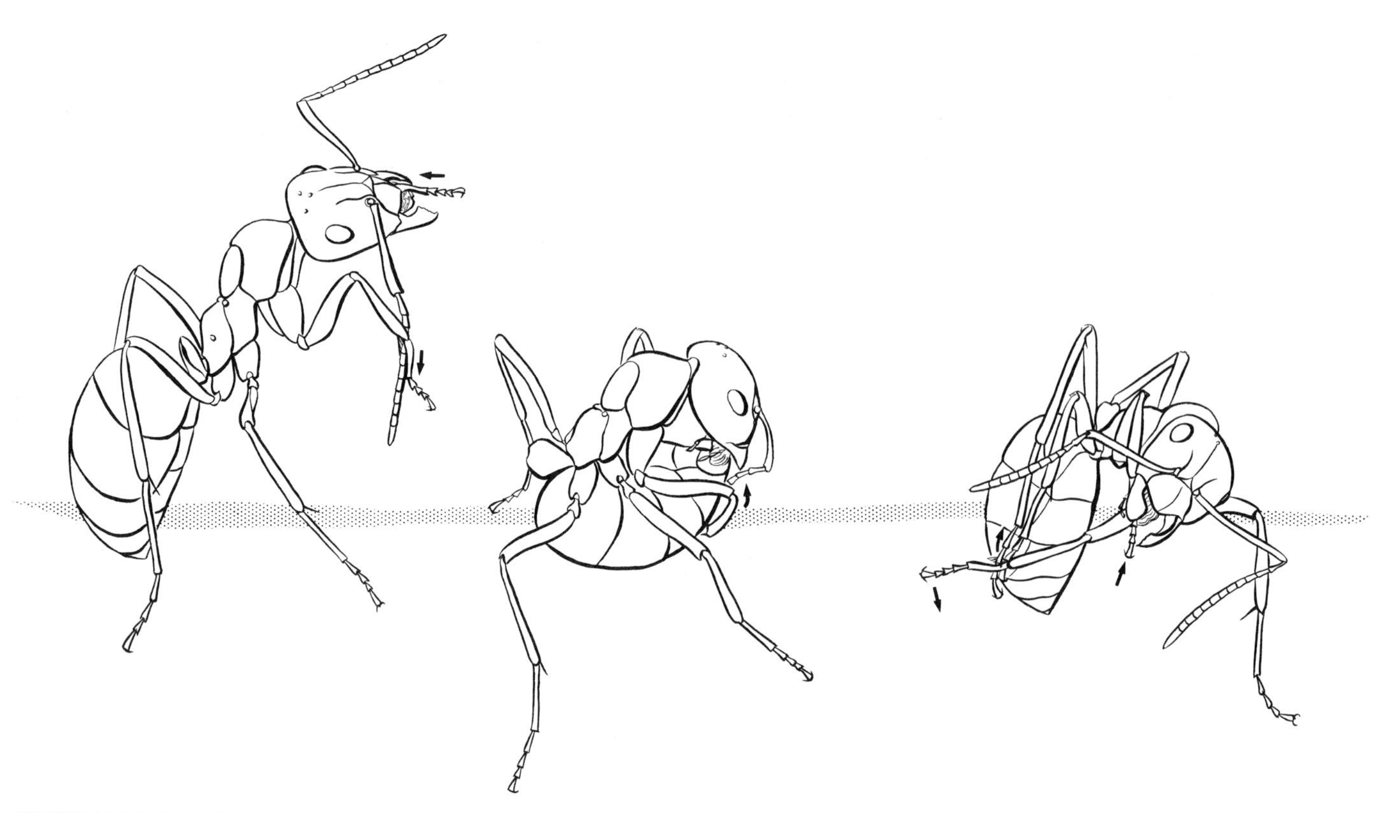

FIGURE 14-10. Three basic movements of self-grooming in the worker ant (*Formica polyctena*): (*left*) drawing an antenna through the comb of the right foreleg while simultaneously cleaning the left foreleg with the tongue; (*middle*) licking the tip of the abdomen; (*right*) rubbing three of its legs together while drawing the right front tarsus over the tongue (original drawing by Turid Hölldobler).

10. The tarsi of both forelegs and one of the middle legs reciprocally clean each other.
11. The tarsi of one foreleg and all three legs on the opposite side reciprocally clean each other. To accomplish this intricate maneuver, which is found only in the ants, the body forms a tripod consisting of the tip of the gaster and the unoccupied middle and hind legs of one side. During the operation one tarsus may be drawn over the mouthparts, as shown in Figure 14-10.
12. The tarsi of the hind legs reciprocally clean each other.
13. The tarsi of the hind legs and one middle leg reciprocally clean each other.

In summary, ants perform most of the self-grooming movements which their rigid exoskeletons make possible. But the relative frequencies of the movements vary enormously. Also, entire movements have been added or dropped, mostly dropped, in the course of evolution from the lower Hymenoptera. The total effect of self-grooming in ants is to collect detritus and secretory materials from the surface of the body and move it anteriorly to the mouth. Here most of it is collected in the infrabuccal pocket and later discarded in the form of waste pellets. But at least some of the material must find its way into the alimentary tracts, with possible effects that have yet to be investigated. Self-grooming movements are also likely to function in the spread of secretory materials over the body. In the one documented example, the queen honeybee transfers 9-ketodecenoic acid back over her body in this fashion.

Many, but not all, social insects also groom nestmates with their glossae (tongues). All ants, honeybees, and termites apparently do it frequently. At least some of the social wasps, for example, *Polistes* (Eberhard, 1969), groom nestmates. On the other hand, the phenomenon is only occasional in the meliponines (Sakagami and Oniki, 1963) and is evidently very rare or absent in the bumblebees (Free and Butler, 1959) and the primitively social halictine bee *Dialictus zephyrus* (Batra, 1966a). The significance of the presence or absence of grooming of nestmates is not at all understood at this time. It may well function, at least in part, in the transfer of pheromones. For example, it is known that the queen substance of the honeybee, 9-ketodecenoic acid, is initially transmitted in this fashion from the queen to workers (Butler, 1954b). Bumblebees, on the other hand, neither groom nestmates nor employ a queen substance.

15 Group Effects and the Control of Nestmates

Group Effects

In *Animal Aggregations* (1931) and *The Social Life of Animals* (1938), Warder Clyde Allee assembled a mélange of examples from the literature and his own experiments to illustrate the diverse ways in which crowding alters the physiology and behavior of organisms. He noted that in some animal species there is an easily perceived optimal density at which survival rate and fecundity are at a maximum. Below or above the optimum, these demographic parameters diminish in a complex and unpredictable manner. Their decline is due to one or more unrelated factors that crop up in bewildering variety: resistance to poison (versus self-poisoning with metabolic products), ability to evade parasites and predators, predilection for cannibalism, probability of finding mates, rate of fixation of lethal and subvital genes through inbreeding, and so on. Crowding increases learning rate in some animals (goldfish, for example), but decreases it in others (cockroaches and parakeets). In a few specialized species, population density even determines the sex of the participating individuals, a trait shown by the cladoceran genus *Moina,* for example, or their phase of polymorphism, as in the migratory locusts.

With but a moment's reflection one realizes that *all* species must have an optimal density, whether weakly or sharply defined, so that, in a sense, Allee's generalization is trivial. However, the eclectic phenomena of social physiology are made much more interesting by the proposition, advanced by Allee and later authors, that aggregating behavior has been designed in evolution to place population densities somewhere near the optimum levels of the respective species. An instructive example is the "Fraser Darling effect" observed in sea birds that nest colonially: the members of larger colonies stimulate each other to commence breeding earlier and over a shorter span of time than do members of small colonies. In the herring gull the effect results in a higher survival rate in the larger colonies. This is because the percentage of mortality in young chicks per unit of time, due chiefly to attacks from predators and adult herring gulls, is approximately a constant. Therefore, the shorter the breeding period, the lower the overall percentage of mortality (Darling, 1938). Darling's effect sets the lower limit of the optimal density. The upper limit is set by food shortages that develop in the sea around the breeding islands (Ashmole, 1963; W. H. Drury, personal communication).

The study of the effects of crowding per se has subsequently proved fruitful in ecology by revealing the kinds of density-dependent behavioral factors through which many, but not all, animal species limit their own population growth (Calhoun, 1962; Lack, 1966; Clark *et al.,* 1967). In France, Grassé and his associates have extended the concept in still another useful direction. For many years they have studied the effects of varying population density on the behavior and physiology of social insects—a wise choice as it turned out since the phenomena are elaborately developed in these most social of all animals. Grassé (1946) proposed to classify all the effects of "social physiology" into two categories: *mass effects,* in which the medium is modified by the population; and *group effects,* in which the members of the population affect one another directly by sensory stimulation. The phrase *effet de groupe* is used constantly by French zoologists; but Grassé's terminology has not caught on elsewhere because, like Allee's earlier exposition of "group behavior," it has been much too amorphously formulated. There is to start with no clear boundary between mass and group effects. Furthermore, it is almost impossible to make a sharp distinction between group effect in the Allee-Grassé sense and the rest of communicative behavior. Like the word trophallaxis, the expression "group effect" can, with little effort, be stretched to become synonymous with communication in the broadest sense.

Even so, I believe we can make at least temporarily good use of the label to cover a particular set of communicative phenomena that are of considerable importance in insect societies. I would therefore like to suggest the following refinement of Grassé's definition: *a group effect is an alteration in behavior or physiology within a species brought about by signals that are directed in neither space nor time.* Alarm signals, trail substances, and sex attractants obviously do not qualify. Most primer pheromones, which act on animals over long periods of time without necessarily evoking a directed response, do qualify. In addition, there exists a wide range of communicative phenomena in social insects which are undirected and long lasting but not necessarily pheromonal in nature. They are the coactions that the French investigators have been intuitively calling group effects. Most are examples of what has been termed social facilitation in the psychological literature, that is, communication which promotes activity rather than inhibiting it. The conception of social facilitation as a discrete phenomenon began in studies of human social psychology. Allport (1924) defined it as "an increase of response merely from the sight or sound of others making the same movement." To complete the terminology, the opposite effect from social facilitation should be labeled social inhibition. Blackith (1957) has described facilitation in the pure sense in social Hymenoptera. When workers of bumblebees return to the nest after a successful foraging trip, they are unable to direct their nestmates to the newfound sources. However, their entry into the colony causes an increase in movement, and a larger number of foragers leave the nest in the period immediately following. Workers of the social wasp genus *Vespula* tend to cluster around the nest entrance just before flying off. When one worker does fly, it is usually followed all at once by a group of others. Spieth (1948) observed that, when foragers of *Polistes* return to the nest with food, they set off a wave of activity that often extends throughout the small colony. Not only do many of the workers proceed to divide the food and pass it back and forth, but individuals that have not shared in the exchange also visit larvae for secretions, start cleaning themselves, or simply walk around. Ants of the species *Lasius emarginatus* excavate the soil and attend larvae at a higher rate when in large groups. When Francfort (1945) separated small groups of this species consisting of four to six workers from larger groups by only a gauze barrier, their activity rate remained high, but it dropped when he inserted a glass barrier. Francfort concluded that the facilitating stimulus is an odor. Walter Hangartner (1969a), in following up Francfort's experiment, discovered that fire ant workers (*Solenopsis geminata*) attempt to dig through porous barriers put up to separate them from other members of their colony. He was able to induce the same effect by substituting tubes containing slightly higher concentrations of carbon dioxide. It is entirely possible that Francfort's result was also due to carbon dioxide or some other general metabolic product, rather than a specialized "facilitation" pheromone.

The relation between activity and group size is not a simple one. Shisan C. Chen (1937a,b) reported that workers of the carpenter ant *Camponotus japonicus aterrimus* placed in groups of two or three in earth-filled jars began digging sooner, moved more earth per ant, and displayed less variation in individual effort than workers placed alone in the same kinds of containers. "Leader" ants, that is, the ants that worked best when alone, had a stimulating effect on nestmates, while the slower "follower" ants had a retarding effect. Leader ants also had a higher metabolic rate, as evidenced by their greater vulnerability to starvation, to drying, and to poisoning by chloroform and ether fumes. Sakagami and Hayashida (1962), in their studies of the slave-making ants *Polyergus samurai* and *Formica sanguinea* and the slave species *F. fusca,* obtained quite different results. When they increased the number of *F. fusca* workers to eight, the average digging performance in both the presence and absence of the slave-makers either remained the same or began to fall off. This is the direct opposite of the accelerating effect reported by Chen. The inhibition seems to be explainable at least in part by the fact that the "elites" (the term given to leader ants by Sakagami and Hayashida following a suggestion of Marguerite Combes [1937]) make up less and less of the sample as the sample size is increased. Stated another way, fewer workers assume the elite role in crowds.

How can we account for the opposing results from the

two experiments? It is possible that the design of the experiments might provide the explanation. Chen used spacious jars in which the ants worked down into a horizontal earthen surface, while Sakagami and Hayashida used narrow (a single centimeter wide) horizontal tubes in which the ants had to crowd past each other to approach vertical earthen surfaces. It is equally possible that *Camponotus* and *Formica* elites differ from each other in their pattern of response, or at least in the way their elitism is invoked. There can be no doubt that elitism is a well-marked phenomenon in ants generally. The work on the lifetime activities of individuals by Dieter Otto and others (see Chapter 8) has shown that workers from the same colony often differ profoundly in their work patterns and in the intensity of their day-to-day activity. Elitism is a common if not universal phenomenon in ants, but the way it is summoned in groups may vary among species.

A diversity of behavior patterns other than nest building are modulated by the size of the group. The aggressiveness of individual social insects increases as the size of the crowd of nestmates around it grows. In the subterranean ant *Acanthomyops claviger,* workers placed in solitude inside artificial nests are nearly insensitive to the natural alarm substances of the species, such as undecane and citronellal, but those placed in the same nests with a few hundred nestmates respond normally to the pheromones (Wilson, unpublished). Similar effects involving a lower threshold of numbers or "critical mass" have been well documented in the honeybee. Martin (1963) found that a minimum of two hundred workers is needed for swarming to be completed. A smaller number of bees can be induced to make the attempt in the company of a fecundated queen, but the effort fails because the group is unable to form a cluster away from the mother hive. Darchen (1957) has shown how the critical mass varies according to the presence or absence of the queen and queen substitutes in the honeybee hive. The following approximate minimal numbers of workers are needed to build combs within an ordinary hive:

50 when a live queen is present;
200 when a freshly dead queen is present;
5,000 when laying workers (but no queen) are present;
10,000 when neither queen nor laying workers are present.

Experiments by Jaycox (1970) demonstrated that a similar stimulating effect is exercised by the queen on the food-collecting activity of the workers. Approximately the same level of foraging activity can be attained by presenting the workers with preparations of synthetic queen substance, *trans*-9-keto-2-decenoic acid. In colonies of the termite *Kalotermes flavicollis,* the amount of liquid food exchanged, both that discharged from the mouth and from the anus, increases with the number of individuals present. The increase is accelerated when the royal couple are also present. In large colonies the royal couple are fed exclusively on saliva produced by stomodeal regurgitation. When isolated with only 30 nymphs in attendance, the couple are also given a small quantity of proctodeal material, and this amount increases sharply if the size of the group is cut still further (Alibert, 1968).

Finally, group size can affect fecundity and longevity in females in ways that cannot always be explained simply as the outcome of differences in nutrition. Free (1957c) found that the degree of ovarial development of bumblebee workers increases with the number of individuals kept together. He ascribed the effect to the social facilitation of feeding on the available food stores. The relation is far from perfect, however, since isolated workers occasionally undergo strong ovarian development. A similar and more effective form of facilitation occurs in the honeybee. When maintained in groups in the absence of the queen, some members of normal-sized colonies always develop their ovaries and even begin to lay eggs, but this does not occur if the workers are kept in complete solitude (Hess, 1942). Grassé and Noirot (1944) have found that ant and termite workers live for shorter periods of time when isolated than when kept in groups, even when they have an unlimited food supply. Maximum longevity is attained when the group size is on the order of ten or greater. When queens of the ant species *Lasius flavus* and *Solenopsis saevissima* are allowed to associate in small groups during colony founding, their survival rate is higher, and they rear the first brood more quickly (Waloff, 1957; Wilson, 1966). Workers of the army ant *Neivamyrmex carolinensis* survive longer in the presence of their queen—a result that can be due either to some unknown salutary effect of the secretions obtained from the queen's body, or the greater cohesion of the workers resulting from the queen's presence, or both (Watkins and Rettenmeyer, 1967). Survivorship is sometimes a function of age. If males of the carpenter ant *Camponotus herculeanus* are isolated in the early part of their life, at a time when they would normally be engaging in liquid food exchange, their lives are shortened. But this is not the case if they are isolated in middle age, at the time they would normally hibernate, or later in the spring, when they are ready to leave on their nuptial flights (Hölldobler, 1966).

In sum, the "group effects" that have been described in social insects are a curiously miscellaneous lot of important but little-understood communicative phenomena. Research will have to be pressed at the neurophysiological and biochemical levels before much knowledge is gained of their causal mechanisms and multiple roles in social regulation.

Dominance Hierarchies and the Evolution of Queen Control

In the primitive insect societies strife and competition prevail, leading to the emergence of a single female, the queen, which physically dominates the rest of the adult members of the colony. In the more advanced insect societies, particularly those in which strong differences between the queen and worker castes exist, the queen also exercises control over the workers but usually in a more subtle manner devoid of overt aggression. The gradual change from what might be regarded as brutish dominance to the more refined modes of queen control is one of the clear-cut evolutionary trends that extend across the social insects as a whole, and one worth tracing in some detail.

Aggression and dominance were first recorded by the Swiss entomologist Pierre Huber (1802) in his pioneering study of bumblebees. He saw that, when a queen lays her eggs, some of the workers try to steal and eat them, but the queen usually repels these intruders with great fury. Similar observations have been made by a long line of subsequent investigators. The Austrian entomologist Eduard Hoffer (1882–1883) discovered that the dominance relationship is orderly and predictable. As he described it: "Punishment is almost always meted out with the legs and mandibles, and the guilty individual never even attempts to defend herself, all her efforts being directed toward making the quickest possible escape. The punishment is sometimes so rough that the poor creature is seriously wounded or even killed." Once egg stealing has been rebuffed for a few hours, the attempts by the workers "become less and less frequent, and finally cease altogether; and these same little insects which previously tried their very best to destroy the newly-laid eggs, now become attentive guardians and nurses of their embryo brothers and sisters; they keep them warm and tenderly provide them with nourishment continuously thereafter."*

The early observers were puzzled by what seemed to them an impulse toward race suicide. Huber offered a surprisingly modern hypothesis that invokes density-dependent control of the kind most recently envisioned as a "social convention" by Wynne-Edwards (1962). He stated, "The bumblebees are the largest insects that feed on honey; and if their number trebled or quadrupled, other insects would not find any nourishment, and perhaps their species would be destroyed." Pérez (1899) viewed egg eating as simply evidence of selfish behavior on the part of the workers and therefore an imperfection in the social order. But there is a more straightforward explanation of the phenomenon, one that involves a function adaptive to the colony as a whole. Lindhard (1912) observed an instance in which a queen of *Bombus lapidarius* extracted the egg of a rival worker and fed it to a queen larva. Also noting that very few if any worker-laid eggs ever survive, he hypothesized that the objects are not meant to develop but instead serve as a kind of "royal food" for prospective queens. If there is any truth in this conclusion, it is consistent with the general occurrence of trophic eggs laid by workers in other kinds of social Hymenoptera (Chapter 14).

The existence of a similar dominance system in the paper wasp *Polistes gallicus* was intimated by the studies of Heldmann (1936a,b), who found that, when two or more females start a nest together in the spring, one comes to function as the egg layer, while the others take up the role of workers. The functional queen feeds more on the eggs laid by her partners than the reverse, thus exercising a form of reproductive dominance. Pardi (1940, 1948, 1951) discovered that the queen establishes her position and controls the other wasps by direct aggressive behavior, and he went on to analyze this form of social organization in detail. Dominance behavior has also been documented in *P. chinensis* by Morimoto (1961a,b), *P. fadwigae* by

*Since this phase of work on the biology of bumblebees has been largely forgotten, even by entomologists, it has become the custom to give the credit for the discovery of dominance hierarchies to Schjelderup-Ebbe (1922) and later students of social behavior in domestic fowl and other vertebrates. But it is clear that the same basic conception had been worked out many years earlier by Hoffer in his studies on bumblebees. Furthermore, Hoffer's observations were confirmed and refined by later investigators, especially Lindhard, well in advance of the vertebrate studies.

Yoshikawa (1963b), and *P. canadensis* and *P. fuscatus* by Mary Jane Eberhard (1969). The relationship among the adult members of the *Polistes* colony is somewhat more elaborate and stereotyped in form than that in *Bombus* colonies. Instead of simple despotism by the queen, there exists in most cases a linear order of ranking involving a principal egg layer, the queen, and the remainder of the associated females, called auxiliaries, who form a graded series in the relative intensities of egg laying, foraging, and comb building (Table 15-1). Dominant individuals receive more food whenever food is exchanged, they lay more eggs, and they contribute less work. They establish and maintain their rank by a series of frequently repeated aggressive encounters. At lowest intensity the exchange is simply a matter of posture; the dominant individual rises on its legs to a level higher than the subordinate, while the subordinate crouches and lowers its antennae. At higher intensities leg biting occurs (Figure 15-1), and at highest intensity the wasps grapple and attempt to sting each other. During the brief fights, the contestants sometimes lose their hold on the nest and fall to the ground. Injuries are rare, although Eberhard once saw a female of *P. fuscatus* killed in a fight. The severest conflicts occur between wasps of nearly equal rank and during the early days of the association when the nest is first being constructed. As time passes, the wasps fit more easily into their roles, and aggressive exchanges become more subdued, less frequent, and eventually purely postural in nature. When the first adult workers eclose from the pupal instar, during later stages of colony development, their relations to the foundresses are invariably subordinate and nonviolent. The workers also form a hierarchy among themselves; those that emerge early tend to dominate their younger nestmates. If the alpha female is removed, the next highest auxiliaries intensify their hostility to each other until one of their number becomes unequivocally dominant; then the exchanges subside to the previous low level. Eberhard (1969) found that the tropical species *P. canadensis* differs from all the temperate species studied to date in that its contests are more violent and result in the losers leaving the nest to attempt nest founding on their own elsewhere. The queens are also less dependent on the indirect technique of egg eating to maintain reproductive control. Thus *P. canadensis* approaches a condition of true despotism, whereas the temperate species so far studied possess a colony organization that can be characterized as an uneasy oligarchy.

The dominant females of *Polistes* colonies maintain their superior reproductive status by three means: they demand and receive the greatest share of food whenever it becomes scarce; they lay the greatest number of eggs in newly constructed brood cells; and they remove and eat the eggs of subordinates when these rivals succeed in laying in empty cells. The size and degree of development of the ovaries is loosely correlated with the rank of the wasp (Pardi, 1948; Deleurance, 1952b; Gervet, 1962). When a female slips in rank, her ovaries also decrease in size. It is tempting to think that subordination leads to decrease in ovarian development simply because the individual receives less food; in other words, she is subjected to "nutritional castration" of the sort originally conceived by Marchal (1896, 1897). The same effect might be achieved by the closely related phenomenon of "work castration," in which the subordinates are forced to ex-

TABLE 15-1. Division of labor according to rank within a group of colony-founding females of the paper wasp *Polistes fuscatus* observed for 26 daylight hours between 18 May and 14 June 1965 (from Eberhard, 1969).

Identification number	Dominance rank	Number of eggs laid	Number of eggs of others eaten	Number of new cells started	Foraging rate (loads per hour observed)	Number of loads received from associates
13	1	9	4	0	.08	25
34	2	5	2	1	.50	20
35	3	0	0	1	1.41	0
28	4	0	0	2	1.56	5
15	5	0	0	8	1.80	3
6	6	0	0	0	1.22	1
18	7	0	0	0	1.50	0

FIGURE 15-1. In this exchange between two colony-founding females of *Polistes fuscatus*, the dominant wasp has risen on its legs above the subordinate one and seized it by a hind leg. The subordinate crouches and lowers its antennae (from Eberhard, 1969).

pend more of their energy reserves in foraging and nest building (Plateaux-Quénu, 1961; Spradbery, 1965). But the matter is far more complex than this. Energy deprivation and ovarian development very likely play some kind of a role, but other factors are at least equally important. When Gervet (1956) chilled the queens of *Polistes gallicus* overnight, to a degree that inhibited their egg production but not their daily activity, they retained their dominant status. They left more brood cells unfilled, however, and this in turn stimulated ovarian development in the subordinates. Deleurance (1948) was able to remove all of the ovaries of dominant *P. gallicus* by surgery, yet even this treatment did not affect their rank. Thus rank seems to determine ovarian development, while the reverse is not the case. Moreover, Gervet's result precludes the possibility that ovarian development is controlled in a straightforward manner by nutrition. What evidently matters most is behavioral control over the empty brood cells. But what determines this more purely psychological phenomenon in turn is not at all clear. Experience seems to have something to do with it, since the first females arriving at new nest sites tend to dominate later arrivals.

It is also true, as noted already, that older workers tend to dominate their younger nestmates.

The despotisms and dominance hierarchies of *Bombus* and *Polistes* have many qualities in common with aggressively organized vertebrate societies. But are they also similar in being based on recognition of individuals and memory of past experiences? This is at least a theoretical possibility in the smaller *Bombus* and *Polistes* colonies, which contain from several to a few tens of individuals. It is still a possible factor in colonies of a few hundred individuals, where the queens and high-ranking auxiliaries comprise a small company of elites who stay close to the active brood cells and are therefore in frequent contact. However, it becomes very difficult to conceive in the largest insect societies. Sakagami (1954) observed mild hostility among worker honeybees when the queen was absent, but he could see no evidence of a regular, *Polistes*-like dominance hierarchy. He argued that since honeybee colonies contain tens of thousands of workers, few of whom live for more than a month, such a complex system of personal relationships is an impossibility.

Sakagami's reasoning is correct to a point, but it does not preclude a looser dominance system based on variation in individual aggressiveness—as opposed to learned, pairwise relationships. In fact, Montagner (1966) has described just such a system in several European species of the social wasp subgenus *Vespula* (*Paravespula*). These insects form large annual colonies, which at the height of the season in late summer consist of a single mother queen and thousands of workers. As in *Polistes*, dominant workers assume an erect stance and subordinate ones crouch low during their encounters. Dominant workers take food, and subordinates regurgitate it to them. Often subordinate individuals regurgitate to each other while both assume crouching postures. Since the workers are sterile, competition in egg laying and egg stealing is not involved. All of the workers defer to the queen in both food exchange and oviposition. The virgin queens, who appear only in late summer, are dominant over the workers. As the numbers of virgin queens increase, and the foraging workers become fewer, there is fierce competition over the dwindling food supply, leading even to the point of fighting. The disruption contributes to the final breakup of the colony in the fall. Montagner, by marking and following individual wasps, showed that the general level of dominance of single workers fluctuates to a limited extent independently of that of its nestmates, resulting in a dynamic equilibrium in the overall frequencies of different categories of encounters during food exchange. As a consequence, the dominance behavior plays a determining role in the division of labor. Montagner did not attempt to ascertain whether the wasps recognize one another individually, but, in view of their large numbers, this seems unlikely. I have watched motion pictures made by Montagner of his colonies with this point in mind, and it seemed to me that the outcomes of the encounters could just as easily be explained as ad hoc resolutions based on the relative aggressiveness (or eagerness) of the pairs of workers at the moment they meet. The *Vespula* case, involving as it does seemingly selfish behavior among members of a sterile worker caste in the presence of the queen, nevertheless raises some difficulties in the genetic theory of social behavior. I have tried to resolve them elsewhere, in Chapter 17.

Except for the peculiar situation in *Vespula* (and a probably similar situation in the closely related genus *Vespa;* see Plateaux-Quénu [1961]) dominance based on overt aggression is virtually absent in the more advanced insect societies with sterile, morphologically distinct worker castes. In the stingless bees, the queen eats worker eggs—indeed, they are a principal component of her diet—but these are presented to her by the workers in the course of brood cell construction in a highly stereotyped fashion (see Chapter 5). At other times workers tend to avoid the queen and will often crouch in what appears to be a submissive posture when they encounter her face to face. But, with the possible exception of vertex tapping in *Melipona quadrifasciata anthidioides,* the queen does not behave in an aggressive manner (Sakagami *et al.,* 1965). Honeybee workers, as just mentioned, are mildly aggressive toward each other in the absence of the queen, but form no structured dominance relationships. Ant and termite workers, so far as I know, never show aggression toward nestmates under normal conditions, and no relationships that could be construed as dominance hierarchies in the usual sense appear to develop.

The relationship between queen and workers in the higher social insects can be characterized in a phrase: gentle despotism. Under the influence of continuous, nonaggressive signals from the queens, the workers contribute selflessly to the upkeep of the queen and to the rearing of her offspring, which it should be remembered are their brothers and sisters. In most cases where this advanced form of queen control has been analyzed, it has turned out to be mediated by one or more pheromones passed from the queen to the worker. The best-known example is queen control in the honeybee, *Apis mellifera.* When the mother queen of a honeybee colony is removed, the workers respond in as short a time as 30 minutes by changing from a state of organized activity to one of disorganized restlessness. In a few more hours they begin to alter one or more worker brood cells into emergency queen cells, within which a new nest queen is eventually produced. A few days later, some of the workers begin to experience increased ovarian development. These combined releaser and primer effects were found by Butler (1954b) to be at least partly due to the removal of a "queen substance," an inhibiting pheromone produced continuously by the queen. Now it appears that at least two inhibitory pheromones are involved in these effects. Moreover, the special treatment accorded the queen is due to at least two additional attractive scents, one of which is manufactured by Koschevnikov's gland at the base of the sting (Butler, 1961; Butler and Simpson, 1965).

One of the inhibitory pheromones present in the queen honeybee is *trans*-9-keto-2-decenoic acid. This substance was characterized in 1960 by Butler, Callow, and Johnston in England and, at the same time, by Barbier and Lederer in France. It is produced entirely in the queen's mandibular glands. The odor of the ketodecenoic acid alone is sufficient to inhibit to some extent both queen-rearing behavior and ovary development in worker bees. It works in conjunction with at least one additional inhibitory scent produced in a part of the body other than the mandibular glands. When experimentally injected into the body cavity of the worker, thus bypassing the external chemoreceptors, it continues to inhibit ovary development but not queen-rearing behavior (Butler and Elaine Fairey, 1963). Lüscher and Walker (1963) found that the corpora allata of workers increase in size for the first few days following removal of the queen, and they hypothesized that the inhibitory pheromones act by suppressing secretion of the "gonadotropic" hormone. Using synthetic material, Gast (1967) then confirmed that the queen substance inhibits development of these endocrine glands. Whether the pheromones produce their effects by direct action on the corpora allata or indirectly through a more circuitous route in the central nervous system remains to be learned.

The queen must dispense on the order of 0.1 μg of 9-ketodecenoic acid per day per worker in order for the inhibitory effect to work on the colony as a whole. Yet at any given moment she carries only about 100 μg of the substance in and on her body (Butler and Patricia Paton, 1962). The question is then raised as to how the queen manages to generate the 2 mg or more a day needed to supply her 20,000–80,000 daughters. We also know that, when the queen is removed, the supply of 9-ketodecenoic acid falls below the threshold level within hours. How is such a relatively stable molecule deactivated so quickly? Possible answers to both questions have been supplied by the recent studies of Norah Johnston, Law, and Weaver (1965). These investigators traced the metabolism of a radioactive form of the pheromone fed to worker bees and found that within 72 hours, 95 percent of it had been converted into inactive substances, consisting principally of 9-ketodecanoic acid, 9-hydroxydecanoic acid, and 9-hydroxy-2-decenoic acid. The metabolic route shown in Figure 15-2 was then postulated. Johnston *et al.* have further hypothesized the existence of a "pheromone cycle." The inactive molecules might be passed back to the queen as part of the regurgitated glandular queen food. The queen could then convert them into the active form by very simple enzymatic processes, resulting in a saving to the queen of a relatively enormous amount of energy required for the complete synthesis of the fatty acid chain.

An additional puzzle was created by the early findings on inhibition by the "queen substance:" How can a normal honeybee colony produce new queens during the annual breeding season in the presence of the mother queen? The solution was obtained by Butler (1960). By appropriate experiments, he first excluded the hypothesis that the attractiveness of "queen substance" is greater for preswarm workers than for workers from nonswarming colonies. Then he determined that the amount of "queen substance" (as measured by the inhibiting power of whole-body extracts) in mother queens from colonies involved in normal reproduction (by means of swarming) is only about a quarter that of mother queens from nonreproducing colonies.

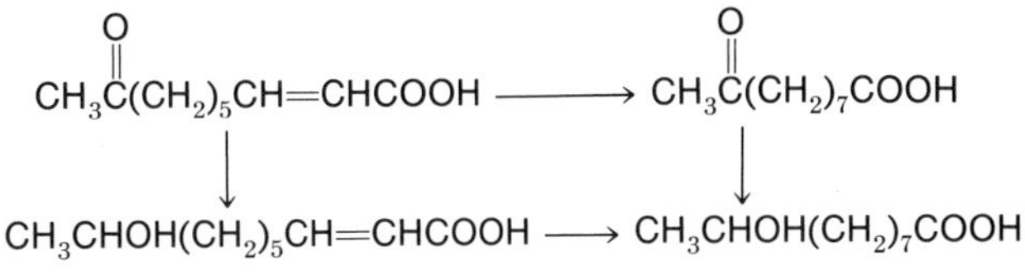

$$\begin{array}{ccc} CH_3\overset{O}{\overset{\|}{C}}(CH_2)_5CH{=}CHCOOH & \longrightarrow & CH_3\overset{O}{\overset{\|}{C}}(CH_2)_7COOH \\ \downarrow & & \downarrow \\ CH_3CHOH(CH_2)_5CH{=}CHCOOH & \longrightarrow & CH_3CHOH(CH_2)_7COOH \end{array}$$

FIGURE 15-2. *Trans*-9-keto-2-decenoic acid and its inactive derivatives produced within the body of the worker honeybee (based on Johnston, Law, and Weaver, 1965).

In view of the fact that the "queen substance" of the honeybee can no longer be equated with 9-ketodecenoic acid alone, Butler and Paton (1962) suggested that the term be discontinued. However, "queen substance" is well entrenched by now, and deservedly so. I would like to suggest that it be continued to loosely designate any pheromone, or set of pheromones, used in queen control.

One additional class of queen substances, therefore, are the caste-inhibitory pheromones produced by the primary reproductive females of the lower termites. These pheromones are decisive in their action—preventing pseudergates from transforming into secondary reproductives—and act over a period of only a few days. Although their chemistry has not yet been elucidated, a great deal is known about their anatomical sources, the means by which they are transmitted, and their varied physiological effects (see Chapter 10).

Queen control over workers and brood also occurs widely in the ants. Its action, which is usually indirect, varies greatly among species. The control involves at least three forms of inhibition. First, the presence of the queen can restrict worker ovarian development, an effect that has been observed thus far in *Leptothorax* (Bier, 1954b), *Myrmica* (Mamsch and Bier, 1966), *Plagiolepis* (Passera, 1965a) and *Formica* (Bier, 1956). These genera represent three of the most divergent subfamilies of ants—Myrmicinae, Dolichoderinae, and Formicinae. Further research may reveal that worker ovary inhibition is even more widespread, if not universal, in the ants. Brian (1958a), in drawing attention to the many cases known of workers laying in the presence of queens, erroneously concluded that in ants "the queen does not inhibit worker oviposition entirely, if at all." We now know that in many such cases the eggs laid by workers in the presence of queens are specialized trophic eggs, incapable of developing and serving no purpose other than to provide protein for the queen and larvae. Passera (1965a) has shown that workers of *Plagiolepis pygmaea* lay both trophic and normal eggs when isolated from queens, but only trophic eggs when a queen is present.

The second potential effect of the queen's presence is to reduce the percentage of larvae that develop into queens, and therefore to increase the percentage that develop into workers. This form of inhibition, so obviously parallel to the caste inhibition of honeybees and termites, has been documented in species of *Monomorium* (Peacock *et al.*, 1954), *Myrmica* (Brian and Carr, 1960), *Plagiolepis* (Passera, 1965b), and *Formica* (Bier, 1956). A third effect of the queen, so far recorded only in *Formica* (Bier, 1956), is the suppression of mating on the part of the virgin queens and workers while they are in the vicinity of the nest.

The various forms of queen control have been ascribed to two very different physiological mechanisms. Cecily Carr (1962) reported that when dead *Myrmica* queens were presented at regular intervals to queenless colonies, their presence suppressed the growth of the larvae—and therefore reduced the likelihood of the larvae transforming into queens. The effect was removed when the corpses were separated from the larvae and nurse workers by a double gauze barrier, which prevented them from being touched or licked. Carr concluded that the inhibition is due to a relatively nonvolatile pheromone. Direct evidence for such a substance was later claimed by Brian and Hibble (1963), who obtained inhibition of larval growth by painting ethanol extracts of queen heads on the larvae and workers and also by feeding it to them in sugar. Brian and Blum (1969) subsequently narrowed this activity to the sterol fraction of the extracts.

In contrast, Bier (1958) postulated that the inhibition of worker ovaries (but not necessarily larval development and caste determination) is due to a shortage of "profertile substances" normally produced in the glands of the workers. This material is considered to be necessary for full ovarian development. When the queen is present, she appropriates a lion's share of it, leaving too little for the workers to develop their ovaries on their own. Mamsch (1967) has presented indirect evidence for the existence of profertile substances in *Myrmica ruginodis.* He found that, when larvae are absent, the queen's power to suppress the ovaries of her workers is decreased. The result is consistent with the notion that the queen, the larvae, and the workers all compete for the profertile substances. It is conceivable that when the larvae are removed, enough of the substances are shunted to the workers to make development of their ovaries possible.

None of this evidence, for either inhibitory pheromones or profertile substances, is convincing. It is still too early to pick between the two hypotheses, or even to exclude the possibility that both forms of queen control exist. Clearly, the problem needs to be resolved by the chemical identification and quantitative assay of whatever pheromones or food substances are involved.

It is likely that queen control in ants will prove very

complicated. This is indicated not only by the growing complexity of the experimental evidence just cited, but also by some peculiar subsidiary phenomena that have been reported in social parasites. Queens of the workerless parasitic species *Myrmecia inquilina,* for example, in some unknown fashion inhibit the production of brood belonging to their host species, *M. nigriceps* and *M. vindex,* even though they tolerate the presence of the host queens and in fact show no trace of overt hostility to them (Haskins and Haskins, 1964). An even stranger relationship exists between *Plagiolepis xene,* another workerless parasite, and its host species *P. pygmaea.* As in the mixed *Myrmecia* colonies, the host and parasite queens live amicably side by side. But the presence of *xene* queens reduces fecundity in *pygmaea* queens, while the presence of *pygmaea* queens raises it in the *xene* queens (Passera, 1966).

In the eusocial halictines *Halictus duplex* (Sakagami and Hayashida, 1958) and *Evylaeus marginatus* (Plateaux-Quénu, 1961), removal of the queen promotes ovarian development and egg laying in the workers. The nature of the control has not been investigated. Also, the existence of queen control in other halictines remains only an untested possibility.

Finally, a subtle form of queen control involving pheromones exists in the vespine wasps. Long ago Marchal (1896, 1897) discovered that workers isolated from the queen undergo increased ovarian development. Montagner (in Spradbery, 1965) found that workers taken from colonies prior to male production in late summer initiate ovarian development within ten days, but during male production they require only two days. These observations suggest that workers are adapted in some way to contribute male-determined eggs of their own. In fact, Montagner (1963a) has noted that the queen tends to remain on the lowest comb where the greatest activity occurs, while males tend to be produced in the uppermost combs. It is conceivable that workers farthest removed from the queen are contributing male-determined eggs. A more direct form of evidence pointing to a true queen substance was finally adduced by Ishay, Ikan, and Bergmann (1965). These Israeli biologists noted that the single mother queen of each large colony of *Vespa orientalis* is very attractive to the workers, who frequently crowd around her and lick her body, especially her head and mouthparts (Figure 15-3). When a queen is removed, the workers undergo a series of striking changes reminiscent of honeybees deprived of queen substance. Within hours they become

FIGURE 15-3. The mother queen of *Vespa orientalis* is attractive to workers in the intervals between ovipositions. The workers arrange themselves in a circle around her and lick her body, especially the head (from Ishay and Schwarz, 1965).

quarrelsome and join in fights that are sometimes severe enough to be fatal. Many leave the nest and roam outside. Normal care of the brood is disrupted, and after a few days some of the workers even begin to lay eggs. Ishay and his associates found that alcoholic extracts from mother queens of *V. orientalis* were sufficient to attract and to calm the workers. When placed in cotton wads, this material was also licked by the wasps. Next, Ikan *et al.* (1969) identified a substance from the head of *V. orientalis* with at least some of the expected properties of the queen substance. This is δ-*n*-hexadecalactone:

$$CH_3(CH_2)_{10}\text{—[six-membered lactone ring: O, =O]}$$

In synthetic form this substance (unlike other substances tested, including other δ-lactones) is attractive to workers. It also induces queenless groups of workers to build queen cells at the end of the growing season, during September and October, a behavioral response that normally occurs only in the presence of an egg-laying queen.

16 Social Homeostasis and the Superorganism

A fundamental property of all life is the ability to maintain physiological steady states. In a properly tuned organism, the constant values attained in pH, in concentrations of dissolved nutrients, in proportions of active organelles, and so on, are always held close to the optimal values for survival and reproduction. This homeostasis, as W. B. Cannon first termed it in 1932 in *The Wisdom of the Body,* is biology's refined analogue of the control systems of advanced man-made machines. Like machine systems, physiological homeostasis is self-regulated by internal feedback loops that increase the values of important variables when they fall below certain levels and throttle them down when they exceed other, higher levels. The efficiency of homeostatic devices can be evaluated by the following standard test used in engineering: first the system is perturbed by altering the environment, then measurements are taken of the deviation of the variable from the optimum and the speed with which it returns. A common variant of the method is to place the system in a constant environment that differs from the optimum. Under this circumstance the property concerned will respond by fluctuating either around the optimum or at some level removed from it. A familiar machine analogue is the slight fluctuation in the temperature of a room containing a thermostat. The two measures of homeostatic efficiency that can be taken in this special kind of environmental alteration are, first, the amount by which the average value deviates from the optimum (the difference between the average room temperature and the thermostat setting), and, second, the magnitude of the fluctuations (the average variation in room temperature). In recent years the analyses and descriptions of steady state systems in physiology have become increasingly mathematical and are now usually incorporated into general control system theory, a branch of engineering that was created during World War II to facilitate the design of automatic weapon systems (Grodins, 1963).

At a higher level, social insects display marked homeostasis in the regulation of their own numbers and of their nest environment. This class of steady-state regulation has been aptly termed social homeostasis by Emerson (1956b). An example, illustrated in Figure 16-1, is the regulation of nest temperature by two kinds of social insects, the honeybee and a species of social wasp in the genus *Vespula,* when their nests are placed in cold air. It can be seen that not only do the honeybees maintain a higher nest temperature but they also permit less fluctuation. On both counts, therefore, the honeybee colony thermostat can be said to be superior to that of the wasps. A particular homeostatic control of this sort can be most succinctly represented by the standard block diagram of systems analysis given in Figure 16-2.

Thermoregulation by Honeybees

The regulation of nest environment is a general capability of the social insects. It is minimal or absent in those more primitive groups that form very small colonies but universal in the groups that form large and complex colonies. Control is achieved in one or the other of two very general ways: through nest structure designed to achieve long-term control over a wide variety of environmental change, or through the short-term behavioral responses of individual colony members to particular environmental perturbations. Both methods are sometimes employed together. For example, many ant and termite

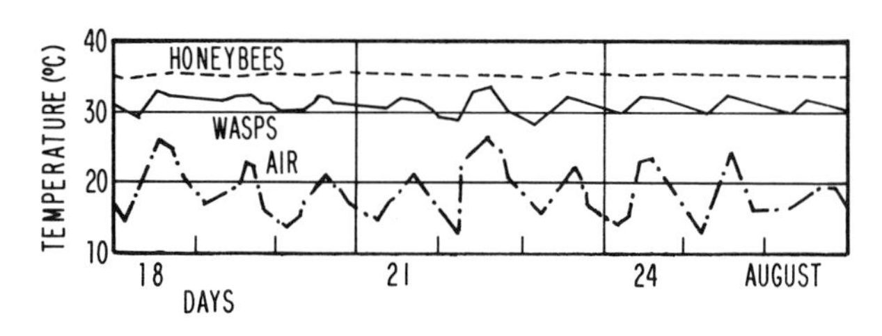

FIGURE 16-1. The nest temperature maintained by colonies of honeybees (*Apis mellifera*) and social wasps (*Vespula vulgaris*) in cold air (redrawn from Kemper and Döhring, 1967).

nests are elaborately constructed to cause automatic temperature and humidity regulation within the brood chambers. In addition, the individual colony members accomplish a fine adjustment by moving the brood to those precise spots within the chambers that contain the closest approach to the preferred microenvironment.

The best understood of all control systems is thermoregulation in the honeybee colony. A favorite subject of melittologists for decades, it has been especially well studied by Gates (1914), Park (1925), Hess (1926), Himmer (1926, 1927), H. F. Wilson and Milum (1927), Corkins (1930), Anderson (1948), and Lindauer (1954b). Its physiological basis has been well reviewed by Ribbands (1953); the behavioral control mechanisms, by Lindauer (1961). The honeybee colony makes an important first step toward thermoregulation by selecting a nest site, such as a hollow tree trunk or artificial hive, that tightly encloses the brood combs and the majority of the adult workers at all times. The workers use various plant gums, collectively referred to as propolis, to seal off all crevices and openings except for a single entrance hole. This procedure not only keeps enemies out, but, just as importantly, it holds in heat and moisture. The precision of temperature control then attained within the nest is astonishing. From late spring to fall, when the workers are foraging and brood is present and growing, the interior temperature of the hive is almost always between 34.5° and 35.5°C—in other words, just below the normal body temperature of man. In winter the temperature of the clustered bees falls below this level, but it is still held very high (between 20° and 30°C) most of the time and is almost never allowed to fall below 17°C (see Figure 16-3). On one remarkable occasion recorded by Gates (1914) the temperature of the adult bee clusters was 31°C at the same time the air temperature outside the hive was −28°C, a difference of 59°C! The ability of the bees to withstand high temperatures is equally impressive. Lindauer (1954b) placed a hive of bees in full sunlight on a lava field near Salerno, Italy, where the surface temperature reached 70°C. As long as he permitted workers to take all the water they wanted from a nearby fountain, they were able to maintain the temperature inside the hive at the desired 35°C.

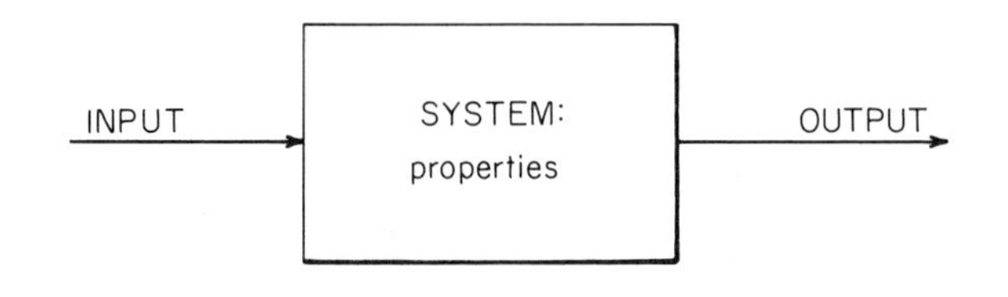

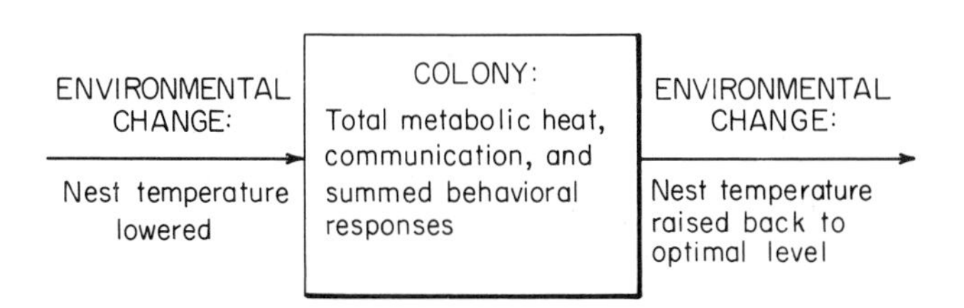

FIGURE 16-2. A control system is analyzed by measuring the alteration (input) imposed on it, the response (output), and, if possible, the internal properties of the mechanism that control the response. The standard block diagram of the upper figure is translated in the lower figure into the properties that are measurable in thermoregulation and other forms of colony homeostasis in the social insects.

How do the worker bees do it? First, they are able to generate a respectable amount of heat as a by-product of metabolism. The amount produced varies greatly according to the age and activity of the individual bees, the humidity and temperature of the hive, and the time of the year. However, under most conditions each bee generates at least 0.1 calorie per minute at 10°C (M. Roth, in Chauvin, 1968a). Presumably a colony of moderate size, containing 20,000 or more workers, is capable of producing thousands of calories per minute.

The honeybee colony makes use of this natural output of heat, which is about average at a rate per gram for insects generally, together with several ingenious behav-

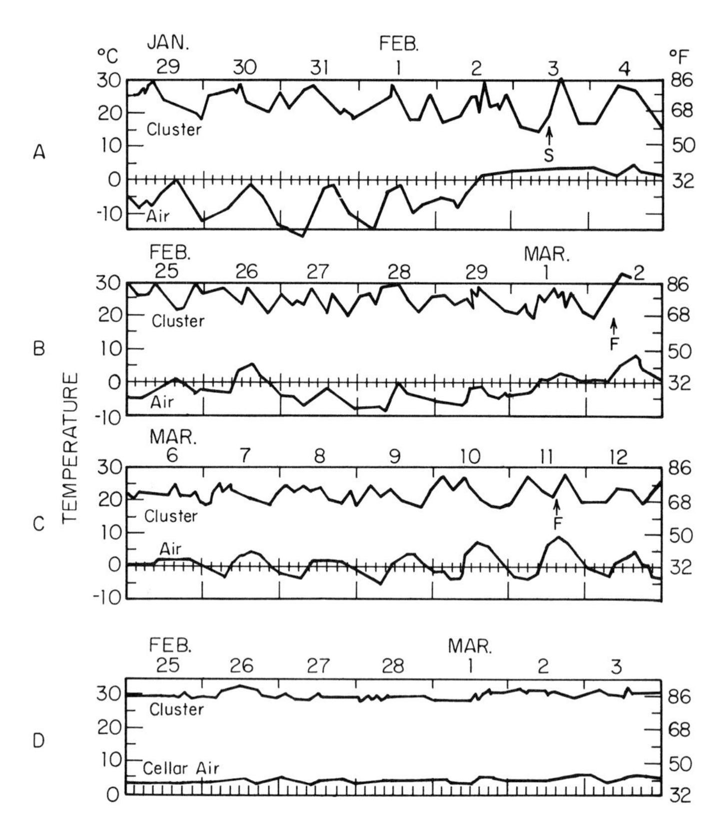

FIGURE 16-3. Portions of a continuous record (*A–C*) of the temperature of honeybee clusters within the hive and of the air outside the hive during winter; (*D*) record of a hive kept in an unheated cellar; after disturbances (*S*) and exercise flights (*F*), the cluster temperature temporarily rose (redrawn from Ribbands, 1953; based on Himmer, 1926).

ioral devices, to hold the hive temperature at the preferred levels. The winter temperature of the hive, as we have seen, is less closely regulated than the summer hive temperature. The mechanisms used in cold weather are first the formation of clusters and second the adjustment of cluster tightness, which is achieved as the outside temperature drops. The workers bunch closer together and the total cluster size correspondingly shrinks (Figure 16-4). Clusters begin to be formed when the hive temperature around the bees falls below 18°C. This raises the temperature surrounding the bodies of the bees to some undetermined level. By the time the hive temperature has dropped to 13°C, and the temperature of the outside air has fallen much lower than that, most of the bees have formed into a very compact cluster that covers part of the brood combs like a warm, living blanket. The outer zone of the cluster is composed of several layers of bees who sit quietly with their heads pointed inward. Those composing the inner zone are more active. They move about restlessly, feed on the honey stores, and from time to time shake their abdomens and breathe more rapidly. Direct measurements have shown that the central bees generate most of the heat, while the outer bees serve as an insulating shell. Together they prevent the temperature of the inner zones of the cluster from falling below 20°C even when the air immediately surrounding the cluster inside the hive approaches the freezing mark.

Temperature control on summer days is even more sophisticated and precise. As summer heat drives the inner hive temperature upward past 30°C or thereabouts, the temperature of the air immediately surrounding the adult workers and brood starts to rise above the preferred 35°C level. At first the workers cool the hive by fanning with their wings to circulate air over the brood combs and then out the nest entrance. When the hive temperature exceeds about 34°C, this simple device no longer suffices. Now water evaporation is added by an elaborate series of behavioral acts. Water is carried into the nest by the

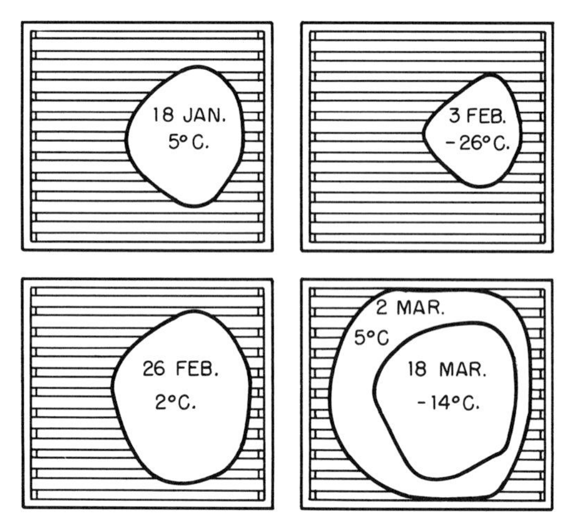

FIGURE 16-4. As the outside temperature falls during the winter, honeybee workers group together more tightly, and the size of the cluster correspondingly shrinks. In these diagrams the outlines of the cluster of a colony are shown on the background of the interior of the hive on various winter days. The dates and outside temperatures are provided inside the outlines (redrawn from Ribbands, 1953; based on H. F. Wilson and Milum, 1927).

workers and distributed over the brood cells, as illustrated in Figure 16-5. Other workers regurgitate droplets onto their tongues and then extend the tongues outward, spreading the water into films from which evaporation is rapid. Other workers fan their wings to drive the moist air away from the brood cells and out of the nest.

The tuning of this social air-conditioning system has been characterized by Lindauer (1961):

One might imagine that each bee would set out to collect water as soon as it detects the overheating. This is not so; not just any bee is able to collect water. Only bees that know the terrain and are experienced foragers can perform this task. Furthermore, it is an advantage when bees are divided into water collectors and water sprinklers. So there is a very strict division of labor between the older flying bees, who collect water outside, and the younger hive bees, who distribute it. Indeed, we found a special mode of communication by means of which the water collectors receive insructions from the hive bees about when to start collecting and when to stop.

Let us begin with the simple case. Let us assume that water collecting is still in progress and the foragers are to be informed whether or not there is need for more water. To transmit this information the hive bees make use of *the short moment when they have contact with the collectors:* this is during water delivery at the entrance hole. As long as overheating exists, the homecoming foragers are relieved of their burdens with great greed; three or four bees at once may rush up to a collector and suck from her the extruded water droplet. This stormy begging informs the collector bee that there is a pressing need for more water. When the overheating begins to subside, however, the hive bees show less interest in the water collectors. The latter now have to run around in the hive themselves, trying to find somewhere a bee that will relieve them of at least part of the water load. The delivery in such cases takes much more time, of course. This rejecting attitude contains the message "Water needs fulfilled," and the water collecting will thus stop, even though the collectors themselves have not been at the brood nests to experience the changed temperature situation.

This delivery time is in fact an accurate gauge of water demand. As shown in [the accompanying figure], with delivery times up to 60 sec, the collecting is continued industriously. Beyond that, however, the eagerness for collecting decreases rapidly, and when delivery takes longer than 3 min collecting practically ceases altogether.

A second point is apparent: When delivery times are very short (up to 40 sec), the water collectors even perform recruiting dances to stimulate hive bees to fly to the water source... These recruiting dances subside somewhat if the delivery time is longer than 40 sec, and later cease completely.

The recruitment of water carriers and the control of water intake at the colony level is thus seen to be a form of mass communication resembling in many respects recruitment by odor trails in ants. In both systems it is only by the summation of large numbers of actions at the level of individual workers that the quantitative needs of the colony as a whole can be measured and filled.

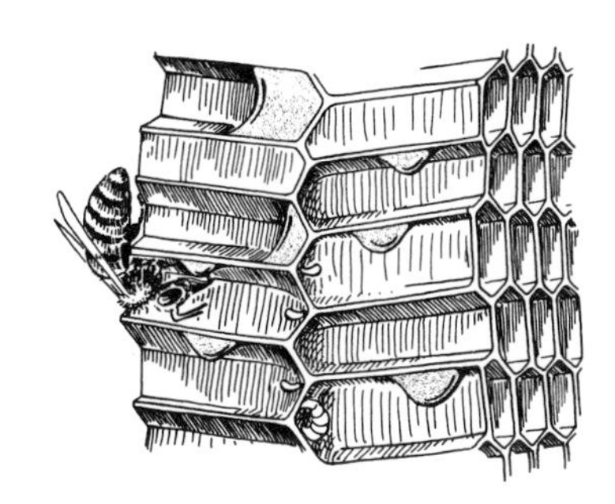

FIGURE 16-5. As the hive becomes overheated, workers distribute droplets of water over the brood cells as shown here. As the water evaporates, it cools the surrounding area. Meanwhile other bees fan the moist air out of the nest (after Lindauer, 1961; from Park, 1925).

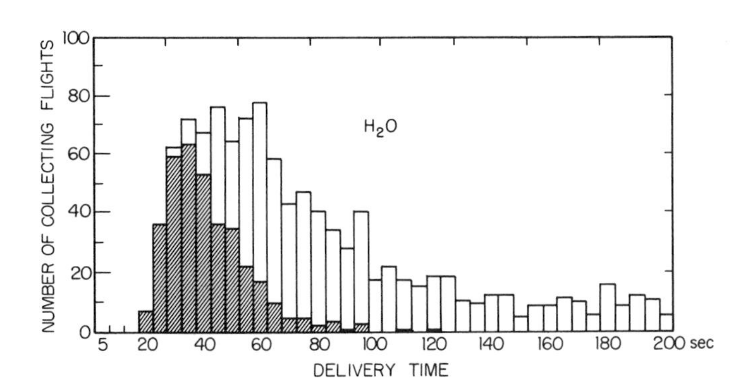

FIGURE 16-6. Need for water in the overheated honeybee hive is communicated to the forager bees by the receiving hive bees who then distribute it over the combs in droplets. Communication consists of the following relation: the more quickly water is accepted by the hive bees, the more readily the forager bees collect additional water. The number of water-collecting flights decreases as the time required for delivery increases. With very short delivery times (20–40 seconds, shown here in shaded columns), the foraging bees are even induced to perform an alerting dance after each collecting flight to recruit new bees to help gather water (redrawn from Lindauer, 1961).

Microclimatic Control by Other Social Insects

Thermoregulation is a common phenomenon throughout the remainder of the social Hymenoptera, but nowhere is it known to be as organized and precise as in the honeybee. Fanning is employed by primitively eusocial bees in the halictine genus *Augochlorella* (Michener, 1969a), as well as by bumblebees (Plath, 1923; Himmer, 1933; Hasselrot, 1960). None of these bees are known to employ water cooling, but their temperature regulation is, nonetheless, efficient. Himmer, for example, found that the inner temperatures of the larger nests of *Bombus agrorum* are kept within one degree of 30°C. Stingless bees of the genus *Melipona* also use fanning to cool overheated nests. The bees place their bodies and beat their wings in such a way as to draw air through the tiny channels of the batumen plates and drive it out through the nest entrance in the manner illustrated in Figure 16-7 (Nogueira-Neto, 1948; Kerr *et al.,* 1967).

FIGURE 16-7. Workers of the stingless bee *Melipona seminigra merrillae* fanning at their nest entrance. This action draws warm air out through the nest opening and cool air into the nest through pores located elsewhere in the nest wall (drawing by João Maria F. de Camargo, in Kerr *et al.,* 1967).

Among the social wasps, colonies of *Polistes, Polybia, Vespa,* and *Vespula* use both fanning and water transport to cool overheated nests (Steiner, 1930; Schwarz, 1931; Himmer, 1932; Weyrauch, 1936; Gaul, 1952; Kemper and Döhring, 1967). In fact, *Polistes* workers show a sequence of two responses to rising temperatures which is remarkably similar to that of the honeybees. According to Steiner, the adult wasps begin to fan with their wings when the temperature at the surface of the brood comb reaches some point around 35°C. Even males take part in this activity. If the temperature continues to rise, adult females, especially the mother queen, begin to collect water and to distribute it in droplets over the nest surface. Despite the fact that the nests of *Polistes* consist of single brood combs exposed to the air, a satisfactory degree of control is attained. During most of the summer days the temperature of the comb surface ranges between 34° and 37.5° and usually stays close to 34.5°, the optimal temperature for brood development. Nests of *Vespula* (*Paravespula*) are kept much cooler. The interior of one nest of *V. vulgaris* monitored by Himmer rose to 34° on the hottest day of the study period, when the air temperature outside reached 36°, and it fell to only 26° when the outside temperature dipped to 9° on the coldest day. During the observation period the outside temperature averaged 18.5°, but the internal nest temperature remained almost always between 29.5° and 32°.

The ants are also able to regulate the environment of their nests, but they differ radically from the remainder of the social Hymenoptera in the means by which they do it. Lacking wings, worker ants are of course not able to ventilate the nests by fanning. And, since they are also unable to commute rapidly to sources of open water, they do not employ droplet evaporation in the manner of the honeybees and social wasps. Instead, they rely heavily on the location and construction of the nest to achieve automatic microclimate regulation. In addition, they are able to move the brood around at will within the chambers of the nest to reach the spots best suited for development. In most ant species all stages of brood are kept in the warmest chambers, in the upper range of 25°–40°C when these temperatures are available, but pupae are usually segregated into the drier parts. Therefore, when choosing microenvironments within the nest, the workers can be said to seek one or the other of at least three preferenda: one for themselves when alone, one for eggs and larvae, and one for pupae. The great majority of bees and wasps, on the other hand, rear their immature stages in fixed brood cells and are not able to transport them to shifting nursery areas in the face of changing local conditions.

As a homeostatic measure of last resort, the colonies of most species of ants are willing to move to new nest sites that provide more favorable microenvironments. During these colony emigrations, queens walk under their own power and the brood is carried. The winged social Hymenoptera are also able to vacate old, unfavorable sites and construct new nests elsewhere; the process is called "absconding" by melittologists. The brood cannot be moved in this case, and it is left behind to die. The adults must then start a new brood from the egg stage. In the meliponine bees, absconding is a rare and especially precarious undertaking. When it occurs, as reported in *Trigona* by Portugal-Araújo (1963), for example, the mother queen must be left behind since she is unable to fly, and the colony must produce a new queen and see her successfully through her nuptial flight in order to survive. Because such restrictions do not apply to ants, it is no surprise to learn that they change nests more frequently than do bees and wasps. There are even a few species of ants, such as *Monomorium pharaonis, Tapinoma melanocephalum, Iridomyrmex humilis,* and *Paratrechina longicornis,* which specialize in occupying flimsy, unstable nest sites and rely heavily on frequent colony emigrations to maintain a favorable brood environment. Captive colonies of these species in the laboratory appear "nervous" to the observer; at the slightest disturbance the workers begin to pick up pieces of brood and walk around. Given access to a better nest chamber, they will lay odor trails and organize emigration within minutes. Such species resemble the "fugitive" or "opportunistic" species of ecological classifications. That is, they depend on finding newly opened sites, exploit them for whatever brief time they are suitable, and then quickly move on as soon as the sites become unfavorable. Other ant species change nest sites only when the old site is drastically altered, and they require hours or days to accomplish it. As a rule, the more elaborate the nest structure of the species, the less mobile the colonies. Harvesting ants of the genus *Pogonomyrmex* and fungus-growing ants of the genus *Atta,* for example, usually emigrate only when the old nest is seriously disturbed.

Ant colonies that nest in the soil benefit from the circumstance that, at depths below a few centimeters, both temperature and humidity are subject to relatively little fluctuation throughout the year. In warm temperate zones especially, the soil microclimate appears to be close to ideal for brood development. Most ant species construct nests primarily in the soil or at the soil surface just beneath the covering layer of leaf litter and humus, and many occupy pieces of rotting wood. The largest number of species in tropical forests, by contrast, utilize small pieces of rotting wood on the ground, a smaller number nest arboreally or in rotting logs, and still fewer nest entirely in the soil (Wilson, 1959d). The largest number of ant species in north temperate areas nest in the soil beneath rocks, a tendency that is especially marked in the boreal coniferous forests that contain the effective northernmost limit of the distribution of ants as a whole. A considerable number of northern species also occupy dead logs and stumps or nest in the open soil.

The thermoregulatory properties of rocks, especially those that are flat and set shallowly into the soil, are obvious. In temperate zones during the spring, the sun warms the rocks and the soil just underneath much more rapidly than the surrounding soil and humus, and it thus allows an earlier initiation of egg laying and brood development than would be possible otherwise. The same property is possessed by the bark of decaying stumps and logs and the frass-filled spaces underneath. In spring, workers, queens, and brood crowd together in such places and only gradually retreat to the inner chambers of the nest at the onset of hot, dry weather. Species that utilize rocks, stumps, or logs are also less vulnerable to low humidities and high temperatures than those that nest exclusively in the soil. Gösswald (1938) found that in Germany only the latter class of species requires 100 percent relative air humidity for indefinite survival.

A more advanced form of microclimatic regulation has been attained by the small minority of ant species that build mounds. True mounds are not to be confused with simple "craters," which are no more than rings of excavated soil around nest entrances. They are symmetrically shaped piles of excavated soil, rich in organic materials, perforated with dense systems of interconnected galleries and chambers that serve as living quarters for the ants, and often thatched with bits of leaves and stems or sprinkled with pebbles or pieces of charcoal. The soil beneath the mounds also contains very extensive galleries and chambers, and normally only a fraction of the colony's population is to be found in the mound itself at any given moment. Mound-building species are found in the myrmicine genera *Atta, Acromyrmex, Myrmicaria, Pogonomyrmex,* and *Solenopsis* in the tropics and warm temperate zones in various parts of the world, in the dolichoderine genus *Iridomyrmex* in Australia, and in the formicine genera *Formica* and *Lasius* in Europe, Asia, and North America. True mounds are encountered in a wide range of environments, but they are most common in habitats subject to extremes of temperature and humidity—bogs, stream banks, coniferous woodland, and

FIGURE 16-8. Two mound nests, probably belonging to the same colony, of the Allegheny mound-building ant *Formica exsectoides* in Pennsylvania. Both mounds have been cut back to the center to reveal the honeycomb-like system of earthen galleries and chambers (from McCook, 1877).

deserts. Pierre Huber (1810) first suggested that the primary function of mounds is microclimatic regulation, and the hypothesis became a subject for investigation by a long line of later European and American investigators. The most intensive analyses have been by Andrews (1927) on *Formica exsectoides* (see Figure 16-8), Wellenstein (1928) on *F. rufa,* Steiner (1929) on members of the *F. rufa* and *F. exsecta* groups, Katô (1939) on *F. truncorum,* Raignier (1948) on *F. polyctena,* and Scherba (1959, 1962) on *F. ulkei.*

All of these studies have confirmed that mound construction provides higher internal nest temperatures during cool weather. Katô, Raignier, and Steiner further reported that, in the *F. rufa* group of species, the mound temperatures, especially 20 to 30 cm beneath the surface of the mound apex, vary less than those of the surrounding air and soil and stay consistently close to the preferenda displayed by the ants. On the other hand, Scherba (1962) discovered a more complex, and in some ways more interesting, situation in *F. ulkei,* a member of the *exsecta* group. In this species the mound temperatures not only remain higher, they vary much more with depth than the surrounding soil, as illustrated in Figure 16-9. The *F. ulkei* mound, in short, provides a steeper temperature gradient than does unworked soil. As a result, workers are able to discriminate finely when selecting brood chambers that match the different preferred temperatures of the various developmental stages of the brood. Moisture is also much more uniform in the *F. ulkei* mounds. Compared with the adjacent soil, the mound nests monitored by Scherba had higher minimum values at 30 cm and lower maximum values at 5 cm and, consequently, a smaller moisture gradient between these two depths. The grand weekly moisture content of the nests was 29.40 ± 0.54 percent at 30 cm and 27.23 ± 0.66 percent at 5 cm. There can be no doubt that the humidity regulation attained by *F. ulkei* in its mounds is adaptive. The workers regularly shift brood up and down through the dense system of chambers and vertical galleries, with the pupae normally being concentrated in the drier upper layers. By means of laboratory experiments, in which workers were permitted to move their brood along artificially produced humidity gradients, Scherba was able to determine the humidity preferenda for larvae and pupae. The results provide a remarkably close match with the moisture contents at different levels of the natural mound nests just cited: 29.58 ± 0.04 percent for larvae and 27.58 ± 0.04 percent for pupae.

In spite of the vast numbers of temperature and hu-

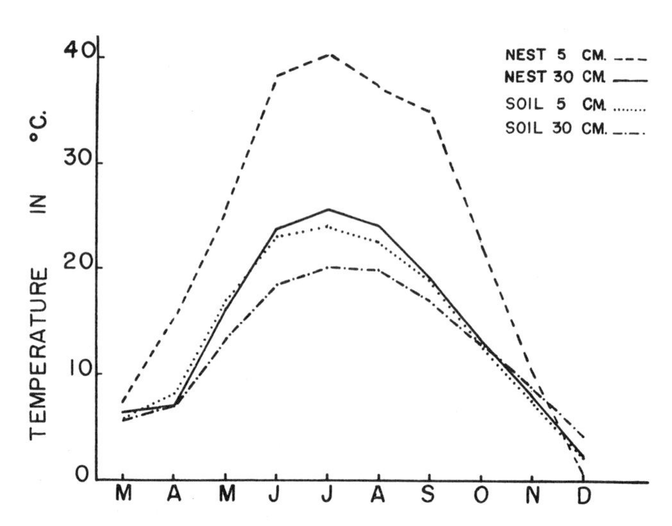

FIGURE 16-9. The seasonal cycle of mean temperature within 20 nests of the mound-building ant *Formica ulkei* and the soil adjacent to the nests. The mound temperatures not only average higher than those in the surrounding soil, they also vary with depth, providing a gradient along which the ants distribute the various developmental stages according to their particular temperature preferenda (from Scherba, 1962).

midity measurements taken in ant mounds by a host of investigators during the past hundred years, we still have only a very limited understanding of the basis for environmental control in these nests. The outer, crust-like layer of the mound seems to reduce loss of heat and moisture, as one would expect. The shape of the mound itself exposes it to more sunlight and enhances warming on cool days, especially in the spring and fall. The mounds of some species of *Lasius* and *Formica* in addition have longer, more gently sloping faces to the south, which increases the amount of exposure still more (Andrews, 1927). For centuries such nests have been used as crude compasses by natives of the Alps. Thatching of the mound surface, a common feature in the nests of many but not all species, appears to reduce erosion of the crust by rain. Perhaps it also provides trapped air spaces that improve insulation. The higher concentration of organic material found in all such nests results in greater moisture retention, but this is offset to some unknown degree by the increased surface area provided by the system of chambers and galleries.

Regardless of which specific features are important in functional design, the mound is certainly no accidental accumulation of excavated earth. It is in a constant state of flux, as workers move materials around to reinforce and to repair the crust and the interior (Cole, 1932; Kloft, 1959a; Chauvin, 1959). When mounds are broken apart or their shape altered experimentally, the ants set to work immediately to restore the original form. Furthermore, some mound-building species regularly open and close the nest entrances in apparent response to humidity or temperature changes. Workers of *Formica rufa* dig large openings on the surface of the mound following rains, and Wellenstein (1928) has suggested that this alteration serves to reduce humidity inside the nest. Unfortunately, these seeming correlations between environmental change and behavioral responses will remain no more than subjects for speculation until a serious micrometereology of ant nests is created, one that utilizes physical models and experiments with both real and simulated nest structures.

In a few cases ants arrange their own bodies to achieve environmental control in a manner that suggests the winter clustering of honeybees. The most impressive example is temperature control within the nomadic-phase bivouacs of the surface-dwelling army ants (Schneirla, Brown, and Brown, 1954; Jackson, 1957). The hundreds of thousands of *Eciton* workers belonging to a single colony form shelters out of nets and chains created from their own bodies, which are hooked together by their tarsal claws and piled layer upon layer to create a single massive cylinder suspended from some log or tree trunk on the rain forest floor. In the center of the mass, where the queen and brood are placed, the temperature is consistently 2–5°C higher than the surrounding air (see Figure 16-10). So far as is known, the effect is attained entirely by the trapping of metabolic heat within the air spaces created by the

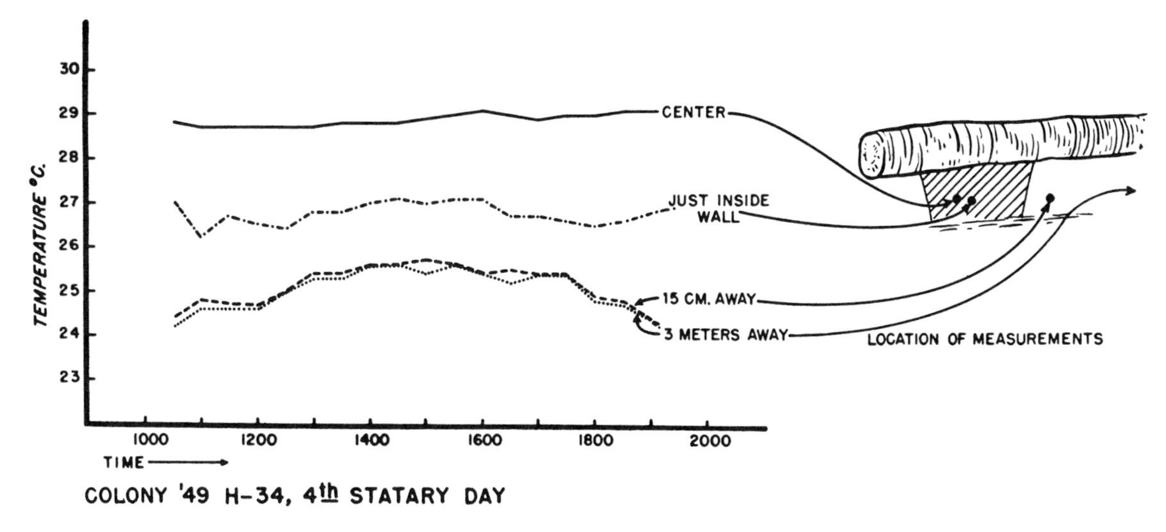

FIGURE 16-10. Thermoregulation within a body mass of army ants (*Eciton hamatum*) in a Central American rain forest. The ants form bivouacs consisting entirely of their own intertwined bodies suspended from a tree trunk or log, as indicated in the right-hand portion of the figure. Within this cluster temperatures are maintained at a level several degrees higher than the surrounding air (from Schneirla, Brown, and Brown, 1954).

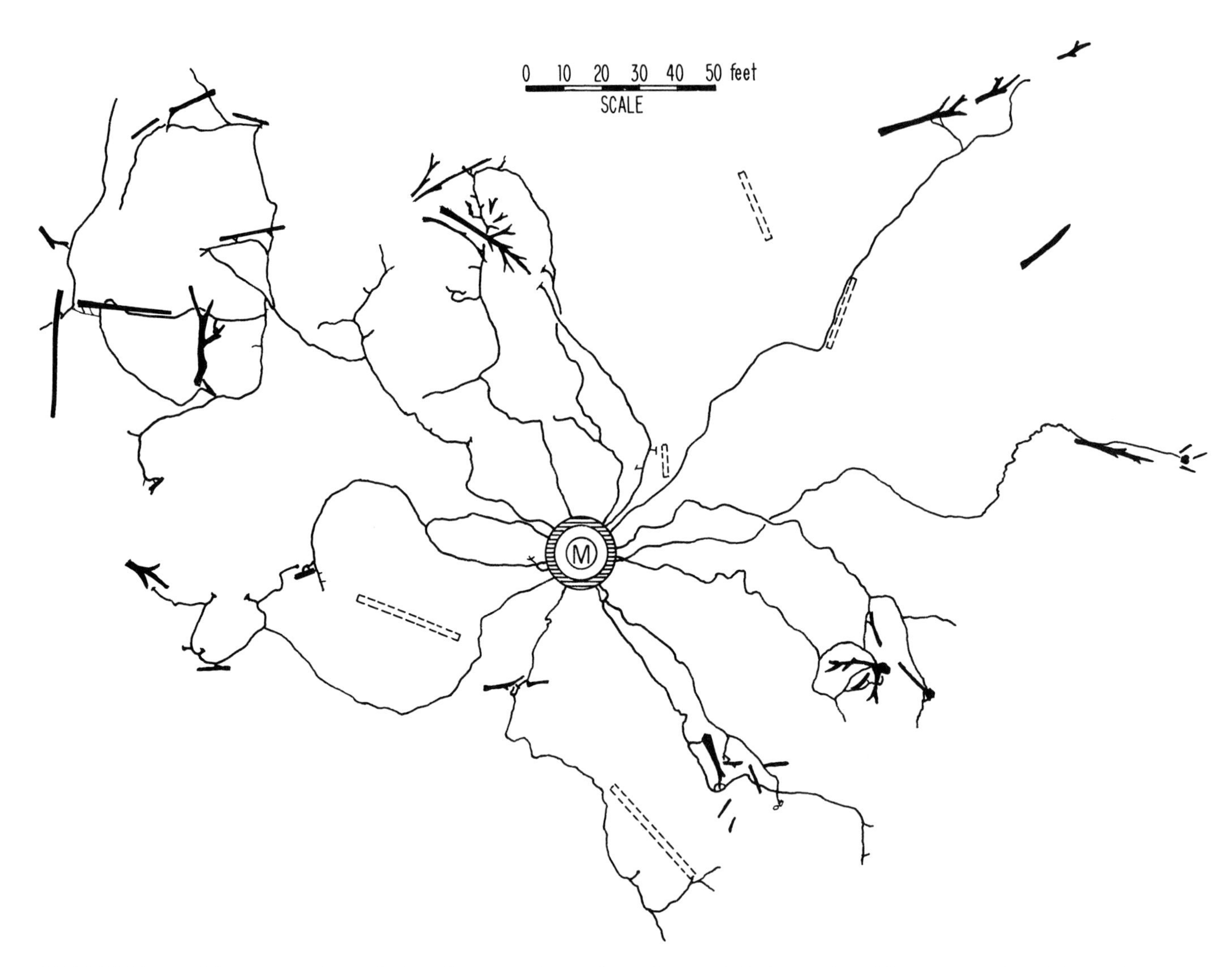

FIGURE 16-11. The nests of the evolutionarily more advanced termites are built from soil and fecal material, and the workers forage out through or over the soil to collect cellulose. This diagram shows the large foraging area, approximately 1.5 acres in extent, utilized by a single colony of *Coptotermes lacteus* in Australia. The bulk of the colony, which contained approximately a million workers, was housed in the mound nest (*M*) in the center of the diagram. Workers foraged through subterranean galleries to dead logs and branches to obtain food; these objects are outlined in black if still solid, or in dashed lines if already reduced to a shell by feeding (modified from Ratcliffe and Greaves, 1940).

intertwined bodies of the ants. A different kind of direct behavioral thermoregulation has been suggested for *Formica rufa* by Zahn (1958). In cool weather workers leave the nests in large numbers to sun themselves in open air close by. Zahn believes that, when the workers return, they significantly raise the nest temperature by radiating heat from their bodies. This *Wärmeträgertätigkeit* was put forth as a new kind of stereotyped social behavior. The amount of heat flux generated by the movement of such workers has not been measured, however; nor can it be certainly identified as anything more than a felicitous by-product of normal foraging activity.

Although the ants present a great array of adaptations in their nest structure, it is the termites that have raised nest architecture to the heights in the service of the colony homeostasis. In fact, the entire history of the termites, from their cryptocercoid roach origins to the appearance of the advanced fungus-growing termitids, can be viewed as a slow escape by means of architectural innovation from a dependence on rotting wood for shelter (Emerson, 1938; Noirot, 1958–1959). Most of the lower termites belonging to the families Kalotermitidae, Hodotermitidae, and Rhinotermitidae live partly or entirely in the wood on which they feed. Galleries are lined and crevices filled with fecal pellets containing large quantities of lignin, and special salivary secretions may be added as a kind of glue. Microenvironmental regulation is evidently not pronounced in such nests. Like the soil-dwelling ants, the primitive termites rely heavily on the natural insulating properties of their nest materials and on their own ability to move from chamber to chamber as local conditions demand. The more advanced termites, on the other hand—from certain rhinotermitid genera such as *Coptotermes* up to the Hodotermitidae and Termitidae—penetrate the soil and rely principally on special structures, usually regularly formed mounds or towers, that are constructed above ground level away from the sources of cellulose (see Figure 16-11).

Lüscher (1956b, 1961a) found that the formed nests of the higher termites vary enormously in their ability to regulate temperature. Three of the African species studied—*Amitermes evuncifer, Thoracotermes macrothorax,* and *Microcerotermes edentatus*—build small nests with thin walls and possess almost no thermoregulatory ability. Oxygen and carbon dioxide are exchanged directly through the cardboard-like outer nest walls. Because of the small size of the nests, this crude method of ventilation suffices. A fourth species, *Cephalotermes rectangularis,* constructs a larger nest with a much thicker wall. The result is greatly improved thermoregulation (illustrated in Figure 16-12), but the *Cephalotermes* pay a price in slower, less efficient ventilation.

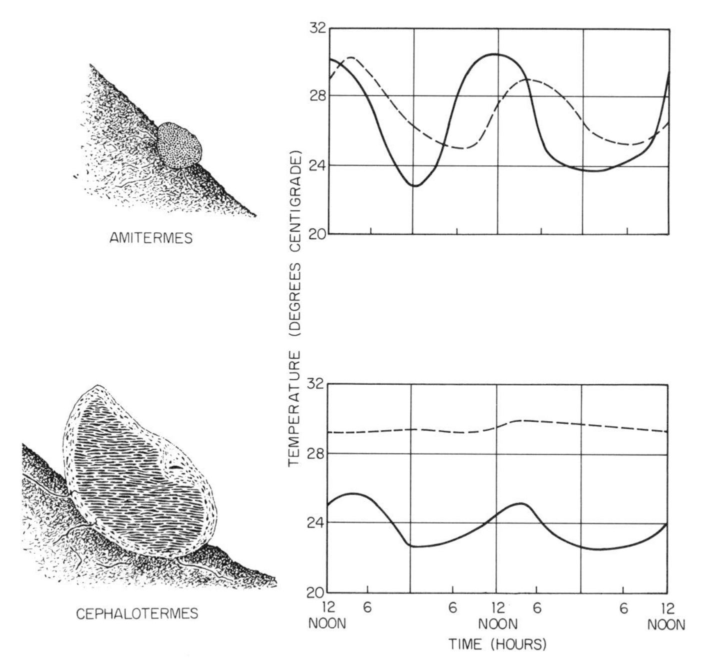

FIGURE 16-12. Longitudinal sections of nests of the African termitids *Amitermes evuncifer* and *Cephalotermes rectangularis.* On the right are temperatures, recorded continuously for 48 hours, of the inside of the nests and of the air outside. The smaller, more simply constructed *Amitermes* nest tracks the outside temperature closely, but the larger and better insulated *Cephalotermes* nest maintains a higher and more nearly constant temperature (modified from Lüscher, 1961a).

The fungus-growing termites of the genus *Macrotermes* have achieved a brilliant advance over *Cephalotermes* and other termites in the degree of nest microclimatic regulation. This has been accomplished entirely by innovations in nest architecture. The entire nest, to put the matter as succinctly as possible, has been turned into an air-conditioning system. The basic principles, first elucidated in full by Lüscher, are summarized in Figure 16-13.

Other Examples of Social Homeostasis

Constancy attained by behavioral and physiological regulation is a feature built into virtually every facet of the life of insect societies. Much of the communication

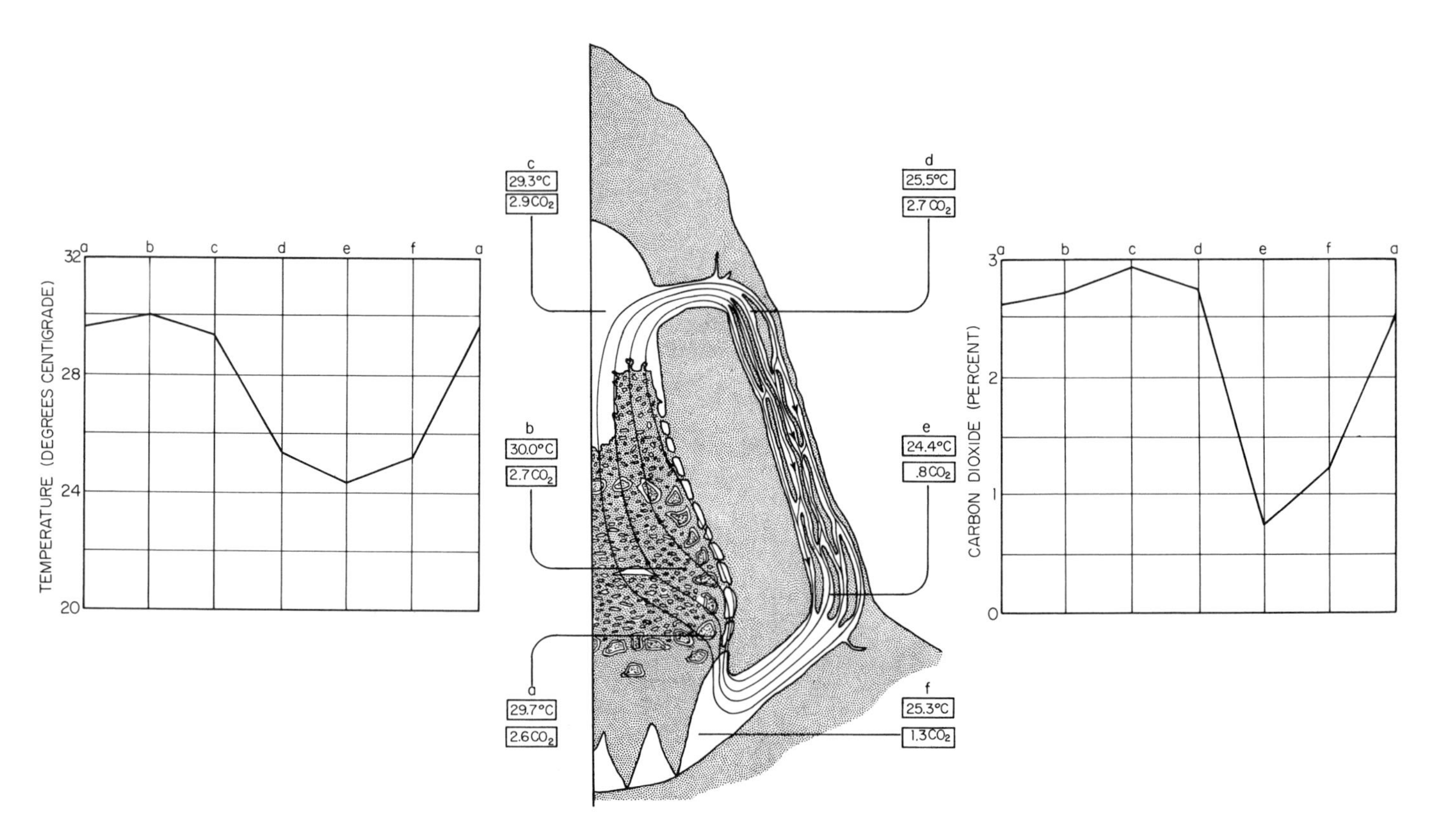

FIGURE 16-13. Air flow and microclimatic regulation in a nest of *Macrotermes natalensis* in Africa. (The diagrammatic longitudinal section of the nest shown here can be compared with the more detailed drawing in Figure 6-11.) At each of the positions indicated, the temperature (in degrees Celsius) is shown in the upper rectangle and the percentage of carbon dioxide appears in the lower rectangle. As air warms in the central core of the nest (*a, b*) from the metabolism of the huge colony inside, it rises by convection to the large upper chamber (*c*) and then out to a flat, capillary-like network of chambers (*d*) next to the outer nest wall. In the outer chambers the air is cooled and refreshed. As this occurs, it sinks to the lower passages of the nest beneath the central core (*e, f*). The curves at the side show how the temperature and carbon dioxide change during circulation. These changes are brought about by the diffusion of gases and the radiation of heat through the thin, dry walls of the ridges (modified from Lüscher, 1961a).

among the colony members is in fact devoted to it. The great diversity of the kinds of controls are illustrated by the following examples:

1. Queens tend to inhibit oviposition by workers and the creation of new, rival queens by an armament of devices ranging from direct aggression to the secretion of inhibitory primer pheromones. In the termites, caste-specific inhibitory pheromones are also produced by reproductive males and possibly also by soldiers. In some ant species and in termites generally, the workers kill off (and in some cases also eat) all supernumerary queens until only one remains in each colony. (Chapters 10 and 15).

2. When colonies of the termite genus *Zootermopsis* are kept on a diet lacking in nitrogen, the workers restore the balance by cannibalism. The intensity of the cannibalism is proportional to the length of the period of starvation. When protein is again added to the diet, the cannibalism ceases (Cook and Scott, 1933). In *Coptotermes lacteus,* surplus virgin reproductives are consumed when it is no longer possible to conduct nuptial flights, in effect rendering these individuals useless for reproductive purposes (Ratcliffe *et al.,* 1952).

3. The number of workers of fire ants (*Solenopsis saevissima*) attracted to a food source along an odor trail increases linearly as the amount of trail pheromone laid

down increases, and the amount of pheromone increases with the number of workers laying trails, which, in turn, is simply the number that are able to feed at the food source. As a result, the number of workers approaching a food source along an odor trail is a linear function of its size, and the number at the food source at any given moment approaches a constant (see Chapter 13). Mass communication of this kind implies homeostatic controls since, by definition, it involves the regulation of numbers of participating individuals. Another closely parallel example is the control of the number of honeybee workers visiting a food source or new nest site by means of the number of workers induced to perform waggle dances.

4. The queen of *Bombus agrorum* (and probably other bumblebee species as well) exercises a subtle form of birth control related to the number of workers available to serve as nurses for the larvae. The queens always construct the egg cells themselves, and most of the cells are placed on top of cocoons containing pupae that are not more than three days old. The number of cells are thus dependent to some extent on the number of workers that will soon be emerging. In addition, the queen lays from 4 to 16 eggs per cell, and the number in any given cell tends to increase in proportion to the number of cocoons underneath the egg cell. The queen somehow gauges the number of future workers that will soon emerge in the vicinity of that particular egg cell, thereby deciding the number of eggs she will lay (Anne D. Brian, 1951).

The Superorganism Concept and Beyond

The idea of social homeostasis leads easily to the visualization of the entire insect colony as a kind of superorganism. In fact, the story of the superorganism concept, from its origin as a philosophical ιδεα sixty years ago to its present sharp decline in contemporary thinking, should prove instructive to historians of science as well as to biologists with a more immediate interest in the subject. During some forty years, from 1911 to about 1950, this concept was a dominant theme in the literature on social insects. Then, at the seeming peak of its maturity it faded, and today it is seldom explicity discussed. Its decline exemplifies the way inspirational, holistic ideas in biology often give rise to experimental, reductionist approaches that supplant them. For the present generation, which is so devoted to the reductionist philosophy, the superorganism concept provided a very appealing mirage. It drew us to a point on the horizon. But, as we worked closer, the mirage dissolved—for the moment at least—leaving us in the midst of unfamiliar terrain, the exploration of which came to demand our undivided attention.

William Morton Wheeler believed in the reality of the superorganism. In his famous essay "The ant colony as an organism" (1911) he stated that "the animal colony is an organism and not merely the analogue of the person." Of course, one has to pay close attention to his definition of organism: "An organism is a complex, definitely coordinated and therefore individualized system of activities, which are primarily directed to obtaining and assimilating substances from an environment, to producing other similar systems, known as offspring, and to protecting the system itself and usually also its offspring from disturbances emanating from the environment. The three fundamental activities enumerated in this definition, namely nutrition, reproduction and protection, seem to have their inception in what we know, from exclusively subjective experience, as feelings of hunger, affection and fear respectively."

It was not very difficult to apply these malleable criteria to the ant colony and other insect societies. Wheeler saw several important qualities of the ant colony that qualified it as an organism:

1. It behaves as a unit.
2. It shows some idiosyncrasies in behavior, size, and structure that are peculiar to the species and other idiosyncrasies that distinguish it from other colonies belonging to the same species.
3. It undergoes a cycle of growth and reproduction that is clearly adaptive.
4. It is differentiated into "germ plasm" (queens and males) and "soma" (workers).

In 1928 Wheeler began calling the social insect colony a superorganism. His ideas on the subject had changed little by that time, although he had begun to conceive in a vague fashion of the phenomenon that was later to be called homeostasis: "We have seen that the insect colony or society may be regarded as a super-organism and hence as a living whole bent on preserving its moving equilibrium and integrity."

Wheeler's view of the superorganism was influenced by earlier, philosophical writings, particularly those of Herbert Spencer, Ernst Haeckel, and G. T. Fechner on emergent evolution and the hierarchical structure of the universe. He was very much a conceptualist in the old scholastic sense, so much so he would have felt at ease with Abelard. By this I mean that he seemed to think of the superorganism as a real, fixed category to which given sets of organisms could be fitted or not.

Other authors dealt with the idea in a more lyrical fashion. Maurice Maeterlinck spoke of the "spirit of the hive" in his *The Life of the Bee* (1901): "Where is this 'spirit of the hive' . . . where does it reside? It disposes pitilessly of the wealth and the happiness, the liberty and life, of all this winged people; and yet with discretion, as though governed itself by some great duty." Later, in *The Life of the White Ant* (1927) and *The Life of the Ant* (1930), Maeterlinck brought this pretty nonsense in line with Wheeler's concept, to which he was openly indebted, and gave it a more scientific veneer. In the early 1920's Eugène N. Marais, a South African, developed the concept of the superorganism from his own observations of termites and expressed the idea in an article in an Afrikaans-language journal *Die Huisgenoot* (Home Companion). This article proved so popular that Marais wrote others, and for a time his work was widely cited by journalists. A book written in Afrikaans, *Die Siel van die Mier* (The Soul of the White Ant), followed in 1933 and was subsequently translated into English and German. During the later years of his life, Marais claimed that his idea had been stolen by Maeterlinck, and his case has more recently been melodramatically championed by Robert Ardrey in the best-selling book *The Territorial Imperative.* But it is all just a tempest in a teapot. Wheeler was clearly the architect of the idea, for what it is worth, and neither Marais nor Maeterlinck had any claim for priority or visible effect on later students of social insects.

Alfred E. Emerson, who followed closely after Wheeler with a long series of strongly philosophical papers on social insects (1938–1962), was the next serious scholar of the subject. He viewed the superorganism somewhat differently from Wheeler. His attitude was, and still is, nominalist in the modern scientific sense, rather than conceptualist, and he regarded the superorganism concept as a primarily heuristic device. The supraorganism, as Emerson preferred to call it from 1950 on, is "strictly analogical," a framework on which to hang factual knowledge of social insects and a provider of clues about undetected phenomena. The clues are there, Emerson (1959) asserted, because the colony is the unit of selection and must develop adaptations that closely parallel those seen in the physiology of individual organisms. In particular, insect societies can be expected to have evolved so as to improve homeostasis within the individual colony: "Analogous similarities between levels, such as the evolution of sterile somatic cells in the organism and sterile castes in the social insect colony, result from natural selection of whole units favoring division of labour, integration and improved homeostasis. Communication through behavioural specializations and trophallactic exchange is characteristic of insect societies. However, improvement of division of labour with consequent improved efficiency, and improvement of integration and co-operation with consequent complexity of organization, are means to an improved self-regulation of optimal nutritional, defensive and micrometerological conditions (homeostasis)."

But, as Emerson himself realized (for example, in 1941), the analogies can only be stretched a little before they lose their heuristic value and become word games. It was not long before even Emerson's liberalized superorganism concept came under heavy critical fire (Novikoff, 1945a,b; Schneirla, 1946b). In retrospect, it is clear that these critics need never have spoken. As I have already stated, the current generation of students of social insects, which was beginning to take its training about this time, saw its future in stepwise experimental work on narrowly conceived problems, and it has chosen to ignore the superorganism concept—at least as an explicitly formulated idea. M. V. Brian, in a recent book on the population biology of social insects (1965b), does not even mention the superorganism concept; nor does he cite the articles just mentioned by Emerson, Wheeler, and others, except for one by Emerson which is mentioned in one sentence as an earlier "review." J. H. Sudd's book on ant behavior (1967) does bring the subject up, but only as a rather disapproving afterthought on the last two pages. Martin Lüscher (1962b) has spoken approvingly of the analogies of organism and superorganism and the similar methods of analyzing the two levels of organization, but chiefly as a philosophical aid in reflecting on his own researches on caste determination and microclimatic regulation in termites. Seldom has so ambitious a scientific concept been so quickly and almost totally discarded.

The superorganism concept faded not because it was wrong but because it no longer seemed relevant. It is not necessary to invoke the concept in order to commence work on animal societies. The concept offers no techniques, measurements, or even definitions by which the intricate phenomena in genetics, behavior, and physiology can be unraveled. It is even difficult to cite examples where the conscious use of the idea led to a new discovery in animal sociology. How, then, did it have the major impact on current research that I have claimed? I said that the idea was inspirational and provided an image that helped channel research in a productive manner. This is true; it is also true that the idea as originally formulated by Wheeler had just that right amount of fact and fancy to generate a mystique. Consider the style in which Wheeler wrote in 1928: "Another more general problem is suggested by the insect society, or colony as a whole, which as I have shown in another place (1911) is so strikingly analogous to the Metozoan body regarded as a colony of cells, or indeed to any living organism as a whole, that the same very general laws must be involved. But the biologist, with his present methods is powerless to offer any solution of the living organism as a whole . . . We can only regard the organismal character of the colony as a whole as an expression of the fact that it is not equivalent to the sum of its individuals but represents a different and at present inexplicable 'emergent level' in the sense of Alexander (1920), Sellars (1922), C. Lloyd Morgan (1923), Parker (1924), Wheeler (1926), Smuts (1926) and others."

Here, then, is the mirage that drew us on. The words "powerless" and "inexplicable" were a challenge leveled at the next generation. Once the challenge had been taken and progress achieved by technical advancements never conceived of by Wheeler, it was natural for the superorganism concept to become déclassé. But it would be wrong to overlook the significant, albeit semiconscious, role this idea has played in the history of the subject.

Finally, it might be asked what vision, if any, has replaced the superorganism concept. Surely there is no new holistic conception, as I have tried to make clear in this book. Certain analogies with other whole systems will prove useful insofar as they call attention to poorly understood organizational processes and the techniques by which they can be analyzed. For example, Meyer (1966) has drawn attention to similarities between the coactions of neurons in the brain and of workers in an insect colony. He has suggested, vaguely but probably correctly, that background noise in both instances is reduced and concerted action achieved in ways that can be elucidated by the general language of cybernetics. In addition to the prospect of a common language and a set of mathematical techniques, there exists among experimentalists a shared faith that characterizes the reductionist spirit in biology generally, that in time all the piecemeal analyses will permit the reconstruction of the full system in vitro. In this case an in vitro reconstruction would mean the full explanation of social behavior by means of integrative mechanisms experimentally demonstrated and the proof of that explanation by the artificial induction of the complete repertory of social responses on the part of isolated members of insect colonies. The proof can be expected to include the following achievements:

—— The rearing of isolated larvae or nymphs and the determination of their caste at the will of the investigator by appropriate manipulation of food, pheromones, hormones, and other caste-biasing factors.

—— The activation of social behavioral responses, including the more intricate and delicate aspects of brood care, by exposure of isolated colony members to synthetic pheromones, sounds, and other stimuli emanating from lifeless dummies or else presented wholly *in vacuo.*

—— The cybernetic simulation of nest building, incorporating into the model only those behavioral elements and sequences of elements actually observed in individual workers, leading in turn to the successful prediction of the responses of isolated workers presented with synthetic nests in various stages of construction.

At the present time we cannot come close to any one of these three accomplishments. Nevertheless, I believe they represent, in what I admit is a loosely defined way, the goals toward which much of contemporary research on social insects must ultimately be directed. Add to this the continuing quest for precise evolutionary, that is, genetic, explanations of the origin of sociality and variations among the species in details of social structure, and one has the exciting modern substitute for the superorganism concept.

17 The Genetic Theory of Social Behavior

The Units of Natural Selection

While composing the *Origin of Species,* Charles Darwin encountered in the social insects the "one special difficulty, which at first appeared to me insuperable, and actually fatal to my whole theory." How, he asked, could the worker caste of insect societies have evolved if they are sterile and leave no offspring? This paradox proved truly fatal to Lamarck's theory of evolution by the inheritance of acquired characters, for Darwin was quick to point out that the Lamarckian hypothesis requires characters to be developed by use or disuse of the organs of individual organisms and then passed directly to the next generation, an impossibility when the organisms are sterile. To save his own theory, Darwin introduced the idea of natural selection operating at the level of the family rather than the single organism. In retrospect, his logic seems impeccable. If some of the individuals of the family are sterile and yet important to the welfare of fertile relatives, as in the case of insect colonies, selection at the family level is inevitable. With the entire family serving as the unit of selection, it is the capacity to generate sterile but altruistic relatives that becomes subject to genetic evolution. To quote Darwin, "Thus, a well-flavoured vegetable is cooked, and the individual is destroyed; but the horticulturist sows seeds of the same stock, and confidently expects to get nearly the same variety; breeders of cattle wish the flesh and fat to be well marbled together; the animal has been slaughtered, but the breeder goes with confidence to the same family." (*The Origin of Species,* 1859: p. 237.) Employing his familiar style of argumentation, Darwin noted that intermediate stages found in some living species connect at least some of the extreme sterile castes, making it possible to trace the route along which they evolved. As he wrote, "With these facts before me, I believe that natural selection, by acting on the fertile parents, could form a species which regularly produce neuters, either all of a large size with one form of jaw, or all of small size with jaws having a widely different structure; or lastly, and this is the climax of our difficulty, one set of workers of one size and structure, and simultaneously another set of workers of a different size and structure . . . " (*The Origin of Species,* 1859: p. 241.)

Darwin was speaking here about the soldiers and minor workers of ants. A modern physiological interpretation of the same phenomenon has already been presented, in Chapter 8.

The point of this argument was lost on a surprising number of subsequent students of social insects. Wasmann (1891) and Raignier (1952), for example, found it impossible to accept natural selection as the explanation of the origin of the slave-making habits in certain ant species because the queen—the sole fundatrix of the colony—does not take part in slave raids. An equally misdirected rebuttal came from Eidmann (1929), who concluded that the only way the brood-care instincts exhibited by worker ants can be transmitted is if the workers themselves lay the eggs that produce the males. Weyer (1929) and K. Hölldobler (1936) not only questioned Eidmann's logic but also showed that in *Formica rufa* at least some of the males came from queen-laid eggs. We now know of one ant genus (*Solenopsis*) in which all of the males must come from the queen, because the worker ovaries are totally degenerate. But of course this is no longer a crucial point.

Modern theory utilizes an advanced form of Darwin's concept of colony-level selection in explanations of the evolution of the worker castes.

Sturtevant (1938) advanced the theory by postulating a hierarchy of three units on which selection can operate simultaneously: namely, the individual, the colony of individuals, and the population of colonies of individuals. If selection operates at the level of the colony, individual altruism can arise. Genetically based phenotypes that sacrifice themselves—by willing exposure to danger or by sterility—can still be favored in natural selection, providing the sacrifice confers sufficiently great fitness on relatives who carry the same genes by virtue of common descent. For the social insects this means that altruistic behavior increases the fitness of the colony as a whole. Put another way, altruism on the part of organisms subsumes selection at the colony (or family) level because there is no other known procedure consistent with conventional selection theory by which altruistic genes can be fixed in populations. The only way such genes might be fixed in the absence of colony-level selection is by genetic drift, which is a most unlikely mechanism in view of the large size of most populations of social insects.

Altruism of Individuals

Altruism is self-destructive behavior performed for the benefit of others. The use of the word altruism in biology has been faulted by Williams and Williams (1957), who suggest that the alternative expression "social donorism" is preferable because it has less gratuitous emotional flavor. Even so, altruism has been used as a term in connection with evolutionary argumentation by Haldane (1932) and rigorous genetic theory by Hamilton (1964), and it has the great advantage of being instantly familiar. The self-destruction can range in intensity all the way from total bodily sacrifice to a slight diminishment of reproductive powers. Altruistic behavior is of course commonplace in the responses of parents toward their young. It is far less frequent, and for our purposes much more interesting, when displayed by young toward their parents or by individuals toward siblings or other, more distantly related members of the same species. Altruism is a subject of importance in evolution theory because it implies the existence of group selection, and its extreme development in the social insects is therefore of more than ordinary interest. The great scope and variety of the phenomenon in the social insects is best indicated by citing a few concrete examples:

— The soldier caste of most species of termites and ants is virtually limited in function to colony defense (Chapters 8 and 10). Soldiers are often slow to respond to stimuli that arouse the rest of the colony, but, when they do, they normally place themselves in the position of maximum danger. When nest walls of higher termites such as *Nasutitermes* are broken open, for example, the white, defenseless nymphs and workers rush inward toward the concealed depths of the nest, while the soldiers press outward and mill aggressively on the outside of the nest. Nutting (personal communication) witnessed soldiers of *Amitermes emersoni* in Arizona emerge from the nest well in advance of the nuptial flights, wander widely around the nest vicinity, and effectively tie up in combat all foraging ants that could have endangered the emerging winged reproductives.

— I have observed that injured workers of the fire ant *Solenopsis saevissima* leave the nest more readily and are more aggressive on the average than their uninjured sisters. Dying workers of the harvesting ant *Pogonomyrmex badius* tend to leave the nest altogether. Both effects may be no more than meaningless epiphenomena, but it is also likely that the responses are altruistic. To be specific, injured workers are useless for most functions other than defense, while dying workers pose a sanitary problem.

— Alarm communication, which is employed in one form or other throughout the higher social groups, has the effect of drawing workers toward sources of danger while protecting the queens, the brood, and the unmated sexual forms.

— Honeybee workers possess barbed stings that tend to remain embedded when the insects pull away from their victims, causing part of their viscera to be torn out and the bees to be fatally injured (Sakagami and Akahira, 1960). A similar defensive maneuver occurs in many polybiine wasps, including *Synoeca surinama* and at least some species of *Polybia* and *Stelopolybia* (Rau, 1933; W. D. Hamilton, personal communication) and the ant *Pogonomyrmex badius*. The fearsome reputation of social bees and wasps in comparison with other insects is due to their general readiness to throw their lives away upon slight provocation.

—— When fed exclusively on sugar water, honeybee workers can still raise larvae—but only by metabolizing and donating their own tissue proteins (Haydak, 1935). That this donation to their sisters actually shortens their own lives is indicated by the finding of de Groot (1953) that longevity in workers is a function of protein intake.

—— Female workers of most social insects curtail their own egg laying in the presence of a queen, either through submissive behavior or through biochemical inhibition (Chapter 15). The workers of many ant and stingless bee species lay special trophic eggs that are fed principally to the larvae and queen.

—— The "communal stomach," or distensible crop, together with a specially modified proventriculus, forms a complex storage and pumping system that functions in the exchange of liquid food among members of the same colony in the higher ants (Eisner, 1957). In both honeybees and ants, newly fed workers often press offerings of ingluvial food on nestmates without being begged, and they may go so far as to expend their supply to a level below the colony average (Gösswald and Kloft, 1960; Lindauer, 1961; Wallis, 1961; Lange, 1967).

These diverse physiological and behavioral responses are difficult to interpret in any way except as altruistic adaptations that have evolved through the agency of natural selection operating at the colony level. The list by no means exhausts the phenomena that could be placed in the same category.

Altruism of Colonies

If selection at the level of the colony can indeed generate the evolution of altruistic behavior on the part of individual members of the colony, it is also conceivable that selection at the level of the whole population of colonies can generate altruistic behavior on the part of entire colonies, or at least of the queens representing colonies. Does such colony altruism exist? Sturtevant (1938) thought that it might in the case of the mound-building ants *Formica exsectoides* and *F. rufa.* After the nuptial flights the queens are usually adopted by colonies of the same species and not necessarily the same colonies that produced them. It seemed reasonable to Sturtevant that "the colony long outlives the queen that originally established it, its fertility being dependent upon a whole series of queens that have no necessary genetic relationship to each other or to the founder." The implication is that populations of colonies dominated by those less disposed to adopt unrelated queens have less chance for survival, and are subject to replacement by genotypes from nearby populations whose colonies are more altruistically inclined.

Williams and Williams (1957) opposed the idea of selection at the level of entire populations. They attempted to refute Sturtevant's hypothesis with the following argument: "The observation of queens of different 'varieties' present in one nest is apparently based on the facultative parasitism found in *F. rufa* and its relatives, and has no direct bearing on selection within a breeding population. The crucial point is the lack of genetic relationship between conspecific queens. If this were a regular occurrence, reproductive competition between the queens would necessarily promote the loss of the sterile castes and would have to be opposed by selection at a higher level. But if the multiple queens were normally sisters, natural selection between these sister-groups might normally favor the production of sterile castes (sibling donorism) within such groups. The essential statistical data on the frequency of unrelated queens in one colony have apparently not been gathered."

Thus, Williams and Williams call attention to the disconcerting fact that selection can also be between sister-groups, a unit somewhat above the level of an individual colony but still well below an entire population. Moreover, they regard adoption behavior as part of the particular scheme of colony multiplication evolved by these species of *Formica.* Many of the new colonies are founded by groups of workers and one or more mated queens which walk away from the home nest to new nest sites. Multiple queens can be thought of as an enabling device for this mode of colony multiplication, and the average number of queens in a single colony, as the balance struck between the accumulation of endogenous queens and their allocation to daughter colonies.

In fact, the cohesive structure of populations of nests in the *Formica exsecta* and *F. rufa* groups renders the line between colony and population indistinct and the existence of "colony altruism" difficult to evaluate. The difficulty is clearly exposed in the case of *F. opaciventris,* a member of the *exsecta* group. Among the 400 mounds of the Moose Island, Wyoming, population of this species, Scherba (1961) determined that each summer only 20 to 25 percent release alates of either sex, while only 3 to 5 percent release queens. Scherba considered this low incidence to be evidence of some integrating mechanism within the entire population of mounds leading to a "reproductive division of labor." Later Scherba (1964a) used marking techniques to confirm a phenomenon that has long been suspected for species of the *Formica exsecta* group: the extensive exchange of workers among mounds by daily visiting. Thus the entire population of mounds can be regarded in at least one sense as harboring a single large colony. Scherba nevertheless concluded that each mound contains a discrete colony since the mounds are regularly spaced and not connected by visually apparent runways. Yet his own data indicate an average worker exchange among mounds of 1.99 ± 0.70 percent over a period of several days, enough it would seem to allow for an extensive amount of physiological integration. Perhaps the condition is intermediate, and the mounds harbor "semi-colonies." It is probably true, as Scherba implies, that the reproductive structure he described evolved by selection at the population level. Yet we cannot conclude from this inference that colony altruism is involved because on Moose Island the colony and the population are very nearly the same.

I have taken a different approach to the question of colony altruism (Wilson, 1963b). Assuming that colony altruism should be most favored by a high population extinction rate, it should follow that the population extinction rate is highest in rare or locally distributed species. In the termites, in which both males and females are derived from diploid zygotes, the effective breeding size of a population ($\tilde{N}$ in the parlance of population genetics; see Li, 1955) is

$$\frac{4N_t}{N_c}\left(\frac{\overline{MQ}}{1+\overline{M}}\right)$$

and in a population of social Hymenoptera, in which the males are derived from unfertilized eggs, the breeding size is

$$\frac{4.5N_t}{N_c}\left(\frac{\overline{MQ}}{1+2\overline{M}}\right)$$

where N_t is the total number of adult individuals, of all castes, in the population of colonies;
N_c is the average number of adult individuals per colony;
$\overline{Q}$ is the average number of queens contributing offspring to individual colonies;
$\overline{M}$ is the average number of males that fertilized each queen (these individuals no longer need be living).

$\tilde{N}$ will be exactly equal to the term given if breeding is panmictic and less than the term if it is not. (I am indebted to Ross Crozier and Warwick Kerr for pointing out the difference stemming from the two kinds of sex determination.)

Examination of these two very similar formulas shows that the most efficient means of enlarging effective population size is by increasing the number of queens. Merely adding a single additional queen to a monogynous colony system, for example, has the effect of doubling the effective population size in both cases. Recent theoretical computations by MacArthur and Wilson (1967) indicate that a doubling of the population size will enormously increase the mean survival time of the populations under a wide range of conditions. Consequently polygyny alone might "rescue" rare populations from quick extinction. Put another way, polygyny could be established in systems of populations small enough to be subject to frequent extinction. Moreover, polygyny in this case implies a toleration among nest queens that is altruistic in nature.

By considering the records of certain groups of rare or very local ant species, namely social parasites and forms mostly limited to bogs and caves, it could be demonstrated that a significantly higher proportion were polygynous in comparison with related species. Monogyny is the rule in other ant species, and we must assume that it is generally adaptive. The inference I draw from this relation is that selection is strong enough at the population level in these rare species, due to the danger of extinction through small population size, to force polygyny on the colonies. But the inference is weak since polygyny also occurs in some ant species that are abundant and widespread, and causal factors other than population-level selection must therefore exist. An alternative explanation is that because of the small number of individuals the average genetic relationship of associated queens would be closer than in larger, more typical populations; hence, altruistic cooperation among queens would be favored due to sister-group selection, a relation that has already been suggested for *Formica rufa* by Williams and Williams. I see no way to decide between these two competing hypotheses on the basis of existing data.

Decrease of colony size, which the formula discloses to

be a less efficient method of increasing effective population size (and which is even more obviously altruistic in nature), did not occur. Finally, the data on rare ant species show no evidence of an increase in exogamy, the tendency for the sexual forms to mate with members of other colonies. On the contrary, the rare species have reduced exogamy. Mating among members of the same colony, which has the effect of turning the colony into something resembling a self-fertilizing hermaphrodite, has long been recognized by myrmecologists as a trait of the rare, parasitic species. The males are commonly apterous or subapterous, mating takes place in or near the nest, and the fecundated, winged queens then either disperse in search of new host colonies or else return to the old. Whether intracolonial mating characterizes other categories of rare species is an open question. Since the trait must cause a decrease in the effective breeding size of the population, just the reverse of what purely logical considerations concerning population size alone dictate, it is necessary to consider other possible advantages of such a design feature. At least one can be deduced: intracolonial mating certainly eliminates the loss of virgin reproductives that normally occurs during dispersal and insures that the queens will be fecundated, however scarce the species. This advantage can easily outweigh disadvantages from inbreeding. Passing from random mating to perfect inbreeding only reduces the effective breeding size of the population by half, a deficit that can be balanced by merely doubling the average number of nest queens. The exact extent of true brother-sister mating is unknown. Because of the additional trait of polygyny in these same species, the offspring of several matings probably breed with each other as a matter of course. In fact, in the only test of this kind of which I am aware, Wesson (1939) did find that the queens and males of the dulotic ant *Harpagoxenus americanus* prefer to mate with unrelated individuals. It is even conceivable that parasitic species are less "adelphogamous" than previously assumed since it has not been established with certainty that true brothers and sisters within polygynous colonies really mate with each other at all. In polygynous colonies the opposite may be true.

To summarize this rather complicated subject, altruism by colonies or by queens representing colonies is an important theoretical possibility with implications ranging beyond the behavior of individual social insects to the organization of whole populations of colonies. The evidence available to test the idea is still indirect in nature, rather intricate, and compromised by the obvious presence of alternative explanations that cannot yet be ruled out. In particular, we still do not know enough about the demarcation of colonies in the more promising species (such as the members of the *Formica exsecta* and *F. rufa* groups) even to distinguish "colonies" from "populations." We also understand very little of the adaptive significance of polygyny, and, more importantly, the degree of kinship of queens that live together in polygynous colonies. A definitive judgment cannot be made until more data are available on these subjects. Meanwhile, some further insight into altruism and the origin of societies can be gained by an examination of certain other aspects of the genetics of social insects, especially the haplodiploid method of sex determination in the hymenopterans.

Sex Determination

In almost all of the species of Hymenoptera thus far studied, fertilized eggs produce females, and unfertilized eggs produce males. This mode of sex determination, known as Dzierzon's rule (it was first demonstrated in the honeybee in 1845 by J. Dzierzon) provides a key piece to the understanding of the evolution of social behavior in the Hymenoptera. It has been modified by modern genetics to the concept of haplodiploidy, which is defined as the mode of sex determination in which males are derived from haploid eggs and females from diploid eggs.

The genetics of haplodiploid sex determination have been most recently reviewed by Kerr (1962b, 1967). The modern period of its study began in 1939–1943 when Whiting developed the multiple allele hypothesis to explain sex determination in the parasitoid wasp *Bracon hebetor* (= *Habrobracon juglandis*). His findings indicate that *Bracon* females are heterozygous for at least one pair of sex alleles located on an undetermined number of loci, such as X_1X_2, X_4X_3, X_5X_4, . . . , while haploid males are hemizygous by virtue of their parthenogenetic origin, yielding X_1, X_2, X_3, X_4, . . . Homozygous diploids, for example X_1X_1, X_2X_2, X_3X_3, . . . are weak males of low viability. Mackensen (1943, 1951) extended the Whiting model to the honeybee, *Apis mellifera,* and noted that what are believed to be the homozygous-diploid male larvae die within four days after hatching. Rothenbuhler (1957, 1958) found several new lines of evidence to support the Mackensen hypothesis, while Laidlaw, Gomes, and Kerr

(1956) and Kerr (1967) used indirect means to estimate the presence of approximately 12 sex alleles.

Further experiments showed that the simple form of the Whiting model cannot be applied to certain other hymenopterans such as *Melipona, Melittobia,* and *Telenomus.* In order to encompass these cases, Cunha and Kerr (1957) proposed a more general model that assumes the existence of a series of male tendency genes (m) and a series of female tendency genes (f), distributed through several chromosomes. The m genes are not additive in effect, and whether they occur in a single dose in hemizygotes or in a double dose in diploids, the total effect can be represented approximately as a constant M in the degree to which they bias the individual toward maleness. The f genes, on the other hand, are postulated to be cumulative in effect, producing a femaleness effect (F) in a hemizygote and ($2F$) in a diploid. Sex would be determined by the inequalities $2F > M =$ females, $M > F =$ male. The conditions of the Whiting model are explained as the outcome of pairs of f genes that have lost their cumulative effect in homozygous condition, but still express it in heterozygous condition as a heterotic effect.

The Whiting state of *Apis mellifera* sex determination is regarded by Kerr (1967) as being a derived condition. The condition is not shared by other members of the Apidae thus far studied, namely species of *Bombus, Melipona, Meliponula,* and *Trigona.* A Whiting state has nonetheless been reported to control caste determination in *Melipona* (Kerr and Nielsen, 1966). Two genes are involved; the double heterozygote $X_1{}^aX_2{}^a$, $X_1{}^bX_2{}^b$ becomes a queen provided the larva receives enough food, while the homozygotes ($X_1{}^aX_1{}^a$, $X_1{}^bX_1{}^b$, and so on) always become workers (see Chapter 9).

The parthenogenetic origin of hymenopteran males means that all alleles will be expressed in a homozygous (or, more precisely, hemizygous) condition. As a result, lethals and subvitals will be exposed each generation and rapidly reduced in the populations by selection, overdominance will be rendered negligible, and total genetic variability in the population will tend to be reduced. However, these negative effects will be true only for genes that are expressed in the male. Those limited in expression to female characters will behave just as though they existed in wholly diploid populations, enjoying the same potential variability and obeying the same equilibrium laws (Kerr, 1967). Another curious effect is that characters that are both under polygenic control and not sex limited should be more variable among males in sibling groups than among females. In fact, under the simplest possible conditions (panmixia and an absence of dominance and epistasis), the theory of polygenic inheritance predicts a genetic variance in males four times that of their biparental sisters. Since most characteristics are under polygenic control, it should, therefore, generally be the case that males are more variable than virgin queens collected from the same colony. Recently Kathleen R. Eickwort (1969b) found this proposition to be true for ten external morphological characters which she measured in the paper wasp *Polistes exclamans.*

After Forel (1874) first demonstrated Dzierzon's Rule to apply to the ant *Formica sanguinea,* Lubbock (1894), Goetsch (1937a), Goetsch and Käthner (1937), Bier (1952), Brian (1953a), and others confirmed it in a variety of higher ant genera and species, while Haskins and Haskins (1950a) established its presence in the very primitive genus *Myrmecia.* But many exceptions have also been noted. Impaternate workers were reported in *Lasius niger* by Reichenbach (1902), and the result confirmed in this species by Crawley (1912) and Bier (1952); in *Atta cephalotes* by Tanner (1892); in *Aphaenogaster rudis* (= *A. fulva aquia*) and *A. lamellidens* by Haskins and Enzmann (1945); in *Oecophylla longinoda* by Ledoux (1950); and in *Formica polyctena* by Otto (1960).

While thelytokous (female-producing) parthenogenesis is now well documented as a minority phenomenon in ants, it is distinguished by some peculiar features whose causes are not well understood. Haskins and Enzmann made a careful attempt to induce virgin queens of *Aphaenogaster* to rear their own young, but they were only partly successful. Of the few individuals brought through to the pupal stage, most were males. In the case of *Aphaenogaster rudis,* only 18 of 100 virgin queens reared brood to the pupal stage, and they produced only 2 females among them; the virgin queens of *A. lamellidens* did only slightly better. Bier found that a similar difficulty in rearing workers from worker-laid eggs of *Lasius niger* is due to the fact that the great majority of larvae coming from such eggs are actually male determined and die at an early age. Eggs that are female determined are viable but rare.

These results can be explained if we realize that thelytokous parthenogenesis in the Hymenoptera can result only if meiosis is modified in some major way. In order to produce a diploid egg without fertilization, it is necessary either for the first, or reductional, meiotic division to be eliminated (apomictic parthenogenesis), or else for the haploid products of the reductional division to fuse secondarily to produce a truly zygotic nucleus (automictic parthenogenesis). The third possible form of parthenogenesis is, of course, generative or haploid parthenogenesis, which involves complete meiosis without subsequent nuclear fusion and is the normal mode of male determination in the Hymenoptera (White, 1954; Suomalainen, 1962). Thelytokous parthenogenesis, whether apomictic or automictic, is evidently a deviation from the norm in most groups of ants, and it consequently produces very variable and uncertain results.

A variable, facultative thelytoky also occurs in honeybees. In the South African strain *Apis mellifera capensis,* the workers of queenless hives normally lay eggs that develop into workers or queens. But in other varieties of *A. mellifera,* thelytoky is a rarity. Mackensen (1943) determined that only about 1 percent of unfertilized eggs develop into females in the three strains he studied. Experiments by Tucker (1958) revealed that the diploidy underlying the thelytoky is automictic in origin. Under certain conditions, and especially when oviposition by the queen is increasing, the second meiotic division of the two products of the first division produces two pronuclei and two polar bodies instead of the usual single pronucleus and three polar bodies. The two pronuclei then undergo syngamy and form a zygote. The mechanics of the first meiotic division evidently insure that heterozygous combinations of the sex alleles are produced, so that a female is produced rather than an inviable diploid male.

Two instances are known in which thelytoky has ceased to be an anomaly and has instead become a fixed feature of the life cycle. According to Ledoux (1949, 1950), apomictic parthenogenesis plays a central role in the intricate life cycle of the African weaver ant *Oecophylla longinoda.* Ledoux's evidence and inferences are summarized in the following sequence:

1. After the nuptial flight, the fecundated queen attempts to establish a colony. She lays eggs 1.0 to 1.2 mm long that develop into workers. In nature, very few of the attempts to start colonies in this way meet success.

2. Many new colonies, on the other hand, are founded by groups of workers, which leave the territory of the mother colony and construct new nests of their own. Some of the workers then lay small eggs 0.6 mm in length, most of which develop into workers and a few into queens. Ledoux hypothesizes that the parthenogenesis in this case is apomictic, resulting in diploid eggs, and that the eggs are small because they are ejected before normal meiosis can begin. The workers also lay a few larger eggs, about 1.1 mm in length, which develop into males. These are interpreted as being typical haploid products of two meiotic divisions, and the parthenogenesis in this case would therefore seem to be generative or haploid.

3. Occasionally a winged queen accompanies workers on a nest-founding expedition. Her eggs are the same size as those of solitary nest-founding queens (1.0–1.2 mm long). If fertilized, they give rise to workers; if unfertilized, to males. Ledoux was unable to find evidence that fertilized eggs laid by queens and raised by workers, that is, eggs laid under optimal trophic conditions, ever produce other queens. Nevertheless, he was not able to exclude this possibility with finality.

In sum, the *Oecophylla longinoda* life cycle appears to contain a true alternation of generations. Queens produce workers, and only workers, from fertilized eggs, and males from unfertilized eggs by haploid parthenogenesis; the workers in turn produce other workers as well as queens from unfertilized eggs by apomictic parthenogenesis, and males from unfertilized eggs by haploid parthenogenesis. This is a nice scheme, but it is unfortunately marred by the results of an independent study of *Oecophylla longinoda* by Way (1954a), who found no evidence of reproduction by workers: "On several occasions when trees which contained nests but no queen were deliberately isolated from the rest of the colony, the ants eventually died out; also in nursery experiments with artificially established colonies, queenless colonies died out unless brood was added from time to time. In a further experiment, batches of the *O. longinoda* nests, several of which contained virgin males and females, were placed in each of forty-eight coconut palms. The ants became established in most of the trees but, after a year, all the colonies had died out except two which were flourishing. One contained an old queen which had probably been present in the introduced nests, the other contained brood of all ages, although the queen was not found. These results may suggest that, in the absence of a fertile queen, neither sexual forms nor workers can reproduce." In view of the important implications of Ledoux's *cycle évolutif,* further experimentation combined with karyological investigation is needed.

It is nevertheless quite possible that Ledoux's schema will prove correct, in spite of Way's negative evidence, since a very similar life cycle has also been proposed for the parasitic *Harpagoxenus americanus* by Wesson (1939) and for four species of *Crematogaster* (*C. auberti, C. scutellaris, C. vandeli, C. skounensis*) by Soulié (1960a). Soulié also found that *Crematogaster* queens can produce only workers, and then solely from fertilized eggs, while workers can create queens by apomictic parthenogenesis. Both queens and workers can produce males by haploid parthenogenesis. The cycle is perfectly alternating, that is, comprised of the sequence queens→workers→queens→ workers→ . . . even though queens and workers representing two or three generations usually coexist in a given colony at the same time. When food is scarce, worker-laid eggs tend to be laid prior to the first meiotic division (apomictic parthenogenesis) and thus to produce queens. Queens of these species of *Crematogaster* are consequently associated with hard times, the direct opposite of the situation found in most other social insects.

It must be kept in mind that the conceptions of the alternation of generations in *Oecophylla* and *Crematogaster* are inferential, based largely on rearing experiments and still unsupported by cytological evidence. But, if correctly construed, they call attention to the possibility of similar phenomena in other groups of the large world ant fauna, of which the life cycles are still largely unexplored. Thelytoky of worker-laid eggs may even be much more widespread than suspected earlier, for the simple reason that it cannot be detected until an explicit search is made for it in laboratory experiments.

What would be the adaptive value in social insects of such an alternation between sexual and asexual generations? It seems likely that worker thelytoky has arisen as an auxiliary means of colony reproduction, permitting colonies first to expand rapidly into new nest sites around the colony nest sites and then to produce winged queens in them under more stringent trophic conditions. This can be viewed as an opportunistic strategy of growth and reproduction. If I have interpreted it correctly, thelytoky might be found to occur most commonly in species that occupy habitats, or nest sites, that fluctuate a great deal in time and space and consequently undergo a high rate of turnover in resident colonies.

The mechanism of sex determination in termites differs radically from that of the social Hymenoptera in that it is not based on haplodiploidy. Like most insect groups, including the orthopteroids with which the termites share a common ancestry, both sexes normally come from fertilized, diploid eggs. Light (1944a) found that, when females of *Zootermopsis* are segregated as virgin nymphs, they later lay eggs that always produce females. A parthenogenetic strain cannot be maintained in the laboratory; the female offspring of the parthenogenetic females die in the early instars. We can guess from this sparse evidence that sex determination in termites is of a conventional genic type, and that parthenogenesis does not play an important role in termite social evolution.

The Implications of Haplodiploidy

One of the most remarkable features of true sociality in insects is that it is very nearly confined to the Hymenoptera. It has originated at least twice in the wasps, more precisely at least once each in the stenogastrine and vespine-polybiine vespids (Evans, 1958) and probably a third time in the sphecid *Microstigmus* (Matthews, 1968); eight or more times in the bees (Michener and Lange, 1958a; Michener, 1962, and personal communication); and at least once or perhaps twice in the ants. These estimates must be qualified in that Michener and his associates have found both semisocial and social species in at least six genera of the Halictidae. It follows that true sociality has originated at least six times in this family alone. In addition, both subsocial and social species are known in the xylocopine genera *Allodape* and *Allodapula,* while the true sociality that characterizes the Apinae appears to represent yet another independent development. In ants, the sharp division between the poneroid and myrmecioid lines, which extends at least all the way back to the Cretaceous Sphecomyrminae, raises the possibility of a dual origin of sociality (Wilson, Carpenter, and Brown, 1967a,b).

In summary, it is evident that eusociality has arisen at least eleven times in the Hymenoptera. Quite probably this lower estimate will increase as our knowledge of hymenopteran biology, especially in tropical bees, increases. Yet, throughout the entire remainder of the Arthropoda, true sociality is known to have originated in only one other living group, the termites. This dominance of the social condition by the Hymenoptera cannot be a concidence. Of the 685,900 living species of insects described to 1948, only 103,000, or 15 percent, belong to the Hymenoptera (Sabrosky, 1952). This has been the situation throughout at least the Cenozoic (Theobald, 1937, in Kerr, 1967). Moreover, sociality is further limited within the Hymenoptera to the aculeate wasps and to their immediate descendants, the ants, which, together, comprise no more than 25,000 known living species.

The tendency of aculeate Hymenoptera to evolve eusocial species can perhaps be ascribed in part to their mandibulate mouthparts, which lend themselves so well to manipulation of objects, or to the penchant of aculeate females for building nests to which they return repeatedly, or to the frequent close relationship between mother and young. These and perhaps some other biological features are prerequisites for the evolution of full sociality. But they are shared in full by many other, species-rich groups of arthropods, including the spiders, earwigs, orthopterans, and beetles, none of which, with the exception of the termites, have achieved full sociality.

Haplodiploidy is also a characteristic of the Hymenoptera shared with few other insect groups. Two authors have independently suggested a connection between haplodiploidy and the frequent occurrence of sociality. Richards (1965) suggested that the control which haplodiploidy grants the female over the sex of her own offspring has eased the way to colonial organization. This is undoubtedly true. The postponement of male production until late in the season is a characteristic of advanced sociality, for example, in the annual Halictidae (Knerer and Plateaux-Quénu, 1967b). At the same time, it is not a characteristic of many other Halictidae that are primitively social but truly social nonetheless. In other words, sex control by the mother is a general feature of higher social evolution but not a prerequisite for the attainment of full sociality.

Hamilton (1964) has created an audacious genetic theory of the origin of sociality which assigns the central role to haplodiploidy in a wholly different way. Working from traditional axioms of population genetics, he first deduced the following principle that applies to any genotype: in order for an altruistic trait to evolve, the sacrifice of fitness by an individual must be compensated for by an increase in fitness in some group of relatives by a factor greater than the reciprocal of the coefficient of relationship to that group. The coefficient of relationship is the equivalent of the average fraction of genes shared by common descent; thus, in sisters it is one-half; in half sisters, one-fourth; in first cousins, one-eighth; and so on. The following example should make the relation intuitively clearer: if an individual sacrifices its life or is sterilized by some inherited trait, in order for that trait to be fixed in evolution it must cause the reproductive rate of sisters to be more than doubled, or that of half-sisters to be more than quadrupled, and so on. The full effects of the individual on its own fitness and on the fitness of all of its relatives, weighted by the degree of relationship to the relatives, is referred to as the "inclusive fitness." This measure can be treated as the equivalent of the classical measure of fitness, which takes no account of effects on relatives; Hamilton's theorem on altruism consists merely of a more general restatement of the basic axiom that genotypes increase in frequency if their relative fitness is greater.

Next Hamilton pointed out that due to the haplodiploid mode of sex determination in Hymenoptera, the coefficient of relationship among sisters is three-fourths; whereas, between mother and daughter, it remains one-half. This is the case because sisters share all of the genes they receive from their father (since their father is homozygous), and they share on the average of one-half of the genes they receive from their mother. Each sister receives one-half of all of its genes from the father and one-half from the mother, so that the average fraction of genes shared through common descent between two sisters is equal to

$$\left(1 \times \frac{1}{2}\right) + \left(\frac{1}{2} \times \frac{1}{2}\right) = \frac{3}{4}$$

Therefore, in cases where the mother lives as long as the eclosion of her female offspring, those offspring may increase their inclusive fitness more by care of their younger sisters than by an equal amount of care given to their own offspring. In other words, hymenopteran species should tend to become social, all other things being equal.

This idea is so simple and starkly mechanical that my own first reaction to it was to reject it out of hand. But the implications, once the proposition is made the basis of evolutionary models, are so extensive and intricate that I soon became absorbed in its possibilities. Making predictions about social behavior on the basis of coefficients of relationship is at the very least a pleasant game and a valuable exercise in heuristics. Hamilton himself gained several new insights that deserve attention quite apart from the theory that generated them. He also presented a fair accounting of the evidence that is not consistent with his theory. Let us now examine the consequences of the Hamilton theory in some detail, evaluate its merits on the basis of existing knowledge, and try to identify the direction that future studies must take in order to provide a surer judgment. Consider first the pedigree and table of coefficients of relationship given in Figure 17-1. Most of the following predictions can be made from this table, together with an elementary knowledge of social insects.

Prediction. True sociality should occur more frequently in groups with haplodiploidy than in those without it.

Evidence. It has already been stressed that eusociality arose at least eleven times within the Hymenoptera, an order characterized by haplodiploidy, but only once (in the termites) in the large number of other insect groups that do not exhibit haplodiploidy. This fact is notably consistent with the Hamilton thesis. On the other hand, eusociality is limited to the aculeates. Why is it not also found in the phytophagan and terebrant Hymenoptera? The answer may be that these groups seldom if ever possess the second essential preadaptation, namely the continuing presence of the mother wasp in close association with her brood.

Outside the Hymenoptera, haplodiploidy occurs in some mites, some thysanopterans, some aleurodids, the iceryine coccids, and the beetle genus *Micromalthus* (White, 1964). Entwistle (1964) has further reported the existence of arrhenotoky in the ambrosia beetle *Xyleborus compactus,* which strongly indicates the operation of a haplodiploid mechanism in this species also. All of these arthropods form close associations of adults and young, and *Xyleborus* belongs to a group of beetles that has evolved the kind of brood care that is widespread in the presocial Hymenoptera (see Chapter 7). Consequently, the absence of eusociality in such a sizable haplodiploid assemblage must be regarded as negative evidence to be weighed against the Hamilton thesis.

Prediction. The basic condition of the model is negated if queens are multiply inseminated by unrelated males.

Suppose a female is mated by n males and they are respectively responsible for proportions $f_1, f_2, \ldots f_n$, where $\sum_{i=1}^{n} f_i = 1$, of her female progeny. The average coefficient of relationship ($\bar{r}$) between daughters is then

$$\frac{1}{2}\left(\frac{1}{2} + \sum_{i=1}^{n} f_i^{\,2}\right)$$

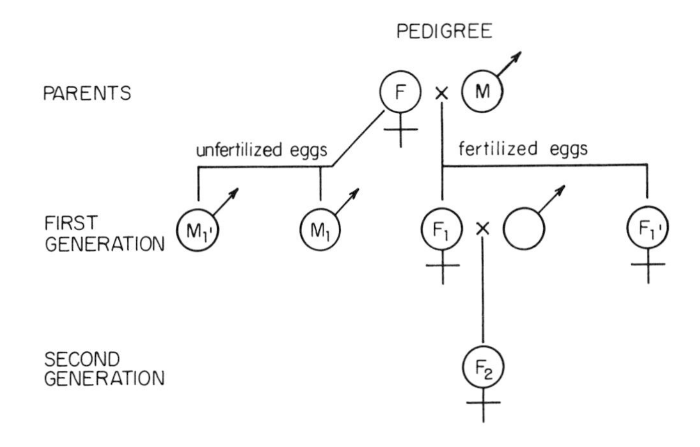

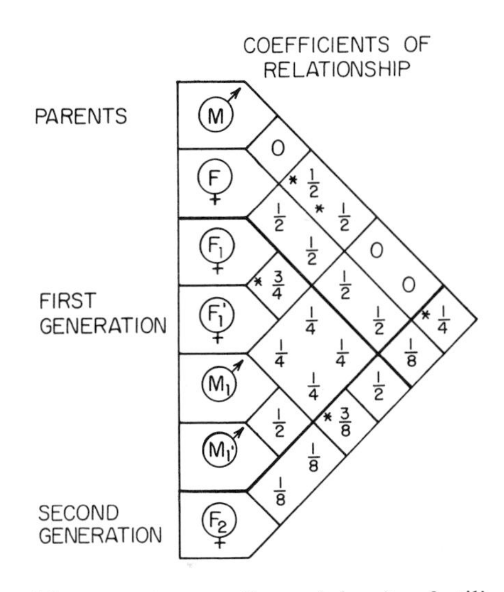

FIGURE 17-1. A hymenopteran pedigree: (*above*) unfertilized eggs laid by the female yield males (M_1 and M_1'), while fertilized eggs yield females (F_1, F_1', and F_2); (*below*) coefficients of relationship for the pedigree. The coefficients are the equivalent of the average fraction of genes held in common between the two individuals by reason of common descent. Asterisks indicate the coefficients that would diminish in cases of polyandrous insemination, assuming the fathership of particular offspring to be unknown (modified from Hamilton, 1964).

In particular, if all males contribute equally, we have

$$\bar{r} = \frac{1}{2}\left(\frac{1}{2} + \frac{1}{n}\right)$$

which is the lowest $\bar{r}$ for any given value of n. If two males contribute equally we have $\bar{r} = 1/2$, which is the same value as between mother and daughter. Furthermore, as the number of contributing males increases the value of $\bar{r}$ can never drop as low as the limiting value

$$\lim_{n\to\infty} \frac{1}{2}\left(\frac{1}{2} + \frac{1}{n}\right) = \frac{1}{4}$$

To put the matter simply, if as few as three or four males inseminate single females about equally as a normal course of events, the coefficient of relationship between sisters will be driven below $\frac{1}{2}$, the value of the bond between mother and daughter. The intrinsic advantage to caring for sisters rather than offspring will have been lost.

Evidence. Some species of social Hymenoptera employ single insemination, but most others that have been studied employ multiple insemination. Queens of the myrmicine ant *Pheidole sitarches* drop quickly to the ground after being mounted by a male, shed their wings and hide shortly after copulation, thus insuring that only one insemination takes place (Wilson, 1957). Queens of the stingless bee *Melipona quadrifasciata* return to the nest from their mating flight with the male genitalia still inserted in their vaginal chamber. Kerr, Zucchi, Nakadaira, and Butolo (1962) found virgin males of this species to contain an average of 1,156,850 spermatozoans, while two recently mated queens contained 950,000 and 1,018,333 spermatozoans respectively in their spermathecae. It thus seems probable that queens of this species are inseminated only once.

Nevertheless, multiple inseminations have been reported more frequently than single inseminations in the social Hymenoptera. Kerr (1961) observed a queen of the myrmicine fungus-growing ant *Mycocepurus goeldii* copulating consecutively with four males. Virgin males of the related fungus-growing species *Atta sexdens rubropilosa* were found to contain 44 to 80 million spermatozoans, while newly mated queens held 206 to 319 million spermatozoans in their spermathecae. Evidently the queens were each inseminated by at least three males. Kannowski (1963) observed queens of *Formica subintegra* mating two to four times. Multiple copulations have also been observed in the formicine species *F. montana* (Kannowski, 1963), *F. rufa* (Marikovsky, 1961), *F. opaciventris* (Scherba, 1961), and *Prenolepis imparis* (Talbot, 1945b). Multiple inseminations in the honeybee *Apis mellifera* have been documented by Triasko (in Kerr *et al.*, 1962), Taber (1954), Taber and Wendel (1958), and Kerr *et al.* (1962). Using genetic markers and sperm counts, these authors have independently estimated that most queens are inseminated by seven to twelve males during their nuptial flights. On the other hand, Alber (1956) noted that, when flights are held in bad weather, the queen may be inseminated only once.

This evidence of the widespread practice of multiple insemination is not favorable to Hamilton's thesis. There are two ways by which the difficulty could be explained away: if the males involved in multiple insemination were closely related to one another so that their sperm would be genetically identical or very similar; or, if multiple insemination were shown to be limited to species with a strong difference between the queen and worker castes so that it could be interpreted as an adaptation acquired secondarily after the worker caste had evolved so far as to be irreversible. The evidence on both aspects is too meager to draw any conclusion at this time. Kerr *et al.* (1962) marked drones of the stingless bee *Trigona postica* from four hives with different pigments and found that individual mating aggregations were comprised of mixtures from all of the hives. This is evidence for genetic outbreeding. On the other hand, it is not known whether the *T. postica* queens mate singly or with multiple males.

Prediction. Males should be more consistently selfish than females toward everyone else in the colony.

From the matrix in Figure 17-1 it can be seen that the males are related to their mother by a degree of $\frac{1}{2}$ but to every other contemporary female of the colony by a degree of $\frac{1}{4}$, while females are related to their mothers by $\frac{1}{2}$, to their sisters by $\frac{3}{4}$, and to their brothers by $\frac{1}{4}$. (For purposes of illustration, the only case considered here is that in which females are inseminated by a single male.)

Evidence. Of course, the selfishness of drone behavior is well known—in our language, the word "drone" has come to designate any lazy, parasitic person. Not only do hymenopteran males contribute virtually nothing to the labor of the colony (see Chapter 8), but they are also highly competitive in begging food from female members of the colony and become quite aggressive in contending with other males for access to females during the nuptial flights.

The one exception to this rule of total egocentricity of which I am aware is the willingness of male ants to regurgitate food to nestmates (Gösswald and Kloft, 1960; B. Hölldobler, 1966). Santschi's observation (1907) of ergatoid *Cardiocondyla* males carrying pupae has not been repeated, and Le Masne (1956a) failed to find evidence of the same behavior in ergatoid males of *Hypoponera.*

I do not find it a serious drawback to this line of evidence, as Hamilton (1964) does, that males of solitary Hymenoptera are also selfish. With the exception of *Pison* and *Trypoxylon,* the males of which have been observed to remain with their mates at the entrance of the hollow-stemmed nests, no case is known of males cooperating in the building and defense of nests. But social behavior of this kind is not to be expected in males of solitary species any more than it is in males of social species. Solitary males are bound to the daughters of their mates by a relationship of $\frac{1}{2}$ (that is, half of the genes come from them), and by an average of less than $\frac{1}{2}$ if their mate has been inseminated by other males as well. Because of the haplodiploid sex determination, they have no relationship at all with the male offspring in the nests. Their fitness would more likely be increased by investing their energy in inseminating as many females as possible.

Prediction. Females should be more altruistic in their behavior toward their sisters and less so toward their brothers and nieces.

Sisters share as many as $\frac{3}{4}$ of their genes, but a female shares only $\frac{3}{8}$ of her genes with her nieces and $\frac{1}{4}$ with her brothers.

Evidence. We have already seen that altruism between sisters is the most fundamental quality of higher hymenopteran societies. It is also true that sisters give short shrift to their brothers. In the European paper wasp *Polistes gallicus,* for example, males easily obtain regurgitated food from workers only when food is generally abundant. In hard times they are ignored and must live on regurgitated material solicited from larvae (Montagner, 1964). Young honeybee drones are at first entirely fed by workers. Later they obtain part of their nourishment on their own from honey cells, and, as they age still further, they begin to be attacked by the workers. This leads to the famous "battle of the drones," which has been described so well by von Frisch (1954): "Now that males have become useless, the workers start plucking and biting those very drones whom up to now they have nursed and fed, pinching them with their firm jaws wherever they can get hold of them. Grasping their feelers or their legs, they try to pull them away from the combs, and to drag them towards the door of the hive. They could not make their meaning clearer. Once they are turned out of the hive, the drones, unable to fend for themselves, are doomed to die of starvation. With great obstinacy they try to force their way back, only to be received again by the workers' biting jaws, and even by their poisonous sting, to which they yield without offering any resistance. For drones do not possess a poisonous sting, nor for that matter the least fighting spirit. Thus they find their inglorious end at the portals of the bee dwelling, driven out and starved, or stung to death, on a fine summer's day." In the case of the stingless bees (*Melipona* and *Trigona*), the males accede more gracefully. After they reach the age of 15 to 20 days, they begin to leave the nest and search for their own food in flowers, usually the same ones visited by their sisters (Warwick E. Kerr, personal communication).

It is an equally notable fact that females in general try to avoid having nieces. This statement is another way of saying that, as a rule, only one queen is tolerated in a colony (monogyny). As Charles Butler so pointedly described the honeybee society in the *Feminine Monarchie,* written shortly after the death of Elizabeth I:

> *For the Bees aborre as well polyarchie, as anarchie, God having shewed in the unto mean express patterne of a perfect monarchie, the most natural and absolute forme of governmet.*

In nonparasitic social Hymenoptera, three common modes of founding new colonies exist: (a) the queen starts a colony by herself, with or without the help of workers, and the colony remains monogynous; (b) more than one queen founds the colony, but subsequently all but one are eliminated; or (c) two or more queens found a colony and subsequently persist together. The first two categories include the great majority of known cases, but the second category deserves emphasis with reference to Hamilton's theory. Laboratory experiments have demonstrated that, in the ants *Lasius flavus* (Waloff, 1957) and *Solenopsis saevissima* (Wilson, 1966), queens that found their colonies in groups have a higher survival rate than queens founding colonies alone. They are also able to rear their brood more quickly. This is a sufficient explanation of why founding queens of these species are often encountered in groups in nature. Once the first brood is reared, however, the behavior of the queens changes radically. In Waloff's experiments the *Lasius flavus* queens were amicable and cooperative until the first pupae appeared, when they began to split into smaller groups. (Artificially grouped *L. niger* queens behaved aggressively from the beginning, and they often fought to the death.) In a single experiment on *Solenopsis saevissima,* conducted by Michael Mullin (personal communication), the queens fought to the death following eclosion of the first workers. Large colonies of the carpenter ants *Camponotus herculeanus* and *C. ligniperda* often contain several queens, but these individuals are intolerant of each other and maintain territories within the far-flung galleries of the individual nest (B. Hölldobler, 1962). A nearly identical situation exists in the large colonies of the Brazilian bumblebee *Bombus atratus.* During the polygynic phase of the life cycle, the queens are intolerant of each other; each rules in her own well-delimited territory and will fight to the death if a rival invades the territory (Ronaldo Zucchi, personal communication).

In other ant genera besides *Camponotus,* oligogyny is occasional in species usually thought of as monogynous. When Headley (1949) excavated 46 nests of *Aphaenogaster rudis,* he found no queens in 6 of the nests, a single queen in each of 38 nests, and two queens each in 2 nests; Talbot (1951) obtained similar proportions in the 71 additional nests of *A. rudis* she excavated. In 20 nests of *Prenolepis imparis,* Talbot (1943) found no queen in 3 of the nests, a single queen in each of 15 nests, and two queens in each of 2 nests. Occasional coexistence of two queens has been recorded in *Atta texana* and *Solenopsis saevissima* (J. C. Moser and M. S. Blum, personal communication). The factors permitting oligogyny, including the possible one of queen territoriality, have not been investigated in most species. Unlike Hölldobler's evidence for *Camponotus,* the data for *Aphaenogaster* and *Prenolepis* do not show an association between oligogyny and larger colony size.

Aggression, competition, and dominance among laying females of the same colony is widespread in the social Hymenoptera. It can be generally interpreted as a mechanism favoring daughters over nieces and less closely related females. When the dominant female of a bumblebee or wasp (*Polistes*) colony is removed, other workers, particularly those with the best-developed ovaries, threaten and fight until one attains a clear-cut dominant position. Aggression among queen honeybees involves complex behavior, including sound signals utilized only for this purpose. When the honeybee queen is removed, only a few of the workers develop the ability to lay eggs; these quarrel among themselves until one or a few become dominant.

No Amazonian *Dames, nor* Indians *more,*
With loyal Awe their Idol *Queen adore.*
Whilst she survives, in concord and content
The Commons live, by no Divisions rent;
But the great Monarch's Death dissolves the Government.

Joseph Warder (1726)

In the halictid bee *Dialictus zephyrus,* which has a relatively low social organization, females commonly substitute their own eggs for those of their sisterly rivals in brood cells (Batra, 1964). Baroni Urbani (1968) has observed a *Polistes*-like dominance of one queen over another in a single colony of the ant *Myrmicina graminicola.* In sum, it is necessary to conclude that in the social Hymenoptera females generally tend to be selfless in the rearing of sisters; when confronted with the choice of rearing nieces or their own daughters, they tend to choose their daughters.

Deviations from this rule inevitably exist. I have already cited cases of oligogyny, possibly associated with a form of territoriality, that would render the situation, in effect, monogynous. While it might thus prove easy to bring some instances of oligogyny in line with the Hamilton thesis, this is quite difficult to do with the many other recorded examples of polygyny. The presence of queens living together in harmony have been reported in some of the primitively social bees and wasps, in the polybiine wasps, and in both primitive and advanced ant genera including *Myrmecia, Amblyopone, Crematogaster, Myrmica, Pheidole, Tetramorium, Monomorium, Tapinoma, Iridomyrmex, Formica, Camponotus,* and many others. These present numerous exceptions to what must be regarded as the relatively weak rule of monogyny and conditional oligogyny in the social Hymenoptera. The mere occurrence of supernumerary queens, however, leaves open the question as to how many are actually contributing eggs to the colony. Multiple egg layers have been recorded in the Polybiini (Richards and Richards, 1951), but in most cases the observations were not close enough to determine the extent of the phenomenon. In other words, it is still very possible for a colony to be morphologically polygynous but physiologically or behaviorally monogynous. An example is already available in *Polistes,* where associated females look alike but are rigidly segregated in rank and function.

Hamilton (1964), in an attempt to account for the occurrence of polygyny, hypothesized that the phenomenon ordinarily involves sisters, or at least close cousins. Such a circumstance would be normal if the "population viscosity" were high, that is, if the insects did not wander far from their birthplace, or if some other circumstance increased the degree of inbreeding. He expresses the following idea: "The geographic distribution of the association phenomenon in *Polistes* is striking (Yoshikawa, 1957). We may state it as a general, though by no means unbroken, rule that northern species approximate to the vespine [solitary] mode of colony foundation and tropical species to the polybiine [associational] . . . The single species *Polistes gallicus* illustrates the tendency well. At the northern edge of its range in Europe its females usually found nests alone. In Italy and Southern France the females found nests in companies; while in North Africa the species is said to found colonies by swarming with workers (Richards and Richards, 1951). We here suggest two hypotheses which could bring these facts into conformity with our general theory. The first posits a general higher viscosity of the tropical populations. This will cause, through inbreeding, all coefficients of relationship to have higher actual values than we could get taking into account only connections through the past one or two generations. And it will also increase the tendency for casual neighbors to be related, which is clearly of potential importance for the association phenomenon . . . The second hypothesis appeals to the lack of marked seasons in the tropics causing a lack of synchronism in the breeding activity of insects. This will tend to cause inbreeding because it scarifies the mating population."

Even if it is true that inbreeding is gratuitously enhanced by a tendency for females to remain close to the parental nest, a phenomenon that has incidentally been documented in *Polistes* by Rau (1940) and Mary Jane West Eberhard (personal communication), the effect is nevertheless inadequate at its best to explain why reproductive females subordinate themselves to other females of the same generation. For even if the associated females were full sisters, the subordinate female would be taking care of nieces with a common reproductive bond of $3/8$, whereas she could be taking care of her own daughters and sharing a common bond of $1/2$. The missing piece of the theory has been supplied by what might be termed the "spinster hypothesis," invented by West (Mrs. Eberhard) (1967). This author points out that nest-founding females of *Polistes* vary greatly in ovarian development and that rank in the dominance hierarchy varies directly with this development. It is further true that most new *Polistes* nests fail. Consequently, the probability of a female with low fertility establishing and bringing a nest through to maturity may simply be so low that it is more profitable, as measured by inclusive fitness, for these high-risk individuals to subordinate themselves to female relatives in foundress associations.

Prediction. Workers should favor their own sons over their brothers.

Females are related to their sons by a degree of $1/2$ but to their brothers by a degree of only $1/4$.

Evidence. This result of the theory is very odd but nevertheless can be reasonably well documented. In 1870 von Siebold, after discovering that virgin females of *Polistes gallicus* lay eggs that turn into males, expressed the opinion that there is a division of labor, with fecundated females producing other females and virgins producing males. Marchal (1896) also reported the presence of laying workers in queenright wasp colonies. Yamanaka (1928), in testing von Siebold's idea, observed that males of *Polistes fadwigae* are produced both by the queen and virgin workers. In *Vespula* (*Paravespula*) the queen tends to remain on the lowest comb of the nest, where most of the worker activity occurs, and, during queen production, most male pupae are found in the uppermost combs. Montagner (1963b) interpreted the pattern as the outcome of the decrease of effectiveness of the queen in inhibiting worker ovariole development, with the result that the workers tend to lay eggs on the combs farthest removed from the queen. The same phenomenon has been observed directly in the bumblebee *Bombus atratus* by Ronaldo Zucchi (personal observation). Worker oviposition is widespread in the ants, from the primitive Myrmeciinae to the advanced Myrmicinae, Dolichoderinae, and Formicinae (Freeland, 1958; Passera, 1965a). By feeding queens of *Myrmica* P^{32} and thus labeling their eggs, Brian (1968) was able to show that the workers lay in the presence of the queens and that most males are derived from worker-laid eggs. And, according to Zucchi, this is also the case in *Bombus atratus.* Thus, although the origin of male-determined eggs is difficult to ascertain in normal colonies, the ease with which workers were implicated when such studies were conducted suggests that the phenomenon is widespread.

Evaluation of the Hamilton theory. In balance, the Hamilton theory of insect sociality seems to me to be consistent with enough evidence, and to account uniquely for enough phenomena, to justify its provisional acceptance. What this means more precisely is that the factor of haplodiploid bias should be taken into account in future evolutionary interpretations and as a guideline in planning some further empirical research. Enough negative evidence nevertheless exists to show that either the bias can be canceled under certain circumstances or that it requires some additional factor, such as high population viscosity or colony inbreeding, in order to operate. The question now before us is whether the factor is always canceled, in spite of its abstract appeal to the theorists, and whether perhaps other, hidden explanations exist for the phenomena that seem to be consistent with the theory.

The best way to test any theory is to continue to bear down relentlessly on its most detailed implications. From the evidence just presented, it is clear that we require deeper investigations on the following subjects concerned with the social Hymenoptera: the extent of both morphological and physiological, that is, true, polygyny among the various groups; the relation of physiological polygyny to population viscosity and to degree of relatedness among the cooperating queens; the number of males that inseminate single queens, especially in the primitively social bees and wasps where virtually no information is currently available; the correlation between the number of inseminations and the degree of female polymorphism; and the parentage of the males. If the new data give a detailed fit to the theory of haplodiploid bias or some modification of it, the theory can eventually be judged good and true. If the data do not fit but nevertheless prove intrinsically valuable and lead to new ideas and investigations (as I believe they will), then the theory will have to be judged as having been merely good.

Selfish Altruism and Altruistic Selfishness

The interaction between selection at the individual level and selection at the colony level is complex, and its effects are often subtle. The evolutionary interpretation of social behavior patterns can be treacherous when pursued in a piecemeal fashion. A behavioral act which on limited view appears to be selfish, or even destructive to others, may in fact be seen to be part of a pattern of coordinated, largely altruistic behavior when reviewed as a part of total colony biology. And another behavioral act, which on first view seems to be altruistic with respect to some other individual, may be revealed as a competitive device among siblings when reviewed in terms of the theory of inclusive fitness. Perhaps I can explain how such paradoxical relations can exist in nature.

We have seen how, in the higher social wasps of the genus *Vespula*, there exists a dominance hierarchy among workers (Montagner, 1966). Liquid food exchange is the medium of the hierarchy, and workers contend, often aggressively, for the privilege of receiving the crop contents of other workers by means of regurgitation. This display of apparently selfish behavior among members of the worker caste is not typical of social insects generally and cannot easily be explained by the theory of natural selection. It is true that the *Vespula* workers lay some eggs of their own which develop into males, and it is at least conceivable that dominant individuals perpetuate the "selfish" genes underlying dominance behavior by contributing more than their share of the eggs. To be sure, dominance hierarchies appear early in the growth of the colony, long before male eggs are laid, and there is only a loose positive correlation between ovarian development and dominance rank. But dominance can still conceivably be interpreted as selfish behavior based on genes favored in the worker-male hereditary lines.

However, a second explanation of the phenomenon, clearly altruistic in content, emerges when the organization of the colony as a whole is examined. Montagner has shown that the dominance hierarchies are the basis of an efficient division of labor among the workers. The "low-ranking" workers are the foragers, who gather the food and nest construction materials and turn them over to higher ranking workers on entering the nest. The highest-ranking workers remain in the nest, attending the larvae and building and repairing the brood cells. Thus the dominance behavior serves as a mechanism that apportions the colony labor, and one can reasonably suppose that it contributes to the fitness of the colony as a whole. A similar hierarchical organization, based on liquid food exchange and associated with ovarian development, has been discovered in the ant *Formica polyctena* by Lange (1967). Unlike the *Vespula*, the *Formica* do not display overt aggression in their interactions.

The *Formica polyctena* case forms a transition between *Vespula* and a more subtle but interesting situation in the honeybee *Apis mellifera*. In the latter species there is also a kind of "dominance hierarchy" of food exchange by which food flows from the foragers to the nurse bees. There is no overt aggression, and most of the bees change their status with increasing age from "dominant" nurses to "subordinate" foragers. Most importantly, workers do not normally contribute to drone production (Ribbands, 1953), so we can discount the selfish gene hypothesis. In short, what appears to be selfish behavior when viewed in a few individuals over a short period of time, is more evidently altruistic behavior when interpreted at the level of the colony over a longer period of time.

Consider next an example of the reverse oxymoron: altruistic selfishness. Brian and Hibble (1963) observed that, in the presence of the queen, workers of *Myrmica ruginodis* give the largest larvae less food and bite them deeply enough to leave cuticular scars. As a result, the growth of the large larvae is slowed, the small larvae grow faster, and fewer larvae reach the critical size required to turn them into queens. Here we have an example of what can be legitimately interpreted as colony regulation, with the workers deferring to the presence of a laying queen by preventing the production of additional queens from among their sisters. But is it really colony regulation, and can it even be viewed as altruistic? When analyzed by means of selection theory, we find that workers are actually increasing their own inclusive fitness by favoring their mother over their sisters. If they favored their sisters to become queens, they would increase the probability of having to rear their own nieces, to whom they are related by degree $\frac{5}{8}$ (assuming brother-sister matings) or less, whereas by favoring their mother they will continue to rear sisters, to whom they are related by degree $\frac{3}{4}$. In this case the simplest explanation involves selection at the level of groups of individuals within the colony rather than at the level of the colony as a whole.

18 Compromise and Optimization in Social Evolution

The evolution of social insects offers an interesting array of the kinds of questions that can only be solved by the modern methods of population biology. We must ask, for example, why so many species of wasps and bees are at a primitive level of sociality, while others have converged with the ants and termites to the apex of eusociality. At a different level, what is the reason that colonies of some species tolerate only one queen, but those of other species many queens? And why have some species evolved elaborate worker polymorphism, while related species have retained strict monomorphism? It will also be eventually necessary to inquire why, within limited taxa, some species employ odor trails and others do not, some have colonies of millions of workers and others less than a hundred, some spread their colonies among multiple nest sites, and others cluster in one unit, some engage in regurgitative food exchange and others do not. Correct answers to evolutionary questions can be given that are either trivial or incisive. The full, meaningful answer must both identify the function of the trait under examination and assess the historical accidents that have caused the trait to evolve to a particular stage at a particular time.

Consider, for example, the matter of the primitively social species: why have they progressed no further? Two extreme possibilities can be envisioned. First, there is what might be termed the "disequilibrium case." This means that the species is still actively evolving toward a higher social level. The situation can arise if social evolution is so slow that the species is embarked on a particular adaptive route but is still in transit. Or it can also be produced, even if social evolution is fast enough, providing extinction rates of evolving species are so high that only a few species ever make it and most are consequently in transit at any given time. Implicit in any disequilibrium hypothesis is the assumption that the high eusocial state, or some particular high eusocial state, is the *summum bonum,* the solitary adaptive peak toward which the species and its relatives are climbing. The opposite extreme is the "equilibrium case," in which species at different levels of social evolution are more or less equally well adapted. There are multiple adaptive peaks—solitary, presocial, eusocial—and evolution toward or away from each is about equally fast, or at least the rate is not strictly correlated with the level of sociality. The equilibrium hypothesis envisions lower levels of sociality as compromises struck by species faced with opposing selection pressures. Now it need only be added that the social levels of some species can probably be characterized as equilibrial with reasonable accuracy and others as disequilibrial to varying degrees. To speak of compromise and equilibrium in a precise manner is to employ the language of optimization theory, a young science that has scarcely begun to be applied to biology as a whole, much less to the study of social insects. The remainder of this chapter will be devoted to a formulation of the elements of the subject, in the belief that it offers the opportunity of opening a whole new era of investigation and understanding in animal behavior.

The germ of the idea with reference to the social insects is to be found in the essays of O. W. Richards (in Richards and Richards, 1951; and Richards, 1953a) on colony growth in the polybiine wasps. It will be remembered from Chapter 3 that the polybiine species vary greatly in colony size; the mature colonies of some never have more than 20 or 30 cells, while others may have many thousands. These differences are correlated with the degree of complexity of nest structure, and possibily other behavioral traits, so it is reasonable to think of colony size as one index of the level of social evolution. But rather than view polybiines as moving upward along the scale of colony size in evolution, Richards raised what he himself termed a "heretical" possibility: that some aspects of social life can be disadvantageous and stagnating. There is a special reason why large colony size might be the wrong strategy for a particular polybiine species: "In Guiana, some of the most dangerous enemies were the Driver ants (Ecitonini) but enemies of various sorts occur everywhere. It is a matter of speculation what are the respective likelihoods of discovery by a searching enemy of one large or many small colonies. Each colony may be regarded as so much locked up capital. There is perhaps more chance of one of the numerous small ones being discovered but less is lost if it is destroyed. On the other hand, for an enemy hunting by scent, a large colony might be detected at a much greater distance though it might also be able to defend itself better. Driver ants appear to be almost irresistable and it seems conceivable that the most desirable size of colony and the length of time for it to reach maturity may be partly determined by the attacks of enemies." (Richards, 1953a.)

Polybiine colonies cannot resist attacks by the ecitonine army ants, which in some places in South America strike as frequently as once every several months on the average. This pressure alone should drive colony size down to the minimum, other factors being equal. On the other hand, attacks by vertebrate predators are also frequent, and they can best be resisted by large colonies. The colony size arrived at in the evolution of the individual polybiine species must be a balance struck to accommodate, at least in part, these two opposing pressures. It is interesting that colonies of stingless bees (Meliponini) are nearly immune to army ant attack, evidently as a consequence of the hard cerumen casements and narrow entrance tunnels in the nest structure and possibly also the chemical defense employed by the workers (Rettenmeyer, 1963a). Stingless bee colonies are also quite large as a rule, and it is not at all out of line to speculate that this trait, along with whatever more advanced social qualities are made possible by it, have been due to the liberation of meliponines from predation by army ants.

The Reversibility of Social Evolution

In his discussion of social evolution in bees, Michener (1964c, 1969a) has reflected further on the concepts of disadvantage and compromise. He has pointed out that, in all of the species of social Hymenoptera so far studied, the rate of colony productivity, measured in new offspring per adult female per unit of time, falls off with an increase in colony size. Typical examples are given in Figure 18-1. Ultimately, this means that, for a population of colonies comprised of a given biomass, the smaller the average colony size, the greater will be the number of sexual forms produced each generation. By the process of natural selection, such a correlate left unopposed should result in the steady reduction of colony size and eventually elimi-

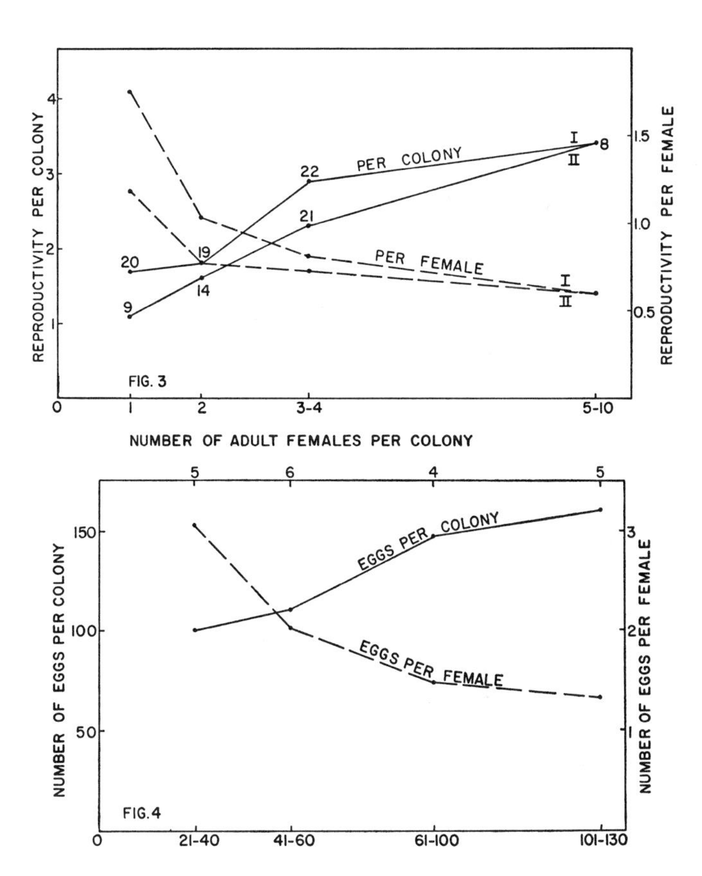

FIGURE 18-1. Part of the demonstration by Michener (1964c) that in the social Hymenoptera the growth rate of the colony measured in new individuals per adult female per unit of time decreases with increase in colony size. (*above*) In the halictine bee *Lasioglossum rhytidophorum*, "reproductivity," a measure based on the numbers of small larvae, eggs, and pollen balls, increases with colony size when the count is taken for the entire colony, but it decreases when the average (per female) count is taken; curves I and II represent different times of the year, while the numbers beside the points represent the number of nests in the sample. (*below*) The same is shown for the wasps *Polybia bistriata* and *P. bicytarella* (data from Richards and Richards, 1951).

nate social behavior altogether. Of course, the trend is opposed. There are three advantages generally conferred by social behavior which can be expected to favor large colony size:

1. *Improved defense.* The nest can be guarded continuously and more efficiently. If the species has a worker caste and the colony size is large enough, individual workers can literally throw their lives away in counterattacks without significantly reducing colony fitness.

2. *Improved homeostasis.* Regulation of temperature and humidity are perfected when enough individuals are present to construct a sufficiently large and complex nest, to contribute their own metabolic heat and water in adequate quantities, and to specialize in behavior for the creation of special mass effects (see Chapter 11).

3. *Improved labor.* Groups of workers can perform tasks, such as the removal of large obstructions during nest building, the capture of larger or more dangerous prey, and so forth, that are impossible for single individuals.

These favorable aspects of social behavior are so clearly dominant in the more advanced social insects as to render Michener's reproductivity effect seemingly insignificant. But this is not at all the case for the primitive eusocial bees. In these insects the degree of nest homeostasis does not exceed that achieved by related solitary forms, and the total of the cooperative labor in a colonial group does not seem to exceed the sum of that accomplished by a group of solitary bees of comparable number. What does matter, according to Michener (1958), is the improved defense against parasitic and predatory arthropods that association with a little group of nestmates provides. Several observers in addition to Michener have witnessed guard bees protecting their nests against mutillid wasps and ants. Lin (1964) was able to demonstrate that groups of *Dialictus zephyrus* females are more effective than solitary individuals in repelling mutillids. Michener and Kerfoot (1967) provided indirect evidence that groups of *Pseudaugochloropsis* females survive longer than solitary ones, but they were only able to conjecture that improved nest defense is responsible.

It follows that, if the reproductivity effect is driving colony size down in evolution and predator-parasite pressure is driving it up, a slackening of predator-parasite pressure should automatically result in a decrease in colony size and perhaps even go so far as to result in the abolishment of social behavior altogether. This is precisely what Michener (1969a) suggests has happened in certain borderline cases in the Halictinae and Ceratinini. An example may be the species of *Exoneurella,* which appear to have reverted to a solitary state from a primitively eusocial state.

Extending this reasoning, we can next inquire whether altered selective pressures ever drive species of termites, ants, and apine bees all the way back from the higher eusocial states. Put another way, is there a point of no return in social evolution? In partial answer it can be pointed out that many of the social parasites in bees, wasps, and ants have reversed an important part of their sociality; that is, they have abandoned the worker caste. But parasites are still completely eusocial by virtue of being wholly dependent on the eusocial colonies of their hosts. Leaving this special case aside, there are reasons for believing that a point of no return does indeed exist. The highest insect societies have lost elements of behavior that would be very difficult to reattain in evolution. The broods of some groups, for example, have become dependent on the adults in a way not seen in solitary insects. Termite eggs will not even hatch unless cared for by adults, and ant larvae and pupae require intimate grooming in every stage of their development and must be assisted at each ecdysis. Also, while Michener's reasoning concerning the preeminence of defense as a factor in the evolution of primitive sociality is probably correct, the other pressures just listed, namely, advantages conferred by homeostasis and cooperative labor, are clearly at work in more advanced insect societies, and it is unlikely that all would ever cease in concert.

The Significance of Monogyny

Had the Queens been more numerous, it would have engaged too large a Circle of Attendants; had they been less, or equal with Regard to Size, it would not so well have answered the different Proportions of Young observable in the several Colonies. So exact are the Wonders of Providence! in nothing superfluous or deficient.

William Gould (1747)

Species of social insects are characterized not only by the kinds of castes they produce but also, within reasonable limits, by the ratios of individuals belonging to each caste within single colonies (Noirot, 1954; Michener, 1961b; Wilson, 1968a). As was shown in Chapters 8-10, a great deal of experimental evidence indicates that the ratios are regulated through the additive inhibition by individuals of a given caste on the production of new members of the same caste. In at least some of the cases, the control is achieved through primer pheromones. Although physiological explanations of caste control are now emerging, little is understood of the adaptive significance of fixed caste ratios. The present level of our understanding can be illustrated by a further consideration of monogyny, the condition of the occurrence of single queens in colonies. Drawing largely from a recent review of the subject by Michener (1964c), I have been able to conceive of four evolutionary explanations of the phenomenon, no two of which are mutually exclusive.

1. As pointed out by Michener, the inverse relation between colony size and average worker productivity implies that maximum average individual productivity is achieved by very young, monogynously founded colonies. This seems to me to be an unlikely selective factor leading to monogyny, however, since what matters most is the probability of colony survival, which in turn depends more on the size and productivity of the colony as a whole than on the average capability of individuals. A large polygynous colony of weak individuals would be superior to a very small, monogynous colony. Other factors, therefore, must drive species toward monogyny.

2. Caste and social behavior would be simple and, hence, might be easier to control if only one queen were present. This hypothesis has a certain immediate plausibility, but it has yet to be supported by any convincing empirical evidence.

3. Weismann (1893) suggested that evolution would proceed most rapidly in monogynous colonies because such colonies expose a more limited number of genotypes to selection. This is true, of course, but it is a second-order factor, rather like mutator genes or dominance modifiers, and its effects could be easily overturned by opposing selection of a more direct kind. Again, I am inclined to believe that other factors are more potent in promoting monogyny.

4. Unless the queens are very closely related, it will be ultimately advantageous for them to eliminate one another or else to avoid sharing the same nest site. Hamilton (1964) has expressed the same idea another way: polygyny ("pleometrosis") should occur only when population viscosity is high enough to insure frequent association of sibling queens and, hence, to promote an altruistic sharing of resources.

I would argue that the fourth hypothesis is not only the simplest explanation available, but it is also adequate to explain most of the current empirical information. As I stressed earlier, in Chapter 17, there is a strong tendency for queens to eliminate rivals except in cases where an association temporarily provides them with an advantage over single queens founding other colonies. Where it does occur, the hostility is usually in marked contrast to the amicable behavior displayed among all the other members of the colony.

There is, nevertheless, a second form of competition that results in monogyny and that cannot be so easily brought in line with the fourth hypothesis. This is the phenomenon of the execution of supernumerary queens by the workers. When the stings of honeybee queens sharing the same nest are amputated, so that they are unable to kill each other, the workers still attack and remove some of the surplus (Darchen and Lensky, 1963). To take another example, the life cycle of stingless bees unfolds in such a way that only one virgin queen at a time can be supplied with a new nest and supporting worker force, and surplus virgin queens are executed by the workers soon after one of their sisters makes a successful nuptial flight (Moure, Nogueira-Neto, and Kerr, 1958). Also, when extra secondary reproductives of *Kalotermes flavicollis* are produced in artificial nests by experimental means, they are quickly killed and eaten by the workers (Lüscher, 1952b). Multiple queens of normally monogynous species of the ant genera *Crematogaster* and *Plagiolepis* can be combined artificially in laboratory colonies, but in time the workers destroy all but one (Baroni Urbani and Soulié, 1962; Soulié, 1964; Passera, 1963b). My own studies have shown a remarkable form of monogyny control on the part of workers of the fire ant *Solenopsis saevissima* (Wilson, 1966). Any number of young, mated queens can be successfully introduced into a queenless *S. saevissima* colony if the entire group is chilled to immobility prior to the introductions. Within a day or two, however, the workers begin to crowd about and to attack individual queens. Their method of execution is to seize these individuals by their appendages, spread-eagle them, and gradually dismember them over a period of hours or days. Figure 18-2 depicts data from one such experiment in which each of five colonies were initially given five queens. In each group the workers soon reduced the number to one. In over twenty such trials involving two to five queens and colonies of several hundred workers, the number of queens

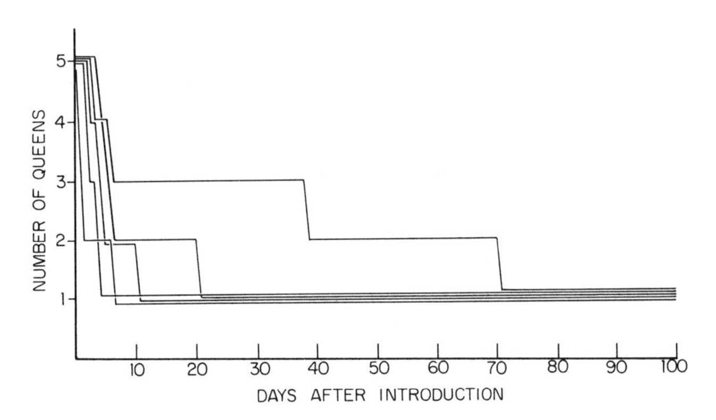

FIGURE 18-2. Reduction of supernumerary queens through execution in experimental colonies of the fire ant *Solenopsis saevissima.* The queens were all young, recently mated individuals collected just after a nuptial flight. The workers came from an old laboratory colony. In each of the five replications, the number of queens was reduced from five to one in no more than 70 days and held at one thereafter for at least 150 days (from Wilson, 1966).

was always reduced to one. Only in a single case, where the colony started with two queens (among three other colonies set up in the same way) were both queens tolerated long enough to contribute significantly to further growth of the colony. No colony made the mistake of executing the last remaining queen. The genetic relationship of the queens to the colonies was not known, but it was probably not very close. All were collected from a very large population over a wide area (around Baton Rouge, Louisiana) so that the chance of queens and colonies originating from the same nest was small. There is as yet no answer to the question of how the workers select queens for execution. It may be on the basis of dissimilarity of body odor, which in turn may or may not have a genetic origin, or it may be on the basis of competitive production of some unknown pheromone. Nor is there yet any clue as to how the workers know enough to save the last queen.

To return to the original proposition, that monogyny evolves by competition among queens, it can be argued that much of the evidence directly supports this view and the remainder is at least compatible with it. Seemingly the least favorable evidence is the phenomenon of queen execution by workers. How could reginicide behavior evolve in workers, except by selection among colonies rather than among queens in single colonies? An answer, not necessarily the correct one, is that the queen-worker complex could evolve so as to have workers remove queens with the least familiar odor, if some of the odor differences were genetic in origin. The reason is that under natural conditions resident queens and associated workers would be genetically very close. In *S. saevissima* and other species where queens are sometimes introduced from the outside, it would be advantageous to the colony as a whole for workers to destroy intruders because the new queens would almost certainly be less closely related. A strange aspect of the *S. saevissima* example is that, although the queens fight among themselves in the early stages of colony founding, in older colonies they completely avoid each other and let the workers handle the executions. It is as though too much is at stake in the undamaged person of the queen within older colonies, and the risk is therefore too great to the whole colony genotype for the queen to engage in further hostility so long as the more expendable workers are able to do it for her.

Whatever the selective pressures that bring monogyny about in the first place, it seems probable that its establishment is conducive to the evolution of what we generally regard as some of the highest forms of social behavior. This inference was drawn in the following way by Haskins (1966). Suppose hostility among queens and queen execution did not exist. Then we would expect most or all social species to be polygynous. The queen-worker ratio in the colony of each species would be a compromise reached between two opposing tendencies. On the one hand, there would be a selfish tendency in females to become fertile and lay eggs of their own; on the other hand, there would be an altruistic tendency for females to become workers in order to increase the total colony fitness. So long as the degree of polygyny was fixed by such an uneasy balance, caste differences would not be great and the development of more elaborate forms of mass behavior dependent on caste differentiation would be inhibited. But once monogyny is fixed (for whatever reason) the species is "free" to evolve in this direction. The polybiine wasps typify the first and more primitive condition, while most of the ants, termites, and apine bees typify the more advanced condition based on monogyny.

The Significance of Worker Inefficiency

Man's most cherished notion about social insects has always been that the workers are highly virtuous, industrious, and efficient. The following encomium from Pierre

André Latreille (1802) is typical of the traditional view: "Peut-on voir une société dont les membres qui la composent aient plus d'amour public? qui soient plus des intéressés? qui aient pour le travail une ardeur plus opiniâtre et plus soutenue? Quel singulier phénomène!"

For the past several years, however, some entomologists have begun to suspect that, contrary to folklore, the members of insect societies do not labor as hard or as efficiently as they might in the interests of colony growth. Otto (1958a) found that workers of the ant *Formica polyctena* (= "*Formica rufa minor*") are idle approximately half of the time. In my own time-budget studies of the harvester ant *Pogonomyrmex badius* the workers were unoccupied most of the time, even during the hours of the day when foraging was at its peak. Lindauer (1961) has commented on the large amount of time honeybee workers spend "loafing." Of a total of 177 observation hours spread over the whole life of one selected bee, 70 were spent doing nothing; another 56, patrolling the nest. Brian (1957d) has identified at least three sources of inefficiency in the rearing of larvae by *Myrmica rubra* colonies: (a) large larvae are so disproportionately favored by the nurse workers that a suboptimal food distribution results; (b) the larvae are grouped together by the workers in such a way as to prevent the most direct servicing by the greatest number of available nurses; and (c) the larvae receive food through a trial-and-error method rather than by transmitting "hunger signals" at a distance. Michener's reproductivity effect described earlier—that in social Hymenoptera the productivity of the average worker (measured by number of brood per worker at a given time) falls off when the number of workers increases—would also seem to reflect inefficiency in organization. Sakagami and Hayashida (1962), in their measurements of work efficiency in *Formica* and *Polyergus*, obtained the surprising result that, as the number of workers available for excavating sand increased, both the proportion of "elite" (hardworking) individuals and the average amount of work per individual decreased. This effect is just the opposite of social facilitation, which has long been assumed—perhaps erroneously—to prevail in ant societies on the basis of Chen's (1937) rather limited experiments with *Camponotus*.

It is easy to conceptualize insect colonies as analogues of simple machines and to invent ways by which worker behavior might be modified to increase worker and reproductive efficiency. But the encompassing and far more complex problem (of which efficiency is only a part) is that of adaptation, and adaptation implies a permissible optimum by which colony fitness can be measured. It is the optimum, or "goal" in machine-oriented terminology, which we understand least. Is it possible that *Myrmica* larvae are arranged in clusters, *Pogonomyrmex* workers are idle a seemingly unconscionable amount of time, and *Formica* workers work less in the presence of many companions because these behavior patterns are adaptive in other ways? Michener (1964c) offers a solution to part of the problem when he says, "It may well be that the large numbers of inefficient workers have their chief selective value, not in the greater reproductivity which they make under ordinary conditions, but in their importance in defense against natural enemies or against major changes in the physical environment." Lindauer (1961) has referred to the loafers in the honeybee colony as "the reserve troops, employed at critical points in the labor market as the necessity arises." Their homeostatic role is especially clearly indicated in thermoregulation and food gathering.

We have just begun to articulate the connections between worker behavior, colony efficiency, and colony fitness in the genetic sense. Three major efforts are needed before a deeper understanding of this fundamental issue can be attained: more complete studies of colony physiology, field studies of colony mortality under natural conditions, and a mathematical theory of genetic fitness constantly extended to keep pace with the first two. Let us now turn to the subject of ergonomics, which provides an entrée into the theoretical analysis of genetic fitness at the colony level.

The Ergonomics of Caste

In Chapters 8–10 it was shown how past studies of caste systems in the social insects have focused on the obvious question of the genetic and physiological mechanisms that control caste determination in the individual insect. In the great majority of bees, and in most and perhaps all ants, wasps, and termites, caste is environmentally determined. The environmental controls are diverse in nature and differ from group to group. They include the biasing influences of yolk nutrients, of various quantitative and qualitative factors in larval feeding, of temperature changes, and of the caste-specific pheromones. There is a second major problem connected with caste that is evolutionary in nature and can be phrased as follows: why do the ratios of the castes (in an entire colony population) vary among species of social insects? This question, like the other questions on optimality and natural selection just considered, is much less obvious than the physiological one and can be considered only in the context of the ecology of the individual species. Much empirical information on ratios exists, but there is little theory. Only recently have insect sociologists begun to handle the data in systematic fashion, as, for example, in the work of Richards and Richards (1951), Lindauer (1961), Hamilton (1964), Brian (1965b), and Wilson (1966, 1968a).

This matter of the presence or absence of a given caste, together with its relative abundance when present, should be susceptible to some form of optimization theory provided we are able to assume selection at the colony level. In fact, colony selection in the advanced social insects does appear to be the one example of group selection that can be accepted unequivocally so long as we are careful to bear in mind that the group in this case is the colony and not the population of colonies. To be sure, it is the queen, the mother of all the workers and second generation reproductives in the colony, which transmits the gametes and is the ultimate focus of selection. In this special sense colony selection differs from group selection in the hypothetical Wynne-Edwards sense, where most or all of the mature individuals of the populations are involved in reproduction (Wynne-Edwards, 1962; G. C. Williams, 1966). But it remains true that the colony is selected as a whole, and its members contribute to colony fitness rather than to individual fitness. It therefore seems a sound procedure to accept colony selection as a mechanism and to press on in search of optimization theory based on the assumption that the mechanism operates generally. For, if selection is mostly at the colony level, workers can be altruistic with respect to the remainder of the colony, and their numbers and behavior can be regulated in evolution to achieve maximum colony fitness. What has been lacking so far is an entrée to the theory of group behavior, a way of abstracting our empirical knowledge of caste and colony ergonomics* into a form that can be used to analyze optimality. In a recent article (Wilson, 1968a) I attempted a first formulation by means of the techniques of linear programming and obtained some surprising but still largely theoretical results. The essential arguments and results are presented here in a simplified form.

First, consider the concept of cost in colony reproduction. As colonies grow, their caste ratios change. Very young colonies founded by single queens typically consist only of the queens and minor workers. As they approach maturity, these same colonies may add medias and soliders. Finally, they produce males and new, virgin queens. Here we will consider ergonomics and cost in the mature colony only. A mature colony is defined as a colony large enough to produce new, virgin queens. Also, for convenience, the category "caste" will include both *physical castes,* such as minor workers and soldiers, and *temporal "castes."* The latter are classes of individuals in those various periods of labor specialization that most individual social insects pass through in the course of their lives. What determines the efficiency of the mature colony is the number of workers in each temporal caste at any given moment. This conception is spelled out in the example given in Figure 18-3.

In the mature colony, depending on the species, the adult force may contain anywhere from a few tens of workers to several millions. The number is a species characteristic. It has been evolved as an adaptation to ultimate limiting factors in the environment. An ultimate limit may be imposed by the constraints of a peculiar kind of nest site to which the species is adapted, by the restricted productivity of some prey species on which the species specializes, or, conversely, by a prey species or competitor so physically formidable as to require a larger worker force as a minimum for survival. These and other ultimate, limiting factors have been documented and discussed in the literature by Brian (1965b). The mature colony, on reaching its predetermined size, can be expected to contain caste ratios that approximate the *optimal mix.* This mix is simply the ratio of castes that can achieve the maximum rate of production of virgin queens and males while the colony is at or near its maximum size.

It is helpful to think of a colony of social insects as operating somewhat like a factory constructed inside a fortress. Entrenched in the nest site and harassed by enemies and capricious changes in the physical environment, the colony must send foragers out to gather food while converting the secured food inside the nest into virgin queens and males as rapidly and as efficiently as possible. The rate of production of the sexual forms is an

*I have suggested the term "ergonomics," borrowed from human sociology (*e.g.,* Murrell, 1965), to identify the quantitative study of the distribution of work, performance, and efficiency in insect societies (Wilson, 1963a).

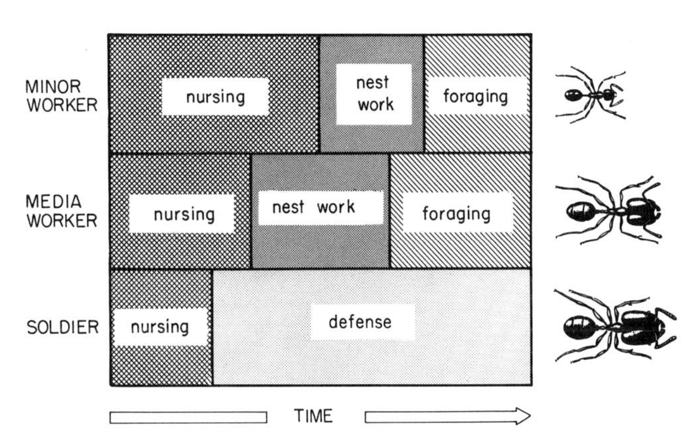

FIGURE 18-3. This diagram visualizes the principal work periods traversed in the life spans of three worker subcastes of a generalized polymorphic ant species. The work periods are those periods in which the indicated task is the one most frequently performed; other tasks may be performed, but less often. The forms of the castes and the sequences of work periods within each caste are based on real species, but the precise durations of the periods are imaginary. In this case each of the eight periods, the total (arbitrary) of periods encountered in all three castes together, is treated as a separate "caste." The optimal mix can be evolved both by varying the relative numbers in each subcaste and the relative time spent in each work period (the ants represented in this figure and in succeeding figures in this chapter belong to the myrmicine genus *Pheidole* and are shown only as an intuitive aid) (after Wilson, 1968a).

important, but not an exclusive, component of colony fitness. Suppose we are comparing two genotypes belonging to the same species. The relative fitness of the genotypes could be calculated if we had the following complete information: the survival rates of queens and males belonging to the two genotypes from the moment they leave the nest on the nuptial flights; their mating success; the survival rate of the fecundated queens; and the growth rates and survivorship of the colonies founded by the queens. Such complete data would, of course, be extremely difficult to obtain. In order to develop an initial theory of ergonomics, however, it is possible to get away with restricting the comparisons to the mature colonies. In order to do this and still retain precision, it would be necessary to take the difference in survivorship between the two genotypes outside the period of colony maturity and reduce it to a single weighting factor. But we can sacrifice precision without losing the potential for general qualitative results by taking the difference as zero. Now we are concerned only with the mature colony, and the production of sexual forms becomes (keeping in mind the artificiality of our convention) the exact measure of colony fitness. The role of colony-level selection in shaping population characteristics within the colony can now be clearly visualized. If, for example, colonies belonging to one genotype contain on the average 1,000 sterile workers and produce 10 new, virgin queens in their entire life span, and colonies belonging to the second genotype contain, on the average, only 100 workers but produce 20 new, virgin queens in their life span, the second genotype has twice the fitness of the first, despite its smaller colony size. As a result, selection would reduce colony size. The lower fitness of the first could be due to a lower survival rate of mature colonies, or to a smaller average production of several forms for each surviving mature colony, or to both. The important point is that the rate of production in this case is the measure of fitness, and evolution can be expected to shape mature colony size and organization to maximize this rate.

The production of sexual forms is determined in large part by the number of "mistakes" made by the mature colony as a whole in the course of its fortress-factory operations. A mistake is made when some potentially harmful contingency is not met—a predator successfully invades the nest interior, a breach in the nest wall is tolerated long enough to dessicate a brood chamber, a hungry larva is left unattended, and so forth. The cost of the mistakes for a given category of contingencies is the product of the number of times a mistake is made times the reduction in queen production per mistake. With this formal definition, it is possible to derive in a straightforward way a set of basic theorems on caste. In the special model, the average output of queens is viewed as the difference between the ideal number made possible by the productivity of the foraging area of the colony and the number lost by failure to meet some of the contingencies. (The model can be modified to incorporate other components of fitness without altering the results.) The evolutionary problem which I postulate to have been faced by social insects can be solved as follows: the colony produces the mixture of castes that maximizes the output of queens. In order to describe the solution in terms of simple linear programming, it is necessary to restate the solution in terms of the dual of the first statement: the colony evolves the mixture of castes that allows it to produce a given number of queens with a minimum quantity of workers. In other words, the objective is to minimize the energy cost.*

The simplest case involves two contingencies whose costs would exceed a postulated "tolerable cost" (above which, selection takes place), together with two castes whose efficiencies at dealing with the two contingencies differ. The inferences to be made from this simplest situation can be extended to any number of contingencies and castes.

The most important step is to relate the total weights, W_1 and W_2, of the two castes in a colony at a given instant to the frequency and importance of two contingencies and the relative efficiencies of the castes at performing the necessary tasks. By stating the problem as the minimization of energy cost (see Wilson, 1968a), the relation can be given in linear form as follows:

Contingency Curve 1

$$W_1 \doteq \frac{\ln F_1 - \ln k_1 x_1}{\alpha_{11} \ln (1 - q_{11})} - \frac{\alpha_{12} \ln (1 - q_{12})}{\alpha_{11} \ln (1 - q_{11})} W_2$$

Contingency Curve 2

$$W_1 \doteq \frac{\ln F_2 - \ln k_2 x_2}{\alpha_{21} \ln (1 - q_{21})} - \frac{\alpha_{22} \ln (1 - q_{22})}{\alpha_{21} \ln (1 - q_{21})} W_2$$

W_1 is the weight of all members belonging to caste 1 in an average colony;

W_2 is the weight of all members belonging to caste 2 in an average colony;

F_1 and F_2 are the highest tolerable costs due to contingencies 1 and 2;

α_{11} is a constant such that $\alpha_{11} W_1$ gives the average number of individual contacts with a contingency of type 1 by members of caste 1 during the existence of the contingency;

α_{12} is a constant such that $\alpha_{12} W_2$ gives the average number of individual contacts with a contingency of type 1 by members of caste 2 during the existence of the contingency;

α_{21} and α_{22} are constants similar to the above two but with reference to contingencies of type 2;

q_{11} is the probability that, on encountering contingency 1, a worker of caste 1 responds successfully;

q_{12} is the probability that, on encountering contingency 1, a worker of caste 2 responds successfully;

q_{21} and q_{22} are the probabilities of the above two but with reference to contingency 2;

x_1 and x_2 are the average costs (in this case, measured in nonproduction of virgin queens) per failure to meet contingencies 1 and 2, respectively;

k_1 and k_2 are the frequencies of contingencies 1 and 2, respectively, for a given period of time.

*Levins (1968) has rederived the same theorems in terms of the opposite dual in order to align them with his general theory of fitness sets. His method has pedagogic advantages, but it is more difficult to relate to the underlying behavioral phenomena.

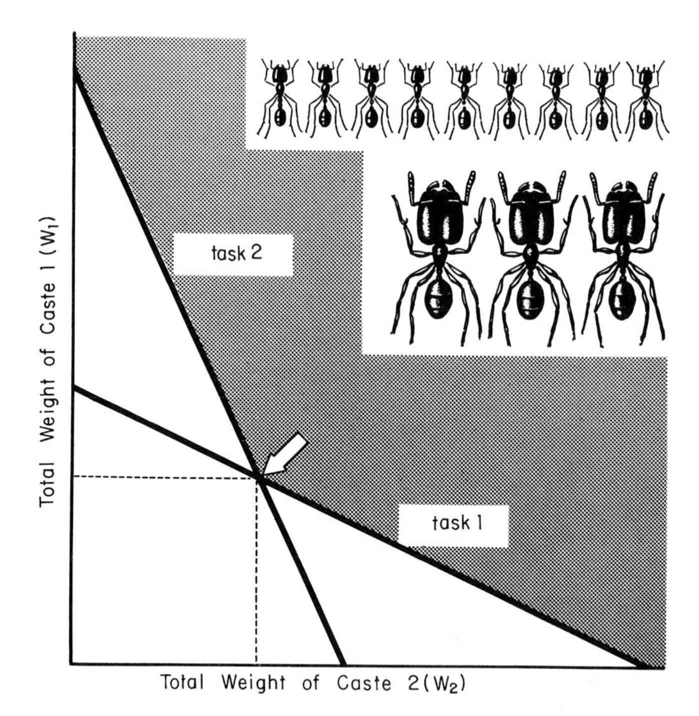

FIGURE 18-4. This diagram shows the general form of the solution to the optimal mix problem in evolution. In this simplest possible case, two kinds of contingencies ("tasks") are dealt with by two castes. The optimal mix for the colony, measured in terms of the respective total weights of all the individuals in each caste, is given by the intersection of the two curves. Contingency curve 1, labeled "task 1," gives the combination of weights (W_1 and W_2) of the two castes required to hold losses in queen production to the threshold level due to contingencies of type 1; contingency curve 2, labeled "task 2," gives the combination with reference to contingencies of type 2. The intersection of the two contingency curves determines the minimum value of $W_1 + W_2$ that can hold the losses due to both kinds of contingencies to the threshold level. The basic model can now be modified to make predictions about the effects on the evolution of caste ratios of various kinds of environmental changes (from Wilson, 1968a).

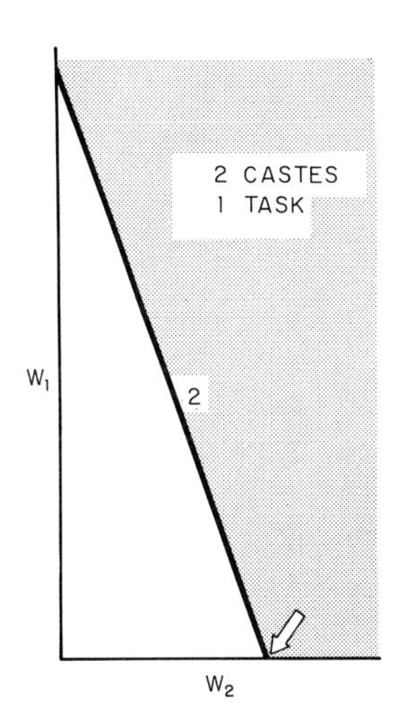

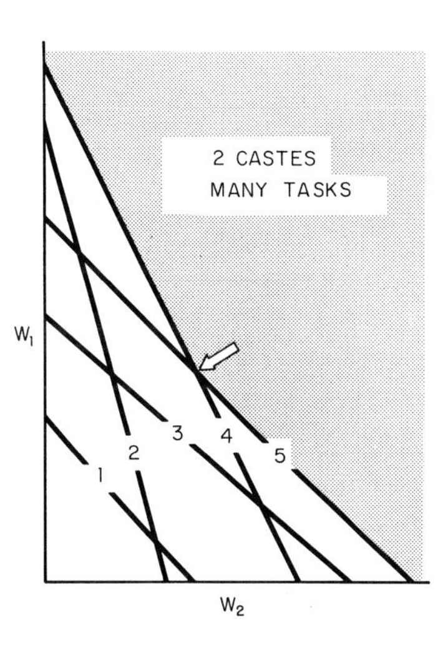

FIGURE 18-5. The diagram on the left shows that, when there are more castes than tasks, the number of castes will be reduced in evolution to equal the number of tasks. The surplus castes removed will be the least efficient ones (in this case, caste 1). The diagram on the right shows that if there are more tasks than castes, the optimal mix of castes will be determined entirely by those tasks, equal or less in number to the number of castes, which deal with the contingencies of greatest importance to the colony (in this case, tasks 4 and 5) (from Wilson, 1968a).

I have presented this amount of detail to illustrate one particular form that contingency curves might take, using conventions that relate to intuitively simple ideas concerning behavior. In fact, no contingency curves of actual species have been drawn. At present, the required steps of defining contingencies and measuring their effects in natural populations are technically formidable. The important point is that under a very wide range of conceivable conditions the contingency curves would be linear or almost linear, or at least could be rendered graphically in linear form.

The optimal mix of castes is the one that gives the minimum summed weights of the different castes while keeping the combined cost of the contingencies at the maximum tolerable level. The manner in which the optimal mix is approached in evolution is envisaged as follows. Any new genotype that produces a mix falling closer to the optimum is also one that can increase its average net output of queens and males. In terms of energetics, the average number of queens and males produced per unit of energy expended by the colony is increased. Even though colonies bearing the new genotype will contain about the same adult biomass as other colonies, their average net output will be greater. Consequently, the new genotype will be favored in colony-level selection, and the species as a whole will evolve closer to the optimal mix.

The general form of the solution to the optimal mix problem is given in Figure 18-4. It has been postulated that behavior can be classified into sets of responses in a one-to-one correspondence to a set of kinds of contingencies. Even if this conception only roughly fits the truth, it is enough to develop a first theory of ergonomics. For example, the graphical presentation in Figures 18-5 and 18-6 shows that so long as the contingencies occur with relatively constant frequencies, it is an advantage for the species to evolve so that in each mature colony there is one caste specialized to respond to each kind of contingency. In other words, one caste should come into being that perfects the appropriate response, even at the expense of losing proficiency in other tasks.

A curious possible effect in the evolution of castes is illustrated in Figure 18-7. This theorem was derived as an answer to the following question: If proliferation and divergence of castes are the expected consequences of selection at the colony level, why have they not reached greater heights throughout the social insects? In fact, these qualities vary greatly from genus to genus and even from species to species. The only answer consistent with the theory is that, as in most evolving systems, the various levels reached by individual species are compromises

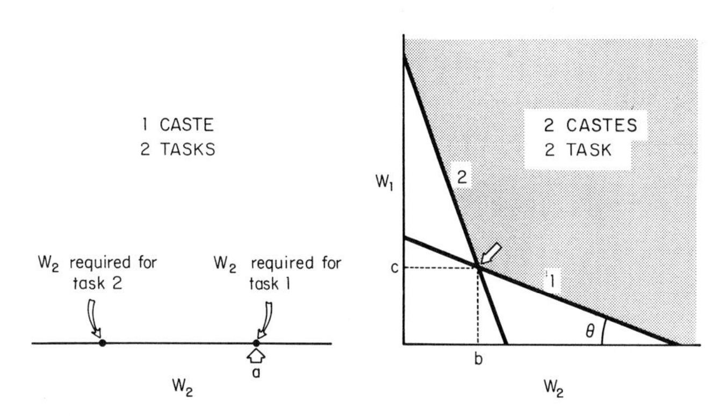

FIGURE 18-6. It is always to the advantage of the species to evolve new castes until there are as many castes as contingencies, and each caste is specialized uniquely on a single contingency. This theorem can be substantiated readily by comparing the two graphs in this figure. With the addition of caste 1 in the right-hand figure, the total weight of workers is changed from a to $b + c$. Since caste 1 specializes in task 1, θ is acute; therefore, $a - b > c$ and $a > b + c$ for all a, b, and c (from Wilson, 1968a).

between opposing selection pressures. The obvious pressure that must oppose proliferation and divergence is fluctuation of the environment. From Figure 18-7 we can see that a long-term change can eliminate a caste if the caste that supersedes it (by taking over its tasks through superior numbers) is not very specialized. In this example, contingency 2 has been increased in frequency (or importance), shifting the contingency curve to the right of the contingency 1 curve intercept of the W_2 axis. Consequently, the number of caste 2 workers required to take care of contingency 2 is also more than enough to take care of contingency 1. The presence of caste 1 now reduces colony fitness, and if the environmental change is of long duration, caste 1 will tend to be eliminated by colony-level selection. In this case the species tracks the environment to acquire a new optimal mix that just happens to eliminate the superseded caste. Thus, if the critical features of the environment are changing at a rate slow enough to be tracked by the species but too fast to permit much specialization of individual castes, both the number and the degree of specialization of castes will be kept low.

At another level, the critical features of the environment may be changing too fast to be tracked genetically, yet too slow to provide each colony with a consistent average for the duration of its life. In this case, a mix of specialized castes would be inferior to a few generalized forms able to adapt to new circumstances.

This form of ergonomic theory also reveals two ways in which the consequences of colony-level selection can be the exact opposite of those stemming from individual-level selection. In Figures 18-8 and 18-9 a relation is

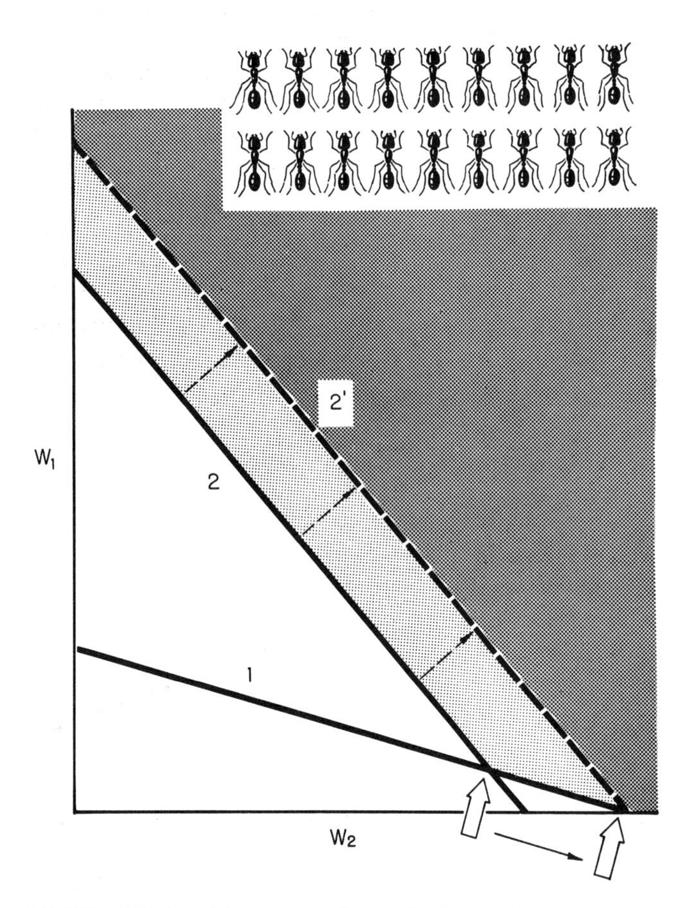

FIGURE 18-7. A long-term change in the environment can cause the evolutionary loss of a caste, even when the task to which the caste is specialized remains as frequent and important as ever (from Wilson, 1968a).

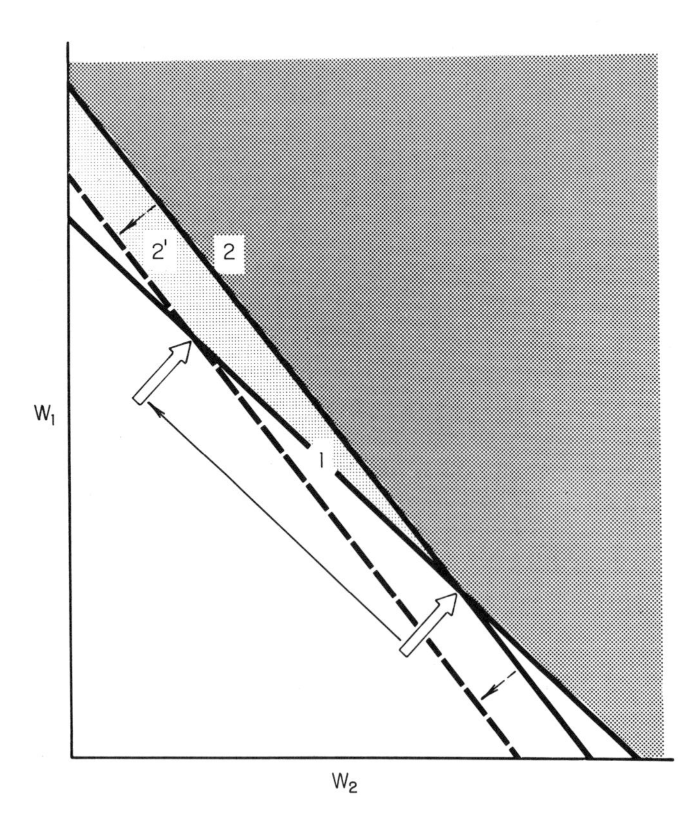

FIGURE 18-8. If the castes are relatively unspecialized, small but long-term changes in the environment will ultimately result in large evolutionary changes in the optimal caste ratios (from Wilson, 1968a).

shown to exist between the prior degree of caste specialization and the magnitude of change in the optimal mix which is invoked by a given change in the environment. The castes represented in Figure 18-8 are relatively unspecialized. Task 2 is shown to become somewhat less common (or less important), resulting in a shift of the contingency curve toward the origin without a change in slope. As a consequence, the optimal mix changes from one comprised predominantly of caste 2 to one comprised predominantly of caste 1. In contrast, the castes represented in Figure 18-9 are highly specialized, and a shift in the contingency curve results in little change in caste ratios. These models lead to the conclusion that species with initially unspecialized castes will have on the average fewer castes and more variable caste ratios, and this effect will be enhanced in fluctuating environments. The more specialized the castes become in evolution, the more entrenched they become, in the sense that they are more likely to be represented in the optimal mix regardless of long-term fluctuations in the environment. Here we have a peculiar theoretical result of colony-level selection, the opposite of individual-level selection. For in classical population genetic theory, which entails individual selection, it is the generalized genotypes and species, and not the specialized ones, that are most likely to survive in the face of long-term fluctuation in the environment.

The second peculiar result of colony-level selection, illustrated in Figure 18-10, involves the relation between the efficiency and the numerical representation of a given caste. If, in the course of evolution, one caste increases in efficiency and the others do not, the proportionate total weight of the improving caste will decrease. In other words, the expected result of colony-level selection is precisely the opposite of that of individual selection, which would be an increase in the more efficient form.

Ergonomic theory will not be easy to test. The required steps of defining contingencies and measuring their effects in natural populations will require closer attention to the

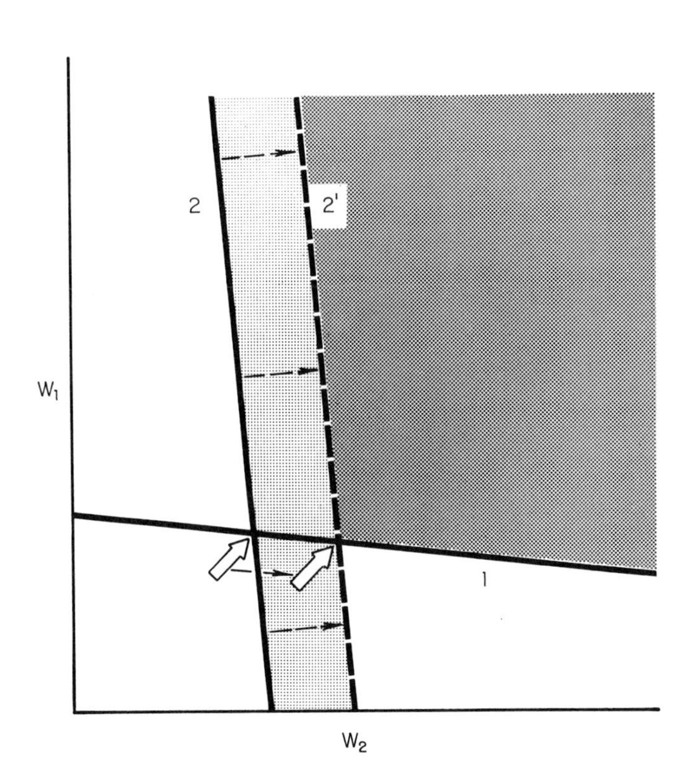

FIGURE 18-9. The more specialized the castes are in aggregate, the less evolutionary change there will be in the optimal mix in the face of long-term environmental change (from Wilson, 1968a).

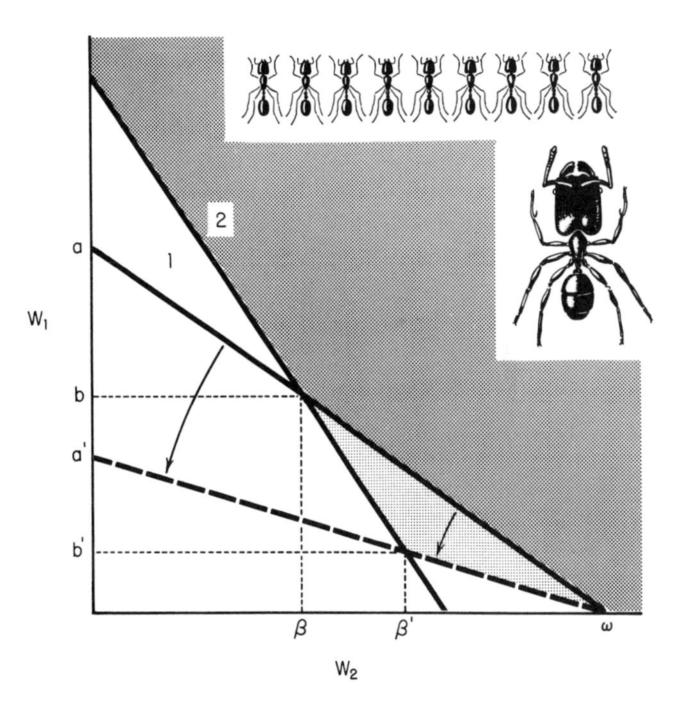

FIGURE 18-10. If one caste increases in efficiency during the course of evolution and the others do not, the proportionate total weight of the improving caste will decrease. This theoretical result of colony-level selection is the opposite of what might be expected from individual-level selection, which tends to increase improving phenotypes (from Wilson, 1968a).

biology of insect colonies than has been attempted in the past. Yet I can see no way of probing very deeply into the evolution of castes except by this means, or at least by comparable studies guided by some other, more clever form of ergonomic theory.

There exists a small amount of indirect empirical evidence relevant to the ergonomic theorems just presented. It is the case, for example, that some phyletic ant lines have lost a caste (the soldier caste) secondarily. Although the theory allows for this possibility, it is not proved by its realization. A second, more suggestive piece of evidence is the fact that physical castes are more frequent in tropical ant faunas than in temperate ant faunas. This rule is consistent with the postulate that castes always tend to proliferate in evolution but are simultaneously being reduced in response to fluctuations in the environment, the degree of response being proportionate to the degree of fluctuation. Third, it is a fact consistent with the theory, but still far from proving it, that the most specialized castes are found primarily in tropical genera and species. The bizarre soldiers of ant genera such as *Paracryptocerus, Pheidole* (*Elasmopheidole*), *Acanthomyrmex, Zatapinoma, Camponotus* (*Colobopsis*), and of termite genera such as *Nasutitermes, Mirotermes, Anacanthotermes,* and *Capritermes* are all but limited to the tropics and subtropics. Polymorphism in temperate ant species, representing the less extreme members of *Pheidole, Solenopsis, Monomorium, Myrmecocystus,* and *Camponotus,* is predominantly of the simpler forms produced by elementary allometry. This climatic correlation is predictable from the theorem that specialization in castes already in existence should increase indefinitely until countered by opposite selective pressures imposed by fluctuations in the environment.

19 Symbioses Among Social Insects

The insect society is a decidedly more open system than the lower units of biological organization such as the organism and the cell. In the course of evolution the tenuous lines of communication among the members of insect colonies have been repeatedly opened and extended to incorporate alien species. Many kinds of ants, for example, adopt aphids, mealybugs, and other homopterans as cattle to provide a steady source of honeydew; a few raid colonies of other species to acquire workers as domestic slaves, or form close alliances with other species to utilize common odor trails and to defend common nest sites. Just as frequently, the lines of communication have been tapped by other alien species which have insinuated themselves into the colony as social parasites. In each of their myriad forms the social symbioses are most strongly developed in the ants, but enough examples exist in other major groups to suggest that certain underlying physiological and behavioral principles are at work in all.

The "Ultimate" Social Parasite

There is no better way to begin a survey of the social symbioses than by considering the most extreme example known, that of the "ultimate" parasitic ant *Teleutomyrmex schneideri.* This remarkable species was discovered by Heinrich Kutter (1950a) in the Saas-Fee, an isolated valley of the Swiss Alps near Zermatt. Its behavior has been studied by Stumper (1950) and Kutter (1969), its neuroanatomy by Brun (1952), and its general anatomy and histology by Gösswald (1953). A second population has been reported from near Briançon in the French Alps by Collingwood (1956). Appropriately, the name *Teleutomyrmex* means "final ant."

The populations of *T. schneideri,* like those of most workerless parasitic ant species (Wilson, 1963b), are small and isolated. The Swiss population appears to be limited to the eastern slope of the Saas-Fee, in juniper-*Arctostaphylos* woodland ranging from 1,800 to 2,300 m in elevation. The ground is covered by thick leaf litter and sprinkled with rocks of various sizes, providing, in short, an ideal environment for ants. The ant fauna is of a typically boreal European complexion, comprised of the following free-living species listed in the order of their abundance (Stumper, 1950): *Formica fusca, F. rufa, Tetramorium caespitum, Leptothorax acervorum, L. nigriceps, Camponotus ligniperda, Myrmica lobicornis, M. sulcinodis, Camponotus herculeanus, Formica sanguinea, F. rufibarbis, F. pressilabris,* and *Manica rubida.* For some unexplained reason this little assemblage is extremely prone to social parasitism. *Formica sanguinea* is a facultative slave-making species, preying on the other species of *Formica. Doronomyrmex pacis,* a workerless parasite living with *Leptothorax acervorum,* was discovered by Kutter as a genus new to science in the Saas-Fee forest in 1945. In addition, Kutter and Stumper have found *Epimyrma stumperi* in nests of *Tetramorium caespitum.* Still more recently, two parasitic *Leptothorax, goesswaldi* and *kutteri,* have turned up in nests of *L. acervorum* (Kutter, 1969).

Teleutomyrmex schneideri is a parasite of *Tetramorium caespitum.* Like so many other social parasites, it is phylogenetically closer to its host than to any of the other members of the ant fauna to which it belongs. In fact, it may have been derived directly from a temporarily free-living offshoot of this species, since *T. caespitum* is the only nonparasitic tetramoriine known to exist at the present time through most of central Europe. It is difficult

FIGURE 19-1. The extreme social parasite, *Teleutomyrmex schneideri,* with its host *Tetramorium caespitum.* The two *Teleutomyrmex* queens sitting on the thorax of the host queen have not yet undergone ovarian development, and their abdomens are consequently flat and unexpanded. One still bears her wings and is, therefore, almost certainly a virgin. The third *Teleutomyrmex* queen, which rides on the abdomen of the host queen, has an abdomen swollen with hyperdeveloped ovarioles. A host worker stands in the foreground (drawing based on a painting by W. Linsenmaier; courtesy of R. Stumper).

to conceive of a stage of social parasitism more advanced than that actually reached by *Teleutomyrmex schneideri.* The species occurs only in the nests of its hosts. It lacks a worker caste, and the queens contribute in no visibly productive way to the economy of the host colonies. The queens are tiny compared with most ants, especially other tetramoriines; they average only about 2.5 mm in total length. They are unique among all known social insects in being ectoparasitic, that is, they spend much of their time riding on the backs of their hosts (Figure 19-1). The *Teleutomyrmex* queens display several striking morphological features that are correlated with this peculiar habit. The ventral surface of the gaster (the large terminal part of the body) is strongly concave, permitting the parasites to press their bodies close to those of their hosts. The tarsal claws and arolia are unusually large, permitting the parasites to secure a strong grip on the smooth chitinous body surface of the hosts. The queens have a marked tendency to grasp objects. Given a choice, they will position themselves on the top of the body of the host queen, either on the thorax or the abdomen. Deprived of the nest queen, they will then seize a virgin *Tetramorium* queen, or a worker, or a pupa, or even a dead queen or worker. Stumper observed a case in which six to eight *Teleutomyrmex* queens simultaneously grasped one *Tetramorium* queen, completely immobilizing her. The mode of feeding by the *Teleutomyrmex* is not known with certainty. The adults are evidently either fed by the host workers through direct regurgitation or else share in the liquid regurgitated to the host queen. In any case, they are almost completely inactive most of the time. They are also very delicate in constitution. When placed in an artificial nest they do not survive more than a day or two, even in the presence of the host workers. The *Teleutomyrmex* adults, especially the older queens, are highly attractive to the host workers, who lick them frequently. According to Gösswald, large numbers of unicellular glands are located just under the cuticle of the thorax, pedicel, and abdomen of the queens; these are associated with glandular hairs and are believed to be the source of a special attractant for the host workers. The abdomens of older *Teleutomyrmex* queens become swollen with fat body and ovarioles, as is shown in Figure 19-1. This physogastry is made possible by the fact that the intersegmental membranes are thicker and more sclerotized than is usually the case in ant queens and can therefore be stretched more. Also, the abdominal sclerites themselves are widely overlapping in the virgin queen, so that the abdomen can be distended to an unusual degree before the sclerites are pulled apart. The ovarioles increase enormously in length, discard their initial orientation, and infiltrate the entire abdomen and even the postpetiolar cavity.

From one to several physogastric queens are found in each parasitized nest, usually riding on the back of the host queen. Each lays an average of one egg every thirty seconds. The infested *Tetramorium* colonies are typically smaller than uninfested ones, but they still contain up to several thousand workers. The *Tetramorium* queens also lay eggs, although none seem capable of developing into sexual forms. Consequently the brood of a parasitized colony consists typically of eggs, larvae, and pupae of *Teleutomyrmex* queens and males mixed with those of *Tetramorium* workers.

The bodies of the *Teleutomyrmex* queens bear the mark of extensive morphological degeneration correlated with their loss of social functions. The labial and postpharyngeal glands are reduced, and the maxillary and metapleural glands are completely absent. The mandibular and tibial glands, on the other hand, are apparently normal.

The integument is thin and less pigmented and sculptured in comparison with that of *Tetramorium;* as a result of these reductions the queens are shining brown, an appearance that contrasts with the opaque blackish brown of their hosts. The sting and poison apparatus are reduced; the mandibles are so degenerate that the parasites are probably unable to secure food on their own; the tibial-tarsal cleaning apparatus is underdeveloped; and, of even greater interest, the brain is reduced in size with visible degeneration in the associative centers. In the central nerve cord, ganglia 9-13 are fused into a single piece. The males are also degenerate. Their bodies, like those of the males of a few other extreme social parasites, are "pupoid," meaning that the cuticle is thin and depigmented, actually greyish in color; the petiole and postpetiole are thick and provided with broad articulating surfaces; and the abdomen is soft and deflected downward at the tip.

In its essentials the life cycle of *Teleutomyrmex schneideri* resembles that of other known extreme ant parasites. Mating takes place within the host nest. The fecundated queens then either shed their wings and join the small force of egg layers within the home nest or else fly out in search of new *Tetramorium* nests to infest. It is probably an easy matter for a *Teleutomyrmex* queen to gain entrance to new nests. Stumper found that the queens could be transferred readily from one *Tetramorium* colony to another, provided the recipient colony originated from the Saas-Fee. However, *Tetramorium* colonies from Luxembourg were hostile to the little parasites. Less surprisingly, ant species from the Saas-Fee other than *Tetramorium caespitum* always rejected the *Teleutomyrmex.*

The Kinds of Social Parasitism in Ants

Social parasitism in ants is complicated, and its study has become virtually a little discipline of entomology in itself. The source of the complexity is first the large number of ant species that have entered into some form of parasitic relationship with each other. Second, at least three major evolutionary routes lead, in a possibly reticulate fashion, to the ultimate stage of permanent, workerless parasitism. Finally, no two species are exactly alike in the details of their parasitic adaptation. Table 19-1 contains a list of the known parasitic ants, together with certain essential data concerning each of them. With this information readily at hand for constant reference, I wish now to present what is deliberately a rather didactic review of the entire subject, attempting to make it as orderly and clear as possible from the outset.

Wasmann (1891) distinguished two classes of consociations, or myrmecobioses as Stumper (1950) later dubbed them, that occur between different species of ants. These are the *compound nests,* in which the two or more species live very close to each other, in some cases even running their nest galleries together, but keep their brood separated; and the *mixed colonies,* in which the brood are mingled and cared for communally.* Compound nests are very common in nature. They reflect relationships that range, depending on the species involved, all the way from the accidental and trivial to total parasitism. Mixed colonies, on the other hand, almost always come about as a result of social parasitism. Forel (1898, 1901b) and Wheeler (1901, 1910) devoted a great deal of attention to compound nests and provided a useful classification of the underlying relationships, complete with a somewhat less useful set of Hellenistic terms to label the various categories. Let us examine this classification briefly. Then we will make more interesting use of it in tracing the evolution of parasitism and other forms of symbioses.

Compound Nests

Plesiobiosis. In this most rudimentary consociation, different ant species nest very close to each other, but engage in little or no direct communication—unless their nest chambers are accidentally broken open, in which case fighting and brood theft may ensue. The less similar the species are to each other morphologically and behaviorally, the more likely they are to cluster together in an accidental, truly "plesiobiotic" relationship. Put the other way, closely related species of ants are the least likely to tolerate each other's presence.

Cleptobiosis. Some species of small ants build nests near those of larger species and either feed on refuse in the host kitchen middens or rob the host workers when they return home carrying food. R. C. Wroughton (quoted by Wheeler, 1910) has described a species of *Crematogaster* in India whose workers "lie in wait for *Holcomyrmex,* returning home, laden with grain, and by threats, rob her

*Forel, in 1898, was actually the first to use the expression "social parasitism."

TABLE 19-1. The known parasitic ants, their hosts, their distribution, and their form of parasitism.

Parasite	Hosts	Nearest related free-living genus	Form of parasitism	Range	Authority
Subfamily Myrmeciinae					
Myrmecia inquilina	*M. nigriceps, M. vindex*	*Myrmecia*	Inquilinism; workerless	Australia	Douglas and Brown (1959); Haskins and Haskins (1964)
Subfamily Pseudomyrmecinae					
Tetraponera ledouxi	*T. anthracina*	*Tetraponera*	Temporary	West Africa	Terron (1969)
Subfamily Myrmicinae					
Myrmica lemasnei	*M. sabuleti*	*Myrmica*	Inquilinism; worker caste present	Pyrénées-Orientales, France	Bernard (1968)
Myrmica myrmecophila	*M. sulcinodis*	*Myrmica*	Inquilinism; workerless	Austria	Bernard (1968)
Myrmica myrmecoxena	*M. lobicornis*	*Myrmica*	Inquilinism; workerless	Switzerland	Kutter (1969)
Sifolinia karavejevi	*M. scabrinodis*	*Pheidole* (?)	Inquilinism; workerless	USSR	Kutter (1969)
Sifolinia laurae	*Myrmica* sp.	*Pheidole* (?)	Inquilinism; workerless	Italy	Emery (1921)
Sifolinia pechi	*M. rugulosa*	*Pheidole* (?)	Inquilinism; workerless	Czechoslovakia	Bernard (1968)
Sommimyrma symbiotica	*M. laevinodis*	?	Inquilinism; workerless	Italy	Kutter (1969)
Symbiomyrma karavajevi	*Myrmica scabrinodis*	?	Inquilinism; workerless	USSR	Arnoldi (1930)
Paramyrmica colax	*Myrmica striolagaster*	*Myrmica* (*M. striolagaster*)	Probably temporary, with worker caste	Texas	Cole (1957)
Manica parasitica	*M. bradleyi*	*Manica*	Inquilinism; workers present	California	Creighton (1950); Wheeler and Wheeler (1968)
Pogonomyrmex anergismus	*Pogonomyrmex rugosus*	*Pogonomyrmex*	Inquilinism; workerless	New Mexico	Cole (1968)
Aphaenogaster tennesseensis	*A. fulva*	*Aphaenogaster*	Temporary	Eastern US	Creighton (1950)
Eriopheidole symbiotica	*Pheidole obscurior*	*Pheidole*	Inquilinism; workerless	Argentina	Kusnezov (1951b)
Epipheidole inquilina	*Pheidole pilifera*	*Pheidole* (*P. pilifera*)	Inquilinism; workers present but scarce	Colorado to Nebraska	Creighton (1950)
Sympheidole elecebra	*Pheidole ceres*	*Pheidole*	Inquilinism; workerless	Colorado	Creighton (1950)
Bruchomyrma acutidens	*Pheidole nitidula*	*Pheidole*	Extreme inquilinism; workerless	Argentina	Bruch (1931)

TABLE 19-1 (*continued*).

Parasite	Hosts	Nearest related free-living genus	Form of parasitism	Range	Authority
Gallardomyrma argentina	*Pheidole nitidula*	*Pheidole*	Extreme inquilinism; workerless	Argentina	Bruch (1932)
Anergatides kohli	*Pheidole megacephala melancholica*	?	Extreme inquilinism; workerless	Congo	Wasmann (1915b); Emery (1922)
Anergates atratulus	*Tetramorium caespitum;* host queen eliminated	?	Extreme inquilinism; workerless	Europe, introduced into US	Wheeler (1910); Creighton (1950)
Teleutomyrmex schneideri	*Tetramorium caespitum*	?	Extreme inquilinism; workerless	Central Europe	Stumper (1950); Kutter (1969)
Strongylognathus (*afer, christophi, huberi, karawajewi, koreanus, kratochvili, rehbinderi, testaceus*)	*Tetramorium caespitum* and *T. jacoti;* host queen tolerated by *S. testaceus*	*Tetramorium*	Degenerate dulosis (*huberi*) or inquilinism (*testaceus*)	Temperate parts of Europe and Asia	Wheeler (1910); Stumper (1950); Pisarski (1966); Kutter (1969)
Harpagoxenus americanus	*Leptothorax longispinosus* and *L. curvispinosus*	*Leptothorax*	Dulosis	Eastern North America	Wesson (1939); Creighton (1950)
Harpagoxenus canadensis	*Leptothorax canadensis*	*Leptothorax*	Dulosis	Eastern North America	Creighton (1950)
Harpagoxenus sublaevis	*Leptothorax acervorum*	*Leptothorax*	Dulosis	Europe	Buschinger (1966); Kutter (1969)
Harpagoxenus zaisanicus	*Leptothorax* sp.	*Leptothorax*	Dulosis (?)	Mongolia	Pisarski (1963)
Leptothorax: buschingeri, goesswaldi, kutteri	*L. acervorum*	*Leptothorax* (*L. acervorum*)	Inquilinism; workerless	Switzerland	Kutter (1967, 1969)
Leptothorax ergatogyna	*L. recedens*	*Leptothorax*	Probably temporary parasitism or facultative inquilinism	France	Bernard (1968)
Myrmoxenus gordiagini	*Leptothorax serviculus*	*Leptothorax*	Dulosis (?); worker caste present	USSR (eastern Siberia)	Ruzsky (1902)
Epimyrma: algeriana, corsica, foreli, gösswaldi, kraussei, ravouxi, stumperi, tamarae, vandeli	*Leptothorax* spp.; queen allowed to live by some species, killed by others	*Leptothorax*	Inquilinism; worker caste present in some species, absent in *E. ravouxi*	Europe	Kutter (1951, 1969); Bernard (1968); Cagniant (1968); Arnoldi (1968)

TABLE 19-1 (*continued*).

Parasite	Hosts	Nearest related free-living genus	Form of parasitism	Range	Authority
Leptothorax duloticus	*L. curvispinosus* and *L. ambiguus*	*Leptothorax*	Dulosis	Eastern US	Wesson (1940b); Talbot (1957)
Leptothorax provancheri (= *L. emersoni*)	*Myrmica brevinodis*	*Leptothorax*	Xenobiosis	North America	Creighton (1950)
Symmyrmica chamberlini	*Manica mutica*	*Leptothorax*	Xenobiosis or inquilinism	Utah	Wheeler (1910)
Chalepoxenus gribodoi	*Leptothorax* spp.; queen survives	*Leptothorax*	Inquilinism; workers present	Central Europe	Kutter (1950b); Bernard (1968)
Doronomyrmex pacis	*Leptothorax acervorum;* queens survive	*Leptothorax*	Inquilinism; workerless	Switzerland	Kutter (1945, 1969)
Formicoxenus nitidulus	*Formica* spp., esp. *nigricans, polyctena,* and *rufa*	*Leptothorax*	Xenobiosis	Europe	Stäger (1925); Stumper (1950)
Monomorium (= *Wheeleriella*) *adulatrix, M.* (= *W.*) *santschii*	*Monomorium salomonis;* host queen is eliminated	*Monomorium*	Inquilinism; workerless	North Africa	Wheeler (1910); Ettershank (1966)
Monomorium (= *Wheeleriella*) *wroughtoni*	*Monomorium salomonis*	*Monomorium*	Inquilinism (?)	India	Forel (1910a); Ettershank (1966)
Monomorium (= *Epixenus*) *andrei, M.* (= *E.*) *biroi,* and *M.* (= *E.*) *creticus*	*Monomorium* (*Xeromyrmex*)	*Monomorium*	Inquilinism; workerless	Syria, Crete	Wheeler (1910); Ettershank (1966)
Monomorium (= *Epoecus*) *pergandei*	*Monomorium minimum;* host queen survives	*Monomorium*	Inquilinism; workerless	Washington, D.C.	Creighton (1950); Ettershank (1966)
Monomorium (= *Phacota*) *noualhieri* and *M.* (= *P.*) *sicheli*	*Monomorium subnitidum*	?	Xenobiosis (?); worker caste present	Spain, Algeria	Wheeler (1910); Ettershank (1966)
Oxyepoecus: bruchi, daguerri, minuta, inquilina	*Pheidole* spp. and *Solenopsis* spp.	*Solenopsis*	Apparently inquilinism; worker caste present	Argentina	Kusnezov (1952); Ettershank (1966)
Monomorium metoecus	*Monomorium minimum*	*Monomorium* (*M. minimum*)	Inquilinism; worker caste present (may be only an aberrant form of the host species)	Southeastern US	Wilson and Brown (1958a)
Solenopsis (= *Labauchena*) *daguerrei*	*Solenopsis saevissima*	*Solenopsis*	Inquilinism; workerless	Argentina, Uruguay	Santschi (1930); Ettershank (1966)

TABLE 19-1 (*continued*).

Parasite	Hosts	Nearest related free-living genus	Form of parasitism	Range	Authority
Solenopsis (=*Labauchena*) *acuminata*	*Solenopsis saevissima* and *S. clytemnestra*	*Solenopsis*	Inquilinism; workerless	Argentina	Borgmeier (1949); Kusnezov (1957b); Ettershank (1966)
Solenopsis (=*Paranamyrma*) *solenopsidis*	*Solenopsis ?clytemnestra*	*Solenopsis*	Inquilinism; workerless	Argentina	Kusnezov (1954, 1957b); Ettershank (1966)
Megalomyrmex symmetochus	*Sericomyrmex amabilis*	*Megalomyrmex*	Xenobiosis	Panama (Central America)	Wheeler (1925)
Megalomyrmex wheeleri	*Cyphomyrmex costatus*	*Megalomyrmex*	Xenobiosis	South America	Weber (1940)
Pseudoatta argentina	*Acromyrmex lundi*	*Acromyrmex* (?)	Inquilinism; workerless	Argentina	Gallardo (1916b); Kusnezov (1954); Weber (personal communication)
Hagioxenus mayri	*Pheidole latinoda*	*Monomorium*	Inquilinism	India	Ettershank (1966)
Hagioxenus schmitzi	*Tapinoma erraticum*	*Monomorium*	Inquilinism; worker caste absent	Israel	Wheeler (1919b); Ettershank (1966)
Xenometra monilicornis	*Cardiocondyla emeryi*	*Cardiocondyla*	Inquilinism; workerless	St. Thomas, West Indies	Wheeler (1910)
Xenometra gallica	*Cardiocondyla elegans*	*Cardiocondyla*	Inquilinism; workerless	France	Bernard (1968)
Crematogaster creightoni and *C. kennedyi*	*C. lineolata;* host queen survives	*Crematogaster*	Inquilinism; workerless	Eastern US	Wheeler (1933b); Creighton (1950)
Crematogaster atitlanica	*C. sumichrasti*	*Crematogaster*	Inquilinism; workerless	Guatemala	Wheeler (1936b)
Kyidris media, K. yaleogyna	*Strumigenys loriae;* host queens survive	?	Inquilinism; workerless	New Guinea	Wilson and Brown (1956)
Strumigenys xenos	*Strumigenys perplexa;* host queens survive	*Strumigenys* (*S. perplexa*)	Inquilinism; workerless	Victoria, Australia	Brown (1955c)
Subfamily Dolichoderinae					
Bothriomyrmex: about 10 spp.	*Tapinoma* spp.; host queen killed	*Tapinoma*	Temporary, during colony formation	Old World	Santschi (1906); Wheeler (1910); Emery (1925); Bernard (1968)
Subfamily Formicinae					
Anoplolepis nuptialis	*A. custodiens*	*Anoplolepis*	Inquilinism; workerless	South Africa	Santschi (1917)
Plagiolepis grassei	*P. pygmaea;* host queen survives	*Plagiolepis* (*P. pygmaea*)	Inquilinism; worker caste rare	Pyrénées-Orientales, France	Le Masne (1956b)
Plagiolepis xene	*P. pygmaea;* host queen survives	*Plagiolepis* (*P. pygmaea*)	Inquilinism; workerless	Hungary to Pyrénées-Orientales	Le Masne (1956b); Passera (1964)

TABLE 19-1 (*continued*).

Parasite	Hosts	Nearest related free-living genus	Form of parasitism	Range	Authority
Aporomyrmex ampeloni	*Plagiolepis vindobonensis;* host queen survives	*Plagiolepis*	Inquilinism; workerless	Austria	Faber (1969)
Aporomyrmex regis	*Plagiolepis* sp.	*Plagiolepis*	Inquilinism; workerless	Daghestan, USSR	Faber (1969)
Camponotus universitatis	*C. aethiops*	*Camponotus*	Inquilinism (?); worker caste present	France, Switzerland	Bernard (1968); Kutter (1969)
Lasius (*Chthonolasius*): 11 species	*Lasius* (*L.*) spp.; host queen eliminated	*Lasius*	Temporary, during colony foundation	Europe, Asia, North America	Wilson (1955a); Cole (1956); Kutter (1969)
Lasius (*Austrolasius*) *reginae*	*Lasius alienus;* host queen killed	*Lasius*	Temporary, during colony foundation	Austria	Faber (1967)
Lasius (*Dendrolasius*): 5 species	*Lasius* spp.; host queen eliminated	*Lasius*	Temporary, during colony foundation	Europe, Asia	Wilson (1955a); Yamauchi and Hayashida (1968)
Acanthomyops latipes, A. murphyi	*Lasius* (*L.*) spp.	*Lasius*	Temporary, during colony foundation	North America	Wing (1968)
Formica (*F.*) "*rufa* group," some but not all of the species, namely: *aquilonia, lugubris, pratensis, rufa, truncorum,* and *uralensis* in Europe; *dakotensis* and *reflexa* in North America.	*Formica fusca* grp. ("subg. *Serviformica*") and *Formica* (*Neoformica*); host queen does not persist	*Formica*	Temporary, during colony foundation; one species (*reflexa*) apparently permanent, with workers	Europe, North America	Wheeler (1910); Creighton (1950); Gösswald (1957); Bernard (1968); Kutter (1969)
Formica (*Coptoformica*), also called the "*exsecta* group" of subgenus *Formica: bruni, exsecta, foreli, forsslundi, goesswaldi, naefi, pressilabris,* and *suecica* in Europe; *exsectoides, opaciventris,* and *ulkei* in North America.	*Formica fusca* group; host queen does not persist	*Formica*	Temporary, during colony foundation	Europe, North America	Wheeler (1910); Creighton (1950); Bernard (1968); Kutter (1969)

TABLE 19-1 (*concluded*).

Parasite	Hosts	Nearest related free-living genus	Form of parasitism	Range	Authority
Formica (*F.*) *microgyna* group: 14 species	*F. fusca* group spp.	*Formica*	Temporary; host queen does not persist	North America	Wheeler (1910); Creighton (1950)
Formica (*Raptiformica*), also called the "*sanguinea* group" of subgenus *Formica*: 2 spp. in Europe and Asia and about 10 spp. in North America	*Formica* spp.	*Formica*	Dulosis	Europe, Asia, North America	Creighton (1950); Bernard (1968); Buren (1968a); Kutter (1969)
Polyergus lucidus, P. rufescens, P. samurai	*Formica* spp.	*Formica*	Dulosis	Europe, Asia, North America	Wheeler (1910); Creighton (1950); Bernard (1968)
Rossomyrmex proformicarum	*Proformica nasuta*	*Formica*	Dulosis	Trans-Caucasus, USSR	Arnoldi (1932)

of her load, on her own private road and this manoeuvre was executed, not by stray individuals, but by a considerable portion of the whole community." Workers of *Conomyrma pyramica* in the southern United States collect dead insects discarded by colonies of *Pogonomyrmex,* including corpses of the *Pogonomyrmex* themselves. My impression in the field has been that some *Conomyrma* colonies obtain a large part of their food in this way, to the point of preventing kitchen middens from building up near the *Pogonomyrmex* nests.

Lestobiosis. Certain small species, most belonging to *Solenopsis* and related genera, stay in the walls of large nests built by other ants or termites and enter the nest chambers of their hosts to steal food and prey on the inhabitants. For example, colonies of the "thief ants" of the subgenus *Solenopsis* (*Diplorhoptrum*), including especially *S. fugax* of Europe and *S. molesta* of the United States, often nest next to larger ant species, stealthily enter their chambers, and prey on their brood. Species of *Carebara* in Africa and tropical Asia frequently construct their nests in the walls of termite mounds and are believed to prey on the inhabitants. A very good and still largely up-to-date review of lestobiosis can be found in an article on ant-termite relations written by Wheeler in 1936.

Parabiosis. In this peculiar form of symbiosis, two or more species use the same nest and sometimes even the same odor trails, but they keep their brood separate. The situation is closely similar to the mixed foraging flocks of birds so prevalent in tropical forests (Moynihan, 1962).

Xenobiosis. This symbiotic state falls just short of a truly mixed colony. One species lives in the walls or chambers of the nests of the other and moves freely among its hosts, obtaining food from them by one means or another, usually by soliciting regurgitation. The brood is still kept separate.

Mixed Colonies

The following phenomena are vital in the later stages of parasitic evolution. In a sense they form categories comparable to those just cited for compound nests, although they are less than ideal because they are not mutually exclusive. Nevertheless, I favor continuing to distinguish them on the grounds that the associated terminology is the familiar one in literature dating back over the past seventy years and, more importantly, the classification can still be relied upon to serve as an adequate guide through the complex relationships as we understand them.

Temporary social parasitism. This symbiosis was first clearly recognized by Wheeler (1904) as a result of his studies of the life cycle of members of the *Formica microgyna* group, especially *F. difficilis.* It has since been discovered in a diversity of genera belonging to several subfamilies. The newly fecundated queen finds a host colony and secures adoption, either by forcibly subduing the workers or by conciliating them in some fashion. The original host queen is then assassinated by the intruder or by her own workers, who somehow come to favor the parasite. With the development of the first parasite brood, the worker force soon becomes a mixture of host and parasite species. Finally, since the host queen is no longer present to replenish them, the host workers die out, and the colony comes to consist entirely of the parasite queen and her offspring.

Dulosis (slavery). Certain ant species have become dependent on workers of other species which they keep as slaves. The slave raids are dramatic affairs in which the slave-making workers go out in columns, penetrate the nests of colonies belonging to other, related species, and bring back pupae to their own nests. The pupae are allowed to eclose, and the workers become fully functional members of the colony. The workers of most slave-making species seldom if ever join in the ordinary chores of foraging, nest building, and rearing of the brood, all of which are left to the slaves.

Inquilinism ("permanent parasitism"). In this final, degenerate stage, the parasitic species spends its entire life cycle in the nests of the host species. Workers may be present, but they are usually scarce and display atrophied behaviors. In many of these species, as for example in *Teleutomyrmex schneideri,* the worker caste has been lost altogether. I am suggesting here the use of the term inquilinism in preference to the somewhat more familiar expression "permanent parasitism" since obligatorily dulotic species are also permanent parasites. Inquilinism and dulosis, on the other hand, form exclusive categories; they are meant to be the streamlined equivalents of Kutter's (1969) "permanent parasitism without dulosis" and "permanent parasitism with dulosis."

The Occurrence of Social Parasitism Throughout the Ants

A rich variety of new parasitic species, representing almost every conceivable evolutionary stage, have been added since the time of Wheeler's classic synthesis in 1910. They continue to be discovered at such a consistently high rate as to suggest that, at this moment, only a small fraction of the total world fauna of social parasites is known. The reason for the slow uncovering of the world fauna seems clear: parasitic species tend to be both rare and locally distributed. As a rule, moreover, the more advanced the stage of parasitism, the rarer the species. Thus, we find (Table 19-1) that temporary social parasites, such as members of the *Formica exsecta* and *Lasius umbratus* groups, are often nearly as widely distributed as their free-living congeners, and a few of the species are also very abundant. Species in which dulosis is weakly developed or even facultative, as, for example, the representatives of the *Formica sanguinea* group, are also relatively abundant and widespread. On the other hand, extreme dulotic species, such as the members of *Strongylognathus, Polyergus,* and *Rossomyrmex,* exist in more restricted, sparser populations. Finally, the extreme workerless parasites are, as a rule, both very rare and very locally distributed. *Anergates atratulus* comes closest to being an exception. It has been collected over a wide area from southern France to Germany, and it has even been accidentally introduced into the United States with its host *Tetramorium caespitum.* Yet everywhere within this range it is still a comparatively rare ant. The great majority of other workerless parasites have been found at only one or two localities and are extremely difficult to locate, even when a deliberate search is made for them in the exact spots where they were first discovered. Usually they give the impression, quite possibly false, of having no more than a toehold on their host populations and of existing close to the edge of extinction.

Most of the known parasitic species have been recorded exclusively from the temperate areas of North America, Europe, and South America. Almost certainly this reflects at least in part the strong bias of ant collectors, most of whom reside in these areas and devote a large part of their lives to a meticulous examination of local faunas. Switzerland, for example, is the present "capital" of parasitic ants for the simple reason that both Auguste Forel and Heinrich Kutter lived there. The United States has benefited similarly from the efforts of W. M. Wheeler, W. S. Creighton, and other gifted resident collectors, while the rich trove of species uncovered in Argentina has been due to three men who spent a large part or all of their lives in the country—Carlos Bruch, Angel Gallardo, and Nicolás Kusnezov. I believe that as the huge and still little-known tropical ant faunas are more carefully worked (there are no resident myrmecologists on the Amazon!), many more parasitic species will come to light. Some extraordinary forms are already known from tropical regions, particularly the extreme workerless *Anergatides kohli* from the Congo and the strange postxenobiotic *Kyidris* parasites from New Guinea. Wheeler (1925) pointed out that females of the numerous species of *Crematogaster* belonging to the subgenera *Atopogyne* and *Oxygyne,* groups widely distributed in Africa, Madagascar, and tropical Asia, have all of the morphological characteristics of northern ants known to be temporary parasites in that they tend to be small and shining and to possess falcate or very oblique mandibles and large postpetioles which are attached broadly to the gasters. The last of these characteristics is usually associated with physogastry, also a common but not diagnostic feature of social parasitism. Emery (1899) has recorded a highly physogastric nest queen of *C.* (*Oxygyne*) *ranavalonae* from Madagascar. At least two species of the Neotropical dolichoderine genus *Azteca* (*aurita* and *fiebrigi*) possess some of these traits. The species of *Rhoptromyrmex,* found in South Africa, Asia, New Guinea, and Australia also possess them (Brown, 1964b). A special study of such species, and any others that can be found to possess various of the "temporary parasite syndrome" of characters, might prove very rewarding to future students of tropical myrmecology.

Even so, the vast differences in quality of sampling from the major parts of the world render the matter inconclusive, and there remains the possibility that life in certain climates and environments actually does predispose ant species toward parasitism. It is true, for example, that a disproportionate number of parasitic species, especially the complete inquilines, occur in mountainous and arid regions. I have already mentioned the extraordinary diversity of parasites found in the little forest of the Saas-Fee. Among numerous other examples that can be cited are the montane species *Epipheidole inquilina, Sympheidole elecebra, Manica parasitica, Paramyrmica colax,* and *Pogonomyrmex anergismus,* which together make up a majority of the known inquiline fauna of North America. Temporary social parasites, along with species that can be tentatively placed in this category by virtue of their morphology, are far more abundant in the colder portions of Europe and North America than in the warm temperate and subtropical portions, even though the faunas of the two climatic zones are otherwise not radically different. Similarly, dulosis is a common phenomenon in the colder parts of Europe and Asia but rare in the warmer parts; and not a single example has ever been reported from the tropical or south temperate zones.

It is conceivable that cooler temperatures facilitate the introduction of parasitic queens in the early evolution of the phenomenon by dulling the responses of the host colonies. I have found, in general, that if ant colonies are first chilled in the laboratory they more likely to adopt queens of their own species which they would otherwise attack and destroy. In nature parasite queens need not wait for winter to utilize this effect. Some degree of chilling, say to 10° or 15°C, occurs commonly during the cool summer nights in mountainous regions, right in the middle of the season of nuptial flights. It should prove instructive to study the effects of various degrees of cooling of potential host colonies on the success of introduction of queens belonging to species at any early stage of inquilinism, such as members of *Epimyrma* and the *Leptothorax goesswaldi* group. Useful information might also be obtained from an analysis of the behavioral effects of cooling on ant groups that most commonly serve as hosts, such as the genera *Leptothorax* and *Formica,* as opposed to those that are relatively immune to social parasitism, such as the genus *Camponotus.*

Whatever the true geographical distribution of social parasitism and its behavioral significance—matters that will be fully clarified only by years of additional field work—there can be no question that the phylogenetic distribution of the phenomenon is very patchy. The more advanced forms of parasitism, namely dulosis and inquilinism, are almost wholly limited to the subfamilies Myrmicinae and Formicinae and are furthermore heavily concentrated in certain genera, including *Pheidole, Myrmica, Leptothorax, Tetramorium, Plagiolepis, Lasius,* and *Formica,* and in the satellite parasitic genera derived from them. A single inquiline (*Myrmecia inquilina*) has been described from the primitive subfamily Myrmeciinae. In view of the relatively small number of species known in the Myrmeciinae (about 120) and the relatively small amount of field study devoted to it to date, parasitism in this group may eventually be found to occur at about the same level of frequency as in the Myrmicinae and Formicinae. The only parasites known with certainty among the Dolichoderinae, on the other hand, are the temporarily parasitic species of *Bothriomyrmex*. This relative immunity is puzzling since the dolichoderines are a relatively large, numerically abundant group of advanced phylogenetic rank. Perhaps the explanation lies in the fact that very few dolichoderine species range into the cooler portions of the North Temperate Zone where parasitic species are most likely to evolve. Yet it is also true that a rich dolichoderine fauna exists in subtropical and temperate Argentina, where many myrmicine parasites have been discovered. No parasitic species of any kind are yet known in the Ponerinae, Cerapachyinae, and Dorylinae. One can speculate almost endlessly on why this is the case. For example, the Ponerinae are primitive (but so are the Myrmeciinae, and in any case many ponerine species form large colonies with advanced social traits). The Dorylinae engage in frequent nest changes (but many parasitic beetles, millipedes, wasps, and other arthropods emigrate with them along their odor trails). I can see no clear correlations between the frequency of social parasitism and geographical distribution, ecology, or social organization that would explain why social parasitism appears to be common in certain major groups of ants and very rare or wholly absent in others. It is even possible that explanations will have to be sought in the evolutionary mechanism itself, as, for example, in the rate of speciation, a point which I will take up shortly in the larger context of a discussion of evolutionary theory.

The Evolution of Social Parasitism in Ants

In 1909 Carlo Emery formulated what is perhaps the single most important generalization concerning social parasitism: "The dulotic ants and the parasitic ants, both temporary and permanent, generally originate from the closely related forms that serve them as hosts." What he meant, of course, was that the parasitic species tend to resemble their host species more closely than any other free-living form. "Emery's rule," as it has been called (Le Masne, 1956b), has continued to hold well for the true inquilines. Taxonomists have stressed that certain of the parasites—for example, *Paramyrmica colax, Epipheidole inquilina, Leptothorax buschingeri,* and *Strumigenys xenos*—really are morphologically more similar to their hosts than to any other known species. At first glance, this relation seems to create a paradox: how can a species generate its own parasite? It might be possible if sympatric speciation occurred, during which certain homogamous mutants or recombinants arise in sufficient number at the same place and time to segregate a distinct breeding population. But current opinion holds that such an event is extremely unlikely (Mayr, 1963). As long ago as 1919 Wheeler experienced difficulty in even conceiving of a mechanism by which it could occur. I have diagramed another and far likelier alternative in Figure 19-2. In what is generally regarded as the prevalent sequence of speciation, a single "parental" species can be divided into two "daughter" species by, first, fragmentation due to geographic barriers and, second, genetic divergence of the populations thus isolated geographically until they acquire intrinsic isolating mechanisms. If and when the newly formed species reinvade one another's ranges, the isolating mechanisms prevent them from interbreeding. And if, in addition, one of the species then becomes specialized as a parasite on the other, the condition of Emery's rule is fulfilled. This model suggests that, all other circumstances

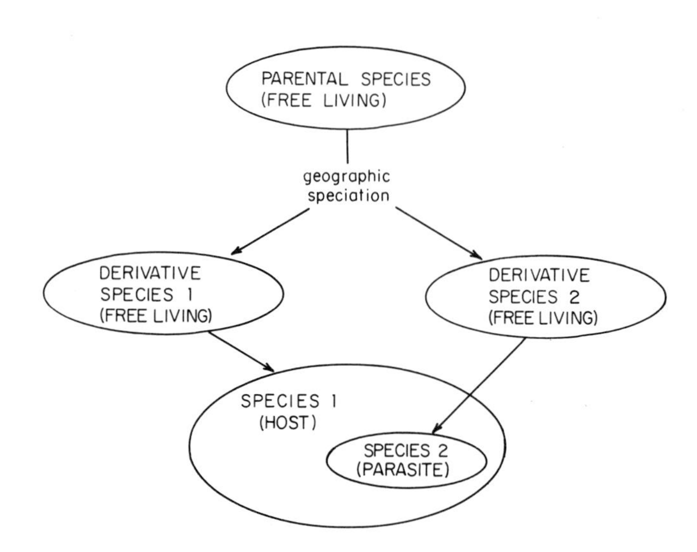

FIGURE 19-2. The means by which a species can originate by geographic speciation and come to live as a social parasite with its closest living relative, in accord with Emery's rule.

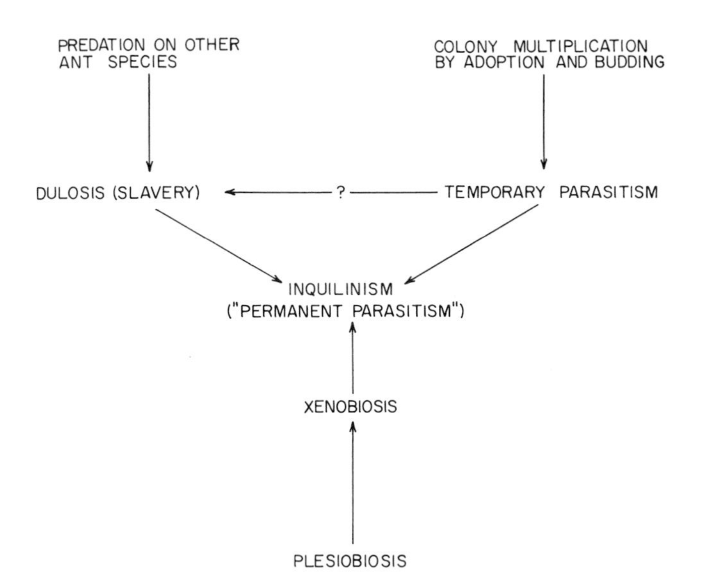

FIGURE 19-3. Hypothesized evolutionary pathways of social parasitism in ants.

being equal, the frequency of parasitism within a given genus should increase as a function of the rate of speciation. A corollary is that the more taxonomically "difficult" the genus, in other words, the larger the percentage of newly formed, indistinctly defined species in it, the higher should be the percentage of parasitism. This prediction does indeed seem to be met by such genera as *Pheidole, Leptothorax, Plagiolepis, Lasius,* and *Formica,* although the correlation through all ant groups taken together is far from perfect. But at least it is clear that the speciation rate is one additional factor that must be considered in future evolutionary analyses of the subject.

Many exceptions to Emery's rule exist. The most notable ones fall into the special categories of xenobiosis and parabiosis, in which the parasitic species typically belongs to a different genus and sometimes even to a different subfamily. The explanation for these two classes of exceptions is simple enough. When members of different genera associate at all, they are not likely to combine their brood, for they tend to be very different in biology and mutually incompatible. Any association achieved will be of a more tenuous sort, involving grooming, food exchange, trail sharing, or combinations of these relations—xenobiosis and parabiosis. The more closely related the two species, the more likely they are to enter into the more intimate forms of parasitism, producing the effect that is generalized in Emery's rule.

How do different ant species come together in symbiosis in the first place? The evolutionary schema presented in Figure 19-3 is a modern extension of one evolved in a long sequence of contributions by Wheeler (1904, 1910), Emery (1909), Escherich (1917), Stumper (1950), and Dobrzański (1965). It differs from earlier arrangements in one principal feature: from the new information concerning the *Kyidris* parasites I have added xenobiosis as a possible alternate route to complete inquilinism. The single most important idea embodied in this diagram is that inquilinism is a convergent phenomenon, reached independently by many different species following one or the other of at least three available pathways in evolution. Also, complete inquilinism is viewed as an evolutionary sink; a return to free life or even to a partially parasitic existence by reversed evolution seems impossible. For convenience I have arranged known cases of social parasitism according to these hypothesized sequences.

The Temporary Parasitism Route

The earliest stages of temporary parasitism are displayed by members of the *Formica rufa* group (Gösswald, 1951a,c; Kutter, 1913, 1969). Several of the members, *F. lugubris, F. polyctena,* and *F. pratensis,* form colonies with multiple queens. New colonies are usually created by budding, or "hesmosis" as it has occasionally been called. Following the nuptial flights the newly fecundated queen normally returns to the home nest, and at some later date she may move to a new site nearby with a group of workers. The new unit thus created is a colony only in the purely spatial sense because it may exchange workers with the mother nest for an indefinite time afterward. Multiplication by budding creates the pattern, so characteristic of these species of *Formica,* of dense aggregations of interconnecting nests that dominate local areas. Occasionally young queens do not find their way back to a nest of their own species. They may then seek adoption in a colony of the *Formica fusca* group. Whenever one of them succeeds in penetrating such an alien nest, the host queen is somehow eliminated; the intruder takes over the role of egg laying exclusively, and eventually the host workers die off. The final result is that the colony consists entirely of the intruder and her offspring. Such temporary parasitism is regarded as a secondary mode of colony founding for these ants since mixed host-parasite colonies are rarely encountered in nature.

However, a closely related species, *F. rufa,* has taken the step of founding its colonies predominantly by temporary parasitism, relying on budding in only a minority of instances. Its host species in Europe include *F. fusca* and *F. lemani.* The *rufa* queen is still a rather inept parasite. On approaching the host colony she does not hide, play dead, conciliate, or display any of the other dissembling tricks ordinarily used by parasitic queens; instead, she plunges right into the nest. Such intrusions frequently result in the death of the queen at the hands of hostile host workers, but enough attempts succeed to maintain *F. rufa* as one of the more abundant and widespread ant species of Europe.

Most European students of *Formica,* starting with Emery (1909), have argued that loss of the ability to found nests in the usual claustral manner, with the resulting dependence on adoption and budding, preadapts members of the *rufa* group to temporary parasitism on other species. Also, the fact that *F. rufa* itself is monogynous (its colonies each tolerate only one egg-laying queen) predisposes this species even further to incursions on other species.

The species of the *exsecta* group of *Formica* (collectively referred to by European writers as the "subgenus *Coptoformica*") have a life cycle very similar to that of the *rufa* group species, except that the queens have become more skillful at penetrating host colonies (Kutter, 1956, 1957). The European species of *F. exsecta,* for example, depend chiefly on homospecific adoption and budding, but a few queens seek colonies of the *fusca* group of species (called the "subgenus *Serviformica*"). The *exsecta* queens stalk the host colonies and either enter the nests by stealth or else permit themselves to be carried in by host workers. The *exsecta* queens are smaller and shinier than those of most members of the *rufa* group, and they seem to be treated with less hostility by the host workers. This is also the case for *F. pressilabris,* a second member of the *exsecta* group found in Europe. Queens approached by host workers lie down and "play dead" by pulling their appendages into the body in the pupal posture. In this position they are picked up by the host workers and carried down into the nests without any outward show of hostility. Later they somehow manage to eliminate the host queen and take over the reproductive role. Similar life histories have been described for the North American species of the *exsecta* group (Wheeler, 1906; Creighton, 1950; Scherba, 1958, 1961) and have been postulated on the basis of limited laboratory experiments to characterize members of the North American *microgyna* group (Wheeler, 1910).

Further subtleties have been developed by the related genus *Lasius.* Apparently all of the species of the subgenera *Austrolasius, Chthonolasius,* and *Dendrolasius* are temporary parasites on members of the subgenus *Lasius.* This relationship is obligatory, not optional as in the *rufa* and *exsecta* groups of *Formica.* The colonies are monogynous for the most part, and homospecific adoption is not practiced. When newly mated queens of *L. umbratus* are searching for a host colony, they first seize a worker in their mandibles, kill it, and run around with it for a while before attempting to penetrate the nest (K. Hölldobler, 1953). Apparently all of the parasitic *Lasius* get rid of the host queens, but the exact means employed are still unknown in most cases. The queens of *L. reginae,* a species recently discovered in Austria by Faber (1967), eliminate their rivals by rolling them over and throttling them (Figure 19-4).

Assassination is also the technique employed by the queens of the dolichoderine species *Bothriomyrmex decapitans* and *B. regicidus* in gaining control of colonies of *Tapinoma* (Santschi, 1906, 1920). These temporary parasites occur in the deserts of North Africa. After the nuptial flight, the *Bothriomyrmex* queen sheds her wings and searches over the ground until she finds a *Tapinoma* nest. She allows herself to be accosted by the aroused *Tapinoma*

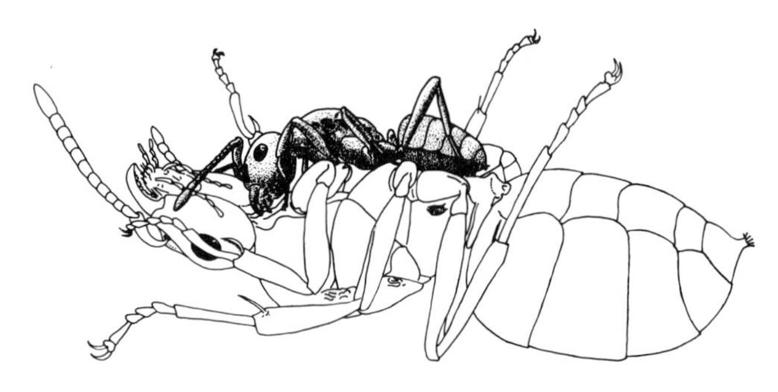

FIGURE 19-4. A newly mated queen of the temporary social parasite *Lasius reginae* has entered a nest of the host species *Lasius alienus* and is strangling the queen (from Faber, 1967).

workers and dragged by them into the interior of their nest. There she takes refuge among the brood or on the back of the *Tapinoma* queen. In time, she settles down for good on the back of the host queen and begins the one act for which she is uniquely specialized: slowly cutting off the head of her victim. When this is accomplished, sometimes only after many hours, the *Bothriomyrmex* takes over as the sole reproductive, and the colony eventually comes to consist entirely of her offspring and herself. A similar mode of entry into host nests is employed by the myrmicine *Monomorium* (= *Wheeleriella*) *santschii,* also a native of North Africa and a permanent workerless parasite of *Monomorium salomonis.* In this case, however, it is the *salomonis* workers who destroy their own queen. They then adopt the *santschii* queen as the sole reproductive in her place (F. Santschi, in Forel, 1906).

The evolutionary transition from temporary parasitism of the *Bothriomyrmex* type to inquilinism of the *Monomorium santschii* type is not a difficult step to imagine. It is, however, the species of the genus *Epimyrma* that display most convincingly the early stages of inquilinism following its evolutionary origin from temporary parasitism (Gösswald, 1933, 1934; Kutter, 1951, 1969). In at least five of the eight known species, a worker caste still exists, but it is relatively scarce and rather similar morphologically to the queen, although still a discrete apterous phase. It apparently never aids the host workers in foraging, nest labor, or brood care. All of the species of *Epimyrma* are parasitic on *Leptothorax.* The queen mates in her home nest, then leaves the nest, sheds her wings, and searches for a new host colony. The mode of entry and subsequent behavior varies greatly among the various species. The queen of the French species *E. vandeli,* upon approaching a *Leptothorax unifasciata* colony, makes repeated hostile approaches to the host workers and "intimidates" them, to use Kutter's expression. If she succeeds in entering the nest, she kills the host queen and secures complete adoption by the rest of the colony. The queen of *E. goesswaldi,* on the other hand, calms the host workers (*L. unifasciata* in Germany) by stroking them with her antennae and lower mouthparts. Once inside the nest, she mounts the host queen from the rear, seizes her around the neck with her saber-shaped mandibles, and kills her. *E. stumperi,* studied in Switzerland by Kutter, uses still another variation to enter the nests of its host, *L. tuberum.* The queen first stalks the host colony with slow, deliberate movements. When approached by the *Leptothorax* workers, she "freezes," crouches down, and seems to feign death. After a time she begins to mount the workers from the rear, strokes their bodies with her foreleg combs, and grooms herself, perhaps thereby passing nest odors back and forth. With this display of sophistication in evidence, it is not surprising to find that queens of *E. stumperi* are able to penetrate host colonies more quickly than the other *Epimyrma* species so far studied. Once inside the nest, the *E. stumperi* queen begins an implacable round of assassination directed at the host queens, of which there are usually at least several in the *L. tuberum* colonies. She mounts each queen in turn, forces her to roll over, then seizes her by the throat with her mandibles. The sharp tips of the mandibles pierce the soft intersegmental membrane of the neck of the victim. The *Epimyrma* maintains her grip for hours or even days, until the *Leptothorax* queen finally dies. Then she moves on to the next queen, and this is repeated until none are left. It is a matter of more than ordinary interest that the *E. stumperi* workers also occasionally mount *Leptothorax* workers and go through an ineffectual rehearsal of the assassination behavior, but without harming their "victims" and with no visible benefit to the parasites. This seems best interpreted as a partial transfer of the queen's behavioral pattern to the vestigial worker caste where it has neither positive nor harmful effects.

But why does the *Epimyrma* queen go to all this trouble? Since the species have already entered a permanently inquiline state, with total dependence on the host workers, it would seem an error to exterminate the host queens, which are, after all, the source of the labor force. However, the *Epimyrma* habit of reginicide cannot be written off simply as an unfortunate vestige from an earlier time when the *Epimyrma* ancestors were temporary parasites. It turns out that, when deprived of their own queens, some of the *Leptothorax* workers begin laying eggs, even in the presence of the *Epimyrma* queen. These develop into workers and thus ensure an indefinite continuation of the worker force. Even so, there is one species, *E. ravouxi,* a parasite of *L. unifasciatus,* which has taken the final step of permitting the host queens to live. *E. ravouxi,* in other words, has moved on into advanced inquilinism, and in this respect it is indistinguishable from other inquiline species whose probable evolutionary history is not nearly so well displayed.

The Dulosis Route

Slavery in ants, particularly as practiced by *Polyergus rufescens* and the species of the *Formica sanguinea* group, has been a favorite subject of myrmecologists in Europe and the United States ever since it was first described by Pierre Huber in 1810. Darwin was fascinated by the phenomenon, and in his book *On the Origin of Species* he offered the first hypothesis of how it originated in evolution. The ancestral *Formica,* he proposed, began by raiding other species of ants in order to obtain their pupae for food. Some of the pupae survived in the storage chambers long enough to eclose as workers, whereupon they were accepted by their captors as nestmates. This fortuitous addition to the work force helped the colony as a whole, and consequently there was an increasing tendency, propelled by natural selection, to raid other colonies solely for the purpose of obtaining slaves. If Darwin's explanation seems at first a bit farfetched, it is only commensurate with the phenomenon itself. Several authors, most notably Erich Wasmann (1905), rejected Darwin's hypothesis on various grounds, chiefly a priori in nature. But the years have brought an increasing amount of confirmation through the discovery of species whose behavior collectively bridges the gap between predator and slave-maker in ever shorter, more plausible steps.

It is a commonplace observation that some ant species are much more territorial than others. The colonies of a few of the most aggressive ones go so far as to extirpate rival species from local areas by attacking their nests and raiding their brood. Behavior of this kind has been recorded not only in the Cerapachyinae and many Dorylinae, which specialize in preying on other ant species, but also in a diversity of higher ant genera, including *Solenopsis, Pheidole, Iridomyrmex,* and *Anoplolepis,* who use predation as an auxiliary technique in territorial aggression. More to the point of our evolutionary reconstruction, such behavior is well marked in a few species of *Formica.* Scherba (1964b) found that in Wyoming the dense, polydomous colonies of *F. opaciventris,* a member of the *exsecta* group, commonly oust colonies of *F. fusca* from their nest sites by laying siege to them and robbing larvae and pupae when they get the chance. When Kutter (1957) placed colonies of *F. naefi,* also a member of the *exsecta* group, near colonies of species belonging to the *fusca* group, the *naefi* attacked their neighbors, penetrated their nest, and carried away both the brood and the adult workers. Kutter is unsure whether such behavior occurs in nature, but he has noted that all larger *naefi* colonies observed in the field contain a few *fusca*-group workers. It seems reasonable to suggest that *naefi* represents the first, truly dulotic stage envisioned in Darwin's hypothesis.

It is also possible that the phenomenon described by Kutter occurs in other ants, or at least in other *Formica.* King and Sallee (1957, 1962) have reported the puzzling existence of natural mixed colonies of *Formica clivia* and *F. fossaceps* that persisted over a period of up to 16 years in Iowa. Both workers and sexuals of the two forms were produced in the nests. King and Sallee considered the possibilities that the two forms are either genetic morphs or distinct species linked in some aberrant and unexplained symbiosis. The field data strongly suggest the second alternative. In laboratory experiments small homospecific groups of workers readily accepted queens of the opposite species combined with alien worker groups. The significance of this permissiveness, and the nature of the interaction of homospecific colonies of the same species in nature, are clearly promising subjects for future study.

The next step in the dulotic progression is best exemplified by *F. sanguinea,* a European species that has been thoroughly studied by Huber (1810), Forel (1874), Wasmann (1891), Dobrzański (1961, 1965), and others. The "sanguinary ants" are very aggressive and territorial, dominating the local spots near their nests that are richest in food. They are "facultative slave-holders," in Wasmann's terminology, since colonies are sometimes found with no slaves present. Also, workers isolated in laboratory nests are able to conduct all of the affairs of colony life, including nest construction, in a competent manner. According to Wheeler (1910) the percentage of slaveless colonies in different populations of *sanguinea* varies enormously, from about 2 percent to 98 percent. The commonest slaves taken belong to the *fusca* group ("*Serviformica*") and include *fusca, lemani,* and *rufibarbis;* less commonly exploited are *gagates, cunicularia, transkaukasica,* and *cinerea,* all of which are also members of the *fusca* group as conceived in the broadest sense. On rare occasions workers of the *rufa* group, in particular *nigricans* and *rufa,* have been found in *sanguinea* nests, but always in the company of *fusca*-group slaves (Bernard, 1968). As a rule, *sanguinea* colonies enslave the *fusca*-group species nearest their nest, and the seeming prefer-

FIGURE 19-5. *Formica sanguinea,* the slave-making "sanguinary ant" of Europe: (*a*) dealated queen; (*b*) pseudogyne, an abnormal female form found in colonies of this (and other) species of *Formica* infested by lomechusine staphylinid beetles; (*c*) worker; (*d*) head of worker, showing the notched clypeus that characterizes members of the *sanguinea* group of Formica. Colonies of this species are not dependent on slaves for survival, and they engage only occasionally in raids. The morphology is not especially adapted for slave making (from Wheeler, 1910).

ences are merely a reflection of local relative abundance of the slave species. Two or even three slave species are sometimes present in a given *sanguinea* nest simultaneously, and the composition of slaves may change from year to year.

The raids of *sanguinea* have been lucidly described by Wheeler (1910):

> The sorties occur in July and August after the marriage flight of the slave species has been celebrated and when only workers and mother queens are left in their formicaries. According to Forel the expeditions are infrequent—"scarcely more than two or three a year to a colony." The army of workers usually starts out in the morning and returns in the afternoon, but this depends on the distance of the *sanguinea* nest from the nest to be plundered. Sometimes the slave-makers postpone their sorties till three or four o'clock in the afternoon. On rare occasions they may pillage two different colonies in succession before going home. The *sanguinea* army leaves its nest in a straggling, open phalanx sometimes a few meters broad and often in several companies or detachments. These move to the nest to be pillaged over the directest route permitted by the often numerous obstacles in their path. As the forefront of the army is not headed by one or a few workers that might serve as guides, but is continually changing, some dropping back while others move forward to take their places, it is not easy to understand how the whole body is able to go so directly to the nest of the slave species, especially when this nest is situated, as is often the case, at a distance of 50 or 100 m . . . When the first workers arrive at the nest to be pillaged, they do not enter it at once, but surround it and wait till the other detachments arrive. In the meantime the *fusca* or *rufibarbis* scent their approaching foes and either prepare to defend their nest or seize their young and try to break through the cordon of *sanguinea* and escape. They scramble up the grass-blades with their larvae and pupae in their jaws and make off over the ground. The sanguinary ants, however, intercept them, snatch away their charges and begin to pour into the entrances of the nest. Soon they issue forth one by one with the remaining larvae and pupae and start for home. They turn and kill the workers of the slave-species only when these offer hostile resistance. The troop of cocoon-laden *sanguinea* straggle back to their nest, while the bereft ants slowly enter their pillaged formicary and take up the nurture of the few remaining young or await the appearance of future broods.

The communicative signals that trigger and orient the raids of colonies belonging to the *sanguinea* group of slave-making ants have recently been identified, at least in part, by Fred E. Regnier and myself (unpublished observations). We found that workers of the American species *Formica rubicunda* readily follow artificial odor trails made from whole body extracts of *rubicunda* workers and applied with a camel's hair brush over the ground in the vicinity of the nest. When the trails were drawn away from the nest opening in the afternoon, at about the time raids are usually conducted, the *rubicunda* workers showed behavior that was indistinguishable from ordinary raiding sorties. They ran out of the nest and along the trail in an excited fashion, and, when presented with colony fragments of a slave species (*F. subsericea*), they proceeded to fight with the workers and to carry the pupae back to their nest. It seems likely that under normal circumstances *rubicunda* scouts lay odor trails from the slave colonies they discover to the home nest, and the raids result when nestmates follow the trails out of the home nest back to the source. This is probably the general mode of communication among slave-making ants. As we shall see shortly, it is the technique employed by the evolutionarily superior amazon ants of the genus *Polyergus,* as well as by the myrmicine slave-makers of the genus *Harpagoxenus.* The tendency of *Formica sanguinea* to fan out into "phalanxes" in their outward march does not conflict with this interpretation; there could be several odor trails involved, around which orientation is less than perfect.

The general biology and raiding behavior of *F. subintegra,* an American member of the *sanguinea* group, have been studied by Wheeler (1910) and by Talbot and Kennedy (1940). The latter investigators, by keeping a chronicle over many summers of a population on Gibraltar Island, in Lake Erie, were able to show that raiding is much more frequent than in *sanguinea.* Some colonies raided almost daily for weeks at a time, striking out in any one of several directions on a given day. Occasionally the forays continued on into the night, in which case the *subintegra* workers remained in the looted nest overnight and returned home the next morning. In other details the raiding behavior resembled that of *sanguinea.* More recently, Regnier and I discovered that each *subintegra* worker possesses a grotesquely hypertrophied Dufour's gland, which contains approximately 700 μg of a mixture of decyl, dodecyl, and tetradecyl acetates. These substances are sprayed at the defending colonies during the slave raids. They act at least in part as "propaganda substances" because they evaporate slowly and help to alarm and to disperse the defending workers.

Little is known about the other nine or so American species of the *sanguinea* group (Creighton, 1950; Buren, 1968a), and their study is likely to reveal new behavioral phenomena related to dulosis. For example, a colony of *F. wheeleri* that I observed in Yellowstone Park, Wyoming (Wilson, 1955b), divided its labor in a remarkable fashion between two species of slaves. *F. neorufibarbis* accompanied the *wheeleri* on a raid (against colonies of *F. fusca* and *F. lasioides* simultaneously) and assisted them in excavating and breaking into the plundered nests. Later, when the mixed nest was excavated for closer examination, the *neorufibarbis* were found to be concentrated in the middle and upper layers. They were very aggressive and joined the *wheeleri* in defending the nest. The workers of the second slave species, *F. fusca,* did not accompany the slave-makers on the raid, and later they made only feeble attempts to defend the nest. Instead, they were found concentrated in the lower layers of the nest close to the brood, and most had their crops distended with liquid food. These circumstances suggest that the *fusca* workers were specializing on food storage and brood care. A deeper significance of the dulotic habit is indicated by this example. It is apparent that the slave-maker colony not only adds to its labor force quantitatively by taking slaves, but it can also incorporate specialists that increase the efficiency of the colony in a fashion analogous to that seen in normal worker polymorphism.

The mode of colony founding by queens of the *F. sanguinea* group has not been observed in nature, and this surprising gap in our information continues to prevent a secure understanding of the evolutionary origins of dulosis. Wheeler (1906) conducted a series of laboratory experiments on the American species *F. rubicunda* which strongly indicate that the queens can function at least facultatively as temporary parasites. When he placed newly dealated (but still virgin) *rubicunda* queens in nests containing workers and brood of *F. fusca,* they responded in an aggressive and effective manner. They advanced on the *fusca* colonies, fighting and killing *fusca* workers that attacked them, then seizing and sequestering the *fusca* pupae, until finally all of the *fusca* workers were dead and the *rubicunda* queens stood guard over the confiscated brood. When new *fusca* workers emerged from the brood pile at a later date, they accepted the *rubicunda* queens and began to lick and to feed them. Viehmeyer (1908) and Wasmann (1908) subsequently repeated Wheeler's experiment with young mated queens of *F. sanguinea* and obtained the same result. The behavior of the intruding queens differs markedly from those belonging to the *fusca* and *microgyna* groups used in parallel experiments. There is no reason to doubt that, at least under certain conditions, the *sanguinea*-group queens do start new colonies by this form of unaided assault on colonies of slave species. Wheeler, in his early writings, and later Santschi (1906) and Wasmann (1908), believed that such temporary parasitism not only characterized the ancestors of the slave-making *Formica* but was a prerequisite for the evolution of the dulotic habit itself. Together they postulated this explanation as an alternative to the Darwinian hypothesis, believing that, once predatory habits evolved in the queen during nest founding, it was far easier for the species to extend such behavior to the worker caste in the form of raiding for slaves. Later, Wheeler (1910) saw the incongruity in his position, namely, that dulosis represents a wholly new behavior pattern that cannot be viewed simply as a variant of the temporarily parasitic mode of colony founding. He concluded, "In my opinion both temporary parasitism and dulosis have arisen independently from the practice of *F. rufa* and *F. sanguinea* of adopting fertilized queens of their own species . . ." This opinion seems about right at the present time. The

Darwin hypothesis is not excluded by the demonstration of temporary social parasitism in the *sanguinea* group; it is, moreover, considerably strengthened by the growing evidence of raiding and accidental dulosis in *F. naefi* and *F. sanguinea,* as I pointed out earlier. We still know very little about whether the various species of the *sanguinea* group rely on temporary parasitism to start new colonies, as opposed to homospecific adoption followed by budding.

The pinnacle of the slave-holding way of life (or nadir if you prefer) is reached in the formicine genus *Polyergus,* a totally dulotic group of species that have evidently been phylogenetically derived from *Formica.* Three species are known: *rufescens* of Europe and North America, *lucidus* of North America, and *samurai* of Japan and eastern Siberia (see Figure 19-6). These "Amazon ants" are nowhere common, but their striking appearance (large size, bright red or black coloration, and shining body surface), the extraordinary degree of their behavioral specialization, and the spectacular qualities of their slave raids have placed them among the most frequently studied of all the ants (Huber, 1810; Forel, 1874; Wheeler, 1910; Emery, 1915; Creighton, 1950; Dobrzańska and Dobrzański, 1960; Gösswald and Kloft, 1960; Beck, 1961; Sakagami and Hayashida, 1962; Yasuno, 1964b; Köhler, 1966; Talbot, 1967; Marlin, 1968). As usual, no one has ever approached Wheeler's ability to distill the important information in the form of a gripping narrative, and I must again defer to him. In the following passage he describes *P. rufescens:*

> The worker is extremely pugnacious, and, like the female, may be readily distinguished from the other Camponotine [formicine—E. O. Wilson] ants by its sickle-shaped, toothless, but very minutely denticulate mandibles. Such mandibles are not adapted for digging in the earth or for handling thin-skinned larvae or pupae and moving them about in the narrow chambers of the nest, but are admirably fitted for piercing the armor of adult ants. We find therefore that the amazons never excavate nests nor care for their own young. They are even incapable of obtaining their own food, although they may lap up water or liquid food when this happens to come in contact with their short tongues. For the essentials of food, lodging and education they are wholly dependent on the slaves hatched from the worker cocoons that they have pillaged from alien colonies. Apart from these slaves they are quite unable to live, and hence are always found in mixed colonies inhabiting nests whose architecture throughout is that of the slave species. Thus the amazons display two contrasting sets of instincts. While in the home nest they sit about in stolid idleness or pass the long hours begging the slaves for food or cleaning themselves and burnishing their ruddy armor, but when outside the nest on one of their predatory expeditions they display a dazzling courage and capacity

FIGURE 19-6. Some principal events in the life of a worker of the slave-making ant *Polyergus rufescens:* (*a*) during a raid the worker attacks a resisting *Formica fusca* worker, piercing its head with its saber-shaped mandibles; (*b*) it carries a captured *fusca* pupa homeward; (*c*) it is fed by a *fusca* slave which has eclosed from a captured pupa (original drawing by Turid Hölldobler).

for concerted action compared with which the raids of *sanguinea* resemble the clumsy efforts of a lot of untrained militia. The amazons may, therefore, be said to represent a more specialized and perfected stage of dulosis than that of the sanguinary ants. In attaining to this stage, however, they have become irrevocably dependent and parasitic. Wasmann believes that *Polyergus* is actually descended from *F. sanguinea,* but it is more probable that both of these ants arose in pretertiary times from some common but now extinct ancestor. The normal slaves of the European amazons are the same as those reared by *sanguinea,* viz: *F. fusca, glebaria, rubescens, cinerea,* and *rufibarbis;* and of these *fusca* is the most frequent. But the ratio of the different components in the mixed nests is the reverse of that in *sanguinea* colonies, there being usually five to seven times as many slaves as amazon workers. The simultaneous occurrence of two kinds of slaves in a single nest is extremely rare, even when the same amazon colony pillages the nests of different forms of *fusca* during the same season. This is very probably the result of the slaves having a decided preference for rearing only the pupae of their own species or variety and eating any others that are brought in. Two slave forms may, however, appear in succession in the same nest. Near Morges, Switzerland, Professor Forel showed me an amazon colony which during the summer of 1904 contained only *rufibarbis* slaves, but during 1907 contained only *glebaria.*

Unlike *sanguinea, rufescens* makes many expeditions during July and August, but these expeditions are made only during the afternoon hours. One colony observed by Forel (1874) made 44 sorties on thirty afternoons between June 29 and August 18. It undoubtedly made many more which were not observed, as Forel was unable to visit the colony daily . . . Forel estimated the number of amazons in the colony at more than 1,000 and the total number of pupae captured at 29,300 (14,000 *fusca,* 13,000 *rufibarbis,* and 2,300 of unknown provenience, but probably *fusca*). The total number for the summer (1873) was estimated at 40,000. This number is certainly above the average, as the amazon colony was an unusually large one. Colonies with only 300 to 500 amazons are more frequent, but a third or half of the above number of pillaged cocoons shows what an influence the presence of a few colonies of these ants must have on the *Formica* colonies of their neighborhood. Of course, only a small proportion of the cocoons are reared. Many of them are undoubtedly injured by the sharp mandibles of the amazons and many are destroyed and eaten after they have been brought home.

The tactics of *Polyergus,* as I have said, are very different from those of *sanguinea.* The ants leave the nest very suddenly and assemble about the entrance if they are not, as sometimes happens, pulled back and restrained by their slaves. Then they move out in a compact column with feverish haste, sometimes, according to Forel, at the rate of a meter in 33 ⅓ seconds or 3 cm. per second. On reaching the nest to be pillaged, they do not hesitate like *sanguinea* but pour into it at once in a body, seize the brood, rush out again and make for home. When attacked by the slave species they pierce the heads or thoraces of their opponents and often kill them in considerable numbers. The return to the nest with the booty is usually made more leisurely and in less serried ranks.

The means by which the *Polyergus* workers are able to mobilize themselves within minutes and run in a compact column straight for the target colony has long been one of the classic problems of entomology. Very recently Mary Talbot (1967) seems to have solved it. While watching *Polyergus lucidus* colonies in Michigan, she noticed that, prior to the onset of each raid, several scout workers explored the surrounding terrain, including the vicinity of the specific nest later raided. By monitoring the *Polyergus* nests carefully, she saw that the beginning of the raid was often signaled by the appearance of a scout returning from the direction of the target nest. As she describes it, "On other days the departure of scouts was less conspicuous, and seldom was one lucky enough to spot a scout coming in. But whenever an ant came in hurriedly from the grass and went directly into the nest, there was an outpouring of ants. It was thus assumed that whenever a sudden emergence occurred it was in response to a messenger arriving with news of a located colony. If this was correct and if the scouting ant, which found a colony, laid down an odor trail on its way home, then the odor must have been quite long lasting, for it sometimes took an ant 30 to 45 minutes to return from a raided nest. It seemed unlikely that a raiding group could be following anything but an odor trail, for it moved rapidly, did not maintain leaders, and usually stopped at exactly the right place."

The next logical step was to try to induce false raids by means of the artificial trail test. Talbot accomplished this in a manner that decisively favored her hypothesis. When she laid down dichloromethane extracts of whole *Polyergus* bodies over the ground along an arbitrary path away from the nest and at the time of day raids normally occur, *Polyergus* workers poured from the nest and followed the trails to the end. Thus Talbot was able to activate the raid swarms at will and lead them to targets of her choosing. Finally, she induced a complete raid on a colony of *Formica nitidiventris* by placing it in a box two meters from a *Polyergus* colony and drawing an artificial *Polyergus* trail to it.

Emery (1911), by employing introduction experiments of the kind invented by Wheeler for studies of temporary parasitism, discovered that the queens of *Polyergus rufescens* act essentially like those of *Epimyrma* and certain other parasitic groups during colony founding. When presented with a colony of *Formica fusca* in the laboratory, the *rufescens* queen works her way into the nest by submissive posturing and secures adoption by the *fusca* workers and queen. Then, after a week or so has elapsed, she kills the *fusca* queen by piercing her head with her sharp mandibles. The frequency with which this mode of colony formation is used in nature is not known. It must occur at least occasionally since single dealated *P. rufescens* queens have been found alone in small *F. fusca* nests on at least two occasions. In addition, according to a recent report by Marlin (1968), colonies of *P. lucidus* can reproduce by budding. Some of the queens return to their home nests following the nuptial flights. Later they accompany workers on a raid and remain behind with a few of them in a plundered *Formica* nest or in some other neighboring nest site.

In 1932 K. V. Arnoldi reported the discovery of a new and equally spectacular kind of formicine slave-making ant. The species, *Rossomyrmex proformicarum,* superficially resembles *Polyergus,* but has evidently been derived from *Formica*-like ancestors in a line separate from that leading to the amazon genus. *R. proformicarum* is locally common in the *Artemisia-Festuca* steppe at several localities in southeastern Russia. It enslaves *Proformica nasuta,* an abundant dweller of xeric habitats which is closely related to the genus *Formica.* The method of raiding is unique. In one instance observed by Arnoldi, the *Rossomyrmex* emerged from their nest and formed into pairs as follows: one worker seized another by its mandibles, whereupon the second worker folded up its legs, tucked under its abdomen, and allowed itself to be carried in the typical formicine fashion. About fifty pairs set off in a loose file in this fashion. After traveling for some 50 m they halted, uncoupled, and wandered about, apparently searching without success for *Proformica* nests. The following day the same colony was encountered in the midst of a raid. When Arnoldi arrived on the scene, a few pairs were still outward bound, but the majority were already pillaging a *Proformica* nest and carrying home the captured pupae. The transporting behavior described in this case is basically the same as that used by *Formica* during colony movements from one nest site to another. The *Rossomyrmex* seem simply to have adapted it to a new function. An intermediate stage in the evolution of the *Rossomyrmex* habit is displayed by the slave-maker *Formica wheeleri,* whose workers have been observed carrying each other back to the home nest following a raid (Wilson, 1955b).

Even more remarkable in another sense is the existence of a phylogenetically independent form of dulosis in the myrmicine genera *Harpagoxenus, Leptothorax,* and *Strongylognathus.* The *Harpagoxenus* case is the most specialized and also by far the best understood. The North American species *H. americanus* has been closely studied by Sturtevant (1927), Creighton (1929), and especially Wesson (1939), whose analysis of the life cycle and behavior is a model of its kind. The *Harpagoxenus* workers are small, blackish brown ants superficially resembling some of the *Leptothorax* species they enslave. Their most distinguishing feature is the presence of "antennal scrobes"—long, deep pits along the sides of the head into which the antennae are folded for protection during the raids. *H. americanus* is a relatively widespread species, existing in very local but dense populations from Ontario south to Virginia and west to Ohio. It enslaves two of the commonest *Leptothorax* species of eastern North America, *L. curvispinosus* and *L. longispinosus.* In the populations studied by Wesson, the ratio of *H. americanus* to local *Leptothorax* colonies was about 1: 15. The mixed colonies contained up to 50 *Harpagoxenus* workers and 300 *Leptothorax* worker slaves. Most colonies were much smaller than this, the medians being about 6 *Harpagoxenus* and 30 *Leptothorax,* respectively.

Following the nuptial flight in early or middle July, the newly fecundated queen sheds her wings and crawls about on the ground or low vegetation in search of a *Leptothorax* colony. On encountering a nest, she begins quite literally to throw the *Leptothorax* adults out. As each worker approaches her in turn, she seizes it by an antenna, drags it out of the nest entrance, and flings it to one side. She avoids attacks on her own body by very rapid, shifting movements. After she has savaged the colony in this manner for a while, the *Leptothorax* queen and workers finally panic and desert the nest. The *Harpagoxenus* queen then appropriates the larvae and pupae left behind.

When *Leptothorax* workers later eclose from the pirated brood, they adopt the *Harpagoxenus* queen without hesitation, and soon afterward she begins to lay eggs. The *Harpagoxenus* workers that develop from these eggs are degenerate in behavior. They spend almost all of their time in the nest grooming each other and "loafing." They are fed with liquid food regurgitated to them by the *Leptothorax* slaves, who also assume the thankless tasks

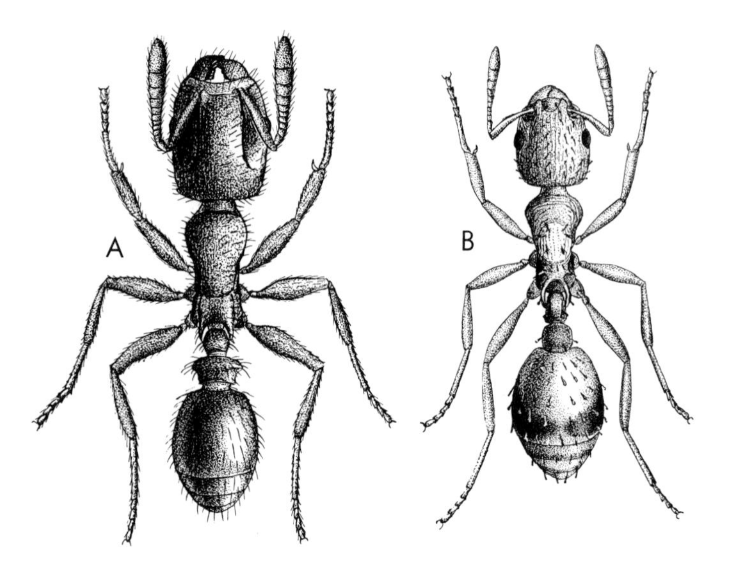

FIGURE 19-7. Workers of *Harpagoxenus americanus* (*A*); and of one of its slave species, *Leptothorax curvispinosus* (*B*) (from Wheeler, 1910).

of foraging, nest construction, and brood care. When the *Harpagoxenus* depart on the slave raids, on the other hand, they reveal themselves to be efficient little fighting machines. The raids are initiated by scouts, who hunt singly for *Leptothorax* colonies in the vicinity of the *Harpagoxenus* nest. When a *Harpagoxenus* scout encounters a *Leptothorax* nest, it normally attempts to penetrate the entrance without hesitation. If the colony is small and weak, it may succeed in capturing it single-handed, after which it begins to transport the *Leptothorax* brood back to its own nest. If, on the other hand, the scout is repulsed, it returns to its nest, excites its nestmates (evidently by release of a pheromone), and soon sets out again for the newly discovered *Leptothorax* nest. This time it lays down a short-lived odor trail which draws out a tight little column of *Harpagoxenus* workers and *Leptothorax* slaves. If this group is still not sufficient to breach the *Leptothorax* nest, some of the *Harpagoxenus* workers return to the home nest and bring out auxiliary columns.

Toward the end of the summer, the raids are transmuted into an unusual form of colony multiplication. An increasing tendency develops for some of the *Harpagoxenus* workers to remain behind in the conquered nests, where they stand guard over the *Leptothorax* brood. When this happens, the expatriates soon lose contact with the home nest, and they are treated as queens by the *Leptothorax* workers who subsequently eclose from the pupae. The eggs laid by the *Harpagoxenus* workers are apparently unfertilized and give rise mostly to males, along with a few workers. Such secondary colonies are very common and even rival in number the primary colonies started by single *Harpagoxenus* queens. Of 32 colonies censused by Wesson in Maryland, Ohio, and Pennsylvania, no less than 16 were populated exclusively by *Harpagoxenus* workers together with their slaves.

The European representative of the genus, *H. sublaevis,* has been studied by Adlerz (1896), Viehmeyer (1921), and Buschinger (1966, 1968). Its life cycle is very similar to that of *H. americanus,* except that some of the workers are inseminated and serve as the usual reproductives; morphologically complete queens are relatively scarce. A slave-making *Leptothorax, L. duloticus,* was discovered in Ohio by Wesson (1940b) and later found in Michigan as well by Talbot (1957). The rather fragmentary information available suggests that *L. duloticus* is basically similar to *Harpagoxenus* in its biology.

Finally, in the Palaearctic myrmicine genus *Strongylognathus* we are privileged to see the transition from dulosis to full inquilinism. The natural history of the genus has been gradually explored over a period of many years by Forel (1874, 1900), Wasmann (1891), Wheeler (1910), Kutter (1923, 1969), Pisarski (1966), and others. *Strongylognathus* is closely related to *Tetramorium,* and its species enslave members of the latter genus. The most favored slave species is *T. caespitum,* one of the most abundant and widespread ant species of Europe. *S. alpinus* has a life cycle more or less typical of the majority of *Strongylognathus* species. It is at an evolutionary level somewhat less advanced than that of *Harpagoxenus* in the one special sense that the behavior of its workers is less degenerate. The workers, like those of most parasitic ant species, do not forage for food or care for the immature stages; nevertheless, they still feed themselves and assist in nest construction. The raids of *alpinus* are notoriously difficult to observe. They occur in the middle of the night and take place, for the most part, along underground galleries. The *alpinus* workers are accompanied by *Tetramorium caespitum* slaves, who, true to the aggressive nature of their species, join in every phase of the raid. Warfare against the target colony is total: the nest queen and winged reproductives are killed, and all of the brood and surviving workers are carried back and incorporated into the mixed colony. This union of adults should not be too surprising when it is recalled that *T. caespitum* colonies, even in the absence of *Strongylognathus,* frequently conduct pitched battles that sometimes terminate in colony fusion. The *S. alpinus* workers are well equipped for lethal fighting. Like many other dulotic and parasitic ant species, they possess saber-shaped mandibles adapted for piercing the heads of their resisting victims (see Figure 19-8). The mode of colony multiplication is not known, but it is at least clear that the host queen is somehow eliminated in the process.

One member of the genus, *Strongylognathus testaceus,* has completed the transition to complete inquilinism. The *Tetramorium* queen is tolerated and lives side by side with the *S. testaceus* queen. There are fewer *testaceus* than host workers, the usual situation found in advanced dulotic species. The *testaceus* workers do not engage in ordinary household tasks and are wholly dependent on the host workers for their upkeep. But the key fact is that they also do not engage in slave raids. Somehow the reproductive ability of the host queen is curtailed. She generates only workers and no reproductives. Only the *S. testaceus* queen is privileged to produce both castes. Nevertheless, the presence of the *Tetramorium* queens permits the mixed colonies to attain great size. Wasmann found one comprised of between 15,000 and 20,000 *Tetramorium* workers and several thousand *Strongylognathus* workers. The brood consisted primarily of queen and male pupae of the inquiline species. It is evident that *S. testaceus* is in a stage of parasitic evolution just a step beyond that occupied by *S. alpinus.* The worker caste of *testaceus* has been retained, and it still has the murderous-looking mandibles dating from the species' dulotic past, but it has evidently lost all of its former functions and is in the process of being reduced in numbers. Probably *S. testaceus* is on the way to dropping the worker caste altogether, a final step that would take the species into the ranks of the extreme inquilines.

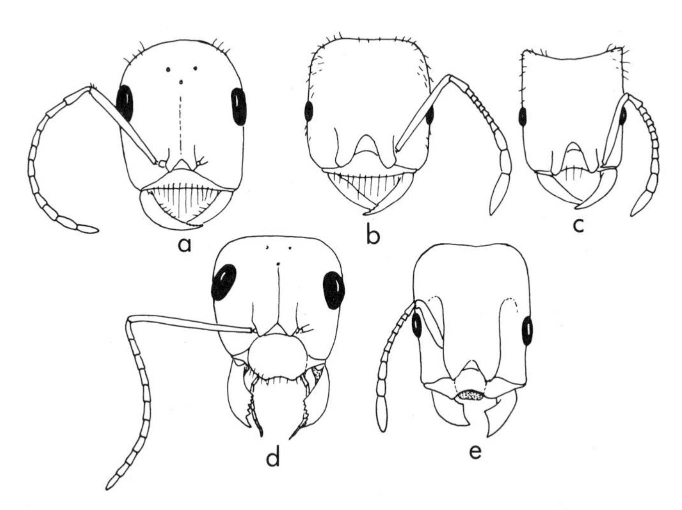

FIGURE 19-8. Heads of workers of five species of dulotic ants, showing varying degrees of modification of the mandibles for fighting during the slave raids: (*a*) *Polyergus rufescens,* (*b*) *Strongylognathus alpinus,* and (*c*) *Strongylognathus testaceus* have saber-shaped mandibles used to pierce the exoskeletons of their victims; (*e*) *Harpagoxenus sublaevis* has sharp, clipper-shaped mandibles used to nip and cut the appendages of opponents; (*d*) *Formica sanguinea,* a facultative slave-maker whose workers still carry on normal work loads in their own nests, has unmodified mandibles with a full set of teeth on the gripping edge (from Kutter, 1969).

Xenobiosis and Trophic Parasitism

The classic example of xenobiosis is the relation of the "shampoo ant" *Leptothorax provancheri* to its host *Myrmica brevinodis* as described by Wheeler (1903, 1910). Species of *Leptothorax* generally form small colonies that nest in tight little places, for example, inside hollow twigs lying on the ground, rotted acorns, or abandoned beetle galleries in the bark of trees. The workers forage singly, and, when they encounter other ants, they usually avoid them by moving in a stealthy, unobtrusive manner. Because of these traits, colonies of *Leptothorax* are often found close to the nests of larger ants, and their workers are able to forage freely among their large neighbors. The trend has been extrapolated into parasitism by *L. provancheri.* This species has been found living only in close association with colonies of *M. brevinodis.* (Wheeler erroneously referred to these two forms by their synonymic names *L. emersoni* and *M. canadensis,* respectively.) Both species occur widely through the northern United States and southern Canada. Colonies of *M. brevinodis* construct their nests in the soil, in clumps of moss, and under logs or stones, especially in wet meadows and bogs. Smaller *L. provancheri* colonies excavate their nests near the surface of the soil and join them to the host nests by means of short galleries. They keep their broods strictly apart. The *Myrmica* are too large to enter the narrow *Leptothorax* galleries, but the *Leptothorax* move freely through the nests of their hosts. The *Leptothorax* workers do not forage for their own food. They depend almost entirely on crop liquid obtained from the host workers, using begging movements to induce the incoming *Myrmica* foragers to regurgitate to them. They also mount the *Myrmica* adults and lick them in what Wheeler has described as "a kind of feverish excitement," to which the hosts respond with "the greatest consideration and affection." Wheeler was under the impression that the *Myrmica* sought the *Leptothorax* in order to obtain a "shampoo," and he believed at first that the relationship might be mutually beneficial. Later (1910) he conceded that the *Leptothorax* are probably no more than parasites. They are, nevertheless, far from being helplessly dependent on *Myrmica brevinodis.* Not only do they construct their own nests and rear their own brood, but they are also able to feed themselves, albeit awkwardly, when isolated in artificial nests in the laboratory.

Formicoxenus nitidulus is a northern European species with habits closely similar to *Leptothorax provancheri.* This reddish little ant closely resembles *Leptothorax* and may have been derived from it in evolution. It is specialized for life inside the large mound nests of *Formica rufa* and *F. pratensis.* It has also been found occasionally in nests of other *Formica* species, including *F. exsecta* and *F. fusca,* and even *Polyergus rufescens* and *Tetramorium caespitum.* The relation of *Formicoxenus* to its hosts has been studied over a period of many years by Forel (1874, 1886), Adlerz (1884), Wasmann (1891), Janet (1897a), Wheeler (1910), Stäger (1925), and Stumper (1949, 1950). The colonies, which contain 100-500 workers and multiple queens, nest exclusively within the walls of the host nests. They excavate their own chambers and keep their brood strictly segregated. Like the *Leptothorax* shampoo ants, they build narrow galleries that open directly into the interior of the *Formica* nests, and from these they periodically emerge to forage among the host workers. But, unlike the *Leptothorax,* they do not lick their hosts. In fact, it has been difficult to observe interactions of any kind between the *Formicoxenus* and the *Formica.* Although Stäger reported regurgitation from *Formica* workers to *Formicoxenus,* Stumper has concluded that this must be uncommon since most of the time the *Formicoxenus* workers appear to keep strictly to themselves. *F. nitidulus* nevertheless displays at least two remarkable adaptations to its commensal existence. First, the males are wingless and highly worker-like in appearance (Figure 19-9). They

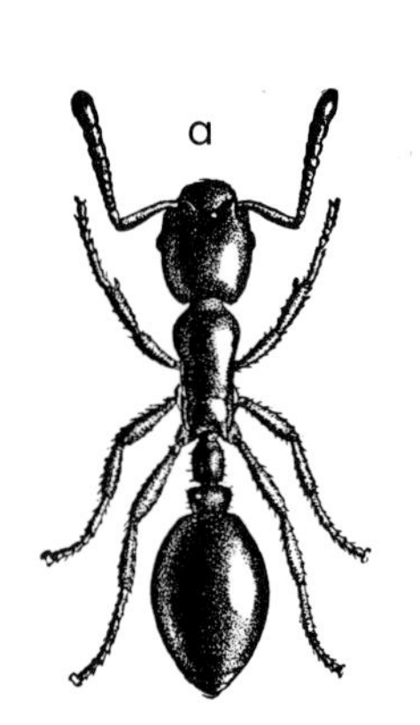

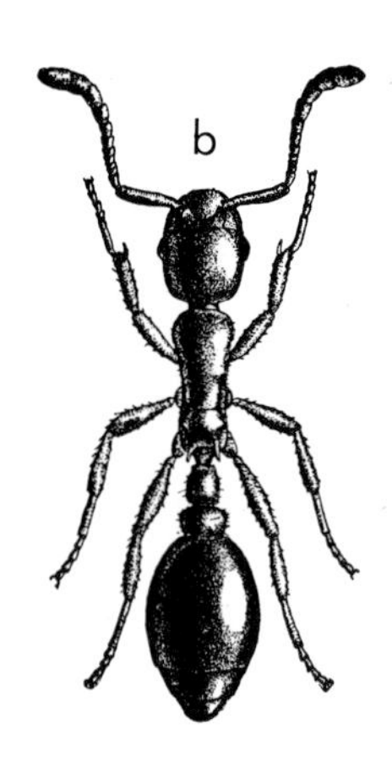

FIGURE 19-9. *Formicoxenus nitidulus,* a European myrmicine ant that lives xenobiotically in nests of *Formica:* (*a*) worker; (*b*) ergatomorphic male (from Wheeler, 1910).

can be distinguished externally only by their longer antennae, which contain one more segment than those of the worker, by an additional abdominal segment, and of course by the extrusible portions of their genitalia. The matings take place on top of the host nests. The second adaptation to life with *Formica* is the ability to emigrate in the columns of the host workers when the latter change nest sites. It has not been determined whether the *Formicoxenus* follow *Formica* trail pheromones in such cases or whether they orient visually with their rather well-developed eyes.

In 1925 Wheeler reported the discovery of a new and thoroughly surprising case of xenobiosis between a myrmicine guest-ant, *Megalomyrmex symmetochus,* and a fungus-growing host species *Sericomyrmex amabilis,* also a myrmicine. The *Sericomyrmex* found modest-sized colonies, comprised of 100–300 workers and a queen, that nest in the wet soil of the laboratory clearing on Barro Colorado Island, Panama. They subsist entirely on a special fungus raised on beds of dead vegetable material. The *Megalomyrmex* form smaller colonies, consisting of 75 adults or less, that live directly among the fungus gardens of the host. Since the *Sericomyrmex* also place their brood in the gardens, the young of both species become mixed to a limited extent. However, the *Megalomyrmex* tend to segregate their brood in little clumps, each of which is closely attended by a few workers, and neither species feeds or licks the brood of the other. The most remarkable fact is that the *Megalomyrmex* appear to subsist exclusively on the fungus. This represents a major dietary shift which must have occurred relatively recently in the evolution of the genus. Because liquid food exchange is either uncommon or completely lacking in fungus-growing ants, the *Megalomyrmex* do not secure nutriment from the *Sericomyrmex* in this way. They do, however, lick the body surfaces of their hosts.

An important common feature of the three examples of xenobiosis just cited is what German writers call *Futterparasitismus,* which can perhaps best be translated into the rather formal expression "trophic parasitism." This is intrusion into the social system of another species in order to steal food. Trophic parasitism does not by itself require a close association of nests or even entry into the host nest by foraging workers. In other words, it can occur apart from xenobiosis. A weak, nonxenobiotic form of such parasitism is exhibited by *Camponotus lateralis* toward *Crematogaster scutellaris* in Europe. Goetsch (1953b) and Kaudewitz (1955) have described instances in which *Camponotus* workers followed the *Crematogaster* odor trails to their feeding grounds and exploited the same food resources during the same time of day. The *Crematogaster* were hostile to the *Camponotus,* which assumed a crouching, conciliatory posture when they met the legitimate users of the trails. Unlike the xenobionts, the *C. lateralis* nest separately. Moreover, the relationship is not obligatory on *lateralis,* since colonies and foraging workers of that species are often found far from *Crematogaster* nests.

On the island of Trinidad, in the West Indies, I discovered an instance of trail sharing that approaches a neutral, or commensalistic relationship (Wilson, 1965b). Each of the several colonies of the formicine *Camponotus beebei* encountered were in close association with a large colony of the dolichoderine *Azteca chartifex,* one of the dominant ant species of the island forests. The *Camponotus* nested in cavities in tree branches near the arboreal carton nests of the *Azteca,* and their workers followed the *Azteca* odor trails down the branches and tree trunks to foraging areas on the nearby ground and weedy vegetation. The diets of the two species were not determined, but, regardless of the degree of similarity, potential interference between the two species was reduced by the existence of opposite daily schedules. The *Camponotus* therefore "borrowed" the *Azteca* trails when the owners were putting them to minimal use. The *Azteca* workers were hostile to the *Camponotus* workers and attacked them on the rare occasions when the latter slowed in their running, but the *Camponotus* were larger and faster and usually easily avoided their hosts without causing any visible disturbance. On a single occasion, a *Camponotus* worker was seen to lead out a tight column of six other *Camponotus* workers, guiding them along by means of a short-lived odor trail. The pheromone was laid directly on top of the *Azteca* odor trail, yet the *Azteca* workers did not fall in line or show any other response to the passage of the *Camponotus* group. Thus it appears that the *Camponotus,* while "eavesdropping" on the *Azteca* odor trails, have reserved a special recruitment trail system of their own which they do not share with their hosts.

An interesting evolutionary question can now be raised: do xenobiosis or the more tenuous forms of trail parasitism ever lead to full inquilinism? The evidence to look for is the coexistence in the same genus of xenobiotic species and fully inquiline species, both of which parasitize other species belonging to the same genus. This is the criterion, it will be recalled, by which inquilinism in *Strongylognathus testaceus* was inferred to be of dulotic origin. So far, no such examples have been found. Even so, some of the traits of *Kyidris,* an inquilinous genus which will be described in a moment, at least suggest the possibility of a xenobiotic origin. Also, *Sifolinia* and *Symmyrmica* are genera related to *Pheidole* or *Leptothorax* that live in an undefined form of symbiosis with *Myrmica* and *Manica* respectively. Should subsequent research reveal them to be inquilines rather than xenobionts, the case for a xenobiotic origin would be strong, especially in view of the fact that both genera are morphologically very close to the xenobionts *Leptothorax provancheri* and *Formicoxenus nitidulus* and share related hosts. Unfortunately, *Sifolinia* and *Symmyrmica* are among the rarest and least known of all ant genera. Their rediscovery and deeper study might prove to have special significance.

Parabiosis

Forel (1898) designated as "parabiosis" the following complex behavior which he discovered in Colombia. Colonies of the arboreal, rain forest ant species *Crematogaster limata parabiotica* and *Monacis debilis* (called *Dolichoderus debilis* var. *parabiotica* by Forel) commonly nest in close association, with the nest chambers kept separate but connected by passable openings. Also, the workers of the two species run together along common odor trails. Wheeler (1921c) confirmed the phenomenon in his

Guyana studies and showed that the two species collect honeydew together from membracids. Wheeler also discovered a similar association between *Crematogaster limata parabiotica* and *Camponotus femoratus.* Both species were observed utilizing common trails and gathering honeydew from jassids and membracids on the same plants, as well as nectar from the same extrafloral nectaries of *Inga.* Not only were the *Crematogaster* and *Camponotus* workers tolerant of each other in this potentially competitive situation, but they were also on quite intimate terms. They "greeted" each other with calm antennation on the trails, and on three occasions Wheeler observed *Camponotus* workers actually regurgitating to individuals of *Crematogaster.*

It has not yet been established whether these examples of parabiosis are mutualistic or parasitic in nature. If parasitic, then parabiosis can of course scarcely be distinguished from xenobiosis; the only differences could be in certain trivial relationships among the workers of the participating species while inside their nests. And at best the distinction between mutualism and parasitism must be a subtle one in such a complicated relationship. The form "*parabiotica*" of *Crematogaster limata* is evidently always associated with other ants. If future taxonomic studies prove it to be a species distinct from the typical *limata,* it is more likely also to be a parasite, for it would then be revealed to be dependent on its associates, while colonies of the "typical" species (*limata*) nest and forage by themselves. But the prima facie case for mutualism seems even stronger at this time. The broods are never mixed, and, as Weber (1943) has pointed out on the basis of his own studies, all of the parabiotic species participate vigorously together in nest defense. There is no evidence that the presence of the *Crematogaster* harms the other species, except possibly by competition for the same food resources. On the contrary, *Camponotus femoratus* maintains flourishing populations in localities where virtually every colony lives in parabiosis with *Crematogaster.*

The Degrees of Inquilinism

Once an ant species enters complete inquilinism, whether by temporary parasitism, dulosis, or xenobiosis, it seems to evolve quickly on down into a state of abject dependence on its host. It acquires some of what can be termed the "inquiline syndrome," a set of characteristics found in varying combinations in all of the relatively specialized inquiline species:

1. The worker caste is lost.
2. The queen is either replaced by an ergatogyne, or ergatogynes appear together with a continuous series of intergrades connecting them morphologically to the queens.
3. There is a tendency for multiple egg-laying queens to coexist in the same host nest.
4. The queen and male are reduced in size, often dramatically so; in some cases (for example, *Teleutomyrmex schneideri, Aporomyrmex ampeloni, Plagiolepis xene*) the queen is actually smaller than the host worker.
5. The male becomes "pupoid": its body is thickened, the petiole and postpetiole become much more broadly attached, the genitalia become more exserted, the cuticle becomes thin and depigmented, and the wings are reduced or lost. The extreme examples of this trend are displayed by *Anergates atratulus, Anergatides kohli,* and *Bruchomyrma acutidens* (see Figure 19-10).
6. There is a tendency for the nuptial flights to be curtailed, and to be replaced by mating activity among nestmates ("adelphogamy") within or near the host nest. Dispersal of the queen afterward is very limited.
7. Probably as a consequence of the curtailment of the nuptial flight just cited, the populations of inquiline species are usually very fragmented and limited in their geographic distribution.
8. The wing venation is reduced.
9. Mouthparts are reduced, with the mandibles becoming smaller and toothless and the palps losing segments. Concomitantly, the inquilines lose the ability to feed themselves and must be sustained by liquid food regurgitated to them by the host workers.
10. Antennal segments are fused and reduced in number.
11. The occiput, or rear portion of the head, of the queen is narrowed.
12. The central nervous system is reduced in size and complexity, usually through reduction of associative centers.
13. The petiole and postpetiole are thickened, especially

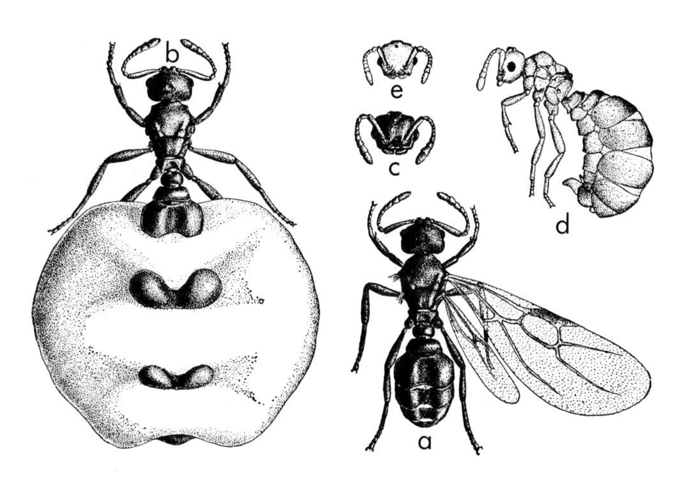

FIGURE 19-10. *Anergates atratulus,* an extreme workerless ant parasite that lives with *Tetramorium caespitum* in Europe: (*a*) virgin queen; (*b*) old queen with a typically physogastric abdomen; (*c*) head of queen, showing the reduced mandibles and narrowed posterior half of the head characteristic of extreme inquilines; (*d, e*) male, a pupoid form of the kind also found in some other extreme inquilinous species (from Wheeler, 1910).

the latter, and the postpetiole acquires a broader attachment to the gaster.

14. A spine is formed on the lower surface of the postpetiole (the *Parasitendorn* of Kutter).
15. The propodeal spines (if present in the ancestral species) "melt," that is, they thicken and often grow shorter, and their tips are blunted.
16. The cuticular sculpturing is reduced or lost altogether over most of the body; in extreme cases the body surface becomes strongly shining.
17. The exoskeleton becomes thinner and less pigmented.
18. Many of the exocrine glands are reduced or lost, a trait already described in some detail in the earlier account of *Teleutomyrmex schneideri.*
19. The queens become highly attractive to the host workers, which lick them frequently. This is especially true of the older, physogastric individuals, and it appears to be due to the secretion of special attractant substances which are as yet chemically unidentified.

One of the most interesting of these trends—namely, reduction of the worker caste—is examined more closely here. The inquiline species *Kyidris yaleogyna* of New Guinea represents the most primitive known level in this evolutionary regression since it retains an abundant and partly functional worker caste (Wilson and Brown, 1956). Colonies of *K. yaleogyna* are parasitic on *Strumigenys loriae,* one of the more abundant and ecologically widespread of the Papuan ants. Both *Kyidris* and *Strumigenys* are members of the tribe Dacetini of the subfamily Myrmicinae, but they are otherwise very different from one another. *Kyidris* is a short-mandibulate form, closer to *Smithistruma* and *Serrastruma* than to the highly distinctive, long-mandibulate *Strumigenys.* Four mixed colonies were discovered nesting in pieces of decaying wood on the floor of rain forests. In each, the *Strumigenys* slightly outnumbered the parasites. One large colony collected *in toto* contained 1,622 workers and 16 dealated queens of *Strumigenys,* in combination with 1,170 workers, 4 dealated queens, 84 alate queens, and 51 males of *Kyidris.* A second colony contained 243 workers and 4 dealated queens of *Strumigenys* and 64 workers, 2 dealated queens, and 31 males of *Kyidris.* These large groups lived in completely harmonious mixtures in which the parasitic nature of the *Kyidris* was only subtly evident. The *Kyidris* workers foraged for food. One group was found attending coccids near the nest. Others engaged in hunting for small insects, but, compared with the *Strumigenys,* they were quite ineffectual. They wandered through the food chambers of the artificial nests like typical restless dacetines, but rarely tried to catch prey. Even when they tried, they usually failed, in sharp contrast to the highly efficient performances of their *Strumigenys* nestmates. One *Kyidris* worker was seen to seize a symphylan, pull it backwards, hold it for about thirty seconds without trying to sting it, and finally release it when it began to struggle. Another seized an entomobryid collembolan, pulled it back vigorously, then lost it when the entomobryan kicked with its furcula. Still another was seen actually to carry an entomobryid at a brisk clip across the food chamber floor; it reached the entrance to the brood chamber only to have a *Strumigenys* meet it and take the insect away. The general impression we gained was that the predatory behavior has regressed, but not completely disappeared, in *Kyidris* workers. In the artificial nests, at least, the *Strumigenys* workers did most of the productive hunting. The *Kyidris* we studied also aided in brood care much less frequently than the *Strumigenys,* and their efforts seemed ineffectual. *Kyidris* workers were never observed in nest construction. They received regurgitated liquid food from

Table 19-2. Characteristics of a host ant species (*Plagiolepis pygmaea*) and two closely related species that live as social parasites with it (based on data from Le Masne, 1956b).

Characteristic	*P. pygmaea* (host)	*P. grassei* (intermediate parasite)	*P. xene* (extreme parasite)
Length of queen (mm)	3.4–4.5	2.0–2.4	1.2–1.3
Development of queen wings	Normal	Normal	Variable, often rudimentary
Condition of worker caste	Normal, abundant; appears before sexual forms	Rare, only one to every 10 fecundated queens; appears after sexual forms	Absent
Condition of male	Normal, with functional wings	Slightly female-like; wings occasionally rudimentary	Very female-like; wings always rudimentary

the *Strumigenys* workers, and sometimes obtained food by inserting their mouthparts between those of two *Strumigenys* exchanging regurgitated material, but they were never seen to offer anything in return. In sum, the New Guinea *Kyidris* appear to represent inquilinism at a very early stage when the worker caste has only begun to reduce its behavioral repertory. Probably the degeneration has proceeded past the point of no return since it is doubtful if *Kyidris* colonies could survive without their hosts.

A somewhat more advanced stage of behavioral decay is shown by the workers of *Strongylognathus huberi,* a European species already mentioned in the previous section on dulosis. The workers are able to conduct raids, they participate in nest building, and they can feed themselves. But, unlike the *Kyidris* workers, they have lost the capacity either to hunt or to care for brood. In the great majority of all other dulotic and inquilinous ant species that still possess a worker caste, the workers appear to have entirely lost the ability to carry on the ordinary functions of nest construction, food gathering, and queen and brood care.

It is little wonder, then, that most truly inquilinous species have taken still one more step in evolution and discarded the worker caste altogether. The stages leading to this final abrogation have been beautifully documented in the genus *Plagiolepis* by Le Masne (1956b) and Passera (1966, 1968). The two species *P. grassei* and *P. xene* are parasitic on the closely related free-living form *P. pygmaea.* In certain key characteristics, namely loss of worker caste, size reduction, and alteration of the male form, *xene* qualifies as an extreme inquiline, while *grassei* occupies an almost exactly intermediate position between it and the free-living *Plagiolepis* (see Table 19-2). The most interesting annectant feature of *grassei* is in the status of the workers. This caste is almost extinct, and it appears in a given host nest only after the winged parasitic sexuals are produced—the reverse of the order that is universal in free-living ant species.

The majority of inquiline species for which adequate information is available permit the host queens to live. This is the situation one would intuitively expect. It seems to make good sense for the parasites to insure themselves a long-lasting supply of host workers. This reasoning is also consistent with the fact that species of *Strongylognathus* which obtain host workers by slave raids also destroy the host queens, while *S. testaceus,* the one species of the genus that does not obtain host workers in this fashion, tolerates the host queens. But what of the minority of inquilines whose presence causes the death of the host queens? We have seen that the queen-killing *Epimyrma* are not necessarily inconvenienced by their act, for the *Leptothorax* workers commence laying and produce other workers. This is not the case, however, for *Anergates atratulus* (Figure 19-10), whose presence evidently induces the *Tetramorium* workers to destroy their own queen. Because *A. atratulus* is such an advanced inquiline, the trait cannot be dismissed as an inconvenient holdover from a dulotic (or temporarily parasitic) past in the unknown *Anergates* ancestor. It is much more plausible that *A. atratulus* has simply adopted a different reproductive strategy from that of other inquilinous species. For some species, in which either the host colonies or the parasites themselves are relatively short-lived, it may be advantageous to get rid of the host queen and invest all of the efforts of the host workers into producing as many parasite queens and males as soon as possible. This is the "big bang" strategy of reproduction, essentially the same as that employed by such fishes as the migratory eels and salmon and such plants as the bamboos (Gadgil and Bossert, 1970). Other parasitic ant species, with longer lives and host colonies that are more stable, would find it advantageous to let the host queens live and to employ the host colony in the production of parasite queens and males at a lower rate—but for a longer period of time. We should bear in mind that even the continuous reproducers inhibit host reproduction to some degree. In general, the production of host sexual forms in the presence of inquilines is a rare event even in those cases where the host queen is permitted to live. Furthermore, Passera (1966) has recently discovered that, in *Plagiolepis pygmaea,* even the ability to produce workers is partially inhibited by the presence of parasitic *P. xene* queens. The theoretical implications raised by these various observations can be best formalized in the following conjecture: *The degree of reproductive repression inflicted by a given inquiline species is such as to maximize the total production of parasite queens and males per host colony under the particular ecological conditions in which the mixed colonies occur.* In order to test this hypothesis and more generally to advance a population theory of social parasitism, data on the population dynamics of both parasitized and unparasitized host colonies is needed.

Social Parasitism in Wasps

There are three basic ways in which one species of aculeate wasp can parasitize another. The first is by direct "parasitoid" attack, wherein the female simply stings the prey into a state of paralysis, lays an egg on it, and departs. When the young larva hatches, it feeds at will on the bulky carcass. The parasitoid habit can be regarded as no more than a slightly modified form of predation since the host is invariably killed in the end, however slow the process. Aculeate parasitoids, such as the species of Mutillidae that prey on bees, have a life cycle similar in essentials to the Terebrantia, or "true" parasitoid wasps (Clausen, 1940; Ferguson, 1962).

The second form of parasitism practiced by some aculeate species is cleptoparasitism, or "cuckoo" reproduction. The German expression for the habit, *Arbeitsparasitismus,* describes it exactly, for in this case the female seeks out a prey already captured and stored by a host female and appropriates it for her own use. Various aspects of the phenomenon have been documented by Wheeler (1919b), Crèvecoeur (1931), Wolf (1951), Evans *et al.* (1953), Olberg (1959), and Gillaspy (1963). Cleptoparasitism is not very common in wasps. It is known so far only from several genera in the Pompilidae (= Psammocharidae) and bembicine and gorytine Sphecidae. The life cycle of the bembicine *Stizoides unicinctus,* studied by the Raus and F. X. Williams and reviewed by Wheeler, is typical. The female digs her way into the nest of a female of *Chlorion thomae,* a member of the Sphecidae, after the latter individual has provisioned it with a cricket, closed the entrance, and departed. The *Stizoides* eats the *Chlorion* egg and lays her own. Members of the closely related genus *Stizus* have radically different habits. They are free living and typical "digger wasps" in their behavior. The females excavate their own burrows in the sand, glue a single egg in the terminal burrow, and progressively feed the hatching larva with crickets or hemipterans (Evans, 1966). As a rule, cleptoparasitic aculeates prey on only one or a very few species of other aculeates belonging to a different genus or even a different family.

The third form of parasitism found in the aculeate wasps is social parasitism, in which the parasitic female intrudes into the nest of another, social species and usurps the position of the queen. The phenomenon appears to be at least as common in the social wasps as in the ants (see Table 19-3). Two of the 16 species of *Vespula* in North America are permanent parasites, and a third is a facultative temporary parasite. Twenty-three, or 10 percent, of the 225 known species of *Mischocyttarus* are permanent parasites, as are 3, and possibly more, of the 130 known *Polistes* species. Social parasitism is, nevertheless, not nearly as diversified as in ants. Where the ants follow at least three basically different kinds of intermediate parasitism (temporary parasitism, dulosis, xenobiosis) that can serve as stepping-stones on the evolutionary road to inquilinism, the wasps have evidently followed only one—temporary parasitism. Furthermore, within each of the evolutionary stages arrayed along this single evolutionary pathway, the wasps show very little behavioral variation.

Taylor (1939), generalizing from his observations on *Vespula squamosa,* was the first to propose an explicit

TABLE 19-3. The socially parasitic aculeate wasps.

Parasite	Hosts	Form of parasitism	Range	Authority
Vespidae, subfamily Vespinae				
Vespa dybowskii	*Vespa crabro, V. xanthoptera*	Facultative temporary parasitism	Siberia, temperate Asia	Sakagami and Fukushima (1957b)
Vespula (*V.*) *austriaca*	*Vespula* (*V.*) *rufa*	Inquilinism (obligatory, permanent parasitism)	Europe, North America	Weyrauch (1937, 1938); de Beaumont (1958)
Vespula (*V.*) *squamosa*	*Vespula* (*V.*) *vidua*	Facultative temporary parasitism	Eastern US, Texas, Mexico, Central America	Taylor (1939); Miller (1961)
Vespula (*Dolichovespula*) *adulterina*	*Vespula* (*Dolichovespula*) *saxonica, V.* (*D.*) *norwegica*	Inquilinism	Europe	Weyrauch (1937, 1938); de Beaumont (1958)
Vespula (*Dolichovespula*) *omissa*	*Vespula* (*Dolichovespula*) *silvestris*	Inquilinism	Europe	Weyrauch (1937, 1938); de Beaumont (1958)
Vespula (*Dolichovespula*) *arctica* [= *V. adulterina?*]	*Vespula* (*Dolichovespula*) *arenaria*	Inquilinism	North America	Taylor (1939); Miller (1961)
Vespidae, subfamily Polistinae				
Polistes atrimandibularis	*P. bimaculatus, P. omissus*	Inquilinism	Europe	Weyrauch (1937, 1938); Scheven (1958)
Polistes semenowi	*P. gallicus, P. nimpha*	Inquilinism	Europe, North Africa	Weyrauch (1937, 1938); Scheven (1958)
Polistes sulcifer	*P. gallicus*	Inquilinism	Europe	Weyrauch (1937, 1938); Scheven (1958)
Polistes macrocephalus	?	Inquilinism (?)	Africa	Bequaert (1940)
Polistes perplexus	?	Inquilinism (?)	Southwestern US	Bequaert (1940)
Polistes sp.	*P. aterrimus*	Inquilinism (?)	Brazil	Rodrigues (1968)
Mischocyttarus flavitarsoides	*M. flavitarsus*	Inquilinism	Western US	Zikán (1949)
Mischocyttarus immarginatoides	*M. immarginatus*	Inquilinism	Central America or Mexico	Zikán (1949)
Mischocyttarus: 21 spp.	*Mischocyttarus* spp.	Inquilinism	Tropical and subtropical Brazil, Paraguay, Peru	Zikán (1949)

evolutionary scheme for the origin of inquilinism in wasps. He suggested that queens belonging to species prone to parasitism first tend to usurp the reproductive position in alien colonies of their own species. Then, with further evolution, they extend this behavior to colonies of other, closely related species. Eventually they come to depend on interspecific parasitism altogether, and, finally, the worker caste is dropped. This is the same sequence believed to have been followed by many phyletic lines of parasitic ants which evolved along the temporary parasitism route. Since Taylor first postulated this scheme, his categories have become well enough documented in the wasps, to the exclusion of other conceivable categories, that they lend considerable strength to the hypothesis. The known cases, arranged according to Taylor's sequence of increasing specialization, are:

1. *Facultative, temporary parasitism within species.* It is generally true that colonies of polistine wasps are territorial and aggressively organized. Moreover, the nests of *Polistes* are frequently founded by two or more queens, who contend for the alpha position. The alpha individual, or "queen" in this purely behavioral sense, dominates the brood cells and lays most or all of the eggs, while the other females are forced to assume the role of workers. Yoshikawa (1955) cited a case in which a *Polistes fadwigae* female arrived at a well-developed nest and usurped the alpha position. Later (1963d) he obtained the same result by detaching *P. fadwigae* nests and suspending them within 5 cm of each other. Under this abnormal circumstance the alpha individuals consistently contended with each other until only one was in a clearly dominant position. Under more natural conditions, when the nest of a foundress *P. fadwigae* female is destroyed, she sometimes tries to usurp the nest of another, more successful female of the same species. The Vespinae display similar tendencies. T. Shida, in a lecture cited by Sakagami and Fukushima (1957b), reported that overwintered queens of *Vespula lewisii* show a strong interest in old nests of the same species. Very likely this brings them into contact with other queens and established colonies. Janet (1903) described a case in which a queen of *Vespa crabro* penetrated an alien nest of her own species, deposed the original queen, and secured adoption by the workers.

2. *Facultative, temporary parasitism between species.* In their study of the life cycle of *Vespa dybowskii,* Sakagami and Fukushima (1957b) found that overwintered queens of this temperate Oriental species are able to start their own colonies independently, but they prefer to enter small established colonies of *Vespa crabro* and usurp the position of the mother queen. Sometimes the queens parasitize colonies of a second free-living species, *V. xanthoptera,* in the same fashion. The parasitism is facilitated by the fact that *V. dybowskii* emerges from hibernation later each spring than the host species, so that vulnerable young host colonies are present in large numbers when the *dybowskii* queens begin searching for a nest site. The usurpation process has not been directly observed, but the confrontation is evidently violent in nature, and the host queen does not survive. The first offspring of the *dybowskii* queen are workers, and they join in the usual chores side by side with the host workers. By the end of the summer the last of the host workers have died of natural causes, and the colony then consists entirely of *dybowskii* workers and the newly emerged *dybowskii* males and virgin queens. The mixed colonies are striking in appearance, because the abdomens of the host workers are brightly banded while those of the *dybowskii* workers are solid black. According to Sakagami and Fukushima, rural Japanese people are aware of the existence of mixed colonies and of the gradual change in the color pattern to a predominantly *dybowskii* type as the season progresses. Taylor (1939), on the basis of a single record of a *Vespula squamosa* queen in a *V. vidua* colony in West Virginia, postulated that *squamosa* is also a facultative temporary parasite. It is perhaps significant that *squamosa* queens emerge from hibernation later than those of *vidua.* But no further conclusions about this case can be drawn until other examples are studied.

3. *Obligatory, temporary parasitism between species.* This stage, so common in the ants, has not yet been documented in the social wasps, even though it seems to be a probable step in the progression to inquilinism.

4. *Obligatory, permanent parasitism between species (inquilinism).* The known examples are listed in Table 19-3. Thanks to the work of de Beaumont and Matthey (1945) and of Scheven (1958) we have a clear picture of this final evolutionary stage of parasitism in the paper-wasp genus *Polistes.* The three parasitic species of Europe, *atrimandibularis, semenowi,* and *sulcifer,* are workerless, and the queens have completely lost the ability to build nests or to care for the young. The degeneration in the behavior of the parasites is even more notable in that the queens of the host species are fully capable of performing such tasks long after they have a worker force to support

FIGURE 19-11. Female heads of *Polistes gallicus* (*left*) and its social parasite *P. sulcifer* (*right*). The parasitic species differs in minor details of head shape and coloration as well as in its more recurved mandibles, a general characteristic of the parasitic members of this genus (from Scheven, 1958).

them. The parasite queens nevertheless closely resemble the host queens in other ways, differing only in minor anatomical features such as coloration and the more recurved condition of the mandibles (Figure 19-11).

At this point it is appropriate to bring up a matter of taxonomic procedure. Some specialists have followed the practice of splitting off the parasitic species of *Polistes* as a distinct genus bearing the name *Sulcopolistes* or *Pseudopolistes,* but this seems to me to be of dubious value. Quite apart from the fact that the morphological differences are minor, there is every probability that workerless inquilinism has been derived independently more than once in *Polistes.* In normal taxonomic procedure a genus is defined as a monophyletic set of similar species set off from other genera by stronger characteristics than those ordinarily used to distinguish species. By either criterion *Sulcopolistes* (= *Pseudopolistes*) does not seem to qualify.

The relatively slight differences in head structure that separate the parasitic species of *Polistes* from their hosts is reflected in the manner in which the queens gain control of the host nests. They simply impose themselves as the top-ranking (alpha) female, in a manner that does not differ significantly from the tactics used by high-ranking host females themselves. The parasite approaches her chief rival and antennates her vigorously while attempting to climb on top of her. Then, if necessary, she bites and stings the host queen into submission. Only minor differences among the three species have been observed in the pattern of movements employed in the attack. The host capitulates by crouching down, pulling in her antennae, and allowing herself to be mounted. During subsequent liquid food exchanges, the parasite always receives rather than donates. She seems to be able to attain her dominant position by greater physical strength and staying power. But, unlike many of the parasitic ants and apparently some of the vespine wasps as well, she does not win by virtue of superior mandibular weaponry; nor does she normally harm the host females physically. On the contrary, the host queens of colonies conquered by *P. atrimandibularis* and *P. semenowi* are permitted to remain on the nest near the parasite queen, albeit at a subordinate rank in the role of workers. The queens of colonies conquered by *P. sulcifer,* on the other hand, always disappear. But even in this case they seem to leave willingly and only rarely are put to death. The parasite invasions usually occur in the first half of June, when the nests are still small, and in many instances prior to the eclosion of the first host workers. This would seem to be the most opportune time to invade. Only the nest-founding queens have to be subdued, and the host workers, already reared by the labor of the nest-founding queens, accept the parasite as their queen when they eclose soon afterward.

It is still uncertain whether inquilinous species of *Polistes* exist in other parts of the world. Very little deliberate search has been made for them. The three species listed in Table 19-3 from Africa, the United States, and Brazil, respectively, are placed in the inquiline class only tentatively, on the basis of morphology and suggestive but inadequate field data. Careful observations of the kind conducted by Scheven on the European species are greatly needed in these regions and other parts of the far-flung range of *Polistes.*

Our information on social parasitism in the closely related New World genus *Mischocyttarus* comes entirely from the report by Zikán (1949), and it is fragmentary and puzzling. Zikán did not present data concerning any one of the 23 parasitic species he described as new to science in this monograph. Instead, he stated that each of the parasites resembles its respective host species more closely than any other member of the genus, and that in each case it differs from the host species by the same set of morphological characteristics: the head relatively more voluminous, the genae wider, the ocelli smaller, the thorax more voluminous and proportionately wider, and the pronotal carinae oriented differently. According to Zikán's formula, then, one needs only to specify the diagnostic characteristics of the host species and add to them the parasitic traits just cited in order to produce the diagnosis for the parasitic species. Zikán created the taxonomic name of each parasitic species by taking the base of the specific name of its host species and adding the suffix *-oides.* Thus *M. fumigatoides* parasitizes *M. fumigatus, M. gracilioides* parasitizes *M. gracilis,* and so forth. Although the arrangement is not very satisfying, it is at least plausible since the parasitic characters cited are of the kind associated with fighting in other aculeate social parasites. Also, the close similarity claimed between host and parasite conforms with Emery's rule, already shown to hold quite well in other aculeate groups.

According to Zikán, the life cycle of the *Mischocyttarus* parasites resembles that of parasitic species of *Polistes.* The parasitic queen invades an established nest of the host and subordinates the alpha female without killing her. She then does nothing but receive care for herself and her own brood. Host brood production ceases, and the host population eventually dies out from normal causes. When the host colony can no longer support them, the parasites are said to leave in search of new host nests. Zikán's entire account is very vague and scanty, and some effort should be made to test its validity.

The permanent social parasites of the Vespinae are less tolerant of the host queens. According to Weyrauch (1937, 1938) and de Beaumont (1958), the parasitic *Vespula* queens invade an established host nest and in some manner dispose of the host queens soon afterward. No one has observed the actual elimination, but it is generally assumed that the parasites kill the host queens in combat. It has at least been established that when two or more queens of one of the species, *V. omissa,* invade a nest, they fight to the death until only one remains. The parasites are physically well endowed to function as assassins. In most cases they are distinguished from the host species by their stronger exoskeletons, the closer fitting of their abdominal segments, and their strongly developed and recurved stings, broader heads, more powerfully built mandibles, and more sharply bidentate clypei (see Figure 19-12). Emery's rule is exemplified very well in these wasps. As shown in Figure 19-13, taxonomists believe that the parasitic species of Europe were derived from the same immediate ancestral stock as the host species, or else a stock very close to it.

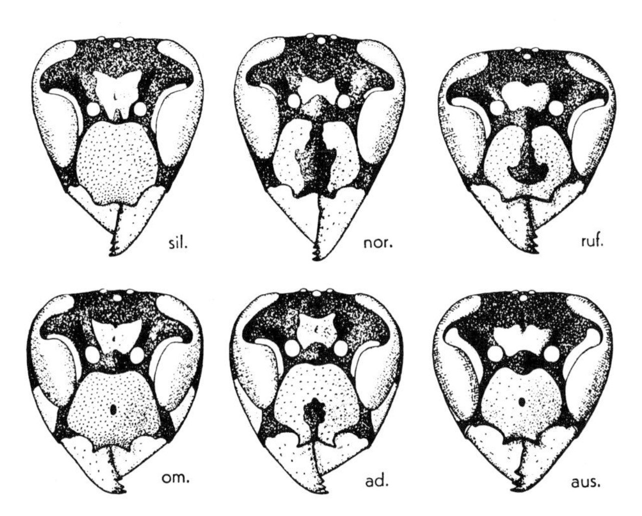

FIGURE 19-12. Queens of three socially parasitic species of the genus *Vespula* compared with those of the host species. The host species in the top row (*silvestris, norwegica, rufa*) are matched with their respective parasites directly below (*omissa, adulterina, austriaca*). Note that each of the parasites differs from its host by its broader head and the more sharply bidentate condition of the clypeus; in color pattern, however, it resembles the host species more closely than either of the two remaining free-living species (from de Beaumont, 1958).

Social Parasitism in Bees

Cleptoparasitism in the bees has been reviewed by Wheeler (1919b), Grütte (1935), and Michener (1944), and documented in more recent years by Ellen Ordway (1964), Knerer and Atwood (1967), Rozen and Michener (1968), and others. It is a very common phenomenon generally and is widespread in such principal families as the Halictidae, Anthophoridae, Megachilidae, and Apidae. In the typical case involving a solitary species, the parasite female simply enters a brood cell already provisioned by a host female and lays an egg of her own. The rightful inhabitant of the cell is then destroyed, either at the outset by the parasite female, or later by the parasite larva itself. *Stelis,* a parasitic megachild genus, contains species that use the more common second method. The larvae are equipped with sharp, falcate mandibles which they use to attack and kill the host larvae; in some cases they also eat their victims.

Unlike the aculeate wasps, the bees do not include any known parasitoids in their ranks. This is to be expected from a group of species that rarely consume animal protein of any kind. On the other hand, members of the meliponine genus *Lestrimelitta* do engage in nest robbing, or "cleptobiosis." Nest robbing, it will be recalled, is common in ants but apparently absent in social wasps. *Lestri-*

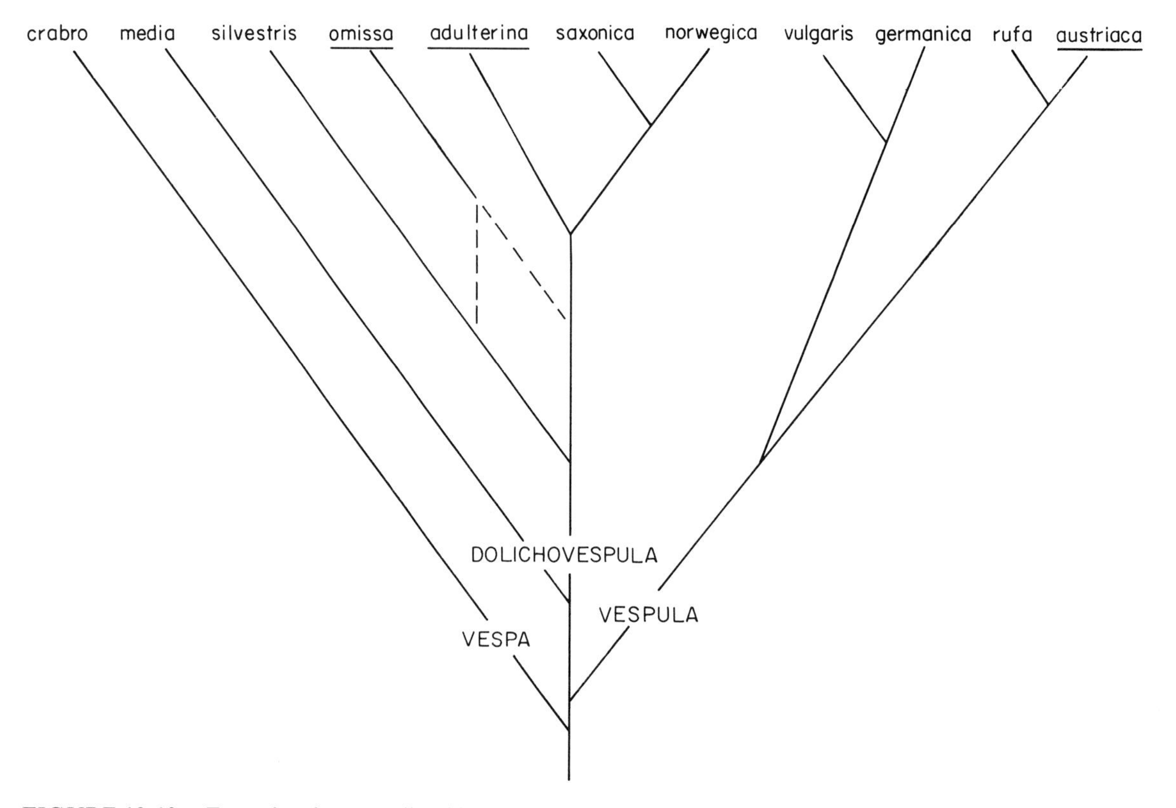

FIGURE 19-13. Emery's rule exemplified in the European Vespinae. In this phylogenetic diagram of the simple branching kind proposed by de Beaumont (1958), the parasitic species (*underlined*) are postulated to be derived from the same immediate stocks as their hosts. The nature of the origin of *omissa* (*dashed lines*), which may have come from either the *silvestris* or *saxonica* phylads, is uncertain.

melitta consists of two Neotropical species, *limao* and *ehrhardti,* and one little-known African species, *cubiceps. L. limao,* a common species from Mexico to Argentina, makes its living by invading nests of *Melipona* and *Trigona* and seizing their stored supplies (Schwarz, 1948; Sakagami and Laroca, 1963). The *Lestrimelitta* lack a scopa (corbicula) on the hind legs, which was evidently lost in evolution as part of their parasitic specialization. Instead, they carry the stolen supplies in their crops and later store them in their own storage pots in the form of honey-pollen mixtures. The invasion is accompanied by the release of a mandibular gland substance with a strong lemon-like odor, the principal component of which has recently been identified by Blum (1966b) as citral. This appears to agitate and drive out the host workers. Sometimes the *L. limao* occupy the plundered nest and, like the dulotic ants of the genera *Harpagoxenus* and *Polyergus,* multiply their own colonies in this fashion. Robbing behavior has also been documented in *L. ehrhardti* by Sakagami and Laroca (1963) and in *L. cubiceps* by Portugal-Araújo (1958). The evolution of the *Lestrimelitta* parasitism is not difficult to imagine. Both *Apis* and nonparasitic meliponines occasionally engage in robbing, both within and between species. The *Lestrimelitta* appear simply to have adopted such behavior as an exclusive way of life.

Nest robbing has recently been reported in the Halictidae by Knerer and Plateaux-Quénu (1967c). In the spring the polygynous hibernating assemblages of young *Halictus scabiosae* females break up, and some of the auxiliaries disperse away from the home nest. Many of these individuals construct new nests of their own. However, others invade the newly founded, monogynous nests of other species, most frequently *Evylaeus nigripes,* and drive out or kill the rightful occupants. It is possible that this second behavior is an evolutionary precursor to the kind of obligatory cleptoparasitism displayed by more specialized halictid genera such as *Sphecodes.*

True social parasitism, in which the intruding female becomes a part of the host society, is limited to two of the more primitively social groups, namely the allodapines and the bumblebees (see Table 19-4). As is the case with their equivalents in the ants and wasps, the inquilinous bees are usually most closely related to the species, or the group of species (the genus), which they parasitize. Thus Emery's rule, which was first articulated for the ants alone,

TABLE 19-4. Socially parasitic bees.

Parasite	Hosts	Form of parasitism	Range	Authority
Xylocopidae Tribe Ceratinini (the "allodapines")				
Braunsapis associata	[a] *Braunsapis unicolor*	Inquilinism	Queensland	Michener (1961c, 1970)
Braunsapis praesumptiosa	*Braunsapis simillima*	Inquilinism	Queensland	Michener (1961c, 1970)
Braunsapis natalica	[a] *Braunsapis grandiceps, B. facialis*	Inquilinism	Natal	Michener (1970)
Braunsapis breviceps	*Braunsapis* sp.	Inquilinism (?)	Malaya	Michener (1966c, 1970)
Allodape greatheadi	*Allodape interrupta*	Inquilinism (?)	Uganda	Michener (1970)
Allodapula guillarmodi	*Allodapula* sp.	Inquilinism (?)	Cape Province	Michener (1970)
Macrogalea mombasae	[a] *Macrogalea candida*	Inquilinism	Kenya, Tanzania	Michener (1970)
Inquilina excavata	[a] *Exoneura variabilis*	Inquilinism	Queensland	Michener (1965b, 1966c, 1970)
Nasutapis straussorum	[a] *Braunsapis facialis*	Inquilinism	Natal	Michener (1970)
Eucondylops konowi	[a] *Allodapula variegata*	Inquilinism	Cape Province, Natal	Michener (1970)
Eucondylops reducta	[a] *Allodapula melanopus*	Inquilinism	Cape Province	Michener (1970)
Apidae subfamily Bombinae				
Bombus distinguendus	[a] *Bombus subterraneus*	Facultative, temporary parasitism	Europe	Lindhard (cited by Free and Butler, 1959)
Bombus terrestris	[a] *Bombus lucorum*	Facultative, temporary parasitism	Europe	Sladen (1912)
Bombus americanorum	[a] *Bombus separatus*	Facultative, temporary parasitism	North America	Frison (1930); Plath (1934)
Bombus affinus	[a] *Bombus terricola*	Facultative, temporary parasitism	North America	Plath (1934)
Bombus hyperboreus	[a] *Bombus polaris*	Obligatory, permanent	North America	K. W. Richards (1971)
Psithyrus spp.: 10 in Europe, 8 in North America	[a] *Bombus spp.*	Inquilinism; workerless	North Temperate Zone	Free and Butler (1959)

[a] Hosts in whose nests the parasites have actually been found. The others are close relatives of the parasites and probably serve as hosts.

proves to have general applicability throughout the social Hymenoptera. The parasites are nevertheless easy to distinguish. Richards (1927a) lists no less than 27 morphological characteristics by which all known species of the parasitic bumblebee genus *Psithyrus* can be separated from the host genus *Bombus,* and he argues that at least 15 of these represent adaptations to the parasitic mode of life. The number of characteristics generally present in parasitic bees are fewer in number and include (Michener, 1961c): (1) reduction or loss of the pollen-collecting scopas; (2) loss of the basitibial and pygidial plates; (3) strong development of the sting; (4) development of an unusually long and mobile, or otherwise modified, apical part of the abdomen; and (5) development of an unusually strong and coarsely punctate integument and of ridges and spines protecting the neck and base of the metasoma. All of these features appear to be either reductions associated with the lack of a need to construct and provision nests or else adaptations for defense against attacking host females. The loss of the scopa (see Figure 19-14) is regarded as the hallmark of parasitism since, without the structure, the females cannot provision their own nests with pollen.

The existence of true inquilines among the allodapine bees has recently been established beyond much doubt, but information on their biology is still fragmentary. From comparisons of many nests collected in Australia, for example, Michener (1965b) concluded that *Inquilina excavata* females somehow enter the stem nests of *Exoneura variabilis* and induce the little colonies to adopt them. The *Exoneura* egg layers are tolerated by the parasites and, in some cases at least, are permitted to lay eggs. *Exoneura* brood are not allowed to mature, however; probably the *Inquilina* females eat the eggs before they hatch. Only *Inquilina* females and males grow up in the infested nests. Eventually the last of the *Exoneura* die off, leaving pure groups of *Inquilina* to occupy the nests temporarily. But the *Inquilina* are not able to carry on by themselves. They have reduced scopas, have never been observed to visit flowers, and evidently must receive all of their food directly from their *Exoneura* hosts. A second allodapine parasite discovered by Michener (1961c), *Allodapula associata,* is unique in that it still possesses well-developed pollen hairs. Since individuals were found in the nests of *A. unicolor* with pollen clinging to the hairs, Michener inferred that the inquilines cooperate in foraging. If this interpretation is correct, *A. associata* represents the earliest stage of inquilinism yet recorded in the bees, one fully comparable with the primitive state exemplified by *Kyidris* in the ants.

FIGURE 19-14. The hind leg of the socially parasitic queen of *Psithyrus rupestris* (*left*) lacks the fringe of hairs (corbicula) on the tibia used in collecting pollen; that of the queen of *Bombus lapidarius* (*right*), a free-living species, possesses it (from de Beaumont, 1958; after Sladen, 1912).

The bumblebee parasites have been much more thoroughly studied, by Kirby (1802), Lepeletier (1832), and Hoffer (1888) and, more recently, by F. W. L. Sladen, O. E. Plath, T. H. Frison, and others. The newer literature has been reviewed by de Beaumont (1958) and Free and Butler (1959). Social parasitism in the bumblebees springs directly from the aggressive organization of their colonies. A strong tendency toward the habit is displayed by some of the free-living species of *Bombus* themselves. The queens of *B. lapidarius* and *B. terrestris,* for example, occasionally invade established nests and try to overthrow the foundress queen. In such cases the foundress and her worker offspring fight back to retain possession, often to the death. Multiple invasions frequently occur; in one extreme case Sladen found the corpses of 20 *B. terrestris* queens piled up in a single nest. Free (1961a) used experimental introductions to observe the invasion process directly. His notes are worth quoting: "On several occasions I introduced the dominant bee of one queenless colony to another. On such occasions the two dominant bees always singled each other out and often attempted to bite and to clasp each other as though to sting, although they never actually did so. However, within a few minutes one of them (mostly the introduced bee) hid in a corner of the nest box. When I returned the introduced bee to its own colony its status there depended on whether it had been dominated in the strange colony. When it had been dominant in the strange colony, it was invariably dominant on return to its own. When it had been beaten in the strange colony, it was nearly always dominated on return to its own, the bee which dominated being that which had previously been the second dominant. These experiments show that although a bee's social status in one of these queenless colonies largely depends on the amount its ovaries are developed, other factors such as memory of previous conflicts, and familiarity with the territory on which it is fighting may also play a part."

Thus the "mental set" of a given bumblebee queen can tip the balance to victory or defeat. It would seem an easy evolutionary step for the queens of intrinsically more aggressive species to move in on colonies belonging to other, relatively docile species. Something like this appears to have occurred repeatedly in the evolution of the bumblebees. The queens of certain species of *Bombus,* including *terrestris* itself, as well as *affinus, americanorum,* and *distinguendus,* occasionally conquer the nests of other *Bombus* species. The host workers then rear the alien brood, which develops into normally functioning workers. As the original host workers gradually die off from natural causes and are replaced by the expanding population of parasites, the colony becomes indistinguishable from one founded by a queen in the usual, independent manner.

In *Psithyrus,* the all-parasitic genus of bumblebees, and in *Bombus hyperboreus,* the queens are permanent inquilines in the nest of free-living *Bombus* and do not produce a worker caste. Their behavior is nevertheless not radically different from the dominant queens of the free-living and temporarily parasitic species of *Bombus.* The young *Psithyrus* queens are fecundated at the end of the summer, and they normally overwinter in cavities in the ground, just like young *Bombus* queens. They emerge later in the following spring than do the queens of the host species and seek out the nests when the host colonies are already well along in development. The behavior of the *Psithyrus* females on entering the nests varies greatly among species and possibly among individuals belonging to the same species as well. In some cases, if we are to believe all of the differing accounts, the parasite queen at first attempts to avoid contact with the inhabitants, hiding under the comb and nesting material. When approached, she uses a calm, deliberate manner to ingratiate herself. The queen of *Psithyrus variabilis,* a species belonging to this conciliator category, responds to attack by drawing her legs close to her body and remaining motionless for long periods of time (Frison, 1926). The heavy sclerites and tough intersegmental membranes of the intruder usually protect her, and as Sladen (1912) has put it, "the workers soon cease to show any hostility towards her. Even the queen grows accustomed to the presence of the stranger, and her alarm disappears, but it is succeeded by a kind of despondency. Her interest and pleasure in her brood seem less, and so depressed is she that one can fancy she has a presentiment of the fate that awaits her. It is by no means a cheerful family, and the gloom of impending disaster seems to hang over it. But while the queen grows more dejected, the *Psithyrus* grows more lively, and takes an increasing interest in the comb, crawling about over it with unwonted alacrity, and examining it minutely."

What the invaded colony experiences, of course, is not exactly a presentiment of doom but simply a demotion within the dominance hierarchy as the *Psithyrus* female assumes the alpha position. The females of some species of *Psithyrus,* according to Free and Butler (1959), achieve the same result by direct assaults on the host colony. A queen using this method grapples with the defending workers and makes preliminary stinging motions, but usually releases them before following through. The treatment gradually subdues the workers. Sometimes the defenders "ball" the *Psithyrus,* swarming over her in such numbers as to cover her with a sphere made up of their own bodies. And occasionally the *Psithyrus* is killed. This happens when a *Bombus* worker is able to sting her in the neck or in one of the few other vulnerable spots on her body. Sladen reported that the queens of two species occurring in England, *Psithyrus rupestris* and *P. vestalis,* always kill the host queens. In contrast, Plath (1922) found that queens of the American species *P. ashtoni* and *P. laboriosus* rarely, if ever, kill the host queens; instead, they are permitted to live in subordinate positions in the nests. Whatever the details, all observers agree that the parasite females at least permit no more *Bombus* to be reared. In the tyrannical manner employed by alpha *Bombus* queens themselves, the parasites try to prevent the *Bombus* foundress and workers from laying eggs in their own cells. When eggs are deposited, the *Psithyrus* promptly destroy them.

Although the taxonomic evidence is still equivocal (Richards, 1927a; Free and Butler, 1959), the species of *Psithyrus* appear in at least some cases to have been derived independently from a common stock with the host species. Richards (1927a, 1953b) has presented a plausible model of how such parasitism might arise in these and other groups of bees in the North Temperate Zone. Starting with the cardinal fact that parasitic queens emerge later in the spring than their hosts, he reasoned that the precondition for parasitism is the existence of two closely related species, one northern in distribution and the other southern. When the southern species penetrates the range of the northern species, it will at first tend to emerge later in the spring than the northern species within the zone of overlap. A second precondition, which has been well documented in some species of *Bombus,* is the tendency for queens to invade colonies of their own species. Given the availability of a closely related species which already has well-developed colonies and which, in the initial stages of range invasion, is more abundant as well, there might be a tendency for the invader to evolve in the direction of interspecific parasitism. And as the parasitism advances from the facultative temporary state to complete inquilinism, the range of the southern invader would be wholly absorbed within that of the host species. Richards' hypothesis could be tested by carefully examining the ranges of many species in various degrees of parasitism to see if the predicted correlation exists. This has not, to my knowledge, been attempted for the bumblebees. In the ants and social wasps at least, where good distributional data are available, the model does not seem to fit well. To take one conspicuously contrary example, *Vespa dybowskii* is a facultative temporary parasite of *V. crabro* and *V. xanthoptera* in the temperate part of Asia; in other words, it is at the earliest stage in the hypothesized scale of evolution. Yet it is located in the northern parts of the ranges of its hosts.

Social Parasitism in Termites

True inquilinism is unknown in termites. Instead, members of three genera, *Ahamitermes, Incolitermes,* and *Termes,* have become what might be termed "nest parasites" on other kinds of termites. That is, they have specialized on living in cavities in the nest walls of their hosts and feeding on the carton material that makes up the nest walls. The alate forms of two of the species, *Incolitermes pumilis* and *Termes insitivus,* even go so far as to enter the host chambers occasionally and to mingle briefly with the host colonies. The situation has no exact parallel in any of the other social insects, but in its extreme form it approaches xenobiosis in ants.

Mound-building termites are strongly vulnerable to this form of parasitism. The mounds are solidly constructed, often the most durable features of the ground environment. They offer conspicuous landmarks for flying alates. They also provide unusually favorable microenvironments for colonization if the alates are able to remain hidden in the mound walls and to avoid contact with the hostile inhabitants. Numerous cases have been reported where colonies of two or more termite species live in close association, and these commonly represent different genera or even different families. Frequently the relationship is exploitative, one species appropriating part of the nest of another. For example, of 150 species studied by Ernst (1960) in Africa, 70 percent were at least occasionally disturbed by other species encroaching on their nests. The vast majority of such associations around the world appear to be fortuitous. Species of the closely related Australian genera *Ahamitermes* and *Incolitermes* have, however, become true obligatory nest parasites (Hill, 1942; Calaby, 1956; Gay, 1956, 1966). The three species of *Ahamitermes, hillii, inclusus,* and *nidicola,* are rare and local in distribution. Each builds its nests exclusively in the large mounds of one or two species of the rhinotermitid *Coptotermes.* The arrangement of the two colonies is intimate, as exemplified in Figure 19-15. The *Ahamitermes,* whose colonies are much less populous than those of the hosts, appear to feed exclusively on the carton material of the host nest. The two species maintain themselves strictly apart. Even when larger *Ahamitermes* colonies occupy a considerable portion of the inner section of the compound nest, they never encroach on the *Coptotermes* nursery or extend downward below the level of the center of the nursery. They construct a small dome to one side of the *Coptotermes* mound, which provides a private escape route for their alate reproductives at the time of the nuptial flight.

The closely related species *Incolitermes pumilis,* only recently removed from *Ahamitermes* and placed in its own genus by Gay (1966), has taken the association with *Coptotermes* one step further. The parasitism is essentially like that of *Ahamitermes,* except that the *Incolitermes* do not build a lateral dome or any other form of escape route to accommodate their alate reproductives. Limited obser-

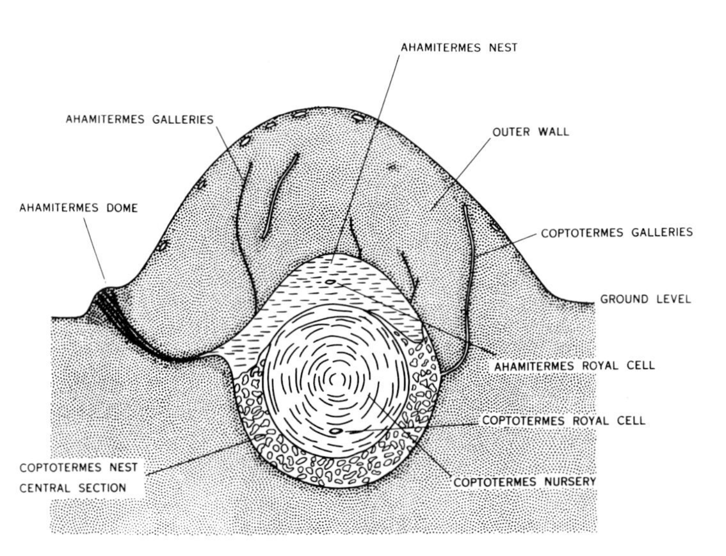

FIGURE 19-15. A diagrammatic cross section of a *Coptotermes acinaciformis* mound occupied by a colony of the parasitic termite *Ahamitermes hillii* (from Calaby, 1956).

vations by Calaby (1956) indicate that at some time prior to the nuptial flight of the *Coptotermes*, the *Incolitermes* alates enter the *Coptotermes* chambers and mingle with the host alates. When the *Coptotermes* alates emerge through the top of the mound for the nuptial flight, the *Incolitermes* depart with them.

A parallel series of adaptations are displayed within the genus *Termes* in South America, Africa, and Australia (Emerson, 1938; Hill, 1942; Skaife, 1954a). Most species are facultative in their relation to other mound builders, frequently but not invariably attaching themselves to the nests of members of such dominant genera as *Amitermes* and *Coptotermes*. According to Emerson, however, *T. inquilinus* has been found only in association with *Constrictotermes cavifrons*, a South American rain-forest species that constructs carton nests on the undersurface of slanting tree trunks. The *T. inquilinus* colonies appear to subsist on the carton material, and, like *Ahamitermes*, they do not mix directly with their hosts. A similar relationship is believed to exist between *Termes fur* and *Constrictotermes cyphergaster*. One tantalizing set of observations published by Hill (1942) suggests that the Australian species *T. insitivus* may be at least as advanced a xenobiont as *Incolitermes*. Nymphs and alates of this rare species have been found in galleries of *Nasutitermes magnus*, mingled with the host colony. The nonreproductive castes of *T. insitivus* have not been found, and the precise nature of the association remains unknown.

Thus with the remotely possible exception of *Termes insitivus* no case of true inquilinism, in which the parasitic species has been integrated into the society of the host species, has ever been demonstrated in the termites. Why does this sharp difference from the social Hymenoptera exist? It cannot be that termite colonies are especially xenophobic and hence incapable of adopting strange species. The truth is that large numbers of species of beetles, flies, and other arthropods have managed to insinuate themselves as adopted members of termite colonies. Some, moreover, are highly modified for a socially parasitic life and are cared for by the workers with the same kind of attention lavished on the termite reproductives. So instead we must look for some feature of the life cycle of termites that makes it unlikely that parasitism is ever attempted. This has proved quite easy to find: termite reproductives seldom if ever attempt to reenter colonies of their species following the nuptial flights. It will be recalled that this preadaptive form of colony-founding behavior is most prominently developed in just those groups of social Hymenoptera, namely certain groups of the ants, the social wasps, and the bumblebees, where parasitism of all kinds is at a peak. To be sure, parasitism is absent in the stingless bees and honeybees, even though the fecundated queens reenter nests of their own species. But in these groups the queen returns specifically to her home colony, and her behavior comprises only one stereotyped element in an unusually complex nest-founding pattern. It therefore seems that the only readily available evolutionary route left open to the termites is xenobiosis derived from nest parasitism. But xenobiosis involves hosts of different genera or even different families and, as has been clearly shown in the ants, it is the least likely means of attaining the final state of inquilinism.

Parasitism between Major Groups of Social Insects

I know of no convincingly documented case of a species belonging to one of the major groups of social insects that lives as a social parasite in the nest of a species belonging to another major group. In other words, bumblebees do not parasitize vespid wasps, vespid wasps do not parasitize ants, and so on. The one possible exception to this rule is *Stigmacros termitoxenus,* a peculiar Australian formicine ant discovered by Wheeler (1936a) in a nest of the termite *Tumulitermes peracutus.* Here is Wheeler's account of the single collection made: "On September 18, 1931, near Mullawa, West Australia, I came upon a colony of diminutive termites nesting under a flat stone in earthen galleries which they had built in a bunch of dry grass. On breaking open one of the galleries I saw several ants of the same size and color as the somewhat more numerous termites and moving about among them. After carefully collecting the occupants of the gallery and making allowance for escaping individuals, the ant-colony was found to comprise only 25 to 30 workers and a single ergatomorphic female (queen). I failed to find any additional ants in the termitary and saw no traces of their brood. The female and more than half the workers attracted my attention because their gasters were enormously distended . . . This distension (physogastry) was not due to liquid stored in their crops but to an unusual accumulation of fat . . . The ants belong to an undescribed species of *Stigmacros,* an exclusively Australian genus of which eleven species are described and of which Mr. John Clark and I have taken quite a number of unpublished forms. Although I have examined hundreds of *Stigmacros* from numerous localities in Eastern, Southern and Western Australia, I have seen no traces of physogastry except in the Mullawa specimens. Since the ants were living in what appeared to be friendly relations with their hosts, I suspect that they are fed by the termite workers and that the physogastry of the female and so many of the workers, like the physogastry of the termite workers and queens, is a result of this feeding." Wheeler's argument is far from convincing, but the phenomenon he suggests is so extraordinary as to warrant a special effort to rediscover and study in detail the biology of *S. termitoxenus.*

20 Symbioses with Other Arthropods

Sphecophiles, Myrmecophiles, Melittophiles, Termitophiles

A remarkable legion of animal species exploit the colonies of social insects in one way or another. Most do so only occasionally, functioning as casual predators or temporary nest commensals. But a great many others are dependent on social insects during part or all of their life cycles. Depending on the identity of the host, such species are referred to as *sphecophiles* (symbionts of social wasps), *myrmecophiles* (symbionts of ants), *melittophiles* (symbionts of social bees), or *termitophiles* (symbionts of termites). Within the scope of obligatory relationships, it is customary to apply broad definitions of these symbiotic categories.* Among the myrmecophiles, for example, should be included the Australian carabid beetle *Sphallomorpha colymbetoides,* the larvae of which excavate cicindellid-like burrows at the edge of meat-ant nests (*Iridomyrmex*) and make their living snatching worker ants (B. P. Moore, personal communication). Also included are the carabids of the genus *Helluomorphoides,* which follow the trails of army ants (*Neivamyrmex*), prey on their brood, but do not live in the bivouacs (Plsek *et al.,* 1969; Topoff, 1969); and perhaps even the tachinid and conopid flies that follow army ant raids in order to parasitize the cockroaches and crickets flushed into the open by the ants (Rettenmeyer, 1961b). On the other hand, I would exclude, somewhat arbitrarily and for purposes of convenience, a large number of species belonging to such groups as the nematodes, squamiferid isopods, polyxenid millipedes, coreid and mirid bugs, psocopterans, and many others, which sometimes abound in and around ant and termite nests but are capable of completing their whole life cycles elsewhere.

Although a respectable list of vertebrate species, including synbranch eels, frogs, lizards, snakes, birds, and small- and medium-sized mammals, occasionally live with social insects or prey upon them, very few are specialized for such an association (Myers, 1929, 1935; Hindwood, 1959; Scherba, 1965). The Rufous Woodpecker, *Micropterus brachyurus,* of tropical Asia hollows its nest cavities in the arboreal carton nests of ants belonging to the genus *Crematogaster.* It feeds on the ants, which are allowed to run freely through the nest cavities without doing visible harm to the birds nesting inside. The Gartered Trogon (*Trogon caligatus*) of South America excavates its nest cavities in the carton nests of polybiine wasps, eating the adult inhabitants as it proceeds. In Trinidad the Green-winged Parakeet (*Forpus passerinus viridissimus*) almost invariably nests in the carton nests of termites. The aardvarks (genus *Orycteropus*) prey on mound-building termites in Africa and construct their burrows in the nests of the insects. Although myrmecophiles, sphecophiles, and termitophiles thus exist among the vertebrates, the vast majority of obligate symbionts are arthropods, and it is among these organisms that the most striking adaptations for life with social insects have taken place. In the following analysis of this subject, particular attention will be

*In this book *symbiosis* is defined in the sense usually employed by American biologists, to include all categories of close and protracted interactions between individuals of different species, rather than in the narrower European sense of an exclusively beneficial interaction. Accordingly, three principal kinds of symbiosis can be recognized: parasitism, in which one partner benefits as the other suffers; commensalism, in which one partner benefits and the other is not affected either way; and mutualism, in which both species benefit.

paid to the arthropod species that have actually insinuated themselves into the social organization of their hosts. The means by which the symbionts have "broken the code" of their hosts is, in my opinion, the single most interesting aspect of the biology of these insects, and one which is only now beginning to yield to experimental investigation. Before approaching this phenomenon, however, it is necessary to review the background of symbiosis more comprehensively, with due attention to the complicated history of the subject.

Erich Wasmann initiated the modern study of arthropod symbionts. Beginning in 1894, he used a simple classification that divides species into five behavioral categories:

1. *Synechthrans.* These arthropods are treated in a hostile manner by the social insects among whom they live. They are scavengers and predators for the most part, and they manage to stay alive by greater speed and agility or the use of defensive mechanisms such as repellent secretions and retraction beneath shell-like body forms.

2. *Synoeketes.* These arthropods, which are also primarily scavengers and predators, are ignored by their hosts, either because they are too swift, as, for example, the *Cyphoderus* collembolans, or else very sluggish and apparently neutral in odor, as, for example, the syrphid fly larvae of the genus *Microdon.*

3. *Symphiles.* Also referred to occasionally as "true" guests or myrmoxenes (in the case of myrmecophiles), these symbionts are accepted to some extent by their hosts as though they were members of the colony. In a few cases the hosts do nothing more than inspect them and carry them about from time to time, but there also exist species that are groomed, fed, and even reared with the hosts' own larvae.

4. *Ectoparasites and endoparasites.* These arthropods are conventional parasites that either live on the body surfaces of their hosts, licking up their oily secretions, stealing food from them, or biting through the exoskeleton and feeding on their blood, or else penetrate the body itself as true parasites or parasitoids. No further account will be taken of the latter category here since it is comprised, for the most part, of species of wasps, flies, mermithid nematodes, and other organisms whose parasitic behavior is not ordinarily distinguishable from that of similar species that prey on nonsocial insects.

5. *Trophobionts.* These are the phytophagous homopterans and lycaenid and riodinid caterpillars that are not dependent on the social insects for food but actually supply their hosts with food in the form of honeydew. In exchange, they receive protection from parasites and predators.

To complete the terminology, a distinction must be made between *ectosymbionts,* those organisms that live on or among their hosts, and *endosymbionts,* which are internal parasites.

Wasmann's scheme was accepted and popularized by Wheeler, who provided thorough and creative reviews in his own work (1910, 1923, and 1928). Wheeler had a special liking for parasitology, and some of his best and most authoritative writing was on this subject. With the accumulation of more detailed information on the behavior of the symbionts in recent years, however, the Wasmannian classification has turned out to be considerably less than perfect. Symphiles, for instance, do not always exist on the charity of their hosts, grooming and soliciting food from them as Wasmann and his contemporaries believed. Many of the species prey on the hosts and their offspring at the same time that they are being treated amicably by them. More importantly, a few myrmecophiles fit more than one category at different times. Certain North American species of the scarab genus *Cremastocheilus,* for example, were recently discovered by Cazier and Mortenson (1965) to be obligate predators of the larvae of the ant genus *Myrmecocystus*—the first recorded case of primarily predatory behavior within the Scarabaeidae. The reactions of the ants toward the *Cremastocheilus* are ambivalent. Under some circumstances a given beetle is treated as a synechthran. Several workers seize it, forcibly haul it out of the nest, and dump it somewhere in the refuse zone. Moments later, the same individual may be greeted by another worker of the same colony, who then attempts to pull it back in the direction of the nest. The *Cremastocheilus* adults are furnished with tufts of golden hairs ("trichomes") at the anterior and posterior corners of the thorax. Cazier and Mortenson believe that these structures dispense a substance which is attractive to the ants and causes them to escort the beetles into the nest. In short, the ants treat the beetles sometimes as synechthrans and sometimes as symphiles. Most of the time, however, the *Cremastocheilus* have the status of synoeketes, that is, they are simply ignored and allowed to wander through the nest without interference. A second, and even more instructive example has been provided by Bert Hölldobler's study of *Amphotis marginata* in Europe. These nitidulid beetles make their living by waiting along the odor trails of the common formicine ant *Lasius fuliginosus* and soliciting food from homeward-bound workers. They accomplish the latter goal with surprising ease, simply by drumming their mouthparts on the labia of the ants in imitation of the tarsation signal used by the ants themselves to initiate food exchange. A worker thus approached is usually fooled into disgorging a drop of liquid from its crop and holding it for a while between its mandibles while the beetle feeds. A short time later, when it somehow becomes aware that something is wrong, it retracts the drop and attacks the *Amphotis*. The beetle is able to protect itself by pulling in its appendages beneath its carapace-like dorsum and sitting tight on the ground. Such fluctuating behavior on the part of the host will probably prove widespread as more examples of imperfectly adapted symbionts are studied in detail. In spite of such occurrences, the Wasmann nomenclature continues to be useful in designating the majority of cases, and it is frequently employed as a kind of shorthand in the literature on social symbioses.

The Diversity of Ectosymbionts

Table 20-1 provides a good idea of the great range of arthropods that have become obligate ectosymbionts. Many thousands of species, representing at least 17 orders, 120 families, and hundreds of genera, are involved in such a relationship. This list is undoubtedly far from exhaustive, despite the fact that it was prepared with the help of many entomologists who specialize in the individual orders and families cited. The literature on myrmecophiles and termitophiles in particular is enormous and growing rapidly each year, much of it consisting of incidental notes buried in taxonomic and ecological studies of selected genera and higher taxa, not all of whose species are symbiotic in their behavior. Among the generalizations that can be made is the obvious one that certain taxa are much more preadapted for life as ectosymbionts than others. The Acarina, for example, are the foremost representatives among ectosymbiont species in terms of sheer numbers. They find easy entry into nests either as scavengers, too docile and small to be evicted by their hosts, or as ectoparasites adapted for life on the body surfaces of the hosts. In the past, mites have been the least studied of all the major symbiotic arthropods, and probably many genera and species remain to be discovered. They are rivaled in diversity by the staphylinid beetles, a family of approximately 28,000 species. There are over 2,000 species in the Aleocharinae alone, the subfamily to which most of the symbiotic forms belong. Staphylinids, like acarines, are predisposed to life in ant and termite nests by virtue of their preference for moist, hidden environments and the role they commonly assume as generalized scavengers and predators. According to Seevers, at least fourteen phyletic lines of staphylinids have produced termitophiles independently, while others have entered into close associations with doryline army ants. These beetles fill a variety of niches within the nests of their hosts, and they have been particularly successful at integrating themselves as symphiles. Both the mites and the staphylinids exemplify the rule that the largest proportion of ectosymbionts, insofar as they can be judged by comparison with their closest relatives, have been derived from subterranean or cryptobiotic, medium-sized arthropods that originally lived as generalized predators and scavengers.

By far the greatest diversity of species of myrmecophiles and termitophiles, measured either per host species or per host colony, are found with host species that form exceptionally large mature colonies. The ultimate of this trend is found in the great faunas of symbionts that live with the doryline army ants, the meat ants of the genus *Iridomyrmex,* and the nasutitermitine and macrotermitine termites, the nests of which normally contain from tens of thousands to millions of inhabitants. An exceptional variety of guests has also been recorded from the large colonies of *Hypoclinea* in tropical Asia and the north temperate species of the *Formica rufa* group. By contrast, very few symbionts are known from nests of species with the smallest mature colony sizes, such as the majority of the ponerine and dacetine ants and the kalotermitid termites.

This rule of population size lends itself readily to theoretical analysis. The insect colony and its immediate environment can be thought of as an island which symbiotic organisms are continuously attempting to colonize. In general, species with the largest mature colony size also enjoy the longest average span of mature colony life. When colony life is long, the probability that symbiotic propagules will penetrate a given colony at some time or other is high. It is also true that if the colony size is large the equilibrial population size of its symbionts will be proportionately large and their species extinction rate (measured as the number of symbiont populations going extinct per colony per unit of time) will be correspondingly low. Finally, the more individuals in a colony, the more diverse the microhabitats presented by its nest and the greater the potential diversity of symbiont species. In sum, large colony size enhances three factors—long colony life, low symbiont extinction rates, and high microhabitat

TABLE 20-1. Ectosymbionts of social insects.

Taxa	Hosts	Biology	Selected references
ISOPODA			
Squamiferidae (*Platyarthrus*)	Many ant genera	Scavenge, occasionally attend aphids	Bernard (1968)
Porcellionidae (*Metaponorthus*)	*Messor*	Feed on stored grain	Bernard (1968)
Schobliidae (*Schoblia, Termitoniscus*)	Termites: *Termes, Macrotermes*	?	Verhoeff (1939)
PSEUDOSCORPIONIDA			
Cheliferidae (*Dasychernes*)	Stingless bees: *Melipona*	Found exclusively in the nests; possibly predaceous on hosts	Salt (1929)
ARANEA			
Clubionidae (*Phrurolithus*)	Ants: *Crematogaster*	?	Emerton (1911)
Clubionidae (*Andromma*)	Termites: *Macrotermes*	?	Fage (1938)
Oonopidae (*Brucharachne, Myrmecoscaphiella*)	Army ants: *Eciton*	?	Fage (1938)
Oonopidae (*Oonops, Dysderina, Gamasomorpha*)	Termites: *Capritermes, Eutermes*	?	Bristowe (1938)
Theridiidae [*Thymoites* (= *Brontosauriella*)]	Termites: *Eutermes*	?	Bristowe (1938)
ACARINA			
Circocyllibanidae, Coxequesomidae, Laelaptidae	Ecitonine army ants	Phoretic on adults	Rettenmeyer (1961c)
Planodiscidae, Scutacaridae	Ecitonine army ants	Phoretic on legs of adults	Rettenmeyer (1961c)
Macrochelidae, Neoparasitidae, Pyemotidae	Ecitonine army ants	Phoretic and ectoparasitic on adults and larvae	Rettenmeyer (1961c)
Possibly: Acaridae, Anoetidae, Ereynetidae, Hypochthoniidae	Ecitonine army ants	?	Rettenmeyer (1961c)
Antennophoridae, Laelaptidae	Nonlegionary ants	Phoretic on adults, live on regurgitated food	Janet (1897b); Wheeler (1910); Bernard (1968)
Laelaptidae	Stingless bees: *Melipona, Trigona*	?	Salt (1929)
Gamasidae, Uropodidae	Stingless bees: *Melipona, Trigona*	Scavengers and ectoparasites	Wheeler (1910); Bernard (1968)
Gamasidae	Stingless bees: *Trigona*	Ectoparasites (?)	Salt (1929)
Scutacaridae (*Acarapis*)	Honeybees	Some species are ectoparasitic on adults	Morgenthaler (1968)
Parasitidae (*Parasitus*)	Bumblebees	?	Free and Butler (1959)
Uropodidae (*Uropoda*)	Termites: *Cornitermes, Eutermes, Glyptotermes*	Scavengers	Hirst (1927)
DIPLOPODA			
Stylodesmidae (*Calymmodesmus, Rettenmeyeria, Yucodesmus*)	Ecitonine army ants	Scavengers, treated as symphiles	Rettenmeyer (1962c)

TABLE 20-1 (*continued*).

Taxa	Hosts	Biology	Selected references
(Possibly some species of *Blaniulus, Nopoiulus, Polydesmus, Polyxenus*)	Nonlegionary ants	Scavengers	Donisthorpe (1927); Manfredi (1949); Bernard (1968)
(*Leuritus, Stenitus, Gasatomus, Tidopterus*)	Termites: *Nasutitermes*	?	Chamberlin (1923)
COLLEMBOLA			
Entomobryidae (Cyphoderinae: *Calobatinus, Cephalophilus, Cyphoda, Cyphoderodes, Cyphoderus, Cyphoderinus, Megacyphoderus, Serroderus,* and others)	Many ant and termite species, as well as stingless bees	Scavengers; some extreme termitophiles lick exudates of the queen, steal regurgitated food, or eat symbiotic fungi.	Salt (1929); Delamare Deboutteville (1948)
Entomobryidae (*Lepidocyrtinus, Lepidoregia*)	Termites	Probably scavengers	Delamare Deboutteville (1948)
Isotomidae (*Isotominella*)	Termites: *Acanthotermes*	Probably scavengers	Delamare Deboutteville (1948)
Oncopoduridae (*Oncopodura*)	Termites	Probably scavengers	Delamare Deboutteville (1948)
THYSANURA			
Nicoletiidae (*Grassiella, Trichatelura*)	Ecitonine army ants	Groom hosts, feed on their prey	Rettenmeyer (1963b)
Nicoletiidae (*Atelura*)	Many ant species	Steal regurgitated food	Janet (1896); Wheeler (1910); Pohl (1957)
Probably other genera: (*Assmuthia, Crypturella, Lepisma, Lepismina, Allatelura, Atopatelura, Braunsina, Ecnomatelura, Metriotelura, Neatelura, Platystylea*)	Ants and termites	?	Escherich (1905); Folsom (1923); Wygodzinsky (1961); Joseph and Mathad (1963)
ORTHOPTERA			
Gryllidae (*Myrmecophila*)	Many ant species	Lick host secretions	Wheeler (1900a, 1910); Bernard (1968)
BLATTARIA			
Attaphilidae (*Attaphila*)	Fungus-growing ants: *Atta, Acromyrmex*	Lick host secretions	Wheeler (1900b, 1910); Roth and Willis (1960); Princis (1960); Moser (1964)
Atticolidae (*Atticola, Myrmecoblatta, Myrmeblattina, Phorticolea*)	Fungus-growing ants	?	Roth and Willis (1960); Princis (1960)
Nothoblattidae (*Nothoblatta*)	Fungus-growing ants	?	Roth and Willis (1960); Princis (1960)
Euthyrrhaphidae (*Sphecophila*)	Wasps: *Polybia*	?	Shelford (1906); Roth and Willis (1960); Princis (1960)
Euthyrrhaphidae (*Tivia*)	Termites	?	Shelford (1907); Rehn (1926); Roth and Willis (1960); Princis (1960)

TABLE 20-1 (*continued*).

Taxa	Hosts	Biology	Selected references
Homoeogamiidae (*Ergaula*)	Termites	?	Roth and Willis (1960); Princis (1960)
Nocticolidae (*Nocticola*)	Termites	?	Roth and Willis (1960); Princis (1960)
Oulopterygidae (*Oulopteryx*)	Stingless bees: *Melipona*	?	Roth and Willis (1960)
HOMOPTERA			
Aphididae, Coccidae, Pseudococcidae, Chermidae, Cercopidae, Membracidae, Fulgoridae, Jassidae	Ants	Trophobiosis	(See special section at end of chapter)
Membracidae	Social wasps: *Brachygastra*	Trophobiosis	(See special section at end of chapter)
Membracidae	Stingless bees: *Trigona*	Trophobiosis	(See special section at end of chapter)
Termitococcidae (*Termitococcus*)	Termites: *Capritermes, Leucotermes*	?	Silvestri (1938); Jakubski (1965)
Jassidae (*Ulopella*)	Termites: *Amitermes*	?	Poisson (1938)
HEMIPTERA			
Termitaphididae (*Termitaphis, Termitaradus*)	Termites: *Heterotermes, Rhinotermes, Coptotermes, Amitermes*	Possibly feed on fungi in nests	Usinger (1942)
Vianaididae (*Vianaida, Anommatocoris, Thaumamannia*)	Ants	Found in ants' nest but not yet proven to be obligate guests	Drake and Davis (1959); R. C. Froeschner (personal communication)
Reduviidae (*Ptilocerus*)	Ants: *Hypoclinea*	Attract workers from odor trails and prey on them	Jacobson (1911)
PSOCOPTERA			
(*Hemiseopsis, Liposcelis, Seopsis,* and others)	Termites: *Macrotermes, Odontotermes,* and others	Unknown; obligatory status not proved (E. L. Mockford, personal communication)	Mockford (1965)
NEUROPTERA			
Chrysopidae (*Nadiva*)	Ants: *Camponotus*	Larvae are symphiles	Weber (1942)
Chrysopidae (*Italochrysa*)	Ants: *Crematogaster*	Larvae stay close to nests and feed on ant larvae	Principi (1946)
Berothidae (*Lomamyia*)	Termites: *Reticulitermes*	Larvae live with the termites and prey on them	E. G. MacLeod (personal communication)
Mantispidae (*Trichoscelia*)	Wasps: Polybiini	?	Brauer (1869)
COLEOPTERA			
Carabidae (*Adelotopus, Nototarus, Philophlaeus, Pseudotrechus, Tachys, Tachyura*)	Nonlegionary ants: *Messor, Iridomyrmex, Lasius, Formica*	?	Lea (1910, 1912); Lindroth (1966); Bernard (1968); T. L. Erwin (personal communication)
Carabidae (*Helluomorphoides*)	Army ants (*Neivamyrmex*)	Follow odor trails, prey on brood	Plsek *et al.* (1969); Topoff (1969)
Carabidae (*Sphallomorpha*)	Ants: *Iridomyrmex*	Burrow near nest, prey on adults	Moore (1964b and personal communication)
Carabidae (*Glyptus, Orthogonius, Rhopalomelus*)	Termites: *Macrotermes* and other higher termites	Larvae live in the nests, are physogastric in later developmental stages	Escherich (1909)

TABLE 20-1 (*continued*).

Taxa	Hosts	Biology	Selected references
Paussidae (probably all species)	Many kinds of ants	Prey on host adults and larvae	Mou (1938); Janssens (1949); Darlington (1950); Le Masne (1961a,b); Reichensperger (1939)
Paussidae (*Physea*)	Leaf-cutting ants (*Atta*)	Apparent symphile	van Emden (1936); Darlington (1950)
Ptiliidae (*Xenopteryx*)	Termites: *Speculitermes*	?	Dybas (1961)
Limulodidae (most or all of the species)	Ants, especially army ants	Groom hosts; possibly other food habits	Lea (1910); Park (1933); Seevers and Dybas (1943); Wilson *et al.* (1954)
Leptinidae (*Leptinus*)	Ants (*Formica*) and bumblebees	?	Park (1929); Cumber (1949b)
Leiodidae (*Attaephilus, Attumbra, Catopomorphus, Dissochaetus, Echinocoleus, Eocatops, Myrmicholeva, Nemadus, Philomessor, Ptomaphaginus, Ptomaphagus, Synaulus*)	Ants: *Leptogenys, Pogonomyrmex, Aphaenogaster, Messor, Formica, Camponotus*	?	Lea (1910); Hatch (1933); Jeannel (1936); Fall (1937)
Leiodidae (*Platycholeus*)	Termites: *Zootermopsis*	?	Hatch (1933)
Leiodidae (*Parabystus, Scotocryptus*)	Stingless bees: *Melipona, Trigona*	Probably scavengers	Wasmann (1904); Salt (1929); Portevin (1937)
Scydmaenidae (many genera and species)	Many ant species	?	Lea (1910, 1912); Costa Lima (1962); Bernard (1968)
Staphylinidae (many tribes and genera)	Doryline and ecitonine army ants	Depending on the species, predators on the hosts or scavengers; at least some also groom the hosts	Wasmann (1917); Patrizi (1948); Paulian (1948); Kistner (1958, 1964, 1965, 1966a,b, 1968a); Seevers (1965); Koblick and Kistner (1965); Akre and Rettenmeyer (1966); Akre and Torgerson (1968)
Staphylinidae (many tribes and genera)	Nonlegionary ants	Some are symphiles. Depending on the species, either predators on the hosts or insects living with them, or scavengers, or inducers of regurgitation, or host-groomers; or a combination of these roles	Lea (1910, 1912); Wheeler (1910); Wasmann (1915a); Donisthorpe (1927); Bernard (1968); Hölldobler (1967a,b, 1969a,b)
Staphylinidae (*Quedius, Velleius*)	Wasps: *Vespa* and *Vespula*	Scavenge on nest refuse and prey on larvae	zur Strassen (1957); Kemper and Döhring (1967)
Staphylinidae (*Belonuchus*)	Stingless bees: *Melipona*	?	Wasmann (1904); Salt (1929)
Staphylinidae (many tribes and genera)	Termites	Some are symphiles. Some steal regurgitated food, others feed on symbiotic fungi. *Termitopullus* bite the host larvae and probably feed on their haemolymph. Food habits of most species are unknown	Lea (1910); Wheeler (1923); Grassé and Poisson (1940); Seevers (1957); Pasteels (1967, 1969); Kistner (1968b-d, 1969)
Pselaphidae (many genera and species)	Many ant species	Predators on hosts	Brauns (1914); Lea (1919); Park (1964); Bernard (1968)

TABLE 20-1 (*continued*).

Taxa	Hosts	Biology	Selected references
Clavigeridae (all species)	Many ant species	Most or all are symphiles that solicit regurgitated food from hosts; also groom hosts	Lea (1910, 1912); Park (1964); Bernard (1968)
Histeridae (*Euxenister, Pulvinister*)	Ecitonine army ants	Groom hosts and prey on host brood	Reichensperger (1924); Akre (1968)
Histeridae (*Abraeus, Chlamydopsis, Hetaerius, Myrmetes, Orectoscelis, Sternocoelis*)	Many nonlegionary ant genera	Feed on hosts and on prey captured by hosts; solicit regurgitated food	Lea (1910, 1919); Wheeler (1910); Bernard (1968)
Histeridae (*Cossyphodister, Notocoelis, Teratosoma, Thaumataerius*)	Termites: *Cornitermes, Syntermes*	?	Mann (1923); Reichensperger (1936)
Scarabaeidae (*Cremastocheilus*)	Several ant genera	Obligate predators on the host larvae; also feed occasionally on the crop contents of dead workers	Cazier and Statham (1962); Cazier and Mortenson (1965)
Scarabaeidae (*Euparixia*)	Ants: *Atta*	?	Woodruff and Cartwright (1967)
Scarabaeidae (*Cryptodus, Euphoria*)	Ants: *Camponotus, Formica, Leptomyrmex*	?	Lea (1910); Wheeler (1910)
Scarabaeidae (*Afroharoldius, Chaetopisthes, Coenochilus, Corythoderus, Novapus, Ryparus, Termitodius, Termitotrox*)	Termites	?	Escherich (1909); Lea (1910); Paulian (1947); Janssens (1949); Martinez (1950); Cartwright and Woodruff (1969)
Karumiidae (all species ?)	Termites	?	Arnett (1964)
? Cantharidae (*Ctenophorellus*)	Termites	?	Silvestri (1920)
Thorictidae (most or all of the species)	Ants: *Messor, Pheidole, Tetramorium, Cataglyphis*	Attack legs or antennae of hosts to be carried into nests, then scavenge on refuse and dead ants	Wheeler (1910); Banck (1927)
Ptinidae (about 12 genera)	Ants: several genera	?	Lawrence and Reichardt (1969)
Lymexylidae (*Atractocerus*)	Termites	?	Escherich (1909)
Nitidulidae (*Amphotis*)	Ants: *Lasius*	Solicit regurgitated food	Wasmann (1892); Donisthorpe (1927); Hölldobler (1968)
Nitidulidae (*Epuraea*)	Bumblebees and social wasps (*Vespula*)	Probably scavengers	Scott (1920)
Nitidulidae (*Brachypeplus, Carpophilus*)	Stingless bees: *Trigona*. Occasional in feral honeybee nests.	?	Lea (1910, 1912)
Rhizophagidae (*Monotoma*)	Ants: *Formica*	?	Bernard (1968)
Cucujidae (*Nepharis, Nepharinus*)	Ants	?	Lea (1910)
Cucujidae (*Nausibius*)	Stingless bees: *Trigona*	?	Wasmann (1904); Salt (1929)
Cryptophagidae (*Emphylus, Catopochrotus*)	Ants: *Crematogaster, Formica*	?	Reitter (1889); Bernard (1968)
Cryptophagidae (*Antherophagus*)	Bumblebees	Adults ride on adults by biting their proboscises; scavengers in nests	Scott (1920); Plath (1934); Cumber (1949b)

TABLE 20-1 (*continued*).

Taxa	Hosts	Biology	Selected references
Cryptophagidae (*Catopochrotides*)	Termites: *Hodotermes*	?	Kieseritzky and Reichardt (1936)
Erotylidae (*Tritomidea*)	Ants: *Amblyopone*	?	Lea (1910)
Erotylidae (*Episcaphula*)	Termites	?	Lea (1910)
Aculagnathidae (*Aculagnathus*)	Ants: *Amblyopone*	?	Oke (1932)
Endomychidae (*Symbiotes, Trochoideus*)	Ants: *Formica, Lasius,* and others	Probably scavengers	Lawrence and Reichardt (1969)
Endomychidae (*Trochoideus*)	Termites	Scavengers	Kemner (1924)
Merophysiidae (*Coluocera*)	Ants	?	Bernard (1968)
Lathridiidae (*Cartodere*)	Ants: several genera	?	Bernard (1968)
? Lathridiidae (*Ceroncinus*)	Termites: *Eutermes*	?	Silvestri (1920)
Colydiidae (*Ditoma, Kershawia, Myrmechixenus, Euclarkia*)	Ants: *Crematogaster, Iridomyrmex,* others	?	Lea (1910, 1919); Clark (1920)
Tenebrionidae (*Cossyphodes, Cossyphodites, Oochrotus,* and others)	Ants: *Aphaenogaster, Messor, Cardiocondyla, Plagiolepis*	?	Basilewsky (1952); Bernard (1968)
Tenebrionidae (*Rhysopaussus, Reichenspergeria, Xenotermes, Ziaelas,* and others)	Termites: *Macrotermes, Odontotermes*	?	Escherich (1909); Hozawa (1914); Wasmann (1921)
Lagriidae (*Lagria*)	Ants: *Myrmecia, Brachyponera*	?	Lea (1910, 1912)
Dacoderidae (*Tretothorax*)	Ants: *Leptogenys, Odontomachus*	?	Lea (1910); Watt (1967)
Melandryidae (*Troctontus*)	Termites: *Microcerotermes*	Larvae are licked and fed regurgitated liquid by host workers	Silvestri (1920); Grassé (1939); Hollande *et al.* (1951)
Rhipiphoridae (*Metoecus*)	Vespine wasps	Predators on wasp larvae	Chapman (1870); Kemper and Döhring (1967)
Coccinellidae (*Coccinella distincta, Hyperaspis*)	Ants: *Formica, Tapinoma*	Feed on aphids and fulgorids kept by the ants	Silvestri (1903); Donisthorpe (1927)
Coccinellidae (*Cleidostethus*)	Stingless bees: *Melipona*	?	Salt (1929)
Clytridae (*Clytra, Hockingia, Saxinis*)	Ants: *Atta, Formica, Camponotus,* and other nonlegionary groups	Larvae feed on vegetable material in nest	Jolivet (1952); Selman (1962); Hocking (1970)
Cryptocephalidae (*Isnus*)	Ants: arboreal species in *Acacia*	?	Selman (1962)
Brenthidae (*Amorphocephalus* and other genera)	Ants: *Camponotus* and probably other genera	Solicit regurgitated food from hosts; also feed hosts by regurgitation	Lea (1910, 1912); Kleine (1925); Le Masne and Torossian (1965)
Curculionidae (*Liometophilus*)	Ants: *Liometopum*	?	Fall (1912)
LEPIDOPTERA			
Lycaenidae (at least 245 species, and probably most of the 3000–4000 species, as a primitive trait)	Many ant genera and species	Ants attend larvae on the food plant. Larvae of some species enter the nest to pupate. Some also feed on the host brood or solicit food from the host workers	Wheeler (1910, 1928); Lamborn (1914); Donisthorpe (1927); Hinton (1951); Clark and Dickson (1956); Downey (1961, 1962)

TABLE 20-1 (*continued*).

Taxa	Hosts	Biology	Selected references
Riodinidae (*Anatole, Hamearis, Nymphidium, Theope;* probably most other genera)	Ants: *Solenopsis, Camponotus,* and other genera	As in Lycaenidae	Hinton (1951); Ross (1966)
Pieridae (several genera; obligatory status not certain)	Ants	Possess attractive glands and are attended by ants on the food plants	Hinton (1951)
Tineidae (*Ardiosteres, Atticonviva, Iphierga, Myrmecozela*)	Ants: *Formica* and other	Feed on nest material and detritus	Hinton (1951)
Tineidae (*Hypophrictoides*)	Ants: *Hypoclinea, Plagiolepis*	Prey on pupae	Roepke (1925)
Tineidae (*Passalactis, Plastopolypus*)	Higher termites	?	Silvestri (1905, 1920); Trägårdh (1907a); Poulton (1936); Hollande *et al.* (1951)
Tineidae (*Melissoblaptes*)	Wasps: *Vespa*	Larvae feed on nest materials	Wheeler (1928)
Cosmopterygidae (*Batrachedra*)	Ants: *Polyrhachis*	Larvae feed on host brood	Hinton (1951)
Cyclotornidae (*Cyclotorna*)	Ants: *Iridomyrmex*	Larvae prey on cicadellids and later on ant larvae as symphilic guests	Dodd (1912)
Pyralidae (*Stenachroia*)	Ants: *Crematogaster*	?	Turner (1913)
Pyralidae (*Pachypodistes*)	Ants: *Hypoclinea*	Larvae feed on nest carton	Hagmann (1907)
Pyralidae (*Wurthia*)	Ants: *Oecophylla, Polyrhachis*	Larvae prey on the host brood	Kemner (1923); Hinton (1951)
Pyralidae (*Galleria*)	Honeybee: *Apis*	Larvae feed on wax combs	Wheeler (1928)
Pyralidae (*Aphomia, Vitula*)	Bumblebees: *Bombus*	Larvae feed on waxen cells, occasionally on refuse or brood also	Frison (1926); Cumber (1949b); Free and Butler (1959); Pouvreau (1967)
Agrotidae (*Epizeuxis*)	Ants: *Formica*	Larvae are scavengers	F. Smith (1941)
DIPTERA			
Psychodidae (*Termitadelphos, Termitodipteron*)	Termites	?	Brues *et al.* (1954)
Culicidae (*Harpagomyia*)	Ants: *Crematogaster*	Adults solicit regurgitated food from workers on the odor trails	Farquharson (1918); Wheeler (1928)
Chironomidae (*Forcipomyia*)	Ants: *Formica*	Larvae prey on the host larvae	E. Séguy in Bernard (1968)
Ceratopogonidae (*Ceratopogon*)	Ants: *Formica*	Larvae live in host nests; obligatory relationship uncertain	Long (1902); E. Séguy in Bernard (1968)
Sciaridae (*Sciara*)	Ants: *Formica, Lasius*	Larvae live in host nests and are groomed by workers	E. Séguy in Bernard (1968)
Sciaridae (*Austrosciara*)	Termites	?	Brues *et al.* (1954)

TABLE 20-1 (*continued*).

Taxa	Hosts	Biology	Selected references
Cecidomyiidae (*Termitomastus*)	Termites: *Anoplotermes*	Physogastric adults; feeding behavior unknown	Silvestri (1920)
Phoridae (many genera and species)	Many ant genera; the flies are especially abundant in and around colonies of ecitonine army ants.	Usually scavengers; some species also share the prey of their hosts or feed on the hosts themselves	Borgmeier (1961–65); Rettenmeyer and Akre (1968); E. Séguy in Bernard (1968)
Phoridae (many genera and species)	Many termite genera	?	Schmitz (1916); Borgmeier (1961–65)
Phoridae (*Melittophora, Pseudohypocera*)	Stingless bees: *Melipona, Trigona*	*Melittophora* are brood commensals; *Pseudohypocera* are scavengers	Salt (1929)
Termitoxeniidae (all genera: *Termitoxenia, Termitomyia, Odontoxenia,* and others)	Higher termites	Physogastric adults; feeding behavior unknown	Bugnion (1913); Schmitz (1940, 1952); Borgmeier (1964a); Delachambre (1965); Kistner (1969)
Thaumatoxenidae (*Thaumatoxena*)	Higher termites	?	Trägårdh (1908); Schmitz (1915); Poisson (1937)
Syrphidae (*Microdon*)	Many ant genera	Scavengers	Wheeler (1910); Andries (1912); E. Séguy in Bernard (1968)
Syrphidae (*Volucella*)	Bumblebees and social wasps	Scavengers	Cumber (1949b); Kemper and Döhring (1967)
Conopidae (*Stylogaster*)	Ecitonine army ants	Hovering females follow raids of ecitonine army ants and parasitize cockroaches flushed out by the ants; they probably also parasitize adult tachinids that follow the ants	Rettenmeyer (1961b)
Tachinidae (*Androeuryops, Calodexia*)	Ecitonine army ants	Females follow ant raids and parasitize cockroaches and crickets flushed out by the ants	Rettenmeyer (1961b)
Anthomyiidae (*Epiplastocerus, Episthetosoma, Paraplastocerus, Plastocerontus, Prosthetosoma, Tetraplastocerus*)	Termites	Larvae, which possess exudatoria, live in termite nests	Silvestri (1920); Hollande *et al.* (1951)
Milichiidae (*Prosoetomilichia*)	Ants: *Hypoclinea*	Adults feed on anal droplets of workers	Jacobson (1909)
Braulidae (*Braula*)	Honeybees: *Apis*	The "bee louse," ectoparasite on the host adults	Imms (1942); Kaschef (1959)

TABLE 20-1 (*concluded*).

Taxa	Hosts	Biology	Selected references
HYMENOPTERA			
Perilampidae (*Echthrodape*)	Allodapine bees: *Braunsapis*	Symphile; external parasite on host pupae	Michener (1969b)
Diapriidae (*Phaenopria, Mimopria, Myrmecopria*)	Ecitonine army ants	Some species are mimetic and run in the host columns	Brues (1902); Wing (1951); Masner (1959)
Diapriidae (*Ashmeadopria, Auxopaedeutes, Bruesopria, Geodiapria, Lepidopria, Loxotropa, Solenopsia, Tetramopria,* and others)	Nonlegionary ants: *Solenopsis* (*Diplorhoptrum*) especially, also *Tetramorium*	*Lepidopria* and *Solenopsia* adults solicit regurgitation; larvae probably endoparasites of host brood	K. Hölldobler (1928); Wing (1951)
Bethylidae, Ceraphronidae, Dryinidae (many genera and species)	Many nonlegionary ant genera and species	Larvae probably internal parasites; some adults are mimetic and may have close relations with host colonies	Donisthorpe (1927); Bernard (1968)

diversity—which reinforce each other to produce a higher diversity and abundance of symbiotic species.

The evidence shows that many species are able to maintain a viable population size only under the protection of large colonies. The great majority of very specialized species are uncommon, and many are very rare and local in distribution. To cite several of many examples, no more than 400 specimens of the aberrant African termitophiles of the genera *Termitomimus* and *Nasutimimus* have ever been collected, despite repeated and intensive search for them in places where they are known to occur (Kistner, 1968d). Myrmecophilous, trail-following millipedes are absent from most army ant colonies, even though the latter contain many thousands and even millions of host individuals. The largest number recorded from one bivouac by Rettenmeyer (1962a), who made a special study of them, was 299, and even this population contained a mixture of two genera and several species. The most abundant insect guests of army ants—and perhaps of any kind of social insects—are phorid flies, but, even when all species are combined, the number of adults in a single bivouac usually does not exceed 4,000 (Rettenmeyer and Akre, 1968). Staphylinids are also very numerous around army ant colonies, but there is seldom if ever more than one beetle per thousand workers, and the average is far lower (Akre and Rettenmeyer, 1966). Using the populations of phorids and staphylinids as rough standards, the relative abundance of the ecitophiles can be judged from the total collections from all colonies made by Rettenmeyer to the year 1962: 8,000 mites, 2,400 phorid flies, 1,100 limulodid beetles, 300 staphylinid beetles, 300 collembolans, 170 thysanurans, 150 millipedes, 140 histerid beetles, and 6 diapriid wasps.

The concept of an insect colony as an island, or, to use slightly more precise language, a partially isolated ecosystem, can be extended to gain a better understanding of certain aspects of the biology of the symbionts. Consider the matter of microhabitat apportionment. The species of mites associated with army ants, according to Rettenmeyer, can be conveniently divided into two main groups, those that live in the refuse piles, or "kitchen middens," of the ants and those that live on the bodies of the ants or within their bivouacs. The latter group, the only true myrmecophiles in this case, have finely apportioned the environment through amazing feats of specialization. Several of the extreme adaptations are illustrated in Figure 20-1. Of special interest are the Circocyllibanidae and Coxequesomidae, which appear to live only on cer-

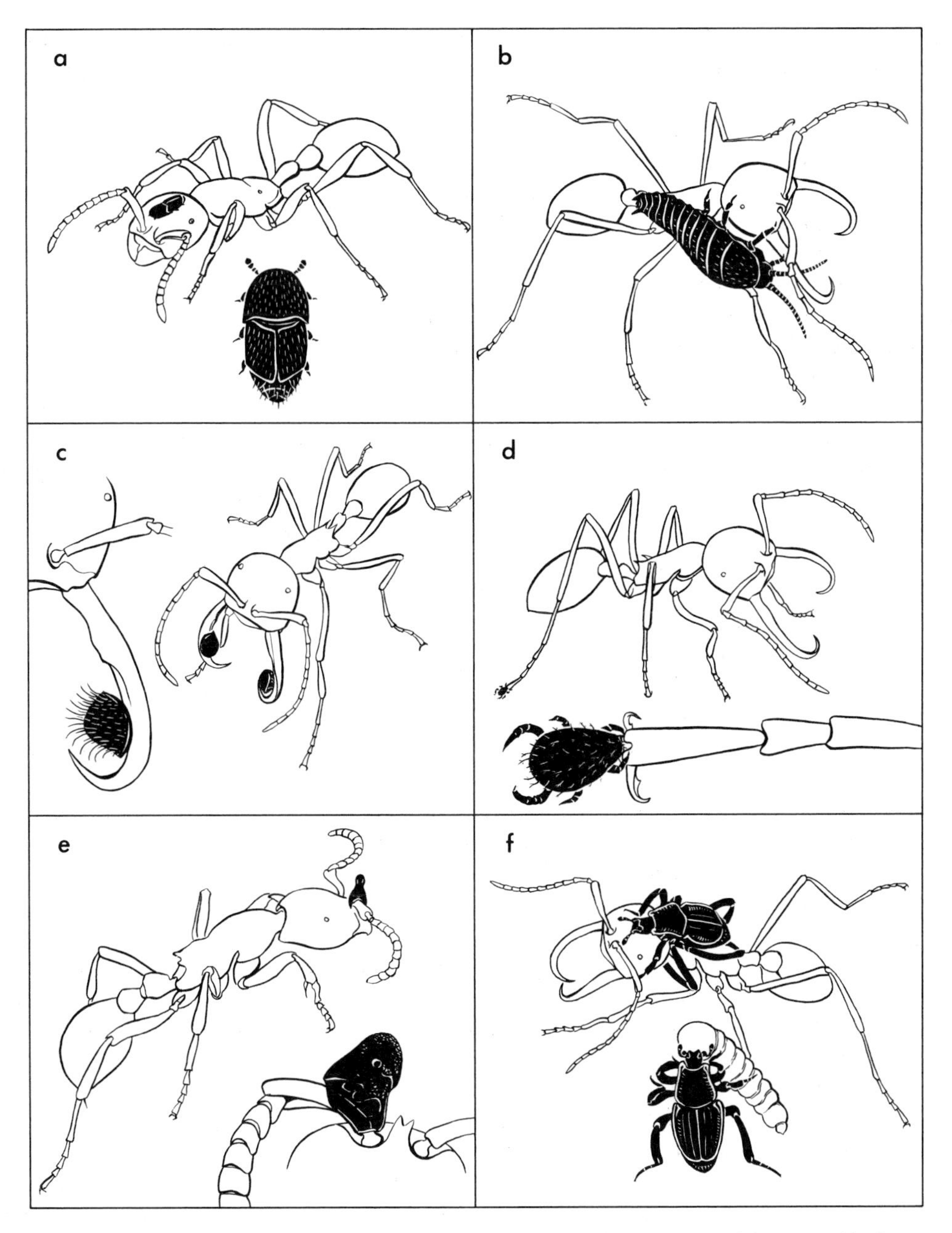

FIGURE 20-1. Six arthropod guests of army ants, which display a few of the many kinds of symbiotic adaptations to these hosts: (*a*) *Paralimulodes wasmanni* is a limulodid beetle that spends most of its time riding on the bodies of host workers (*Neivamyrmex nigrescens*); (*b*) *Trichatelura manni* is a nicoletiid silverfish that scrapes and licks the body secretions of its hosts (*Eciton* spp.) and shares their prey; (*c*) the *Circocylliba* mite shown here belongs to a species specialized for riding on the inner surface of the mandibles of major workers of *Eciton;* (*d*) *Macrocheles rettenmeyeri* is another mite species which is normally found attached in the position shown, serving as an extra "foot" for workers of *Eciton dulcius;* (*e*) *Antennequesoma* is a mite genus highly specialized for attachment to the first antennal segment of army ants; (*f*) the histerid beetle *Euxenister caroli* grooms the adults of *Eciton burchelli* and feeds on their larvae (original drawings by Turid Hölldobler, based on photographs by C. W. Rettenmeyer and R. D. Akre and sketches of *Paralimulodes* by Wilson *et al.,* 1954).

tain parts of the host bodies, such as the mandibles, scapes, or coxae. Even more extraordinary is the total adaptation of *Macrocheles rettenmeyeri.* The mite feeds on blood taken from the terminal membranous lobe (arolium) of the hind tarsus of a large media or soldier of its exclusive host species, *Eciton dulcius.* In the process it allows its entire body to be used by the host ant as a substitute for the terminal segment of the foot. Rettenmeyer (1962b) described its behavior: "The mite inserts its chelicerae into the membrane, but it is not known if the palpi assist in holding to the ant. Presumably the mite feeds in this position and may be considered a true ectoparasite rather than a strictly phoretic form . . . Army ants characteristically form clusters by hooking their tarsal claws over the legs or other parts of the bodies of other workers. Temporary nests or bivouacs of *Eciton dulcius crassinode* may have clusters several inches in diameter hanging in cavities in the ground. In small laboratory nests, when a worker hooked the leg with the mite onto the nest or another worker, the hind legs of the mite usually served in place of the ant's tarsal claws. The hind legs of the mite were never seen in a straight position but were always curved. In all cases the entire hind legs of the mite served as claws, not just the mite's claws. No difference was noted in the behavior of the ant whether its own claws or the hind legs of the mite were hooked onto some support. When no other part of the ant's leg was touching any object, adequate support was provided by the mite's legs for at least five minutes." A second species, *M. dibanos,* was observed to parasitize *Eciton vagans* in the same manner.

Mites, being much smaller than their hosts, function primarily as ectoparasites or scavengers within the army ant nests. Many of the insects, on the other hand, either scavenge or obtain part of their nutriment by licking the oily secretions of the bodies of the hosts, a specialized behavior referred to by Wheeler and some later authors as "strigilation." Other insects feed on the booty brought in by the host workers, or else prey on the hosts themselves, especially the defenseless larvae and pupae. A few species combine strigilation with prey sharing or predation on the hosts. The nutritive value of strigilation is open to some question, at least in the case of the ecitophiles. Both Rettenmeyer and Akre have suggested that it also serves to transfer the host odor to the parasites, a logical hypothesis but one that will be difficult to test. As illustrated in Figure 20-1, the insect guests vary greatly in size and in the locations they occupy with respect to the host workers. They also vary in the spatial positions they occupy within both the bivouac as a whole and in the ant columns during colony emigrations.

The Ectosymbiotic Adaptations

It is apparent that, in the course of adapting to social insects as hosts, many of the symbionts have undergone considerable evolution in their food habits. The nearest relatives of the nitidulid *Amphotis marginata,* to take one example, are sap feeders. If we assume that the ancestors of *A. marginata* had a similar diet, it is understandable but no less impressive that as a myrmecophile it was able to shift to the honeydew (a sap product) supplied to it indirectly by its ant hosts. *Amorphocephalus coronatus* and other myrmecophilous brenthids have probably made a similar transition from some form of phytophagous behavior, while *Cremastocheilus* has changed from what was almost certainly a herbivorous diet to become the only known fully carnivorous scarabaeid. Judging from the behavior of most other cockroaches and crickets, *Attaphila* and *Myrmecophila* shifted from herbivorous or detritus feeding habits to their present apparent dependence on the strigilation of host secretions. The same may be true of the ecitophilous thysanurans, which now feed on both the body secretions of their hosts and the fresh prey brought in by them. The adults of the beetle family Aculagnathidae possess unique elongate, sucking mouthparts. Their feeding behavior is still unknown, but it will almost certainly be found to represent a major departure from the ancestral mode. It is equally apparent that, when the extreme symbiotic staphylinids are analyzed more carefully, especially the more extreme members of the Corotocini, other examples of dietary shifts will be discovered.

As an accompaniment to the deep penetration of insect colonies, symbionts of the various groups have tended independently to undergo one or more of a relatively small set of evolutionary changes in morphology and behavior. These convergent traits, which have been documented and analyzed in steadily increasing detail by entomologists from Wasmann, Wheeler, Janet, Escherich, and Donisthorpe to Seevers, Kistner, Pasteels, Rettenmeyer, and their co-workers, provide an understanding of the minimal adaptations required for ectosymbiosis. In the following summary, first the morphological and then the behavioral traits will be discussed with particular reference to the ways in which they promote the integration of symbionts with their host colonies.

Lighter body coloration. The more completely adapted myrmecophiles and termitophiles tend to be lighter in color than their free-living relatives. Among the symphilic beetles, a peculiar dark blood-red coloration is very common and is often combined with an oily body surface. The loss of coloration, including that leading to the symphilic red, may be nothing more than accommodation to a more subterranean existence, or part of the widespread parasitic syndrome of morphological regression to be described later.

Appeasement substances and trichomes. Many of the better-integrated symphiles dispense attractive substances to their hosts from epidermal glands. These materials are eagerly licked up by the host workers and even by other symbionts living with the colony. They are generally interpreted as devices for gaining acceptance and keeping favor with the hosts, and for this reason I suggest that they can be conveniently referred to as "appeasement substances." Bert Hölldobler (1970) has referred to the behavioral phenomenon as *Besänftigung,* that is, appeasement or calming, and the glands that produce the effect as *Besänftigungsdrusen.* Our best information on morphology comes from Pasteel's excellent study (1968) of the epidermal glandular system of species of the corotocine staphylinid genus *Termitella,* which live as symphiles in the nests of *Nasutitermes.* In addition to the "primary glandular system," which is shared with other staphylinids and functions principally to manufacture defensive and lubricating secretions, *Termitella* adults possess a "secondary glandular system" for the production of appeasement substances. The secondary system consists principally of the voluminous, paired poststernal glands, which are illustrated in Figure 20-2. These organs, which have evidently evolved *de novo* in the termitophilous forms, produce secretions which are licked by the termite hosts. The unicellular glands of type 1 are hypertrophied in *Termitella* and also produce attractants.

Similar novel glands exist in the myrmecophilous Staphylinidae (B. Hölldobler, 1971). As shown in Figure 20-3, the glands used in communication by *Atemeles* are at a different location from those of *Termitella,* and they probably have been independently derived in evolution. When a beetle approaches the nest of a host colony to seek adoption, it wanders around until it encounters a

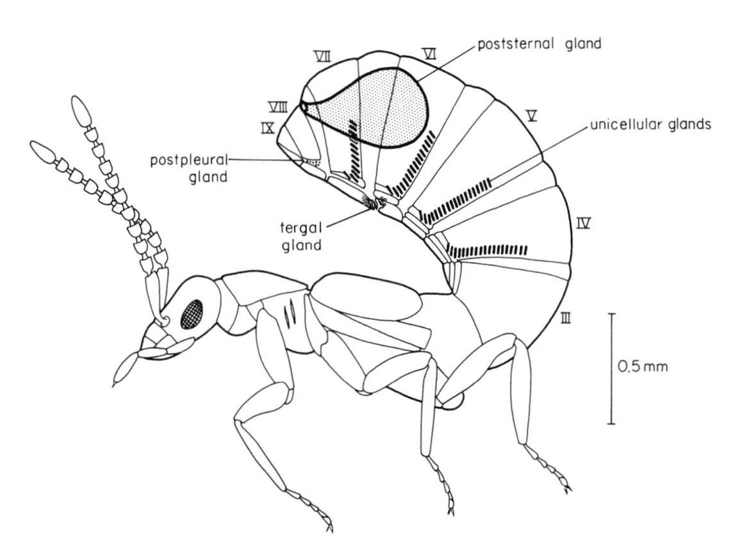

FIGURE 20-2. Side view of a termitophilous staphylinid of the genus *Termitella,* showing the location of the principal epidermal glands of the abdomen. The primary system includes the tergal gland, which produces a defensive secretion, unicellular glands of "type 1" located along the edges of sternites IV to VII, and the postpleural gland. The secondary system consists principally of the poststernal glands; the left member of the pair is shown here. The poststernal glands and enlarged unicellular glands of type 1 produce substances attractive to the termite hosts (redrawn from Pasteels, 1968).

worker. Then it turns to present its appeasement gland, located at the tip of the abdomen. The secretion of the gland is at least partly proteinaceous, and it contains no appreciable amounts of carbohydrates. The ant feeds on the material, seeming to grow calmer in the process. Then it moves over to the "adoption glands," which it also licks. After this second repast, the ant transports the beetle into the nest.

A large percentage of the symphilic beetles and at least one proctotrupid wasp considered to be symphilic possess peculiar tufts of red or golden yellow hairs called trichomes or trichodes. These brush-like structures aid in the dissemination of appeasement substances. The host workers, and sometimes even other symbionts, lick and gnaw the hairs and glandular openings at their bases. Wasmann (quoted by Wheeler, 1910) summarized the location of the trichomes: "On the sides and base of the portions of the abdomen not covered by the wing-cases (*Lomechusa* group of Staphylinids, many Clavigerids); on the tip of the abdomen (*Lomechusa* group), or pygidium

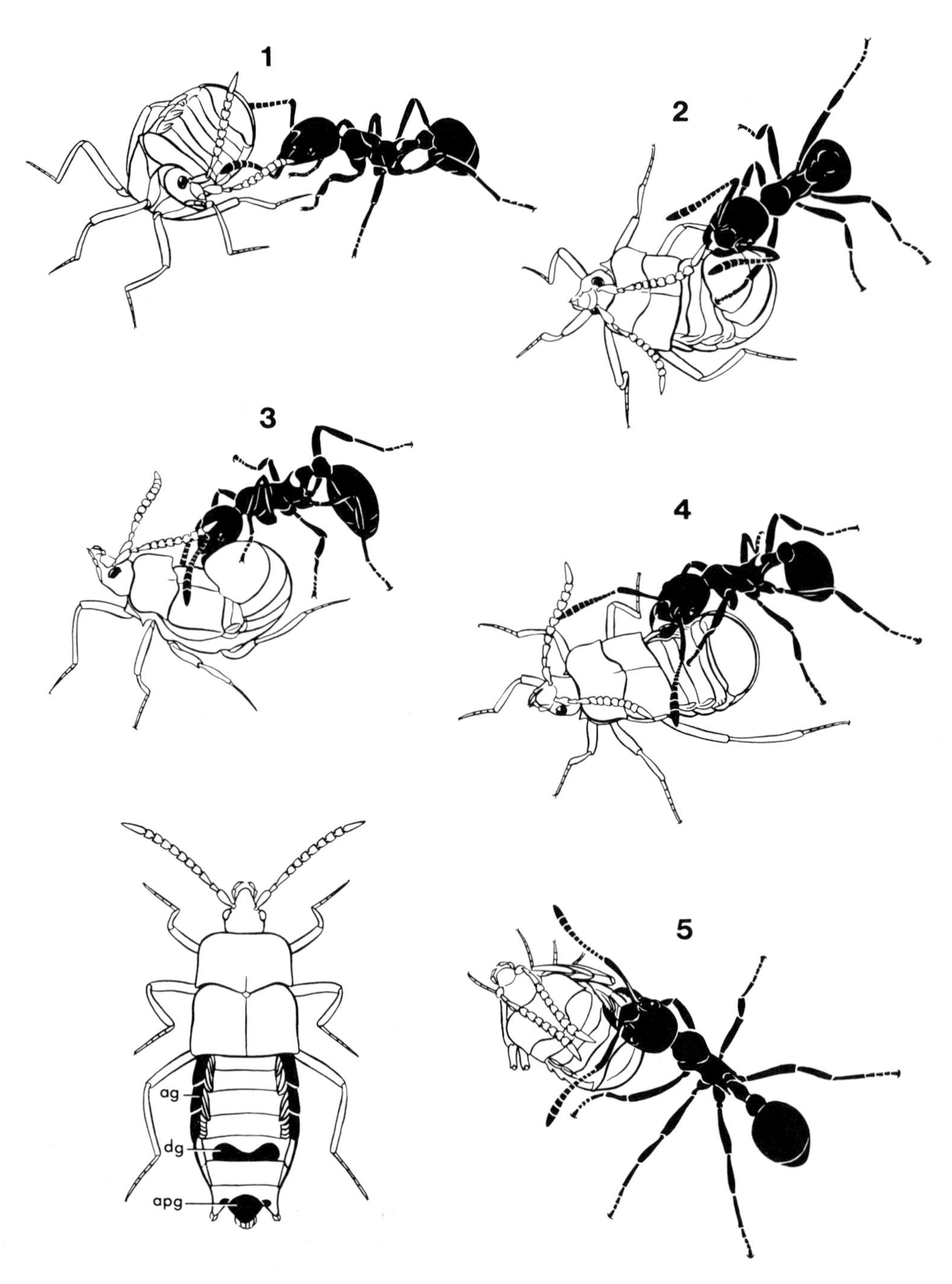

FIGURE 20-3. Symphily in the staphylinid beetle *Atemeles pubicollis.* The figure at the lower left indicates the location of the three principal abdominal glands of the parasite: (*ag*) adoption glands; (*dg*) defensive glands; (*apg*) appeasement gland. The beetle presents its appeasement gland to a worker of *Myrmica* that has just approached it (1). After licking the gland opening (2), the worker moves around to lick the adoption glands (3, 4), after which it carries the beetle into the nest (5) (after Hölldobler).

(many *Paussus*); at the tips of the wing-cases (many Clavigerids, *Chaetopisthes* among termitophilous Scarabaeids); on the sides of the wing-cases (many *Paussus*); at the posterior corners or edges of the prothorax (*Pleuropterus* and many *Paussus* among Paussids, *Lomechon* among Silphids, *Corythoderus* and *Chaetopisthes* among termitophilous Scarabaeids, *Tylois* among Histerids, many *Thorictus* among Thorictids); on the anterior corners of the prothorax (*Napochus termitophilus*); on the much elevated sides of the prothorax (*Teratosoma* among Histerids, *Gnostus* among Gnostids); in a median transverse groove on the prothorax (many *Paussus*); on the neck, between the head and prothorax (the myrmecophilous *Napochus* among the Scydmaenidae, *Tetramopria* among Proctotrupids); on a perforated horn on the vertex (several *Paussus*); on the front (*Pogonoxenus* among Tenebrionids); on the antennal club (many *Paussus*); and even on the coxae and tips of the femora (*Lomechusa*)." (See Figure 20-4.)

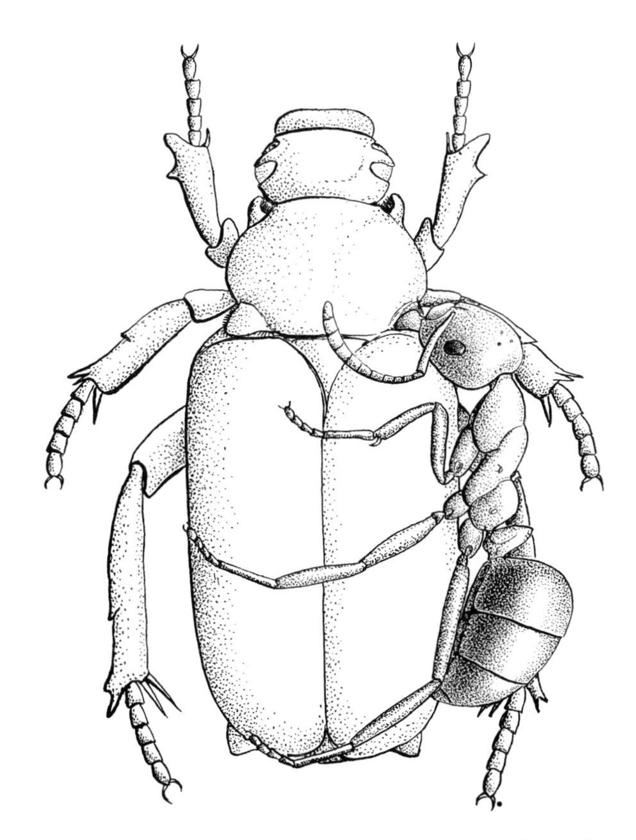

FIGURE 20-4. A worker of *Formica integra* gnawing at the thoracic trichomes of the scarab beetle *Cremastocheilus castaneae* (from Wheeler, 1910).

We do not know whether the appeasement substances are simple nutrients, or secondary phagostimulants, or more specialized compounds that mimic the attractant pheromones of the hosts. If they prove to be nutrients or phagostimulants, the further problem exists of how the symphiles keep from being consumed entirely. When chemical identifications are achieved in accompaniment with new behavioral studies, they are likely to further illuminate not only symbiosis but also some of the basic problems of communication in social insects.

Mimicry, physogastry, exudatoria, glandular antennae. Among the battalions of parasitic staphilinids which march with the army ants are many species that strikingly resemble their hosts. This myrmecoid appearance is found almost nowhere else in the Staphylinidae. It has originated many times over through modifications of the abdomen and thorax which then create an ant-like "petiole." In addition, there is a strong tendency to resemble the hosts in the overall slender body form, in color, and even in the sculpturation of the body surface. Seevers (1965) showed that petioles have been created in no less than seven ways in various groups of the Staphylinidae, with several of the modifications having been chosen by two or more groups independently. In some phyletic lines the anterior abdominal segments have simply been pinched into tubes or cones to make a narrowed articulation with the thorax. In others the petiole is constructed entirely either from the tergite of the second abdominal segment, the sides of which have been extended ventrally to form a partial or complete tube, or from elements of the third abdominal segment. In still others the large metepimera of the thorax are prolonged ventrally and joined to create a fused, tubular structure that articulates posteriorly with the abdomen. Three of these basic types are represented in Figure 20-5. Although the ant-like species are accompanied in the nests by staphylinids of a more generalized body shape, their large numbers and the remarkable degree of evolutionary convergence they encompass leave little doubt that some kind of mimicry is involved. The question is, who is being fooled? Wasmann (1903 and contained references), in first documenting the phenomenon from museum specimens, was persuaded that the form of the body, including the false petiole, is tactile mimicry that deceives the host ants. He further believed that the coloration is visual mimicry that deceives birds

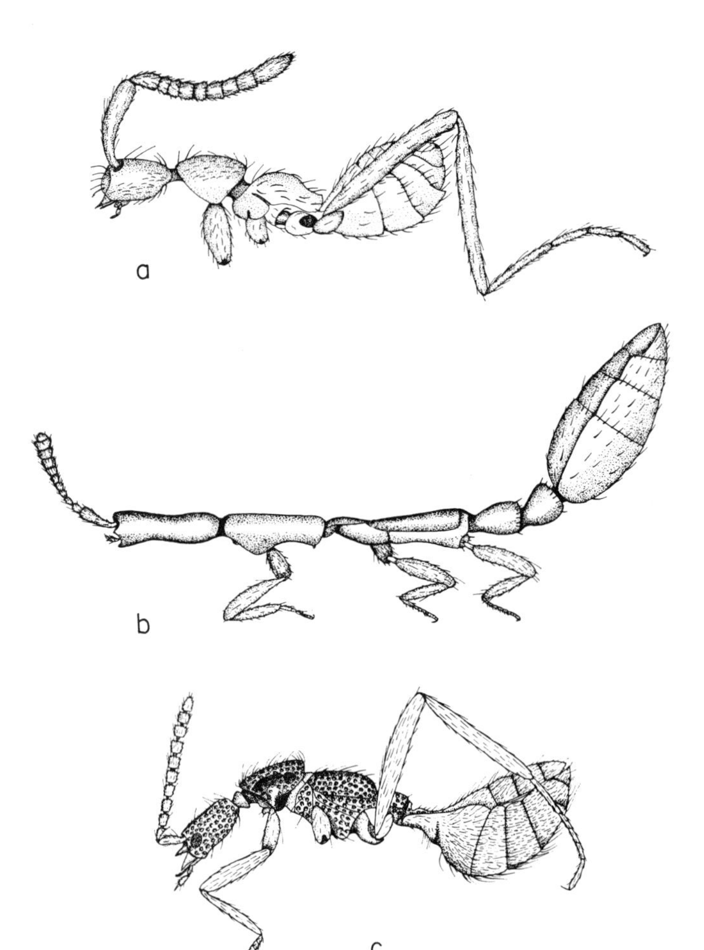

FIGURE 20-5. Examples of mimetic staphylinid beetles that live with army ants: (*a*) *Myrmeciton antennatum,* a guest of New World army ants (*Labidus*) which has a generalized "petiole" formed simply by a constriction of the anterior abdominal segments; (*b*) *Mimanomma spectrum,* a guest of African driver ants (*Anomma*), in which the sclerites of each of the two anterior abdominal segments have fused together to form petiolar segments; (*c*) *Ecitosius gracilis,* a guest of New World army ants (*Neivamyrmex*) in which the petiole is formed from the third abdominal segment (from Seevers, 1965).

and other predators which might otherwise be able to pick the beetles out of the columns of running ants. The same interpretation has been adopted by Kistner (1966a,b, 1969) on the basis of his own field observations of staphylinid guests of African driver ants. Species of *Dorylomimus, Dorylonannus, Dorylogaster,* and *Mimanomma* run with the ants in their columns. When a host worker encounters one of the beetles, it touches it lightly with its antennae. Kistner states, "This action is identical with that of an ant when it meets another ant. Typically, the antenna rubs along the ant and lingers at the petiole, then both ants move on. I have interpreted this palpation as a signal which tells the blind ants that the passing insect is another ant. I have further interpreted this with regard to the Dorylomimini, that the constriction in the abdomen is such that the same signal is evoked when palpated by the ants and both ants and myrmecophiles go their separate ways. Thus the morphological constriction permits the myrmecophile to function within the colony as though it were an ant." I find the Wasmann-Kistner explanation unconvincing. We know that chemical identification is paramount in ants generally. When the surface odor of a worker ant is disturbed only slightly, the ant is swiftly attacked by its own sisters even though its morphology remains unchanged. Having watched the antennal greetings of many kinds of ants in life and on slow-motion film, it has not seemed to me that the workers routinely inspect one another's midsection, although I have not examined driver ants and cannot dispute Kistner's impression that it occurs in these particular insects. Although the hypothesis of tactile mimicry must not be dismissed out of hand, an alternative hypothesis is equally promising. It is possible that predators watching the ant columns for edible morsels might be fooled by both the color of the beetles and their shape as well. Thus all of the mimicry apparent to taxonomists may be visual and directed at predators outside the host colony.

If morphological mimicry on the part of symbionts is indeed visual in its form of transmission, we should expect it to be absent in cases where the animals spend their entire lives underground. Turning to the termitophiles, we find evidence which in my view strongly supports this interpretation. The vast majority of the termitophilous staphylinids and other insects do not in the least resemble termites. A common modification is that shown by *Termitella* (Figure 20-2) in which the abdomen is capable of being curled forward over the back. This action does not increase the resemblance of the body to that of a termite worker. On the contrary, the resulting double-tiered structure lessens the resemblance, especially when the beetle is viewed from the side. What the action does do is to bring the appeasement substances of the abdominal glands to the front of the termite, where they can be brought into play before the host workers have much chance to examine the beetles during face-to-face encounters. However, there is a complication to the staphylinid story, one that involves a wholly different form of adaptation. This is physogastry, the exceptional growth of the abdomen which occurs in the adults of many of the termitophilous species. In extreme cases, one of which is illustrated in Figure 20-6, the abdomen is bloated to a degree exceeding even that found in the physogastric termite queens. The abdominal sclerites are pushed far apart and come to comprise only an insignificant portion of the body surface. The abdomen has been turned permanently up and over the remainder of the body, so that its tip points obliquely upward or straight forward. Studies by Emerson, Seevers, and Kistner have revealed that the physogastry is due entirely to postimaginal growth, just as it is in the termite queens. The young adults have the typical staphylinid form when they first emerge from the pupal stage and add abdominal tissue only gradually thereafter. Trägårdh (1907b) found that in *Termitomimus* the physogastry is the result primarily of the hypertrophy of the fat bodies and secondarily of the enlargement of the ovaries. He was impressed by the superficial resemblance of the bloated abdomen of this beetle to a termite worker—hence, the Greek suffix *-mimus* given in the scientific name of the beetle. Trägårdh actually went so far as to suggest that the resemblance is functional mimicry. This idea was ignored for the most part by other entomologists because at about the same time Wasmann and Holmgren propounded their hypothesis that physogastry in termite queens serves to generate fatty exudates attractive to workers. Wheeler (1918, 1923) extended this notion and argued that the physogastric staphylinids are simply imitating the queen in order to gain the same attention from the workers. This viewpoint acquired considerable impetus when Wheeler (1928) later observed termite workers carrying and licking physogastric adults of *Spirachtha* in South America. Consequently, the notion of chemical mimicry has taken precedence over that of visual or tactile mimicry in the case of the physogastric staphylinids. There nevertheless exist three genera belonging to the tribe Corotocini (*Spirachtha, Spirachthodes,* and *Coatonachthodes*) in which the physical resemblance of the abdomen to a termite worker is too close to allow the hypothesis of morphological mimicry to be dropped entirely. Adults belonging to these aberrant forms have three or four pairs of finger-like appendages ("exudatoria") growing from the sides of the inverted portion of the abdomen in about the positions one would expect to find legs and antennae if the abdomen were to be viewed as a termite pseudobody. The *ne plus ultra* of this phenomenon is shown by Kistner's *Coatonachthodes ovambolandicus,* the species illustrated in Figure 20-6. Only a single specimen is known, and it was found as a

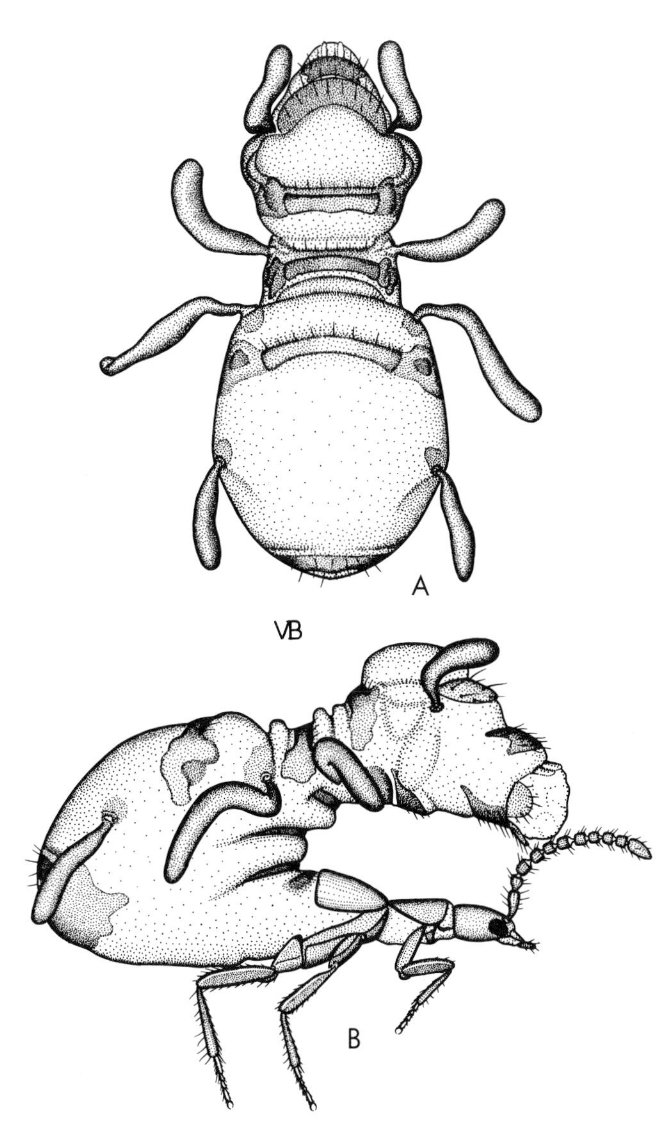

FIGURE 20-6. *Coatonachthodes ovambolandicus,* a physogastric staphylinid beetle that lives with *Fulleritermes* in Africa (*A,* top view; *B,* side view). The swollen upper part of the abdomen resembles the body of a termite worker, complete with four pairs of "exudatoria" that resemble termite appendages (from Kistner, 1968c).

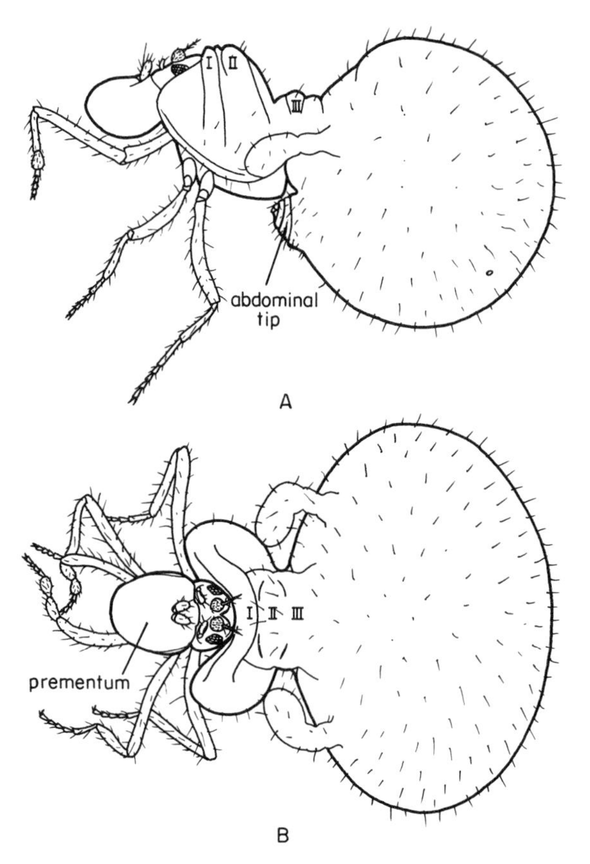

FIGURE 20-7. The termitoxeniid fly *Cheiloxenia obesa,* a guest of the termite *Allodontermes giffardii* in Africa (*A*, side view; *B*, top view). This is the extreme case of physogastry known among the termitophilous Diptera. The anterior segments of the abdomen are numbered in this drawing to show the extent of their displacement (redrawn from Delachambre, 1966).

guest in a colony of *Fulleritermes contractus* in Southwest Africa. One must admit that the resemblance of its abdomen to the entire body of a termite is uncanny. The effect is even stronger in photographs of the preserved specimen, since the exoskeleton of the pseudobody is darkened in just the right places to give the impression of the existence of a pair of mandibles and of constrictions separating the pseudocaput, pseudothorax, and pseudoabdomen. There is no way at present to decide whether this resemblance is mere coincidence, the kind of random correspondence that might be expected to appear somewhere among the hundreds of termitophilous staphylinids, or whether it is the most nearly perfect example yet uncovered of an evolutionary trend to achieve morphological mimicry by means of physogastry. Any significant advance in our understanding of physogastry in general must await biochemical analyses and an experimental testing of the several possible variants of the mimicry hypothesis. Physogastry also occurs in the fly families Cecidomyiidae and Termitoxeniidae, which belong to the suborders Nematocera and Brachycera, respectively. Future comparisons of the behavior and physiology of these phylogenetically disparate groups with each other and with the beetles should prove instructive. Delachambre (1966) has recently described what may be the ultimate fly species in *Cheiloxenia obesa.* This little termitoxeniid, a guest of *Allodontermes giffardii* in Africa, is convergent in many respects to the extreme corotocine staphylinids, although it does not resemble a termite worker in overall appearance (Figure 20-7). The first two segments of the physogastric abdomen overlap the thorax and hind margin of the head like a thick blanket. To the rear the main portion of the abdomen swells into a massive globe that sends out a single pair of lateral "exudatoria." The tip of the abdomen, unlike that of the physogastric staphylinids, projects beneath the remainder of the body. The prementum and middle segments of the fore tarsi are also swollen, neither feature being paralleled in the beetles. Looking at this bizarre creature, and the other extreme termitoxeniids and staphylinids as well, one gets the impression that the physogastry is designed to project fatty tissue, and the appeasement substances that may be contained in it, in all directions from the body of the termitophile. In other words, it is possible that the abdomen has been reshaped into a kind of conciliatory shielding device.

The exudatoria may or may not serve a similar function of attraction and appeasement toward termite hosts. Wheeler (1928) actually saw workers of *Nasutitermes* licking the exudatoria and other parts of the bodies of *Spirachtha mirabilis* in a nest in Guyana. Silvestri (1920) described a remarkable series of other termitophilous insect groups from West Africa where larvae have well-developed lateral exudatoria, including *Troctontus* (Coleoptera: Melandryidae); *Ceroncinus* (Coleoptera: ?Lathridiidae); *Ctenophorellus* (Coleoptera: ?Cantharidae); *Epiplastocerus, Plastocerontus, Prosthetosoma,* and *Tetraplastocerus* (Diptera: Anthomyiidae); and *Plastopolypus* (Lepidoptera: Tineidae). Few if any of these species are physogastric as well, suggesting that either the exudatory organs can supplant the abdomen in its role of physogastry or else they serve a different function. In fact, two alternative functions have been suggested. Silvestri believed that the organs produce defensive secretions, which can be brought to bear against host termites in the event of an attack. Hollande, Cachon, and Vaillant (1951), on the other hand, concluded from a careful histological study of *Troctontus, Plastopolypus,* and a variety of termitophilous fly larvae that the organs are sensory in function. They noted that the appendages, which they preferred to call *dactylonèmes* instead of exudatoria, do not contain special glandular tissue and openings in most of the species; instead, they are richly supplied with sensory units, especially sensilla coeloconica. Whether this much more informed opinion is correct, or whether the appendages serve various functions in different termitophile groups, it is at least clear that there is a strong selection favoring the development of these novel organs since they are of independent origin in each of the families that possesses them. It may be useful to note also that superficially similar organs exist in larvae of the ant genus *Pachysima,* which is not associated with termites in any way. The function of the exudatoria in these insects is also a complete mystery (see Chapter 14).

Finally, a strange glandular swelling of the antennae occurs in at least two groups of myrmecophilous beetles, the Paussidae and ectrephine Ptinidae. The club-like form of the antennae of the myrmecophilous Clavigeridae, Pselaphidae, and Endomychidae, and of the termitophilous Tenebrionidae may represent the same phenomenon. Mou (1938) has shown that paussid antennae contain considerable amounts of glandular tissue, although the behavioral significance of the trait seems not to have been explicitly investigated. The following observation by Clark (1923) on Australian myrmecophilous ptinids is very suggestive. He watched adults of *Enasiba tristis* and *E. conifer* in laboratory nests containing a colony of *Iridomyrmex conifer* and noted that occasionally "one, and sometimes two ants were seen to attach themselves to the antennae, and appeared to be getting great satisfaction by nibbling and licking the apical joints, stroking the beetle meanwhile with their antennae. None of the ants were seen to attach themselves to the fascicles in the prothorax."

Limuloid body form. A great many of the staphylinid and limulodid beetle guests of army ants possess a protective body form radically different from the mimetic type (Wasmann, 1920; Seevers, 1965). The overall shape superficially resembles that of the Horseshoe Crab *Limulus:* the pronota and elytra of the beetles are expanded and carapace-like, covering all or most of the head, mouthparts, appendages, and even the abdomen, which is partly retractile. Wasmann, Wheeler, and others have interpreted the limuloid body type to be a defensive design against predators and the host ants themselves. Wasmann referred to it often as the *Trutztypus,* despite the fact that *Schutztypus* is more appropriate to its putative function. A limuloid form also occurs in some symbiotic flies. The most extreme example known, and surely one of the most bizarre morphological modifications of any kind in all the insects, is that of *Thaumatoxena.* The adults look more like little crustaceans than the true dipterans they are (Figure 20-8).

Morphological regression. Although certain organs have been acquired *de novo* by the extreme symbionts, for example, trichomes, exudatoria, and some of the exocrine

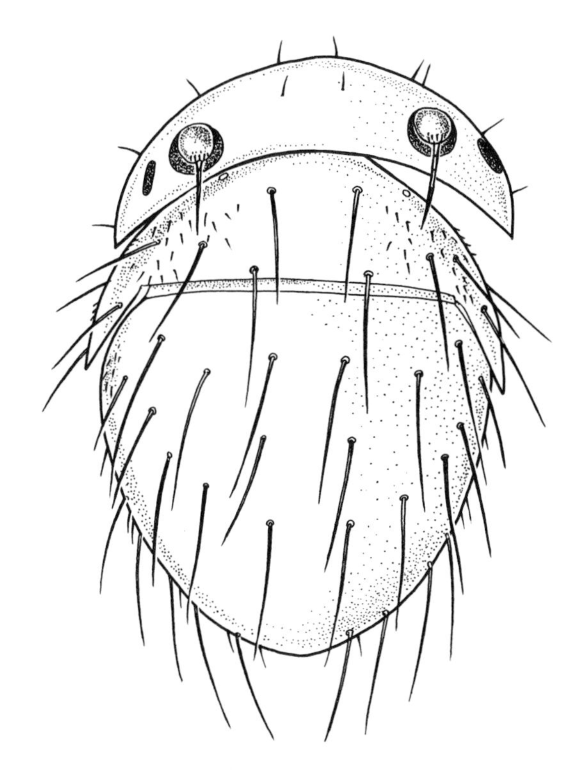

FIGURE 20-8. An adult of the phoroid fly *Thaumatoxena wasmanni,* showing the extreme limuloid body form associated with life in termite colonies (redrawn from Trägårdh, 1908).

glands, the prevailing feature of morphological specialization has been the reduction or loss of parts. Flightlessness, accompanied by shortening or even loss of the wings, is common. Reduction or loss of the eyes is also widespread. Other general trends include reduction in antennal and tarsal segmentation and, at least in some of the liquid-feeding symphilic beetles, degeneration of the mouthparts. As one might expect, the greatest amount of regression occurs in the symphilic species that have become most closely integrated with the host society. Regression is also relatively advanced in species that have adapted to a more subterranean existence in order to live with their hosts. Seevers (1965) conducted a special survey of the phenomenon in the staphylinids that live with army ants and noted the following recurring traits: eyelessness, winglessness, elytral fusion, elytral reduction, fusion of cephalic sclerites, extreme fusion of thoracic sclerites, reduction in number of antennal and tarsal segments, and loss of paratergites and parasternites. A complete loss of elytra has occurred in *Mimeciton,* a condition that may be unique among all of the adult nonlarviform beetles. Seevers found that the eyes have been lost independently in at least ten of the phyletic lines of Aleocharinae that live with driver ants. Important regressive changes have also occurred in the intestines of the adults of myrmecophilous silverfish, which subsist entirely on liquid meals obtained from their ant hosts (Pohl, 1957).

Integrative behavior. A gradually increasing degree of behavioral integration with the host society constitutes the most obvious of the evolutionary pathways along which progress in symbiosis can be measured. Multiple devices exist for achieving integration. They vary greatly from species to species, in ways that are equally dependent on the ancestry of the symbionts and on the ecology and behavior of the hosts. The explicit analysis of this phenomenon has scarcely begun. A model is provided by the recent study of ecitophilous staphylinids by Akre and Rettenmeyer (1966). These authors started by making a distinction between "generalized" species, having the typical appearance of most Staphylinidae and with no obvious modifications for life with ants, and "specialized" species, which include the mimetic forms. The specialized staphylinids are further distinguished by the possession of rigid abdomens held in a single position. Paradoxically, these insects are also less common than the nonmimetic forms in the host colonies, although, of course, this trait was not used in the analysis. Akre and Rettenmeyer were able to compile an impressive list of consistent behavioral differences between the two kinds of species, which is reproduced in Table 20-2. All represent steps on the part of the specialized species to adapt more closely to the host ants and to their way of life. It seems likely that biologists, by accumulating a number of such dichotomous lists for various taxa of symbionts, especially for those living with host species which themselves occupy very different niches, will be able to gain deeper insight into both the varieties of symbiotic existence and the communicative devices invented to achieve them.

This brings us to the matter of communication between hosts and symbionts—the "breaking of the host code" by the guests of the ants and termites. How do such alien creatures manage this? The puzzlement of early investigators was neatly expressed by Wheeler when he said, "Were we to behave in an analogous manner we should live in a truly Alice-in-Wonderland society. We should delight in keeping porcupines, alligators, lobsters, etc., in our homes, insist on their sitting down to table with us and feed them so solicitously with spoon victuals that our children would either perish of neglect or grow up as hopeless rhachitics." This problem has begun to yield somewhat to experimental investigation in the past several years, especially when use has been made of the same techniques that earlier proved successful in the analysis of primary communication among the social insects. It seems logical also to apply the same behavioral classification to both cases, and this procedure will be used in the following account of symbiotic adaptations.

1. *Trail following.* That the guests of army ants can orient by the odor trails of the hosts without further cues provided by the hosts themselves has been known for a long time. Some of the phorids bring up the "rear guard" of emigration columns, sprinting deftly along each twist and turn of the newly vacated trails. Similar unaided trail following can be observed in millipedes and other symbionts when these animals are traveling in ant columns but have entered a stretch momentarily devoid of ants (Rettenmeyer, 1963b). Akre and Rettenmeyer (1968) investigated this phenomenon systematically by exposing a large variety of guests of ecitonine army ants to natural odor trails laid over the floor of a laboratory arena. They found that nearly all of the species tested, including a random sample of Staphylinidae, Histeridae, Phoridae, Limulodidae, Thysanura, and Diplopoda, were able to orient accurately by means of the trails alone (see Figure

TABLE 20-2. Behavioral differences between morphologically "generalized" and "specialized" Staphylinidae associated with army ants (from Akre and Rettenmeyer, 1966).

Generalized species	Specialized species
1. Live around periphery of bivouacs or in refuse deposits	1. Live within bivouacs; rarely seen in refuse deposits or outside bivouacs (except on emigrations)
2. Found at start of emigration before brood is carried; may be found in smaller numbers during rest of emigration; also found at end of emigration, including after ants have disappeared from column	2. Most frequently found when first part of brood is carried; may be present throughout time when brood is seen; rare or absent at other times
3. Run along edges of emigration columns and short distances from columns; sometimes in centers of columns	3. Run in centers of emigration columns
4. Sometimes found at bivouac sites shortly after ants have emigrated	4. Absent or rare at bivouac sites shortly after ants have emigrated
5. Do not ride on brood or booty carried by workers in raid or emigration columns	5. May ride on booty and brood in emigration columns
6. Not carried by ants (excluding attacking workers)	6. Carried by workers (few species only)
7. Frequently attacked by ants in nest	7. Usually not attacked by ants in nest
8. Attack adult ants but usually kill only injured or weak workers	8. Do not attack adult ants
9. Can tolerate a wider range of ecological conditions; live with a few or no ants for a day to weeks	9. Can tolerate only a narrow range of ecological conditions; usually die within a few hours of removal from colony
10. Never groom living workers; sometimes appear to feed on surfaces of dead workers; workers do not groom staphylinids	10. Often groom surfaces of living workers; workers tolerate grooming and may groom staphylinids
11. Never rub their legs on workers	11. May rub their legs on workers
Examples: *Microdonia, Tetradonia, Ecitana, Ecitodonia*	Examples: *Ecitomorpha, Ecitophya, Probeyeria, Ecitosius*

20-9). In some instances the guests were more sensitive to the trail pheromones than the very ants that secreted them. In choice tests, the ecitophiles generally preferred trails laid by their own host species to those of other army ants, and sometimes they were even repelled by the trails of the wrong species. However, a few instances were recorded in which the ecitophiles chose the trail of nonhost species. None of the symbionts showed an ability to distinguish the trails of its own host colony in competition with trails laid by other colonies belonging to the same species. In case any doubt remains that at least some symbionts are actually guided by the pheromone itself, the experiments of Moser (1964) on *Attaphila fungicola* can be cited. This little cockroach lives with the fungus-growing ant *Atta texana,* the workers of which depend strongly on odor trails for orientation during foraging expeditions and their infrequent colony emigrations. That *Attaphila* might respond to the ant pheromone had been suggested earlier by observations of *A. schuppi* following trails of *Acromyrmex* in the field in Brazil (Bolivar, 1905). Moser demonstrated that *A. fungicola* will follow the pheromone when it is taken directly from the poison glands of the host workers and laid along artificial trails in the laboratory. He also noted that the cockroaches do not depend solely on this ability to disperse from nest to nest because individuals have been observed riding on the backs of *Atta texana* queens during the nuptial flights of these ants.

2. *Attraction and appeasement.* A general characteristic of advanced symphiles is their ability to attract host workers. As the earlier discussion of mimicry and physogastry stressed, the attraction appears to be linked to the continuing process of appeasement within the host nests. It is also the means by which some symphilic species induce their hosts to bring them into the nests in the first place. If the communication is simple and direct enough, it should be possible to extract the attractants, to place them on olfactorily inert dummies, and, by presenting the

FIGURE 20-9. A symbiotic millipede (*Calymmodesmus* sp.) is seen running along the odor trail of an emigration column of the army ant *Labidus praedator* (photograph courtesy of C. W. Rettenmeyer).

dummies to host workers, to induce the response in the absence of other sensory cues. This step has been accomplished by Bert Hölldobler (1970), who demonstrated that the scent of *Atemeles* larvae can be transferred to dummies without losing its attractiveness to the host ant workers. The technique can be employed as a bioassay that could lead eventually to the purification and chemical identification of the attractant. When it succeeds, we will undoubtedly confirm that the symphilic substances are more than just elementary nutritive substances or even analogues of the natural host pheromones. Several authors have spoken of a narcotizing effect of symphilic substances. Park (1964), for example, noted cases in which worker ants licked the trichomes of adult clavigerids for long periods of time, after which they were "so overwhelmed by this trichome stimulant that they became temporarily disoriented and less sure of their footing." Jacobson (1911) has described a remarkable case in which an outright predator also uses an intoxicating attractant. The reduviid bug *Ptilocerus ochraceus* of Java feeds to a large extent on *Hypoclinea biturberculata* (placed by Jacobson and his contemporaries in the genus *Dolichoderus*), one of the most abundant ants of southeastern Asia.

> The way in which the bugs proceed to entice the ants is as follows. They take up a position in an ant-path or ants find out the abodes of the bugs, and attracted by their secretions visit them in great numbers. On the approach of an ant of the species *Dolichoderus bituberculatus* the bug is at once on the alert; it raises halfway the front of the body, so as to put the trichome in evidence. As far as my observations go the bugs only show a liking for *Dolichoderus bituberculatus;* several other species of ants, *e.g. Crematogaster difformis* Smith and others, which were brought together with them, were not accepted; on the contrary, on the approach of such a stranger, the bug inclined its body forwards, pressing down its head; the reverse therefore of the inviting attitude taken up towards *Dolichoderus bituberculatus.* In meeting the latter the bug lifts up its front legs, folding them in such a manner that the tarsi nearly meet below the head. The ant at once proceeds to lick the trichomes, pulling all the while at the tuft of hairs, as if milking the creature, and by this manipulation the body of the bug is continually moved up and down. At this stage of the proceedings the bug does not yet attack the ant; it only takes the head and thorax of its victim between its front legs, as if to make sure of it; very often the point of the bug's beak is put behind the ant's head, where this is joined to the body, without, however, doing any injury to the ant. It is surprising to see how the bug can restrain its murderous intention as if it was knowing that the right moment had not yet arrived. After the ant has indulged in licking the tuft of hair for some minutes the exudation commences to exercise its paralyzing effect. That this is only brought about by the substance which the ants extract from the trichome, and not by some thrust from the bug, is proved by the fact, that a great number of ants, after having licked for some time the secretion from the trichome, leave the bug to retire to some distance. But very soon they are overtaken by the paralysis, even if they have not been touched at all by the bug's proboscis. In this way a much larger number of ants is destroyed than actually serves as food to the bugs, and one must wonder at the great profligacy of the ants, which enables them to stand such a heavy draft on the population of one community. As soon as the ant shows signs of paralysis by curling itself up and drawing in its legs, the bug at once seizes it with its front legs, and very soon it is pierced and sucked dry.

A drawing of *Ptilocerus ochraceus* is reproduced in Figure 20-10. The bug's circumspect behavior quite possibly has adaptive value. *Hypoclinea bituberculata* is a member of the Dolichoderinae, a subfamily that has developed powerful defensive secretions which render them immune to most predators. *H. bituberculata* is probably equipped in this fashion since its colonies are able to maintain large foraging columns in exposed areas during the day. We can guess that in the course of its evolution *Ptilocerus ochraceus* has simply discovered an efficient method for anesthetizing its prey, thereby avoiding the chemical sprays. The example of *Ptilocerus* also has a bearing on a once famous controversy about evolution. Erich Wasmann was not only the pioneering student of myrmecophiles and termitophiles, but also a dedicated Jesuit philosopher who seemed compelled to use his knowledge of these insects to disprove the theory of evolution by natural selection. He believed that the ants and termites take care of their guests because of certain "symphilic instincts," which are modifications of their general philoprogenitive instincts. The hosts were envisaged as applying a special kind of "amical selection" to breed the more peculiar characteristics such as the trichomes and exudatoria, much as human beings deliberately select certain strains of domestic animals for their desirable traits. But, since the symbionts are mostly parasites who harm their hosts, sometimes to the point of jeopardy, such selection could not be Darwinian—it must all be part of the Creator's plan for a balanced and harmonious universe. W. M. Wheeler, Wasmann's chief adversary in the more theoretical discussions, and Karl Escherich were firm in pointing out that the conventional Darwinian explanation holds equally well in this case, if only we accept the possibility that parasites are capable of perverting the otherwise adaptive responses of their hosts utilized in colony com-

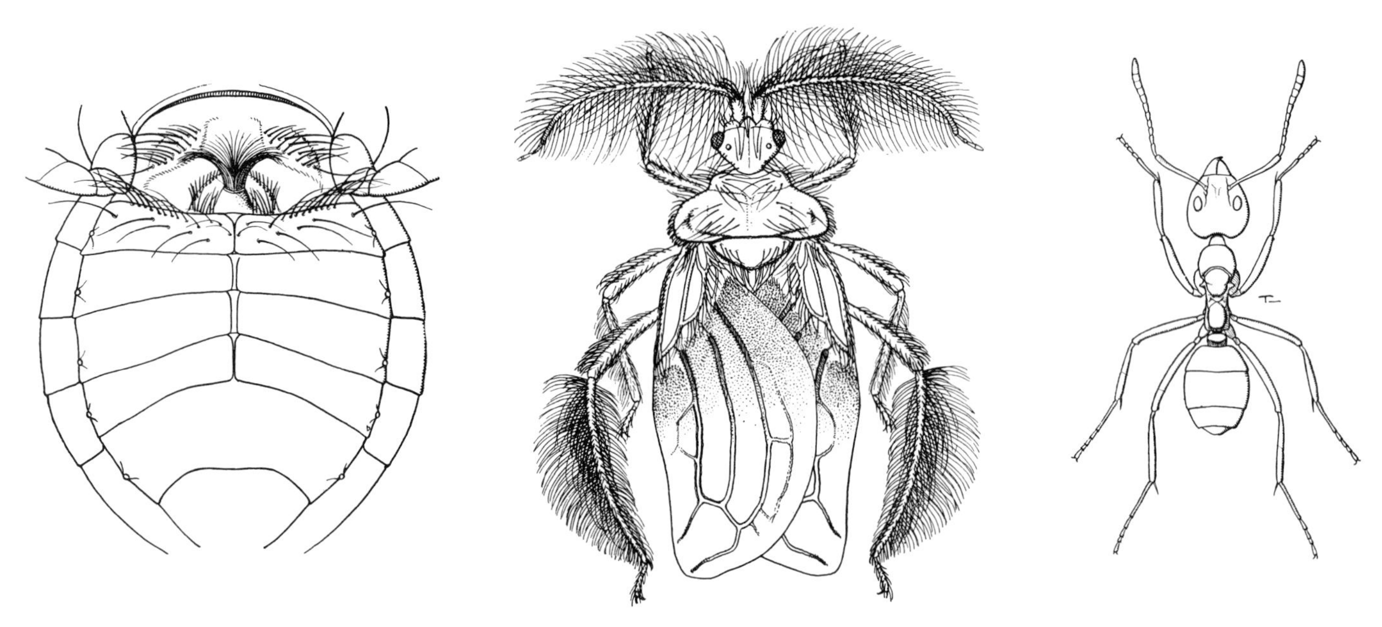

FIGURE 20-10. *Ptilocerus ochraceus* and the ant (*Hypoclinea bituberculata*) on which it feeds. At the left is the undersurface of the bug's abdomen with the trichomes that dispense a tranquilizing attractant to the ants (from China, 1928).

munication. *Ptilocerus ochraceus* offers strong evidence for this particular interpretation since the ant victims could in no way have deliberately selected the bugs for their trichomes and poisonous secretions, even in the unlikely event that an overpowering parental instinct drove them to the *Ptilocerus* abattoirs in the first place. Also, the experimental evidence on the kinds of releasing stimuli employed by symbionts to intrude into the host systems can best be explained as adaptations by the symbionts to their hosts, rather than the reverse.

3. *Regurgitation.* Several species of the mite genus *Antennophorus,* studied intensively in Europe many years ago by Janet (1897b), Wasmann (1902), and Karawajew (1906), occur only in nests of the ant genus *Lasius.* They ride on the bodies of the ants, shifting their positions when two or more are present on the same ant so as to produce a balanced load. When three mites are present, for example, two ride on either side of the abdomen, and the third, on the undersurface of the head. When four are present, two position themselves on either side of the abdomen and head, respectively. The *Antennophorus* live on food regurgitated by the *Lasius,* either imbibing it as it passes between workers or soliciting it directly by stroking the heads of the workers with the long, antenna-like forelegs, in apparent imitation of the tactile signals used by the ants themselves. The same capacity to gull ants into regurgitation is possessed by the lomechusine staphylinid beetles *Atemeles, Lomechusa,* and *Xenodusa* (Wasmann, 1915a), by beetles belonging to *Amphotis* and *Claviger* (Donisthorpe, 1927; Park, 1964), and even by adult mosquitoes (*Harpagomyia*) and phorid flies (*Metopina*) (Farquharson, 1918; K. Hölldobler, 1928). This remarkable vulnerability on the part of the ants suggests that there must be some simple trick involved in the soliciting procedure. Recently, Bert Hölldobler (1967a,b, 1970) has nicely demonstrated the nature of the minimal tactile signal required. The most susceptible worker ant is one that has just finished a meal and is searching for nestmates with which to share its crop contents (see Chapter 14). In order to gain its attention, a nestmate (or myrmecophile) has only to tap its body lightly with antennae or forelegs. This causes the donor to turn and face the individual that gave the signal. If tapped lightly and repeatedly on the labium, it will regurgitate. Other ants ordinarily use their fore tarsi for this purpose, while myrmecophiles use either their tarsi or antennae. The

larvae of *Atemeles* and *Lomechusa,* according to Hölldobler, curve the front part of their bodies upward and push their labia against those of the host workers (Figure 20-12). Even these clumsy imitations are enough if the donors are heavily laden with crop liquid. Figure 20-11 shows the whole sequence of the interplay between an adult *Atemeles* and its host.

Most of the more casual observations of regurgitation between host and symbionts suggest that the exchange is exploitative, meaning that the liquid flows exclusively from the host to the symbiont. Hölldobler's (1967b) experiments with radioactive tracers have proved this to be the case for the larvae of *Atemeles* and *Lomechusa* that live in nests of *Formica.* However, an exception has been reported in the case of *Amorphocephalus coronatus.* Adults of this brenthid beetle live with *Camponotus* in southern Europe. According to Le Masne and Torossian (1965), they receive food from some of the host workers and regurgitate part of it back to other host workers. This is the first reported example of which I am aware that could be construed as altruistic behavior on the part of symphilic beetles. The Lycaenidae and honeydew-excreting Homoptera, which we will examine at greater length shortly, donate some of their own food to the hosts in exchange for protection on their feeding grounds. The *Amorphocephalus,* in contrast, are said to give food back that was originally received from the hosts. The implication is an important one—that a symbiont can be integrated as a colony unit to the extent of showing altruistic behavior. For this reason regurgitation in *Amorphocephalus* and other liquid-feeding symbionts deserves closer study in the future.

4. *Recognition.* The key problem facing a symbiont attempting to live as a symphile is to win initial acceptance from the host workers, at least as a neutral object and, preferably, as a full nestmate. Some symphiles do achieve complete adoption as colony members. They are antennated, groomed, and fed like other workers or host larvae. Somehow they acquire the distinctive odors of the host species they victimize. Rettenmeyer (1961a) has suggested the interesting possibility that the guests of army ants pick up the odor in an active fashion while grooming their hosts. It is certainly true that the grooming behavior of symphilic staphylinids and histerids cannot be attributed entirely to the ingestion of edible cuticular materials. Akre and Torgerson (1968) have described the elaborate grooming rituals of a staphylinid guest of *Neivamyrmex:* "While *Probeyeria* and *Ecitophya* straddled their host across the longitudinal axis of the body, *Diploeciton* assume a position parallel to, but slightly to one side of and on top of, the ant. To position itself, a beetle grasps with its mandibles the scape of an antenna of an ant close to its base. It then positions its body parallel to the body of the ant and uses the first and third legs on the lower

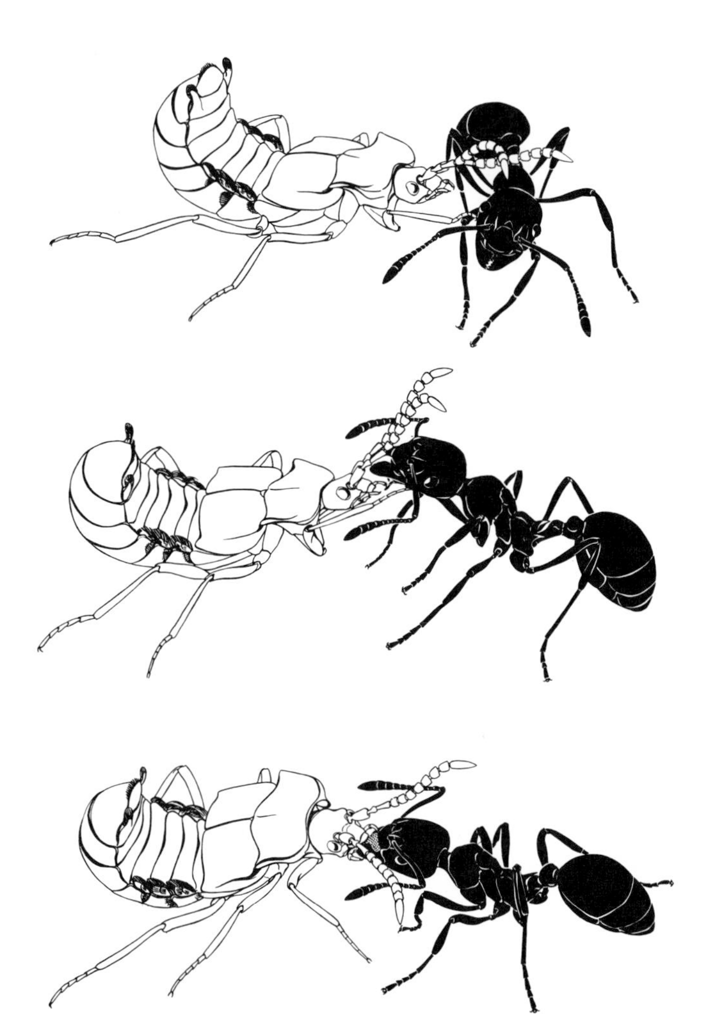

FIGURE 20-11. The soliciting of regurgitation by the staphylinid *Atemeles pubicollis.* The beetle gains the attention of the ant worker (*Myrmica* sp.) by tapping it on the side with its antennae (*top*). The ant turns (*center*) and then is induced to regurgitate when the beetle taps its fore tarsi on her labium (*bottom*) (from B. Hölldobler, 1970).

side of the body to brace against the substrate. The three legs on the other side of the body then straddle the ant. The mesothoracic leg on the bottom curls under and around the thorax of the ant. This places the sternum of the beetle's thorax against the side of the thorax of the ant as though riding 'sidesaddle.' In this position the beetle rubs the ant with its legs. The mesothoracic lower leg rubs the bottom of the thorax and the upper legs rub on the dorsal area of the ant; the prothoracic leg usually rubs the head of the ant, the mesothoracic leg rubs on the thorax and gaster, while the metathoracic leg is used sparingly to rub the gaster of the ant. The rubbing strokes are rather slow and alternate between stroking the body of the ant and the staphylinid's own body. The front leg is rubbed on the head and thorax, both middle legs are rubbed on the elytra and the globular portion of the myrmecoid abdomen, while the metathoracic leg was rubbed only rarely on the abdomen." The stroking movements of the staphylinids and histerids calm the ants and in some instances, according to Akre and Torgerson, seem to paralyze them partially. However, this is not another case of a fatal seduction of the kind worked by *Ptilocerus,* for the ecitophiles do not attack their adult hosts. They feed only on the brood, which, interestingly enough, are not recipients of the grooming ritual.

A still rarer feat for symbionts is to dupe the host workers into treating them as particular immature stages in the host brood development. This the larvae of the lomechusine staphylinids have accomplished with distinction (Wasmann, 1915a). These parasites are grub shaped and able to move about only short distances with the use of their short legs. They are treated by their *Formica* hosts like their own larvae and allowed to lie among the host brood, which they proceed to consume voraciously. The lomechusine grubs are also adept at soliciting regurgitated food from the ants (Figure 20-12). Hölldobler (1967b) has discovered that a substance identified by the workers with their own brood can be separated from the bodies of the *Atemeles.* When he extracted the parasite larvae in acetone and soaked dummies in the mixture, the dummies became temporarily attractive to the ant workers and were treated as pieces of brood. Thus it appears possible that nothing more than production of the right "pseudopheromone" and one or two crude signaling movements can suffice to insert a parasite into the heart of the host colony. The adaptations of the lomechusines appear not to have gone very far beyond this point. Wasmann found that members of the genus *Lomechusa* suffer heavy mortality at the pupal stage for the very reason that the ants keep on treating them as brood. A mature larva of *Formica* must be buried in soil in order to spin its cocoon and pupate inside. After this is accomplished, the nurse workers dig it up, clean it off, and place it with other cocooned pupae in one of the drier brood chambers. If the whole procedure is practiced on a *Lomechusa* larva, the result is fatal. The larva needs to be buried all right, but afterwards the pupa cannot tolerate exhumation and transport to the dry chambers. The species itself is able to survive only because a small percentage of the *Lomechusa* pupae are overlooked by the *Formica* workers and allowed to remain covered by soil. An additional population control, noted by Hölldobler (personal communication), is cannibalism among the *Lomechusa* larvae.

We have seen how guests of the ecitonine army ants are able to identify and track down their host species by following the odor trails. Such behavior is not universal among symbionts. When Hölldobler (1969a) gave *Atemeles pubicollis,* a guest of nonlegionary ants in Europe, an opportunity to follow trails of their hosts, they failed to respond. They also paid no attention to the exits of laboratory nests containing host colonies so long as the test arenas were kept in still air. But when a weak air current was drawn first over the colonies and then in the direction of the beetles, the *Atemeles* ran upwind. This set of stimuli closely approximates the natural condition a beetle would find itself in while searching for host nests over the surface of the ground. Hölldobler also discovered that the odor preference of the *A. pubicollis* adults changes with age. Wasmann had learned many years previously, and Hölldobler confirmed, that the beetles migrate from nests of *Formica* to those of *Myrmica* six to ten days after they eclose from the pupae. The following spring the beetles return to nests of *Formica,* where they reproduce. According to Hölldobler, the physiological basis of the shift is quite elementary: after eclosion, the beetles are attracted to air laden with *Myrmica* scent in preference to that of *Formica,* but, when they emerge from hibernation, their preference switches back to *Formica.* The adaptive basis may be the availability of larvae, which are the chief prey of the beetles. *Myrmica* maintains larvae in its nest throughout the late fall, winter, and early spring, but *Formica* do not. However, during the summer the *Formica* colonies provide a richer source of larvae.

FIGURE 20-12. *Above:* A larva of the staphylinid *Atemeles pubicollis* feeds on a larva of one of its host ants (*Formica*). *Below:* A *Formica* worker regurgitates to an *Atemeles* larva. Glands believed to be the source of a false brood identification odor are located in pairs on the dorsal surface of each of the body segments of the *Atemeles* larva (from B. Hölldobler, 1971).

5. *Alarm.* Although the relationship of most myrmecophiles and termitophiles to their hosts appears to be wholly parasitic, the *Amorphocephalus* case should alert us to the possibility that symbionts can sometimes contribute in a positive manner to the communication system of the colony. Blum (1969) reports that an inquilinous Japanese species of the staphylinid genus *Zyras* produces a scent that resembles citronellal, one of the alarm pheromones of its ant host *Lasius spathepus.* The *Zyras* substance induces a strong alarm response in the worker ants. It is not known whether the beetle directs its substance at the hosts, or whether it discharges only in the presence of common enemies in a fashion geared to the needs of the colony communication system.

Why Do Social Wasps and Bees Have Fewer Guests?

We now come to the interesting question of why obligate symbionts are not as highly evolved in the nests of social wasps and bees as they are in the nests of ants and termites. Not only are fewer species associated with bees and wasps, but their adaptations for symbiotic life are generally far less pronounced. The chief exceptions are mites, beetles, and flies that live as scavengers and brood commensals. Some of the species living with stingless bees possess limuloid or turtle-like body forms (see Figure 20-13). Most of the other species seem to be quite generalized in form and behavior and are probably either attacked by their hosts or at best treated indifferently. I know of only a single example of a true symphile among these insects, by which I mean a species that communicates with the hosts on their own terms and is integrated to some degree into their society. This is *Echthrodape africana,* a perilampid wasp whose larvae are ectoparasites on the pupae of allodapine bees of the genus *Braunsapis* in Africa. According to Michener (1969b), the *Echthrodape* larvae are apparently moved around the nest by the host adults and placed with host brood in comparable stages of development. The feces of both the host and parasite larvae are also cleaned out by the adult *Braunsapis.*

The general lack of sophistication among sphecophiles and melittophiles seems unlikely to be due to some basic peculiarity in the social behavior of the hosts. Social wasps and bees employ most of the same fundamental communication forms as do ants and termites—trophallaxis, grooming, progressive care of larvae, nestmate recognition

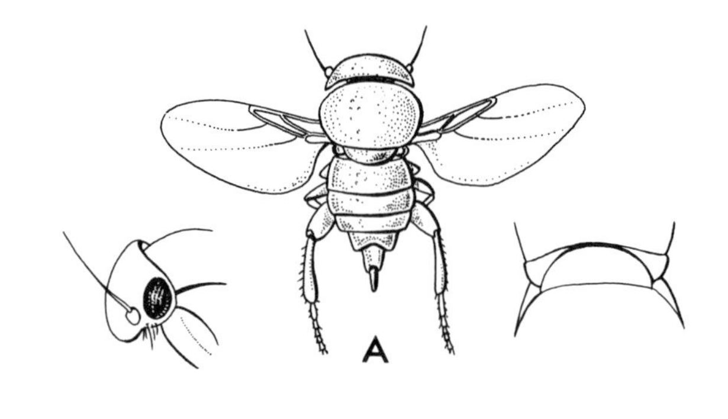

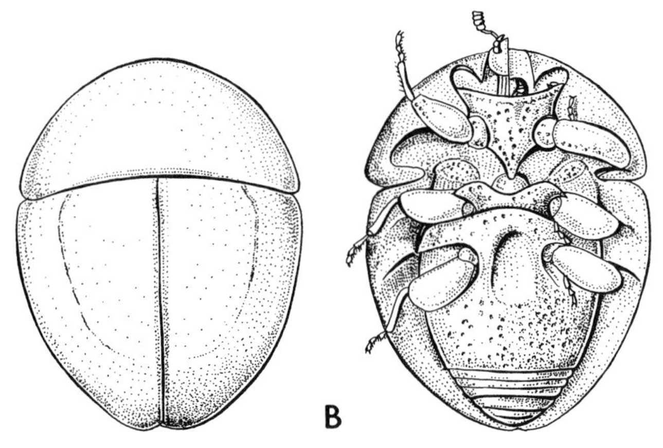

FIGURE 20-13. Guests of meliponine bees that have developed defensive body forms: (*A*) *Melittophora salti,* a phorid fly with short legs, sparse pilosity, and overlapping body parts that protect the softer intersegmental zones; (*B*) *Cleidostethus meliponae,* a tiny, aberrant coccinellid beetle with completely retractable head and appendages (redrawn from Salt, 1929).

and territorial defense, alarm, and even trail following. Much more impressive are the great differences that exist between the army ants and the higher termites, the two groups with the largest and most complex symbiont faunas. Bees and wasps keep their brood in cells and do not transport them from place to place. However, this does not seem likely to exert a great influence on the development of guests that live on regurgitated food or booty brought in by host workers or even, for that matter, on the host brood itself. No really important differences in colony size exist. Some of the higher Polybiini form colonies of thousands of adults. The mature colonies of the honeybees and some meliponines contain tens of thousands of workers, among the largest found in the social insects and at least comparable in size to some colonies of ants that support a rich symbiotic fauna. Nor are the nests of social wasps and bees particularly short-lived or unstable. The colonies of some species of melipo-nines and polybiines, for example, often remain in the same positions for years.

However, the colonies of wasps and bees appear to me to be distinguished by one important feature that does provide a plausible explanation for the scarcity of symbi-onts. Both groups tend to construct compact, tightly sealed nests in arboreal locations. There must be relatively few potential arthropod guest species that are preadapted for the penetration of such nests. The right combination would seem to be some degree of specialization for ar-boreal life and a preference for dark, tight spaces, perhaps together with a tolerance for higher temperatures and lower humidities. These preferences are displayed by only a small minority of arthropod species, including those adapted for life in tree holes, standing dead branches, and the deeper layers of bark. Even the ones thus preadapted to enter wasp and bee nests must pass some formidable additional obstacles—thick envelopes of carton or wax sometimes reinforced by cement-like propolis and narrow, tightly guarded nest entrances often lined with sticky or repellent substances. Also, detritus on which would-be scavengers can feed is scarce inside the nests because the workers of many species simply take most of it to the nest entrance and eject it. Also, as Salt (1929) has pointed out, bees produce smaller amounts of refuse in the first place, because the pollen and nectar that make up their diets are highly concentrated food sources. By comparison ant and termite nests are relatively open systems, rich in refuse and embedded in environments not greatly different from their own interiors. Carton, wax, and uncongenial sealing materials are seldom used as principal building materials. A great array of arthropod species, in fact the largest single component of the entire arthropod fauna of the world, lives in soil and rotting vegetable matter of the kind that forms the matrix of most ant and termite nests. Op-portunities abound for penetration into the many cham-bers and galleries that go unguarded from time to time. It seems reasonable, therefore, to suggest that the differ-ences in symbiont diversity between the major groups of social insects has originated in evolution as a statistical outcome of the initial differences in the opportunities for invasion presented by their nests.

The Trophobionts

As aphids feed on the phloem sap of plants, they pass a sugar-rich liquid through their gut and back out through the anus in only slightly altered form. During the passage of this "honeydew," as much as one-half of the free amino acids are absorbed by the gut (according to measurements made in *Tuberolachnus salignus* by Mittler, 1958), sugars are partly absorbed and converted into glucosucrose, melezitose, and higher oligosaccharides, while other classes of substances, including organic acids, B-vitamins, and minerals, are probably partially taken up as well. The remainder of the complex mixture is excreted. To process a large volume of phloem sap and discard the excess as honeydew evidently costs the aphid less in calories than a more nearly total extraction from smaller quantities of sap. The amounts of honeydew produced by individuals are often prodigious. First instar nymphs of Mittler's *Tuberolachnus* excreted droplets at the rate of seven drop-lets per hour, each droplet containing 0.065 μl, and the total output per aphid was 133 percent of the aphid's weight every hour. Other species that have been analyzed are slightly more modest, their hourly output ranging from 1.9 to 13.3 percent of body weight per hour (Auclair, 1963). When unaided by ants, the aphid disposes of the droplets by flicking them away with its hind legs or cauda or by expelling them by contracting its rectum or entire abdomen. The honeydew then falls upon the vegetation and ground below. Similar substances are excreted by several other groups of sap-feeding Homoptera, including scale insects (Coccidae), mealybugs (Pseudococcidae), jumping plant lice (Chermidae = Psyllidae), treehoppers (Jassidae, Membracidae), leafhoppers (Cicadellidae), froghoppers or spittle insects (Cercopidae), and members of the "lantern-fly" family (Fulgoridae). Sometimes honeydew accumulates in large enough quantities to be usable by man. The manna "given" to the Israelites in the Old Testament account was almost certainly the ex-cretion of the coccid *Trabutina mannipara,* which feeds on tamarisk. The Arabs still gather the material, which they call "man." In Australia, chermid honeydew is collected as food by the aborigines. Referred to as "sugar-lerp," up to three pounds can be harvested by one person in a single day. It is no surprise, therefore, to find that ants also gather honeydew of all kinds. Many, perhaps most, species collect it from the ground and vegetation where it falls. But many others have developed the capacity to solicit the honeydew directly from the homopterans themselves. A great majority of the members of the three phylogenetically most advanced subfamilies, the Myrmicinae, Dolichoderinae, and the Formicinae, attend homopterans to some extent. To use one last term from Wasmann, the ants can be said to have entered into *trophobiosis* with the homopterans. As trophobionts, the homopterans differ from other ectosymbionts in a basic way: they obtain their own food and pass some of it on to their hosts. The extent of utilization of honeydew in the diet varies greatly among ant species. In the highly predaceous myrmicine *Daceton armigerum,* it is only an occasional event (Wilson, 1962b). The diet of the common wood ant of Europe (*Formica rufa*), according to one study by Wellenstein (1952), consists of about 62 percent honeydew. Extreme trophobiosis is practiced by certain subterranean ants believed to be totally dependent on root aphids and coccids. The best-known examples are two formicine genera: *Acanthomyops,* which is restricted to North America (Wheeler, 1910; Wing, 1968), and *Acropyga,* studied in Central and South America by Bünzli (1935) and Weber (1944). In neither case has it been established beyond doubt that the colonies live entirely off their "cattle," but the circumstantial evidence is strong that they can do so. Honeydew may or may not suffice by itself. As Way (1963) has pointed out, the data are not adequate to determine the matter. Experiments by Lange (1960b) and Ayre (1960) showed that colonies of *Formica polyctena* decline slowly when fed with pure bee honey made from honeydew. It is nevertheless quite possible that the honeydew obtained directly from root homopterans is nutritionally richer. It is further possible that *Acanthomyops* and *Acropyga* obtain extra protein by eating some of their homopterans. Cropping of homopterans does occur under at least some circumstances. For example, Way (1954b) found that laboratory colonies of the weaver ant *Oecophylla longinoda,* when presented with excessive members of its trophobiotic coccid, killed and removed individuals until the population had reached the level required for a sufficient but not excessive outflow of honeydew.

The means by which ants obtain honeydew from their aphids was first fully described by Pierre Huber in 1810, and his account is accurate enough to provide a paradigm of this form of behavior. The ants involved belonged to the common European formicine species *Lasius niger:*

> A thistle branch was covered with brown ants. I watched the latter for quite a while in an effort to determine the precise moment at which they released the secretion. I found that the liquid very rarely exuded by itself and that the aphids, if separated from the ants, discharged it to a distance with a sudden jerking movement. Why did all the ants on the stems have their abdomens apparently distended with liquid? I was able to answer this question by watching a single ant, whose exact procedure can be described as follows. I saw her first walk over some of the aphids without pausing or disturbing them. But she soon halted near one of the smallest individuals and appeared to caress its abdomen, stroking it alternately with one and then the other antenna. I was surprised to see the liquid emitted from the aphid's body, and that the ant seized and drank the droplet at once. She then applied her antennae to another, much larger aphid, which voided a larger drop of the nutritious liquid. The ant moved forward to receive it and then moved on to a third aphid, which was caressed in the same manner. The liquid was voided immediately and then imbibed by the ant. She moved on; a fourth aphid, probably already exhausted, refused to respond to her solicitations, and the ant, probably realizing she had nothing to expect, deserted this individual for a fifth one, from which I saw her obtain a further supply of food. A few such repasts are quite sufficient, and the satiated ant returns to the nest . . . I witnessed this remarkable procedure thousands and thousands of times; it was always followed by the ants when they wished to obtain food from the aphids. If the latter are neglected too long, they discharge the honeydew on the leaves, where the returning ants find and collect it prior to approaching the insects from which it originated. On the other hand, if the ants visit the aphids zealously, the latter seem to comply with their desires by hurrying up the evacuation of the liquid. This is indicated by the diameter of the droplet. At such times they do not eject the ant-manna at a distance but, in a manner of speaking, retain it and hand it over to their attendants. It sometimes happens that the ants are so numerous on a particular plant that they exhaust the aphids with which it is covered. Under these circumstances they stroke the bodies of the aphids in vain and are forced to wait until these insects have pumped up a fresh supply of sap from the stems. The aphids are by no means parsimonious, and if they have anything to give, never fail to respond to the solicitation of the ants. I have repeatedly seen the same aphid yield several drops in succession to different ants that seemed very eager for the syrup.

Typical aphid tending by the related genus *Formica* is illustrated in Figure 20-14. The interaction appears to be essentially similar among all the homopterans and the ant species that attend them. Moreover, the mutualism is probably of ancient origin. Wheeler (1914) has described finding aphids associated with *Iridomyrmex goepperti* in a block of Baltic amber, of Oligocene age. *Iridomyrmex,* like many of the other ant genera common in the Baltic amber and other early Tertiary fossil deposits, has survived relatively unchanged in external morphology, and its living species still attend honeydew-producing homopterans.

The protection by the ants permits the homopterans to

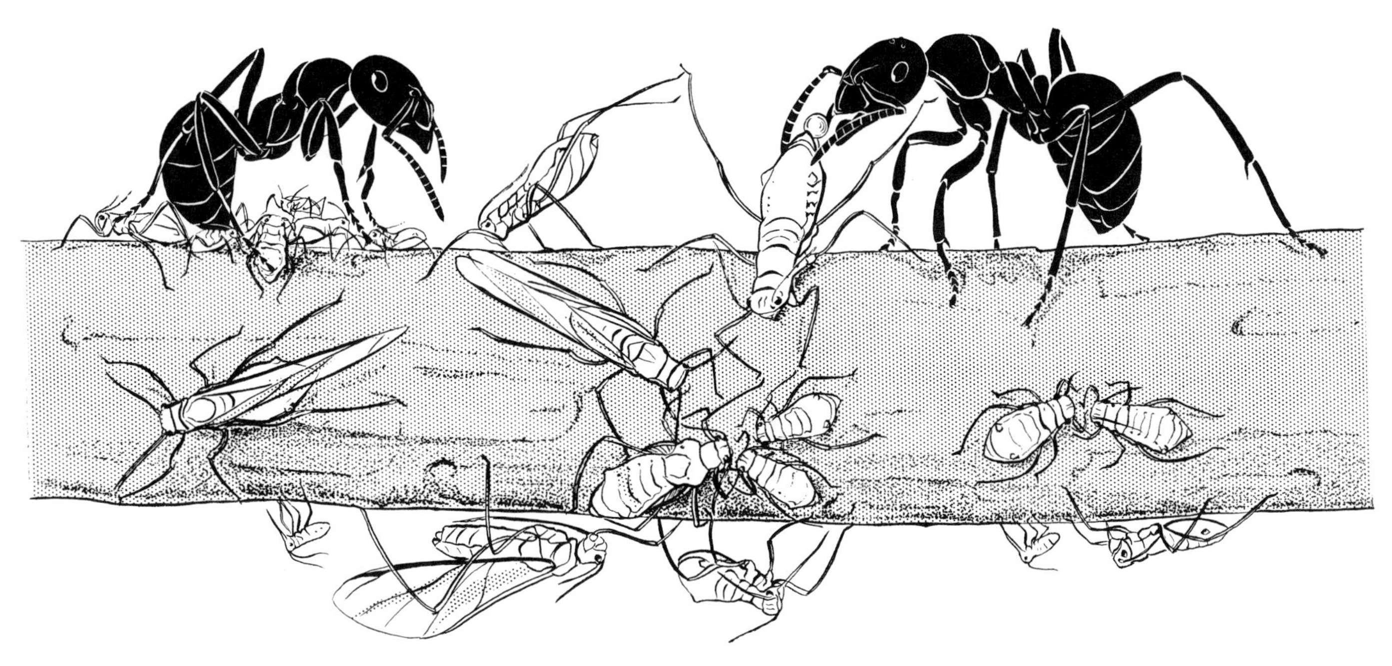

FIGURE 20-14. Workers of *Formica polyctena* attending aphids (*Lachnus roboris*) (original drawing by Turid Hölldobler).

develop larger, more stable populations and increases their overall rate of dispersal. Because of this important fact, economic entomologists have conducted intensive research on all conceivable aspects of the mutualism, and a large and specialized literature on the subject has grown through the years. Reviews that mark successive stages in the development of the subject and yet are only partly overlapping in empirical content have been provided by Büsgen (1891), Wheeler (1910), Jones (1929), Herzig (1937), Nixon (1951), and Zwölfer (1958), and, most recently and authoritatively, by Auclair (1963) and Way (1963). Of paramount interest is our improving knowledge of the special adaptations that both of the insect partners have evolved with the apparent sole function of implementing the mutualism. The following brief account is drawn chiefly from the accounts by Wheeler and Way.

Modifications of homopterans to ants. Symbiotic homopterans ease out the droplets of honeydew when solicited by ants, rather than ejecting them at a distance. Individuals of the black bean aphid (*Aphis fabae*) show the following typical specialized responses in the presence of ants: the abdomen is raised slightly, the hind legs are kept down instead of being lifted and waved as in unattended aphids, and the honeydew droplet is emitted slowly and held on the tip of the abdomen while it is being consumed by the ants. If a droplet is not accepted, the aphid will often withdraw it back into the abdomen. The required stimulus for honeydew excretion is a simple, mechanical one. According to Mordwilko (1907), many of the symbiotic aphid species can be induced to emit a droplet by merely brushing the abdomen with some delicate object in imitation of the caressing movements of the ant's antennae and forelegs. I have done this successfully myself with giant myrmecophilous coccids in New Guinea using one of my own hairs, and then collected the material directly from the insects and tasted it. Constant attentions from the ants of this sort increases the production of honeydew by aphids by as much as a factor of three.

The extreme myrmecophilous aphids have evolved to the status of little more than domestic "cattle." They have reduced or lost the usual defensive structures found in free species, including the defensive abdominal spouts called cornicles that secrete a quickly hardening wax, the dense cloaks of flocculent wax filaments secreted by special hypodermal glands, the sclerotized exoskeletons, and the modifications of the legs for jumping. On the other hand they have acquired a new trophobiotic organ, a circlet of stiff hairs around the anus that serves to hold the honeydew droplet while it is being eaten by the ant. Long anal hairs found on certain pseudococcids evidently serve the

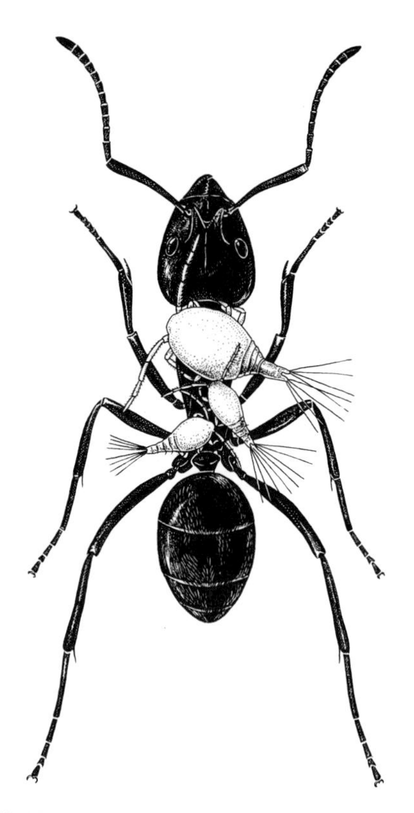

FIGURE 20-15. Javan pseudococcids of the genus *Hippeococcus* escape from danger by climbing onto the backs of their host ants and allowing themselves to be carried to safety. Their legs and tarsi are apparently specially modified for this purpose. Here three individuals are shown being carried by a worker of *Hypoclinea gibbifer* (modified and redrawn from Reyne, 1954).

same purpose (see Figure 20-15). Kloft (1959b) has made the intriguing suggestion that the rear of the aphid's abdomen resembles the head of an ant worker, and the kicking of the aphid's hind legs represents an imitation of the antennal movements of an ant. The stimuli presented induce the visiting ant workers to mistake the aphid for a donor ant in the special way that any animal makes a mistake when confronted with the small but vital set of releasers out of context. The solicitation that follows, according to Kloft, is just a slight perversion of the ordinary trophallaxis occurring between sister workers. While very original, this is not a persuasive argument. Coccids and mealybugs are attended with equal fervor and precision, yet their appearance and behavior make them appear wholly different from their ant hosts. The resemblances between the aphids and the heads of ants are at best subtle ones that might easily be termed coincidence.

Special note must be taken of the extraordinary transport behavior shown by the genus *Hippeococcus*. These pseudococcids, described from Java as a new genus by Reyne (1954), are kept by ants of the genus *Hypoclinea* in their underground nests and on trees and shrubs nearby. When they are disturbed the small *Hippeococcus* climb onto the bodies of the ants or else are gathered by the ants in their mandibles. The ant riding is made possible by long, grasping legs and flat, sucker-like tarsi (see Figure 20-15).

Experiments by El-Ziady and Kennedy (1956) have revealed that the presence of ants of the genera *Lasius* and *Paratrechina* delays the production of alates in populations of species of *Aphis* and hence postpones their dispersal and increases their population density. Several physiological models have been proposed to explain the result, but critical experimentation to choose among them is still lacking. The important point remains that ants are capable, accidentally or otherwise and with or without cooperative adaptations on the part of the aphids, of increasing the population densities of their charges and thus increasing their honeydew yield.

The life cycles of homopterans, which Zwölfer and others have documented so thoroughly, crucially affect the success of the trophobiosis. The ideal trophobiont is one which feeds on a single species of food plant and whose life cycle is not tightly synchronized, so that stages capable of producing honeydew are available throughout the year. The evidence shows that not only do taxa possessing these properties as original traits become trophobiotic more frequently, but some species also acquire or enhance them secondarily in evolution after the association with ants has begun.

In passing I believe it will be useful to negate what has long threatened to turn into a serious canard in the literature of mimicry and symbiosis. In 1891 Edward B. Poulton reported that leafhoppers imitate the cut pieces of leaves carried by the parasol ants of the genus *Atta*. Poulton published a figure showing nymphs of a membracid species (believed to belong to the genus *Stegaspis*) walking among columns of homeward bound *A. cephalotes* workers. There is no doubt that the flat, green upper body of the treehopper resembles a piece of leaf carried by the ants, and that the lower part of their bodies and appendages could pass for the burdened ants themselves. Although I have not encountered any mention of this case in subsequent literature, it has been frequently discussed among naturalists in conversation. Both Wheeler and William Beebe, for example, believed in it. And, if true, it would be, as Poulton said, an example of protective mimicry "more wonderful in its detail and complexity than any which has been heretofore described." However, the evidence is very flimsy and dubious. The membracids were not seen in the column of ants; Poulton's drawing is an admitted fiction. They were collected from a bush in Guyana, which may or may not have been visited by *Atta*. Species of *Atta* are not known to attend homopterans in any case, and there is no obvious reason why membracids would ever need to leave their host plant in the company of the ants. Nevertheless, it would be interesting if some entomologist could rediscover these insects and study their behavior in the field to settle the matter once and for all.

Modifications of ants to Homoptera. The fact that ants do not normally attack and eat their homopteran associates is evidence by itself that behavioral evolution has occurred which accommodates them to the mutualism. Their mode of soliciting honeydew is essentially the same as that used to initiate regurgitation within the colony. But the fact that it is directed at such alien arthropods seems to indicate the existence of a second major behavioral adaptation on the part of the ants. Some species of ants go so far as to care for the homopterans inside their nests. In the early studies of S. A. Forbes and F. M. Webster, for example, it was discovered that the eggs of the corn-root aphid (*Aphis maidiradicis*) are kept by colonies of the ant *Lasius neoniger* in their nests throughout the winter. The following spring the newly hatched nymphs are transported to the roots of nearby food plants. If the host plants are uprooted, the ants move them to undisturbed root systems in the vicinity. During the late spring and summer, some of the aphids transform into alates and disperse on their own. After they have settled and begun to feed, they may be adopted by other ant colonies in whose territories they happen to have fallen (Forbes, 1906). When eggs are tended in this way, they are often mixed with the host brood. Also, when the nest is disturbed, the ants pick up their homopterans and transport them to safety in a manner indistinguishable from the rescue of their own brood. Ant workers chase potential predators and parasites away from the homopterans but, again, in the same way that they protect their nests and inert masses of food. Generalizations of this kind led Herzig (1937) and Nixon (1951) to question whether the ant-homopteran symbiosis is really an advanced mutualism involving extensive coadaptation. Herzig went so far as to suggest that most of the behavior of the ants toward aphids can be explained as a compromise between two opposing primitive motivations: avoidance because of unpalatability of their flesh, and attraction because of their honeydew. But now there is sufficient evidence to show that the ants do respond toward the aphids in specialized ways that can only represent an adaptation on their part. It has been established beyond doubt that workers of at least some ant species carry their homopteran guests to the appropriate part of the food plant, and at the correct stage of the trophobionts' development. Such behavior has been documented, for example, in the case of *Acropyga* and its root coccids by Bünzli (1935), in *Oecophylla* and *Saissetia* by Way (1954b), and in *Lasius* and *Stomaphis* by Goidanich (1959). Even more impressive is the fact that the queens of certain ant species carry coccids in their mandibles during the nuptial flight. This habit, which has no parallel in the behavior of non-coccidophilous ants, has been observed in a species of *Cladomyrma* in Sumatra by Roepke (1930), in *Acropyga paramaribensis* in Surinam by Bünzli (1935), and in an unidentified formicine (possibly *Acropyga*) in China by W. L. Brown (1945).

Ants are not the only organisms that attend homopterans. According to Salt (1929), stingless bees of the genus *Trigona* collect honeydew directly from membracids in Brazil, and at least one species "tickles" the treehoppers to induce flow. Belt, in *The Naturalist in Nicaragua* (1874), observed social polybiine wasps of the genus *Brachygastra* attending membracids. "The wasp stroked the young hoppers, and sipped up the honey when it was extruded, just like the ants. When an ant came up to a cluster of leaf-hoppers attended by a wasp, the latter would not attempt to grapple with its rivals on the leaf, but would fly off and hover over the ant; then when its little foe was well exposed, it would dart at it and strike it to the ground." Silvanid beetles of the Neotropical genera *Coccidotrophus* and *Eunausibius* attend pseudococcids in the hollow leaf petioles of *Tachigalia* (Wheeler, 1928). A few lycaenid butterflies milk homopterans while in the adult or larval stage. The adults of *Allotinus horsfieldi*, for example, solicit honeydew from aphids by stroking them with their forelegs (Bingham, 1907). Larvae of *Lachno-*

cnema bibulus depend entirely on honeydew solicited from membracids and jassids and possess modifications in the forelegs that are apparently adapted to this single purpose (Hinton, 1951). Certain flies (*Revellia quadrifuscata*) solicit honeydew from membracids by drumming on the backs of the homopterans with their fore tarsi (Andrews, 1930).

It is also true that trophobionts occur in other groups besides the Homoptera. Green (1900) recorded a case of a hemipteran bug, *Coptosoma* sp. (Plataspididae), being attended in Ceylon by *Crematogaster* workers. A majority of the lycaenid and riodinid butterflies are closely associated with ants in the same way (Wheeler, 1910; Lamborn, 1914; Donisthorpe, 1927; Hinton, 1951). According to Hinton, myrmecophily appears to be a primitive trait among the living lycaenids; its absence in some species is considered to be due to a secondary loss in evolution. Lycaenid larvae are visited by ant workers while they feed on their host plants. By stroking their backs with their antennae and forelegs, the ants induce the larvae to discharge a secretion from an unpaired gland located on the dorsum of the seventh abdominal segment, and they imbibe the liquid like honeydew. Other, paired glands on the eighth abdominal segment seem in some species to produce a volatile attractant that draws the ants to the caterpillars in the first place, but they do not emit a liquid during the milking sequence. A typical episode is recorded by A. L. Rayward (in Donisthorpe, 1927) of an interaction between caterpillars of the Chalk-hill Blue (*Agriades coridon*) and workers of the ant *Lasius flavus* in England. The caterpillars, which were feeding on *Hippocrepis,* were literally covered by the ants, twenty being observed on and around one individual alone. When a caterpillar was isolated and a single *flavus* worker placed near it, the ant "at once began to run to and fro about the body of the larva, waving its antennae excitedly. In a few minutes it found its way to the gland on the seventh abdominal segment, and stroked it with a rapid movement of the antennae and first pair of legs. This action was repeated several times, when suddenly the gland was distended, and one, two, or more tiny beads of a crystal-clear fluid were slowly expelled, and were greedily sucked up by the ant." In return, the lycaenid and riodinid caterpillars are protected from predators and parasites by their attendants. Sometimes the ants go so far as to construct crude earthen "cattle-sheds" around the larvae of the same kind sometimes built for honeydew-producing homopterans (Ross, 1966).

The relationship has become clearly obligatory on many if not all of the lycaenids attended by ants. If the dorsal gland exudates are not removed regularly by ants, the caterpillars soon die (Hinton, 1951). Furthermore, a few of the species have evolved beyond trophobiosis into other forms of symbiosis with the ants. For example, the larva of the Large Blue (*Lycaena arion*), another species found in England, feeds on wild thyme and is attended by ants until it reaches the third instar. Then it crawls down onto the ground and wanders about until it meets a worker of the ant genus *Myrmica.* After the ant milks it of some of its abdominal secretion, the larva deforms its body into a striking new shape: it retracts its head and swells its thoracic segments up while at the same time constricting its abdominal segments, giving the body a hunched, tapered look (Figure 20-16). Apparently this serves as a signal to the ant. Perhaps the altered body form releases an attractant or imitation ant-larval substance. Whatever the case, the ant now picks the caterpillar up and transports it back to its nest. Once ensconced in the brood chambers, the caterpillar turns carnivore, feeding heavily on the helpless host ant larvae. When it reaches full ma-

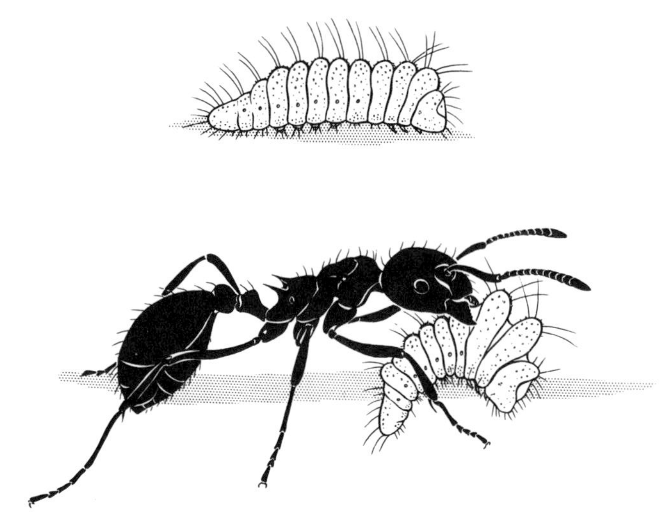

FIGURE 20-16. The adoption procedure of the third instar larva of the Large Blue, *Lycaena arion.* The individual above is searching for a host ant and still has the typical shape of a lycaenid larva. The larva at the bottom has been milked by the *Myrmica* worker, and, after hunching its body up, it is being transported back to the nest by the ant (redrawn from Frohawk, 1915).

turity, it pupates and overwinters in the nest, emerging as an imago the following June.

The caterpillar of the widespread Asiatic and Australian lycaenid *Liphyra brassolis* feeds on larvae of the green weaver-ant *Oecophylla smaragdina* (Dodd, 1902). It is covered by a tough shell, apparently as protection against its hosts, and when it pupates it remains within the larval exoskeleton like a cyclorrhaph dipteran. When the butterfly hatches, it is covered by a dehiscent cloak of white, gray, and brown scales. Then, as the *Oecophylla* attempt to seize the butterfly during its egress from the nest, they get only mandibles and antennae full of the scales. According to Hinton (1951), a similar adaptation has been reported in several other lycaenid species as well as in the South American pyralid *Pachypodistes goeldi.*

21 The Population Dynamics of Colonies

The first nineteen chapters of this book summarized what is usually regarded as the substance of insect sociology—those tangled patterns of life cycle, caste, and communication that make up the unique adaptations of social existence. In Chapters 17 and 18, which dealt with genetic theory and optimization, we examined the underlying mechanism of natural selection. Throughout, partial explanations of social phenomena were contrived from the evidence concerning proximate causation in physiology and behavior combined with the theory of ultimate causation through colony-level selection. But now it must be recognized that a major piece is still missing. This is the population ecology of colonies, a subject the full implications of which have only begun to be envisioned, and then cloudily, by students of social insects within the past twenty years. In Figure 21-1, I have tried to schematize the relationship of this new discipline to what has gone before. The idea can be expressed essentially as follows. The superficial aspects of caste, communication, and other social phenomena represent adaptations that are fixed by natural selection at the colony level. Natural selection at the colony level is the differential survival and reproduction of sets of very closely related genotypes. In order to understand survival and reproduction fully, it is necessary to go beyond insect sociology into population and community ecology. A conscientious review of those subjects has already been provided by M. V. Brian (1965b), and the reader should see his book for many details important in themselves but not essential to explanations of social phenomena that will not be repeated here. This is particularly true of the topics of predation, food supply, and community organization. What I would like to attempt now is to make explicit the connections between population dynamics and social phenomena, as conceived in Figure 21-1, and to discuss at some length the most relevant ideas and data from population ecology. This approach has revealed serious shortcomings in both theory and factual information. The theory of population ecology is inadequate for most conventional animal populations—see, for example, the pessimistic appraisals by Slobodkin (1961) and Watt (1968)—and it is largely undeveloped for the special case of social insect colonies. Conversely, when existing data on insect colonies are assembled in a form applicable to existing theory, they are usually very incomplete. The best hope is that, by examining this important subject in a systematic manner, the most fruitful paths for future research can be identified.

The Survivorship of Colony Members

The attrition of individuals in colonies of social insects is intense. In the army ant *Eciton hamatum*, the rate of loss is about 1,000 workers every day, or somewhat less than 1 percent of the entire force. The attrition rate of the household ant *Monomorium pharaonis* is 2 percent, and that of bumblebees (*Bombus agrorum* and *B. humilis*) is 4 percent per day. A colony of honeybees containing 60,000 workers in June loses them at the rate of 1,800, or 3 percent, every day. During the first 14 days of adult life, workers of the hornet *Vespa orientalis* suffer a mortality of 8.8 percent if confined to the nest and 42.5 percent if allowed to forage outside (Ishay *et al.,* 1967). The mortality of immature stages is also very high. Anne D. Brian (1951, 1952) noted that, in the colonies of *Bombus agrorum* which she observed, 71 percent of the eggs hatched, 73

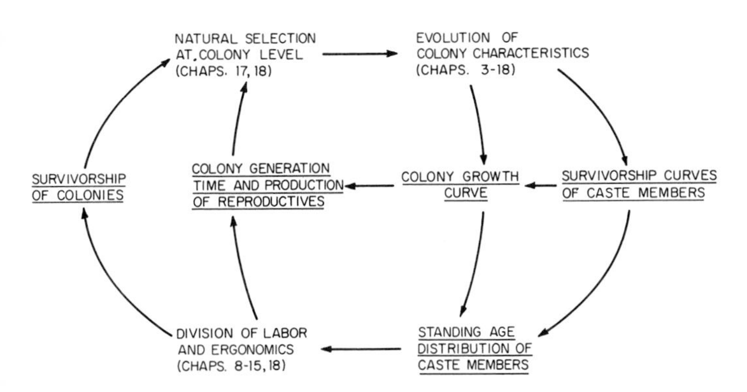

FIGURE 21-1. The causal relationships of principal phenomena in the biology of social insects. The phenomena of population dynamics which are reviewed in chapter 21 are underlined.

percent of the larvae reached the pupal stage, and 90 percent of the pupae yielded adults; thus, only 47 percent of the eggs survived to the adult stage. M. V. Brian (1951b) obtained similar results in laboratory colonies of the ant *Myrmica ruginodis* (= *M. "rubra"*): 50 percent of the eggs hatched, 74 percent of the larvae reached the pupal stage, and 89 percent of the pupae eclosed, resulting in a final survivorship, from egg to adult, of 33 percent.

Students of animal demography have found it convenient to recognize the four basic types of survivorship curves represented in Figure 21-2. The few such curves that have been worked out in the social insects vary markedly among the species in interesting ways that can be related to their social organization. In species of *Bombus*, for example, the worker caste lacks temporal division of labor. Individuals can take up any task at any age, and most begin foraging while still young. Since the heaviest mortality in social insects generally occurs among foraging workers, it is not surprising to find that the *Bombus* survivorship curve is type III. M. V. Brian (1965b) has in fact shown that Anne Brian's mortality data on *B. humilis* adult workers fits the negative exponential curve

$$N_t = N_o e^{-\mu t}$$

where N_o is the number of newly eclosed workers, N_t the number surviving after t days, and the mortality constant μ is approximately 0.04 per day. Honeybee workers, in contrast, follow a well-defined program of tasks during their adult life (see Chapter 9) and do not normally begin foraging until the tenth day or later. Sakagami and Fukuda (1968) found that the survivorship curve of these insects is type I. In the protected environs of the nest, mortality is very low, but when workers begin leaving the hive to go on foraging trips they also start perishing at a very high rate (see Figure 21-3). This adherence to the type I curve is shared with relatively stable, affluent human populations, but it is relatively rare elsewhere in the animal kingdom. It is literally true that this demographic property originated in honeybees as a result of their higher social status. There is another aspect of honeybee mortality that has a bearing on the evolutionary theory of senescence. Although most flying bees eventually die from predation or accidents, individuals protected from such misfortunes die of physiological senescence anyway. Even under the most favorable conditions, few

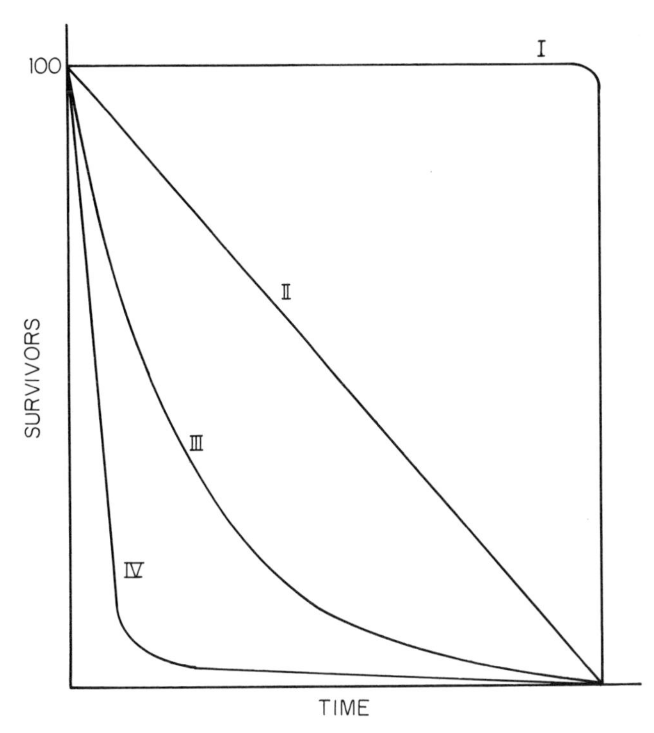

FIGURE 21-2. The four elementary types of survivorship curves in animal populations. Type I: Mortality is concentrated in the old animals, often as a result of physiological senescence. Type II: Mortality takes a constant number of animals per unit time regardless of the age and size of the population. Type III: Mortality takes a constant proportion of the surviving animals. Type IV: Mortality is concentrated on the youngest animals (redrawn from Slobodkin, 1961; based on Pearl and Miner, 1935, and Deevey, 1947).

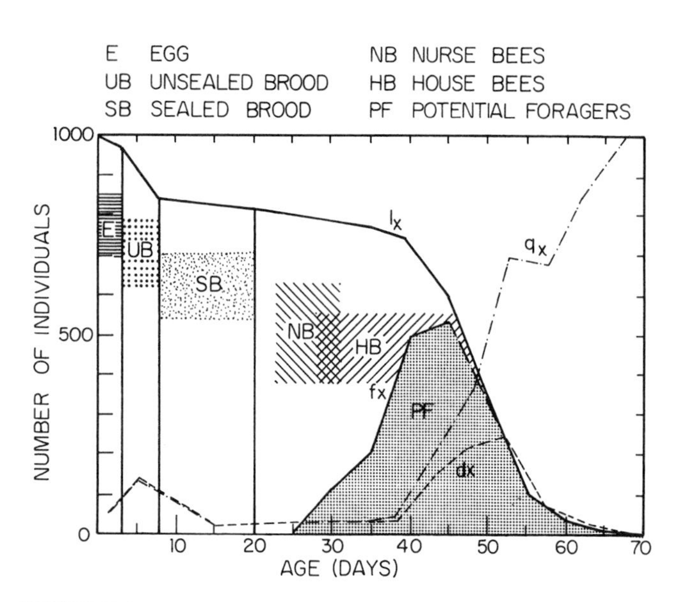

FIGURE 21-3. The survivorship curve of honeybee workers is of type I, as in advanced human societies. This curve is based on bees born in early summer. Those born in the fall and which overwinter also conform to the type I curve, but enjoy a much longer average life span: (d_x) number dying during respective age interval out of 1,000 born; (q_x) number dying per 1,000 alive at respective age interval (redrawn from Sakagami and Fukuda, 1968).

workers in a queenright colony in summer conditions can live to an adult age of greater than 50 days. Rockstein (1950a), for example, found only 11 workers alive from an original marked group of 2,700 at the end of an observation period of 51 days. Anna Maurizio (1950) found no survivors at all after 38 days from an original group of over a thousand, even among samples that had been protected by caging. Hodge (1894), Helen Pixell-Goodrich (1920), and Rockstein (1950b) have reported that as worker bees age, their brain cells decline in number and become increasingly vacuolated. In brains of bees measured by Rockstein, the average number of cells in medial transects was 522 in newly emerged individuals; this number decreased to 445, 434, and 369, successively, in bees 2, 4, and 6 weeks old. These findings are consistent with the theory of Medawar (1957), who argued that senescence is the result of an evolutionary programming of the physiology of individual organisms to postpone internal breakdown until past the age that most of the population would be killed off by external causes. Developmental events are arranged to give the maximum vitality to the age groups that have not yet been killed off to a significant extent, even if part of the physiological consequence is an increased rate of physiological breakdown (senescence) in older age groups. In other words, as L. B. Slobodkin has neatly put it, causes of mortality attract each other. In the case of the honeybee, the rapid mortality of foraging workers due to accident has seemingly "attracted" mortality by senescence, which the honeybee queen makes up by a moderately high oviposition rate—on the order of a thousand or more eggs a day.

Existing data are insufficient to permit deeper insight into the evolution of colonial survivorship curves. Also, we lack information about the derivative property of standing age distribution curves among the workers of insect colonies. These data are needed to complete the analysis of regular temporal polyethism (see Chapter 9). The gap could be filled in dramatic fashion if some reproducible morphological and biochemical criteria of insect age were to be discovered. Michener *et al.* (1955) have used the degree of wear in the wings and mandibles in rough demographic estimates of halictid bee populations, while Bassindale (1955) has employed changes in abdominal coloration to estimate the ages of meliponine bees. These and other methods that depend on the behavioral and physiological idiosyncrasies of the individual colony members can have only limited success. I have made several attempts to find measurable, behavior-free aging processes in ants, but have never hit upon anything sufficiently reliable. The problem poses an interesting and important challenge to insect physiologists.

Meanwhile, the best information on aging that can be extracted from existing literature, aside from the survivorship curves just described, are longevity records of individual colony members. A large fraction of all of the available data are provided in Table 21-1. The following key generalizations can be made from these records:

1. Mother queens live, as expected, much longer than workers in all groups of social insects. A few are among the longest-lived of all insects—surviving for as much as ten or even fifteen years.

2. No correlation is yet evident between the degree of sociality of a species and the longevity of its members. The primitively social halictids and bumblebees, for example, are as long-lived as the highly advanced honeybees.

3. The visible correlations of longevity are with two other factors: members of annual species live for shorter

TABLE 21-1. Longevities of individual social insects.[a]

Species	Caste	Locality	Average longevity or range	Maximum recorded longevity	Authority	Comments
WASPS						
Polistes fadwigae	Worker	Japan	2½ mos.	?	Yoshikawa (1963c)	
Mischocyttarus drewseni	Queen	Brazil	66 days	44–88 days	R. L. Jeanne (personal communication)	Based on 4 individuals
Mischocyttarus drewseni	Worker	Brazil	21 days	2–97 days	R. L. Jeanne (personal communication)	Based on 282 individuals
Vespa vulgaris	Worker	England	3–4 wks.	?	Ritchie (1915)	
Vespa orientalis	Worker	Israel	46 days	64 days	Ishay *et al.* (1967)	Based on 113 individuals (confined to nest)
ANTS						
Myrmica laevinodis (= "*rubra*")	Worker	England	?	2 yrs.	Brian (1951b)	
Aphaenogaster rudis	Queen	United States	8.7 yrs.	4.6–13 yrs.	Haskins (1960)	Based on 11 queens kept in laboratory nests
Aphaenogaster rudis	Worker	United States	?	>3 yrs.	Fielde (1904b)	
Lasius alienus	Queen	France	?	9.25 yrs.	Janet (1904)	Based on a single queen in a laboratory nest
Stenamma westwoodi	Queen	England	?	17 or 18 yrs.	Donisthorpe (1936)	Based on a single queen in a laboratory nest
Monomorium pharaonis	Queen	England	?	39 wks.	Peacock and Baxter (1950)	
Monomorium pharaonis	Worker	England	?	9–10 wks.	Peacock and Baxter (1950)	
BEES						
Dialictus versatus	Queen	United States	11 mos.	?	Michener (1969a)	Annual species
Dialictus versatus	Worker	United States	3 wks.	?	Michener (1969a)	Annual species
Evylaeus nigripes	Queen	France	12–15 mos. (range)	—	Plateaux-Quénu (1965)	Annual species
Evylaeus marginatus	Queen	France	3–5 yrs. (range)	—	Plateaux-Quénu (1962)	Perennial species
Allodape angulata	Queens and workers	South Africa	<1 yr.	<1 yr.	Skaife (1953)	Annual species
Melipona quadrifasciata	Queen	Brazil	?	3 yrs., 1 mo.	Kerr *et al.* (1962)	One record only

[a] In Hymenoptera only adult life spans are given.

TABLE 21-1 (*continued*).

Species	Caste	Locality	Average longevity or range	Maximum recorded longevity	Authority	Comments
Trigona xanthotricha	Worker	Brazil	?	94 days	Kerr and Santos Netos (1956)	Many
Bombus humilis	Worker	England	17.5 days	?	A. D. Brian in M. V. Brian (1965b)	Based on survival rates of wild bees; negative exponential survival curve
Apis mellifera	Worker (summer)	Germany	32.1 days	55 days	Rösch (1925)	13
Apis mellifera	Worker (summer)	Italy	?	38 days	Maurizio (1950)	Many
Apis mellifera	Worker (summer)	England	33.6 ± 0.2 days	40 days	Ribbands (1952)	47
TERMITES						
Kalotermes flavicollis	Pseudergates	France	?	2 yrs.	Grassé (1949)	
Neotermes tectonae	Pseudergates	Java	3–5 yrs.	?	Kalshoven (1930)	
Incisitermes minor	Primary queen and male	California	?	10–12 yrs.	Harvey (1934)	
Reticulitermes lucifugus	Worker	Europe	?	>5 yrs.	Buchli (1958)	
Reticulitermes lucifugus	Soldier	Europe	?	>5 yrs.	Buchli (1958)	
Reticulitermes lucifugus	Replacement reproductive	Europe	?	>7 yrs.	Buchli (1958)	
Coptotermes acinaciformis	Workers, soldiers	Australia	?	about 2 yrs.	Gay *et al.* (1955)	
Coptotermes lacteus	Workers, soldiers	Australia	?	about 2 yrs.	Gay *et al.* (1955)	
Cyclotermes sp.	Primary queen and male	India	?	12 yrs.	Beeson (1941)	
Cubitermes ugandensis	Worker	Africa	196–339 days	about 339 days	Williams (1959b)	
Macrotermes bellicosus	Queen	Africa	?	10 yrs.	F. A. Fenton (in Snyder, 1956)	

terms than those of perennial species, a hardly surprising result but one which does not necessarily proceed a priori (witness the very short lives of honeybee workers, who live in potentially immortal colonies); also, there is a connection with phylogeny, in that wasps and bees appear to have shorter lives than ants and termites, but for reasons yet to be fathomed.

The Regulation of Colony Growth

There is some cause to believe, from both theoretical considerations and fragmentary empirical evidence, that the increase in the numbers of members of individual insect colonies conforms at least approximately to the same laws that govern the growth of nonsocial animal populations. Actual measurements show that the colony increases exponentially in its initial growth phase, but that as it approaches "maturity," meaning the stage in which virgin queens are produced, its growth rate tapers off, bringing the colony population in the end to an asymptotic limit or else setting it into decline. Before examining the evidence in detail, let us review the elementary theoretical background that can give it meaning. For an "ideal" animal population, in which constant space and resources are available and the age distribution is in a steady state, the increase in individuals can be approximately described by the logistic equation,

$$\frac{dN}{dt} = rN\left(\frac{K - N}{K}\right)$$

where $N =$ the number of individuals in the population at any given point in time;
$K =$ the "carrying capacity of the environment," or the maximum number of individuals that can be supported by that portion of the environment in which the population is contained;
$r =$ the "intrinsic rate of population increase," equal to the instantaneous birth rate minus the instantaneous death rate.

The solution to this equation yields a sigmoid curve of increasing numbers (N) with the passage of time, and a point of inflection (at which dN/dt reaches its maximum) at $K/2$. Although adherence to the elementary logistic or some simple modification thereof has not been rigorously demonstrated in the growth of social insect colonies, the growth curves are frequently sigmoid in shape. The best-analyzed example, from Nolan's data on honeybees, is given in Figure 21-4. Another, well-known example is the increase in the number of nest craters in a colony of *Atta sexdens* measured by Autuori and analyzed by Bitancourt. Each crater is a ring of excavated earth around a nest entrance, and the number of nest entrances was assumed by Bitancourt, perhaps reasonably, to be a linear function of the number of workers in the colony. Growth data from colonies of *Neotermes tectonae* and *Lasius alienus* collected respectively by L. G. E. Kalshoven and Charles Janet have also been fitted to logistic curves by Bodenheimer (1937). Both Bodenheimer and Brian calculated intrinsic rates of increase (r) from these and similar, but more fragmented, data on other social insect groups. Brian (1965b) summarizes the findings as follows: the ant *Myrmica* has the lowest colony growth rate, with $r = 0.001$/individual/day; the bee *Halictus* and termite *Neotermes* are next with 0.003; *Apis* (0.03), *Bombus* (0.03), and *Vespula* (0.05) are considerably higher; while the polybiine wasps, at 0.1, have a much higher value than all the rest. These numbers, although scattered and very few, seem to suggest that the major groups of social insects differ from each other in the basic property of colony growth rate, with ants and termites being low, bees intermediate, and wasps high.

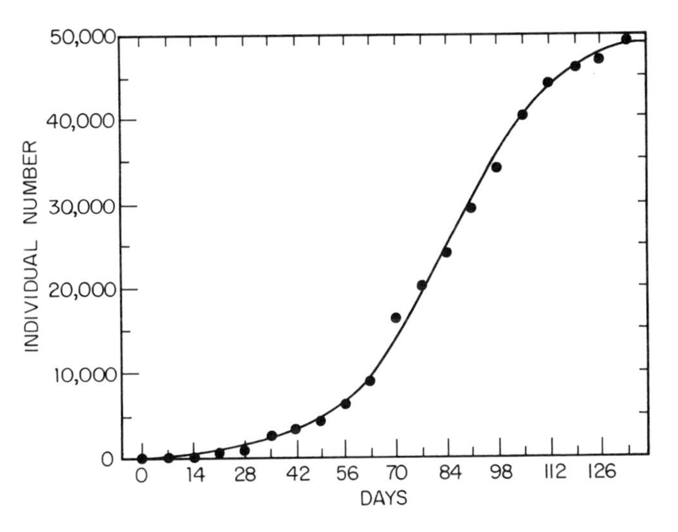

FIGURE 21-4. Population growth of a colony of the Cyprian race of the honeybee. All life stages are included except the initial population of a single mother queen and about 5,000 overwintered workers (redrawn from Sakagami and Fukuda, 1968; based on data from Nolan, 1928, with more recent corrections for estimated mortality).

The importance of a fit to the logistic curve, aside from providing a first crude picture of the full course of population growth, lies in its implication of density-dependent negative feedback in the later stages of colony growth. The identification and measurement of the factors comprising the feedback should become one of the principal targets of population studies in the social insects, as it already is in the analysis of other animal species. Only when this information becomes available will more exact models of colony population growth be possible. In the meantime, the first of the density-dependent negative factors to receive close attention in theoretical studies has been the emission of the reproductive forms. The generating idea is very simple: the more workers there are, the larger the proportion of individuals who turn into males and virgin queens. And since these reproductive individuals are both expensive to make and counterproductive purely in terms of calories, their manufacture puts a drag on the further production of both workers and other reproductives. The process will inevitably reduce the population growth rate to zero. By constructing models that make reasonable and, for the most part, readily verifiable assumptions, it has been possible to arrive at some unexpected and interesting results concerning the regulation of growth rate and final colony size. Richards and Richards (1951), for example, employed a simple iterative algebraic model to simulate growth in colonies of social wasps. The following parameters were recognized:

d = developmental time in days, egg to adult;
N = number of individuals in the colony, all castes;
P = number of founder workers;
r = average number of brood cells added to colony each day per worker;
s = average length of worker life in days;
t = time elapsed in days since founding of the colony.

A fraction $1/s$ of the worker force dies off each day. It follows that during the development of the first brood, meaning up to time d, the number of individuals in the colony on any given day will be

$$N_t = P\left(1 - \frac{1}{s}\right)^t + \sum_{x=1}^{t} rP\left(1 - \frac{1}{s}\right)^t$$

where the first term is the number of the original workers remaining after t days, and the second term is the cumulative number of cells the wasps have built during t days. Each cell is assumed to contain one immature individual, so that the second term represents the number of the immature individuals, and the first term represents all of the surviving adults. Beyond d, the time at which the first brood begins to eclose into adults (which themselves begin to add brood cells), the model building becomes a complicated and laborious numerical exercise. Richards and Richards have nevertheless carried it through to show the following results for all values of real parameters known to them. First, the colony will grow exponentially. Second, and less obviously, the ratio of larvae to workers will start relatively high, a condition that tends to cause the larvae to be fed less and to develop into other workers. But after the beginning of the emergence of the first brood, the ratio will drop sharply and remain low thereafter, a balance conducive to the production of queens. Finally, the production of queens will slow the growth of the colony and may stop it altogether or even reverse it. An example of the first stage of colony growth, according to the Richards' model, is given in the accompanying table. A similar model, with mostly comparable results also applicable to social wasps, has been presented by Lövgren (1958).

M. V. Brian (1957a, 1965b) also developed more general models with the rate and timing of the emission of the reproductive forms taken as the principal population limiting factor. His most important result is the discovery that small changes in this single factor can produce radical variation in entire patterns of colony growth. In the following account I have changed the notation and method of presentation considerably to make the ideas clearer to the general reader. Brian's parameters are the following:

a,b,c = fitted constants;
λ = birth rate of the worker population;
μ = death rate of the worker population;
N_w = number of workers in the colony;
N_I = number of immature individuals (eggs, larvae, pupae) giving rise to workers later;
N_J = number of immature individuals (eggs, larvae, pupae) giving rise to virgin queens in the next generation.
s = a constant with value < 1.

The rate of increase of the workers in a colony should be equal to some function of the number of workers present minus two quantities: the death rate of the workers and some function of the number of queens being simultaneously produced. In symbols,

$$\frac{dN_w}{dt} = \lambda N_w^{\,s} - \frac{N_w}{c\lambda N_w^{\,s}} - \mu N_w$$

The unimpeded population birth rate is given as $\lambda N_w^{\,s}$, with $s < 1$, since empirical studies on ants of the genus *Myrmica* (by Brian, 1953b) have shown that larval growth increases logarithmically with the number of nurse workers. The rate of production of queens is given by Brian in the above equation as a function of the ratio of workers to new worker larvae being born, an assumption also in at least rough accordance with the known facts about *Myrmica.* If for simplicity we take s to equal 0.5, the growth equation integrates to

$$N_{w(t)} = \frac{\left(\lambda - \frac{1}{c\lambda}\right)^2}{\mu^2}[1 - e^{-\mu t/2}]^2$$

TABLE 21-2. Growth of a hypothetical wasp colony (according to the model of Richards and Richards, 1951). Original worker number is 88; developmental time, 30 days (egg stage 6 days, larval and pupal stages 12 days each); rate of cell addition, 0.5 cell per worker per day; and average worker longevity, 45 days.

	Day											
Number	1	2	15	16	20	25	30	31	32	45	60	90
Cells	44	87	555	584	690	800	885	921	977	3350	8972	74787
Larvae	0	0	360	395	438	378	318	306	294	1060	3880	35554
Workers	88	86	60	58	50	40	30	72	113	555	885	9302
Ratio of larvae to workers	0	0	6.0	6.8	8.8	9.4	10.6	4.2	2.6	1.9	4.4	3.8

which rises sigmoidally toward the limit $[(\lambda - 1/c\lambda)/\mu]^2$ as t becomes large.

It should also be approximately true that

$$\frac{dN_J}{dt} = bN_w$$

in other words, that queen larvae, which receive special attention and are present in smaller numbers, should increase as a linear function of the number of workers. Further, the rate of increase of worker larvae should be some monotonic function, perhaps logarithmic, of the number of workers present, minus approximately the rate of increase of the queen larvae. In symbols,

$$\frac{dN_I}{dt} = aN_w{}^s - \frac{dN_J}{dt}$$
$$= aN_w{}^s - bN_w$$

By varying the value of b slightly, very different solutions are obtained (Figure 21-5). It remains to be seen whether Brian's formulations contain a sufficient set of parameters to be predictive. His model does have the virtue of being the first to provide simulations of strongly differing life cycles which do, in fact, occur in nature. For example, the "explosive" form of queen emission is exemplified by the annual life cycles of temperate polistine and vespine wasps (Figure 21-6), as well as those of many species of halictid bees and bumblebees. Periodically oscillating steady states occur in honeybees, stingless bees, army ants, and other groups that multiply by colony fission. In these cases the oscillation is intensified by the departure of workers with one or more members of the newly increased queen force. Most ant species with seasonal nuptial flights conducted en masse are mildly oscillatory. Others that release their sexual forms gradually and at a low rate, or not at all, might approach the condition of a nonoscillating steady state. We are mostly ignorant concerning the extent to which the cycles are controlled by endogenous versus exogenous factors. It is probable that the basic parameters of the Richards and Brian models are subject to significant alteration by environmental change. In the warm parts of New Zealand, for example, colonies of introduced

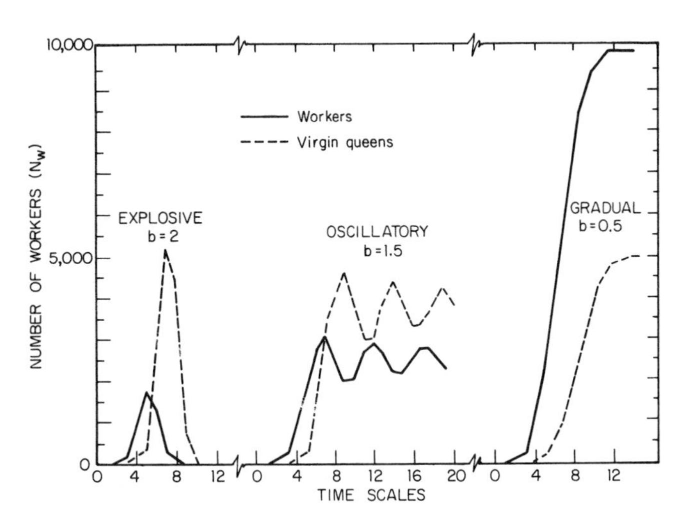

FIGURE 21-5. Growth of populations of social insects with all parameters held constant except b, the coefficient of production of queens. As b is decreased from 2 to 0.5, the growth curve changes from the "explosive" form, which terminates in the death of the colony following emission of the virgin queens, through an oscillatory steady state, to a gradual approach to a steady state (modified from Brian, 1965b).

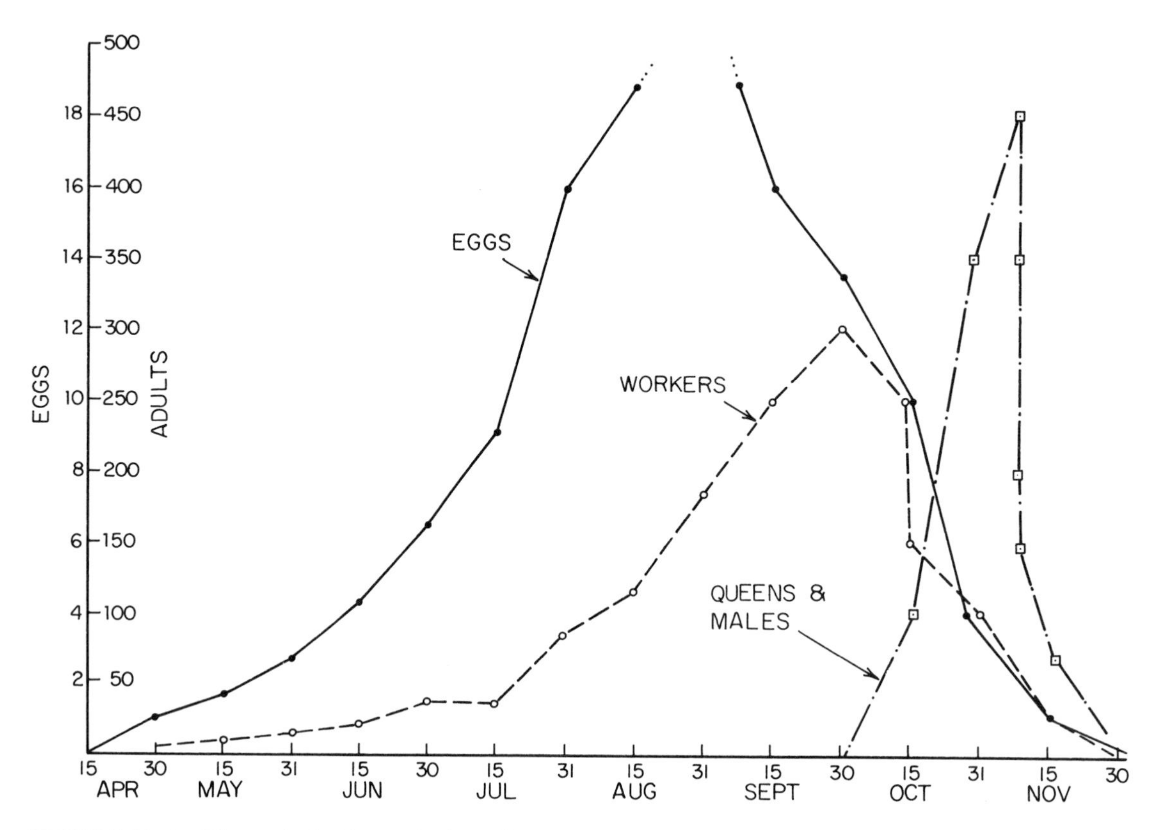

FIGURE 21-6. The explosive form of colony life cycle is exemplified by the hornet *Vespa orientalis,* in which worker production is sacrificed for massive production of queens and males at the end of the summer. As shown in this record of the history of a single colony, the final result is a rapid and terminal decline of the colony as a whole (redrawn from Ishay, Bytinski-Salz, and Shulov, 1967).

species of *Bombus, Polistes,* and *Vespula* are frequently able to overwinter and, by continuing to grow on into a second or even third summer, attain much larger sizes than the annual colonies of the same species in the country of their origin (Cumber, 1949c, 1951). In other words, the explosive phase of the colony cycle is postponed or eliminated.

Table 21-3 gives most of the available data on colony size in various species of social insects. These raw numbers tell us nothing directly about the growth rates or factors limiting mature colony size, but they do permit some inferences when comparisons are made among major groups. Such conclusions lead easily to more fundamental physiological questions to be taken up shortly:

1. There exists only a weak correlation between average colony longevity (see also Table 21-4) and mature colony size. Within the social wasps, for example, the annual colonies of the temperate *Vespa* and *Vespula* species are fully as large as those of the perennial Polybiini. Within the Polybiini, on the other hand, the very short-lived colonies of *Mischocyttarus* and *Polistes* also have the smallest colony size. By the same token it is only the perennial species of social bees, in the genera *Apis, Melipona,* and *Trigona* particularly, that mount colonies with tens of thousands of workers.

2. There is no clear relation between climate and colony size. In the ants, if anything, temperate species tend to have somewhat larger colonies on the average. This is due to a special ecological effect connected with restrictions in nest site, which follows.

3. There is a strong correlation between preferred nest site and mature colony size. The social wasps, which

TABLE 21-3. Number of adults in colonies of social insects.

Species	Locality	Number of adults	Number of adults (range)	Authority	Number of colonies censused
WASPS					
Polistinae					
Protopolybia minutissima	South America	51, 99, 100, 693	?	Richards and Richards (1951)	4
P. pumila	South America	7,237	?	Richards and Richards (1951)	1
Brachygastra scutellaris	South America	863 and 876	?	Richards and Richards (1951)	2
B. mellifica	Texas to Central America	10,000 and 15,000	?	Schwarz (1929), Naumann (1968)	2
Metapolybia cingulata	South America	37, 54, 57, 106	?	Schwarz (1929), Naumann (1968)	4
Polybia atra	South America	1528	?	A. Hase (in Bodenheimer, 1937)	1
P. micans	South America	207	?	Richards and Richards (1951)	1
P. rejecta	South America	2780	?	Richards and Richards (1951)	1
P. occidentalis	South America	130 and 227	?	Richards and Richards (1951)	2
P. parvula	South America	148 and 868	?	Richards and Richards (1951)	2
P. bistriata	South America	73[a]	29–129	Richards and Richards (1951)	14
P. bicyttarella	South America	81[a]	33–143	Richards and Richards (1951)	9
P. catillifex	South America	6, 17, 24, 41	?	Richards and Richards (1951)	4
Pseudochartergus fuscatus	South America	210	?	R. L. Jeanne (1970c)	1
Apoica thoracica	South America	236	?	Richards and Richards (1951)	1
A. albomacula	South America	58	?	Richards and Richards (1951)	1
Stelopolybia pallens	South America	168[a]	25–390	Richards and Richards (1951)	4
Stelopolybia testacea	South America	6,466	?	R. L. Jeanne (1970c)	1
Ropalidia spp.	Africa	?	to 500 or 600	Carl (1934)	Many
Mischocyttarus drewseni (incipient colonies omitted)	South America	22[a]	12–37	R. L. Jeanne (personal communication)	7
Polistes carnifex	South America	2, 9	?	Richards and Richards (1951)	2
P. canadensis	South America	16[a]	6–24	Richards and Richards (1951)	5
P. crinitus	West Indies	5, 5, 20	?	Richards and Richards (1951)	3
Vespinae					
Vespa crabro	Germany	~100[a]	to 400	Kemper and Döhring (1961)	8
Vespula cuneata	North Carolina	539	?	Manee (1915)	1
Vespula diabolica	United States	115	?	Duncan (1924)	1

[a] Average number; other data pertain to individual colonies.

TABLE 21-3 (*continued*).

Species	Locality	Number of adults	Number of adults (range)	Authority	Number of colonies censused
Vespula germanica	New England	146, 2,090	?	Baerg (1921)	2
Vespula germanica	New England	648	?	Wyman (1860)	1
Vespula germanica	Germany	1,348	?	Pechlaner (1904)	1
Vespula maculata	United States	105	?	Betz (1932)	1
Vespula maculata	United States	494	?	Rau (1929)	1
Vespula sylvestris	Germany	95	to 180	Kemper and Döhring (1961)	Several
Vespula vulgaris	England	5,207	?	Crawshay (1905)	1
Vespula vulgaris	England	1,197	?	Ritchie (1915)	1
Vespula vulgaris	Germany	~1,000[a]	to 2,847	Kemper and Döhring (1961)	Several
Sphecidae					
Microstigmus comes	Central America	?	1–18	Matthews (1968b)	Many
ANTS					
Ponerinae					
Amblyopone pallipes	Quebec	12[a]	4–35	Francoeur (1965)	6
Prionopelta opaca	New Guinea	20	?	Wilson (1959d)	1
Platythyrea parallela	New Guinea	50	?	Wilson (1959d)	1
Rhytidoponera araneoides	New Guinea	~50[a]	?	Wilson (1959d)	3
R. laciniosa	New Guinea	100, 150	?	Wilson (1959d)	2
Gnamptogenys macretes	New Guinea	40	?	Wilson (1959d)	1
Leptogenys bituberculata	New Guinea	300	?	Wilson (1959d)	1
L. diminuta	New Guinea	90–400	?	Wilson (1959d)	4
L. purpurea	New Guinea	500, 2,000	?	Wilson (1959d)	2
Mesoponera papuana	New Guinea	10	?	Wilson (1959d)	1
Diacamma rugosum	New Guinea	30, 50	?	Wilson (1959d)	2
Cryptopone motschulskyi	New Guinea	20	?	Wilson (1959d)	1
Ectomomyrmex striatulus	New Guinea	10, 20	?	Wilson (1959d)	2
Myopias sp. 1	New Guinea	60	?	Wilson (1959d)	1
M. sp. 2	New Guinea	40, 70	?	Wilson (1959d)	2
M. sp. 3	New Guinea	30	?	Wilson (1959d)	1
Cerapachyinae					
Cerapachys opaca	New Guinea	100	?	Wilson (1959d)	1
C. polynikes	New Guinea	20	?	Wilson (1959d)	1
Dorylinae					
Eciton burchelli	Central America	?	150,000–700,000	See Chapter 4	
E. hamatum	Central America	?	100,000–500,000	See Chapter 4	
Neivamyrmex nigrescens	United States	?	80,000–140,000	See Chapter 4	
Aenictus currax	New Guinea	>100,000	?	Wilson (1959d)	1
A. laeviceps	Philippine Islands	?	60,000–110,000	See Chapter 4	
Anomma wilverthi	Africa	?	2×10^6–22×10^6	See Chapter 4	
Myrmicinae					
Myrmica ruginodis	Scotland	1,216[a]	305–2,855	Brian (1950)	12
Myrmica schencki emeryana	United States	255[a]	35–561	Talbot (1945a)	36
Pogonomyrmex badius	United States	~5,000[a]	?	Golley and Gentry (1964)	Many
P. barbatus	United States	12,358	?	Wildermuth and Davis (1931)	1
Aphaenogaster rudis	Michigan, United States	326[a]	26–2,079	Talbot (1951)	72

TABLE 21-3 (*continued*).

Species	Locality	Number of adults	Number of adults (range)	Authority	Number of colonies censused
A. rudis	Ohio, United States	280[a]	11–950	Headley (1949)	46
A. treatae	United States	682[a]	65–1,662	Talbot (1954)	30
A. dromedarius	New Guinea	100	?	Wilson (1959d)	1
Stenamma diecki	Quebec	41[a]	5–103	Francoeur (1965)	7
Leptothorax curvispinosus	United States	84[a]	8–368	Headley (1943)	38
L. longispinosus	United States	47[a]	2–142	Headley (1943)	97
Ancyridris sp.	New Guinea	10, 15	?	Wilson (1959d)	2
Pristomyrmex sp.	New Guinea	100	?	Wilson (1959d)	1
Adelomyrmex biroi	New Guinea	10	?	Wilson (1959d)	1
Acidomyrmex melleus	New Guinea	>5,000	?	Wilson (1959d)	2
Vollenhovia brachycera	New Guinea	150	?	Wilson (1959d)	1
Myrmecina transversa	New Guinea	100	?	Wilson (1959d)	1
Tetramorium caespitum	England (1963)	14,068[a]	2,603–29,571	Brian *et al.* (1967)	23
T. caespitum	England (1964)	7,881[a]	1,395–30,943	Brian *et al.* (1967)	26
Meranoplus spinosus	New Guinea	150	?	Wilson (1959d)	1
Cardiocondyla thoracica	New Guinea	70	?	Wilson (1959d)	1
C. paradoxa	New Guinea	50	?	Wilson (1959d)	1
Crematogaster elegans	New Guinea	300	?	Wilson (1959d)	1
C. subtilis	New Guinea	>5,000	?	Wilson (1959d)	1
C. dohrni artifex	India	56,947	?	Ayyar (1937)	1
C. dohrni rogenhoferi	India	5,690	?	Roonwal (1954)	1
Strumigenys bajarii	New Guinea	400	?	Wilson (1959d)	1
S. frivaldszkyi	New Guinea	15	?	Wilson (1959d)	1
S. mayri	New Guinea	100	?	Wilson (1959d)	1
S. loriai	New Guinea	300, 500	?	Wilson (1959d)	2
Rhopalothrix biroi	New Guinea	50	?	Wilson (1959d)	1
Dacetinops cibdela	New Guinea	10	?	Wilson (1959d)	1
Pheidole sp.	New Guinea	150	?	Wilson (1959d)	1
Pheidologeton sp.	New Guinea	>3,000	?	Wilson (1959d)	1
Sericomyrmex amabilis	Central America	?	to 300	Wheeler (1925)	Many
S. urichi	Central and South America	?	200–1,691	Weber (1967)	Many
Atta colombica	Central America	Between 10^6 and 2.5×10^6	?	M. Martin (personal communication)	1
Myrmeciinae					
Myrmecia gulosa	Australia	188, 1,586	?	Haskins and Haskins (1950a)	2
M. pilosula	Australia	553, 862	?	Haskins and Haskins (1950a)	2
M. vindex	Australia	109, 272	?	Haskins and Haskins (1950a)	2
Dolichoderinae					
Leptomyrmex fragilis	New Guinea	350	?	Wilson (1959d)	1
Iridomyrmex scrutator	New Guinea	?	500–>3,000	Wilson (1959d)	1
Formicinae					
Pseudolasius breviceps	New Guinea	200, 500	?	Wilson (1959d)	2
Acropyga sp.	New Guinea	>1,000	?	Wilson (1959d)	1
Paratrechina pallida	New Guinea	500	?	Wilson (1959d)	1
P. sp. 1	New Guinea	200	?	Wilson (1959d)	1

TABLE 21-3 (*continued*).

Species	Locality	Number of adults	Number of adults (range)	Authority	Number of colonies censused
P. sp. 2	New Guinea	150	?	Wilson (1959d)	1
Prenolepis imparis	United States	1,582[a]	48–2,208	Talbot (1943)	11
Formica exsectoides	United States	41,366 and 238,510	?	Cory and Haviland (1938)	2
F. pallidefulva nitidiventris	United States	2,352[a]	541–7,050	Talbot (1948)	24
F. "incerta"	United States	714[a]	107–1,668	Talbot (1948)	24
F. rufa	Europe	?	to >100,000	Brian (1965b)	Many
Camponotus papua	New Guinea	300	?	Wilson (1959d)	1
C. confusus	New Guinea	200	?	Wilson (1959d)	1
C. vitreus	New Guinea	>4,000	?	Wilson (1959d)	1
C. pennsylvanicus	United States	?	1,943–2,500	Pricer (1908)	Many
Calomyrmex laevissimus	New Guinea	250	?	Wilson (1959d)	1
Polyrhachis hirsutula	New Guinea	150	?	Wilson (1959d)	1
P. limbata	New Guinea	100	?	Wilson (1959d)	1
P. debilis	New Guinea	~325[a]	300–350	Wilson (1959d)	3
P. rufiventris	New Guinea	200	?	Wilson (1959d)	1
P. omymyrmex	New Guinea	60	?	Wilson (1959d)	1
BEES					
Halictidae					
Halictus scabiosae	Switzerland	3[a]	1–6	Batra (1966b)	7
Dialictus versatus	United States	?	5–105	Michener (1966b)	29
Dialictus zephyrus	United States	?	to 45	Batra (1966a)	Many
Evylaeus marginatus	France	?	200–400	Plateaux-Quénu (1962)	Many
Pseudaugochloropsis costaricensis	Central America	5[a]	2–7	Michener and Kerfoot (1967)	5
P. nigerrima	Central America	3[a]	2–6	Michener and Kerfoot (1967)	4
Anthophoridae					
Exoneura variabilis	Australia	?	1–6	Michener (1965b)	Many
Apidae					
Bombus spp.	England	?	100–400 (according to species)	Free and Butler (1959)	Many
Bombus medius	Mexico	2,183	?	Michener and LaBerge (1954)	1
Melipona spp.	South America	?	500–4,000 (according to species)	Nogueira-Neto (1951)	Many
Trigona spp.	South America	?	300–80,000 (according to species)	Nogueira-Neto (1951)	Many
Trigona spp.	South America	?	to 100,000	Michener (1969a)	Many
Apis mellifera	Holland	22,579	?	Swammerdam (1737–38)	1
Apis mellifera	World-wide	?	20,000–80,000	Ribbands (1953)	Many
TERMITES[b]					
Mastotermitidae					
Mastotermes darwiniensis	Australia	?	Several thousands to >1 million	Hill (1942)	?

[b] Nymphs included in population counts of termite colonies.

TABLE 21-3 (*continued*).

Species	Locality	Number of adults	Number of adults (range)	Authority	Number of colonies censused
Kalotermitidae					
Cryptotermes brevis	United States	~300[a]	?	McMahan (1966)	?
C. havilandi	United States	?	>3,000	Wilkinson (1962)	?
Incisitermes minor	California	?	1,000–2,750	Harvey (1934)	Many
I. minor	Arizona	9,200	?	Nutting (1969)	1
Marginitermes hubbardi	United States	?	to 2,267	Nutting (1969)	?
Paraneotermes simplicicornis	United States	?	to 1,394	Nutting (1966a)	?
Pterotermes occidentis	United States	760 and 2,911	?	Nutting (1966b)	2
Hodotermitidae					
Zootermopsis angusticollis	United States	?	to 8,000	Castle (1934)	Many
Z. laticeps	United States	2,367	?	Nutting (1969)	1
Rhinotermitidae					
Coptotermes acinaciformis	Australia	?	$>1.25 \times 10^6$	Greaves (1964)	Many
C. frenchi	Australia	?	$>7.5 \times 10^5$	Greaves (1964)	Many
C. lacteus	Australia	?	6×10^5–1.1×10^6	Gay and Greaves (1940)	1
Termitidae					
Amitermes hastatus	South Africa	?	to 50,000	Skaife (1955)	Many
Apicotermes desneuxi	West Africa	11,638[a]	to 78,200	Bouillon (1964)	36
Macrotermes natalensis	West Africa	~2,000,000[a]	?	Lüscher (1955)	Many
Macrotermes spp.	West Africa	?	to "several millions"	Grassé (1949)	Many
Microcerotermes arboreus	West Africa	5,876	?	Emerson (1938)	1
Nasutitermes exitiosus	Australia	?	4.84×10^5–8.07×10^5	Holdaway *et al.* (1935)	4
N. surinamensis	South America	~3,000,000	?	Emerson (1938)	?
N. sp.	Jamaica	631,878	?	Andrews (1911)	1
Odontotermes obesus	India	90,961	?	Gupta (1953)	1
Trinervitermes geminatus	West Africa	?	19,000–52,000	Sands (1965)	Many

construct their own nests of carton and in most cases suspend them from arboreal supports, generally have smaller mature colonies than the soil-dwelling ants and termites. Among these insects it is the exceptional ground-nesting species belonging to the genus *Vespula* that form the largest colonies. Among the rain forest ants of New Guinea, those that nest in rotting logs and other pieces of rotting wood on the ground (almost all Ponerinae and Cerapachyinae and the majority of Myrmicinae) form smaller colonies than those living in less restricted nest sites, such as the open soil of the forest floor (*Acidomyrmex, Pheidologeton, Leptomyrmex, Pseudolasius, Acropyga,* most *Paratrechina*), open air at the ground surface (Dorylinae), and various sectors of the tree canopy (most *Crematogaster, Iridomyrmex, Camponotus,* and *Polyrhachis*). A similar relation exists within the termites. It is the soil-dwelling species of *Mastotermes, Coptotermes,* and Termitinae (particularly *Macrotermes*) that develop truly enormous colonies. The bees offer a curious exception. The soil-dwelling halictines form much smaller colonies than the Meliponini and *Apis,* most of which build their own nests in trees, but, as we have seen, most halictines also form short-lived annual colonies.

4. There is only a weak correlation with other evolutionary trends manifested within the same major phyletic groups. *Mastotermes darwiniensis,* the most primitive living termite species, forms huge colonies comparable to those of the advanced fungus-growing macrotermitines. The species of *Myrmecia,* one of the most primitive ant genera, have larger colonies than do most members of the ad-

TABLE 21-4. Longevities of established colonies of social insects.

Species	Location	Longevity: Average	Longevity: Maximum	Authority	Comments
WASPS					
Polistes spp.	North Temperate Zone	<6 mos.	6 mos.	See Chapter 3	Annual life cycle
Mischocyttarus spp.	Brazil	6 mos.	7.5 mos.	R. L. Jeanne (personal communication)	No seasonal synchrony
Vespula spp.	North Temperate Zone	<6 mos.	6 mos.	See Chapter 3	Annual life cycle
ANTS					
Formica pratensis	Switzerland	?	>56 yrs.	Forel (1928)	Replacement reproductives make the colony potentially immortal
F. rufa	England	?	>60 yrs.	C. Darwin (1859)	Replacement reproductives make the colony potentially immortal
F. ulkei	United States	?	>25 yrs.	Dreyer (1942)	Replacement reproductives make the colony potentially immortal
BEES					
Halictus spp. and *Lasioglossum* spp.	North Temperate Zone	<6 mos.	6 mos.	(See Chapter 5)	Annual life cycle
Evylaeus marginatus	France	?	6 yrs.	Plateaux-Quénu (1962)	Perennial life cycle
TERMITES					
Neotermes tectonae	Java	10 yrs.	15 yrs.	Kalshoven (1930)	Longevity potentially greater when supplementary reproductives arise
Neotermes castaneus	United States	?	>24 yrs.	Emerson (1939)	Based on one laboratory colony; colonies may produce supplementaries
Cyclotermes sp.	India	?	12 yrs.	C. F. C. Beeson (in Snyder, 1956)	Royal pair are not replaced by supplementaries, and colony dies with them
Amitermes hastatus	South Africa	?	25 yrs.	Skaife (1955)	
Cubitermes fungifaber	West Africa	?	>5 yrs.	Noirot (1969b)	Colonies live only "5 years after the egress of the nest from the soil"
Nasutitermes triodiae	Australia	?	>63 yrs.	Hill (1942)	
Macrotermes spp.	Africa	?	>80 yrs.	Grassé (1949)	

vanced subfamilies. One interesting rule does stand out, nonetheless: the most elaborate forms of social behavior occur in species with large, perennial colonies. These include the waggle dance of *Apis,* the fungus gardening of the Attini and Macrotermitinae, and the legionary behavior of the Dorylinae and certain genera of Ponerinae.

5. The great variation in colony size among species belonging to the same taxonomic group (see, for example, the data for polybiine wasps and myrmicine ants) attests

to the capacity of this population trait to evolve with relative speed. We have seen that small alterations in such physiological parameters as mean worker life span and thresholds in queen determination of larvae can bring about major differences in mature colony size. Given this potency to adapt colony size to local environmental conditions at the species level, we should feel encouraged to investigate which of the conditions are critical in evolution.

In undertaking evolutionary interpretation, it is necessary to stress the distinction between proximate and ultimate causation. In the case of the polybiine wasps, O. W. and Maud J. Richards attempted to explain the curtailment of population growth as the result of worker turnover and emission of sexual forms. Theirs was an explanation of proximate causation. They also suggested that the parameters of worker turnover and emission of sexual forms are set in the course of evolution in such a way as to maximize colony fitness. This is an explanation of ultimate causation. The Richardses went on to suggest that the optimal size of wasp colonies is the balance struck to accommodate two kinds of predation: on the one hand, if the colonies are made small they lose less "capital" in the form of immature stages when they are raided by army ants; if, on the other hand, they are *too* small they are more vulnerable to vertebrates, which (unlike army ants) can be intimidated by the adult wasps. Consequently, the optimal colony size could be determined, all other factors being equal, by the balance determined between the frequency and severity of attacks by army ants and the frequency and severity of attacks by vertebrates. It needs to be added that small colony size is probably only a concomitant of the main feature that confers the real selective advantage, namely short colony developmental time. The crucial feature is the capacity to produce and disperse females in a short enough time to make an attack by army ants on any given colony improbable. This will have the result of reducing the average colony size. The "capital," that is, immature stages, produced per worker per unit of time is actually larger in small colonies and therefore cannot be the principal factor as envisaged by the Richardses.

At least two kinds of ultimate factors besides predation have been inferred in other population studies. I have cited already the restricting nature of the rotting-wood nest sites in the ant species of New Guinea rain forests. They are the most favored class of nest sites in the rain forest faunas of the tropical Oriental and Australian regions, and, apparently as a consequence, a very large percent of the ant species have small mature colony sizes. O. W. Richards (1927b) found that among the subarctic European species of *Bombus,* the "mature" colony size is relatively small. That is, fewer workers are produced in each colony prior to the onset of queen production. The implication is that, in the brief summers available to the bees, it is necessary to reduce mature colony size in order to hasten colony development and to produce a sufficiently large crop of queens. Whatever advantages accrue to larger colony size are overridden by this single overwhelming stress factor.

Up to this point we have been examining the negative factors of colony growth, whose proximate manifestation is the mortality of colony members and the emission of reproductive forms. Let us now consider the input side of the ledger, namely the oviposition rate of the queens and laying workers. In the differential equations of colony growth the rate of colony growth was written as λN, where λ is the birth coefficient and N is the total number of adult workers in the colony. In nonsocial animal populations λN means simply that the rate of increase is equal to the number of breeding individuals times the average birth rate per individual. In a colony of social insects, however, few if any of the workers breed. Is it therefore valid to regard the colony birth rate as a linear function of the number of workers? The answer depends on which life stage and which type of social insect are considered. If we mean increase of adult individuals in the colony, and if the oviposition rate of the queen finds no limit within the range of N measured, λN should give a close approximation because the rate at which new adults are created will be expected to increase linearly as the number of workers available to care for the brood increases. But if it is the increase in egg production that matters, new measures are required. The number of eggs being laid cannot be expected to be a simple multiple of the number of workers present, even if the queens have an effectively unlimited capacity to increase egg production. In fact Brian (1957a) showed that in *Myrmica ruginodis* the oviposition rate of a queen increases slightly with an increase in the worker force. To be exact, the number of eggs laid in three weeks at 20°C was $37.5 + 0.249\ N_w$, where N_w was the number of workers. This result was confirmed, at least qualitatively, in the experiments of Mamsch (1967). Subsequently Brian found that in the related species *M. rubra* the queen reaches the limit of her egg-laying capacity when accompanied by 40 workers. Between 72 and 84 percent of colonies in nature have more than this number of workers per queen. The workers are therefore underemployed and more prone to devote their energies to the rearing of new queens from the available larvae.

In social insects generally the potential oviposition rate of the individual queen has tended to be raised by one or the other of four evolutionary changes: an increase in the number of ovarioles, a lengthening of the ovarioles, an increase in the egg maturation rate, and a reduction in the size of the egg (Hagan, 1954; Iwata and Sakagami, 1966). These trends are correlated with the mature size of the colony. For example, queens of Halictidae have the basic aculeate hymenopteran character of three ovarioles per ovary, while *Apis mellifera* queens have an average of 160 per ovary. In the Meliponini, which like *Apis* produce large colonies, there are only four ovarioles per ovary, but each ovariole is very long, coiled through the abdomen, and packed with eggs. In the ants the number of ovarioles is closely correlated with colony size. This relation is exemplified by the following series arranged in order of ascending colony size, with the number of ovarioles given in parentheses: *Leptothorax congruus* (3), *Aphaenogaster famellica* (10–11), *Camponotus obscuripes* (30–34), *Crematogaster laboriosa* (18–25), *Lasius niger* (61–70), *Eciton burchelli* (1,200–1,300).

The evolutionary increase in oviposition capacity has nevertheless not been adequate to free the social wasps from this particular constraint. In the short-cycle polybiine wasps, where data were pooled from 33 colonies belonging to several species, Richards and Richards (1951) found that the rate of cell construction depends on the capacity of the queens to lay eggs, rather than the reverse. If the developmental time is accepted as 21 days, the number of cells constructed per worker per day is $0.3986 + 0.0119\,q$, where q is the number of egg-laying queens in the nest. According to Spradbery (1965), the nest queens of *Vespula,* of which there is only one to a colony, reach the limit of their oviposition capacity at the peak of colony growth late in the summer, and their eggs are often supplemented by worker-laid eggs afterwards.

It is of the greatest advantage for newly founded colonies to build up their worker strength quickly. In all social insects the highest mortality occurs in the founding stages and is due chiefly to the inability of young colonies to withstand predation. In *Neotermes tectonae,* to take one of the better-documented examples, colonies are very likely to be destroyed by ants until they acquire 50 to 75 members (Kalshoven, 1930). It follows that, in the engineering of the early stages of colony development, a premium must be set on rapid increase. The incipient colony is analogous to a pioneer population of solitary organisms. In both cases the need is to build numerical strength to a point where extinction of the entire colony becomes improbable; in other words, the young colony must maximize its rate of increase. This idea can be expressed in a more explicit form by use of the Euler equation,

$$\int_0^{\infty} e^{-rx} l_x m_x \, dx = 1$$

where l_x is the probability that the colony member survives to age x, m_x is the number of offspring it creates (or in the case of workers, helps to create), and r is the rate of increase. For the sake of simplicity, the age distribution is made stable and the average number of offspring created up to the moment of colony extinction is set at unity. The colony, to put the argument entirely in terms of the equation, should try to maximize r. Lewontin (1965) has employed the Euler equation to consider effects that joint changes in the survivorship and fecundity schedules would have on r in an expanding population of nonsocial organisms. As I have pointed out (Wilson, 1966), his results have some promising applications to the analysis of social insects. First, Lewontin deduced that rapid development would be selected more efficiently in the evolutionary increase of r than would high total fecundity. This result is at least consistent with the following observed facts about social insects: (1) it is the rule for new colonies to be started by single queens, the collaboration of two or more queens apparently in most species not adding sufficiently to r to make polygyny advantageous; (2) the first broods develop more quickly than later broods; (3) first-brood workers are smaller than those of later broods (thus favoring a quicker start in population growth). Lewontin's second result of use is that for a given total production of viable offspring, r-selection will favor lower fecundity and lower survival rates. This inference is in accord with the following additional facts about social insects: (1) queen-laid eggs are sacrificed freely in the feeding of the first-brood larvae; and (2) workers of incipient colonies are much less aggressive than those of mature colonies and hence far less likely to throw their lives away, but they nevertheless have shorter natural life spans. All of the conditions listed, except monogyny, are removed when the colonies approach the carrying capacity of the environment and a high r is no longer required.

The Survivorship of Colonies

The growth of individual colonies is characterized in most species by a rapid increase in numbers over the first part of the life cycle which is repeated faithfully each generation. Colony growth is typically subjected to intense r-selection. The growth of the population of colonies, however, is far less likely to incorporate such episodes. The extent to which they do will have a wholly different set of effects on the age structure and longevity of colonies as well as on their production of reproductive forms.

In order to see this two-layered relationship clearly, it is necessary to recall that in the social insects the unit of selection is the colony. More accurately, it is the queen, since in a loose sense the colony is only the somatic extension of this individual. In the discrete-generation case, fitness can be measured as the replacement rate $\int_0^\infty L_x M_x \, dx$, where L_x is the probability that a queen survives to age x and M_x is the rate of production of new queens at age x. For a persistent species, which in social insects is comprised of a population of colonies, the replacement rate for all genotypes taken together over a long period of time is equal to unity. But, if the species is evolving (and all almost certainly are, all the time), the various genotypes will have replacement rates not equal to unity. The components of colony fitness, then, can be classified according to whether they contribute to the survivorship of queens (L_x) or to the production of new queens (M_x). The growth rate of colonies and colony size are important only insofar as they increase these two major components of fitness. This explains, in a very general way, why in some species the societies grow slowly, why mature colony size is quite small, why rate of queen production is very low, or a combination of all these things. A low queen (colony) M_x is simply compensated for by a high queen (colony) L_x, or vice versa. The same physiological and behavioral characteristics that lower fecundity can be expected to raise survivorship.

But, having said this much, it must be added that we know so little about the relation of behavior and physiology to life and fecundity schedules as to make theoretical extensions treacherous. One promising empirical generalization that can be made is that, when colonies are started without the assistance of workers by a single queen, a group of queens, or by a royal pair (in termites), the colony survivorship curve is of type IV. This is an abstraction of the commonly observed fact that a very large number of reproductive individuals are liberated in the nuptial flights and all but a tiny fraction die soon afterward. For example, of 69 incipient colonies belonging to two species of *Polistes* studied by Yoshikawa (1954), only two colonies, or 3 percent of the total, survived to maturity. Similarly heavy early mortality occurs in the hornets and yellow jackets of the subfamily Vespinae (Scott, 1944; Duncan, 1939) and in the halictid bees (Batra, 1966a). Batra's survivorship curves, which comprise one of the few sets of data of this kind available in the social insects, are reproduced in Figure 21-7. Autuori (1956) found an average of 2,902 virgin queens per nest in the fungus-growing ant *Atta sexdens,* 1,688 in *A. bisphaerica,* and 978 in *A. laevigata,* all available for nuptial flights; while Wildermuth and Davis (1931) estimated that 80,000 to 1,000,000 of *Pogonomyrmex barbatus* were released from an 80-acre alfalfa field in one year. At most, a few dozen such individuals survive the nuptial flight and start incipient colonies. Because a single mature colony lives for years, an

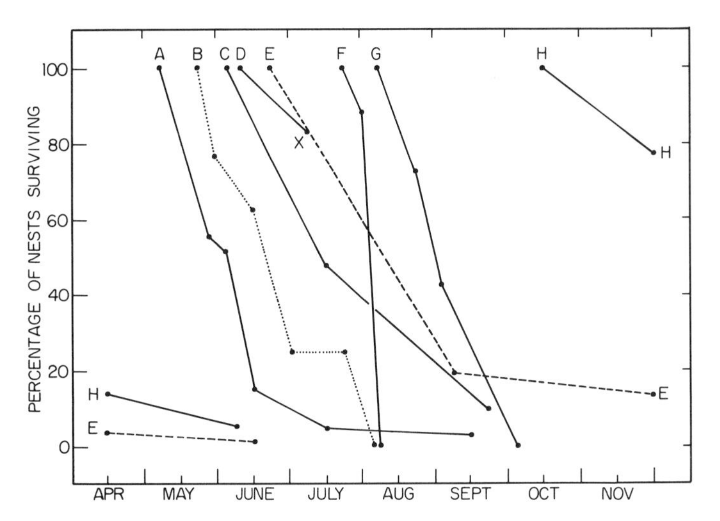

FIGURE 21-7. Colony survivorship curves of the halictid bee *Dialictus zephyrus.* Each curve represents a population of colonies in a different locality in Kansas (redrawn from Batra, 1966a).

average of less than one of the new colonies per year, certainly no more than 0.1 percent of the original virgin queens together with their subsequent offspring, can be expected to survive to maturity in a stable population of colonies. Colony-level mortality of this severity is commonplace throughout the ants and termites, reaching its extreme in the higher termites. According to Ratcliffe, Gay, and Greaves (1952), one mature colony of *Nasutitermes exitiosus* produces as many as 44,000 new winged reproductives each year and over a million during its lifetime. Since half the latter number, say 500,000, are virgin primary queens, it follows that in a stable population of *exitiosus* colonies an average of only one royal pair, or less than 10^{-5} of the total produced, ever succeeds in producing a mature colony of their own. In both termites and ants, most of the emerging queens succumb to predators, especially birds and worker ants, in a few hours during and immediately following the nuptial flights. Of those that survive, few are able to find suitable nesting sites and eventually are destroyed by predators or workers of alien colonies as they wander about in exposed situations. Studies by Brian (1952b) on the British ant fauna indicate that the quantity of nest sites is a primary limiting factor of population density and exercises its effect mostly at the time of the nuptial flight. Noirot (1958–1959) concluded from his field studies in Africa that termite densities are limited both through competition among termite colonies for nest sites and predation by ants, especially in the early stages of colony growth. It follows that selection is heaviest, and hence evolution potentially fastest, at the beginning of the colony life cycle. As future studies unfold, we should expect to discover many new complex adaptations in the young queen in this period of her life, especially as they relate to habitat searching and the avoidance of other social insects (Wilson and Hunt, 1966).

I have stressed that the average replacement rate of a persistent species will be unity if measured over a sufficiently long period of time. The same may or may not be true of local populations monitored over short periods of time. In the case of weedy or "fugitive" species, that is, species that exploit newly opened habitats by means of greater dispersal power or faster reproduction, populations are probably either increasing or decreasing at a fairly steep rate at any given moment. Other, more stable species have populations in which growth will be relatively slight. Excellent examples of stable species are found in certain mound-building ants of the genus *Formica.* Through censuses taken by E. A. Andrews, E. N. Cory, W. A. Dreyer, Elizabeth Haviland, R. L. King, R. M. Sallee, G. Scherba, and Mary Talbot in the United States over a period of several decades, quantitative records are available that are unusually complete for social insects. In populations of *F. exsectoides, F. opaciventris,* and *F. ulkei* there has been in each case an apparent gradual "aging" process, a definite decrease in the proportion of small mounds and a concomitant increase in the proportion of larger mounds (Scherba, 1963). Nest densities ranged between 0.01 and 0.04 per square meter with little change over a period of years. The average longevity of a *F. ulkei* nest was estimated by Dreyer (1942) to be about 20–25 years. New queens are added through adoption, making the age of the colony independent of the age of the queen. According to Scherba, a Wyoming population of *F. opaciventris* remained steady during a three-year observation period at a density of 0.04 nests per square meter, a nest birth rate of 5 to 13 percent, and a nest death rate of 8 to 9 percent. In general, mortality was highest among the new nests.

The relatively long time required to reach maturity and the long average life of colonies thereafter promote short-term stability in the population of colonies. Colonies of the ant species *Myrmica rubra* and *M. ruginodis,* to cite an extreme example, require eight to ten years of growth before they begin producing queens in Scotland (Brian, 1957a). The long potential life of mother queens of insect societies is well known and has already been documented in Table 21-1. The potential longevity of individual colonies must be at least equally great. The meager data bearing on this point are summarized in Table 21-4. The colonies of various species can be classified very simply into two types: the mortal and the potentially immortal. The mortal colony has only one or a very few queens, and new queens are not added when the original ones die. The colony lasts no longer than one worker's lifetime beyond the death of the last queen. Examples include many and perhaps most species of ants, halictid bees, and social wasps. Potentially immortal colonies are found in three classes of species: (1) those that have multiple queens and adopt new ones as a matter of course (*Formica exsecta* group, some polybiine wasps); (2) those that reproduce by fission, in which new queens arise periodically and depart with a portion of the worker force, thus paralleling the budding process of "immortal" hydras and some other invertebrates (examples in the social insects include army

TABLE 21-5. Size and caste composition of colonies of various ages of the termite *Cubitermes severus* (from Paulette Bodot, 1969).

Population	Age		
	Young (pop. <10,000)	Adult (pop. 10,000–40,000)	"Senile" (pop. >40,000)
Number of colonies observed	25	45	12
Average population	5,600	30,000	50,000
Workers (percent)	47	63	85
Larvae (percent)	52	36	14
Soldiers (percent)	0.85	0.75	1.15

ants, honeybees, stingless bees); and (3) those that automatically create replacement reproductives when the original royal members die (some ants, for example *Monomorium pharaonis* and species of *Oecophylla,* and most termites). It is curious that at least some termite species have apparently surrendered the capacity for colony immortality in the course of their evolution. The primitive termites, including *Mastotermes darwiniensis* and the Kalotermitidae, Hodotermitidae, and Rhinotermitidae, as well as most members of the advanced family Termitidae, all have the power to produce supplementary reproductives. However, the trait appears to have been lost in the termitid genus *Apicotermes* and its close relatives (Noirot, 1969a). In one species of termitine, *Cubitermes severus,* adultoid supplementary reproductives occur, but colonies appear to age and die anyway. In Table 21-5 are given Paulette Bodot's data on *C. severus* colonies of different ages, including those considered to be in a state of senility. In the adult period the colony regularly produces new winged reproductive forms, which are released in nuptial flights. In the "senile" period this output diminishes, then stops, and at the very peak of its numerical strength the colony population begins to decline. From the juvenile through the senile periods, there is a steady reduction in the proportion of larvae to workers, while the percentage of soldiers remains approximately constant. Skaife (1954a) has reported a similar intrinsic decline in colonies of the Cape black-mound termite, *Amitermes hastatus,* even though this species has a limited capacity to produce supplementaries. I believe it is premature to generalize from this limited evidence, as Noirot has done, that colony senescence is universal in the higher termites. Nevertheless, great interest in the matter accrues from the paradoxical fact that it occurs at all. For if the seizure of a nest site and the race to maturity are the most critical steps in the life cycle of termite colonies, with success being crowned only once every ten thousand or hundred thousand attempts, what could possibly be the adaptive value of surrendering supplementary reproduction and giving in to senescence?

Whatever the statistics of colony longevity, it remains generally true that stable species of social insects have populations consisting mostly of mature colonies varying little in the number of queens and workers. This principle has been upheld by numerous data collected by Headley (1943) on the ants *Aphaenogaster rudis, Leptothorax curvispinosus,* and *L. longispinosus;* by Talbot (1943, 1951) on *Aphaenogaster rudis* and *Prenolepis imparis;* by Bouillon (1964) on the termite *Apicotermes desneuxi;* by Bodot (1964) on species of *Cubitermes;* and by Sands (1965) on *Trinervitermes geminatus.* Their censuses are still too limited to allow a broad extension to social insects as a whole. However, my own subjective impression while collecting hundreds of ant species in both the North and South Temperate Zones and tropics around the world has been that, as a rule, mature or nearly mature colonies outnumber the younger stages.

Once a habitat is populated by mature colonies of social insects, the total numbers of individual insects can vary without radically altering the number of colonies. In fact, the colonial organization of the populations can be expected to serve as a homeostatic device in damping fluctuations in the numbers of individual insects. The reason is that it is possible for a cutback in numbers of individuals, even a drastic one, to reduce the average size of colonies without changing the number of colonies. Thus

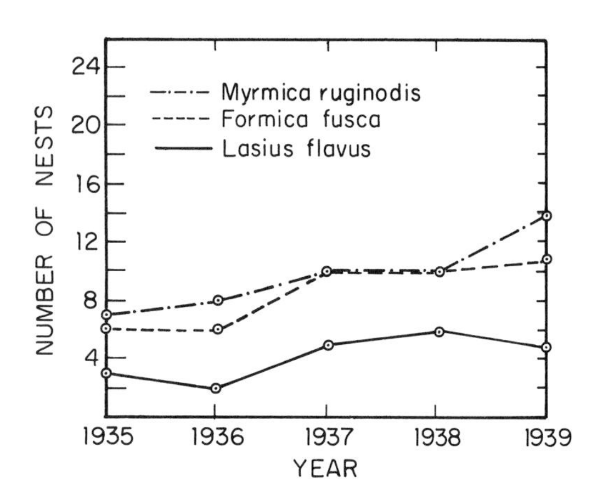

FIGURE 21-8. Fluctuation in the number of nests of three species of ants in a small area of bracken heath in Yorkshire, England (redrawn from Pickles, 1940).

the reduction does not extinguish the species, nor does it even change its distribution within the local environment. When conditions ameliorate, the colonies can serve as nuclei in the rapid restoration of the populations of individuals. This inference is supported by the data of Pickles (1940), who for a period of four years kept careful records of both the nest populations and biomasses of ant species in a small area of bracken heath in northern England. As shown in Figures 21-8 and 21-9, the number of nests of the three species increased gradually by a factor of approximately two, while the numbers of individuals were fluctuating in a much stronger and less concordant fashion. The most interesting example is that of *Formica fusca.* Although 1939 saw the numbers of individuals of this species descending to low levels, the number of nests had actually risen, so that the chances of the species vanishing from the study area remained very remote even during this bad year.

The principal conclusions that can be drawn from the information just presented are that colony mortality is very heavy at the earliest stages, that mature colonies are relatively very persistent, and that the populations of colonies are therefore probably changing in density only very slowly—at least compared with nonsocial insect species—in all but the least stable habitats. An important consequence of these generalizations is that the mere production of new queens contributes less to queen (= colony) fitness than it otherwise would. At least equally sensitive to evolution are the behavioral and growth characteristics that contribute to colony survivorship. Thus, it is not surprising to find cases like that of *Formica opaciventris,* which releases queens from only 3 to 5 percent of the mounds yearly (Scherba, 1961), or of *Eciton burchelli* and *E. hamatum,* whose colonies multiply only slowly by simple annual budding (Schneirla, 1956b). Each of these species has elaborate behavioral adaptations that promote colony survivorship and, hence, high population stability. Population stability is not always of high selective advantage, as can be seen in the common occurrence of fugitive species in most groups of animals. Where it has evolved, however, the following relation can be predicted: the greater the stability of the population of colonies, the lower the productivity of new queens and the higher the survival rate of new colonies. A difficult but important task of the future will be the closer measurement of population stability and the correlation of this variable with other characteristics of colony growth and behavior.

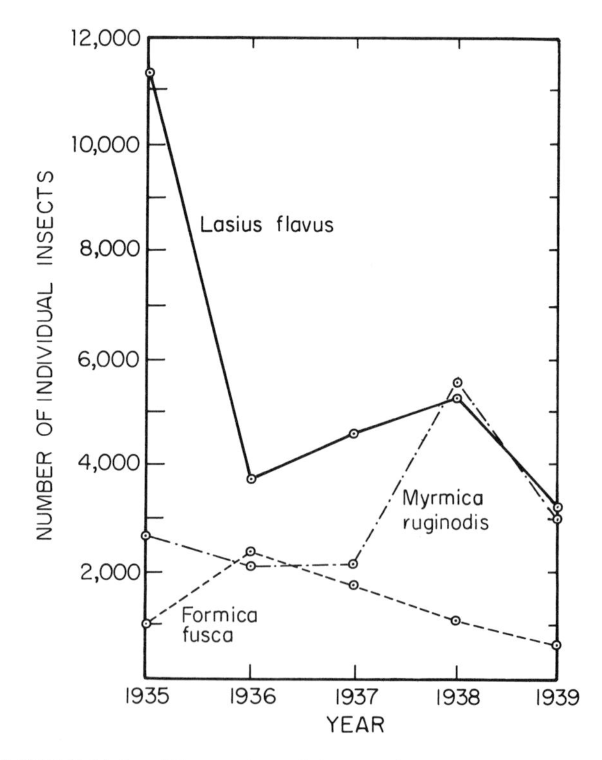

FIGURE 21-9. Fluctuation of the numbers of individuals in the same three species of ants shown in figure 21-8. The numbers given for each species are the combined totals for larvae, pupae, and adults (redrawn from Pickles, 1940).

Competition and Territoriality

The worst enemies of social insects are other social insects. Ants in particular are their chief predators. A great many studies on termite ecology, including those of Bugnion (1922), Wheeler (1936a), Weber (1949a), and R. M. C. Williams (1959a), have confirmed the statement of Emile Hegh (1922) that "Of all the enemies of termites, the ants are the most active and effective." Whole genera of tropical ants—*Centromyrmex, Acanthostichus, Megaponera, Ophthalmopone, Termitopone, Carebara*—are specialized for feeding on termites. The driver ants of the genus *Anomma,* according to Paulette Bodot (1964), are the chief predators of the great mound-building colonies of *Macrotermes* in Africa. Since the colonies of both the victims and the huntresses each contain millions of workers, the depredations must reach truly titanic dimensions. And Auguste Forel was equally correct in stating that "The greatest enemies of ants are other ants, just as the greatest enemies of men are other men." All of the Cerapachyinae so far as known and many of the Ponerinae and Dorylinae are specialized predators on other ant species. A great many less-specialized species in a considerable array of genera, including such dominant elements as *Ectatomma, Myrmica, Solenopsis, Lasius,* and *Formica,* regularly include other ants in their diets. Almost half of the prey of *Formica subnitens,* for example, consist of other ant species (Ayre, 1959). The chief predators of social wasps in tropical America are ants, especially army ants of the genus *Eciton* (Richards and Richards, 1951; R. L. Jeanne, personal communication), while in tropical Asia *Aenictus* army ants, as well as wasps of the genus *Vespa,* regularly attack other social wasps (F. X. Williams, 1919).

Predation in the social insects grades imperceptibly into fierce competition for nest sites, food, and territory. So ubiquitous and overt are these interactions that social insects, especially ants, have served as a principal source of documentation for general competition theory in biology (Brian, 1955b, 1965b; Wilson, 1971). Almost all modern ecologists would agree with the following definition offered by Miller (1967) or some variation of it: "the active demand by two or more individuals of the same species (intraspecies competition) or members of two or more species at the same trophic level (interspecies competition) for a common resource or requirement that is actually or potentially limiting." This definition is consistent with the assumptions of the Lotka-Volterra equations, which still form the basis of much of the mathematical theory of competition. It also matches the intuitive conception held by most modern ecologists concerning the underlying behavioral processes. Competition normally arises only when populations become crowded enough for a shortage to develop in one or more resources. When it does come into play, it reduces population growth, and, if permitted to increase unimpeded, it will eventually reduce the growth rate to zero. Much of the interference between insect colonies consists of open aggression, which includes elements of predation. This generalization can be exemplified almost anywhere in the world by putting down moist cubes of sugar and observing the interaction of the ant species attracted to them. As a matter of fact, I often do this experiment as a form of amusement. On the streets of San Juan, Puerto Rico, to take one of many examples, anywhere from one to six species are attracted to the same sugar bait. *Paratrechina longicornis,* the swift-running *hormiga loca* (crazy ant) found in tropical cities around the world, is an example of a class of species I have come to call "opportunists." The workers are very adept at locating food and are often the first to arrive at newly placed sugar baits. They fill their crops rapidly and hurry to recruit nestmates with odor trails laid from the rectal sac of the hind gut. But they are also very timid in the presence of competitors. As soon as more aggressive species begin to arrive in force, the *longicornis* draw back and run excitedly in search of new, unoccupied baits. The species survives by arriving and recruiting early enough to preempt a share of the food, without however exposing the workers to any great risk in battle. A second class of species, called "extirpators," recruit by odor trails and fight it out with competitor species. Examples include species of *Pheidole,* the fire ant *Solenopsis geminata,* and *Camponotus sexguttatus.* All have a well-developed soldier caste which, in the case of *Pheidole* and *Solenopsis* at least, plays a key role in the fighting. Injury and death among the combatants are commonplace. Finally, a third class of species ("insinuators"), exemplified by *Tetramorium simillimum* and species of *Cardiocondyla,* rely on small size and stealthy behavior to reach the sugar baits. The colonies are small and lack a soldier caste. When a *Cardiocondyla* scout worker discovers food, it recruits only one nestmate at a time by means of tandem running. The small numbers of workers reaching the baits by this means are usually able to ease their way to the edge of the sugar baits through crowds of "extirpator" workers without eliciting aggressive responses. The three categories of behavior are not completely rigid. The opportunistic and insinuator species sometimes fight, and the extirpator species usually retreat when they arrive tardily at baits already dominated by another extirpator species.

Aggressive behavior at sugar baits is a manifestation of the most prevalent form of territorial behavior in ants. Under normal circumstances the extirpator species dominate their natural persistent carbohydrate sources, which are pieces of decaying fruit and groups of aphids, coccids, fulgorids, membracids, and other honeydew-yielding homopterous insects located on plants near the nests (Elton, 1932; Morisita, 1941; Talbot, 1943; Brian, 1955b; Tsuneki and Adachi, 1957; Yasuno, 1965). In an ecological study of Polish ants, Dobrzańska (1958) noticed two different techniques by means of which various species partition their foraging grounds. In certain formicines such as *Lasius niger* and members of the *rufa* and *sanguinea* groups of *Formica,* persistent trails lead to plants containing honeydew-secreting insects. Each colony controls a set of plants or portions of plants which it defends not only from other colonies of its own species but also from other ant species. Individual workers tend to return consistently to the same spots on their foraging trips, and some stand guard there for hours or even days at a time. The second technique is utilized by such myrmicines as *Leptothorax acervorum, Myrmica scabrinodis,* and *Tetramorium caespitum.* Workers hunt singly for food sources and when successful recruit nestmates rapidly by means of odor trails. This second method of partitioning is clearly adapted to the exploitation of large but temporary food sources, while the first method is ideal for stable sources, regardless of whether they are large or small. Some ant species utilize both methods. Species of fire ants, particularly *Solenopsis geminata* and *S. saevissima,* as well as the species of *Tetramorium,* have highly efficient recruitment trail systems, but they also attempt to take permanent possession of persistent food sources when they are encountered.

Although ant workers of species employing both types of partition react aggressively to alien ants at the food sites, those relying principally on fixed territories—the first "technique" just described—are, as a rule, more aggressive. Species that use neither of the methods are the least aggressive of all ants. In the Kayano grassland of Hokkaido, for example, Yasuno (1965) found that *Formica exsecta* and *F. truncorum,* which tend to occupy permanent territories, are more aggressive than *Camponotus herculeanus* and *Myrmica ruginodis* and able to replace them at food sites. The latter two species in turn dominate *Formica japonica,* a nonterritorial and opportunistic species. The workers of *japonica* range widely from their nests in search of small food items and unoccupied food sites. Occasionally nests are located in empty spaces between the territories of other species, in which case the foraging pattern of the *F. japonica* becomes more compacted.

Careful studies on a variety of myrmicine and formicine ant species by Talbot (1943), Yasuno (1965), and Brian, Hibble, and Kelly (1966) have shown that, while the defended feeding territories are persistent throughout the life of the colony, they shift, amoeba-like, in size and shape from season to season and even, in the extreme case of *Prenolepis imparis,* from day to day. One excellent example involving five species in the Kayano grassland is illustrated in Figure 21-10. A second quality of hostility at food sites is that it tends to be the most intense among alien colonies of the same species, and progressively less so the greater the taxonomic difference between the inter-

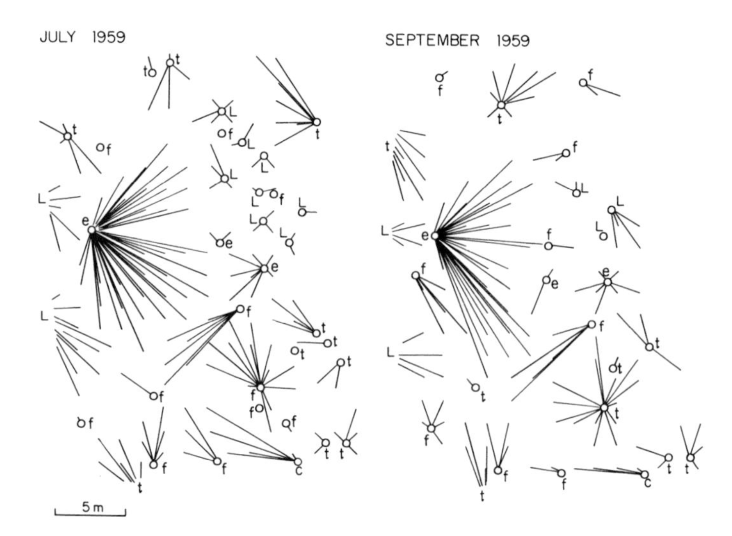

FIGURE 21-10. The feeding territories of colonies of five ant species in a stand of pine seedlings in the Kayano grassland, Hokkaido, Japan. Each nest is indicated by a small open circle, and lines are drawn between it and the pine seedlings occupied by the colony. These data document changes in nest site by the smaller colonies and the changing compass of the territories by colonies of all sizes: (*C*) *Camponotus herculeanus;* (*e*) *Formica exsecta;* (*f*) *Formica japonica;* (*L*) *Lasius niger;* (*t*) *Formica truncorum* (redrawn from Yasuno, 1965).

acting species (Dobrzańska and Dobrzański, 1962; Brian, 1965b). This effect is due in part to a passive attitude that workers of some species take to members of other species. In other cases workers meeting members of very different species on foraging trips merely dash away, and this avoidance may be continued, for a while at least, even when enemies are encountered at food sites. But when members of the same or closely related species are encountered, there is a predilection to pause and investigate the stranger by antennal contact, an intimacy that often results in fighting. Taxonomically remote species also are less likely to encounter each other at food sites because of a tendency to hunt in different microhabitats at different times of the day. This phylogenetic rule of species interference helps to explain the patterns of species replacement on Pacific islands described by Wilson and Taylor (1967) and Wilson and Hunt (1967). On the larger islands such as Savai'i, Upolu, and Tahiti, with areas of a thousand square kilometers or more, two or more species belonging to the same genus frequently coexist in large numbers. An exception involves the larger, more aggressive species of *Pheidole. P. fervens,* a widespread Indo-Australian species, is unknown from Samoa at the present time but it is a dominant ant in the Society Islands. *P. megacephala,* a pantropical species of African origin, has the reverse distribution: it is dominant on Upolu (Samoa) but rare or absent in the Society Islands. *P. oceanica,* another Indo-Australian element, replaces *megacephala* on Savai'i (Samoa) and occurs on Upolu only on the western side facing Savai'i; it is, furthermore, relatively uncommon in the Society Islands. Elsewhere in Polynesia the complementarity among the three species is maintained. *Fervens* occurs on Tonga and Pitcairn; it is only occasional on the Marquesas and unknown in Hawaii. *Megacephala* is absent from Tonga and Pitcairn but is dominant on the Marquesas and in Hawaii. On smaller islands, the most closely related species of certain other genera, such as *Cardiocondyla, Solenopsis,* and *Paratrechina,* often display complementary patterns. On the smallest islands, members of different genera sometimes replace each other, as, for example, the larger species of *Pheidole* and *Solenopsis.*

The intensity of territorial defense varies among ant species along a gradient that appears to represent an evolutionary progression (Brian, 1965b). Colonies of the primitive genus *Myrmecia,* together with most ponerines, defend only the immediate nest area. The workers forage singly and normally capture only prey items that can be returned to the nest single-handedly. They also gather nectar from flowers in a solitary fashion resembling that of wasps and nonsocial bees. They do not recruit other workers to such food sites; nor do they fight to defend them. Defense of food sites evidently appeared in the course of evolution when ants began to utilize persistent carbohydrate sources. The simplest known case is that of *Prenolepis imparis,* whose workers defend pieces of rotting fruit up to distances of 1.5 m from their nest entrances. They show a generalized hostility to alien *P. imparis* workers, to those of other ant species, and in fact to all insects that attempt to approach their pieces of fruit. But the defense is limited strictly to the fruit, and foraging workers from as many as three colonies may search the same ground simultaneously without interference (Talbot, 1943). The next step in the evolutionary progression seems to be represented by the wood ant *Formica rufa.* Workers from the huge colonies that characterize this species travel along regular pathways as long as 200 m to visit several forest trees. By virtue of their large numbers and exceptionally aggressive behavior, they keep the trees mostly clear of other insects. Certain other ant species, notably *Pheidole megacephala, Solenopsis saevissima,* and *Iridomyrmex humilis,* go one step further than *Formica rufa.* They monopolize not only the proven food sources but also most or all of the remaining foraging area. Rival colonies are attacked and eliminated, especially those that have similar nest site preferences and whose worker castes are of comparable physical size. Finally, the extreme of territorial behavior is displayed by a few species specialized for living in obligate ant plants (see Chapter 4). Their workers are extremely aggressive in manner, totally monopolize the tree in which they live, and attack almost every other animal species encountered.

As a rule, colonies of social insects belonging to the same species repel each other by one means or other, with the result that populations of mature colonies are "overdispersed"—spaced so that the distances between them are too uniform to have been randomly set. The documentation of this phenomenon in ants is extensive (Brian, 1956b,c, 1965b; Talbot, 1943, 1954; Wilson, 1959d; Yasuno, 1964a, 1965). Of special interest is the discovery by Waloff and Blackith (1962) that nests of *Lasius flavus,* a species known from direct behavioral observations to be territorial, are randomly distributed in areas of low population densities but overdispersed in areas of high densities. A different pattern altogether is displayed by

Formica japonica, which, as we have already seen, is a nonterritorial, "opportunistic" species. According to Yasuno (1964a), colonies at the Kayano grassland are overdispersed where the species occurs in uniform but low-density stands, and aggregated where they occur in a species-rich zone along the border of an adjacent forest. It seems probable that competition for food spaces the colonies out in the former habitat, while pressure from other, more aggressive species forces them together along the forest border.

Spacing can be achieved among territorial species not only by the aggressive interaction of developed colonies but also, at the very beginning, by the individual actions of the nest-founding queens. In a study of habitat selection in Scottish ants, Brian (1952a) demonstrated a remarkable "rank order" among species in the appropriation of favored nest sites by competing queens. In the cool, moist woodland of western Scotland, ant colonies are limited to sunny places where higher temperatures persist long enough to permit the rearing of brood. As newly mated ant queens enter rotting pine stumps, they move to the south side of the vertical surface just beneath the bark. When individuals belonging to the same species encounter one another, they group together or space out at very short intervals. When queens of different species meet, however, they space out at much greater distances. Under such conditions, *Formica fusca* occupies the southern face of the stump, which is the warmest area, while *Myrmica ruginodis* (= *M.* "*rubra*") moves, for the most part, to the east face, which is the second warmest. *M. scabrinodis* takes what is left. The tiny *Leptothorax acervorum* avoids conflict altogether by occupying galleries in the core of the stump too narrow to admit the other species. Brian was then able to demonstrate by a laboratory experiment, the results of which are given in Figure 21-11, that the segregation of the species is due, at least in large part, to the repulsion of the *Myrmica* by the *Formica* from the favored southerly position. In later stages of their population growth, the colonies of the different species continue to occupy different positions, and in some cases they remain there into maturity.

Spacing within species often involves a more destructive interference among the queens and colonies. It is a common observation that ant queens and young colonies are destroyed by other colonies belonging to the same species. Large numbers of newly mated queens of *Formica fusca* are captured and killed as they run past the nest entrances

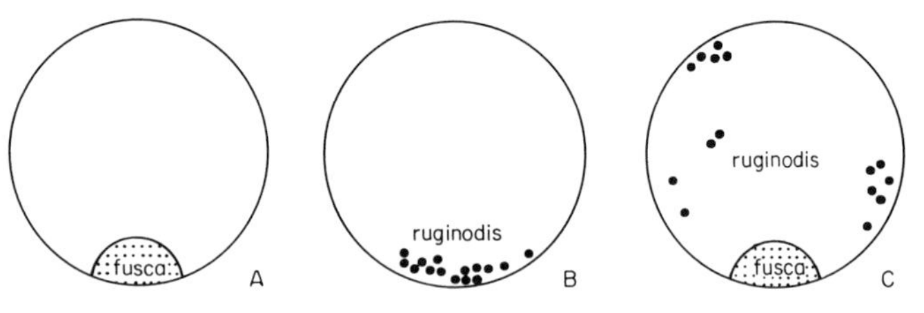

FIGURE 21-11. M. V. Brian's experimental demonstration of behavioral exclusion leading to species segregation in ant queens. Queens were placed in glass vessels heated to 30°C at one side (the bottom of the diagram) and their positions recorded ten minutes later. (*A, B*) When placed with members of their own species alone, both the *Formica fusca* queens (located as a group in the stippled areas) and the *Myrmica ruginodis* queens (shown individually as dots) invariably chose the warmest part of the chambers; (*C*) when mixed together, the *M. ruginodis* were displaced to the cooler part. This segregation closely resembles that observed in dead stumps under natural conditions (redrawn from Brian, 1952b).

(Donisthorpe, 1915); the same fate terminates a large percentage of the colony-founding queens of *Iridomyrmex detectus* and *Solenopsis saevissima* (Wilson, unpublished observations). Queens of *Myrmica* and *Lasius* are harried by ant colonies, including those belonging to their own species, and finally they are either driven from the area or killed (Brian, 1955b, 1956b,c; E. O. Wilson and G. L. Hunt, unpublished observations). It is also frequently true that colony-founding ant queens and juvenile colonies are more abundant where mature colonies are scarce or absent. Brian, who has studied this effect in the British fauna in some detail, discovered a striking inverse correlation in various habitats between the density of adult colonies and of foundress queens of *Formica* and *Myrmica.* Similar dispersing effects have been recorded in other social insects. In stable habitats of southwestern Australia, mature colonies of the termite *Coptotermes brunneus* are spaced about 90 m apart. In the intervening areas colony-founding queens are caught and destroyed. Also, the mature colonies compete intensely for the limited foraging space in the few available trees (Greaves, 1962). A similar pattern of strong territoriality has been described in *Hodotermes mossambicus* by Nel (1968). It is true of termites generally that when more than one couple belonging to the same species succeed in founding colonies together, they coexist peaceably or even combine forces for a while. But within a few months at most fighting and cannibalism ensue, until finally only a single couple—and, hence, one effective colony—survives (Nutting, 1969). Colonies of the paper wasp *Polistes fadwigae* located 3.5 m apart steal and eat each other's larvae. If they are brought by the experimenter to within 5 cm of each other, the dominant females fight until a new dominance order is achieved, and the colonies fuse (Yoshikawa, 1963b). Honeybee workers from different colonies fight at the same food dishes when the sugar supply begins to be used up (Kalmus, 1941). Under more natural conditions, honeybee colonies placed together have been shown, by use of radioactive tagging, to restrict each other's foraging areas as a function of the degree of crowding (Levin, 1961; Levin and Glowska-Konopacka, 1963).

Territorial fighting among mature colonies of both the same and differing species is common but not universal in ants. It has been recorded in very diverse genera of which the following form only a partial list: *Pseudomyrmex, Myrmica, Pogonomyrmex, Leptothorax, Solenopsis, Pheidole, Tetramorium, Iridomyrmex, Azteca, Anoplolepis, Oecophylla, Formica, Lasius, Camponotus.* The most dramatic battles known within species are those conducted by the common pavement ant *Tetramorium caespitum.* First described by the Reverend Henry C. McCook (1879a) from observations in Penn Square, Philadelphia, these "wars" can be witnessed in abundance on sidewalks and lawns in towns and cities of the eastern United States throughout the summer. Masses of hundreds or thousands of the small dark brown workers lock in combat for hours at a time, tumbling, biting, and pulling each other, while new recruits are guided to the melee along freshly laid odor trails. Although no careful study of this phenomenon has been undertaken, it appears superficially to be a contest between adjacent colonies in the vicinity of their territorial boundaries. Curiously, only a minute fraction of the workers are injured or killed.

Territorial wars between colonies of differing ant species occur only occasionally in the cold temperate zones. Colonies of *Myrmica* and *Formica,* for example, sometimes overrun and capture nest sites belonging to other species in the same genus (Brian, 1952a; Scherba, 1964b). Intense aggression is, on the other hand, very common in the tropics and warm temperate zones. Certain pest species, particularly *Pheidole megacephala,* the fire ant *Solenopsis saevissima,* and the so-called Argentine ant, *Iridomyrmex humilis,* are famous for the belligerency and destructiveness of their attacks on native ant faunas wherever they have been accidentally introduced by human commerce (M. R. Smith, 1936b; Wilson and Brown, 1958b; Haskins, 1939; Haskins and Haskins, 1965; Wilson and Taylor, 1967). They even go so far as to eliminate some of the species, especially those closest to them taxonomically and ecologically. In the case of *I. humilis,* only the smallest, least aggressive ant species remain unaffected. It is possible that, given enough time, evolutionary change will result in the "taming" of such species and their increasingly harmonious assimilation into the native faunas. Wheeler (annotations to Réaumur, 1926) reminds us that in the early 1500's a stinging ant appeared in such huge numbers as to cause the near abandonment of the early Spanish settlements on Hispaniola and Jamaica. The colonists of Hispaniola called on their patron saint, St. Saturnin, to protect them from the ant and conducted religious processions through the village streets to exorcise the pest. What was evidently the same species, which came to be known by the Linnaean name *Formica omnivora,* appeared in plague proportions in Barbados, Grenada, and Martinique in the 1760's and 1770's. The legislature of Grenada offered a reward of £ 20,000 for anyone who could devise a way of exterminating the pest, but to no avail. It now appears that *F. omnivora* was none other than the familiar fire ant, *Solenopsis geminata,* which at the present time is a peaceable and only moderately abundant member of the West Indies fauna.

Iridomyrmex humilis and *Pheidole megacephala* are wholly incompatible species, with *humilis* the usual if not invariable winner in such subtropical localities as Madeira, Oahu, and Bermuda. The *Iridomyrmex* populations penetrate new environments on foot, with raiding columns of workers clearing the way and pioneer communities of workers and queens following into the freshly opened nest sites. New populations, on the other hand, spring from little groups of workers and queens that are carried inadvertently in human freight and baggage. The gradual replacement of *Pheidole megacephala* by *Iridomyrmex humilis* on Bermuda, which began with the establishment of *humilis* around 1953, has been especially well documented by Haskins and Haskins (1965) and Crowell (1968). Fighting and replacement among both native and introduced ant species is a frequent and conspicuous event in cultivated land in the tropics. It has been observed at length in *Pheidole, Anoplolepis,* and *Oecophylla* in coconut groves of Tanzania by Way (1953) and in citrus groves of South Africa by Steyn (1954), and in these three genera plus *Iridomyrmex* in coconut groves of the Solomon Islands by E. S. Brown (1959). Here is Brown's account of a typical event at Lunga, on the island of Guadalcanal:

> *Pheidole megacephala* was advancing rapidly into *Oecophylla* territory, and at the periphery of its area of spread its 'advanced guard' was installed on the bases of the trunks of a number of palms which still had *Oecophylla* colonies in the crowns. On the bases of the trunks of many of the palms were some old earthen runways of termites, and *Pheidole*, being small, had concealed itself in these; where there were small holes in the wall of the runway, *Pheidole* were stationed with heads facing outwards. Meanwhile larger numbers of *Oecophylla* had descended from the crown on to the base of the trunk within an inch of the *Pheidole* position, and there was, more or less, a state of deadlock; the *Oecophylla* ants were unable to get access to the *Pheidole*, while the latter were in insufficient numbers to attack. When an individual of *Pheidole* ventured out of its shelter it was immediately set upon by a number of *Oecophylla* and killed . . . That most such battles eventually turned in favor of *Pheidole* is evident from the rapid territorial advances it was making at this period.

Anoplolepis longipes follows a more direct battle plan, exemplified in the following accounts by Brown of encounters with *Oecophylla smaragdina* and *Iridomyrmex myrmecodiae:*

> [The] invading *Anoplolepis* ants move on to the base of the trunk, which evokes the descent of large numbers of *Oecophylla* to ring the trunk in defensive formation just above them. It then becomes a ding-dong struggle, the dividing line between the two species sometimes moving up or down a few feet from day to day; any ant wandering alone into the other species' territory is usually surrounded and overcome. Eventually one species will get the better of the other, but this may not happen for several days or weeks . . .

> *Anoplolepis* had advanced on to the base of the trunk of a palm occupied by *Iridomyrmex*, which had descended in force from the trunk and formed a formidable phalanx of countless individuals, almost completely covering the surface of the trunk over about 2 ft. of its length. After a few days this defensive formation was still intact, but had retreated higher up the trunk; eventually it was driven from the trunk altogether, and later *Anoplolepis* took possession of the crown.

I do not wish to leave the impression that all competitive interactions between ant species involve brutish frontal assaults by fighting workers. Of course such events are the ones that attract attention and are most frequently recorded. When competition among other species pairs less selectively chosen are examined more closely, alternative and more subtle techniques are revealed. Brian (1952a,b) found that the takeover of nest sites by various species of Scottish ants is usually gradual and may involve any of several methods. *Myrmica scabrinodis,* for example, seizes nests of *M. ruginodis* either by direct siege, causing total evacuation of the *ruginodis,* by gradual encroachment of the nest, chamber by chamber, or by occupation following greater tenacity in the face of adverse physical conditions, particularly severe cold, that drive the other species away. In the course of ecological studies in the Dry Tortugas, Florida, Richard Levins and I found that *Pheidole megacephala* is able to displace *Solenopsis globularia* at food dishes in most instances not only by direct aggressive behavior but also by its greater capacity to maintain calm during the encounters. When workers of the two species meet, they usually dash away from each other in panic, temporarily losing contact with the odor trails that hold nestmates together as a force. By assembling on their own odor trails more quickly following the encounters, the *megacephala* are able to reorganize more efficiently and to mobilize forces strong enough to repel the still scattered *globularia* workers. In general, overt forms of aggression have been seen most frequently in the following circumstances: (1) where the species have been recently introduced by man and hence are "unfamiliar" with their competitors in the evolutionary sense; (2) where the species are strongly territorial; (3) where the habitats have been artificially simplified, as in cultivated land and urban environments, so that the species brought into juxtaposition are ones which seldom encounter each other in their more complex native environments. It is to be expected that, as behavioral studies become more exacting in the future, the less direct forms of contest competition within and among social insect species will be relatively better documented. It also seems certain that examples of "scramble" competition will be revealed, involving the mutual appropriation of common resources in short supply even in the absence of direct physical encounters.

The Control of Colony Density and Species Diversity

Interference between colonies belonging to the same species has the important effect of increasing the numbers of competing species that can coexist. This result is predicted in the graphical models devised by Gause and Witt (1935) on the basis of the Lotka-Volterra competition equations. In words, the Gause-Witt theory states that, if two species interfere with one another to any extent, one will always replace the other unless the following condition is met: the population densities of the two species must be self-limiting in such a way that they will stop increasing before the other species becomes extinct. The most familiar way in which such an equilibrial coexistence can come into being is if the two species occupy sufficiently different niches. Then one will tend to reach a limit in the part of the habitat to which it is specialized and reaches maximal densities before it is able to crowd out the second species in the part of the habitat optimal for the second species. This special case has become so familiar in ecological writing as to be frequently referred to as *the* "Gause hypothesis," "Gause's law," and so forth, but the Gause-Witt model embraces other potential mechanisms as well. Consider, for example, the possibility that the population density of each species is under the control of a parasite specialized for feeding on it. This, too, could easily lead to stable coexistence of the two prey species—as well as of the two parasitic species. It follows that any density-dependent control peculiar to a species will contribute to the stable coexistence of competing species. In this light the work of Pontin (1960, 1961, 1963) takes on a particular significance. Pontin made careful studies of the ecology of two related species of formicine ants, *Lasius flavus* and *L. niger,* with special reference to the ways in which each affects the survival and reproduction of the other. In calcareous grassland near Wytham, England, the two species are dominant species, and their colonies are intermingled at saturated densities. In order to measure the consequences of interaction, Pontin first placed newly mated *Lasius* queens in tubes with openings large enough to admit workers but too small to permit the escape of the queens, and seeded them within the territories of mature colonies. He found that the queens were attacked and destroyed preferentially by workers of their own species. Studies were then made of the relation between productivity of new queens and the distance between the nest of the colony and the nearest nests belonging to both species. In a related experiment, colonies of *L. flavus* were transplanted to new positions in a circle around nests of *L. niger* in order to increase the competitive pressure on them. The results showed conclusively that queen production is reduced more by intraspecific than by interspecific interference. Therefore, through both the depression of the production rate of new queens and their destruction following the nuptial flights, each of the two species controls its own population densities to a greater extent than those of its competitor. This effect fulfills, at least in principle, the essential condition of the Gause-Witt equilibrium. The behavioral basis of the effect can only be guessed. Perhaps the reason why workers attack alien queens of their own species preferentially is that, as Brian (1956b) has claimed to be the case in *Myrmica,* they tend to be repulsed at a distance by the odors of both queens and workers of alien species. Such a response, which serves primarily as an adaptation to avoid injurious conflict, could not be extended to members of the same species without interfering with normal communication within the colonies.

Interference between mature colonies seems to be reduced by innate ecological differences between the two species. *Lasius niger* is a versatile ant that nests in rotting stumps, beneath stones, or in the open soil, and forages both below and above ground and up onto low vegetation. *L. flavus* is a primarily subterranean species that builds mounds in the open soil. Where the two species live together in the Wytham grassland, *niger* inserts itself in suitable nest sites between the *flavus* mounds in the manner illustrated in Figure 21-12. By competition for space and food (and limited predation on *flavus,* which is not reciprocated) *niger* depresses the queen production of *flavus.* Symmetrically, *flavus* takes away space and food from *niger;* and it also interferes with *niger* by using stones as props for the mounds of excavated earth, thus covering them and denying them to the *niger* for use as nest covers. But the degree of interference between the two species seems to be superimposed on a larger degree of interference among colonies belonging to the same species. The latter, intraspecific interference is not only enough to stabilize the populations of colonies, but it is

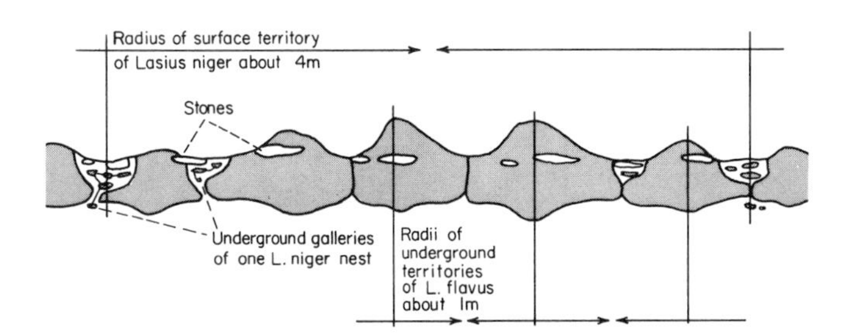

FIGURE 21-12. A diagrammatic vertical section through interspersed nests of the competing ant species *Lasius flavus* and *L. niger.* The territories of *L. flavus* are stippled (redrawn from Pontin, 1961).

sufficiently greater than interspecific interference to permit the permanent coexistence of the two species.

Let us next consider both the theoretical argument and the evidence from the *Lasius* example in an evolutionary context. In order for species to coexist, it is necessary that each of them be sufficiently different to reach their equilibrial densities before eliminating their competitors, and the usual way this occurs is through differences in critical dimensions of the "niche," namely, those parameters of habitat, nest site, diet, foraging periodicity, and other factors capable of limiting populations. Now when the ranges of two species first meet, it may be that the species are already so different that competition is negligible, and the ranges come to overlap with no difficulty. But if interference is considerable, it will be of adaptive value for the species to diverge ecologically in the zone of overlap. In the case of social insects such "character displacement" (Brown and Wilson, 1956) will have a dividend measurable in increased colony survival or queen production or both. In view of the kind of competition revealed by Pontin's analysis it is appropriate that one of the first and best-documented examples of ecologically based character displacement occurs in the genus *Lasius* (Wilson, 1955a). *L. flavus,* one of the protagonists in Pontin's study, occurs throughout the temperate portions of Europe, Asia, and North America. Over most of this range it occupies a wide array of principal habitats, including grassland and both deciduous and coniferous woodland with various degrees of shading and drainage. In the forested portion of eastern North America, however, it encounters a very closely related species, *L. nearcticus.* There it is more restricted in habitat choice, being limited chiefly to open woodland and fields, while *nearcticus* occupies the darker, moister woodland habitats. Where they occur together, the two species can be distinguished by no less than five morphological characters, including differences in eye size, color, maxillary palp development, antennal length, and head shape. At least the first three of these reflect a greater adaptation on the part of *flavus* for a less subterranean existence. From the Great Plains of North America, where *nearcticus* is left behind, westward to the Pacific coast and beyond across Asia and Europe, *flavus* displays both a wider ecological range and a greater variation in each of the morphological characters sufficient to incorporate the *nearcticus* as well as the eastern *flavus* traits. In short, where *flavus* overlaps the range of *nearcticus,* it is displaced to more open habitats, and its populations display morphological characteristics correlated with specialization to these habitats.

A case of character displacement in actual progress has been recorded in fire ants by Wilson and Brown (1958b). The South American species *Solenopsis saevissima* was introduced into the port of Mobile, Alabama, around 1918, and in the 1940's began a rapid expansion that was to extend its range over most of the southern United States by 1970. The species builds up very dense populations in open habitats, but is largely absent from woodland. *S. xyloni,* a closely related native species which also favors open environments, has been mostly eliminated from its old range in the southern United States in only twenty years. *S. geminata,* on the other hand, has been only partially displaced by *saevissima;* whereas previously *geminata* occurred in both open and woodland environments, now it is limited mostly to woodland where the *saevissima* do not penetrate. There has been a concurrent morphological change in the *geminata* population inside the *saevissima* range. Previously the open environment was occupied chiefly by a reddish color form and the woodland by a dark brown color form; with the advent of *saevissima,* the reddish color form has been mostly eliminated.

The evolutionary phenomenon of character displacement may or may not be of general occurrence in social insects. It can usually be detected only by extensive studies of geographic variation and so far has been recorded only in several ant genera, including *Odontomachus, Rhytidoponera, Pristomyrmex, Solenopsis,* and two species groups of *Lasius.* It deserves closer study because of the likelihood that interspecific competition sets constraints not only on the distribution, ecology, and morphology of particular social insect species, but on their social characteristics as well. A species restricted to pieces of dead wood or some other cramping nest site will, by interspecific competition, tend to evolve a smaller mature colony size and lower its production rate of sexual forms. If the reduction is great enough, it may abandon odor trails as a form of communication. Elsewhere (Wilson, 1959e, 1961) I have presented combined biogeographic and ecological evidence suggesting that just such a restriction has occurred commonly in the evolution of the Pacific ant fauna.

Once the nest site of an individual colony has been selected, and if the area of its foraging ground is fixed by the colonies in residence around it, the productivity of the colony will probably depend on the food yield of the territory. It follows that in truly territorial species the production rate of new queens in mature colonies will increase as a function of the size of the territory. The relation has been verified in the two analyses in which the proposition has been tested: that of Pontin (1961) on *Lasius* and that of Brian, Hibble, and Kelly (1966) on *Myrmica.* This does not mean that the populations of colonies are necessarily food limited. On the contrary, Brian (1956b,c) has provided evidence that in the west of Scotland the density of colonies is controlled by the density of available nest sites. In England as a whole, which has only a marginal environment for ants, "nest sites need to have high insolation, to be moist but not too wet, to be soft enough for excavation but mechanically stable, and they need to be within reach of plants to supply sugar and water through Homoptera and protein through the many small insects that feed on the plants and their litter." The constraining condition of having to nest in a place with few plants and yet still be near places with many plants is offset by the extreme abundance of insect food during the spring and summer. In Brian's study area in western Scotland, much of the land is covered by higher vegetation of one form or another. Consequently nest sites are in short supply and the ant colonies compete heavily for them. When Brian laid out slabs of rocks, which make ideal nest sites, in an open, food-rich area populated by *Myrmica ruginodis,* the number of colonies increased. In a parallel study, he found that the gradual growth of trees in a glade, which increased shading and reduced the number of warm, dry nest sites on the ground, resulted in a reduction in the number of ant colonies. The same general conclusion was independently drawn by Gösswald (1951b) on the basis of his long-term studies of the same ant genera in Germany. He pointed, for example, to the curious fact that dense ant populations characteristically occur around quarries, where fragments of stones lying on the ground provide an unusual number of nest sites. The abundance of nest sites appears to be a common primary control of colony density throughout the social insects. It is manifestly the principal factor in the meliponine bees of tropical America (Moure, Nogueira-Neto, and Kerr, 1958), and in ant and termite species that are restricted to pieces of dead wood in tropical forests (Noirot, 1958–59; Wilson, 1959d).

Actually, the factors controlling the density of populations of social insects are probably what ecologists refer to as intercompensatory. This means that, in a given environment at a given time, one factor is usually limiting, and, if it were removed, the population would increase until a second factor became limiting, and so on. If nest sites became unlimited in a food-rich area, the colony density would increase until food became scarce. If food were then presented in unlimited amounts, the populations would presumably increase until territorial behavior (to be sure, centered around smaller territories) stabilized the population at a new, still higher level. This simple sequence is based on only a few empirical observations, mostly those of Brian, and it is speculative when applied to social insects as a whole. It is altogether probable that other sequences and other factors are at work. It is even likely that different populations belonging to the same species equilibrate under different schedules of controls. One suggestive example has been provided by Medler (1957). When he set out nest boxes for bumblebees in northern Wisconsin, where these insects are already abundant, he found that the population increased, but not so in southern Wisconsin, where bumblebees are scarce. Medler concluded that in northern Wisconsin nest sites are limiting, but that in southern Wisconsin, which is more agricultural, food is scarce and sporadic enough to supplant nest sites as the limiting factor. This whole subject invites new experimental work, and the mostly sedentary and perennial natures of social insects makes them ideally suited for it.

To return finally to the matter of the determination of species diversity, current theory teaches that the number of species is a function of the diversification of those factors that are primary in controlling population density. Climate also plays a role, to be sure. As a rule, the more severe and fluctuating the climate, the fewer the species. Thus, the only social insects on Iceland are *Vespula germanica* and *Bombus jouellus,* plus one or two introduced ants restricted to heated buildings. For equally obvious reasons, social wasps, apine bees, and termites are scarce in deserts. The degree of geographic isolation is also important; Hawaii, for example, has no native species of ants or termites. But in most parts of the world that are neither greatly isolated nor uncongenial to social insect life the number of coexisting species is probably a function of the degree to which the population-limiting factors are diversified. In some environments, for example, the Dutch forests analyzed by Westhoff and Westhoff-De Joncheere (1942), this means the complexity of the habitat insofar as it determines variation in the kinds of available nest sites. In other environments, to a degree yet to be measured by any kind of field research, it would mean that the variety of available prey and other food items is crucial.

The Dispersal of Colonies

Despite the great diversity of their species and their overwhelming numerical preponderance on all the continents of the world, the social insects have seldom been able to colonize the oceanic islands. Prior to the arrival of man, ants did not range east of Samoa and Tonga in the Pacific; nor did they reach the islands of the mid-Atlantic (Wilson and Taylor, 1961, 1967). As few as eight stocks make up the existing native fauna of New Zealand, while the basic fauna of the West Indies is only slightly richer (Brown, 1958; Wheeler and Mann, 1914). Almost no bee genera containing social species reach New Zealand; the single known exception is a stock belonging to *Lasioglossum.* Native allodapines and apids are absent from all of Melanesia and Polynesia east of the Solomon Islands as well (Michener, 1965a). A few termite genera, mostly members of the relatively primitive Kalotermitidae and Rhinotermitidae, have colonized oceanic islands, but the vast majority are wholly restricted to the continents (Emerson, 1952).

The limitation in the dispersal ability of these insects is rooted in their social way of life. Several specific examples will make this relation clear. The American myrmicine ant *Pheidole sitarches* conducts mass nuptial flights, during which the males form aerial swarms just above open spaces in woodland and the queens fly to them to acquire a partner. Each queen mates only once. As soon as a male alights on her back, she falls to the ground, allows the copulation to proceed for a few minutes, then rejects the male, sheds her wings, and begins to crawl over the ground in search of a nest site (Wilson, 1957). As a consequence, each new generation disperses only as far as the nuptial swarms can be organized away from the parental nests, plus the very short distance walked by the dealated queens. In many other ant species the queen continues flying after copulation, so that a single individual can serve as a propagule to be carried by the wind. On small islands in the Florida Keys from which entire arthropod faunas were removed by methyl bromide fumigation, new ant populations were always started by individual queens who flew or were blown across the water (Simberloff and Wilson, 1969). The maximum distances that can be covered by such journeys can be inferred from other kinds of distributional data on the Florida Keys fauna. All of the ten arboreal ant species occurring in identical habitats in the vicinity of Key West have dispersed over the 5-mile water gap to the Marquesas group, but only one (*Pseudomyrmex elongatus*) has crossed the additional 40 miles to colonize the Dry Tortugas. The fire ant *Solenopsis saevissima,* which spreads by the dissemination of mated queens from nuptial flights high in the air, extended its range by approximately 5 miles a year during the period of maximal expansion in the 1940's and 1950's (Wilson and Brown, 1958b).

According to Johnson (1966) alate termites appear to be specialized for dispersal within a zone of relatively calm air near the ground. The observed dispersal distances, compiled by Nutting (1969) from the literature and his own studies, are accordingly short: *Kalotermes flavicollis,* a few dozen meters; *Cryptotermes havilandi,* 1–45 m; *Incisitermes minor,* 120 m; *Zootermopsis angusticollis,* about 350 m; *Anacanthotermes ochraceus,* about 100 m; *Reticulitermes lucifugus,* 10–200 m; *Macrotermes* sp., a few kilometers; *Odontotermes angustatus,* more than a single kilometer; *Amitermes minimus,* several hundred meters. The maximum distance reached by individuals in a large sample, say of a million or more, has not been measured. Harvey (1934) believed that *Incisitermes minor* alates could reach 2 km or farther, and Glick (1939) recovered small numbers of alates of *Reticulitermes virginicus* from airplane-towed traps at altitudes of as high as 1,000 m. But even if long-distance dispersal were accomplished by a very small fraction of the alate population, it is unlikely that they would be the source of a new population since colonies can be founded only by a cooperating female and male. The probability must be very remote that one-millionth or even one-thousandth of a population of alates, scattered through an arc of terrain many kilometers from the parental nests, could locate each other and form into bisexual pairs in the brief interval of time before predators eliminated them.

Even more formidable obstacles to long-distance dispersal face the apine bees. Honeybee colonies move from place to place in swarms, and new sites are chosen by scout bees who communicate the locations through waggle dancing. The combination of the two constraints make jumps of more than a few kilometers improbable. Meliponine bees emigrate to new sites through a still more complicated chain of events, consisting of the discovery of new sites by scouts, construction of the nests by workers who fly back and forth between the old and new sites, and, finally, emigration of the newly mated queen. All of this usually takes place over a few tens or hundreds of meters at most. According to Moure, Nogueira-Neto, and Kerr (1958), the known record is held by a colony of *Trigona* (*Plebeia*) *emerina* which was introduced into Louisiana. It traveled 365 m in an attempt to start a new nest in the strange environment.

In a curious fashion, some ant species have traded dispersal ability for potential colony immortality. One of the most striking examples is provided by *Iridomyrmex humilis.* In this species there are no clear colony boundaries; each population is comprised of but one huge, widely dispersed colony with a great many relatively small queens. Nuptial flights are extremely rare, matings normally occur within the nest, and the newly fecundated queens stay where they are. Under natural conditions the species therefore spreads very slowly. The same combined traits—unicolonial or at most weakly multicolonial populations, multiple nest queens, and low dispersal rates—are shared with *Pseudomyrmex "flavidula"* (R. Leuthold, personal communication), *Myrmica ruginodis* "subsp. *microgyna*" (Brian, 1965b), *Formica pallidefulva "incerta"* (Talbot, 1948), and *F. polyctena* (Raignier, 1948; Gösswald, 1951a). An intriguing additional aspect of the phenomenon is that each of these species is accompanied in at least part of its range by one or more sibling species, forms which are extremely similar in morphology but contrast strongly in behavior and population characteristics. They are multicolonial; each colony contains only one or a very few, relatively large queens; and the queens conduct normal nuptial flights terminating in claustral colony foundation. In short, they possess the primitive formicid traits. Such pairs (or sets of three or more) of unicolonial and multicolonial sibling species may be widespread in the ants. Without making any particular search for them, I have noted their apparent presence in the *Crematogaster minutissima* and *Formica neorufibarbis* groups in the United States, and Marikovsky (1963) has described what may be still another example in the *Formica sanguinea* populations of Siberia. Brian's analysis of the *Myrmica ruginodis* case is the most informative to date. He regards the two forms, the multicolonial *"macrogyna"* and the unicolonial *"microgyna"* as subspecies, but the sum of the information he has published about them—differences in queen size, life cycle, and habitat preference; different, albeit overlapping, geographical distributions; and rejection of *microgyna* queens by *macrogyna* colonies—strongly suggests that they are in fact distinct species. Otherwise, Brian's evolutionary interpretation of the case seems to be sound and might prove applicable to the other examples as well. He points out that *microgyna* occurs principally in species-poor environments in northwestern Britain, while *macrogyna* occurs widely but in scattered localities in species-rich environments to the south. In these species-rich environments, the possibility of colony and even population-of-colonies extinction through competitive exclusion is higher, so that macrogynous queens capable of starting new colonies in temporarily vacant sites are needed. Where competitive pressure is less, the species can afford to gamble on greater colony longevity. Accordingly, the colonies can be fused into single units, and newly mated queens can safely return to the parental nests. And since colonies are longer lived, fewer empty nest sites will be available at any given time, making the dissemination of queens away from the parental nests less profitable. The result will be a dual pressure for the unicolonial species to suspend claustral nest founding altogether. The adoption of unicolonialism, with its increased degree of inbreeding and reliance on budding as the principal means of reproduction, is an evolutionary step that parallels the adoption of apomixis and vegetative reproduction in nonsocial organisms. Both permit the rapid growth of populations in habitats that are relatively free of competitors but sparsely distributed.

22 The Prospect for a Unified Sociobiology

When the same parameters and quantitative theory are used to analyze both termite colonies and troops of rhesus macaques, we will have a unified science of sociobiology. Can this really ever happen? As my own studies have advanced, I have been increasingly impressed with the functional similarities between insect and vertebrate societies and less so with the structural differences that seem, at first glance, to constitute such an immense gulf between them. Consider for a moment termites and macaques. Both are formed into cooperative groups that occupy territories. The group members communicate hunger, alarm, hostility, caste status or rank, and reproductive status among themselves by means of something on the order of 10 to 100 nonsyntactical signals. Individuals are intensely aware of the distinction between groupmates and nonmembers. Kinship plays an important role in group structure and has probably served as a chief generative force of sociality in the first place. In both kinds of society there is a well-marked division of labor, albeit with a much stronger reproductive component on the part of the insects. The details of organization have been evolved by an evolutionary optimization process of unknown precision, during which some measure of added fitness was given to individuals with cooperative tendencies—at least toward relatives. The fruits of cooperativeness depend upon the particular conditions of the environment and are available to only a minority of animal species during the course of their evolution.

From the specialist's point of view this comparison may at first seem facile—or worse. But it is out of such deliberate oversimplification that the beginnings of a general theory are made. The formulation of a theory of sociobiology constitutes, in my opinion, one of the great manageable problems of biology for the next twenty or thirty years. Let me try to guess part of its future outline and some of the directions in which it is most likely to lead animal behavior research. As I suggested in the previous chapter on population biology, the evolution of social behavior can be fully comprehended only through an understanding, first, of demography, which yields the vital information concerning population growth and age structure, and, second, of the genetic structure of the populations, which tells us what we need to know about effective population size in the genetic sense, the coefficients of relationship within the societies, and the amounts of gene flow between them. The principal goal of a general theory of sociobiology should be an ability to predict features of social organization from a knowledge of these population parameters combined with information on the behavioral constraints imposed by the genetic constitution of the species. It will be a chief task of evolutionary ecology, in turn, to derive the population parameters from a knowledge of the evolutionary history of the species and of the environment in which the most recent portion of that history has unfolded. This sequential relation between evolutionary studies, ecology, and sociobiology is represented in Figure 22-1.

Notice that the word "theory" is given a special meaning in this schema. It is conceived as being hypothetico-deductive, or neo-Cartesian in nature, which means that it is built from models designed explicitly to test and extend our basic assumptions. This form of ratiocination has, of course, worked exceedingly well in the physical sciences and is just now beginning to have an influence throughout evolutionary biology. It proceeds roughly in three steps: First, empirical knowledge, both actual and

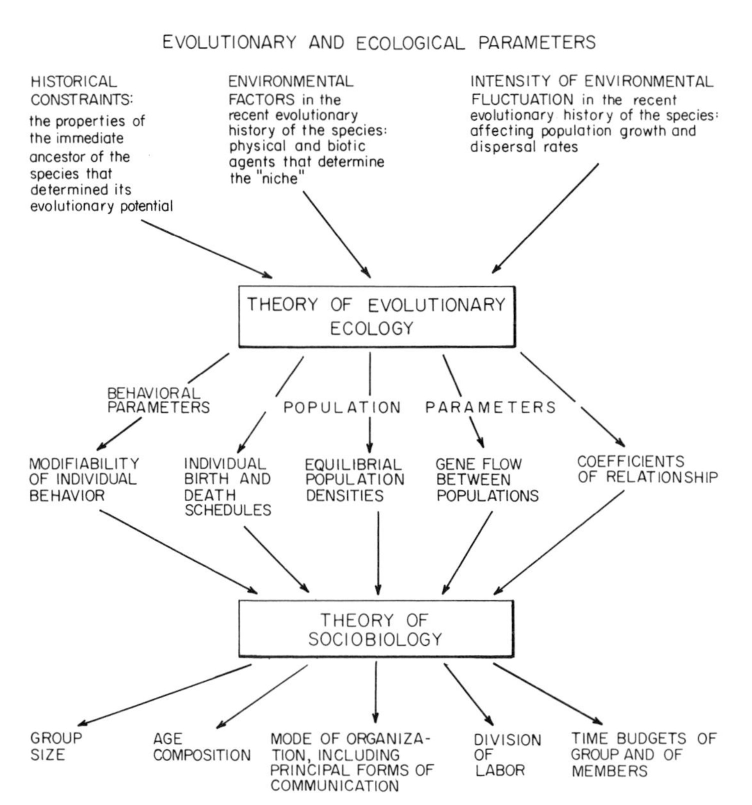

FIGURE 22-1. A diagram of the connections that can be made between phylogenetic studies, ecology, and sociobiology.

desired, is organized into concepts which in their most desirable form are sets of measured variables. What we know, or think we know, is then expressed in precisely defined relations among the concepts. These are the postulates of our knowledge. In the second step of the hypothetico-deductive method the consequences of the postulates are deduced by means of models. New predictions about the real world are thereby contrived, and, in the third and final step, they are examined through experimentation and analysis in order to test and revise the original concepts and postulates. The predictions, insofar as they can be verified, provide postulates for the next, subordinate level of theory. This is the procedure that can best be followed to link evolutionary theory and ecology to population biology and thence to sociobiology.

Evolutionary biology is now in the process of building fragments of such a theory by approximately the means just outlined. The role of historical constraints has been examined by a number of recent authors in, for example, the volumes edited by Baker and Stebbins (1965), Waddington (1968, 1969), and Lewontin (1968). Levins (1968) has deliberately set out to develop deductive theories of the niche and population genetics in fluctuating environments. At the next level, MacArthur and Wilson (1967) have derived some of the population parameters and linked them to dispersal and distribution, while Hamilton (1964) has produced the first comprehensive theory of altruistic behavior based on first principles of population genetics. An increasing number of experimental biologists are devoting themselves to the study of behavioral genetics, which holds the key to the population consequences of behavioral modifiability (Parsons, 1967). The remaining lacunae of evolutionary biology are nevertheless wide and deep, and relevant experimental investigation is in its earliest stages.

In the future development of sociobiology, the insect societies will play a key role. They are so remote in phylogenetic origin from vertebrate societies as to resemble creations of some other world. They provide the comparative material every good theoretical scheme needs. Where the theory predicts convergence of a functional trait, its validity can be put to a convincingly stringent test by comparing insects and vertebrates. It would be appropriate, therefore, to close this brief attempt at prophesy by identifying the basic differences that visibly exist, not between insects and vertebrates as organisms, but between the societies they form. If we exclude man, with his unique language and revolutionary capacity for cultural transmission, the best-organized societies of vertebrates can be distinguished by a single trait so overriding in its consequences that the other characteristics seem to flow from it. This is personal recognition among the members of the group. As a rule each adult animal knows and bears some particular relationship to every other member. Status is extremely important. Where dominance hierarchies exist, elaborate signaling is employed to implement them. Parent-offspring relationships are specific to individuals, tightly binding, and relatively long in duration. Within primate societies, personal groupings often ascend to the level of cliques, and, in the case of the macaques, to the "uncle" and "aunt" forms of parental care. Group members spend large amounts of their time and energy in establishing and maintaining these multifarious personal bonds. Associated with the personal form of communication is a prolonged period of socialization in the young. By constant experience, much of it obtained in the form of play, the young animal learns its personal relationship to its parents and establishes an early status among its age peers. Some division of labor exists: there are weakly defined "specialists" in defense, leadership, and foraging. But it is not based upon morphologically defined castes, and it does not include reproductive neuters.

In contrast, the insect societies are, for the most part, impersonal. The small, relatively primitive colonies of bumblebees and *Polistes* wasps are based on dominance hierarchies, and individuals appear to recognize one another to a limited extent. In other kinds of social insects, however, personalized relationships play little or no role. The sheer size of the colonies and the short life of the members combine to make it inefficient, if not impossible, to establish individual bonds. The average adult honeybee, for example, lives for only about six weeks in the midst of a rapidly changing population of up to 80,000 workers. The army ant worker has only a few weeks or months to become acquainted with a million or more nestmates. In the more advanced societies even the relationship of the workers to the queen is impersonal. She is "recognized" by a small set of pheromones, which can be extracted and absorbed into dummy substitutes of inert material. Where songbirds have individual calls that are identified by territorial neighbors, and mammals use individual scent marks, the members of an insect colony employ signals that are for the most part uniform throughout the species. The one known exception is the colony odor, which is acquired at least in part from food and nesting material and is used to distinguish nestmates, all of them together, from members of other colonies. Socialization is minimal in insects, and play is apparently absent.

Yet the insect colony, as a unit, can equal or exceed the accomplishments of the nonhuman vertebrate society as a unit. One need only think of the giant air-conditioned nests of the macrotermitines, the organized foraging expeditions of the army ants, and the swift recruitment of honeybees in response to pheromones and waggle dancing to see the strength of this generalization. The insects do it with programmed divisions of behavior among castes. Their great innovation in evolution was the reproductive neuter, which fixed the colony as the unit of natural selection and removed the limits on the amount of caste differentiation that could occur among the colony members. Once this happened, it was possible to fashion entire organisms to perform difficult, highly specialized tasks and to combine them in mixtures capable of matching the feats of single vertebrates. Had vertebrates gone the same route, the results would have been much more spectacular. The individual bird or mammal, possessing a brain vastly larger than that of an insect, might have been shaped into a comparably better specialist. The vertebrate society then could have evolved into an unimaginably more complex, efficient unit—at the price, of course, of independent action on the part of its members. Vertebrates have instead remained chained to the cycle of individual reproduction. This forever enhances freedom on the part of the individual at the expense of efficiency on the part of society. The dilemma of mankind is that technology and population growth have propelled us to the point where we could perhaps operate better as a society with termite-like altruism and regimentation, yet we cannot and must not forsake the primate individuality that brought us to the threshold of civilization in the first place.

The optimistic prospect for sociobiology can be summarized briefly as follows. In spite of the phylogenetic remoteness of vertebrates and insects and the basic distinction between their respective personal and impersonal systems of communication, these two groups of animals have evolved social behaviors that are similar in degree of complexity and convergent in many important details. This fact conveys a special promise that sociobiology can eventually be derived from the first principles of population and behavioral biology and developed into a single, mature science. The discipline can then be expected to increase our understanding of the unique qualities of social behavior in animals as opposed to those of man.

This great attribute, which signifies unbounded wisdom, induces us to admire those laws by which providence rules the insect societies and reserves to herself their exclusive direction; and it shows us that in delivering man to his own guidance, she has subjected him to a great and heavy responsibility. If natural history only serves to prove this truth, it will have attained the most dignified end of which science may boast—that of endeavoring to ameliorate the human species by the examples it lays before us.

Pierre Huber (*1810*)

Glossary

Included in this list are some terms used generally in entomology as well as most of those restricted to insect sociology. The glossary was designed to make it possible to read the book with no more than an elementary background in biology and, specifically, to eliminate the need to refer to textbooks in general entomology.

Active space. The space within which the concentration of a pheromone (or any other behaviorally active substance) is at or above threshold concentration. The active space of a pheromone is, in fact, the signal itself.

Aculeate. Pertaining to the Aculeata, or stinging Hymenoptera, a group including the bees, ants, and many of the wasps.

Adoption substance. A secretion presented by a social parasite that induces the host insects to accept it as a member of their colony.

Adult transport. The carrying or dragging of one adult social insect by a nestmate, usually during colony emigrations. In ants, adult transport is a very frequent and stereotyped form of behavior.

Adultoid reproductive. In the higher termites, a supplementary reproductive that is a fully developed imago and is thus morphologically indistinguishable from the primary reproductive it replaces.

Age polyethism. The regular changing of labor roles by colony members as they age.

Aggregation. A group of individuals, comprised of more than just a mated pair or a family, that have gathered in the same place but do not construct nests or rear offspring in a cooperative manner (opposed to colony, *q.v.*).

Alarm-defense system. Defensive behavior which also functions as an alarm signaling device within the colony. Examples include the use by certain ant species of chemical defensive secretions that double as alarm pheromones.

Alarm pheromone. A chemical substance exchanged among members of the same species that induces a state of alertness or alarm in the face of a common threat.

Alarm-recruitment system. A communication system that rallies nestmates to some particular place to aid in the defense of the colony. An example is the odor trail system of lower termites, which is used to recruit colony members to the vicinity of intruders and breaks in the nest wall.

Alitrunk. The mesosoma of the higher Hymenoptera, including the true thorax and (fused anteriorly to the thorax) the first abdominal segment.

Allodapine. A ceratinine bee belonging to *Allodape* or one of a series of closely related genera, all of which are either eusocial or socially parasitic. Excluded from this informal taxonomic category is *Ceratina,* the only other major living genus of the tribe Ceratinini.

Allometric growth. See allometry.

Allometry. Any size relation between two body parts that can be expressed by $y = bx^a$, where a and b are fitted constants. In the special case of isometry, $a = 1$, and the relative proportions of the body parts therefore remain constant with change in total body size. In all other cases ($a \neq 1$) the relative proportions change as the total body size is varied (see Chapter 8).

Altruism. Self-destructive behavior performed for the benefit of others.

Ambrosia. The fungus cultivated by wood-boring scolytoid beetles. Sometimes the term is used specifically to designate the part of the fungus that grows out into the burrows and is eaten by the beetles.

Ambrosia beetle. A wood-boring scolytoid beetle that cultivates fungus ("ambrosia") for food (see also fungus-growing beetle).

Antennation. Touching with the antennae. The movement can serve as a sensory probe or as a tactile signal to another insect.

Apiary. A place where honeybees are kept. Specifically, a group of hives.

Appeasement substance. A secretion presented by a social parasite that reduces aggression in the host insects and aids the parasite's acceptance by the host colony.

Apterous neoteinic. In termites, the same as ergatoid reproductive.

Army ant. A member of an ant species that shows both nomadic and group-predatory behavior. In other words, the nest site is changed at relatively frequent intervals, in some cases daily, and the workers forage in groups (same as legionary ant).

Arolium. A cushion-like pad located between the tarsal claws and comprising part of the pretarsus.

Arrhenotoky. The production of males from unfertilized eggs.

Astelocyttarous. Pertaining to nests, and especially wasp nests, in which the comb is attached directly to a support and lacks pillars.

Basitarsus. The proximal or basal segment of the tarsus.

Batumen. A protective layer of propolis or hard cerumen (sometimes vegetable matter, mud, or various mixtures) that encloses the nest cavity of a colony of stingless bees.

Bivouac. The mass of army ant workers within which the queen and brood find refuge.

Blastogenesis. The origin of different caste traits from variation in either the ovarian environment of the egg or the nongenetic contents of the egg (opposed to genetic control of caste and trophogenesis).

Brachypterous neoteinic. In termites, the same as nymphoid reproductive.

Brood. The immature members of a colony collectively, including eggs, nymphs, larvae, and pupae. In the strict sense eggs and pupae are not members of the society, but they are nevertheless referred to as part of the brood.

Brood cell. A special chamber or pocket built to house immature stages.

Budding. The same as colony fission.

Callow workers. Newly eclosed adult workers whose exoskeleton is still relatively soft and lightly pigmented.

Calyptodomous. Pertaining to nests, especially wasp nests, in which the combs are surrounded by an envelope.

Caste. Broadly defined, as in ergonomic theory (Chapter 18), any set of individuals of a particular morphological type, or age group, or both, that performs specialized labor in the colony. More narrowly defined, any set of individuals in a given colony that are both morphologically distinct and specialized in behavior.

Cerumen. A brown mixture of wax and propolis used for nest construction by social bees.

Chain transport. The relaying of food from one worker to another in the course of transporting it back to the nest.

Cladogram. A diagram showing nothing more than the sequence in which groups of organisms are interpreted to have originated and diverged in the course of evolution.

Claustral colony founding. The procedure during which queens (or royal pairs in the case of termites) seal themselves off in cells and rear the first generation of workers on nutrients obtained mostly or entirely from their own storage tissues, including fat bodies and histolysed wing muscles.

Cleptobiosis. The relation in which one species robs the food stores or scavenges in the refuse piles of another species, but does not nest in close association with it.

Cleptoparasitism. The parasitic relation in which a female seeks out the prey or stored food of another female, usually belonging to a different species, and appropriates it for the rearing of her own offspring.

Colony. A group of individuals, other than a single mated pair, which constructs nests or rears offspring in a cooperative manner (opposed to aggregation, *q.v.*).

Colony fission. The multiplication of colonies by the departure of one or more reproductive forms, accompanied by groups of workers, from the parental nest, leaving behind comparable units to perpetuate the "parental" colony. This mode is referred to occasionally as hesmosis in ant literature and sociotomy in termite literature. Swarming in honeybees can be regarded as a special form of colony fission.

Colony odor. The odor found on the bodies of social insects which is peculiar to a given colony. By smelling the colony odor of another member of the same species, an insect is able to determine whether it is a nestmate (see nest odor and species odor).

Comb (of cells or cocoons). A layer of brood cells or cocoons crowded together in a regular arrangement. Combs are a characteristic feature of the nests of many species of social wasps and bees.

Commensalism. Symbiosis in which members of one species are benefited while those of the other species are neither benefited nor harmed.

Communal. Applied to the condition or to the group showing it in which members of the same generation cooperate in nest building but not in brood care.

Communication. Action on the part of one organism (or cell) that alters the probability pattern of behavior in another organism (or cell) in an adaptive fashion.

Compound nest. A nest containing colonies of two or more species of social insects, up to the point where the galleries of the nests anastamose and the adults sometimes intermingle but where the broods of the species are still kept separate (see mixed nest).

Copularium. The first chamber built by a colony-founding couple of termites.

Corpora allata. Paired endocrine organs located just behind the brain; the source of juvenile hormone.

Dealate. Referring to an individual that has shed its wings, usually after mating; used both as an adjective and a noun.

Dealation. The removal of the wings by the queens (and also males in the termites) during or immediately following the nuptial flight and prior to colony foundation.

Developmental cycle. The period from the birth of the egg to the eclosion of the adult insect (applied to social wasp studies by Richards and Richards, 1951).

Dichthadiiform ergatogyne. A member of an aberrant reproductive caste, limited to army ants, which is characterized by the possession of a wingless alitrunk, a huge gaster, and an expanded postpetiole.

Dimorphism. In caste systems, the existence in the same colony of two different forms, including two size classes, not connected by intermediates.

Diphasic allometry. Polymorphism in which the allometric regression line, when plotted on a double logarithmic scale, "breaks" and consists of two segments of different slopes whose ends meet at an intermediate point (see Chapter 8).

Disc. See imaginal disc.

Division of labor. See polyethism.

Dominance hierarchy. The physical domination of some members of a group by other members, in relatively orderly and long-lasting patterns. Except for the highest and lowest ranking individuals, a given member will dominate one or more of its companions and be dominated in turn by one or more of the others. The hierarchy is initiated and sustained by hostile behavior, albeit sometimes of a subtle and indirect nature.

Dorylophile. An obligatory guest of one of the army ants belonging to the tribe Dorylini.

Driver ants. African legionary ants belonging to the genus *Anomma* and, less frequently, other members of the tribe Dorylini.

Drone. A male social bee, especially a male honeybee or bumblebee.

Dulosis. The relation in which workers of a parasitic (dulotic) ant species raid the nests of another species, capture brood (usually in the form of pupae), and rear them as enslaved nestmates.

Ecitophile. An obligatory guest of one of the army ants belonging to the tribe Ecitonini, especially of the genus *Eciton* itself.

Eclosion. Emergence of the adult (imago) from the pupa; less commonly, the hatching of an egg.

Ectosymbiont. A symbiont that associates with the host colonies during at least part of its life cycle in some relationship other than internal parasitism.

Elite. Referring to a colony member displaying greater than average initiative and activity.

Emery's rule. The rule that species of social parasites are very similar to their host species and therefore presumably closely related to them phylogenetically.

Emigration. The movement of a colony from one nest site to another.

Envelope. A sheath of carton or wax surrounding the nest of a social insect, especially that of a social wasp.

Epigaeic. Living, or at least foraging, primarily above ground (opposed to hypogaeic).

Epinotum. See propodeum.

Ergatogyne. Any form morphologically intermediate between the worker and the queen.

Ergatoid male. See ergatomorphic male.

Ergatoid reproductive. A supplementary reproductive termite without a trace of wing buds, usually larval in external form, and with a distinctively rounded head (same as third-form reproductive, tertiary reproductive, and apterous neoteinic).

Ergatomorphic male. An individual with normal male genitalia and a worker-like body.

Ergonomics. The quantitative study of work, performance, and efficiency.

Ethocline. A series of different behaviors observed among related species and interpreted to represent stages in a single evolutionary trend.

Eusocial. Applied to the condition or to the group possessing it in which individuals display all of the following three traits: cooperation in caring for the young; reproductive division of labor, with more or less sterile individuals working on behalf of individuals engaged in reproduction; and overlap of at least two generations of life stages capable of contributing to colony labor. This is the formal equivalent of the expressions "truly social" or "higher social" which are commonly used with less exact meaning.

Exploratory trail. An odor trail laid more or less continuously by the advance workers of a foraging group. This kind of communication is used regularly by army ants (opposed to recruitment trail).

Exudatoria. Finger-like appendages found on the larvae of certain ant species and on a variety of termitophiles. According to the hypothesis of W. M. Wheeler (see Chapter 14) the exudatoria produce secretions attractive to the ant or termite workers.

Facilitation. The same as social facilitation; see under group effect.

First-form reproductive. See primary reproductive.

Formicarium. Same as formicary.

Formicary. A nest of ants. The term is also commonly applied to an ant mound or an artificial nest used in the laboratory to house ants.

Fungus-growing beetle. Any beetle that utilizes symbiotic fungi for food. The term applies not only to the "ambrosia beetles" of the Scolytoidea but also to certain bark-inhabiting scolytids and other kinds of beetles (some Anobiidae, Curculionidae, etc.) that are associated with symbiotic fungi.

Gaster. A special term occasionally applied to the metasoma, or terminal major body part, of ants.

Genus. A set of similar species of relatively recent common ancestry.

Gongylidium. A swollen hyphal tip of the symbiotic fungi cultured by attine ants. The ants pick the gongylidia as food. A group of gongylidia is referred to as a staphyla.

Grooming. The licking of the body surfaces of nestmates. Self-grooming also occurs, in which individuals clean their own bodies both by licking and stroking with the legs.

Group effect. An alteration in behavior or physiology within a species brought about by signals that are directed in neither space nor time. A simple example is social facilitation, in which there is an increase of an activity merely from the sight or sound (or other form of stimulation) coming from other individuals engaged in the same activity.

Group-predation. The hunting and retrieving of living prey by groups of cooperating animals. A behavior pattern best developed in army ants.

Guest. A social symbiont.

Gymnodomous. Pertaining to nests, especially wasp nests, that lack an envelope.

Gynandromorph. An individual containing patches of both male and female tissue.

Gynergate. A female containing patches of tissue of both the queen and worker castes.

Haplodiploidy. The mode of sex determination in which males are derived from haploid eggs and females from diploid eggs.

Harvesting ants. Ant species that store seeds in their nests. Many taxonomic groups have developed this habit independently in evolution.

Hemimetabolous. Undergoing development which is gradual and lacks a sharp separation into larval, pupal, and adult stages. Termites, for example, are hemimetabolous (opposed to holometabolous).

Hesmosis. Same as colony fission.

Heterogony. Same as allometry.

Hive aura. See nest odor.

Hive odor. See nest odor.

Holometabolous. Undergoing a complete metamorphosis during development, with distinct larval, pupal, and adult stages. The Hymenoptera, for example, are holometabolous (opposed to hemimetabolous).

Homeostasis. The maintenance of a steady state, especially a physiological or social steady state, by means of self-regulation through internal feedback responses.

Homopteran. A member of, or pertaining to, the insect order Homoptera, which includes the aphids, jumping plant lice, treehoppers, spittlebugs, whiteflies, and related groups.

Honeybee. A member of the genus *Apis.* Unless qualified otherwise, a honeybee is more particularly a member of the domestic species *A. mellifera,* and the term is usually applied to the worker caste.

Honey pot (*honeypot*). A container made by stingless bees or bumblebees from soft cerumen and used to store honey.

Hornet. A large wasp of the family Vespinae, particularly a member of the genus *Vespa* or (in the United States) the bald-faced hornet *Vespula* (*Dolichovespula*) *maculata.*

Hypogaeic. Living primarily underground (subterranean) or at least beneath cover such as leaf litter, stones, and dead bark (cryptobiotic).

Imaginal disc. A relatively undifferentiated tissue mass occurring in the body of a larva which is destined to develop later into an adult organ.

Imago. The adult insect. In termites, the term is usually applied only to adult primary reproductives.

Ingluvial. Pertaining to the crop, the distensible middle portion of the foregut in which (in many species) liquid food is stored.

Inquilinism. The relation in which a socially parasitic species spends the entire life cycle in the nests of its host species. Workers are either lacking or, if present, are usually scarce and degenerate in behavior. This condition is sometimes referred to loosely as "permanent parasitism."

Insect society. In the strict sense, a colony of eusocial insects (ants, termites, eusocial wasps, or bees). In the broad sense adopted in this book, any group of presocial or eusocial insects.

Insect sociology (*insect sociobiology*). The study of social behavior and population characteristics related to social behavior in insects.

Instar. Any period between molts during the course of development.

Involucrum. A sheath of soft cerumen surrounding the brood chamber in a nest of stingless bees (Meliponini).

Isometry. The condition in which the sizes of two body parts remain constant relative to each other as total body size is increased (isometry is a special case of allometry, *q.v.*).

Labium. The lower "lip," or lowermost mouthpart-bearing segment of insects, located just below the mandibles and the maxillae.

Larva. An immature stage which is radically different in form from the adult; characteristic of the holometabolous insects, including the Hymenoptera. In the termites, the term is used in a special sense to designate an immature individual without any external trace of wing buds or soldier characteristics.

Legionary ant. See army ant.

Lestobiosis. The relation in which colonies of a small species nest in the walls of the nests of a larger species and enter the chambers of the larger species to prey on brood or to rob the food stores.

Macrocephalic female. In the social halictine bees, a larger female possessing a disproportionately large head. Such individuals are usually the egg layers of the colony.

Major worker. A member of the largest worker subcaste, especially in ants. In ants the subcaste is usually specialized for defense, so that an adult belonging to it is often also referred to as a soldier (see media worker and minor worker).

Mandibulate soldier. A soldier which has large mandibles used in colony defense.

Mass communication. The transfer of information among groups of individuals of a kind that cannot be transmitted from a single individual to another. Examples include the spatial organization of army ant raids, the regulation of numbers of worker ants on odor trails, and certain aspects of the thermoregulation of nests.

Mass provisioning. The act of storing all of the food required for the development of a larva at the time the egg is laid (opposed to progressive provisioning).

Media worker. In polymorphic ant series involving three or more worker subcastes, an individual belonging to the medium-sized subcaste(s) (see minor worker and major worker).

Melittology. The scientific study of bees.

Melittophile. An organism that must spend at least part of its life cycle with bee colonies.

Mesosoma. The middle of the three major divisions of the insect body. In most insects it is the strict equivalent of the thorax, but in higher Hymenoptera it includes the propodeum.

Metasoma. The hindmost of the three principal divisions of the insect body. In most insect groups it is the strict equivalent of the abdomen. In the higher Hymenoptera it is composed only of some of the abdominal segments, since the first segment (the "propodeum") is fused with the thorax and has therefore become part of the mesosoma.

Minima. In ants, a minor worker.

Minor worker. A member of the smallest worker subcaste, especially in ants. Same as minima (see media worker and major worker).

Mixed nest. A nest containing colonies of two or more species of social insects, in which mixing of both the adults and brood occurs (see compound nest).

Molt (*moult*). The casting off of the outgrown skin or exoskeleton in the process of growth. Also the cast-off skin itself. The word is further used as an intransitive verb to designate the performance of the behavior.

Monogyny. The existence of only one functional queen in the nest (opposed to polygyny).

Monomorphism. The existence within a species or colony of only a single worker subcaste.

Monophasic allometry. Polymorphism in which the allometric regression line has a single slope; in ants the use of the term also implies that the relation of some of the body parts measured is nonisometric (see Chapter 8).

Mound nest. A nest at least part of which is constructed of a mound of soil or carton material that projects above the ground surface. The architecture of the mound is often elaborate, specific in plan to the species, and evidently adapted to contribute to microclimate control within the nest.

Multicolonial. Pertaining to a population of social insects which is divided into colonies that recognize nest boundaries (opposed to unicolonial).

Mutualism. Symbiosis that benefits the members of both of the participating species.

Mycangium. Any one of a variety of special pocket-shaped receptacles used to carry symbiotic fungi. Found in the scolytoid beetles.

Myrmecioid complex. One of the two major taxonomic groups of ants; the name is based on the subfamily Myrmeciinae, one of the constituent taxa. It should not be confused with the subfamily Myrmicinae, which belongs to the poneroid complex (see Chapter 4).

Myrmecodomatia. Structures of higher plants that appear to have evolved, in the course of mutualistic evolution, to serve as dwelling places for ants.

Myrmecology. The scientific study of ants.

Myrmecophile. An organism that must spend at least part of its life cycle with ant colonies.

Myrmecophytes. Higher plants that live in obligatory, mutualistic relationship with ants.

Nanitic worker. The dwarf workers produced from either the first ant broods or later ant broods that have been subjected to starvation. Nanitic workers occur in both monomorphic and polymorphic species.

Nasus. The snout-like organ possessed by soldiers of some species in the Nasutitermitinae. The nasus is used to eject poisonous or sticky fluid at intruders.

Nasute soldier. A soldier termite possessing a nasus (*q.v.*).

Necrophoresis. Transport of dead members of the colony away from the nest.

Necrophoric behavior. Same as necrophoresis.

Neoteinic. A supplementary reproductive termite. Used either as a noun or as an adjective (*e.g.,* neoteinic reproductive).

Nest odor. The distinctive odor of a nest, by which its inhabitants are able to distinguish the nest from those belonging to other colonies or at least from the surrounding environment. In some cases the insects, *e.g.,* honeybees and some ants, can orient toward the nest by means of the odor. It is possible that the nest odor is the same as the colony odor (*q.v.*) in some cases. The nest odor of honeybees is often referred to as the hive aura or hive odor.

Nest parasitism. The relation, found in some termites, in which colonies of one species live in the walls of the nests of a second, host species and feed directly on the carton material of which they are constructed.

Nest robbing. Same as cleptobiosis.

Nomadic phase. The period in the activity cycle of an army ant colony during which the colony forages more actively for food and moves frequently from one bivouac site to another. At this time the queen does not lay eggs, and the bulk of the brood is in the larval stage (opposed to statary phase.)

Nomadism. The relatively frequent movement by an entire colony from one nest site to another.

Nuptial flight. The mating flight of the winged queens and males.

Nymph. In general entomology, the young stage of any insect species with hemimetabolous development. In termites, the term is used in a slightly more restricted sense to designate immature individuals which possess external wing buds and enlarged gonads and which are capable of developing into functional reproductives by further molting.

Nymphoid reproductive. A supplementary reproductive termite bearing wing buds (same as second-form reproductive, secondary reproductive, and brachypterous neoteinic).

Ocellus. One of the three simple eyes of adult insects, located on or near the center line of the dorsal surface of the head. The ocelli should be distinguished from the laterally placed compound eyes.

Odor trail. A chemical trace laid down by one insect and followed by another. The odorous material is referred to either as the trail pheromone or the trail substance.

Oligogyny. The occurrence in a single colony of from two to several functional queens. A special case of polygyny.

Ommatidium. One of the basic visual units of the insect compound eye. The ommatidia are bounded externally by the facets that together make up the glassy, rounded outer surface of the eye.

Oophagy. Egg cannibalism; the eating by a colony member of its own eggs or those laid by a nestmate.

Ovariole. One of the egg tubes which, together, form the ovary in female insects.

Palpation. Touching with the labial or maxillary palps. The movement can serve as a sensory probe or as a tactile signal to another insect.

Parabiosis. The utilization of the same nest and sometimes even the same odor trails by colonies of different species, which nevertheless keep their brood separate.

Parasitism. Symbiosis in which members of one species exist at the expense of members of another species, usually without going so far as to cause their deaths.

Parasitoid. A parasite that slowly kills the victim, this event occurring near the end of the parasite's larval development. The term is also used as an adjective.

Parasocial. See presocial.

Parecium. The air space surrounding the fungus garden in the nest of a macrotermitine termite.

Partially claustral colony founding. The procedure during which the queen founds the colony by isolating herself in a chamber but still occasionally leaves to forage for part of her food supply.

Patrolling. The act of investigating the nest interior. Worker honeybees are especially active in patrolling and are thereby quick to respond as a group to contingencies when they arise in the nest (see Chapter 9).

Pedicel. The "waist" of the ant. It is made up of either one segment (the petiole) or two segments (the petiole plus the postpetiole).

Permanent social parasitism. See inquilinism.

Petiole. The first segment of the "waist" of aculeate Hymenoptera; this is, in fact, the second abdominal segment since the first abdominal segment (propodeum) is fused to the thorax.

Pheromone. A chemical substance, usually a glandular secretion, which is used in communication within a species. One individual releases the material as a signal and another responds after tasting or smelling it.

Phragmocyttarous. Pertaining to nests, and especially wasp nests, in which combs are attached laterally to the inner surface of a bag-like envelope.

Phragmosis. The condition in which the head or tip of the abdomen is truncated and is used as a living plug for the nest entrance. Occurs in ants and termites, usually in the soldier caste.

Physogastry. The swelling of the abdomen to an unusual degree due to the hypertrophy of fat bodies, ovaries, or both.

Piping. The sound emitted by young honeybee queens after their emergence. It induces return calls ("quacking") from other virgin queens still in the royal cells and stimulates swarming behavior by the workers.

Pleometrosis. Same as polygyny (*q.v.*).

Plesiobiosis. The close proximity of two or more nests, accompanied by little or no direct communication between the colonies inhabiting them.

Pollen pot. A container made by stingless bees from soft cerumen and used to store pollen.

Pollen-storers. Bumblebee species that store pollen in abandoned cocoons. From time to time the adult females remove the pollen from the cocoons and feed it into a larval cell in the form of a liquid mixture of pollen and honey (opposed to pouch-makers).

Polydomous. Pertaining to single colonies that occupy more than one nest.

Polyethism. Division of labor among members of a colony. In social insects a distinction can be made between caste polyethism, in which morphological castes are specialized to serve different functions, and age polyethism, in which the same individual passes through different forms of specialization as it grows older.

Polygyny. The coexistence in the same colony of two or more egg-laying queens. When multiple queens found a colony together, the condition is referred to as primary polygyny. When supplementary queens are added after colony foundation, the condition is referred to as secondary polygyny. The coexistence of only two or several queens is sometimes called oligogyny (opposed to monogyny).

Polymorphism. In social insects, the coexistence of two or more functionally different castes within the same sex. In ants it is possible to define polymorphism somewhat more precisely as the occurrence of nonisometric relative growth occurring over a sufficient range of size variation within a normal mature colony to produce individuals of distinctly different proportions at the extremes of the size range.

Poneroid complex. One of the two major taxonomic groups of ants; the name is based on the subfamily Ponerinae, one of the constituent taxa (see Chapter 4).

Postpetiole. In certain ants, the second segment of the "waist." This is in fact the third abdominal segment, since the first abdominal segment (propodeum) is fused to the thorax.

Pouch-makers. Bumblebee species that build special wax pouches adjacent to groups of larvae and fill them with pollen (opposed to pollen-storers).

Presocial. Applied to the condition or to the group possessing it in which individuals display some degree of social behavior short of eusociality. Presocial species are either subsocial, *i.e.,* the parents care for their own nymphs and larvae, or else parasocial, *i.e.,* one or two of the following three traits are shown: cooperation in care of young, reproductive division of labor, overlap of generations of life stages that contribute to colony labor.

Pretarsus. The terminal segment of the leg, consisting usually of a pair of lateral claws and the arolium.

Primary reproductive. In termites, the colony-founding type of queen or male derived from the winged adult.

Proctodeum. The hindgut and Malpighian tubules of an insect.

Progressive provisioning. The act of providing the larva with meals at intervals during its development (opposed to mass provisioning).

Propodeum. In higher Hymenoptera, the first abdominal segment when it is fused with the alitrunk. Same as epinotum.

Propolis. A collective term for the resins and waxes collected by bees and brought to their nests for use in construction and in sealing fissures in the nest wall.

Pseudergate. A special caste found in the lower termites, comprised of individuals who have either regressed from nymphal stages by molts that reduced or eliminated the wing buds, or else were derived from larvae by undergoing nondifferentiating molts. Pseudergates serve as the principal elements of the "worker" caste, but remain capable of developing into other castes by further molting.

Pterothorax. The wing-bearing portion of the thorax of winged insects, *i.e.,* the meso- and metathorax.

Pupa. The inactive instar of the holometabolous insects (including the Hymenoptera) during which development into the final adult form is completed.

Quacking. The calls emitted by virgin honeybee queens in their cells in response to the "piping" sounds of the first virgin queen to emerge in the same hive.

Quasisocial. Applied to the condition or to the group showing it in which members of the same generation use the same composite nest and cooperate in brood care.

Queen. A member of the reproductive caste in semisocial or eusocial species. The existence of a queen caste presupposes the existence also of a worker caste at some stage of the colony life cycle. Queens may or may not be morphologically different from workers.

Queen control. The inhibitory influence of the queen on the reproductive activities of the workers and other queens.

Queenright. Referring to a colony, especially a honeybee colony, that contains a functional queen.

Queen substance. Originally, the set of pheromones by which the queen honeybee continuously attracts and controls the reproductive activities of the workers. Nowadays the term is commonly used in a narrower sense to designate *trans*-9-keto-2-decenoic acid, the most potent component of the pheromone mixture. In the present book it is suggested that it be defined more broadly in line with the original usage as any pheromone or set of pheromones used by a queen to control the reproductive behavior of the workers or other queens.

Recruitment trail. An odor trail laid by a single scout worker and used to recruit nestmates to a food find, a desirable new nest site, a breach in the nest wall, or some other place where the assistance of many workers is needed (opposed to exploratory trail).

Relative growth. The relative increase of one body part with respect to another as total body size is varied. Allometry (*q.v.*) is a special form of relative growth.

Replacement reproductive. Same as supplementary reproductive.

Replete. An individual ant whose crop is greatly distended with liquid food, to the extent that the abdominal segments are pulled apart and the intersegmental membranes are stretched tight. Repletes usually serve as living reservoirs, regurgitating food on demand to their nestmates.

Retinue. A group of workers, not necessarily permanent or even long lasting in composition, who closely attend the queen.

Royal cell. In honeybees, the large, pitted, waxen cell constructed by the workers to rear queen larvae. In some species of termites, the special cell in which the queen is housed.

Royal jelly. A material supplied by workers to female larvae in royal cells which is necessary for the transformation of larvae into queens. Royal jelly is secreted primarily by the hypopharyngeal glands and consists of a rich mixture of nutrient substances, many of them possessing a complex chemical structure (see worker jelly).

Sclerite. A portion of the body wall bounded by sutures.

Secondary reproductive. In termites, the same as nymphoid reproductive.

Second-form reproductive. In termites, the same as nymphoid reproductive.

Self-grooming. See grooming.

Semisocial. Applied to the condition or to the group showing it in which members of the same generation cooperate in brood care and there is also a reproductive division of labor, *i.e.,* some individuals are primarily egg layers and some are primarily workers.

Sensillum. In insects, a simple sense organ or one of the structural units of a compound sense organ.

Slavery. Same as dulosis.

Social facilitation. See under group effect.

Social homeostasis. The maintenance of steady states at the level of the society either by control of the nest microclimate or by the regulation of the population density, behavior, and physiology of the group members as a whole.

Social insect. In the strict sense, a "true social insect" is one that belongs to a eusocial species; in other words, it is an ant, a termite, or one of the eusocial wasps or bees. In the broad sense, a "social insect" is one that belongs to either a presocial or eusocial species.

Social parasitism. The coexistence in the same nest of two species of social insects, of which one is parasitically dependent on the other. The term can also be applied loosely to the relation between symphiles and their social insect hosts.

Society. A group of individuals belonging to the same species and organized in a cooperative manner. Some amount of reciprocal communication among the members is implied.

Sociobiology. The study of all aspects of communication and social organization.

Sociotomy. Same as colony fission.

Soldier. A member of a worker subcaste specialized for colony defense.

Species odor. The odor found on the bodies of social insects which is peculiar to a given species. It is possible that the species odor is merely the less distinctive components of a larger mixture comprising the colony odor (*q.v.*).

Sphecology. The scientific study of wasps.

Sphecophile. An organism that must spend at least part of its life cycle with wasp colonies.

Staphyla. A group of gongylidia, the swollen hyphal tips produced by fungi that live in symbiosis with attine ants.

Statary phase. The period in the activity cycle of an army ant colony during which the colony is relatively quiescent and does not move from site to site. At this time the queen lays the eggs and the bulk of the brood is in the egg and pupal stages (opposed to nomadic phase).

Stelocyttarous. Pertaining to nests, and especially wasp nests, in which the combs are attached to the support by pillars.

Sternite. A ventral sclerite; in other words, a portion of the body wall bounded by sutures and located in a ventral position (opposed to tergite).

Stigmergy. The guidance of work performed by individual colony members by the evidences of work previously accomplished rather than by direct signals from nestmates.

Stingless bee. A bee belonging to the apine tribe Meliponini.

Stochastic theory of mass behavior. The theory that transition probabilities in the behavior of individual social insects are programmed to produce optimal mass responses of the colony as a whole, that the probabilities have been determined by selection at the colony level, and that they represent a sensitive adaptation to the particular environmental conditions in which the species has existed during recent evolutionary time.

Stomodeum. The foregut of an insect.

Storage pots. Containers made of soft cerumen for the storage of food in the nests of social bees. Some of the pots constructed by meliponine bees contain only honey (honey pots) and others only pollen (pollen pots).

Straight run. The middle run made by a honeybee worker during the waggle dance and the element that contains most of the symbolical information concerning the location of the target outside the hive. The dancing bee makes a straight run, then loops back to the left (or right), then makes another straight run, then loops back in the opposite direction, and so on—the three basic movements together form the characteristic figure-eight pattern of the waggle dance.

Stridulation. The production of sound by rubbing one part of the body surface against another.

Strigilation. The licking of secretions from the body of another animal.

Subsocial. Applied to the condition or to the group showing it in which the adults care for their nymphs or larvae for some period of time (see also presocial).

Superorganism. Any society, such as the colony of a eusocial insect species, possessing features of organization analogous to the physiological properties of a single organism. The insect colony, for example, is divided into reproductive castes (analogous to gonads) and worker castes (analogous to somatic tissue); it may exchange nutrients by trophallaxis (analogous to the circulatory system), and so forth.

Supersedure. In honeybees, the replacement of the resident queen, usually an old or sickly individual, with a new queen reared by the workers. The process is distinct from colony multiplication by swarming.

Supplementary reproductive. A queen or male termite that takes over as the functional reproductive after the removal of the primary reproductive of the same sex. Supplementary reproductives are adultoid, nymphoid, or ergatoid in form (*q.v.*).

Surface pheromone. A pheromone with an active space restricted so close to the body of the sending organism that direct contact, or something approaching it, must be made with the body in order to perceive the pheromone. Examples include the colony odors of many species.

Swarming. In honeybees, the normal method of colony reproduction, in which the queen and a large number of workers depart suddenly from the parental nest and fly to some exposed site. There they cluster while scout workers fly in search of a suitable new nest cavity. In ants and termites, the term "swarming" is often applied to the mass exodus of reproductive forms from the nests at the beginning of the nuptial flight.

Symbiont. An organism that lives in symbiosis with another species.

Symbiosis. The intimate, relatively protracted, and dependent relationship of members of one species with those of another. The three principal kinds of symbiosis are commensalism, mutualism, and parasitism (*q.v.*).

Symphile. A symbiont, in particular a solitary insect or other kind of arthropod, which is accepted to some extent by an insect colony and communicates with it amicably. Most symphiles are licked, fed, or transported to the host brood chambers, or treated to a combination of these actions.

Synechthran. A symbiont, usually a scavenger, a parasite, or a predator, which is treated with hostility by the host colony.

Synoekete. A symbiont that is treated with indifference by the host colony.

Tandem running. A form of communication, used by the workers of certain ant species during exploration or recruitment, in which one individual follows closely behind another, frequently contacting the abdomen of the leader with its antennae.

Tarsation. Touching with the tarsi, especially the touching of another insect as a tactile signal.

Tarsus. The foot of an insect, the one- to five-segmented appendage attached to the tibia, or lower leg segment.

Temporal polyethism. Same as age polyethism (*q.v.*).

Temporary social parasitism. Parasitism in which a queen of one species enters an alien nest, usually belonging to another species, kills or renders infertile the resident queen, and takes her place. The population of the colony then becomes increasingly dominated by the offspring of the parasite queen as the host workers die off from natural causes.

Tergite. A dorsal sclerite; in other words, a portion of the body wall bounded by sutures and located in a dorsal position (opposed to sternite).

Termitarium. A termite nest. Also, an artificial nest used in the laboratory to house termites.

Termitology. The scientific study of termites.

Termitophile. An organism that must spend at least part of its life cycle with termite colonies.

Tertiary reproductive. In termites, the same as ergatoid reproductive.

Thelytoky. The production of females from unfertilized eggs.

Third-form reproductive. In termites, the same as ergatoid reproductive.

Trail parasitism. See trophic parasitism.

Trail pheromone. A substance laid down in the form of a trail by one animal and followed by another member of the same species.

Trail substance. Same as trail pheromone.

Trichome. A tuft of long, often yellow or golden hairs associated with glandular areas on the body surfaces of many myrmecophilous beetles. The hairs are believed to aid in the dissemination of attractants.

Triphasic allometry. Polymorphism in which the allometric regression line, when plotted on a double logarithmic scale, "breaks" at two points and consists of three segments. In ants, the two terminal segments usually have slight to moderately high slopes and the middle segment has a very high slope (see Chapter 8).

Trophallaxis. The exchange of alimentary liquid among colony members and guest organisms, either mutually or unilaterally. In stomodeal trophallaxis the material originates from the mouth; in proctodeal trophallaxis it originates from the anus.

Trophic egg. An egg, usually degenerate in form and inviable, which is fed to other members of the colony.

Trophic parasitism. The intrusion of one species into the social system of another, as, for example, by utilization of the trail system, in order to steal food.

Trophogenesis. The origin of different caste traits from differential feeding of the immature stages (opposed to genetic control of castes and blastogenesis).

Unicolonial. Pertaining to a population of social insects in which there are no behavioral colony boundaries (opposed to multicolonial).

Waggle dance. The dance whereby workers of various species of honeybees (genus *Apis*) communicate the location of food finds and new nest sites. The dance is basically a run through a figure-eight pattern, with the middle, transverse line of the eight containing the information about the direction and distance of the target (see Chapter 13).

Worker. A member of the nonreproductive, laboring caste in semisocial and eusocial species. The existence of a worker caste presupposes the existence also of royal (reproductive) castes. In termites, the term is used in a more restricted sense to designate individuals in the family Termitidae which completely lack wings and have reduced pterothorax, eyes, and genital apparatus.

Worker jelly. A secreted material supplied by workers to larvae in regular brood cells that causes the larvae to develop into workers (see royal jelly).

Xenobiosis. The relation in which colonies of one species live in the nests of another species and move freely among the hosts, obtaining food from them by regurgitation or other means but still keeping their brood separate.

Yellow jacket (*yellow-jacket* or *yellowjacket*). In the United States, any one of a number of ground-nesting, light-colored wasps of the genus *Vespula.*

Bibliography

Abushama, F. T., 1966. Electrophysiological investigations on the antennal olfactory receptors of the damp-wood termite *Zootermopsis angusticollis*. *Entomologia Experimentalis et Applicata,* 9(3): 343–348.

Adam, Brother, 1951. Introduction of queens. *Proceedings of the 14th International Beekeeping Congress,* 10: 1–5.

Adlerz, G., 1884. Myrmecologiska studier. I. *Formicoxenus nitidulus* Nyl. *Öfversigt af Kongl. Vetenskaps-Akademiens Förhandlingar,* 8: 43–64.

———, 1886. Myrmecologiska studier. II. Svenska myror och deras lefnadsförhå llanden. *Bihang Till K. Svenska Vetenskaps-Akademiens Handlingar,* 11(18): 1–320.

———, 1896. Myrmecologiska studier. III. *Tomognathus sublaevis* Mayr. *Bihang Till K. Svenska Vetenskaps-Akademiens Handlingar,* 21(4): 1–76.

Ahmad, M., 1950. The phylogeny of termite genera based on imago-worker mandibles. *Bulletin of the American Museum of Natural History,* 95(2): 37–86.

Ahrens, W., 1934. Zur Kenntnis der Homologien akzessorischer Geschlechtsdrüsen bei Insekten. *Zoologischer Anzeiger,* 108(7–8): 187–195.

Akre, R. D., 1968. The behavior of *Euxenister* and *Pulvinister,* histerid beetles associated with army ants. *Pan-Pacific Entomologist,* 44(2): 87–101.

———, and C. W. Rettenmeyer, 1966. Behavior of Staphylinidae associated with army ants (Formicidae: Ecitonini). *Journal of the Kansas Entomological Society,* 39(4): 745–782.

———, and C. W. Rettenmeyer, 1968. Trail-following by guests of army ants (Hymenoptera: Formicidae: Ecitonini). *Journal of the Kansas Entomological Society,* 41(2): 165–174.

———, and R. L. Torgerson, 1968. The behavior of *Diploeciton nevermanni,* a staphylinid beetle associated with army ants. *Psyche, Cambridge,* 75(3): 211–215.

Alber, M. A., 1956. Multiple mating. *British Bee Journal,* 83: 134–135, 84: 6–7, 18–19.

Alexander, R. D., 1961. Aggressiveness, territoriality, and sexual behavior in field crickets (Orthoptera: Gryllidae). *Behaviour,* 17(2–3): 130–223.

———, T. E. Moore, and R. E. Woodruff, 1963. The evolutionary differentiation of stridulatory signals in beetles (Insecta: Coleoptera). *Animal Behaviour,* 11(1): 111–115.

Alibert, Mme J., 1968. Influence de la société et de l'individu sur la trophallaxie chez *Calotermes flavicollis* Fabr. et *Cubitermes fungifaber* (Isoptera). *Colloques Internationaux du Centre National de la Recherche Scientifique, Paris, 1967,* No. 173: 237–288.

Allee, W. C., 1931. *Animal aggregations: A study in general sociology.* University of Chicago Press, Chicago. ix + 431 pp.

———, 1938. *The social life of animals.* W. W. Norton and Co., Inc., New York. 293 pp.

Allen, M. Delia, 1959. The "shaking" of worker honeybees by other workers. *Animal Behaviour,* 7(3–4): 233–240.

———, 1965. The role of the queen and males in the social organization of insect communities. *Symposium of the Zoological Society of London,* 14: 133–157.

Allport, F. H., 1924. *Social psychology.* Houghton Mifflin Co., Boston. xiv + 453 pp.

Anderson, E. J., 1948. Hive humidity and its effect upon wintering of bees. *Journal of Economic Entomology,* 41(4): 608–616.

Andrews, E. A., 1911. Observations on termites in Jamaica. *Journal of Animal Behavior,* 1(3): 193–228.

———, 1927. Ant-mounds as to temperature and sunshine. *Journal of Morphology and Physiology,* 44(1): 1–20.

———, 1930. Honeydew reflexes. *Physiological Zoölogy,* 3(4): 467–484.

Andries, Maria, 1912. Zur Systematik, Biologie und Entwicklung von *Microdon* Meigen. *Zeitschrift für Wissenschaftliche Zoologie,* 103(2): 300–361.

Arnett, R. H., Jr., 1964. Notes on Karumiidae (Coleoptera). *Coleopterists' Bulletin,* 18(3): 65–68.

Arnold, G., 1915–1926. A monograph of the Formicidae of South Africa [6 pts., plus an appendix]. *Annals of the South African Museum,* 14(1–6): 1–766, 23(2): 191–295.

———, 1947. A key to the African genera of the Apidae. *Journal of the Entomological Society of Southern Africa,* 9(2): 193–218.

Arnoldi, K. V., 1930. Studien über die Systematik der Ameisen. VI. Eine neue parasitische Ameise, mit Bezugnahme auf die Frage nach der Entstehung der Gattungsmerkmale bei den parasitären Ameisen. *Zoologischer Anzeiger,* 91(9–12): 267–283.

———, 1932. Biologische Beobachtungen an der neuen paläarktischen Sklavenhalterameise *Rossomyrmex proformicarum* K. Arn., nebst einigen Bemerkungen über die Beförderungsweise der Ameisen. *Zeitschrift für Morphologie und Ökologie der Tiere,* 24(2): 319–326.

———, 1968. Important additions to the myrmecofauna (Hymenoptera, Formicidae) of the USSR, with some new descriptions [in Russian]. *Zoologichyeskiy Zhurnal, Moscow,* 47(12): 1800–1822.

Ashmole, N. P., 1963. The regulation of numbers of tropical oceanic birds. *Ibis,* 103b: 458–473.

Auclair, J. L., 1963. Aphid feeding and nutrition. *Annual Review of Entomology,* 8: 439–490.

Autrum, H., 1936. Über Lautäusserungen und Schallwahrnehmung bei Arthropoden. I. Untersuchungen an Ameisen. Eine allgemeine Theorie der Schallwahrnehmung bei Arthropoden. *Zeitschrift für Vergleichende Physiologie,* 23(3): 332–373.

———, 1959. Nonphotic receptors in lower forms. *In* J. Field, H. W. Magoun, V. E. Hall, eds., *Handbook of Physiology, Section 1: Neurophysiology, Vol. I.* American Physiological Society, Washington, D.C. pp. 369–385.

———, and W. Schneider, 1948. Vergleichende Untersuchungen über den Erschütterungssinn der Insekten. *Zeitschrift für Vergleichende Physiologie,* 31(1): 77–88.

———, and Marieluise Stoecker, 1950. Die Verschmelzungsfrequenzen des Bienenauges. *Zeitschrift für Naturforschung,* Abt. B, 5(1): 38–43.

———, and Vera von Zwehl, 1964. Die spektrale Empfindlichkeit einzelner Sehzellen des Bienenauges. *Zeitschrift für Vergleichende Physiologie,* 48(4): 357–384.

Autuori, M., 1956. La fondation des sociétés chez les fourmis champignonnistes du genre "Atta" (Hym. Formicidae). *In* M. Autuori *et al., L'instinct dans le comportement des animaux et de l'homme.* Masson et Cie, Paris. pp. 77–104.

Ayre, G. L., 1959. Food habits of *Formica subnitens* Creighton (Hymenoptera: Formicidae) at Westbank, British Columbia. *Insectes Sociaux,* 6(2): 105–114.

———, 1960. Der Einfluss von Insektennahrung auf das Wachstum von Waldameisenvölkern. *Naturwissenschaften,* 47(21): 502–503.

———, 1963. Response to movement by *Formica polyctena* Forst. *Nature, London,* 199: 405–406.

Ayyar, P. N. K., 1937. A new carton-building species of ant in South India, *Crematogaster dohrni artifex,* Mayr. *Journal of the Bombay Natural History Society,* 39(2): 291–308.

Baerg, W. J., 1921. Notes on the nest and the population of a colony of *Vespa diabolica. Journal of Economic Entomology,* 14(6): 509–510.

Baker, H. G., and G. L. Stebbins, eds., 1965. *The genetics of colonizing species.* Academic Press, New York. xv + 588 pp.

Banck, L. J., 1927. *Contributions to myrmecophily. I. An anatomical-histological and experimental-biological study of Thorictus foreli Wasm.* Saint-Paul, Fribourg. 83 pp. [Cited by Plath, 1935].

Barbier, M., 1968. Biochimie de l'abeille. *In* R. Chauvin, ed., *Traité de biologie de l'abeille, Vol. I.* Masson et Cie, Paris. pp. 378–409.

———, and E. Lederer, 1960. Structure chimique de la 'substance royale' de la reine d'abeille (*Apis mellifica L.*). *Compte Rendu de l'Académie des Sciences, Paris,* 250(26): 4467–4469.

———, E. Lederer, and T. Nomura, 1960. Synthèse de l'acide céto-9 décène-2-trans oïque ('substance royale') et de l'acide céto-8 nonène-2-trans oïque, *Compte Rendu de l'Académie des Sciences, Paris,* 251(10): 1133–1135.

Baroni Urbani, C., 1965. Sull'attivata' di foraggiamento notturna del *Camponotus nylanderi* Em. *Insectes Sociaux,* 12(3): 253–263.

———, 1968. Domination et monogynie fonctionnelle dans une société digynique de *Myrmecina graminicola* Latr. *Insectes Sociaux,* 15(4): 407–411.

———, and J. Soulié, 1962. Monogynie chez la fourmi *Cremastogaster scutellaris* [Hymenoptera Formicoidea]. *Bulletin de la Société d'Histoire Naturelle de Toulouse,* 97(1–2): 29–34.

Barr, Barbara A., 1969. Sound production in Scolytidae (Coleoptera) with emphasis on the genus *Ips. Canadian Entomologist,* 101(6): 636–672.

Basilewsky, P., 1952. Les Cossyphodidae de l'Afrique Noire (Coleoptera Heteromeroidea Tenebrionaria). *Publicações Culturais da Companhia de Diamantes de Angola,* 14: 7–16.

Bassindale, R., 1955. The biology of the stingless bee *Trigona* (*Hypotrigona*) *gribodoi* Magretti (Meliponidae). *Proceedings of the Zoological Society of London,* 125(1): 49–62.

Bates, H. W., 1863. *The naturalist on the River Amazons,* 2 vols. John Murray, London. ix + 351, vi + 423.

Batra, L. R., 1963. Ecology of ambrosia fungi and their dissemination by beetles. *Transactions of the Kansas Academy of Science,* 66(2): 213–236.

———, 1966. Ambrosia fungi: Extent of specificity to ambrosia beetles. *Science,* 153: 193–195.

Batra, Suzanne W. T., 1964. Behavior of the social bee, *Lasioglossum zephyrum,* within the nest (Hymenoptera: Halictidae). *Insectes Sociaux,* 11(2): 159–185.

———, 1966a. The life cycle and behavior of the primitively social bee, *Lasioglossum zephyrum* (Halictidae). *Kansas University Science Bulletin,* 46(10): 359–422.

———, 1966b. Nesting behavior of *Halictus scabiosae* in Switzerland (Hymenoptera, Halictidae). *Insectes Sociaux,* 13(2): 87–92.

———, 1966c. Nests and social behavior of halictine bees of India (Hymenoptera: Halictidae). *Indian Journal of Entomology,* 28(3): 375–393.

———, 1968. Behavior of some social and solitary halictine

bees within their nests: A comparative study (Hymenoptera: Halictidae). *Journal of the Kansas Entomological Society,* 41(1): 120–133.

Baumgärtner, H., 1928. Der Formensinn und die Sehschärfe der Bienen. *Zeitschrift für Vergleichende Physiologie,* 7: 56–143.

Bazire-Benazet, Madeleine, 1957. Sur la formation de l'oeuf alimentaire chez *Atta sexdens rubropilosa* Forel, 1908 (Hym. Formicidae). *Compte Rendu de l'Académie des Sciences, Paris,* 244(9): 1277–1280.

Beamer, R. H., 1930. Maternal instinct in a Membracid (*Platycotis vittata*) (Homop.). *Entomological News,* 41(10): 330–331.

Beaumont, J. de, 1958. Le parasitisme social chez les Guêpes et les Bourdons. *Mitteilungen der Schweizerischen Entomologischen Gesellschaft,* 31(2): 168–176.

———, and R. Matthey, 1945. Observations sur les *Polistes* parasites de la Suisse. *Bulletin de la Société Vaudoise des Sciences Naturelles,* 62(263): 439–454.

Beck, H., 1961. Vergleichende Untersuchungen ueber einige Verhaltensweisen von *Polyergus rufescens* Latr. und *Raptiformica sanguinea* Latr. *Insectes Sociaux,* 8(1): 1–11.

Becker, G., 1969. Rearing of termites and testing methods used in the laboratory. *In* K. Krishna and F. M. Weesner, eds., *Biology of termites, Vol. I.* Academic Press, New York. pp. 351–385.

Bee Research Association, 1967. Bibliography on honeybees other than *Apis mellifera. Bee World,* 48(1): 8–15, 18.

Beebe, W., 1919. The home town of the army ants. *The Atlantic Monthly,* 124(4): 454–464.

Beeson, C. F. C., 1941. *The ecology and control of forest insects of India and neighboring countries.* Dehra Dun, India. 1007 pp.

Beig, D., and S. F. Sakagami, 1964. Behavior studies of the stingless bees, with special reference to the oviposition process, II. *Melipona seminigra merrillae* Cockerell. *Annotationes Zoologicae Japonenses,* 37(2): 112–119.

Beling, Ingeborg, 1929. Über das Zeitgedächtnis der Bienen. *Zeitschrift für Vergleichende Physiologie,* 9(2–3): 259–338.

Belt, T., 1874. *The naturalist in Nicaragua.* John Murray, London. xvi + 403 pp.

Bequaert, J., 1918. A revision of the Vespidae of the Belgian Congo based on the collection of the American Museum Congo Expedition, with a list of Ethiopian Diplopterous wasps. *Bulletin of the American Museum of Natural History,* 39(1): 1–384.

———, 1922. Ants of the American Museum Congo Expedition. A contribution to the myrmecology of Africa. IV. Ants in their diverse relations to the plant world. *Bulletin of the American Museum of Natural History,* 45(1): 333–583.

———, 1928. A study of certain types of Diplopterous wasps in the collection of the British Museum. *Annals and Magazine of Natural History,* (10)2(7): 138–176.

———, 1933. The Nearctic social wasps of the subfamily Polybiinae (Hymenoptera; Vespidae). *Entomologica Americana* (n.s.), 13(3): 87–148.

———, 1935. Presocial behavior among the Hemiptera. *Bulletin of the Brooklyn Entomological Society,* 30(5): 177–191.

———, 1938. A new *Charterginus* from Costa Rica, with notes on *Charterginus, Pseudochartergus, Chartergus, Pseudopolybia, Epipona* and *Tatua* (Hymenoptera, Vespidae). *Revista de Entomologia, Rio de Janeiro,* 9(1–2): 99–117.

———, 1940. An introductory study of *Polistes* in the United States and Canada with descriptions of some new North and South American forms (Hymenoptera; Vespidae). *Journal of the New York Entomological Society,* 48(1): 1–31.

———, 1944a. The social Vespidae of the Guianas, particularly of British Guiana. *Bulletin of the Museum of Comparative Zoology, Harvard,* 94(7): 249–304.

———, 1944b. A revision of *Protopolybia* Ducke, a genus of neotropical social wasps (Hymenoptera, Vespidae). *Revista de Entomologia, Rio de Janeiro,* 15(1–2): 97–134.

Bergström, G., and J. Löfqvist, 1968. Odour similarities between the slave-keeping ants *Formica sanguinea* and *Polyergus rufescens* and their slaves *Formica fusca* and *Formica rufibarbis. Journal of Insect Physiology,* 14(7): 995–1011.

Berland, L., 1942. Les *Polistes* de France [Hym. Vespidae]. *Annales de la Société Entomologique de France,* 111: 135–148.

Bernard, F., 1951. Adaptations au milieu chez les fourmis sahariennes. *Bulletin de la Société d'Histoire Naturelle de Toulouse,* 86(1–2): 88–96.

———, 1968. *Les fourmis (Hymenoptera Formicidae) d'Europe occidentale et septentrionale. Faune de l'Europe et du Bassin Méditerranéen, No. 3.* Masson et Cie, Paris. 411 pp.

Bernstein, S., and Ruth A. Bernstein, 1969. Relationship between foraging efficiency and the size of the head and component brain and sensory structures in the red wood ant. *Brain Research,* 16(1): 85–104.

Bethe, A., 1898. Dürfen wir den Ameisen und Bienen psychische Qualitäten zuschreiben? *Pflügers Archiv für die Gesamte Physiologie,* 70: 15–100.

Betrem, J. G., 1960. Ueber die Systematik der *Formica rufa*-Gruppe. *Tijdschrift voor Entomologie,* 103(1–2): 51–81.

Betz, Barbara J., 1932. The population of a nest of the hornet *Vespa maculata. Quarterly Review of Biology,* 7(2): 197–209.

Bevan, C. W. L., A. J. Birch, and H. Caswell, 1961. An insect repellant from black cocktail ants. *Journal of the Chemical Society,* Pt. 1: 488.

Bier, K., 1952. Zur scheinbaren Thelytokie der Ameisengattung *Lasius. Naturwissenschaften,* 39(18): 433.

———, 1953. Beziehungen zwischen Nährzellkerngrösse und Ausbildung ribonukleinsäurehaltiger Strukturen in den Oocyten von *Formica rufa rufo-pratensis minor* Gösswald. *Verhandlungen der Deutschen Zoologischen Gesellschaft, Freiburg, 1952,* pp. 369–374.

———, 1954a. Über den Saisondimorphismus der Oogenese von *Formica rufa rufo-pratensis minor* Gössw. und dessen Bedeutung für die Kastendetermination. *Biologisches Zentralblatt,* 73(3–4): 170–190.
———, 1954b. Über den Einfluss der Königin auf die Arbeiterinnenfertilität im Ameisenstaat. *Insectes Sociaux,* 1(1): 7–19.
———, 1956. Arbeiterinnenfertilität und Aufzucht von Geschlechtstieren als Regulationsleistung des Ameisenstaates. *Insectes Sociaux,* 3(1): 177–184.
———, 1958. Die Regulation der Sexualität in den Insektenstaaten. *Ergebnisse der Biologie,* 20: 97–126.
Bingham, C. T., 1897. *Hymenoptera. Vol. I. Wasps and bees. In* W. T. Blanford, ed., *The fauna of British India, including Ceylon and Burma.* Taylor and Francis, London. xxix + 579 pp.
———, 1903. *Hymenoptera. Vol. II. Ants and cuckoo-wasps. In* W. T. Blanford, ed., *The fauna of British India, including Ceylon and Burma.* Taylor and Francis, London. xix + 506 pp.
———, 1907. *Butterflies, Vol. II. In* C. T. Bingham, ed., *The fauna of British India, including Ceylon and Burma.* Taylor and Francis, London. viii + 480 pp.
Birukow, G., 1954. Photo-Geomenotaxis bei *Geotrupes silvaticus* Panz. und ihre zentralnervöse Koordination. *Zeitschrift für Vergleichende Physiologie,* 36(2): 176–211.
Bitancourt, A. A., 1941. Expressão matematica do crescimento de formigueiros de "*Atta sexdens rubropilosa*" representado pelo aumento do numero de olheiros. *Archivos do Instituto Biológica, São Paulo,* 12(16): 229–236.
Blackith, R. E., 1957. The analysis of social facilitation at the nest entrance of some Hymenoptera. *Physiologia Comparata et Oecologia,* 4: 388–402.
———, 1958a. Visual sensitivity and foraging in social wasps. *Insectes Sociaux,* 5(2): 159–169.
———, 1958b. An analysis of polymorphism in social wasps. *Insectes Sociaux,* 5(3): 263–272.
Blatter, E., 1928. Myrmecosymbiosis in the Indo-Malayan flora. *Journal of the Indian Botanical Society,* 7(3–4): 176–185.
Blest, A. D., 1960. The evolution, ontogeny and quantitative control of the settling movements of some New World saturniid moths, with some comments on distance communication by honey-bees. *Behaviour,* 16(3–4): 188–253.
Blum, M. S., 1966a. The source and specificity of trail pheromones in *Termitopone, Monomorium* and *Huberia,* and their relation to those of some other ants. *Proceedings of the Royal Entomological Society of London,* 41(10–12): 155–160.
———, 1966b. Chemical releasers of social behavior. VIII. Citral in the mandibular gland secretion of *Lestrimelitta limao* (Hymenoptera: Apoidea: Melittidae). *Annals of the Entomological Society of America,* 59(5): 962–964.
———, 1969. Alarm pheromones. *Annual Review of Entomology,* 14: 57–80.
———, and C. A. Portocarrero, 1964. Chemical releasers of social behavior. IV. The hindgut as the source of the odor trail pheromone in the neotropical army ant genus *Eciton. Annals of the Entomological Society of America,* 57(6): 793–794.
———, and G. N. Ross, 1965. Chemical releasers of social behaviour. V. Source, specificity, and properties of the odour trail pheromone of *Tetramorium guineense* (F.) (Formicidae: Myrmicinae). *Journal of Insect Physiology,* 11(7): 857–868.
———, and E. O. Wilson, 1964. The anatomical source of trail substances in formicine ants. *Psyche, Cambridge,* 71(1): 28–31.
———, J. C. Moser, and A. D. Cordero, 1964. Chemical releasers of social behavior. II. Source and specificity of the odor trail substances in four attine genera (Hymenoptera: Formicidae). *Psyche, Cambridge,* 71(1): 1–7.
———, F. Padovani, and E. Amante, 1968. Alkanones and terpenes in the mandibular glands of *Atta* species (Hymenoptera: Formicidae). *Comparative Biochemistry and Physiology,* 26(1): 291–299.
———, S. L. Warter, and J. G. Traynham, 1966. Chemical releasers of social behaviour—VI. The relation of structure to activity of ketones as releasers of alarm for *Iridomyrmex pruinosus* (Roger). *Journal of Insect Physiology,* 12(4): 419–427.
———, R. M. Crewe, J. H. Sudd, and A. W. Garrison, 1969. 2-Hexenal: Isolation and function in a *Crematogaster* (*Atopogyne*) sp. *Journal of the Georgia Entomological Society,* 4(4): 145–148.
Blüthgen, P., 1926, 1931. Beiträge zur Kenntnis der indomalayischen Halictus- und Thrinchostoma-Arten. (Hym. Apidae, Halictini). *Zoologische Jahrbücher, Abt. Syst., Ökol. Geogr.,* 51(4–6): 376–698, 61(3): 285–346.
———, 1961. Die Faltenwespen Mitteleuropas (Hymenoptera, Diploptera). *Abhandlungen der Deutschen Akademie der Wissenschaften zu Berlin,* 2: 1–251.
Boch, R., 1957. Rassenmässige Unterschiede bei den Tänzen der Honigbiene (*Apis mellifica* L.). *Zeitschrift für Vergleichende Physiologie,* 40(3): 289–320.
———, and D. A. Shearer, 1962. Identification of geraniol as the active component in the Nassanoff pheromone of the honey bee. *Nature, London,* 194: 704–706.
———, and D. A. Shearer, 1964. Identification of nerolic and geranic acids in the Nassanoff pheromone of the honey bee. *Nature, London,* 202: 320–321.
———, D. A. Shearer, and B. C. Stone, 1962. Identification of iso-amyl acetate as an active component in the sting pheromone of the honey bee. *Nature, London,* 195: 1018–1020.
Bodenheimer, F. S., 1937. Population problems of social insects. *Biological Reviews* (Cambridge Philosophical Society), 12(4): 393–430.
Bodot, Paulette, 1964. Études écologiques et biologiques des termites dans les savanes de Basse Côte d'Ivoire. *In* A. Bouillon, ed., *Études sur les termites africains.* Masson et Cie, Paris. pp. 252–262.

———, 1969. Composition des colonies de termites: ses fluctuations au cours du temps. *Insectes Sociaux,* 16(1): 39–53.

Boeckh, J., 1967. Reaktionsschwelle, Arbeitsbereich und Spezifität eines Geruchsrezeptors auf der Heuschreckenantenne. *Zeitschrift für Vergleichende Physiologie,* 55(4): 378–406.

———, K. E. Kaissling, and D. Schneider, 1965. Insect olfactory receptors. *Cold Spring Harbor Symposia on Quantitative Biology,* 30: 263–280.

Bohart, R. M., 1949. Notes on North American *Polistes* with descriptions of new species and subspecies (Hymenoptera, Vespidae). *Pan-Pacific Entomologist,* 25(3): 97–103.

Bolivar, I., 1905. Les blattes myrmécophiles. *Mitteilungen der Schweizerischen Entomologischen Gesellschaft,* 11(3): 134–141.

Bonnet, C., 1779. Observation XLIII. Sur un procédé des fourmis. *Oeuvres d'Histoire Naturelle et de Philosophie,* 1: 535–536.

Borgmeier, T., 1949. Formigas novas ou pouco conhecidas de Costa Rica e da Argentina (Hymenoptera, Formicidae). *Revista Brasileira de Biologia,* 9(2): 201–210.

———, 1955. Die Wanderameisen der Neotropischen Region (Hym. Formicidae). *Studia Entomologica,* 3: 1–716.

———, 1959. Revision der Gattung *Atta* Fabricius (Hymenoptera, Formicidae). *Studia Entomologica* (n.s.), 2(1–4): 321–390.

———, 1961. Weitere Beitraege zur Kenntnis der neotropischen Phoriden, nebst Beschreibung einiger *Dohrniphora*-Arten aus der indo-australischen Region (Diptera, Phoridae). *Studia Entomologica* (n.s.), 4(1–4): 1–112.

———, 1962. Versuch einer Uebersicht ueber die neotropischen Megaselia-Arten, sowie neue oder wenig bekannte Phoriden verschiedener Gattungen (Diptera, Phoridae). *Studia Entomologica* (n.s.), 5(1–4): 289–488.

———, 1963. Revision of the North American phorid flies. Pt. I. The Phorinae, Aenigmatiinae and Metopininae, except *Megaselia* (Diptera, Phoridae). *Studia Entomologica* (n.s.), 6(1–4): 1–256.

———, 1964a. A generic revision of the Termitoxeniinae, with the description of a new species from Burma (Diptera, Phoridae). *Studia Entomologica* (n.s.), 7(1–4): 73–95.

———, 1964b. Revision of the North American phorid flies. Pt. II. The species of the genus *Megaselia,* subgenus *Aphiochaeta* (Diptera, Phoridae). *Studia Entomologica* (n.s.), 7(1–4): 257–416.

———, 1965. Revision of the North American phorid flies. Pt. III. The species of the genus *Megaselia,* subgenus *Megaselia* (Diptera, Phoridae). *Studia Entomologica* (n.s.), 8(1–4): 1–160.

Bossert, W. H., and E. O. Wilson, 1963. The analysis of olfactory communication among animals. *Journal of Theoretical Biology,* 5: 443–469.

Bouillon, A., 1964. Étude de la composition des sociétés dans trois espèces d'*Apicotermes* Holmgren (Isoptera, Termitinae). *In* A. Bouillon, ed., *Études sur les termites africains.* Masson et Cie, Paris. pp. 181–196.

Boven, J. K. A. van, 1958. Allometrische en biometrische Beschouwingen over het Polymorfisme bijenkele Micrensoorten (Hym. Formicidae). *Verhandelingen van de Koninklijke Vlaamse Academie voor Wetenschappen, Letteren en Schone Kunsten van België, Klasse Wetenschappen,* 56: 1–134.

Branner, J. C., 1910. Geologic work of ants in tropical America. *Bulletin of the Geological Society of America,* 21: 449–496.

Brauer, F., 1869. Beschreibung der Verwandlungsgeschichte der *Mantispa styriaca* Poda und Betrachtungen über die sogenannte Hypermetamorphose Fabre's. *Verhandlungen der Zoologisch-Botanischen Gesellschaft in Wien,* 19: 831–840.

Brauns, H., 1902. *Eucondylops* n. g. Apidarum (Hym.). *Zeitschrift für Systematische Hymenopterologie und Dipterologie,* 2: 377–380.

———, 1914. Descriptions of some new species of myrmecophilous beetles from southern Rhodesia. *Proceedings of the Rhodesia Scientific Association,* 13(3): 32–42.

———, 1926. A contribution to the knowledge of the genus *Allodape,* St. Farg. & Serv. Order Hymenoptera; Section Apidae (Anthophila). *Annals of the South African Museum,* 23(3): 417–434.

Brian, Anne D., 1951. Brood development in *Bombus agrorum* (Hym., Bombidae). *Entomologist's Monthly Magazine,* 87: 207–212.

———, 1952. Division of labour and foraging in *Bombus agrorum* Fabricius. *Journal of Animal Ecology,* 21(2): 223–240.

———, 1954. The foraging of bumble bees. Pt. I. Foraging behaviour; Pt. II. Bumble bees as pollinators. *Bee World,* 35(4): 61–67, (5): 81–91.

Brian, M. V., 1950. The stable winter population structure in species of *Myrmica. Journal of Animal Ecology,* 19(2): 119–123.

———, 1951a. Caste determination in a myrmicine ant. *Experientia,* 7(5): 182–183.

———, 1951b. Summer population changes in colonies of the ant *Myrmica. Physiologia Comparata et Oecologia,* 2(3): 248–262.

———, 1952a. Interaction between ant colonies at an artificial nest-site. *Entomologist's Monthly Magazine,* 88: 84–88.

———, 1952b. The structure of a dense natural ant population. *Journal of Animal Ecology,* 21(1): 12–24.

———, 1953a. Oviposition by workers of the ant *Myrmica. Physiologia Comparata et Oecologia,* 3(1): 25–36.

———, 1953b. Brood-rearing in relation to worker number in the ant *Myrmica. Physiological Zoölogy,* 26(4): 355–366.

———, 1955a. Studies of caste differentiation in "*Myrmica rubra*" L. 2. The growth of workers and intercastes. *Insectes Sociaux,* 2(1): 1–34.

———, 1955b. Food collection by a Scottish ant community. *Journal of Animal Ecology,* 24(2): 336–351.

———, 1956a. Studies of caste differentiation in *Myrmica rubra* L. 4. Controlled larval nutrition. *Insectes Sociaux,* 3(3): 369–394.

———, 1956b. The natural density of *Myrmica rubra* and

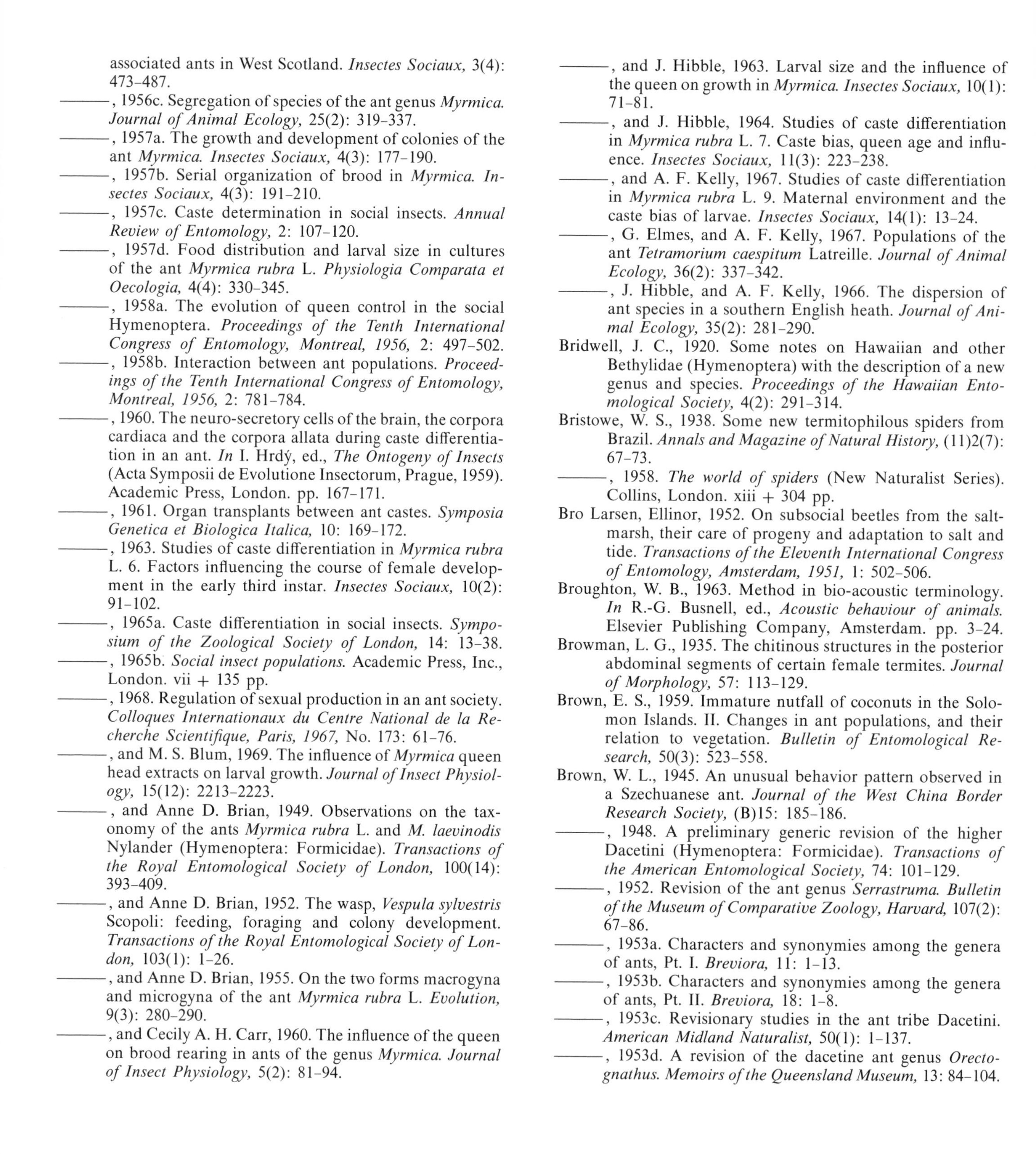

associated ants in West Scotland. *Insectes Sociaux,* 3(4): 473–487.

———, 1956c. Segregation of species of the ant genus *Myrmica. Journal of Animal Ecology,* 25(2): 319–337.

———, 1957a. The growth and development of colonies of the ant *Myrmica. Insectes Sociaux,* 4(3): 177–190.

———, 1957b. Serial organization of brood in *Myrmica. Insectes Sociaux,* 4(3): 191–210.

———, 1957c. Caste determination in social insects. *Annual Review of Entomology,* 2: 107–120.

———, 1957d. Food distribution and larval size in cultures of the ant *Myrmica rubra* L. *Physiologia Comparata et Oecologia,* 4(4): 330–345.

———, 1958a. The evolution of queen control in the social Hymenoptera. *Proceedings of the Tenth International Congress of Entomology, Montreal, 1956,* 2: 497–502.

———, 1958b. Interaction between ant populations. *Proceedings of the Tenth International Congress of Entomology, Montreal, 1956,* 2: 781–784.

———, 1960. The neuro-secretory cells of the brain, the corpora cardiaca and the corpora allata during caste differentiation in an ant. *In* I. Hrdý, ed., *The Ontogeny of Insects* (Acta Symposii de Evolutione Insectorum, Prague, 1959). Academic Press, London. pp. 167–171.

———, 1961. Organ transplants between ant castes. *Symposia Genetica et Biologica Italica,* 10: 169–172.

———, 1963. Studies of caste differentiation in *Myrmica rubra* L. 6. Factors influencing the course of female development in the early third instar. *Insectes Sociaux,* 10(2): 91–102.

———, 1965a. Caste differentiation in social insects. *Symposium of the Zoological Society of London,* 14: 13–38.

———, 1965b. *Social insect populations.* Academic Press, Inc., London. vii + 135 pp.

———, 1968. Regulation of sexual production in an ant society. *Colloques Internationaux du Centre National de la Recherche Scientifique, Paris, 1967,* No. 173: 61–76.

———, and M. S. Blum, 1969. The influence of *Myrmica* queen head extracts on larval growth. *Journal of Insect Physiology,* 15(12): 2213–2223.

———, and Anne D. Brian, 1949. Observations on the taxonomy of the ants *Myrmica rubra* L. and *M. laevinodis* Nylander (Hymenoptera: Formicidae). *Transactions of the Royal Entomological Society of London,* 100(14): 393–409.

———, and Anne D. Brian, 1952. The wasp, *Vespula sylvestris* Scopoli: feeding, foraging and colony development. *Transactions of the Royal Entomological Society of London,* 103(1): 1–26.

———, and Anne D. Brian, 1955. On the two forms macrogyna and microgyna of the ant *Myrmica rubra* L. *Evolution,* 9(3): 280–290.

———, and Cecily A. H. Carr, 1960. The influence of the queen on brood rearing in ants of the genus *Myrmica. Journal of Insect Physiology,* 5(2): 81–94.

———, and J. Hibble, 1963. Larval size and the influence of the queen on growth in *Myrmica. Insectes Sociaux,* 10(1): 71–81.

———, and J. Hibble, 1964. Studies of caste differentiation in *Myrmica rubra* L. 7. Caste bias, queen age and influence. *Insectes Sociaux,* 11(3): 223–238.

———, and A. F. Kelly, 1967. Studies of caste differentiation in *Myrmica rubra* L. 9. Maternal environment and the caste bias of larvae. *Insectes Sociaux,* 14(1): 13–24.

———, G. Elmes, and A. F. Kelly, 1967. Populations of the ant *Tetramorium caespitum* Latreille. *Journal of Animal Ecology,* 36(2): 337–342.

———, J. Hibble, and A. F. Kelly, 1966. The dispersion of ant species in a southern English heath. *Journal of Animal Ecology,* 35(2): 281–290.

Bridwell, J. C., 1920. Some notes on Hawaiian and other Bethylidae (Hymenoptera) with the description of a new genus and species. *Proceedings of the Hawaiian Entomological Society,* 4(2): 291–314.

Bristowe, W. S., 1938. Some new termitophilous spiders from Brazil. *Annals and Magazine of Natural History,* (11)2(7): 67–73.

———, 1958. *The world of spiders* (New Naturalist Series). Collins, London. xiii + 304 pp.

Bro Larsen, Ellinor, 1952. On subsocial beetles from the salt-marsh, their care of progeny and adaptation to salt and tide. *Transactions of the Eleventh International Congress of Entomology, Amsterdam, 1951,* 1: 502–506.

Broughton, W. B., 1963. Method in bio-acoustic terminology. *In* R.-G. Busnell, ed., *Acoustic behaviour of animals.* Elsevier Publishing Company, Amsterdam. pp. 3–24.

Browman, L. G., 1935. The chitinous structures in the posterior abdominal segments of certain female termites. *Journal of Morphology,* 57: 113–129.

Brown, E. S., 1959. Immature nutfall of coconuts in the Solomon Islands. II. Changes in ant populations, and their relation to vegetation. *Bulletin of Entomological Research,* 50(3): 523–558.

Brown, W. L., 1945. An unusual behavior pattern observed in a Szechuanese ant. *Journal of the West China Border Research Society,* (B)15: 185–186.

———, 1948. A preliminary generic revision of the higher Dacetini (Hymenoptera: Formicidae). *Transactions of the American Entomological Society,* 74: 101–129.

———, 1952. Revision of the ant genus *Serrastruma. Bulletin of the Museum of Comparative Zoology, Harvard,* 107(2): 67–86.

———, 1953a. Characters and synonymies among the genera of ants, Pt. I. *Breviora,* 11: 1–13.

———, 1953b. Characters and synonymies among the genera of ants, Pt. II. *Breviora,* 18: 1–8.

———, 1953c. Revisionary studies in the ant tribe Dacetini. *American Midland Naturalist,* 50(1): 1–137.

———, 1953d. A revision of the dacetine ant genus *Orectognathus. Memoirs of the Queensland Museum,* 13: 84–104.

———, 1954. Remarks on the internal phylogeny and subfamily classification of the family Formicidae. *Insectes Sociaux,* 1(1): 21–31.
———, 1955a. A revision of the Australian ant genus *Notoncus* Emery, with notes on the other genera of Melophorini. *Bulletin of the Museum of Comparative Zoology, Harvard,* 113(6): 471–494.
———, 1955b. Ant taxonomy. *In* E. L. Kessel, ed., *A century of progress in the natural sciences, 1853–1953.* California Academy of Sciences, San Francisco. pp. 569–572.
———, 1955c. The first social parasite in the ant tribe Dacetini. *Insectes Sociaux,* 2(3): 181–186.
———, 1957. Predation of arthropod eggs by the ant genera *Proceratium* and *Discothyrea. Psyche, Cambridge,* 64(3): 115.
———, 1958. A review of the ants of New Zealand (Hymenoptera). *Acta Hymenopterologica, Tokyo,* 1(1): 1–50.
———, 1959. The neotropical species of the ant genus *Strumigenys* Fr. Smith: Group of *gundlachi* (Roger). *Psyche, Cambridge,* 66(3): 37–52.
———, 1960a. Contributions toward a reclassification of the Formicidae. III. Tribe Amblyoponini (Hymenoptera). *Bulletin of the Museum of Comparative Zoology, Harvard,* 122(4): 145–230.
———, 1960b. Ants, acacias and browsing mammals. *Ecology,* 41(3): 587–592.
———, 1964a. The ant genus *Smithistruma:* A first supplement to the world revision (Hymenoptera: Formicidae). *Transactions of the American Entomological Society,* 89: 183–200.
———, 1964b. (Revision of *Rhoptromyrmex*). *Pilot Register of Zoology,* Card Nos. 11–19.
———, 1965. Contributions to a reclassification of the Formicidae. IV. Tribe Typhlomyrmecini (Hymenoptera). *Psyche, Cambridge,* 72(1): 65–78.
———, 1967. A new *Pheidole* with reversed phragmosis (Hymenoptera: Formicidae). *Psyche, Cambridge,* 74(4): 331–339.
———, 1968. An hypothesis concerning the function of the metapleural glands in ants. *American Naturalist,* 102: 188–191.
———, 1969. *Strumigenys lopotyle* species nov. *Pilot Register of Zoology,* Card No. 27.
———, and W. W. Kempf, 1960. A world revision of the ant tribe Basicerotini. *Studia Entomologica* (n.s.), 3(1–4): 161–250.
———, and W. W. Kempf, 1967. *Tatuidris,* a remarkable new genus of Formicidae (Hymenoptera). *Psyche, Cambridge,* 74(3): 183–190.
———, and E. O. Wilson, 1956. Character displacement. *Systematic Zoology,* 5(2): 49–64.
———, and E. O. Wilson, 1959a. The evolution of the dacetine ants. *Quarterly Review of Biology,* 34: 278–294.
———, and E. O. Wilson, 1959b. The search for *Nothomyrmecia. Western Australian Naturalist,* 7: 25–30.
Bruch, C., 1931. Notas biológicas y sistematicas acerca de "Bruchomyrma acutidens" Santschi. *Revista del Museo de La Plata,* 33: 31–55.
———, 1932. Descripción de un género y especie nueva de una hormiga parásita (Formicidae). *Revista del Museo de La Plata,* 33: 271–275.
———, 1936. Notas sobre el "Camuatí" y las avispas que lo construyen. *Physis, Buenos Aires,* 12(43): 125–135.
Brues, C. T., 1902. New and little-known guests of the Texan legionary ants. *American Naturalist,* 36: 365–378.
———, A. L. Melander, and F. M. Carpenter, 1954. Classification of insects. *Bulletin of the Museum of Comparative Zoology, Harvard,* 108: 1–917.
Brun, R., 1914. *Die Raumorientierung der Ameisen und das Orientierungsproblem im allgemeinen.* Gustav Fischer, Jena. viii + 234 pp.
———, 1923. Vergleichende Untersuchungen über Insektengehirne, mit besonderer Berücksichtigung der pilzhutförmigen Körper (Corpora pedunculata Dujardini). *Schweizer Archiv für Neurologie und Psychiatrie,* 13: 144–172.
———, 1952. Das Zentralnervensystem von *Teleutomyrmex schneideri* Kutt. ♀ (Hym. Formicid.). *Mitteilungen der Schweizerischen Entomologischen Gesellschaft,* 25(2): 73–86.
Bruns, H., 1954a. Wann und in welchem Umfang wird die Kl. Fichtenblattwespe (*Lygaeonematus abietum* Htg.) von der Roten Waldameise (*Formica rufa* L.) eingetragen? *Forstwirtschaftliche Centralblatt,* 73: 35–40.
———, 1954b. Beobachtungen zum Verhalten (insbesondere Tagesrhythmus) der Roten Waldameise (*Formica rufa*) während des Nahrungserwerbes. (Vorläufige Mitteilung). *Zeitschrift für Tierpsychologie,* 11(1): 151–154.
Buchli, H. H. R., 1950. Recherche sur la fondation et le développement des nouvelles colonies chez le termite lucifuge (*Reticulitermes lucifugus* Rossi). *Physiologia Comparata et Oecologia,* 2: 145–160.
———, 1958. L'origine des castes et les potentialites ontogéniques des termites européens du genre *Reticulitermes* Holmgren. *Annales des Sciences Naturelles,* (11)20(18): 263–429.
Buckingham, Edith N., 1911. Division of labor among ants. *Proceedings of the American Academy of Arts and Sciences,* 46(18): 425–507.
Buddenbrock, W. von, 1917. Die Lichtkompassbewegungen bei den Insekten, insbesondere den Schmetterlingsraupen. *Sitzungsberichte der Heidelberger Akademie der Wissenschaften,* Abt. B, 1: 3–26.
Büdel, A., and E. Herold, eds., 1960. *Biene und Bienenzucht.* Ehrenwirth Verlag, München. xii + 379 pp.
Bugnion, E., 1913. *Termitoxenia.* Étude anatomo-histologique. *Annales de la Société Entomologique de Beligique,* 57: 23–44.
———, 1922. La guerre des fourmis et des termites, la genèse des instincts expliquée par cette guerre. Appendix in A.

Forel, *Le monde social des fourmis du globe comparé à celui de l'homme, Vol. 3.* Librairie Kundig, Geneva. pp. 173-225.

———, 1928. The war between the ants and the termites. A study of the origin of instinct. *In* A. Forel, *The social world of the ants compared with that of man, Vol. 2* (tr. C. K. Ogden). G. P. Putnam's Sons, Ltd., London. pp. 353-404.

———, 1930. Les pièces buccales, le sac infrabuccal et le pharynx des fourmis. *Bulletin. Société Royale Entomologique d'Égypte* (n.s.), 14(2-3): 85-210.

Bullock, T. H., and G. A. Horridge, 1965. *Structure and function in the nervous systems of invertebrates. Vols. I and II.* W. H. Freeman, San Francisco. xx + 1719 pp.

Bünzli, G. H., 1935. Untersuchungen über coccidophile Ameisen aus den Kaffeefeldern von Surinam. *Mitteilungen der Schweizerischen Entomologischen Gesellschaft,* 16(6-7): 453-593.

Burdick, D. J., and M. S. Wasbauer, 1959. Biology of *Methocha californica* Westwood (Hymenoptera: Tiphiidae). *Wasmann Journal of Biology,* 17(1): 75-88.

Buren, W. F., 1958. A review of the species of *Crematogaster,* sensu stricto, in North America (Hymenoptera: Formicidae), Pt. I. *Journal of the New York Entomological Society,* 66: 119-134.

———, 1968a. Some fundamental taxonomic problems in *Formica* (Hymenoptera: Formicidae). *Journal of the Georgia Entomological Society,* 3(2): 25-40.

———, 1968b. A review of the species of *Crematogaster,* sensu stricto, in North America (Hymenoptera, Formicidae). Pt. II. Descriptions of new species. *Journal of the Georgia Entomological Society,* 3(3): 91-121.

Buschinger, A., 1966. Untersuchungen an *Harpagoxenus sublaevis* Nyl. (Hym., Formicidae). I. Freilandbeobachtungen zu Verbreitung und Lebensweise. *Insectes Sociaux,* 13(1): 5-16.

———, 1968. Untersuchungen an *Harpagoxenus sublaevis* Nyl. (Hymenoptera, Formicidae). III. Kopula, Koloniegründung, Raubzüge. *Insectes Sociaux,* 15(1): 89-104.

Büsgen, M., 1891. Der Honigtau: Eine biologische Studie an Pflanzen und Pflanzenläusen. *Zeitschrift für Naturwissenschaften,* 25: 339-428.

Butenandt, A., B. Linzen, and M. Lindauer, 1959. Über einen Duftstoff aus der Mandibeldrüse der Blattschneiderameise *Atta sexdens rubropilosa* Forel. *Archives d'Anatomie Microscopique et de Morphologie Expérimentale,* 48: 13-19.

Butler, C. G., 1954a. *The world of the honeybee* (New Naturalist Series). Collins, London. xiv + 226 pp.

———, 1954b. The method and importance of the recognition by a colony of honeybees (*A. mellifera*) of the presence of its queen. *Transactions of the Royal Entomological Society of London,* 105(2): 11-29.

———, 1960. The significance of queen substance in swarming and supersedure in honey-bee (*Apis mellifera* L.) colonies. *Proceedings of the Royal Entomological Society of London,* (A)35: 129-132.

———, 1961. The scent of queen honeybees (*A. mellifera* L.) that causes partial inhibition of queen rearing. *Journal of Insect Physiology,* 7(3-4): 258-264.

———, 1964. Pheromones in sexual processes in insects. *Symposium of the Royal Entomological Society of London,* 2: 66-77.

———, 1965. Sex attraction in *Andrena flavipes* Panzer (Hymenoptera: Apidae), with some observations on nest-site restriction. *Proceedings of the Royal Entomological Society of London,* (A)40: 77-80.

———, 1966a. Mandibular gland pheromone of worker honeybees. *Nature, London,* 212: 530.

———, 1966b. Bee Department. *Report of the Rothamsted Experimental Station for 1966:* 208-215.

———,1967. Insect pheromones. *Biological Reviews* (Cambridge Philosophical Society), 42(1): 42-87.

———, 1969. Some pheromones controlling honeybee behaviour. *Proceedings of the Sixth Congress of the International Union for the Study of Social Insects, Bern, 1969,* pp. 19-32.

———, and D. H. Calam, 1969. Pheromones of the honey bee—the secretion of the Nassanoff gland of the worker. *Journal of Insect Physiology,* 15(2): 237-244.

———, and Elaine M. Fairey, 1963. The role of the queen in preventing oogenesis in worker honeybees. *Journal of Apicultural Research,* 2(1): 14-18.

———, and J. B. Free, 1952. The behaviour of worker honeybees at the hive entrance. *Behaviour,* 4(4): 262-292.

———, and Patricia N. Paton, 1962. Inhibition of queen rearing by queen honey-bees (*Apis mellifera* L.) of different ages. *Proceedings of the Royal Entomological Society of London,* (A)37: 114-116.

———, and J. Simpson, 1965. Pheromones of the honeybee (*Apis mellifera* L.). An olfactory pheromone from the Koschewnikow gland of the queen. *Vědecké Práce Výzkumných Ústavů Zemědělských Včelařskéhov v Dole, 1965,* pp. 33-36.

———, and J. Simpson, 1967. Pheromones of the queen honeybee (*Apis mellifera* L.) which enable her workers to follow her when swarming. *Proceedings of the Royal Entomological Society of London,* (A)42: 149-154.

———, R. K. Callow, and J. R. Chapman, 1964. 9-Hydroxydec-*trans*-2-enoic acid, a pheromone stabilizing honeybee swarms. *Nature, London,* 201: 733.

———, R. K. Callow, and Norah C. Johnston, 1961. The isolation and synthesis of queen substance, 9-oxodec-*trans*-2-enoic acid, a honeybee pheromone. *Proceedings of the Royal Society,* (B)155: 417-432.

———, D. J. C. Fletcher, and Doreen Watler, 1969. Nest-entrance marking with pheromones by the honeybee, *Apis mellifera* L., and by a wasp, *Vespula vulgaris* L. *Animal Behaviour,* 17(1): 142-147.

Butler, Charles, 1609. *The feminine monarchie. On a treatise concerning bees, and the due ordering of them.* Joseph Barnes, Oxford.

Buttel-Reepen, H. von, 1915. *Leben und Wesen der Bienen.* Friedr. Vieweg und Sohn, Braunschweig. xiv + 300 pp.

Buysson, R. du, 1903. Monographie des guêpes ou *Vespa. Annales de la Société Entomologique de France,* 72(2): 260–288.

———, 1905. Monographie des Vespides du genre *Nectarina. Annales de la Société Entomologique de France,* 74(4): 537–566.

———, 1906. Monographie des Vespides appartenant aux genres *Apoica* et *Synoeca. Annales de la Société Entomologique de France,* 75(3): 333–362.

———, 1909. Monographie des Vespides du genre *Belonogaster. Annales de la Société Entomologique de France,* 78(2): 199–270.

Cagniant, H., 1968. Description d'*Epimyrma algeriana* (nov. sp.) (Hyménoptères Formicidae, Myrmicinae), fourmi parasite. Représentation des trois castes. Quelques observations biologiques, écologiques et éthologiques. *Insectes Sociaux,* 15(2): 157–170.

Calaby, J. H., 1956. The distribution and biology of the genus *Ahamitermes* (Isoptera). *Australian Journal of Zoology,* 4(2): 111–124.

Cale, G. H., 1946. Chapters xiii–xv. *In* R. A. Grout, ed., *The hive and the honeybee.* Dadant & Sons, Hamilton, Illinois. pp. 300–391.

Calhoun, J. B., 1962. Population density and social pathology. *Scientific American,* 206(2): 139–148.

Callow, R. K., J. R. Chapman, and Patricia N. Paton, 1964. Pheromones of the honeybee: Chemical studies of the mandibular gland secretion of the queen. *Journal of Apicultural Research,* 3(2): 77–89.

Cannon, W. B., 1939. *The wisdom of the body,* rev. ed. W. W. Norton and Co., Inc., New York. 333 pp.

Carl, J., 1934. *Ropalidia montana* n. sp. et son nid. Un type nouveau d'architecture vespienne. *Revue Suisse de Zoologie,* 41(4): 675–691.

Carne, P. B., 1966. Primitive forms of social behaviour, and their significance in the ecology of gregarious insects. *Proceedings of the Ecological Society of Australia,* 1: 75–78.

Carney, W. P., 1970. Laboratory maintenance of carpenter ants. *Annals of the Entomological Society of America,* 63(1): 332–334.

Carpenter, F. M., 1930. The fossil ants of North America. *Bulletin of the Museum of Comparative Zoology, Harvard,* 70(1): 1–66.

Carr, Cecily A. H., 1962. Further studies on the influence of the queen in ants of the genus *Myrmica. Insectes Sociaux,* 9(3): 197–211.

Carthy, J. D., 1950. Odour trails of *Acanthomyops fuliginosus. Nature, London,* 166: 154.

———, 1951a. The orientation of two allied species of British ant. I. Visual direction finding in *Acanthomyops* (*Lasius*) *niger. Behaviour,* 3(4): 275–303.

———, 1951b. The orientation of two allied species of British ant. II. Odour trail laying and following in *Acanthomyops* (*Lasius*) *fuliginosus. Behaviour,* 3(4): 304–318.

Cartwright, O. L., and R. E. Woodruff, 1969. Ten *Rhyparus* from the Western Hemisphere (Coleoptera: Scarabaeidae: Aphodiinae). *Smithsonian Contributions to Zoology,* No. 21: 1–20.

Castle, G. B., 1934. The damp-wood termites of western United States, genus *Zootermopsis* (formerly, *Termopsis*). *In* C. A. Kofoid *et al.,* ed., *Termites and termite control,* 2nd ed. University of California Press, Berkeley. pp. 273–310.

Cavill, G. W. K., and Phyllis L. Robertson, 1965. Ant venoms, attractants, and repellents. *Science,* 149: 1337–1345.

———, P. J. Williams, and F. B. Whitfield, 1967. α-Farnesene, Dufour's gland secretion in the ant *Aphaenogaster longiceps* (F. Sm.). *Tetrahedron Letters,* 23: 2201–2205.

Cazier, M. A., and M. A. Mortenson, 1965. Bionomical observations on myrmecophilous beetles of the genus *Cremastocheilus* (Coleoptera: Scarabaeidae). *Journal of the Kansas Entomological Society,* 38(1): 19–44.

———, and Marjorie Statham, 1962. The behavior and habits of the myrmecophilous scarab *Cremastocheilus stathamae* Cazier with notes on other species (Coleoptera: Scarabaeidae). *Journal of the New York Entomological Society,* 70(3): 125–149.

Chamberlin, R. V., 1923. On four termitophilous millepeds from British Guiana. *Zoologica, New York,* 3(21): 411–421.

Chapman, T. A., 1870. Some facts towards a life-history of *Rhipiphorus paradoxus. Annals and Magazine of Natural History,* (4)6(34): 314–326.

Chauvin, R., 1959. Contribution à l'étude de la construction du dôme chez *Formica rufa,* II. *Insectes Sociaux,* 6(1): 1–11.

———, 1960a. Facteurs d'asymétrie et facteurs de régulation dans la construction du dôme chez *Formica rufa,* IV. *Insectes Sociaux,* 7(3): 201–205.

———, 1960b. Les substances actives sur le comportement à l'intérieur de la ruche. *Annales de l'Abeille,* 3(2): 185–197.

———, 1964. Expériences sur l'apprentissage par équipe du labyrinthe chez *Formica polyctena. Insectes Sociaux,* 11(1): 1–20.

———, ed., 1968. *Traité de biologie de l'abeille,* 5 vols.: *Vol. I, Biologie et physiologie générales,* xvi + 547; *Vol. II, Systéme nerveux, comportement et régulations sociales,* viii + 566; *Vol. III, Les produits de la ruche,* viii + 400; *Vol. IV, Biologie appliquée,* viii + 434; *Vol. V, Histoire, ethnographie et folklore,* viii + 152. Masson et Cie, Paris.

———, 1968a. Énergétique (calorimétrie) des abeilles d'après les travaux de M. Roth (1964–1965). *In* R. Chauvin, ed. (*q.v.*), *Traité de biologie de l'abeille, Vol. I.* pp. 245–261.

———, 1968b. Les voies nerveuses principales du cerveau. Eléments de physiologie cérébrale et musculaire. *In*

R. Chauvin, ed. (*q.v.*), *Traité de biologie de l'abeille, Vol. II.* pp. 49–99.
———, 1968c. Chimiosensibilité chez l'abeille. *In* R. Chauvin, ed. (*q.v.*), *Traité de biologie de l'abeille, Vol. II.* pp. 122–145.
———, 1968d. *Animal societies from the bee to the gorilla.* (English tr. by G. Ordish). Hill and Wang, New York. 281 pp.
———, and J. Denis, 1965. Une araignée sociale du Gabon. *Biologia Gabonica,* 1(2): 93–99.
———, and C. Noirot, eds., 1968. *L'effet de groupe chez les animaux.* Colloques Internationaux du Centre National de la Recherche Scientifique, Paris, 1967, No. 173. 390 pp.
Chen, S. C., 1937a. Social modification of the activity of ants in nest-building. *Physiological Zoölogy,* 10(4): 420–436.
———, 1937b. The leaders and followers among the ants in nest-building. *Physiological Zoölogy,* 10(4): 437–455.
China, W. E., 1928. A remarkable bug which lures ants to their destruction. *Natural History Magazine,* 1(6): 209–213.
Clark, G. C., and C. G. C. Dickson, 1956. The honey gland and tubercles of larvae of the Lycaenidae. *Lepidopterists' News,* 10(1–2): 37–43.
Clark, J., 1920. Notes on Western Australian ant-nest beetles. *Journal and Proceedings of the Royal Society of Western Australia,* 6(2): 97–104.
———, 1923. A new myrmecophilous beetle. *Journal and Proceedings of the Royal Society of Western Australia,* 9(1): 44–46.
———, 1925. The ants of Victoria. *Victorian Naturalist,* 42(3): 58–64, (6): 135–144.
———, 1934. Notes on Australian ants, with descriptions of new species and a new genus. *Memoirs of the National Museum of Victoria,* 8: 5–20.
———, 1943. A revision of the genus *Promyrmecia* Emery (Formicidae). *Memoirs of the National Museum of Victoria,* 13: 83–149.
———, 1951. *The Formicidae of Australia. Volume I. Subfamily Myrmeciinae.* Commonwealth Scientific and Industrial Research Organization, Melbourne, Australia. 230 pp.
Clark, L. R., P. W. Geier, R. D. Hughes, and R. F. Morris, 1967. *The ecology of insect populations in theory and practice.* Methuen and Company, Ltd., London. xiii + 232 pp.
Clausen, C. P., 1940. *Entomophagous insects.* McGraw-Hill Book Co., New York. x + 688 pp.
Cleveland, L. R., 1926. Symbiosis among animals with special reference to termites and their intestinal flagellates. *Quarterly Review of Biology,* 1(1): 51–60.
———, S. R. Hall, Elizabeth P. Sanders, and Jane Collier, 1934. The wood-feeding roach *Cryptocercus,* its Protozoa, and the symbiosis between Protozoa and roach. *Memoirs of the American Academy of Arts and Sciences,* 17(2): 185–342.
Cockerell, T. D. A., 1909. Descriptions of Hymenoptera from Baltic amber. *Schriften der Physikalisch-Ökonomischen Gesellschaft zu Königsberg,* 50: 1–25.
Cohen, W. E., 1933. An analysis of termite (*Eutermes exitiosus*) mound material. *Journal of the Council for Scientific and Industrial Research, Australia,* 6(3): 166–169.
Cole, A. C., 1932. The rebuilding of mounds of the ant, *Pogonomyrmex occidentalis,* Cress. *Ohio Journal of Science,* 32(3): 245–246.
———, 1956. Studies of Nevada ants. II. A new species of *Lasius* (*Chthonolasius*) (Hymenoptera: Formicidae). *Journal of the Tennessee Academy of Science,* 31(1): 26–27.
———, 1957. *Paramyrmica,* a new North American genus of ants allied to *Myrmica* Latreille (Hymenoptera: Formicidae). *Journal of the Tennessee Academy of Science,* 32(1): 37–42.
———, 1968. *Pogonomyrmex harvester ants: A study of the genus in North America.* University of Tennessee Press, Knoxville. x + 222 pp.
———, and J. W. Jones, 1948. A study of the weaver ant, *Oecophylla smaragdina* (Fab.). *American Midland Naturalist,* 39(3): 641–651.
Collingwood, C. A., 1956. A rare parasitic ant (Hym., Formicidae) in France. *Entomologist's Monthly Magazine,* 92: 197.
———, 1958. The ants of the genus *Myrmica* in Britain. *Proceedings of the Royal Entomological Society of London,* (A)33: 65–75.
Combes, Marguerite, 1937. Existence probable d'une élite non différenciée d'aspect, constituant les véritables ouvrières chez les *Formica. Compte Rendu de l'Académie des Sciences, Paris,* 204(22): 1674–1675.
Cook, S. F., and K. G. Scott, 1933. The nutritional requirements of *Zootermopsis* (*Termopsis*) *angusticollis. Journal of Cellular and Comparative Physiology,* 4(1): 95–110.
Cordero, A. D., 1963. An unusual behavior of the leafcutting ant queen *Acromyrmex octospinosa* (Reich). *Revista de Biologia Tropicale,* 11(2): 221–222.
Corkins, C. L., 1930. The metabolism of the honeybee colony during winter. *Bulletin, Wyoming Agricultural Experiment Station,* 175: 1–54.
Cory, E. N., and Elizabeth E. Haviland, 1938. Population studies of *Formica exsectoides* Forel. *Annals of the Entomological Society of America,* 31(1): 50–56.
Costa Lima, A. da, 1962. Micro-coleoptero representante da nova subfamilia Plaumanniolinae (Col., Ptinidae). *Revista Brasileira de Biologia,* 22(4): 413–418. [Transferred to Scydmaenidae by Lawrence and Reichardt, 1966].
Coville, F. V., 1890. Notes on bumble-bees. *Proceedings of the Entomological Society of Washington,* 1(4): 197–202.
Crampton, G. C., 1923. A comparison of the terminal abdominal structures of an adult alate female of the primitive termite *Mastotermes darwinensis* with those of the roach *Periplaneta americana. Bulletin of the Brooklyn Entomological Society,* 18(3): 85–93.
Crawley, W. C., 1912. Parthenogenesis in worker ants, with

special reference to the colonies of *Lasius niger,* Linn. *Transactions of the Entomological Society of London,* 1911(4): 657–663.

Crawshay, G. A., 1905. A large community of *Vespa vulgaris. Entomologist's Monthly Magazine,* 41: 8–10.

Creighton, W. S., 1929. Further notes on the habits of *Harpagoxenus americanus. Psyche, Cambridge,* 36(1): 48–50.

———, 1950. The ants of North America. *Bulletin of the Museum of Comparative Zoology, Harvard,* 104: 1–585.

———, 1953. New data on the habits of *Camponotus* (*Myrmaphaenus*) *ulcerosus* Wheeler. *Psyche, Cambridge,* 60(2): 82–84.

———, 1963. Further studies on the habits of *Cryptocerus texanus* Santschi (Hymenoptera: Formicidae). *Psyche, Cambridge,* 70(3): 133-143.

———, and R. H. Crandall, 1954. New data on the habits of *Myrmecocystus melliger* Forel. *The Biological Review, City College of New York,* 16(1): 2–6.

———, and Martha P. Creighton. 1959. The habits of *Pheidole militicida* Wheeler (Hymenoptera: Formicidae). *Psyche, Cambridge,* 66(1–2): 1–12.

———, and R. E. Gregg, 1954. Studies on the habits and distribution of *Cryptocerus texanus* Santschi (Hymenoptera: Formicidae). *Psyche, Cambridge,* 61(2): 41–57.

Crèvecoeur, A., 1931. Le maraudage occasionnel. Tendance au cleptôparasitisme chez divers Psammocharidae (Hym.). *Mémoires de la Société Entomologique de Belgique,* 23(5): 183–187.

Crewe, R. M., and M. S. Blum, 1970. Identification of the alarm pheromones of the ant *Myrmica brevinodis. Journal of Insect Physiology,* 16(1): 141–146.

———, J. M. Brand, D. J. C. Fletcher, and S. H. Eggers, 1970. The mandibular gland chemistry of some South African species of *Crematogaster* (Hymenoptera: Formicidae). *Journal of the Georgia Entomological Society,* 5(1): 42–47.

Crowell, K. L., 1968. Rates of competitive exclusion by the Argentine ant in Bermuda. *Ecology,* 49(3): 551–555.

Crozier, R. H., 1968. Cytotaxonomic studies on some Australian dolichoderine ants (Hymenoptera: Formicidae). *Caryologia,* 21(3): 241–259.

Cruz-Landim, Carminda da, 1963. Evolution of the wax and scent glands in the Apinae (Hymenoptera: Apidae). *Journal of the New York Entomological Society,* 71: 2–13.

———, and A. Ferreira, 1968. Mandibular gland development and communication in field bees of *Trigona* (*Scaptotrigona*) *postica* (Hymenoptera: Apidae). *Journal of the Kansas Entomological Society,* 41(4): 474–481.

Cumber, R. A., 1949a. The biology of humble-bees, with special reference to the production of the worker caste. *Transactions of the Royal Entomological Society of London,* 100(1): 1–45.

———, 1949b. Humble-bee parasites and commensals found within a thirty mile radius of London. *Proceedings of the Royal Entomological Society of London,* (A)24: 119–127.

———, 1949c. An overwintering nest of the humble-bee *Bombus terrestris* (L.) (Hymenoptera, Apidae). *New Zealand Science Review,* 7: 96–97.

———, 1951. Some observations on the biology of the Australian wasp *Polistes humilis* Fabr. (Hymenoptera: Vespidae) in North Auckland (New Zealand), with special reference to the nature of the worker caste. *Proceedings of the Royal Entomological Society of London,* (A)26: 11–16.

———, 1954. The life-cycle of humble-bees in New Zealand. *New Zealand Journal of Science and Technology,* (B)36(1): 95–107.

Cunha, A. B. da, and W. E. Kerr, 1957. A genetical theory to explain sex determination by arrhenotokous parthenogenesis. *Forma et Functio,* 1(4): 33–36.

Curtis, Helena, 1968. *Biology.* Worth Publishers, Inc., New York. xvii + 854.

Dade, H. A., 1962. *Anatomy and dissection of the honeybee.* Bee Research Association, London. xi + 158 pp.

Darchen, R., 1957. La reine d'*Apis mellifica,* les ouvrières pondeuses et les constructions cirières. *Insectes Sociaux,* 4(4): 321-325.

———, 1965. Éthologie d'une araignée sociale, *Agelena consociata* Denis. *Biologia Gabonica,* 1(2): 117–146.

———, and J. Lensky, 1963. Quelques problèmes soulevés par la création de sociétés polygynes d'abeilles. *Insectes Sociaux,* 10(4): 337–357.

Darling, F. F., 1938. *Bird flocks and the breeding cycle.* Cambridge University Press, Cambridge. x + 124 pp.

Darlington, P. J., 1950. Paussid beetles. *Transactions of the American Entomological Society,* 76: 47–142.

Darwin, C. R., 1859. *On the origin of species by means of natural selection, or the preservation of favoured races in the struggle for life,* 1st ed. John Murray, London. ix + 502 pp.

Daumer, K., 1956. Reizmetrische Untersuchung des Farbensehens der Bienen. *Zeitschrift für Vergleichende Physiologie,* 38(5): 413–478.

———, 1958. Blumenfarben, wie sie die Bienen sehen. *Zeitschrift für Vergleichende Physiologie,* 41(1): 49–110.

Deevey, E. S., 1947. Life tables for natural populations of animals. *Quarterly Review of Biology,* 22(4): 283–314.

Delachambre, J., 1965. Recherches sur les Termitoxeniidae (Diptera). I. Description d'une nouvelle espèce Africaine et sa larve. *Insectes Sociaux,* 12(3): 273–283.

———, 1966. Recherches sur les Termitoxeniidae (Diptera). II. Description de deux nouvelles espèces Africaines. *Insectes Sociaux,* 13(2): 105–115.

Delamare Deboutteville, C., 1948. Recherches sur les Collemboles termitophiles et myrmécophiles (Écologie, Éthologie, Systématique). *Archives de Zoologie Expérimentale et Générale,* 85(5): 261-425.

Deleurance, É.-P., 1948. Le comportement reproducteur est indépendant de la présence des ovaires chez *Polistes* (Hyménoptères Vespides). *Compte Rendu de l'Académie des Sciences, Paris,* 227(17): 866–867.

———, 1950. Sur le mécanisme de la monogynie fonctionnelle

chez les *Polistes. Compte Rendu de l'Académie des Sciences, Paris,* 230(8): 782–784.
———, 1952a. Étude du cycle biologique du couvain chez *Polistes.* Les phases "couvain normal" et "couvain abortif." *Behaviour,* 4(2): 104–115.
———, 1952b. Le polymorphisme social et son déterminisme chez les Guêpes. *Colloques Internationaux du Centre National de la Recherche Scientifique, Paris, 1950.* No. 34: 141–155.
———, 1955a. L'influence des ovaires sur l'activité de construction chez les *Polistes* (Hyménoptères Vespides). *Compte Rendu de l'Académie des Sciences, Paris,* 241(16): 1073–1075.
———, 1955b. Contribution à l'étude biologique des *Polistes* (Hyménoptères Vespides). II. Le cycle évolutif du couvain. *Insectes Sociaux,* 2(4): 285–302.
———, 1957. Contribution à l'étude biologique des *Polistes* (Hyménoptères Vespides). I. L'activité de construction. *Annales des Sciences Naturelles,* (11)19(1–2): 91–222.
Deligne, J., 1965. Morphologie et fonctionnement des mandibules chez les soldats des Termites. *Biologia Gabonica,* 1(2): 179–186.
Délye, G., 1957. Observations sur la fourmi saharienne *Cataglyphis bombycina* Rog. *Insectes Sociaux,* 4(2): 77–82.
Desneux, J., 1904. Isoptera, Fam. Termitidae. *Genera Insectorum* (Wytsman), fasc. 25: 1–52.
Dethier, V. G., 1953. Vision. *In* K. D. Roeder, ed., *Insect physiology.* John Wiley & Sons, Inc., New York. pp. 488–522.
———, 1957. Communication by insects: physiology of dancing. *Science,* 125: 331–336.
———, 1963. *The physiology of insect senses.* Methuen and Co., Ltd., London. ix + 266 pp.
Dobrzańska, Janina, 1958. Partition of foraging grounds and modes of conveying information among ants. *Acta Biologiae Experimentalis, Warsaw,* 18: 55–67.
———, 1959. Studies on the division of labour in ants genus *Formica. Acta Biologiae Experimentalis, Warsaw,* 19: 57–81.
———, 1966. The control of the territory by *Lasius fuliginosus* Latr. *Acta Biologiae Experimentalis, Warsaw,* 26(2): 193–213.
———, and J. Dobrzański. 1960. Quelques nouvelles remarques sur l'éthologie de *Polyergus rufescens* Latr. *Insectes Sociaux,* 7(1): 1–16.
———, and J. Dobrzański, 1962. Quelques observations sur les luttes entre différentes espèces de fourmis. *Acta Biologiae Experimentalis, Warsaw,* 22(4): 269–277.
Dobrzański, J., 1956. Badania nad zmystem czasu u mrówek. [Investigations on time sense in ants]. *Folia Biologica, Kraków,* 4(3–4): 385–397.
———, 1961. Sur l'éthologie guerrière de *Formica sanguinea* Latr. (Hyménoptère, Formicidae). *Acta Biologiae Experimentalis, Warsaw,* 21: 53–73.
———, 1965. Genesis of social parasitism among ants. *Acta Biologiae Experimentalis, Warsaw,* 25(1): 59–71.
———, 1966. Contribution to the ethology of *Leptothorax acervorum* (Hymenoptera: Formicidae). *Acta Biologiae Experimentalis, Warsaw,* 26(1): 71–78.
Dodd, F. P., 1902. Contribution to the life-history of *Liphyra brassolis,* Westw. *Entomologist,* 35: 153–156.
———, 1912. Some remarkable ant-friend Lepidoptera of Queensland. *Transactions of the Entomological Society of London,* 1911(3): 577–590.
Dodson, C. H., 1966. Ethology of some bees of the tribe Euglossini (Hymenoptera: Apidae). *Journal of the Kansas Entomological Society,* 39(4): 607–629.
Doflein, F., 1905. Beobachtungen an den Weberameisen (*Oecophylla smaragdina*). *Biologisches Centralblatt,* 25(15): 497–507.
Dönhoff, E., 1855. Über das Herrschen verschiedener Triebe in verschieden Lebensaltern bei den Bienen. Reprinted in *Beiträge zur Bienenkunde.* Pfenningstorff, Berlin. [Cited by Ribbands, 1953].
Donisthorpe, H. St. J. K., 1915. *British ants, their life-history and classification.* William Brendon and Son, Ltd., Plymouth, Eng. xv + 379 pp.
———, 1927. *The guests of British ants, their habits and life-histories.* George Routledge & Sons, Ltd., London. xxiii + 244 pp.
———, 1936. The oldest insect on record. *Entomologist's Record and Journal of Variation,* 48(1): 1–2.
Dostal, Brigitte, 1958. Riechfähigkeit und Zahl der Riechsinneselemente bei der Honigbiene. *Zeitschrift für Vergleichende Physiologie,* 41(2): 179–203.
Douglas, A., and W. L. Brown, 1959. *Myrmecia inquilina* new species: the first parasite among the lower ants. *Insectes Sociaux,* 6(1): 13–19.
Downey, J. C., 1961. Myrmecophily in the Lycaenidae (Lepidoptera). *Proceedings of the North-Central Branch, American Association of Economic Entomologists,* 16: 14–15.
———, 1962. Myrmecophily in *Plebejus* (*Icaricia*) *icarioides* (Lepid.: Lycaenidae). *Entomological News,* 73(3): 57–66.
Drake, C. J., and N. T. Davis, 1959. The morphology, phylogeny, and higher classification of the family Tingidae, including the description of a new genus and species of the subfamily Vianaidinae (Hemiptera: Heteroptera). *Entomologica Americana* (n.s.), 39: 1–100.
Dreyer, W. A., 1942. Further observations on the occurrence and size of ant mounds with reference to their age. *Ecology,* 23(4): 486–490.
Ducke, A., 1910. Révision des guêpes sociales polygames d'Amérique. *Annales Historico-Naturales Musei Nationales Hungarici,* 8(2): 449–544.
———, 1914. Über Phylogenie und Klassifikation der sozialen Vespiden. *Zoologische Jahrbücher, Abt. Syst. Ökol. Geogr. Tiere,* 36(2–3): 303–330.
Duncan, C. D., 1924. *Dolichovespula diabolica* Sauss. and its supposed variety *fernaldi* Lewis (Hymenoptera, Vespidae). *Pan-Pacific Entomologist,* 1(1): 40–42.
———, 1939. A contribution to the biology of North American

vespine wasps. *Stanford University Publications, Biological Sciences,* 8(1): 1–272.

Dupraw, E. J., 1965. Non-Linnaean taxonomy and the systematics of honeybees. *Systematic Zoology,* 14(1): 1–24.

Dybas, H. S., 1961. A new genus of feather-wing beetles from termite nests in Bolivia (Coleoptera: Ptiliidae). *Fieldiana, Zoology,* 44(8): 57–62.

Eason, E. H., 1964. *Centipedes of the British Isles.* Frederick Warne and Co., Ltd., London. x + 294 pp.

Eberhard, Mary Jane West, 1969. The social biology of polistine wasps. *Miscellaneous Publications, Museum of Zoology, University of Michigan,* 140: 1–101.

Ehrhardt, Sophie, 1931. Über Arbeitsteilung bei *Myrmica-* und *Messor-*Arten. *Zeitschrift für Morphologie und Ökologie der Tiere,* 20(4): 755–812.

Eibl-Eibesfeldt, I., and Eleonore Eibl-Eibesfeldt, 1967. Das Parasitenabwehren der Minima-Arbeiterinnen der Blattschneider-Ameise (*Atta cephalotes*). *Zeitschrift für Tierpsychologie,* 24: 278–281.

Eickwort, Kathleen R., 1969a. Separation of the castes of *Polistes exclamans* and notes on its biology (Hym.: Vespidae). *Insectes Sociaux,* 16(1): 67–72.

———, 1969b. Differential variation of males and females in *Polistes exclamans. Evolution,* 23(3): 391–405.

Eidmann, H., 1928. Weitere Beobachtungen über die Koloniegründung einheimischer Ameisen. *Zeitschrift für Vergleichende Physiologie,* 7(1): 39–55.

———, 1929. Die Koloniegründung von *Formica fusca* L. nebst Untersuchungen über den Brutpflegeinstinkt von *Formica rufa* L. *Zoologischer Anzeiger* (*Wasmann-Festband*), 82: 99–114.

———, 1935. Zur Kenntnis der Blattschneiderameise *Atta sexdens* L., insbesondere ihrer Ökologie, Teil I. *Zeitschrift für Angewandte Entomologie,* 22(2): 185–241.

Eisner, T., 1957. A comparative morphological study of the proventriculus of ants (Hymenoptera: Formicidae). *Bulletin of the Museum of Comparative Zoology, Harvard,* 116(8): 439–490.

———, and W. L. Brown, 1958. The evolution and social significance of the ant proventriculus. *Proceedings of the Tenth International Congress of Entomology, Montreal, 1956,* 2: 503–508.

———, and G. M. Happ, 1962. The infrabuccal pocket of a formicine ant: A social filtration device. *Psyche, Cambridge,* 69(3): 107–116.

Elton, C. S., 1932. Territory among wood ants (*Formica rufa* L.) at Picket Hill. *Journal of Animal Ecology,* 1(1): 69–76.

El-Ziady, Samira, and J. S. Kennedy, 1956. Beneficial effects of the common garden ant, *Lasius niger* L., on the black bean aphid, *Aphis fabae* Scopoli. *Proceedings of the Royal Entomological Society of London,* (A)31: 61–65.

Emden, F. van, 1936. Eine interessante zwischen Carabidae und Paussidae vermittelnde Käferlarve. *Arbeiten über Physiologische und Angewandte Entomologie aus Berlin-Dahlem,* 3: 250–256.

Emerson, A. E., 1929. Communication among termites. *Transactions of the Fourth International Congress of Entomology, Ithaca, 1928,* 2: 722–726.

———, 1938. Termite nests—a study of the phylogeny of behavior. *Ecological Monographs,* 8(2): 247–284.

———, 1939. Populations of social insects. *Ecological Monographs,* 9(3): 287–300.

———, 1941. Biological sociology. *Journal of the Scientific Laboratories of Denison University,* 36: 146–155.

———, 1949. The organization of insect societies. *In* W. C. Allee *et al.*, eds., *Principles of animal ecology.* W. B. Saunders Co., Philadelphia. pp. 419–435.

———, 1952. The biogeography of termites. *Bulletin of the American Museum of Natural History,* 99(3): 217–225.

———, 1955. Geographical origins and dispersions of termite genera. *Fieldiana, Zoology,* 37: 465–521.

———, 1956a. Ethospecies, ethotypes, taxonomy, and evolution of *Apicotermes* and *Allognathotermes* (Isoptera, Termitidae). *American Museum Novitates,* 1771: 1–31.

———, 1956b. Regenerative behavior and social homeostasis in termites. *Ecology,* 37(2): 248–258.

———, 1959. Social insects. *Encyclopaedia Britannica,* 20: 871–878.

———, 1962. Vestigial characters, regressive evolution and recapitulation among termites. *In* Proceedings of the New Delhi Symposium, 1960, *Termites in the humid tropics.* UNESCO, Paris. pp. 17–30.

———, 1965. A review of the Mastotermitidae (Isoptera), including a new fossil genus from Brazil. *American Museum Novitates,* 2236: 1–46.

———, 1967. Cretaceous insects from Labrador. 3. A new genus and species of termite (Isoptera: Hodotermitidae). *Psyche, Cambridge,* 74(4): 276–289.

———, 1969. A revision of the Tertiary fossil species of the Kalotermitidae (Isoptera). *American Museum Novitates,* 2359: 1–57.

———, and F. A. Banks, 1957. Five new species and one redescription of the Neotropical genus *Armitermes* Wasmann (Isoptera, Termitidae, Nasutitermitinae). *American Museum Novitates,* 1841: 1–17.

Emerton, J. H., 1911. New spiders from New England. *Transactions of the Connecticut Academy of Arts and Sciences,* 16: 383–407.

Emery, C., 1891. Le formiche dell'ambra siciliana nel Museo Mineralogico dell'Universita di Bologna. *Memorie della R. Accademia delle Scienze dell'Istituto di Bologna,* (5)1: 567–591.

———, 1894. Die Entstehung und Ausbildung des Arbeiterstandes bei den Ameisen. *Biologisches Centralblatt,* 14(2): 53–59.

———, 1895. Die Gattung *Dorylus* Fab. und die systematische Eintheilung der Formiciden. *Zoologische Jahrbücher, Abt. Syst. Ökol. Geogr. Tiere,* 8(5): 685–778.

———, 1896. Le polymorphisme des fourmis et la castration alimentaire. *Compte Rendu des Séances du Troisième Congrès International de Zoologie, Leyde, 1895,* pp. 395–410.

———, 1897. Descriptions de deux fourmis [Hymén.]. *Bulletin de la Société Entomologique de France,* 1897(1): 12–14.
———, 1898. Beiträge zur Kenntniss der paläarktischen Ameisen. *Öfversigt af Finska Vetenskapssocietetens Förhandlingar,* 20: 124–151.
———, 1899. Glanures myrmécologiques [Hymén.]. *Bulletin de la Société Entomologique de France,* 1899(2): 17–20.
———, 1902. An analytical key to the genera of the family Formicidae, for the identification of the workers (tr. W. M. Wheeler). *American Naturalist,* 36: 707–725.
———, 1909. Über den Ursprung der dulotischen, parasitischen und myrmekophilen Ameisen. *Biologisches Centralblatt,* 29(11): 352–362.
———, 1910–1925. Formicidae: Dorylinae, Ponerinae, Dolichoderinae, Myrmicinae, Formicinae. *In* P. Wytsman, *Genera Insectorum,* fasc. 102 (1910); fasc. 118 (1911); fasc. 137 (1912); fasc. 174 (1921–1922); fasc. 183 (1925). V. Verteneuil & L. Desmet, Brussels.
———, 1911. Beobachtungen und Versuche an *Polyergus rufescens. Biologisches Centralblatt,* 31(20): 625–642.
———, 1915. Histoire d'une société expérimentale de *Polyergus rufescens. Revue Suisse de Zoologie,* 23(9): 385–400.
———, 1921. Quels sont les facteurs du polymorphisme du sexe féminin chez les fourmis? *Revue Générale des Sciences Pures et Appliquées,* 32: 737–741.
———, 1922. Hymenoptera. Fam. Formicidae. Subfam. Myrmicinae. *Genera Insectorum,* fasc. 174: 1–397.
———, 1925. Les espèces européennes et orientales du genre *Bothriomyrmex. Bulletin de la Société Vaudoise des Sciences Naturelles,* 56: 5–22.
Entwistle, P. F., 1964. Inbreeding and arrhenotoky in the ambrosia beetle *Xyleborus compactus* (Eichh.) (Coleoptera: Scolytidae). *Proceedings of the Royal Entomological Society of London,* (A)39: 83–88.
Erhardt, H.-J., 1962. Ablage übergrosser Eier durch Arbeiterinnen von *Formica polyctena* Förster (Ins., Hym.) in Gegenwart von Königinnen. *Naturwissenschaften,* 49(22): 524–525.
Ernst, E., 1959. Beobachtungen beim Spritzakt der *Nasutitermes*-Soldaten. *Revue Suisse de Zoologie,* 66(2): 289–295.
———, 1960. Fremde Termitenkolonien in *Cubitermes*-Nestern. *Revue Suisse de Zoologie,* 67(2): 201–206.
Esch, H., 1963. Über die Auswirkung der Futterplatzqualität auf die Schallerzeugung im Werbetanz der Honigbiene (*Apis mellifica*). *Verhandlungen der Deutschen Zoologischen Gesellschaft, Wien, 1962,* pp. 302–309.
———, 1967a. The evolution of bee language. *Scientific American,* 216(4): 96–104.
———, 1967b. Die Bedeutung der Lauterzeugung für die Verständigung der stachellosen Bienen. *Zeitschrift für Vergleichende Physiologie,* 56(2): 199–220.
———, 1967c. The sounds produced by swarming honey bees. *Zeitschrift für Vergleichende Physiologie,* 56(4): 408–411.
———, Ilse Esch, and W. E. Kerr, 1965. Sound: An element common to communication of stingless bees and to dances of the honey bee. *Science,* 149: 320–321.
Escherich, K., 1902. Biologische Studien über algerische Myrmekophilen, zugleich mit allgemeinen Bemerkungen über die Entwicklung und Bedeutung der Symphilie. *Biologisches Centralblatt,* 22(20–22): 638–663.
———, 1905. Das System der Lepismatiden. *Zoologica, Stuttgart,* 18(43): 1–164.
———, 1906. *Die Ameise. Schilderung ihrer Lebensweise.* Friedrich Vieweg und Sohn, Braunschweig. xx + 232 pp.
———, 1909. *Die Termiten oder weissen Ameisen.* Werner Klinkhardt, Leipzig. xii + 198 pp.
———, 1917. *Die Ameise. Schilderung ihrer Lebensweise.* 2nd ed. Friedrich Vieweg und Sohn, Braunschweig. xvi + 348 pp.
Ettershank, G., 1966. A generic revision of the world Myrmicinae related to *Solenopsis* and *Pheidologeton* (Hymenoptera: Formicidae). *Australian Journal of Zoology,* 14: 73–171.
———, 1967. A completely defined synthetic diet for ants (Hym., Formicidae). *Entomologist's Monthly Magazine,* 103: 66–67.
Evans, H. E., 1958. The evolution of social life in wasps. *Proceedings of the Tenth International Congress of Entomology, Montreal, 1956,* 2: 449–457.
———, 1964. Observations on the nesting behavior of *Moniaecera asperata* (Fox) (Hymenoptera, Sphecidae, Crabroninae) with comments on communal nesting in solitary wasps. *Insectes Sociaux,* 11(1): 71–78.
———, 1966. *The comparative ethology and evolution of the sand wasps.* Harvard University Press, Cambridge, Mass. xvi + 526 pp.
———, 1969. Three new Cretaceous aculeate wasps (Hymenoptera). *Psyche, Cambridge,* 76(3): 251–261.
———, and Mary Jane West Eberhard, 1970. *The wasps.* University of Michigan Press, Ann Arbor. vi + 265 pp.
———, C. S. Lin, and C. M. Yoshimoto, 1953. A biological study of *Anoplius apiculatus autumnalis* (Banks) and its parasite, *Evagetes mohave* (Banks). *Journal of the New York Entomological Society,* 61(2): 61–78.
Ezhikov, T., 1934. Individual variability and dimorphism of social insects. *American Naturalist,* 68: 333–344.
Faber, W., 1967. Beiträge zur Kenntnis sozialparasitischer Ameisen. 1. *Lasius* (*Austrolasius* n. sg.) *reginae* n. sp., eine neue temporär sozialparasitische Erdameise aus Österreich (Hym. Formicidae). *Pflanzenschutz-Berichte,* 36(5–7): 73–107.
———, 1969. Beiträge zur Kenntnis sozialparasitischer Ameisen. 2. *Aporomyrmex ampeloni* nov. gen., nov. spec. (Hym. Formicidae), ein neuer permanenter Sozialparasit bei *Plagiolepis vindobonensis* Lomnicki aus Österreich. *Pflanzenschutz-Berichte,* 39(3–6): 39–100.
Fabre, J. H., 1918. *The sacred beetle and others* (tr. A. T. de Mattos). Dodd, Mead and Co., New York. xxvi + 425 pp.
Fage, L., 1938. Quelques Arachnides provenant de fourmilières

ou de termitières du Costa Rica. *Bulletin du Muséum d'Histoire Naturelle, Paris,* (2)10(4): 369–376.

Fall, H. C., 1912. Four new myrmecophilous Coleoptera. *Psyche, Cambridge,* 19(1): 9–12.

———, 1937. The North American species of *Nemadus* Thom., with descriptions of new species (Coleoptera, Silphidae). *Journal of the New York Entomological Society,* 45(3–4): 335–340.

Farish, D. J., 1969. Grooming in the Hymenoptera (Insecta). Ph.D. Thesis, Department of Biology, Harvard University. 176 pp.

Farquharson, C. O., 1918. *Harpagomyia* and other Diptera fed by *Cremastogaster* ants in S. Nigeria. *Proceedings of the Entomological Society of London,* 1918 (pt. I): xxix–xxxix.

Faure, J. C., 1940. Maternal care displayed by mantids (Orthoptera). *Journal of the Entomological Society of Southern Africa,* 3: 139–150.

Fenton, F. A., 1952. *Field crop insects.* Macmillan Company, New York. ix + 405 pp.

Ferguson, W. E., 1962. Biological characteristics of the mutillid subgenus *Photopsis* Blake and their systematic values (Hymenoptera). *University of California Publications in Entomology,* 27(1): 1–92.

Fernández-Morán, H., 1958. Fine structure of the light receptors in the compound eyes of insects. *Experimental Cell Research,* suppl. 5: 586–644.

Fiebrig, K., 1907. Nachtrag zu: Eine Wespen zerstörende Ameise aus Paraguay. *Zeitschrift für Wissenschaftliche Insektenbiologie,* 3(5–6): 154–156.

Fielde, Adele M., 1901. A study of an ant. *Proceedings of the Academy of Natural Sciences of Philadelphia,* 53(2): 425–449.

———, 1903. Artificial mixed nests of ants. *Biological Bulletin, Marine Biological Laboratory, Woods Hole,* 5(6): 320–325.

———, 1904a. Power of recognition among ants. *Biological Bulletin, Marine Biological Laboratory, Woods Hole,* 7(5): 227–250.

———, 1904b. Tenacity of life in ants. *Biological Bulletin, Marine Biological Laboratory, Woods Hole,* 7(6): 300–309.

———, 1905. The progressive odor of ants. *Biological Bulletin, Marine Biological Laboratory, Woods Hole,* 10(1): 1–16.

———, and G. H. Parker, 1904. The reactions of ants to material vibrations. *Proceedings of the Academy of Natural Sciences of Philadelphia,* 56(2): 642–650.

Fink, D. E., 1915. The eggplant lace-bug. *Bulletin, United States Department of Agriculture,* 239: 1–7.

Finke, Ingrid, 1958. Zeitgedächtnis und Sonnenorientierung der Bienen. Lehramtsarbeit Naturwissenschaftliche Fakultät der Universität München. [Cited by von Frisch, 1967a].

Fischer, W., 1957. Untersuchungen über die Riechschärfe der Honigbiene. *Zeitschrift für Vergleichende Physiologie,* 39(6): 634–659.

Flanders, S. E., 1945. Is caste differentiation in ants a function of the rate of egg deposition? *Science,* 101: 245–246.

———, 1957. Regulation of caste in social Hymenoptera. *Journal of the New York Entomological Society,* 65: 97–105.

Fletcher, D. J. C., and J. M. Brand, 1968. Source of the trail pheromone and method of trail laying in the ant *Crematogaster peringueyi. Journal of Insect Physiology,* 14(6): 783–788.

Florey, E., 1954. Experimentelle Erzeugung einer "Neurose" bei der Honigbiene. *Naturwissenschaften,* 41(7): 171.

Folsom, J. W., 1923. Termitophilous Apterygota from British Guiana. *Zoologica, New York,* 3(19): 383–402.

Forbes, J., 1954. The anatomy and histology of the male reproductive system of *Camponotus pennsylvanicus* DeGeer (Formicidae Hymenoptera). *Journal of Morphology,* 95(3): 523–555.

Forbes, S. A., 1906. The corn root-aphis and its attendant ant. (*Aphis maidi-radicis* Forbes and *Lasius niger* L., var. *americanus* Emery). *Bulletin, United States Department of Agriculture, Division of Entomology,* 60: 29–39.

Forel, A., 1874. *Les fourmis de la Suisse.* Société Helvétique des Sciences Naturelles, Zurich. iv + 452 pp. (1920, revised and corrected. Imprimerie Coopérative, La Chaux-de-fonds. xvi + 333 pp.)

———, 1878. Études myrmécologiques en 1878 (première partie) avec l'anatomie du gésier des fourmis. *Bulletin de la Société Vaudoise des Sciences Naturelles,* 15(80): 337–392.

———, 1886. Études myrmécologiques en 1886. *Annales de la Société Entomologique de Belgique,* 30: 131–215.

———, 1896. Ants' nests. *Report of the Smithsonian Institution for 1894,* pp. 479–505.

———, 1898. La parabiose chez les fourmis. *Bulletin de la Société Vaudoise des Sciences Naturelles,* 34(130): 380–384.

———, 1900. Fourmis du Japon. Nids en toile. *Strongylognathus huberi* et voisins. Fourmilière triple. *Cyphomyrmex wheeleri.* Fourmis importées. *Mitteilungen der Schweizerischen Entomologischen Gesellschaft,* 10(7): 267–287.

———, 1901a. Die psychischen Fähigkeiten der Ameisen und einiger anderer Insekten. *Verhandlungen des V. Internationalen Zoologen-Congresses zu Berlin, 1901,* pp. 141–169.

———, 1901b. Fourmis termitophages, Lestobiose, *Atta tardigrada,* sous-genres d'*Euponera. Annales de la Société Entomologique de Belgique,* 45(12): 389–398.

———, 1902. Beispiele phylogenetischer Wirkungen und Rückwirkungen bei den Instinkten und dem Körperbau der Ameisen als Belege für die Evolutionslehre und die psychophysiologische Identitätslehre. *Journal für Psychologie und Neurologie,* 1: 99–110.

———, 1906. Moeurs des fourmis parasites des genres *Wheeleria* et *Bothriomyrmex. Revue Suisse de Zoologie,* 14(1): 51–69.

———, 1908. *The senses of insects* (tr. of 1910b by M. Yearsley). Methuen and Co., London. xv + 324 pp.

———, 1910a. Glanures myrmécologiques. *Annales de la Société Entomologique de Belgique,* 54: 6–32.

———, 1910b. *Das Sinnesleben der Insekten.* E. Reinhardt, Munich.
———, 1921–1923. *Le monde social des fourmis du globe comparée à celui de l'homme,* 5 vols. Libraire Kundig, Geneva. xxxvii + 948 pp.
———, 1928. *The social world of the ants compared with that of man,* 2 vols. (tr. C. K. Ogden). G. P. Putnam's Sons, Ltd., London. xlv + 551, xx + 445 pp.
Forrest, Helen F., 1963. Three problems in invertebrate behavior. III. The production of audible sound by common ants and its possible uses in communication, with special reference to stridulation. Ph.D. Thesis, Rutgers University, New Brunswick, New Jersey. pp. 221–350.
Francfort, R., 1945. Quelques phénomènes illustrant l'influence de la fourmilière sur les fourmis isolées. *Bulletin de la Société Entomologique de France,* 50(2): 95–96.
Francke-Grosmann, Helene, 1956. Hautdrüsen als Träger der Pilzsymbiose bei Ambrosiakäfern. *Zeitschrift für Morphologie und Ökologie der Tiere,* 45(3): 275–308.
———, 1967. Ectosymbiosis in wood-inhabiting insects. *In* S. M. Henry, ed., *Symbiosis, Vol. II.* Academic Press, Inc., New York. pp. 141–205.
Francoeur, A., 1965. Écologie des populations de fourmis dans un bois de chênes rouges et d'érables rouges. *Naturaliste Canadien,* 92(10–11): 263–276.
Frank, A., 1941. Eigenartige Flugbahnen bei Hummelmännchen. *Zeitschrift für Vergleichende Physiologie,* 28: 467–484.
Franklin, H. J., 1912–1913. The Bombidae of the New World. *Transactions of the American Entomological Society,* 38(3–4): 177–486, 39(2): 73–200.
Free, J. B., 1955a. The behaviour of egg-laying workers of bumblebee colonies. *British Journal of Animal Behaviour,* 3(4): 147–153.
———, 1955b. The division of labour within bumblebee colonies. *Insectes Sociaux,* 2(3): 195–212.
———, 1955c. Queen production in colonies of bumblebees. *Proceedings of the Royal Entomological Society of London,* (A)30: 19–25.
———, 1956. A study of the stimuli which release the food begging and offering responses of worker honeybees. *British Journal of Animal Behaviour,* 4(3): 94–101.
———, 1957a. The food of adult drone honeybees (*Apis mellifera*). *British Journal of Animal Behaviour,* 5(1): 7–11.
———, 1957b. The transmission of food between worker honeybees. *British Journal of Animal Behaviour,* 5(2): 41–47.
———, 1957c. The effect of social facilitation on the ovarial development of bumble-bee workers. *Proceedings of the Royal Entomological Society of London,* (A)32: 182–184.
———, 1958. The defence of bumblebee colonies. *Behaviour,* 12(3): 233–242.
———, 1959. The transfer of food between the adult members of a honeybee community. *Bee World,* 40(8): 193–201.
———, 1960. The distribution of bees in a honey-bee (*Apis mellifera* L.) colony. *Proceedings of the Royal Entomological Society of London,* (A)35: 141–144.
———, 1961a. The social organization of the bumble-bee colony. A lecture given to The Central Association of Bee-keepers on 18th January, 1961. North Hants Printing and Publishing Co., Ltd., Fleet, Hants, England. 11 pp.
———, 1961b. Hypopharyngeal gland development and division of labour in honey-bee (*Apis mellifera* L.) colonies. *Proceedings of the Royal Entomological Society of London,* (A)36: 5–8.
———, 1965. The allocation of duties among worker honeybees. *Symposium of the Zoological Society of London,* 14: 39–59.
———, 1969. Influence of the odour of a honeybee colony's food stores on the behaviour of its foragers. *Nature, London,* 222: 778.
———, and C. G. Butler, 1959. *Bumblebees* (New Naturalist Series). Collins, London. xiv + 208 pp.
Freeland, J., 1958. Biological and social patterns in the Australian bulldog ants of the genus *Myrmecia. Australian Journal of Zoology,* 6(1): 1–18.
Freisling, J., 1938. Die Bauinstinkte der Wespen (Vespidae). *Zeitschrift für Tierpsychologie,* 2(1): 81–98.
Friese, H., 1909. Die Bienen Afrikas nach dem Stande unserer heutigen Kenntnisse [Dr. Leonhard Schultze, Forschungsreise in Südafrika II(2)]. *Jenaische Denkschriften,* 14: 85–475.
———, 1923. *Die europäischen Bienen (Apidae). Das Leben und Wirken unserer Blumenwespen.* de Gruyter, Berlin and Leipzig. vii + 456 pp.
Frisch, K. von, 1914. Der Farbensinn und Formensinn der Biene. *Zoologische Jahrbücher, Abt. Allgem. Zool. Physiol. Tiere,* 35(1–2): 1–188.
———, 1919. Über den Geruchsinn der Biene und seine blütenbiologische Bedeutung. *Zoologische Jahrbücher, Abt. Allgem. Zool. Physiol. Tiere,* 37(1–2): 1–238.
———, 1921. Über den Sitz des Geruchsinnes bei Insekten. *Zoologische Jahrbücher, Abt. Allgem. Zool. Physiol. Tiere,* 38(1): 1–68.
———, 1934. Über den Geschmackssinn der Biene. *Zeitschrift für Vergleichende Physiologie,* 21(1): 1–156.
———, 1940. Die Tänze und das Zeitgedächtnis der Bienen im Widerspruch. *Naturwissenschaften,* 28(5): 65–69.
———, 1949. Die Polarisation des Himmelslichtes als orientierender Faktor bei den Tänzen der Bienen. *Experientia,* 5(4): 142–148.
———, 1950. Die Sonne als Kompass im Leben der Bienen. *Experientia,* 6(6): 210–221.
———, 1954. *The dancing bees: An account of the life and senses of the honey bee* (tr. Dora Ilse). Methuen and Company, Ltd., London. xiv + 183 pp.
———, 1965. *Tanzsprache und Orientierung der Bienen.* Springer-Verlag, Berlin. vii + 578 pp. [Tr. to English by L. E. Chadwick, 1967a, as *The dance language and orientation of bees.*]

———, 1967a. *The dance language and orientation of bees* (tr. Chadwick). Belknap Press of Harvard University Press, Cambridge. xiv + 566 pp.

———, 1967b. Honeybees: Do they use direction and distance information provided by their dancers? *Science,* 158: 1072–1076.

———, and R. Jander, 1957. Über den Schwänzeltanz der Bienen. *Zeitschrift für Vergleichende Physiologie,* 40(3): 239–263.

———, and G. A. Rösch, 1926. Neue Versuche über die Bedeutung von Duftorgan und Pollenduft für die Verständigung im Bienenvolk. *Zeitschrift für Vergleichende Physiologie,* 4(1): 1–21.

Frison, T. H., 1926. Contribution to the knowledge of the interrelations of the bumblebees of Illinois with their animate environment. *Annals of the Entomological Society of America,* 19(2): 203–235.

———, 1930. A contribution to the knowledge of the bionomics of *Bremus americanorum* (Fabr.) (Hymenoptera). *Annals of the Entomological Society of America,* 23(4): 644–665.

Frohawk, F. W., 1915. Further observations on the last stage of the larva of *Lycaena arion. Transactions of the Entomological Society of London,* 1915 (pt. II): 313–316.

Fulton, B. B., 1924. Some habits of earwigs. *Annals of the Entomological Society of America,* 17(4): 357–367.

Gadgil, M., and W. H. Bossert, 1970. Life historical consequences of natural selection. *American Naturalist,* 104: 1–24.

Gallardo, A., 1916a. Las hormigas de la República Argentina. Subfamilia Dolicoderinas. *Anales del Museo Nacional de Historia Natural de Buenos Aires,* 28: 1–130.

———, 1916b. Notes systématiques et éthologiques sur les fourmis attines de la République Argentine. *Anales del Museo Nacional de Historia Natural de Buenos Aires,* 28: 317–344.

———, 1918. Las hormigas de la República Argentina. Subfamilia Ponerinas. *Anales del Museo Nacional de Historia Natural de Buenos Aires,* 30: 1–112.

———, 1920. Las hormigas de la República Argentina. Subfamilia Dorilinas. *Anales del Museo Nacional de Historia Natural de Buenos Aires,* 30: 281–410.

———, 1932. Las hormigas de la República Argentina. Subfamilia Mirmicinas. *Anales del Museo Nacional de Historia Natural de Buenos Aires,* 37: 37–170.

Gary, N. E., 1961. Queen honey bee attractiveness as related to mandibular gland secretion. *Science,* 133: 1479–1480.

Gast, R., 1967. Untersuchungen über den Einfluss der Königinnensubstanz auf die Entwicklung der endokrinen Drüsen bei der Arbeiterin der Honigbiene (*Apis mellifica*). *Insectes Sociaux,* 14(1): 1–12.

Gates, B. N., 1914. The temperature of the bee colony. *Bulletin, United States Department of Agriculture,* 96: 1–29.

Gaul, A. T., 1941. Experiments in housing vespine colonies, with notes on the homing and toleration instincts of certain species. *Psyche, Cambridge,* 48(1): 16–19.

———, 1947. Additions to Vespine biology. III. Notes on the habits of *Vespula squamosa* Drury (Hymenoptera, Vespidae). *Bulletin of the Brooklyn Entomological Society,* 42(3): 87–96.

———, 1948. Additions to Vespine biology. V. The distribution of labor in the colonies of hornets and yellowjackets. *Bulletin of the Brooklyn Entomological Society,* 43(3): 73–79.

———, 1952. Additions to Vespine biology. IX. Temperature regulation in the colony. *Bulletin of the Brooklyn Entomological Society,* 47(3): 79–82.

Gause, G. F., and A. A. Witt, 1935. Behavior of mixed populations and the problem of natural selection. *American Naturalist,* 69: 596–609.

Gay, F. J., 1956. New species of termites from Australia. *Proceedings of the Linnean Society of New South Wales,* 80(3): 207–213.

———, 1966. A new genus of termites (Isoptera) from Australia. *Journal of the Entomological Society of Queensland,* 5: 40–43.

———, 1967. A world review of introduced species of termites. *Bulletin, Commonwealth Scientific and Industrial Research Organization* (Melbourne), 286: 1–88.

———, and T. Greaves, 1940. The population of a mound colony of *Coptotermes lacteus* (Frogg.). *Journal of the Council for Scientific and Industrial Research, Australia,* 13(2): 145–149.

———, T. Greaves, F. G. Holdaway, and A. H. Wetherly, 1955. Standard laboratory colonies of termites for evaluating the resistance of timber, timber preservatives, and other materials to termite attack. *Bulletin, Commonwealth Scientific and Industrial Research Organization* (Melbourne), 277: 1–60.

Gerstung, F., 1891–1926. *Der Bien und seine Zucht,* 7 eds. Pfenningsdorf, Berlin. [Cited by Ribbands, 1953].

Gervet, J., 1956. L'action des températures différentielles sur la monogynie fonctionnelle chez les *Polistes* (Hyménoptères Vespides). *Insectes Sociaux,* 3(1): 159–176.

———, 1962. Étude de l'effet de groupe sur la ponte dans la société polygyne de *Polistes gallicus. Insectes Sociaux,* 9(3): 231–263.

———, 1964. Le comportement d'oophagie différentielle chez *Polistes gallicus* L. (Hymen. Vesp.). *Insectes Sociaux,* 11(4): 343–382.

Ghent, A. W., 1960. A study of the group-feeding behaviour of larvae of the Jack Pine Sawfly, *Neodiprion pratti banksianae* Roh. *Behaviour,* 16(1–2): 110–148.

Ghent, R. L., and N. E. Gary, 1962. A chemical alarm releaser in honey bee stings (*Apis mellifera* L.). *Psyche, Cambridge,* 69(1): 1–6.

Gillaspy, J. E., 1963. The genus *Stizoides* (Hymenoptera: Sphecidae: Stizini) in North America, with notes on the Old World fauna. *Bulletin of the Museum of Comparative Zoology, Harvard,* 128(7): 369–391.

Glick, P. A., 1939. The distribution of insects, spiders and mites in the air. *Technical Bulletin, United States Department of Agriculture,* 673: 1–60.

Glöckner, W. E., 1957. Ueber Schnuervesuche an Formiciden waehrend der Metamorphose. *Insectes Sociaux,* 4(2): 83–90.

Goetsch, W., 1937a. Koloniegründung und Kastenbildung im Ameisenstaat. *Forschungen und Fortschritte,* 14: 223.

———, 1937b. Die Entstehung der "Soldaten" im Ameisenstaat. *Naturwissenschaften,* 25(50): 803–808.

———, 1946. Vitamin "T," ein neuartiger Wirkstoff. *Österreichische Zoologische Zeitschrift,* 1: 49–57.

———, 1947. Der Einfluss von Vitamin T auf Gestalt und Gewohnheiten von Insekten. *Österreichische Zoologische Zeitschrift,* 1: 193–247.

———, 1951. Ergebnisse und Probleme aus dem Gebiet neuer Wirkstoffe. *Österreichische Zoologische Zeitschrift,* 3: 140–174.

———, 1953a. *Die Staaten der Ameisen,* 2nd ed. Springer-Verlag, Berlin. viii + 152 pp.

———, 1953b. *Vergleichende Biologie der Insekten-Staaten.* Geest & Portig K.-G., Leipzig. viii + 482 pp.

———, and H. Eisner, 1930. Beiträge zur Biologie Körnersammelnder Ameisen, II. Teil. *Zeitschrift für Morphologie und Ökologie der Tiere,* 16(3–4): 371–452.

———, and Br. Käthner, 1937. Die Koloniegründung der Formicinen und ihre experimentelle Beeinflussung. *Zeitschrift für Morphologie und Ökologie der Tiere,* 33(2): 201–260.

Goidanich, A., 1959. Le migrazioni coatte mirmecogene dello *Stomaphis quercus* Linnaeus, Afide olociclico monoico omotopo (Hemiptera Aphidoidea Lachnidae). *Bollettino dell'Istituto di Entomologia della Università degli Studi di Bologna,* 23: 93–131.

Goldsmith, T. H., 1958. The visual system of the honeybee. *Proceedings of the National Academy of Sciences of the U.S.A.,* 44(2): 123–126.

———, 1962. Fine structure of the retinulae in the compound eye of the honey-bee. *Journal of Cell Biology,* 14(3): 489–494.

———, and H. R. Fernandez, 1966. Some photochemical and physiological aspects of visual excitation in compound eyes. *In* C. G. Bernhard, ed., *The functional organization of the compound eye.* Pergamon Press, Oxford, Eng. pp. 125–143.

Goll, W., 1967. Strukturuntersuchungen am Gehirn von *Formica. Zeitschrift für Morphologie und Ökologie der Tiere,* 59(2): 143–210.

Golley, F. B., and J. B. Gentry, 1964. Bioenergetics of the southern harvester ant, *Pogonomyrmex badius. Ecology,* 45(2): 217–225.

Gonçalves, L. S., 1969. A study of orientation information given by one trained bee by dancing. *Journal of Apicultural Research,* 8(3): 113–132.

Gösswald, K., 1932. Ökologische Studien über die Ameisenfauna des mittleren Maingebietes. *Zeitschrift für Wissenschaftliche Zoologie,* 142(1–2): 1–156.

———, 1933. Weitere Untersuchungen über die Biologie von *Epimyrma gösswaldi* Men. und Bemerkungen über andere parasitische Ameisen. *Zeitschrift für Wissenschaftliche Zoologie,* 144(2): 262–288.

———, 1934. Die Grundzüge der stammesgeschichtlichen Entwicklung des Ameisenparasitismus, neu beleuchtet durch die Entdeckung einer weiteren parasitischen Ameise. *Entomologische Beihefte aus Berlin-Dahlem,* 1: 57–62.

———, 1938. Über den Einfluss von verschiedener Temperatur und Luftfeuchtigkeit auf die Lebensäusserungen der Ameisen. *Zeitschrift für Wissenschaftliche Zoologie,* 151(3): 337–381.

———, 1951a. *Die rote Waldameise im Dienste der Waldhygiene. Forstwirtschaftliche Bedeutung, Nutzung, Lebensweise, Zucht, Vermehrung und Schutz.* Metta Kinau Verlag, Lüneburg. 160 pp.

———, 1951b. Zur Ameisenfauna des Mittleren Maingebietes mit Bemerkungen über Veränderungen seit 25 Jahren. *Zoologische Jahrbücher, Abt. Syst. Ökol. Geogr. Tiere,* 80(5–6): 507–532.

———, 1951c. Versuche zum Sozialparasitismus der Ameisen bei der Gattung *Formica* L. *Zoologische Jahrbücher, Abt. Syst. Ökol. Geogr. Tiere,* 80(5–6): 533–582.

———, 1953. Histologische Untersuchungen an den arbeiterlosen Ameise *Teleutomyrmex schneideri* Kutter (Hym. Formicidae). *Mitteilungen der Schweizerischen Entomologischen Gesellschaft,* 26(2): 81–128.

———, 1957. Über die biologischen Grundlagen der Zucht und Anweiselung junger Königinnen der Kleinen Roten Waldameise nebst praktischen Erfahrungen. *Waldhygiene,* 2: 33–53.

———, and K. Bier, 1955. Beeinflussung des Geschlechtsverhältnisses durch Temperatureinwirkung bei *Formica rufa* L. *Naturwissenschaften,* 42(5): 133–134.

———, and K. Bier, 1957. Untersuchungen zur Kastendetermination in der Gattung *Formica.* 5. Der Einfluss der Temperatur auf die Eiablage und Geschlechtsbestimmung. *Insectes Sociaux,* 4(4): 335–348.

———, and W. Kloft, 1956. Untersuchungen über die Verteilung von radioaktiv markiertem Futter im Volk der Kleinen Roten Waldameise (*Formica rufopratensis minor*). *Waldhygiene,* 1: 200–202.

———, and W. Kloft, 1960. Neuere Untersuchungen über die sozialen Wechselbeziehungen im Ameisenvolk, durchgeführt mit Radio-Isotopen. *Zoologische Beiträge,* 5(2–3): 519–556.

Gotwald, W. H., and W. L. Brown, 1966. The ant genus *Simopelta* (Hymenoptera: Formicidae). *Psyche, Cambridge,* 73(4): 261–277.

Gould, W., 1747. *An account of English ants.* A. Millar, London. xv + 109 pp.

Grabensberger, W., 1933. Untersuchungen über das Zeitge-

dächtnis der Ameisen und Termiten. *Zeitschrift für Vergleichende Physiologie,* 20(1-2): 1-54.
Graham, K., 1967. Fungal-insect mutualism in trees and timber. *Annual Review of Entomology,* 12: 105-126.
Grandi, G., 1961. Studi di un entomologo sugli imenotteri superiori. *Bollettino dell'Istituto di Entomologia della Università degli Studi di Bologna,* 25: 1-659.
Grassé, P.-P. 1939. Rapports d'une larve de Coléoptère termitophile (*Troctontus appendiculatus* Silv.) avec ses hôtes. *Compte Rendu de l'Académie des Sciences, Paris,* 208(11): 831-832.
———, 1942. L'essaimage des Termites: Essai d'analyse causale d'un complex instinctif. *Bulletin Biologique de la France et de la Belgique,* 76(4): 347-382.
———, 1946. Sociétés animales et effet de groupe. *Experientia,* 2(3): 77-82.
———, 1949. Ordre des Isoptères ou Termites. *In* P.-P. Grassé, ed., *Traité de Zoologie, Vol. IX.* Masson et Cie, Paris. pp. 408-544.
———, 1959. La reconstruction du nid et les coordinations interindividuelles chez *Bellicositermes natalensis* et *Cubitermes* sp. La théorie de la stigmergie: Essai d'interprétation du comportement des termites constructeurs. *Insectes Sociaux,* 6(1): 41-83.
———, 1967. Nouvelles expériences sur le termite de Müller (*Macrotermes mülleri*) et considérations sur la théorie de la stigmergie. *Insectes Sociaux,* 14(1): 73-102.
———, and R. Chauvin, 1944. L'effet de groupe et la survie des neutres dans les sociétés d'insectes. *Revue Scientifique,* 82(7): 461-464.
———, and C. Noirot, 1945. La transmission des Flagellés symbiotiques et les aliments des Termites. *Bulletin Biologique de la France et de la Belgique,* 79(4): 273-292.
———, and C. Noirot, 1946a. La production des sexués néoténiques chez le termite à cou jaune (*Calotermes flavicollis* F.): inhibition germinale et inhibition somatique. *Compte Rendu de l'Académie des Sciences, Paris,* 223(21): 869-871.
———, and C. Noirot, 1946b. Le polymorphisme social du termite à cou jaune (*Calotermes flavicollis* F.). La production des soldats. *Compte Rendu de l'Académie des Sciences, Paris,* 223(22): 929-931.
———, and C. Noirot, 1947. Le polymorphisme social du termite à cou jaune (*Calotermes flavicollis* F.). Les faux-ouvriers ou pseudergates et les mues régressives. *Compte Rendu de l'Académie des Sciences, Paris,* 224(3): 219-221.
———, and C. Noirot, 1948. Sur le nid et la biologie du *Sphaerotermes sphaerothorax* (Sjöstedt), termite constructeur de meules sans champignons. *Annales des Sciences Naturelles,* (11)10(2): 149-166.
———, and C. Noirot, 1951a. Nouvelles recherches sur la biologie de divers Termites champignonnistes (Macrotermitinae). *Annales des Sciences Naturelles,* (11)13(3): 291-342.
———, and C. Noirot, 1951b. La sociotomie: migration et fragmentation de la termitière chez les *Anoplotermes* et les *Trinervitermes. Behaviour,* 3(2): 146-166.
———, and C. Noirot, 1958a. La meule des termites champignonnistes et sa signification symbiotique. *Annales des Sciences Naturelles,* (11)20(2): 113-128.
———, and C. Noirot, 1958b. La société de *Calotermes flavicollis* (Insecte, Isoptère), de sa fondation au premier essaimage. *Compte Rendu de l'Académie des Sciences, Paris,* 246(12): 1789-1795.
———, and C. Noirot, 1958c. Construction et architecture chez les termites champignonnistes (Macrotermitinae). *Proceedings of the Tenth International Congress of Entomology, Montreal, 1956,* 2: 515-520.
———, and C. Noirot, 1959. L'évolution de la symbiose chez les Isoptères. *Experientia,* 15(10): 365-372.
———, and C. Noirot, 1961. Nouvelles recherches sur la systématique et l'éthologie des termites champignonnistes du genre *Bellicositermes* Emerson. *Insectes Sociaux,* 8(4): 311-359.
———, and R. Poisson, 1940. Recherches sur les insectes termitophiles. I. Une nouvelle espèce de *Termitodiscus* [Col. Staphylinidae] et son éthologie. *Bulletin de la Société Entomologique de France,* 45(8): 82-90.
Grassi, B., and Andrea Sandias, 1893-1894. Costituzione e sviluppo della società dei Termitidi. Osservazioni sui loro costumi con un'Appendice sui Protozoi Parassiti dei Termitidi e sulla famiglia delle Embidine. *Atti dell'Accademia Gioenia di Scienze Naturali,* (4)6(13): 1-75, (4)7(1): 1-76.
———, and Andrea Sandias, 1896. The constitution and development of the society of termites: Observations on their habits; with appendices on the parasitic Protozoa of Termitidae, and on the Embiidae. *Quarterly Journal of Microscopical Science,* 39(155): 245-322. [English tr., 1893-1894.]
Greaves, T., 1962. Studies of foraging galleries and the invasion of living trees by *Coptotermes acinaciformis* and *C. brunneus* (Isoptera). *Australian Journal of Zoology,* 10(4): 630-651.
———, 1964. Temperature studies of termite colonies in living trees. *Australian Journal of Zoology,* 12(2): 250-262.
Green, E. E., 1900. Note on the attractive properties of certain larval Hemiptera. *Entomologist's Monthly Magazine,* 36(2nd ser., 11): 185.
Gregg, R. E., 1942. The origin of castes in ants with special reference to *Pheidole morrisi* Forel. *Ecology,* 23(3): 295-308.
Grodins, F. S., 1963. *Control theory and biological systems.* Columbia University Press, New York. vii + 205 pp.
Groot, A. P. de, 1953. Protein and amino acid requirements of the honeybee (*Apis mellifica* L.). *Physiologia Comparata et Oecologia,* 3(2-3): 197-285.
Grütte, E., 1935. Zur Abstammung der Kuckucksbienen (Hymenopt. Apid.) *Archiv für Naturgeschichte,* (n.f.), 4(4): 449-534.

Gupta, S. D., 1953. Ecological studies of termites. Pt. I. Population of the mound-building termite, *Odontotermes obesus* (Rambur) (Isoptera: Family Termitidae). *Proceedings of the National Institute of Sciences of India,* 19(5): 697-704.

Haas, A., 1946. Neue Beobachtungen zum Problem der Flugbahnen bei Hummelmännchen. *Zeitschrift für Naturforschung,* 1(10): 596-600.

———, 1952. Die Mandibeldrüse als Duftorgan bei einigen Hymenopteren. *Naturwissenschaften,* 39(20): 484.

Hagan, H. R., 1954. The reproductive system of the army-ant queen, *Eciton* (*Eciton*). Pt. I. General anatomy. *American Museum Novitates,* 1663: 1-12.

Hagen, H. A., 1858. Monographie der Termiten. *Linnaea Entomologica* (*Z. Ent. Ver. Stettin.*), 12: 1-461.

Hagen, K. S., 1962. Biology and ecology of predaceous Coccinellidae. *Annual Review of Entomology,* 7: 289-326.

Hagen, V. W. von, 1939. The ant that carries a parasol. *Natural History,* 43(1): 27-32.

Hagmann, G., 1907. Beobachtungen über einen myrmekophilen Schmetterling am Amazonenstrom. *Biologisches Centralblatt,* 27(11): 337-341.

Hahn, E., 1958. Untersuchungen über die Lebensweise und Entwicklung der Maulwurfsgrille (*Gryllotalpa vulgaris* Latr.) im Lande Brandenburg. *Beiträge zur Entomologie,* 8(3-4): 334-365.

Haldane, J. B. S., 1932. *The causes of evolution.* Longmans, Green and Co., Ltd., London. vii + 235 pp.

———, 1955. Animal communication and the origin of human language. *Science Progress,* 43(171): 385-401.

———, and Helen Spurway, 1954. A statistical analysis of communication in "*Apis mellifera*" and a comparison with communication in other animals. *Insectes Sociaux,* 1(3): 247-283.

Hamilton, W. D., 1964. The genetical evolution of social behavior, I and II. *Journal of Theoretical Biology,* 7: 1-52.

Hammann, Eleonore, 1957. Wer hat die Initiative bei den Ausflügen der Jungkönigin, die Königin oder die Arbeitsbienen? *Insectes Sociaux,* 4(2): 91-106.

Hangartner, W., 1967. Spezifität und Inaktivierung des Spurpheromons von *Lasius fuliginosus* Latr. und Orientierung der Arbeiterinnen im Duftfeld. *Zeitschrift für Vergleichende Physiologie,* 57(2): 103-136.

———, 1969a. Carbon dioxide, releaser for digging behavior in *Solenopsis geminata* (Hymenoptera: Formicidae). *Psyche, Cambridge,* 76(1): 58-67.

———, 1969b. Structure and variability of the individual odor trail in *Solenopsis geminata* Fabr. (Hymenoptera, Formicidae). *Zeitschrift für Vergleichende Physiologie,* 62(1): 111-120.

———, 1969c. Trail laying in the subterranean ant, *Acanthomyops interjectus. Journal of Insect Physiology,* 15(1): 1-4.

———, and S. Bernstein, 1964. Über die Geruchsspur von *Lasius fuliginosus* zwischen Nest und Futterquelle. *Experientia,* 20(7): 392-393.

———, J. M. Reichson, and E. O. Wilson, 1970. Orientation to nest material by the ant, *Pogonomyrmex badius* (Latreille). *Animal Behaviour,* 18(2): 331-334.

Hansson, Å., 1945. Lauterzeugung und Lautauffassungsvermögen der Bienen. *Opuscula Entomologica,* suppl. 6: 1-124.

Hardin, G., 1956. Meaninglessness of the word protoplasm. *Scientific Monthly, New York,* 82(3): 112-120.

Harris, W. V., 1961. *Termites, their recognition and control.* Longmans, Green & Co., Ltd., London. xii + 187 pp.

———, and W. A. Sands, 1965. The social organization of termite colonies. *Symposium of the Zoological Society of London,* 14: 113-131.

Hartig, T., 1844. Ambrosia des *Bostrychus dispar. Allgem. Forst-u. Jagtzg.,* 13: 73-74. [Cited by Graham, 1967].

Harvey, P. A., 1934. Life history of *Kalotermes minor. In* C. A. Kofoid *et al.,* eds., *Termites and termite control,* 2nd ed. University of California Press, Berkeley. pp. 217-233.

Haskell, P. T., 1961. *Insect sounds.* H. F. & G. Witherby, Ltd., London. viii + 189 pp.

Haskins, C. P., 1939. *Of ants and men.* Prentice-Hall, Inc., New York. vii + 244 pp.

———, 1941. Note on the method of colony foundation of the ponerine ant *Bothroponera soror* Emery. *Journal of the New York Entomological Society,* 49: 211-216.

———, 1960. Note on the natural longevity of fertile females of *Aphaenogaster picea. Journal of the New York Entomological Society,* 68: 66-67.

———, 1966. [Discussion of E. O. Wilson, "Behaviour of social insects"]. *Symposium of the Royal Entomological Society of London,* 3: 93-94.

———, and E. V. Enzmann, 1938. Studies of certain sociological and physiological features in the Formicidae. *Annals of the New York Academy of Sciences,* 37(2): 97-162.

———, and E. V. Enzmann, 1945. On the occurrence of impaternate females in the Formicidae. *Journal of the New York Entomological Society,* 53: 263-277.

———, and Edna F. Haskins, 1950a. Notes on the biology and social behavior of the archaic ponerine ants of the genera *Myrmecia* and *Promyrmecia. Annals of the Entomological Society of America,* 43(4): 461-491.

———, and Edna F. Haskins, 1950b. Note on the method of colony foundation of the ponerine ant *Brachyponera* (*Euponera*) *lutea* Mayr. *Psyche, Cambridge,* 57(1): 1-9.

———, and Edna F. Haskins, 1951. Note on the method of colony foundation of the ponerine ant *Amblyopone australis* Erichson. *American Midland Naturalist,* 45(2): 432-445.

———, and Edna F. Haskins, 1955. The pattern of colony foundation in the archaic ant *Myrmecia regularis. Insectes Sociaux,* 2(2): 115-126.

———, and Edna F. Haskins, 1964. Notes on the biology and social behavior of *Myrmecia inquilina.* The only known myrmeciine social parasite. *Insectes Sociaux,* 11(3): 267-282.

———, and Edna F. Haskins, 1965. *Pheidole megacephala* and *Iridomyrmex humilis* in Bermuda—equilibrium or slow replacement? *Ecology,* 46(5): 736-740.

———, and R. M. Whelden, 1954. Note on the exchange of ingluvial food in the genus *Myrmecia. Insectes Sociaux,* 1(1): 33–37.

Hasselrot, T. B., 1960. Studies on Swedish bumblebees (genus *Bombus* Latr.) their domestication and biology. *Opuscula Entomologica,* suppl. 17: 1–192.

Hatch, M. H., 1933. Studies on the Leptodiridae (Catopidae) with descriptions of new species. *Journal of the New York Entomological Society,* 41(1–2): 187–239.

Hauschteck, Elisabeth, 1963. Chromosomes of Swiss ants. *Proceedings of the Eleventh International Congress of Genetics, The Hague,* 1: 140.

Haydak, M. H., 1935. Brood rearing by honeybees confined to a pure carbohydrate diet. *Journal of Economic Entomology,* 28(4): 657–660.

———, 1943. Larval food and development of castes in the honeybee. *Journal of Economic Entomology,* 36(5): 778–792.

———, 1945. The language of the honeybee. *American Bee Journal,* 85: 316–317.

Headley, A. E., 1943. Population studies of two species of ants, *Leptothorax longispinosus* Roger and *Leptothorax curvispinosus* Mayr. *Annals of the Entomological Society of America,* 36(4): 743–753.

———, 1949. A population study of the ant *Aphaenogaster fulva* ssp. *aquia* Buckley (Hymenoptera, Formicidae). *Annals of the Entomological Society of America,* 42(3): 265–272.

Hean, A. F., 1943. Notes on maternal care in thrips. *Journal of the Entomological Society of Southern Africa,* 6: 81–83.

Heath, H., 1927. Caste formation in the termite genus *Termopsis. Journal of Morphology and Physiology,* 43: 387–425.

———, 1928. Fertile termite soldiers. *Biological Bulletin, Marine Biological Laboratory, Woods Hole,* 54(4): 324–326.

Hebling, N. J., W. E. Kerr, and Florence S. Kerr, 1964. Divisão de trabalho entre operárias de *Trigona* (*Scaptotrigona*) *xanthotricha* Moure. *Papéis Avulsos do Departamento de Zoologia, São Paulo,* 16(13): 115–127.

Hecker, H., 1966. Das Zentralnervensystem des Kopfes und seine postembryonale Entwicklung bei *Bellicositermes bellicosus* (Smeath.) (Isoptera). *Acta Tropica,* 23(4): 297–352.

Hegh, E., 1922. *Les termites.* Imprimerie Industrielle et Financière, Brussels. 756 pp.

Hein, G., 1954. Der Rucktanz als wesentlicher Bestandteil der Bienentänze. *Experientia,* 10(1): 23–24.

Heldmann, G., 1936a. Ueber die Entwicklung der polygynen Wabe von *Polistes gallica* L. *Arbeiten über Physiologische und Angewandte Entomologie aus Berlin-Dahlem,* 3: 257–259.

———, 1936b. Über das Leben auf Waben mit mehreren überwinterten Weibchen von *Polistes gallica* L. *Biologisches Zentralblatt,* 56(7–8): 389–401.

Heran, H., 1952. Untersuchungen über den Temperatursinn der Honigbiene (*Apis mellifica*) unter besonderer Berücksichtigung der Wahrnehmung strahlender Wärme. *Zeitschrift für Vergleichende Physiologie,* 34(2): 179–206.

———, 1959. Wahrnehmung und Regelung der Flugeigengeschwindigkeit bei *Apis mellifica* L. *Zeitschrift für Vergleichende Physiologie,* 42(2): 103–163.

———, 1968. Régulation thermique et sens thermique chez l'abeille. *In* R. Chauvin, ed. (*q.v.*), *Traité de biologie de l'abeille, Vol. II.* pp. 173–199.

Hertz, Mathilde, 1930. Die Organisation des optischen Feldes bei der Biene, II. *Zeitschrift für Vergleichende Physiologie,* 11(1): 107–145.

———, 1935. Zur Physiologie des Formen- und Bewegungssehens. III. Figurale Unterscheidung und reziproke Dressuren bei der Biene. *Zeitschrift für Vergleichende Physiologie,* 21(4): 604–615.

Herzig, J., 1937. Ameisen und Blattläuse. *Zeitschrift für Angewandte Entomologie,* 24: 367–435.

Hess, Gertrud, 1942. Über den Einfluss der Weisellosigkeit und des Fruchtbarkeitsvitamins E auf die Ovarien der Bienenarbeiterin. *Beihefte zur Schweizerischen Bienenzeitung,* 1(2): 33–110.

Hess, W. R., 1926. Die Temperaturregulierung im Bienenvolk. *Zeitschrift für Vergleichende Physiologie,* 4(4): 465–487.

Heyde, Kaethe, 1924. Die Entwicklung der psychischen Fähigkeiten bei Ameisen und ihr Verhalten bei abgeänderten biologischen Bedingungen. *Biologisches Zentralblatt,* 44(11): 623–654.

Heymons, R., 1929. Über die Biologie der Passaluskäfer. *Zeitschrift für Morphologie und Ökologie der Tiere,* 16(1–2): 74–100.

Hill, G. F., 1942. *Termites* (*Isoptera*) *from the Australian Region.* Commonwealth of Australia, Council for Scientific and Industrial Research. 479 pp.

Himmer, A., 1926. Der soziale Wärmehaushalt der Honigbiene. I. Die Wärme im nichtbrütenden Wintervolk. *Erlanger Jahrbuch für Bienenkunde,* 4: 1–51.

———, 1927. Der soziale Wärmehaushalt der Honigbiene. II. Die Wärme der Bienenbrut. *Erlanger Jahrbuch für Bienenkunde,* 5: 1–32.

———, 1932. Die Temperaturverhältnisse bei den sozialen Hymenopteren. *Biological Reviews* (Cambridge Philosophical Society), 7(3): 224–253.

———, 1933. Die Nestwärme bei *Bombus agrorum* F. *Biologisches Zentralblatt,* 53(5–6): 270–276.

Hinde, R. A., 1966. *Animal behaviour: A synthesis of ethology and comparative psychology.* McGraw-Hill Book Co., New York. x + 534 pp.

Hindwood, K. A., 1959. The nesting of birds in the nests of social insects. *Emu,* 59(1): 1–36.

Hingston, R. W. G., 1929. *Instinct and intelligence.* Macmillan Company, New York. xv + 296 pp.

Hinton, H. E., 1944. Some general remarks on sub-social beetles, with notes on the biology of the staphylinid, *Platystethus arenarius* (Fourcroy). *Proceedings of the Royal Entomological Society of London,* (A)19: 115–128.

———, 1951. Myrmecophilous Lycaenidae and other Lepidoptera—a summary. *Proceedings of the South London*

Entomological and Natural History Society, 1949–50, pp. 111–175.

Hirashima, Y., 1961. Monographic study of the subfamily Nomiinae of Japan (Hymenoptera, Apoidea). *Acta Hymenopterologica, Tokyo,* 1(3): 241–303.

Hirst, S., 1927. Acarina. On a gamasid mite (*Uropoda* (*Uroobovella*) *samoae,* sp. n.) occurring on the termite *Calotermes* (*Glyptotermes*) *xantholabrum* Hill. In *Insects of Samoa. Pt. 8. Terrestrial Arthropoda other than insects,* fasc. 1: 25–27. British Museum (Natural History), London.

Hocking, B., 1970. Insect associations with the swollen thorn acacias. *Transactions of the Royal Entomological Society of London,* 122(7): 211–255.

Hodge, C. F., 1894. Changes in the ganglion cells from birth to senile death: Observations on man and honey-bee. *Journal of Physiology,* 17: 129–134.

Hodgson, E. S., 1955. An ecological study of the behavior of the leaf-cutting ant *Atta cephalotes. Ecology,* 36(2): 293–304.

Hoffer, E., 1882–1883. *Die Hummeln Steiermarks. Lebensgeschichte und Beschreibung Derselben,* two pts. Leuschner & Lubensky, Graz. 92 pp., 98 pp. [Behavioral descriptions are in the first part, published in 1882.]

———, 1888. Die Schmarotzerhummeln Steiermarks. *Mitteilungen des Naturwissenschaftlichen Vereins für Steiermark, 1888,* pp. 84–88.

Hoffman, Irmgard, 1960. Untersuchungen über die Herkunft von Komponenten des Königinnenfuttersaftes der Honigbiene. *Zeitschrift für Bienenforschung,* 5: 101–111.

Hoffman, W. E., 1924. Biological notes on *Lethocerus americanus* (Leidy). *Psyche, Cambridge,* 31(5): 175–183.

Holdaway, F. G., F. J. Gay, and T. Greaves, 1935. The termite population of a mound colony of *Eutermes exitiosus* Hill. *Journal of the Council for Scientific and Industrial Research, Australia,* 8(1): 42–46.

Hollande, A., J. Cachon, and F. Vaillant, 1951. Recherches sur quelques larves d'insectes termitophiles (Muscidae, Calliphoridae, Oestridae, Tineidae, Melandryidae). *Annales des Sciences Naturelles,* (11)13(4): 365–396.

Hölldobler, B., 1962. Zur Frage der Oligogynie bei *Camponotus ligniperda* Latr. und *Camponotus herculeanus* L. (Hym. Formicidae). *Zeitschrift für Angewandte Entomologie,* 49(4): 337–352.

———, 1964. Untersuchungen zum Verhalten der Ameisenmännchen während der imaginalen Lebenszeit. *Experientia,* 20(6): 329.

———, 1966. Futterverteilung durch Männchen in Ameisenstaat. *Zeitschrift für Vergleichende Physiologie,* 52(4): 430–455.

———, 1967a. Verhaltensphysiologische Untersuchungen zur Myrmecophilie einiger Staphylinidenlarven. *Verhandlungen der Deutschen Zoologischen Gesellschaft, Heidelberg, 1967,* pp. 428–434.

———, 1967b. Zur Physiologie der Gast-Wirt-Beziehungen (Myrmecophilie) bei Ameisen. I. Das Gastverhältnis der *Atemeles-* und *Lomechusa*-Larven (Col. Staphylinidae) zu *Formica* (Hym. Formicidae). *Zeitschrift für Vergleichende Physiologie,* 56(1): 1–21.

———, 1968. Der Glanzkäfer als "Wegelagerer" an Ameisenstrassen. *Naturwissenschaften,* 55(8): 397.

———, 1969a. Host finding by odor in the myrmecophilic beetle *Atemeles pubicollis* Bris. (Staphylinidae). *Science,* 166: 757–758.

———, 1969b. Orientierungsmechanismen des Ameisengastes *Atemeles* (Coleoptera, Staphylinidae) bei der Wirtssuche. *Verhandlungen der Deutschen Zoologischen Gesellschaft, Würzburg, 1969,* pp. 580–585.

———, 1970. Zur Physiologie der Gast-Wirt-Beziehungen (Myrmecophilie) bei Ameisen. II. Das Gastverhältnis des imaginalen *Atemeles pubicollis* Bris. (Col. Staphylinidae) zu *Myrmica* und *Formica* (Hym. Formicidae). *Zeitschrift für Vergleichende Physiologie,* 66(2): 215–250.

———, 1971. Communication between ants and their guests. *Scientific American,* 224(3): 86–93.

———, and U. Maschwitz, 1965. Der Hochzeitsschwarm der Rossameise *Camponotus herculeanus* L. (Hym. Formicidae). *Zeitschrift für Vergleichende Physiologie,* 50(5): 551–568.

Hölldobler, K., 1928. Zur Biologie der diebischen Zwergameise (*Solenopsis fugax*) und ihrer Gäste. *Biologisches Zentralblatt,* 48(3): 129–142.

———, 1936. Beiträge zur Kenntnis der Koloniengründung der Ameisen. *Biologisches Zentralblatt,* 56(5–6): 230–248.

———, 1953. Beobachtungen über die Koloniengründung von *Lasius umbratus umbratus* Nyl. *Zeitschrift für Angewandte Entomologie,* 34(4): 598–606.

Hollingsworth, M. J., 1960. Studies on the polymorphic workers of the army ant *Dorylus* (*Anomma*) *nigricans* Illiger. *Insectes Sociaux,* 7(1): 17–37.

Holmgren, N., 1909. Termitenstudien. Part 1. Anatomische Untersuchungen. *Kungliga Svenska Vetenskapsakademiens Handlingar* (n.f.), 44(3): 1–215.

———, 1913. Termitenstudien. 4. Versuch einer systematischen Monographie der Termiten der orientalischen Region. *Kungliga Svenska Vetenskapsakademiens Handlingar* (n.f.), 50(2): 1–276.

Horridge, G. A., 1968. *Interneurons: Their origin, action, specificity, growth, and plasticity.* W. H. Freeman, San Francisco. xxiii + 436 pp.

Houston, T. F., 1970. Discovery of an apparent male soldier caste in a nest of a halictine bee (Hymenoptera: Halictidae), with notes on the nest. *Australian Journal of Zoology,* 18(3): 345–351.

Howse, P. E., 1964. The significance of the sound produced by the termite *Zootermopsis angusticollis* (Hagen). *Animal Behaviour,* 12(2–3): 284–300.

———, 1966. Air movement and termite behaviour. *Nature, London,* 210: 967–968.

———, 1968. On the division of labour in the primitive termite *Zootermopsis nevadensis* (Hagen). *Insectes Sociaux,* 15(1): 45–50.

———, 1970. *Termites: A study in social behaviour.* Hutchinson University Library, London. 150 pp.

Hoyle, G., 1965. Neural control of skeletal muscle. *In* M. Rockstein, ed., *The physiology of Insecta, Vol. II.* Academic Press, Inc., New York. pp. 407–449.

Hozawa, S., 1914. Note on a new termitophilous Coleoptera found in Formosa (*Ziaelas formosanus*). *Annotationes Zoologicae Japonenses,* 8(3–4): 483–488.

Hubbard, H. G., 1897. The ambrosia beetles of the United States. *Bulletin, United States Department of Agriculture* (n.s.), 7: 9–30.

Huber, François, 1792. *Nouvelles observations sur les abeilles, adressées à M. Charles Bonnet.* Barde, Manget and Co., Geneva. 368 pp. [1808. English tr., 2nd ed. John Anderson, Edinburgh. xxv + 314 pp.]

Huber, Franz, 1965. Neural integration (central nervous system). *In* M. Rockstein, ed., *The physiology of Insecta, Vol. II.* Academic Press, Inc., New York. pp. 333–406.

Huber, Jakob, 1905. Über die Koloniengründung bei *Atta sexdens. Biologisches Centralblatt,* 25(18): 606–619, (19): 625–635.

Huber, Pierre, 1802. Observations on several species of the genus *Apis,* known by the name of Humble-bees, and called Bombinatrices by Linnaeus. *Transactions of the Linnaean Society of London,* 6: 214–298.

———, 1810. *Recherches sur les moeurs des fourmis indigènes.* J. J. Paschoud, Paris. xvi + 328 pp.

Hummel, H., and P. Karlson. 1968. Hexansäure als Bestandteil des Spurpheromons der Termite *Zootermopsis nevadensis* Hagen. *Hoppe-Seyler's Zeitschrift für Physiologische Chemie,* 349: 725–727.

Hurd, P. D., 1955. The aculeate wasps. *In* E. L. Kessel, ed., *A century of progress in the natural sciences 1853–1953.* California Academy of Sciences, San Francisco. pp. 573–575.

Hüsing, I. O., 1968. Anatomie générale et cytologie. *In* R. Chauvin, ed. (*q.v.*), *Traité de biologie de l'abeille, Vol. II.* pp. 45–48.

Huxley, J., 1930. *Ants.* Jonathan Cape and Harrison Smith, New York. 113 pp.

———, 1932. *Problems of relative growth.* Dial Press, New York. xix + 276 pp.

Ihering, H. von, 1894. Die Ameisen von Rio Grande do Sul. *Berliner Entomologische Zeitschrift,* 39(3): 321–446.

———, 1896. Zur Biologie der sozialen Wespen Brasiliens. *Zoologischer Anzeiger,* 19(516): 449–453.

———, 1898. Die Anlage neuer Colonien und Pilzgärten bei *Atta sexdens. Zoologischer Anzeiger,* 21(556): 238–245.

Ihering, R. von, 1904. As Vespas sociaes do Brasil. *Revista do Museu Paulista,* 6: 97–309.

Ikan, R., E. D. Bergmann, J. Ishay, and S. Gitter, 1968. Proteolytic enzyme activity in the various colony members of the Oriental hornet, *Vespa orientalis* F. *Life Sciences,* 7(18): 929–934.

———, R. Gottlieb, E. D. Bergmann, and J. Ishay, 1969. The pheromone of the queen of the Oriental hornet, *Vespa orientalis. Journal of Insect Physiology,* 15(10): 1709–1712.

Imms, A. D., 1919. On the structure and biology of *Archotermopsis,* together with descriptions of new species of intestinal Protozoa and general observations on the Isoptera. *Philosophical Transactions of the Royal Society,* (B)209: 75–180.

———, 1942. On *Braula coeca* Nitsch and its affinities. *Parasitology,* 34(1): 88–100.

Ishay, J., 1964. Observations sur la biologie de la guêpe orientale *Vespa orientalis* F. *Insectes Sociaux,* 11(3): 193–206.

———, 1965. Entwicklung und Aktivität im Nest von *Vespa orientalis* F. *Deutsche Entomologische Zeitschrift* (n.f.), 12(4–5): 397–419.

———, and R. Ikan, 1969. Gluconeogenesis in the Oriental hornet *Vespa orientalis* F. *Ecology,* 49(1): 169–171.

———, and J. Schwarz, 1965. On the nature of the sounds produced within the nest of the Oriental hornet, *Vespa orientalis* F. *Insectes Sociaux,* 12(4): 383–387.

———, H. Bytinski-Salz, and A. Shulov, 1967. Contributions to the bionomics of the Oriental hornet (*Vespa orientalis* Fab.). *Israel Journal of Entomology,* 2: 45–106.

———, R. Ikan, and E. D. Bergmann, 1965. The presence of pheromones in the Oriental hornet, *Vespa orientalis* F. *Journal of Insect Physiology,* 11(9): 1307–1309.

Iwata, K., 1967. Report of the fundamental research on the biological control of insect pests in Thailand. II. The report on the bionomics of subsocial wasps of Stenogastrinae (Hymenoptera, Vespidae). *Nature and Life in Southeast Asia* (Tokyo), 5: 259–293.

———, and S. F. Sakagami, 1966. Gigantism and dwarfism in bee eggs in relation to the modes of life, with notes on the number of ovarioles. *Japanese Journal of Ecology,* 16(1): 4–16.

Jackson, W. B., 1957. Microclimatic patterns in the army ant bivouac. *Ecology,* 38(2): 276–285.

Jacobs-Jessen, Una F., 1959. Zur Orientierung der Hummeln und einiger anderer Hymenopteren. *Zeitschrift für Vergleichende Physiologie,* 41(6): 597–641.

Jacobson, E., 1909. Ein Moskito als Gast und diebischer Schmarotzer der *Cremastogaster difformis* Smith und eine andere schmarotzende Fliege. *Tijdschrift voor Entomologie,* 52: 158–164.

———, 1911. Biological notes on the hemipteron *Ptilocerus ochraceus. Tijdschrift voor Entomologie,* 54: 175–179.

Jacoby, M., 1937. Das räumliche Wachsen des *Atta*-Nestes vom 50. bis zum 90. Tage (Hym. Formicidae). *Revista de Entomologia, Rio de Janeiro,* 7(4): 416–425.

———, 1944. Observações e experiências sôbre *Atta sexdens rubropilosa* Forel visando facilitar seu combate. *Boletim do Ministério da Agricultura, Industria e Comercio, Rio de Janeiro,* May 1943: 1–55.

———, 1952. Die Erforschung des Nestes der Blattschneider-Ameise *Atta sexdens rubropilosa* Forel (mittels des

Ausgussverfahrens in Zement), Teil I. *Zeitschrift für Angewandte Entomologie,* 34(2): 145–169.
Jakubski, A. W., 1965. *A critical revision of the families Margarodidae and Termitococcidae (Hemiptera, Coccoididea).* Trustees British Museum (Natural History), London. 187 pp.
Jamieson, C. A., and G. H. Austin, 1958. Preference of honeybees for sugar solutions. *Proceedings of the Tenth International Congress of Entomology, Montreal, 1956,* 4: 1059–1062.
Jander, R., 1957. Die optische Richtungsorientierung der Roten Waldameise (*Formica rufa* L.). *Zeitschrift für Vergleichende Physiologie,* 40(2): 162–238.
———, 1963. Insect orientation. *Annual Review of Entomology,* 8: 95–114.
Jander, Ursula, 1966. Untersuchungen zur Stammesgeschichte von Putzbewegungen von Tracheaten. *Zeitschrift für Tierpsychologie,* 23: 799–844.
Janet, C., 1896. Sur les rapports des Lépismides myrmécophiles avec les fourmis. *Compte Rendu de l'Académie des Sciences, Paris,* 122(13): 799–802.
———, 1897a. *Rapports des animaux myrmécophiles avec les fourmis.* H. Ducourtieux, Limoges. 98 pp.
———, 1897b. Sur les rapports de l'*Antennophorus uhlmanni* Haller avec le *Lasius mixtus* Nyl. *Compte Rendu de l'Académie des Sciences, Paris,* 124(11): 583–585.
———, 1898a. *Système glandulaire tégumentaire de la Myrmica rubra. Observations diverses sur les fourmis.* G. Carré et C. Naud, Paris. 28 pp.
———, 1898b. Sur une cavité du tégument servant, chez les Myrmicinae, à étaler au contact de l'air, un produit de sécrétion. *Compte Rendu de l'Académie des Sciences, Paris,* 126(16): 1168–1171.
———, 1903. *Observations sur les guêpes.* C. Naud, Paris. 85 pp.
———, 1904. *Observations sur les fourmis.* Ducourtieux et Gout, Limoges. 68 pp.
———, 1907. *Anatomie du corselet et histolyse des muscles vibrateurs, après le vol nuptial, chez la reine de la fourmi (Lasius niger).* Ducourtieux et Gout, Limoges. 149 pp.
Janssens, André, 1949. Un Scarabaeinae termitophile nouveau du Congo belge. *Revue de Zoologie et de Botanique Africaines,* 42(1–2): 183–184.
Janssens, Emile, 1949. Sur la massue antennaire de *Paussus* Linné et genres voisins. *Bulletin de l'Institut Royal des Sciences Naturelles de Belgique,* 25(22): 1–9.
Janzen, D. H., 1966. Coevolution of mutualism between ants and acacias in Central America. *Evolution,* 20(3): 249–275.
———, 1967. Interaction of the bull's-horn acacia (*Acacia cornigera* L.) with an ant inhabitant (*Pseudomyrmex ferruginea* F. Smith) in eastern Mexico. *Kansas University Science Bulletin,* 47(6): 315–558.
———, 1969a. Birds and the ant X acacia interaction in Central America, with notes on birds and other myrmecophytes. *Condor,* 71(3): 240–256.
———, 1969b. Allelopathy by myrmecophytes: The ant *Azteca* as an allelopathic agent of *Cecropia. Ecology,* 50(1): 147–153.
Jay, S. C., 1964a. Rearing honeybee brood outside the hive. *Journal of Apicultural Research,* 3(1): 51–60.
———, 1964b. Starvation studies of larval honey bees. *Canadian Journal of Zoology,* 42(3): 455–462.
Jaycox, E. R., 1970. Honey bee queen pheromones and worker foraging behavior. *Annals of the Entomological Society of America,* 63(1): 222–228.
Jeanne, R. L., 1970a. Social biology of the neotropical wasp *Mischocyttarus drewseni.* Ph.D. Thesis, Department of Biology, Harvard University. 170 pp.
———, 1970b. Chemical defense of brood by a social wasp. *Science,* 168: 1465–1466.
———, 1970c. Descriptions of the nests of *Pseudochartergus fuscatus* and *Stelopolybia testacea,* with a note on a parasite of *S. testacea* (Hymenoptera, Vespidae). *Psyche, Cambridge,* 77(1): 54–69.
Jeannel, R., 1936. Monographie des Catopidae. *Mémoires du Muséum Nationale d'Histoire Naturelle, Paris* (n.s.), (A) 1: 1–433.
Jerdon, T. C., 1854. A catalogue of the species of ants found in Southern India. *Annals and Magazine of Natural History,* (2)13(73): 45–56, (74): 100–110.
Johnson, C. G., 1966. A functional system of adaptive dispersal by flight. *Annual Review of Entomology,* 11: 233–260.
Johnson, R. A., 1954. The behavior of birds attending army ant raids on Barro Colorado Island, Panama Canal Zone. *Proceedings of the Linnean Society of New York,* Nos. 63–65: 41–70.
Johnston, Norah C., J. H. Law, and N. Weaver, 1965. Metabolism of 9-ketodec-2-enoic acid by worker honeybees (*Apis mellifera* L.). *Biochemistry,* 4: 1615–1621.
Jolivet, P., 1952. Quelques données sur la myrmécophilie des Clytrides (Col. Chrysomeloidea). *Bulletin de l'Institut Royal des Sciences Naturelles de Belgique,* 28(8): 1–12.
Jones, C. R., 1929. Studies on ants and their relation to aphids. *Bulletin, Colorado State University Agricultural Experiment Station,* 341: 1–96.
Jones, S., 1935. A note on the distribution, oviposition and parental care of *Scutigerella unguiculata* Hansen var. *indica* Gravely. *Journal of the Bombay Natural History Society,* 38(1): 209–211.
Jonescu, C. N., 1909. Vergleichende Untersuchungen über das Gehirn der Honigbiene. *Zeitschrift für Naturwissenschaften,* 45: 111–180.
Jordan, R., 1936. Beobachtungen der Arbeitsteilung im Hummelstaate (*Bombus muscorum*). *Archiv für Bienenkunde,* 17: 81–91.
Joseph, K. J., and S. B. Mathad, 1963. A new genus of termitophilous *Atelurinae* (Thysanura: Nicoletidae) from India. *Insectes Sociaux,* 10(4): 379–386.
Jucci, C., 1932. Sulla presenza di batteriociti nel tessuto adiposo dei termitidi. *Archivio Zoologico Italiano,* 16: 1422–1429.
Kaestner, A., 1968. *Invertebrate zoology. Vol. 2. Arthropod rela-*

tives, Chelicerata, Myriapoda (tr. and adapted from the German by H. W. Levi and Lorna R. Levi). Interscience Publishers, John Wiley & Sons., New York. ix + 472 pp.

Kaiser, P., 1954. Über die Funktion der Mandibeln bei den Soldaten von *Neocapritermes opacus* (Hagen). *Zoologischer Anzeiger,* 152(9–10): 228–234.

———, 1956. Die Hormonalorgane der Termiten im Zusammenhang mit der Entstehung ihrer Kasten. *Mitteilungen aus dem Hamburgischen Zoologischen Museum und Institut,* 54: 129–178.

Kaissling, K.-E., and M. Renner, 1968. Antennale Rezeptoren für Queen Substance und Sterzelduft bei der Honigbiene. *Zeitschrift für Vergleichende Physiologie,* 59(4): 357–361.

Kalmus, H., 1937. Vorversuche über die Orientierung der Biene im Stock. *Zeitschrift für Vergleichende Physiologie,* 24(2): 166–187.

———, 1941. Defence of source of food by bees. *Nature, London,* 148: 228.

———, and C. R. Ribbands, 1952. The origin of the odours by which honeybees distinguish their companions. *Proceedings of the Royal Society,* (B)140: 50–59.

Kalshoven, L. G. E., 1930. *Bionomics of Kalotermes tectonae Damm. as a base for its control* [in Dutch]. H. Veenman & Zonen, Wageningen. 154 pp.

———, 1936. Onze Kennis van de Javaansche Termieten. *Handelingen. Nederlandsch-Indisch Natuurwetenschappelijk Congres,* 7: 427–435. [Cited by Sands, 1969.]

Kannowski, P. B., 1963. The flight activities of formicine ants. *Symposia Genetica et Biologica Italica,* 12: 74–102.

Karawajew, W., 1906. Weitere Beobachtungen über Arten der Gattung *Antennophorus. Mémoires de la Société des Naturalistes de Kiew,* 20(2): 209–229.

Karlson, P., and A. Butenandt, 1959. Pheromones (ectohormones) in insects. *Annual Review of Entomology,* 4: 39–58.

———, M. Lüscher, and H. Hummel, 1968. Extraktion und biologische Auswertung des Spurpheromons der Termite *Zootermopsis nevadensis. Journal of Insect Physiology,* 14(12): 1763–1771.

Kaschef, A.-H., 1959. The sensory physiology and behaviour of the honeybee louse *Braula coeca* Nitzsch (Diptera, Braulidae). *Insectes Sociaux,* 6(4): 313–342.

Kaston, B. J., 1936. The senses involved in the courtship of some vagabond spiders. *Entomologica Americana* (n.s.), 16(2): 97–166.

———, 1965. Some little known aspects of spider behavior. *American Midland Naturalist,* 73(2): 336–356.

Katô, M., 1939. The diurnal rhythm of temperature in the mound of an ant, *Formica truncorum truncorum* var. *yessenni* Forel, widely distributed at Mt. Hakkôda. *Science Reports of the Tôhoku Imperial University,* 14(1): 53–64.

Kaudewitz, F., 1955. Zum Gastverhältnis zwischen *Cremastogaster scutellaris* Ol. mit *Camponotus lateralis bicolor* Ol. *Biologisches Zentralblatt,* 74(1–2): 69–87.

Kelner-Pillault, S., 1969. Abeilles fossiles ancetres des apides sociaux. *Proceedings of the Sixth Congress of the International Union for the Study of Social Insects, Bern, 1969,* pp. 85–93.

Kemner, N. A., 1923. Hyphaenosymphilie, eine neue, merkwürdige Art von Myrmekophilie bei einem neuen myrmekophilen Schmetterling (*Wurthia aurivillii* n. sp.) aus Java beobachtet. *Arkiv für Zoologi,* 15(15): 1–28.

———, 1924. Über die Lebensweise und Entwicklung des angeblich myrmecophilen oder termitophilen Genus *Trochoideus* (Col. Endomych.), nach Beobachtungen über *Trochoideus termitophilus* Roepke auf Java. *Tijdschrift voor Entomologie,* 67: 180–194.

Kemper, H., and Edith Döhring, 1961. Zur Frage nach der Volksstärke und der Vermehrungspotenz bei den sozialen Faltenwespen Deutschlands. *Zeitschrift für Angewandte Zoologie,* 48(2): 163–197.

———, and Edith Döhring, 1967. *Die sozialen Faltenwespen Mitteleuropas.* Paul Parey, Berlin. 180 pp.

Kempf, W. W., 1951. A taxonomic study on the ant tribe Cephalotini (Hymenoptera: Formicidae). *Revista de Entomologia, Rio de Janeiro,* 22(1–3): 1–244.

———, 1958. New studies of the ant tribe Cephalotini (Hym. Formicidae). *Studia Entomologica* (n.s.), 1(1–2): 1–168.

———, 1959. A revision of the Neotropical ant genus *Monacis* Roger (Hym., Formicidae). *Studia Entomologica* (n.s.), 2(1–4): 225–270.

———, 1963. A review of the ant genus *Mycocepurus* Forel (Hym. Formicidae). *Studia Entomologica* (n.s.), 6(1–4): 417–432.

———, 1965. A revision of the Neotropical fungus-growing ants of the genus *Cyphomyrmex* Mayr. Part II. Group of *rimosus* (Spinola) (Hym. Formicidae). *Studia Entomologica* (n.s.), 8(1–4): 161–200.

Kennedy, J. S., ed., 1961. *Insect polymorphism. Symposium of the Royal Entomological Society of London,* 1: 1–115.

Kerr, W. E., 1950a. Evolution of the mechanism of caste determination in the genus *Melipona. Evolution,* 4(1): 7–13.

———, 1950b. Genetic determination of castes in the genus *Melipona. Genetics,* 35(2): 143–152.

———, 1961. Acasalamento de rainhas com vários machos em duas espécies da Tribu Attini (Hymenoptera, Formicoidea). *Revista Brasileira de Biologia,* 21(1): 45–48.

———, 1962a. Tendências evolutivas na reprodução dos himenópteros sociais. *Arquivos do Museu Nacional, Rio de Janeiro,* 52: 115–116.

———, 1962b. Genetics of sex determination. *Annual Review of Entomology,* 7: 157–176.

———, 1967. Genetic structure of the populations of Hymenoptera. *Ciência e Cultura, São Paulo,* 19(1): 39–44.

———, 1969. Some aspects of the evolution of social bees (Apidae). *In* T. Dobzhansky, M. K. Hecht, and W. C. Steere, eds., *Evolutionary biology, Vol. 3.* Appleton-Century Crofts, New York. pp. 119–175.

———, and N. J. Hebling, 1964. Influence of the weight of

worker bees on division of labor. *Evolution,* 18(2): 267–270.

———, and Vilma Maule, 1964. Geographic distribution of stingless bees and its implications (Hymenoptera: Apidae). *Journal of the New York Entomological Society,* 72: 2–18.

———, and R. A. Nielsen, 1966. Evidences that genetically determined *Melipona* queens can become workers. *Genetics,* 54(3): 859–866.

———, and G. R. dos Santos Netos, 1956. Contribuição para o conhecimento da bionomia dos Meliponini. 5. Divisão de trabalho entre as operarias de *Melipona quadrifasciata quadrifasciata* Lep. *Insectes Sociaux,* 3(3): 423–430.

———, A. Ferreira, and N. Simões de Mattos, 1963. Communication among stingless bees—additional data (Hymenoptera: Apidae). *Journal of the New York Entomological Society,* 71: 80–90.

———, A. C. Stort, and M. J. Montenegro, 1966. Importância de alguns fatôres ambientais na determinação das castas do gênero *Melipona. Anais da Academia Brasileira de Ciências,* 38(1): 149–168.

———, R. Zucchi, J. T. Nakadaira, and J. E. Butolo, 1962. Reproduction in the social bees (Hymenoptera: Apidae). *Journal of the New York Entomological Society,* 70: 265–276.

———, S. F. Sakagami, R. Zucchi, V. de Portugal-Araújo, and J. M. F. de Camargo, 1967. Observações sôbre a arquitetura dos ninhos e comportamento de algumas espécies de abelhas sem ferrão das vizinhanças de Manaus, Amazonas (Hymenoptera, Apoidea). *Atas do Simpósio sôbre a Biota Amazônica, Conselho Nacional de Pesquisas, Rio de Janeiro,* 5(Zool.): 255–309.

Keys, J. H., 1914. Some further remarks on *Aëpophilus bonnairei,* Sign. *Entomologist's Monthly Magazine,* 50: 284–285.

Kieseritzky, V., and A. Reichardt, 1936. Zur Frage über die Selbständigkeit der Familie Catopochrotidae, im Zusammenhang mit der Entdeckung einer neuen Gattung. *Trudȳ Zoologicheskogo Instituta, Leningrad,* 3: 693–697.

Kiil, V., 1934. Untersuchungen über Arbeitsteilung bei Ameisen (*Formica rufa* L., *Camponotus herculeanus* L. und *C. ligniperda* Latr.). *Biologisches Zentralblatt,* 54(3–4): 114–146.

Kincaid, T., 1963. The ant-plant, *Orthocarpus pusillus,* Bentham. *Transactions of the American Microscopical Society,* 82(1): 101–105.

King, G. E., 1933. The larger glands in the worker honey-bee. A correlation of activity with age and with physiological functioning. Abstract of Ph.D. Thesis, Graduate School of the University of Illinois (1928). 20 pp.

King, R. L., and R. M. Sallee, 1957. Mixed colonies in ants: third report. *Proceedings of the Iowa Academy of Science,* 64: 667–669.

———, 1962. Further studies on mixed colonies in ants. *Proceedings of the Iowa Academy of Science,* 69: 531–539.

———, and F. Walters, 1950. Population of a colony of *Formica rufa melanotica* Emery. *Proceedings of the Iowa Academy of Science,* 57: 469–473.

Kirby, W., 1802. *Monographia Apum Angliae,* 2 vols. J. Raw, Ipswich, Eng. 258 pp., 388 pp.

Kirkpatrick, T. W., 1957. *Insect life in the tropics.* Longmans, Green and Co., London. xiv + 311 pp.

Kistner, D. H., 1958. The evolution of the Pygostenini (Coleoptera Staphylinidae). *Annales du Musée Royal du Congo Belge, Sér. 8, Zool.,* 68: 5–198.

———, 1964. New species of the genus *Dorylocratus* with notes on their behavior (Coleoptera: Staphylinidae). *Pan-Pacific Entomologist,* 40(4): 246–254.

———, 1965. A revision of the species of the genus *Phyllodinarda* Wasmann with notes on their behavior. *Pan-Pacific Entomologist,* 41(2): 121–132.

———, 1966a. A revision of the African species of the aleocharine tribe Dorylomimini (Coleoptera: Staphylinidae). II. The genera *Dorylomimus, Dorylonannus, Dorylogaster, Dorylobactrus,* and *Mimanomma,* with notes on their behavior. *Annals of the Entomological Society of America,* 59(2): 320–340.

———, 1966b. A revision of the myrmecophilous tribe Deremini (Coleoptera: Staphylinidae). Part I. The *Dorylopora* complex and their behavior. *Annals of the Entomological Society of America,* 59(2): 341–358.

———, 1968a. Revision of the myrmecophilous species of the tribe Myrmedoniini. Part II. The genera *Aenictonia* and *Anommatochara*—their relationship and behavior. *Annals of the Entomological Society of America,* 61(4): 971–986.

———, 1968b. A taxonomic revision of the termitophilous tribe Termitopaedini, with notes on behavior, systematics, and post-imaginal growth (Coleoptera: Staphylinidae). *Miscellaneous Publications of the Entomological Society of America,* 6(3): 141–196.

———, 1968c. Revision of the African species of the termitophilous tribe Corotocini (Coleoptera: Staphylinidae). I. A new genus and species from Ovamboland and its zoogeographic significance. *Journal of the New York Entomological Society,* 76(3): 213–221.

———, 1968d. Revision of African species of the termitophilous tribe Corotocini (Coleoptera, Staphylinidae). II. The genera *Termitomimus* Trägårdh and *Nasutimimus* new genus and their relationships. *Coleopterists' Bulletin,* 22(3): 65–93.

———,1969. The biology of termitophiles. *In* K. Krishna and F. M. Weesner, eds. (*q.v.*), *Biology of termites, Vol. I.* pp. 525–557.

Kleber, Elisabeth, 1935. Hat das Zeitgedächtnis der Bienen biologische Bedeutung? *Zeitschrift für Vergleichende Physiologie,* 22(2): 221–262.

Kleine, R., 1925. Die Myrmekophilie der Brenthiden. *Zoologische Jahrbücher, Abt. Syst. Geogr. Biol. Tiere,* 49(3): 197–228.

Kloft, W., 1959a. Zur Nestbautätigkeit der Roten Waldameisen. *Waldhygiene,* 3/4: 94–98.

———, 1959b. Versuch einer Analyse der trophobiotischen Beziehungen von Ameisen zu Aphiden. *Biologisches Zentralblatt,* 78(6): 863–870.

———, 1960a. Die Trophobiose zwischen Waldameisen und Pflanzenläusen mit Untersuchungen über die Wechsel wirkungen zwischen Pflanzenläusen und Pflanzengeweben. *Entomophaga,* 5: 43–54.

———, 1960b. Wechselwirkungen zwischen pflanzensaugenden Insekten und den von ihnen besogenen Pflanzengeweben, Teil I. *Zeitschrift für Angewandte Entomologie,* 45(4): 337–381.

Kneitz, G., 1964. Saisonales Trageverhalten bei *Formica polyctena* Foerst. (Formicidae, Gen. *Formica*). *Insectes Sociaux,* 11(2): 105–129.

Knerer, G., 1969. Brood care in halictine bees. *Science,* 164: 429–430.

———, and C. E. Atwood, 1964. Further notes on the genus *Evylaeus* Robertson (Hymenoptera: Halictidae). *Canadian Entomologist,* 96(7): 957–962.

———, and C. E. Atwood, 1966a. Polymorphism in some Nearctic halictine bees. *Science,* 152: 1262–1263.

———, and C. E. Atwood, 1966b. Nest architecture as an aid in halictine taxonomy (Hymenoptera: Halictidae). *Canadian Entomologist,* 98(12): 1337–1339.

———, and C. E. Atwood, 1967. Parasitization of social halictine bees in southern Ontario. *Proceedings of the Entomological Society of Ontario* (1966), 97: 103–110.

———, and Cécile Plateaux-Quénu, 1966a. Sur l'importance de l'ouverture des cellules à couvain dans l'évolution des Halictinae (Insectes Hyménoptères) sociaux. *Compte Rendu de l'Académie des Sciences, Paris,* 263(21): 1622–1625.

———, and Cécile Plateaux-Quénu, 1966b. Sur le polymorphisme des femelles chez quelques Halictinae (Insectes Hyménoptères) paléarctiques. *Compte Rendu de l'Académie des Sciences, Paris,* 263(22): 1759–1761.

———, and Cécile Plateaux-Quénu, 1966c. Sur la polygynie chez les Halictinae (Insectes Hyménoptères). *Compte Rendu de l'Académie des Sciences, Paris,* 263(25): 2014–2017.

———, and Cécile Plateaux-Quénu, 1967a. Sur la production continue ou périodique de couvain chez les Halictinae (Insectes Hyménoptères). *Compte Rendu de l'Académie des Sciences, Paris,* 264(4): 651–653.

———, and Cécile Plateaux-Quénu, 1967b. Sur la production de mâles chez les Halictinae (Insectes, Hyménoptères) sociaux. *Compte Rendu de l'Académie des Sciences, Paris,* 264(8): 1096–1099.

———, and Cécile Plateaux-Quénu, 1967c. Usurpation de nids étrangers et parasitisme facultatif chez *Halictus scabiosae* (Rossi) (Insecte Hyménoptère). *Insectes Sociaux,* 14(1): 47–50.

Koblick, Tonya Ann, and D. H. Kistner, 1965. A revision of the species of the genus *Myrmechusa* from tropical Africa with notes on their behavior and their relationship to the Pygostenini (Coleoptera: Staphylinidae). *Annals of the Entomological Society of America,* 58(1): 28–44.

Koch, A., 1938. Symbiosestudien III: Die intrazellluare Bakteriensymbiose von *Mastotermes darwiniensis* Froggatt (Isoptera). *Zeitschrift für Morphologie und Ökologie der Tiere,* 34(4): 584–609.

Köhler, F., 1955. Wache und Volksduft im Bienenstaat. *Zeitschrift für Bienenforschung,* 3: 57–63.

———, 1966. Untersuchungen zur Orientierung der Raubzüge der Amazonenameise *Polyergus rufescens* Latr. *Insectes Sociaux,* 13(4): 305–309.

König, J. C., 1779. Naturgeschichte der sogenannten weissen Ameisen. *Besch. Berlin. Ges. Naturf. Freunde,* 4: 1–28. [Cited by Snyder, 1956].

Körner, Ilse, 1939. Zeitgedächtnis und Alarmierung bei den Bienen. *Zeitschrift für Vergleichende Physiologie,* 27(4): 445–459.

Krafft, B., 1966a. Étude du comportement social de l'Araignée *Agelena consociata* Denis. *Biologia Gabonica,* 2(3): 235–250.

———, 1966b. Premières recherches de laboratoire sur le comportement d'une araignée sociale nouvelle "*Agelena consociata* Denis." *Revue du Comportement Animal* (Crepin-Leblond, Paris), 1(4): 25–30.

———, 1967. Thermopreferendum de l'araignée sociale *Agelena consociata* Denis. *Insectes Sociaux,* 14(2): 161–182.

Kratky, E., 1931. Morphologie und Physiologie der Drüsen im Kopf und Thorax der Honigbiene (*Apis mellifica* L.) *Zeitschrift für Wissenschaftliche Zoologie,* 139: 120–200.

Krishna, K., 1961. A generic revision and phylogenetic study of the family Kalotermitidae (Isoptera). *Bulletin of the American Museum of Natural History,* 122(4): 303–408.

———, and Frances M. Weesner, eds., 1969. *Biology of termites, Vol. I.* Academic Press, Inc., New York. xiii + 598 pp.

———, and Frances M. Weesner, eds., 1970. *Biology of termites, Vol. II.* Academic Press, Inc., New York. xiv + 643 pp.

Krombein, K. V., ed., 1958. Hymenoptera of America north of Mexico. Synoptic catalog. *Agriculture Monograph,* 2(1st suppl.): 1–305.

———, and B. D. Burks, eds., 1967. Hymenoptera of America north of Mexico. Synoptic catalog. *Agriculture Monograph,* 2(2nd suppl.): 1–584.

Kröning, F., 1925. Über die Dressur der Biene auf Töne. *Biologisches Zentralblatt,* 45(8): 496–507.

Kühn, A., 1927. Über den Farbensinn der Bienen. *Zeitschrift für Vergleichende Physiologie,* 5(4): 762–800.

Kullmann, E., 1968. Soziale Phaenomene bei Spinnen. *Insectes Sociaux,* 15(3): 289–297.

Kusnezov, N., 1951a. "Dinergatogina" en *Oligomyrmex bruchi* Santschi (Hymenoptera Formicidae). *Revista Sociedad Entomologica Argentina,* 15: 177–181.

———, 1951b. Un caso de evolucion eruptiva. *Eriopheidole symbiotica* nov. gen. nov. sp. (Hymenoptera, Formicidae). *Memorias del Museo de Entre Rios, Paraná, Republica Argentina,* 29(Zool.): 7–31.

———, 1952. Acerca de las hormigas simbióticas del género *Martia* Forel (Hymenoptera, Formicidae). *Acta Zoologica Lilloana del Instituto "Miguel Lillo", Tucumán, Republica Argentina,* 10: 717–722.

———, 1954. Un genero nuevo de hormigas (*Paranamyrma solenopsidis* nov. gen. nov. sp.) y los problemas relacionados (Hymenoptera, Formicidae). *Memorias del Museo de Entre Rios, Paraná, Republica Argentina,* 30(Zool.): 7–21.

———, 1957a. Numbers of species of ants in faunae of different latitudes. *Evolution,* 11(3): 298–299.

———, 1957b. Die Solenopsidinen-Gattungen von Südamerika (Hymenoptera, Formicidae). *Zoologischer Anzeiger,* 158(11–12): 266–280.

Kutter, H., 1913. Ein weiterer Beitrag zur Frage der sozialparasitischen Koloniegründung von *F. rufa* L. Zugleich ein Beitrag zur Biologie von *F. cinerea. Zeitschrift für Wissenschaftliche Insektenbiologie,* 9(6–7): 193–196.

———, 1923. Die Sklavenräuber *Strongylognathus huberi* For. ssp. *alpinus* Wheeler. *Revue Suisse de Zoologie,* 30(15): 387–424.

———, 1945. Eine neue Ameisengattung. *Mitteilungen der Schweizerischen Entomologischen Gesellschaft,* 19(10): 485–487.

———, 1950a. Über eine neue, extrem parasitische Ameise. 1. Mitteilung. *Mitteilungen der Schweizerischen Entomologischen Gesellschaft,* 23(2): 81–94.

———, 1950b. Über zwei neue Ameisen. *Chalepoxenus insubricus* u. *Epimyrma stumperi. Mitteilungen der Schweizerischen Entomologischen Gesellschaft,* 23(3): 337–346.

———, 1951. *Epimyrma stumperi* Kutter (Hym. Formicid.). 2. Mitteilung. *Mitteilungen der Schweizerischen Entomologischen Gesellschaft,* 24(2): 153–174.

———, 1956. Beiträge zur Biologie palaearktischer *Coptoformica* (Hym. Form.). *Mitteilungen der Schweizerischen Entomologischen Gesellschaft,* 29(1): 1–18.

———, 1957. Zur Kenntnis schweizerischer Coptoformicaarten (Hym. Form.). 2. Mitteilung. *Mitteilungen der Schweizerischen Entomologischen Gesellschaft,* 30(1): 1–24.

———, 1963. Miscellanea myrmecologica II. B. Künstliche Allianzkolonien. *Mitteilungen der Schweizerischen Entomologischen Gesellschaft,* 36(4): 324–329.

———, 1964. Miscellanea myrmecologica III. B. Künstliche Allianzkolonien (Fortsetzung). *Mitteilungen der Schweizerischen Entomologischen Gesellschaft,* 37(3): 131–137.

———, 1967. Beschreibung neuer Sozialparasiten von *Leptothorax acervorum* F. (Formicidae). *Mitteilungen der Schweizerischen Entomologischen Gesellschaft,* 40(1–2): 78–91.

———, 1969. Die sozialparasitischen Ameisen der Schweiz. *Neujahrsblatt Herausgegeben von der Naturforschenden Gesellschaft in Zürich,* 171: 1–62.

Kuwabara, M., 1957. Bildung des bedingten Reflexes von Pavlovs Typus bei der Honigbiene, *Apis mellifica. Journal of the Faculty of Science, Hokkaido University,* ser. 6(Zool.), 13(1–4): 458–464.

Lacher, V., 1964. Elektrophysiologische Untersuchungen an einzelnen Rezeptoren für Geruch, Kohlendioxyd, Luftfeuchtigkeit und Temperatur auf den Antennen der Arbeitsbiene und der Drohne (*Apis mellifica* L.). *Zeitschrift für Vergleichende Physiologie,* 48(6): 587–623.

———, 1967. Verhaltensreaktionen der Bienenarbeiterin bei Dressur auf Kohlendioxid. *Zeitschrift für Vergleichende Physiologie,* 54(1): 74–84.

Lack, D., 1966. *Population studies of birds.* Clarendon Press, Oxford, Eng. 341 pp.

Laidlaw, H. H., Jr., F. P. Gomes, and W. E. Kerr, 1956. Estimation of the number of lethal alleles in a panmictic population of *Apis mellifera* L. *Genetics,* 41(2): 179–188.

Lamborn, W. A., 1914. On the relationship between certain West African insects, especially ants, Lycaenidae and Homoptera. *Transactions of the Entomological Society of London, 1913* (Pt. III): 436–498; Appendix, 499–524.

Lange, R., 1958. Die deutschen Arten der *Formica rufa*-Gruppe. *Zoologischer Anzeiger,* 161(9–10): 238–243.

———, 1960a. Über die Futterweitergabe zwischen Angehörigen verschiedener Waldameisenstaaten. *Zeitschrift für Tierpsychologie,* 17(4): 389–401.

———, 1960b. Modellversuche über den Nahrungsbedarf von Völkern der Kahlrückigen Waldameise *Formica polyctena* Först. *Zeitschrift für Angewandte Entomologie,* 46(2): 200–208.

———, 1967. Die Nahrungsverteilung unter den Arbeiterinnen des Waldameisenstaates. *Zeitschrift für Tierpsychologie,* 24(5): 513–545.

Lappano, Eleanor Rita, 1958. A morphological study of larval development in polymorphic all-worker broods of the army ant *Eciton burchelli. Insectes Sociaux,* 5(1): 31–66.

Latreille, P. A., 1802. *Histoire naturelle des fourmis.* Théophile Barrois père, Libraire, Paris. xvi + 445 pp.

Law, J. H., E. O. Wilson, and J. A. McCloskey, 1965. Biochemical polymorphism in ants. *Science,* 149: 544–546.

Lawrence, J. F., and H. Reichardt, 1966. The systematic position of *Plaumanniola* Costa Lima (Coleoptera: Scydmaenidae). *Coleopterists' Bulletin,* 20(2): 39–42.

———, and H. Reichardt, 1969. The myrmecophilous Ptinidae (Coleoptera), with a key to Australian species. *Bulletin of the Museum of Comparative Zoology, Harvard,* 138(1): 1–27.

Lea, A. M., 1910. Australian and Tasmanian Coleoptera inhabiting or resorting to the nests of ants, bees, and termites. *Proceedings of the Royal Society of Victoria* (n.s.), 23(1): 116–230.

———, 1912. Australian and Tasmanian Coleoptera inhabiting or resorting to the nests of ants, bees and termites. Supplement. *Proceedings of the Royal Society of Victoria* (n.s.), 25(1): 31–78.

———, 1919. Notes on some miscellaneous Coleoptera, with

descriptions of new species, Pt. V. *Transactions and Proceedings of the Royal Society of South Australia,* 43: 166–261.

Lecomte, J., 1952. Hétérogénéité dans le comportement agressif des ouvrières d'*Apis mellifica. Compte Rendu de l'Académie des Sciences, Paris,* 234(8): 890–891.

———, 1956. Über die Bildung von 'Strassen' durch Sammelbienen, deren Stock um 180° gedreht wurde. *Zeitschrift für Bienenforschung,* 3: 128–133.

Ledoux, A., 1949. Le cycle évolutif de la fourmi fileuse (*Oecophylla longinoda* Latr.). *Compte Rendu de l'Académie des Sciences, Paris,* 229(3): 246–248.

———, 1950. Recherche sur la biologie de la fourmi fileuse (*Oecophylla longinoda* Latr.). *Annales des Sciences Naturelles,* (11)12(3–4): 313–461.

———, 1958. Biologie et comportement de l'Embioptère *Monotylota ramburi* Rims.-Kors. *Annales des Sciences Naturelles,* (11)20(4): 515–532.

Lee, J., 1938. Division of labor among the workers of the Asiatic carpenter ants (*Camponotus japonicus* var. *aterrimus*). *Peking Natural History Bulletin,* 13(2): 137–145.

Legewie, H., 1925. Zur Theorie der Staatenbildung. I. Teil und II. Teil. Die Biologie der Furchenbiene *Halictus malachurus* K. *Zeitschrift für Morphologie und Ökologie der Tiere,* 3(5): 619–684, 4(1–2): 246–300.

Le Masne, G., 1953. Observations sur les relations entre le couvain et les adultes chez les fourmis. *Annales des Sciences Naturelles,* (11)15(1): 1–56.

———, 1956a. La signification des reproducteurs aptères chez la fourmi *Ponera eduardi* Forel. *Insectes Sociaux,* 3(2): 239–259.

———, 1956b. Recherches sur les fourmis parasites *Plagiolepis grassei* et l'évolution des *Plagiolepis* parasites. *Compte Rendu de l'Académie des Sciences, Paris,* 243(7): 673–675.

———, 1961a. Recherches sur la biologie des animaux myrmécophiles. III. L'adoption des *Paussus favieri* Fairm. par une nouvelle société de *Pheidole pallidula* Nyl. *Compte Rendu de l'Académie des Sciences, Paris,* 253(15): 1621–1623.

———, 1961b. Recherches sur la biologie des animaux myrmécophiles. IV. Observations sur le comportement de *Paussus favieri* Fairm., hôte de la fourmi *Pheidole pallidula* Nyl. *Annales de la Faculté des Sciences de Marseille,* 31: 111–130.

———, 1965. Les transports mutuels autour des nids de *Neomyrma rubida* Latr.: Un nouveau type de relations interspécifiques chez les fourmis? *Comptes Rendus du Ve Congrès de l'Union Internationale pour l'Étude des Insectes Sociaux, Toulouse, 1965,* pp. 303–322.

———, and Annie Bonavita, 1967. Colony-founding according to archaic type (with repeated foraging) in the ant *Neomyrma rubida* Latr. *Proceedings of the Tenth International Ethological Conference, Stockholm, 1967,* mimeographed summary.

———, and C. Torossian, 1965. Observations sur le comportement du Coléoptère myrmécophile *Amorphocephalus coronatus* Germar (Brenthidae) hôte des *Camponotus. Insectes Sociaux,* 12(2): 185–194.

Lengerken, H. von, 1939. *Die Brutfürsorge- und Brutpflegeinstinkte der Käfer.* Akademische Verlagsgesellschaft M.B.H., Leipzig. viii + 285 pp.

———, 1951. Zur Brutbiologie des Spanischen Mondhornkäfers (*Copris hispanus* L.). *Biologisches Zentralblatt,* 70(9–10): 418–432.

———, 1954. *Die Brutfürsorge- und Brutpflegeinstinkte der Käfer,* 2nd ed. 383 pp.

Lepeletier de Saint-Fargeau, A. L. M., 1832. Observations sur l'ouvrage intitulé: "Bombi Scandinaviae Monographicè Tractati, *etc.,* a Gustav. Dahlbom. Londini Gothorum, 1832"; auxquelles on a joint les caractères des genres *Bombus* et *Psithyrus,* et la description des espèces qui appartiennent au dernier. *Annales de la Société Entomologique de France,* 1(4): 366–382.

Leuthold, R. H., 1968a. A tibial gland scent-trail and trail-laying behavior in the ant *Crematogaster ashmeadi* Mayr. *Psyche, Cambridge,* 75(3): 233–248.

———, 1968b. Recruitment to food in the ant *Crematogaster ashmeadi. Psyche, Cambridge,* 75(4): 334–350.

Levieux, J., 1966. Note préliminaire sur les colonnes de chasse de *Megaponera foetens* F. (Hyménoptère Formicidae). *Insectes Sociaux,* 13(2): 117–126.

Levin, M. D., 1961. Interactions among foraging honey bees from different apiaries in the same field. *Insectes Sociaux,* 8(3): 195–201.

———, and S. Glowska-Konopacka, 1963. Responses of foraging honeybees in alfalfa to increasing competition from other colonies. *Journal of Apicultural Research,* 2(1): 33–42.

Levins, R., 1968. *Evolution in changing environments: Some theoretical explorations.* Princeton University Press, Princeton, New Jersey. ix + 120 pp.

Lewontin, R. C., 1965. Selection for colonizing ability. *In* H. G. Baker and G. L. Stebbins, eds., *The genetics of colonizing species.* Academic Press, Inc., New York. pp. 79–94.

———, ed., 1968. *Population biology and evolution.* Syracuse University Press, Syracuse, New York. vii + 205 pp.

Li, C. C., 1955. *Population genetics.* University of Chicago Press, Chicago, Illinois. xi + 366 pp.

Light, S. F., 1944a. Parthenogenesis in termites of the genus *Zootermopsis. University of California Publications in Zoology,* 43(16): 405–412.

———, 1944b. Experimental studies on ectohormonal control of the development of supplementary reproductives in the termite genus *Zootermopsis* (formerly *Termopsis*). *University of California Publications in Zoology,* 43(17): 413–454.

———, and Frances M. Weesner, 1951. Further studies on the production of supplementary reproductives in *Zooter-*

mopsis (Isoptera). *Journal of Experimental Zoology,* 117(3): 397–414.

Lin, N., 1964. Increased parasitic pressure as a major factor in the evolution of social behavior in halictine bees. *Insectes Sociaux,* 11(2): 187–192.

Lind, Nancy K., 1970. Studies in the exocrinology of ants. Ph.D. Thesis, Department of Biology, Harvard University. 122 pp.

Lindauer, M., 1952. Ein Beitrag zur Frage der Arbeitsteilung im Bienenstaat. *Zeitschrift für Vergleichende Physiologie,* 34(4): 299–345.

———, 1954a. Dauertänze im Bienenstock und ihre Beziehung zur Sonnenbahn. *Naturwissenschaften,* 41(21): 506–507.

———, 1954b. Temperaturregulierung und Wasserhaushalt im Bienenstaat. *Zeitschrift für Vergleichende Physiologie,* 36(4): 391–432.

———, 1955. Schwarmbienen auf Wohnungssuche. *Zeitschrift für Vergleichende Physiologie,* 37(4): 263–324.

———, 1957. Sonnenorientierung der Bienen unter der Äquatorsonne und zur Nachtzeit. *Naturwissenschaften,* 44(1): 1–6.

———, 1960a. Die Sinne der Biene. *In* A. Büdel and E. Herold, eds., *Biene und Bienenzucht.* Ehrenwirth Verlag, Munich. pp. 28–47.

———, 1960b. Time-compensated sun orientation in bees. *Cold Spring Harbor Symposium on Quantitative Biology,* 25: 371–377.

———, 1961. *Communication among social bees.* Harvard University Press, Cambridge, Mass. ix + 143 pp.

———, and W. E. Kerr, 1958. Die gegenseitige Verständigung bei den stachellosen Bienen. *Zeitschrift für Vergleichende Physiologie,* 41(4): 405–434.

———, and W. E. Kerr, 1960. Communication between the workers of stingless bees. *Bee World,* 41: 29–41, 65–71.

———, and H. Martin, 1963. Über die Orientierung der Biene im Duftfeld. *Naturwissenschaften,* 50(15): 509–514.

———, and H. Martin, 1968. Die Schwereorientierung der Bienen unter dem Einfluss des Erdmagnetfeldes. *Zeitschrift für Vergleichende Physiologie,* 60(3): 219–243.

———, and J. O. Nedel, 1959. Ein Schweresinnesorgan der Honigbiene. *Zeitschrift für Vergleichende Physiologie,* 42(4): 334–364.

Lindhard, E., 1912. Humlebien som Husdyr. Spredte Traek af nogle danske Humlebiarters Biologi. *Tidsskrift for Landbrukets Planteavl,* 19: 335–352.

Lindroth, C. H., 1966. The ground-beetles (Carabidae, excl. Cicindelinae) of Canada and Alaska, Pt. 4. *Opuscula Entomologica,* suppl., 29: 409–648.

Long, W. H., 1902. New species of *Ceratopogon. Biological Bulletin, Marine Biological Laboratory, Woods Hole,* 3(1–2): 3–14.

Lövgren, B., 1958. A mathematical treatment of the development of colonies of different kinds of social wasps. *Bulletin of Mathematical Biophysics,* 20(2): 119–148.

Lubbock, J., 1894. *Ants, bees, and wasps: A record of observations on the habits of the social Hymenoptera.* D. Appleton & Co., New York. xix + 448 pp.

Lucas, H., 1885. [Communication sur les *Myraptera scutellaris* White (Hyménoptères)]. *Annales de la Société Entomologique de France,* (6)5. *Bulletin Entomologique Sèance du 11 Mars:* LIV.

Lukoschus, F., 1955. Die Bedeutung des innersekretorischen Systems fuer die Ausbildung epidermaler Kastenmerkmale bei der Honigbiene (*Apis mellifica* L.). *Insectes Sociaux,* 2(3): 221–236.

———, 1956. Zur Kastendetermination bei der Honigbiene. *Zeitschrift für Bienenforschung,* 3(8): 190–199.

Lund, A. W., 1831. Lettre sur les habitudes de quelques fourmis du Brésil, addressée à M. Audouin. *Annales des Sciences Naturelles,* (1)23: 113–138.

Lüscher, M., 1951. Beobachtungen über die Koloniegründung bei verschiedenen afrikanischen Termitenarten. *Acta Tropica,* 8(1): 36–43.

———, 1952a. New evidence for an ectohormonal control of caste determination in termites. *Transactions of the Ninth International Congress of Entomology, Amsterdam, 1951,* 1: 289–294.

———, 1952b. Die Produktion und Elimination von Ersatzgeschlechtstieren bei der Termite *Kalotermes flavicollis* Fabr. *Zeitschrift für Vergleichende Physiologie,* 34(2): 123–141.

———, 1953. The termite and the cell. *Scientific American,* 188(5): 74–78.

———, 1955. Der Sauerstoffverbrauch bei Termiten und die Ventilation des Nestes bei *Macrotermes natalensis* (Haviland). *Acta Tropica,* 12(4): 289–307.

———, 1956a. Die Entstehung von Ersatzgeschlechtstieren bei der Termite *Kalotermes flavicollis* Fabr. *Insectes Sociaux,* 3(1): 119–128.

———, 1956b. Die Lufterneuerung im Nest der Termite *Macrotermes natalensis* (Haviland). *Insectes Sociaux,* 3(2): 273–276.

———, 1958. Von der Gruppe zum "Staat" bei Insekten. *In* F. E. Lehmann, ed., *Gestaltungen sozialen Lebens bei Tier und Mensch.* Francke Verlag, Bern. pp. 48–65.

———, 1960. Hormonal control of caste differentiation in termites. *Annals of the New York Academy of Sciences,* 89(3): 549–563.

———, 1961a. Air-conditioned termite nests. *Scientific American,* 205(1): 138–145.

———, 1961b. Social control of polymorphism in termites. *Symposium of the Royal Entomological Society of London,* 1: 57–67.

———, 1962a. Hormonal regulation of development in termites. *Symposia Genetica et Biologica Italica,* 10: 1–11.

———, 1962b. Sex pheromones in the termite superorganism. *General and Comparative Endocrinology,* 2(6): 615.

———, 1963. Functions of the corpora allata in the development of termites. *Proceedings of the Sixteenth Interna-*

tional Congress of Zoology, Washington, D.C., 1963, 4: 244–250.

———, 1964. Die spezifische Wirkung männlicher und weiblicher Ersatzgeschlechtstiere auf die Entstehung von Ersatzgeschlechtstieren bei der Termite *Kalotermes flavicollis* (Fabr.). *Insectes Sociaux,* 11(1): 79–90.

———, 1969. Die Bedeutung des Juvenilhormons für die Differenzierung der Soldaten bei der Termite *Kalotermes flavicollis. Proceedings of the Sixth Congress of the International Union for the Study of Social Insects, Bern, 1969,* pp. 165–170.

———, and B. Müller, 1960. Ein spurbildendes Sekret bei Termiten. *Naturwissenschaften,* 47(21): 503.

———, and A. Springhetti, 1960. Untersuchungen über die Bedeutung der Corpora Allata für die Differenzierung der Kasten bei der Termite *Kalotermes flavicollis* F. *Journal of Insect Physiology,* 5(3–4): 190–212.

———, and I. Walker, 1963. Zur Frage der Wirkungsweise der Königinnenpheromone bei der Honigbiene. *Revue Suisse de Zoologie,* 70(2): 304–311.

Lyford, W. H., 1963. Importance of ants to brown podzolic soil genesis in New England. *Harvard Forest Paper* (Petersham, Massachusetts), no. 7: 1–18.

Maa, T., 1953. An inquiry into the systematics of the tribus Apidini or honeybees (Hym.). *Treubia,* 21(3): 525–640.

MacArthur, R. H., and E. O. Wilson, 1967. *The theory of island biogeography.* Princeton University Press, Princeton, New Jersey. xi + 203 pp.

Macgregor, E. C., 1948. Odour as a basis for orientated movement in ants. *Behaviour,* 1(3–4): 267–296.

Mackensen, O., 1943. The occurrence of parthenogenetic females in some strains of honeybees. *Journal of Economic Entomology,* 36(3): 465–467.

———, 1951. Viability and sex determination in the honey bee (*Apis mellifera* L.). *Genetics,* 36(5): 500–509.

Maeterlinck, M., 1905. *The life of the bee* (English tr. by A. Sutro). Dodd, Mead, and Co., New York. 427 pp.

———, 1927. *The life of the white ant.* (English tr. by A. Sutro). George Allen and Unwin, Ltd., London. 213 pp.

———, 1930. *The life of the ant.* (English tr. by B. Miall). George Allen and Unwin, Ltd., London. 192 pp.

Maidl, F., 1934. *Die Lebensgewohnheiten und Instinkte der staatenbildenden Insekten.* Fritz Wagner, Vienna. x + 823 pp.

Malyshev, S. I., 1960. The history and the conditions of the origin of instinct in ants (Hymenoptera, Formicoidea) [in Russian]. *Horae Societatis Entomologicae Unionis Soveticae,* 47: 5–52.

———, 1968. *Genesis of the Hymenoptera and the phases of their evolution* (tr. from Russian by B. Haigh; ed. by O. W. Richards and B. Uvarov). Methuen & Co., Ltd., London. viii + 319 pp.

Mamsch, E., 1967. Quantitative Untersuchungen zur Regulation der Fertilität im Ameisenstaat durch Arbeiterinnen, Larven und Königin. *Zeitschrift für Vergleichende Physiologie,* 55(1): 1–25.

———, and K. Bier, 1966. Das Verhalten von Ameisenarbeiterinnen gegenüber der Königin nach vorangegangener Weisellosigkeit. *Insectes Sociaux,* 8(4): 277–284.

Manee, A. H., 1915. Observations in Southern Pines, North Carolina (Hym., Col.). *Entomological News,* 26(6): 265–268.

Manfredi, P., 1949. Miriapodi mirmecofili. *Natura, Milano,* 40(3–4): 82–83.

Mani, M. S., and S. N. Rao, 1950. A remarkable example of maternal solicitude in a thrips from India. *Current Science,* 19(7): 217.

Mann, W. M., 1923. New genera and species of termitophilous Coleoptera from northern South America. *Zoologica, New York,* 3(17): 323–366.

Marais, E. N., 1933. *Die Siel van die Mier.* J. L. van Schaik, Pretoria. 182 pp.

Marak, G. E., and J. J. Wolken, 1965. An action spectrum for the fire ant (*Solenopsis saevissima*). *Nature, London,* 205: 1328–1329.

Marchal, P., 1896. La reproduction et l'évolution des guêpes sociales. *Archives de Zoologie Expérimentale et Générale,* (3)4: 1–100.

———, 1897. La castration nutriciale chez les Hyménoptères sociaux. *Compte Rendu de la Société de Biologie, Paris,* 1897: 556–557.

Marikovsky, P. I., 1961. Material on sexual biology of the ant *Formica rufa* L. *Insectes Sociaux,* 8(1): 23–30.

———, 1962. On some features of behavior of the ants *Formica rufa* L. infected with fungous disease. *Insectes Sociaux,* 9(2): 173–179.

———, 1963. The ants *Formica sanguinea* (Latr.) as pillagers of *Formica rufa* (Lin.) nests. *Insectes Sociaux,* 10(2): 119–128.

Markin, G. P., 1970. Food distribution within laboratory colonies of the Argentine ant, *Iridomyrmex humilis* (Mayr). *Insectes Sociaux,* 17(2): 127–157.

Markl, H., 1962. Borstenfelder an den Gelenken als Schweresinnesorgane bei Ameisen und anderen Hymenopteren. *Zeitschrift für Vergleichende Physiologie,* 45(5): 475–569.

———, 1965. Stridulation in leaf-cutting ants. *Science,* 149: 1392–1393.

———, 1966a. Schwerkraftdressuren an Honigbienen. I. Die geomenotaktische Fehlorientierung. *Zeitschrift für Vergleichende Physiologie,* 53(3): 328–352.

———, 1966b. Schwerkraftdressuren an Honigbienen. II. Die Rolle der schwererezeptorischen Borstenfelder verschiedener Gelenke für die Schwerekompassorientierung. *Zeitschrift für Vergleichende Physiologie,* 53(3): 353–371.

———, 1967. Die Verständigung durch Stridulationssignale bei Blattschneiderameisen. I. Die biologische Bedeutung der Stridulation. *Zeitschrift für Vergleichende Physiologie,* 57(3): 299–330.

———, 1968a. Conservation de l'équilibre et sens de la pesanteur chez l'abeille. *In* R. Chauvin, ed. (*q.v.*), *Traité de biologie de l'abeille, Vol. II.* pp. 146–172.

———, 1968b. Die Verständigung durch Stridulationssignale bei Blattschneiderameisen. II. Erzeugung und Eigenschaften der Signale. *Zeitschrift für Vergleichende Physiologie,* 60(2): 103–150.

———, and M. Lindauer, 1965. Physiology of insect behavior. *In* M. Rockstein, ed., *The physiology of Insecta, Vol. II.* Academic Press, Inc., New York. pp. 3–122.

Marko, P., I. Pecháň, and J. Vittek, 1964. Some phosphorus compounds in royal jelly. *Nature, London,* 202: 188–189.

Marlin, J. C., 1968. Notes on a new method of colony formation employed by *Polyergus lucidus lucidus* Mayr (Hymenoptera: Formicidae). *Transactions of the Illinois State Academy of Science,* 61(2): 207–209.

Martin, H., 1964. Zur Nahorientierung der Biene im Duftfeld. Zugleich ein Nachweis für die Osmotropotaxis bei Insekten. *Zeitschrift für Vergleichende Physiologie,* 48(5): 481–533.

———, 1965. Leistungen des topochemischen Sinnes bei der Honigbiene. *Zeitschrift für Vergleichende Physiologie,* 50(3): 254–292.

———, and M. Lindauer, 1966. Sinnesphysiologische Leistungen beim Wabenbau der Honigbiene. *Zeitschrift für Vergleichende Physiologie,* 53(3): 372–404.

Martin, M. M., and N. A. Weber, 1969. The cellulose-utilizing capability of the fungus cultured by the attine ant *Atta colombica tonsipes. Annals of the Entomological Society of America,* 62(6): 1386–1387.

———, R. M. Carman, and J. G. MacConnell, 1969. Nutrients derived from the fungus cultured by the fungus-growing ant *Atta colombica tonsipes. Annals of the Entomological Society of America,* 62(1): 11–13.

———, J. G. MacConnell, and G. R. Gale, 1969. The chemical basis for the attine ant-fungus symbiosis. Absence of antibiotics. *Annals of the Entomological Society of America,* 62(2): 386–388.

Martin, P., 1963. Die Steuerung der Volksteilung beim Schwärmen der Bienen. Zugleich ein Beitrag zum Problem der Wanderschwärme. *Insectes Sociaux,* 10(1): 13–42.

Martinez, A., 1950. Lamellicornia neótropica. II. Una nueva especie del género *Termitodius* Wasmann (Aphodiinae, Col.). *Arthropoda, Buenos Aires,* 1: 167–173.

Maschwitz, U., 1964. Gefahrenalarmstoffe und Gefahrenalarmierung bei sozialen Hymenopteren. *Zeitschrift für Vergleichende Physiologie,* 47(6): 596–655.

———, 1966a. Das Speichelsekret der Wespenlarven und seine biologische Bedeutung. *Zeitschrift für Vergleichende Physiologie,* 53(3): 228–252.

———, 1966b. Alarm substances and alarm behavior in social insects. *Vitamins and Hormones,* 24: 267–290.

———, K. Koob, and H. Schildknecht, 1970. Ein Beitrag zur Funktion der Metathoracaldrüse der Ameisen. *Journal of Insect Physiology,* 16(2): 387–404.

Masner, L., 1959. A revision of ecitophilous diapriid-genus *Mimopria* Holmgren (Hym., Proctotrupoidea). *Insectes Sociaux,* 6(4): 361–367.

Matsumura, F., H. C. Coppel, and A. Tai, 1968. Isolation and identification of termite trail-following pheromone. *Nature, London,* 219: 963–964.

Matthews, R. W., 1968a. *Microstigmus comes: Sociality in a sphecid wasp. Science,* 160: 787–788.

———, 1968b. Nesting biology of the social wasp *Microstigmus comes* (Hymenoptera: Sphecidae, Pemphredoninae). *Psyche, Cambridge,* 75(1): 23–45.

Maurizio, Anna, 1950. The influence of pollen feeding and brood rearing on the length of life and physiological condition of the honeybee. *Bee World,* 31(1): 9–12.

Mayr, E., 1963. *Animal species and evolution.* Belknap Press, Harvard University Press, Cambridge, Mass. xiv + 797 pp.

Mazokhin-Porshnyakov, G. A., 1968. The learning ability of insects and their capacity to generalize optical stimuli [in Russian]. *Revue d'Entomologie de l'URSS,* 47: 362–379.

———, 1969. Die Fähigkeit der Bienen, visuelle Reize zu generalisieren. *Zeitschrift für Vergleichende Physiologie,* 65: 15–28.

McClure, H. E., 1932. Incubation of bark bug eggs (Hemiptera: Aradidae). *Entomological News,* 43(7): 188–189.

McCluskey, E. S., 1958. Daily rhythms in male harvester and Argentine ants. *Science,* 128: 536–537.

———, 1963. Rhythms and clocks in harvester and Argentine ants. *Physiological Zoölogy,* 36(3): 273–292.

———, 1965. Circadian rhythms in male ants of five diverse species. *Science,* 150: 1037–1039.

———, 1967. Circadian rhythms in female ants, and loss after mating flight. *Comparative Biochemistry and Physiology,* 23(2): 665–677.

———, and C. E. Carter, 1969. Loss of rhythmic activity in female ants caused by mating. *Comparative Biochemistry and Physiology,* 31(2): 217–226.

McCook, H. C., 1877. Mound-making ants of the Alleghenies, their architecture and habits. *Transactions of the American Entomological Society,* 6: 253–296.

———, 1879a. Combats and nidification of the pavement ant, *Tetramorium caespitum. Proceedings of the Academy of Natural Sciences of Philadelphia, 1879,* 31: 156–161.

———, 1879b. *The natural history of the agricultural ant of Texas. A monograph of the habits, architecture, and structure of Pogonomyrmex barbatus.* Academy of Natural Sciences, Philadelphia. 208 pp.

———, 1882. *The honey ants of the Garden of the Gods, and the occident ants of the American Plains.* J. B. Lippincott & Co., Philadelphia. 188 pp.

McGurk, D. J., Jennifer Frost, E. J. Eisenbraun, K. Vick, W. A. Drew, and J. Young, 1966. Volatile compounds in ants: Identification of 4-methyl-3-heptanone from *Pogonomyrmex* ants. *Journal of Insect Physiology,* 12(11): 1435–1441.

———, Jennifer Frost, G. R. Waller, E. J. Eisenbraun, K. Vick,

W. A. Drew, and J. Young, 1968. Iridodial isomer variation in dolichoderine ants. *Journal of Insect Physiology,* 14(6): 841–845.

McKittrick, Frances A., 1965. A contribution to the understanding of cockroach-termite affinities. *Annals of the Entomological Society of America,* 58(1): 18–22.

McMahan, Elizabeth A., 1966. Food transmission within the *Cryptotermes brevis* colony (Isoptera: Kalotermitidae). *Annals of the Entomological Society of America,* 59(6): 1131–1137.

Medawar, P. B., 1957. *The uniqueness of the individual.* Methuen and Company, London. 191 pp.

Medler, J. T., 1957. Bumblebee ecology in relation to the pollination of alfalfa and red clover. *Insectes Sociaux,* 4(3): 245–252.

Medugorac, I., and M. Lindauer, 1967. Das Zeitgedächtnis der Bienen unter dem Einfluss von Narkose und von sozialen Zeitgebern. *Zeitschrift für Vergleichende Physiologie,* 55(4): 450–474.

Meidell, O., 1934. Fra dagliglivet i et homlebol. *Naturen,* 58(3): 85–95, (4): 108–116.

Menozzi, C., 1930. Formiche della Somalia Italiana meridionale. *Memorie della Società Entomologica Italiana,* 9: 76–131.

Menzel, R., 1968. Das Gedächtnis der Honigbiene für Spektralfarben. I. Kurzzeitiges und langzeitiges Behalten. *Zeitschrift für Vergleichende Physiologie,* 60(1): 82–102.

Meyer, J., 1966. Essai d'application de certains modèles cybernétiques à la coordination chez les insectes sociaux. *Insectes Sociaux,* 13(2): 127–138.

Michener, C. D., 1944. Comparative external morphology, phylogeny, and a classification of the bees (Hymenoptera). *Bulletin of the American Museum of Natural History,* 82(6): 157–326.

———, 1953a. Comparative morphological and systematic studies of bee larvae with a key to the families of hymenopterous larvae. *Kansas University Science Bulletin,* 35(8): 987–1102.

———, 1953b. Problems in the development of social behavior and communication among insects. *Transactions of the Kansas Academy of Science,* 56(1): 1–15.

———, 1954. Bees of Panamá. *Bulletin of the American Museum of Natural History,* 104(1): 1–175.

———, 1958. The evolution of social behavior in bees. *Proceedings of the Tenth International Congress of Entomology, Montreal, 1956,* 2: 441–447.

———, 1960. Observations on the behaviour of a burrowing bee (*Amegilla*) near Brisbane, Queensland (Hymenoptera, Anthophorinae). *Queensland Naturalist,* 16(3–4): 63–67.

———, 1961a. Observations on the nests and behavior of *Trigona* in Australia and New Guinea (Hymenoptera, Apidae). *American Museum Novitates,* 2026: 1–46.

———, 1961b. Social polymorphism in Hymenoptera. *Symposium of the Royal Entomological Society of London,* 1: 43–56.

———, 1961c. Probable parasitism among Australian bees of the genus *Allodapula* (Hymenoptera, Apoidea, Ceratinini). *Annals of the Entomological Society of America,* 54(4): 532–534.

———, 1962. Biological observations on the primitively social bees of the genus *Allodapula* in the Australian region (Hymenoptera, Xylocopinae). *Insectes Sociaux,* 9(4): 355–373.

———, 1964a. Evolution of the nests of bees. *American Zoologist,* 4(2): 227–239.

———, 1964b. The bionomics of *Exoneurella,* a solitary relative of *Exoneura* (Hymenoptera: Apoidea: Ceratinini). *Pacific Insects,* 6(3): 411–426.

———, 1964c. Reproductive efficiency in relation to colony size in hymenopterous societies. *Insectes Sociaux,* 11(4): 317–341.

———, 1965a. A classification of the bees of the Australian and South Pacific regions. *Bulletin of the American Museum of Natural History,* 130: 1–362.

———, 1965b. The life cycle and social organization of bees of the genus *Exoneura* and their parasite, *Inquilina* (Hymenoptera: Xylocopinae). *Kansas University Science Bulletin,* 46(9): 317–358.

———, 1966a. Interaction among workers from different colonies of sweat bees (Hymenoptera, Halictidae). *Animal Behaviour,* 14(1): 126–129.

———, 1966b. The bionomics of a primitively social bee, *Lasioglossum versatum* (Hymenoptera: Halictidae). *Journal of the Kansas Entomological Society,* 39(2): 193–217.

———, 1966c. Parasitism among Indoaustralian bees of the genus *Allodapula* (Hymenoptera: Ceratinini). *Journal of the Kansas Entomological Society,* 39(4): 705–708.

———, 1969a. Comparative social behavior of bees. *Annual Review of Entomology,* 14: 299–342.

———, 1969b. Immature stages of a chalcidoid parasite tended by allodapine bees (Hymenoptera: Perilampidae and Anthophoridae). *Journal of the Kansas Entomological Society,* 42(3): 247–250.

———, 1969c. African genera of allodapine bees (Hymenoptera: Anthophoridae: Ceratinini). *Journal of the Kansas Entomological Society,* 42(3): 289–293.

———, 1970. Social parasites among African allodapine bees (Hymenoptera, Anthophoridae, Ceratinini). *Journal of the Linnean Society, London, Zoology,* 49(3): 199–215.

———, and W. B. Kerfoot, 1967. Nests and social behavior of three species of *Pseudaugochloropsis* (Hymenoptera: Halictidae). *Journal of the Kansas Entomological Society,* 40(2): 214–232.

———, and W. E. LaBerge, 1954. A large *Bombus* nest from Mexico. *Psyche, Cambridge,* 61(2): 63–67.

———, and R. B. Lange, 1958a. Distinctive type of primitive social behavior among bees. *Science,* 127: 1046–1047.

———, and R. B. Lange, 1958b. Observations on the behavior of Brasilian halictid bees. V. *Chloralictus. Insectes Sociaux,* 5(4): 379–407.

———, and R. B. Lange, 1959. Observations on the behavior of Brazilian halictid bees (Hymenoptera, Apoidea). IV. *Augochloropsis,* with notes on extralimital forms. *American Museum Novitates,* 1924: 1–41.

———, E. A. Cross, H. V. Daly, C. W. Rettenmeyer, and A. Wille, 1955. Additional techniques for studying the behavior of wild bees. *Insectes Sociaux,* 2(3): 237–246.

———, R. B. Lange, J. J. Bigarella, and R. Salamuni, 1958. Factors influencing the distribution of bees' nests in earth banks. *Ecology,* 39(2): 207–217.

Miller, C. D. F., 1961. Taxonomy and distribution of Nearctic *Vespula. Canadian Entomologist,* 93, suppl. 22: 1–52.

Miller, E. M., 1969. Caste differentiation in the lower termites. *In* K. Krishna and F. M. Weesner, eds. (*q.v.*), *Biology of termites, Vol. I.* pp. 283–310.

Miller, R. S., 1967. Pattern and process in competition. *Advances in Ecological Research,* 4: 1–74.

Milliron, H. E., 1961. Revised classification of the bumblebees—a synopsis (Hymenoptera: Apidae). *Journal of the Kansas Entomological Society,* 34(2): 49–61.

Millot, J., and M. Vachon, 1949. Ordre des scorpions. *In* P.-P. Grassé, ed., *Traité de zoologie, Vol. 6.* Masson et Cie, Paris. pp. 386–436.

Milojevic, B. D., 1940. A new interpretation of the social life of the honey bee. *Bee World,* 21(4): 39–41.

Milum, V. G., 1955. Honey bee communication. *American Bee Journal,* 95(3): 97–104.

Minnich, D. E., 1932. The contact chemoreceptors of the honey bee, *Apis mellifera* Linn. *Journal of Experimental Zoology,* 61(3): 375–393.

Mitchell, T. B., 1960. Bees of the eastern United States, Vol. I. *Technical Bulletin, North Carolina Agricultural Experiment Station,* 141: 1–538.

Mittelstaedt, H., 1950. Physiologie des Gleichgewichtssinnes bei fliegenden Libellen. *Zeitschrift für Vergleichende Physiologie,* 32(5): 422–463.

Mittler, T. E., 1958. Studies on the feeding and nutrition of *Tuberolachnus salignus* (Gmelin) (Homoptera, Aphididae). III. The nitrogen economy. *Journal of Experimental Biology,* 35(3): 626–638.

Mockford, E. L., 1957. Life history studies on some Florida insects of the genus *Archipsocus* (Psocoptera). *Bulletin of the Florida State Museum, Biological Sciences,* 1(5): 253–274.

———, 1965. Some South African Psocoptera from termite nests. *Entomological News,* 76(7): 169–176.

Moeller, A., 1893. Die Pilzgärten einiger südamerikanischer Ameisen. *Botanische Mitteilungen aus den Tropen,* 1893(6): 1–127.

Montagner, H., 1963a. Contribution à l'étude de dèterminisme des castes chez les Vespides. *Compte Rendu de la Société Biologie, Paris,* 157(1): 147–150.

———, 1963b. Étude préliminaire des relations entre les adultes et le couvain chez les guêpes sociales du genre *Vespa,* au moyen d'un radio-isotope. *Insectes Sociaux,* 10(2): 153–165.

———, 1964. Étude du comportement alimentaire et des relations trophallactiques des mâles au sein de la société de guêpes, au moyen d'un radio-isotope. *Insectes Sociaux,* 11(4): 301–316.

———, 1966. Le mécanisme et les conséquences des comportements trophallactiques chez les guêpes du genre *Vespa.* Thèses, Faculté des Sciences de l'Université de Nancy, France. 143 pp.

———, 1967. Comportements trophallactiques chez les guêpes sociales. Film produced by Service du Film de Recherche Scientifique; 96, Boulevard Raspail, Paris. No. B2053.

———, and G. Courtois, 1963. Données nouvelles sur le comportement alimentaire et les échanges trophallactiques chez les guêpes sociales. *Compte Rendu de l'Académie des Sciences, Paris,* 256(19): 4092–4094.

Moore, B. P., 1964a. Volatile terpenes from *Nasutitermes* soldiers (Isoptera, Termitidae). *Journal of Insect Physiology,* 10(2): 371–375.

———, 1964b. Australian larval Carabidae of the subfamilies Broscinae, Psydrinae and Pseudomorphinae. *Pacific Insects,* 6(2): 242–246.

———, 1966. Isolation of the scent-trail pheromone of an Australian termite. *Nature, London,* 211: 746–747.

———, 1968. Studies on the chemical composition and function of the cephalic gland secretion in Australian termites. *Journal of Insect Physiology,* 14(1): 33–39.

———, 1969. Biochemical studies in termites. *In* K. Krishna and F. M. Weesner, eds. (*q.v.*), *Biology of termites, Vol. I,* pp. 407–432.

Mordwilko, A., 1907. Die Ameisen und Blattläuse in ihren gegenseitigen Beziehungen und das Zusammenleben von Lebewesen überhaupt. *Biologisches Centralblatt,* 27(7): 212–224, (8): 233–252.

Morgenthaler, O., 1968. Les maladies infectieuses des ouvrières. *In* R. Chauvin, ed. (*q.v.*), *Traité de biologie de l'abeille, Vol. IV.* pp. 324–395.

Morimoto, R., 1960. Experimental study on the trophallactic behavior in *Polistes* (Hymenoptera, Vespidae). *Acta Hymenopterologica, Tokyo,* 1(2): 99–103.

———, 1961a. On the dominance order in *Polistes* wasps, I (Studies on the social Hymenoptera in Japan, XII). *Science Bulletin of the Faculty of Agriculture, Kyushu University,* 18(4): 339–351.

———, 1961b. On the dominance order in *Polistes* wasps, II (Studies on the social Hymenoptera in Japan, XIII). *Science Bulletin of the Faculty of Agriculture, Kyushu University,* 19(1): 1–17.

Morisita, M., 1941. Interrelations between *Formica fusca japonica* and other species of ants on a tree. *Kontyû,* 15: 1–9.

Morse, R. A., 1963. Swarm orientation in honeybees. *Science,* 141: 357–358.

———, and N. E. Gary, 1961. Colony response to worker bees confined with queens (*Apis mellifera* L.). *Bee World,* 42(8): 197–199.

———, and F. M. Laigo, 1969. *Apis dorsata* in the Philippines (including an annotated bibliography). *Monograph of the Philippine Association of Entomologists, Inc.* (University of the Philippines, The College, Laguna, P. I.). No. 1: 96 pp.

———, D. A. Shearer, R. Boch, and A. W. Benton, 1967. Observations on alarm substances in the genus *Apis. Journal of Apicultural Research,* 6(2): 113–118.

Mosconi, P. B., 1958. Osservazioni istologiche sul complesso endocrino e neurocrino di *Mastotermes darwiniensis* Froggatt. *Symposia Genetica,* 6: 77–90.

Moser, J. C., 1963. Contents and structure of *Atta texana* nests in summer. *Annals of the Entomological Society of America,* 56(3): 286–291.

———, 1964. Inquiline roach responds to trail-marking substance of leaf-cutting ants. *Science,* 143: 1048–1049.

———, 1967. Mating activities of *Atta texana* (Hymenoptera, Formicidae). *Insectes Sociaux,* 14(3): 295–312.

———, and M. S. Blum, 1963. Trail marking substance of the Texas leaf-cutting ant: Source and potency. *Science,* 140: 1228.

———, R. C. Brownlee, and R. Silverstein, 1968. Alarm pheromones of the ant *Atta texana. Journal of Insect Physiology,* 14(4): 529–535.

Moskovljevíc-Filipovíc, V. C., 1956. [The influence of normal and modified social structures on the development of the pharyngeal glands and the work of the honeybee (*Apis mellifera* L.)]. *Monogr. Srpska Akad. Nauka.,* 262: 1–101. [Cited by Free, 1965].

Mou, Y. C., 1938. Morphologische und histologische Studien über Paussidendrüsen. *Zoologische Jahrbücher, Abt. Anat. Ontog. Tiere,* 64(2–3): 287–346.

Moure, J. S., 1961. A preliminary supra-specific classification of the Old World meliponine bees (Hymenoptera, Apoidea). *Studia Entomologica* (n.s.), 4(1–4): 181–242.

———, 1967. A check-list of the known euglossine bees (Hymenoptera, Apidae). *Atas do Simpósio sôbre a Biota Amazônica Conselho Nacional de Pesquisas, Rio de Janeiro,* 5(Zool.): 395–415.

———, P. Nogueira-Neto, and W. E. Kerr, 1958. Evolutionary problems among Meliponinae (Hymenoptera, Apidae). *Proceedings of the Tenth International Congress of Entomology, Montreal, 1956,* 2: 481–493.

Moynihan, M., 1962. The organization and probable evolution of some mixed species flocks of Neotropical birds. *Smithsonian Miscellaneous Collections,* 143(7): 1–140.

Muesebeck, C. F. W., K. V. Krombein, and H. K. Townes, eds., 1951. Hymenoptera of America north of Mexico. *Agriculture Monograph,* 2: 1–1420.

Müller, J., 1826. *Zur vergleichenden Physiologie des Gesichtssinnes des Menschen und der Tiere.* Leipzig.

Müller, W., 1886. Beobachtungen an Wanderameisen (*Eciton hamatum* Fabr.). *Kosmos,* 18: 81–93.

Muma, M. H., and W. J. Gertsch, 1964. The spider family Uloboridae in North America north of Mexico. *American Museum Novitates,* 2196: 1–43.

Murrell, K. F. H., 1965. *Ergonomics: Man in his working environment.* Chapman and Hall, London. xix + 496 pp.

Myers, J. G., 1929. The nesting-together of birds, wasps and ants. *Proceedings of the Entomological Society of London,* 4: 80–88.

———, 1935. Nesting associations of birds with social insects. *Transactions of the Royal Entomological Society of London,* 83(1): 11–22.

Naarman, Hiltrud, 1963. Untersuchungen über Bildung und Weitergabe von Drüsensekreten bei *Formica* (Hymenopt. Formicidae) mit Hilfe der Radioisotopenmethode. *Experientia,* 19(8): 412–413.

Naumann, M. G., 1968. A revision of the genus *Brachygastra* (Hymenoptera: Vespidae). *Kansas University Science Bulletin,* 47(17): 929–1003.

Nedel, J. O., 1960. Morphologie und Physiologie der Mandibeldrüse einiger Bienen-Arten (Apidae). *Zeitschrift für Morphologie und Ökologie der Tiere,* 49(2): 139–183.

Neese, V., 1968. Le sens de la vue. *In* R. Chauvin, ed. (*q.v.*), *Traité de biologie de l'abeille, Vol. II.* pp. 101–121.

Nel, J. J. C., 1968. Aggressive behaviour of the harvester termites *Hodotermes mossambicus* (Hagen) and *Trinervitermes trinervoides* (Sjöstedt). *Insectes Sociaux,* 15(2): 145–156.

Nixon, G. E. J., 1951. *The association of ants with aphids and coccids.* Commonwealth Institute of Entomology, London. 36 pp.

Nixon, H. L., and C. R. Ribbands, 1952. Food transmission within the honeybee community. *Proceedings of the Royal Society,* (B)140: 43–50.

Nogueira-Neto, P., 1948. Notas bionômicas sobre Meliponíneos (Hymenoptera, Apoidea). I. Sobre a ventilação dos ninhos e as construções com ela relacionadas. *Revista Brasileira de Biologia,* 8(4): 465–488.

———, 1950. Notas bionômicas sobre Meliponíneos (Hymenoptera, Apoidea). IV. Colônias mistas e questões relacionadas. *Revista de Entomologia, Rio de Janeiro,* 21(1–2): 305–367.

———, 1951. Stingless bees and their study. *Bee World,* 32(10): 73–76.

———, 1970. *A criação de abelhas indígenas sem ferrão* (*Meliponinae*). Editora Chácaras e Quintais, São Paulo. 365 pp.

Noirot, C., 1952. Les soins et l'alimentation des jeunes chez les termites. *Annales des Sciences Naturelles,* (11)14(2–4): 405–414.

———, 1954. Le polymorphisme des termites supérieurs. *Année Biologique,* 30(11–12): 461–474.

———, 1955. Recherches sur le polymorphisme des termites supérieurs (Termitidae). *Annales des Sciences Naturelles,* (11)17(3–4): 399–595.

———, 1956. Les sexués de remplacement chez les termites supérieurs (Termitidae). *Insectes Sociaux,* 3(1): 145–158.

———, 1958–1959. Remarques sur l'écologie des termites. *Annales de la Société Royale Zoologique de Belgique,* 89(1): 151–169.

———, 1969a. Glands and secretions. *In* K. Krishna and F. M. Weesner, eds. (*q.v.*), *Biology of termites, Vol. I.* pp. 89–123.

———, 1969b. Formation of castes in the higher termites. *In* K. Krishna and F. M. Weesner, eds. (*q.v.*), *Biology of termites, Vol. I.* pp. 311–350.

———, and Cécile Noirot-Timothée, 1965a. Organisation de la glande sternale chez *Calotermes flavicollis* F. (Insecta, Isoptera). Étude au microscope électronique. *Compte Rendu de l'Académie des Sciences, Paris,* 260(23): 6202–6204.

———, and Cécile Noirot-Timothée, 1965b. La glande sternale dans l'évolution des termites. *Insectes Sociaux,* 12(3): 265–272.

Nolan, W. J., 1924. The division of labour in the honeybee. *North Carolina Beekeeper,* (October): 10–15. [Cited by Free, 1965].

———, 1925. The brood-rearing cycle of the honeybee. *Bulletin, United States Department of Agriculture,* 1349: 1–56.

———, 1928. Seasonal brood-rearing activity of the Cyprian honeybee. *Journal of Economic Entomology,* 21(2): 392–403.

Noll, J., 1931. Untersuchungen über die Zeugung und Staatenbildung des *Halictus malachurus* Kirby. *Zeitschrift für Morphologie und Ökologie der Tiere,* 23(1–2): 285–368.

Nørgaard, E., 1956. Environment and behaviour of *Theridion saxatile. Oikos* (Acta Oecologica Scandinavica), 7(2): 159–192.

Norris, Maud J. (Mrs. O. W. Richards), 1968. Some group effects on reproduction in locusts. *Colloques Internationaux du Centre National de la Recherche Scientifique, Paris, 1967,* No. 173: 147–161.

Novikoff, A. B., 1945a. The concept of integrative levels and biology. *Science,* 101: 209–215.

———, 1945b. Continuity and discontinuity in evolution. *Science,* 102: 405–406.

Nunberg, Marian, 1951. Contribution to the knowledge of prothoracic glands of Scolytidae and Platypodidae (Coleoptera). *Annales Musei Zoologici Polonici,* 14(18): 261–265.

Nutting, W. L., 1966a. Colonizing flights and associated activities of termites. I. The desert damp-wood termite *Paraneotermes simplicicornis* (Kalotermitidae). *Psyche, Cambridge,* 73(2): 131–149.

———, 1966b. Distribution and biology of the primitive dry-wood termite *Pterotermes occidentis* (Walker) (Kalotermitidae). *Psyche, Cambridge,* 73(3): 165–179.

———, 1969. Flight and colony foundation. *In* K. Krishna and F. M. Weesner, eds. (*q.v.*), *Biology of termites, Vol. I.* pp. 233–282.

Odhiambo, T. R., 1959. An account of parental care in *Rhinocoris albopilosus* Signoret (Hemiptera-Heteroptera: Reduviidae), with notes on its life history. *Proceedings of the Royal Entomological Society of London,* (A)34: 175–185.

———, 1960. Parental care in bugs and non-social insects. *New Scientist,* 8: 449–451.

Ohaus, F., 1909. Bericht über eine entomologische Studienreise in Südamerika. *Stettiner Entomologische Zeitung,* 70(1): 3–139.

Oke, C., 1932. Aculagnathidae. A new family of Coleoptera. *Proceedings of the Royal Society of Victoria* (n.s.), 44(1): 22–24.

Ökland, F., 1930. Studien über die Arbeitsteilung und die Teilung des Arbeitsgebietes bei der Roten Waldameise (*Formica rufa* L.). *Zeitschrift für Morphologie und Ökologie der Tiere,* 20(1): 63–131.

———, 1934. Utvandring og overvintring hos den røde skogmaur (*Formica rufa* L.). *Norsk Entomologisk Tidsskrift,* 3(5): 316–327.

Olberg, G., 1959. *Das Verhalten der solitären Wespen Mitteleuropas* (*Vespidae, Pompilidae, Sphecidae*). VEB Deutscher Verlag der Wissenschaften, Berlin. xiii + 402 pp.

Opfinger, Elisabeth, 1949. Zur Psychologie der Duftdressuren bei Bienen. *Zeitschrift für Vergleichende Physiologie,* 31(4): 441–453.

Ordway, Ellen, 1964. *Sphecodes pimpinellae* and other enemies of *Augochlorella. Journal of the Kansas Entomological Society,* 37(2): 139–152.

———, 1965. Caste differentiation in *Augochlorella* (Hymenoptera, Halictidae). *Insectes Sociaux,* 12(4): 291–308.

Ormerod, E. L., 1868. *British social wasps: An introduction to their anatomy and physiology, architecture, and general natural history.* Longmans, Green, Reader, and Dyer, London. xi + 270 pp.

Orösi-Pál, Z., 1956. Az épitö alkalom hatása a viaszmirigy müködéséne. *Méhészet, Budapest,* 4: 105. [Cited by Free, 1965].

Osanai, M., and H. Rembold, 1968. Entwicklungsabhängige mitochondriale Enzymaktivitäten bei den Kasten der Honigbiene. *Biochimica et Biophysica Acta,* 162: 22–31.

Otto, D., 1958a. Über die Arbeitsteilung im Staate von *Formica rufa rufo-pratensis minor* Gössw. und ihre verhaltensphysiologischen Grundlagen: Ein Beitrag zur Biologie der Roten Waldameise. *Wissenschaftliche Abhandlungen der Deutschen Akademie der Landwirtschaftswissenschaften zu Berlin,* 30: 1–169.

———, 1958b. Über die Homologieverhältnisse der Pharynx- und Maxillardrüsen bei Formicidae und Apidae (Hymenopt.). *Zoologischer Anzeiger,* 161(9–10): 216–226.

———, 1960. Zur Erscheinung der Arbeiterinnenfertilität und Parthenogenese bei der Kahlrückigen Roten Waldameise (*Formica polyctena* Först.), (Hym.). *Deutsche Entomologische Zeitschrift* (n.f.), 7(1–2): 1–9.

Pain, Janine, 1961a. Sur la phérormone des reines d'abeilles et ses effets physiologiques. *Annales de l'Abeille,* 4(2): 73–158.

———, 1961b. Absence du pouvoir d'inhibition de la phérormone I sur le développement ovarien des jeunes ouvrières d'abeilles. *Compte Rendu de l'Académie des Sciences, Paris,* 252(15): 2316–2317.

———, 1964. Premières observations sur une espèces nouvelle d'araignées sociales (*Agelena consociata* Denis). *Biologia Gabonica,* 1(1): 47–58.

Painter, T. S., 1969. The origin of the nucleic acid bases found in the royal jelly of the honeybee. *Proceedings of the National Academy of Sciences of the U.S.A.,* 64(1): 64–66.

Pandazis, G., 1930. Über die relative Ausbildung der Gehirnzentren bei biologisch verschiedenen Ameisenarten. *Zeitschrift für Morphologie und Ökologie der Tiere,* 18(1–2): 114–169.

Pardi, L., 1940. Ricerche sui Polistini. I. Poliginia vera ed apparente in *Polistes gallicus* (L.). *Processi Verbali della Società Toscana di Scienze Naturali in Pisa,* 49: 3–9.

———, 1942. Ricerche sui Polistini. V. La poliginia iniziale di *Polistes gallicus* (L.). *Bollettino dell'Istituto di Entomologia della Università degli Studi di Bologna,* 14: 1–106.

———, 1948a. Beobachtungen über das Interindividuelle Verhalten bei *Polistes gallicus* (Untersuchungen über die Polistini, No. 10). *Behaviour,* 1(2): 138–172.

———, 1948b. Dominance order in *Polistes* wasps. *Physiological Zoölogy,* 21(1): 1–13.

———, 1951. Studio della attività e della divisione di lavoro in una società di *Polistes gallicus* (L.) dopo la comparsa delle operaie (Ricerche sui Polistini, XII). *Archivio Zoologico Italiano,* 36: 363–431.

Park, Orlando, 1929. Ecological observations upon the myrmecocoles of *Formica ulkei* Emery, especially *Leptinus testaceus* Mueller. *Psyche, Cambridge,* 36(3): 195–215.

———, 1933. Ecological study of the ptiliid myrmecocole, *Limulodes paradoxus* Matthews. *Annals of the Entomological Society of America,* 26(2): 255–261.

———, 1964. Observations upon the behavior of myrmecophilous pselaphid beetles. *Pedobiologia,* 4(3): 129–137.

Park, Wallace, 1925. The storing and ripening of honey by honeybees. *Journal of Economic Entomology,* 18(2): 405–410.

Parsons, P. A., 1967. *The genetic analysis of behaviour.* Methuen and Company, London. ix + 174 pp.

Passera, L., 1963a. Le cycle évolutif de la fourmi *Plagiolepis pygmaea* Latr. (Hyménoptères, Formicoïdea, Formicidae). *Insectes Sociaux,* 10(1): 59–69.

———, 1963b. Les relations sociales chez la fourmi *Plagiolepis pygmaea* Latr. (Hyménoptères, Formicoidea, Formicidae). *Insectes Sociaux,* 10(2): 103–110.

———, 1964. Données biologiques sur la fourmi parasite *Plagiolepis xene* Stärcke. *Insectes Sociaux,* 11(1): 59–70.

———, 1965a. Inhibition de la ponte des ouvrières par les reines chez la fourmi *Plagiolepis pygmaea* Latr. *Comptes Rendus de la Ve Congrès Union Internationale pour l'Étude des Insectes Sociaux, Toulouse, 1965,* pp. 293–302.

———, 1965b. Rôle des reines sur le devenir des larves du sexe femelle chez la fourmi *Plagiolepis pygmaea* Latr. (Hym. Formicidae). *Compte Rendu de l'Académie des Sciences, Paris,* 260(26): 6979–6981.

———, 1966. Fécondité des femelles au sein de la myrmécobiose *Plagiolepis pygmaea* Latr.—*Plagiolepis xene* Star. (Hyménoptères, Formicidae). *Compte Rendu de l'Académie des Sciences, Paris,* 263(21): 1600–1603.

———, 1968. Observations biologiques sur la fourmi *Plagiolepis grassei* Le Masne Passera parasite social de *Plagiolepis pygmaea* Latr. (Hym. Formicidae). *Insectes Sociaux,* 15(4): 327–336.

———, J. Bitsch, and C. Bressac, 1968. Observations histologiques sur la formation des oeufs alimentaires et des oeufs reproducteurs chez les ouvrières de *Plagiolepis pygmaea* Latr. (Hymenoptera Formicidae). *Compte Rendu de l'Académie des Sciences, Paris,* 266(24): 2270–2272.

Pasteels, J. M., 1965. Polyéthisme chez les ouvriers de *Nasutitermes lujae* (Termitidae Isoptères). *Biologia Gabonica,* 1(2): 191–205.

———, 1967. Contribution à l'étude systématique des Staphylins termitophiles du Gabon (Coleoptera Staphylinidae, Aleocharinae). *Revue de Zoologie et de Botanique Africaines,* 76(1–2): 43–69.

———, 1968. Le système glandulaire tégumentaire des Aleocharinae (Coleoptera, Staphylinidae) et son évolution chez les espèces termitophiles du genre *Termitella. Archives de Biologie, Liège,* 79(3): 381–469.

———, 1969. Les glandes tégumentaires des staphylins termitophiles. III. Les Aleocharinae des genres *Termitopullus* (Corotocini, Corotocina), *Perinthodes, Catalina* (Termitonannini, Perinthina), *Termitusa* (Termitohospitini, Termitusina). *Insectes Sociaux,* 16(1): 1–26.

Patrizi, S., 1948. Contribuzioni alla conoscenza delle formiche e dei mirmicofili dell'Africa orientale. V. Note etologiche su *Myrmechusa* Wasmann (Coleoptera Staphylinidae). *Bollettino dell'Istituto di Entomologia della Università degli Studi di Bologna,* 17: 168–173.

Paulian, R., 1947. Un Termitotrox [Col. Scarabaeidae] de Côte d'Ivoire. *Bulletin de la Société Entomologique de France,* 52(8): 134–136.

———, 1948. Observations sur les Coléoptères commensaux d'*Anomma nigricans* en Côte d'Ivoire. *Annales des Sciences Naturelles,* (11)10(1): 79–102.

Pavan, M., and G. Ronchetti, 1955. Studi sulla morfologia esterna e anatomia interna dell'operaia di *Iridomyrmex humilis* Mayr e ricerche chimiche e biologiche sulla iridomirmecina. *Atti della Società Italiana di Scienze Naturali, Milano,* 94(3–4): 379–477.

Peacock, A. D., and A. T. Baxter, 1950. Studies in Pharaoh's ant, *Monomorium pharaonis* (L.). 3. Life history. *Entomologist's Monthly Magazine,* 86: 171–178.

———, F. L. Waterhouse, and A. T. Baxter, 1955. Studies in pharaoh's ant, *Monomorium pharaonis* (L.). 10. Viability in regard to temperature and humidity. *Entomologist's Monthly Magazine,* 91: 37–42.

———, D. W. Hall, I. C. Smith, and A. Goodfellow, 1950. The biology and control of the ant pest *Monomorium pharaonis* (L.). *Miscellaneous Publications, Department of Agriculture for Scotland,* 17: 1–51.

———, I. C. Smith, D. W. Hall, and A. T. Baxter, 1954. Studies in Pharaoh's ant, *Monomorium pharaonis* (L.). 8. Male production by parthenogenesis. *Entomologist's Monthly Magazine,* 90: 154–158.

Pearl, R., and J. R. Miner, 1935. Experimental studies on the duration of life. XIV. The comparative mortality of certain lower organisms. *Quarterly Review of Biology,* 10(1): 60–79.

Pearse, A. S., Marguerite T. Patterson, J. S. Rankin, and G. W. Wharton, 1936. The ecology of *Passalus cornutus* Fabricius, a beetle which lives in rotting logs. *Ecological Monographs,* 6(4): 455–490.

Pechlaner, E., 1904. Zum Nestbau der *Vespa germanica. Verhandlungen der Zoologisch-Botanischen Gesellschaft in Wien,* 54: 77–79.

Pérez, J., 1899. *Les abeilles.* Librairie Hachette et Cie, Paris. viii + 348 pp.

Perrelet, A., and F. Baumann, 1969. Evidence for extracellular space in the rhabdome of the honeybee drone eye. *Journal of Cell Biology,* 40(3): 825–830.

Petersen, B., 1968. Some novelties in presumed males of Leptanillinae (Hym., Formicidae). *Entomologiske Meddelelser,* 36: 577–598.

Pflumm, W., 1969. Stimmungsänderungen der Biene während des Aufenthalts an der Futterquelle. *Zeitschrift für Vergleichende Physiologie,* 65: 299–323.

Pickens, A. L., 1932. Observations on the genus *Reticulitermes* Holmgren. *Pan-Pacific Entomologist,* 8(4): 178–180.

Pickles, W., 1940. Fluctuations in the populations, weights and biomasses of ants at Thornhill, Yorkshire, from 1935 to 1939. *Transactions of the Royal Entomological Society of London,* 90(17): 467–485.

Piéron, H., 1904. Du rôle du sens musculaire dans l'orientation de quelques espèces de fourmis. *Bulletin de l'Institut Général Psychologique,* 4(2): 168–186.

Pietschker, H., 1911. Das Gehirn der Ameise. *Zeitschrift für Naturwissenschaften,* 47: 43–114.

Pisarski, B., 1963. Nouvelle espèce du genre *Harpagoxenus* Forel de la Mongolie (Hym. Form.). *Bulletin de l'Académie Polonaise des Sciences, Classe II* (*Zool.*), 11(1): 39–41.

———, 1966. Études sur les fourmis du genre *Strongylognathus* Mayr (Hymenoptera, Formicidae). *Annales Zoologici, Warszawa,* 23(22): 509–523.

Pixell-Goodrich, Helen L. M., 1920. Determination of age in honey-bees. *Quarterly Journal of Microscopical Science,* 64(254): 191–206.

Plateaux-Quénu, Cécile, 1960. Nouvelle preuve d'un déterminisme imaginal des castes chez *Halictus marginatus* Brullé. *Compte Rendu de l'Académie des Sciences, Paris,* 250(26): 4465–4466.

———, 1961. Les sexués de remplacement chez les insectes sociaux. *Année Biologique,* 37(5–6): 177–216.

———, 1962. Biology of *Halictus marginatus* Brullé. *Journal of Apicultural Research,* 1: 41–51.

———, 1965. Sur le cycle biologique de *Halictus nigripes* Lep. *Compte Rendu de l'Académie des Sciences, Paris,* 260(8): 2331–2333.

———, 1967. Tendances évolutives et degré de socialisation chez les Halictinae [Hym., Apoidea]. *Annales de la Société Entomologique de France* (n.s.), 3(3): 859–866.

Plath, O. E., 1922. Notes on *Psithyrus,* with records of two new American hosts. *Biological Bulletin, Marine Biological Laboratory, Woods Hole,* 43(1): 23–44.

———, 1923. Observations on the so-called trumpeter in bumblebee colonies. *Psyche, Cambridge,* 30(5): 146–154.

———, 1934. *Bumblebees and their ways.* Macmillan Co., New York. xvi + 201 pp.

———, 1935. Insect societies. *In A handbook of social psychology.* Clark University Press, Worcester, Mass. pp. 83–141.

Plowright, R. C., and S. C. Jay, 1966. Rearing bumble bee colonies in captivity. *Journal of Apicultural Research,* 5(3): 155–165.

———, 1968. Caste differentiation in bumblebees (*Bombus* Latr.: Hym.). I. The determination of female size. *Insectes Sociaux,* 15(2): 171–192.

Plsek, R. W., J. C. Kroll, and J. F. Watkins, 1969. Observations of carabid beetles, *Helluomorphoides texanus,* in columns of army ants and laboratory experiments on their behavior. *Journal of the Kansas Entomological Society,* 42(4): 452–456.

Pohl, L., 1957. Vergleichende anatomisch-histologische Untersuchungen an *Lepisma saccharina* Linné und der Myrmecophilen *Atelura formicaria* Heyden (Beitrag zur Myrmecophilie, erster Abschnitt). *Insectes Sociaux,* 4(4): 349–363.

Poisson, R., 1937. A propos de *Thaumatoxena wasmanni* Breddin et Börner 1904. Insecte Diptère commensal des Termites. *Bulletin de la Société Entomologique de France,* 42(13–14): 201–208.

———, 1938. *Ulopella termiticola,* nov. gen., n. sp., type nouveau d'Ulopinae commensal des Termites [Hym. Hom. Jassidae]. *Bulletin de la Société Entomologique de France,* 43(1–2): 13–17.

Pontin, A. J., 1960. Field experiments on colony foundation by *Lasius niger* (L.) and *L. flavus* (F.) (Hym., Formicidae). *Insectes Sociaux,* 7(3): 227–230.

———, 1961. Population stabilization and competition between the ants *Lasius flavus* (F.) and *L. niger* (L.). *Journal of Animal Ecology,* 30(1): 47–54.

———, 1963. Further considerations of competition and the ecology of the ants *Lasius flavus* (F.) and *L. niger* (L.). *Journal of Animal Ecology,* 32(3): 565–574.

Portevin, G., 1937. Liodides nouveaux des collections du Muséum. *Revue Française d'Entomologie,* 4(1): 31–36.

Portier, P., and M. Duval, 1929. Recherches sur la teneur en gaz carbonique de l'atmosphère interne des fourmilières. *Compte Rendu de la Société de Biologie, Paris,* 102(35): 906–908.

Portugal-Araújo, V. de, 1958. A contribution to the bionomics of *Lestrimelitta cubiceps* (Hymenoptera, Apidae). *Journal of the Kansas Entomological Society,* 31(3): 203–211.

———, 1963. Subterranean nests of two African stingless bees (Hymenoptera: Apidae). *Journal of the New York Entomological Society,* 71: 130–141.

Poulton, E. B., 1891. On an interesting example of protective mimicry discovered by Mr. W. L. Sclater in British Guiana. *Proceedings of the Zoological Society of London, 1891,* pp. 462–464.

———, 1936. Further references to termitophilous tineid larvae of the genus *Passalactis* Meyr. *Proceedings of the Royal Entomological Society of London,* (A)11: 98.

Pouvreau, A., 1965. Sur une méthode d'élevage des bourdons (*Bombus* Latr.) a partir de reines capturées dans la nature. *Annales de l'Abeille,* 8(2): 147–159.

———, 1967. Contribution à l'étude morphologique et biologique d'*Aphomia sociella* L. (Lepidoptera, Heteroneura, Pyralidoidea, Pyralididae), parasite des nids de bourdons (Hymenoptera, Apoidea, *Bombus* Latr.). *Insectes Sociaux,* 14(1): 57–72.

Pricer, J. L., 1908. The life history of the carpenter ant. *Biological Bulletin, Marine Biological Laboratory, Woods Hole,* 14(3): 177–218.

Principi, Maria Matilde, 1946. Contributi allo studio dei Neurotteri Italiani. IV. *Nothochrysa italica* Rossi. *Bollettino dell'Istituto di Entomologia della Università degli Studi di Bologna,* 15: 85–102.

Princis, K., 1960. Zur Systematik der Blattarien. *Eos, Madrid,* 36(4): 427–449.

Pukowski, Erna, 1933. Ökologische Untersuchungen an *Necrophorus* F. *Zeitschrift für Morphologie und Ökologie der Tiere,* 27(3): 518–586.

Quénu (Plateaux-Quénu), Cécile, 1957. Sur les femelles d'été de *Halictus scabiosae* (Rossi) (Insecte Hyménoptère). *Compte Rendu de l'Académie des Sciences, Paris,* 244(8): 1073–1076.

Quilico, A., P. Grünanger, and M. Pavan, 1960. Sul componente odoroso del veleno del formicide *Myrmicaria natalensis* Fred. *Proceedings of the Eleventh International Congress of Entomology, Vienna, 1960,* 3: 66–68.

Raffy, Anne, 1929. L'atmosphère interne des fourmilières contient-elle de l'oxyde de carbone? *Compte Rendu de la Société Biologie, Paris,* 102(35): 908–909.

Raignier, A., 1948. L'économie thermique d'une colonie polycalique de la fourmi des bois (*Formica rufa polyctena* Foerst.). *La Cellule,* 51(3): 279–368.

———, 1952. *Vie et moeurs des fourmis.* Payot, Paris.

———, and J. Van Boven, 1955. Étude taxonomique, biologique et biométrique des *Dorylus* du sous-genre *Anomma* (Hymenoptera Formicidae). *Annales de Musée Royal du Congo Belge,* n.s. 4, Sciences Zoologiques, 2: 1–359.

Ratcliffe, F. N., and T. Greaves, 1940. The subterranean foraging galleries of *Coptotermes lacteus* (Frogg.). *Journal of the Council for Scientific and Industrial Research, Australia,* 13(2): 150–160.

———, F. J. Gay, and T. Greaves. 1952. *Australian termites. The biology, recognition, and economic importance of the common species.* Commonwealth Scientific and Industrial Research Organization, Melbourne. 124 pp.

Rau, P., 1929. The nesting habits of the bald-faced hornet, *Vespa maculata. Annals of the Entomological Society of America,* 22(4): 659–675.

———, 1930. Animosity and tolerance in several species of *Polistes* wasps. *Journal of Comparative Psychology,* 10(3): 267–286.

———, 1933. *The jungle bees and wasps of Barro Colorado Island* (*with notes on other insects*). Privately published, Phil Rau, Kirkwood, St. Louis Co., Missouri. 324 pp.

———, 1940. Co-operative nest-founding by the wasp, *Polistes annularis* Linn. *Annals of the Entomological Society of America,* 33(4): 617–620.

Rayment, T., 1932. The stingless bees of Australia. *Victorian Naturalist,* 49(1): 9–15.

———, 1935. *A cluster of bees.* Endeavour Press, Sydney. 752 pp.

———, 1951. Biology of the reed-bees, with descriptions of three new species and two allotypes of *Exoneura. Australian Zoologist,* 11(4): 285–313.

———, 1955. Dimorphism and parthenogenesis in halictine bees. *Australian Zoologist,* 12(2): 142–153.

Réaumur, René Antoine Ferchault de, 1742. *Mémoires pour servir á l'histoire des insectes, Vol. VI.* Imp. Royale, Paris. 608 pp.

———, 1926. *The natural history of ants* (from an unpublished manuscript in the Archives of the Academy of Sciences of Paris, written sometime between October 1742 and January 1743; Tr. into English by W. M. Wheeler, with annotations). Alfred A. Knopf, New York. xvii + 280 pp.

Regnier, F. E., and E. O. Wilson, 1968. The alarm-defence system of the ant *Acanthomyops claviger. Journal of Insect Physiology,* 14(7): 955–970.

———, and E. O. Wilson, 1969. The alarm-defence system of the ant *Lasius alienus. Journal of Insect Physiology,* 15(5): 893–898.

Rehn, J. A. G., 1926. Zoological results of the Swedish expedition to Central Africa 1921. Insecta. 18. Blattidae. *Arkiv för Zoologi,* 18A(18): 1–24.

Reichenbach, H., 1902. Ueber Parthenogenese bei Ameisen, und andere Beobachtungen an Ameisenkolonien in künstlichen Nestern. *Biologisches Centralblatt,* 22(14–15): 461–465.

Reichensperger, A., 1924. Neue südamerikanische Histeriden als Gäste von Wanderameisen und Termiten, II Teil. *Revue Suisse de Zoologie,* 31(4): 117–152.

———, 1936. Beitrag zur Kenntnis der Myrmecophilen- und Termitophilenfauna Brasiliens und Costa Ricas, IV (Col. Hist. Staphyl. Pselaph.). *Revista de Entomologia, Rio de Janeiro,* 6(2): 222–242.

———, 1939. Beiträge zur Kenntnis der Myrmecophilen- und Termitophilenfauna Braziliens und Costa Ricas, VI (Col. Hist. Staph.). *Revista de Entomologia, Rio de Janeiro,* 10(1): 97–137.

Reichle, F., 1943. Untersuchungen über Frequenzrhythmen bei Ameisen. *Zeitschrift für Vergleichende Physiologie,* 30: 227–251.

Reitter, E., 1889. Zwei neue Coleopteren-Gattungen aus Transkaukasien. *Wiener Entomologische Zeitung,* 8(9): 289–292.

Rembold, H., 1965. Biologically active substances in royal jelly (tr. A. Dietz). *Vitamins and hormones,* 23: 359–382.

———, 1969. Biochemie der Kastenentstehung bei der Honigbiene. *Proceedings of the Sixth Congress of the International Union for the Study of Social Insects, Bern, 1969,* pp. 239–246.

———, and Gisela Hanser, 1964. Über den Weiselzellenfuttersaft der Honigbiene. VIII. Nachweis des determinierenden Prinzips im Futtersaft der Königinnenlarven. *Hoppe-Seyler's Zeitschrift für Physiologische Chemie,* 339: 251–254.

Renner, M., 1960. Das Duftorgan der Honigbiene und die physiologische Bedeutung ihres Lockstoffes. *Zeitschrift für Vergleichende Physiologie,* 43(4): 411–468.

———, and Margot Baumann, 1964. Über Komplexe von subepidermalen Drüsenzellen (Duftdrüsen?) der Bienenkönigin. *Naturwissenschaften,* 51(3): 68–69.

Rettenmeyer, C. W., 1961a. Arthropods associated with Neotropical army ants with a review of the behavior of these ants (Arthropoda: Formicidae: Dorylinae). Ph.D. Dissertation, University of Kansas. xv + 605 pp.

———, 1961b. Observations on the biology and taxonomy of flies found over swarm raids of army ants (Diptera: Tachinidae, Conopidae). *Kansas University Science Bulletin,* 42(8): 993–1066.

———, 1961c. Behavior, abundance and host specificity of mites found on Neotropical army ants (Acarina; Formicidae: Dorylinae). *Proceedings of the Eleventh International Congress of Entomology, Vienna, 1960,* 1: 610–612.

———, 1962a. The diversity of arthropods found with Neotropical army ants and observations on the behavior of representative species. *Proceedings of the North Central Branch, American Association of Economic Entomologists,* 17: 14–15.

———, 1962b. Notes on host specificity and behavior of myrmecophilous macrochelid mites. *Journal of the Kansas Entomological Society,* 35(4): 358–360.

———, 1962c. The behavior of millipeds found with Neotropical army ants. *Journal of the Kansas Entomological Society,* 35(4): 377–384.

———, 1963a. Behavioral studies of army ants. *Kansas University Science Bulletin,* 44(9): 281–465.

———, 1963b. The behavior of Thysanura found with army ants. *Annals of the Entomological Society of America,* 56(2): 170–174.

———, 1970. Insect mimicry. *Annual Review of Entomology,* 15: 43–74.

———, and R. D. Akre, 1968. Ectosymbiosis between phorid flies and army ants. *Annals of the Entomological Society of America,* 61(5): 1317–1326.

Reyne, A., 1954. *Hippeococcus* a new genus of Pseudococcidae from Java with peculiar habits. *Zoologische Mededelingen, Leiden,* 32(21): 233–257.

Rhein, W. von, 1933. Über die Entstehung des weiblichen Dimorphismus im Bienenstaate. *Wilhelm Roux' Archiv für Entwicklungsmechanik der Organismen,* 129(4): 601–665.

———, 1956. Über die Ernährung der Arbeitermade von *Apis mellifica* L., insbesondere in der Altersperiode. *Insectes Sociaux,* 3(1): 203–212.

Ribbands, C. R., 1952. Division of labour in the honeybee community. *Proceedings of the Royal Society,* (B)140: 32–43.

———, 1953. *The behaviour and social life of honeybees.* Bee Research Association, Ltd., London. 352 pp.

———, 1955. The scent perception of the honeybee. *Proceedings of the Royal Society,* (B)143: 367–379.

———, 1956. The scent language of honey bees. *Report of the Smithsonian Institution for 1955,* pp. 369–377.

———, 1965. The role of recognition of comrades in the defence of social insect communities. *Symposium of the Zoological Society of London,* 14: 159–168.

———, and Nancy Speirs, 1953. The adaptability of the homecoming honeybee. *British Journal of Animal Behaviour,* 1(2): 59–66.

———, H. Kalmus, and H. L. Nixon, 1952. New evidence of communication in the honeybee colony. *Nature, London,* 170: 438–440.

Richard, G., 1950. Le phototropisme du termite à cou jaune (*Calotermes flavicollis* Fabr.) et ses bases sensorielles. *Annales des Sciences Naturelles,* (11)12(3–4): 485–603.

———, 1969. Nervous system and sense organs. *In* K. Krishna and F. M. Weesner, eds. (*q.v.*), *Biology of termites, Vol. 1.* pp. 161–192.

Richards, K. W., 1971. Biology of Arctic bumblebees. *Quaestiones Entomologicae* (in press).

Richards, O. W., 1927a. The specific characters of the British humblebees (Hymenoptera). *Transactions of the Entomological Society of London,* 75(2): 233–268.

———, 1927b. Some notes on the humble-bees allied to *Bombus alpinus,* L. *Tromsø Museums Årshefter,* 50(6): 1–32.

———, 1945. A revision of the genus *Mischocyttarus* de Saussure (Hymen., Vespidae). *Transactions of the Royal Entomological Society of London,* 95(7): 295–462.

———, 1946. Observations on *Bombus agrorum* (Fabricius) (Hymen., Bombidae). *Proceedings of the Royal Entomological Society of London,* (A)21: 66–71.

———, 1953a. The care of the young and the development of social life in the Hymenoptera. *Transactions of the Ninth International Congress of Entomology, Amsterdam, 1951,* 2: 135–138.

———, 1953b. *The social insects.* Harper Torchbooks, New York. 219 pp.

———, 1962. *A revisional study of the masarid wasps* (*Hymenoptera, Vespoidea*). Trustees of the British Museum, London. vii + 294 pp.

———, 1965. Concluding remarks on the social organization of insect communities. *Symposium of the Zoological Society of London,* 14: 169–172.

———, and Maud J. Richards, 1951. Observations on the social wasps of South America (Hymenoptera Vespidae). *Transactions of the Royal Entomological Society of London,* 102(1): 1–170.

Ritcher, P. O., 1958. Biology of Scarabaeidae. *Annual Review of Entomology,* 3: 311–334.

Ritchie, J., 1915. Some observations and deductions regarding the habits and biology of the common wasp. *Scottish Naturalist,* 1915(47): 318–331.

Roberts, R. B., and C. H. Dodson, 1967. Nesting biology of two communal bees, *Euglossa imperialis* and *Euglossa ignita* (Hymenoptera: Apidae), including description of larvae. *Annals of the Entomological Society of America,* 60(5): 1007–1014.

Rockstein, M., 1950a. Longevity in the adult worker honeybee. *Annals of the Entomological Society of America,* 43(1): 152–154.

———, 1950b. The relation of cholinesterase activity to change in cell number with age in the brain of the adult worker honeybee. *Journal of Cellular and Comparative Physiology,* 35(1): 11–23.

Rodrigues, Vilma Maule, 1968. Estudo sôbre vespas sociais do Brasil (Hymenoptera-Vespidae). D.Sc. Thesis, Faculdade de Filosofia, Ciências e Letras de Rio Claro, Universidade de Campinas. 113 pp.

Roepke, W., 1925. Eine neue myrmekophile Tineïde aus Java: *Hypophrictoïdes dolichoderella. Tijdschrift voor Entomologie,* 68: 175–194.

———, 1930. Ueber einen merkwürdigen Fall von "Myrmekophilie," bei einer Ameise (*Cladomyrma* sp.?) auf Sumatra beobachtet. *Miscellanea Zoologica Sumatrana,* 45: 1–3.

Roonwal, M. L., 1954. On the structure and population of the nest of the common Indian tree ant, *Crematogaster dohrni rogenhoferi* (Mayr) (Hymenoptera, Formicidae). *Journal of the Bombay Natural History Society,* 52(2–3): 354–364.

———, 1962a. Recent developments in termite systematics (1949–60). *In* Proceedings of the New Delhi Symposium, 1960, *Termites in the humid tropics.* UNESCO, Paris. pp. 31–50.

———, 1962b. Biology and ecology of Oriental termites. No. 5. Mound structure, nest and moisture-content of fungus combs in *Odontotermes obesus,* with a discussion on the association of fungi with termites. *Records of the Indian Museum,* 58(3–4): 131–150.

Rösch, G. A., 1925. Untersuchungen über die Arbeitsteilung im Bienenstaat. 1. Teil. Die Tätigkeiten im normalen Bienenstaate und ihre Beziehungen zum Alter der Arbeitsbienen. *Zeitschrift für Vergleichende Physiologie,* 2(6): 571–631.

———, 1927. Über die Bautätigkeit im Bienenvolk und das Alter der Baubienen. Weiterer Beitrag zur Frage nach der Arbeitsteilung im Bienenstaat. *Zeitschrift für Vergleichende Physiologie,* 6(2): 264–298.

———, 1930. Untersuchungen über die Arbeitsteilung im Bienenstaat. 2. Teil. Die Tätigkeiten der Arbeitsbienen unter experimentell veränderten Bedingungen. *Zeitschrift für Vergleichende Physiologie,* 12(1): 1–71.

Ross, G. N., 1966. Life-history studies on Mexican butterflies. IV. The ecology and ethology of *Anatole rossi,* a myrmecophilous metalmark (Lepidoptera: Riodinidae). *Annals of the Entomological Society of America,* 59(5): 985–1004.

Roth, L. M., and E. R. Willis, 1960. The biotic associations of cockroaches. *Smithsonian Miscellaneous Collections,* No. 141, vi + 470 pp.

Rothenbuhler, W. C., 1957. Diploid male tissue as new evidence on sex determination in honey bees *Apis mellifera* L. *Journal of Heredity,* 48(4): 160–168.

———, 1958. Genetics and breeding of the honey bee. *Annual Review of Entomology,* 3: 161–180.

———, J. M. Kulinčević, and W. E. Kerr, 1968. Bee genetics. *Annual Review of Genetics,* 2: 413–438.

Roubaud, E., 1911. The natural history of the solitary wasps of the genus *Synagris. Report of the Smithsonian Institution for 1910,* pp. 507–525.

———, 1916. Recherches biologiques sur les guêpes solitaires et sociales d'Afrique. La genèse de la vie sociale et l'évolution de l'instinct maternel chez les vespides. *Annales des Sciences Naturelles,* (10)1: 1–160.

Rozen, J. G., and C. D. Michener, 1968. The biology of *Scrapter* and its cuckoo bee, *Pseudodichroa* (Hymenoptera: Colletidae and Anthophoridae). *American Museum Novitates,* 2335: 1–13.

Ruppli, E., and M. Lüscher, 1964. Die Elimination überzähliger Ersatzgeschlechtstiere bei der Termite *Kalotermes flavicollis* (Fabr.) (Vorläufige Mitteilung). *Revue Suisse de Zoologie,* 71(3): 626–632.

Ruzsky, M., 1902. Neue Ameisen aus Russland. *Zoologische Jahrbücher, Abt. Syst. Ökol. Geogr. Tiere,* 17(3): 469–484.

Sabrosky, C. W., 1952. How many insects are there? *In* A. Stefferud, ed., *Insects: The yearbook of agriculture 1952.* U. S. Department of Agriculture, Washington, D.C. pp. 1–7.

Sakagami, S. F., 1954. Occurrence of an aggressive behaviour in queenless hives, with considerations on the social organization of honeybee. *Insectes Sociaux,* 1(4): 331–343.

———, 1960. Ethological peculiarities of the primitive social bees, *Allodape* Lepeltier and allied genera. *Insectes Sociaux,* 7(3): 231–249.

———, 1966. Techniques for the observation of behaviour and social organization of stingless bees by using a special hive. *Papéis Avulsos do Departamento de Zoologia, São Paulo,* 19(12): 151–162.

———, and Y. Akahira, 1960. Studies on the Japanese honeybee, *Apis cerana cerana* Fabricius. VIII. Two opposing adaptations in the post-stinging behavior of honeybees. *Evolution,* 14(1): 29–40.

———, and H. Fukuda, 1968. Life tables for worker honeybees. *Researches on Population Ecology,* 10(2): 127–139.

———, and K. Fukushima, 1957a. Reciprocal thieving found in *Polistes fedwigae* Dalla Torre (Hymenoptera, Vespidae). *Journal of the Kansas Entomological Society,* 30(4): 140.

———, and K. Fukushima, 1957b. *Vespa dybowskii* André as a facultative temporary social parasite. *Insectes Sociaux,* 4(1): 1–12.

———, and K. Fukushima, 1957c. Some biological observations on a hornet *Vespa tropica* var. *pulchra* (Du Buysson), with special reference to its dependence on *Polistes* wasps (Hymenoptera). *Treubia,* 24(1): 73–82.

———, and K. Fukushima, 1961. Female dimorphism in a social halictine bee, *Halictus* (*Seladonia*) *aerarius* (Smith) (Hymenoptera, Apoidea). *Japanese Journal of Ecology,* 11(3): 118–124.

———, and K. Hayashida, 1958. Biology of the primitive social bee, *Halictus duplex* Dalla Torre. I. Preliminary report on the general life history. *Annotationes Zoologicae Japonenses,* 31(3): 151–155.

———, and K. Hayashida, 1960. Biology of the primitive social bee, *Halictus duplex* Dalla Torre. II. Nest structure and immature stages. *Insectes Sociaux,* 7(1): 57–98.

———, and K. Hayashida, 1961. Biology of the primitive social bee, *Halictus duplex* Dalla Torre. III. Activities in spring solitary phase. *Journal of the Faculty of Science, Hokkaido University,* ser. 6 (Zool.), 14(4): 639–682.

———, and K. Hayashida, 1962. Work efficiency in heterospecific ant groups composed of hosts and their labour parasites. *Animal Behaviour,* 10(1–2): 96–104.

———, and K. Hayashida, 1968. Bionomics and sociology of the summer matrifilial phase in the social halictine bee, *Lasioglossum duplex. Journal of the Faculty of Science, Hokkaido University,* ser. 6 (Zool.), 16(3): 413–513.

———, and R. Ishikawa, 1969. Note préliminaire sur la répartition géographique des bourdons japonais, avec descriptions et remarques sur quelques formes nouvelles ou peu connues. *Journal of the Faculty of Science, Hokkaido University,* ser. 6 (Zool.), 17(1): 152–196.

———, and S. Laroca, 1963. Additional observations on the habits of the cleptobiotic stingless bees, the genus *Lestrimelitta* Friese (Hymenoptera, Apoidea). *Journal of the Faculty of Science, Hokkaido University,* ser. 6 (Zool.), 15(2): 319–339.

———, and C. D. Michener, 1962. *The nest architecture of the sweat bees* (*Halictinae*): *A comparative study of behavior.* University of Kansas Press, Lawrence. 135 pp.

———, and Y. Oniki, 1963. Behavior studies of the stingless bees, with special reference to the oviposition process. I. *Melipona compressipes manaosensis* Schwarz. *Journal of the Faculty of Science, Hokkaido University,* ser. 6 (Zool.), 15(2): 300–318.

———, and F. L. Wain, 1966. *Halictus latisignatus* Cameron: A polymorphic Indian halictine bee with caste differentiation (Hymenoptera, Halictidae). *Journal of the Bombay Natural History Society,* 63(1): 57–73.

———, and K. Yoshikawa, 1968. A new ethospecies of *Stenogaster* wasps from Sarawak, with a comment on the value of ethological characters in animal taxonomy. *Annotationes Zoologicae Japonensis,* 41(2): 77–84.

———, and R. Zucchi, 1965. Winterverhalten einer neotropischen Hummel, *Bombus atratus,* innerhalb des Beobachtungskastens: Ein Beitrag zur Biologie der Hummeln. *Journal of the Faculty of Science, Hokkaido University,* ser. 6 (Zool.), 15(4): 712–762.

———, and R. Zucchi, 1967. Behavior studies of the stingless bees, with special reference to the oviposition process. VI. *Trigona* (*Tetragona*) *clavipes. Journal of the Faculty of Science, Hokkaido University,* ser. 6 (Zool.), 16(2): 292–313.

———, and R. Zucchi, 1968. Oviposition behavior of an Amazonian stingless bee, *Trigona* (*Duckeola*) *ghilianii. Journal of the Faculty of Science, Hokkaido University,* ser. 6 (Zool.), 16(4): 564–581.

———, S. Laroca, and J. S. Moure, 1967. Two Brazilian apid nests worth recording in reference to comparative bee sociology, with description of *Euglossa melanotricha* Moure sp. n. (Hymenoptera, Apidae). *Annotationes Zoologicae Japonensis,* 40(1): 45–54.

———, Maria J. Montenegro, and W. E. Kerr, 1965. Behavior studies of the stingless bees, with special reference to the oviposition process. V. *Melipona quadrifasciata anthidioides* Lepeletier. *Journal of the Faculty of Science, Hokkaido University,* ser. 6 (Zool.), 15(4): 578–607.

Salt, G., 1929. A contribution to the ethology of the Meliponinae. *Transactions of the Entomological Society of London,* 77(2): 431–470.

Sands, W. A., 1956. Some factors affecting the survival of *Odontotermes badius. Insectes Sociaux,* 3(4): 531–536.

———, 1957. The soldier mandibles of the Nasutitermitinae (Isoptera, Termitidae). *Insectes Sociaux,* 4(1): 13–24.

———, 1960. The initiation of fungus comb construction in laboratory colonies of *Ancistrotermes guineensis* (Silvestri). *Insectes Sociaux,* 7(3): 251–263.

———, 1965. Mound population movements and fluctuations in *Trinervitermes ebenerianus* Sjöstedt (Isoptera, Termitidae, Nasutitermitinae). *Insectes Sociaux,* 12(1): 49–58.

———, 1969. The association of termites and fungi. *In* K. Krishna and F. M. Weesner, eds. (*q.v.*), *Biology of termites, Vol. I.* pp. 495–524.

Santschi, F., 1906. A propos des moeurs parasitiques temporaires des fourmis du genre *Bothriomyrmex. Annales de la Société Entomologique de France,* 75: 363–392.

———, 1907. Fourmis de Tunisie capturées en 1906. *Revue Suisse de Zoologie,* 15(2): 305–334.

———, 1911. Observations et remarques critiques sur le mécanisme de l'orientation chez les fourmis. *Revue Suisse de Zoologie,* 19(13): 303–338.

———, 1917. Fourmis nouvelles de la Colonie du Cap, du Natal et de Rhodesia. *Annales de la Société Entomologique de France,* 85: 279–296.

———, 1920. Fourmis du genre *Bothriomyrmex* Emery (Systématique et moeurs). *Revue Zoologique Africaine,* 7(3): 201–224.

———, 1923. L'orientation sidérale des fourmis, et quelques considérations sur leurs différentes possibilités d'orientation. *Mémoires de la Société Vaudoise des Sciences Naturelles,* 1(4): 137–176.

———, 1930. Un nouveau genre de fourmi parasite sans ouvrières de l'Argentine. *Revista Sociedad Entomologico Argentina,* 13: 81–83.

Saussure, H. de, 1853–1858. *Monographie des guêpes sociales ou de la tribu des vespiens, ouvrage faisant suite à la monographie des guêpes solitaires.* V. Masson, Paris. cxcix + 256 pp.

Savage, T. S., 1847. On the habits of the "Drivers" or Visiting Ants of West Africa. *Transactions of the Entomological Society of London,* 5(1): 1–15.

Schaller, F., 1956. Die Endosymbiose und Brutpflege der Erdwanze *Brachypelta aterrima* (Heteropt., Cydnidae). *Verhandlungen der Deutschen Zoologischen Gesellschaft, Hamburg, 1956,* pp. 118–123.

Schedl, K. E., 1958. Breeding habits of arboricole insects in Central Africa. *Proceedings of the Tenth International Congress of Entomology, Montreal, 1956,* 1: 183–197.

Scherba, G., 1958. Reproduction, nest orientation and population structure of an aggregation of mound nests of *Formica ulkei* Emery ("Formicidae"). *Insectes Sociaux,* 5(2): 201–213.

———, 1959. Moisture regulation in mound nests of the ant, *Formica ulkei* Emery. *American Midland Naturalist,* 61(2): 499–508.

———, 1961. Nest structure and reproduction in the mound-building ant *Formica opaciventris* Emery in Wyoming. *Journal of the New York Entomological Society,* 69: 71–87.

———, 1962. Mound temperatures of the ant *Formica ulkei* Emery. *American Midland Naturalist,* 67(2): 373–385.

———, 1963. Population characteristics among colonies of the ant *Formica opaciventris* Emery (Hymenoptera: Formicidae). *Journal of the New York Entomological Society,* 71: 219–232.

———, 1964a. Analysis of inter-nest movement by workers of the ant *Formica opaciventris* Emery (Hymenoptera: Formicidae). *Animal Behaviour,* 12(4): 508–512.

———, 1964b. Species replacement as a factor affecting distribution of *Formica opaciventris* Emery (Hymenoptera: Formicidae). *Journal of the New York Entomological Society,* 72: 231–237.

———, 1965. Observations on *Microtus* nesting in ant mounds. *Psyche, Cambridge,* 72(2): 127–132.

Scheven, J., 1958. Beitrag zur Biologie der Schmarotzerfeldwespen *Sulcopolistes atrimandibularis* Zimm., *S. semenowi* F. Morawitz und *S. sulcifer* Zimm. *Insectes Sociaux,* 5(4): 409–437.

Schjelderup-Ebbe, T., 1922. Beiträge zur Sozialpsychologie des Haushuhns. *Zeitschrift für Psychologie,* 88(3–5): 225–252.

Schmid, J., 1964. Zur Frage der Störung des Bienengedächtnisses durch Narkosemittel, zugleich ein Beitrag zur Störung der sozialen Bindung durch Narkose. *Zeitschrift für Vergleichende Physiologie,* 47(6): 559–595.

Schmidberger, J., 1837. Der Apfel-Borkenkäfer. *Bostrichus dispar.* Autor. (*Apate dispar.* Fabr.). *In* Kollar, V., ed., *Naturgeschichte der schädlichen Insecten in Beziehung auf Landwirthschaft und Forstcultur.* k. k. Landwirthschafts-Gesellschaft, Vienna. pp. 261–270.

Schmidt, Anneliese, 1938. Geschmacksphysiologische Untersuchungen an Ameisen. *Zeitschrift für Vergleichende Physiologie,* 25(3): 351–378.

Schmidt, G. H., 1964. Aktivitätsphasen bekannter Hormondrüsen während der Metamorphose von *Formica polyctena* Foerst. (Hym. Ins.). *Insectes Sociaux,* 11(1): 41–57.

Schmidt, R. S., 1955. Termite (*Apicotermes*) nests—important ethological material. *Behaviour,* 8(4): 344–356.

Schmitz, H., 1915. Die Wahrheit über *Thaumatoxena* Breddin et Börner. Neue Beiträge zur Kenntnis der myrmecophilen und termitophilen Phoriden Nr. 1. *Zoologischer Anzeiger,* 45(12): 548–564.

———, 1916. Wissenschaftliche Ergebnisse einer Forschungsreise nach Ostindien . . . VI. Neue termitophile Dipteren aus den Familien der Termitoxeniiden und Phoriden. *Zoologische Jahrbücher, Abt. Syst. Ökol. Geogr. Tiere,* 39(2): 211–266.

———, 1940. Zum Ausbau der Systematik der Termitoxeniidae (Dipt.). *Proceedings of the Sixth International Congress of Entomology, Madrid, 1935,* pp. 9–15.

———, 1952. Das fehlende Bindeglied zwischen den Dipterenfamilien Phoridae und Termitoxeniidae in Afrika gefunden. *Transactions of the Ninth International Congress of Entomology, Amsterdam, 1951,* 1: 113–116.

Schneider, D., 1965. Chemical sense communication in insects. *Symposium of the Society for Experimental Biology,* 20: 273–297.

Schneirla, T. C., 1933a. Studies on army ants in Panama. *Journal of Comparative Psychology,* 15(2): 267–299.

———, 1933b. Some important features of ant learning. *Zeitschrift für Vergleichende Physiologie,* 19(3): 439–452.

———, 1938. A theory of army-ant behavior based upon the analysis of activities in a representative species. *Journal of Comparative Psychology,* 25(1): 51–90.

———, 1940. Further studies on the army-ant behavior pattern: Mass-organization in the swarm-raiders. *Journal of Comparative Psychology,* 29(3): 401–460.

———, 1941. Social organization in insects, as related to individual function. *Psychological Review,* 48(6): 465–486.

———, 1943. The nature of ant learning. II. The intermediate stage of segmental maze adjustment. *Journal of Comparative Psychology,* 35(2): 149–176.

———, 1944. The reproductive functions of the army-ant queen as pace-makers of the group behavior pattern. *Journal of the New York Entomological Society,* 52: 153–192.

———, 1946a. Ant learning as a problem in comparative psychology. *In* P. Harriman, ed., *Twentieth century psychology.* Philosophical Library, New York. pp. 276–305.

———, 1946b. Problems in the biopsychology of social organization. *Journal of Abnormal and Social Psychology,* 41(4): 385–402.
———, 1949a. Army-ant life and behavior under dry-season conditions. 3. The course of reproduction and colony behavior. *Bulletin of the American Museum of Natural History,* 94(1): 3–81.
———, 1949b. Problems in the environmental adaptation of some new-world species of doryline ants. *Anales del Instituto de Biología, Universidad de Mexico,* 20(1–2): 371–384.
———, 1952. Basic correlations and coordinations in insect societies with special reference to ants. *Colloques Internationaux du Centre National de la Recherche Scientifique, Paris, 1950,* No. 34: 247–269.
———, 1953a. The army-ant queen: Keystone in a social system. *Bulletin de la IIIe Congrès Union Internationale pour l'Étude des Insectes Sociaux,* 1: 29–41.
———, 1953b. Modifiability in insect behavior. *In* K. D. Roeder, ed., *Insect physiology.* John Wiley & Sons, Inc., New York. pp. 723–747.
———, 1956a. The army ants. *Report of the Smithsonian Institution for 1955,* pp. 379–406.
———, 1956b. A preliminary survey of colony division and related processes in two species of terrestrial army ants. *Insectes Sociaux,* 3(1): 49–69.
———, 1957a. A comparison of species and genera in the ant subfamily Dorylinae with respect to functional pattern. *Insectes Sociaux,* 4(3): 259–298.
———, 1957b. Theoretical consideration of cyclic processes in doryline ants. *Proceedings of the American Philosophical Society,* 101(1): 106–133.
———, 1958. The behavior and biology of certain Nearctic army ants. Last part of the functional season, southeastern Arizona. *Insectes Sociaux,* 5(2): 215–255.
———, 1961. The behavior and biology of certain Nearctic Doryline ants—sexual broods and colony division in *Neivamyrmex nigrescens. Zeitschrift für Tierpsychologie,* 18(1): 1–32.
———, 1963. The behaviour and biology of certain Nearctic army ants: Springtime resurgence of cyclic function—southeastern Arizona. *Animal Behaviour,* 11(4): 583–595.
———, 1965. Cyclic functions in genera of legionary ants (Subfamily Dorylinae). *Proceedings of the Twelfth International Congress of Entomology, London, 1964,* pp. 336–338.
———, and R. Z. Brown, 1950. Army-ant life and behavior under dry season conditions. 4. Further investigation of cyclic processes in behavioral and reproductive functions. *Bulletin of the American Museum of Natural History,* 95(5): 263–353.
———, and R. Z. Brown, 1952. Sexual broods and the production of young queens in two species of army ants. *Zoologica, New York,* 37(1): 5–32.
———, and G. Piel, 1948. The army ant. *Scientific American,* 178(6): 16–23.
———, R. Z. Brown, and Frances C. Brown, 1954. The bivouac or temporary nest as an adaptive factor in certain terrestrial species of army ants. *Ecological Monographs,* 24: 269–296.
Schön, A., 1911. Bau und Entwicklung des tibialen Chordotonalorgans bei der Honigbiene und bei Ameisen. *Zoologische Jahrbücher, Abt. Anat. Ontog. Tiere,* 31(3): 439–472.
Schöne, H., 1961. Complex behavior. *In* T. H. Waterman, ed., *The physiology of Crustacea, Vol. II.* Academic Press, Inc., New York. pp. 465–520.
Schorr, Hildegard, 1957. Zur Verhaltensbiologie und Symbiose von *Brachypelta aterrima* Först. (Cydnidae, Heteroptera). *Zeitschrift für Morphologie und Ökologie der Tiere,* 45(6): 561–602.
Schreiner, J., 1906. Die Lebensweise und Metamorphose des Rebenschneiders oder grossköpfigen Zwiebelhornkäfers (*Lethrus apterus* Laxm.) (Coleoptera, Scarabaeidae). *Horae Societatis Entomologicae Rossicae,* 37(3–4): 197–208.
Schricker, B., 1965. Die Orientierung der Honigbiene in der Dämmerung zugleich ein Beitrag zur Frage der Ocellenfunktion bei Bienen. *Zeitschrift für Vergleichende Physiologie,* 49(5): 420–458.
Schubart, O., 1934. Tausendfüssler oder Myriapoda. I. Diplopoda. *Tierwelt Deutschlands,* 28: 301.
Schwarz, E. A., 1904. [Brief communication: Socialism in Arachnidae, *Uloborus republicanus* Simon]. *Proceedings of the Entomological Society of Washington,* 6(3): 147–148.
Schwarz, H. F., 1929. Honey wasps. *Natural History,* 29(4): 421–426.
———, 1931. The nest habits of the diplopterous wasp *Polybia occidentalis* variety *scutellaris* (White) as observed at Barro Colorado, Canal Zone. *American Museum Novitates,* 471: 1–27.
———, 1948. Stingless bees (Meliponidae) of the Western Hemisphere. *Bulletin of the American Museum of Natural History,* 90: xvii + 546 pp.
Schwarz, R., 1955. Über die Riechschärfe der Honigbiene. *Zeitschrift für Vergleichende Physiologie,* 37(3): 180–210.
Scott, H., 1920. Notes on the biology of some inquilines and parasites in a nest of *Bombus derhamellus* Kirby; with a description of the larva and pupa of *Epuraea depressa* Illig. (= *aestiva* Auctt.: Coleoptera, Nitidulidae). *Transactions of the Entomological Society of London, 1920,* pp. 99–127.
———, 1929. On some cases of maternal care displayed by cockroaches and their significance. *Entomologist's Monthly Magazine,* 65: 218–222.
———, 1944. Notes on the season of 1943. *Entomologist's Monthly Magazine,* 80: 1–4.
Sebeok, T. A., 1965. Animal communication. *Science,* 147: 1006–1014.
Seevers, C. H., 1957. A monograph on the termitophilous Staphylinidae (Coleoptera). *Fieldiana, Zoology,* 40: 1–334.
———, 1965. The systematics, evolution and zoogeography of

staphylinid beetles associated with army ants (Coleoptera, Staphylinidae). *Fieldiana, Zoology,* 47(2): 137–351.
———, and H. Dybas, 1943. A synopsis of the Limulodidae (Coleoptera): A new family proposed for myrmecophiles of the subfamilies Limulodinae (Ptiliidae) and Cephaloplectinae (Staphylinidae). *Annals of the Entomological Society of America,* 36(3): 546–586.
Sekiguchi, K., and S. F. Sakagami, 1966. Structure of foraging population and related problems in the honeybee, with considerations on the division of labour in bee colonies. *Report of the Hokkaido National Agricultural Experiment Station,* 69: 1–65.
Selman, B. J., 1962. Remarkable new chrysomeloids found in the nests of arboreal ants in Tanganyika (Coleoptera: Clytridae and Cryptocephalidae). *Annals and Magazine of Natural History,* (13)5: 295–299.
Sernander, R., 1906. Entwurf einer Monographie der europäischen Myrmekochoren. *Kungliga Svenska Vetenskapsakademiens Handlingar,* 41(7): 1–410.
Shannon, C. E., and W. Weaver, 1949. *The mathematical theory of communication.* University of Illinois Press, Urbana. 117 pp.
Shear, W. A., 1970. The evolution of social phenomena in spiders. *Bulletin of the British Arachnological Society,* 1(5): 65–76.
Shearer, D. A., and R. Boch, 1965. 2-Heptanone in the mandibular gland secretion of the honey-bee. *Nature, London,* 206: 530.
Shelford, R., 1907a. Studies of the Blattidae, V–VII. *Transactions of the Entomological Society of London, 1906,* Pt. 4: 487–519.
———, 1907b. Blattoidea, in Sjöstedt's "Kilimandjaro-Meru Expedition", No. 17: 13–48. [Cited by Wheeler, 1928].
Shuel, R. W., and S. E. Dixon, 1960. The early establishment of dimorphism in the female honeybee, *Apis mellifera* L. *Insectes Sociaux,* 7(3): 265–282.
Siebold, C. T. von, 1870. Ueber die Parthenogenesis der *Polistes gallica. Zeitschrift für Wissenschaftliche Zoologie,* 20(2): 236–242.
Silvestri, F., 1903. Contribuzioni alla conoscenza dei Mirmecofili. I. Osservazioni su alcuni mirmecofili dei dintorni di Portici. *Annuario del Museo Zoologico della R. Università di Napoli* (n.s.), 1(13): 1–5.
———, 1905. Contribuzione alla conoscenza dei termitidi e termitofili dell'Eritrea. *Redia,* 3: 341–359.
———, 1920. Contribuzione alla conoscenza dei termitidi e termitofili dell'Africa occidentale. II. Termitofili, Parte Seconda. *Bollettino del Laboratorio di Zoologia Generale e Agraria della Facoltà Agraria in Portici,* 14: 265–319.
———, 1938. Ridescrizione del genere *Termitococcus* Silv. con una specie nouova del Brasile e descrizione di un nuovo genere affine. *Bollettino del Laboratorio di Zoologia Generale e Agraria della Facoltà Agraria in Portici,* 30: 32–40.
Simberoff, D. S., and E. O. Wilson, 1969. Experimental zoogeography of islands: The colonization of empty islands. *Ecology,* 50(2): 278–296.
Simon, E., 1891. Observations biologiques sur les Arachnides. *Annales de la Société Entomologique de France,* 60: 5–14.
Simpson, J., 1960. The functions of the salivary glands of *Apis mellifera. Journal of Insect Physiology,* 4(2): 107–121.
———, 1963. Queen perception by honey bee swarms. *Nature, London,* 199: 94–95.
———, 1964. The mechanism of honey-bee queen piping. *Zeitschrift für Vergleichende Physiologie,* 48(3): 277–282.
———, 1966. Repellency of the mandibular gland scent of worker honey bees. *Nature, London,* 209: 531–532.
———, and Sarah M. Cherry, 1969. Queen confinement, queen piping and swarming in *Apis mellifera* colonies. *Animal Behaviour,* 17: 271–278.
Skaife, S. H., 1953. Subsocial bees of the genus *Allodape* Lep. & Serv. *Journal of the Entomological Society of Southern Africa,* 16(1): 3–16.
———, 1954a. The black-mound termite of the Cape, *Amitermes atlanticus* Fuller. *Transactions of the Royal Society of South Africa,* 34(1): 251–271.
———, 1954b. Caste differentiation among termites. *Transactions of the Royal Society of South Africa,* 34(2): 345–353.
———, 1955. *Dwellers in darkness.* Longmans, Green and Co., Ltd., London. x + 134 pp.
———, 1961. *The study of ants.* Longmans, Green and Co., Ltd., London. vi + 178 pp.
Skellam, J. G., M. V. Brian, and J. R. Proctor, 1959. The simultaneous growth of interacting systems. *Acta Biotheoretica,* 13(2–3): 131–144.
Skwarra, Elisabeth, 1929. Die Ameisenfauna des Zehlaubruches. *Schriften der Physikalisch-Okonomischen Gesellschaft zu Königsberg,* 66(2): 3–174.
———, 1934. Ökologie der Lebensgemeinschaften mexikanischer Ameisenpflanzen. *Zeitschrift für Morphologie und Ökologie der Tiere,* 29(2): 306–373.
Sladen, F. W. L., 1896. Humble bees. *British Bee Journal,* 24: 37, 47–48.
———, 1901. A scent-producing organ in the abdomen of the bee. *Gleanings in Bee Culture,* 29: 639.
———, 1902. A scent-producing organ in the abdomen of the worker of *Apis mellifica. Entomologist's Monthly Magazine,* 38: 208–211.
———, 1912. *The humble-bee, its life-history and how to domesticate it, with descriptions of all the British species of Bombus and Psithyrus.* Macmillan and Co., Ltd., London. xiii + 283 pp.
Slobodkin, L. B., 1961. *Growth and regulation of animal populations.* Holt, Rinehart, and Winston, New York. viii + 184 pp.
Smeathman, H., 1781. Some accounts of the termites which are found in Africa and other hot climates. *Philosophical Transactions of the Royal Society,* (B)71: 139–142.
Smith, Falconer, 1941. A note on noctuid larvae found in ants' nests (Lepidoptera; Hymenoptera: Formicidae). *Entomological News,* 52(4): 109.
———, 1942. Polymorphism in *Camponotus* (Hymenoptera-Formicidae). *Journal of the Tennessee Academy of Science,* 17(4): 367–373.

Smith, M. R., 1936a. The ants of Puerto Rico. *University of Puerto Rico, Journal of Agriculture,* 20: 819–875.

———, 1936b. Distribution of the Argentine ant in the United States and suggestions for its control or eradication. *Circular, United States Department of Agriculture,* 387: 1–39.

———, 1942. The legionary ants of the United States belonging to *Eciton* subgenus *Neivamyrmex* Borgmeier. *American Midland Naturalist,* 27(3): 537–590.

———, 1943. A generic and subgeneric synopsis of the male ants of the United States. *American Midland Naturalist,* 30(2): 273–321.

———, 1947. A generic and subgeneric synopsis of the United States ants, based on the workers (Hymenoptera: Formicidae). *American Midland Naturalist,* 37(3): 521–647.

Snodgrass, R. E., 1956. *Anatomy of the honey bee.* Comstock Publishing Associates, Cornell University Press, Ithaca, New York. xiv + 334 pp.

Snyder, T. E., 1949. Catalog of the termites (Isoptera) of the world. *Smithsonian Miscellaneous Collections,* 112: 1–490.

———, 1956. Annotated, subject-heading bibliography of termites 1350 B.C. to A.D. 1954. *Smithsonian Miscellaneous Collections,* 130: 1–305.

———, 1968. Second supplement to the annotated, subject-heading bibliography of termites 1961–1965. *Smithsonian Miscellaneous Collections,* 152(3): 1–188.

———, and J. Zetek, 1924. Damage by termites in the Canal Zone and Panama and how to prevent it. *Bulletin, United States Department of Agriculture,* 1232: 1–26.

Soulié, J., 1960a. Des considérations écologiques peuvent-elles apporter une contribution à la connaissance du cycle biologique des colonies de *Cremastogaster* (Hymenoptera-Formicoidea). *Insectes Sociaux,* 7(3): 283–295.

———, 1960b. La "sociabilité" des *Cremastogaster* (Hymenoptera-Formicoidea). *Insectes Sociaux,* 7(4): 369–376.

———, 1961. Quelques notes éthologiques sur la vie dans le nid chez deux espèces méditerranéennes de *Cremastogaster* (Hymenoptera-Formicoidea). *Insectes Sociaux,* 8(1): 95–98.

———, 1964. Le contrôle par les ouvrières de la monogynie des colonies chez *Sphaerocrema striatula* (Myrmicidae, Cremastogastrini). *Insectes Sociaux,* 11(4): 383–388.

Spangler, H. G., 1967. Ant stridulations and their synchronization with abdominal movement. *Science,* 155: 1687–1689.

Spencer, H., 1894. Weismannism once more. *Contemporary Review,* 1894, pp. 592–608.

Spieth, H. T., 1948. Notes on a colony of *Polistes fuscatus hunteri* Bequaert. *Journal of the New York Entomological Society,* 56: 155–169.

Spradbery, J. P., 1965. The social organization of wasp communities. *Symposium of the Zoological Society of London,* 14: 61–96.

Stäger, R., 1923. Resultate meiner Beobachtungen und Versuche an *Aphaenogaster testaceo-pilosa* Lucas, *spinosa* Emery, var. *nitida* Emery. *Zeitschrift für Wissenschaftliche Insektenbiologie,* 18(12): 351–356.

———, 1925. Das Leben der Gastameise (*Formicoxenus nitidulus* Nyl.) in neuer Beleuchtung. *Zeitschrift für Morphologie und Ökologie der Tiere,* 3(2–3): 452–476.

———, 1931. Über das Mitteilungsvermögen der Waldameise beim Auffinden und Transport eines Beutestückes. *Zeitschrift für Wissenschaftliche Insektenbiologie,* 26(4–6): 125–137.

———, 1935. Über Verkehrs- und Transportverhältnisse auf den Strassen der Waldameise. *Revue Suisse de Zoologie,* 42(3): 459–460.

———, 1939. Neue Beobachtungen und Versuchsanstellungen mit Ameisen. *Mitteilungen der Naturforschenden Gesellschaft in Bern, 1938,* pp. 1–15.

Stahel, G., and D. C. Geijskes, 1939. Ueber den Bau der Nester von *Atta cephalotes* L. und *Atta sexdens* L. (Hym. Formicidae). *Revista de Entomologia, Rio de Janeiro,* 10(1): 27–78.

Steche, W., 1957. Gelenkter Bienenflug durch "Attrappentänze." *Naturwissenschaften,* 44(22): 598.

Steiger, U., 1967. Über den Feinbau des Neuropils im Corpus pedunculatum der Waldameise. Elektronenoptische Untersuchungen. *Zeitschrift für Zellforschung,* 81: 511–536.

Steiner, A., 1929. Temperaturuntersuchungen in Ameisennestern mit Erdkuppeln, im Nest von *Formica exsecta* Nyl. und in Nestern unter Steinen. *Zeitschrift für Vergleichende Physiologie,* 9(1): 1–66.

———, 1930. Neuere Ergebnisse über den sozialen Wärmehaushalt der einheimischen Hautflügler. *Naturwissenschaften,* 18(26): 595–600.

Steyn, J. J., 1954. The pugnacious ant (*Anoplolepis custodiens* Smith) and its relation to the control of citrus scales at Letaba. *Memoirs of the Entomological Society of Southern Africa,* 3: iii, 1–96.

Stöckhert, E., 1923. Über Entwicklung und Lebensweise der Bienengattung *Halictus* Latr. und ihrer Schmarotzer (Hym.). Zugleich ein Beitrag zur Stammesgeschichte des Bienenstaates. *Konowia,* 2: 48–64, 146–165, 216–247.

Strassen, R. zur, 1957. Zur Oekologie des *Velleius dilatatus* Fabricius eines als Raumgast bei *Vespa crabro* Linnaeus lebenden Staphyliniden (Ins. Col.). *Zeitschrift für Morphologie und Ökologie der Tiere,* 46(3): 243–292.

Stuart, A. M., 1960. Experimental studies on communication in termites. Ph.D. Thesis, Department of Biology, Harvard University. 95 pp.

———, 1963a. Origin of the trail in the termites *Nasutitermes corniger* (Motschulsky) and *Zootermopsis nevadensis* (Hagen), Isoptera. *Physiological Zoölogy,* 36(1): 69–84.

———, 1963b. Studies on the communication of alarm in the termite *Zootermopsis nevadensis* (Hagen), Isoptera. *Physiological Zoölogy,* 36(1): 85–96.

———, 1964. The structure and function of the sternal gland

in *Zootermopsis nevadensis* (Isoptera). *Proceedings of the Zoological Society of London,* 143(1): 43–52.

———, 1967. Alarm, defense, and construction behavior relationships in termites (Isoptera). *Science,* 156: 1123–1125.

———, 1969. Social behavior and communication. *In* K. Krishna and F. M. Weesner, eds. (*q.v.*), *Biology of termites, Vol. I.* pp. 193–232.

———, 1970. The role of chemicals in termite communication. *In* J. W. Johnston, D. G. Moulton, and A. Turk, eds., *Advances in chemoreception. Vol. I. Communication by chemical signals.* Appleton-Century-Crofts, New York. pp. 79–106.

———, and P. Satir, 1968. Morphological and functional aspects of an insect epidermal gland. *Journal of Cell Biology,* 36(3): 527–549.

Stumper, R., 1921. Études sur les fourmis. III. Recherches sur l'éthologie du *Formicoxenus nitidulus* Nyl. *Bulletin de la Société Entomologique de Belgique,* 3: 90–97.

———, 1949. Études myrmécologiques. IX. Nouvelles observations sur l'éthologie de *Formicoxenus nitidulus* Nyl. *Bulletin de la Société des Naturalistes Luxembourgeois* (n.s.), 43: 242–248.

———, 1950. Les associations complexes des fourmis. Commensalisme, symbiose et parasitisme. *Bulletin Biologique de la France et de la Belgique,* 84(4): 376–399.

———, 1956. Sur les sécrétions des fourmis femelles. *Compte Rendu de l'Académie des Sciences, Paris,* 242: 2487–2489.

———, 1961. Radiobiologische Untersuchungen über den sozialen Nahrungshaushalt der Honigameise *Proformica nasuta* (Nyl). *Naturwissenschaften,* 48(24): 735–736.

Sturdza, S. A., 1942. Beobachtungen über die stimulierende Wirkung lebhaft beweglicher Ameisen auf träge Ameisen. *Bulletin de la Section Scientifique de l'Académie Roumaine,* 24: 543–546.

Sturtevant, A. H., 1927. The social parasitism of the ant *Harpagoxenus americanus. Psyche, Cambridge,* 34(1): 1–9.

———, 1938. Essays on evolution. II. On the effects of selection on social insects. *Quarterly Review of Biology,* 13(1): 74–76.

Sudd, J. H., 1957. Communication and recruitment in Pharaoh's ant, *Monomorium pharaonis* (L.). Animal Behaviour, 5(3): 104–109.

———, 1960. The foraging method of Pharaoh's ant, *Monomorium pharaonis* (L.). *Animal Behaviour,* 8(1–2): 67–75.

———, 1962. The source and possible function of the odour of the African stink-ant, *Paltothyreus tarsatus* F. (Hym., Formicidae). *Entomologist's Monthly Magazine,* 98: 62.

———, 1963. How insects work in groups. *Discovery, London,* (June 1963): 15–19.

———, 1967. *An introduction to the behaviour of ants.* Arnold, Ltd., London. viii + 200 pp.

Sumichrast, Fr., 1868. Notes on the habits of certain species of Mexican Hymenoptera presented to the American Entomological Society. *Transactions of the American Entomological Society,* 2: 39–44.

Suomalainen, E., 1962. Significance of parthenogenesis in the evolution of insects. *Annual Review of Entomology,* 7: 349–366.

Swammerdam, J., 1737–1738. *Biblia naturae.* Leyden.

Sykes, W. H., 1835. Descriptions of new species of Indian ants. *Transactions of the Entomological Society of London,* 1(2): 99–107.

Szabó-Patay, J., 1928. A kapus-hangya. *Természettudományi Közlöny, Budapest,* 1928: 215–219.

Szlep, Raja, and T. Jacobi, 1967. The mechanism of recruitment to mass foraging in colonies of *Monomorium venustum* Smith, *M. subopacum* ssp. *phoenicium* Em., *Tapinoma israelis* For. and *T. simothi* v. *phoenicium* Em. *Insectes Sociaux,* 14(1): 25–40.

Taber, S., 1954. The frequency of multiple mating of queen honey bees. *Journal of Economic Entomology,* 47(6): 995–998.

———, and J. Wendel, 1958. Concerning the number of times queen bees mate. *Journal of Economic Entomology,* 51(6): 786–789.

Talbot, Mary, 1943. Population studies of the ant, *Prenolepis imparis* Say. *Ecology,* 24(1): 31–44.

———, 1945a. Population studies of the ant *Myrmica schencki* ssp. *emeryana* Forel. *Annals of the Entomological Society of America,* 38(3): 365–372.

———, 1945b. A comparison of flights of four species of ants. *American Midland Naturalist,* 34(2): 504–510.

———, 1948. A comparison of two ants of the genus *Formica. Ecology,* 29(3): 316–325.

———, 1951. Populations and hibernating conditions of the ant *Aphaenogaster* (*Attomyrma*) *rudis* Emery (Hymenoptera: Formicidae). *Annals of the Entomological Society of America,* 44(3): 302–307.

———, 1954. Populations of the ant *Aphaenogaster* (*Attomyrma*) *treatae* Forel on abandoned fields on the Edwin S. George Reserve. *Contributions from the Laboratory of Vertebrate Biology of the University of Michigan,* 69: 1–9.

———, 1957. Population studies of the slave-making ant *Leptothorax duloticus* and its slave *Leptothorax curvispinosus. Ecology,* 38(3): 449–456.

———, 1967. Slave-raids of the ant *Polyergus lucidus* Mayr. *Psyche, Cambridge,* 74(4): 299–313.

———, and C. H. Kennedy, 1940. The slave-making ant, *Formica sanguinea subintegra* Emery, its raids, nuptial flights and nest structure. *Annals of the Entomological Society of America,* 33(3): 560–577.

Tanner, J. E., 1892. *Oecodoma cephalotes.* Second paper. *Trinidad Field Naturalists' Club,* 1: 123–127.

Taylor, L. H., 1939. Observations on social parasitism in the genus *Vespula* Thomson. *Annals of the Entomological Society of America,* 32(2): 304–315.

Taylor, R. W., 1967. A monographic revision of the ant genus *Ponera* Latreille (Hymenoptera: Formicidae). *Pacific Insects Monograph,* 13: 1–112.

Terron, G., 1969. Mise en évidence du parasitisme temporaire de *Tetraponera anthracina* Santschi par *Tetraponera ledouxi* nov. spec. (Hym. Formicidae, Promyrmicinae). *Annales de la Faculté des Sciences du Cameroun,* 3: 113–115.

Tevis, L., 1958. Interrelations between the harvester ant *Veromessor pergandei* (Mayr) and some desert ephemerals. *Ecology,* 39(4): 695–704.

Thompson, Caroline B., 1919. The development of the castes of nine genera and thirteen species of termites. *Biological Bulletin, Marine Biological Laboratory, Woods Hole,* 36(6): 379–398.

———, 1922. The castes of *Termopsis. Journal of Morphology,* 36(4): 495–531.

Thorpe, W. H., 1956. *Learning and instinct in animals,* 1st ed. Methuen and Co., Ltd., London. viii + 493 pp. (1963. 2nd ed. Methuen and Co., Ltd., London. x + 558 pp.)

Thurm, U., 1964. Mechanoreceptors in the cuticle of the honey bee: Fine structure and stimulus mechanism. *Science,* 145: 1063–1065.

Tillyard, R. J., 1931. The wing-venation of the order Isoptera. I. Introduction and the family Mastotermitidae. *Proceedings of the Linnaean Society of New South Wales,* 56(4): 371–390.

Topoff, H. R., 1969. A unique predatory association between carabid beetles of the genus *Helluomorphoides* and colonies of the army ant *Neivamyrmex nigrescens. Psyche, Cambridge,* 76(4): 375–381.

Torossian, C., 1959. Les échanges trophallactiques proctodéaux chez la fourmi *Dolichoderus quadripunctatus* (Hyménoptère-Formicoidea). *Insectes Sociaux,* 6(4): 369–374.

———, 1961. Les échanges trophallactiques proctodéaux chez la fourmi d'Argentine: *Iridomyrmex humilis* (Hym. Form. Dolichoderidae). *Insectes Sociaux,* 8(2): 189–191.

———, 1965. [Comment]. *Compte Rendu de la Ve Congrès de l'Union Internationale pour l'Étude des Insectes Sociaux, Toulouse, 1965,* p. 301.

Townsend, G. F., and R. W. Shuel, 1962. Some recent advances in apicultural research. *Annual Review of Entomology,* 7: 481–500.

Trägårdh, I., 1907a. Notes on a termitophilous tineid larva. *Arkiv för Zoologi,* 3(22): 1–7.

———, 1907b. Description of *Termitomimus,* a new genus of termitophilous, physogastric Aleocharini, with notes on its anatomy. *Zoologiska Studier Tillagnade Prof. T. Tullberg, Uppsala.* pp. 172–190.

———, 1908. Contributions to the knowledge of *Thaumatoxena* Bredd. & Börn. *Arkiv för Zoologi,* 4(10): 1–12.

Treat, A. E., 1958. Social organization in the Moth Ear Mite (*Myrmonyssus phalaenodectes*). *Proceedings of the Tenth International Congress of Entomology, Montreal, 1956,* 2: 475–480.

Tretzel, E., 1961. Biologie, Ökologie und Brutpflege von *Coelotes terrestris* (Wider) (Araneae, Agelenidae). II. Brutpflege. *Zeitschrift für Morphologie und Ökologie der Tiere,* 50(4): 375–542.

Tsuneki, K., 1953. On colour vision in two species of ants, with special emphasis on their relative sensitivity to various monochromatic lights. *Japanese Journal of Zoology,* 11(1): 187–221.

———, and Y. Adachi, 1957. The intra and interspecific influence relations among nest populations of four species of ants. *Japanese Journal of Ecology,* 7: 166–171.

Tucker, K. W., 1958. Automictic parthenogenesis in the honey bee. *Genetics,* 43(3): 299–316.

Tulloch, G. S., 1932. A gynergate of *Myrmecia. Psyche, Cambridge,* 39(1–2): 48–51.

———, 1935. Morphological studies of the thorax of the ant. *Entomologica Americana* (n.s.), 15(3): 93–131.

Turner, A. J., 1913. Studies in Australian Lepidoptera, Pyralidae. *Proceedings of the Royal Society of Queensland,* 24: 111–163.

Usinger, R. L., 1942. Revision of the Termitaphididae (Hemiptera). *Pan-Pacific Entomologist,* 18(4): 155–159.

Vachon, M., 1952. *Études sur les scorpions.* Institut Pasteur d'Algérie, Alger. 482 pp.

Vecht, J. van der, 1957. The Vespinae of the Indo-Malayan and Papuan areas (Hymenoptera, Vespidae). *Zoologische Verhandelingen,* 34: 1–83.

———, 1962. The Indo-Australian species of the genus *Ropalidia* (*Icaria*) (Hymenoptera, Vespidae) (2nd pt.). *Zoologische Verhandelingen,* 57: 1–72.

———, 1966. The East-Asiatic and Indo-Australian species of *Polybioides* Buysson and *Parapolybia* Saussure. *Zoologische Verhandelingen,* 82: 1–42.

———, 1967. Bouwproblemen van sociale wespen. *Verhandelingen der K. Nederlandsche Akademie van Wetenschappen, Afdeeling Natuurkunde,* 76(4): 59–68.

Velthuis, H. H. V., and J. van Es, 1964. Some functional aspects of the mandibular glands of the queen honeybee. *Journal of Apicultural Research,* 3(1): 11–16.

Verhoeff, K. W., 1939. Zur Kenntnis der Schöbliiden. *Zoologischer Anzeiger,* 125(5–6): 135–137.

Verlaine, L., 1931. L'instinct et l'intelligence chez les Hyménoptères. XII. Les collectivités d'Abeilles sont-elles gouvernées par des traditions? *Mémoires de la Société Entomologique de Belgique,* 23(5): 191–222.

Verron, H., 1963. Rôle des stimuli chimiques dans l'attraction sociale chez *Calotermes flavicollis* (Fabr.). *Insectes Sociaux,* 10(2): 167–184, (3): 185–296, (4): 297–335.

Viana, M. J., and J. A. Haedo Rossi, 1957. Primer hallazgo en el hemisferio sur de Formicidae extinguidos y catalogo mundial de los Formicidae fosiles. *Ameghiniana* (*Buenos Aires*), 1(1–2): 108–113.

Vick, K. W., W. A. Drew, E. J. Eisenbraun, and D. J. McGurk, 1969. Comparative effectiveness of aliphatic ketones in eliciting alarm behavior in *Pogonomyrmex barbatus* and *P. comanche. Annals of the Entomological Society of America,* 62(2): 380–381.

Viehmeyer, H., 1908. Zur Koloniegründung der parasitischen Ameisen. *Biologisches Centralblatt,* 28(1): 18–32.

———, 1921. Die mitteleuropäischen Beobachtungen von *Harpagoxenus sublevis* Mayr. *Biologisches Zentralblatt,* 41(6): 269–278.

Voss, Christiane, 1967. Über das Formensehen der roten Waldameise (*Formica rufa*-Gruppe). *Zeitschrift für Vergleichende Physiologie,* 55(3): 225–254.

Vosseler, J., 1905. Die ostafrikanische Treiberameise (Siafu). *Pflanzer* (Dar-es-Salam), 19: 289–302.

Vowles, D. M., 1950. Sensitivity of ants to polarized light. *Nature, London,* 165: 282–283.

———, 1954a. The orientation of ants. I. The substitution of stimuli. *Journal of Experimental Biology,* 31(3): 341–355.

———, 1954b. The orientation of ants. II. Orientation to light, gravity, and polarized light. *Journal of Experimental Biology,* 31(3): 356–375.

———, 1954c. The function of the corpora pedunculata in bees and ants. *British Journal of Animal Behaviour,* 2(3): 116.

———, 1955. The structure and connexions of the corpora pedunculata in bees and ants. *Quarterly Journal of Microscopical Science,* 96(2): 239–255.

———, 1961. Neural mechanisms in insect behaviour. *In* W. H. Thorpe and O. L. Zangwill, eds., *Current problems in animal behaviour.* Cambridge University Press, London. pp. 5–29.

Waddington, C. H., ed., 1968. *Towards a theoretical biology. 1. Prolegomena.* Aldine Publishing Co., Chicago. ii + 234 pp.

———, ed., 1969. *Towards a theoretical biology. 2. Sketches.* Aldine Publishing Co., Chicago. ii + 351 pp.

Wagner, W., 1906. Psycho-biologische Untersuchungen an Hummeln mit Bezugnahme auf die Frage der Geselligkeit im Tierreiche, Teil I und II. *Zoologica, Stuttgart,* 19(46): 1–239.

Wahl, O., 1932. Neue Untersuchungen über das Zeitgedächtnis der Bienen. *Zeitschrift für Vergleichende Physiologie,* 16(4): 529–589.

———, 1933. Beitrag zur Frage der biologischen Bedeutung des Zeitgedächtnisses der Bienen. *Zeitschrift für Vergleichende Physiologie,* 18(4): 709–717.

Wallis, D. I., 1961. Food-sharing behaviour of the ants *Formica sanguinea* and *Formica fusca. Behaviour,* 17(1): 17–47.

———, 1962. Aggressive behaviour in the ant, *Formica fusca. Animal Behaviour,* 10(3–4): 267–274.

Waloff, N., 1957. The effect of the number of queens of the ant *Lasius flavus* (Fab.) (Hym., Formicidae) on their survival and on the rate of development of the first brood. *Insectes Sociaux,* 4(4): 391–408.

———, and R. E. Blackith, 1962. The growth and distribution of the mounds of *Lasius flavus* (Fabricius) (Hym: Formicidae) in Silwood Park, Berkshire. *Journal of Animal Ecology,* 31(3): 421–437.

Walsh, C. T., J. H. Law, and E. O. Wilson, 1965. Purification of the fire ant trail substance. *Nature, London,* 207: 320–321.

Warburg, O., 1892. Ueber Ameisenpflanzen (Myrmekophyten). *Biologisches Centralblatt,* 12(5): 129–142.

Warder, J., 1726. *The true Amazons: Or, the Monarchy of Bees,* 6th ed. John Pemberton, London. xxiv + 112 pp.

Wasmann, E., 1891. *Die zusammengesetzten Nester und gemischten Kolonien der Ameisen.* Aschendorffschen Buchdruckerei, Münster i. W. vii + 262 pp.

———, 1892. Zur Biologie einiger Ameisengäste. *Deutsche Entomologische Zeitschrift,* 1892(2): 347–351.

———, 1894. *Kritisches Verzeichniss der myrmecophilen und termitophilen Arthropoden. Mit Angabe der Lebensweise und mit Beschreibung neuer Arten.* Felix L. Dames, Berlin. xvi + 231 pp.

———, 1899. Die psychischen Fähigkeiten der Ameisen. *Zoologica, Stuttgart,* 26: 1–133.

———, 1902. Zur Kenntnis der myrmecophilen *Antennophorus* und anderer auf Ameisen und Termiten reitender Acarinen. *Zoologischer Anzeiger,* 25(661): 66–76.

———, 1903. Zum Mimicrytypus der Dorylinengäste. *Zoologischer Anzeiger,* 26(704): 581–590.

———, 1904. Contribuição para o estudo dos hospedes de abelhas brazileiras. *Revista do Museu Paulista,* 6: 482–487.

———, 1905. Nochmals zur Frage über die temporär gemischten Kolonien und den Ursprung der Sklaverei bei den Ameisen. *Biologisches Centralblatt,* 25(19): 644–653.

———, 1908. Weitere Beiträge zum sozialen Parasitismus und der Sklaverei bei den Ameisen. *Biologisches Centralblatt,* 28(8): 257–271.

———, 1915a. Neue Beiträge zur Biologie von *Lomechusa* und *Atemeles,* mit kritischen Bemerkungen über das echte Gastverhältnis. *Zeitschrift für Wissenschaftliche Zoologie,* 114(2): 233–402.

———, 1915b. Anergatides Kohli, eine neue arbeiterlose Schmarotzerameise vom oberen Kongo (Hym., Form.). *Entomologische Mitteilungen,* 4(10–12): 279–288.

———, 1917. Neue Anpassungstypen bei Dorylinengästen Afrikas (Col., Staphylinidae). *Zeitschrift für Wissenschaftliche Zoologie,* 117(2): 257–360.

———, 1920. *Die Gastpflege der Ameisen, ihre biologischen und philosophischen Probleme.* Verlag von Gebrüder Borntraeger, Berlin. xvii + 176 pp.

———, 1921. Ueber einige indische Rhysopaussinen (Col., Tenebrionidae). *Tijdschrift voor Entomologie,* 64: 14–30.

Watkins, J. F., 1964. Laboratory experiments on the trail following of army ants of the genus *Neivamyrmex* (Formicidae: Dorylinae). *Journal of the Kansas Entomological Society,* 37(1): 22–28.

———, and T. W. Cole, 1966. The attraction of army ant workers to secretions of their queens. *Texas Journal of Science,* 18(3): 254–265.

———, and C. W. Rettenmeyer, 1967. Effects of army ant queens on longevity of their workers (Formicidae: Dorylinae). *Psyche, Cambridge,* 74(3): 228–233.

———, T. W. Cole, and R. S. Baldridge, 1967. Laboratory studies on interspecies trail following and trail preference of army ants (Dorylinae). *Journal of the Kansas Entomological Society,* 40(2): 146–151.

Watt, J. C., 1967. The families Perimylopidae and Dacoderidae (Coleoptera, Heteromera). *Proceedings of the Royal Entomological Society of London,* (B)36: 109–118.

Watt, K. E. F., 1968. *Ecology and resource management. A quantitative approach.* McGraw-Hill Book Co., New York. xii + 450 pp.

Way, M. J., 1953. The relationship between certain ant species with particular reference to biological control of the coreid, *Theraptus* sp. *Bulletin of Entomological Research,* 44(4): 669–691.

———, 1954a. Studies of the life history and ecology of the ant *Oecophylla longinoda* Latreille. *Bulletin of Entomological Research,* 45(1): 93–112.

———, 1954b. Studies on the association of the ant *Oecophylla longinoda* (Latr.) (Formicidae) with the scale insect *Saissetia zanzibarensis* Williams (Coccidae). *Bulletin of Entomological Research,* 45(1): 113–134.

———, 1963. Mutualism between ants and honeydew-producing Homoptera. *Annual Review of Entomology,* 8: 307–344.

Weaver, N., 1955. Rearing of honeybee larvae on royal jelly in the laboratory. *Science,* 121: 509–510.

———, 1957. Effects of larval age on dimorphic differentiation of the female honey bee. *Annals of the Entomological Society of America,* 50(3): 283–294.

———, 1966. Physiology of caste determination. *Annual Review of Entomology,* 11: 79–102.

———, Elizabeth C. Weaver, and J. H. Law, 1964. The attractiveness of citral to foraging honeybees. *Progress Report of the Texas Agricultural Experiment Station, Texas A & M University,* 2324: 1–7.

———, Norah C. Johnston, Ronna Benjamin, and J. H. Law, 1968. Novel fatty acids from the royal jelly of honeybees (*Apis mellifera,* L.). *Lipids,* 3(6): 535–538.

Weber, N. A., 1940. The biology of the fungus-growing ants. Pt. VI. Key to *Cyphomyrmex,* new Attini and a new guest ant. *Revista de Entomologia, Rio de Janeiro,* 11(1–2): 406–427.

———, 1941. The biology of the fungus-growing ants. Pt. VII. The Barro Colorado Island, Canal Zone, species. *Revista de Entomologia, Rio de Janeiro,* 12(1–2): 93–130.

———, 1942. A neuropterous myrmecophile, *Nadiva valida* Erichs. *Psyche, Cambridge,* 49(1–2): 1–3.

———, 1943. Parabiosis in Neotropical "ant gardens." *Ecology,* 24(3): 400–404.

———, 1944. The Neotropical coccid-tending ants of the genus *Acropyga* Roger. *Annals of the Entomological Society of America,* 37(1): 89–122.

———, 1946. The biology of the fungus-growing ants. Part IX. The British Guiana species. *Revista de Entomologia, Rio de Janeiro,* 17(1–2): 114–172.

———, 1947. A revision of the North American ants of the genus *Myrmica* Latreille with a synopsis of the Palearctic species, I. *Annals of the Entomological Society of America,* 40(3): 437–474.

———, 1948. A revision of the North American ants of the genus *Myrmica* Latreille with a synopsis of the Palearctic species, II. *Annals of the Entomological Society of America,* 41(2): 267–308.

———, 1949a. New ponerine ants from Equatorial Africa. *American Museum Novitates,* 1398: 1–9.

———, 1949b. The functional significance of dimorphism in the African ant, *Oecophylla. Ecology,* 30(3): 397–400.

———, 1950. A survey of the insects and related arthropods of Arctic Alaska, Pt. I. *Transactions of the American Entomological Society,* 76(3): 147–206.

———, 1955. Pure cultures of fungi produced by ants. *Science,* 121: 109.

———, 1956. Treatment of substrate by fungus-growing ants. *Anatomical Record,* 125(3): 604–605.

———, 1957a. Fungus-growing ants and their fungi: *Cyphomyrmex costatus. Ecology,* 38(3): 480–494.

———, 1957b. Dry season adaptations of fungus-growing ants and their fungi. *Anatomical Record,* 128(3): 638.

———, 1957c. Weeding as a factor in fungus culture by ants. *Anatomical Record,* 128(3): 638.

———, 1958. Evolution in fungus-growing ants. *Proceedings of the Tenth International Congress of Entomology, Montreal, 1956,* 2: 459–473.

———, 1966. Fungus-growing ants. *Science,* 153: 587–604.

———, 1967. Growth of a colony of the fungus-growing ant *Sericomyrmex urichi* (Hymenoptera: Formicidae). *Annals of the Entomological Society of America,* 60(6): 1328–1329.

Wehner, R., and M. Lindauer, 1966. Zur Physiologie des Formensehens bei der Honigbiene. I. Winkelunterscheidung an vertikal Orientierten Streifenmustern. *Zeitschrift für Vergleichende Physiologie,* 52(3): 290–324.

Weil, E., 1958. Zur Biologie der einheimischen Geophiliden. *Zeitschrift für Angewandte Entomologie,* 42(2): 173–209.

Weir, J. S., 1958a. Polyethism in workers of the ant *Myrmica. Insectes Sociaux,* 5(1): 97–128.

———, 1958b. Polyethism in workers of the ant *Myrmica,* Pt. II. *Insectes Sociaux,* 5(3): 315–339.

———, 1959a. Egg masses and early larval growth in *Myrmica. Insectes Sociaux,* 6(2): 187–201.

———, 1959b. The influence of worker age on trophogenic larval dormancy in the ant *Myrmica. Insectes Sociaux,* 6(3): 271–290.

———, 1959c. Changes in the retro-cerebral endocrine system of larvae of *Myrmica,* and their relation to larval growth and development. *Insectes Sociaux,* 6(4): 375–386.

Weismann, A., 1893. The all-sufficiency of natural selection. *Contemporary Review,* 64: 309–338, 596–610.

Weiss, B. A., and T. C. Schneirla, 1967. Inter-situational transfer in the ant *Formica schaufussi* as tested in a two-phase single choice-point maze. *Behaviour,* 28(3–4): 269–279.

Weiss, K., 1953. Versuche mit Bienen und Wespen in farbigen Labyrinthen. *Zeitschrift für Tierpsychologie,* 10(1): 29–44.

———, 1957. Zur Gedächtnisleistung von Wespen. *Zeitschrift für Vergleichende Physiologie,* 39(6): 660–669.

Wellenstein, G., 1928. Beiträge zur Biologie der roten Waldameise (*Formica rufa* L.) mit besonderer Berücksichtigung klimatischer und forstlicher Verhältnisse. *Zeitschrift für Angewandte Entomologie,* 14: 1–68.

———, 1952. Zur Ernährungsbiologie der Roten Waldameise (*Formica rufa* L.). *Zeitschrift für Pflanzenkrankheiten,* 59: 430–451.

Wellington, W. G., 1957. Individual differences as a factor in population dynamics: the development of a problem. *Canadian Journal of Zoology,* 35(3): 293–323.

Wenner, A. M., 1961. Division of labor in a honey bee colony—a Markov process? *Journal of Theoretical Biology,* 1: 324–327.

———, 1962. Communication with queen honey bees by substrate sound. *Science,* 138: 446–448.

———, 1964. Sound communication in honeybees. *Scientific American,* 210(4): 117–123.

———, and D. L. Johnson, 1967. [Reply to K. von Frisch, 1967b]. *Science,* 158: 1076–1077.

———, P. H. Wells, and D. L. Johnson, 1969. Honey bee recruitment to food sources: Olfaction or language? *Science,* 164: 84–86.

———, P. H. Wells, and F. J. Rohlf, 1967. An analysis of the waggle dance and recruitment in honey bees. *Physiological Zoölogy,* 40(4): 317–344.

Werringloer, Anneliese, 1932. Die Sehorgane und Sehzentren der Dorylinen nebst Untersuchungen über die Facettenaugen der Formiciden. *Zeitschrift für Wissenschaftliche Zoologie,* (A)141(3): 432–524.

Wesson, L. G., 1939. Contribution to the natural history of *Harpagoxenus americanus* (Hymenoptera: Formicidae). *Transactions of the American Entomological Society,* 65: 97–122.

———, 1940a. An experimental study on caste determination in ants. *Psyche, Cambridge,* 47(4): 105–111.

———, 1940b. Observations on *Leptothorax duloticus. Bulletin of the Brooklyn Entomological Society,* 35(3): 73–83.

———, and R. G. Wesson, 1940. A collection of ants from southcentral Ohio. *American Midland Naturalist,* 24(1): 89–103.

West (Eberhard), Mary Jane, 1967. Foundress associations in polistine wasps: dominance hierarchies and the evolution of social behavior. *Science,* 157: 1584–1585.

———, and R. D. Alexander, 1963. Sub-social behavior in a burrowing cricket *Anurogryllus muticus* (De Geer). Orthoptera: Gryllidae. *Ohio Journal of Science,* 63(1): 19–24.

Westhoff, V., and J. N. Westhoff-De Joncheere, 1942. Verspreiding en nestoecologie van de mieren in de nederlandsche bosschen. *Tijdschrift over Plantenziekten,* 48(5): 138–212.

Westwood, J. O., 1838. *The entomologist's text book.* W. S. Orr, London. x + 432 pp.

Weyer, F., 1929. Die Eiablage bei *Formica rufa*-Arbeiterinnen. *Zoologischer Anzeiger,* 84(9–10): 253–256.

———, 1935. Epithelerneuerung im Mitteldarm der Termiten während der Häutung. *Zeitschrift für Morphologie und Ökologie der Tiere,* 30(4): 648–672.

Weygoldt, P., 1969. *The biology of pseudoscorpions.* Harvard University Press, Cambridge, Mass. xiv + 145 pp.

Weyrauch, W. K., 1929. Experimentelle Analyse der Brutpflege des Ohrwurmes *Forficula auricularia* L. *Biologisches Zentralblatt,* 49(9): 543–558.

———, 1935a. Wie entsteht ein Wespennest? I. Teil. Beobachtungen und Versuche über den Papierbereitungsinstinkt bei *Vespa, Dolichovespula* und *Macrovespa. Zeitschrift für Morphologie und Ökologie der Tiere,* 30(3): 401–431.

———, 1935b. *Dolichovespula* und *Vespa.* Vergleichende Übersicht über zwei wesentliche Lebenstypen bei sozialen Wespen. Mit Bezugnahme auf die Frage nach der Fortschrittlichkeit tierischer Organisation. *Biologisches Zentralblatt,* 55(9–10): 484–524.

———, 1936. Das Verhalten sozialer Wespen bei Nestüberhitzung. *Zeitschrift für Vergleichende Physiologie,* 23(1): 51–63.

———, 1937. Zur Systematik und Biologie der Kuckuckswespen *Pseudovespa, Pseudovespula* und *Pseudopolistes. Zoologische Jahrbücher, Abt. Syst. Ökol. Geogr. Tiere,* 70(3–4): 243–290.

———, 1938. Nachtrag zu meiner Arbeit über Pseudovespinen und Pseudopolistinen. *Zoologischer Anzeiger,* 121(1–2): 33–37.

Wheeler, G. C., and Jeanette Wheeler, 1951. The ant larvae of the subfamily Dolichoderinae (Hymenoptera, Formicidae). *Proceedings of the Entomological Society of Washington,* 53(4): 169–210.

———, and Jeanette Wheeler, 1953. The ant larvae of the subfamily Formicinae, Pts. I and II. *Annals of the Entomological Society of America,* 46(1): 126–171, (2): 175–217.

———, and Jeanette Wheeler, 1956. The ant larvae of the subfamily Pseudomyrmecinae (Hymenoptera: Formicidae). *Annals of the Entomological Society of America,* 49(4): 374–398.

———, and Jeanette Wheeler, 1960a. The ant larvae of the subfamily Myrmicinae. *Annals of the Entomological Society of America,* 53(1): 98–110.

———, and Jeanette Wheeler, 1960b. Techniques for the study of ant larvae. *Psyche, Cambridge,* 67(4): 87–94.

———, and Jeanette Wheeler, 1964a. The ant larvae of the subfamily Ponerinae: supplement. *Annals of the Entomological Society of America,* 57(4): 443–462.

———, and Jeanette Wheeler, 1964b. The ant larvae of the subfamily Dorylinae: supplement. *Proceedings of the Entomological Society of Washington,* 66(3): 129–137.

———, and Jeanette Wheeler, 1965. The ant larvae of the subfamily Leptanillinae (Hymenoptera, Formicidae). *Psyche, Cambridge,* 72(1): 24–34.

———, and Jeanette Wheeler, 1968. The rediscovery of *Manica parasitica* (Hymenoptera: Formicidae). *Pan-Pacific Entomologist,* 44(1): 71–72.

Wheeler, W. M., 1900a. The habits of *Myrmecophila nebrascensis* Bruner. *Psyche, Cambridge,* 9(294): 111–115.

———, 1900b. A new myrmecophile from the mushroom gardens of the Texan leaf-cutting ant. *American Naturalist,* 34: 851–862.

———, 1901. The compound and mixed nests of American

ants. *American Naturalist,* 35: 431–448, 513–539, 701–724, 791–818.

———, 1903. Ethological observations on an American ant (*Leptothorax emersoni* Wheeler). *Journal für Psychologie und Neurologie,* 2(1–2): 1–31.

———, 1904. A new type of social parasitism among ants. *Bulletin of the American Museum of Natural History,* 20(30): 347–375.

———, 1905. Worker ants with vestiges of wings. *Bulletin of the American Museum of Natural History,* 21(24): 405–408.

———, 1906. On the founding of colonies by queen ants, with special reference to the parasitic and slave-making species. *Bulletin of the American Museum of Natural History,* 22(4): 33–105.

———, 1907a. The fungus-growing ants of North America. *Bulletin of the American Museum of Natural History,* 23(31): 669–807.

———, 1907b. The polymorphism of ants, with an account of some singular abnormalities due to parasitism. *Bulletin of the American Museum of Natural History,* 23(1): 1–93.

———, 1908. Honey ants, with a revision of the American Myrmecocysti. *Bulletin of the American Museum of Natural History,* 24(20): 345–397.

———, 1910. *Ants: Their structure, development and behavior.* Columbia University Press, New York. xxv + 663 pp.

———, 1911. The ant-colony as an organism. *Journal of Morphology,* 22(2): 307–325.

———, 1914. The ants of the Baltic amber. *Schriften der Physikalisch-Ökonomischen Gesellschaft zu Königsberg,* 55: 1–142.

———, 1915. Some additions to the North American ant-fauna. *Bulletin of the American Museum of Natural History,* 34(12): 389–421.

———, 1916a. The marriage-flight of a bull-dog ant (*Myrmecia sanguinea* F. Smith). *Journal of Animal Behavior,* 6(1): 70–73.

———, 1916b. The Australian ants of the genus *Onychomyrmex. Bulletin of the Museum of Comparative Zoology, Harvard,* 60(2): 45–54.

———, 1918. A study of some ant larvae with a consideration of the origin and meaning of social habits among insects. *Proceedings of the American Philosophical Society,* 57: 293–343.

———, 1919a. A new paper-making *Crematogaster* from the southeastern United States. *Psyche, Cambridge,* 26(4): 107–112.

———, 1919b. The parasitic Aculeata, a study in evolution. *Proceedings of the American Philosophical Society,* 58(1): 1–40.

———, 1921a. Observations on army ants in British Guiana. *Proceedings of the American Academy of Arts and Sciences,* 56(8): 291–328.

———, 1921b. A study of some social beetles in British Guiana and of their relations to the ant-plant Tachigalia. *Zoologica, New York,* 3(3): 35–126.

———, 1921c. A new case of parabiosis and the "ant gardens" of British Guiana. *Ecology,* 2(2): 89–103.

———, 1922. Ants of the American Museum Congo Expedition. A contribution to the myrmecology of Africa. VII. Keys to the genera and subgenera of ants. VIII. A synonymic list of the ants of the Ethiopian region. IX. A synonymic list of the ants of the Malagasy Region. *Bulletin of the American Museum of Natural History,* 45(1): 631–710, 711–1004, 1005–1055.

———, 1923. *Social life among the insects.* Harcourt, Brace and Company, New York. vii + 375 pp.

———, 1925. A new guest-ant and other new Formicidae from Barro Colorado Island, Panama. *Biological Bulletin, Marine Biological Laboratory, Woods Hole,* 49(3): 150–181.

———, 1927a. The physiognomy of insects. *Quarterly Review of Biology,* 2(1): 1–36.

———, 1927b. *Emergent evolution and the social.* Kegan Paul, Trench, Trubner, and Co., Ltd., London. 57 pp.

———, 1928. *The social insects: Their origin and evolution.* Kegan Paul, Trench, Trubner and Co., Ltd., London. xviii + 378 pp.

———, 1933a. *Colony-founding among ants, with an account of some primitive Australian species.* Harvard University Press, Cambridge, Mass. x + 179 pp.

———, 1933b. A second parasitic *Crematogaster. Psyche, Cambridge,* 40(2): 83–86.

———, 1934. A second revision of the ants of the genus *Leptomyrmex* Mayr. *Bulletin of the Museum of Comparative Zoology, Harvard,* 77(3): 69–118.

———, 1936a. Ecological relations of ponerine and other ants to termites. *Proceedings of the American Academy of Arts and Sciences,* 71(3): 159–243.

———, 1936b. A singular *Crematogaster* from Guatemala. *Psyche, Cambridge,* 43(2–3): 40–48.

———, 1937. *Mosaics and other anomalies among ants.* Harvard University Press, Cambridge, Mass. 95 pp.

———, 1942. Studies of Neotropical ant-plants and their ants. *Bulletin of the Museum of Comparative Zoology, Harvard,* 90(1): 1–262.

———, and I. W. Bailey, 1920. The feeding habits of pseudomyrmine and other ants. *Transactions of the American Philosophical Society* (n.s.), 22: 235–279.

———, and W. M. Mann, 1914. The ants of Haiti. *Bulletin of the American Museum of Natural History,* 33(1): 1–61.

White, M. J. D., 1954. *Animal cytology and evolution,* 2nd ed. University Press, Cambridge, Eng. xiv + 454 pp.

———, 1964. Cytogenetic mechanisms in insect reproduction. *Symposium of the Royal Entomological Society of London,* 2: 1–12.

Whiting, P. W., 1938. Anomalies and caste determination in ants. *Journal of Heredity,* 29(5): 189–193.

———, 1939. Sex-determination and reproductive economy in *Habrobracon. Genetics,* 24(1): 110–111.

———, 1943. Multiple alleles in complementary sex determination of *Habrobracon. Genetics,* 28(5): 365–382.

Wilde, J., 1615. *De Formica, Liber Unus.* Ambergae apud Schönfeld. [Copy in the library of E. O. Wilson.]

Wildermuth, V. L., and E. G. Davis, 1931. The red harvester ant and how to subdue it. *Farmers' Bulletin, United States Department of Agriculture,* 1668: 1–12.

Wilkinson, W., 1962. Dispersal of alates and establishment of new colonies in *Cryptotermes havilandi* (Sjöstedt) (Isoptera, Kalotermitidae). *Bulletin of Entomological Research,* 53(2): 265–286.

Williams, C. B., 1964. *Patterns in the balance of nature and related problems in quantitative biology.* Academic Press, London. vii + 324 pp.

Williams, E. C., 1941. An ecological study of the floor fauna of the Panama rain forest. *Bulletin of the Chicago Academy of Sciences,* 6(4): 63–124.

Williams, F. X., 1919. Philippine wasp studies. II. Descriptions of new species and life history studies. *Bulletin of the Experiment Station, Hawaiian Sugar Planters' Association, Entomology Series,* 14: 19–184.

———, 1928. Studies in tropical wasps—their hosts and associates (with descriptions of new species). *Bulletin of the Experiment Station, Hawaiian Sugar Planters' Association, Entomology Series,* 19: iv + 179.

Williams, G. C., 1966. *Adaptation and natural selection. A critique of some current evolutionary thought.* Princeton University Press, Princeton, New Jersey. x + 307 pp.

———, and Doris C. Williams, 1957. Natural selection of individually harmful social adaptations among sibs with special reference to social insects. *Evolution,* 11(1): 32–39.

Williams, R. M. C., 1959a. Flight and colony foundation in two *Cubitermes* species (Isoptera: Termitidae). *Insectes Sociaux,* 6(2): 203–218.

———, 1959b. Colony development in *Cubitermes ugandensis* Fuller (Isoptera: Termitidae). *Insectes Sociaux,* 6(3): 291–304.

Willinck, A., 1952. Los véspidos sociales argentinos, con exclusión del género *Mischocyttarus* (Hym., Vespidae). *Acta Zoológica Lilloana del Instituto "Miguel Lillo," Tucumán, Argentina,* 10: 105–151.

———, 1953. Las especies Argentinas de *"Mischocyttarus"* de Saussure (Hym., Vespidae). *Acta Zoológica Lilloana del Instituto "Miguel Lillo," Tucumán, Argentina,* 14: 317–340.

Willis, E. O., 1967. The behavior of bicolored antbirds. *University of California Publications in Zoology,* 79: 1–127.

Wilson, E. O., 1953. The origin and evolution of polymorphism in ants. *Quarterly Review of Biology,* 28(2): 136–156.

———, 1954. A new interpretation of the frequency curves associated with ant polymorphism. *Insectes Sociaux,* 1(1): 75–80.

———, 1955a. A monographic revision of the ant genus *Lasius. Bulletin of the Museum of Comparative Zoology, Harvard,* 113(1): 1–201.

———, 1955b. Division of labor in a nest of the slave-making ant *Formica wheeleri* Creighton. *Psyche, Cambridge,* 62(3): 130–133.

———, 1957. The organization of a nuptial flight of the ant *Pheidole sitarches* Wheeler. *Psyche, Cambridge,* 64(2): 46–50.

———, 1958a. Observations on the behavior of the cerapachyine ants. *Insectes Sociaux,* 5(1): 129–140.

———, 1958b. The beginnings of nomadic and group-predatory behavior in the ponerine ants. *Evolution,* 12(1): 24–31.

———, 1958c. Studies on the ant fauna of Melanesia. I. The tribe Leptogenyini. II. The tribes Amblyoponini and Platythyreini. *Bulletin of the Museum of Comparative Zoology, Harvard,* 118(3): 101–153.

———, 1958d. A chemical releaser of alarm and digging behavior in the ant *Pogonomyrmex badius* (Latreille). *Psyche, Cambridge,* 65(2–3): 41–51.

———, 1959a. Studies on the ant fauna of Melanesia VI. The tribe Cerapachyini. *Pacific Insects,* 1(1): 39–57.

———, 1959b. Communication by tandem running in the ant genus *Cardiocondyla. Psyche, Cambridge,* 66(3): 29–34.

———, 1959c. Source and possible nature of the odor trail of fire ants. *Science,* 129: 643–644.

———, 1959d. Some ecological characteristics of ants in New Guinea rain forests. *Ecology,* 40(3): 437–447.

———, 1959e. Adaptive shift and dispersal in a tropical ant fauna. *Evolution,* 13(1): 122–144.

———, 1961. The nature of the taxon cycle in the Melanesian ant fauna. *American Naturalist,* 95: 169–193.

———, 1962a. Chemical communication among workers of the fire ant *Solenopsis saevissima* (Fr. Smith). 1. The organization of mass-foraging. 2. An information analysis of the odour trail. 3. The experimental induction of social responses. *Animal Behaviour,* 10(1–2): 134–164.

———, 1962b. Behavior of *Daceton armigerum* (Latreille), with a classification of self-grooming movements in ants. *Bulletin of the Museum of Comparative Zoology, Harvard,* 127(7): 403–422.

———, 1963a. The social biology of ants. *Annual Review of Entomology,* 8: 345–368.

———, 1963b. Social modifications related to rareness in ant species. *Evolution,* 17(2): 249–253.

———, 1964. The true army ants of the Indo-Australian area (Hymenoptera: Formicidae: Dorylinae). *Pacific Insects,* 6(3): 427–483.

———, 1965a. Chemical communication in the social insects. *Science,* 149: 1064–1071.

———, 1965b. Trail sharing in ants. *Psyche, Cambridge,* 72(1): 2–7.

———, 1966. Behaviour of social insects. *Symposium of the Royal Entomological Society of London,* 3: 81–96.

———, 1968a. The ergonomics of caste in the social insects. *American Naturalist,* 102: 41–66.

———, 1968b. Chemical systems. *In* T. A. Sebeok, ed., *Animal communication, techniques of study and results*

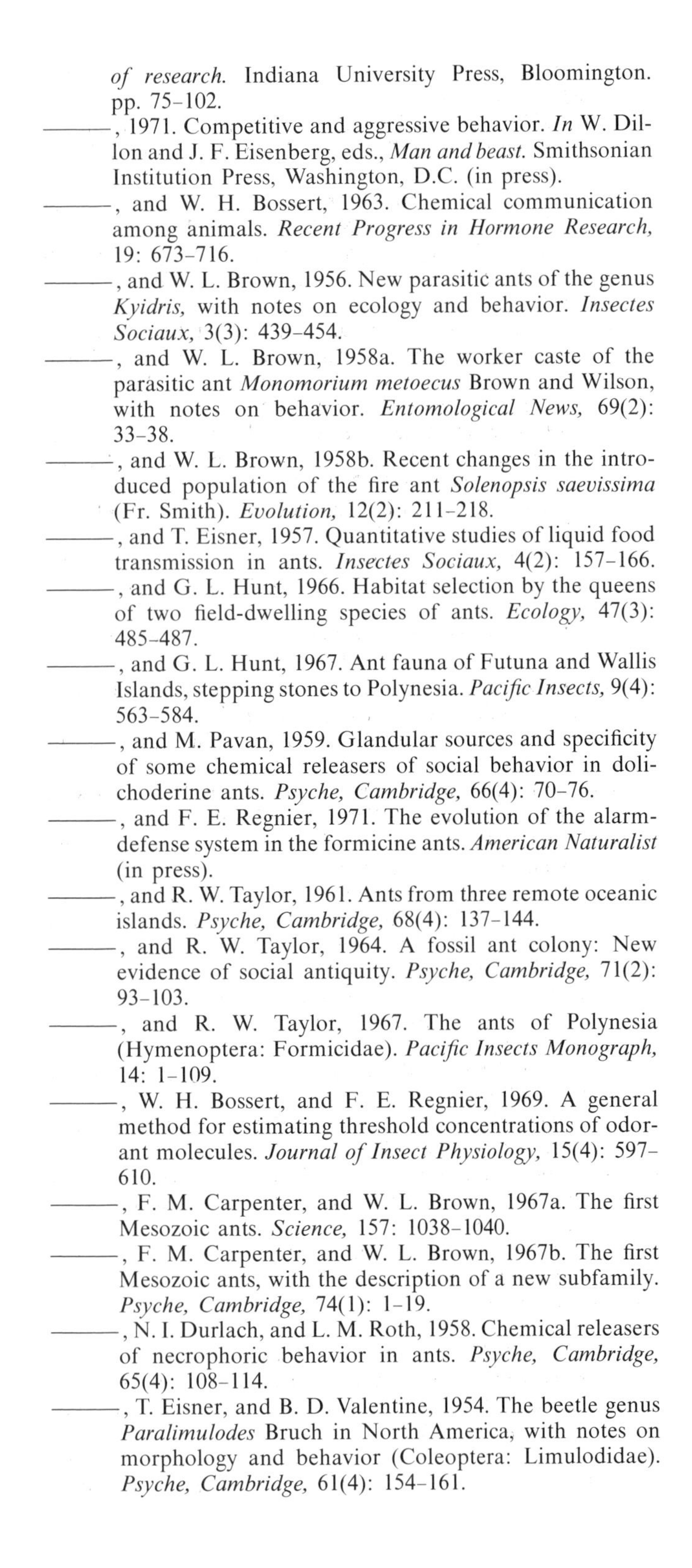

of research. Indiana University Press, Bloomington. pp. 75–102.

———, 1971. Competitive and aggressive behavior. *In* W. Dillon and J. F. Eisenberg, eds., *Man and beast.* Smithsonian Institution Press, Washington, D.C. (in press).

———, and W. H. Bossert, 1963. Chemical communication among animals. *Recent Progress in Hormone Research,* 19: 673–716.

———, and W. L. Brown, 1956. New parasitic ants of the genus *Kyidris,* with notes on ecology and behavior. *Insectes Sociaux,* 3(3): 439–454.

———, and W. L. Brown, 1958a. The worker caste of the parasitic ant *Monomorium metoecus* Brown and Wilson, with notes on behavior. *Entomological News,* 69(2): 33–38.

———, and W. L. Brown, 1958b. Recent changes in the introduced population of the fire ant *Solenopsis saevissima* (Fr. Smith). *Evolution,* 12(2): 211–218.

———, and T. Eisner, 1957. Quantitative studies of liquid food transmission in ants. *Insectes Sociaux,* 4(2): 157–166.

———, and G. L. Hunt, 1966. Habitat selection by the queens of two field-dwelling species of ants. *Ecology,* 47(3): 485–487.

———, and G. L. Hunt, 1967. Ant fauna of Futuna and Wallis Islands, stepping stones to Polynesia. *Pacific Insects,* 9(4): 563–584.

———, and M. Pavan, 1959. Glandular sources and specificity of some chemical releasers of social behavior in dolichoderine ants. *Psyche, Cambridge,* 66(4): 70–76.

———, and F. E. Regnier, 1971. The evolution of the alarm-defense system in the formicine ants. *American Naturalist* (in press).

———, and R. W. Taylor, 1961. Ants from three remote oceanic islands. *Psyche, Cambridge,* 68(4): 137–144.

———, and R. W. Taylor, 1964. A fossil ant colony: New evidence of social antiquity. *Psyche, Cambridge,* 71(2): 93–103.

———, and R. W. Taylor, 1967. The ants of Polynesia (Hymenoptera: Formicidae). *Pacific Insects Monograph,* 14: 1–109.

———, W. H. Bossert, and F. E. Regnier, 1969. A general method for estimating threshold concentrations of odorant molecules. *Journal of Insect Physiology,* 15(4): 597–610.

———, F. M. Carpenter, and W. L. Brown, 1967a. The first Mesozoic ants. *Science,* 157: 1038–1040.

———, F. M. Carpenter, and W. L. Brown, 1967b. The first Mesozoic ants, with the description of a new subfamily. *Psyche, Cambridge,* 74(1): 1–19.

———, N. I. Durlach, and L. M. Roth, 1958. Chemical releasers of necrophoric behavior in ants. *Psyche, Cambridge,* 65(4): 108–114.

———, T. Eisner, and B. D. Valentine, 1954. The beetle genus *Paralimulodes* Bruch in North America, with notes on morphology and behavior (Coleoptera: Limulodidae). *Psyche, Cambridge,* 61(4): 154–161.

———, T. Eisner, G. C. Wheeler, and Jeanette Wheeler, 1956. *Aneuretus simoni* Emery, a major link in ant evolution. *Bulletin of the Museum of Comparative Zoology, Harvard,* 115(3): 81–99.

Wilson, H. F., and V. G. Milum, 1927. Winter protection for the honeybee colony. *Research Bulletin, Agricultural Experiment Station, University of Wisconsin,* 75: 1–47.

Wing, M. W., 1951. A new genus and species of myrmecophilous Diapriidae with taxonomic and biological notes on related forms (Hymenoptera). *Transactions of the Royal Entomological Society of London,* 102: 195–210.

———, 1968. Taxonomic revision of the Nearctic genus *Acanthomyops* (Hymenoptera: Formicidae). *Memoirs, Cornell University Agricultural Experiment Station,* 405: 1–173.

Wolf, E., 1934. Das Verhalten der Bienen gegenüber flimmernden Feldern und bewegten Objekten. *Zeitschrift für Vergleichende Physiologie,* 20(1–2): 151–161.

Wolf, H., 1951. Die parasitische Lebensweise der Grabwespengattung *Nysson* Latr. *Nachrichten des Naturwissenschaftlichen Museums der Stadt Aschaffenburg,* 33: 77–80.

Woodruff, R. E., and O. L. Cartwright, 1967. A review of the genus *Euparixia* with description of a new species from nests of leaf-cutting ants in Louisiana (Coleoptera: Scarabaeidae). *Proceedings of the United States National Museum,* 123: 1–21.

Wright, S., 1945. Tempo and mode in evolution: A critical review. *Ecology,* 26(4): 415–419.

Wygodzinsky, P., 1961. A new genus of termitophilous Atelurinae from South Africa (Thysanura: Nicoletiidae). *Journal of the Entomological Society of Southern Africa,* 24(1): 104–109.

Wykes, G. R., 1952. The preferences of honeybees for solutions of various sugars which occur in nectar. *Journal of Experimental Biology,* 29(4): 511–519.

Wyman, J., 1860. Observations on the habits of a species of hornet (*Vespa*), which builds its nest in the ground. *Proceedings of the Boston Society of Natural History,* 7: 411–418.

Wynne-Edwards, V. C., 1962. *Animal dispersion in relation to social behaviour.* Oliver and Boyd, Edinburgh and London. xi + 653 pp.

Yamanaka, M., 1928. On the male of a paper wasp, *Polistes fadwigae* Dalla Torre. *Science Reports of the Tôhoku Imperial University, Sendai, Japan,* ser. 4 (Biol.), 3(3): 265–269.

Yamauchi, K., and K. Hayashida, 1968. Taxonomic studies on the genus *Lasius* in Hokkaido, with ethological and ecological notes (Formicidae, Hymenoptera). I. The subgenus *Dendrolasius* or Jet Black Ants. *Journal of the Faculty of Science, Hokkaido University,* ser. 4 (Zool.), 16(3): 396–412.

Yarrow, I. H. H., 1955a. The type species of the ant genus *Myrmica* Latreille. *Proceedings of the Royal Entomological Society of London,* (B)24: 113–115.

———, 1955b. The British ants allied to *Formica rufa* L. (Hym.,

Formicidae). *Transactions of the Society for British Entomology,* 12(1): 1–48.

Yasuno, M., 1964a. The study of the ant population in the grassland at Mt. Hakkôda. II. The distribution pattern of ant nests at the Kayano grassland. *Science Reports of the Tôhoku University, Sendai, Japan,* ser. 4 (Biol.), 30(1): 43–55.

———, 1964b. The study of the ant population in the grassland at Mt. Hakkôda. III. The effect of the slave making ant, *Polyergus samurai,* upon the nest distribution pattern of the slave ant, *Formica fusca japonica. Science Reports of the Tôhoku University, Sendai, Japan,* ser. 4 (Biol.), 30(2): 167–170.

———, 1965. Territory of ants in the Kayano grassland at Mt. Hakkôda. *Science Reports of the Tôhoku University, Sendai, Japan,* ser. 4 (Biol.), 31(3): 195–206.

Yoshikawa, K., 1954. Ecological studies of *Polistes* wasps. I. On the nest evacuation. *Journal of the Institute of Polytechnics, Osaka City University,* ser. D (Biol.), 5: 9–17.

———, 1955. A polistine colony usurped by a foreign queen. Ecological studies of *Polistes* wasps, II. *Insectes Sociaux,* 2(3): 255–260.

———, 1962a. Introductory studies on the life economy of polistine wasps. I. Scope of problems and consideration on the solitary stage. *Bulletin of the Osaka Museum of Natural History,* 15: 3–27.

———, 1962b. Introductive studies on the life economy of polistine wasps. II. Superindividual stage. 1. Incipient behaviors. *Japanese Journal of Ecology,* 12(5): 187–190.

———, 1962c. Introductory studies on the life economy of polistine wasps. VI. Geographical distribution and its ecological significances. *Journal of Biology, Osaka City University,* 13: 19–43.

———, 1962d. Introductory studies on the life economy of polistine wasps. VII. Comparative consideration and phylogeny. *Journal of Biology, Osaka City University,* 13: 45–64.

———, 1963a. Introductory studies on the life economy of polistine wasps. II. Superindividual stage. 2. Division of labor among workers. *Japanese Journal of Ecology,* 13(2): 53–57.

———, 1963b. Introductory studies on the life economy of polistine wasps. II. Superindividual stage. 3. Dominance order and territory. *Journal of Biology, Osaka City University,* 14: 55–61.

———, 1963c. Introductory studies on the life economy of polistine wasps. III. Social stage. *Journal of Biology, Osaka City University,* 14: 63–66.

———, 1963d. Introductory studies on the life economy of polistine wasps. IV. Analysis of social organization. *Journal of Biology, Osaka City University,* 14: 67–85.

———, 1963e. Introductory studies on the life economy of polistine wasps. V. Three stages relating to hibernation. *Journal of Biology, Osaka City University,* 14: 87–96.

———, R. Ohgushi, and S. F. Sakagami, 1969. Preliminary report on entomology of the Osaka City University 5th Scientific Expedition to Southeast Asia 1966. With descriptions of two new genera of stenogastrine wasps by J. van der Vecht. *Nature and Life in Southeast Asia* (Tokyo), 6: 153–182.

Zahn, M., 1958. Temperatursinn, Wärmehaushalt und Bauweise der Roten Waldameisen (*Formica rufa* L.). *Zoologische Beiträge,* 3(2): 127–194.

Zerrahn, Gertrud, 1934. Formdressur und Formunterscheidung bei der Honigbiene. *Zeitschrift für Vergleichende Physiologie,* 20(1–2): 117–150.

Zikán, J. F., 1949. O gênero *Mischocyttarus* Saussure, (Hymenoptera, Vespidae), com a descrição de 82 espécies novas. *Boletim Parque Nacional do Itatiáia, Rio de Janeiro,* 1: 1–251.

Zimmermann, K., 1930. Zur Systematik der palaearktischen *Polistes* (Hym. Vesp.). *Mitteilungen aus dem Zoologischen Museum in Berlin,* 15(3–4): 609–621.

———, 1931. Studien über individuelle und geographische Variabilität paläarktischer *Polistes* und verwandter Vespiden. *Zeitschrift für Morphologie und Ökologie der Tiere,* 22(1): 173–230.

Zuberi, H. A., 1963. L'anatomie comparée du cerveau chez les termites en rapport avec le polymorphisme. *Bulletin Biologique de la France et de la Belgique,* 97(1): 147–208.

Zwölfer, H., 1958. Zur Systematik, Biologie und Ökologie unterirdisch lebender Aphiden (Homoptera, Aphidoidea) (Anoeciinae, Tetraneurini, Pemphigini und Fordinae). Teil IV (Ökologische und systematische Erörterungen). *Zeitschrift für Angewandte Entomologie,* 43(1): 1–52.

Index

. . . humming-bird filled with baked almonds, surrounded by a Spring linnet, which, in turn, was enveloped by an English snipe. These the carcass of a stuffed goose surrounded, covering which were two canvas-back ducks . . . the whole placed within the bosom of a Chicago goose. Soaked in raisin wine for six days, then larded, and smoked three weeks over burning sandalwood, it was at last placed on the spit and roasted with pig-pork drippings."

The day after Ralston's Bank of California closed its doors on clamoring investors, August 26, 1875, in a financial scandal that rocked the State, Ralston's body was found floating in the Bay. The estate passed to his partner, William Sharon, whose daughter's marriage was the last of Belmont's social flings. The great mansion became successively a young ladies' seminary, a private insane asylum, and finally, the COLLEGE OF NOTRE DAME, opened here in 1923 by the Sisters of Notre Dame de Namur.

SAN CARLOS, 24.2 *m.* (21 alt., 21,370 pop.), a town of homes among oaks, named for Lt. Juan Manuel de Ayala's *San Carlos*, first vessel to enter the Golden Gate, was the seat of Rancho de las Pulgas (the fleas). "The ranch had been well-named by the matter-of-fact Spanish," wrote Gertrude Atherton in later years. "I may add that it was no breach of decorum to speak of fleas in California, nor even to scratch."

The flower beds in the neighborhood supply eastern markets with carloads of chrysanthemums, asters, roses, violets, lilies, irises, anemones, gardenias, acacia, heather, and peach blossoms.

REDWOOD CITY, 26.2 *m.* (10 alt., 50,000 pop. est.), seat of San Mateo County, spreads over land some miles eastward from the docks and piers, canneries, and salt works of its deep-water bay frontage. Redwood Slough once extended inland to the mouth of Redwood Creek, now the center of town, where the Mexican rancheros shipped from the Embarcadero de las Pulgas. As shipment of redwood lumber hauled from the forested ridges to the west began in 1850, Embarcadero became a busy shipbuilding, wagonmaking, and blacksmithing center. Each time that San Francisco burned to the ground the demand for lumber swelled. The redwood business prompted the renaming of the town in 1858, despite founder S. M. Mezes' attempts to substitute his own name for the more euphonious Embarcadero.

The country estates of ATHERTON, 28.7 *m.* (52 alt., 7,717 pop.), are so heavily wooded that little of the town beyond its shady lanes winding under great live oaks can be seen from the highway. Besides its mansions, it has the old railroad station where the Peninsula's first steam carriages stopped.

The first estate here was Faxon Dean Atherton's mile-square Valparaiso Park, laid out in 1860. The family mansion (no longer standing) was later described by his daughter-in-law, Gertrude, as "a large comfortable house with two bath rooms—few houses boasted more than one —and a wing for the servants . . . About the house was a continuous bed of Parma violets whose fragrance greeted one when passing the deer park. (The deer generally died, homesick for their forests on the mountains.)" Faxon Atherton, "in his early youth, had adventured as far as Chile in search of his fortune. He made it in hardware. Not long after his arrival he married Dominga de Goni . . ." Mrs. Atherton, her daughter-in-law wrote, "had hopes of making a true Atherton out of me, and I sometimes wonder she did not . . . 'Ladies in Spain do not write,' she said to me when I began to betray symptoms; and it was quite twelve years after I published my first novel before the painful subject that I wrote at all was mentioned by any of the family in my presence . . ." Life in Valparaiso Park for Gertrude Atherton seemed to be a long series of summer afternoons spent with neighbors "on the wide verandah, sewing, embroidering, exchanging recipes, gossiping. I often wondered if life anywhere else in the whole wide world were as dull."

When James L. Flood, the former San Francisco saloonkeeper who rose to sudden riches by speculation in Virginia City mines, began in 1878 to build the scrollwork-festooned extravaganza of gables, cupolas, and porticos that he called Linden Towers, "the impertinent invasion of . . . the Bonanza millionaires" threw the country aristocracy into a furore, and "for weeks the leading topic on the verandah was whether or not the Floods should be called upon when they moved in." The "colossal white house . . . looked more like a house on a wedding cake than something to live in." In the end, "for business reasons, impressed upon them by their husbands, the women did call." When the Floods returned the call at Valparaiso Park—Mrs. Flood wearing "a flowing dark blue silk wrapper, discreetly ruffled, and 'Miss Jennie' a confection of tourquoise-green flannel trimmed with deep flounces of Valenciennes lace!"—Gertrude Atherton fancied "they went away . . . with the pleasant feeling of superiority that only multi-millions can give." The trappings of Linden Towers—the sterling silver soap dishes initialed J.F., the statues and tapestries, the marble fireplaces and carved rosewood panels—went on the auction block in 1934 when the mansion was torn down; all that remains of the JAMES L. FLOOD ESTATE on Middlefield Rd. are the lodge, iron gateways, and a wall.

MENLO PARK, 29.9 *m.* (63 alt., 26,957 pop.), on US 101 connects with State 84, which crosses the Upper Bay on Dumbarton Bridge to Newark (9,884 pop.) in Alameda County. It developed in 1863 around an estate named by two Irishmen for their home in Ireland, was chosen in 1871 by Milton S. Latham, Governor and U. S. Senator, for his stately, pillared mansion. When the Duke of Manchester passed through on his way around the world, he was escorted here by train to meet the country fashionables. The company was gathered in the drawing room—the women in their Paris gowns and the men in their evening best eagerly awaiting their first sight of a real duke when the English butler announced him. "And then," as Gertrude Atherton told the story, "the duke strode in, and they nearly fainted. He wore boots that reached his thighs and a red flannel shirt! . . .

Histor

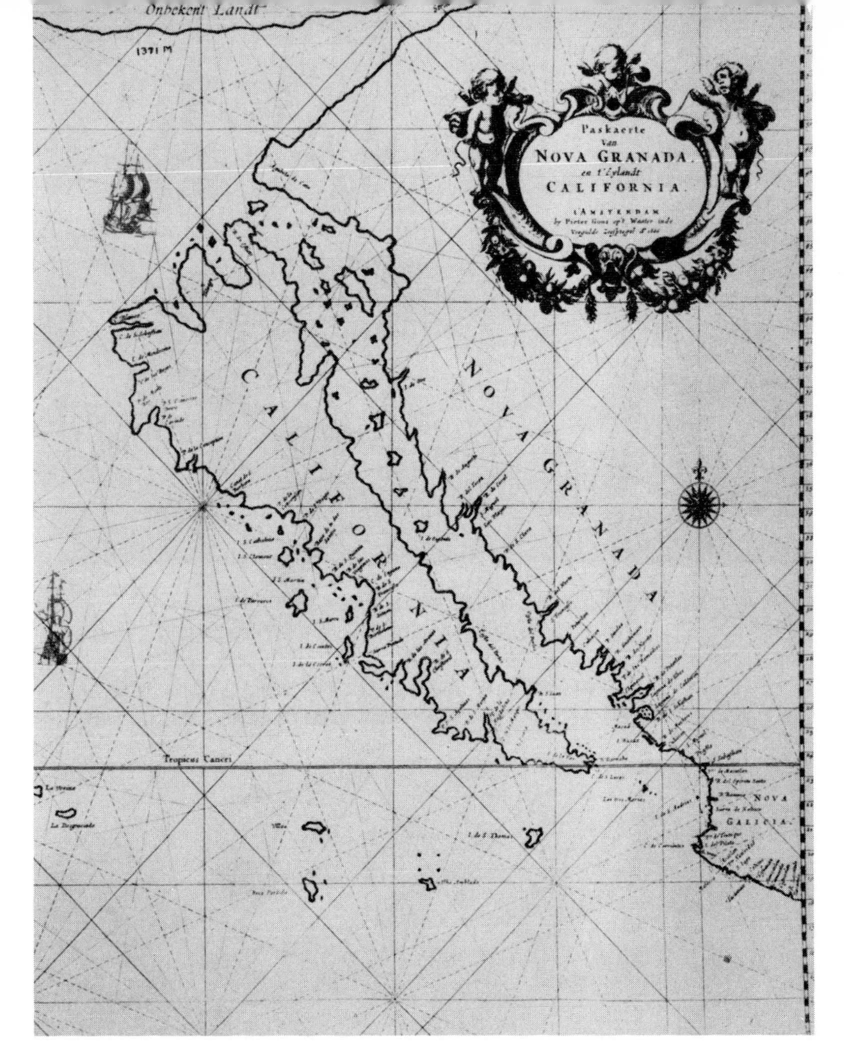

MAP OF CALIFORNIA DRAWN IN 1666

A VIEW OF SUTTER'S MILL, AND CULLOMA VALLEY

WORKING AT SUTTER'S MILL, 1850,
TWENTY FIVE FEET FROM WHERE GOLD WAS DISCOVERED

RUSSIAN CHURCH, FORT ROSS, 1812

HORNITOS

SAN FRANCISCO IN 1849

Courtesy of Bancroft Library, University of California

PRAIRIE SCHOONER, BROUGHT FROM OHIO
TO YOLO COUNTY BY JOHN BEMMERLY IN 1849.

ON TO THE GOLD FIELDS

LYNCH LAW, 1856

STAGE COACH AND TRAIN, CISCO, 1869

PONY EXPRESS, HIGHWAYMEN IN PURSUIT

Poor man, he was terribly mortified, and explained to his suave and smiling host that all he knew of California he had gleaned from the stories of Bret Harte, and had provided himself with what he believed to be the regulation Western costume . . ." The MILTON S. LATHAM HOUSE, damaged in the 1906 earthquake and never repaired, stands among gardens, fountains, stables, and carriage houses on Ravenswood Avenue.

Visible above liveoaks and eucalypti for miles is the PALO ALTO (tall tree), 30.6 *m.,* whose branches shaded the camp of the Gaspar de Portolá expedition, November 6-11, 1769. The "barbarous heathen" of the region whom the chronicler, Fray Crespi, described as "very affable, mild, and docile, and very generous" believed that the Great Spirit dwelt in the tree and held their councils beneath it. When the expedition led by Don Fernando de Rivera y Moncada to hunt for a mission site on the Bay passed by five years later, Fray Palou marked the spot as suitable; and the Juan Bautista de Anza Party, bound for San Francisco in March 1776, found the cross he had erected. The tree in those days was double-trunked, but one of the trunks fell across the creek in the late 1890's. A reproduction of it appears on the seal of Leland Stanford Jr. University. Before the World War, Stanford students used to mount guard over the tree nightly just before their annual "Big Game" with the University of California, to prevent raiding parties of California students from cutting it down.

PALO ALTO, 31.1 *m.* (63 alt., 57,800 pop. 1965 est.), is a college town with gardens along avenues lined with live oaks and pepper trees. It owes its community theater, its advanced schools, and its distinguished residents to the influence of Stanford University. It was the home of Herbert Hoover, 31st President of the United States. The Hoover House is at 623 Miranda Rd. Palo Alto has developed an important Public Library of its own and a Community Center that has adapted Spanish colonial style to the surroundings, which are embellished with magnificent oak trees.

When the university was opened in 1891, Timothy Hopkins was still raising hay on the town site; he gave up his land in 1894 with the proviso that if liquor were ever sold on it, the land would revert to its original owners. To quench their thirst, early Stanford students traveled southward to the village of Mayfield (now part of Palo Alto), so named allegedly, because a recorder's office clerk mistook the first letter of Hayfield for an "M"; there they found no less than 27 saloons along the noisy main street in which to roar their drinking songs.

Right from Palo Alto on wide, palm-bordered Palm Drive (University Ave.) to the sandstone pillars guarding the entrance to the 9,000-acre campus of STANFORD UNIVERSITY (formerly Leland Stanford Jr. University). The campus, nicknamed The Farm, covers the former Palo Alto Stock Farm where Leland Stanford (one of the "Big Four," early railroad builders and financiers), who became Governor (1861-62) and U. S. Senator (1885-93), raised thoroughbred horses. Construction of the university buildings was begun in May, 1887, with a $30,000,000 endowment fund established by Stanford and his wife as a memorial to their only child, who had died at the age of 16. The university was opened to students in 1891. The institution had more women students than men until Mrs. Stanford limited the number of women to 500 (whereupon Stanford co-eds were dubbed the "500"); but women now form a large part of nearly 10,000 students.

Dominating the campus and visible for miles is HOOVER TOWER, 285 ft. tall, completed 1941, which houses the invaluable collections of the HOOVER INSTITUTION ON WAR, REVOLUTION AND PEACE, begun by Herbert Hoover. The tower has a 35-bell carillon presented in honor of Hoover's relief work during World War I by the government of Belgium.

On Lomita Dr. to the LELAND STANFORD JUNIOR MUSEUM (*open 10-5, adm. 25¢*), a reproduction in concrete of the museum at Athens, with mosaic panels and bronze entrance doors added. The south wing is devoted to anthropological, archeological, and paleontological exhibits; notable are the prehistoric Indian artifacts, baskets of the Klamath, Haida carvings in soft slate, and the David Starr Jordan drawings of fish of the Miocene period. Also in the south wing are the Timothy Hopkins Korean Collection, including costumes, household goods, screens, and carved furniture; the Di Cesnola Collection of Greek and Roman pottery and glass, excavated at Cyprus in 1865-76; and copies of Baron Rothschild's Tanagra figurines. The west wing houses the Ikeda Collection of Japanese and Chinese ceramics and bronzes, including an old peachblow vase, and an Imperial yellow Chinese vase. The north wing has a collection of early Californiana, including the first passenger locomotive used in California (1864), and the "last spike" of the Central Pacific Railroad, driven by Governor Stanford. Also here are early California tools and arms, Moro and Igorote weapons from the Philippines, and Chinese robes and ceramics. On the second floor is the Stanford collection of guns, jewelry, clothing and toys.

Left from Lomita Dr., 0.3 *m.* to the OUTDOOR AUDITORIUM landscaped with trees and shrubs to make it look as natural as possible. The STANFORD STADIUM, 0.6 *m.*, a great pit, holds more than 90,000 spectators. On alternate years the Big Game—between Stanford and the University of California—is held here, jamming the countryside with parked cars and traffic.

At the end of Palm Dr. is the UNIVERSITY QUADRANGLE, 1 *m.* The 15 buildings of the outer quadrangle and the 12 buildings of the inner quadrangle, built of buff sandstone and red tile are connected by long colonnades enclosing green lawns. The MEMORIAL CHURCH, built in 1900 by Mrs. Stanford in memory of her husband, is a buff sandstone edifice of cruciform shape in modified Moorish Romanesque design. The mosaic of the facade is a duplicate of one imported from Italy but destroyed in the 1906 earthquake. The brilliant stained glass windows depict Biblical incidents. Marble columns support the dome. The reredos is a mosaic reproduction of Da Vinci's *Last Supper*. JORDAN HALL, NW. corner of the Quadrangle, has among natural history exhibits an unusual collection of fishes. On Lasuen St., NE. corner of the Quadrangle, is the THOMAS WELTON STANFORD ART GALLERY, the gift of Thomas Welton Stanford, the university founder's brother; it houses a collection of paintings and reproductions of classic statuary and holds frequent temporary exhibits; the most costly painting is a portrait by Sir Joshua Reynolds.

The MAIN LIBRARY on Lasuen St., facing the Quadrangle, is the central unit in the system of 40 school, departmental and special collections, totaling more than 1,500,000 catalogued and 2,000,000 uncatalogued books and pamphlets organized for use.

MEMORIAL HALL dedicated 1937, contains plaques honoring Stanford students and faculty who died in the World Wars. It has an auditorium for 1,700.

LAURENCE FROST AMPHITHEATER given by the parents of a member of the Class of 1935, is used for concerts and art festivals and seats 8,000.

Right on the Searsville Rd. about 0.3 *m.* to the FOOD RESEARCH INSTITUTE, organized under a Carnegie Corporation grant, which studies the production, distribution, and consumption of food. A well-equipped laboratory facilitates the primary job of research and the secondary one of teaching.

Highly important in the Stanford curriculum is nuclear engineering, which has been developed in the decades since World War II. One of the most important installations in any educational institution is the STANFORD LINEAR ACCELERATOR CENTER, built under contract with the Atomic Energy Commission at a cost of $114,000,000, fully operable in 1966. It was planned to house a two-mile accelerator on a 500-acre site near Sand Hill Road and produce a beam with energy of from 10 to 20 billion electron volts. The device extends from the foothills to the university campus. It is housed in a concrete box, 10,000 feet long, 10 by 11 feet in cross section, placed under 25 feet of soil to absorb radiation. Above it klystron tubes accelerate the electron beam. The klystron tube was developed at Stanford in 1957.

The HARRIS J. RYAN ENGINEERING LABORATORY is devoted to nuclear engineering and includes a pool-type reactor, completed 1959. This is part of the SCIENCE QUADRANGLE, the focal point of which is the Physics Lecture Hall, built 1957 and financed by royalties from the Klystron tube, primary component of radar and electron linear accelerators, invented by Stanford physicists led by Russel H. Varian, for whom the Laboratory of Physics is named. The High-Energy Physics Laboratory operates the billion-volt electron linear accelerator, a huge atom smasher.

STANFORD MEDICAL CENTER, stands on 56 acres and houses the School of Medicine and Palo Alto-Stanford Hospital. The SCHOOL OF EDUCATION building dates from 1938, the MUSIC BUILDING from 1957. The University's President lives in the LOU HENRY HOOVER HOUSE, former campus home of the Hoovers and donated by Mr. Hoover.

The SCHOOL OF ENGINEERING has among its seven departments Aeronautics and Astronautics, centered in the DANIEL GUGGENHEIM AERONAUTIC LABORATORY and the WILLIAM FREDERICK DURAND LABORATORY. The AIR SCIENCE program prepares cadets as Air Force officers, using ROTC courses, with jet aircraft flights as part of their preparation. Free world and totalitarian military systems are studied, as well as the uses of U. S. air power in national security.

The Stanford Guide Service is located at the Information Center at the main entrance to the Quadrangle and in Hoover Tower. Tours of the campus are conducted daily at 2 p.m. from the Information Center. Tours of the Medical Center begin each day at 3 p.m. in the Hospital lobby. Student guides are available for special group tours if arranged ten days in advance with the Office of University Relations, Stanford University, Stanford, California (Telephone: 321-2300, extension 2862). The Information Center has an assortment of maps, postcards, slides and other descriptive materials. Open daily 10-5.

Parking is restricted except nights, Saturdays, Sundays, and University holidays. Visitor parking is permitted in specially marked visitor zones, time-limit zones, off-campus vehicle zones, and unmarked zones.

On the rolling hills of the campus' southeastern outskirts is (R), 32.6 *m.,* an abandoned tunnel, a round red-brick tower, and an arched bridge in a tangle of weeds. An almost legendary figure—"The Frenchman," Peter Coutts, had his Matadero (slaughterhouse) Ranch here in 1874, with a race track, orchards, vineyards, a cottage, dovecotes, thoroughbred horses, and Ayrshire cattle. When he began developing a shrubbery-fringed lake with green islands connected by bridges, neighbors wondered what he was doing; and when he began digging tunnels into the hills, their speculations grew into wild rumors. Coutts was a fugitive, people said, and he was building a tunnel to flood the countryside from his lake in case his enemies found him. Suddenly in 1880 he disappeared. Although many of his neighbors never knew it, he had merely returned to France to resume his life as Paulin Caperon, Parisian banker and publisher, interrupted nine years earlier when he had been forced to flee to Switzerland because of his newspaper's opposition to Napoleon III's policy in the Franco-Prussian War.

US 101 cuts now through the fruit trees that sweep in row on row across SANTA CLARA VALLEY. The broad plain was so thickly studded with great oaks that to Capt. George Vancouver in 1792, it looked like "a park which had originally been planted with true old English oak." Now in the spring, from the foothills of its mountain walls—the Mount Hamilton Range on the east and the Santa Cruz Mountains on the west—it looks more like an expanse of snowdrifts because of the orchards white with blossoms. The almonds flower first, in late February, and following them in succession until early April, the peaches, cherries, pears, apricots, prunes and apples. In summer an army of wandering fruit pickers invades the valley—an army as large as the host of visitors in blossom time. The trays of prunes and apricots drying in the hot sun cover acres. The millions of prune trees stem from the tiny orchard of prune trees of the French type, *le petit pruneau d'Agen,* planted here in 1856 by Louis Pellier.

At 35.1 *m.* is the junction with a paved road.

At 35.1 *m.* is the junction with a paved road. Right on this road is LOS ALTOS, 2 *m.* (200 alt., 19,696 pop.), noted for its gardens and country club, where houses among oaks overlook the valley orchards. FOOTHILL COLLEGE, established 1958, had 8,268 students and a faculty of 281 in 1964-65.

MOUNTAIN VIEW, 37 *m.* (76 alt., 30,889 pop.) is an industrial and residential center, with packing houses and canneries, where the pick of surrounding orchards and berry patches is brought for processing. Before the railroad arrived in 1864, the Mountain View House with its wide verandah where men sat in tilted chairs with their feet on the railing, chewing tobacco and arguing, was a main stop for the four-horse Concord coaches on the San Francisco-San Jose line.

At 40.6 *m.* is the junction with State 9. Left on State 9 is SUNNYVALE, 0.4 *m.* (95 alt., 52,898 pop.), once the seat of the tract purchased by Martin Murphy, Jr. from Rancho Pastoría de las Borregas (shepherds of the lambs). Among fig trees grown from mission cuttings stands the two-story white MARTIN MURPHY, JR. HOUSE, put together with lumber cut to size in Boston and shipped round the Horn when the Murphy family settled here in 1849. A room with an altar was set aside for mass, marriages and christenings, which were celebrated by priests from the Mission Santa Clara.

Sunnyvale is the site of the International Science Center for Research and Development. Between 1950 and 1960 Sunnyvale annexed 41,447 people from surrounding areas.

Right on State 9 is CUPERTINO, 3 *m.* (3,664 pop.), a crossroads town in flat orchard lands, settled in the 1850's by squatters who banded together when the owners of Rancho Quito tried to chase them off. In the 1880's several sea captains retired and built prim New England cottages here.

From the foothills, SARATOGA, 7.5 *m.* (100-800 alt., 14,891 pop.), looks down on orchards; its annual Blossom Festival draws throngs. It began life as McCarthysville, laid out by Martin McCarthy at the foot of the mountain toll road. It was rechristened Saratoga in 1863 because, like the New York watering place of that name, it was near springs.

The last flair of open-house hospitality in the bonanza tradition took place at Senator James Duval Phelan's VILLA MONTALVO, on the Saratoga-Los Gatos Highway at the eastern edge of Saratoga. The great tile-roofed two-story California-Spanish villa overlooks terraces approached by a long driveway and mounted by marble steps leading to a wide veranda. Roundabout are a guest house, a 20-car garage, an oval pool reflecting wistaria-draped pergolas, an amphitheater, and a fountain dedicated to Ordanez de Montalvo, who in 1620 was the first to use the name California. The doors were opened to guests in 1914. Phelan entertained legislators, opera stars, and the literary elite with luncheons on the terraces, open-air banquets at barbecue pits, and dinners in the patio where screaming macaws vied with a string orchestra in the balcony. After his death in 1930 the villa was bequeathed to the San Francisco Art Association "for the development of art, literature, music, and architecture in promising students." Only a caretaker disturbs the quiet today.

State 9 winds into the Santa Cruz Mountains to CONGRESS SPRINGS, 8.5 *m.,* in a wooded canyon, where capitalists who wanted to make the spot a private vacation place for themselves opened Congress Hall in 1866.

From the summit (2,650 alt.) of Castle Rock Ridge, at the junction with State 5 (*see TOUR 1b*), 14.4 *m.,* the highway descends in curves to the entrance of BIG BASIN REDWOODS STATE PARK, 26.8 *m.* (*lodge, campground*) in Big Basin, a 11,553-acre reserve of massive, cinnamon-brown redwoods lifting spire-shaped green crests as high as 300 feet. In the shadowy depths, tangled with Ceanothus, madrone, oak, toyon, and Tumion, the floor is carpeted with ferns and shrubs. The park was the first set aside for preservation of the redwoods. Its formation in 1901 followed a long campaign carried on in the editorial columns of the Redwood City *Times and Gazette* by Ralph Smith and continued after his death in 1887 by Andrew P. Hill, early photographer of the redwoods.

GOVERNOR'S CAMP, 28.7 *m.* (*post office, inn and cottages, store*), so named in remembrance of visits by two Governors of California and a Governor of Utah in 1901 and 1902, is the starting point for the Redwood Trail, looping through the park past trees of curious formation. One of the tallest is the ANIMAL TREE, deformed by burls at its base. Even taller are the FATHER OF THE FOREST and the MOTHER OF THE FOREST, which towers 320 feet, although its top is broken off.

State 9 follows Boulder Creek to its confluence with Bear Creek and the San Lorenzo River at the settlement of BOULDER CREEK, 37.9 (500 pop.), a colony of summer homes that began as a redwood lumber camp.

The way leads down the valley to BEN LOMOND, 41.8 *m.* (458 pop.), which grew up in the 1880's at the base of Ben Lomond Mountain, named by a Scottish immigrant.

FELTON, 44.9 *m.* (350 pop.), stands within Rancho Zayante, which Isaac Graham bought in 1841—allegedly with the $36,000 indemnity he got for the outrage of having been arrested at Monterey as a dangerous foreigner and shipped to Mexico.

Left from Felton 1 *m.* crossing a COVERED BRIDGE (1892) to MOUNT HERMON, a mountain resort. At 4 *m.* is the junction with State 17 (*see below*).

State 9 continues to the junction with a dirt road, 46.5 *m.;* L. here 0.5 *m.*

to the BIG TREES, a county park of Sequoia sempervirens. The GIANT (306 feet), the JUMBO (250 feet), and the CATHEDRAL GROUP are among the tallest. The cavernous opening of the hollow FRÉMONT TREE (285 feet), large enough to hold 50 people, is persistently reported to have sheltered Capt. John C. Frémont and his aides in 1846. Frémont's comment, when he visited the grove in later years, was: "It is a good story, let it stand."

State 9 skirts the bank of the San Lorenzo River to its junction with State 1 (*see TOUR 1b*), 51.7 *m.*, on the edge of Santa Cruz (*see TOUR 1b*).

Under the branches of the ARMISTICE OAK (R), 42.6 *m.*, Santa Clara Valley's last anti-Yankee uprising ended January 8, 1847, when Francisco Sánchez and his men gave up their arms to the gringos. The armistice followed a skirmish nearby on January 2—the "Battle of the Mustard Stalks," later dignified as the Battle of Santa Clara—in which the Californians lost four dead.

SANTA CLARA, 45.9 *m.* (94 alt.) is one of the two cities of Santa Clara County that have profited greatly by the movement of population into the area, the other being San Jose. In 1950 Santa Clara had 11,702 people; in 1960, 58,880; by 1964 the State Dept. of Finance estimated 81,800.

Located 48 *m.* south of San Francisco, this growing city is served by the Southern Pacific and Western Pacific railroads, by Pacific Airlines from San Jose Airport, and the major highways of Nimitz Freeway and US 101. Its principal products are canned foods, dried fruit, especially prunes; fiberglas, and electronics. Santa Clara completed a new $1,500,000 city hall in 1964 and a new library in 1965. It houses 66,000 volumes. The Kaiser Hospital, costing $5,500,000, was completed in 1965. The first orchards in the neighborhood were set out by the padres of Mission Santa Clara de Asís, founded January 12, 1777 by the Rio Guadalupe. Padre Palou wrote: "Besides an abundance of water in the river, there are several springs which fill the ditches made to carry the water to the fields for irrigation." The "abundance of water" proved the mission's undoing, for the river overflowed into it. On higher ground the padres built a new church (1781-84), only to have it come crashing about their heads in the earthquakes of 1812 and 1818. The third church was dedicated in 1822.

On March 19, 1851, Father John Noboli, S.J., began adapting the mission to the requirements of Santa Clara College (chartered 1855), now the UNIVERSITY OF SANTA CLARA. The church, which became the college chapel, was so badly damaged by the earthquakes of 1865 and 1868 that it finally had to be demolished. When the frame church that replaced it burned in 1926, a fifth church reproducing in concrete the adobe third was built to house relics rescued from the ashes. Among its most sacred mementoes is the memorial to Fray Maguín de Catala, "The Holy Man of Santa Clara," reputed to have prophesied the conquest of California by the United States, the discovery of gold, and the destruction of San Francisco by earthquake. All that remains of the third MISSION SANTA CLARA is a remnant of the adobe cloisters in the olive-shaded rear garden. The three bells, dated 1789, 1799, and 1805, were replaced by a new set through a gift from Alphonso XIII of Spain.

The Spanish-style buildings of the University form a series of quadrangles. The UNIVERSITY LIBRARY exhibits old missals, vestments, breviaries, paintings, and chairs, rescued from the various churches. In the THEATER, students enact a Passion Play every fifth year. The University of Santa Clara is best known for the meteorological and astronomical observations of its scientists. Weather predictions of the late Father Ricard, called "The Padre of the Rains," were so often accurate that farmers used to telephone for the next day's forecast. The university has developed strong schools of law, engineering and business administration. In 1962 women were enrolled for the first time. It had 3,895 students in 1965.

Along the Alameda (tree-lined road) now followed by US 101, once stretched three lanes of willow trees with dovecotes in their branches. On festival days the doves were unloosed with trailing ribbons fastened to their feet to fly in front of little girls scattering wild flowers.

At SAN JOSE, 49.6 *m.* (85 alt., 300,000 pop.) (*see SAN JOSE*), is the junction with US 101-Alt. (*see above*).

1. Left from San Jose on E. Santa Clara St., which becomes Alum Rock Ave., to the junction with Mount Hamilton Rd., 6.5 *m.;* L. here on Alum Rock Ave. 2 *m.* to ALUM ROCK PARK (*picnicking, swimming, dancing, and other facilities*), San Jose's 716-acre park, spreading up oak-studded Penitencia Creek canyon, with footpaths and bridle trails. The park, named for an "alum rock" in the lower canyon, has many mineral springs.

Right from Alum Rock Ave. on Mount Hamilton Rd. and up the mountainside, making 365 turns in 5 miles. In January and February, when the first flowers are blooming in the valley, the peak is often still snow-capped, dazzlingly white against a blue sky. The summit of MOUNT HAMILTON, 18.8 *m.* (4,029 alt.), a mile-long ridge, commands far reaching views.

LICK OBSERVATORY of the University of California is located on the west peak of Mount Hamilton (4,029 alt.) and is reached from San Jose by the Mount Hamilton Stage Line, taxicabs and private cars on a winding road nearly 19 miles long. Distinctive among the group of buildings are the domes housing the telescopes, of which the 120-inch reflecting telescope is second in size only to the 200-inch instrument on Palomar Mountain. James Lick, a real estate speculator who settled in San Francisco in 1848, at the age of 80 designated $700,000 for the erection of an observatory that was to become a department of the University of California. The observatory was completed after his death and turned over to the university in 1888. It then had a 12-inch refractor and a 36-inch refractor, the most powerful instrument in use at that time.

In 1895 a 36-inch reflecting telescope was added, and in 1939 came a twin astrographic reflector containing two 20-inch lenses of 12-foot focal length each. The great 120-inch reflecting telescope, installed in 1959, is the principal instrument for astrophysical investigations at Lick.

A refracting telescope is a large spyglass with a glass lens at the upper end usually having two or more components. In using the 36-inch refractor eye-pieces magnifying 100 to 2,000 times are employed and at times the moon's surface appears as if only 120 miles away. In a reflecting telescope, on the other hand, parallel light rays from a star are passed into an image by a concave mirror mounted at the lower end of the tube. It brings all the wave-lengths from ultraviolet to infrared into a common focus. The photographic plate can record objects quite invisible to the naked eye. When the 120-inch reflector was completed in 1959 the total cost was $2,800,000. The pyrex glass blank for the reflector was cast by the Corning Glass Co. The mirror weighs 4½ tons and with its supports is mounted at the lower end of an open steel tube weighing 45 tons.

The astrograph is a gigantic camera that photographs an area of the sky 6 degrees on a side on a plate 17 inches square. A complete set of 1,246 exposures of the sky begun in 1947 was completed in 1954. A second set will be started in 1967 in order to record the changes in position.

In orderly motion about the sun there are nine principal planets with their satellites, 30,000 or more minor planets, and hundreds of comets. The major planets have been extensively photographed at Lick. Four of Jupiter's satellites were discovered here, and the rotation periods of Saturn and Neptune were determined. Here the transparency of the thin atmosphere of Mars has been carefully analyzed in the light of different colors. Lick photographs of the dark surface markings have shown slow changes over the years. Excellent photographs of the moon have been obtained and made available to specialists. More than 30 new comets have been discovered by Lick Observatory.

Lick Observatory is open daily from 1 p.m. to 5 p.m., except on University administrative holidays or in periods of heavy snowfall. Admission is free. The sky may be viewed through the 12-inch and 36-inch telescopes on Fridays between 7 p.m. and 9 p.m. standard time, by those who have requested entrance two weeks in advance by mail with stamped, return envelope. The number of visitors is limited to 250. If weather is threatening visitors may call the Observatory for verification; phone 258-5061, San Jose. The Mount Hamilton Stage Bus Line leaves Greyhound depot Mondays through Saturdays 9:30 a.m. to 3:30 p.m.; Friday evening trips can be arranged by reservation. Motor cars take the Santa Clara St. (Alum Rock Ave.) turnoff from US 101 bypass in San Jose for the narrow and winding Mount Hamilton Road.

State 17 turns L. from the junction with Stevens Creek Rd. on San Jose-Los Gatos Rd. to LOS GATOS, 10 *m.* (412 alt., 9,036 pop.), within confines of the former Rancho Rinconada de los Gatos (little corner of the cats). The heights above were formerly infested with wildcats.

As State 17 rises through the canyon past pleasant foothill homes, it passes the cloistered Jesuit SACRED HEART NOVITIATE (L), rising like a medieval monastery from terraced St. Joseph's Hill. From the summit of the range at INSPIRATION POINT (1,600 alt.), the road drops down the western slopes to the junction with State 1 (*see TOUR 1b*), 32.8 *m.*, at the northern edge of SANTA CRUZ (*see TOUR 1b*).

At 51.5 *m.* on US 101, Almaden Rd. follows Arroyo de los Alamitos to NEW ALMADEN, 12.3 *m.* (473 alt.), a long abandoned mining town. The reddish hillsides are scarred with abandoned shafts, flumes, and chimneys, for New Almaden in its day produced more quicksilver than any other place in America. As early as 1824 the Robles brothers,

Secundino and Teodoro, and Antonio Sunol had followed the Indian trail to the hill of cinnabar here where the short, dark Olhones of the Santa Clara Valley delighted to smear themselves with the powdered red pigment; but finding neither gold nor silver, they traded their claim for a rancho. A shrewd Mexican engineer, Andrés Castillero, gained title to the deposits in 1845 and began recovering mercury with whalers' try-pots for retorts. When gold miners found a sudden need for quicksilver to use in gold recovery processes after 1848, long pack mule trains loaded with heavy flasks of mercury began following the trail from New Almaden to the port of Alviso on San Francisco Bay. The 20-room CASA GRANDE of the mine superintendent, at the head of the street (L), with wide balconies, two-foot-thick walls, and hand-carved, inlaid fireplaces, was one of the great mansions of its time. The mines were worked until after the turn of the century, when the price of mercury fell. New Almaden's peace was undisturbed until its revival as a week-end resort in the late 1920's.

As US 101 crosses lower Santa Clara Valley's orchard stretches, passing through fruit-shipping centers with small populations—COYOTE, 60.8 *m.,* MADRONE, 66.7 *m.,* MORGAN HILL, 68.7 *m.* (3,151 pop.), and SAN MARTIN, 72.4 *m.*—it cuts across the vast domain of the Murphy family. An Irish immigrant, Martin Murphy in 1844 bought an 8,927-acre ranch, and his descendants expanded his holdings so that by 1883 they owned 3,000,000 acres in three states. In the Diablo Mountains, 14 *m.* east of Morgan Hill is HENRY W. COE STATE PARK, 2,200 ft. alt., 12,161 acres. Here are ANDERSON RESERVOIR, 1,600 acres, and COYOTE RESERVOIR, east of Gilroy, 688 acres. Near San Jose are STEVENS CREEK PARK and RESERVOIR, 415 acres, and COYOTE CREEK PARK. Other reservoirs in Santa Clara County are LEXINGTON, GUADALUPE, ALMADEN, CALERO and CHESBRO, lying between Los Gatos and Morgan Hill. Toward the Coast are UVAS CANYON PARK, 580 acres, UVAS RESERVOIR, and MOUNT MADONNA PARK, west of Gilroy, 3,093 acres.

GILROY, 78.6 *m.* (190 alt., 7,348 pop.), is headquarters for ranchers of the lower Santa Clara Valley's truck gardens, orchards, dairy, hog, and poultry farms. A Scotchman, John Cameron, took the name Gilroy when he jumped ship at Monterey in 1814. He acquired Rancho San Ysidro by marrying Ygnacio Ortega's daughter. At Gilroy is the junction with State 152.

East from Gilroy State 152 crosses Merced County and connects with US 99 south of Chowchilla. About halfway is LOS BANOS (170 alt., 5,272 pop.). Twelve miles west of Los Banos is the new SAN LUIS DAM AND RESERVOIR, a Federal-State joint project to cost over $670,000,000. The reservoir will hold 2,100,000 acre-ft. of water from the Sacramento-San Joaquin Delta. Pumps will lift the water which will serve the Central Valley project through the Delta-Mendota Canal and the State Water Project. Eight generators will have a capacity of 424,000 kw.

Right from Gilroy on paved State 152, up the slopes of the Santa Cruz Mountains to the junction with a winding dirt road, 8.5 *m.*; R. here 3 *m.* to MOUNT MADONNA COUNTY PARK on the summit of MOUNT MADONNA (1,897 alt.); the terraces, fountains, and trees, around the mansion built here by the cattle king Henry Miller are being overrun by a wilderness of fern, madrone, manzanita, and live-oak. Henry Miller was born Heinrich Kreiser in Germany; in 1858—the year in which he formed a partnership with Charles Lux—he took the name of a chance acquaintance whose ticket he had used on the voyage to California. For 58 years the partners went on acquiring lands and cattle until the "Kingdom of Miller and Lux" included more than 1,000,000 acres, in California, Oregon, Nevada and Arizona. The vast herds—all marked with the Miller and Lux Double-H brand—were driven all the way to market in San Francisco from one feeding and resting place on the firm's property to another.

From the summit of Hecker Pass (1,300 alt.), 11 *m.,* State 152 descends the western slopes to WATSONVILLE (*see TOUR 1b*), 17 *m.,* at the junction with State 1 (*see TOUR 1b*).

At 86.5 *m.* is the junction with a narrow paved road.

Left on this road is SAN JUAN BAUTISTA, 3.7 *m.* (200 alt., 1,046 pop.), a ranching community with galleried adobe dwellings at the base of the Gabilan Range. The gnarled trees of its plaza still have the hitching rings that once tethered the horses of caballeros. Every year during the St. John's Day weekend the inhabitants don Spanish costumes for a two-day fiesta with a barbecue, a parade, dancing, and pageantry in the OPEN-AIR THEATER to commemorate the founding on St. John's Day, 1797, and the completion on the same day 15 years later of MISSION SAN JUAN BAUTISTA, fifteenth and largest of the California missions, built with Indian labor.

The mission church, cruciform in shape, with a squat corner belfry, stands beside a low quadrangle with a sweeping red tile roof, facing the Plaza. The massive walls are of adobe up to the 20-foot level and of brick above. In the interior, designed with unusual restraint, the murals have disappeared beneath coats of whitewash. On the floor are the original clay tiles, worn smooth. The ceiling is modern, but the heavy beams are those that Indian laborers set in place and bound with rawhide thongs. Beside the old high altar, fashioned of redwood and painted with vegetable dyes by Indian artisans, are chairs, candelabra, an altar rail and figures of saints brought from Mexico. The niched reredos in back was carved by Thomas Doak, California's first Yankee settler, in return for board and lodging. A little rear room contains two baptismal fonts carved in 1797 from limestone. In other rooms are embroidered ceremonial robes, old books, tools, religious paraphernalia, and hand-inscribed music on parchment. The old music box, said to have been given by Capt. George Vancouver, was carried to Indian encampments by Padre Felipe Arroyo on muleback in an effort to tempt erring converts back to the mission with its tinkling eighteenth-century tunes. Behind the mission, under gnarled olive trees that date from the mission's founding, more than 4,000 Indian victims of the white man's diseases lie buried in a big trench.

San Juan's oldest building, now protected along with others as part of the San Juan Bautista State Monument, is the PLAZA HOTEL, W. side of the Plaza, probably built about 1792 as a one-story, tiled adobe and enlarged with a second story of wood in 1856 when Angelo Zanetta opened it as a hotel. A stopping place for as many as 11 stage lines on the San Francisco-Los Angeles route, it made a name for its wines and cuisine. The CASTRO HOUSE next door, a typical two-story tile-roofed early California mansion with an overhanging balcony around its upper floor, was built in 1825 by Gen. Jose Castro, afterwards acting governor (1835) and *comandante general* of the northern California forces (1846). In 1849 it was purchased by Patrick Breen, a Donner Party survivor, and used as an inn. As a guest here in later years, Helen Hunt Jackson began to write *Ramona*—until the Breens' house manager discovered she was not a Catholic and put her out. On the south side of the Plaza is the ZANETTA HOUSE, a two-story edifice with the usual second-floor balcony, where gay balls were held in the sixty's and seventy's when its owner was Angelo Zanetta.

Above San Juan on PAGAN HILL a gigantic cross of concrete marks the site of a great wooden cross erected by the mission fathers to ward off evil when they discovered their Indian neophytes secretly practicing pagan religious rites on the spot. Above it rises GABILAN PEAK (3,169 alt.), highest in the range, now part of FRÉMONT PEAK STATE PARK, where Lt. Col. John C. Frémont retreated with his forces March 6, 1846, defying orders by Mexican officials to leave the country. For three days the flag of the United States flew over the log fort he erected, until Gen. Jose Castro led forces from San Juan whereupon Frémont, outnumbered, withdrew to Sutter's Fort.

Right on State 156 east to HOLLISTER (6,071 pop.), north on State 156 connecting with State 152, east to Pacheco Pass. Between here and LOS BANOS (6,090 pop.) the Bureau of Reclamation has been building the huge San Luis Dam and Reservoir. The dam rises 384.6 ft. and the cost, State and Federal, is placed at $671,542,000. *See also page 383.*

Before reaching Los Banos State 152 crosses the DELTA MENDOTA CANAL, which carries water 120 miles from the Delta to Mendota Pool, where the water irrigates land formerly served by the San Joaquin River. This makes possible storage of river water behind Friant Dam and Millerton Lake for diversion to the east side of the valley.

The main side route turns R. from Hollister on State 25 down the San Benito Valley to TRES PINOS, 19.1 *m.,* among thoroughbred ranches.

The Rancho Ciénega de los Paicines of which PAICINES, 24.1 *m.,* is headquarters, is still intact; on its thousands of acres are alfalfa patches, prune orchards, and herds of pure-bred, short-horn cattle. The rugged fastnesses of the badlands roundabout were long the haunt of the bandit gang led by Tiburcio Vásquez, and here at Paicines' crossroads store Vásquez made his last robbery. Vásquez' career had begun in 1854, when as a youth of 15 he stabbed a constable during a quarrel at a Monterey fandango and escaping into the hills joined a band of horse thieves. For years, sheriffs, posses and vigilantes were unable to bring him to justice for robberies and murder. According to modern standards his monetary rewards were trifling, for his men never divided more than $2,000 among themselves. The robbery at Paicines in August 1873 was his undoing; he left three dead men behind—a deaf man who had not heard his orders, a Portuguese who had not understood the language he was speaking, and a hotel keeper who had stood in the way of a bullet sent through his door when he refused to open it. When the neighbors became angry the gang fled south, trailed by sheriffs. A colleague cuckolded by Vásquez betrayed his hideout. Brought back to Hollister in May 1874, he was tried and hanged at San Jose March 19, 1875. The calm and dignity that he showed before witnesses of the hanging bore out his proudest boast—that he was "muy caballero."

At 43.3 *m.* is the junction with an improved road; R. here 5 *m.* to the eastern entrance of PINNACLES NATIONAL MONUMENT (*guides, cabins, campgrounds*), 10,000 acres of crags, rocks, pinnacles, and forests among the Chalones peaks of the Gabilan Range. The weird, theatrically colored rock formations of dark red conglomerate found nowhere else in North America, have been carved into long twisting caverns, narrow gorges between precipitous walls with spire-like abutments, and great terraced domes fluted with vertical grooves. The spires rise from 600 to 1,000 feet. The walls of the BRIDAL CHAMBER, a semi-circular amphitheater overgrown with thickets of live-oak and wild cherry, rise a sheer 500 feet.

State 25 continues southward to the junction with a paved road, 54.8 *m.;* R. here to KING CITY (*see below*), 69.8 *m.,* on US 101.

US 101 twists over the slopes of a low divide and across the broad plains near the mouth of the Salinas Valley, so named for the Salinas

River's salt pools along its lower course. The sandy river bed, virtually dry in summer when the river flows partly underground, stretches southeastward a hundred miles between the Santa Lucia Range on the west and the Gabilan and Diablo Ranges on the east. In the early days, wild oats and yellow mustard grew so tall on the Salinas Plain that horsemen traveling across it had to stand on their saddles to see ahead. Over the valley bottoms and foothill slopes roved the herds of a long string of ranchos. The valley is still cattle country—although most of the bottom land is covered now with rows of lettuces and sugar beets, grain and alfalfa fields, berry patches and orchards—and many of the ways of the rancho era linger on.

California's biggest rodeo, held every July in the SALINAS RODEO FIELD (L), 104.6 *m.,* keeps alive a tradition dating from pre-American times, when the rodeo held first place among the events that enlivened the dull everyday routine. Once a year the whole countryside joined in rounding up the range cattle into corrals, where they were slaughtered —not for meat, because there was no way of shipping it, but for hides and tallow. After the work was done, there were horse races, bull and bear fights, and singing and dancing. The trick-riding exhibitions of today's rodeos would have provided few thrills for the men of those days, who grew up on their horses. One of their stunts was to pick up a coin from the ground while riding at a dead run; another was to snatch off the head of a chicken, buried up to its neck in the ground, while galloping past; and another, to carry a trayful of brimming glasses at a gallop and deliver it without spilling a drop.

SALINAS, 105.6 *m.* (44 alt., 28,957 pop.), seat of Monterey County, is still a cattle-raising center, as it was in the days when nothing stood here but the buildings of one of the Ranchos del Rey, where the stock of the mission at Carmel and of the presidio at Monterey were pastured. The town dates from 1856, when Deacon Elias Howe erected the Half-Way House, an inn, store, and county meeting hall. In 1869 Salinas had a population of 600—including 10 Chinese, 3 Negroes, and 50 Mexicans—and the first newspaper in the section, the Salinas *Standard,* then a year old. Today its bustling modern business district is the market center for dairy farms, truck gardens, and lettuce fields.

The Salinas area has more than 50,000 acres in lettuce. In and about the town cluster the ice-making plants and packing sheds that prepare the crop for shipment to eastern markets. More than 2,000 field workers move in to plant, cultivate, and harvest the spring, summer, and fall crops. In 1936, when lettuce workers struck for higher wages and better working conditions, Salinas burst into the Nation's headlines as the scene of tear-gas battles between State Highway Patrol officers and strikers. Highlight of this strike was the mobilization which followed a report to the Highway Patrol that a Communist advance on Salinas was under way. Red flags proving the statement were taken from the highway and rushed to Sacramento. Airplanes were sent up to reconnoiter. Embattled growers prepared to defend life and property. In the meantime, an indignant highway commission requested that the flags placed as markers on roadsides by its workmen be returned to serve their purpose of warning motorists.

An unusual Salinas industry is the AMERICAN RUBBER PRODUCERS GUAYULE PLANT, opened in 1931 with a daily production capacity of 15,000 pounds, which manufactures rubber for automobile tires from the Mexican guayule shrub. It was established here after 14 years of experimentation had determined that the valley offered proper conditions for raising the plant. After four years' growth the plant is harvested and its trunk and roots are cleaned, chopped, and ground into fine particles. The rubber is skimmed from the fiber, refined, then dried and compressed. Since 1926 thousands of acres have been planted with guayule.

Left from Salinas on a paved road is NATIVIDAD, 5.7 *m.* (161 alt.), a bustling station on the Coast Stage Line in the fifty's now falling into ruin since the railroad and the highway were laid out to the west. The two-story adobe CASA DE JOAQUIN GOMEZ overlooks it from the Gabilan foothills. Here one winter night in 1847 young William T. Sherman and a companion from Monterey hitched their horses "and went in just as Gómez was about to sit down to a tempting supper of stewed hare and tortillas . . . The allowance, though ample for one, was rather short for three," wrote Sherman, "and I thought the Spanish grandiloquent politeness of Gómez, who was fat and old, was not over-cordial . . . I was helped to a dish of rabbit, with what I thought to be an abundant sauce of tomato. Taking a good mouthful, I felt as though I had taken liquid fire; the tomato was *chile colorado,* or red pepper, of the purest kind. It nearly killed me, and I saw Gómez' eyes twinkle for he saw that his share of supper was increased."

The chile was perhaps Gómez' way of retaliating for the losses suffered by his fellow countrymen in the Battle of Natividad, fought here November 16, 1846, when 150 Californians mustered to attack 60 or 70 Yankees on their way with 300 horses to join Frémont at Monterey. A scouting party sent ahead by the Yankees was ambushed and several of the men were wounded or killed; one of its members, a Delaware Indian, who brought news of the battle to Alcalde Walter Colton at Monterey, "was attacked by three Californians—one of whom he shot with his rifle, another he killed with his tomahawk, and the third fled." In the sharp skirmish that followed when the main detachment of Americans came up, the Californians inflicted rather serious damage on their opponents before retiring. The Americans saved their horses, but at the cost of four or five wounded and as many killed.

At 108 *m.* on US 101 is the junction with a paved road.

Right on this road to SPRECKELS, 2 *m.*, a company town around a large sugar beet factory (*visitors 9:30-1:30; guides; plant closed Dec.-July*). It was built in 1899 by Claus Spreckles. California's sugar beet crop leads the nation and is worth about $48,000,000 a year.

As US 101 cuts southeast in an arrow-straight line through trim rows of yellow-green lettuce heads, blue-green alfalfa patches, carrot fields and flower-seed farms, the valley begins to narrow. The pepper-tree-shaded village of CHUALAR (a place abounding in wild pigweed), 116.2 *m.*, stands among lettuce fields. GONZALES, 122.4 *m.* (127 alt., 2,138 pop.), is a dairy center; thousands of cows in surrounding pastures supply its milk-condensing plant. SOLEDAD (Sp., solitude), 130.2 *m.* (189 alt., 2,837 pop.), is a farming and stock-raising center.

At 132.3 *m.* is the junction with a paved road.

Right on this road 1 *m.* to the junction with a paved road; R. here 2 *m.* to the adobe remains, now in process of restoration of MISIÓN DE NUESTRA SENORA DE LA SOLEDAD (Our Lady of Solitude), whose founders named it more fittingly than they knew. The thirteenth of the missions, it was founded October 9, 1791, by Fray Fermín Francisco de Lasuen; the thatched adobe chapel was finished in 1797. It was so impoverished after its secularization in 1835 that Padre Vincente Sarria, who refused to leave, died on the altar steps before mass one Sunday morning of starvation.

The main side route continues on a graveled road to PARAISO HOT SPRINGS, 5.8 *m.* (1,160 alt.), a resort in the Santa Lucia Mountains, where the mission padres used to come to drink the sparkling mineral waters. "Eternal Paradise" was the name they gave a 20-acre tract here, granted to them in 1791.

US 101 parallels the sandy river bed of the Salinas River, choked with cottonwoods and willows. The jagged peaks of the Pinnacles National Monument (*see above*) appear (L), in sharp contrast to the rounded brown Gabilan Mountains. Long files of eucalypti, planted as windbreaks, crisscross the fields.

Around GREENFIELD, 139.6 *m.* (287 alt., 150 pop.), stretch the alfalfa fields, green the year round, for which it is named, and seed farms, gooseberry patches, and orchards.

At 150.6 *m.* is the junction with an oiled road.

Right on this road through rolling grain and cattle country, where the hot summer sun bakes out a pungent smell of dry grass, tarweed, and dust, to the summit of Jolon Grade, 10.3 *m.* (1,343 alt.), and down into sequestered Jolon Valley. Over the rough route in a rattling spring wagon came George Atherton and his wife Gertrude with their child and its nurse and their Chinese cook in the eighty's sent here by George's father, Faxon Dean Atherton, to evict squatters from his 43,000-acre Rancho Milpitas. At JOLON, 17.4 *m.* (978 alt.), "a straggling village on the edge of a ranch," as Gertrude wrote—now a mere remnant of a town with little left but a country store, church, and school—they stayed in a room that had neither a fireplace nor a window at the two-story ADOBE HOTEL, built in 1870 to house stagecoach travelers. (It is still standing, its upper balcony shaded by an ancient grapevine.) The next morning, "armed to the teeth," George set out with two sheriffs to evict the squatters.

Right from Jolon on a dirt road 6.5 *m.* to MISIÓN SAN ANTONIO DE PADUA, where "the squatters had herded their families and livestock . . . while they went off to seek a warmer hospitality elsewhere" when Gertrude Atherton visited the place several days later. "It was a strange sight. The church and yard were crowded with women, children, sheep, and goats. Winter was approaching and it was already very cold, but the immensely fat Mexican women wore but a single calico garment. The brown children, playing with the goats, were stark naked. It was no warmer in the tottering church and the first rain would add to their miseries."

When Fray Junípero Serra founded San Antonio de Padua, third of the missions, July 14, 1771, in an oak-dotted glen, he began ringing the bells hung from an oak branch, crying: "Hear, O Gentiles! Come! Oh come to the holy Church of God! Come, Oh come, and receive the Faith of Christ!" A lone Indian appeared to watch him celebrate mass at the rustic altar. The mission grew rapidly in wealth, until its fine horses, its 50,000 cattle and 50,000 sheep began to tempt horse and cattle thieves; a stone mill, operated with water brought for miles through a stone zanja, was built to grind its wheat. All that remains today is the crumbling brick-faced church (1810-21) with its arched portals and bell opening flanked by low bell towers and the crumbling colonnade with clinging rose vines stretching off to two burnt-brick wine presses above a wine cellar. A modern shingle roof protects the interior, where benches, an altar and organ, and some old wooden statues have been placed. Every year St. Anthony's Day, June 13, is celebrated with masses and a fiesta.

A long bridge carries US 101 across the sandy wash of the Salinas River to KING CITY, 152.4 *m.* (338 alt., 2,937 pop.), chief marketing center of the southern Salinas Valley, lying at the edge of barren, wind-swept hills. Along the sidewalks are still a few old hitching posts, used occasionally by cowboys in from the ranges. King City, named for one of its first settlers, was founded in 1868. Its surrounding farms produce pink beans, fruit, grain, and cattle. At King City is a junction with the paved road leading to Pinnacles National Monument (*see above*).

As US 101 crosses the Salinas Valley's lower reaches, the blue and purple of the Santa Lucia Mountains dominate the landscape. Since Mexican days SAN LUCAS, 161.7 *m.* (408 alt., 75 pop.), has had a tradition of raising thoroughbred horses. As the highway continues through cattle ranges between low, bare hills, the valley narrows. SAN ARDO, 172.2 *m.* (458 alt., 111 pop.), lies among rolling acres of wheat field and pasture. Along the river bank through low hills US 101 continues to BRADLEY, 187.4 *m.* (552 alt., 110 pop.), which ships diatomaceous earth.

The village of SAN MIGUEL, 198.3 *m.* (615 alt., 1,000 pop.), surrounded by almond orchards, clusters near MISIÓN SAN MIGUEL ARCANGEL, sixteenth of the missions and probably the most unspoiled, founded July 25, 1797, on St. James Day but dedicated to St. Michael because another mission had already been named for St. James. The mission buildings include the church (1816-18) and a row of low buildings all opening on a court bordered by a corridor with arches of varying shape and size. The inside walls of the chapel are decorated with simple designs—fret borders, painted columns and a low dado, in imitation of marble, the work of Indians under Spanish direction.

West of San Miguel in San Luis Obispo County between US 101 and the Coast is Lake Nacimiento, a large reservoir fed by the Nacimiento River from the Santa Lucia Mountains. This river and the San Antonio feed the Salinas River, which flows into the Salinas Reservoir.

The rolling land around PASO ROBLES, 207.2 *m.* (721 alt., 6,900 pop.), is still dotted with the great spreading oaks that led the first Spanish explorers to name the place El Paso de los Robles. Its almond trees are more famous now than its oaks, for Paso Robles is surrounded by a large almond acreage; the trees blossom in February, turning the countryside into one of those scenes so often photographed for State advertising literature. Long before Paso Robles was a town, the spot was known for its hot springs, by which the padres of Mission San Miguel placed a wall of rude logs to impound a pool. Even grizzly bears are said to have resorted to its warm waters for bathing; old-timers told of a grizzly who made regular nocturnal trips to cure a lame leg, grasping with forepaws an overhanging limb while he dipped his posterior with evident enjoyment. The MUNICIPAL BATH HOUSE is built over the main sulphur spring and the hotel mud-bath house and hot sulphur pool over the original lithia springs.

ATASCADERO (Sp., bog), 218.3 *m.* (849 alt., 12,000 pop.), rimmed by a semi-circle of wooded hills, was conceived in 1913 as a model community, designed on paper to conform to uniform architectural standards and laid out on a scale large enough to accommodate a metropolis—so large that its scattered residences could never get close enough together to compose a town in the usual sense. Following the failure of the enterprise, its grandiose rococo civic center was converted into a school, and its small-scale farms went to seed, until eventually the town found its salvation in chicken and turkey raising.

Atascadero is the junction with US 466, which runs west to State 1 north of Morro Bay.

On a knoll near the cattle-shipping center of SANTA MARGARITA, 225.8 *m.* (998 alt., 211 pop.), is the ASISTENCIA OF SANTA MARGARITA, an outpost of Mission San Luís Obispo, of which Capt. Alfred Robinson wrote in 1831: "It was divided into store rooms for different kinds of grain, and apartments for the accommodation of the mayordomo, servants, and wayfarers. At one end was a chapel, and snug lodging-rooms for the priest . . . and the holy friars of the two missions occasionally met there to acknowledge to each other their sins."

Left from Santa Margarita on State 178, paved, up Pozo Valley, through the fruit and dairy centers of POZO, 18.5 *m.,* and LA PANZA, 34.3 *m.,* and across the high plateau of Carisa Plain (1,400 alt.). In the southwestern part of the plain rises a lonely sandstone butte enclosing a great oval cavity with overhanging walls, entered by a narrow opening on the east, which Spanish settlers called the PAINTED ROCK. Along the walls for a length of 60 feet are a number of rude aboriginal paintings.

To the north is LA PANZA SUMMIT, 2,700 ft.; to the south FREEBORN MTN., 3,311 ft.; and MT. MACHESNA, 4,054 ft. This is the heart of the great LOS PADRES NATIONAL FOREST, which takes in a part of the Sierra Madre and extends from Monterey County, through San Luis Obispo, Santa Barbara and into Ventura Counties. (Forest headquarters, Federal Bldg., Santa Barbara.)

US 101 twists up the wild, wooded slopes of the Santa Lucia Mountains to CUESTA PASS (1,570 alt.), 125.7 *m.,* in sharp hairpin curves. Over this pass, in November 1846, John C. Frémont led his small army at night in a pelting rain to attack the town of San Luis Obispo, believing it to be heavily garrisoned. Only the sleeping villagers were there when the gallant Frémont rode down the main street and made them prisoners. Frémont reported that he had "brilliantly captured San Luis Obispo without bloodshed." A historian reviewing the incident remarked: "It detracted somewhat from its brilliancy that there was nothing to surrender, and nobody authorized to do it if there had been, and the army was uncomfortably wet."

At the southern base of the Santa Lucia Mountains is SAN LUIS OBISPO, 243.3 *m.* (200-300 alt., 20,386 pop., 1960; 24,100, 1963 est.), seat of San Luis Obispo County. It grew up around MISIÓN SAN LUIS OBISPO DE TOLOSA (St. Louis, Bishop of Toulouse), Chorro and Montgomery Sts., which is said to have been so named because two of the pyramidal volcanic peaks in the neighborhood suggested a bishop's mitre. When Fray Junípero Serra established the mission, fifth of the chain, September 1, 1772, he found the Indians of the place friendly because soldiers of the Gaspar de Portolá expedition three years before had killed with their carbines and spears several of the great bears that were terrorizing the Indian rancherías. Padre José Cavaller began his work with four soldiers and two Lower California Indians for company, his provisions limited to 50 pounds of flour, three pecks of wheat, and one barrel of brown sugar. The burning of the tule-thatched mission roofs three times in a row led the friars to experiment with making tiles, which proved so sucessful that they became the fashion after 1784 for all California missions. Most noted of Padre Cavaller's successors was the rugged individualist, Fray Antonio Martínez who, in 1818, drilled and equipped Indian troops to help fight the pirate, Hippolyte de Bouchard, and got up out of a sick-bed to lead them. Later Fray Martínez grew friendly with Yankee skippers engaged in smuggling goods into California, often entertaining them with his best wines. The mission was formed in 1844 into a pueblo that was incorporated in 1856 as a city, but the settlement kept the attributes of a sleepy Mexican village until 1894, when the railroad pierced the mountains from Santa Margarita. Many streets follow winding trails beaten by Spanish horsemen.

The church of SAN LUIS OBISPO DE TOLOSA has been restored and holds services. Its museum has relics of Fra Serra's day. Also of interest are the County Historical Museum, Ah Louis' Chinese store of 1874, and Sinsheimer Bros. store, bringing back the interior furnishings and stock of nearly 100 years ago.

CALIFORNIA STATE POLYTECHNIC COLLEGE, founded 1901, had 10,347 students in 1965. It has branches at Pomono and San Dimas. Salinas Reservoir served by the Salinas River is located east of the Santa Lucia Range and provides irrigation for this area of low rainfall.

At San Luis Obispo is the junction with State 1 (*see TOUR 1c*).

Section c. SAN LUIS OBISPO to LOS ANGELES; 203.6 m. US 101

This section of US 101, heavily traveled both by private motor cars and by trucks, swings briefly to the coast south of San Luis Obispo and then weaves inland across a broad river valley and into typical barren hills. Returning to the shore of the Pacific, which it follows for many miles, it swings inland again over low hills and descends to the San Fernando Valley. Its way into Los Angeles is disfigured with roadside stands and advertisements.

Southwest of SAN LUIS OBISPO, 0 *m.,* US 101 passes through a small valley and reaches the Pacific, where it skirts the beach of San Luis Obispo Bay for several miles.

PISMO BEACH, 12.2 *m.* (35 alt., 1,782 pop.), well advertised by posters and banners, is a busy seaside resort during the summer, popular for its Pismo clams. This variety is seldom found elsewhere to any extent, and unrestricted digging had made them almost extinct until State regulations conserved them.

US 101 leaves the sea and runs through Arroyo Grande Valley, where much of the once rich topsoil, improperly cultivated, has been washed away. In October, 1934, CCC workers began erecting check dams and contour ditches to catch the silt and restore fertility to 7,000 acres. Banks were seeded with a grass mixture, mostly red barley, to hold the soil. Thousands of trees and shrubs were planted.

ARROYO GRANDE, 16.2 *m.* (114 alt., 3,291 pop.), once merely a stagecoach station, sprang into life in 1877 because of a land rush. Surrounding it are cattle and grain ranches.

NIPOMO, 24.6 *m.* (114 alt., 750 pop.), is on one of the few early land grants that have remained in the hands of one family.

SANTA MARIA, 32.6 *m.* (204 alt., 20,027 pop. 1960; 34,000 est., 1964), is the market center in Santa Maria Valley. Grain, beans, and seed are the leading crops. Trees and shrubs line most of the town's streets, though until the late 1860's the area was so desolate that adjoining ranchos scorned the sandy flats. The first building, erected in 1871, was a small general store to accommodate farmers who had begun irrigating the land. Santa Maria owes much of its more recent development to large oil districts nearby. The town has notably wide streets because four pioneer farmers, laying out the townsite, wanted plenty of space to turn their eight-mule teams.

For miles the countryside along US 101 is planted with beans. The lanes of eucalypti along the road lead to LOS ALAMOS, 49.4 *m.* (569 alt., 600 pop.).

BUELLTON, 66.4 *m.* (490 alt., 107 pop.), is called the Mission Cross Roads because it lies on the route between two missions.

Left from Buellton on a paved road to the MISSION SANTA YNEZ, 4 *m.,* on the outskirts of the hamlet of SOLVANG. The nineteenth mission, established September 17, 1804, by Fra Estevan Tapis and three other brothers, its prosperity reached a peak in 1820, when it owned 12,000 head of stock. After 1850 the buildings began to crumble. They are of adobe, low, painted white, and roofed with red tile, with an arched colonnade in front. The walls are five or six feet thick, to support the big, hand-hewn beams of the roof. In the mission are vessels of beaten copper, parchment volumes of church music, bound mission records, and carved wooden crucifixes.

US 101 crosses the Santa Ynez River, 66.9 *m.,* and winds through heavily wooded country to Nojugui Pass (900 alt.) in the Santa Ynez Mountains. At places the hills are almost solid rock with massive blocks towering over the highway. Sycamores with mottled gray trunks grow along creek beds.

LAS CRUCES, 75.9 *m.* (339 alt.), at a junction with State 1 (*see TOUR 1c*) is a postoffice for numerous farmers.

The lovely GAVIOTA (seagull) PASS (200 alt.), between Las Cruces and the ocean, was named by the Gaspar de Portolá expedition when one of his soldiers shot a seagull here in 1769. The shore at the lower end was once a favorite spot for the landing of goods smuggled in to evade the high tariffs imposed by the Mexican Government. Yankee ships anchored in the small coves and sent their cargoes inland for sale to missions and landholders.

US 101 runs on a narrow shelf above the sea to GOLETA, 102.4 *m.* (50 alt., 519 pop.), named for a schooner built by Captain William G. Dana of Nipomo in 1828.

Passing between small farms, the highway is lined for several miles with poplars and bordered at intervals by walnut orchards. In the purplish blue Santa Ynez Mountains (L), the peak of La Cumbre (3,985 alt.) stands out.

SANTA BARBARA, 108.4 *m.* (*see SANTA BARBARA*).

US 101 follows narrow and unattractive streets. An alternate route follows broad palm-lined Cabrillo Boulevard, skirting the bay, passing the yacht harbor and the bird refuge lagoon, and connecting with US 101 at the eastern city limits.

In the low wooded hills to the south is the millionaire colony of MONTECITO, 112.6 *m.* (250 alt.). Signs marked "Private" bar the paved roads leading to the estates of such living trademarks as Stetson, Fleischmann, Armstrong, Pillsbury, and Du Pont. (*Adm. to many estates on special tours conducted periodically by Santa Barbara Garden Club.*) The four-lane highway is divided by a parkway overgrown with iceplant, a flat-leaved plant (*genus Mesembryanthemum*), found along the southern coast from Marin County to the Mexican Border, with stout stems and large leaves encrusted by shining beads of a gummy substance that glistens like frost in the sunlight.

SUMMERLAND, 114.9 *m.*, is the site of the world's first offshore oil field. Drilling began in 1896 from wooden piers built out into the tidelands and spread along the beach. Today modern equipment is in operation on platforms built in the sea.

Near the ocean just east of CARPINTERIA, 118.4 *m.* (12 alt., 4,998 pop.), are several pits of asphalt, used by the natives to waterproof their houses and canoes. A spreading Torrey pine (L), some 60 feet high, was transplanted from the island of Santa Cruz about 1900.

Right from Carpinteria on Linden Avenue into the CARPINTERIA STATE BEACH PARK, 0.5 *m.* Under a roof about 300 yards (L) is the dead trunk of a grapevine, planted in 1846. Before the vine died, in 1916, the circumference of its trunk measured nine feet; it covered a quarter of an acre and yielded eight tons of grapes each year.

Leaving the flat plains, US 101 mounts the narrow pass of RINCON CAPE, 121.4 *m.* In 1838, would-be Governor Alvarado sent out a hundred men with three cannons against the slightly larger army of Governor Carrillo, Mexican appointee, and chased him into the Mission San Buenaventura. The total casualty was a Carrillo man, killed by a sniper from the Mission belfry. Alvarado became governor.

An innovation in underwater drilling was completed here in 1961 when Richfield Oil Corp., operating from a vessel, capped a well and installed pipes in 55 ft. of water nearly a mile off shore. Also off Rincon Richfield has built an island of rubble to support drilling and storage tanks. A causeway connects it with the shore.

VENTURA (San Buenaventura, 48 alt., 29,114 pop. 1960; 40,432 est., 1965) exporting center for oil and agricultural products of Ventura County is expanding as a resort. Nearby is San Buenaventura Beach State Park. Located on Santa Barbara Channel, Ventura exports through Port Hueneme, 12 *m.* south. It was visited by Cabrillo (1542) and Portola. County farm products include fruits, nuts, vegetables and beef cattle; oil production is around 30,000,000 bbl. annually. Ventura Marina (opened 1963) is expected to berth more than 2,000 boats.

Two Norfolk pines towering above the belfry mark MISSION SAN BUENAVENTURA (*small fee*), Main and Figueroa Sts. This outpost, ninth of the California missions, was the last founded by Father Serra, leader of the California missionary campaign, who planted his huge cross on the hill to the rear and consecrated the ground in 1782. The mission was twice damaged by fire in 1791-92, and by earthquake in 1812; little of the original structure remains. The low tile roofed structure with its heavy buttresses and large corner tower is largely a restoration. In keeping with Franciscan simplicity the facade has little ornamental detail. The tower with its double, arcaded belfry is topped with a stripe-ribbed dome. Perhaps the most distinguishing feature of the structure is its bold and simple mass. Mission relics are shown in a small building opening onto the patio, in which is an ancient stone olive-crusher. The wooden bells that once hung in the belfry are now in the museum. Each is made of a single block of wood with metal plates inside, against which the wooden clappers struck. Such bells are common in certain parts of Mexico.

The PIONEER MUSEUM (*free*), in the VENTURA COUNTY COURTHOUSE, N. California and Poli Streets, has a collection including feather flowers made by elderly women in Vermont, a small scale model of the mission, Eskimo harpoons, model of South Sea Island canoes, and stone bowls of the Canalinos. In front of the Courthouse is a large cement STATUE OF JUNIPERO SERRA, erected by the Federal Art Project.

US 101 passes over the Ventura River delta, a district of walnut trees and orange groves blocked by long windbreaks of tamarisk, eucalyptus, and other evergreens. The windbreaks are favorite nesting spots of western mocking-birds, slim and graceful, with slaty black wings and tail, and a prominent white wing patch.

At 137.6 *m.* is the junction with State 126, paved.

Left on the Freeway through Santa Clara Valley, a fertile strip of fields and citrus orchards along the Santa Clara River, sheltered between the San Rafael Range on the north and the Sierra San Fernando on the south. The valley is dotted with oil wells, still pumping oil from the oldest field in California; from 1866 until discovery of the Puente Hill district in 1880, it yielded nearly all the oil produced in the State.

At 5.8 *m.* is the junction with a paved road; R. on this road to the farmers' village of SATICOY (Ind., I have found it), 1 *m.* (149 alt., 400 pop.), founded in 1861 as SATICOY SPRINGS, where a number of Chumash Indians headed by the chieftainess Pomposa were living as late as 1870. The springs were their traditional gathering place for ceremonials.

Oil fields vary with lemon orchards on the approach to SANTA PAULA, 13.2 *m.* (288 alt., 15,050 pop. est., 1964). Originally the site of Mupu village, where lived Indians who helped build Mission San Buenaventura. Orange and lemon trees were introduced in quantity in 1875. Exploitation of oil resources of the area began in 1883; in 1887 the first refinery was built in Santa Paula. In 1890 the pioneers formed the Union Oil Co. Its original office is now the OIL MUSEUM.

The countryside around FILLMORE, 24.3 *m.* (460 alt., 4,808 pop.), produces oil, lemons, and oranges. The town lies at the mouth of wooded Sespe Canyon, a favorite resort for trout fishermen and game hunters.

The home place of Helen Hunt Jackson's *Ramona* was CAMÚLOS RANCHO, 33.7 *m.*, the Moreno Rancho of the novel, which Mrs. Jackson observed with such an eye for detail during a two-hour visit in 1882 that she was able to describe it later with complete fidelity. It remains as it was when she saw it, "one of the best specimens to be found in California of the representative house of the half-barbaric, half-elegant, wholly-generous, and free-handed life led by the Mexican men and women of degree in the early part of this century. . . ." Camúlos Rancho was the heart of the 2,000-acre Rancho San Francisco granted in 1839 to St. Antonio del Valle, on which the first California gold was discovered—March 9, 1842. The rancho passed to his son, Don Ygnacio del Valle, whose wife, Señora Doña Ysabel, was the original of *Ramona's* Señora Morena, although Mrs. Jackson never saw her, and to his grandson, Reginald F. del Valle, the original of the boy Felipe. The severely simple whitewashed adobe house (1852) encloses on three sides a patio with graveled walks between flower beds, rose and cypress hedges, and a fountain, overlooked by wide verandas. The oldest part of the house is the front wing, containing the chief apartments, where Mrs. Jackson placed the rooms of Ramona and Father Salvierderra in the novel; the wing on the west and the *cocina* (kitchen) on the north were added ten years later. The "greater part of the family life went on" on the verandas, wrote Mrs. Jackson. "All the kitchen work, except the actual cooking, was done here, in front of the kitchen doors and windows. Babies slept, were washed, sat in the dirt, and played on the veranda. The women said their prayers, took their naps, and wove their lace there. . . . The herdsmen and shepherds smoked there, lounged there, trained their dogs there." The family chapel, near the house, is a tiny frame structure with an altar containing *santas* (saints) brought from Spain. Nearby are an orange grove with an old fountain, two ancient bells hanging from a wooden frame, a brick winery and a grape-arbor, and an olive-mill. In the family graveyard, a square white vault contains the grave of Ygnacio del Valle.

At 43.6 *m.* is the junction with US 99 (*see TOUR 3d*).

At 139.1 *m.* is a junction with Mulligan Alley, a dirt road.

Right here to the OLIVAS ADOBE, 1.4 *m.*, once the dwelling of Don Raymundo Olivas, who took a leading part in suppressing a native uprising. The house is now the Old Adobe Gun Club.

South of MONTALVO, 140.7 *m.* (93 alt., 2,028 pop.), a lonely railroad station and cluster of seedy frame buildings that ships large quantities of walnuts, US 101 crosses the Santa Clara River, in the willows of which Tiburcio Vásquez is said to have hidden while evading, in accredited will-o'-the-wisp fashion, the punitive efforts of the combined constabulary of five counties.

At EL RIO, 143.2 *m.,* a few frame houses squatting around a store in the midst of fields and orchards, is the junction with US 101-Alt. (*see TOUR 2B*). The general store of its founder is still standing (L), with the black letters on its brick wall reading "New Jerusalem—Simon Cohn, Proprietor."

Tree-shaded CAMARILLO, 151.4 *m.* (150 alt., 2,359 pop.), was named for an early ranchero.

Right from Camarillo, a paved road leads to CAMARILLO STATE HOSPITAL, 4.6 *m.,* a large group of white buildings in California-Spanish style. Off US 101 is OXNARD AIR FORCE BASE (1,609 pop.). On the seashore at Point Mugu is a U.S. NAVAL AIR MISSILE TEST CENTER.

US 101 continues inland, following the old Camino Real into the higher inland valleys, cultivated, irrigated, hemmed by mountains, and cut by streams. From the summit of Conejo Grade (778 alt.) is a view of the patterned mountains and valley drained by the Santa Clara River.

At 157.5 *m.* is the junction with a paved road.

Right on this road to HIDDEN VALLEY, 5.5 *m.,* a mountain-hemmed district of grazing lands approached through walnut groves and a narrow pass.

SHERWOOD FOREST, 9 *m.,* takes its name from the scattered clusters of oaks on the northern shore of SHERWOOD LAKE. Large houses cling to the steep hillsides, including the home of the former movie czar, Will Hays. The signs, "Private Road," forbid circling the lake, but the view from the hillside reveals the body of water below. This was the location for parts of the movies *Robin Hood* and *Tarzan of the Apes.* East of the lake settlement is the dam that creates the lake and a slight divide from which the road turns L. into Potrero Rd. and to a junction with US 101 at 13 *m.*

The two-story white clapboarded house (R) with green jalousies, in NEWBURY PARK, 158.4 *m.* (700 alt., 135 pop.), was once a stage coach station.

THOUSAND OAKS, 161.9 *m.* (2,934 pop.), started as a collection of tourist accommodations around the GOEBEL LION FARM, which supplied many animals used by movie studios. This area has numerous associations with motion picture making. On the coast off State 1 is the LEO CARRILLO BEACH STATE PARK.

US 101 continues through huge oak trees, whose acorns supplied the Oak Grove Indians with their chief foodstuff—acorn meal. Since these trees took root two centuries ago, the water level has dropped to such an extent that no young growth can take hold.

At 164.9 *m.* is the junction with Potrero Road (*see above*) and at 168.9 *m.* is the junction with a paved road.

Right on this second road to PARAMOUNT MOTION PICTURE SETS, 1 *m.,* make-believe western cow town, Spanish haciendas, African native huts, and Colonial villages sprawling in false-front arrays in the valley below the road.

MALIBU LAKE, 3 *m.,* small, irregular, and deep in the crotch of steep and grotesquely-shaped mountains, is surrounded with homes and gardens. The MALIBU MOUNTAIN CLUB (R) is a favored resort of the cinema world.

East of the CALABASAS POST OFFICE, 174.9 *m.,* is the LOS ANGELES PET CEMETERY, approached by a private gravel road up a gentle slope of the Simi Hills. At this point is a view of the San Fernando Valley, a fertile farming area 25 miles long.

CALABASAS, 177.1 *m.* (928 alt., 150 pop.), is now a collection of tourist facilities but once was southern California's tough town. On the outskirts, in an area of orange and walnut groves, US 101 crosses the extraordinary city limits of Los Angeles.

The highway now moves into the desirable residential sections of the San Fernando Valley, the fastest growing part of Los Angeles. Just inside the city limits is the expanding community of WOODLAND HILLS (37,000 pop., 1963). This is at the intersection of State 27 which comes down from the north and proceeds to the coast at TOPANGA BEACH. About 6 *m.* north of Woodland Hills on State 27 is Chatsworth (33,100 pop., 1963) and nearby is Chatsworth Reservoir. Two *m.* north of Woodland Hills is CANOGA PARK (79,100 pop., 1963), site of Atomics International, pioneer in nuclear reactors. State 27 then moves outside the city limits to TOPANGA, in the Topanga Canyon area (*see page 416*) not yet embraced in the wide arms of Los Angeles.

TARZANA, 182.4 *m.* (800 alt., 20,000 pop. est. 1965), is an unincorporated community inside the city limits of Los Angeles bounded by Reseda, Encino, Woodland Hills and Mulholland Drive. Originally a poultry and berry farm called Runnymede III, it took the name of Tarzana in 1928 by permission of Edgar Rice Burroughs, author, whose Tarzana Ranch occupied part of the site. The ranch has been subdivided. East of Tarzana are the outskirts of Hollywood and other closely populated parts of the City of Los Angeles.

At the intersection of Radford Avenue, 191.1 *m.* can be seen the buildings used for many years by REPUBLIC PICTURES, principally for the production of western motion pictures.

At 193 *m.* is a junction with Lankershim Boulevard. Left on the Boulevard to UNIVERSAL CITY, an independent municipality of 410 acres, filled with numerous studio buildings and the tall MCA TOWER, headquarters for Universal motion picture, radio and television facilities. See description of studios and tours under HOLLYWOOD.

South of the junction US 101 follows Cahuenga Boulevard and descends through Cahuenga Pass to HOLLYWOOD, 195.5 *m.,* a district of Los Angeles.

Section d. LOS ANGELES *to* MEXICAN BORDER; *141.4 m.* US 101; Interstate 5

Leaving Los Angeles, this section of US 101 becomes identical with Interstate 5 and moves through oil fields, truck gardens, and citrus groves. It returns to the sea at Doheny Park and skirts a barren coastline edged by brush-clad hills, passing occasional resort towns, south to San Diego and the Mexican border.

US 101 follows N. Spring St., in the center of LOS ANGELES, 0 *m.,* then, after several turns follows Whittier Blvd. to MONTEBELLO, 8.9 *m.* (600 alt., 32,097 pop.), surrounded by an incongruous mixture of flower gardens and oil fields. It has numerous nurseries engaged in the wholesale ornamental shrub and flower business, and holds the annual State Flower and Horticultural Show. Thousands of barrels of oil are pumped daily from wells in the vicinity.

Right from Montebello on First Street, which becomes Bluff Road, to the SITE OF THE BATTLE OF SAN GABRIEL, 1 *m.,* where on January 8, 1847, the decisive engagement in the conquest of California by the Yankees was fought. Gen. Stephen W. Kearny and Commodore Robert F. Stockton, with a company of 600, engaged a band of several hundred Californians under Gen. José Flores. Typical of the skirmishes throughout the United States conquest of California, this battle lasted less than an hour, and the total number killed on each side was two, the number injured, eight. The occupation of Los Angeles followed the engagement.

At 9.3 *m.* on US 101 is the junction with Rosemead Boulevard.

Left on Rosemead Boulevard to San Gabriel Boulevard; L. on San Gabriel to the FIRST SITE OF MISSION SAN GABRIEL, 1.5 *m.,* established in 1771 by padres from Mission San Diego; moved five miles north in 1776.

US 101 runs eastward across the San Gabriel River, passing the PIO PICO MANSION, 11.9 *m.* (R), a State monument, built by Pío Pico, last Mexican Governor of California.

East of the San Gabriel River, 12 *m.* north of Long Beach on US 101 is DOWNEY (120 alt., 82,505 pop. 1960, 92,000 est. 1964), named for John Gately Downey, early governor. His adobe house, restored, is located on Norwalk Blvd. The city was incorporated in 1956. Here are the Space & Information Systems Div. of North American Aviation, employing 19,000, and a plant of Aerojet-General Corp.

Adjoining Downey is SOUTH GATE, industrial center (58,000 pop., est. 1964), 9 *m.* from Los Angeles on Long Beach Freeway and State 10. Bounded by South Gate and Compton on State 7 is LYNWOOD (35,000 pop., est. 1965).

LOS NIETOS, 2.5 *m.* (159 alt., 1,240 pop.), sprawls in the midst of orchards, fields and truck gardens. Three-fifths of the residents are of Spanish or Mexican descent.

A thick forest of derricks identifies SANTA FE SPRINGS, 3.6 *m.* (16,342 pop.) one of the most productive oil fields in California. The Santa Fe Railroad purchased the newly platted town of Fulton Sulphur Springs in 1886 and renamed it. The Los Angeles Pioneer Oil Company abandoned its lease when experimental wells here failed to produce oil.

The NORWALK STATE HOSPITAL (L), 4.6 *m.*, was located on this site in 1916 after several other sites had been rejected. One of them was Signal Hill, which, shortly afterward, became one of the State's richest oil developments.

From Downey US 101 (Santa Ana Freeway) continues to NORWALK, 17 *m.* (97 alt., 86,739 pop., 1960; 93,506 est. 1964). This "skyrocketing city," carved from Rancho Los Coyotes in 1870 and named for a Connecticut town, was incorporated 1957. One-third of its residents are in industry and 83% of its houses are owner-occupied. CERRITOS COLLEGE, est. 1956, enrolled 8,746 in 1965. The Library has 64,000 volumes.

BUENA PARK (46,401 pop., 1960) is south of Norwalk. About 2 *m.* south on State 39 is KNOTT'S BERRY FARM, with a Ghost Town, Mining Camp, Trails Museum, etc. *Open daily, 11 a.m.-9 p.m.* Opposite is the ALLIGATOR FARM, with live reptiles. MOVIELAND WAX MUSEUM, opened 1962, exhibits lifesized replicas of movie stars in original costumes.

WHITTIER (250 alt., 33,663 pop., 1960; 68,383 after annexation of East Whittier, 1963) at the southwest slope of the Puente Hills was named for the poet John G. Whittier when settled by Iowa Quakers, 1887. WHITTIER COLLEGE, liberal arts, had 2,235 students in 1965. Whittier was long the home of Richard E. Nixon. Near US 101 is WHITTIER STATE SCHOOL, where boys are given vocational training.

Avocados and oranges are the principal fruits raised in the Puente Hills. LA HABRA, 17.9 *m.* (325 alt., 25,136 pop.) is near an oil field, which delivers oil by pipeline to San Pedro. LA PUENTA, 6 *m.* n. on State 39, another new city, had 24,723 pop. in 1960.

At 21.2 *m.* is a junction with Imperial Highway. Left 2 *m.* to BREA (363 alt., 8,487 pop.), oil industry center.

FULLERTON, 25.9 *m.* (161 alt.) Orange county oil-producing and industrial center, increased from 13,948 pop. in 1950 to 56,180 in 1960 and an est. 73,672 in 1964. Hughes Aircraft Co., employs 4,800; Beckman Instruments, 3,000; Hunts Foods, 1,440. CALIFORNIA STATE COLLEGE AT FULLERTON is one of the newest of the State's eighteen; opened 1959, it enrolls more than 5,000. Modern buildings stand on a campus of 235 acres. The six-story LETTERS AND SCIENCE BUILDING cost $6,000,000.

ANAHEIM, 27.9 *m.* (165 alt.), founded by Germans in 1857 as a communal settlement, was named Ana from the Santa Ana River and *Heim,* German for home. In 1950 it had 14,556 people and was a center of orange production. In July, 1955, the Walt Disney corporation built DISNEYLAND, an elaborate amusement center costing $17,000,000. In 1960 the U. S. Census reported 104,184 pop.; in 1965 the promoters claimed 150,000. Easily accessible by US 101 and Harbor Blvd., Anaheim acquired numerous motels and convention attractions. It has Disneyland Hotel (450 r.), Charter House Hotel, Jolly

Roger Inn, Fantasy Motel, Anaheim Motor Lodge and others, some quite spacious.

Anaheim's new STADIUM, cost $24,000,000, home of the Angels baseball team, was opened April 9, 1966, with 40,735 present. In the opening game the Angels lost to the San Francisco Giants. Sid Ziff wrote in the *Los Angeles Times:* "It's a compact Dodger stadium. Everything is closer to the playing field."

Disneyland appeals to nostalgia for the old West and the 1890's, and to make-believe. It is divided into sections: Tomorrowland (trip to the moon); Frontierland (stage-coach, steamboat days); Fantasyland (castles, fairytale characters); Adventureland (tropical jungles); Good Old Nineties (Main Street, Town Square, street cars, railroad). A full-size steamboat, the *Mark Twain,* and a square-rigger, *Columbia,* move through lagoons. The parking place has space for 10,000 cars.

Admission, $1.75 for adults; $1.20 for teenagers up to 17; 60¢ for children 11 and under. Buses from Los Angeles; helicopter from Los Angeles International Airport.

US 101 meets Chapman Ave., 30.9 *m.* Left on Chapman to ORANGE, 1.8 *m.* (176 alt., 26,444 pop. 1960; 67,170, est. 1965). Originally a distributing point for oranges, it has now experienced an "explosive industrial boom" and set aside 1,500 acres for industries. It has a new Library of 60,000 volumes and a Civic Center. In 1868 A. B. Chapman and Andrew Glassell received 40 acres in payment of attorney's fees. They called the town Richland, and in 1875 changed it to Orange. A legend says Chapman played cards with three associates to determine whether it should be Orange, Lemon, Olive or Almond and Orange won. CHAPMAN COLLEGE, founded 1861, moved to Orange from Los Angeles, 1954. It has a symphony orchestra, three choral groups and PURCELL ART GALLERY.

The orange, introduced into Spain from China, was brought into the State from Lower California by the Franciscan padres. The first orchard of any size was planted at San Gabriel Mission near Los Angeles about 1805. In 1873 two small navel orange trees were planted at Riverside (*see TOUR 13b*). In 1876, when the first Valencia seedlings arrived from London, the citrus industry was already established.

Two varieties of orange—the navel and the Valencia—which ensure fruit ripening throughout the year, are grown to the virtual exclusion of the few other existing varieties. The Valencia production of which is favored over the navel, is a summer-ripening variety grown in the cooler coastal regions; it hangs in full color on the tree in spring along with the new blossoms. The navel is a winter-ripening fruit grown in the warmer inland areas. It is seedless, more highly colored than the Valencia, and is distinguished from all other varieties by the navel formation of the skin at the blossom end, which gives the fruit its name.

Citrus growing in California is probably the most intensively developed crop-culture in the world. Every tree in the important groves of the State is of pedigreed stock. Careful records of the performance of each tree are kept, and only those known to produce high quality fruit are used.

Seed for the root stock is selected from some hardy, strong-rooted, disease-resisting variety—usually the sour-orange—and planted in lath-house beds. (Sweet orange rootstock has been found best for lemons.) The rootstock seedling tree is removed from the seed-bed after a year or more of growth, when it is about twelve inches high, and replanted in an outdoor nursery, where it grows a year or two longer before being budded. In this operation, a twig bearing a healthy bud is cut from a heavy-yielding parent tree that is an inherent bearer of a stable type and inserted in a slit made in the bark of the young seedling about four inches above the ground. When the grafted bud has grown into a branch, the seedling's own top is cut off, and the grafted branch is trained to form another top for the tree, which is ready for transplanting in the orchard the following year. It grows another three years before it is ready to bear commercially, making a minimum of six years from the sprouting of the sour-orange seed. Meanwhile control of the tree's shape is continued by pruning. Bearing commercially in its sixth year, an orange tree does not reach full bearing age until its tenth year. It continues to increase in size and yield for fifty or more years if well cared for.

The war against citrus pests is costly and difficult, since the use of tree medicines varies according to the soil, tree, and climatic types and conditions. Citrus red spider, most destructive and hardest to control, and the various types of scale insects are the commonest pests. About once a year the orchard trees are covered with a canvas tarpaulin and fumigated to kill scale. Insecticides of various types are sprayed upon the trees to eliminate other pests.

On every cold night in winter the grower must be ready with his orchard heaters. The heaters long in general use are oil-burning stack pots, which are placed between the tree rows, one to a tree. With the broadcast of a frost warning the watchman in charge of an orchard stays up all night, keeping crews ready to light the heaters with gasoline torches resembling an engineer's long-spouted oil-can. The burners must be watched and regulated at intervals. Where the smudge-pot heating method is used, a thick blanket of black smoke produced by the fuel protects the trees from frost. Threat of frost is greatest about an hour before sunrise. During a cold period everything within miles—clothing, furniture, faces—is covered with the greasy soot.

During southern California's rainless summers the orchards must receive a 48-hour irrigation at three to five week intervals, depending upon the type of soil and the climate. Orchard lysimeters are used for frequent tests of the soil to measure the percolation of water through it, for the crop is as easily ruined by too much water as by too little. Water is usually carried to the edge of the orchards by an underground iron or concrete conduit, and brought above ground from this by a concrete stand pipe that empties directly into the rows of furrows passing between the trees.

Fertilization of the orchard is another problem thoroughly studied by citrus experts. The growing of a "cover crop," such as vetch or clover, which is plowed under in the spring, adds nitrogen to the soil. Stable manure and various other organic fertilizers are used. Inorganic fertilizers are necessary in some soils to add minerals or to counteract soil acidity. As in the base of insecticides, unless orchard conditions are correctly diagnosed and the proper materials, quantities, and methods used, harm to tree or fruit or both results.

Finally, the trees in windy areas must be protected by the planting of orchard-bordering rows of tall-growing evergreens. The eucalyptus is commonly used because it grows rapidly, resists disease, and is adapted to the climate, but it is a heavy surface feeder, retarding the growth of nearby orchard trees; the Monterey cypress grows less rapidly, but is also less injurious to adjacent orchard growth; the athel is even more injurious to the grove than the eucalyptus, but because of its excessively rapid growth it is used as a preliminary windbreak while slower-growing trees are maturing.

As picking time approaches a few typical fruits from the trees are tested and analyzed as to sugar content, and the time of picking determined thereby. Nearly every California citrus orchardist belongs to a growers' cooperative association (most of them to the giant California Fruit Growers' Exchange), and the fruit of these orchard owners is harvested by specially-trained picking-crews sent out by the association's nearest packing house. As in the growing of the orange, the utmost care is used in its handling from the moment it is cut from the tree until it reaches its market. The skin of the orange, although it is a perfect seal for the fruit, is very easily harmed by scratching or bruising, and blue mold may set in before the market is reached, or even before the fruit can be packed. Therefore, the orange "picker"—who does no actual picking but nips the stem with a pair of clippers close enough to the fruit that no sharp protrusion is left to scratch against other oranges—wears soft gloves and has a sack specially designed with a buttoned flap that releases the fruit from the bottom into the field box. (In gathering lemons, the picker also carries a ring and cuts only the fruit which is too large to pass through it; lemons are thus picked by size every month while still a dark green color.)

After being picked, oranges are allowed to stand a day or two at the packing house. During this time some of the moisture in the rind evaporates, making the skin less liable to injury in handling. The fruit is soaked in deep trays of warm water, then passed through long rows of revolving brushes that wash away traces of dirt, then rinsed in cold water, and finally passed over rollers or more brushes under a blast of air that dries them for the grading table. As they roll on canvas belts past the trained operators (usually women), they are sorted into grades according to certain standards of quality and appearance. Many packing houses now employ a recently invented citrus fluoroscope, with which frost damage, granulation and other internal imperfections are detected by X-ray. The size or the condition of the skin of an orange does not determine the inner quality of the fruit. Often, in the season when Valencias are on the market, oranges with a greenish tinge on the rind are mistaken for immature fruit. Actually, fruit in this condition is in its maximum stage of maturity.

After citrus fruits have been graded, stamped in vegetable dye with the name of the grade, and separated according to size and grade, they are wrapped in tissue paper and packed in boxes in a symmetrical pattern that allows a certain number of oranges of a certain size to be packed in each box. There are ten principal sizes of oranges: 100's, packed 100 to a box, are the largest; 344's, packed 344 to a box, are the smallest. In the bigger packing houses, the crates are thoroughly cooled before shipment in cars that are iced in summer and sometimes heated in winter. In recent years the large growers' cooperative associations have begun to use cull—below merchantable grade—fruit in various by-products, such as canned juice, juice concentrate, orange oils and acids, citrate of lime, lemon oil, and citrus pectin. Research laboratories conducted by the associations are studying methods of putting the citrus fruits to other uses.

The non-profit growers' cooperatives have contributed considerably to the reduction of the high cost of orange growing in California by facilitating picking, hauling, packing, and marketing and standardizing costs all along the line. The fruit of the grower-members is sold in a pool and the proceeds are divided among the orchardists on the basis of the quality, quantity, and grades of fruit each has put into the pool.

EL MODENA, 4.5 *m.* (250 alt., 510 pop.), is a small trading center on the outskirts of Orange.

East of El Modena, Chapman Avenue becomes County Park Road, which with its various continuations rises into the cactus-covered foothills of the Santa Ana Mountains. From this road other roads follow canyons into recreational areas.

At 7.8 *m.* is a junction with Santiago Canyon Road; L. (straight ahead) 1 *m.* to IRVINE PARK, a recreational area with oak-shaded picnicking and camping facilities. SANTIAGO RESERVOIR AND DAM, 3 *m.* (700 alt.), 160 feet high and 1,400 feet long, is a private reserve for irrigation purposes, Orange County's largest water conservation structure.

The main side route continues southeastward on Santiago Canyon Road to a junction with Silverado Canyon Road, 14.3 *m.* Left (straight ahead) to ROME SHADY BROOK, 3.5 *m.* (*accommodations, and small swimming pool*). Numerous summer homes line the canyon and hiking and horseback trails run through the surrounding CLEVELAND NATIONAL FOREST, a reserve of 390,000 acres formed in 1910 to insure watershed protection; millions of gallons of water are piped from its springs and rivers to supply the surrounding country.

Santiago Canyon Road, the main side road, continues southeastward to a junction with Modjeska Grade Rode, 18.9 *m.*; L. here 0.7 *m.* to the FOREST OF ARDEN, once the home of Mme. Helene Modjeska, the Polish actress. The large white house across the creek is well hidden among high trees and other greenery. In 1876 Modjeska came here with a group of refugee Polish artists. The colony failed. Modjeska learned English in a few months, turned to the American stage and began a new career that carried her to her greatest triumphs. Named for a scene in a favorite play, her home became a Mecca for well-known artists and actors. Shortly before her death in 1909, she moved to Modjeska Island.

EL TORO, 27.8 *m.* (144 alt.), is the home of AIRCRAFT, FLEET MARINE FORCE, PACIFIC, headquarters for all Fleet Marine Aviation Units on the West Coast and the only staging and training base for Marine Aviation personnel on the Coast. The Marine Corps Air Facility is a helicopter unit. Personnel is 12,000 uniformed, 1,100 civilian. With a plant investment of $74,000,000 the base has an annual payroll of about $74,000,000 and expends $14,000,000.

SANTA ANA, 34.9 *m.* (135 alt., 100,350 pop. 1960; 130,272, est. 1964), founded 1869, is the seat of Orange County, ten miles from Newport Beach, in the valley discovered on St. Anne's Day. Once a citrus center it now has more than 200 diversified industries, including electronics, pharmaceuticals, metal products and beet sugar refining. Its Civic Center, Eighth and Ross Sts., recently added the new PUBLIC LIBRARY, modern in style. CHARLES W. BOWERS MEMORIAL MUSEUM contains relics of Spanish and Indian days in California; also Madame Helena Modjeska's memorabilia. SANTA ANA COLLEGE (Junior) had 4,625 students in 1964. The MUNICIPAL BOWL, opened 1963, seats 10,000.

The old Courthouse was the favorite objective of eloping couples in the period before California demanded a three-day interval between marriage license and ceremony. Members of the motion picture colony took advantage of Santa Ana's hospitable parsons, who performed marriages at all hours.

Large pepper trees line the main street of TUSTIN, 37.9 *m.* (122 alt., 2,006 pop.), which rivaled Santa Ana before the extension of the railroad to the latter city.

South of Tustin is RED HILL (325 alt.), known to the Spanish pioneers as Cerrito de las Ranas (hill of the frogs). It was a landmark to Indians, missionaries, and Spanish and Mexican rancheros; and served as a direction finder for early map makers. The twin peaks of MODJESKA (5,481 alt.), and SANTIAGO (5,680 alt.), on the eastern horizon, are the highest in the Santa Ana Mountains.

At 48.9 *m.* is the junction with the paved El Toro Rd. (*see above*).

SAN JUAN CAPISTRANO, 57.9 *m.* (103 alt., 1,200 pop.), grew up around Mission San Juan Capistrano, and is populated with the descendants of early Mexican settlers. The village declared war on Mexico, during the early mission days, because of harsh treatment inflicted upon the Indians by Mexican officials.

The much-pictured ruins of MISSION SAN JUAN CAPISTRANO (*adm. 25¢*) are dear to the hearts of California mission romanticists. The church is in much the same condition to which the 1812 and 1918 earthquakes reduced it; few of the subsidiary structures have been restored. The mission was formally dedicated on Nov. 1, 1776, by Father Junípero Serra, and named for St. John of Capistrano, the Crusader. Construction was begun in 1797 and completed in 1806. The church was built in the form of a cross, 180 feet long and 90 feet wide; in its day it was one of the most beautiful of the California missions, with an arched roof, seven domes, and a tall campanario (belfry) that could be seen for 10 miles. An official record says that a Mexican sculptor was sent by the Franciscans to carve the stone arches, cornices, and doorways.

The church was occupied only six years and three months. In 1812 an earthquake wrecked the roof, the cloisters, the nave, five of the domes, and leveled the tall campanario, killing 29 persons. Enforcement of the Secularization Act severed the Indians from the mission settlements. In 1865 a rebuilding of the walls with adobe bricks was followed by a heavy rainstorm that reduced the building to mud.

The mission is widely known for the "Capistrano swallows" that have built their homes in the ruins from Spanish Colonial times. It is said that for nearly a century the swallows have left the mission on St. John's Day, October 23, and returned again in the spring on St. Joseph's Day, March 19. Only once have they been late, delayed by storm at sea, but they arrived only four hours behind their schedule. In the fall of 1936 elaborate preparations were made by the National Broadcasting Company to give listeners-in the fluttering sounds of their departure, but the birds muffed the chance, disappearing before necessary connections could be made.

Melodramatic tales of the supernatural woven about Capistrano are too well patterned to do the relic justice. These generally involve a dark-robed and rope-girdled Franciscan ghost that reputedly walks, then melts into the garden twilight, and of bells that once rang in the mission campanario despite the motionless bell ropes, coiled as usual on their pegs. The climax—recited *con agitato*—came with the discovery that a young Indian girl, a neophyte, had expired at the very moment of the reverberations of the self-tolling bells.

Within the entrance to the mission enclosure is a small garden beside the ruins of the church. The curious diamond-shaped tiles on the floor of the sanctuary were made on the hillside north of the mission. Opening on the garden is a museum with old Spanish vestments, sheepskin and parchment bound manuscripts, Mexican and Spanish paintings and statues, an ancient confessional, Indian frescoes, and a golden altar.

Left from San Juan Capistrano on State 74 to SAN JUAN HOT SPRINGS, 12.4 *m.,* a spot of green in a narrow canyon formed by precipitous chaparral-covered mountains. Steaming sulphur waters pour into various-sized baths and concrete pools belonging to a resort.

The thick-walled adobe CAPISTRANO MISSION TRADING POST, 60.4 *m.,* overlooking the ocean was built in 1820. It was headquarters for an extensive trade in hides and other commodities between the mission and the Yankee clippers.

At DOHENY PARK, 60.9 *m.* (15 alt., 549 pop.), is the junction with US 101 Alt. (*see TOUR 2B*). Doheny Palisades (L), high above the surf-level road, was developed as a restricted residential area by E. L. Doheny, the oil magnate. The precipitous sea cliffs are attractively landscaped, terraced with ice plant and bougainvillea.

SAN CLEMENTE (St. Clement), 64.4 *m.* (5 alt., 8,527 pop.), sometimes referred to as the "Spanish Village," was founded in 1925 by a Los Angeles realtor. The unvarying white stucco of all buildings and houses in this model village is surmounted by equally unvarying red tile roofs. In 1955 residential lots were priced at $1,000 to $1,500. After the Freeway of Interstate 5 was built the lots sold for $4,000 to $6,000. Building permits rose from $3,716,000 in 1960 to $13,920,000 in 1963.

A pueblo-style gateway leads into SAN CLEMENTE STATE PARK, 66.3 *m.* (*camping, picnicking facilities*) on the cliffs overlooking the Pacific, and a broad beach.

A tiny hexagonal toolhouse (L) marks LAS FLORES, 78.4 *m.,* a station of the Atchison, Topeka & Santa Fe Railroad serving the Santa Margarita Rancho. The stockyards and chute for loading cattle into railroad cars have almost become a landmark of the past, for trucks are now the favored method of transporting cattle.

At 86.4 *m.* is the junction with a private dirt road.

Left here through a wire gate to the headquarters of the RANCHO SANTA MARGARITA, 9.5 *m.,* in an adobe hacienda erected by Pío Pico in 1837, now a ranch office and residence. The large one-story house stands near the Santa Margarita River. The present owner, a corporation, engages in diversified farming and ranching. Controlled by Mission San Luís Rey, the rancho was returned to the Indians in 1835. In the same year, through political manipulations, the influential Picos acquired the property. Here they lived until 1862, when Andrés sold his share, 66,500 acres, to his brother Pío for $1,000 plus a San Diego residence. Pío—according to the story—shortly afterward acquired the adjoining 44,000-acre Las Flores Rancho from its Indian owners. Pío Pico was an inveterate gambler. It is said that he rode to the races with a mule loaded with silver coin for betting. His gambling led him to sell the rancho to John Forster for $14,000. Forster called himself Don Juan Forster, married Pico's sister Ysidora, and lived here 30 years, allowing only Spanish to be spoken.

OCEANSIDE, 87.4 *m.* (45 alt., 28,800 pop., 1962), is a distributing center for fruit and farm products, with a fine beach augmented by harbor facilities for small craft, including 600 berths for boats. It is at the terminus of State Highways 76 and 78. OCEANSIDE-CARLSBAD COLLEGE (Junior) on State Route 196 had 1,578 students in 1964.

The Miss Southern California contest is held on Oceanside Beach annually in June. North of Oceanside is CAMP PENDLETON, training base of First Division, U. S. Marine Corps.

A 90-acre corner of Camp Pendleton has been set aside for the nation's largest nuclear generating plant of SAN ONOFRE, a pressurized water reactor, planned to produce 395,000 kilowatts for Southern California Edison Co. and San Diego Gas & Electric Co. The new plant will use 72 tons of uranium dioxide producing heat equivalent to 18 million bbl. of fuel oil and lasting 3½ years. About 10,000 visitors have been registered annually at the Nuclear Information Center. (*free*)

The white campanario of MISSION SAN LUIS REY, 4.7 *m.*, is a pleasing part of the landscape of the hilly valley and fixes the position of the church in the cloistered establishment. The mission was founded in 1798, completed in 1802, and dedicated to San Luís, Rey de Francia. It was designed to serve a tribe of the Shoshone later known as the Luiseno. Fra Antonio Peyri, one of the founders, led the mission until secularization in 1829, after which it declined. American troops occupied it during the Mexican War.

Architecturally, San Luís Rey is regarded as one of the most impressive of all the California missions. Its style was a composite of Spanish-Moorish and Mexican, built of adobe faced with brick. The brick-red delineaments of the facade and belfry and the cloisters of the long monastery running west of the church form a well proportioned whole. Farther west are remains of cloisters—rows of plaster-chipped arches high above the spread of the valley. Remains of the extensive adobe wall that once surrounded the mission are near the highway.

The church has been greatly restored. The murals on the interior columns and beams are not vivid in color, but rich with small detail and paganistic patterning. The emblem of the Third Order of St. Francis, a cross and the stigmata (five wounds of Christ), is over the arched entrance to the mortuary chapel. Restoration began in 1892. It is now a Franciscan seminary.

A few miles up the San Luis River is SAN ANTONIO DE PALA, an *asistencia* of the mission, built 1816 by Fra Peyri, abandoned 1846, restored since 1959. The buildings are adobe surmounted by red tiles. The bell-tower holds two bells, one above the other.

CARLSBAD, 90.4 *m.* (42 alt., 9,253 pop.), in a winter vegetable growing district, ships green vegetables, flowers, and bulbs. CARLSBAD BEACH STATE PARK, immediately south of town, possesses a bathing beach, picnic grounds, and campsites.

Discovery of a well of mineral water led early settlers to name the town after Karlsbad, Bohemia. The ARMY AND NAVY ACADEMY prepares cadets for college. Carlsbad is on the Freeway based on US 101, which permits travel through Los Angeles to Ventura without a stoplight. Buena Vista Lagoon is the site of the Maxton Brown Bird Sanctuary, with 125 varieties of waterfowl.

ENCINITAS, 99.4 *m.* (85 alt., 2,786 pop.), settled 1854 by Germans, is in a flower producing area. The Midwinter Flower Show is held in March.

SOLANA BEACH, 103.4 *m.* (65 alt.), is in an area that grows flowers and citrus fruits.

DEL MAR, 105.9 *m.* (100 alt., 3,124 pop.), is the scene in August of the annual San Diego County Fair.

Left from Del Mar to the SAN DIEGO COUNTY FAIRGROUNDS AND RACE TRACK. Horse racing is conducted by the Del Mar Turf Club, founded 1937 by Bing Crosby and Pat O'Brien and operated for Boys, Inc., estab- 1954 by Clint Murchison and the late Sid Richardson. Known for Futurity and Debutante stakes for 2-year-olds. Season is 42 days from end of July to mid-September; average daily attendance 12,000. In 1964 473,000 attended and wagered $41,383,677.

OSUNA ADOBE, left, 4.8 *m.,* house of the first alcade of San Diego under Mexico, who fought in the Battle of San Pasqual, was long the country home of Bing Crosby. RANCHO SANTA FE occupies most of the old Osuna Rancho. It is a landscaped residential area with an attractive civic center, luxury inn and golf course. Douglas Fairbanks once owned a large citrus ranch here called Rancho Zorro.

US 395 crosses LAKE HODGES east of HODGES DAM, 11.7 *m.,* part of the San Diego water system.

ESCONDIDO, 18 *m.* is at the junction with US 395. (*See TOUR 6c, page 526.*)

On the coast, west of US 101 is TORREY PINES PARK, with the spectacular Torrey Pines of irregular shape with needles in clusters of five, 5 to 7 inches long. At 109.3 turn left to CAMP MATTHEWS of the U. S. Marine Corps and its Rifle Range.

Right from US 101 to the coastline at LA JOLLA and SCRIPPS INSTITUTION OF OCEANOGRAPHY, founded 1892 by E. W. Scripps, publisher, and his half-sister, Ellen Browning Scripps, a division of the University of California since 1912. Notable for its physical, chemical, geological and geophysical studies of oceans and its biological research. Dr. Roger Revelle is director. Museum and aquarium, open 8 to 6. Studies of the earth's crust under water have produced much valuable data. Action of the sea on the human body has concerned Scripps in recent years. Six research vessels of which the Argo, 213 ft. long and 2,079 tons, is the largest, plumb the ocean's depths. After studying conflicting data Scripps established that the Mariana Trench, southeast of Guam, is the deepest at 35,810 ft. and the Tonga Trench, east of the Tongo Islands, is second at 35,435 ft. Scripps operates a Floating Instrument Platform (FLIP), a laboratory with room for twelve, which can assume a vertical position of 300 ft. in the water with 85% under water.

Sealab II, underwater laboratory of the U. S. Navy, with Scripps scientists aboard, in 1965 made three visits to the ocean floor off La Jolla. The 12 by 57 ft. steel cylinder, carrying 10 men, remained 200 ft. below 15, 30 and 45 days, getting electricity by cable from the shore. Sealab II was made by San Francisco Naval Shipyard.

The beach below the institution and the rocky stretch to the north harbor most of the marine life characteristic of this section. Perhaps the most common fish seen along shore is the wriggling, fast-swimming tidepool fish or sculpin, which darts about in rocky pools. The sting ray or stingaree, a flat fish with a long, whiplike tail, is occasionally encountered in quiet water at very low tide, particularly on muddy bottoms. Its sharp, barbed spine can inflict a painful wound.

The beach south of the institution is a favorite spawning ground of grunion, a small, slender smelt that comes up on the sandy shore to lay its eggs a few minutes after the high tides of March, April, May, and June. On the second, third, and fourth nights after the full of the moon during these months, the female grunion rides in on the advancing wave, and then squirms and flops back into the wash of the next wave. In the brief time she is out of the water—about 30 seconds—the grunion digs tail first into the sand for about half the depth of her body and deposits her eggs. Two weeks later, at the time of the next high tide, the waves loosen the ripened eggs from the sand. As the eggs are freed the baby grunion hatch and are washed back into the sea.

The spawning habits of this striped smelt provide a popular southern California sport. About a quarter of an hour after high tide on the nights the grunion run, the long sandy beaches are crowded with people gathering the fish by moonlight, bonfires, and flashlights. Some nights, during the height of the season, thousands of grunion are visible at one time. No license is required, but the fish must be taken by hand in the brief seconds it is out of the water.

LA JOLLA, 3.9 *m.*, a suburb of San Diego, occupies a small rocky promontory popular with easterners who have built lovely homes and gardens on top of its sheer cliffs overlooking the sea. The name is variously explained as a corruption of *la joya* (the jewel), and *la hoya* (the hollow). The seven LA JOLLA CAVES are accessible from Coast Blvd. In the contrast of light and darkness from within the caves the openings to the sea show strange forms, sometimes human in aspect.

La Jolla's most popular spot is the cove, with a small beach beneath the cliffs and a miniature marine garden off the rocks.

Prominent in the town's development was Ellen Browning Scripps, who endowed the Scripps Institution of Oceanography and the La Jolla Museum of Art.

PACIFIC BEACH, 7.4 *m.* (65 alt., 1,626 pop.), is a resort suburb. MISSION BEACH, 8.9 *m.* (25 alt., 1,092 pop.), a narrow strand between Mission Bay and the ocean is popular for paddle-boarding. OCEAN BEACH, 10.4 *m.* (20 alt., 4,012 pop.), is a residential and resort area that has an indoor salt water plunge. A State fish and game refuge is on the northern edge of town, along Mission Bay.

The route winds back to a junction with US 101 (Pacific Highway) at 13.5 *m.*

US 101 continues south to SAN DIEGO, 123.8 *m.* and connects with US 395 (*see TOUR 6c*) and US 80 (*see TOUR 14b*).

Right (straight ahead) from San Diego on Pacific Blvd. to the CORONADO FERRY (*fee for auto*), 0.1 *m.*, and across San Diego Bay (0.7 *m.*) to CORONADO, 1.5 *m.* (25 alt., 18,039 pop.), a residential and resort community named for the Coronado Islands. The town's most notable feature is the HOTEL DEL CORONADO, a well-maintained but rambling-towered and ornate building designed by Stanford White. It is noted for the beauty of its patio with the immense flame-colored bougainvillea vine. A toll bridge from San Diego to Coronado has been authorized by the State.

From the beach part of the CORONADO ISLANDS are seen, some 20 miles to the south in Mexican territory. This group of three is the southernmost of the California Channel Islands, all rough, barren, and without fresh water. They had no permanent Indian population and are at present uninhabited, though sportsmen frequently visit them for the excellent big-game fishing. Discovered by the Cabrillo expedition in 1542, the islands are named for the Spanish explorer Coronado.

South of Coronado on State 75 is "Tent City," conspicuously lacking in tents, but having an abundance of straight-row frame shacks; and the SILVER STRAND, a narrow bar of sand and sand growth, some 8 miles long and in places less than half a city block wide over which the highway leads to Mexico.

SILVER STRAND STATE PARK, 5.8 *m.*, has campgrounds and picnic tables. At the south end of the strand is the city of IMPERIAL BEACH, 9.3 *m.* (19,186 pop. 1965, est.).

East of San Diego Bay the Harbor Drive follows the shore. Farther back the Montgomery Freeway combines the routes of US 101 and State 5, serving National City and Chula Vista and joined at Palm City by State 75, which has come down the west side of San Diego Bay.

NATIONAL CITY (32,771 pop. 1960) on San Diego Bay adjoins San Diego on the south and is served by the Santa Fe, San Diego & Arizona Eastern, and Montgomery Freeway. It is meatpacker for Imperial Valley livestock. The U. S. Naval Station with 2,000 employees, has maintenance of the Pacific Reserve Fleet.

CHULA VISTA (Sp., beautiful view), (74 alt., 50,669 pop. 1964), on the Bay, in the center of flourishing truck gardening, is largely residential. Rohr Corp. is the principal aircraft and missile engine producer employing nearly 5,000. Fiesta de la Luna, a flower festival is held annually in July.

At 137.5 *m.* is the junction with State 75 (*see above*).

The southern boundary of SAN YSIDRO, 140.9 *m.* (70 alt., 2,000 pop.), one of the principal ports of entry from Mexico, is the INTERNATIONAL BOUNDARY, 141.4 *m.*, between the United States and Mexico. Over the line is Tijuana, with its popular race track and Mexican shops.

Tour 2A

Junction with US 101—Lakeport—Middletown—Calistoga—St. Helena—Napa—Vallejo—Junction with US 40; 124.9 *m.* State 20 and State 29.

Roadbed partly oil surfaced, partly paved; easy grades
Accommodations adequate.

Into the wooded glens of hermit-like Lake County—never penetrated by a railroad—lead State 20 and 29. From the rocky hillsides gush mineral springs, and in the sheltered, oak-dappled valleys, clearings are green with bean vines on poles, pear and walnut orchards. The road hugs the shore of hill-encompassed Clear Lake, largest sheet of fresh water lying wholly within the State. From narrow defiles it climbs over the slopes of soaring Mount St. Helena. The vineyard-mantled slopes of widening Napa Valley border it as it continues to the mouth of the Napa River on San Pablo Bay.

State 20 branches eastward from US 101 (*see TOUR 2a*), 0 *m.*, at a point 2 miles north of Ukiah, along the course of upper tributaries of the Russian River through rising foothills. It skirts twin BLUE LAKES, 14.5 *m.*, (*summer resorts*) whose deep blue waters mirror steep-walled wooded heights. At the trading and resort village of UPPER LAKE, 22.9 *m.* (1,350 alt., 400 pop.), the route turns R. from State 20 on State 29.

The placid reaches of CLEAR LAKE are embraced by undulating, oak-forested hill slopes. The Pomo Indians who dwelt on the islands, ferrying across on tule *balsas* (rafts), had a legend about the region that told how the mighty chief Konochti was so angry when his enemy, the chief Kahbel, asked for his daughter Lupiyomi in marriage that he took up his stand at the Narrows—where a long tongue of land almost cuts the lake in two—to do battle. Across the water the two chiefs hurled at each other the huge boulders that strew the mountainside until Kahbel lay dead and Konochti, dying, sank back to form great MOUNT KONOCHTI (4,200 alt.), which towers between the two arms of the lake east of the Narrows. The Pomos dwelt here undisturbed, chipping their spearheads out of black obsidian and weaving the plumes of the blue quail into their baskets, until white men came, trappers and cattle herders, in the 1840's. Before the end of the decade they had run afoul of the white man's justice when United States soldiers, to punish them for slaying two of their fellow countrymen, surrounded them on an island and shot them down. They forgave the white man, however, and signed a Treaty of Peace and Friendship with him in August 1851—a treaty giving up their lands, for which they received a gift of "Ten head of beef cattle, three sacks of bread and sundry clothing."

The resort and farming center of LAKEPORT, 31.1 *m.* (1,350 alt., 2,303 pop.), seat of Lake County, once known as Forbestown for William Forbes, who deeded 40 acres of his land for a county seat in 1861, looks over the great expanse of the upper lake toward the hazy hills of the far shore.

Hidden from the lake by Mount Konochti's orchard-mantled slopes is KELSEYVILLE, 39 *m.* (1,500 alt., 200 pop.), called Peartown by its boosters, where Salvador and Antonio Vallejo built a log cabin for the mayordomo and the ten vaqueros who herded the cattle of their Rancho Laguna de Lu-Pi-Yo-Mi over the ranges. In 1847 they drove their cattle away and sold their land to Benjamin Kelsey, a man named Stone, and others. When Kelsey and Stone had an adobe built here, they earned the enmity of their Indian laborers by their unkindness, and in revenge the Indians killed them both in 1849.

At 44.1 *m.* is the junction with a paved road.

Left on this road to the farming center of LOWER LAKE, 9.3 *m.* (1,350 alt., 870 pop.), dating from 1858, and L. from Lower Lake on State 53, 0.8 *m.*, to CACHE CREEK, near the outlet to the lake, where a dam built in 1866 by the Clear Lake Water Company raised the lake level until it flooded houses, farm lands and orchards. When repeated appeals for redress went unheeded, the farmers on Sunday morning, November 15, 1868, after a prayer service led by their pastor, began to destroy the dam—a job that required two days and two nights. It was never rebuilt.

State 29 continues through narrow, twisting valleys past hot springs resorts to MIDDLETOWN, 64.2 *m.* (1,300 alt., 450 pop.), serving the farming, dairying, and poultry-raising region of the Loconomi Valley. It was named Middletown because it lies exactly midway between Lower Lake and Calistoga (*see below*).

Along St. Helena Creek State 29 winds up a long grade at the base of MOUNT ST. HELENA (4,343 alt.), which Robert Louis Stevenson called "the Mont Blanc of the Coast Range." At the summit of the grade (2,960 alt.), 73.2 *m.,* is the ROBERT LOUIS STEVENSON MONUMENT (R), a pillar on a rock base holding an open book carved from Scotch granite. It stands on the site of the bunk-house where Stevenson spent his honeymoon with his bride, Fannie Van de Grift Osbourne, in the summer of 1880. "The mountain," Stevenson observed in the notes for *The Silverado Squatters* which he wrote here, ". . . feeds in the springtime many splashing brooks. From its summit you must have an excellent lesson in geography. . . . Three counties, Napa, Lake, and Sonoma, march across its cliffy shoulders. Its naked peak stands nearly 4,500 feet above the sea. Its sides are fringed with forest, and the soil, where it is bare, glows warm with cinnabar." This is now Robert Louis Stevenson Memorial State Park.

At the head of upper Napa Valley's vineyard stretches is its trading center, CALISTOGA, 82.2 *m.* (365 alt., 1,514 pop.), the watering place laid out by the enterprising Mormon, Samuel Brannan, in 1859, on the site which the Indians called Colaynomo (oven place), knowing that underground heat warmed its hot springs. Planning a popular spa, he built a hotel and 20 cottages and christened his development Calistoga —a compound from the names California and Saratoga. At Larkmead Winery champagne is produced in the Hanns Kornell Cellars by the bottle-fermenting method. Napa County Historical Museum is here.

Right from Calistoga on State 28, 0.9 *m.* to the junction with a paved road; L. here to the PETRIFIED FOREST, 5 *m.,* a tract one mile long and one-quarter mile wide covered mostly by silicified and opalized redwoods. The trees lie in two tiers with their tops pointing away from Mount St. Helena, indicating that lava from that mountain killed them and preserved them in stone form. The conversion from wood to stone was so perfect that the texture and fiber are completely preserved. Most of the trees are broken but the pieces retain their relative positions. Many are of great size, the largest having been uncovered in 1919 at a depth of 90 feet. The Queen of the Forest is 80 feet long and 12 feet in mean diameter; The Monarch, 126 feet long and 8 feet in mean diameter.

At 86.7 *m.* the highway passes the field (L) in which stands the old BALE MILL built in 1846 by the English settler Dr. Edward Bale, nephew-in-law of Gen. Mariano Vallejo and owner of the Carne Humana (Sp., human flesh) Rancho. The vine-covered water wheel, 40 feet in diameter, dwarfs the weather-beaten mill—a small, two-story, gabled-roofed frame building, dignified by the false front typical of its era. The mill was restored by the Native Sons of Napa County.

ST. HELENA, 90.2 *m.* (299 alt., 2,722 pop.), has many old wine cellars. Beside State 29 on the northern outskirts is the BERINGER BROTHERS WINERY (*open*), in whose underground storge cellars, hewn from solid limestone hundreds of thousands of gallons of wine are aging in great casks. There are fourteen wineries inside the limits of St. Helena and twenty-eight within a radius of six miles. North of St. Helena the Christian Brothers maintain aging cellars in a large stone winery building, where 2,000,000 gallons of wines are stored in oaken casks. At the southern end of St. Helena Louis M. Martini Winery operates two cellars and 900 acres of vineyards. Others in the area are Freemark Abbey, Mont La Salle, Souverain, Hanns Kornell, Chas. Krug, Leon Brendel and Sutter Home wineries. All are open to the public.

Left from St. Helena on a paved road 7.3 *m.* to the PACIFIC UNION COLLEGE, conducted by the Seventh Day Adventist Conference, which trains about 1,300 ministers, doctors, nurses, and teachers on the crater of an extinct volcano. The mountain has been drilled to a depth of 600 feet, where water at a temperature of 97° was tapped.

Southeast of St. Helena the highway passes RUTHERFORD, 94.2 *m.* (183 alt., 350 pop.), and OAKVILLE, 96.2 *m.* (153 alt., 219 pop.), small wine-producing centers.

YOUNTVILLE, 99.6 *m.* (150 alt., 360 pop.), stands on what was once the southern section of the 11,814-acre Rancho Caymus, granted in 1836 to George C. Yount, who came to California from North Carolina with the Wolfskill party. Two miles north of here, Yount built the first white habitation in Napa Valley. Yount's Kentucky log blockhouse was unquestionably the first in California. In the walls of the 18-foot square lower room portholes were placed for defense. A year later Yount replaced the blockhouse with a narrow adobe structure, 100 feet long; the walls of this, too, he pierced with portholes.

War veterans have been cared for in the VETERANS HOME (R) on the southern outskirts of Yountville since 1881. Opened for disabled veterans of the Mexican War and the Grand Army of the Republic, the institution was privately operated until taken over by the State in 1897.

NAPA, 108.6 *m.* (20 alt., 26,650 pop.), seat of Napa County, serves the ranchers and grape-growers of the valley, tans leather and manufactures leather goods, ships valley products by barge down the Napa River. In spite of its bustling air, its neon lights, and automobiles, Napa has an old, settled air, emphasized by venerable shade trees and buildings of another era. Its courthouse square is an oasis of peace. Napa's history goes back to 1832, when the first settlers arrived, though the town did not come into existence until 1848, when Nathan Coombs bought the site from holders of the original Spanish grant. During the 1850's timber from valley mills was hauled to Napa and shipped by water to San Francisco. Napa became a mining center when the Silver Rush of 1858 brought many prospectors. In the 1860's vineyards multiplied. The first vineyards had been planted by settlers who received cuttings from the Missions at Sonoma and San Rafael. Eight miles northwest of Napa the vineyards of the Christian Brothers surround their Mont LaSalle Winery and Novitiate.

Northeast of Napa by Highway 37 (or by 29 to Rutherford, thence via 128), to LAKE BERRYESSA, 25 m. long, 3 m. wide, 275 ft. deep, with a shoreline of 160 m. This fishing and recreation area was formed after 1957 when the Bureau of Reclamation built MONTICELLO DAM. It is 304 ft. tall, contains 335,000 cu. yds. of concrete and cost $10,000,000.

The land east of Napa belonged to the Rancho Tulocay, where Cayetano Juarez stocked his herds in 1837 and built the first of the two JUAREZ ADOBES (L), 110 *m.,* in 1840. The second and larger one was built in 1845. During the Bear Rebellion at Sonoma (*see TOUR 2b*) of which he was alcalde at the time, Juarez distinguished himself by swimming nine miles to escape capture.

State 29 continues southward to VALLEJO, 123 *m.* (10 alt., 62,700 pop. 1963 est.), situated where the Napa River flows into San Pablo Bay. The town is built on low hillsides, its streets reaching down to the water's edge. Far-seeing Gen. Mariano G. Vallejo, undertaking a pilgrimage with his young bride in the late 1830's across the 99,000-acre domain of his Rancho Suscol (inhabited in those days only by Indians and wild beasts), declared that here he would found a city to bear his name. When the Yankee conquerors set up their State of California in 1850, Vallejo welcomed the new regime with an offer of 156 acres for a State capital. Here he proposed to endow a university, erect a museum, churches, schools, and asylums, and lay out public parks. Since his was the highest bid, the new government accepted it, and a town was laid out. But Vallejo's ambitions had proved too great a strain for his purse. When the State Legislature assembled on January 5, 1852, they found their promised State building still unfinished. Disgruntled at being compelled to sit on boxes and barrels and to spend the night on the river steamer *Empire,* they moved away precipitately on January 12 to Sacramento, only to leave there with equal haste when a flood sent them scuttling. Vallejo finally was inaugurated as the seat of government with an elaborate Christmas ball, and in January 1853, the peripatetic government returned; but on February 4, it left for good, first going to Benicia and finally to Sacramento. Vallejo himself petitioned to be released from his obligation. The town flourished because of the purchase in 1853 of nearby Mare Island for a United States navy yard, opened 1854 by Commander, later Admiral, David G. Farragut. Vallejo has the CALIFORNIA MARITIME ACADEMY and is redeveloping approx. 125 acres between the main business district and the waterfront. Plans include a freeway and a civic center with new library, city hall and auditorium. The city has constructed a marina that ultimately will berth 1,000 boats at the north end of the waterfront.

The galley *San Carlos,* first Spanish ship to sail through the Golden Gate, sighted the low-lying island, and Capt. Don Juan Pérez de Ayala christened it Isla Plana (flat island). Later the crude cattle ferry that plied across Carquinez (Ind., great serpent) Strait was caught in a squall and overturned by the stampeding animals; some of them swam shore, among them General Vallejo's prized white mare, which he found a few days later, peacefully grazing on the island. Overjoyed, he named the place La Isla de la Yegua (the island of the mare). So fond was Vallejo's young wife of the view from the island shore that she once said she wished it were hers. "My dear, it is yours," replied the General; and it was acknowledged as her property. The island was sold for $7,000 in 1850, and a year later for $17,000.

MARE ISLAND NAVAL SHIPYARD covers in part the narrow spit of land between Napa River Channel and San Pablo Bay and uses 2,700 acres, and controls 5,067 additional acres of reclaimable tidelands. It has four shipbuilding ways, four drydocks, up to 740 ft. long; 724 buildings, and vast facilities for constructing, repairing, overhauling and outfitting ships of the United States Navy, providing test work and research, as well as service and material to other defense units. By 1964 Mare Island had built or was building 13 nuclear-powered submarines. When the keel for the nuclear-powered polaris missile submarine Mariano G. Vallejo was laid in 1964 it was the 507th vessel to be built at the shipyard. During World War II the yard launched 392 ships and overhauled 1,207.

Among the extraordinary skills and facilities at Mare Island are nuclear reactor technology, SUBROC missile system installation, a large-scale computer, a polymer laboratory, a vertical boring mill, ball valve manufacture for nuclear submarines, radioactive testing, ultrasonic testing capacity, and others. Civilian employment averages 10,000; military, more than 4,000. The U. S. Naval Schools Command has more than 1,300 military personnel in the Guided Missile School, Naval Tactical Data System School, and Cryptographic Repair School. The military has available housing, barracks, dispensary, public school, athletic and chapel facilities. Ships and administration buildings are open to visitors on Navy Day, October 27.

At 124.8 *m.* on State 29 is the junction with a narrow paved road.

Right on this road 0.3 *m.* to MORROW COVE, a grassy hollow framed by tall eucalypti, facing Carquinez Strait and the Contra Costa (Sp., opposite coast) County Hills.

State 29 terminates at the junction with US 40 (*see TOUR 9b*), 124.9 *m.* The Carquinez Strait Bridge (1927; parallel span, 1958) carries US 40 across the strait to Contra Costa County.

Tour 2B

Junction with US 101—Oxnard—Santa Monica—Long Beach—Doheny Park; 113.6 *m.* US 101-Alt., State 1.

Branching southward from US 101 just south of the Santa Clara River crossing, US 101 A (State 1) leaves the beet fields, orange groves and walnut orchards of the river delta and swings out to the shore of the Pacific. As the route continues along the smooth, curving beach, a rampart of rain-cut cliffs on the inland side hides the brush-covered Santa Monica Mountains paralleling the coast. The highway swings around Santa Monica Bay, cuts across the neck of the Palos Verdes peninsula, and continues down the coast, passing a long string of seaside resorts.

South of a junction with US 101, 7 miles south of Ventura, State 1 leads to OXNARD, a city that began as a single beet sugar refinery in open fields. In 1960 Oxnard had 40,265 pop.; in November, 1964, by special U. S. Census, 58,269. In 1897 three Oxnard brothers built the factory, which operated until 1958. The city was incorporated in 1903. World War II brought U. S. Naval Construction Battalion Center at PORT HUENEME, 3 *m.* away. The U. S. Naval Air Missile Test Center, now headquarters for Pacific Missile Range, was established 1946 at Point Mugu. Oxnard Air Force Base was opened 1952. The three bases employ 7,650 civilians in addition to Service personnel.

SILVER STRAND and HOLLYWOOD-BY-THE-SEA, 5.5 *m.,* and HOLLYWOOD BEACH, 6 *m.,* are seaside resorts at the end of sand dunes still pointed out as a film "location" for Rudolph Valentino's *The Sheik.*

Across lagoon lands, scarred with white patches of alkali and planted here and there with oats, State 1 travels seaward. From the beetling bluffs of Point Mugu it winds above breakers, on a ledge cut from rocky cliffs so steep that road surveyors had to be lowered by ropes in places.

MALIBU (Ind., deer), 36.8 *m.,* favorite beach-spot of cinema celebrities, is well-protected against sight-seers and autograph hunters by uniformed guards stationed at all entrances. Its dwellings, some of elaborate scale, sit facing the beach with their backs turned aloofly to the highway.

South of Malibu, on a promontory a mile from the road, is (L) the great white RINDGE MANSION, of the California Spanish type. It was never completed by its builder—wealthy, eccentric May K. Rindge. The 16,350-acre Rindge estate, Rancho Malibu, was purchased in 1892 by Fred H. Rindge, son of a Massachusetts manufacturer, and pastured with great herds and flocks. In 1905, before he could realize his dream of building a California Riviera here, he died, leaving the estate to his wife and three children. His widow began a 30-year series of bitter lawsuits, violence, and strategy to prevent the building of railroads and highways on her vast property. She hired armed, mounted guards to patrol her boundaries, built high wire fences with barred and chained gates, plowed county-built highways under, turned droves of hogs upon cuts for new roads or planted them with alfalfa; and during 1915-17, when her gates were systematically smashed and her guards overpowered every week by crowds of farmers and travelers trying to get through to Santa Monica, she dynamited her roads. She brought suits to oust squatters and to punish trespass, libel, and defamation of character. Her battle with the State against its plan to run the coast highway through her beach land was carried to the Supreme Court, where she lost. Her son brought her into court with the charge that her fights were dissipating the estate at the rate of more than $1,000,000 a year. When in 1929, four years after leasing the land that is now Malibu, she started building her mansion, she soon found herself without cash. In 1938 the Federal District Court awarded control of the estate to a trustee-corporation authorized to bring order to its finances.

The J. PAUL GETTY MUSEUM at Malibu (*open Wednesdays and Saturdays, 3-5 p.m. by appointment; free*) has marbles from the Elgin Collection, Lansdowne Hercules, Louis XV and Louis XVI galleries with Boucher tapestries and 18th century French furniture. Works by Titian, Tintoretto, Rubens, Gainsborough, other masters 15th to 17th centuries.

South of MALIBU PIER, 37.9 *m.* are expensive beach homes and cottages, ranging in architectural types from trim shingle-roofed New England cottages to ingeniously elaborate French châteaux. Many of the dwellings were built by motion-picture celebrities when the seaside film colony began to develop but sold when Malibu proved more fashionable.

At 41.9 *m.* is the junction with paved State 27.

Left here through TOPANGA CANYON, a deep gorge in the Santa Monica Mountains, where a four-day brush fire in 1938 burned many of the old sycamores, and scores of mountain cabins. At TOPANGA SPRINGS, 7.7 *m.*, are picnic grounds. TOPANGA SUMMIT (1,560 alt.), 9.7 *m.*, is a splendid viewpoint at the edge of a sheer precipice (*telescopes, 10¢*).

Just beyond Topanga the route crosses the city line of Los Angeles. It proceeds past elaborate cliffside homes and strips of cottage-dotted beach at PACIFIC PALISADES.

At 53.2 *m.* is the junction with Sunset Blvd.

Left on Sunset Blvd., 1 *m.* to the ADOLF BERNHEIMER ORIENTAL GARDENS and HOUSE, a 7-acre showplace with landscaped grounds and bronze decorations from China and Japan. (*Admission charged.*)

Facing Santa Monica Bay from a high mesa is SANTA MONICA, 48.5 *m.* (100 alt., 83,249 pop.), a residential-resort city where summer visitors swell the population to more than 100,000. Along the bluffs for 3 miles stretches green, tree-shaded PALISADES PARK. Many fine houses along the beach were built by movie stars. SANTA MONICA CITY COLLEGE on Pico Blvd., enrolls more than 10,000 students. Will Rogers is commemorated both by WILL ROGERS BEACH and by WILL ROGERS STATE PARK, near Sunset Blvd., 186 acres, comprising his house and stables and a collection of memorabilia.

The land here was once part of two ranchos, "the place called San Vicente," and "the place called Santa Monica," over whose boundaries the Sepulveda, Reyes and Marquez families quarreled for many years. In the 1860's people of Los Angeles began making buggy excursions to the beach for swimming, beach bonfires, and Saturday dances in a big tent. Across the plain where the town now stands, ox-teams hauled *brea* (tar) from La Brea tar pits to a little wharf for loading on San Francisco-bound steamers.

East on Santa Monica Blvd. (State 2) to CENTURY CITY, 14 *m.* west of the center of Los Angeles, an urban complex built since 1963 on 260 acres originally owned by Twentieth Century-Fox Film Corp., now owned by Aluminum Company of America (ALCOA). Dominating it is a luxury hotel, the CENTURY PLAZA, opened June, 1966, designed by Minoru Yamasaki, and operated by the International Hotel Corp., a 20-story structure built in a curve, with convention halls and parking below ground. The civic area, planned by Welton Beckett & Associates and leased by Twentieth Century-Fox, has CENTURY SQUARE, a shopping center on an elevated mall; twin 13-story office buildings and a number of 27-story apartment houses. The former studio lot, once part of the ranch of Tom Mix, has become a luxurious residential and office center at a reported investment of $500,000,000.

Santa Monica is at the junction with US 66 (*see TOUR 12c*). US 101A leaves the ocean and turns R. on Lincoln Blvd.

At 49.8 *m.* is the junction with Ocean Park Blvd.

Left on Ocean Park Blvd. to the DOUGLAS AIRCRAFT FACTORY (*closed to public*), 1.6 *m.,* adjoining an airport and testing ground on Clover Field (*open to visitors*). Clover Field was the starting and finishing place of the U. S. Army "'round the world" flight in 1929.

At 52.1 *m.* is the junction with Venice Blvd.

Right here to VENICE, 0.7 *m.* (20 alt., 39,200 pop. 1963) pleasure town with an elaborate amusement section on the beach and pier. In the early 1900's Abbott Kinney, a middle-west manufacturer, set about creating a Venice on the tidal flats. He built 15 miles of concrete canals radiating from a central artery and planned to line their banks with Italian Renaissance houses. A few such structures were built, gondolas with singing gondoliers were trolled about the waterways, and lectures, "art" exhibits, and Chautauqua meetings were held in an attempt to make the town the western cultural center. Kinney provided transportation to bring prospective lot-buyers from Los Angeles and other places. They came in large numbers, but unfortunately they preferred the bathing beach to the lecture hall. In a final desperate effort Kinney brought in the divine Sara Bernhardt to play *Camille* in an auditorium at the end of a pier. When that, too, failed, he quietly surrendered and imported sideshow freaks, tent-shows, and flip-flop entertainment devices. And the boom began.

Since 1958 construction of the small-craft harbor of MARINA DEL REY adjoining Venice Beach south of Washington Blvd. started a new resort. There will be slips for 6,000 boats and storage facilities for 4,000.

US 101 climbs into Del Rey Hills to the junction with 83rd St., 54.7 *m.* Left to LOYOLA UNIVERSITY, 1 *m.* At 55 *m.* is the junction with Manchester Ave. Left on Manchester 3 *m.* to:

INGLEWOOD (150-245 ft. alt., 63,790 pop. 1960; 84,100, 1964, est.) an independent city with Los Angeles on west, north and east. Northwest is HUNTINGTON PARK (29,920 pop.); south are LENNOX (31,224 pop.), and HAWTHORNE (33,035 pop.). Inglewood is an industrial and residential community served by San Diego Freeway and Imperial Highway.

Inglewood, incor. 1908, grew up on the site of the Rancho Aguaje de la Centinela (spring of the sentinel), granted in 1844 to Ignacio Machado, who prized it so little that he traded it for a Los Angeles adobe and threw in two barrels of wine to make it a bargain. The land, passing into the hands of Yankees in 1857, was acquired in 1873 by Daniel Freeman, who laid out a town and sold lots. Although the ADOBE DEL RANCHO AGUAJE DE LA CENTINELA, a simple house of sun-dried clay with deeply-recessed windows and vine-clad corridors, is all that remains of Inglewood's Spanish past, the town celebrates "Centinela Days" with pageantry every August.

HOLLYWOOD PARK (*racing daily except Sun. and Mon. June 1-July 31*) is the $2,500,000 ultra-modern home of southern California's mid-summer racing season. The 315-acre plant includes a one-mile oval track; a white concrete-and-steel double-decked grandstand seating 12,000 people; a clubhouse with terraces, lounges, and boxes for 8,000; an enclosed paddock equipped with indirect lighting; and a projection screen that flashes pictures of photographic finishes. The stables have 1,200 stalls and 300 tack- and feed-rooms. Within the 3,700-acre infield are three lakes, covering 10 acres, stocked with ducks, geese, and swans. Opened in June 1938, the track is owned and operated by the Hollywood Turf Club, which in turn is owned by a group of motion picture celebrities.

LOS ANGELES INTERNATIONAL AIRPORT, 57.1 *m.,* one of the great aircraft terminals of the nation, has 1,000 landings and take-offs daily. (*See LOS ANGELES.*) Nearby are the large plants of North American Aviation, Inc., Douglas Aircraft Co. and Northrup Co.

At 58.3 *m.* is the junction with paved El Segundo Blvd.

Right on El Segundo Blvd. is EL SEGUNDO, 1.3 *m.* (35 alt., 16,223 pop.), surrounded by huge black oil tanks, concrete reservoirs, and steel-mast lightning rods, which grew up around the Standard Oil Company's second (*segundo*) California refinery. Today aircraft industries predominate, North American Aviation employing 19,000. Hughes Aircraft Co. 5,700 and Aerospace Corp. 4,700.

At 60.2 *m.* is the junction with Manhattan Beach Blvd.

Right on Manhattan Beach Blvd. is MANHATTAN BEACH, 0.7 *m.* (46 alt., 1,891 pop.), a family seaside resort (*bath-houses and fishing tackle*) near the sandy strand of MANHATTAN BEACH STATE PARK, municipal regulations forbid noisy amusements.

HERMOSA (Sp., beautiful) BEACH, 61.1 *m.* (43 alt., 16,115 pop.), is a quiet town of frame bungalows and seaside cottages.

REDONDO BEACH, 62.6 *m.* (16 alt., 46,986 pop., 1960), at the southern end of Santa Monica Bay, is situated on the site of the Dominguez Rancho of 1784. By the development of King Harbor, a $30,000,000 project, Redondo Beach obtained marina facilities of unusual extent, with 1,400 moorings. Pirate's Isle is a 53,000 sq. ft. salt water swimming pool kept at 80° F. Here are located the Space Technology Laboratories of Thompson Ramo Wooldridge, Inc. dealing with space communications, radar systems, computers, and applied aerodynamics, employing more than 7,300 scientific personnel. The city has easy access to Pacific Coast Highway and San Diego Freeway.

Left from Redondo Beach on Torrance Blvd. is TORRANCE, 4.1 *m.* (75 alt., 100,991 pop. 1960), founded in 1911 by Jared Sidney Torrance, Pasadena utilities magnate, who studied the best city plans in many parts of the world before he laid out his industrial-residential community. The oil fields, machine and tool shops, steel plant, and other factories are in a section apart from the quiet tree-shaded trees lined with houses. In Torrance Columbis Steel Co., now U.S. Steel, built a large plant.

At 64.7 *m.* on US 101A is the junction with Palos Verdes Dr.

At PORTUGUESE BEND Wayfarer's Chapel of the Church of the New Jerusalem commemorates Emanuel Swedenborg.

Right here to PALOS VERDES, 0.6 *m.,* a residential district and site of a State College. On Palos Verdes Drive South, jutting out to the sea is MARINELAND of the PACIFIC, described as a 3-ring sea circus, where whales, sea lions and porpoises perform and thousands of fish are on display. Reached from Los Angeles center via Crenshaw, Hawthorne or Sepulveda Blvds.

Right from Palos Verdes on Palos Verdes Drive West to the junction with a dirt road, 2 *m.;* here 0.1 *m.* to the slim white tower of POINT VICENTE LIGHTHOUSE (*open 2-4 Tues. and Thurs.*).

Palos Verdes Dr. continues, skirting wide views of the Pacific, to the junction with Twenty-fifth St. 2 *m.;* straight ahead on Twenty-fifth St. to the junction with Paseo del Mar, 0.6 *m.*

Right on Paseo del Mar is FORT MACARTHUR UPPER RESERVATION, 0.9 *m.,* its great guns hidden from sight, guarding the approach to Los Angeles Harbor from the edge of ocean bluffs.

Crossing the ubiquitous Los Angeles city limits, Twenty-fifth St. continues to the junction with Pacific St. in SAN PEDRO (*see below*), 0.8 *m.*

US 101A, State 11, the Harbor Freeway and a number of other Los Angeles highways converge on the coast at SAN PEDRO, a unit of the PORT OF LOS ANGELES, tenth in tonnage in the United States. San Pedro is an independent community within the City of Los Angeles, the other units of the Port being Wilmington and Terminal Island, with an available waterfront 28 miles long. In 1964 San Pedro had an estimated population of 68,700. It is located on the southeast slopes of the Palos Verdes Peninsula, 24 *m.* south of Los Angeles City Hall. It became a part of the city in 1909.

When Yankee Capt. William Shaler's *Lelia Byrd,* anchored here in 1805—nearly three centuries after the bay's discovery by Juan Rodríguez Cabrillo—the place was an open roadstead bordered by mud flats, serving Mission San Gabriel for an embarcadero. Although foreign trade was forbidden by law, settlers—and the padres—were quite willing to barter their hides and tallow for contraband manufactured goods. Even after San Pedro was officially recognized as a port in 1826, smuggling was continued to evade custom duties. In 1852 the harbor became a port of call for its first boat line, when the *Sea Bird* began scheduled trips to and from San Francisco. When Richard Henry Dana returned here a quarter century after the visit described in *Two Years Before the Mast,* however, freight was still lightered "up the creek" from the bay through a winding channel at this "most desolate place on the California Coast."

The first Government harbor improvement, made in 1877, was the deepening of the channel to a depth of 16 feet. In 1892 Congress appropriated money for harbor development. In 1899 began the titanic job of creating a man-made harbor which by 1939 had cost the City of Los Angeles and the Federal Government a total of nearly $60,000,000.

The GOVERNMENT BREAKWATER, extending 2.11 miles from the tip of Point Fermin to a concrete lighthouse, built 1899-1910 of sandstone and granite blocks, protects the OUTER HARBOR. As incoming vessels round the breakwater, they are piloted up the channel to the 1,600-foot-wide TURNING BASIN and brought up to the docks. The FORT MACARTHUR LOWER RESERVATION at the southern end of Pacific Ave., guards the West Channel from the edge of steep bluffs, bristling with anti-aircraft service guns, railroad guns with 40-foot barrels, and the derricks that feed them heavy shells. Behind the piers and warehouses along the docks is a belt of fish canneries and industrial plants.

The city, State, and Federal buildings of the SAN PEDRO CIVIC CENTER, bounded by Harbor Blvd., Beacon, 6th and 9th Sts., stand along a palm-lined parkway on a low bluff near the center of the business district. Over the foothill slopes, rising above, ramble the homes of the residence area. At the landward end of the breakwater is CABRILLO BEACH PARK, end of Stephen M. White Dr. (*open 6 a.m.-12 p.m.*). The CABRILLO BEACH MUSEUM (*open 9-5 in winter, 10-6 in summer*) has a large aquarium and collection of shells and mounted fish. POINT FERMIN PARK, on Paseo del Mar at the tip of the Peninsula (*picnicking*), is a 27-acre expanse of trees and lawns with winding flower-bordered footpaths on the bluffs overlooking the Pacific. In the park is the POINT FERMIN LIGHTHOUSE, built in 1876.

The VINCENT THOMAS BRIDGE, opened November, 1963, connects San Pedro with TERMINAL ISLAND. This suspension bridge is the third largest in California, surpassed only by the Golden Gate and the Oakland Bay bridges. It is 6,060 ft. long with approaches and has a central span of 1,500 ft. and side spans of 500 ft. each. Clear height above water is 185 ft., height of main towers is 365 ft., and 105,000 tons of concrete were poured. Plans call for connecting the bridge to Harbor Freeway at Pacific Ave. The bridge cost $21,000,000. About 9,500 motor cars crossed the bridge daily by May, 1965.

WILMINGTON, 72.6 *m.* (13 alt., 36,300 pop. est. 1963), with its business streets, warehouses, and shipping offices lies on the flat ground facing Terminal Island across the San Pedro Inner Harbor. When the harbor and island were just a slough and a sand bar, this was marshy land, with the fish-smelling huts of an Indian village squatting on the mud. In 1857 Phineas Banning bought a 2,400-acre parcel of old Rancho San Pedro, laid out a town called New San Pedro (now Wilmington) and began development of the harbor, and ran stages here to meet weekly steamers from San Francisco, Los Angeles' first landing-place for ocean-going freight. Wilmington grew with the building of a railroad, linking city with port, in 1869. A few years later the Southern Pacific Railroad built across the mud flats to deep water at San Pedro. In 1906 Los Angeles inched up to the harbor by annexing a "shoestring strip" to the very gates of Wilmington, and in 1909 added the town.

BANNING PARK (*free*), E. M St. and Banning Blvd., is a 20-acre tract, once the heart of the Phineas Banning estate, landscaped with a quiet, old-fashioned charm. A circular walk, lined with a white picket fence and rows of eucalyptus trees, leads to the old BANNING HOUSE preserved by the Parks Dept. of the City of Los Angeles. (*Tours Sundays, April-October.*)

DRUM BARRACKS, 1031 Cary Ave., built in 1861, was originally a troop barracks and a supply depot for the U. S. Army of the Southwest, 1862-1868. It was named for Gen. Richard Colton Drum, father of Gen. Hugh Drum, U.S.A.

The SANTA CATALINA ISLAND TERMINAL, at the foot of Avalon Blvd., is the embarcadero for Santa Catalina Island (*see TOUR 2c*). The CALIFORNIA YACHT CLUB, on Yacht St., is a white frame building topped by a tower resembling a lighthouse.

Wilmington is at the junction with US 6 (*see TOUR 7b*), which unites with US 101A to Long Beach.

South of the FISHING VILLAGE is FISH HARBOR, a small basin protected by a sea wall. It is headquarters for 2,300 fishermen who put out to sea each day in 1,200 vessels, returning in the evening with their catch which, if luck is good, loads the vessels almost gunwales down. Big boats—such as the tuna clippers—travel in fleets, some going as far as Panama or the Galapagos. Sardine fishers find schools of fish on moonless nights by their phosphorescent glow. Fish Harbor has the nation's largest fish packing plants for tuna, mackerel and sardines.

LONG BEACH, 77.5 *m.* (*See LONG BEACH*).

At 84.1 *m.* is a junction with a beach road. Left on this road to SEAL BEACH, a resort town (6,994 in 1960; 18,000, est. 1966).

US 101 crosses two inlets of Anaheim Bay to a tongue of land between a lagoon and the ocean, favored for beach cabins. At 78.7 *m.* is SUNSET BEACH; at 79 *m.* is BOLSA BAY, where tule-grown marshes invite duck hunting. (*Speedboats, rowboats, available.*)

At 90.8 *m.* is HUNTINGTON BEACH (50,290 pop., 1960; 66,000 est. 1966), which boomed when oil was found there in 1920. It now has 1,691 active wells and several oil refineries. Derricks stand everywhere and offshore drilling is pursued. The Municipal Pier is open for fishing. The GOLDEN WEST COLLEGE (junior) was ready for students in 1966. The city sponsors the All Southland Salute to Santa Claus band review, with 50 bands taking part.

At 96.4 *m.* is a junction with State 55. Left on this road to COSTA MESA, incor. 1953, which had 37,550 in 1960 and claimed 68,600 in 1965. This is in the fast-growing coastal area of Orange County. It has Orange Coast College (junior), with 15,970 students in 1965, and is developing a new Civic Center. Industries are moving into this sector.

Right on State 55 over an inlet of NEWPORT BAY to NEWPORT BEACH, 1 *m.*, a fast-growing recreation area comprising the communities of BALBOA PENINSULA, BALBOA ISLAND, EAST BLUFF, DOVER SHORES, LIDO ISLE, NEWPORT HEIGHTS, WEST NEWPORT, CORONA DEL MAR, CAMEO SHORES, WESTCLIFF and LIDO SANDS. In 1960 it had 26,564 residents; in 1965 36,650, expanding in the summer to about 60,000. The boat harbor has room for 7,500 vessels. The Pacific Coast Regatta is held in the spring and a water carnival, Tournament of Lights, is a midsummer feature. The harbor, originally a hideout for smugglers, became a busy port after 1872, but silt from the Santa Ana River made huge dredging operations necessary. Houses are on leased land, owned by the city and the Irvine Co., basis of which was the Irvine Ranch of 93,000 acres stretching for eight miles from Newport to Laguna Beach.

Three miles inland from Corona del Mar and the ocean is the new campus of UNIVERSITY OF CALIFORNIA, IRVINE, halfway between Santa Ana Freeway and Pacific Coast Highway (State 1). The Irvine Co. donated 1,000 acres, to which the State added 510. Construction began in 1964 of a community-college plan designed by Wm. L. Pereira & Associates, around which the Irvine Co. expects to develop 10,000 acres for a new city. The COLLEGE OF ARTS, LETTERS & SCIENCE was ready for students in 1966, and a 75,000 vol. library, assembled at the San Diego campus, was installed. A 300-ft. bell tower will be erected by private donations.

The Irvine segment of the University of California is an example of the ability of the educational system of California to adapt itself to the needs of the community. With the tremendous increase in population came the rush of thousands of young people to the gates of the colleges and universities. With funds available it became possible for the University to lay out a campus, provide a physical plant, equip it and allocate instructors ready to teach 1,000 students at opening. The core of the system is the liberal arts college, where undergraduates can major in the humanities, fine arts, social sciences, biological sciences and physical sciences. Two professional schools and an institute were to be ready at the same time: the Graduate School of Administration, dealing with business, public and educational administration, and the School of Engineering, with an upper division and graduate program, emphasizing electronics. An Institute of Environmental Planning deals with research in resource and urban planning. The initial faculty was placed at 105. Plans call for expansion of the library to 250,000 volumes by 1970, 500,000 by 1975. While these facilities are being developed students at Irvine are given easy access to collections at the Los Angeles, Riverside and San Diego libraries of the University of California.

LAGUNA BEACH, 104.9 *m.* (17 alt., 9,288 pop. 1960) a resort favored by artists, has taken the lead in mounting annual cultural events that draw thousands of spectators. Its FESTIVAL OF ARTS, founded 1932, reproduces famous paintings and sculpture with living characters, employing local residents. Using its 2,400-seat amphitheater, it grosses $250,000 a season with a net profit of $75,000, half of which goes to the city and half to other cultural activities, including the FESTIVAL OF OPERA, and CIVIC BALLET, the SCHOOL OF ART AND DESIGN, and the LAGUNA CHORALE. There are also a Chamber Music Society and the COMMUNITY PLAYERS, for whom a new theater is projected. The LAGUNA BEACH ART ASSOCIATION GALLERY exhibits paintings by members.

THREE ARCHES, 110.4 *m.,* a cliff beach colony, is named after rocks on the seashore.

Overlooking DANA COVE is the hamlet of DANA POINT, 112.4 *m.* (175 alt.). The cliffs rose "twice as high as our royal-mast-head," observed Richard Henry Dana when the *Pilgrim* anchored here in 1835. "The shore is rocky, and directly exposed to the southeast, so that vessels are obliged to slip and run for their lives on the first sign of the gale." When loading hides were thrown to the beach from the cliffs.

DOHENY PARK (*see TOUR 2d*), 113.6 *m.,* is at the junction with US 101 (*see TOUR 2d*).

Tour 2C

Wilmington to Avalon, Santa Catalina Island, 27 *m.* by boat. Also via Catalina Airlines from Los Angeles International Airport and Avalon Transport from Long Beach.

Twenty-seven miles southwest of Los Angeles Harbor in the rolling wastes of the Pacific Ocean rise the mist-blown peaks of Santa Catalina Island. Twenty-two miles long and from one-half to eight miles wide, the island looks from the mainland as if a section of the Coast Range had been transplanted to the open ocean. In its numerous valleys the foliage is luxuriant; the rugged higher slopes are covered with chaparral. Abounding with tuna, swordfish, yellowtail, white sea bass, rock bass, barracuda, mackerel, bonita, whitefish, and sheepshead, the waters about the island are popular with sport fisherman.

Santa Catalina Island has one incorporated city, a $2,000,000 casino, a network of roads and trails, a $1,000,000 water system, a small industrial center, and numerous highly commercialized tourist attractions.

The island was discovered by Juan Rodríguez Cabrillo, Portuguese navigator who, seeking the mythical Strait of Anian under orders from the Spanish Crown, put into the small, placid bay now called Avalon on October 7, 1542. He named the island San Salvador, for his flagship. On November 28, 1602, the islanders saw the Spanish King's next emissary, Sebastián Viscaíno, enter the harbor with his three white-winged ships. He gave the island its present name in honor of St. Catherine. In 1811 a Russian vessel, seeking the prized sea otters, landed in the bay and slaughtered many of the Indians. Until 1821, when Mexico freed herself from Spain and lifted the Spanish ban on foreign trade in California, Santa Catalina was the base for unlawful trading operations with the mainland.

Although gold had been discovered on Santa Catalina in 1834, it was not until 1863 that several prospectors "struck it rich," starting a gold rush; some 100,000 feet of claims were staked and filed in the Los Angeles County Recorder's office, and indefatigable prospectors even ran their mine tunnels under the ocean floor. The boom was cut short by three developments: a new island owner, José María Covarrubias, bought the property in 1855 and vociferously objected to the freebooting activities of the prospectors; a pirate scare frightened the Federal Government; and last but not least the gold ran out.

The Government became worried when it was learned that Confederate sympathizers were planning to use Santa Catalina as a base for pirating the gold-carrying ships in the coastal trade, and Northern troops were hurriedly sent in with orders to evacuate all private citizens by February 1, 1864.

From then until 1919, when William Wrigley, Jr., bought the property, the various owners attempted subdivisions, promotion of pleasure resorts, and a silver mining venture. But it was Wrigley who had the formula needed to make the island a commercial success and who had the millions to create a romantic, palm-studded town on the bay. He is buried in a mausoleum of white stone quarried on the island, located at the head of Avalon Canyon, 1 *m.* southwest of Bird Park.

Boats for Santa Catalina leave from the CATALINA TERMINAL, foot of Avalon Blvd., Wilmington. More than 250,000 persons travel to the island by water annually.

AVALON, 27 *m.* (0 alt., 2,200 pop.), the main settlement and center of resort and sports activities of SANTA CATALINA ISLAND, lies at the mouth of a large canyon along the crescent-shaped shore of Avalon Bay on the northwestern side of the island. The buildings are the usual miscellaneous collection of stucco-covered and frame structures in the Californian version of what Spanish design should be. Crescent Avenue, the principal street, follows the curve of the bay, widening in the downtown area to the size of a plaza. Palm and olive trees, set in stone boxes in the center of the street, shade low settees and stone benches; grassy squares and sparkling fountains, strolling señoritas in spangled skirts and strumming troubadours in velvet costumes, round out the scene. There are three hotels operated by the Santa Catalina Island Company, and privately operated hotels, apartment houses, and bungalow courts.

Most of the island's tourist attractions are in Avalon or near it. Regularly scheduled boat and motorcar trips to the remoter points of interest leave from Avalon, as do the various speedboat, glass-bottom boat, and flying fish trips (*see below*).

The AVALON BOARDWALK skirts the harbor from Pebbly Beach. With many of the island's major attractions fronting on it, it is the principal recreation center.

EL ENCANTO, in the heart of Avalon, is a plaza surrounded by small shops and cafés.

The AVALON CASINO (*open 2-4 p.m. daily*), on the boardwalk, rises from the northwest promontory of Avalon Bay. The $2,000,000 white, circular building is styled in an adaptation of Moorish design. On the lower floor is a large motion picture theater. Five ramps give access to the second-story ballroom. A cocktail lounge with a 100-foot-long bar has walls covered with fantastic fish murals.

The SANTA CATALINA AIRPORT lies at the head of a deep canyon at HAMILTON BEACH, northwest of Avalon. Regular flights daily from Los Angeles via Catalina Airlines and Long Beach via Avalon Air Transport.

The AVALON GREEK THEATER, east end of Crescent Ave., is used chiefly by civic and fraternal organizations.

The SANTA CATALINA ISLAND VISITORS' COUNTRY CLUB (*greens fee*), Sumner Ave. and Fremont St., is a rambling stucco clubhouse crowning a knoll and surrounded by an 18-hole golf course. There are also tennis courts and a 9-hole pitch-and-put course.

The CATALINA BASEBALL PARK, Fremont St. and Avalon Blvd., is the spring-training headquarters of the Chicago Cubs.

The AVALON BIRD PARK (*open 8-6*), in Avalon Canyon (*reached only by bus, leaving corner Crescent and Metropole Aves. every one-half hr. from 7-7*), is a 20-acre home for more than 8,000 birds of 650 varieties, mostly from foreign countries.

The GLIDDEN INDIAN MUSEUM (*open 9-5*), on the hill overlooking Avalon, is a one-story yellow and tan frame structure exhibiting Indian artifacts and skeletons found on the island.

PEBBLY BEACH, at end of the Boardwalk, 1.5 miles east of Avalon, is an industrial district; here are a large pottery plant, a furniture factory, and a stone quarry. Mexican employees live in a small village near the shore.

EAGLES NEST LODGE, 10 miles from Avalon on the Avalon-Isthmus motor road (*on the Isthmus Auto Tour; see below*), is headquarters for goat, boar, and quail hunters, and an overnight stop for hikers and riding parties. The one-story rustic wooden lodge is named for nearby

EAGLE MOUNTAIN, on whose 1,000-foot pinnacle scores of eagles nest. All hunting is conducted on supervised trips with guides (license). Boar and goats are hunted the year round from horseback with deer rifles. Large numbers of wild mountain goats roam the island interior. Some historians credit Cabrillo with having abandoned some goats on the island in 1542; others contend that they were left by Father Torquemada of the Vizcaino expedition of 1602.

The ISTHMUS lies in the northwestern section of Santa Catalina Island, 14 miles by water, 22 miles by motor road, from Avalon (*reached on the 'Round the Island Cruise and Isthmus Auto Tour; see below*), at a point where the island was almost cut in two when the land sagged during some cataclysm in the distant past. The terrain here is a flat mesa between bordering mountains. Fronting PAPEETE BEACH and ISTHMUS COVE is a settlement of one-story, thatched bungalettes for accommodation of a fluctuating summer population. The South Sea Island effect has been sought for all structures, including the store. Both *Rain,* silent film, and *Sadie Thompson,* sound film, were produced here. Midway between Isthmus Cove and Catalina Harbor stands the former UNITED STATES GOVERNMENT BARRACKS, now living quarters for employees of the island operating company; it was built to house Union troops.

CONDUCTED TOURS

(*All boats leave from Avalon Pier; auto trips from Avalon Plaza*)

1. The GLASS BOTTOM BOAT TRIP (*fare $1.80, children $1*), visits the marine gardens that extend 17 miles along the protected north shore.

2. The SEAL ROCK TRIP (*fare $1.25, children $1*), is a one and one-half hour, 11-mile cruise in a 60-foot boat along the jagged lee shore to a sunny, wave-lashed cluster of rocks just off NORTHEAST POINT, where several hundred sea lions live.

3. The EVENING FLYING FISH TRIP (*Apr.-Oct.*) is a night ride behind a 45-million candlepower searchlight, into the beam of which glide thousands of flying fishes, their highly colored "wings" (fins) iridescent in the glare.

4. The AVALON SPEEDBOAT TRIP is a 50-mile-an-hour dash from Avalon to a point about 1 mile beyond the Santa Catalina Airport and return.

5. The ISTHMUS BOAT TRIP (*May.-Oct.*) is a 3-hour 28-mile round trip cruise from Avalon Bay to Isthmus Cove along the northern coast. A 1-hour stop for lunch and sightseeing is made at the Isthmus.

6. The 'ROUND THE ISLAND CRUISE (*10:30 a.m. Sun. only, April to Oct.*).

7. The ISTHMUS AUTO TOUR follows the Old Stage Road over the Summit (1,520 alt.), past Middle Ranch Valley, and Eagles Nest Lodge, to the Isthmus.

8. The STARLIGHT DRIVE (*April to Oct.*) is a 55-minute 7-mile evening trip in an open bus. The sights are set to music by a Spanish-costumed guitar player who goes along.

9. The SKYLINE DRIVE and interior is a motor trip of nearly four hours. (*Fare $4.75, children $2.50*).

Tour 3

(Ashland, Ore.) — Yreka — Redding — Red Bluff — Sacramento — Stockton — Fresno — Bakersfield — Los Angeles — Ontario — Redlands — Coachella — Brawley — Calexico — (Mexicali, Mexico); US 99 and 99 W.; Interstate 5.

Oregon Line to Mexican Border, 880.6 *m.*

Paved roadbed throughout, open all season.
Southern Pacific R.R. parallels route between Oregon Line and Bakersfield, Saugus and Los Angeles, Banning and Mecca, Brawley and Calexico; Santa Fe Ry. parallels route between Oakland and Bakersfield.
Accommodations plentiful; numerous camping sites in northern counties.

US 99 presents a complete cross section of California. From the rugged wall of the Siskiyous on the north, it winds down barren river canyons, round Mount Shasta, and along the twisting Sacramento River between steep, evergreen-forested slopes. Southward it parallels the two great rivers, the Sacramento and the San Joaquin, that drain the far-reaching plains of the Central Valley, stretching for 750 miles in an unbroken sweep. From the valley's southern end it climbs the arid, brush-clad Tehachapi Mountains and descends to the fertile Southern California valleys where citrus groves and truck gardens extend for mile on mile. Its course then turns southeast through the sagebrush reaches of a desert trough between the San Jacinto and San Bernardino Mountains, past the Salton Sea, into irrigated farmlands of the sun-scorched Imperial Valley. The highway passes smoking lumber mills among bare, stump-dotted slopes; mines and quarries on scarred mountain sides and dredges scooping river gravel; herds cropping broad pastures and far-spreading acres of hay and grain; fruit-laden orchards around packing houses, truck garden plots, and groves of date palms. The climate varies sharply, from the bracing cold of mountain highlands to the sultry heat of Imperial Valley below sea level.

Section a. OREGON LINE to SACRAMENTO; 298.7 m. US 99-99 W. Interstate 5

Northern California's oldest road in continuous use—the Oregon-California Trail—roughly paralleled today's concrete roadbed. Starting as a faint pathway blazed through the wilderness by venturesome scouts and trappers from 1827 on, it was followed in the 1840's by early immigrants with their ox-drawn wagons, driving their cattle before them. It became a well-defined pack trail in the next decade, thronged with gold seekers, mule trains, ox-teams and covered wagons. In the late 1850's, the first turnpike between Portland and Sacramento was opened, and as the tide of travel mounted, outposts of civilization sprang up along the way. On the campsites where fur hunters had stopped for rest, inns were built. As freighters and stagecoaches supplanted mule trains, stage stations were opened; in 1886-87, when the Southern Pacific pushed its tracks from Redding to Oregon along the route, these became railroad stations and, finally, villages and towns.

South of the Oregon Line, 0 *m.,* crossed 22.8 miles south of Ashland, Oregon, stretch the wild slopes of the Siskiyou Mountains. As the highway descends, the lava mass of BLACK MOUNTAIN (5,270 alt.) looms up (L). Above it towers snow-crowned Mount Shasta (*see below*), seen for nearly 100 miles southward at intervals.

HORNBROOK, 8.2 *m.* (2,115 alt., 300 pop.), lies (L) downhill across Cottonwood Creek, its old brick business structures, store buildings with wooden awnings, and handful of shanties shadowed by shaggy black walnut trees.

To CAMP LOWE, 10.4 *m.,* where the rippling Klamath River (L) swings in to the highway, come anglers for the fine steelhead trout and salmon fishing.

At 14.7 *m.* is the junction with State 96, which follows the Klamath River westward.

Right on State 96, along the canyon of the lower Klamath, winding through one of California's most primitive regions. Until 1850, when a party of miners followed the river from the ocean, panning for gold at every bar, its course—variously known as Clamitte, Klamet, Indian Scalp, and Smith River—was unknown and incorrectly shown on the maps.

At 2 *m.* is a boundary of Klamath National Forest, an area of 1,291,619 acres, crisscrossed by pine-clad ridges and cut by deep canyons. Almost untouched by logging operations, the reserve guards about 15 billion feet of sugar and ponderosa pine and Douglas fir. It is dotted with hydraulic and drift mines, dredges, wing dams, and stamp mills.

BROWN'S RESORT, 16 *m.,* is a stopping place for fishermen and hunters.

At 33 *m.* is the junction with a dirt road; L. here 3 *m.* to SCOTT BAR (1,687 alt.), in the deep gorge of Scott River, a ghost town of dilapidated buildings where prospectors led by John Scott panned for gold in 1850. Driven out by Indians, they spread news of a rich strike that brought miners flocking in. By 1863 William H. Brewer found the placers worked out, the population departed, and half the houses empty.

The road winds on a ledge cut from steep canyon sides to SPRING FLAT, 16 *m.* (Forest Service Camp). Here trails start into the MARBLE MOUNTAIN WILDERNESS AREA, a wildly picturesque 237,527-acre tract, comprising, among other tall outcroppings Kings Castle, a monumental pile (7,396 alt.) of limestone and marble streaked with a vein of glistening white. Of the other peaks, Boulder Peak (8,317 alt.) is the highest. Near their summits are about 50 small crystal-clear lakes, stocked with trout, and a dozen or more streams with 250 miles of fishing waters. Deer, bear, and mountain lion roam the region. The Forest Service maintains five camps:

Paradise Lake, Marble Valley, Sky High Valley, Spirit Lake, and Upper Cabin.

On the main route is HAMBURG, 35 *m.* (1,580 alt.) with a general store and tourist cottages. A bustling mining town up to 1861, it had disappeared two years later, its buildings swept away by a flood.

At SEIAD VALLEY (*cabins; restaurant*), 45 *m.* (1,383 alt., 50 pop.), on a fertile little flat, which Brewer found to be "a delightful spot . . . an oasis in the desert," a New York farmer settled in 1854 and amassed a fortune raising potatoes that he sold for 15¢ a pound to the miners.

HAPPY CAMP, 65 *m.* (1,088 alt., 566 pop.), a popular place for year-round fishing and a point of departure for trails into the wilderness (*guides, packers, and pack trip supplies*), is the District Forest Supervisor's office.

At PICK-AW-ISH CAMP, 74 *m.*, Indians gather yearly in the dark of the August moon for a three-day festival. Among the ceremonies are the Brush, Deerskin, and Coyote Dances and the Working of the Earth.

SOMES BAR, 106 *m.* (580 alt., 50 pop.), where the Salmon River joins the Klamath, began as a mining town. It is a favorite spot for year-round fishing for king salmon and steelhead trout and a starting point for trails into the western part of the Marble Mountain Primitive Area.

SIX RIVERS NATIONAL FOREST lies west of the Coast Range in Del Norte and Humboldt Counties. It occupies 305,000 acres or more than 476 square miles in Humboldt County. The U. S. Forest Service maintains camp and recreation facilities near Orleans, Bluff Creek and Weitchpec; on Highway 299 in the Willow Creek area and on State 36 across the Van Duzen River east of Bridgeville

WEITCHPEC, 130 *m.* (367 alt., 64 pop.), is at the confluence of the Trinity River and the Klamath on the northern boundary of the HOOPA INDIAN RESERVATION, which occupies approximately 100 square miles of the Hoopa, or Trinity, Valley. The reservation was established in 1865, following a treaty of peace (1864) that ended 12 years of bitter warfare between the Hoopas (Hupalos) and white settlers. Soldiers were stationed at the reservation until 1892. About 500 Indians, many of mixed blood, support themselves here by farming, stock raising, and occasional work outside the reservation.

At HOOPA, 141.4 *m.* (100 pop.), only town on the reservation, are stores, administrative offices, and elementary and high schools.

Highway crosses the reservation's southern boundary at 144.8 *m.* to the junction with US 299 (*see TOUR 8b*), 153.4 *m.*

At the southern end of KLAMATH BRIDGE, 14.8 *m.*, a graceful concrete span, the Shasta River, foaming over its stony bed far below (R), joins the Klamath. The highway twists through the tortuous gorge of the Shasta on a ledge blasted from the craggy walls, bridging the river three times more within the next three miles. Highest of the spans is steel and concrete PIONEER BRIDGE, 16.7 *m.*, arching the canyon 252 feet above the river.

At 21 *m.* is the junction with a dirt road.

Right on this road is HAWKINSVILLE, 0.3 *m.*, with its red church steeple rising above a handful of weather-beaten shanties. In 1851 it was a flourishing center of trade. Here began a string of miners' cabins that followed Yreka Creek southward for 3 miles.

The road winds on to HUMBUG CREEK, 5.6 *m.*, supposedly so named because a company of prospectors on the way there in 1851 met others returning who insisted that the rumors of gold were "all humbug" but paying no heed, pushed on and made a rich strike. Joaquin Miller, who, as a youth, had a cabin on the creek bank, drew a gloomy picture of the region: "It lay west of the city (Yreka), a day's ride down in a deep, densely timbered cañon, out of sight of Mount Shasta, out of sight of everything—even the sun . . ."

Beyond Humbug Creek is the junction with a dirt road; L. here 1.4 *m.* to the SITE OF HUMBUG CITY, chief of the mining camps. A wild place it was, "a sort of Hades," said Miller, "a savage Eden, with many Adams walking up and down and plucking of every tree, nothing forbidden here; for here, so far as it would seem, are neither laws of God nor man." The principal saloon was the Howlin' Wilderness, an immense log cabin with a log fire always burning in the huge fireplace, where so many fights broke out that the common saying was, "We will have a man for breakfast tomorrow."

YREKA, 23.1 *m.* (2,624 alt., 4,759 pop.), seat of Siskiyou County, is rimmed by sheltering slopes. Surrounding the compact business district along Miner Street are blocks of gracious old white houses on streets shaded by locust trees. After Abraham Thompson in March, 1851, struck it rich at Black Gulch Camp, 2,000 men flocked in within less than six weeks. The place was known as Thompson's Dry Diggings; then, when a town was laid out in May, as Shasta Butte City; and finally, in 1852, as Yreka, which is thought to be a corruption of Wai-ri-ka (Ind., mountain). It had its share of the early Indian troubles, once barely escaping massacre when an Indian woman, Klamath Peggy, traveled 20 miles of rough mountains to warn the citizens of an impending attack. Approaching along devious trails over the brush-covered hills, the Klamath warriors found Yreka strongly guarded by sentries and withdrew. For years Klamath Peggy lived in Yreka, fearing the vengeance of her kinsmen; finally Yreka's people pensioned her.

In Yreka's HALL OF RECORDS, of painted stucco (in a locust-shaded square at Fourth and Lane Sts.), is the SISKIYOU COUNTY GOLD EXHIBIT of ores and nuggets, a small fraction of the millions that the region has yielded. At the KLAMATH NATIONAL FOREST SUPERVISOR'S HEADQUARTERS (*open 8:30-4:30*), Broadway and Miner St., is a huge relief map of Klamath National Forest.

At 25.1 *m.* is the junction with a paved road.

Right here 0.1 *m.* to the SITE OF GREENHORN, marked today by a nondescript cluster of frame structures. In the early 1850's, a company of miners had failed to find gold here, and when a greenhorn asked where to mine, they directed him to their abandoned claim; the joke turned on them when he made a rich strike. Here was the battlefield of the "Greenhorn War" of 1855, which broke out when the Greenhorn men, angered over diversion of the water for their claims from Greenhorn Creek, cut the Yreka Flats Ditch. A Yreka court promptly issued an injunction. They defied it. When Yreka officers countered by arresting one of them, the Greenhorn men marched on Yreka and freed him. But the court decision stood, and the Yreka Flats Ditch Association continued using water from Greenhorn Creek.

Along Yreka Creek the road runs over rugged slopes to the summit of FOREST HOUSE MOUNTAIN (4,159 alt.), and along Moffett Creek into lovely SCOTT VALLEY, named for John Scott, who discovered gold at Scott's Bar.

At 15.3 *m.* is the junction with a dirt road; R. here 7.6 *m.*, along McAdams Creek to its confluence with Deadwood and Cherry Creeks, to the SITE OF DEADWOOD, second only to Yreka among the region's mining centers in 1854-55. Here Joaquin Miller wrote his first poem, an epithalamium for the marriage of Deadwood's cook to a Yreka lady; he recited it at the reception for the pair.

FORT JONES, 17 *m.*, on the main side road (2,747 alt., 302 pop.), once known by a variety of names: Wheelock, for the man who built a hotel here in the 1850's; Scottsburg, for the surrounding Scott Valley; and Ottiewa, for a group of the Shasta Indians, acquired its present name from the camp the First United States Dragoons maintained here between 1852 and 1858 for protection against Indian raids.

GREENVIEW, 22 *m.* (2,812 alt., 94 pop.), is a starting point for hiking and horseback trips west into the valleys of the Salmon Mountains, where gold mining began in the 1850's.

ETNA, 29 *m.* (2,941 alt., 379 pop.), in the heart of rich farm lands, grew up about a flour mill, named the Rough and Ready, which began competition in 1856 with the neighboring Etna Mills, built two years earlier. As settlements sprang up about the two mills, rivalry waxed hot between them; but in 1863 the older fell behind when its post office was shifted to its neighbor. Rough and Ready Mills even took over the loser's name, giving up its own, since the town Rough and Ready in Nevada County had prior claims.

Southwest of Etna the road, now unpaved, climbs a 15 percent grade to the summit of SALMON MOUNTAIN (5,969 alt.), which affords a panorama of the Scott River Valley. Down another 15 percent grade it winds along North Russian Creek to FINLEY CAMP, 48 *m.* (2,600 alt.), on the North Fork of the Salmon River.

SAWYER'S BAR, 54 *m.* (2,171 alt., 100 pop.), began as a mining camp. It has a Roman Catholic Church built in the early days of whip-sawed lumber.

Left from Sawyer's Bar 5 *m.* on a dirt road with a 20 percent grade to the junction with a dirt road.

(1) Right here 0.5 *m.* to BLACK BEAR MINE, where quartz has been extracted since the days when mill machinery had to be carried over the mountains on muleback or on ox-drawn sleds.

(2) Left here on a narrow mountain road is CECILVILLE, 20 *m.* (2,350 alt., 20 pop.), an Indian village. This is a point of entry to the SALMON-TRINITY ALPS PRIMITIVE REGION (*see TOUR 8b*). A Forest Service trail (R) follows the South Fork of the Salmon River into the wilderness.

West of Sawyer's Bar the main side road continues along the North Fork of Salmon River. At its confluence with the South Fork is FORKS OF SALMON, 70 *m.* (1,242 alt., 100 pop.), where in June 1850 the first prospectors, pushing into the Salmon Mountains along the South Fork, found rich diggings. From Forks of Salmon, the gold-seekers spread along the North Fork and over the divide into Scott Valley, until the river was dotted with camps.

At 87 *m.* is the junction with State 96 (*see above*).

Through the narrow neck of the valley US 99 veers (L) between low undulating hills.

At 32.3 *m.* is the junction with a dirt road.

Left on this road, which runs through fields of blue-green alfalfa, dotted with trim, freshly painted farmhouses and barns amid clumps of trees, is

GRENADA, 1 *m.* (2,561 alt., 300 pop.), center of a prosperous farming and dairying district.

Here the highway swerves to cut in a straight line across the Shasta Valley's level, monotonous sweep of hayfield and grazing land with scattered farmhouses half-hidden among bunched trees.

Along the roadside (L) are low, smoothly rounded, conical hills of volcanic origin, visible throughout the length of Shasta Valley. Indian legend tells how the Great Spirit whose dwelling place was Mount Shasta, wanting à home close to his own for his only daughter, set the Indians to work building Shastina, the smaller peak that juts up from the western slope below the main crater. They carried dirt and rock in great baskets until one morning the Great Spirit saw that his daughter's dwelling place, already larger than he had planned, would soon be as great as his own. Instantly he commanded the Indians to stop work. Each emptied the dirt from his basket wherever he stood; and there, for each basketful of earth, was left a little, moundlike hill.

Around GAZELLE, 41.5 *m.* (2,758 alt., 116 pop.), a handful of frame houses, weather-beaten barns, and white-painted business buildings among bushy black walnut trees, stretch grazing lands.

At 48.9 *m.* is the junction with a paved road.

Left on this road is EDGEWOOD, 1.3 *m.* (2,953 alt., 250 pop.), a dairying center on the site picked by the first travelers on the California-Oregon trail for a stopping place, where William and Jackson Brown built a log cabin in 1851.

WEED, 52.1 *m.* (3,466 alt., 4,000 pop.), at the junction with US 97 (*see TOUR 3A*), is a lumber town, bleak and raw looking, in a hill-rimmed hollow. From the logged-over slopes, dotted with scrubby timber and blackened stumps, the brush sweeps down to encroach on weather-beaten houses and rickety fences. Along the railroad sidings (L) beyond the grimy business district, spread great lumber mills with vast rows of stacked pine boards. In clearings at the edge of the brush huddle desolate, unpainted company shacks, barracklike rooming houses, and company stores.

South of Weed US 99 winds over rounded hills through a wasteland of brush sparsely dotted with pines. The dense forests of pine, spruce, and fir that clothed these lower slopes of Shasta before 1880 were recklessly denuded. The lumber companies, when they had cut the most accessible timber, burned off the land to destroy the slash and debris, killing the young trees and seedlings.

The highway cuts around the eastern flanks of BLACK BUTTE (6,344 alt.), a volcanic cone that lifts its triple-crested peak 3,000 feet above the plain, reaching an elevation of 3,937 feet at 58 *m.*

MOUNT SHASTA CITY, 61.3 *m.* (3,554 alt., 1,923 pop.), was settled in the 1850's, near the base of the immense, snow-swept peak for which it was named. Roundabout spread the lush green meadows of Strawberry Valley, sweeping up the slopes to the belt of dark green conifers at the edge of the snowfields. To early travelers it was Sisson's, in honor of J. H. Sisson, pioneer postmaster and hotel keeper, who guided parties up Shasta's slopes. Only in 1924, as the tide of tourist travel grew, did the town become Mount Shasta City.

1. Left from Mount Shasta City on Alma St. to the Shasta Trail, 3 *m.*, where the trip can be continued by foot or, in part, on horseback. (*Season: July-Sept. For amateurs, round trip from lodge to summit requires 8 to 10 hours; record is 2 hours 24 minutes. Climbers should wear calked shoes, heavy clothing, colored glasses and carry food for three meals and water, an alpenstock, and a gunnysack or piece of carpet for sliding the mountain in descent.*) The trail, narrow, rough, and dusty, but compensating for its discomforts by ever changing vistas, winds through stretches of chaparral, in springtime fragrant with wildflowers. Then it ascends through stands of sugar and yellow pine and, at higher elevations, through whitebark pine and Shasta fir. On the edge of the Fir Zone, is the SHASTA ALPINE LODGE (7,992 alt.), 8 *m.*, a stone rest house built at Horse Camp by the Sierra Club. (*Wood for fireplace and blankets from caretaker.*) A start from the lodge at 2 a.m. allows time to reach THUMB ROCK to watch the sunrise. The trail continues from the edge of the Alpine Zone's belt of storm-beaten dwarf pines at 9,000 feet and over stretches of bare, rough, brown lava to the snowfields and ice of the summit, 12 *m.*, of MOUNT SHASTA, where a vast panorama of tumbled mountains and valleys spreads on every side.

"Lonely as God and white as a winter moon," wrote Joaquin Miller, "Mount Shasta starts up sudden and solitary from the heart of the great black forest . . ." Sixth highest mountain in California, it is more impressive than the highest, Mount Whitney.

Mt. Shasta is in SHASTA NATIONAL FOREST, which, with the adjoining TRINITY NATIONAL FOREST comprises wild areas of 2,036,836 acres, where game abounds, hunting, fishing and camping grounds are available, and industries are permitted. Hotels, motels and lodges are plentiful. MOUNT SHASTA RECREATION AREA surrounds Lake Shasta and is open for winter sports.

From the higher crevices, steam hisses, and nearby, molten sulphur bubbles out: "last feeble expression," wrote Muir, "of the mighty power that lifted the entire mass of the mountain from the volcanic depths far below the surface of the plain." He went on: "Shasta is a fire mountain, an old volcano, gradually accumulated and built up . . . by successive eruptions of ashes and molten lava which . . . grew outward and upward like the trunk of a knotty, bulging tree." The material came from two craters, that of Shasta itself and a smaller one, Shastina, on the western slope. Then came the glacial winter and with it "a down-crawling mantle of ice upon a fountain of smouldering fire, crushing and grinding its brown, flinty lavas . . ." The tremendous burden of ice shattered the summit and ground deep grooves in the slopes. Then, as the glacial period drew to an end, the ice melted and broke up, leaving tumbled heaps and rings of moraine on the mountainsides. Even today, the great downspreading ice streams of five glaciers—Hotlum, Bolam, Whitney, Wintun, and Konwakiton—grip the eastern and northeastern flanks above the 10,000-foot level, feeding numberless creeks.

Although it dominates the landscape for a hundred miles, Shasta was unknown to white men until Peter Skene Ogden discovered it February 14, 1827. Credit for the first ascent goes to Capt. E. D. Pearce, a merchant of Yreka, who made the climb alone in September 1854. The first ascent by scientists was made in September 1862, when the California State Geological Survey party, led by Josiah Dwight Whitney, followed today's trail to the top and made observations. They discovered that others had preceded them, for at the summit, said Prof. William H. Brewer, Whitney's assistant, "was a liberal distribution of 'California conglomerate,' a mixture of tin cans and broken bottles, a newspaper, a Methodist hymn book, a pack of cards, an empty bottle, and other evidences of a bygone civilization." Near the summit a blizzard trapped John Muir and his companion, Jerome Fay, in April 1875; finding refuge in the hot spring, they were forced to lie for 13 hours—scalded on one side, all but frozen on the other.

2. Right from Mount Shasta City on Alma Street to the STATE FISH HATCHERY (*feeding time 1:30-2:30 p.m.*), 1 *m.* Here, in 17-acre landscaped grounds, shaded by elms, cedars, and cottonwoods, are 50 rearing ponds. In the clear depths gleam the silvery blue-gray trout: Loch Leven, Eastern Brook, Mackinaw, and Rainbow. In five trim white buildings are rows of wooden hatching troughs, where young fish dart back and forth. The plant is capable of hatching 8,000,000 annually.

3. Right from Mount Shasta on Ream Avenue, which soon becomes a dirt road winding up scenic mountain slopes to CASTLE LAKE (*camping*), 10 *m.*, a little forest-girt body of water, more than a mile above sea level, hidden among the recesses back of Castle Crags (*see below*).

US 99 runs past lumber mills, lumberyards, and log ponds. From the green meadows of Strawberry Valley it climbs into sparsely forested rolling country.

At 63.7 *m.* is the junction with State 89 (*see TOUR 5a*).

The highway crosses a high tableland bordered by mountain ridges and descends a long grade to a junction at 67.5 *m.* with a paved road.

Right here to SHASTA SPRINGS, 0.2 *m.* (2,556 alt.). High above the narrow canyon of the Sacramento, on a little plateau carpeted with green lawns, stand trim green and white cottages. Here bubbling Shasta water is bottled and shipped all over the country.

Left 0.3 *m.* from Shasta Springs to MOSS BRAE FALLS. Fairylike falls differing from all others in California, they spill in feathery sprays over banks of fern.

At 69 *m.* is the junction with a paved road.

Right on this road beyond a stone gateway to SHASTA RETREAT, 0.3 *m.* (2,554 alt.), where neat cottages cluster in a thicket of pines and California black oaks.

A high arching concrete span carries US 99 above the Sacramento River into DUNSMUIR, 70.1 *m.* (2,308 alt., 2,873 pop.). Hemmed by mountains, the little city perches on a narrow shelf along the winding canyon bed, its business buildings lining the highway for a mile and a half. Up the side streets, climbing steeply, frame houses cling to pine-forested slopes. Through the heart of town and southward stretch railroad shops and yards of a Southern Pacific division point. Thronged in season by hunters and fishermen, Dunsmuir is the supply center for a region abounding with fish and game.

The entrance (L) to CASTLE CRAGS STATE PARK PICNIC AREA, 74.7 *m.*, is at the northern end of CASTLE CRAGS STATE PARK, 50 square miles of wilderness surrounding the jagged ridge of silver-gray granite that lifts crags (R) above the brow of the wooded hill. Among stately pines are camp sites by stocked streams and lakes.

Left from the entrance to the Picnic Area on a dirt road, across the Sacramento River to the railroad station of CASTLE CRAGS, 0.4 *m.*, the site of Lower Soda Springs in the green meadows at the mouth of Soda Creek. On his way south from the Rogue River in 1843, Lansford Hastings once camped here with 16 others and built, it was said, the old fort of pine logs, Hastings' Barracks. First permanent settler was "Mountain Joe" Doblondy, a guide of Frémont's; he tilled the soil, built houses, kept a hotel, guided travelers up the California-Oregon Trail past Shasta, and fought the Indians. Mountain Joe's fabulous tales lured miners in the spring of 1855, but they were unwarranted. In anger the miners left—only after they had killed or driven away the fish and game on which the Indians lived. Modoc warriors swooped down on the little settlement and burned it in reprisal. They were pursued into Castle Crags and, after a battle, driven out.

At 76.3 *m.* is a junction with a dirt road.

Right on this road, along the bank of Castle Creek, to the base of CASTLE CRAGS, 3.5 *m.* This gigantic pile of gray-white granite, frequently tinged with pale rose, rears its jagged spires a sheer 6,000 feet above forested slopes. The crags were first called Castillo del Diablo. Here, on June 26, 1855, settlers of the region fought the battle with the Modoc who had burned the settlement at Lower Soda Springs. Joaquin Miller's version of it, garnished with boasts of his own exploits, is a colorful if not an eyewitness account; the testimony, all except his own, indicates that he was still a schoolboy in Oregon at the time it occurred. As Miller tells it, the Indians were traced to their hiding place by the flour they had spilled along the way as they made off with their loot. Mountain Joe gathered a company with recruits from the mining camps at Portuguese Flat and Dog Creek. Its leader, Judge R. R. Gibson, who had married the daughter of the Shasta chief, won over the Shasta as allies. Under the highest crag in the northwest corner of BATTLE ROCK, most prominent of the spires, the 29 whites, with Shasta allies of about the same number, fought face to face with the Modoc until they forced them to withdraw, leaving many of their warriors dead. Miller's fanciful tale of his own exploits tells how he was carried, wounded, down the mountainside in a big buckskin bag tied to the back of a wrinkled squaw, and how Mountain Joe cared for his wounds in a camp by the riverbank.

CASTELLA, 76.5 *m.* (1,947 alt., 500 pop.), spreads its straggling outskirts along the highway, but the town itself stands (L) on a paved road that winds down to the pleasant green meadows along the rocky river bed.

The highway winds along the deep canyon of the Sacramento, the roadside edged with pines and glossy yellow-green black oaks in vivid contrast. At 81.2 *m.* begins the climb into the lofty mountainous region that stretches ahead for 35 miles.

SHI-LO-AH MINERAL SPRINGS, (*cabins, trailer-camp, store*), 85.5 *m.*, cascade down the slope into a rough stone basin. The water, rich in sodium compounds and sulphureted hydrogen, lathers without soap, and three times daily changes in color from clear to yellow to deep green.

Along the winding gorge between steep, thick-forested slopes the highway twists and turns to DOG CREEK at 95 *m.* where the canyon opens out (R) to reveal a superbly magnificent vista of rugged and impressive mountainsides. At 96 *m.* is the small village of DELTA. From here the highway runs along the western edge of the northwestern branch of Shasta Lake (see below), and then crosses this northwestern branch of the lake and continues to the Pitt River branch, which impounds the waters of Pitt River, at c. 105 *m.* Here the highway crosses the Pitt River branch on the world's highest double-deck bridge, 560 feet above the river bed. The bridge carries both the highway and the railroad tracks.

At 119 *m.* is CENTRAL VALLEY (pop. 2,202). In Central Valley is the junction with a paved road.

Take latter (R) 6 *m.* to SHASTA DAM, impounding, among others, the Sacramento, Pitt and McCloud Rivers.

Shasta Dam is the chief unit of the Central Valley Irrigation Project which utilizes the northern waters of the Central Valley of California as well as the waters of the San Joaquin River and its tributaries in the more southerly section of the valley.

Shasta Dam is the second highest and second biggest dam in mass content in the United States. It is 602 ft. tall, 3,500 ft. long. It impounds the waters of a huge reservoir known as Shasta Lake. The latter has an area of 29,500 acres. The Central Valley Irrigation Project has been estimated to cost upward of $170,000,000. Thirty-seven miles of railroad tracks had to be relocated to make way for the lake, and a number of small settlements were submerged by the waters of the lake which is the second largest in California. The dam was completed in 1945. *See also page 554.*

In the vicinity of Shasta Dam were a number of copper mines. Mining there started in 1896. The fumes from the smelters killed vegetation for miles around. KENNET, former mining town, is now covered by the Lake. Its pioneer bar has been restored in Sunset Inn, Central Valley.

At 124 *m.* is the junction (L) with US 299 (*see TOUR 8a*), which unites with US 99 between this point and Redding. For 160 miles southward to Carquinez Straits the Sacramento Valley plain extends between sheltering mountain ridges.

On a flat half encircled by the swirling Sacramento River is REDDING, 127 *m.* (557 alt., 52,773 pop.), seat of Shasta County since 1888, a plywood center, dotted with hotels and small industrial plants, at the northern end of the Sacramento Valley. Although it stands within Pierson B. Reading's Rancho Buena Ventura, northernmost Mexican land grant, it owes its name not to Reading but to B. B. Redding, Central Pacific R.R. land agent. The town is a shipping point for the vast fruit-growing, farming, and mining district roundabout. Through it pour sportsmen bound for hunting and fishing. Headquarters of the Shasta-Trinity National Forests is at 1615 Continental St.

Redding is at the southern junction with US 299 (*see TOUR 8b*).

Right from Redding on an unpaved highway to IGO, 12 *m.* (1,081 alt., 40 pop.), and ONO, 17 *m.* (900 alt., 17 pop.), twin mining camps. A miner known as McPherson, one of the first to build a substantial house here, had a small son who would put on his hat whenever his father started for the mines and say, "I go." "Oh, no," was always the answer. When one camp acquired the name Igo, its neighbor consequently was called Ono.

At 134.5 *m.,* on the southern bank of Clear Creek, is a junction with a dirt road.

Right on this road to READING'S BAR, 11 *m.,* where in March 1848, only three months after Marshall's discovery at Coloma, Pierson Reading and his Indians found the first gold in Shasta County. By October 1849, 400 men were here digging. When a prospector who had arrived with one pack horse built a hotel, the settlement became One Horse Town, and, as time passed, simply Horsetown. It grew to a village of 1,000 inhabitants, with stores, hotels, 14 saloons, a Roman Catholic Church—even a newspaper, the *Northern Argus,* established in 1857. But fire leveled the town in 1868, and it never rose again.

ANDERSON, 140.3 *m.* (433 alt., 4,492 pop.), sprawls over weed-grown fields among clumps of oaks and willows, its dingy business blocks facing each other across the highway and railroad tracks. Growing up on the American Ranch, bought in 1856 by Elias Anderson, the settlement soon became a stop for travelers and the starting point of a trail to the Trinity mines.

In the fifties COTTONWOOD, 145.3 *m.* (435 alt., 150 pop.), now a scattered village of rutted winding streets, frame shanties, and red brick and frame false-front stores with wooden awnings, was a miners' trade center.

Left from Cottonwood on the main street, which becomes a bumpy, gravel-strewn road as it winds off into fields and pastures, to the junction with a dirt road, 3.2 *m.;* R. here to the junction with a second dirt road, 3.7 *m.;* L. through a gate to a fork in the road, 4.2 *m.;* L. here, through another gate at 4.7 *m.;* and over stretches of cattle land to the READING ADOBE, 5.7 *m.,* a one-story oblong structure with crumbling walls of reddish adobe brick propped by boards and a brick chimney rising from a peaked, shingled roof. Built in 1846, this was the bunkhouse of .Pierson Barton Reading's buckaroos. Here too stood a smokehouse, a barn, and Reading's two-story mansion where the pioneers of northern California—Sutter, Bidwell, Lassen, and Frémont—often gathered. Reading's 26,000-acre Rancho Buena Ventura, northernmost Mexican grant in California, stretched along the Sacramento; he took possession in 1845. When Shasta County was organized in 1850, the county seat was fixed here, where it remained till its removal to Shasta the next year. Here were grown the first cotton in the State and the first olives in northern California. Reading died in 1868; his grave is on a slight rise nearby.

At 159.5 *m.* on US 99 is a junction with a paved road.

Left on this road to the ADOBE HOUSE OF GEN. WILLIAM B. IDE, 0.9 *m.,* a State Historical Monument and museum of pioneer objects. Built by the first and only president of the short-lived Bear Flag Republic in 1849, this house stands sheltered by ancient oaks in a hedged garden overlooking the Sacramento. Here Ide established a ferry, which continued in operation until the 1870's. A small one-story structure, whose adobe bricks show through peeling whitewash, the house has been enlarged with a frame addition and lean-to porch.

RED BLUFF, 160.6 *m.* (309 alt., 7,202 pop.), seat of Tehama County, spreads along low bluffs above the Sacramento, not far from the cliff of reddish sand and gravel for which Ide's Rancho de la Barranca Colorado (red bluff ranch) was named. Red Bluff's settlers, who first called their town Leodocia, never found the gold for which they were looking, although it was discovered in adjoining sections. Their wealth came from wheatfields and vineyards. A neat little town of quiet, tree-lined streets, with old brick business buildings, Red Bluff is the chief trading center for the upper Sacramento Valley.

In the 1850's the Sacramento, flowing past Red Bluff between tree-bordered banks, was churned into foam by the paddle wheels of steamers puffing up from Sacramento. By wagon and pack train their cargoes went westward to the Trinity mines and the Coast Range diggings. In those days the saloons, taverns, and corrals were alive with hordes of boatmen, packers, gamblers, and miners. As the great valley plains were ploughed and the river water diverted into irrigation ditches, the Sacramento dwindled. It is no longer the State's mightiest avenue of traffic, although it is still navigable for 225 miles at high water, and even now, with the San Joaquin River, forms an inland waterway that ranks high in tonnage of water-borne commerce. Red Bluff now ships farm and forest products, lumber, livestock and fruit. The Red Bluff Roundup, a rodeo, takes place in April.

Of Red Bluff's former river traffic, the only remaining evidence is a neglected cluster of one-story false-front buildings, once shipping offices, overlooking the river. The old two-story red brick CITY HALL, at Pine and Washington Sts., will soon be replaced by a new building. In a green square bordered by Pine and Washington Sts., is the modern TEHAMA COUNTY COURTHOUSE. The JOHN BROWN FAMILY HOUSE, at 135 Main St., is an undistinguished little clapboarded cottage with shingled roof and porch. In 1864 Red Bluff's citizens, ardent admirers of the abolitionist crusader, raised money to aid John Brown's widow and three daughters, who lived here until 1870.

The WILLIAM B. IDE MEMORIAL MUSEUM, exhibits relics of the early settlers and objects picked up along the Lassen Trail, and occupies the Ide Adobe State Monument one mile north of Red Bluff.

Red Bluff is at the junctions with US 99E (*see TOUR 3B*) and State 36 (*see TOUR 6A*). South of Red Bluff the route follows US 99W.

South of Red Bluff the level plains of the Sacramento Valley spread like the prairie of the Midwest, rimmed by distant foothills. The route runs through grazing lands dotted with oaks and through great grain-fields crisscrossed by miles of fences. At intervals are the little towns, half hidden in clumps of trees.

CORNING, 178.6 *m.* (277 alt., 3,006 pop.), stands in a belt of olive groves. The Sevillano olive, grown here in huge quantities, is one of the largest and best varieties; olive oil refining and olive canning are the chief industrial activities. The scattered frame houses are shaded by umbrella trees and long files of shaggy palms.

US 99 crosses an irrigation canal into ORLAND, 192.1 *m.* (259 alt., 2,534 pop.), where modern schools and business buildings are shaded by umbrella and black walnut trees. The Reclamation Service

Orland Irrigation Project was the first organized in the State after the passage of the Wright Irrigation Act in 1887.

Below Orland stretch grainfields where wheat and barley were first grown on a large scale in the late 1850's. As thousands of acres of virgin land were sown, California's output of wheat became as valuable as her gold. By 1872 the Glenn-Colusa area produced 1,000,000 sacks of grain yearly. Overlord of the region was Dr. Hugh J. Glenn, who in 1867 purchased the 7,000-acre Rancho Jacinto and added to it until his holdings totaled 55,000 acres, of which 45,000 were in wheat. To put in the crop, 108 mule teams were required; they were accompanied by cook houses and feed and water wagons. From 1874 until his death in 1883, Glenn was the leading grain farmer in the United States. Afterwards poor crops and low prices led to the subdivision of the ranch. During the early 1890's, as wheat yield decreased throughout the region, grain farming was linked with stock raising and the big ranches were broken up into small tracts.

WILLOWS, 208.6 *m.* (139 alt., 4,139 pop.), seat of Glenn County, was so named because the clump of willows bordering the creek was once the only landmark between the river and the foothill settlements. This is headquarters for Mendocino Forest. The Sacramento Valley Irrigation Company pumps water from the Sacramento River. Sheep raising is profitable.

US 99 cuts into the heart of a great rice-growing belt. At harvest time the lush, water-soaked fields are a yellow-green sea of bending, plumed stalks. Rice was first grown in Colusa County in 1911.

MENDOCINO NATIONAL FOREST, west of US 99 covers more than 867,000 acres and includes mountains up to 8,500 ft., YOLLA BOLLY WILDERNESS AREA, and LAKE PILLSBURY.

At MAXWELL, 224.6 *m.* (95 alt., 506 pop.), ramshackle frame and brick buildings face the warehouses bordering the railroad tracks.

Right from Maxwell on a paved road, past prune and lemon groves covering low hills, to the junction with a dirt road; 6.1 *m.;* L. here 0.2 *m.* past a decrepit farmhouse and around the base of a craggy hill to an old STONE CORRAL, an enclosure of rough stone. Concrete pillars at the entrance bear the inscriptions: "Erected by John W. Steele 1855." Here were rounded up the herds of Granville P. Swift that grazed throughout the valley tended by Indian vaqueros. Swift, a tall Kentuckian, came to California in 1844 with the Kelsey party, first band of settlers to come directly overland; a leader in every important controversy between Yankees and Californians, he headed one of the three companies in the Bear Flag Revolt and served as Captain of Company C of the California Battalion under Frémont. After the Mexican War he became a miner and then a stock raiser.

Amid far-spreading rice fields is WILLIAMS, 233.8 *m.* (84 alt., 851 pop.), laid out by W. H. Williams in 1876, today a supply and shipping point. In 1914 when a band of about 50 I. W. W.'s created much excitement and hostility by marching through the county, the town of Williams calmly provided them with breakfast and gave them $60 for cleaning up its cemetery.

Left from Williams on State 20 is COLUSA, 9.8 *m.* (58 alt., 5,318 pop.), seat of Colusa County. The name comes from Ko-ru (scratcher), head village of the Ko-ru-si tribe, on whose ruins Colusa stands; among these Indians it was the privilege of a bride to begin her honeymoon by scratching her husband's face. Colusa was founded soon after Col. Charles D. Semple purchased two square leagues of land here from John Bidwell. Although snags, sandbars, and sharp turns made the river dangerous, he persisted in his efforts to develop river commerce until Colusa was connected with Sacramento and San Francisco by regular steamboat service. Up the Old River Road, which followed the west bank of the river from Colusa to Shasta City, ran stage coaches and freight wagons. River traffic continued until a railroad was built in 1876. After that a line of barges that towed wood to Sacramento was the only important river line until 1901, when the Farmers' Transportation Company established regular service with boats once a week. Notable for its dignified simplicity is the COLUSA COUNTY COURTHOUSE, a tall white edifice with huge pillars flanking the entrance.

ARBUCKLE, 244.6 *m.* (139 alt., 1,000 pop.), is a trade town. C. H. Locke, observing that oak trees here grew immense crops of acorns, planted 21 acres of almonds in 1892. Today more than 7,000 acres of almond orchards surround the town.

At 268.7 *m.* on US 99 is the junction with a paved road.

Left on this road is YOLO (Ind., place of rushes), 0.4 *m.* (78 alt., 296 pop.), an old-fashioned town with rutted streets and old houses, picket fences, and many walnut trees. From 1857 to 1861 it was the seat of Yolo County.

WOODLAND, 274.2 *m.* (63 alt., 13,524 pop., 1960; 16,201, 1966, est.), is a busy retail center with food processing industries. Woodland's first settler arrived in 1853; two years later a blacksmith shop was set up, about which soon clustered stores and saloons. Experiments in irrigation were begun in the vicinity with the diversion of Cache Creek in 1856. First named Yolo City and nicknamed "By Hell" for an early saloonkeeper's favorite oath, Woodland acquired its present name —suggested by the grove of huge oaks in which it stood—when the post office was opened in 1859. The large, handsome, two-story YOLO COUNTY COURTHOUSE, its central portico with Corinthian columns and a balustraded cornice, stands in a landscaped square between Second and Third Streets.

At 283.8 *m.* is the western junction with US 40 (*see TOUR 9b*), which unites eastward with US 99 (*see TOUR 9b*).

SACRAMENTO, 298.7 *m.* *q.v.*

At Sacramento is the eastern junction with US 40 (*see TOUR 9a*) and the northern junction with US 50 (*see TOUR 10a*). US 40 is being designated INTERSTATE 80.

Section b. SACRAMENTO to BAKERSFIELD; 265.5 m. US 99

This section of US 99 runs through the heart of the great Central Valley, a desert of almost unbelievable fertility under irrigation. But irrigation demands unremitting toil; the omission of a single quarterly watering throughout might kill every tree and cultivated plant on the vast valley floor. Farming here is not farming as Easterners know it; most of the ranches are food factories, with superintendents and foremen, administrative headquarters and machine sheds. Even the owners of small ranches usually concentrate on a single crop; and they must send to the store if they want as much as one egg. In addition to the permanent employees the valley uses a great deal of seasonal labor that forms a constant problem.

In 1965 the Federal law terminating the use of Mexican migrant labor (Braceros) on the farms of California began to be felt and a crisis resulted in some areas because there were not enough farmhands to harvest the crops. Efforts to recruit domestic help from outside the state were discouraging; many of the youths hired for work in the fields could not stand the routine and left. The State of California is working to produce an adequate labor supply, in part by importing aliens under terms of the Immigration and Naturalization law.

South of SACRAMENTO, 0 *m.,* US 99, which is united with US 50 to Stockton, runs between orchards, vineyards, and dairy farms.

OLD ELK GROVE, 16 *m.* now a scattering of houses and an old cemetery, was founded in 1850 by James Hall, who built a hotel that burned in 1857.

Left here on a dirt road to modern ELK GROVE, 1 *m.* (49 alt., 2,205 pop.), which sprang up after the burning of the hotel. The town lies in the midst of a Tokay grape district with one of the oldest wineries in California.

The area around GALT, 23 *m.* (46 alt., 2,600 pop.), which was laid out in 1869, produces poultry and dairy products, figs and grain.

LODI, 30 *m.* (51 alt., 26,700 pop. 1964), is center of Lodi District, featuring sherry and Tokay, the latter from Flame Tokay grapes, first planted in 1870's. Guild Wine Co., Acampo Winery and 18 others have an annual producing capacity of 50,000,000 gallons. The site was picked by the Central Pacific Railway in 1860 as Mokelumne Station, name changed 1873 to Lodi (pronounced Low-dye) presumably after a race horse. It has a Grape and Wine Festival in September. Canning, fruit processing, light industry flourish. COMANCHE DAM, on the Mokelumne River 15 m. east of Lodi, provides a recreational area for the District.

STOCKTON, 43 *m.,* is at the junction with US 50 (*see TOUR 10b*).

US 99 moves south across a flat expanse covered with great walnut orchards, truck farms, melon tracts, vineyards, and pastures.

MANTECA (Sp., *mantequilla,* butter), 56.2 *m.* (40 alt., 8,242 pop. 1960; 11,500 est. 1965), began in 1870 as Cowell's Station on the Central Pacific Ry. Its subsequent development as the shipping center for great quantities of dairy products led to the adoption of its present name. Grapes and sugar beets are also grown nearby.

Left from Manteca on State 120 to OAKDALE, 21.2 *m.* (150 alt., 4,980 pop.), a dairying center set on a plateau above the Stanislaus River overlooking irrigated valley land. Dairying is the principal industry, supplemented by diversified horticulture and almond growing. An annual Almond Blossom Festival is held in the spring. In the robust 1870's and 1880's Oakdale, on the road to the mines, was raucous at times. It takes its name from the live oaks that surrounded the site when the town was founded in 1871.

At 33.2 *m.* is the junction with an oiled road; L. on this road 0.8 *m.* to KNIGHT'S FERRY (200 alt., 80 pop.), through whose mud-bogged streets Mark Twain and Bret Harte trudged in the days when they were nobodies. In the early 1850's Knight's Ferry was temporarily renamed Dentville, for Lewis and John Dent, brothers-in-law of Ulysses S. Grant. One of the town's cherished landmarks is an old-fashioned COVERED BRIDGE over the Stanislaus River reputedly designed by Grant in 1854 while visiting them. Lewis Dent became Grant's aide-de-camp during the Civil War and Minister to Chile during his presidency. Among the buildings dating from the gold rush days are the MASONIC TEMPLE, the FIRE HOUSE with its dilapidated hosecart, and the adobe ruins of the COURT HOUSE, built in 1861 as a hotel, which served as a court house during the town's reign (1862-72) as seat of Stanislaus County.

At 48.2 *m.* is the junction with State 49-108 (*see TOUR 4b*) which unites with State 120 between this point and a junction at 55.1 *m.* (*see TOUR 4b*).

At PRIEST'S STATION, 64.9 *m.* (2,500 alt.), a busy supply depot for early gold miners, the old Priest Hotel was a stopping point on the route to the upper reaches of the Stanislaus River. The dances held in its 50-foot ballroom were gala events of the region.

BIG OAK FLAT, 66 *m.* (2,950 alt., 317 pop.), was founded as Savage Diggings by James Savage, who settled here with his five Indian wives and a retinue of red-skinned servants in 1850. The present name was derived from an oak, 11 feet in diameter at the base, that stood in the center of town until it was felled for the gold that clung to its roots; an arch, with remnants of the tree embedded in its framework, marks the site. Several pioneer structures remain, including the ODD FELLOWS HALL.

GROVELAND, 67.9 *m.* (2,850 alt., 300 pop.), once known as First Garrote for an execution by that method carried out here during the gold rush, is an attractive rural village. Reminders of the past are the iron doors and stone walls of the WELLS FARGO OFFICE and the GROVELAND HOTEL with its second-story balconies.

SECOND GARROTE, 69.9 *m.*, took its name from the HANGMAN TREE, still standing (L), from whose branches criminals of mining days are said to have been hanged. Although historians find no evidence that Bret Harte ever saw this region, two old prospectors, Chaffee and Chamberlain, who lived in the BRET HARTE CABIN (R), are locally claimed to be the originals of his *Tennessee's Partner.*

State 120 continues to rise into the mountains to the junction with a paved road, 85.9 *m.* (*see YOSEMITE PARK TOUR 3*).

Around RIPON, 62.5 *m.* (68 alt., 1,894 pop.), radiate large vineyards. Chief industrial plant in the district is the SCHENLEY DISTILLERY (R). A large road sign proclaims the sprawling, white, peaked-roof building as the "World's Largest Exclusive Brandy Distillery."

Ripon initiated "the world's largest out-door rummy game," sponsored by the American Legion post; the first out-door game was held in 1932, with 600 rummy, whist and bridge players, surrounded by as many onlookers, seated at tables set up in the town's main street. The 1938 card party, which attracted 900 participants, was staged in PORTUGUESE GROVE. Players and spectators alike paid $1 admission. The proceeds, minus the prize money, went to charities.

US 99 crosses the STANISLAUS RIVER, 63.2 *m.,* one of the fabulous streams of the gold rush era, whose pastoral beauty in this region was described by Bret Harte in his *Down on the Stanislaus.*

SALIDA, 65.5 *m.* (70 alt., 600 pop.), a fruit, alfalfa, and grain shipping point, is a center for dove and quail hunting.

MODESTO (Sp. modest) 71.9 *m.* (91 alt., 43,250 pop.), seat of Stanislaus County, is on the Tuolumne River. It was laid out in 1870 by the Central Pacific; when the railroad wished to name it after W. C. Ralston, San Francisco banker, the latter objected, and the officials commemorated his modesty in the town name. It has canneries and packing plants that serve the agricultural area. East on State 132 is TURLOCK LAKE STATE PARK. The Modesto and Turlock Irrigation Districts are served by water impounded by DON PEDRO DAM, 200 ft. high, with a reservoir of 290,000 acre-feet. In 1964 the Federal Power Commission authorized a new dam and reservoir, which will inundate the present dam, be 555 ft. high, 1,900 ft. long and impound more than 2,000,000 acre-feet of water.

In the neighborhood of CERES, 76 *m.* (88 alt., 4,406 pop.), checkered with vineyards and fig, peach, apricot, and pear orchards, the date gardens are of particular interest—new groves with low feathery palms as well as older ones whose trees are so tall that adjustable platforms, resembling painters' scaffolds, are fastened around them to facilitate the picking and pollinating. The latter process, which is always done by hand (late March or April), affects the size of the seed, the size of the fruit, and even the time of its ripening. Experiments in which pollen from different varieties of male palms was applied to separate strands of inflorescence on the same female palm (by means of glassine bags) demonstrated that the growers could control the ripening time by as much as 30 days.

US 99 crosses and recrosses a network of canals on its way to TURLOCK, 84.6 *m.* (101 alt., 10,100 pop.), in the midst of hundreds of small farms—all depending upon the vast Turlock Irrigation District water since 1901. The farms produce dairy and poultry products and a variety of crops: alfalfa, sweet potatoes, watermelons, peaches, apricots, grapes, and grain. The town celebrates an August melon carnival. STANISLAUS STATE COLLEGE, opened 1960, had 796 students in 1964.

US 99 crosses the MERCED RIVER, 92.8 *m.,* which rises in Lakes Merced and Tenaya in Yosemite National Park. The stream was named by Lieut. Gabriel Moraga, a Spanish army officer, who explored the valley in 1813. Weakened by thirst, the party drank greedily of the waters; in gratitude they named it River of Our Lady of Mercy (*de las Mercedes*).

LIVINGSTON, 94.4 *m.* (131 alt., 2,500 pop.), trade center of a sweet potato belt, also ships raisins, grapes, peaches, and alfalfa.

Left from Livingston 4 m. to the 1,100-acre VALLEY AGRICULTURAL CORPORATION VINEYARD, producing Thompson seedless grapes. The vineyards are enclosed by a rabbit-proof fence.

ATWATER, 100.9 *m.* (153 alt., 11,100 pop.), profits by the proximity of CASTLE AIR FORCE BASE, which normally employs 6,000.

BUHACH, 102.4 *m.* (155 alt., 100 pop.), is the trading center for a district of Portuguese farmers who specialize in growing pyrethrum (feverfew), powdered blossoms of which are used in insecticides.

MERCED, 108.3 *m.* (167 alt., 20,068 pop. 1960), seat of Merced County, is the principal motor gateway to Yosemite National Park via State 140. It is in a hay and cotton producing area and has a cottin gin, cement factories and potteries. A rodeo is held in June and a District Fair in October.

Left from Merced on G St., which becomes State 140; at 0.5 *m.* is the junction with Snelling Rd.; L. here to a junction at 6 *m.;* R. 1.5 *m.* to 700-acre LAKE YOSEMITE, a summer resort. On Snelling Rd. is SNELLING, 19 *m.* (252 alt., 324 pop.), once a mining town and until 1872 the seat of Merced County.

On State 140 is PLANADA, 10 *m.* (167 alt., 1,560 pop.), called Geneva when platted in 1912 as a model town.

State 140 ascends 1,900 feet in 27 miles through the foothills.

Since 1854 MARIPOSA (butterfly), 80 *m.* (1,962 alt.), has been the seat of Mariposa County. In summer the town does a brisk tourist business with Yosemite-bound traffic. County population remains near 5,000. The COURTHOUSE, a two-story, square, white-frame building, topped by a slender, square clock tower, is the oldest in California; it was built in 1854, fastened together with wooden pegs. The MARIPOSA GAZETTE OFFICE (1854) has yellowed newspaper files that hold much of California's early history as one of the United States.

Mariposa is at the junction with State 49 (*see TOUR 4b*).

West of Mariposa State 140 mounts steadily, passing camps and resorts to EL PORTAL, 72 *m.,* the principal entrance to Yosemite National Park.

CHOWCHILLA, 125.2 *m.* (240 alt., 4,500 pop.), is in a district of dairying, hog and poultry raising, cotton, fruit, and grain growing. Its Union High School Marching Band has become famous. The Chowchilla River is referred to locally as the region's Mason and Dixon Line; legend has it that Union soldiers marching south from Stockton during the Civil War were ordered to load their guns when they reached it.

US 99 passes through olive groves and apricot orchards to BERENDA, 133.6 *m.* (255 alt.), a small collection of frame dwellings. Beyond the MADERA COUNTY FAIRGROUNDS AND RACE TRACK (R) 139.4 *m.,* it spans the Fresno River.

MADERA, 140.4 *m.* (272 alt., 14,430 pop.), seat of Madera County, was laid out in 1876 by the California Lumber Company. Today it is a center for cotton, grapes, alfalfa, dairy products and livestock grown on the well irrigated area between the city and the San Joaquin River, boundary line of Fresno County. In 1944 the Madera Irrigation District was formed with FRIANT DAM as the key unit of the Central Valley project of the U. S. Bureau of Reclamation. The MADERA CANAL brings water to more than 250,000 acres, and a 153-mile canal serves the lower San Joaquin Valley.

MILLERTON LAKE NATIONAL RECREATION AREA, 21 *m.* east of Madera on State 145, was formed by Friant Dam. Its shoreline is 43 miles long. The former town of Millerton moved to Sacramento.

HERNDON, 152 *m.* (296 alt.), is bowered in fig orchards.

FRESNO, 161 *m.* (*see* FRESNO).

1. Left from Fresno on Kearney Blvd., to KEARNEY PARK, 7 *m.,* an experimental farm and park operated by the University of California. Five thousand acres are devoted to experimental work; the picturesque 240-acre park in the center of the tract is a recreational area.

2. Left from Fresno on State 180 (Ventura Ave.), to a junction at 13.2 *m.;* R. here 2 *m.* to SANGER (370 alt., 8,533 pop.), trade center of a region producing semi-tropical fruits, grapes, plums, and various field crops.

State 180 continues eastward, crosses the fertile Kings River Valley, and begins to rise into the Sierra foothills. The highway passes through GENERAL GRANT GROVE (*see SEQUOIA AND KINGS CANYON NATIONAL PARKS*), where guides are available for trips into adjacent areas. The highway, winding in a generally northward direction through the park, bends abruptly R. when it reaches the canyon of the South Fork of Kings River, 76 *m.* The road has been pushed on up the canyon year by year over what was merely a pack trail. Connoisseurs of mountain scenery have long exalted this route as one of the most beautiful in the world. Everything that the Sierra can offer in the way of breath-taking contrast and dramatic beauty is here—tiny peaceful meadows below towering polished domes, dozens of cascades, any one of which would be considered a crowning glory in an Eastern State, and multi-colored crags against skies unbelievably blue.

3. Left from Fresno on State 168 (Fresno St.), paved in the lower regions, oiled in the mountains, is CLOVIS, 11 *m.* (360 alt., 7,704 pop.), the trade center for a wine-producing, lumbering, and dairying hinterland.

ACADEMY, 22 *m.* (580 alt.), has granite quarries. State 168 then mounts rapidly to TOLLHOUSE, 32 *m.,* at the upper end of the circuitous Tollhouse Road. SHAVER LAKE (5,200 alt.), 48 *m.,* lies star-shaped in a pine-rimmed depression of the Sierras; in its waters are lake, rainbow, Lochleve, big mouth, black bass, and blue gills. HUNTINGTON LAKE 70 *m.* (7,000 alt.), 7 miles long, is fed by melting snows from the 14,000-foot peaks that rise from its shores. One of California's foremost summer and winter mountain resorts, the lake is also part of a gigantic hydro-electric project. From Florence Lake, formed by a great multiple-arch dam across the South Fork of the San Joaquin River 14 miles northeast, water is diverted through a tunnel carved through solid rock to Huntington Lake, where it is held by three concrete dams, each about 1,000 feet long and 200 feet high. From here the water shoots 2,000 feet down a 45-degree incline to the turbines of the Big Creek power house. Along the lake are various bathing beaches and at the tiny resort centers on the shores guides and horses are available for trips into the High Sierra. The highway continues further into the mountains to various summer camps.

South of Fresno trim vineyards border US 99 for unbroken miles.

MALAGA, 166.3 *m.* (293 alt., 125 pop.), is a wine center specializing in muscat grapes, a species imported from Malaga, Spain, in 1852. Sheep raising is the next most important activity of the area.

FOWLER, 170.3 *m.* (290 alt., 1,892 pop.), is an important horse and mule market. It has several whisky warehouses, grape processing plants, and fruit packing houses.

SELMA, 176.3 *m.* (305 alt., 7,100 pop.), is another packing center in the muscat grape belt. Some of the grapes are converted into muscatel wine; some are seeded and dried to become raisins; others—those that come from the vine in large, full clusters—are dried on the stem for the fancy Christmas trade. Although considered by some of better flavor than the Thompson Seedless, the muscat is second in commercial importance because an extra seeding process is required. Most of the drying is done in September; the muscats are spread on large squares of brown paper to stand in the sun for 2 or 3 days.

In the foothills around KINGSBURG, 181 *m.* (300 alt., 3,093 pop.), are many peach orchards, vineyards, and orange groves. The population, still 90 per cent Swedish in descent, is made up of the children and grandchildren of a colony of Michigan Swedes that settled here in the 1870's.

KINGS RIVER, crossed at 182.8 *m.,* was discovered in 1805 by Spaniards who piously named it El Rio de los Santos Reyes (River of the Holy Kings).

At GOSHEN JUNCTION, 196.3 *m.,* is the junction with State 198, paved.

1. Right on this road to HANFORD, 13.2 *m.* (246 alt., 10,133 pop.), seat of Kings County and the trade center of ranchers who specialize in stock-raising and dairying. Around the 50-year-old, two-story tan-brick courthouse, and the squat, granite-block jail an ambitious Civic Center has been built, including a reinforced-concrete Community Auditorium.

The Mussell Slough feud, upon which Frank Norris based his novel, *The Octopus,* came to a climax here. A group of angry ranchers had organized as the Settlers' League to fight the Southern Pacific Railroad, which under an Act of Congress was taking possession of every odd-numbered section of land along its newly built line. In May 1880 they did battle with the sheriff's forces here. Five ranchers and two deputies were killed, and 17 league members went to the jail, which is still standing.

ARMONA, 18.2 *m.* (239 alt., 250 pop.), is a shipping point for spinach, grapes, and other fruit. LEMOORE, 23 *m.* (226 alt., 2,561 pop.), is a dairying and fruit growing trade center.

COALINGA, 55 *m.* (162 alt., 6,500 pop.) is a supply point for oil fields. In 1928 new sources were discovered in the Kettleman Hills. COALINGA COLLEGE (Junior) had 797 students in 1964.

PRIEST VALLEY (2,373 alt., 67 pop.), 46 *m.,* near the crest of the Coast Range, was first explored by old Ben Williams, the trapper, in 1849. Not long after that William Galman and Captain Walker came upon a priest and 100 Indians resting here after rounding up wild horses. The valley was named because of this encounter, but it is better known as the place where Joaquin Murrieta, California's most notorious bandit, was killed in 1853. The spot where Murrieta made his last stand is said to be the Arroya Cantova, at the head of a small nearby canyon.

Unjust and brutal treatment at the hands of some Yankee miners started Murrieta on a career that terrorized the mining regions and stage routes. He began by fighting invaders of his native land but soon became a bandit. He was reputed to have killed every man of the group that originally mistreated him. The State offered $5,000 reward for his capture in 1852. In July 1853, Capt. Harry Love of Santa Clara, with a posse of 20 State Rangers, left San Jose to find him. La Molinera, Murrieta's former wife, told Captain Love that he would probably be hiding here. To mislead spies, Love and his squad pretended they were on their way to Los

Angeles, but at night doubled back to the arroyo, where they surprised Murrieta and his gang. Three-Fingered Jack, Murrieta's lieutenant, scrambled away through the underbrush with Love in chase. Love shot him between the eyes. Murrieta jumped on an unsaddled horse. Henderson, one of the Rangers, followed and shot the horse. Murrieta, stunned, began stumbling away on foot. One shot from Henderson's pistol brought him to one knee, another killed him. A man named Bill Byrnes, who had once been Murrieta's partner in a monte game, cut off his head, which was packed in salt and taken back to San Jose to prove that the bandit was dead. This head and the hand of Three-Fingered Jack were put in jars of alcohol and exhibited all over the State. They disappeared during the San Francisco earthquake and fire of 1906. Murrieta's exploits were celebrated in the picture *Robin Hood of El Dorado,* in which Warner Baxter played the title role.

SAN LUCAS, 71 *m.* (*see TOUR 2b*) is at a junction with US 101 (*see TOUR 2b*).

2. Left from Goshen Junction on State 198 to VISALIA, 6.4 *m.* (333 alt., 15,791 pop., 1960; 18,600, 1965, est.) was founded in 1852 by Nathaniel Vise, a bear hunter, who combined his own surname with his wife's given name, Sallie, to form Visalia. Modern Visalia is quite unlike the town of the 1870's and 1880's, chiefly interested in the cattle business and noted for the skill of its saddlemakers. It was particularly sympathetic to the enraged immigrants who were living on the sections of land given by Congress to the Southern Pacific Railroad. Towns like Visalia, where the inhabitants' interests were tied up with those of the ranchers rather than the railroad, gave encouragement to the revolt and refuge to hunted rebels.

The College of the Sequoias (Junior) has more than 3,000 students.

Visalia, seat of Tulare County, recently built a new Courthouse at a cost of $3,500,000. as well as a new City Hall. It is served by the Southern Pacific, the Santa Fe, and the Orange Belt stages. United Air Lines maintains daily flights from the Municipal Airport.

Under the branches of the ELECTION TREE (R), 14 *m.,* a party under the command of Maj. James D. Savage held an election on July 19, 1852, by which Tulare County was formed. The oak was actually the county seat until Visalia was founded.

The KAWEAH RIVER comes into view at 20 *m.* and parallels the road for 19 miles eastward. Its North and South Forks unite at THREE RIVERS, 28.7 *m.* (825 alt., 18 pop.).

LAKE KAWEAH is formed by Terminus Dam in the Kaweah River, 20 *m.* east of Visalia on State 198. The dam is rolled earthfill, 250 ft. tall, 2,375 ft. long, creating a reservoir of 150,000 acre-feet.

At 29.9 *m.* is the junction with a graveled road; L. here 1 *m.* to KAWEAH, now little but a name. It was founded in 1891 by former members of the International Workingman's Association who, in 1885, had made plans for a socialistic community. The members had shocked the countryside by filing individual claims to what was Sequoia National Park a few years later (*see SEQUOIA AND KINGS CANYON NATIONAL PARKS*). The group, which formally took the name of Kaweah Co-operative Commonwealth Colony in 1886, had ambitious plans that included the founding of a town to be called Avalon. Their difficulties over land claims were but the beginning of their troubles, however, and the plan was finally abandoned.

At 33.9 *m.* is the junction with another dirt road; R. here 25.5 *m.,* cutting across part of Sequoia National Park, to MINERAL KING (*summer lodge*). This tiny settlement (7,831 alt.), at the base of SAWTOOTH PEAK (12,340 alt.), was founded in 1873 by three spiritualists; here they staked the White Chief Lode silver claim, named in honor of the Indian spirit control who they said had guided them to it. By 1875 they had interested enough capital to enable them to build a toll road to the claim. There was a brief inrush of miners but the silver could not be recovered in paying quantities and in 1888 a snowslide destroyed much of the property.

At 37.5 *m.* on State 198 is the Ash Mountain entrance to Sequoia National Park (*see SEQUOIA PARK TOUR*).

TAGUS RANCH, 199.7 *m.,* on State 99 4 *m.* south of Visalia airport, is the 200-acre nucleus of what was once a 7,000-acre orchard property. Guided tours over the area are provided May to September for guests of its motel.

TULARE, 205.2 *m.* (287 alt., 14,265 pop.), has been a hard-luck town. Founded in 1872 as division headquarters and a railroad repair center by the Southern Pacific, it was just beginning to get on its feet in 1883 when a fire destroyed a large part of it. Rebuilt by 1886, it was mowed down again by fire. For 19 years the Southern Pacific pay train rolled in regularly to distribute $40,000 in $20 gold pieces. Then in 1891 a double blow was dealt: the railroad shops were moved to Bakersfield, division headquarters to Fresno. When the pay train stopped rolling the citizens had to turn to the land and develop the orchards and vineyards that make it an important shipping center.

The WHILTON MUSEUM, in the Hotel Tulare, has more than 1,000 mounted native birds and mammals, and an extensive collection of California wild flowers.

Left from Tulare on Tulare St. to LINDSAY, 15.5 *m.* (380 alt., 5,500 pop.), a foot-hill town in a citrus and olive growing belt. The vast acres keep busy several citrus-packing houses, olive canneries, and plants making olive oil. The area is at its loveliest in blossom time. The Orange Blossom Festival takes place annually in April.

Fields of cotton, white-tufted in the fall, border US 99 on its way to TIPTON, 215.5 *m.* (272 alt., 967 pop.), a shipping point for cotton, milk, and poultry.

East from Tipton 16 *m.* to PORTERVILLE (393 alt., 9,275 pop.), named in 1861 as an Overland Mail station on the Tule River. Citrus packing center; distributor for fruits, grapes, cotton. Site of STATE HOSPITAL FOR REHABILITATION. Intersection of State 65 and 190.

LAKE SUCCESS, formed by dam in the Tule River 5 *m.* east of Porterville on State 190, covers 75,000 acre-feet. The rolled earthfill dam has a maximum height of 142 ft., length 3,490 ft.

PIXLEY, 221.5 *m.* (270 alt., 1,200 pop.), dating from the railroad building era, was named for Frank Pixley, founder of the early San Francisco weekly, *The Argonaut.*

EARLIMART, 227 *m.,* is a center for cotton oil.

DELANO, 234.9 *m.* (319 alt., 13,250 pop. est. 1964), is a grain and fruit shipping point.

McFARLAND, 241 *m.* (350 alt., 400 pop.), is a cotton ginning point. Founded in 1877, its first settlers opposed whisky drinking. All land deeds contained a clause prohibiting the selling of liquor. Test cases nullified the claus in 1933-1934.

South of FAMOSA, 246.3 *m.* (423 alt., 110 pop.), shipping point for cotton, corn and dairy products, US 99 unites with US 466 (*see TOUR 11b*), for 19 miles.

The massed rigs of the Kern Front oil field stand (L), 261 *m.*, beyond a strip of fruit orchards; farther on its derricks merge with those of the Round Mountain and Kern River fields.

BAKERSFIELD, 265.5 *m.* (420 alt., 56,848 pop. 1960; 63,500 est. 1963), seat of Kern County, spreads along the Kern River in the southern end of the San Joaquin Valley. It is 291 *m.* south of San Francisco and 113 *m.* north of Los Angeles on US 99 and a junction for US 99 and Highways 466, 178, 119 and 50. It is the distribution center for the rich Kern County products, which include oil and natural gas from 13,500 wells valued at $200,000,000 annually, cotton worth $80,000,000; livestock and poultry, $44,000,000, and grapes and wine $30,000,000. It is served by the Southern Pacific, two major airlines and five bus lines. The new Civic Center in modern style is dominated by the City Hall and the County Administration and Courts Building, and the Civic Auditorium (1962), which seats 7,250. BAKERSFIELD COLLEGE, a two-year course with more than 3,000 students, has a 150-acre campus, on which stands MEMORIAL STADIUM, seating 20,000. Of historical interest are Kern County Museum and Pioneer Village, in which have been assembled the original city hall and pioneer stores and houses.

Between the various oil fields spread the ranches for which the town is a trade and shipping center—a checkerboard of cotton and alfalfa, pastures, vineyards, orchards, and apiaries. Bakersfield was named for Col. Thomas Baker, who arrived in 1862 to direct a reclamation project and remained to lay out the townsite in 1869. At that time the mining town of Havilah was the county seat, but after the enterprising new community organized a Bakersfield Club, an Agricultural Society, and a Cotton Growers' Association, the business men of Havilah began moving in. The editor of the Havilah *Courier,* leading newspaper in Kern County in the 1870's, followed his advertisers with type and press. In 1873, when Bakersfield became the county seat, most of the other citizens of Havilah came in too.

The discovery of gold in Kern River Canyon in 1885 invested Bakersfield with the color and vigor of the earlier Mother Lode boom towns. It assumed all the roughness and toughness of the camps of the unrestricted 1850's; its streets were filled with swaggering miners and gamblers, the sound of gunshots was frequently heard. In 1889 fire destroyed most of the old buildings; rebuilding resulted in modernization. Then came the discovery of oil in the Kern River fields in 1899, and Bakersfield again saw rough and tumble boom days. Today the downtown district (R. of US 99) is metropolitan in its variety of shops, cafés, department stores, theaters, and office buildings. In its streets great motor vans of potatoes, lettuce, and grapes are as familiar a sight as trucks loaded with oil well casing, drilling equipment, and derrick parts.

Bakersfield is at the junction with US 466 (*see TOUR 11b*).

Left in Bakersfield on State 178 (Nile St.) 7.5 *m.,* to a junction with Alfred Harrell Blvd.; L. here 6.5 *m.* to KERN RIVER STATE PARK, a 345-acre recreational reserve on high bluffs overlooking the Kern River. The moonlight view from these bluffs is regarded as exceptionally beautiful, even in this region where scenic grandeur is expected of the surroundings. The park has a small zoo and a pheasant and quail breeding farm for stocking the hunting areas.

At 11 *m.* on State 178 the route enters KERN RIVER CANYON, a gorge walled by steep cliffs; at their base the Kern River (L) tumbles, twisting and cascading, to the valley floor.

MIRACLE HOT SPRINGS, 28 *m.,* is a health resort. The road passes through BODFISH, a mining camp during the 1880 gold rush. Right 8 *m.* on Caliente Road to HAVILAH, old mining town.

ISABELLA DAM, at the junction of State 178 and 155, impounds the waters of Kern River in LAKE ISABELLA, which since 1954 has developed as a recreation area with more than 1,000,000 visitors annually, who come to camp, sail or fish for catfish, black bass and sunfish. The lake has a 38-mile shoreline.

Left from ISABELLA, an old mining site, 4 *m.* to KERNVILLE, (2,569 alt., 175 pop.), which came to life as Whiskey Flat during the 1885 gold rush, when a man named Hamilton opened a saloon, Kernville's first "building," which was merely a plank laid across two whisky barrels. When Whiskey Flat became a prosperous mining center in its own right it took its present name. That it was a camp of quick action and nervous trigger fingers is evidenced by the name of the old cemetery south of the town, still called Gunmen's Row, because so many local men who died with their boots on are buried in it. Kernville, still engaged in mining, is also the trade center for cattle raisers, and an outfitting point for hunters and fishermen. The characteristically "western" or "frontier" appearance of its streets has attracted many a motion picture company on location.

State 178 continues upward through magnificently beautiful country along the south fork of Kern River, following the route blazed by Joseph Walker in 1834. Walker, attached to the party of Captain Bonneville, who was looking over the Oregon country, had been detailed to lead an expedition to explore the country around and beyond the Great Salt Lake; this party, the first to cross the Sierra, spent the winter around San Francisco, then turned south through the San Joaquin Valley and up along Kern River, under guidance of two Indians. They crossed what is now called WALKER PASS (5,248 alt.), 76.6 *m.,* and descended into Indian Wells Valley, where State 178 meets State 14. Left here 11 *m.* to BROWN, on US 395 (*see TOUR 6c*).

Section c. BAKERSFIELD to LOS ANGELES; 109.5 m. US 99

US 99 shoots south across the floor of the San Joaquin Valley in a course that bends only once. Reaching the towering Tehachapi Mountains, it rises swiftly through a jagged canyon into a lofty, mountainous country. From the mountains it descends into the sheltered Santa Clara Valley, climbs into wooded hills and descends again to San Fernando Valley.

South of BAKERSFIELD, 0 *m.,* US 99, bordered by great ash trees and ragged palms, cuts through a cotton- and oil-producing region. In the fall, the distant oil derricks form a strange background for large fields of cotton, white with bursting bolls. Far ahead the road's straight strip is visible to the foot of the range. Grazing lands, dotted with wandering cattle and bands of sheep, stretch away on both sides, with patches of wild flowers in untilled fields.

East of US 99 5 *m.* to LAMONT (6,177 pop. 1960), and 10 *m.* to ARVIN (5,440 pop. 1961), distributors for farm crops and livestock.

On rain-eroded heights (R) perch oil derricks as a 5-mile grade begins the winding ascent of GRAPEVINE CANYON, deep fissure in the TEHACHAPI RANGE, once the home of Yokut and Shoshone. Through this gap in 1772 came Don Pedro Fages, *comandante militar* of Alta California, in search of two deserters. Descending through the canyon, Fages caught his deserters—the first known explorers of the valley. From GRAPEVINE, 31 *m.* (1,700 alt., 15 pop.) (*fuel, restaurant, and cabin accommodations*), is an excellent view of both the canyon and the San Joaquin Valley. A telescope offers a better view—for 5¢ or 10¢, depending upon the length of the peep. Rising steadily, US 99 reaches the summit of the grade, 35.1 *m.*

At 35.2 *m.* is a junction with a dirt road.

Right on this road to FORT TEJON, 0.2 *m.*, a state historical monument. On August 10, 1854, Lt. Col. E. F. Beale established this fort in the wilderness as protection for travelers over the mountain trail and as an administrative post for regulating affairs of the surrounding Indians. Ten years after its establishment, on September 11, 1864, it officially expired. In 1858 a strange procession wound into the clearing of Fort Tejon—a camel train imported from the Near East in an attempt to provide the Army with transportation in the deserts of the South and West—and a year later part of the train returned. The camels were of little use, since on long marches they foundered because their tender feet were not adapted to the rocky soil. In 1864 all that remained in California were auctioned off at Benicia. Some entered the circus, some packed freight, and some, turned loose, frightened the wits out of desert prospectors for many years. Another event of 1858 was the arrival of the first stagecoach of the Butterfield Overland Mail on its way to San Francisco.

On the trunk of an oak tree, 200 feet north of the fort, a Bakersfield party on an outing late in the last century discovered carved on the tree the words:

PETER LEBECK
KILLED
BY
A X BEAR
OCTr. 17
1837

Digging at the base of the tree, they disinterred the skeleton of Peter Lebecque, young voyageur of the Hudson Bay Company. On his way south through this wild land with one or two companions, he had sighted and shot a grizzly as it stood beneath this tree. Believing the animal dead and approaching to obtain the pelt, he was caught by the reviving animal, clawed, and crushed to death.

US 99 climbs through a wooded valley to LEBEC, 38.5 *m.* (3,575 alt., 30 pop.), where shimmering CASTAIC LAKE is visible (L). Ever since an earthquake in 1924, the lake has been a mineral-laden puddle of alkaline water only a few feet deep. There is a tale that white men drowned the inhabitants of a small Indian village in its waters, and that years later the mineralized bodies of men, women, and children bobbed back to the surface. The Indians had been suspected of murdering a cook and a boy at Fort Tejon.

South of CHANDLERS, 40.5 *m.* (*gas station and garage*), US 99 ascends to the summit of TEJON PASS, 41.3 *m.* (4,182 alt.).

GORMAN, 42.7 *m.* (3,774 alt., 68 pop.), was named for Private Gorman of Fort Tejon who, on his discharge from service in 1864, was one of three soldiers to take up homesteads in this region.

US 99 intersects the Grapevine grade, 42.8 *m.,* its former route, considered the final word in road building at the time of its construction in 1919. The Ridge Route, completed in October 1933, follows the course used by the early stagecoaches. An engineering achievement, it cut 9.6 miles from the earlier corkscrew road; but even the new route is not without hazards, for its smooth surface and long downgrades tempt motorists to excessive speed, particularly hazardous because of the crawling lines of diesel-powered trucks.

CASTAIC, 69.7 *m.* (1,008 alt., 61 pop.), is in the fork of US 99 and the old Grapevine road. US 99, winding through the narrow Castaic Valley, emerges into the Santa Clara Valley. Four miles from Castaic State 126 meets US 99.

A spindling steel bridge crosses SAN FRANCISQUITO CREEK, 74 *m.* On the night of March 13, 1928, the St. Francis Dam far up the San Francisquito Canyon broke without warning. Down the canyon roared billions of gallons of Los Angeles' water supply, crushing houses and drowning 600 persons. The rushing mass of water destroyed ten bridges, several miles of highway, part of the Los Angeles Aqueduct, one power plant, several hundred homes, and more than 10,000 acres of crops. Immense chunks of the dam, some as big as houses, are still seen, scattered up and down the canyon.

US 99 rises into the foothills of the SANTA SUSANA MOUNTAINS (R).

At 78.4 *m.* is a junction with a road, oiled for 1 mile.

Right on this road at 2.6 *m.* is a one-way private lane, leading to the FIRST OIL WELL IN CALIFORNIA, 4.6 *m.* Here as early as 1835 vaqueros found cattle from the San Fernando Mission bogged down in beds of pitch. In 1875 a well drilled by manpower to a depth of 30 feet brought to the surface oil subsequently piped to the Newhall Refinery six miles distant. The well is still producing.

US 99 enters San Fernando Valley and is joined by State 14 (US 6).

The black trunk of the LOS ANGELES AQUEDUCT, 84.4 *m.,* snakes down the slope alongside an open spillway near the highway (L). The 233-mile aqueduct, built at a cost of $23,000,000, converted the parched San Fernando Valley into the rich "market basket of Los Angeles," but it also ruthlessly turned towns and farms of fertile Owens River Valley high in the Sierra to semidesert.

SYLMAR, 86.1 *m.,* is a Southern Pacific loading station for an olive-packing plant.

At 86.3 *m.* is a junction with a paved road.

Left on this road to the OLIVE VIEW SANATORIUM, 1.5 *m.*, a tuberculosis hospital supported jointly by Los Angeles County and the State. Behind a front of modern Spanish buildings, a series of neat but barracks-like structures houses a thousand tubercular patients.

The VETERANS' ADMINISTRATION FACILITY, 4 *m.*, is a hospital for tubercular ex-service men. Its sand-colored, red-roofed buildings along the base of the foothills spread over 632 acres of landscaped grounds.

SAN FERNANDO, 88 *m.* (1,066 alt., 17,165 pop.), maintains its independence of Los Angeles, whose territory entirely surrounds it, by means of a municipally owned water system supplied by artesian wells.

Right from San Fernando on State 118 to MISIÓN SAN FERNANDO REY DE ESPANA, 1.6 *m.* (*adm. 50¢*). The Franciscans' seventeenth mission in order of time, it was established Sept. 8, 1797, by Fathers Fermín Lasuen and Francisco Dumetz on a site selected primarily for its natural advantages. By December 1806, the first chapel was completed. Before secularization took place in 1835, Mission San Fernando had become one of the most prosperous in California. Everything has been restored.

MEMORY GARDEN in BRAND PARK (*playgrounds*), directly opposite the convent, is one of the few restored mission gardens. In one section is the mission's fountain, built in 1812-14 in imitation of one in Cordova, Spain.

SAN FERNANDO VALLEY STATE COLLEGE, at nearby NORTHRIDGE (25,500 pop.), enrolls 12,000 students annually.

BURBANK, 96.5 *m.* (560 av. alt., 90,155 pop. 1960; 96,500 est. 1965) occupies 16.7 square miles north of Hollywood, west of Glendale and south of the Verdugo Mountains. It is an independent community with the city of Los Angeles on three sides. It is served by the Southern Pacific Railway and crossed by the Golden State and Ventura Freeways. A city of many detached dwellings, its population zoomed from 34,337 in 1940 with the coming of industry. The land was originally the Rancho de la Providencia, owned by Dr. David Burbank, from whom a land development company bought 9,000 acres. Among its first industries were First National Pictures and Moreland Motor Trucks. Today Warner Bros. Pictures, Inc., which succeeded First National, occupies the world's largest motion picture and television studios at 4000 Warner Blvd. Other studios are those of Walt Disney Productions, 2400 W. Alameda Ave., the Columbia Studio Ranch, 3701 Oak St., and Color City of National Broadcasting Co., 3000 W. Alameda Ave., for which see HOLLYWOOD.

Corporate offices of Lockheed Aircraft Corp. are at 2555 N. Hollywood Way. The Lockheed California Co. A-1 plant occupies nearly 4,000,000 sq. ft. Here and at B-1 plant are built the Orion anti-submarine missile the F-104 Starfighter, and other aircraft and experimental devices for military use, commercial transport, ocean systems, missile support and space programs. Manufacturing equipment includes an 8,000-ton triple-action Birdsboro press, so large that a building had to be constructed around it. This is one of the four major California units of Lockheed, which has plants and bases at 35 separate locations in 15 states. Also in Burbank is Lockheed Air Terminal, Inc., at 2627 N. Hollywood Way, which has runways of 6,000 ft. and 6,900 ft.

GLENDALE, 99.7 *m.* (440 alt., 119,442 pop. 1960; 131,379, 1964, est.), in the Verdugo Hills, is an independent residential suburb of Los Angeles, incor., 1906. A modern city with 15 parks, it has GLENDALE COLLEGE, offering two-year courses and evening classes; a CIVIC AUDITORIUM, the VERDUGO SWIM STADIUM, the BRAND LIBRARY, a symphony orchestra and an art association. The site of Glendale is a part of the former Rancho San Rafael, first (1784) Spanish land grant in California. Sharp Yankee traders, who invaded the region following annexation by the United States, deviously secured possession of most of the rancho. The townsite was plotted in 1886-87. The town languished until 1902. Its reawakening brought the Pacific Electric Railroad from Los Angeles.

CASA ADOBE DE SAN RAFAEL, 1340 Dorothy Dr., stands in a two-acre city park. The building was erected between 1864 and 1872 by the one-time sheriff of Los Angeles County, Tomas Sánchez, whose wife, María Sepulveda, inherited part of the vast Rancho San Rafael. Married at the age of 13, she bore 21 children in this house. Cooking utensils of her time are preserved within, together with other relics. The park is planted with orange, fig, lemon, avocado, tangerine, and eucalyptus trees. LA CASA DE CATALINA VERDUGO was the last of five adobes built on Rancho San Rafael. Across the road is a huge oak beneath which Gen. Andrés Pico made his last camp before surrendering to Gen. John C. Frémont on Jan. 13, 1847.

FOREST LAWN MEMORIAL PARK, Forest and Glendale Aves. (*open 9 a.m. to 5:30 p.m.*), is an elaborately developed graveyard that has had wide publicity. It was planned in 1917 by Dr. Hubert L. Eaton as "a great park, devoid of misshapen monuments . . . a place where lovers new and old shall love to stroll." It is notable for its exact reproduction of great works of art, five churches and a GREAT MAUSOLEUM containing eleven terraces of art masterpieces and more than 300 stained glass windows. The management says: "Forest Lawn is the only place in the world where all of Michelangelo's work may be seen together." Leonardo da Vinci's *Last Supper* in stained glass dominates the MEMORIAL COURT OF HONOR. The *Crucifixion,* a painting 195 ft. long, 45 ft. tall, is presented with hourly lectures. Reproductions include Ghiberti's 15th century bronze doors, Donatello's *St. George,* Michelangelo's *David,* Thorvaldson's *Christus,* Ward's *Washington.* The churches are often used for weddings. The five churches are a New England meeting house reproducing one where Longfellow worshiped; the Little Church of the Flowers, inspired by Stoke Poges; the Church of the Recessional, duplicating Kipling's church at Rottingdean; the Wee Kirk of the Heather, with stained glass windows telling the story of Annie Laurie. Among celebrities entombed here are Gutzon Borglum, Carrie Jacobs Bond, Theodore Dreiser, Robert A. Millikan, Clark Gable, Carole Lombard and Aimee Semple McPherson.

LOS ANGELES, 109.5 *m.* Here is the junction with US 60-70 (*see TOUR 13b*) with which US 99 is united eastward to Ontario.

Section d. POMONA to BEAUMONT, 59.4 m. US 99-70; Interstate 10

The wide concrete roadbed of US 99-70 in this section runs smoothly over miles of flatland, bordered by far-stretching orange groves and acres and acres of close-cropped grape vines.

US 99-70 branches south from US 60 (*see TOUR 13b*) at Pomona.

ONTARIO, 6.2 *m.* (980 alt., 46,617 pop., 1961; 61,250 est., 1964), originally desert land crossed by the Mojave Trail from Arizona to the Mission San Gabriel, was settled by George and William Chaffey from Ontario, Canada in 1882 as a fruit region with a mutual water system for irrigation. Their broad central Euclid Ave., 7 *m.* long, is planted with eucalyptus, pepper and grevillias. Oranges, lemons, peaches and grapes were the chief crops. Visitors gaped at its famous gravity-mule-traction, whereby two mules pulled a street car up a 2% grade for 8 *m.* and then were placed on a flatcar trailer and car and mules descended by gravity. Since World War II the area has developed industries, including Lockheed Aircraft Service, Lockheed Air Terminal, Inc., General Electric, Southern California Aircraft Corp., Aerojet General, employing 3,000. Ontario International Airport is served by three airlines and is the base for California Air National Guard's 163rd Fighter Intercepter Group. The Orange Products Division of Sunkist Growers, Inc., processes 1,500 tons of fruit every 24 hours.

The 5,000-acre GUASTI VINEYARD, 2.2 *m.* (*open 9-5 weekdays*), extends from the foothills of the San Gabriel Mountains deep into the valley floor. For five miles the highway runs between rows of vines stretching far away between furrows of sandy loam. Here are more than 500 varieties of grapes, 25 of which are raised on commercial scale. The annual crop is slightly more than 20,000 tons. The large, modern winery produces both dry and sweet wines and one of the few California "champagne" wines. During harvesting season, from August to December, an average of 400 tons of grapes are stemmed and crushed daily. The stems are returned to the soil of the vineyards in the form of fertilizers.

GUASTI, 9.7 *m.* (952 alt., 600 pop.), at the center of the Guasti vineyards, is recognizable (R) by the long, reddish-colored administration building and the corrugated iron and brick distillery. The small community bears the name of the Italian immigrant, Secundo Guasti, who in 1902 set out the first grapevines here. Eventually he had 5,000 acres and operated as the Italian Vineyard Co. The Guasti family controlled until 1945 when it sold to Garrett & Co. The Garretts had their first vineyards in North Carolina in 1835 and came to the Cucamonga region of California in 1911. The third generation now controls 7,000 acres and three wineries in California and others in New York and North Carolina.

Steep MOUNT SLOVER, 24.1 *m.*, with craggy rock outcroppings, juts up sharply in a white mist of smoke and dust. Here the corrugated iron roofs and smokestacks of a large cement plant climb the hillsides. The whole hill is gradually being carted away and ground up for building and road construction.

COLTON, 27 *m.* (847 alt., 18,666 pop.), is an industrial and railroad center not far from the site of Rancho Jumuba, a Mission San Gabriel stock ranch established before 1819. About a mile away José María Lugo built a house in 1842, when he, his brothers, and Diego Sepulveda were granted the valley. It was on Jumuba that Fort Benson was erected in the late 1850's by Jerome Benson, who raised earthworks and loaded a cannon with rocks to protect his land against seizure by the Mormons who dominated the region. Colton was named for a Southern Pacific official when the railroad arrived in 1875. It is served also by the Union Pacific main line, a branch of the Santa Fe, and Pacific Electric to Los Angeles port. Wyatt and Virgil Earp once lived here; their father was a justice of the peace and Virgil was town marshal.

Colton is at the junction with US 395 (*see TOUR 6e*), which intersects Interstate 10 and combines with US 66 eastward.

From the SANTA ANA RIVER, 28.1 *m.,* a winding trickle in a broad, brush-grown bed, the highway strikes through a jungle of auto courts, roadside cafés, and fruit and vegetable stands; through fields with scattered trees and ranch houses where cattle graze; past barns, fence-posts, and windmills. Orange trees reappear, clustering thick and leafy along the route.

REDLANDS, 32.9 *m.* (1,351 alt., 34,000 pop.), deriving its name from the red soil of the region, was originally a navel orange shipping center. Lying at the eastern boundary of the southern citrus belt, it became the packing and distributing point for citrus fruits. Since 1950 it has diversified its industry, the largest being Lockheed Propulsion Co., a subsidiary of Lockheed Aircraft Corp., which occupies a 700-acre site. It is engaged in development of solid and hybrid rockets and thrust vector control systems, with utilization of gas turbine, ram-jet, rocket, ion and nuclear engines.

Incor. 1888, its first newspaper was called the Citrograph. US 99, US 70, Interstate 10 and State 38 serve Redlands and US 395 is 6 *m.* west. Greyhound and Metropolitan Bus Lines stop here and there is bus service to Southern Pacific and Santa Fe main line points. NORTON AIR FORCE BASE occupies 1,910 acres northwest of the city limits, employing 3,000 military and 10,800 civilian personnel. It is headquarters for the Ballistics Systems Division of the Air Force Command. Also near the base are Space Technological Laboratories and Aerospace Corp. Redlands Municipal Airport has a 4,500 runway for light aircraft.

The REDLANDS COMMUNITY MUSIC ASSOCIATION has been an important influence for music appreciation in this area since 1922. It sponsors concerts and operas at the PROSELLIS BOWL giving 17 free performances of operas and concerts in July and August. Average attendance at the opera season is 38,000. The FOOTLIGHTERS, a little theater group, gives four plays a year. The University sponsors concerts by noted artists.

The UNIVERSITY OF REDLANDS, on University Hill, E. Colton Ave. and University St., occupies 33 buildings on a verdant 100-acre campus. Founded in 1907 by the Southern California Baptist Convention, it had 1,667 students and a faculty of 123 in 1965. In the Greek Theater students hold their annual Zanja Fiesta, named for the old water ditch of mission days, that now carries a stream between the audience and the actors. A 725-acre tract owned by the university in the nearby San Bernardino Mountains is used as a geological station and as a retreat for students.

The A. K. SMILEY PUBLIC LIBRARY has more than 100,000 volumes and 44,000 documents. Its annual circulation reaches 340,395. On the grounds is the LINCOLN MEMORIAL, donated by Robert Watchorn in memory of a son killed in World War I. Murals by Dean Cornwell depict scenes from Lincoln's life. Lincoln relics are exhibited.

The SAN BERNARDINO ASISTENCIA (Sp., chapel), Barton Rd. and Mountain View Ave., is the reproduction of an outpost of San Gabriel Mission that served as both a chapel for the Indians and headquarters for Rancho San Bernardino. As early as 1819 the padres of San Gabriel had built an adobe station at Guachama, the Indian village at this spot. Abandoned after secularization of the missions, it remained virtually deserted until 1842, when the Mexican Government granted a large part of the rancho, including the chapel, to the Lugo family of Los Angeles. In 1852 the Lugos sold the entire 37,000 acres to a colony of Mormons, who used the station as a tithing house until 1857. It was razed to make way for orange trees; oranges have never grown well on this spot—according to legend because an Indian medicine man put a curse on the land. The present *asistencia* was built by the San Bernardino Historical Society, which brought an expert adobe brick- and tile-maker from Mexico to mold the materials as they were fashioned in mission days.

US 99 winds past hedges of roses through Reservoir Canyon. Small green orchards of walnut, pecan, and cashew trees alternate with hayfields. The crest of the San Bernardino Mountains (L) rises above in the distance. Over creek beds and past small farms, the route leads across the flat floor of valley land surrounded by rolling shrub-covered hills.

BEAUMONT, 59.4 *m.* (2,559 alt., 4,432 pop.) (*see TOUR 13b*), is at the junction with US 60 (*see TOUR 13b*), with which US 99 is united for 52.1 miles.

Section e. INDIO to MEXICAN BORDER; 95.4 m. US 99

US 99 in this section traverses a region of sharp contrasts. Its northern end lies through the Coachella Valley, an intensely cultivated area; it cuts through a long, desolate stretch of desert in the region of the Salton Sea; and then it returns to rich farm land as it passes through the Imperial Valley (*see TOUR 14a*).

US 99 branches south from its junction with US 60-70 (*see TOUR 13a*). 0 *m.* at a point 2 miles south of Indio.

COACHELLA, 1.5 *m.* (—72 alt., 4,854 pop.), is a sprawling village in the heart of the Coachella Valley, which extends from San Gorgonio Pass to the northern shores of the Salton Sea between the Little San Bernardino Mountains (L) and the San Jacinto and Santa Rosa Mountains (R). The valley produces dates, grapefruit, cotton, alfalfa, fruit, and vegetables, irrigated by water from deep wells. Like many of the valley towns, Coachella has its Mexican section of ramshackle houses, fronting dust-covered side roads, and inhabited by agricultural workers (*see TOUR 14a*).

Left from Coachella on State 111, the North Shore road, which runs below sea level for its entire length; it is noted for its desert scenery and geological curiosities. The region is at its best in the early winter mornings, when mirages are most common, or in late afternoon when the coloring is most pronounced. In summer the temperature often rises to 125° F.

THERMAL, 3 *m.* (—127 alt., 400 pop.), in a grape and date producing area, is well named, for the days are hot almost all the year, though evenings are fairly comfortable.

MECCA, 10.9 *m.* (—197 alt., 800 pop.), a town of date gardens shimmering in the heat, was named because of its association with dates and desert.

Left from Mecca on Shaver's Canyon Road 3 *m.* to the junction with Painted Canyon Road; L. here 3.5 *m.* through PAINTED CANYON (*no water*) in the vivid Mecca Hills. At 3 *m.* on Shaver's Canyon Road is the junction with a foot trail; straight ahead on this trail 2 *m.* to PAINTED CANYON PICNIC GROUND, where the steep 500-foot walls are a multicolored blend of ochres, reds, lavenders, purples, ashy dark green, and brilliant scarlet. At 4 *m.* on Shaver's Canyon Road is a junction with Dos Palmas Spring Road; R. here 4 *m.* to a junction with a spur road; L. again 4 *m.* to HIDDEN SPRINGS, an oasis in the OROCOPIA (Sp., plenty of gold) MOUNTAINS. Here the Cahuilla Indians once gathered and held powwows.

On State 111 is CALEB'S SIDING, 15.4 *m.*, a railroad loading point. Right here 1 *m.* on a dirt road to the TORRES MARTINEZ RESERVATION, the home of 215 Indians.

South of Caleb's Siding the highway swings toward the Orocopia Mountains and the shore of Salton Sea (*see main route*). In the desert (L) bordering the Salton Sea a few smoke trees are visible. From a distance the slate-gray branches of these trees, which become 10 to 15 feet in height, resemble the long narrow pillar of smoke made by an Indian campfire. When the trees bloom in June their large, powder-blue blossoms give the appearance of the cloud of smoke made by a bonfire.

DATE PALM BEACH, 20.1 *m.*, and SALTON BEACH, 24.1 *m.* (—22 alt.), are small resorts on the shore of the Salton Sea (*camping, swimming, boating*). Some mullet can be caught in the very salty sea, but little else. There is little vegetation besides the olive and buff mesquite and creosote bushes, which cover the desert as it stretches eastward to the gully-slashed slopes of the CHOCOLATE MOUNTAINS.

A solitary CAHUILLA INDIAN ROCK MOUND is visible at 29.1 m., rising out of the desert (L) a mile from the road. On its surface are many petroglyphs. Cut in the solid rock is the likeness of an Indian head. Travertine, or petrified scale (calcium carbonate), clinging to the flat surface of the rocks, shows that it was once below the water line of prehistoric Lake Cahuilla.

South of BOMBAY BEACH, 35.8 *m.*, another small resort, the route swings out into the desert around the southern end of the sea, crossing many little gullies.

NILAND, 53.8 *m.* (— 30 alt., 1,815 pop.), is at the northern end of Imperial Valley (*see TOUR 14a*), surrounded by some of the largest ranches in the valley; on one ranch alone are 4,000 grapefruit, 17,000 orange, 6,000 lemon, and 2,000 tangerine trees. NILAND MARINA, for water sports, is located 11 *m.* north of Niland west of Route 111. At HOT MINERAL SPA, 20 *m.* north of Niland, baths in steaming pools are available.

At 56.8 *m.* is the junction with a dirt road; R. 5 *m.* to the MUD POTS, 20 acres of boiling mud craters, geysers, and mineral springs, bubble up from the depths of the earth into small basins, shooting out jets of steam and sulphurous gases. As the hot grayish mud spouts out, it builds a rim around the edge of each geyser and slowly builds up the edge of the basin.

Fishermen and sailors frequent the northeast shore of Salton Sea, served by State 111. RED HILL MARINA is located north of Calipatria. Mullet Island, farther north, once popular, is visited chiefly by hunters. In the 25-mile SALTON SEA STATE PARK visitors may camp at the water's edge at BOMBAY BEACH, SALTON BEACH, MECCA BEACH and DESERT BEACH.

A short distance by foot trail southwest of the Mud Pots are the INDIAN PAINT POTS, where water from oxide springs has formed motley incrustations.

CALIPATRIA, 60.1 *m.* (—183 alt., 2,542 pop.), is the center of an area where green peas are the principal crop. Alfalfa is also important. A local mill grinds the alfalfa into meal for cattle fodder.

Near Calipatria are several duck preserves and gun clubs. Best duck hunting is found on the southern shores of the Salton Sea in the tule marshes between the New and Alamo Rivers (*blinds and boats available*).

The highway crosses NEW RIVER, 65.8 *m.*, a channel marked by high, reddish bluffs, slashed in the loose alluvial soil of the Imperial Valley in 1905-1907 when the Colorado River broke loose on its last rampage. The break occurred when the irrigation engineers opened a breach in the river wall in 1905 just south of the Mexican boundary, in order to supply valley canals with the water held back by an accumulation of silt. The gap was cut far enough ahead of flood time for safety, but before gates could be installed a premature flood raced down the river. Rebuffed by a natural levee at the tip of the Gulf, it backed through the new opening into the Salton Sink. Within two years the lake covered an area extending from Mecca to Niland, reached a length of 45 miles and a maximum width of 17 miles. The railroad company had to shift 67 miles of track to a higher level. Just as the railroad was rolling down its sleeves, the river again broke its banks and poured disastrous quantities of water into the basin. The company again set to work, built a 90-foot trestle across the break, commandeered equipment from 1,200 miles of track, and dumped 3,000 carloads of rock into the opening. Finally, in February 1907, the break was plugged, and the railroad's ledgers showed another $1,000,000 in the red. Suit was filed against the Federal Government for the $3,000,000 spent, and in 1930 the company received a check for $1,012,665.

South of New River, State 111 is bordered by the emerald green of alfalfa fields, alternating with pasture lands, corn and melon patches, and fruit groves. At intervals the route is lined with eucalyptus trees, planted as windbreaks.

BRAWLEY, 73.8 *m.* (—115 alt., 13,752 pop.), (*see below*), is at the junction with US 99.

At 9.5 *m.* on US 99 is the junction with a dirt road.

Right on this road, crossing the All-American Canal to the FISH TRAPS, 2.5 *m.*, piles of rock two or three feet high. The handiwork of the ancient Mountain Cahuilla, they were devised in the remote period when a great Indian village stood on the shores of the prehistoric Lake Cahuilla.

They are easily discerned among the boulders and consist of circular walls of rock with cleverly arranged openings.

The SALTON SEA, (L), 14 *m.* (—244 alt. at surface), a gourd-shaped body of water 30 miles long and from 8 to 14 miles wide, was only a vast, sandy depression when discovered in 1853 by Professor W. P. Blake, who made the first governmental survey of Imperial Valley. In 1905 the Colorado River overflowed into the Imperial Valley (*see above*) and poured into the Salton Sink, filling it to a depth of 83 feet and a length of 45 miles. When this flood was checked in 1907, it left the lowest area still filled with the present Salton Sea—a lake with no outlets. The present depth of the sea is kept approximately constant, despite evaporation, by waters draining from the irrigation ditches into the New and Alamo Rivers, which empty into the sea.

The lake replaces the final remnant of prehistoric Lake Cahuilla, created when the silt from the Colorado River dammed back the waters at the head of the Gulf of California. The old lake dried up gradually, leaving behind in the San Jacinto Mountains, 1,000 feet above the level of the Salton Sea, enormous beds of fossil sharks' teeth and oyster shells. Bits of fragile conch shell today glisten everywhere in the desert sands of this region. A printer's error in setting *conchilla* (Sp., small sea shell) gave Coachella Valley its unusual name.

The southwest shoreline of Salton Sea, extending 30 *m.* northwest to Travertine Rock, has become a booming recreation area. Largest development is SALTON CITY, with two marinas, motels, the Salton Bay Yacht Club with a $500,000 plant, and the annual 550-mile boat race. Next on the way north come SALTON SEA BEACH, SUN DIAL BEACH and DESERT SHORES.

CORAL POINT is the start of the petrified coral scale of the ancient beach line, which cuts an even whitish stripe along the sides of the Santa Rosa Mountains. Despite the heat, pink and white oleanders flourish along the highway and the green (in winter) or gold (in summer) strands of dodder industriously ensnare occasional trees.

TRAVERTINE ROCK, (R), 23.4 *m.,* a geological relic of Lake Cahuilla, rises from the desert a few score yards from the highway. The once submerged lower half of this great mound of rocks is covered with a layer of travertine (calcium carbonate), a scaly, petrified shell formation; the upper part, of travertine rock, shows its core of dark granite and many petroglyphs, or rock carvings, incised ages ago by the ancestors of the Mission Indians.

The Fish & Game Commission of Imperial County warns that Salton Sea is relatively shallow and subject to turbulence during windstorms. Check weather reports of local radio stations before embarking. Boats must carry Coast Guard approved life-jackets or cushions; fire extinguisher; anchor and line; bailer or bucket; distress flares; lights. California fishing licence plus warm-water stamp required. Limits of catch: 6 corvina, 20 sargo.

At 47.3 *m.* is the junction with State 78 (*see TOUR 6C*).

KANE SPRINGS, 48.8 *m.* (— 150 alt.), is the oldest known water hole on the Colorado Desert (*service station, restaurant*), long a camping ground of desert explorers and Indians. One of the most prevalent of local myths concerns a Spanish galleon that sailed into the northernmost arm of the prehistoric Gulf of California, to be abandoned there with its fabulous cargo of gold. As the sea dried up, the hapless ship sank beneath shifting dunes. In 1890 an oldtimer appeared at Kane Springs asserting he had seen the ancient ship nearby almost covered by a dune. Searchers, however, failed to find it. The probable inspiration for the legend was a boat built in 1862 by a Colorado River mining company, transported part way across the desert by ox team, and then abandoned because of the difficulty of the journey from San Gorgonio Pass to the Colorado River.

South of Kane Springs a long sand mound, or desert wall (R), protects the highway from floods during the rare but sudden thunderstorms. As the road turns abruptly east, sagebrush, creosote, and the grayish green greasewood give way to trees and bamboo grass.

WESTMORELAND, 62.3 *m.* (— 150 alt., 1,400 pop.), founded in 1910, boasts of having the widest main street in the valley and no cemetery. In the community live Yankees, Negroes, Indians (Asiatic), Mexicans, Chinese, and Japanese. As in other valley towns, the population is swelled during the harvest season by itinerant agricultural workers.

For Westmoreland or Benson Boat landing take US 99 to Vendel's Corner, turn right 4 *m.* to Santon Sea.

Through fields of grain, vegetables, and cotton, vast melon patches, and endless rows of winter lettuce, runs a network of irrigation canals. Long lines of sugarcane stand in green files along the edges of ditches. Intense heat prevails in this region throughout the summer months, with temperatures ranging up to 125°, but the torrid heat is made tolerable by the extreme dryness of the desert air. In winter the temperature varies as much as 60° between day and night.

BRAWLEY, 70.2 *m.* (— 115 alt., 13,752 pop. 1962), largest town in Imperial Valley, is the center of highly profitable cattle-feeding operation. Herds are shipped in during the fall, fed scientifically and shipped out before the hot summer months. The industry has seen great expansion since 1950. One of the largest feed lots handles 25,000 head a year. The desert area around Brawley is said to feed approximately 78% of the beef in California, and in an average year 600,000 head of cattle valued at $61,000,000 are fed there. The area also is responsible for growing 135,000 acres of alfalfa, which had six to seven cuttings a year.

Brawley celebrates its distinction by the annual CATTLE CALL and IMPERIAL VALLEY RODEO, which draws more than 50,000 visitors in November. An arena was built for this purpose.

South of Brawley US 99 cuts through one of the valley's most productive areas. Here grow the best cantaloupes, white head lettuce, and alfalfa. Contrary to usual dairying methods, cows are milked in the open and their fodder—baled hay—remains stacked and uncovered in the fields.

IMPERIAL, 79.9 *m.* (—67 alt., 3,000 pop.), is the oldest town in the valley. This is headquarters for Imperial Irrigation District, which covers more than 900,000 acres and diverts water from the Colorado River at Imperial Dam, servicing more than 3,000 miles of irrigation canals and 7,000 miles of drainage. The District controls All-American Canal and the first fifty miles of the Coachella branch of the Canal. More than 435,000 acres were irrigated in 1962 at a rate to consumers of $2 per acre-foot (covering one acre with 1 ft. of water).

The IMPERIAL COUNTY AIRPORT, 80.4 *m.,* is a flagstop for transcontinental airlines; opposite it are the IMPERIAL COUNTY FAIR GROUNDS. The California Midwinter Fair is held in March.

At 84.4 *m.* is a junction with US 80, to El Centro. (*Page 639*)

HEBER, 90.4 *m.* (—9 alt., 991 pop.), is a shopping and shipping point for the southern valley. Founded in 1901 by the Imperial Land Company some distance eastward and named Paringa, the town was moved two years later following completion of the Southern Pacific Railroad survey and renamed for a president of the California Development Company.

CALEXICO, 95.4 *m.* (5 alt., 7,992 pop.), a border town, is a port of entry to the United States. Here Mexican travel permits are available (*no permit necessary if visitor returns same day he leaves*). Calexico is on a former 160-acre tract owned by George Chaffey, one of the promoters of the first irrigation projects in the valley. The tent city of the Imperial Land Company was the first Calexico. A press agent of this company is said to have coined the names of the twin cities of Calexico and Mexicali (just over the border) by combining syllables from the words California and Mexico. Before the repeal of the Volstead Act, Calexico was chiefly a week-end town visited by Imperial Valley workers in pursuit of the bright lights, dance halls, and saloons of the adjacent Mexican town. In recent years Calexico has thrown off its tawdry border aspect and carried out civic improvements, and Mexicali's bars and cantinas have lost the lure they once held for pleasure-bent valley inhabitants.

ROCKWOOD HALL, on the International Boundary Line, is an old adobe building, amid date palms, once used by Charles R. Rockwood, the man most closely associated with the development of Imperial Valley. The structure, restored in 1932, contains two assembly halls.

Calexico is a focal point for tourism; within reach of one-day trips are the American Sahara, the All-American Canal, and the Colorado River, a Butterfield stage depot, and the Yuma territorial prison. Across the border Mexicali offers many attractions. A city of about 175,000, it is the capital of Baja California, an area with many country clubs and opportunities for fishing, golf, swimming, and spectator sports such as bull fighting.

The All-American Canal crosses New River west of Calexico in two 15½ ft. diameter pipes with a capacity of 2,700 cubic ft. per second. Up to 585,500 gallons of water per second pass this point during a peak period.

Tour 3A

(Klamath Falls, Ore.)—Dorris—Macdoel—Weed; US 97
Oregon Line to Weed, 57.4 *m.*

Paved roadbed throughout.
Accommodations limited.

US 97 follows a southwesterly course through the central section of northern California, a thinly populated country, high, rugged, and semiarid. The northern section of the route runs through a high mountain valley, walled by the Siskiyou Range. South of the valley the course swings more sharply westward and follows the foothills at the northwestern base of Mount Shasta through rough, lava-scarred country. The region, one of farming, lumbering, and stock raising, was settled late, after the advent of roads and railroad. Lumbering is the major source of income. Here, as throughout most of Siskiyou County, are stands of yellow and sugar pine, red fir, and cedar. This section has recently become popular for hunting and fishing.

US 97 crosses the Oregon Line, 0 *m.,* 21 miles southwest of Klamath Falls, Ore., and runs through Butte Valley. The valley, 17 miles in length and approximately 10 miles wide, once cradled a great lake, long since dried up. When the Southern Pacific Railroad was built through this region, a colony of Dunkards settled here. Coming from Iowa and states further east, they were unacquainted with the principles of dry farming, and their first crops failed. In time, however, they learned to irrigate from the water sources that underlie the valley.

DORRIS, 3.5 *m.* (4,238 alt., 500 pop.), a trade center for farmers and vacationists who visit the area, was one of the Dunkard colonies. Lower Klamath Lake lies east; also a National Wildlife Refuge.

US 97 continues southwestward. To the west is the basin known as Meiss, or Butte Lake, in ancient times part of the lake that covered the valley.

MACDOEL, 14.5 *m.* (4,258 alt., 150 pop.), also a Dunkard colony, is now a business center for the Butte Valley Irrigation District. The whine of saw mills and the ring of sharp-bitted loggers' axes resound in the high clear air.

US 97 was rerouted south of Macdoel late in 1938 to skirt, rather than cross, DEER MOUNTAIN (5,380 alt.), whose height (L) is unimpressive. For several miles the highway traverses a desolate, sagebrush-covered region.

US 97 enters the rugged, lava-strewn region that fans out from the base (L) of snow-covered Mount Shasta (*see TOUR 3a*).

At 45.6 *m.* is the junction with a rough dirt road.

Right on this road, through pine- and juniper-forested wastes, 2.3 *m.* to the junction with a dirt road; R. here 2 *m.* to the base of SHEEP ROCK (5,500 alt.), a forbidding pile of weather-scarred crags resembling battlements. In early days, several square miles of level surface nearby, dotted with patches of grass, was a winter pasture for thousands of wild Mount Shasta sheep. From the higher ridges they came down to the warm lava crags and plateaus. Naturalist John Muir marveled when he heard how they were seen leaping from an almost perpendicular lava headland 150 feet high ". . . without evincing any extraordinary concern, hugging the rock closely and controlling the velocity of their . . . movements by striking at short intervals and holding back with their cushioned, rubber feet . . ." Long hunted by tribesmen, the sheep disappeared altogether after wanton slaughter by white settlers.

The main side road continues to a junction with a narrow dirt road, 3.3 *m.*; L. here 0.3 *m.* to the end of the road, where a faintly discernible foot trail marked by paint-daubed rocks leads 0.3 *m.* through junipers and over hummocks of lava to a great hole walled with lava crags—the entrance to PLUTO'S CAVE. The trail descends to a great archway that leads into a high-ceilinged cavern. Opposite (R), through another arch is another great well. The rough floor slopes down to the opening of an enormous tunnel, roofed by a jagged ceiling that arches 50 to 60 feet above a floor piled high with tumbled rocks. A short distance beyond the opening is a third large well; beyond another vast arched gateway the cave winds into the depths of the earth,—for nearly a mile it is 30 to 40 feet wide and 50 to 60 feet high. Within its black depths early explorers found owls and bats, the bones and horns of wild sheep, and blackened ashes of fires.

US 97 swings round the reddish brown, craggy hump of YELLOW BUTTE (R), dips to the valley floor, and climbs again into rubble-scarred foothills at the base of Mount Shasta. At 51.9 *m.* there is a view (R) of Little Shasta Valley. The route crosses rolling hills where stands of pine and cedar grow amid stretches of cutover land, and reaches WEED, 57.4 *m.* (3,466 alt., 4,000 pop.) (*see TOUR 3a*), at the junction with US 99 (*see TOUR 3a*).

Tour 3B

Red Bluff—Marysville—Roseville; 123.4 *m.* US 99E

Southern Pacific R.R. parallels route between Marysville and Los Molinos; Sacramento Northern R.R. between Marysville and Chico. Accommodations plentiful; hotels and auto camps in larger towns. Paved highways throughout; open all seasons.

US 99E runs between the river and the highlands through the great plains of the eastern Sacramento Valley. Far eastward, paralleling the route, the land slopes upward to the dim blue foothills of the Sierra Nevada. Level grainfields—green with young wheat and barley in season, or yellow with stubble—and gently rolling stretches of grazing land, dotted with groves of oaks, spread on either side. In the valley lands, where water is plentiful, orchards fringe the highway for miles, white and pink with blossoms in spring. Many of the quiet valley towns, half buried among trees, preserve traces of their mid-nineteenth century origin.

Branching east from US 99, 0 *m.* (*see TOUR 3a*) in RED BLUFF, US 99E crosses the tree-fringed bed of the placid Sacramento River. Beyond orchards and fields, far in the east above the foothills, towers Lassen Peaks (*see TOUR 5a*).

At 2.6 *m.* is the junction with State 36 (*see TOUR 6A*).

At 14.8 *m.* is the junction with a paved road.

Right on this road crossing the Sacramento River to TEHAMA, 1.3 *m.* (218 alt., 190 pop.), until 1857 the seat of Tehama County, a quiet village of dusty, tree-shaded byways, surrounding a frame steepled church and false-front buildings with wooden awnings. Settled in 1847 when Robert H. Thomas built an adobe here on his Rancho de los Saucos (ranch of the elder trees), Tehama became a busy freighting and trading center and the chief ferry crossing between Marysville and Shasta until Red Bluff outrivaled it as a river town.

LOS MOLINOS (the mills), 15.3 *m.* (211 alt., 200 pop.), is headquarters for small dairy, poultry, and orchard tracts.

The SITE OF BENTON CITY, 21.3 *m.,* is marked by a concrete monument and a dugout, formerly a cellar. Peter Lassen laid out a town here in 1847 on his 26,000-acre Rancho Bosquejo. To round up settlers Lassen went to Missouri, in honor of whose expansionist senator, Thomas H. Benton, he named the place. He returned in the summer of 1848 with the first group to come overland to settle in the upper Sacramento Valley. With him he brought a charter, granted by the Grand Lodge of Missouri, for the first Masonic Lodge in California, Western Star Lodge, No. 2. But the discovery of gold depopulated Lassen's embryo city and the lodge was moved to Shasta in 1851.

At 22 *m.* is the junction with a paved road.

Right on this road is VINA, 0.8 *m.* (206 alt., 300 pop.), center of Senator Leland Stanford's former 55,000-acre grape-producing Vina Ranch, established in 1881. Given to Stanford University on his death in 1892, the ranch has been subdivided into fruit, nut, and garden tracts.

US 99E continues through far-stretching grainfields and farmlands. At 37.9 *m.* is the junction with a paved road.

Left here to RICHARDSON SPRINGS, 9.6 *m.* (*hotel and cottages*), an all-year resort in the foothills near mineral waters.

CHICO (Sp., little), 40.8 *m.* (193 alt., 14,757 pop.), a trade center since Gen. John Bidwell planted his orchards here, stands in a farm belt producing fruit, grain, and almonds. In the late 1840's John Bidwell, a member of the first overland party to cross the Sierra Nevada, combined the Rancho Arroyo Chico and Rancho de Farwell into the Rancho Chico which became renowned for its great and varied productivity and its miles of tree-arched avenues. Bidwell maintained an experimental orchard of 1,800 acres, which at the time of his death contained 400 varieties of fruit. A pioneer in raisin growing and olive oil manufacture, he began wine making in 1864 or 1865, but after two years plowed up his vineyards. He later ran for President of the United States on the Prohibition Party ticket. When in 1860 he laid out the town of Chico, he offered free lots to any who would build on his townsite. Before the end of a decade Chico was a city of 2,000, boasting hotels and churches—and even saloons and gambling houses, despite Bidwell's advocacy of temperance.

A memorial to Chico's founder is the granite monument on the SITE OF THE BIDWELL ADOBE, built for him by the Maidu Indians in 1852. Beyond, in a parklike preserve, is the two-story, stone BIDWELL MANSION with its broad verandas and huge central tower. The mansion is now a state monument.

CHICO STATE COLLEGE, has a five-year course and more than 5,000 students. The college was begun in 1887 on a 10-acre plot donated by Bidwell.

The INDIAN VILLAGE, where lived the Maidu (known as the Bidwell Indians), extends along Sacramento Avenue (R) on the northwest outskirts of town. When Bidwell arrived in the late 1840's, the Maidu were wild as deer, the men going about wholly naked, the women clad only in skirts of grass. Under the régime of Bidwell, who built houses, a school and a church for them, they worked as ranch hands.

CHICO TEEN CENTER, opened 1959 in Silver Dollar Fair Grounds, permits teen-agers to act adult roles.

Donated by Bidwell's widow, 2,400-acre BIDWELL PARK (*picnicking and other facilities*), at the east end of Fourth St., winds for 10 miles along Big Chico Creek, a narrow strip luxuriantly green with vines, oaks, and sycamores, which served as the Sherwood Forest of *The Adventures of Robin Hood,* filmed here in 1937. The HOOKER OAK, approached by South Drive, was named for the British botanist, Sir Joseph Hooker, who in 1877 adjudged it the world's largest oak tree; it has been estimated to be 1,000 years old, and rises 101 feet. From its massive trunk, 28 feet in circumference, giant branches spread over an area 147 feet in diameter.

At 42.5 *m.* is the junction with a paved road.

Left here to the UNITED STATES PLANT INTRODUCTION STATION, 2 *m.*, (159 alt.) a 240-acre farm where plants brought from many parts of the world are studied, acclimated and, if possible, developed into proved species for American soils.

The highway cuts through plum and almond orchards to the roadside hamlet of DURHAM, 47.7 *m.* (159 alt., 1,500 pop.). A tract of 6,300 acres here, purchased by the State, was subdivided in 1918 and sold to settlers, many of them war veterans, under supervision of the State Land Settlement Board.

NELSON, 54.4 *m.* (121 alt., 200 pop.), in the midst of a grain-growing area where the first wheat was planted in the early 1850's is named for Capt. A. D. Nelson, early wheat grower.

RICHVALE, 58.4 *m.* (121 alt., 100 pop.), has grown up in the heart of California's leading rice-growing region, where the U. S. Department of Agriculture established in 1908 a rice-experiment station that tested 275 varieties. In 1912, 1,000 acres about Richvale were planted. The rice thrived and spread throughout the region. The side-roads westward run through swampy fields covered with the rippling green of top-heavy rice plants. Low irrigation dikes weave through them in intricate patterns.

At 61.5 *m.* is the junction with a paved highway.

Left here to OROVILLE, 7.9 *m.* (205 alt., 6,115 pop.) at the junction with State 70, US 40 Alt. (*See Tour 6B*).

Deep among peach trees is LIVE OAK, 76.9 *m.* (74 alt., 800 pop.).

Right from Live Oak on a paved road to the base of SUTTER BUTTES, 7 *m.*, jutting up from the flat valley floor. Of the four jagged peaks, SOUTH BUTTE (2,132 alt.) is the highest. The buttes are the eroded remains of an ancient crater that formerly rose twice their present height. The Indians hereabouts were driven from their homelands by horse and cattle thieves who found hiding places in the buttes. Just before the Bear Flag Revolt, (*see TOUR 2A*) Capt. John C. Frémont and his expedition camped here.

The road continues around the base of the buttes.

YUBA CITY, 87.5 *m.* (50 alt., 13,200 pop., 1963), on the Feather River, seat of Sutter County, was laid out in 1849 when John A Sutter deeded the site to Saml. Brannan, *et al.* Called "the peace bowl of the world," the area processes cling peaches, cherries, plums, walnuts, almonds. Feather River Cooperative is the country's largest prune dryer. The largest walnut orange is near Yuba City. Irrigation benefits 75 per cent of the farms, which produce grain and peas in winter and beans and corn in summer. Opposite the old-fashioned HALL OF RECORDS in its tree-shaded square, 229 B Street, a giant walnut tree planted about 1878, rises 100 feet, its branches spreading over a 104-foot area from a trunk 15½ feet in circumference at the 4-foot level.

Yuba City is at the junction with State 20 and across the Feather River has access to US 40 Alt., and State 70.

US 99E crosses the Feather River to MARYSVILLE, 88.6 *m.* (61 alt., 10,394 pop.), seat of Yuba County, at the confluence of the Feather and the Yuba Rivers. Here in 1842 Theodore Cordua built a trading post that became a way station on the Oregon-California Trail. In the winter of 1848-49 Cordua sold out to men who opened a general store and laid out a townsite. Moving spirit behind the town's early growth was Stephen J. Field, later appointed by Lincoln to the United States Supreme Court, who arrived in 1849 and three days later was elected mayor, chiefly because he had bought 200 lots for $16,250. The night of his election the town was named Marysville for Mary Murphy Covillaud, a survivor of the Donner party and wife of one of the owners of the townsite. As head of navigation of Feather River Marysville became the port by which thousands of miners moved to the Yuba and Feather River diggings. By 1854 twenty trucking outfits with 400 wagons and 4,000 mules operated out of the town.

Hydraulic mining for gold choked the Yuba River and raised its bed 70 feet, making necessary huge earthen levies to protect Marysville from inundation. LAKE ELLIS, in the heart of Marysville, was built by diverting water from the Yuba River. Part of the projected county water development is the Bullards Bar Reservoir, to be created by Marysville Dam on the Yuba River, with a shore line of forty miles extending within ten miles of the existing Englebright Dam. Camp Far West, near Beale AFB, acquired a lake when a new dam was built on the Bear River. The Virginia Ranch Dam (Brown's Valley) was completed in 1964 at a cost of $4,900,000. A new YUBA COUNTY COURT HOUSE costing nearly $3,000,000, has been completed.

Historic relics of the gold mining era abound in old brick buildings in Marysville, and in mining sites nearby. Aaron Museum, 704 D St., built 1856, displays documents and mining equipment. Stephen J. Field's house, 630 D St., was built of adobe. The John C. Fall house, 706 G St. (1855) has a free hanging staircase. Ramirez House, 222 Fifth St., built about 1850 by Jose Ramirez, a Chilean, using South American hardwood and 30 in. walls, outside staircases, marble floors.

YUBA COLLEGE (Junior) in 1962 opened at its new 160-acre site on Beale Road; it had 2,992 students in 1964.

In the vicinity are Beale Air Force Base, 14 *m.* east of Maryville; home of a Strategic Air Command; Yuba Consolidated Gold Fields, 10 *m.* east, and Upper Narrows Dam, on Yuba River, a recreation area.

BEALE AIR FORCE BASE, 14 *m.* east of Marysville, is headquarters for the 14th Strategic Aerospace Division of the Strategic Air Command, and for the 456th Strategic Aerospace Wing (SAC). The Wing has three tactical squadrons, manning intercontinental bombers, refuelling with jet aerial tankers and handling Titan I intercontinental ballistic missiles. Missile launch sites are maintained at Lincoln, Sutter Buttes and Chico. Beale AFB has 4,500 military and 355 civilian personnel with an annual payroll of $16,800,000.

Left on this road is HAMMONTON, 9.8 *m.* (135 alt., 836 pop.), which sprang up in 1905 when the Yuba Consolidated Goldfields began dredging in the Yuba River bottoms; it was named for the company's W. P. Hammon, builder of the first successful dredge. As many as seven giant dredges, costing from $1,000,000 to $3,000,000 each, operated here in landlocked artificial basins, scooping up earth in front, washing out the gold, and moving on, trailed by piles of sand, rock, and gravel.

WHEATLAND, 101.6 *m.* (90 alt., 500 pop.), is a village of weather-beaten tree-shaded houses among hop fields. During the picking season, thousands of migratory workers camp on the fringes of the

Industry, Commerce and Transportation

GOLDEN GATE BRIDGE, SAN FRANCISCO

SAN FRANCISCO-OAKLAND BAY BRIDGE

SAN PEDRO HARBOR, TERMINAL ISLAND

MONTEREY BAY FISHING FLEET

FREIGHTER AND LOADING CRANE, OAKLAND HARBOR

WATER FRONT, SAN DIEGO

SOLID PROPELLANT ROCKET MOTOR TEST,
LOCKHEED, BEAUMONT

HARBOR FREEWAY TO LONG BEACH, LOS ANGELES

UNION PASSENGER STATION, LOS ANGELES

TELEVISION CITY, CBS, HOLLYWOOD

WARNER BROS. STUDIOS, BURBANK

)RTHERN CALIFORNIA SAWMILL SITE

(E DALTONS ROBBING A KANSAS BANK,
)RRIGANVILLE MOVIE RANCH

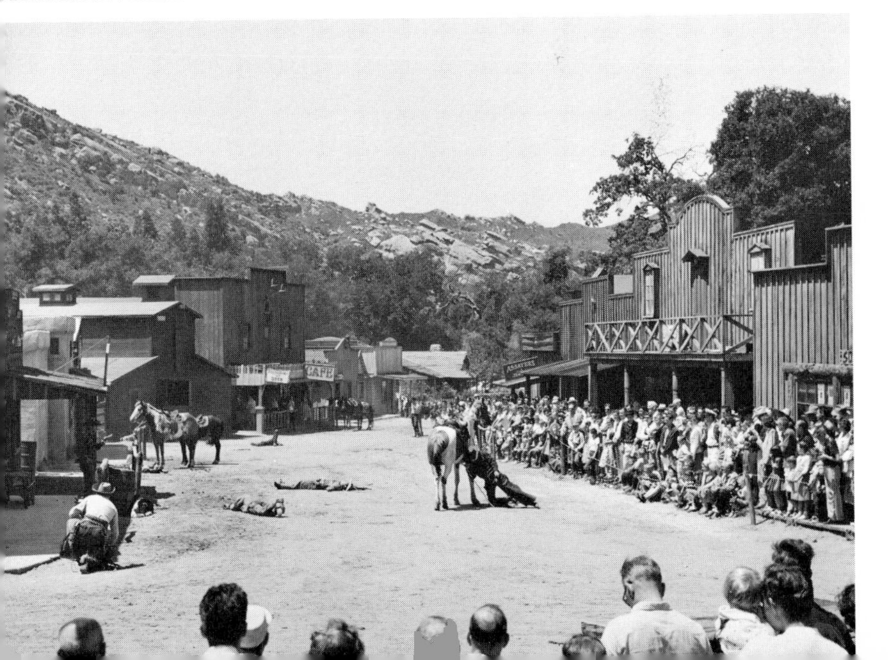

HEIGHT OF SEASON, SANTA ANITA RACE TRACK

RAMS AND COLTS GAME IN COLISEUM, LOS ANGELES

fields, one of which employs as many as 4,000; the Wheatland hop pickers' "riot" of 1913, California's first important strike by field workers, grew out of their protest against miserable living conditions. William Johnson's ranch here, settled in 1844-45, was the first settlement reached by the Argonauts who crossed the Sierra over the Donner's Pass branch of the California trail; in the winter of 1846-47 seven of the Donner party came here to seek help for those still snowbound at Donner Lake (*see TOUR 9A*).

SHERIDAN, 104.9 *m.* (116 alt., 198 pop.), a desolate village of ramshackle frame houses in the midst of an unshaded plain, was first called Union Shed, because of a great shed built in 1857 to shelter freight teams from summer heat and winter rains.

LINCOLN, 112.8 *m.* (163 alt., 3,197 pop.), a trading point for a wide grain- and fruit-growing area, since the 1870's has made pottery. The smokestacks and many-windowed, iron-roofed buildings of the huge pottery and terra cotta works are near the western edge of town. Deposits of glass sand and a lignite coal mine have been discovered nearby.

Softly undulating land, patched with clumps of oaks, sweeps away from the highway as it continues to ROSEVILLE, 123.4 *m.* (6 alt., 13,421 pop.), which clusters about the extensive Southern Pacific shops and yards, its shady side streets of frame dwellings lined with olive, maple, and fig trees. Roseville ships plums, berries, almonds, and grapes chilled in its huge icing plants, in large quantities.

Roseville is at the junction with US 40 (*see TOUR 9A*).

Tour 3C

Greenfield (S. of Bakersfield)—Taft—Maricopa—Ojai—Ventura; 121.5 *m.,* US 399; State 33.
Route paralleled between Ojai and Ventura by Southern Pacific R.R.

Accommodations plentiful except in mountains.
Paved roadbed. Care should be taken in slide areas, particularly during and after rains.

US 399 crosses the flat, torrid floor of the San Joaquin Valley, leaving broad farming and grazing lands for a forest of oilwell derricks. Ascending the Coast Range, the highway twists for almost 70 miles through mountain heights and descends into peaceful Ojai Valley.

West of GREENFIELD, 0 *m.,* at a junction with US 99 (*see TOUR 3C*), is PANAMA, 3.2 *m.,* dependent on dairying.

US 399 follows a straight course through a land devoted to cattle raising. In the summer, it seems to dance in the shimmering heat waves.

At 17.3 *m.* is a junction with a dirt road.

Right on this dirt road to the STATE ZOOLOGICAL PARK, 3.1 *m.*, where herds of elk are occasionally seen grazing from beyond the boundaries.

Passing the Elk Hills (R), dotted with the derricks of the ELK HILLS NAVAL OIL RESERVE, US 399 enters barren Buena Vista Valley, 22.1 *m.* Southwest of the bleached Buena Vista Hills, 27.4 *m.*, is Midway Valley, black with oil derricks that rise above a land the hue of wood ashes. This valuable West Side Oil Field (also called the Midway Sunset Field), dating from 1899, was the homeland of what was probably the most poverty-stricken race on the North American continent. The wealth is far from exhausted, partly because the four companies—Standard, Union, Associated, and General Petroleum—by mutual agreement and their monopoly of pipe lines—control both the spacing of wells and the production rate of the other 200 producing companies in the field.

FORD CITY, 29.6 *m.* (1,000 alt., 3,926 pop.), has a bleak and treeless business street. Adding to the importance and size of the business section, is Standard Oil's CAMP 11C (1,200 pop.), lying west of the community. In the boom days of the field, Ford City was an unnamed tent city; its plethora of Model-T Fords led to the name.

TAFT, 30.3 *m.* (1,000 alt., 3,822 pop.), the metropolis of the West Side Oil Field, is bleakly typical of the oil fields. The town has three men to each woman among the inhabitants, a very busy Supply Row for field equipment, and a monthly population turnover of transient labor that frequently amounts to 25 per cent of the total population. Although most of Taft's 20 bars operate all night because of the three shifts in the fields, life is otherwise quiet. The town has several civic, social, and religious organizations, and in a region where early settlers purchased water by the quart, school facilities have recently been improved by a $50,000 plunge for children.

Only the early history of Taft is spectacular. In 1908 the Southern Pacific and Santa Fe decided to operate a joint line—the Sunset (now owned by the Southern Pacific)—to carry prospective investors to and from a settlement camp in the expanding field. When this camp burned down, a battle started between the Southern Pacific, which owned land north of the tracks, and J. S. Jameson, who owned land to the south. The Southern Pacific erected a town, Moron, and Jameson pushed a rival community, Jameson. At the height of the rivalry, when the outcome of the battle hung upon the location of a proposed post office, a fire obliged the Southern Pacific by destroying Jameson. Moron was chosen as the site of the post office, its name changed to honor President Taft. Later Taft was incorporated and rebuilt Jameson became South Taft.

US 399 follows Kern St. through the residential areas of SOUTH TAFT and TAFT HEIGHTS into an arid, derrick-sentineled countryside. Gradually, the highway rounds the outflanking Temblor Range.

MARICOPA, 38 *m.* (850 alt., 648 pop.), is another bleak and dreary product of the oil fields, a community of treeless streets and sun-baked houses.

US 399 ascends GROCER GRADE of the Temblor Range, where blanched desert land is replaced by the browns and blacks of mountain rock and the green of chaparral.

STUBBLEFIELD GULCH, 49 *m.*, widens into the oak-decked floor of Cuyama Valley, a green region of isolated farms. Cuyama River runs along the highway through the narrowing, southern section of the valley, between hills increasingly rugged.

A boundary of the SANTA BARBARA NATIONAL FOREST is crossed at 60.5 *m.* Only 80,000 of the 1,772,555 acres are strictly timberland; the reserve is primarily for protection of the watershed. Fishing and hunting in this unfrequented region are excellent. Streams here are kept well stocked with trout; deer, quail, wild pigeon, fox, and even mountain lion are among the game.

Continuing upward through slashed, oak-grown hills past CUYAMA PEAK (5,880 alt.), US 399 loses the accompanying murmur of the Cuyama River and begins a steep and twisted ascent to Pine Mountain Crest, 75.8 *m.* (5,300 alt.). Visible (L) is REYES PEAK (7,488 alt.). Behind is a tumultuous terrain of rugged green hills; ahead, scarred country. The highway reaches a trough cut in the mountains by Sespe Creek and follows the course of the trout-stocked waters as its banks narrow and heighten. Oaks lean from precipitous gorge sides.

Crossing a last summit (3,700 alt.), 91.8 *m.*, the highway follows an erratic course, sweeping through the three WHEELER GORGE TUNNELS, 100.8 *m.*, concrete-lined burrowings, through solid mountain sides.

The pleasant countryside is characterized by its many vacation resorts. Bridle trails mark the hillsides. Hot and cold mineral springs and plunges and rustic resort cabins are passed. MATILIJA CANYON, 104.1 *m.*, is abloom through the summer with the lovely matilija-poppies. These flowers, 4 and 5 inches in diameter, whose white petals and yellow centers appear to have been cut from crepe paper, blossom from bushes that reach heights of 5 and 6 feet.

Ojai Valley, 106.1 *m.*, 12 miles long and 3 miles wide, is a rolling, fertile land protected from inclemencies of weather by the Topatopa Mountains on the north and the Sulphur Mountains on the south. It is dotted with luxuriant oaks and sycamores and checkered with fruit orchards.

At 108.3 *m.* is a junction with a paved road.

Left on this road is OJAI, 1 *m.* (743 alt., 4,495 pop.), the residential heart of the valley. Originally, this affluent municipality was called Nordhoff in honor of Charles Nordhoff, whose column in the New York *Herald* in the 1870's and booster book aided the California real estate boom that created the town. (Charles Nordhoff was grandfather of the co-author of the *Bounty* series.) Although in 1888 two Yale men, S. D. and W. L. Thacher, established nearby an expensive preparatory school for the sons of the wealthy, Nordhoff remained primarily a rustic village until

1916. In that year, however, great changes occurred. E. D. Libbey, whose 500-acre estate and Spanish-style castle adjoined the village, persuaded the residents to remodel structures in Spanish style and to change the name of the community. The climate was advertised; resort hotels sprang up; retired men of means came here and took up ranching; ranchers took up culture; the tourists multiplied. The valley population of about 3,500 is swelled during winter and spring months by an influx of tourists and vacationists. Culture is encouraged by art classes every Thursday evening, Sanskrit instruction every afternoon, and open forums maintained by the University of California. Supplementing these activities are those of numerous organizations—choral, histrionic, terpsichorean, orchestral, and others. The Rosicrucians and the adherents of the American Inner School (headquarters of the Theosophist Society) on Krotona Hill form one social group; the memberships of the Jack Boyd Club and the Ojai Valley Country Club form others. The Ojai Valley Tennis Club in the last week of April conducts a State-wide tennis tournament, during which even the schools are closed. So constant are Ojai activities that the city has earned the title, "a miniature Santa Barbara."

South of the Ojai junction, US 399 sweeps between ARBOLADA (R), the former Libbey estate—now a residential subdivision—and the OJAI VALLEY COUNTRY CLUB and GOLF COURSE (L), built with its prize-winning architectural design at an expense of $250,000 and given to the club by Mr. Libbey.

South of MIRROR LAKE (R), 110.4 *m.,* a marshy abode of wild ducks, US 399 leaves Ojai Valley behind. Oak-covered hills again close in upon the highway and the accompanying Ventura River spread.

FRESNO CANYON, 115.8 *m.,* permits a view of FOSTER PARK (L), a wooded mountain side improved by a great, outdoor, concrete AUDITORIUM, the gift of the region's wealthy citizens. South of the derricks of the VENTURA OIL FIELD, 118.8 *m.,* the highway crosses the broadening Ventura River Valley to the junction with US 101 in VENTURA, 121.5 *m.* (*see TOUR 2C*).

Tour 4

Junction with State 24—Sierraville—Sierra City—Auburn—Placerville—Sonora—Mariposa; 301.8 *m.* State 49.

Roadbed narrow and unpaved between State 24 and Sierraville, sometimes impassable in winter; paved elsewhere.
Hotels in larger towns; camping grounds plentiful.

When James Wilson Marshall found a few flakes of gold on January 24, 1848, in the tailrace of a mill he was building on the American River, he started a mass movement into California. More than 100,000 gold seekers poured into El Dorado within the next two years in a feverish search for riches, half depopulating some eastern villages and causing a labor shortage. Some men made a comfortable stake but more were constantly on the move in an effort to find richer strikes—and in the end had nothing to show for their frantic activities. Even though they panned as much as $100 worth of gold a day for a time, they spent it all in gambling or for food, drink, and shelter, retailed at fantastic prices. Sooner or later perhaps half of the gold hunters went back to their eastern homes; some remained to hunt ceaselessly for the fabulous strike that was never made; others moved on to Nevada, Oregon, Idaho, Colorado, and other places where gold and silver were found in the next decades; and a considerable number settled down in California, speculating in land, founding towns, and establishing businesses of one kind and another. A few moved to high positions in the new State.

State 49 runs through the very heart of the California gold country—including the Mother Lode. Within a decade millions of dollars' worth of gold were taken out of the streams and hills, the bulk of it going to comparatively few men. In this brief span, during which mining camps mushroomed on every river bar and wandering prospectors swarmed the hills by the thousands, a civilization sprang up overnight, endured briefly, and fell in ruins—the full-blown but short-lived civilization of the wide-open, riproaring gold towns. After its decade of glory, as placer mines gave out and placer miners wandered away, the gold region lost its fine flush of feverish enthusiasm. Gone was the day of the roving prospector on muleback, whose stock in trade was a pick and a pan. To unlock the riches in the years that followed, tremendous labors were required and millions in capital. The giant mining companies with their vast and elaborate equipment, sank shafts deep to underlying quartz, built networks of tunnels, stamp mills to crush the ore, wing dams, and ditches, lifted rivers from their beds, and brought in hydraulic monitors that washed whole mountains away.

Along State 49 are strewn the relics of these labors and of the men who performed them; decaying shanties of the "pick and pan" men, abandoned hillside shafts of the quartz mines, high-piled débris of the hydraulic workings. Some of the gold rush towns have disappeared completely, others are mere heaps of rubbish. Even in his day Mark Twain could write of ghost towns: "You will find it hard to believe that here stood at one time a fiercely flourishing little city, of 2,000 or 3,000 souls, with its newspaper, fire company, brass band, volunteer militia, bank, hotels, noisy Fourth of July processions and speeches, gambling-halls crammed with tobacco smoke, profanity, and rough-bearded men of all nations and colors, with tables heaped with gold dust—streets crowded and rife with business—town lots worth $400 a front foot—labor, laughter, music, swearing, fighting, shooting, stabbing—a bloody inquest and a man for breakfast every morning—and now nothing but lifeless, homeless solitude. In no other land, in modern times, have towns so absolutely died and disappeared, as in the old mining regions of California."

A few towns, kept alive by continuing or sporadic operations, remain to delight those who take pleasure in pioneer relics. Three or four of them nurture their picturesque aspects for the benefit of visitors, even to the point of encouraging some "quaint character" who can grind out the tales to enchant the antiquarian. The surviving settlements are more or less alike. Each grew up along a road or a stream; hence few have even a single straight street. In each the building that has best weathered the years is usually the office of the Wells Fargo Company, which shipped out the gold; it had to be sturdy to survive attack from the considerable number of gold seekers who preferred others to do the backbreaking digging and panning for them. It also had to be reasonably fireproof because of the frequent fires resulting from the careless, rough and tumble life. In many towns some old stores remain because they, too, were strongly built; like the express offices they handled gold and were subject to attack by bandits, as well as by hilarious, devil-may-care celebrants. A characteristic of the buildings erected with the need of defense in mind is the iron shutters—heavy, full length iron doors that protected all openings, windows and entrances alike. These doors were usually brightly painted; the blues, reds, and greens are now faded pastels. Symbol of aristocracy is the "ancient" building constructed of lumber "brought round the Horn"—the number is smaller than local pride admits. These stand behind ailanthus and locust trees, sometimes in good condition. Another survivor in some places is the fraternal association hall; fittingly, the Independent Order of Odd Fellows was particularly strong in the gold country.

The region still yields gold; in the Mother Lode region alone 175 lode mines and 130 placer mines were active in 1937. Smoke began to rise from old stacks of hillside workings after the 1933 rise in the price of gold again made operations profitable. The 1930's also brought thousands of unemployed men flocking in to try their luck, as did the forty-niners fleeing the economic collapse of the 1840's. Along State 49, near the ruins of crude cabins built by the men who first panned the creeks were shacks and tents of "snipers," who worked the river gravels for what gold they could find, which was rarely more than enough to keep them in coffee and beans. Even the more experienced miners average only $1.53 a day, and the creeks could be worked only for an average of 86 days a year.

Near State 49 between Auburn and Mariposa runs the Mother Lode itself, but the whole region covered by this route claims the name. Geologically, the Mother Lode is a long narrow strip in the Sierra foothills, a mile wide and 120 miles long. Great masses of quartz crop out at intervals, seamed and laced with gold—precious yellow metal inlaid in rock. Because of the more or less continuous deposits, there was a belief in the fifties that this was an unbroken ribbon, set with gold, like a chain of massive nuggets. The streams in the region and the ancient river beds were filled with auriferous gravel, the first gold taken by the early miners.

In the region pierced by this route young Bret Harte (*see LITER-*

ATURE) picked up the material that helped set the pattern for what Stanley Vestal has called "the histrionic West." Westerns—novels, magazines, and movies—maintain the sentimental tradition first accepted and given prestige by New England Brahmin literary patronage of such stories as *Outcasts of Poker Flat*.

Section a. JUNCTION WITH STATE 24 to EL DORADO; 168.4 m. State 49.

State 49 branches southwest from State 24 (*see TOUR 6B*), 0 *m.*, 1 mile west of Chilcoot, across the level mountain-hemmed Sierra Valley.

LOYALTON, 11.5 *m.* (4,949 alt., 837 pop.), named during the Civil War by mountaineers who were loyal Union men, is a lumbering center. An ordinance unusual in an early Western town forbade the sale of liquor within the "city limits," with the result that Loyalton became California's second largest city; its incorporators spread the town as widely as possible to discourage lumberjacks from walking beyond the city limits for a drink.

SIERRAVILLE, 24.5 *m.* (4,950 alt., 180 pop.), at the junction with State 89 (*see TOUR 5b*), a crossroads center where four stage lines once met, lies in the Sierra Valley, where farming began in 1853. In 1881 a fire swept and destroyed practically the entire business section, which, however, was soon rebuilt.

West of Sierraville on State 49 is SATTLEY, 28.5 *m.* (4,992 alt., 30 pop.), at the foot of the steep and winding climb over the mountain through YUBA PASS (6,700 alt.), through which winter storms occasionally sweep, marooning travelers as they did the Duchess and John Oakhurst in Bret Harte's tale of the outcasts.

At 42.5 *m.* is a junction with a dirt road.

Right here into a rugged wilderness, where resorts and camps cluster along almost two score trout-stocked lakes fed by melting snows, to GOLD LAKE, 6.9 *m.* In 1850 one Stoddard led a group of miners here in search of a lake whose shores, he asserted, were covered with pure gold. When they found no nuggets lying about, the miners almost lynched Stoddard on the spot. The expedition, however, led to discovery of the rich Plumas County deposits. At LONG LAKE, 10.4 *m.*, is ELWELL PUBLIC CAMP, starting point for the climb to MOUNT ELWELL (7,846 alt.), which offers fine views of the mountains.

By the North Fork of the Yuba River lies SIERRA CITY, 46.8 *m.* (4,100 alt., 250 pop.), where in 1850 settlers panned the river gravel for gold. From the jagged SIERRA BUTTES (8,600 alt.), rising sheer above it, an avalanche roared down to destroy the first town of shacks and tents; but Sierra City lived to make its fortune from the buttes. Honeycombed with the tunnels of quartz mines by 1852, they yielded gold through the 1850's and 1860's. Among the richest claims was the Monumental Mine, where a 100-pound gold nugget valued at $25,000 was found August 18, 1860. Still producing is the Buttes

Mine, opened in 1850; others, such as the Young America, Colombo, and Gold Point, have taken a new lease on life since 1933.

But the distant clangor of the stamp mills fails to offset the ghostly air of the deserted buildings along the main street. At its head (L) stands an empty frame hotel, built of "resawn" lumber, with broken windowpanes and sagging beams. At the far end is another old hotel where old-timers tell their tales; and opposite it is a butcher shop with the sign: "Fresh Meat Tuesdays, Thursdays and Saturdays." In the FIRE HOUSE (R) is a two-wheeled cart, brought round the Horn in the 1850's.

An imposing structure (R) is the BUSCH BUILDING (1871), which has two stories of brick and a balcony and third story of wood. Over one of the doorways are the initials "E.C.V.," standing for "E. Clampus Vitus," the hilarious "Incomparable Confraternity" that swept through the mining country. It was organized in Pennsylvania in 1847 and transplanted to California before 1853. The organization was one of the biggest hoaxes of a country where hoaxes were the order of the day. The initiations were masterpieces of ingenious and humorous torture, and only those who survived them understood what E.C.V. stood for. So powerful was the sway of the mysterious brotherhood as developed by the miners that newcomers found they could not conduct business in mining towns until they had joined it. When Lord Sholto Douglas brought his theatrical group to Marysville, he found the first night audience too small to pay for the rent of the theater. Enlightened by a friendly miner, he applied for membership, and thereafter the miners flocked to his company's presentations.

State 49 runs westward along the North Fork of the Yuba River, where slopes rise abruptly, scarred by the streams of the hydraulic giants that tore away earth and gravel.

DOWNIEVILLE, 59.4 *m.* (3,000 alt., 640 pop.), seat of Sierra County, is in a little basin walled by sheer mountainsides. At the western end, State 49 runs through JERSEY FLAT, with cottages—some of them built in the 1860's—in bright gardens. Across the river is narrow, locust-bordered Main Street where on Saturday nights, when Downieville had a population of 5,000, rows of horses were tethered to the posts that held up the wooden awnings and men thronged the boardwalks lighted only by the rays of kerosene lamps shining through saloon windows. During the week, heavily loaded pack trains and lumbering stagecoaches choked the dusty street, while miners crowded the stores to buy grub at fabulous prices. The first Yankee woman to arrive in town was escorted through Main Street by most of the male citizens; Signora Elise Biscaccianti, famous pianist of her day, was carried in on men's shoulders, as was her piano.

In September 1849, Frank Anderson arrived, the first man to pan gravel here, and in November, Major William Downie and his motley band—a Kanaka by the name of Jim Crow, an Irishman, an Indian, and 10 Negro sailors. Downie found the river filmed with ice the first morning he went out to pan, but the gravel was so rich in gold that he put up crude cabins and set his men to sifting the snow-covered bars. When provisions began to run low, he sent nine of the men, including Jim Crow, with newly dug gold to buy food and supplies in the lower country. Only Jim Crow came back, and he not till spring, when he found Downie and his companions on the verge of starvation. A crowd of would-be miners followed the Kanaka back to The Forks. Camps sprang up on every bar and flat. At one of them, Tin Cup Diggings, each of three owners filled a tin cup with gold dust every day. There is a story that Jim Crow, boiling a 14-pound salmon, found flakes of gold at the bottom of the pot. Gold continued to turn up in surprising places. A woman who kept a tent restaurant, sweeping her dirt floor one day, saw yellow particles and investigated, then folded her tent, for it was pitched over a gold mine. On Durgan's Flat, across the river, $80,000 worth of gold was taken out during the first half of 1850; in 11 days Frank Anderson and three companions took $12,900 worth from a claim 60 feet square. Still enough remained to make it worthwhile for the Chinese, who came in later years, patiently to lift the heavy stones by hand, clearing out the bed of the North Yuba.

The SIERRA COUNTY COURTHOUSE was erected in 1855 on Durgan's Flat; it has 12-paned windows, lofty ceilings, and an old-time courtroom. In its yard is an arrastre, or mill wheel, used by pioneers to crush gold quartz. Behind it, at the base of PIETY HILL (nearly every mining town had its Piety Hill) is the GALLOWS that served for executions after 1857. The wooden building (1864) in front of the courthouse houses the Downieville *Mountain Messenger,* whose first issue (1853) came off the old Washington hand press now in the basement. Where Downieville's movie house stands today was the theater in which Edwin Booth, Lola Montez, Lotta Crabtree, and other famous troopers played to entertainment-starved audiences who showered them with coins and gold dust pokes worth more than a king's gift after a command performance. Relics of the days when gold dust served for currency are preserved in the MEROUX MUSEUM (*open Sun. and holidays; keys with storekeeper weekdays*) on Main St., typical of the earliest substantial mining camp construction, with walls of flat stones laid horizontally. In the COSTA STORE (L), on Main St., built in 1853 of uncemented shale with walls 4 feet thick at the base and 27 inches at the top, are glass-enclosed scales that still weigh gold dust and nuggets. During the summer the store buys from $50 to $60 worth a day from "snipers." Under the overhanging roof of the ST. CHARLES HOTEL (L), are benches and chairs polished smooth by countless pairs of blue jeans. Except for a row of locust trees in the front of it, the St. Charles, with a balcony running the length of the upper story, looks much as it did when James McNulty built it in 1853.

Superb mountain peaks look down on old GOODYEAR'S BAR, 63.3 *m.,* where the tumbling, ice-cold waters of Goodyear's Creek join the North Yuba. Miles and Andrew Goodyear arrived here in the summer of 1849 and built the solidest cabin in the region. Miles Goodyear had come west in 1836 as a stripling in the little missionary party led by Dr. Marcus Whitman on his way to establish a mission along the Columbia River; when the missionary stubbornly insisted on taking a wagon through from Fort Hall, young Goodyear issued his ultimatum. If the cart went on, he would leave. The cart continued, and Goodyear went off into the wilderness alone. In time he took an Indian wife and started farming near the Great Salt Lake—the first white settler in the region. When the Mormons arrived in 1847, Miles cleared out for another wilderness in California. When he died in the fall of 1849 his brother buried him in a gold rocker until he could take the bones down to Benicia.

Lively camps sprang up nearby: Ranse Doddler, Hoodoo, St. Joe's and Cutthroat Bars, Nigger Slide, and Kennedy's Ranch. During the early 1850's the North Yuba was diverted from its channel so that gold in the bedrock could be taken out; and each summer, after winter floods had washed out the makeshift dams of the preceding season, the tremendous labor of rebuilding them was repeated.

At this point is a junction with an improved mountain road.

Left on this road, which follows the route of the stage road built from Marysville in the early 1860's, to the present-day GOODYEAR'S BAR, 0.2 *m.* (3,200 alt., 60 pop.), a handful of scattered houses. Through pine and cedar, the road climbs to MOUNTAIN HOUSE, 5.9 *m.* (5,641 alt.), where a splendid panorama of wooded mountains and deeply chiseled canyons, with glimpses of the river far below, opens out northward. The inn that stood here was a stopping place as early as 1850, when 2,500 mules carried supplies in pack trains over narrow and precipitous trails to Downieville.

At FOREST CITY, 8.7 *m.* (4,500 alt., 343 pop.), high-gabled houses cling to the walls of the ravine up which climbs the main street. Here in the summer of 1852, a company of sailors found gold. Brownsville they called the thriving town that grew up, but the name was not good enough. Some favored the poetic Yomana, Indian name for a nearby bluff; others, the more practical Forks of Oregon Creek. The final decision was to name the town for the first woman to reside there; but it was neither the first, nor yet the second, who appropriated the honors, but the third, Mrs. Forest Mooney, aggressive feminine journalist.

On a series of terraces connected by winding streets perches ALLEGHANY, 12 *m.* (4,500 alt., 519 pop.), where all the citizens depend on gold mining for a living. Gold was discovered on the creek below in May, 1850, by one of several parties of Hawaiians (Kanakas sent out by a certain Captain Ross, reputed son of King Kamehameha). At the lower end of Alleghany is the BRADBURY HOUSE, where lived Thomas J. Bradbury. Working the nearby Sixteen-to-One Mine, which he discovered in 1908, he found 20 years later that the neighboring Tightner Mine was tapping the same vein, which had its apex in his own backyard. The two mines, together with the nearby Twenty-one, were consolidated as the Original Sixteen-to-One Mine, which still produces.

State 49 winds along the North Yuba, where tributary creeks spill down precipitous ravines. Ahead the way leads up GOLD RIDGE, whose slopes are luxuriantly green with yellow pine, fir, oak, and madrone. A lone cemetery, all but obliterated by the returning forest, a desolate cabin, or an ancient apple tree, mark the sites of once thriving gold camps: Galena, Young's, Railroad, Depot, and Hell's Hills,

Celestial and Oak Valleys, Dad's Gulch, Indian Springs, and Pike City.

At DEPOT HILL, 76.1 *m.,* is the JOUBERT HYDRAULIC MINE (L), which has produced millions in its many years of continuous operation by one family. A gashed hill, a metal pipe line, and hydraulic monitor (or long nozzle), long sluice boxes, a muddy stream red with detritus from the mine are the signs of its activities.

At 81 *m.* is a junction with a mountain road.

Left on this road, which runs between picket fences enclosing weather-worn clapboarded houses, to CAMPTONVILLE, 0.3 *m.* (2,900 alt., 75 pop.), a hamlet whose spreading elms and old-fashioned flower gardens emphasize its Yankee origin. Though the town was moved twice to make way for hydraulic operations and several times almost destroyed by fire, the old houses behind picket fences on the outskirts escaped its vicissitudes; most of the other buildings were restored along the old lines after the fire of 1907. At the western end of town stands the PELTON MONUMENT (R), with a model of the Pelton Water Wheel surmounting a pedestal; Lester A. Pelton's invention of the wheel here in 1878 was an important step in the development of hydroelectric power.

At 82.6 *m.* on State 49 is a junction with an improved road.

Right here to BULLARD'S BAR DAM, 7.6 *m.* (1,500 alt.), which impounds the North Yuba's waters to operate generators in the power-house at its base. The bar was named for a Dr. Bullard, who wandered into town after a shipwreck off the California coast on his way from Brooklyn to the Sandwich Islands. Nothing but names remain as memorials of nearby camps: Foster's Ferry, Stoney, Rock Island, Succor, Slate Range, Cut-eye Foster's, Kanaka, Winslow, Negro, Missouri, Condemned, Frenchmen's, and others.

NORTH SAN JUAN, 90.6 *m.* (1,900 alt., 135 pop.), in spite of its name, was as much a Yankee settlement as any other town along the Mother Lode. The most likely explanation of the Spanish name is that Christian Kientz, a Mexican War veteran, saw a resemblance to San Juan de Ulloa in Mexico in the hill (L) where he discovered gold in 1853, and named it accordingly. A population of 10,000 gathered here when the town was headquarters for the rich hydraulic workings on San Juan Ridge. Here was the main office of the first long distance telephone line in the West, strung up in 1878 for 60 miles, between Milton and French Corral, to connect the hydraulic mining centers.

Along Main Street, bordered with poplars, walnut trees, and locusts, stand decaying, red brick business houses and frame buildings with shake roofs. The two-story brick ODD FELLOWS HALL (L) was dedicated in 1860. In the old frame NATIONAL HOTEL (R), silk-hatted mining company officials from San Francisco discussed gold production with superintendents of the nearby hydraulic mines. Next door (R) is the TOWN HALL, with arched doorway, massive iron doors and shutters, and fine cornice work; from its second-story iron balcony, the town band used to give concerts. An old brick store (L), operating now with electric refrigerators and cash registers, preserves the primitive register in which steel balls, placed in the groove corresponding to the amount of purchase, rolled down, rang a bell, and caused a marker to spring up.

At 93.1 *m.* is the junction with a graveled road.

Right here to FRENCH CORRAL, 4.5 *m.* (1,700 alt., 138 pop.), where a Frenchman built a corral for his mules in 1849; it is the oldest town in San Juan Ridge. The school house (L) was a hotel in the 1850's. Behind the solid brick walls and iron doors and shutters of the WELLS FARGO EXPRESS OFFICE (R) millions in gold were guarded. The MILTON MINING AND WATER COMPANY OFFICE, now a grocery store, was the lower terminus of the Ridge telephone line. One of the Edison instruments, made in Boston in 1876, is preserved here.

At 94.2 *m.* is a junction with a mountain road.

Left here up SAN JUAN RIDGE, where the rich quartz veins of the mountain peaks, during an age long process of erosion, were washed into a preglacial river bed, which geologic upheavals lifted high and dry above today's drainage system. Nowhere in California was hydraulic mining undertaken on so vast and spectacular a scale as here. The network of flumes and canals, more than 300 miles in length, cost not less than $5,000,000. The ridge is almost deserted today. The Sawyer Decision of Jan. 23, 1884, following passage of the Federal Anti-Debris Act the year before, closed all the hydraulic mines in the State, ending a long and bitter warfare by farmers of the valley lowlands against the inundation of their fields by the thousands of tons of silt the rivers carried down from the mines.

TYLER, 5.6 *m.* (2,550 alt.), called Cherokee in the 1850's, has a handful of weathered houses, a church, a school house, and the inevitable "diggin's."

At NORTH COLUMBIA, 9.1 *m.* (3,000 alt., 156 pop.), many of the old homes are still occupied, although the superintendent's office across the mining ditch, the pretentious Eureka Lake and Yuba Canal Company office building, the machine shop where gold was once retorted, and the blacksmith shop, have long been deserted.

A picture of decay is LAKE CITY, 12.5 *m.* (3,300 alt.), where two or three decrepit houses and a forlorn hotel (1855), its balcony sagging drunkenly, huddle by the grassy depression in the pasture which was once the "lake."

Neat gardens still bloom at the doors of the old homes in NORTH BLOOMFIELD, 15.5 *m.* (3,200 alt., 90 pop.), along the broad, locust-lined main street. In the MALAKOFF MINE OFFICE gold was reduced to bars for transportation to the San Francisco Mint. The largest single bar weighed a quarter of a ton and was valued at $114,000. At the edge of town (L) rise the exquisitely molded pinnacles and minarets, "touched with vivid colors like a place enchanted," of the MALAKOFF MINE, most colossal hydraulic excavation on San Juan Ridge.

The name of SNOW TENT, 19.5 *m.* (4,250 alt.), recalls some canvas station—snow-bound in winter—along the old stage route.

Mail is brought into GRANITEVILLE, 28 *m.* (4,900 alt., 150 pop.), by sleigh or on snowshoes in winter. Some deep quartz mines are still worked here.

At BOWMAN LAKE, 34.4 *m.* (5,500 alt.), a massive dam of granite, part of the Nevada Irrigation District system, stands today where a dam was first built in 1868 and twice razed, in 1872 and 1876.

NEVADA (Sp., snowy) CITY, 106 *m.* (2,450 alt., 3,000 pop.), seat of Nevada County, is headquarters of Tahoe National Forest.

Here were the rich placer diggings of the 1850's and 1860's, the hydraulic excavations of the 1870's and the later deep quartz mines.

James W. Marshall, the man who started the gold rush with his discovery at Coloma, found gold at Deer Creek in the summer of 1848 but moved on, searching for richer fields. Within two years 10,000 miners were working every foot of ground within a radius of three miles. Near the rich gravel beds of the LOST HILL section (L), uncovered early in 1850, a mushroom town, Coyoteville, sprang up, named for the coyote or tunnel method of mining; during the two years it lasted, the gravel banks are said to have yielded $8,000,000 in gold dust and nuggets. Forerunner of Nevada City was Beer Creek Diggings or Caldwell's Upper Store—so called for a log cabin store kept by Dr. A. B. Caldwell.

On the main thoroughfare, Broad Street, a prominent land mark is the three-story NATIONAL HOTEL (R), an important stopping place in the 1860's and 1870's, when five or six stagecoaches daily lumbered down the street, to stop with a jolt in front of it. Inside, at the long bar over which $1,000 is said to have passed every day, travelers washed the dust out of their throats with a Pisco punch made from Peruvian brandy, a bonanza cocktail, or the miner's standby, Bourbon whisky. On Commercial Street, the UNION HOTEL (R) stands on a site occupied continuously by a hostelry ever since the spring of 1850, when Madam Penn built the first boarding house in town; this industrious lady often took her turn at the gold rocker in the ravine beyond Coyote Street where the Stampses, first family in town, settled in October 1849. The ASSAYER'S OFFICE (R), with its original furniture and iron safe, is run by the son of the original owner. Although $27,-000,000 in gold is said to have passed through the office, it was never held up or robbed. Here the first samples of ore from the Comstock Lode in Nevada were assayed in 1858.

State 49 strikes southward through green meadows and gardens and over gentle hills. In the meadows where GRASS VALLEY, 110.3 *m.* (2,400 alt., 3,817 pop.), was to spring up, an immigrant party of 1849 found their half-starved cattle, which had strayed from camp on the heights, feasting on the green grass. In August 1849 a band of 20 men, led by a Dr. Saunders, built and wintered in the first cabin on BADGER HILL (L), where homes and gardens spread today. The rich quartz mines which were to make Grass Valley one of the outstanding gold towns were not tapped until 1850. On pine- and fir-clad GOLD HILL, at the eastern outskirts of town (R), stands the KNIGHT MONUMENT, commemorating the first discovery of gold-bearing quartz in California. George Knight's find of 1850 initiated an industry that still produces wealth long after the depletion of the placers and the close of hydraulic operations. Between 1850 and 1857 the GOLD HILL MINE alone produced $4,000,000. The EMPIRE MINE on Ophir Hill, opened in 1850, and the NORTH STAR MINE on Lafayette Hill, opened in 1851, have yielded $80,000,000 worth of gold. Others still in operation are the GOLDEN CENTER and the

IDAHO-MARYLAND, opened in 1863. An immense outlay of capital and labor is represented in the stamp mills, cyanide tanks, and shafts. The quartz veins, one nearly two miles long, are honeycombed with mine workings. The Empire mine is worked to the 7,000 level, 3,620 feet wholly below the shaft collar at 928 feet below sea level. The deepest mine in the district is the North Star, where the 9,800 feet level is 1,600 feet below sea level.

Grass Valley's wide, paved Main Street with its sidewalks, filling stations, and neon signs, has a modern prosperous air. But wooden awnings over the sidewalks remain, and the old buildings, some with new fronts, retain unchanged interiors. More redolent of old times are the side streets, where the homes of two famous early day residents stand. At the corner of Mill and Walsh Sts. is the HOME OF LOLA MONTEZ, to which the famous beauty, who was born María Dolores Porris Gilbert and lived to become Countess of Lansfeld, returned from her triumphs in Europe and America. Here she lived in retirement from 1852 to 1854, with a pet bear, some dogs, and a husband. The husband she later divorced because he killed her bear when it clawed and bit him. The beautiful and daring Lola brought with her a reputation that made her the talk of two continents, fame for public performances of the Spider Dance, and friendships with the great. In her modest cottage—since altered by the addition of a second story—she entertained at soirees, held mainly for the benefit of the younger miners, that became the talk of the Mother Lode.

Nearby is the childhood HOME OF LOTTA CRABTREE (*see THE THEATER*), Lola's protégée. Here, in the two-story house at 220 Mill St., where her mother boarded miners, Lotta met La Montez. The days she spent with the glamorous Lola, learning dance steps, and ballads and riding horseback through the woods, turned her into a dancer and actress. Beginning her career as a child entertainer, she traveled from camp to camp with her mother. At the age of 16 she scored a triumph in San Francisco, and, traveling to the Atlantic Coast, repeated it.

Right from Grass Valley on State 20 to ROUGH AND READY, 4.4 *m.* (1,900 alt., 145 pop.). Here gold was discovered by a group of Wisconsin men who arrived in September, 1849, in a dozen covered wagons that bore the name of the town-to-be painted on their canvas sides. The leader of the group was Captain Townsend, who had served under Gen. Zachary Taylor, old "Rough and Ready" of Mexican War fame. Popular legend says it was the distaste of the Wisconsin men for newcomers from New England that prompted them, at a mass meeting called by E. F. Brundage in 1850, to organize the independent State of Rough and Ready, adopt a constitution, and announce their secession from the Union; thus they would have more freedom to deal with New Englanders as they felt they should be dealt with. A more likely explanation, however, is their rebellious objection to the Federal taxes on miners. Today the Rough and Ready diggin's are half hidden by chaparral; up to the edge of town, where aged apple trees shade weather beaten houses, stretches a green meadow. The white painted frame ROUGH AND READY HOTEL (L), part of which dates from 1853, is now the store and post office.

AUBURN, 134.6 *m.* (1,360 alt., 6,000 pop.), spills over hill and hollow, encircled by orchard-covered knolls. Its winding streets, where old-fashioned white houses sit back among maples and walnuts, are dappled with a leafy lacework of sun and shadow. Long before the orchards were planted, this was a mining camp, for Claude Chana and his Indians mined gold here in May 1848. First called Wood's Dry Diggings, the camp was renamed a year later by miners who had come from Auburn, N. Y., with Stevenson's Volunteer Regiment. Prospectors poured in during 1849, until a network of trails radiated to camps in the hills and ravines; these became turnpikes choked with stagecoaches, mule teams, and freight wagons, where highwaymen often lay in wait for hold-ups. The gold gave out, but Auburn's decline was circumvented by the advent of the railroad in 1865, and by the planting, in the 1880's and 1890's, of the foothill orchards. The flavor of gold days lingers in Old Town, which lies in a hollow at the head of Auburn Ravine, overlooked by the arcaded dome of the handsome, tan brick PLACER COUNTY COURTHOUSE. Here narrow streets twist uphill under wide-branching trees, past crumbling brick buildings with sagging iron doors and shutters and over-hanging balconies. Still open for business are the I. O. O. F. HALL and the PLACER COUNTY BANK. The Auburn *Herald* has been published continuously since 1852.

At Auburn is a junction with US 40 (*see TOUR 9a*).

South of Auburn the rolling countryside gives way to deep-cut gorges twisting between mountain ridges. The highway, wide and well-paved, twists in and out through rough countryside conveying some idea of the hardships its early prospectors endured.

At 138 *m.* is a junction with an improved road.

Left here, over the route of the old turnpike up FOREST HILL DIVIDE, over which in the 1850's a constant stream of traffic poured. An arrow on a marker at 3.5 *m.* points to LIME ROCK (R) across the canyon of the North Fork, where a woman confederate used to signal to highwaymen lying in wait for approaching stages.

At intervals are the sites of the early stage stations: GRIZZLY BEAR HOUSE, 4.5 *m.* (1,600 alt.), BUTCHER'S RANCH, 6.5 *m.*, SHERIDAN'S, 7 *m.*, MILE HILL TOLL HOUSE, 9.3 *m.*, and SPRING GARDEN, 11.2 *m.* (2,400 alt.), where fresh vegetables were served as a luxury.

FOREST HILL, 18.5 *m.* (3,200 alt., 200 pop.), is in one of the State's most productive cement tunnel-mining districts. The FOREST HOUSE (L), which replaced the first store, a brush shanty of 1850, and two brick stores with iron doors and shutters of the gold-rush period, still serve men from the mines. Into the mountain to depths of from 200 to 5,000 feet penetrate the shafts of big mines, among them the Dardanelles, the Rough and Ready, and the Jenny Lind.

Left from Forest Hill 3 *m.* to YANKEE JIM'S (2,650 alt., 16 pop.), where only a cluster of cottages and the acres laid waste by hydraulic mining remain as evidence of the days when this was the trading center for camps in the neighboring canyons. Just north is SHIRT TAIL CANYON, which acquired its name in the summer of 1849, when a miner was discovered busily panning the stream clad in nothing but his shirt. Tributary to Shirt Tail are other canyons, cutting across the rough highland country that might have become Donner County, if a movement of 1869 had succeeded. The proposed new county would have included such settlements as Ground Hog's Glory, Hell's Delight, Miller's Defeat, Ladies' Canyon, Devil's Basin, Hell's Half-acre, and Bogus Thunder.

On the main side road is MICHIGAN BLUFF, 25.6 *m.* (3,500 alt., 39 pop.), on the brink of a ridge 2,000 feet above the Middle Fork of the American River. Here Leland Stanford, who became one of the "Big Four" of railroad fame, kept a store from 1853 to 1855. Most of the frame and brick buildings that lined the locust- and poplar-bordered street are ruins.

Crossing the North Fork of the American River at LYON'S BRIDGE (650 alt.), State 49 winds nearly 1,000 feet up the east side of the canyon to the diminutive settlement of COOL, 141.8 *m.* (1,525 alt., 10 pop.).

Left from Cool on an improved road is GREENWOOD, 7.5 *m.* (1,650 alt., 385 pop.), where the Greenwoods, a father and two sons, found gold and built their cabin in the spring of 1848. Discovering that hunting was more profitable than mining, they gave up their claims to supply prospectors with venison.

On Oregon Creek, where a group of Oregonians discovered gold in 1849, is GEORGETOWN 13 *m.* (2,275 alt., 679 pop.), at the northern end of the true Mother Lode. Growlersburg, it was called in 1850, when George Phipps led a company of sailors in; in honor of him the present title was adopted two years later. In its prime Georgetown was the trading center of 10,000 miners from almost a hundred camps, among them Mamaluke Hill, Sailors' Slide, Divine Gulch, Spanish Dry Diggings, and Volcanoville. The main street, with locust trees, uneven boardwalks, stone and brick buildings—built soon after a fire in 1856—remains much as it was.

A monument to Alexander Bayley's blasted dreams is the BAYLEY HOUSE (R), 145 *m.,* built as an inn. Bayley lavished $20,000 on this three-story brick mansion with its Colonial porticoes and terraced garden. He opened the house with a grand ball on May 15, 1862. Today it stands a wilderness because Bayley guessed wrong; he thought the overland railroad would be carried through here.

In COLOMA, 155 *m.* (850 alt., 400 pop.), on the south fork of the American River is GOLD DISCOVERY STATE PARK marking the SITE OF SUTTER'S SAWMILL (*see SACRAMENTO*), where on Jan. 24, 1848, James W. Marshall, while shutting off the water, happened to notice flakes of gold. Marshall's discovery was not the first in California—six years earlier gold deposits had been found and worked in Placerita Canyon near Los Angeles—but his was the discovery that spread the gold lust over the world. From every town and rancho in California men swarmed in until by summer almost every foot of ground had been staked out. With the influx of fresh thousands from the East, the gold hunters began to push out north and south.

Marshall's discovery won him little but hard luck. When he took his samples of gold to Sacramento to be tested, he was laughed at. It remained for another man, Sam Brannan, a San Francisco publisher, to tell the world, and for yet others to claim the profits. They swarmed in to prospect Marshall's claims and posted armed guards to keep him off them. When he appealed to the courts, friends of the trespassers sat as judge and jury; even his attorneys sold him out and joined the opposition. For ten years Marshall was spied upon, threatened, swindled—and all this time hated so violently that he could scarcely earn a living. Turning to the lecture platform, he spread his story throughout the country; but by the time he returned to California, twenty years after his discovery, it was too late. Larger interests had consolidated all the richer veins of gold, and California justice was confirming them in the ownership of lands, forests, and mines squeezed from the early settlers. In 1872 public opinion forced from the Legislature an appropriation of $200 a month for two years for Marshall, but at the next session this was cut in half. Marshall died in abject poverty on Aug. 10, 1885, and was buried within sight of the spot where he made his find.

Little is left today of the Coloma the forty-niners knew—only a few houses, fallen stone walls, and the ever present locust trees. Half hidden by matted vines and Chinese heaven trees are the CHINA STORE and the JAIL (R), both built of stone and both guarded by iron doors and shutters.

Right from Coloma on a road that leads uphill to the MARSHALL MONUMENT, 0.5 *m.*, a State park of 18 acres. The bronze MARSHALL STATUE, high on a granite pedestal, was erected by the Native Sons of the Golden West in 1890 over Marshall's grave. The MARSHALL CABIN, where he lived from 1848 to 1868, has been restored.

South of Coloma, State 49 passes through fruit orchards. So scarce was fruit in the early 1850's that $800 was once bid for one small plot containing four apple trees. The lack of fruit, green vegetables, and fresh meat in the miners' diet often threatened them with scurvy. Only their strenuous physical work enabled them to digest their steady fare of beans, sowbelly (salt pork), doughy saleratus bread—or sometimes, by way of variety, flapjacks—and coffee brewed from beans crushed in a sack between two stones and used over and over again. Real bread sold for a dollar a slice in hotels, and butter for one slice cost a dollar extra. Potatoes, when they first appeared in the camps, sometimes brought a dollar apiece. Only after 1855, when men began to bring their wives to the camps, did greens become plentiful, for the women planted small kitchen gardens.

PLACERVILLE, 163.3 *m.* (1,848 alt., 4,439 pop.), seat of El Dorado County, was born when men who found the land about Coloma well occupied plodded on to virgin territory. In July 1848, after a tour of the new mines in the Sierra Nevada foothills, Gov. Richard B. Mason reported to Washington, D. C., that William Daylor and Perry McCoon, ranchers from the Sacramento Valley, with their Indian retainers, had struck it rich at a place called "Old Dry Diggin's"; they had taken $17,000 from one small ravine in a week's time. Soon thousands were swarming up and down the gulches around what became known as the Ravine City. The lure of easy wealth brought not only miners but a small army of plundering outlaws known as the Owls.

One night three of them held up a Frenchman named Cailloux at the point of a knife, ransacked his cabin, and stole 50 ounces of gold dust. Captured the next morning, the three culprits were speedily tried and hanged from the great oak, known thereafter as the hang tree, which stood at the corner of Main and Coloma Sts. Punishment for later malefactors was as swift. The street became the alcalde's courtroom, and the crowd the jury for such men as "Irish Dick" Crone, Bill Brown, and others who were summarily hanged. Thus Hangtown acquired the name by which it has been known (unofficially) ever since.

From the first a strategic point on the overland trail and the Coloma Road, Placerville had boomed into a town of 2,000 people by 1850. By 1854 it was a serious contender with San Francisco and Sacramento in wealth and population, surpassed by them only in number of votes. Disastrous fires that swept most of the town—three times in 1856 and again in 1864 and 1865—were only temporary setbacks; its importance as a stopping place on a transcontinental route assured its prosperity. Through Placerville passed the Overland Mail, the Pony Express, and the overland telegraph. When the rush to the Comstock Lode in Nevada broke, Placerville became the chief station on the way from the west. Through it poured a stream of ponderous wagons drawn by six-mule teams, bearing merchandise and provisions to the Washoe mines. During these turbulent days men who were later to become industrial giants began their careers here: Mark Hopkins, later a railroad magnate, who set up shop on the muddy main street with a wagonload of groceries from Sacramento; Philip D. Armour of meat-packing fame, who ran a small butcher shop in the center of town; and John Studebaker, who laid the foundation for a great automobile industry by his success in building miners' wheelbarrows.

Along the winding banks of rock-lined Hangtown Creek, where the first comers settled, twists Placerville's Main Street from which scores of narrow lanes and crooked side streets lead up the ravines and hillsides past neat clapboarded cottages in tangled old gardens. Now a hotel is the IVY HOUSE (L), a rambling three-story brick edifice with wide two-story verandas; it housed the Placerville Academy from 1861 to the 1890's. Also of brick and ivy clad is the METHODIST CHURCH, on a corner of Main St. and Cedar Ravine, dedicated Sept. 8, 1861, during the pastorate of Adam Bland. At the lower end of Main St. is the OLD HANGTOWN BELL, which hung for years in the Plaza, where it was rung to summon the vigilantes or sound fire alarms. At ST. PATRICK'S CHURCH (L), near the edge of town, an old silver bell, the gift of the miners, still calls the faithful to worship; built in 1865, the church stands on the site of the frame structure built in 1852.

Placerville is at a junction with US 50 (*see TOUR 10a*), with which it unites southward to El Dorado.

South of Placerville, State 49 heads into variegated country, climbing up steep slopes, coasting along high ridges where rivers hundreds of feet below churn over sands once rich in gold and shoots down into valleys. In places it winds through rolling hills, green in spring, burned almost white by the sun in summer.

By WEBER'S CREEK, 165.2 *m.* (1,550 alt.), is the SITE OF WEBERVILLE, where in 1848 Capt. Charles M. Weber, founder of Stockton, mined with the help of the Indian chief, José Jesús, and his twenty-five retainers. Setting up his store at this spot, Weber exchanged beads and cloth for the gold brought by the Indians, who found they were on the losing end of the bargain only after he had built up a respectable fortune.

DIAMOND SPRINGS, 176.3 *m.* (1,791 alt., 860 pop.), now a supply center for lumbering interests, was in 1849 a camp on the Carson Pass Emigrant Trail and remained so until the placer and quartz mines in the outlying hills had been worked out. One of the many sandstone buildings that lined the main street was the GOLDEN WEST HOTEL, built in 1856. At about the same time, the lumber for the CALIFORNIA HOTEL (L) was brought round the Horn.

EL DORADO (Sp., the golden), 168.4 *m.* (1,610 alt., 417 pop.), began life with the less romantic name of Mud Springs, inspired by the nearby watering place, which the cattle of immigrants had trampled into a boggy quagmire. The frame HILL HOTEL (R), built in 1852, still holds together, but the UNION CHURCH on the hill (L) is only a restored version of the structure of 1853. The roofless shells and rusted iron shutters of three old business buildings, stand (L), overgrown by ailanthus trees.

At El Dorado is the western junction with US 50 (*see TOUR 10a*).

Section b. EL DORADO to MARIPOSA; 133.4 m. State 49.

South of EL DORADO, 0 *m.*, State 49 continues through the Mother Lode country.

On the SITE OF LOGTOWN, 2.5 *m.*, once a humming camp, are the ruins of a stamp mill, where quartz was crushed to extract the gold. At 7.3 *m.* is the MONTEZUMA MINE, from which probably about $1,000,000 in gold was taken.

Between El Dorado and HUSE BRIDGE over the Cosumnes River, 10.1 *m.*, once stretched a continuous line of stamp mills. Exploring miners went on digging up every foot of ground for gold, until not one of the buildings was left standing.

PLYMOUTH, 14.6 *m.* (900 alt., 500 pop.), amid rolling oak-dotted fields, preserves little evidence of its former wealth. The PLYMOUTH CONSOLIDATED includes all the deep mines in this country, which produced about $15,000,000 worth of gold.

Left from Plymouth on an improved road to FIDDLETOWN, 6.4 *m.* (1,700 alt., 250 pop.), settled by Missourians in 1849. "They are always fiddling. Call it Fiddletown," said one Missouri patriarch; and so the name, which Bret Harte memorialized in *An Episode of Fiddletown,* was chosen. A certain Judge Purinton, whose home, the PURINTON HOUSE, is a local landmark, was so embarrassed at being known as "the man from Fiddletown" in the business circles of Sacramento and San Francisco, that he had the name changed to Oleta in 1878. In later years, however, the pride of the townspeople in the original name revived and it has been officially restored.

On Dry Creek is DRYTOWN, 18.1 *m.* (700 alt., 150 pop.), which, far from being dry, had 26 saloons in its day. In the surrounding gulches—Blood, Murderer's, and Rattlesnake—as much as $100 was sometimes washed out of a single pan. Some of the inhabitants were as wild and woolly as the names of the gulches indicate. The recent re-opening of some of the quartz mines—the AMADOR MOTHER LODE and the FREMONT—keeps 80 to 90 men employed. Closed since 1926, the EXCHANGE HOTEL (R), with its old pump and watering trough, is plastered with posters advertising chewing tobacco, circuses, and an Italian fiesta. The restored TOWN HALL on the hill (L), having served at various times as a church, dance hall, and a residence, is now (1939) a community center.

AMADOR CITY, 21 *m.* (1,100 alt., 171 pop.), at the bottom of a gulch on Amador Creek, was a camp in 1849. Here, on the south side of the creek, was the Ministers' Claim, where in 1851 the Rev. Mr. Davidson and three other preachers made the first quartz discovery in the region. The three reverend gentlemen made quite a good thing out of their claim on weekdays, and preached for the good of the miners' souls on Sundays. Quartz mines, such as the famous KEYSTONE (L)—adjoining the original Amador—yielded fortunes; in 1869, when the Bonanza vein was discovered, the first month's crushing paid $40,000. Rows of decaying frame, brick, and stone buildings line the main street, but prospectors still haunt the neighborhood, for millions in gold may yet be taken from the ledge—they hope.

In SUTTER CREEK, 23.2 *m.* (1,200 alt., 1,013 pop.), which Gen. John Augustus Sutter, in search of a site for his sawmill, passed up in favor of Coloma, many miners decided to settle down after the first gold fever had died. Its main street meanders past neat little cottages among lawns, flowers, and trees and the scattered dumps of quartz mines. Among the older brick buildings, besides MASONIC HALL and the KEYES BUILDING, is the HAYWARD OFFICE, headquarters of Alvinza Hayward, who went broke many times trying to make the quartz finds of 1851 pay. Shafts and tunnels had to be bored through solid rock, the art of timbering mastered to prevent cave-ins, and equipment shipped in at high rates from the East—it all required hundreds of thousands of dollars. Many men in the district were ruined financially, but Hayward kept on developing his mines until he had an income of $50,000 a month. One of his holdings, the CENTRAL EUREKA MINE, opened in 1869, is still the best producer in the neighborhood. Between 1859 and 1872 Leland Stanford, too, had to pour so much cash into his holding, the LINCOLN MINE, formerly the Union, that he once offered to sell for $5,000. Dissuaded by his superintendent, Robert Downs, he held it, made a fortune out of it, and sold it for $400,000. The money went into building the Central Pacific Railroad, from which venture Stanford garnered part of the millions that later endowed Stanford University.

At MARTEL, 25.8 *m.* (1,500 alt., 150 pop.), is a junction with an oiled road.

Right on this road is IONE, 9.4 *m.* (300 alt., 950 pop.), a camp that the first miners called Bedbug and Freeze Out. Later, when churches, homes, schools, and stores replaced the miners' tents, the citizens decided that neither term would do for a post office address. They chose the present name in honor of one of the ladies in Edward Bulwer-Lytton's *The Last Days of Pompeii.* The PRESTON INDUSTRIAL SCHOOL, established here in 1889, is a State reform school for boys.

JACKSON, 28.6 *m.* (1,250 alt., 2,000 pop.), seat of Amador County, boasts of having some of the deepest mines on the North American continent. No ghost town, Jackson is full of life and bustle, its main street crowded with stores, cafés, bars, and movie houses, for the big quartz mines on the northern outskirts, which have been the town's mainstay for 80 years, are still active. The ARGONAUT and the KENNEDY mines, whose deepest shafts strike a mile below the surface, have produced since the early 1850's. Up to 1931 the Argonaut had yielded $17,391,409 worth of gold.

From 1848 to 1851 Jackson was important as a stopping point on the branch of the Carson Pass Emigrant Trail, which met roads here from Sacramento and Stockton. At a spring on the banks of Jackson Creek, teamsters hauling freight between Drytown, and Mokelumne Hill, below, and miners on their way through the Mother Lode used to break their journeys. So convivial were these overnight stops that piles of bottles collected and gave the town its first name, Bottileas. In 1850, having grown to respectable size, it borrowed the name of an energetic citizen, Colonel Jackson; but the rough-and-ready habits of its unruly infancy persisted. More than 10 men were strung up from a great oak that stood on Main Street. As impatient with the formalities of government as it was with those of the courts, Jackson made itself county seat in characteristically aggressive style. With plenty of gold in their pouches, a group of Jacksonians called at Double Springs, the county seat in 1851, and invited all the county officers to refreshments at the local saloon. During the festivities, a pair of Jackson men loaded all of the county archives, seals, and paraphernalia into a buggy and whisked them off to Jackson. Despite the subsequent fury of the county clerk, Jackson remained the county seat. Not Jackson, however, but Mokelumne Hill enjoyed the last laugh; in 1852 the citizens of that town won the seat from Jackson. The overwhelming number of ballots cast in the election—several times as many as there were county residents—was due to unusual industry; mounted men rode all over the county, voting at every camp and town. A decade later Jackson again won the county seat.

Jackson's Main Street, narrow and winding, passes between rows of iron-shuttered stone buildings with overhanging balconies. On

Courthouse Hill stands the brick AMADOR COUNTY COURTHOUSE of the 1850's. Opposite is the WOMAN'S CLUBHOUSE with its Native Daughters' Room, where the Native Daughters of the Golden West organized in 1886. Through a residence section with schools and churches, the road winds down to the little GREEK ORTHODOX CHURCH 'midst tombstones and cypress trees.

At Jackson is a junction with State 8 (*see TOUR 10a*).

State 49 continues south over open, rolling hills. The GINOCCHIO STORE (L), 30.4 *m.,* a roofless shell with stone walls and iron shutters, is all that remains of BUTTE CITY (1,100 alt.), a mining town in the 1850's. Forgotten miners lie in a forlorn graveyard on a hill (R). Nearby towers the conical bulk of JACKSON BUTTE (2,348 alt.), an isolated volcanic peak.

BIG BAR BRIDGE, 31.8 *m.,* spans the Mokelumne River, where in the fall of 1848 Col. J. D. Stevenson and men of his discharged regiment mined gold. Soon camps were on every bar and flat along the stream. Because the miners were an unruly bunch, prone to brandish their Colts at the least dispute, Colonel Stevenson drew up what was probably the first code of miners' laws in California. Here in 1850 the crude Whale Boat Ferry carried passengers across the river; it was replaced by a bridge that yielded as much in tolls as did many a mine. The old GARDELLA INN, at the northern end, and KELTON'S, at the southern end of the bridge are now dwellings. Under water lie the sites of former camps, inundated by PARDEE RESERVOIR, visible in the distance (R), which supplies the cities on the eastern shore of San Francisco Bay, 150 miles away, with water.

High on the divide between the Mokelumne and Calaveras Rivers perches MOKELUMNE HILL, 38.4 *m.* (1,500 alt., 595 pop.), over river beds that yielded rich gold-bearing gravels. Settled in 1848, "Mok Hill," or "The Hill," grew into one of the Mother Lode's biggest and liveliest towns. It is said that at one time one man was killed each week-end for 17 weeks, and at another, five men in one week. From 1857 to 1866 Mokelumne Hill was the seat of Calaveras County. Nearby is FRENCH HILL, scene of the "French War" skirmishes of 1851, in which American miners drove off Frenchmen and appropriated their claims. Here too is NIGGER HILL, named for the Negro who drifted into town one day in the 1850's and asked the miners what they did for a living. All he had to do, they told him, was dig holes in the ground and lift out nuggets. For a joke, they sent him to a hilltop that had been prospected unsuccessfully many times. Within two days he returned, casually exhibiting a poke of gold dust and some nuggets; the jokers promptly dropped whatever they were doing and ran to stake out claims on the hill.

State 49 passes the walls of a gray ruin roofed only by the green of ailanthus trees, and climbs the hill to the I.O.O.F. HALL (L), one of California's first three-story buildings. Next door are the GOLDEN EAGLE HOTEL and STURGES STORE, both built in 1854. On the SITE OF CHINATOWN (R), long ago destroyed by fire, a thousand Chinese once burned punk in their joss houses, bought rice in their shanty stores, and—according to old-timers—sold Chinese slave girls.

On a vacant lot (L) between a garage and barber shop is the SITE OF THE ZUMWALT SALOON, where, according to legend, the desperado, Joaquín Murrieta, sat one evening playing cards, unrecognized. The talk got round to him, as it often did. Flushed with courage and Bourbon, a young miner named Jack slapped a sack of gold on the table and cried, "Here's $500 which says that I can kill that —— —— Murrieta, if I ever come face to face with him!" Onto the table leaped Murrieta, a pistol in each hand, "I am Murrieta. Now is your chance!" No one made a move. The man with a price on his head calmly strode out of the saloon and rode away.

A large ailanthus casts its shadow over the balconies of two stone buildings (L)—the LEDGER HOTEL and the FORMER CALAVERAS COURTHOUSE, now the hotel annex. Built in 1856 as the Hotel de l'Europe, the Ledger has been renovated, but still has its original stone facade and windows, black walnut bar and balustraded staircase. Dating from the early 1850's is a little white church (R), built with gold collected by a godly and determined lady who stopped the men as they came back from the mines at night, holding out a miner's pan for contributions of gold dust and nuggets. From Stockton Hill on the southern outskirts of town rise the ruins of the great three-story HEMMINGHOFFEN-SUESDORF BREWERY, like some ancient abbey, with crumbling, roofless walls, shattered archways, and great vaults choked with fallen stone blocks.

1. Right from Mokelumne Hill on an oiled road is FOSTERIA, 5.3 *m.* (1,500 alt.), at the head of Rich Gulch, where Senator William M. Gwin once operated the famous GWIN MINE.

To CAMPO SECO, 11.9 *m.* (700 alt., 216 pop.), the Penn Copper Mine brought wealth in the 1860's. Ruined stone buildings, a two-story clapboard house, and a cemetery are relics of the past.

2. Left from Mokelumne Hill on a graveled road to a junction at 1.3 *m.;* R. here to the sites of two old mining camps: JESÚS MARÍA, 3.7 *m.* (1,500 alt.), where an old adobe dwelling built by Mexican miners is now a granary; and WHISKEY SLIDE, 7.4 *m.*, where the ruined JOHN NOCE HOUSE of the 1850's exhibits a stone chimney, still intact.

South of Mokelumne Hill, State 49 runs through CHILE GULCH, where American and Chilean miners fought the "Chilean War" in December 1849. The Chileans' practice of acquiring claims in the names of their peons so angered the Yankees that they passed laws against it. In resistance to the law the Chileans, led by a certain Dr. Concha, drove the Americans out of the gulch. The incident embroiled the United States in a diplomatic dispute with Chile, which was settled in favor of the Americans. But the gulch remained the scene of violence; a monument in an old graveyard keeps alive the memory of two young Massachusetts men who "were Cruelly Murdered at the Chilean Gulch, July 18, 1851 by three Mexican assassins for the sake of Gold."

The highway crosses the NORTH FORK OF THE CALAVERAS RIVER, 41.5 *m.,* so-called because Spanish explorers found Indian skulls (*calaveras*) along the banks of its lower course.

SAN ANDREAS, 44 *m.* (1,000 alt., 1,082 pop.), seat of Calaveras County, was first a Mexican town, as a few scattered adobe dwellings built in 1848 indicate. Along the main street—thinly disguised in their modern dress of paint, stucco, chromium, and plate glass—are the buildings of the early Yankee era. Documents dating back to 1866 are preserved in the courthouse; relics and historic records, among them a red sash of Joaquín Murrieta's, in the CALAVERAS CHAMBER OF COMMERCE MUSEUM (L) at the north end of town. San Andreans assert that here in the barroom of the Metropolitan Hotel—which burned down in 1926—occurred the event memorialized in Mark Twain's *The Jumping Frog.* They also assert—and so does many another gold country town—that here Joaquín Murrieta began his career of murder and banditry.

South of San Andreas the long ridge of BEAR MOUNTAIN (2,831 alt.), rises (R) as State 49 winds through dry meadows overgrown with clusters of oaks.

At SAN ANTONIO CREEK, 49.4 *m.,* the few decaying buildings that remain of FOURTH CROSSING, a stop-over point of stagecoaches appears (R). The tallest of them is the HERRICK HOUSE, once a hotel; next to it are the REDDICK HOUSE and the remains of an old stable. Across the street is the dance hall.

ALTAVILLE, 55.2 *m.* (1,525 alt., 320 pop.), where the old inn (L) and the PRINCE STORE (R) date from the 1850's, was known as Cherokee Flat when Bret Harte made it the setting of his poem, "To the Pliocene Skull." Here in February 1866, 130 feet down in the shaft of the Mattison Mine, on the slopes of nearby Bald Hill, was unearthed the Calaveras skull. A storm of controversy arose. Unquestionably Pliocene, declared the geologists, were the mud, gravel, and sand of the stratum in which it had turned up; but the anthropologists, threatened with the upset of every theory about the age of man on earth, were not so sure. Amused at the discomfiture of J. D. Whitney, State geologist, the public hailed the affair as a huge hoax, perpetrated by enemies he had made when he interfered with plans of local financiers for selling wild-cat stocks. Nevertheless, Whitney, in a final summary that appeared 13 years later, maintained the skull's authenticity; the last official word on the subject declared the skull to be prehistoric, if not Pliocene. Old-timers still insist, however, that the affair was a practical joke—that at about the time it was found, Dr. Kelly, the Angels Camp dentist, discovered that the skull of an old skeleton he kept in his office was missing.

ANGELS CAMP, 56.2 *m.* (1,500 alt., 915 pop.), was named not for the heavenly host but for one of Col. J. D. Stevenson's very earthly volunteers, a man called Angel, who found gold on the creek running through town, in the summer of 1848. The man who made Angels Camp famous was Mark Twain. Local people insist that it was in their HOTEL ANGELS barroom that Twain one winter's night heard the story on which he based *The Jumping Frog.* Every year in May, Angels Camp holds a Jumping Frog Jubilee. In the main event frogs, after rigid inspection to prevent loading with buckshot as in Twain's tale, compete for a first prize of $500. Models of frogs appear in the windows of stores and hotels. One restaurant advertises a jumping frog pie—made of prunes and raisins; and local businessmen have organized a Frog Boosting Club.

Angels Camp was also the locale of Bret Harte's story, *Mrs. Skaggs' Husbands.* And here, in a little theater since burned down, appeared Edwin Booth. Here, too, gold was reputedly discovered by Bennager Raspberry, whose name is honored by Raspberry Lane; while hunting, he jammed the ramrod in his gun, and when he shot it out to dislodge it, it stuck in the earth at the roots of a manzanita bush; pulling it out, he found a piece of glittering quartz in a rich vein, from which he took $7,700 in the first three days. Relics of mining days survive on the main street, despite such evidence of progress as neon signs and Venetian blinds. The STICKLE STORE (L), now a five-and-ten, was built in 1857 and the CALAVERAS HOTEL (R) at about the same time. At the southern city limits (L) is a model of the UNDERSHOT WATER WHEEL used to operate an *arrastre* (mill) for crushing gold ore.

Left from Angels Camp on State 4, a paved road, is VALLICITO (Sp., little valley), 4.8 *m.* (1,800 alt., 160 pop.). Mounted on the branches of a big oak tree (R) is an old bell—brought around the Horn—that used to summon the miners of this once lively camp to Sunday services.

DOUGLAS FLAT, 7.3 *m.* (2,000 alt., 85 pop.), once produced enough gold to make hundreds rich for a brief period. A bank as well as a store was the well-preserved stone-walled, iron-shuttered GILLEADO BUILDING (L).

At MURPHYS, 9.1 *m.* (2,200 alt., 600 pop.), the brothers Murphy, John and Daniel, were the first to discover gold. The town has the usual Joaquín Murrieta legend: here the ubiquitous bad man is said to have been a three-card monte dealer in 1851 and to have begun his bloody career when his brother, unjustly accused of horse stealing, was hanged, and he himself flogged. Outwardly much as they were in the 1850's, Murphys' store, saloons, and business houses sprawl along the main street in the shade of elms and Chinese heaven trees. In front of them in the evenings, their chairs tilted back against the walls, loll men in blue jeans and broad felt hats.

The MITCHLER HOTEL (L) was built in 1856 by John L. Sperry to accommodate important visitors when Murphys became the gateway to the Calaveras Big Tree Grove. Inside and out, the Mitchler—which Bret Harte is said to have described in *A Night in Wingdam*—remains just as it was. At the windows hang the old iron shutters; along the second-story balcony runs a fine iron railing. Embedded in the trunks of the heaven trees that shade the front are bits of the wire cable once used for hitching horses, and in the doorway are the bullet marks left from the time when a stranger shot the town's current bad man. Inside, horsehair sofas and red plush furnish the best suite. The iron safe standing behind the bar preserves old ledgers with the names of such notables as Horatio Alger, Jr., Charles E. Bolton (Black Bart), Thomas Lipton, Ulysses S. Grant, Henry Ward Beecher, John Jacob Astor, Jr., John Pierpont Morgan, and Will Rogers, who made a movie here in 1934.

Right from Murphys 1.1 *m.* on a paved road to MERCER'S CAVE, a series of 20 caverns filled with stalactite and stalagmite formations. SHEEP

RANCH, 8.6 *m.* (2,400 alt., 75 pop.), was a quartz mining camp where George Hearst, father of newspaper magnate William Randolph Hearst, laid the foundations for his fortune. An old frame hotel with double balconies still stands.

State 4 continues northwestward through rolling mountain country, cut by gorges. The CALAVERAS BIG TREES, 24.9 *m.*, the first grove of *Sequoia gigantea* discovered, were found by a hunter who was stalking deer for a mining camp in 1854. He hurried back and told of seeing trees taller than the masts of clipper ships and thicker than houses.

The road mounts higher into the Sierra through BIG MEADOWS, 41.7 *m.* (6,600 alt.), and crosses the crest of the Sierra Range at EBBETS PASS, 67.2 *m.* (8,800 alt.), discovered in the 1840's. State 4 comes to a junction with State 89 at 77.9 *m.*; L. here on State 89. MARKLEEVILLE, 81.9 *m.* (5,500 alt., 469 pop.), in a high mountain valley on an early immigrant route, was settled in the 1850's. It later furnished timber for the mines of the Comstock Lode.

At 88.5 *m.* is junction with State 88–89 at Woodford; R. on the highway to cross the Nevada line 21.2 *m.* south of Carson City, Nev.

The main route turns L. on State 88-89 to a point at 95 *m.* where State 88 swings sharply L. and State 89 goes R. (*for both routes, see TOUR 10a*).

By State 49 on ALBANY FLAT, 58.5 *m.*, stands the half-ruined ROMAGGI BUILDING, a store, saloon, and hotel patronized by travelers on the road to the wild camp, a favorite haunt of Joaquín Murrieta and his gang, on the Arroyo de los Muertos (creek of the dead). A sprawling, two-story edifice, with an outside stairway, cellars, and courtyard walls, it was built of rock and adobe in 1852.

The somnolent, half-deserted village of CARSON HILL, 60.1 *m.* (1,400 alt., 50 pop.), looks like a ghost of the old days. CARSON HILL, rising (L) above the village, was the richest diggings in the whole Mother Lode. In Carson Creek James H. Carson, led to the spot by friendly Indians, found gold in August 1848. The 180 ounces of gold that he panned in 10 days were the first of more than $20,000,000 worth that diggings here have produced. A great gash in the hillside shows the position of the MORGAN MINE, where 15 miles of tunnels pierce the hill and one shaft bores 3,500 feet below sea level. Here the largest nugget ever found in the United States, a solid mass of gold weighing 195 pounds and valued at $43,534, was taken out Nov. 22, 1854. The story—another time-worn veteran of the mining country—is that a man called Hance, chasing a stray mule in the vicinity, discovered the mine when he saw an outcropping of quartz in which yellow metal gleamed; with a rock he knocked off a chunk of gold weighing 14 pounds. When Hance sold his claim to Col. Alfred Morgan and five associates, the fight over a rich prize began; miners said that Morgan and his partners had each exceeded the limits of land one man could claim. A hoodlum named Billy Mulligan, backed by a tough bunch recruited from the San Francisco "Hounds," stepped in, seized part of the claim, and held it for a time. After several years of violence, the courts put Morgan in possession, but litigation continued until the entire property had come into the hands of another man, James G. Fair, of Comstock Lode fame.

MELONES (Sp., melons), 63.5 *m.* (700 alt., 75 pop.), in the days when it boasted a population of 5,000, was one of the toughest camps. Melones took its name from a nearby camp, where Mexican miners found coarse gold in the shape of melon seeds. Earlier it was called Slumgullion, for the thick, sticky mud along the river banks; and before that, Robinson's Ferry, for the man who ferried travelers across the river. Robinson sold the ferry in 1856 to Harvey Wood, who acquired a county franchise to operate it for 50 years. Within six weeks after the Morgan mine was opened, the ferry made $10,000 for him, and as time went on, it continued to earn so much that he could spend $40,000 on the toll roads approaching it.

State 49 winds up a ravine to a junction with a paved road at 65.6 *m.*

Left here up JACKASS HILL, where the men who packed supplies in to the miners used to stop overnight, tethering their donkeys. Mark Twain, who spent 5 months here in 1864-5, wrote: ". . . a flourishing city of two or three thousand population had occupied this grassy dead solitude during the flush times of twelve or fifteen years before, and where our cabin stood had once been the heart of the teeming hive . . ." On the summit is MARK TWAIN'S CABIN, 1 *m.*, restored except for the chimney and fireplace; a monument of quartz rock commemorates Twain's stay here as a guest of the Gillis brothers, at a time when he had to flee San Francisco because of debts. He used to sit under the oak before the door and smoke his pipe. It was here that he met the original of Dick Baker in *Roughing It,* Dick Stoker, of whom he wrote: "One of my comrades there—another of those victims of eighteen years of unrequited toil and blighted hopes—was one of the gentlest spirits that ever bore its patient cross in a weary exile: grave and simple Dick Baker, pocket miner of Dead-Horse Gulch."

In MORMON GULCH, where Mormon miners once pitched camp, stands TUTTLETOWN, 66.6 *m.* (1,500 alt., 164 pop.), which became a trade center and pack-mule stop on the old Slumgullion road, after Judge A. A. H. Tuttle built his log cabin here in August 1848. Mark Twain used to come down Jackass Hill to trade at the SWERER STORE (L), a stone structure erected in 1852. In the same year was built the right wing of the old frame TUTTLETOWN HOTEL (L), whose drinking trough is now a goldfish pond.

At 67.8 *m.* is a junction with a paved road.

Right on this road to the RAWHIDE MINE, 2.5 *m.*, where $6,000,000 has been taken from deep quartz veins. Rawhide Flat at this point gives a good view of TABLE MOUNTAIN (L), celebrated in Bret Harte's tales. This flat-topped volcanic mountain, a half-moon stretching for 60 miles, and the region contiguous to the Stanislaus River constitute the true Bret Harte country. As a young man, late in 1855, Harte passed through the region hunting for a camp that would hire him as a teacher. Here he gained the impressions that he wove into the sentimental tales that impressed stay-at-home Easterners of the *Atlantic Monthly* circle and made him famous.

At 70.7 *m.* on State 49 is a junction with a paved road.

Right on this road to SHAWS FLAT, 1 *m.* (2,100 alt., 15 pop.), where Mandeville Shaw planted an orchard in 1849 and where Tarleton Caldwell later planted a garden and black walnut trees. Here in 1855 were discovered the gold deposits of an ancient river channel, so rich that the yield of Caldwell's claim alone was estimated at $250,000. Tunnels were sunk far into Table Mountain; such men as James G. Fair and John B. Stetson got their start here. The camp was called Whimtown because of the many whims (windlasses) used in hoisting ore from the shafts. The MISSISSIPPI HOUSE (R), at the crossroads, was once a hotel, store, and post office. Still in service are the old bar and the patched bar chairs. From the ceiling, covered with stained muslin, hang kerosene lights; on the walls are guns, sconces, horn spoons, and other mining relics.

At 71.1 *m.* is a junction with a paved road.

Left on this road is COLUMBIA, 1.8 *m.* (2,200 alt., 200 pop.), a ghost of what was once the "Gem of the Southern Mines," the richest, noisiest, fastest growing, most spectacularly wicked camp in the Mother Lode. Roundabout lie 300 acres of weirdly shaped, brush-mantled crags, laid bare when the top soil was sluiced away to uncover the rich gravel beneath. From these acres have come $87,000,000 in gold. A party of Mexican miners, driven from their claims by gringos, found new diggings here in March, 1850; with backbreaking labor they carried ore in sacks to the streams and there pounded it in mortars and washed it. In the same month Dr. Thaddeus Hildreth, his brother, and three others, camping here overnight, were drenched by a sudden downpour. As they waited for their blankets to dry out the next morning, March 27, one of the party, John Walker, tried his luck at prospecting. He struck it rich, and within the first half week the Yankees were taking out 15 pounds of gold a day. By mid-April, Hildreth's Diggings had 6,000 residents, the advance guard of an army that was to swell to 15,000. In the meantime, however, the settlement had a serious setback to overcome, for the imposition of a $20 tax on foreign-born miners and the drying up of the streams in summer combined to drive away all but one of the 6,000 inhabitants. Recovering speedily from this handicap, it forged ahead to become a full-fledged metropolis, with 143 faro games, 30 saloons, 4 banks, 27 produce stores, 3 express offices—and an arena for bull-and-bear fights, which, described by Horace Greeley in the New York *Tribune,* is said to have given Wall Street its best-known phrases.

Columbia, one of the best preserved of the ghost towns, has more ancient and fewer modern buildings than any other spot in the Mother Lode. On Kennebec Hill at the southern outskirts, among cypresses, rises the square brick tower of ST. ANN'S CHURCH, built in 1856, which has altar murals painted by James Fallon. Main Street, following the contour of what was Matelot (Fr., sailor) Gulch, has wide-branching shade trees, brick and stone facades, old iron doors and shutters, fancy wrought iron railings, and shade-covered sidewalk canopies. At its southern end in FALLON'S HOTEL (R), where artist James Fallon's father did the honors in the 1860's and 1870's. The old red brick WELLS FARGO OFFICE (*adm. 25¢*) is now a museum; it contains old account books showing what prices once were; sugar $3 a pound, molasses $5 a gallon, flour $1.50 a pound, onions $1 a pound, sardines and lobsters $4 a can, candles 50¢ apiece, and special miners' knives $30 each. The onetime saloon, the STAGE DRIVERS' RETREAT (*adm. free*), Main and Fulton Sts. (L), is now the Golden Nugget Club of old-timers and a museum; the bar remains, and the old piano, brought around the Horn. Across the street (R) is the KNAPP GROCERY STORE, with heavy iron doors barred, which was saved from fire in 1857 when two barrels of vinegar were thrown on the flames. In the FIREHOUSE (R) are two old fire engines and an old fire hose of riveted buffalo hide; Papeets, the older of the two, was built for King Kamehameha of the Sandwich Islands, brought from New York to San Francisco, and sold to Columbia in 1859. The PIONEER SALOON (R), Main and Jackson Sts., built in 1858, housed a saloon and gambling hall on the ground floor and a fandango hall in the cellar. Beyond is the ruined CHINA STORE, half-hidden behind ailanthus trees. The PRESBYTERIAN CHURCH (1864), Jackson and Gold Sts., replacing a church of 1853, is at the edge of the former plaza from which the bell now hanging in the church tower used to summon miners to fires, mass meetings, and hangings. On the top of the hill at the northern outskirts is the two-story red-brown schoolhouse, opened for its first term April 14, 1862. Beside it is a graveyard with marble headstones bearing old-fashioned epitaphs.

SONORA, 73.5 *m.* (1,850 alt., 2,725 pop.), seat of Tuolumne County, is on seven small hills. Though it had little room to spread between the wooded hillsides of its canyon, it has long since outstripped its old-time rival, Columbia, and boasts of churches, schools, and like civic improvements. In 1848 this was Sonorian Camp, settled by Mexicans from the state of Sonora. By November of that year half of the population was rotting with scurvy. A town government was organized to establish a hospital, which treated the victims with raw potatoes at $1.50 a pound and lime juice at $5 a bottle. During 1849 men of several nations arrived, until, on Saturday nights, the narrow streets were packed with as many as 5,000 miners. In June 1850, the new "gringo" government, unappreciative of Sonora's pioneers, imposed a $20 tax on foreign-born miners. Alarmed at the Yankee treachery, Mexicans, Chileans, and others gathered for discussion. Fearing an attack, the Americans organized a mass meeting under the American flag, and headed by fife and drum, with banners waving and guns popping, they marched in a mob on the camp. The Mexicans, however, also fearing an attack, were pulling down their tents and fleeing to hunt new diggings. When the excitement subsided, Sonora found itself with but a fifth of its former population.

Not Yankees but Chileans in 1851 discovered the richest pocket mine in the Mother Lode, the BIG BONANZA (prosperity) on PIETY HILL (R). Yankees, however, garnered its greatest riches, when three partners, who had acquired the mine at little cost in the 1870's, hit a deposit of almost pure gold after working it for some years. In one day they took out $160,000 and within a week, $500,000. The Big Bonanza still produces.

The acquisitive instincts of Sonora's settlers were perhaps best shown by a certain Judge Sullivan. In the first case tried before him he collected three ounces of gold from the defendant, a Mexican, for stealing a certain Mr. Smith's leggings, and one ounce from the plaintiff himself for bothering the judge with so trivial a complaint as the loss of the leggings. Again, when a wealthy citizen haled a penniless man into court for stealing a mule, Judge Sullivan decreed that the wealthy man should pay both fine and costs, since "the court could not be expected to sit without remuneration."

On the narrow side streets are adobe facades with overhanging balconies, mellowing frame houses in old gardens, and heavy, ivy-clad stone walls. ST. JAMES CHURCH (R), built in 1860, rears a cross-tipped spire above cypress trees and lilac bushes. The one-story, red-brick BAUMAN BREWERY on Washington Street opposite the high school, once supplied beer to 40 saloons. The two-story GUNN ADOBE (R), at the southern edge of town, was built by Mexican laborers for

Dr. Lewis C. Gunn in 1850. With new wings and balcony, it is now the Hotel Italia. The hotel parlor was Gunn's office as first county recorder and his print shop, where on July 4, 1850, he started the first newspaper published in the mining region, the Sonora *Herald*.

At Sonora is a junction (L) with State 108 (*see TOUR 6b*), which here unites with State 49.

State 49-108 winds into JAMESTOWN, 77.4 *m.* (1,500 alt., 814 pop.), which has retained some of its old facades, balconies, and sidewalk roofs. Called Jimtown by the miners, it was named for Col. George F. James, a San Francisco lawyer who tried his luck at mining here in 1848 but fought with other miners and left. Its prosperity sprang partly from the rich mines of TABLE MOUNTAIN—where the HUMBUG yielded $4,000,000 in its day, some of it in nuggets the size of hens' eggs.

The WOODS' CROSSING, 78.4 *m.,* marked by a monument of gold quartz, was a busy camp in 1848 and 1849, after the Reverend James Woods and his party discovered gold in the creek.

CHINESE CAMP, 84.3 *m.* (1,400 alt., 237 pop.), which Bret Harte wrote about as "Salvado," is only a handful of brick and adobe buildings. No trace is left of the 5,000 Chinese who mined for gold here in the 1850's. Their very origin is a mystery; some say that a ship's captain left his vessel in San Francisco Bay and brought his Chinese crew here to mine, others, that they were employed by English prospectors. Only 2 miles west, on a plain at the foot of Table Mountain, was fought the first tong war in California. On Sept. 26, 1856, armed with pikes, shovels, pitchforks, daggers, clubs, and a few muskets, 900 men of the Yan Wo Tong and 1200 of the Sam Tu Tong battled. After several hours of yelling and clashing, both sides withdrew. Since the warriors had never before used muskets, the total casualties were four killed and four wounded. At Chinese Camp is a junction with State 120 (*see TOUR 3b*), which unites westward with State 49.

In WOOD'S CREEK CANYON, 87.4 *m.,* are the hoists, dumps, and stamp mill of the EAGLE SHAWMUT MINE, producer of $5,000,000 in gold.

JACKSONVILLE, 88.7 *m.* (900 alt., 63 pop.), honors Col. Alden Jackson who discovered gold here in June 1849. Fire has destroyed all of the old buildings except the JACKSONVILLE HOTEL (L), built in the 1850's.

At 91.2 *m.* is the western junction with State 120 (*see TOUR 3b*).

From the MOCCASIN CREEK POWER HOUSE (L), 93.9 *m.,* with workers' bungalows on green lawns, State 49 climbs the steep sides of Moccasin Creek ravine, skirting views of range after range of mountains, green in the foreground, blue in the distance.

COULTERVILLE, 104.7 *m.* (1,675 alt., 380 pop.), is among almost treeless mountains covered with chaparral and sagebrush. Fortunes came out of these bare-looking slopes during the wild years when 3,000 Yankees and more than 1,000 Chinese tramped the streets of

Coulterville, then called Banderita, for the small red bandannas used by the first Mexican settlers. Little traffic passes through now; once in a while a stray cow wanders through the street, or a pack mule loaded with prospector's trappings; men dressed in blue jeans and work shirts stroll about, unhurried. In the plaza, fringed by old stone and adobe edifices, is an oak from which many a bad man, according to cherished local legend, dangled at the end of a rope. Along the main street, catalpa and umbrella trees shade brick and stone buildings.

BAGBY, 115.9 *m.* (1,000 alt.) formerly Benton Mills, is the site of one of John C. Fremont's enterprises. The foundations of his mill and dam are visible. From here the Merced River flows northwest between Horseshoe Bend Mtn. and Hunter Valley Mtn. to form LAKE McCLURE, which will have a surface area of 7,127 acres at full capacity. The new EXCHEQUER DAM displaces the older dam and rises 879 ft. above sea level. It has a single turbine generator unit capable of 80,100 kilowatts. Six miles downstream is McSWAIN DAM, elevation 425 ft., length 1,500 ft.; its powerhouse will use the discharge from new Exchequer plant. Cost of the two installations is estimated at $31,500,000, of which the Federal Government is expected to contribute $9,500,000.

Crossing Merced River the highway ascends the steep slopes of HELL HOLLOW, a climb affording spectacular views. At the PACIFIC MINE, 118.7 *m.,* gold is being taken from a quartz vein. A turnout at 120.1 *m.* offers an unusual view. The river twists and turns a thousand feet below; to the north and west a seemingly endless procession of ranges melts into haze.

Now almost deserted, BEAR VALLEY, 121.6 *m.* (2,100 alt.), was built by Frémont as headquarters for his rich Mariposa mines. A jagged stone ruin (R), a tree growing within its walls, is FRÉMONT'S COMPANY STORE opened in 1851. Across the street are the FOUNDATIONS OF THE FRÉMONT HOUSE, where Frémont and his wife, Jessie, lived for a short time. Frémont's 44,000-acre Rancho de las Mariposas (ranch of the butterflies) was in the San Joaquin Valley when he bought it for $3,000 in 1847. Discovery of gold the year after made the gold-bearing foothills eastward so much more desirable that Frémont "floated" his lands in that direction—and the United States courts confirmed him in his ownership here, though the place was 50 miles east of the original claim. The move put Frémont in line to become a rich man for there seemed to be wealth hidden in the quartz ledges here. From a place within pistol shot of Bear Valley's main street, $200,000 was taken within four months in 1851. Frémont, however, had many troubles. First were the miners who swarmed the new diggings with the assertion that Mexican grants covered only agricultural and grazing rights, not minerals. A small war, with the aid of militia sent by a compliant governor, was required to oust them. Then followed lengthy litigation in the face of hostile public sentiment, piling up court costs and lawyers' fees; on top of that was a ruinous outlay for stamp mills, tunnels, and shafts; and finally, ore did not persist below the surface. Later Frémont said, "When I came to California I hadn't a cent. Now I owe two million dollars!"

Beside the road at 126.1 *m.* are the ruined walls of flat rocks and adobe of a TRABUCCO STORE (R), trading post of the early 1850's for old MOUNT OPHIR, 126.2 *m.* An obscure trail leads through the brush to the ruined mint, where hexagonal gold slugs worth $50 each were coined in the early 1850's.

MOUNT BULLION, 128.2 *m.* (2,100 alt., 16 pop.), was Princeton, until a more picturesque name was suggested by the nearby mountain, which was named for Frémont's father-in-law, Senator Thomas Hart Benton—nicknamed Old Bullion because of his many political battles about currency. Here more than 2,000 miners used to call for their mail. The low, iron-shuttered adobe (L) is the MARRE STORE, which supplied the miners in the 1860's. Just south of Mount Bullion is the old PRINCETON MINE; first owned by Frémont, the Princeton produced millions in gold.

Right from Mount Bullion on a graveled road to HORNITOS (Sp., little ovens), 13 *m.* (1,000 alt., 62 pop.), which got its name from the appellation given the tombs of its early Mexican settlers; two of them, shaped like square bake ovens, lie in ruins on the hillside below the weather-beaten Roman Catholic Church. The camp here in 1849 was known for its fiestas, fandango halls, gambling, and shooting frays. In the ruined GHIRARDELLI STORE, the "chocolate king" of later days began building his fortune as a merchant. Facing the plaza is (R) the old FANDANGO HALL in which the miners sported when they came in flush from the "diggin's." The two-story HORNITOS HOTEL, on an upper street, was built in 1860. The jail, built of stone with a strong iron door and one little window near the ceiling, was the scene of an early tragedy. A Chinese, after having been constantly annoyed by an American miner, lost his patience and shot his tormentor. He was lodged here and when attempts to lynch him failed, a man handed the prisoner some tobacco through the window; as the Chinese stretched out his hand, several of the miners seized it, pulled him against the window, fastened a rope around him, and tore him to pieces.

MARIPOSA (*see TOUR 3b*), 133.4 *m.*, is at a junction with State 140 (*see TOUR 3b*).

Tour 5

Junction with US 99—Bartle—McArthur Memorial (Burney Falls) State Park—Lassen Volcanic National Park—Quincy—Truckee; 277.7 *m.*, State 89.

Roadbed paved between US 99 and US 299, Manzanita Lake and Lake Almanor, and Quincy and Blairsden; elsewhere dirt or gravel. Accommodations at convenient intervals: lakeside hotels, resorts, camps.

State 89 cuts southeastward across northeastern California, seeming to seek out high and inaccessible splendors. It plunges through the Cascade Range, comes to a chaotic region, where lava has made no man's lands of the forests, and emerges between ridges of the Sierra Nevada, whose granite crags rise sheer from steel-blue lakes. The mark of man is plain along the route: forests thinned by lumbermen, carved mountainsides and piles of tailings left by miners, lakes created in hydroelectric developments. The road flows smoothly through the Feather River region, passes old mining towns, mountain stockfarms where cowboys ride herd, and reaches lakes and meadows where emigrant parties fought the Sierra snows—now a playground of summer and winter sports.

Section a. JUNCTION WITH US 99 *to* MORGAN SPRINGS; *139 m., State 89.*

State 89 branches east from the junction with US 99 (*see TOUR 3a*), 0 *m.*, 2 miles south of Mount Shasta City.

SNOWMAN HILL, 5.2 *m.*, rises (R) within 200 yards of the highway from Shasta's lower slopes. Glistening with snow from December to March, the hill is popular with winter sports enthusiasts, who gather for ski tournaments, snow frolics, and tobogganing.

McCLOUD, 10.6 *m.* (3,254 alt., 1,000 pop.), is a typical lumber town, loud with the whine of the great buzz saws in its mills, where huge logs become sweet-smelling yellow lumber. The town preserves, under altered spelling, the name of the Scot, Alexander Roderick McLeod, who in 1827 led southward the first band of Hudson's Bay Company trappers to penetrate northern California. They narrowly escaped starvation and death when severe weather forced them to cache their furs and traps and seek a warmer climate.

Left from McCloud is a road leading through the once heavily timbered McCloud Flat country, built up of glacial silt from Mount Shasta.

MUD CREEK, 3.4 *m.*, is in the area devastated by the mud flow from Konwakiton Glacier in August, 1924. Breaking loose from the creek banks, the avalanche of water and sediment spread over the country, leaving it a desolate waste. The glacial sediment, deposited in many places to a depth of 15 feet, overflowed into the McCloud River and was carried southward for several hundred miles.

At 9.6 *m.* is the junction with a dirt road; R. here 0.5 *m.*, to the McCloud Ice Caves. Entrance is by a 12-foot ladder through a cave-in. The caverns extend east and west for nearly three-tenths of a mile. In the west cave the glittering wall of ice remains until midsummer.

BARTLE, 32 *m.* (3,970 alt., 10 pop.), is a station on the McCloud Logging Railroad (a common carrier if passengers are willing to ride in the caboose). Guns and other paraphernalia unearthed here in 1874 were believed to be part of the McLeod party's cache. Headwaters of the McCloud River rise in this vicinity.

At Bartle is the junction with the dirt Lava Beds National Monument Rd. (*see TOUR 8A*).

CAYTON, 48 *m.* (3,300 alt., 26 pop.), and CLARK CREEK, 52 *m.* (3,000 alt., 12 pop.), lie in a valley meadow between FORT MOUNTAIN and SOLDIER MOUNTAIN (L) and RED MOUNTAIN (R).

At 54 *m.* State 89 enters McARTHUR MEMORIAL (BURNEY FALLS) STATE PARK, a 335-acre recreational area surrounding BURNEY FALLS, where Lost River rushes from its underground channel to plunge in a double cataract—divided by a rocky island—128 feet into the gorge. Pine needles cushion the trail that winds down the eastern slope of the canyon. As the steady roar of the falls grows louder, a fine, cool mist, heady and invigorating, bathes the pine-scented air. The trail leads to the edge of the wide, dark-watered pool at the base of the falls, seen streaming in twin ribbons of white foam against purple rock and blue sky. From the steep, moss-covered banks on the far side of the gorge grow giant pines, green and straight. Many smaller waterfalls spring from the massed lava, bathing the walls with thin streams of transparent silver. At the foot of the falls, Lost River dashes along its boulder-strewn course to LAKE BRITTON, a mile distant, where a dam impounds its waters and those of the Pit River. The lake, kept stocked with bass, is a favorite with sportsmen. There is excellent trout fishing in Burney Creek, a mile above the falls.

At 61 *m.* is the junction with US 299 (*see TOUR 8a*).

The road, striking into Hat Creek Canyon, follows the winding course of this stream southward past HAT CREEK POST OFFICE (L), 83 *m.* (3,300 alt.).

SUBWAY CAVE (L), 81 *m.* (*flashlights required for exploration*), extends through an old lava flow for several hundred feet, with a flat, level floor and a ceiling 6 to 25 feet high.

The road crosses the northeastern boundary of LASSEN VOLCANIC NATIONAL PARK at 99 *m.*, within Lassen National Forest. The forest, a 1,306,727-acre reserve of rugged mountains, high plateaus, forests, and lakes at the northern end of the Sierra Nevada and the southern tip of the Cascade Range, was set aside in 1906. The park, a mountainous expanse of 106,933 acres surrounding the dormant volcano, Lassen Peak, was created in 1916.

Routes to the Park: By automobile, to Manzanita Lake from Redding, on State 44, 52 *m.;* from Mt. Shasta City on State 89, 100 *m.* To Southwest entrance from Red Bluff, on State 36, 52 *m.*, from Susanville, 69 *m.* By bus: from Red Bluff and Susanville to Mineral, all year; from Redding to Manzanita Lake, June 15 to Sept. 15. By train: Southern Pacific to Redding, Western Pacific to Keddie. Nearest commercial airports at Chico and Redding. Overnight accommodations June 10 to Sept. 20. Cabins available all winter at Mineral and elsewhere. Consult Lassen National Park Co., Manzanita Lake, Calif.

The weird, lava-devastated acres of the park are full of evidence of historic and phehistoric volcanic upheavals—sheer, jagged cliffs; great irregularly shaped rocks; fumaroles; boiling lakes, and steaming mud pots. In winter, when the park is a rolling blanket of snow, the gas and steam vents present a paradoxical picture of boiling coldness.

Just within the park, at 99.7 *m.,* State 89 joins with State 44.

Right on State 44 (*stage service to Lassen Volcanic National Park four times weekly*) is VIOLA, 7 *m.* (4,390 alt., 65 pop.). The road drops gradually to SHINGLETOWN, 18 *m.* (2,700 alt., 320 pop.) and then precipitously to MILLVILLE, 37 *m.* (512 alt., 65 pop.), so named because the first grist mill in the county was opened there.

REDDING (*see TOUR 3a*), 49.5 *m.,* is at the junction with US 99.

State 89 becomes the Lassen Peak Loop, and passes between pine fringed MANZANITA LAKE (R) and REFLECTION LAKE (L) —popular for fishing, swimming, and boating—which mirror in their placid surfaces the towering heights of Lassen Peak and Chaos Crags.

At the eastern end of Manzanita Lake is the settlement of MANZANITA LAKE (*post office, campgrounds, lodge, cabins, store*), 100.5 *m.* (5,845 alt.), an all-year resort. At EDUCATIONAL HEADQUARTERS here are presented lectures on the park's flora and fauna, geology, and natural history. MANZANITA LAKE LODGE, operated by Lassen National Park Co. has accommodations. Hotel bungalows, housekeeping cabins, should be reserved June 20 to Sept. 10. The company's bus makes one round trip daily June-September, to bus depot at Redding. The MAY LOOMIS MEMORIAL MUSEUM has exhibits of volcanic specimens.

The CHAOS JUMBLES, 102.2 *m.,* covering a wide stretch on both sides of the highway, are the tops of the Chaos Crags which broke off and tumbled down.

The road swings northward in a hairpin curve which carries it briefly outside the park's boundaries and then southward. It climbs through a pass between the sloping sides (R) of RAKER CREEK (7,466 alt.), and the domes (L) of CHAOS CRAGS (8,458 alt.), old lava plugs, forced up about 200 years ago. Chaos Crags are aptly named, for nowhere in the park is there a scene of wilder confusion. It is hard to express the feeling of wonder—of being transported into some older world—engendered by their gigantic disarray. The pinkish lava domes are magnificent piles of angular pointed blocks heaped in almost inconceivable disorder among projecting pinnacles. Encircling them are enormous banks of pointed talus, some 1,000 feet high. The crags were formed in three stages of volcanic activity—first, the piling up of a series of cinder cones by explosive eruptions; then the pushing up of steep-sided domes of viscous lava; and finally, the tearing down of the domes by new explosions that hurled vast masses of lava fragments over the vicinity. The remnant of a cinder cone formed in the first stage, its crater 600 feet wide and 60 feet deep, still appears on the southern slope of the crags.

HOT ROCK (R), 107.2 *m.,* a large, black boulder of lava, is said to have retained its warmth for a week after it had been blasted from the volcano.

Through EMIGRANT PASS, 109 *m.,* an old trail once ran along the west fork of Hat Creek.

The DEVASTATED AREA, 109.7 *m.*, is a great V-shaped section, starting where the lava and mud flowed down the northeast slope of Lassen Peak, and widened as it swept toward the park boundary. The mud, lava, and heat of the 1915 eruptions denuded the region of all vegetation. Volcanic ash and ejecta, either carried by the mud and lava flows or tossed by terrific blasts, blanket the entire area. Stripped of foliage, the trunks of some large trees reach polelike toward the sky. Nature is slowly reforesting; small trees have taken root in the bleakness.

At SUMMIT LAKE RANGER STATION, 111.9 *m.* (R), a ranger is glad to supply information and advice on almost any subject concerning Lassen Park—from how to build a campfire to the hibernating habits of bear.

SUMMIT LAKE, 112.2 *m.* (R), is a good camping spot (*two campgrounds; fishing, hiking*). Trails radiate to the CHAIN-OF-LAKES district in the eastern section of the park.

As the highway continues south, WHITE MOUNTAIN (R), a dome volcano of the same type as Lassen Peak, named for the color of its soil and rock, rises close to the road.

At KINGS CREEK MEADOWS, 117.2 *m.* (7,400 alt.), are public campgrounds (*campfire programs nightly during season*).

Right from this point on a foot trail to KINGS CREEK FALLS, 1.3 *m.*

At 121.8 *m.* is the summit (8,512 alt.) of the Lassen Peak Loop (*parking space*).

Right from the summit on a foot trail 2.5 *m.* to LASSEN PEAK (10,453 alt.) (*round trip 3 hours*), the only volcano recently active in the United States. One of a long succession of volcanoes—Mount Rainier, Mount Hood, Mount Shasta, and ancient Mount Mazama—Lassen Peak, standing at the point where a southerly spur of the Cascade Range joins the Sierra Nevada, is the only one not yet extinct.

The peak was named San José by the Spanish soldiers under Capt. Luís Argüello, who sighted it in 1820. It was renamed for Peter Lassen, Danish pioneer who in 1848 led an immigrant band off the Overland Trail in a supposed short-cut to the Sacramento Valley. The travelers suffered such hardships in crossing the craggy region around the peak that their route was dubbed "Lassen's Folly."

The great cone, a good example of the rather rare dome type of volcano, was formed by stiff, viscous lava thrust up through the vent in a bulging domelike form around the crater of an older, gently sloping volcano of the shield type. As the dome rose, a rock mantle (talus) of broken-off fragments sheathed its slopes. Unlike most dome volcanoes, Lassen has a funnel-shaped crater formed by escaping gases, so violent that at times they still shoot forth volcanic "ash." Once an oval-shaped bowl 1,000 feet across and 360 feet deep, it is now nearly choked up with rough lava.

For two centuries Lassen had slumbered, wrapped in a snowy mantle; suddenly on May 30, 1914, it began blasting forth cold lava, cinders, and smoke. The first eruption opened a new vent in the old crater, in which melting snow, trickling down, was converted into steam. For a year mild explosions—more than 150 in all—continued, throwing out chunks of cold lava that were not warm enough to melt snow and clouds of dust that were carried by the winds as far as 15 miles southward. Not until May, 1915, did the first glowing lava appear, streaming from a notch in the crater's rim down the northeastern slope in a 1,000-foot-long tongue. As the snow melted, mud flows swept 20-ton boulders 5 to 6 miles into Hat Creek and Lost Creek valleys. Three days later, a scorching blast felled trees 3 miles away, while smoke and volcanic ash spiraled into the sky 5 miles above the crater. The volcano's force was virtually spent by the end of 1915, although it occasionally spewed forth steam and ash during the next 2 years. After a final series of violent outbursts in May and June, 1917, it subsided into its former peace, having erupted about 300 times in all. Although not yet extinct, it is not expected to erupt again for many years.

LAKE HELEN (8,000 alt.), 122.5 *m.* (R), named for Helen Tanner Brodt, first white woman to climb Lassen Peak, lies within the serrated crater rim of ancestral Mount Tehama, once the dominating volcano of the region. The giant crater at the peak of its sloping cone, which had ceased erupting before the Ice Age, was destroyed by violent explosions, which left behind a great bowl, or caldera, encircled by remnant peaks—Brokeoff Mountain, Mount Diller, and Black Butte.

At 122.6 *m.* is the junction with a foot trail.

Left here 1.3 *m.* to BUMPASS HELL (*round trip about 2½ hours*), a small inferno of mud springs, geysers, boiling pools, steam and hot gas vents (fumaroles). The sulphurous gases of the steam vents, which send up clouds of vapor in the cool air of morning and evening, change the lavas into yellow, olive green, or red earthy materials, or into white clay-like substances.

LAKE EMERALD (*fishing prohibited*), 123 *m.* (R), reflects in its crystal-clear waters the jagged relics of the ancient crater rim. The trout gleaming in its clear depths snap at morsels of food tossed by visitors.

The SULPHUR WORKS, 127.7 *m.*, visible from the road, is an area of steam vents, boilers, and mud pots which show that the lava beneath the surface has not entirely cooled. Below the roadway (R) are natural cauldrons of bubbling hot mud. A wooden platform allows good views, but the slippery footing makes closer exploration dangerous. Above the road (L) are smaller pools filled by scalding sulphur springs, whose chemical content is apparent both from the odor and from the yellows, greens, and reds of the rock and soil over which the waters spill. A shady creek, ice cold, tumbles over its rocky bed only a few yards from the steaming pools.

Lassen Peak Loop twists downward, skirting the lower slopes (R) of Eagle Peak (9,211 alt.), Mount Diller (9,086 alt.), Diamond Peak (7,969 alt.), Brokeoff Mountain (9,232 alt.), and (L) of Black Butte (8,208 alt.)—all remnants of ancient Mount Tehama's crater rim. The highest, Brokeoff Mountain, has one side jaggedly split away. On Diamond Peak, steam vents in the old crater wall are visible.

At 129.1 *m.* is the SULPHUR WORKS CHECKING STATION (*open 6 a.m.-10 p.m.*), and at 131 *m.* the road crosses the southern boundary of Lassen Volcanic National Park through a stone gateway erected in honor of Rep. John E. Raker, who fathered the establishment and development of the park.

At 134 *m.* the route turns L. to MORGAN SPRINGS (*see TOUR 6A*), 139 *m.* at the junction with State 36 (*see Tour 6A*); R. here. State 89 and State 36 are united for 19 miles (*see TOUR 6A*).

Section b. JUNCTION WITH STATE 36 to TRUCKEE; 119.1 m. State 89.

State 89 branches southward from its junction with State 36 (*see TOUR 6A*), 0 *m.*, 3 miles southwest of Chester (*see TOUR 6A*).

LAKE ALMANOR (L), 2.4 *m.*, an island-studded sheet of water 45 square miles in extent, is one of California's largest reservoirs used for hydroelectric development. The name, despite its Spanish sound, was coined from syllables of the names Alice, Martha, and Elinore—daughters of the pioneer Earl family, whose father was president of the Great Western Power Company. Peter Lassen's trail passed through the site of Lake Almanor, then called Big Meadows. About midway of its length, at 30.8 *m.*, is the ALMANOR INN. On the west bank are two small towns, ALMANOR and PRATTVILLE.

At LAKE ALMANOR DAM, 14.4 *m.*, the highway crosses the North Fork of the Feather River, and climbs to a point from which there is a view of the Sacramento Valley, 50 miles to the west.

Circling eastward again, State 89 enters historic country—the district of the Northern Mines, where gold rush immigrants found the precious metal in its original quartz veins, not scattered over hillsides by glaciers and carried to river bars by the streams' action, as in so many other mining regions. Profitable up to the present time have been some of these mines, accounting for part of the prosperity of GREENVILLE, 24.4 *m.* (3,500 alt., 520 pop.), and CRESCENT MILLS, 29.4 *m.* (3,500 alt., 160 pop.). Here fertile agricultural lands reach into the mountains.

There is a gentle climb to INDIAN FALLS, 32.7 *m.* (4,000 alt., 45 pop.).

At 35.7 *m.* is the junction with US 40-Alt. (*see TOUR 6B*), which unites with State 89 for 33 miles.

State 89 branches southward from its junction with US 40-Alt. at BLAIRSDEN, 68.7 *m.*

At 73.1 *m.* is CLIO (4,581 alt.), a settlement that could not at first decide what to call itself. Several names, all unsatisfactory, had been suggested, when the keeper of the general store became aware of the name of the heating stove at which he was looking. "Clio!" he cried. "There's a name for our town." And Clio it became.

CALPINE, 84.1 *m.* (5,000 alt., 600 pop.), lies in the agricultural area of Sierra Valley, one of the few level regions in the county. It is a fruitful valley, a mile high.

At 87.7 *m.* is the junction with State 49 (*see TOUR 4a*), which unites with State 89 for 5 miles. Left here on State 89-49.

State 89 branches southward from its junction with State 49 at SIERRAVILLE (*see TOUR 4a*), 92.7 *m.*

At Little Truckee River, 102.8 *m.*, is a junction with an improved road.

Right here, 7 *m.*, over an early route, to WEBER LAKE (6,769 alt.), which mirrors in one square mile the beauty of the forested mountains that encompass it.

At 106.8 *m.* on State 89 is a junction with an improved mountain road.

Right here 2.9 *m.* to INDEPENDENCE LAKE (7,000 alt.), 3 miles long and very deep, from which flows one of the main branches of the Little Truckee River. It was named by the international charmer, Lola Montez —for whom lofty MOUNT LOLA (9,167 alt.), to the north is named—on July 4, 1853, when she came from Grass Valley (*see TOUR 4a*) with a picnic party.

A typical lumber camp is HOBART MILLS, 112.1 *m.* (5,925 alt., 516 pop.), with its immense sawmills, surrounded by miles of logged-over country.

At the confluence of ALDER and PROSSER CREEKS, 114.6 *m.*, the California Emigrant Trail came in from the Northeast. Somewhere in this region the families of George and Jacob Donner in their slow-moving wagons (*see TOUR 9a*), lagging behind the rest of the party because of a broken axle, were halted by falling snow at the end of October 1846. Encamped in crude huts of canvas and boughs banked with snow, cut off by 5 miles of drifts from their fellow travelers at Donner Lake, they tried to hold off starvation. Here the body of George Donner was buried by Gen. Stephen W. Kearny's men in June 1847.

TRUCKEE, 119.1 *m.* (5,818 alt., 1,000 pop.) (*see TOUR 9a*), is at the junction with US 40 (*see TOUR 9*).

Tour 6

(Lakeview, Ore.)—Alturas—Susanville—(Reno, Nev.)—Bridgeport —Bishop—Independence—Randsburg—San Bernardino—San Diego; US 395, St 74, St 71 and US 395.

Oregon Line to San Diego, **824.9** *m.*

Paved roadbed throughout; open all seasons; snow during winter in northern section. Southern Pacific R.R. parallels route between Oregon Line and Raven-

dale; Western Pacific R.R. between Litchfield and Susanville and between Doyle and Nevada Line; Southern Pacific R.R. between Nevada Line and Inyokern.
Accommodations limited except in larger towns.

Though running so far inland behind the Sierra Nevada that it dips for some distance through Nevada, US 395 is nonetheless the shortest route between southern California and the Pacific Northwest.

Section a. OREGON LINE TO NEVADA LINE; 213.2 m. US 395.

This section of US 395 runs through northeastern California across the level sagebrush carpeted floors of a chain of upland valleys, walled by tumbled mountain sides. Contrasting sharply with the flat sweep of brushland are sawtoothed mountain ranges. This is a dry, desolate land with harsh beauty of coloring—the yellow-gray, dun, and gray green of brush and plains; the brownish green of pine and juniper; the red brown of crags.

Here civilization seems remote. On the isolated fields and pastures, which the first log-cabin settlers guarded with shotguns from hostile Indians, the wilderness of sagebrush still encroaches. Something of the early West still appears in the spurred cowboys in 10-gallon hats, who herd their steers off the highway to make room for motorists, and in such relics as the creaking pump by the well in the yard and the rail fences.

US 395 crosses the Oregon Line 14 miles south of Lakeview, Ore., at PINE CREEK, 0 *m.* (4,900 alt.), where along rutted dirt streets sagging frame houses sprawl among willows and cottonwoods.

US 395 skirts the base (L) of the WARNER MOUNTAINS (5,000-10,000 alt.) for 60 miles. The higher peaks of the range are snow-clad the year round. In monotonous succession their undulating, brush-grown slopes, bristling near the peaks with scrubby juniper and crags, overshadow the highway. Goose Lake Valley (R) sweeps into the distance. A tawny fringe of pastures and hayfields slopes down to the blackish mud flats of dry GOOSE LAKE, once 10 miles wide and 28 to 40 miles long. The lake was dry in 1849 when immigrants drove their oxcarts across its bed; but the heavy rains of following years refilled it, and later settlers moved freight across it in boats. When it again went dry in 1924, the old road over the lake bottom appeared, with traces of campfires and abandoned wagons.

At 9.5 *m.* is the junction with a dirt road.

Left on this road to BUCK CREEK RANGER STATION, at the entrance to MODOC NATIONAL FOREST, 0.6 *m.*, 1,500,594 acres. Here, amid every type of mountain country—from 10,000-foot peaks to level pine-clad plateaus—streams and lakes abound in trout, and the forests in game fowl and Rocky Mountain mule deer. Throughout the Warner Mountain sector, the Forest Service maintains a string of public camps.

The road emerges from pine woods at 2.7 *m.* to wind over the rock-strewn sagebrush wastes of Fandango Valley and up the craggy, juniper-wooded slopes of FANDANGO MOUNTAIN (L), to the summit of FAN-

DANGO PASS (6,100 alt.), 7.8 *m.* Through the pass from the desert country eastward, in the early 1850's, came an immigrant wagon train. Passing the summit, the travelers looked down upon the little valley and the shining waters of Goose Lake. Finding game, grass and sweet water, they camped. In their joy at what they thought was safe arrival at the end of their journey, the company began to dance a hilarious fandango; suddenly, Indian warriors, watching from the forest, swooped down without warning and massacred them.

From the summit of Fandango Pass appears 65-mile-long Surprise Valley, with a chain of three dry lakes: Upper, Middle, and Lower. The valley owes its name to the surprise of a party of Nevadans, who came upon it unexpectedly while pursuing Indians in 1861.

At 11.5 *m.* is the junction with a rough, graveled road; 1. L. here 5.3 *m.* across the fields and pastures of the valley to FORT BIDWELL, 16.8 *m.* (4,740 alt., 462 pop.), overshadowed by the Warner Mountains and the great barren Bidwell Mountains. Here, in 1866, was established a fort, named for John Bidwell, where cavalrymen were stationed to hold the Indians in check. Abandoned as a military post in 1892, it became a Government school for descendants of once hostile tribesmen. In 1930 the barracks were torn down. The Indians remain, housed in neat white cottages clustered round the schoolhouse and hospital on the western edge of town.

2. Right from the junction along dry Upper Lake 21 *m.* to CEDARVILLE (*see below*).

Along the line of US 395 creaking prairie schooners rolled in the late 1840's. Here ran the Applegate Cut-off. In June and July 1846, Jesse and Lindsey Applegate, with the help of 13 others, opened the route that bears their name, and that autumn piloted about 100 wagons over it from Fort Hall, Idaho, into Oregon. The Applegate Cut-off passed between Middle and Upper Lakes, along the western edge of Surprise Valley, through Fandango Pass, down into Goose Lake Valley, and southward along the lakeshore. From Goose Lake's southern end, it struck northwestward across the lava beds past Tule and Clear Lakes into the Klamath country of southern Oregon.

Crossing Lassen Creek, US 395 climbs the burned-over slopes of towering Sugar Hill. As it rounds a bend at 14.3 *m.*, the whole stretch of Goose Lake springs suddenly into view. The highway descends to the valley floor. Among fields with scattered barns and corrals are orchards where excellent apples grow.

DAVIS CREEK, 22 *m.* (4,995 alt., 84 pop.), amid pastures spotted with clumps of willows, is a mere handful of ramshackle houses and abandoned store buildings.

The highway leaves Goose Lake Valley at 28.4 *m.*, and winds up over a rocky waste of sagebrush and juniper to dip into the canyon of the PIT RIVER (R), 31.4 *m.*, a creek winding over brush-grown bottoms through craggy slopes. Along the canyon bed, like a half-ruined wall of crumbling stone winding in and out around the edge of the hilltops, rise the rocky palisades that fringe the vast plateau lying to the west. Over this tableland, known as the Devil's Garden, grows one of the largest unbroken stands of western juniper, a tract of 300,000 acres.

At 37.1 *m.* is the junction with a paved road.

Left on this road, over a desolate mesa, to the cottonwood bordered canyon of THOMAS CREEK, 6 *m.*, along which the road climbs to BONNER GRADE, 8.3 *m.*, named for John H. Bonner, who built this road over the Warner Mountains in 1869.

Deep among pine and fir is the Forest Service CEDAR PASS PUBLIC CAMP, 9.6 *m.* Beyond a bend at 10.3 *m.*, great beetling crags, blackish and tinged with rose, jut pinnaclelike; for one mile these rocky masses overhang the highway.

CEDAR PASS SUMMIT (6,350 alt.), 11.2 *m.*, affords a view of tumbled peaks and ridges wooded with bristling conifers. In graceful curves the road winds downward along Cedar Creek. As pines give way to junipers at lower levels, far stretching Surprise Valley appears ahead, and beyond is the dim blue line of mountains in Nevada.

CEDARVILLE, 17.9 *m.* (4,768 alt., 400 pop.), is an old-fashioned village with wide, tree-lined dirt streets, along which stand old white clapboarded houses. One block R. of Main St., in a grove of tall cottonwoods stands the CRESSLER AND BONNER TRADING POST, a tiny cabin of hand-hewn logs with a shake roof, worn and weathered. Its builder, one Townsend, was killed by Indians soon after he built it in 1865; his widow sold it to William T. Cressler and John H. Bonner, who set up the first store in Modoc County, a trading post that carried on a thriving business with immigrants and later with the settlers of Surprise Valley.

The highway winds on over rolling country, across willow-fringed Pit River and into the level valley of the South Fork of the Pit.

At 42.8 *m.* is the junction with US 299 (*see TOUR 8a*) which, between this point and Alturas, is united with US 395.

ALTURAS, 43.5 *m.* (4,446 alt., 2,819 pop.) (*see TOUR 8a*), is at the junction with US 299 (*see TOUR 8a*).

South of Alturas are pastures and scattered farmhouses. Farther southward is a rolling, dreary waste of sagebrush. East of US 395 is the MODOC NATIONAL WILDLIFE REFUGE.

At 60.1 *m.* is the junction with a dirt road.

Right on this road 1.7 *m.* to a junction with a dirt road; R. here and through a wooden gate 0.3 *m.* to a fork in the road; L. here, through another wooden gate, 1.6 *m.* to a farmhouse at the base of the hills. Here a wagon track, impassable except on foot, leads (L) through the barnyard and across swampy fields 1 *m.* to the mouth of CROOK'S CANYON, where tiny Hilton Creek gurgles among boulders overgrown with sagebrush and juniper.

Beyond, at the base of a rock-crowned hill, is the INFERNAL CAVERNS BATTLE-GROUND. Over the pleasant valley meadows six white marble headstones mark the graves of six of the men killed here September 26, 1867: two sergeants and four privates of the Twenty-third Infantry and the First Cavalry. The battle occurred when a band of about 100 Shoshones, Paiutes, and Pits who had terrorized the country round about were cornered by Gen. George Crook and a force of 65 soldiers. In a seemingly impregnable fortress of caves and rocks the Indians took their stand. After two days of fighting, they were forced out, but only after eight of Crook's men had been killed and 14 wounded.

After passing LIKELY, 63.1 *m.* (4,500 alt., 75 pop.), in the midst of wide-spreading pastures, US 395 climbs to a low pass between rugged South Fork Mountain (R) and Tule Mountain (7,136 alt.) (L), to twist over South Fork Mountain's lower slopes and wind along the shores of Tule Lake Reservoir (L), then descend again to valley plains.

MADELINE, 76.6 *m.* (5,300 alt., 92 pop.), stands at the upper end of a dreary sagebrush waste. About the railroad station huddle a bleak tavern, gas stations, and a handful of shanties.

Across the vast plain the highway continues southward to RAVENDALE, 98.1 *m.* (5,300 alt., 19 pop.), a group of cattle corrals and old houses with peeling paint. Then it enters a rolling, juniper-dotted wilderness and, at 107.2 *m.,* cuts into a wild region of rocky ridges, broken and jagged. Along the edge of the low tableland (R) stands a rocky embankment like a breastwork of stone piled by hand.

Green and fresh, Secret Valley appears suddenly like an oasis far below beyond the crags; but its meadows soon give way to another sagebrush waste.

As US 395 winds down from the pass, 128.2 *m.,* Honey Lake Valley, walled by the lofty peaks of the Sierra Nevada (R), appears in a widening panorama of fields and pastures, clumps of trees and rooftops. A valley 40 miles long and 20 miles wide, it was long the only settled section of Lassen County. Its first dwellers named it for the honeydew on trees and shrubs, from which the Indians made a sort of molasses. Here lived two Indian tribes; along the base of the Sierra, the poverty-stricken Washo, numbering only 900 in 1859; over the rest of the valley, the Pah Ute, 6,000 or 7,000 in number. The Pah Ute proved friendly to white men, visiting their houses and trading furs. Their chief, Old Winnemucca, in 1856 signed a treaty with the whites that provided for punishment to Indians for thieving and to white men for molesting squaws.

Across the sluggish, tule-bordered Susan River, deep among willows, the highway cuts into lush fields, past farmhouses and windmills in clumps of bushy cottonwoods. Here sprawls old-fashioned STANDISH, 138.7 *m.* (4,000 alt., 115 pop.), with a steepled church and belfry-topped schoolhouse, and a string of ancient stores and houses.

At 147.6 *m.* is the junction with State 36 (*see TOUR 6A*).

JOHNSTONVILLE, 148.1 *m.* (4,100 alt., 13 pop.), where tree-shaded houses gather about a schoolhouse, was the home of pioneer Robert Johnston. It once bore the name Toadtown because toads covered the ground during heavy rainstorms.

US 395 here swings sharply southeast. Overshadowed by the rugged, forested Sierra is JANESVILLE, 155.1 *m.* (4,236 alt., 200 pop.). Along the highway for a mile in forest clearings, straggle red, sagging frame houses, abandoned stores, and a little white schoolhouse with the ever-present belfry. Here in 1857 Malcolm Bankhead raised his two-story house of logs and a blacksmith shop. The town was named for his wife, when the post office was opened in 1864.

A few widely scattered houses, a store, and gas station are all that remain of BUNTINGVILLE, 157.2 *m.,* named for A. J. Bunting who opened a store here in 1878.

South of Buntingville, beyond the strip of farmland and pasture glimpsed between black oaks and pines (L) appears the vast, blackish bottom of dry HONEY LAKE.

MILFORD, 167.3 *m.* (4,180 alt., 69 pop.), is a few old buildings, on the edge of the orchards and pastures.

DOYLE, 185.7 *m.* (4,260 alt., 68 pop.), amid sagebrush, is hemmed in by desolate slopes. The route continues southeastward between mountain walls, on one side rock ribbed and barren, forested with green on the other.

STOY JUNCTION, 205.4 *m.,* marks the junction with US 40-Alt. (*see TOUR 6B*).

More and more rugged grow the rocky walls of the Sierra, as the highway runs through an arid expanse to reach the glistening white sheet of dry ALKALINE LAKE and the NEVADA LINE, 213.2 *m.,* at a point 15.6 miles west of Reno, Nev., and 83.2 miles north of a second crossing of the California Line.

Section b. *NEVADA LINE to BISHOP; 131.5 m.* *US 395.*

Skirting the eastern slopes of the Sierra Nevada, this section of US 395 winds through rugged, thinly settled country, warm and leafy in summer, bitterly cold and often piled with impenetrable snowdrifts in winter. Geologically, this region is young. The great cinder cones of Mono Craters—some reaching an elevation of 9,000 feet—and numerous bubbling hot springs are evidences of a great igneous field not yet burned out; it is estimated that volcanic activity may have taken place here as recently as 500 years ago. Sheep and cattle graze in the valleys, but the rugged back country is seldom penetrated save by sportsmen who come to hunt deer and to fish in the numerous high, cold lakes.

US 395 crosses the Nevada Line, 0 *m.,* 76 miles south of Reno (*see NEVADA GUIDE*). TOPAZ LAKE (L), on the line, a tranquil body of water encompassed by pinon-strewn hills, was formed in 1920-21 when the course of Walker River changed and inundated about 1,800 acres of a flourishing ranch.

After skirting Topaz Lake, US 395 passes between SLINKARD'S CANYON, 3.5 *m.,* and its broad residual delta (L), and traverses Antelope Valley.

WALKER RIVER, 4.2 *m.,* is the valley's principal source of water. In 1827 Jedediah Smith, traveling what was to become the Sonora Trail over the Sierra in the first crossing of that range by a white man, trod the banks of this stream on his eastward trek to Salt Lake City. In 1833-34 the trapping and exploring party of Capt. Joseph Walker reached the river, followed it to its source, and then battled through the defiles of the Sierra for 23 days before entering the central plains of California. This was the first westward passage of the mountains. In 1841 the Bidwell-Bartleson party, first of the overland settlers, followed this same river, and a decade later the trail had become a well-defined road leading into the central gold fields.

COLEVILLE, 9.4 *m.* (5,750 alt., 50 pop.), a tourist and farm trading hamlet, lines both sides of the highway beneath scattered shade trees.

WALKER CANYON, 15 *m.*, cut by the willow-lined river, is at the lower end of Antelope Valley.

Part of TOIYABE NATIONAL FOREST is crossed at 16 *m.* Giant pines lean over Walker River, and cottonwoods and willows lend greenery to the jagged, crumbling walls of granite that tower above the road.

SHINGLE MILL PUBLIC CAMP, 20.3 *m.*, is a pine-sheltered clearing beside the river.

CHRIS FLATS PUBLIC CAMP (*stoves and tables*), 24.2 *m.*, lies on a sandy bar of the riverbank beneath cottonwoods and pines.

At 28.4 *m.* is the junction with State 108, paved.

Right on State 108, which ascends the Sierra Nevada, crossing it through SONORA PASS (9,642 alt.), 15.2 *m.* Hordes of fortune hunters swarmed over the route after the Sonora-Mono road to the Nevada silver mining country was cleared in the 1860's. The road descends past a string of outing camps to DARDANELLE (*cabins*), 31.2 *m.* (5,775 alt., 10 pop.), lying near the base of the striking serrated mountain ridge (R) for which it is named.

The TUOLUMNE COUNTY PUBLIC CAMP (L) is at 32.5 *m.*

The way leads down the crest of the narrow ridge between the Stanislaus River (R) and the beautiful North Fork of the Tuolumne (L).

COW CREEK RANGER STATION, 49.5 *m.* (5,750 alt.), consists of a few tree-shaded buildings.

At the mountain resort of STRAWBERRY, 53.5 *m.* (5,240 alt.), is the junction with a dirt road; L. here 2 *m.* to PINECREST, a mountain camp on the southern shore of STRAWBERRY LAKE, near ELEANOR LAKE. The lakes are guarded by granite cliffs surmounted by tier upon tier of somber, dark green pines.

LONGBARN (*hotel*), 63.9 *m.* (5,000 alt., 25 pop.), is in a summer and winter sports area. The snow-blanketed slopes round-about are thronged in season with skiers and tobogganers. Close by is the 1,800-foot toboggan slide of the Tuolumne Club.

CONFIDENCE, 69.7 *m.* (4,250 alt., 63 pop.), trading village of the surrounding mining and lumbering area, was named for the nearby Confidence Mine, that justified its name by producing $4,250,000 in gold.

At 75.7 *m.* is the junction with a dirt road; L. here 1 *m.* to SOULSBYVILLE, (3,000 alt., 320 pop.), a mining town dating back to 1856.

At 80.7 *m.* is the junction with a dirt road; L. on this road 7 *m.* is TUOLUMNE (2,600 alt., 2,000 pop.), a mill center of sugar pine lumbering, lying in a circular basin on Turnback Creek.

SONORA, 83.2 *m.* (1,925 alt., 2,725 pop.), (*see TOUR 4b*), is at the junction with State 49 (*see TOUR 4b*).

FALES HOT SPRINGS, 31.8 *m.*, were discovered in 1867 by Archibald Samuel Fales, who crossed the plains in 1851. Fales had been a prospector, and a driver and guard on the Bodie-Sacramento Stage Line; but rheumatism necessitated his settling in the mud. Around his mud bath there sprang up a saloon, a freight depot, sheep pens, and accompanying evidences of civilization. The present establishment is a rustic collection of buildings around a concrete pool whose waters, laden with sodium, calcium, and magnesium, have a temperature of 180 degrees.

US 395 continues its ascent through hills that begin to lose their rotundity and rise in spires of crumbling rock.

DEVIL'S GATE, 33.1 *m.* (7,549 alt.), a great perpendicular mass of cracked green rock, rises sheer above both sides of the highway.

HUNTOON PUBLIC CAMP, (*stoves*) 40.5 *m.*, is (L) beside a small mountain stream, a tributary of Walker River.

Forsaking the course of the creek, the route crosses the southern line of TOIYABE NATIONAL FOREST, 42.1 *m.*, and approaches secluded Bridgeport Valley, where small bands of cattle graze.

BRIDGEPORT, 45.7 *m.* (6,473 alt., 200 pop.), is in a valley ringed by the snowy peaks of the Sierra and low, pinon-covered hills. Dilapidated, partly crushed houses, long since abandoned, evidence the weight of winter snows. Bridgeport owes its existence to a geographical misunderstanding on the part of the early inhabitants of Mono County, who, in 1864, were amazed to learn that their county seat, Aurora, was in Nevada. Hurriedly the county papers were transferred from Aurora to a tiny roadside camp that later became Bridgeport. The town grew prosperous in the days when gold was plentiful, and built its gingerbread courthouse. Today, with gold scarce, it subsists upon what mining activity remains, upon the surrounding farm trade, and tourist patronage.

The ornate MONO COUNTY COURTHOUSE, a two-story, white structure (L) ornamented with red-trimmed windows, has appeared frequently in motion pictures. A cannon on its lawn is celebrated locally as having been a repository for "hooch" during prohibition days.

South of Bridgeport US 395 again climbs into the mountains. Green foliage covers the bottoms of the canyons; snow-capped peaks glisten in the distance; pinon and pines give verdant ornamentation to mountainsides scarred with green and black rock formations of increasing prominence.

At 53.1 *m.* is the road to BODIE HISTORICAL STATE PARK.

Left on this road to a junction with a dirt road, 4.1 *m.;* L. on this road is BODIE, 13.7 *m.* (8,374 alt., 125 pop.). From 1876 to 1880 a mining town with 12,000 inhabitants, Bodie is today an aggregation of shacks clustered below the great mill and mine (R) on the mountain. Gold was first discovered in the Mono region in 1852; the Mono Trail was blazed from Big Oak Flat through the present Yosemite Park, and in 1859 W. S. Body found gold here. But in 1864 the Sonora Pass wagon road was opened, removing much of the Mono Trail traffic, in 1870 the Aurora mine failed, and it was not until 1876 that fortune again smiled with the unexpected development of the Bodie Mine, a wildcat venture. Population figures leaped. Saloons and gambling halls prospered; drunken miners threw 20-dollar gold pieces in the streets for small boys to spike with their tops. Occasionally guns blazed in the streets; the phrase, "Bad man from Bodie," became popular. By 1883 all the mines but two had closed, and in 1887 even they were consolidated. Finally, on June 24, 1932, fire destroyed most of what remained. The ROSECLIP MINE on the hill above (R) reopened in October, 1936.

Ascending the long grade, US 395 winds through hills clothed with meager bunch grass, desert sage, and needle grass.

The SITE OF DOGTOWN (L), 53.7 *m.*, several hundred yards from the highway, is marked by a few remnants of the stores and houses of the early mining town, probably the earliest settlement in the entire region. Its life was short, for in 1859 the population, lured by more attractive prospects, moved en masse to neighboring Monoville.

VIRGINIA LAKE JUNCTION, 59.7 *m.* (8,138 alt.), at the summit of the grade, is a gas station, store, and a group of tourist cabins.

MONO LAKE, 69 *m.*, though it appears fresh and inviting, is actually a briny deep, in which nothing but one small species of salt-water shrimp and the larvae of one tiny black fly can live. Its waters are so impregnated with soluble alkaline materials that storms, often violent in this region, sometimes pile the shores with a soapy foam several feet thick. In the 87 square miles of water of the lake more than 19 different chemicals are held in solution. Rising above volcanic PAOHA ISLAND in the lake's center, there is visible on clear, still days a misty vapor from the island's hot spring. Another extinct crater, NEGIT ISLAND, also juts above the blue surface of the lake. Above it occasionally soar terns, locally called "Mono pigeons," and sea-gulls that have flown inland from the Pacific. The lake has achieved a certain measure of fame in literature and legend. Mark Twain tells in *Roughing It* of a dog that attained a running speed of 250 miles an hour after taking a swim in the lake. A more modern tale is of a long-haired dog that emerged with nothing left but its bark.

The waters of many streams flow into the new Mono Basin Aqueduct and help to slake the thirst of metropolitan Los Angeles. The drainage from five streams, the Mill, Rush, Walker, Leevining, and Parker, is now diverted by conduits leading to Grants Lake, six miles south of Mono Lake, where it is impounded. From this Grants Lake Dam another conduit leads the water to a 12-mile tunnel emerging near a tributary of the Owens River, whence it flows down to the Los Angeles Aqueduct at Aberdeen.

MONOVILLE, 69.1 *m.* (6,419 alt., 20 pop.), a resort settlement on the shore of Mono Lake, is a summer supply center for campers and miners. Monoville thrived in the wild boom days but, unlike most mining towns, history records no bloodcurdling deeds of its early period; it was always, apparently, as peaceful as it is today. It was still fighting to be the county seat in 1864, and served for 50 years as the starting point for those seeking for "the lost cement mines." According to local legend, these mines were discovered by three German brothers on their way to California in the 1850's and were near the headwaters of the Owens River. In the vein of cement they found "lumps of gold set like raisins in a pudding"; but a bitter winter drove the brothers from their find, killing two and leaving the third insane. It was his babbled story of treasure that started the prolonged treasure hunt.

LEEVINING, 71.8 *m.* (7,000 alt., 80 pop.), derives its name from nearby Leevining Canyon, named for Lee Vining, who in 1852 led a party of prospectors to the first gold strike there.

At 72.1 *m.* is the junction with State 120, the Tioga Pass route.

South of Leevining is a view (L) of a group of the MONO CRATERS, 73.1 *m.* Four of them, resembling gigantic ash heaps, are immediately and easily discernible. Actually there are 20, forming a crescent-shaped range, along the base of which US 395 proceeds for 10 miles through Pumice Valley, covered with fine silt pumice of past volcanic eruptions. Here are REVERSED CREEK RECREATION AREA, Bald Mtn., 9,045 ft. and Glass Mtn., 11,127 ft.

The AEOLIAN BUTTES (L), 78.1 *m.,* scarred and crumbled steps of red rock, are the oldest volcanic formations on Mono Basin.

At 99.7 *m.* is the junction with a dirt road.

Right on this road is MAMMOTH POST OFFICE, 4 *m.,* a few small houses and stores. At 4.2 *m.* is the junction with a dirt road; R. here 3.5 *m.* to an EARTHQUAKE FAULT (R), and to the DEVIL'S POST PILE NATIONAL MONUMENT, 11.5 *m.,* a rectangular area 2.5 miles long and 0.5 miles wide, extending along both sides of the Middle Fork of the San Joaquin River and containing clear evidence of recent volcanic activity. The 40-foot cliff of columnar basalt rearing above the turbulent San Joaquin is composed of columns that are nearly perfect prisms. Two feet in diameter and almost vertical, they fit together like the cells of a honeycomb. Most of the prisms are pentagonal, though there are some with four or six sides. Littering the base of the cliff are fragments of these prisms shaken down by some long-past earthquake.

Ten miles north are BANNER PEAK (12,953 alt.) and MOUNT RITTER (13,153 alt.); just south of the latter are the MINARETS, pinnacles of dark granite.

South of the Devil's Post Pile are the RAINBOW FALLS. Here the San Joaquin makes a 140-foot perpendicular drop into a box canyon, creating a scene of mist-sprayed grandeur.

At 8 *m.* on the main side road is the junction with a dirt road; R. or L. here into the vast region of lakes and resorts known collectively as MAMMOTH LAKES (8,931 alt.). These lakes, the joy of fishermen and vacationists, were once the scene of one of California's feverish gold rushes. In the summer of 1878 the town of Mammoth had a population of 125; in 1879 it had 2,500. A string of settlements had sprung up in the canyon and two semi-weekly newspapers flourished. But the exceptionally bitter winter of 1879-80 caused most of the miners to abandon their claims, and of those who remained three were frozen to death. Although Mammoth had a brief return of prosperity in 1880, the mines were nearly exhausted, and the next year saw the final decline.

At 103.1 *m.* is the junction with a dirt road.

Right on this road to CONVICT LAKE, 2.5 *m.* (7,583 alt.), a pellucid sheet of blue, fronted by a rustic resort for fishermen and backed by the imposing height of Mount Morrison (12,245 alt.). On September 17, 1871, 29 convicts escaped from the Nevada State Penitentiary and made their way to this lake, where they attempted to winter. Six of the desperadoes started south, and on their way met and murdered William Poor, a mail carrier. Outraged citizens in Aurora and Benton organized a posse, which set out for the convict stronghold. The posse reached the lake on September 24, and engaged in a gun battle in which Robert Morrison (for whom the distant peak is named) was killed. Several of the convicts escaped, but were later captured; two were hanged at Bishop; another was returned to the penitentiary.

The semiarid undulations of Long Valley stretch away (L), and in the far background rise the WHITE MOUNTAINS (14,242 alt.).

The ascent of SHERWIN GRADE is begun at 119.1 *m.* Over bare hills the highway climbs to the summit, 119.7 *m.* (6,430 alt.), from which is a view, ahead and below, of vast Owens Valley and the silver line of Owens River. From the summit the road descends in tortuous hairpin curves, frequently flanked (R) by sheer drops.

US 395, flanked (R) by the dull hues of the TUNGSTEN HILLS, traverses greener country and proceeds over the floor of Owens Valley between rows of poplars and oaks. Occasionally hayfields and cornfields border the road.

BISHOP, 131.5 *m.* (4,147 alt., 2,875 pop.), is the largest town in Owens Valley, and the chief business center of the area. It is a supply point for stockraising and mining interests. (Much tungsten is mined in the vicinity.) Sportsmen know it as a gateway and outfitting point for camping, fishing, and packing trips into the Sierra Nevada, which rise (R) more than 10,000 feet above the town.

Bishop was named for Bishop Creek, which in turn was named for Samuel A. Bishop, a Fort Tejon stockman, who drove the first herd of cattle into Owens Valley in 1861. He built two rough pine cabins on his St. Francis Ranch that were besieged by Paiutes during the Indian uprisings of the 1860's, and several times thereafter until Fort Independence (*see below*) was established in 1862. By 1863 settlers had come into Owens Valley in increasing numbers, and the village sprang up 3 miles northeast of Bishop's ranch.

At the northern end of Bishop is the junction with US 6 (*see TOUR 7a*), which unites with US 395 for 123 miles.

Right from Bishop on State 168 along Bishop Creek to the junction with South Fork Rd., 11 *m.;* L. 7.5 *m.* to SOUTH LAKE (9,750 alt.), source of the South Fork of Bishop Creek and popular for trout fishing. The main stream of Bishop Creek, which runs from the Lake Sabrina region in the high Sierra to Owens River at Bishop, is 14 miles in length. Its North, Middle, and South Forks penetrate a mountain area of more than 200 lakes. Dropping an average of 400 feet per mile, the main stream is equipped with seven power stations, the highest at an elevation of 8,000 feet. Its waters are also used for irrigation.

The Paiute inhabited this region before the arrival of white men. Among the most primitive of American Indians—neither planting, tilling, nor cultivating the soil—none the less they used the waters of this creek to aid the growth of their wild food plants. They watered the plots of wild seed, shrubs, and tubers in the spring by damming the creek and running a ditch to each field. In the fall the dam was destroyed and the waters allowed to return to the main channel.

On State 168 is LAKE SABRINA, 15 *m.* (9,150 alt.), headwaters of the Middle Fork of Bishop Creek, and noted for its rainbow and golden trout. Five peaks towering 13,000 feet are visible, including Mt. Darwin, 13,830 ft.

Section c. BISHOP *to* BROWN 124.7 *m.* US 395-US 6.

This section of US 395 penetrates a land of contrasts—cool crests and burning lowlands, fertile agricultural regions and untamed deserts.

It is a land where Indians made a last stand against the invading white man, where bandits sought refuge from early vigilante retribution; a land of fortunes—past and present—in gold, silver, tungsten, marble, soda, and borax; and a land esteemed by sportsmen because of scores of lakes and streams abounding with trout and forests alive with game. The highway follows the irregular eastern base of the towering Sierra Nevada, past the highest peak in any of the States—Mount Whitney—at the western approach to Death Valley, the Nation's lowest, and hottest, area.

South of Bishop, 0 *m.*, US 6-395 traverses the length of flat Owens Valley, an 85-mile trough. Although little rain falls here, the land has great fertility. The Owens River, receiving the waters of swift tributaries, supplies unlimited amounts of irrigation water. Much of the farm land, however, has been acquired by the city of Los Angeles as watershed adjuncts to the Los Angeles Aqueduct, which taps Owens River.

CROWLEY LAKE, a reservoir of the Los Angeles Dept. of Water & Power, is stocked with trout. WHITMORE HOT SPRINGS, 10 *m.* beyond Crowley, also is a Department activity.

BIG PINE, 16.5 *m.* (4,002 alt.), is a village catering to motorists and the outfitting of fishing, packing, and hiking parties.

Right from Big Pine on Big Pine Creek Rd. to GLACIER LODGE (*rates reasonable*), 12 *m.*, and to a hiking trail at 14 *m.*; straight ahead on this trail to PALISADES GLACIER, 20 *m.*

TINEMAHA RESERVOIR (L), 24 *m.*, is a collecting basin of the Los Angeles Aqueduct System.

US 395 cuts through POVERTY HILLS, 24.3 *m.*, between scattered masses of volcanic outcroppings. Lava deposits appear with increasing frequency, and at 30.3 *m.* (L) is an extensive lava bed.

ABERDEEN, 31.6 *m.* (3,814 alt.), is the starting point of the original LOS ANGELES AQUEDUCT, 233 miles long. It took five years to build (1908-1913), and cost $23,000,000. More than 40 miles of tunnel were bored through solid rock, the longest section measuring 26,780 feet. The completed aqueduct includes 142 separate tunnels aggregating almost 52 miles in length, 12 miles of steel siphons about 10 feet in diameter, 24 miles of open, unlined conduit, 39 miles of open cement-lined conduit, and 98 miles of covered conduit. The gravity system is used throughout.

At 42.5 *m.* is the junction with a paved road and with a dirt road.

1. Right on paved road to MOUNT WHITNEY STATE FISH HATCHERY (*open 9-5 weekdays*), 1.4 *m.* whose red stone buildings occupy a parklike tract. Between two and one-half and three and one-half million trout fry are distributed annually from the hatchery.

2. Left on dirt road 0.1 *m.* to the RUINS OF FORT INDEPENDENCE. A decaying, one-story log cabin, once the officers' quarters, is all that remains of the military camp that played a part in quelling the Inyo County Indian uprisings in the 1860's. The fort was established July 4, 1862, and maintained until 1877.

INDEPENDENCE, 45.8 *m.* (3,925 alt., 408 pop.), seat of Inyo County, is an outfitting point for pack trips into the Sierra by way of Kearsarge Pass and a shipping point for farm products.

The EASTERN CALIFORNIA MUSEUM (*open 9-6 weekdays*), in the basement of the INYO COUNTY COURTHOUSE, houses a collection of antiquated firearms used in the Indian wars, old sidesaddles, candle and bullet molds, files of newspapers and historical documents, Indian relics of California and Nevada, flora and minerals of the region, and a collection of photographs of Indian pictographs.

On the crest of the INYO MOUNTAINS (L) is an inaccessible monolith of weather-worn and storm-beaten granite known as the PAIUTE MONUMENT. Although 80 feet high, of a roughly conical shape, it appears from Independence as a slender, heaven-pointing finger.

By MANZANAR (Sp., apple orchard), 51.9 *m.* (3,872 alt., 153 pop.), is the JOHN SHEPHERD RANCH HOUSE (R), built in 1873 of materials brought by wagon from San Pedro, 250 miles away.

The ALABAMA HILLS (R), 56 *m.*, riven by glacial pressure in the Triassic period and rent by earth movements in subsequent ages, rise beside the highway. At the foot of the hills and paralleling the highway runs an EARTHQUAKE FAULT, along which the land dropped from 4 to 12 feet in the temblor of 1872, which took 26 lives.

LONE PINE, 61.6 *m.* (3,728 alt., 1,310 pop.), dating from the early 1850's, caters to tourists and outfits for trips to Mount Whitney. Within packing distance are more than 1,000 mountain lakes stocked with trout.

Right from Lone Pine on State 190, through HUNTERS FLAT (*hunting in season*), 10 *m.*, abounding with deer, bear, sage hen, grouse, and quail, to WHITNEY PORTAL, 14 *m.*, a pack station, marking the end of the motor road.

Straight ahead on a pack trail to OUTPOST CAMP, 18 *m.* (10,300 alt.), which lies just below the timber line in a setting of towering cliffs and green meadows. Lone Pine Creek tumbles over a series of falls (L) as the trail ascends past MIRROR LAKE, 19 *m.* Above stand the last stunted trees, outposts of the timber line. The trail now enters an area of mighty granite walls and boulder-strewn basins where, strangely, some of the loveliest of Alpine wild flowers grow. Upward the trail winds through fields of perpetual snow to the summit of MOUNT WHITNEY (14,495 alt.), 24 *m.* The mighty peak was named by the noted geologist, Clarence King (1842-1901), in honor of Prof. J. D. Whitney, leader of the California Geologic Survey party which in 1864 ascertained that this mountain was the loftiest peak in the United States, exclusive of Alaska. The first attempt to scale Mount Whitney was made by King in 1871, but, confused by storm clouds, he climbed Mount Langley by mistake. Unaware of his error, he published an account claiming first ascent of Mount Whitney. In 1873 W. A. Goodyear climbed Mount Langley, discovered King's marker, and published news of the mistake. King hurried west from New York and on September 19, 1873, climbed the true Mount Whitney. But others had deprived him by one month of the honor of making the first ascent. On August 18, 1873, A. H. Johnson, C. D. Begole, and John Lucas had reached the summit and given the mountain the name Fisherman's Peak, which for a time threatened to supersede the earlier name. The summit is a nearly level area of 3 to 4 acres, sloping slightly westward.

At 63.6 *m.* is the southern junction with State 190.

Left on State 190 to a MARBLE QUARRY, 8 *m.* (L), whose product is noted for its coloring and quality. KEELER, 13 *m.* (3,610 alt., 300 pop.), a desert settlement, has been for many years the headquarters of a plant engaged in extracting soda-ash from Owens Lake.

Left from Keeler on a dirt road 6 *m.* to CERRO GORDO (9,217 alt.), producing silver, lead, and zinc.

At 30.4 *m.* on State 190 is the junction with a dirt road; R. here 6.2 *m.* to DARWIN (4,479 alt., 80 pop.), an old mining town whose population at one time numbered 1,500. The town bears the name of Dr. Darwin French, who in 1860, while searching for the fabulous lost Gunsight Mine, discovered silver and lead ledges in the Coso RANGE, 12 miles south.

State 190 crosses Panamint Valley and ascends the Panamint Range to the western boundary of DEATH VALLEY NATIONAL MONUMENT, 63 *m.* (*see DEATH VALLEY TOUR 1*).

BARTLETT, 70.6 *m.* (3,690 alt., 70 pop.), center of operations of one of the largest borax mining companies in the United States, is by OWENS LAKE, 17 miles long and 10 miles wide which was named for Richard Owen, a member of John C. Frémont's 1845 expedition. Into it from the north flows Owens River. The level of the lake was once much higher; well-defined shore lines mark the continued recession of the waters. Originally it was an inland sea, created in the Ice Age by slow-moving glaciers; as the ice cap slowly receded melting torrents roared into the lower areas and formed inland seas, filled with salts, sodas, and other chemicals leached from the rocks. Through the centuries the waters, no longer replenished by melting ice streams, evaporated and left great arid beds—the dry lakes of today's desert.

CARTAGO, 82.7 *m.* (3,680 alt., 30 pop.), is a railroad village at the southern end of Owens Lake. An employees' town (L), once considered a model housing project, has been virtually abandoned since the lime, soda, and borax plant closed several years ago.

OLANCHA, 85.6 *m.* (3,649 alt., 75 pop.), is named for the Olancha tribe which formerly inhabited the region. In the background (R) is OLANCHA PEAK (12,135 alt.).

UPPER HAIWEE RESERVOIR, 91.4 *m.*, connects one mile south with LOWER HAIWEE RESERVOIR, the two forming one of the largest collecting basins in the Los Angeles aqueduct system.

US 395 crosses lava masses past foothill slopes (R) scarred by great misshapen boulders of volcanic substance.

LITTLE LAKE, 111.5 *m.* (3,172 alt., 23 pop.), is a tourist settlement named for a 10-acre, privately owned lake that gleams beside the road.

At 123.8 *m.* US 6 (*see TOUR 7b*) swings R.; the route turns sharply L. on US 395.

BROWN, 124.7 *m.* (2,400 alt., 51 pop.), is a railroad town on the Southern Pacific-Randsburg Line.

Section d. *BROWN to JUNCTION US 66; 107.5 m. US 395*

This section of US 395 continues through desert country tapped by gold and silver mines.

South of BROWN, 0 *m.,* is INYOKERN, 9.6 *m.* (2,442 alt., 25 pop.), which in summer swelters in the heat of flat Indian Wells Valley. On the lone hotel, a weather-beaten frame structure, is a sign reading: "This is the hotel, believe it or not." At Inyokern is the junction with the Trona Road (*see DEATH VALLEY NATIONAL MONUMENT*).

The Rand Mountains, dark, rounded elevations, loom up straight ahead; the highway begins its ascent of the long grade up the range.

At 34.3 *m.* is the junction with an asphalt paved road.

Right on this road is RANDSBURG, 1 *m.* (3,523 alt., 443 pop.), a mining camp in the Rand Quadrangle, long a very productive mining district, named for the Rand gold-mining district in the Transvaal, South Africa. The town's unpainted weather-worn cabins sprawl over the rocky slopes on the northern flank of Rand Mountain, like those of a typical movie-set mining town. The single street divides a collection of ancient frame houses scattered about at random. On the slopes are the tin and galvanized iron structures of the mines, each complemented by a pile of tailings. Gold was first discovered in Goler Wash, nine miles northwest of Randsburg, in 1893, and dry-washing camps soon sprang up in Last Chance and Red Rock Canyons and at Summit Diggings. The richest mine was the Yellow Aster, on the side of Rand Mountain, which by 1925 had yielded about $10,000,000 in gold. The ore in the Yellow Aster Mine, which is now a "glory hole," 1,200 feet deep, contains enough tungsten to pay all operating expenses, leaving the gold as clear profit. (A glory hole is an open excavation without shafts or tunnels, enabling its operators to recover a large quantity of ore at minimum expense.) The ore taken from this pit is crushed in the stamp mill and its minute particles of gold are recovered by a complex chemical process known as cyaniding.

From JOHANNESBURG, 35.2 *m.* (3,536 alt., 106 pop.), named for the chief city in the Transvaal, most of the glory of mining days has vanished, though a few miners still hopefully work old properties.

At RED MOUNTAIN, 37 *m.* (3,530 alt., 200 pop.), prospector Hamp Williams is said to have stumbled on California's richest silver mine by accident. Overtaken by a snowstorm one day in the winter of 1919, he sought shelter in a pit dug years before by another gold seeker. While crouching there he picked up a fragment of rock that appeared to contain metal; an assay subsequently showed it to be "horn silver," rich in value. Williams staked out a claim, later developed into the Big Kelly Mine and now known as the Rand Silver Mine. Today Red Mountain is cutting in on the social life of Johannesburg and Randsburg with its honky-tonks, dance halls, and drinking places, where miners of the Rand and men from Trona come to spend their leisure hours. As in a movie set of a rough-and-tumble border town, its buildings perch above the road level, the raucous jangle of mechanical pianos coming through the swinging doors of the saloons.

ATOLIA, 40.4 *m.* (3,600 alt., 150 pop.), is a far cry from the booming town that supplied the Allies in the World War with tungsten from the Atolia Tungsten Mine ever since new discoveries in China pushed the price of tungsten too low to justify operation at a profit.

US 395 sweeps across the vast Mojave Desert. No settlement, not even a gasoline station, marks the route until it approaches KRAMER JUNCTION, 64.2 *m.* (2,482 alt.), a filling station and auto camp. The road winds upgrade through the low Kramer Hills and levels out again across the Mojave. It passes through a region of great silence, of sagebrush and creosote bush stretching interminably. Near the southern end of the straightaway, occasional green, irrigated tracts break the tawny monotone of sand.

ADELANTO, 95.7 *m.* (2,900 alt., 300 pop.), is the center of an alfalfa-growing and poultry-raising district, reclaimed by the Mojave River.

The highway mounts steadily to the level sandy plateau of Baldy Mesa and sweeps across it toward the bulwark of the San Bernardino Mountains.

At 107.5 *m.* is the junction with US 66 (*see TOUR 12b*), which unites with US 395 to San Bernardino (*see TOUR 12b*).

Section e. SAN BERNARDINO to SAN DIEGO; 134.8 m. US 395, St 74, St 71 and US 395.

Between a high inland plateau at the mouth of Cajon Pass and the coast, US 395 runs due southward through the oldest citrus producing area in southern California and across low, barren ranges to the coast.

SAN BERNARDINO, 0 *m.* (*see TOUR 12b*), is at the junction with US 66 (*see TOUR 12b*).

At 3 *m.* US 99 (*see TOUR 3d*) unites westward with US 395 to COLTON (*see TOUR 3d*), where US 395 branches L.

HIGHGROVE, 5 *m.* (500 pop.), is a trading post for the surrounding citrus ranches.

In RIVERSIDE, 10.2 *m.* (*see TOUR 13b*), is a junction with US 60 (*see TOUR 13b*), which unites with US 395 southeastward to a junction at 14.7 *m.*, where US 60 turns R.

BOX SPRINGS, 13.5 *m.*, is a filling station and garage.

At 17.1 *m.* is a junction with a paved road.

Left on this road to MARCH FIELD, 2 *m.*, an Army flying field (*open daylight hours*), with a 3,550-foot runway outlined with green lights, a paved landing mat of 450,000 square feet, 200 permanent buildings on the field, and eight miles of paved roads. The field was established in 1917 as a training school, named for Peyton C. March, Jr., son of the Chief of Staff and a victim of World War I. Since transfer of training activities to Texas in 1931, March Field has been occupied by tactical units of the Army Air Corps.

PERRIS, 26.1 *m.* (1,456 alt., 2,950 pop.), platted in 1885-1886 and incorporated in 1911, succeeded an older settlement, Pinecate (Sp., stink bug), 1.5 miles south on the Sante Fe Railway. The 12 rich gold mines that once were very profitable have been worked out.

PERRIS RESERVOIR will become the southern end of the State Water Project, which will deliver water from Plumas and Butte counties for 300 miles to Riverside County, which the water will reach in 1972.

Left from Perris on State 74 through acres of peach and apricot orchards, to HEMET, 16.2 *m.* (1,600 alt., 5,416 pop.), in the center of the Indian country where artists of ancient races have left rock paintings and carvings on boulders in the canyons.

1. Right from Hemet on State Street 2 *m.* to the RAMONA BOWL, a natural amphitheater where the Ramona Pageant, an adaptation of Helen Hunt Jackson's novel *Ramona* is presented yearly in April and May. Scores of metates, or corn-grinding holes, in the vicinity of the bowl, remain to mark the SITE OF THE PAHSITNAH VILLAGE, once the largest in the area. The road southward continues over St. John grade to a junction with State 79 (*see TOUR 6C*) at 24 *m.*

2. Left from Hemet 2.5 *m.* on San Jacinto Ave. to SAN JACINTO (1,550 alt., 2,553 pop.), at the base of the Lakeview Mountains and MOUNT RUDOLPH (2,629 alt.). San Jacinto, which occupies the site of Jusispah Village, one of the seven Indian rancherias of the San Jacinto Valley, was founded by Procco Akimo, a Russian exiled from his native land during the Tsarist terrors of the 1870's. The second non-stop transpolar flight from Russia to the United States terminated safely in a cow pasture 3 miles west of San Jacinto on July 8, 1937, when the fliers missed their goal, March Field.

GILMAN HOT SPRINGS, 7 *m.* (1,600 alt.), in the San Jacinto Mountains, is a resort with Indian lodges, bungalows, bath house, swimming pool, and recreation pavilion.

East of Hemet on State 74 is a junction with a dirt road, 21.8 *m.*

Left on this road 0.4 *m.* to a junction with another dirt road; L. here to SOBOBA HOT SPRINGS and the SOBOBA INDIAN RESERVATION, 4.9 *m.* (1,725 alt.), where about 125 Indians live in neat homes around a large hospital. One of the oldest Indian legends of this region relates that Tuchaipai, the Great Spirit, gave the people their choice of living forever, of dying temporarily, or of dying forever. They argued and deliberated for days while the smoke of their council fires rose high from the hills. Then they called in the Fly, who was growing very tired of hearing them talk. So he said to them: "Oh, you men, what are you talking so much about? Just tell the Great Spirit that you want to die forever and never come back again." And this is what the simple Indians told Tuchaipai, so that now when they die they can never come back again. But the Fly was sorry he had told them such a foolish thing, and ever since he has rubbed his hands together in apology.

At 34.9 *m.* on State 74 is a junction with a paved road.

Left on this road 4 *m.* to IDYLLWILD (5,300 alt., 376 pop.), (*rates reasonable*), a mountain resort with a large rustic lodge and cabins (*guides available*). The mountain slopes heavily covered with pines and big-cone spruce, rise northward to SAN JACINTO PEAK (10,300 alt.) in MT. JACINTO STATE PARK. The Indians say that a powerful demon, called Tahquitz, lives atop San Jacinto. Sometimes he stalks through the canyons wailing and howling through the night for victims—and this is what makes bad weather.

At 32 *m.* the Idyllwild Rd. meets US 60-70-99 (*see TOUR 13b*). State 74 continues to KEEN CAMP POST OFFICE, 35.9 *m.* (4,936 alt.), in a region of Jeffrey pines, with occasional sugar pines, firs, and yellow pines.
At 69.9 *m.* State 74 meets State 111 (*see TOUR 13b*).

The main Tour 6 now follows St 74 south (R) from Perris.

GOOD HOPE MINE (R), 31 *m.*, the most productive gold mine in the district, has an electrically operated mill with daily capacity of 100 tons.

ELSINORE, 38 *m.* (1,286 alt., 2,432 pop.), is a resort on the northern shores of LAKE ELSINORE, a land-locked basin with no outlet whose waters, fed by subterranean springs, have been receding in recent years; around it is a State Park.

According to one of the two stories of the town's naming, it was named for Elsinore in Shakespeare's *Hamlet*. According to the other, corroborated by local tradition, two thirsty travelers, a Yankee and a Spaniard, came upon the lake after a dusty trek. The impetuous Spaniard rushed eagerly into the water up to his armpits and drank greedily. The cautious Yankee inquired, "How does it taste, senor?" The Spaniard, having by this time become aware of the strong mineral content of the water, replied in disgust, "Like 'ell, senor!" In the Elsinore Valley, rich in ores, an 11-pound nugget, largest ever found in southern California, was picked up. Airplanes land at Elsinore's landing field; amphibians and seaplanes, in the lake. Water sports, boating, and speedboat regattas are popular the year round and duck hunting in season.

Right from Elsinore on State 71 to a junction with State 74, 2.4 *m.*; L. on State 74 to the OLYMPIC FIELDS NUDIST COLONY, 7 *m.* which attained public notice in the late 1930's as the West's pioneer nudist colony, founded "on the proposition that the desire for release from clothing is inherent and normal, and that a periodic release from the too-insistent and destructive pressure of our high-speed social order is beneficial mentally, emotionally, and physically."

In Elsinore the main Tour 6 follows St 71 south (L).

South of Elsinore, the route skirts Lake Elsinore and passes into the fertile Temecula Valley, once the wheatfield of Mission San Luis Rey. It now produces beef cattle and dairy products; hay, grain, and alfalfa; olives and walnuts.

At 51 *m.* is a junction with Hawthorne St., a paved road.

Left here to MURRIETA HOT SPRINGS, 3 *m.* (1,309 alt., 153 pop.), a resort (*all accommodations; rates reasonable*).

At 53 *m.* is junction with US 395 which main Tour 6 follows south (R).

TEMECULA (Ind., the rising sun), 56.3 *m.* (1,003 alt., 255 pop.), a trading post for the residents of Temecula Valley, is a dwindling handful of wind-worn, dusty buildings—fewer today than in 1882 when the town was founded on the railroad connecting San Bernardino with San Diego. The tracks were washed out by torrential rains in 1892 and never replaced.

The inhabitants claim for Temecula the dubious distinction that the wind blows every afternoon. Some old timers insist that the natives are so accustomed to this strong wind that they will not come out of doors on the rare windless days.

Joe Winkel of Temecula, frequently mentioned in Herriman's cartoon, "Krazy Kat," was a real person, who settled here in 1902 and opened a bar that attracted newspapermen from coast cities. The saloon, still standing opposite the bank, was in old days the mecca for booted and spurred cowhands who brought great droves of beef stock to the railroad. With the advent of truck and trailer transportation, the cowpuncher abandoned his chaps for mechanic's overalls—but not his visits to Winkel's bar.

At 58.1 *m.* is a junction with State 79 (*see TOUR 6C*).

At 58.4 *m.* is a junction with a dirt road.

Left here, through PECHANGA CANYON, to PALA (Ind., water), 9 *m.* (411 alt., 260 pop.), trading center for the PALA INDIAN RESERVATION. Pala grew up around LA ASISTENCIA DE SAN ANTONIO DE PALA, established by Fr. Antonio Peyri for the Mission San Luis Rey. The restored chapel (*small adm. fee*) is (L) near the northern boundary of the reservation. Of particular interest are the detached bell tower, some vivid Indian murals, and a statue of St. Anthony carved in wood more than a century ago by a Mexican. Erected in 1816, Pala was deserted by its padres in 1829; its buildings, already in ruins before 1836, were almost demolished by flood in 1916. The Pala Reservation has a small village of portable wooden shelters. At the annual celebrations are included the Corpus Christi Fiesta, the San Antonio Fiesta, and a fiesta held on August 25th, which features a rodeo. American music and gambling have largely replaced tribal customs.

Left from Pala, 1 *m.*, to a KUNZITE MINE. This rare stone has been found only here and on the island of Madagascar.

RINCON, 17.8 *m.*, on the main side road, is an Indian trading post; the route continues L. here. Right 1.8 *m.* from Rincon, across SAN LUIS REY RIVER, is the RINCON INDIAN RESERVATION (800 alt., 170 pop.). A 3-day fiesta is held annually on August 24.

At 23.1 *m.* is a junction of the main side road with the paved Highway to the Stars; L. on this road 6.9 *m.* to CRESTLINE, a small campground. Left from Crestline 3.2 *m.* on a dirt road to PALOMAR STATE PARK.

PALOMAR MOUNTAIN OBSERVATORY of the California Institute of Technology occupies a 720-acre plateau on PALOMAR MOUNTAIN (6,126 alt.) 34.7 *m.* north of Crestline by the Highway to the Stars. The Rockefeller Boards provided it in 1928 to supplement the facilities of Mount Wilson Observatory of the Carnegie Institution of Washington. It is the seat of the 200-inch reflecting telescope, most powerful instrument in the world and also has 48-inch and 18-inch Schmidt wide-angle telescopes. The mirror of the 200-inch telescope weighs 16 tons. Originally it had a focal length of 660 inches, reaching stars of the 23rd magnitude; in 1965 this was enlarged to 1,800 inches, reaching 24th magnitude stars.

On the Rincon Rd. at 38.9 *m.* is the junction with State 79 (*see Tour 6C*).

RAINBOW, 63 *m.* (1,051 alt., 56 pop.), on US 395 is in a tiny upland valley. From this point the route climbs the winding, chaparral lined Red Mountain Grade through rugged country.

At 66 *m.* is a junction with an oiled road.

Left on this road to LIVE OAK PARK, 3 *m.*, a natural wooded area set aside for public recreation (*picnic ground; open-air dance hall; playground*).

FALLBROOK, 70.5 *m.* (700 alt., 10,116 pop.), was settled in the 1880's. Because of the soil and climatic advantages, this is an ideal citrus-growing region. In late years avocado orchards have been planted and poultry farms established.

Right from Fallbrook is DE LUZ, 9.9 *m.* (146 alt., 180 pop.), a mountain community in an area noted for deer hunting; quail, doves, and rabbits are also plentiful.

BONSALL, 78 *m.* (172 alt., 213 pop.), is a small trading center in a dairying and farming area.

At OCEANSIDE JUNCTION, 80.5 *m.*, is a junction with Mission Road (*see TOUR 2d*).

A post office and store were established at VISTA, 85.1 *m.* (330 alt.), in 1890 when the Oceanside-Escondido branch of the Santa Fe Railway was completed. Development of the town began in 1926, when water was brought to the district from Lake Henshaw. In 1950 it had 1,705 people; in 1960, 14,795; in 1965, 21,429 est.

At 85.1 *m.* is a junction with State 78.

Right on State 78 to a junction with an oiled highway, 0.1 *m.;* R. here to the old COUTS ADOBE HOUSE of the RANCHO GUAJOME (Ind., home of the frog), built by Lt. Cave J. Couts in 1853. Four wings surround a large central patio; the roofs are still covered with the original dull red tiles. Cave J. Couts, Jr., who inherited the property, maintains it in the style of a Mexican hacienda, employing Indian servants. Helen Hunt Jackson described this rancho in *Ramona.*

SAN MARCOS, 91.5 *m.* (570 alt., 3,049 pop.), at the confluence of three valleys, is on the former Rancho Los Vallecitos de San Marcos. Industries have been moving in. Librascope Division of General Precision Co. makes guidance computers for aircraft and rockets, here.

ESCONDIDO, (*hidden valley*), 97.9 *m.* (650 alt., 16,177 pop., 1960; 26,161, 1965, est.), at a junction of US 395, State 76 and 78, is a growing residential area in the heart of vineyards and citrus orchards. Has an annual avocado show and a grape day. There are two notable wineries, Borra and Ferrara, and the largest lemon-packing plant in the State. Motels on US 395.

The town of Escondido was laid out in 1885 from the former Wolfskill Ranch, which comprised 13,000 acres of the original Rancho Rincon del Diablo.

The outstanding social event of the year in Escondido is the Grape Day celebration (*September*). During the carnival, several tons of grapes are distributed to visitors, politicians orate in the public park, bands play, and at night, church meetings and dances end the festival. This holiday originated in 1905, to celebrate the final liquidation of irrigation district bonds which had been oppressive to the farmers for 20 years. When the bondholders agreed, in 1904, to accept 50 per cent of the face value of the securities, the indebtedness was paid. On September 9, 1905, a gala mass meeting was held and the redeemed bonds were ceremoniously burned before a crowd of 2,000 cheering farmers.

Left from Escondido on State 78 through the farming and poultry regions of Santa Maria Valley.

A monument marks the SAN PASQUAL BATTLEFIELD, 7.3 *m.*, where, in December 1846, Brig. Gen. Stephen W. Kearny attacked the Californian army under General Andrés Pico. The Californians escaped from the conflict with only a few slightly wounded, but 19 of Kearny's men were killed and 17 wounded. It is said that Kearny lost the battle through carelessness, and that there was little excuse for his unprepared attack.

State 78 passes through San Pasqual Valley and ascends San Pasqual Grade.

RAMONA, 18.5 *m.* (1,440 alt., 400 pop.), founded in 1886, is in a region of poultry farms; turkeys are the specialty.

WITCH CREEK, 30.9 *m.*, is a trading post for a small farming district.

East of Witch Creek, State 78 traverses Santa Maria Valley.

At SANTA ISABEL, 34.1 *m.*, State 78 unites eastward with State 79 for 7.9 *m.* to JULIAN (*see TOUR 6C*).

BANNER, 49.1 *m.*, once a boom town, is now a camp and picnic ground (*adm. 25¢ and 50¢*) at the foot of Banner Grade. The region was filled with gold-mad miners soon after gold was discovered in Julian in 1870. The most important mines at Banner were the Redman and the Golden Chariot.

SCISSORS CROSSING, 54.1 *m.*, in the San Felipe Valley, was so named because of a pattern made by converging roads. Pedro Fages, who camped in this valley and gave it its present name, visited Indian villages here in 1782. Scissors Crossing is at the junction with a dirt road (*see TOUR 14b*).

ANZA-BORREGO DESERT STATE PARK, 54.5 *m.* (452 alt.) includes three distinct desert areas and extends from the northern boundary of San Diego County to within a few miles of the Mexican border. In the northeast corner CLARK'S DRY LAKE is used by the Univ. of Maryland for a radio telescope and has two miles of antennas for detecting radio emissions from outer space. On a dirt road off State 78 are BORREGO and BORREGO SPRINGS, the latter an expanding residential center.

Anza-Borrego State Park lies north of State 78, 92 *m.* northeast of San Diego. It has a total of 488,000 acres in San Diego and Imperial Counties. There are 96 campsites and 34 picnic installations, and a motel, restaurant and stores within easy driving distance.

SENTENAC CANYON BRIDGE and GRAPEVINE MOUNTAIN are at 58.7 *m.* At 60.3 *m.* is the junction with Borrego Valley Rd.

In PALM CANYON PLAYGROUND, 15.7 *m.*, are many specimens of the Washingtonia palm tree, California's only native palm tree. Botanists say that this tree is "a residual remnant of the days of the saber-toothed tiger, ancestral horse, camel and ground sloth, whose fossils have been found here." Perhaps 500,000 years ago, this canyon was a palm-lined shore of the Gulf of California. The first grove of palms is an hour's hike (L) up Palm Canyon from the campground.

On State 78 at 78.3 *m.* is a junction with a dirt road.

Left on this road 3.7 *m.* to a GYPSUM MINE in the FISH CREEK MOUNTAINS. At the mine a trail leads several hundred yards through an unnamed canyon to an area covered with FOSSIL FOOTPRINTS, by a prehistoric watering hole where mastodons came to drink.

At 91.3 *m.* is the junction with US 99 (*see TOUR 3e*).

LAKE HODGES, 103 *m.*, is crossed on a concrete bridge. (*Boats rented; bass, trout, crappie, perch.*)

BIG STONE PARK, 111.9 *m.* (*picnic grounds; admission fee*), is in a region of giant boulders. It was a favorite Indian campground, and for 50 miles around rock paintings are found.

The road winds to the CREST OF POWAY GRADE, at 113.9 *m.* (525 alt.), along chaparral covered slopes. (*Caution: limited visibility on curves.*)

At 126 *m.* is a junction with a dirt road.

Left on this road to MISIÓN SAN DIEGO DE ALCALA (*admission fee*), which stands on a small knoll overlooking Mission Valley. Junípero Serra, on July 16, 1769, erected a crude hut on Presidio Hill in San Diego, which was the first of the 21 missions built by the Franciscans in California. In 1774, the mission was transferred to this place, where, on the night of November 5, 1775, an Indian attack destroyed the structure and caused the death of Fr. Luís Jaumé. The garrison, stationed at Presidio Hill, slept through the engagement. Jaumé is buried under the altar of the church.

The mission was rebuilt and dedicated in 1780. Prosperous years followed. Barracks were erected for soldiers, and grain houses, corrals, and dormitories for the Indian converts. A dam, built 5 miles up the river, was 220 feet long, with a gateway 12 feet high in the center.

An earthquake in 1803 destroyed the church, but by 1813 the buildings had been restored and enlarged. Though Mission San Diego was not one of the wealthiest, it owned, in 1830, more than 8,000 head of cattle, 1,000 horses and mules, 16,000 sheep, and controlled 1,506 Indians. With the passage of the Bill of Secularization in 1834, which returned mission property to public use, the mission declined. Richard Henry Dana visited the mission in 1835 and described it in his *Two Years Before the Mast.*

The U. S. Land Commission returned 22 acres of the mission grounds to the Church, that area on which the church stands; but neglect and weathering soon reduced the structure to ruin. The present building, a restoration, was erected in 1931. A parochial school is conducted in the modern buildings beside the mission. In the museum are many relics, among them old baptismal, marriage, and death records, the first pages in the handwriting of Junípero Serra.

The SAN DIEGO RIVER, 126.3 *m.*, like many southern California streams, is dry the greater part of the year. Occasionally, however, heavy rains cause it to flood the floor of Mission Valley.

At 128.6 *m.* is a junction with US 80, El Cajon Blvd. (*see TOUR 14b*).

SAN DIEGO, 134.8 *m.* (*see SAN DIEGO*), is at the junction with US 101 (*see TOUR 2d*).

Tour 6A

Junction with US 395—Susanville—Chester—Mineral—Red Bluff; 112.3 *m.* State 36.

Roadbed paved. Daily bus service between Reno, Nev., and Redding. Accommodations only in larger towns.

This route climbs the rugged gorge of the Susan River, deep among pines, to cross the Sierra Nevada through Fredonyer Pass. It leads through the forested uplands around Lake Almanor, where power houses and lumber mills tap the water and timber resources of the mountains. From Mineral, southern gateway to Lassen Volcanic National Park, it descends through corridors of luxuriant fir and pine to the Central Valley's hot, grassy stretches.

State 36 branches west from its junction with US 395 (*see TOUR 6a*), 0 *m.,* at the northern edge of Johnstonville and crosses level stretches, where willow-fringed creeks meander over lush, haystack-dotted fields. Along this route, paralleling the sluggish, tule-bordered Susan River (L), ran one of the most traveled trails into California, the Noble Emigrant Route, surveyed through Noble's Pass across the Sierra in 1852. By 1855 most of the northern California immigration was following it. From the Humboldt River route through Nevada, it entered California around Honey Lake (*see TOUR 6a*), followed the north bank of the Susan River, and then diverged to continue around the north side of Lassen Peak (*see TOUR 5a*) through Noble's Pass to Shasta (*see TOUR 8b*), whose merchants had put up $2,000 for a wagon road. At Isaac Roop's trading post in Susanville a register was kept; in the period from August 2 to October 4, 1857, it recorded 99 trains passing through the valley, with 306 wagons and carriages, 665 horses and mules, 16,937 cattle, 835 men, 254 women, and 390 children.

SUSANVILLE, 4.7 *m.* (4,195 alt., 5,598 pop.), from a bench above the Susan River commands a view of far-stretching Honey Lake Valley (*see TOUR 6a*). The valley lies in the shadow of low, timbered hills (R) and the lofty, pine-forested ridges of the Sierra (L). Main Street, a broad boulevard sweeping up a slight rise into the heart of town, has a brisk, traffic-thronged air of modernity despite the booted men in wide-brimmed hats who lounge about and the pastures that reach to the very edge of the business district. Susanville's busy air is accentuated by the smokestacks and water towers of the great box factory and lumber mills at its edge. Lumber, cattle, and hay are the town's stock in trade. Above the downtown section and the straggling houses in the hollows around the mills rise the locust- and cottonwood-bordered streets of Susanville's west end, which takes to the heights.

To the site of Susanville in the spring of 1853, alone and penniless, came Isaac N. Roop on horseback from Shasta. He staked out a claim and built a cabin. In June of the year following, came Peter Lassen and a handful of prospectors; they dug a ditch and struck gold. By the end of 1857, besides log cabins with shake roofs and fireplaces, the settlement had one house of boards and at least one cook stove. There was little law, less Gospel; horse thieves made their rendezvous here and shooting scrapes became frequent. For amusement there was gambling everywhere, much whisky drinking, and once in a while a dance.

Susanville's past lives in the names of its streets, which honor its pioneers. Its oldest building, FORT DEFIANCE, stands on Weatherlow St., facing over the valley. This was the cabin that Roop built in July

1854. Here he stored his merchandise and supplies, and to it he brought water from Piute Creek in a half-mile ditch. In its day the cabin even had the honor of serving as capitol of a territory, which its founders called Nataqua (Ind., woman). Nataqua's boundaries hemmed in 50,000 square miles, from the northeast corner of California to 25 miles south of Lake Tahoe. So far away were California on one hand and the Utah Mormon settlements on the other that the valley's settlers determined to set up their own government. On April 26, 1856, they created Nataqua—and while they were at it, the town of Susanville—meeting at Roop House and electing Isaac Roop recorder and Peter Lassen surveyor. But Nataqua was short-lived. A year later the Honey Lake men were joining with settlers to the east in a demand for the establishment of the Territory of Nevada. In the same month, August 1857, California asserted jurisdiction when Plumas County created Honey Lake Township. Grumbling, some of the settlers paid their taxes, but not the 40 or 50 pioneers who had endured the early hardships and the Indian fights. In rebellion they set up a local government in 1858, and the year after helped elect Isaac Roop Provisional Governor of what became, in 1861, Nevada Territory, with Honey Lake Valley included as Roop County.

When California stuck to its claims, the stage was set for the Sagebrush War; and Roop House acquired its new name of Fort Defiance. When Probate Judge John S. Ward and Sheriff William Hill Naileigh, of Roop County, refused to refrain from exercising authority, as the Plumas County Court ordered, Sheriff E. H. Pierce, of Plumas County, arrested Naileigh. He sent Deputy Byers to arrest Ward too, but Isaac Roop and seven mounted men blocked the way with shotguns. Forced to give up Naileigh, because snow in the mountains prevented taking him back to Plumas County, Sheriff Pierce crossed the mountains alone. With a posse of 90 men he returned February 13, 1863, to find an armed force of from 75 to 100 fortified in Roop's cabin. Negotiations proved fruitless. On February 15 Pierce and his men occupied and fortified a barn 200 yards away. When one of them, who went out to bring in timber, was fired on and wounded, the battle began; and for five hours both parties blazed away at each other. An armistice followed. Assured by the men in the fort that they would burn down the town around him unless he surrendered, Sheriff Pierce agreed to stop on condition that each party disband and all officers cease functioning. News of the compromise reached a party of reinforcements from Quincy, seat of Plumas County, who were dragging a small cannon over the mountains through the snow. They turned about and dragged it back. The boundary dispute was settled when the California-Nevada Line was run northward from Lake Tahoe, east of Honey Lake Valley; but hard feelings persisted until the California Legislature, on April 1, 1864, created Lassen County with its seat at Susanville.

Left from Susanville on Weatherlow St. to a fork at 3.1 *m.;* L. here to another fork at 4.2 *m.;* R. here through bristling pines and across peaceful

Elysian Valley where Peter Lassen arrived in June 1855, with a party of prospectors, and found gold. Going back in October 1855, he brought tools, plows and farm implements, cows, oxen, and horses. With his friend, Isadore Meyerwitz, he settled on a 1-mile tract and built a long low cabin on the south side of Lassen Creek.

At 5.9 *m.* is a turnstile (L), the entrance to the grove of pines that shade LASSEN'S GRAVE. Two monuments overlook the pleasant pastures. The old one, erected by the Masons, June 24, 1862, a column of gray stone, fissured and crumbled, topped with an urn, bears the inscription: "In Memory of Peter Lassen, the Pioneer, who was killed by the Indians, April 26, 1859. Aged 66 years." On the modern granite obelisk beside it is the same inscription, and the words: "Erected in Honor of Peter Lassen by the people of the Northern Counties of the State of California."

State 36 winds up the jagged, pine-fringed gorge of the Susan River. At 12.7 *m.* it climbs away from the river, scaling the Sierra to the summit of FREDONYER PASS (5,750 alt.), 18.6 *m.*, and descends the long western slope.

At 27.4 *m.* is the junction with a paved road.

Left on this road is WESTWOOD, 1.2 *m.* (5,082 alt., 1,209 pop.), a company town huddled around the giant mills of the Red River Lumber Company. In July 1938 Westwood was the scene of a bitter labor dispute. When 600 members of the C.I.O. union, the International Woodworkers of America, called a strike in protest against a 17½ percent wage reduction, the lumber company closed its mills, throwing out of work an equal number of "company union" men—members of the Industrial Employees Union. The strike led to a struggle between the two groups of employees, with the C.I.O. men picketing the plant. On July 14 the I.E.U. members, aided by Westwood residents, came to blows with the C.I.O. workers. While some 2,000 men fought with fists, picks, axes, and rifles, deputy sheriffs turned high-pressure fire hose on the pickets. The C.I.O. men were defeated, and a "kangaroo" court drove them and their families out of town. The plant opened the next day. About a week later 300 C.I.O. men voted to return to work when the Red River Company promised to restore their seniority rights.

CHESTER, 40.9 *m.* (4,280 alt., 1,553 pop.), is a lumber town on the northwestern shore of Lake Almanor (*see TOUR 5b*), where power company employees make their homes.

Right from Chester on a dirt road 0.5 *m.* to the junction with a dirt road.

1. Right on this road 11 *m.* to JUNIPER LAKE CAMPGROUND (*undeveloped*), on the western shore of Juniper Lake, in the southeastern part of Lassen Volcanic National Park (*see TOUR 5a*). JUNIPER LAKE RESORT (*tents and cabins, saddle and pack horses, rowboats and motorboats; 25¢ charge per car for parking and $1 per car for passing over privately owned land to Horseshoe Lake*), 12 *m.*, is the starting point for trips into the lake-dotted lava wastes of eastern Lassen Volcanic National Park. Through this region extends the Chain-of-Lakes; Juniper Lake at the base of the ancient volcano, Mount Harkness; Horseshoe Lake, whose waters flow partly into the Feather and partly into the Pit River; sandy-edged Snag Lake, whose clear depths reveal the remains of trees growing at the south end before it was dammed by lava; and Butte Lake, edged by rugged lava shores, near the eastern base of Prospect Peak. Among others are Twin, Echo, Swan, Rainbow, and Chester Lakes.

2. Left from the junction on a dirt road into Warner Valley. LEE CAMP (*hotel, store, gasoline; saddle and pack horses; guide service*), is

at 12 *m.* and KELLY CAMP (*hotel, store, gasoline; saddle and pack horses; guide service*) at 14 *m.* The road crosses the southern boundary of Lassen Volcanic National Park (*see TOUR 5a*) at 14.4 *m.* A RANGER STATION (*information and maps*) is at 15.6 *m.*

DRAKESBAD RESORT (*tents and cabins; saddle horses and mounted guide service; 25¢ charge for non-patrons to visit any of Drakesbad property*), 18.1 *m.,* on Hot Springs Creek below Flatiron Ridge, is a center for exploration of the remote southeastern sector of Lassen Volcanic National Park. Just southward is BOILING SPRINGS LAKE, a bubbling cauldron of volcanic origin encircled by steaming mud pots. To the west, the solfataras of the DEVIL'S KITCHEN hiss in the lava-walled "Canyon of a Thousand Smokes," through which flows a tiny stream called the Little Styx. The lava-scarred region northeast is studded with volcanic peaks, accessible by trail.

At 43.9 *m.* is the junction with State 89 (*see TOUR 5a*), which unites with State 36 between this point and Morgan Springs.

The road climbs through fine stands of fir and pine to DEER CREEK PASS (5,000 alt.), 46.8 *m.,* and descends to MORGAN SPRINGS (*campground*), 63.5 *m.* (4,786 alt.).

Morgan Springs is at the junction with State 89 (*see TOUR 5a*).

In MINERAL (*cabins, lodge, golf course, garage*), 72.5 *m.* (4,800 alt., 50 pop.), a small vacation town on the edge of a wide mountain meadow fringed with dense stands of pine and fir, are the ADMINISTRATIVE HEADQUARTERS of Lassen Volcanic National Park (*see TOUR 5a*).

LASSEN CAMP (*cabins, saddle and pack horses, guide service*) is at 79.5 *m.*

The road winds through wooded gorges, where here and there a mountain stream flashes silver, skirting the base of an occasional cliff. Called in this section Ponderosa Way, it plunges through the deep shade of ponderosa pines. The sky is a narrow slit of blue above the treetops. Between the tree trunks, thickly covered with green moss on the north side, the sun falls in bright shafts. Moss-covered logs, felled when the road was cut through, lie along the way.

On either side at 97 *m.* rise great cliffs. As the road descends into the lower foothills, fields are strewn with masses of broken dark gray volcanic rock. Huge purplish-hued buttes rise in the distance. The flat-topped craters of several extinct volcanoes appear.

The road winds down from the foothills and strikes southward along the course of the Sacramento River to its junction with US 99E (*see TOUR 3B*) 109.5 *m.;* R. here. From this point State 36 and US 99E are united to RED BLUFF (*see TOUR 3a*), 112.3 *m.,* at the junction with US 99 (*see TOUR 3a*).

Tour 6B

Junction with US 395—Portola—Blairsden—Quincy—Rich Bar—Oroville—Marysville—Knights Landing—Woodland; 203.1 *m.* State 70—US Alt. 40.

Paved roadbed throughout; open at all seasons.
Route paralleled by Western Pacific R.R. between Portola and Marysville.
Accommodations plentiful in large towns, mountain resorts open about May 1 to October 1.

State 70 climbs over the Sierra Nevada through Beckwourth Pass, lowest crossing in central and northern California, into the mountain-walled farmlands of Sierra Valley, largest in the Sierra. It winds between thick-forested slopes along the spectacular canyon of the Feather River, dashing over its turbulent course down the range's long western decline through the heart of one of the State's oldest placer mining regions. From the orange grove belt below the foothills, it levels out across the Sacramento Valley, paralleling the Feather River southward through far-reaching peach orchards and beet fields.

State 70 branches west from its junction with US 395 (*see TOUR 6a*), 0 *m.*, 11 miles north of the Nevada Line, into the barren brown foothills of the Sierra's eastern slope.

The road climbs easily to BECKWOURTH PASS, 2.2 *m.* (5,220 alt.), the lowest entrance to the State north of the desert gateways in southern California. Its discovery in the spring of 1850 was the outstanding achievement of the remarkable adventurer and trapper, James P. Beckwourth, son of a Revolutionary War officer and a slave mother. Born in Charlottesville, Virginia, in 1798, Beckwourth tired early of an apprenticeship to a Missouri blacksmith and went westward. His rovings had already carried him into the Rockies with Gen. W. H. Ashley's fur trading expedition of 1824 and had kept him some years among the Crows. Later in Florida he served in the U. S. Army and in New Mexico traded in furs, before he came to California to share in the revolution of 1846. After further exploits in New Mexico, he returned to California and discovered this pass.

"This, I saw at once"—he told his admiring biographer, T. D. Bonner,—"would afford the best wagon-road into the American Valley approaching from the eastward, and I imparted my news to three of my companions. . . . They thought highly of the discovery, and even proposed to associate with me in opening the road . . . I made known my discovery to a Mr. Turner, proprietor of the American Ranch, who entered enthusiastically into my views: it was a thing, he said, he had never dreamed of before. If I could but carry out my plans, and divert travel into that road he thought I should be a made man for life.

Therefore he drew up a subscription list, setting forth the merits of the project and showing how the road could be made practicable to Bidwell's Bar, and thence to Marysville. . . ."

Over the wagon road, opened in 1851, Beckwourth guided wagon trains through the pass. A dark-skinned man, dressed in leather coat and moccasins, who wore his hair in two long braids twisted with colored cloth, riding without a saddle, he made a vivid impression on a traveler in one train, 11-year-old Ina Coolbrith (later California's first poet laureate), in whose honor massive, evergreen-girdled MOUNT INA COOLBRITH (8,311 alt.), south of the pass, is named.

"In the spring of 1852," he told Bonner, "I established myself in Beckwourth Valley. . . . My house is considered the emigrant's landing place, as it is the first ranch he arrives at in the golden state." Here he dedicated to Bonner the autobiography on which his reputation as a liar is based, meanwhile consuming great quantities of rum. The more the two cronies drank, the more Indians Jim would remember having killed single-handed. When he was thoroughly "likkered up"—so the story goes—he would shout, "Paint her up, Bonner! Paint her up!" But fame was short for Beckwourth. His very name was distorted to Beckwith on early maps. Only recently has the correct spelling been restored.

State 70 descends from Beckwourth Pass to CHILCOOT, 6 *m.* (4,995 alt., 85 pop.), a shipping point for Sierra Valley, where farming and stock raising are carried on almost a mile above sea level.

At 7.1 *m.* is the junction with State 49 (*see TOUR 4a*). BECKWOURTH, 17 *m.* (4,874 alt.), was once an active trade center of a lumbering area. State 70 now follows closely the Middle Fork of the FEATHER RIVER, past BECKWOURTH PEAK (7,248 alt.).

PORTOLA, 23 *m.* (4,834 alt., 1,876 pop.), a lumber and railroad center, the largest community in this Plumas Mountain region, is a new and somewhat raw looking town, named for the Portola Fiesta in San Francisco in 1909, commemorating Gaspar de Portola, California's first Spanish governor.

East of Portola State 70 follows Castle Canyon, hemmed in closely by rugged buttes. Those on the south, BECKWITH BUTTES, bear Jim Beckwourth's name incorrectly spelled.

BLAIRSDEN, 34 *m.* (4,500 alt., 262 pop.), is a center for winter sports in the upper Feather River recreation area. Here is a junction with State 89 (*see TOUR 5b*), which unites northwest with State 70 for 33 miles.

Left from Blairsden on a dirt road to MOHAWK, 1.4 *m.,* where there are sulphur springs (*hotel open all year*).

State 70-89 continues northwest from Blairsden, following the Middle Fork of the Feather River, here a dashing stream which swells to river-like proportions only in winter spate. El Rio de las Plumas (the river of the feathers) was the name that Capt. Luis Arguello and his band of Spanish explorers gave it when they came upon its lower regions in 1820 and found them strewn with wildfowl feathers or—according to another surmise—feathery willow pollen.

At 35 *m.* the FEATHER RIVER INN (*American plan; open June 1 to Oct. 1*) nestles among the pines. At SLOAT, 42.5 *m.* (4,115 alt.), a ranch house is open all year. SPRING GARDEN, 46 *m.* (3,965 alt., 125 pop.), is a summer resort.

At 55 *m.* is a junction with a dirt road.

Left on this road up steep grades to ONION VALLEY PASS (*trout fishing*), 17.2 *m.* (6,500 alt.). In its descent the road follows ridges between the Feather and Yuba drainage systems. At 30.1 *m.* is LA PORTE (5,000 alt.), an old mining town (*meals and lodging*). Between here and Strawberry Valley there is fine hunting for deer and quail and good trout fishing—particularly for those prepared to pack into the backcountry off this road. At 43.1 *m.* (3,650 alt.) is STRAWBERRY VALLEY (*meals and lodging*), once a rich mining area. West of RACKERBY, 57.3 *m.* (250 pop.), another old mining camp, near the upper waters of South Honcut Creek, the roadbed is paved; it rejoins State 24 at 78.3 *m.*

QUINCY, 59 *m.* (3,407 alt., 2,723 pop.), seat of Plumas County, is pleasantly situated in the American Valley; its white houses, seen against green meadows and pine clad slopes, give it the look of a Vermont village. The town was established in 1854 by H. J. Bradley, who, wanting to make it the county seat, erected a frame building in the rear of his hotel and offered it free for county use. There are both mining and farming in the vicinity and winter sports in season.

Left from Quincy a paved road leads up Spanish Creek 8 *m.* to MEADOW VALLEY (*cabins and meals May 1 to Nov. 1*) at the base of SPANISH PEAK, and 17 *m.* to BUCKS (*same accommodations*) on BUCKS LAKE (5,071 alt.).

At 65 *m.* on State 70 is KEDDIE (3,223 alt., 150 pop.), a junction on the Indian Valley Railroad and a shipping point for the surrounding fertile mountain valley.

At 67 *m.* is the northern junction with State 89 (*see TOUR 5b*); L. here, through the deep canyon of the North Fork of the Feather, down the long reaches of the Sierra's western slope. Arrow-straight pines and firs mantle the canyon walls.

PAXTON, 69 *m.* (3,080 alt., 30 pop.), formerly called Soda Bar because of its mineral springs, was an early placer mining town.

At TWAIN, 73 *m.* (2,909 alt., 145 pop.), is a mountain tavern on the far side of the river (L).

The highway serpentines down the narrow canyon to RICH BAR, 82 *m.* (2,502 alt., 52 pop.). Rich Bar—the Barra Rica of early Mexican placer miners—quite lived up to its name. Single pans of "dirt" produced from $100 to $1500, and altogether more than $3,000,000 in gold dust were sifted from its gravel.

At BELDEN, 87 *m.* (2,306 alt., 110 pop.), is a junction with a dirt road.

Right on this steep and winding road 14.4 *m.* to LONGVILLE (*meals and lodging*), on YELLOW CREEK (4,609 alt.). This and several other tributary streams that meet the Feather River below Belden afford fine fishing; there is also good grouse, deer, and bear hunting here.

Southwest of Belden the road, the railroad, and the river weave in and out down the canyon's twisting course. The canyon, green with somber pines, widens in places to a yawning gulf, narrows in others to a narrow cleft. The railroad climbs up the steep walls to the heights above and dips again to the river bed. Over its jagged, boulder-piled channel the river spills in snowy, foaming rapids.

East (L) of WORKMAN'S BAR (2,128 alt.), a number of peaks rise about 7,000 feet. At TOBIN, 94.5 *m.* (2,006 alt., 16 pop.), the highway makes one of its many crossings of the river over a concrete bridge. Opposite MERLIN, 99 *m.* (1,756 alt.), State 70 winds along the granite sides of GRIZZLY PEAK. In two places the road passes through tunnels bored through the solid rock of the bluff.

At PULGA (Sp., flea), 108 *m.* (1,380 alt.), the State highway crosses the river on a graceful arch span 350 feet long, high above the railroad bridge, and 170 feet above the churning water. Pulga, at the mouth of Flea Valley, was formerly a populous gold-mining camp called Big Bar before the railroad humbled it with a new name.

South of Pulga State 70 leaves the river canyon for the forested hills of Jarboe Pass and meets the river again at LAS PLUMAS, 120 *m.* (562 alt., 80 pop.). The water for the power plant at Las Plumas comes through a tunnel from Intake, a station 5 miles north, thus avoiding the big bend in the river. The highway once again crosses the Feather River at 130.8 *m.*, then leaves it to traverse rolling hills covered with oaks and digger pines.

OROVILLE, 136 *m.* (205 alt., 6,115 pop. 1960), a tree-shaded town at the base of the Sierra foothills, sprang up in the winter of 1849-50 as Ophir City, a tent town, when gold was discovered here. Discoveries elsewhere depopulated the place in 1852, but it boomed again when a canal brought water in 1856 to dry diggings nearby. Oroville in that year became the seat of Butte County, a city of 4,000, fifth largest in California, boasting horse races, two theaters, 65 saloons. In the 1870's gold lured thousands of Chinese and Oroville's Chinatown was California's largest in 1872.

California dredging began on the Feather River here in 1898, with the floating of the first successful bucket elevator dredge, built by W. P. Hammon and Thomas Couch. The dredges, as many as 44 operating at one time, extracted nearly $30,000,000 in gold within 20 years. One company offered to move and rebuild Oroville if permitted to wash out the gold-bearing gravel below it.

New distinction came to Oroville when Governor Brown on August 13, 1962, signed a bill appropriating $121,000,000 for construction of OROVILLE DAM AND RESERVOIR on the Feather River six miles northeast of Oroville. The plans called for a dam 735 ft. tall, nine feet taller than Hoover Dam (726 ft.), and a reservoir covering 15,500 acres, with a shoreline of 167 miles.

The state of California has constructed an observation point at a site directly above the construction area. The overlook has a paved parking area for more than 250 automobiles and modern facilities. The main building has large windows through which construction can be viewed by tourists.

When mining died down, it was discovered that Oroville stood in a thermal belt suitable for olive groves and citrus and deciduous fruit orchards. Mrs. Ehman of Oroville developed a commercial process for pickling ripe olives, and as a result the city gained one of the largest ripe olive canning plants.

1. Left from Oroville on Berry Creek Road 5.9 *m.* to the junction with a paved road; R. here 15.7 *m.* is FORBESTOWN (2,800 alt.), called a lively mining center for four decades after B. F. Forbes founded it in 1850, now deserted in its mountain cove, a ghost town of heaped debris, old foundations and crumbling structures with fallen roofs.

At 9.4 *m.* on Berry Creek Rd., a stone monument (L), near the South Fork of the Feather River, marks BIDWELL'S BAR, where a camp sprang up soon after John Bidwell found gold there July 4, 1848. By 1853 it had 2,000 inhabitants. Digging went on everywhere—in the streets and under houses. Here the river, lifted from its bed and carried in a flume for miles, yielded rich gold-bearing gravel. Fluming operations reached their height in 1856-57. Meanwhile the diggings were being exhausted, and the whole population stampeded to Oroville, where a new boom was on. Across the river gorge hangs the first suspension bridge erected in California, its 407-foot cables fastened to anchors embedded in rock. The cables were brought around the Horn and the bridge was opened in 1856. Opposite the old red-brick toll house at its southern end is the OLD MOTHER ORANGE TREE, planted by Judge Joseph Lewis in 1856. Its seeds, which brought $1 an ounce, were planted in northern California's first citrus belt. Along the southern sandy shore (L) is BIDWELL'S BAR PARK (*swimming; picnic tables*).

2. Right from Oroville on a paved road crossing the Feather River 1 *m.* to a junction with a paved road; L. here 1 *m.* to THERMALITO (194 alt.), where northern California's first important orange grove was planted in 1886. In the surrounding "thermal belt," oranges ripen even earlier than in southern California.

The main side route turns R. over wooded slopes to TABLE MOUNTAIN, 6.6 *m.,* a mesalike eminence edged with deeply furrowed palisades. In the 1850's its sides were pierced with 35 tunnels by gold miners. Across the top, bright with wild flowers in spring, the road winds over wastelands scarred with patches of blackish rubble.

CHEROKEE, 11 *m.* (1,400 alt., 243 pop.), is a ramshackle hamlet sprawling beyond stone walls and picket fences, in a crag-strewn, pine-wooded hollow. Its name comes from the Cherokee Indians who settled here in 1853. They were driven from their lands in Georgia when white land grabbers discovered gold there; later they left their second home in Arkansas to follow the trappers' route that came to be known as the Cherokee Trail west to California. A mining center in the 1860's, Cherokee gained brief fame with the discovery of a few diamonds in the placer diggings. To the south rises a spur of Table Mountain with deeply scarred sides, site of the abandoned hydraulic SPRING VALLEY GOLD MINE; 100 miles of ditches and pipe lines built about 1870 carried Feather River water here. The mine's operations, spreading layers of sand over valley farm lands below, stirred bitter opposition among the farmers.

State 70 traverses a flat fertile valley, paralleling the Feather River (R).

At 147.7 *m.* is the junction with a paved road (*see TOUR 3B*).

MARYSVILLE (*see TOUR 3B*), 165 *m.*, is at the junction with US 99E (*see TOUR 3B*) with which State 20 is united to YUBA CITY, 166 *m.* (*see TOUR 3B*); L. at Yuba City.

State 70 runs southward through acres of peach orchards.

At 168 *m.* is the junction with a paved road.

Left here, through low-lying fertile orchards, to NICOLAUS, 19.7 *m.* (33 alt., 89 pop.), an old river town.

On State 99 is TUDOR, 174 *m.* (44 alt., 38 pop.), in the heart of one of the State's largest clingstone peach areas; the Phillip cling peach was originated here.

Left from Tudor on a paved road 1.2 *m.* to the junction with the Garden Highway; R. here 1.6 *m.* to the SITE OF HOCK FARM. By the roadside (L) stands the front wall, of rusty iron plates riveted together, of the old fort, which in 1842 stood with the farm on the banks of the Feather River. Named for an Indian village on the Feather, Hock Farm was John Augustus Sutter's chief stock ranch and the first white settlement in Sutter County. After the loss of Sutter's Fort (*see SACRAMENTO*) and most of his fortune, he retired here. From 1850 to 1868, he kept open house in a handsome mansion among gardens, orchard, and vineyard.

At 178 *m.* the route crosses the SUTTER BY-PASS, part of a drainage system built to control the Sacramento River in flood season.

ROBBINS, 186.3 *m.* (49 alt., 200 pop.), is a fruit shipping town.

At 194 *m.* the highway crosses the SACRAMENTO RIVER.

Right from the northern end of the bridge on an improved road to KIRKVILLE, 9.4 *m.* (13 alt., 20 pop.) and COLES LANDING, 14.3 *m.* (17 alt.), busy shipping towns in early steamboat days.

At the southern end of the bridge is KNIGHTS LANDING, 195.3 *m.* (48 alt., 600 pop.), an early shipping point that still retains some of the river town atmosphere of the 1860's. Moldering false-front frame stores, a little askew, with wooden awnings over the board-walks, stretch for two blocks along the river. Dr. William Knight was granted land here through marriage with a Mexican wife. In 1843, two years after he immigrated from New Mexico with the Workman-Howland party, he built a dwelling on Indian "Yoday" Mound: a rude hut of willow poles bound with rawhide, covered with walls of river tules plastered with mud. In the dense woods roundabout roamed grizzly bears. Laid out in 1849 as Baltimore, the place came to be known by Knight's name because of the ferry he had established.

State 70-99 cuts through extensive sugar-beet fields to a junction at 200.5 *m.* with an improved road.

Left on this road 0.8 *m.* to the SPRECKELS SUGAR FACTORY (*open to visitors 9-9 workdays*). In the whirring, light-flooded interior of the factory, pervaded with a syrupy sweet smell of sugar, sugar beets are fed by conveyor belt to a complex apparatus of washers, scales, slicers, diffusion tanks, purifiers, filter presses, evaporators, vacuum pans, centrifugal machines, and driers; built at a cost of $2,500,000, the plant was opened in

August 1937. The founder of the company, German immigrant Claus Spreckels, who came to California in the 1850's, started the State's sugar beet industry when he built a plant in 1888 at Watsonville.

WOODLAND (*see TOUR 3a*), 203.1 *m.* (63 alt., 13,524 pop.), is at the junction with US 99W (*see TOUR 3a*).

Tour 6C

Junction with US 395—Warner Hot Springs—Santa Ysabel—Julian—Junction with US 80; 81.7 *m.* State 79.

Limited accommodations.
Roadbed oiled or paved throughout.

From Temecula Valley this route climbs southeast through a mountain pass to a broad mountain basin in the highland area, and swings south through a fertile mountain valley to a famous gold mining district of pioneer days.

State 79 branches west from US 395, 0 *m.,* through lands formerly belonging to Mission San Luis Rey (*see TOUR 2c*). On the south bank of the Temecula River, which parallels the road, is PECHANGA CEMETERY, an old Indian burial ground in which the prototype of Alessandro, Indian hero of Helen Hunt Jackson's *Ramona,* is buried. This Indian, Juan Diego, was shot by an American in 1877 because of a supposed horse theft.

The old WOLFE TRADING POST (R), 1.9 *m.* (*apply at office of Pauba ranch superintendent for adm.*), once a trading post and tavern kept by Louis Wolfe, is now a storeroom and sleeping quarters for the ranch laborers. Mrs. Wolfe furnished many stories of the Temecula Indian evictions (*see below*) to Helen Hunt Jackson.

At 15.5 *m.* is a junction with a dirt road.

Right on this road to the RUINS OF THE AGUANGA STAGE STATION, 0.3 *m.,* a wooden building that served the Butterfield Overland Stage route. The first stage reached Aguanga in October 1858, and stopped twice a week thereafter until 1861 (*see TOUR 14a*).

In the small cemetery (L), 15.5 *m.* Jacob Bergman and members of his family are buried. Bergman drove the first Butterfield Stage through the valley. His grandson, who lives near the Aguanga post office, has a collection of relics and mementos of the stage line.

AGUANGA, 17.3 *m.,* a post office, store, and service station, was a junction of the old trails, one of which, the San Bernardino-Sonora Road, went northeast to the San Gorgonio Pass and through the San

Bernardino Valley; the other, known as the Colorado Road, went northwest through Temecula.

At 17.4 *m.* is a junction with the Hemet Rd. (*see TOUR 6e*).

The OAK GROVE STAGE STATION, 23.4 *m.,* a well-preserved adobe building, is now a tavern. Oak Grove Valley has some of the finest oaks in southern California. OAK GROVE PUBLIC CAMP GROUND, 23.8 *m.* (*tables and ovens*), is in CLEVELAND NATIONAL FOREST. Opposite the camp is a ranger station.

At the entrance to the 44,000-acre WARNER'S RANCH, 29.4 *m.,* is a hollow (L), known as DEADMAN'S HOLE, named by a Butterfield stage driver who found a dead man beside the spring. Old timers in the San Diego back country have many tales of murders at this spot.

The ranch land was a pivotal center for three Indian peoples, the Luiseno and Cahuilla, of the Shoshonean linguistic stock, and the Diegueno of Yuman origin. The particular group that inhabited this area were the Cupeno, a Yuman tribe. (Their name is a Spanish corruption of the name of their village, Kupa.) To the west were the Luiseno, and to the north and east, extending far into the Colorado Desert, were the Cahuilla. When the Spanish padres made their first trip into the valley, on an expedition from Mission San Diego, they named it Valle de San José. They found 10 villages scattered about the valley, the largest of which was Kupa, at the hot springs, later called Agua Caliente (hot water). The valley was used jointly by the Mission San Luís Rey and the Mission San Diego. In 1836, after secularization, Silvestre de la Portilla received the valley as a grant, but abandoned his interests, and in 1844, Juan José Warner, a Connecticut Yankee born Jonathan Trumbull, was granted the entire Valle de San José. He had come to California in 1831 with the Jackson party and had taken Mexican citizenship papers. After the secularization of the missions the Indians drifted back to their ancestral lands and shifted for themselves. The Cupeno continued to live by the hot springs, where they often lay in the waters and muds throughout cold nights. During Warner's ownership, they were allowed to remain at their long-established ranchería because he needed their labor, though he often complained of their raids upon his stock. Paid three dollars a month, the usual wage at that time for Indians, they were occasionally stimulated with applications of the lash, an established custom.

Warner's Ranch was the first civilized stop west of the perilous Colorado Desert. Kearny and his army rested here on their way from Fort Leavenworth, Kansas, to join Commodore Stockton at San Diego, in 1846. Warner moved to Los Angeles after an Indian uprising in 1851, and by 1861 he had lost his interest in the ranch. In 1880, John G. Downey, a former governor of California, owned the property. The Indians' rights to occupy the land became a matter of controversy. A suit was taken to the Supreme Court of California, which decided against the Indians. In the meantime, the Sequoya League, aided by the reports of Helen Hunt Jackson, succeeded in having the Government investigate their plight. A commision bought 3,438 acres in the Pala

Valley, near the Pala Mission Chapel, for a reservation, and in 1903 the Indians were forcibly removed to it. They have lived peacefully in this new territory, though they are placed among Indians of a different linguistic stock. Warner's Valley is now the property of the San Diego Water Company and the western section of the old rancho has been covered by Lake Henshaw (*see below*). Much of the area is leased to private cattle raisers.

WARNER HOT SPRINGS, 38.3 *m.*, is a mineral springs resort (*cabins, campgrounds, hotel*). Its Spanish-type buildings are among elms and locust trees. To the east is LOS COYOTES INDIAN RESERVATION.

At 41.9 *m.* on State 79 is a junction with a dirt road.

Left here to the HEADQUARTERS OF WARNER'S RANCH, 1 *m.*, and to a well-preserved BUTTERFIELD STAGE STATION, 2 *m.*

MORETTIS, 46.3 *m.*, is a service station.

Right from Morettis to LAKE HENSHAW, 4 *m.*, a reservoir of the San Diego water system, built in 1922.

At 50.5 *m.* in Santa Ysabel Valley, is a junction with a dirt road.

Left on this road to the VOLCAN INDIAN SCHOOL, 0.5 *m.*, on the SANTA YSABEL INDIAN RESERVATION.

At 51.8 *m.* on State 79 is a junction with a dirt road.

Right here to the MESA GRANDE INDIAN RESERVATION, 7 *m.*

SANTA YSABEL CHAPEL (L), 52.2 *m.*, a concrete building, was built in 1924 on the site of an adobe *asistencia* (chapel) erected in 1822 by the padres of Mission San Diego. This branch of the mission never equaled the importance of Pala Chapel, but a mission report of 1822 mentioned a granary, several houses, a cemetery, and 540 baptized Indians. After secularization, Santa Ysabel fell into ruin, though some Indians still continued to attend services. Even after the chapel was completely obliterated, the bells continued to be rung from a framework of logs on which they had been mounted; later they mysteriously disappeared. No Indians live here because the acquisitive whites have pushed them farther into the rough back country.

The SANTA YSABEL CAMPO SANTO (cemetery), 52.3 *m.*, has been used since the 1820's. Nearby are the *enramadas* (arbors) made of green boughs brought from the surrounding hills, where the Indians gather twice a year for traditional festivals.

SANTA YSABEL, 53.8 *m.* (2,983 alt.), is a small trading center (*all accommodations*) at a junction with State 78 (*see TOUR 6e*), with which State 79 unites briefly.

WYNOLA, 56.8 *m.*, is a filling station and an elementary school.

At 60.8 *m.* is a junction with a paved road.

Right on this road to PINE HILLS RESORT, 2.5 *m.* (*lodge and cabin accommodations*). Nearby is CAMP MARSTON, a Y.M.C.A. camp for boys, named in honor of the donor of the tract, George W. Marston.

JULIAN, 61.4 *m.* (4,129 alt.), a boom town during the 1870's, is now a farmers' trade center, surrounded by pine-wooded resort areas. The town was established by the Bailey brothers and their cousins, the Julian brothers, who moved into the region in 1869, shortly after placer gold was discovered near the site of the present town.

In 1870 13-year-old Billy Gorman, who had arrived in Julian with his family from Texas, found a piece of white rock flecked with yellow. He took the quartz to his father, who opened a mine that was to become the George Washington, commemorating the date of the discovery—Washington's Birthday. As news of the strike spread, gold-seekers, gamblers, and their women flocked in, and for a decade the place was enlivened by quarrels, pistol shootings, and stabbings. Twenty-mule teams and wagons were used to haul ore to San Diego. About 1880, when gold had already become scarce in Julian, the big strike at Tombstone, Ariz., was made, and most Julian residents deserted the town. It has been estimated that more than $15,000,000 in gold were taken from the district.

At 61.7 *m.* is the junction with State 78 (*see TOUR 6e*).

At 66.7 *m.* is a junction with a county road.

Right on this road to the 840-acre INAJA (Ind., my springs) INDIAN RESERVATION, 7 *m.* (33 pop.), and the 80-acre COSMIT RESERVATION, 7.7 *m.* (unpopulated). This land, remarkable for its unspoiled beauty, is used mainly for grazing.

CUYAMACA RANCHO STATE PARK, 69.7 *m.,* is a resort area around CUYAMACA LAKE, part of San Diego's water supply system. In pioneer days it was known as La Laguna Que Se Seca (the lake that dries up). Adjoining on the west are CAPITAN GRANDE INDIAN RESERVATION and EL CAPITAN RESERVOIR.

At 81.7 *m.* is a junction with US 80 (*see TOUR 14b*), at a point 3 miles east of Descanso.

Tour 7

(Tonopah, Nev.)—Bishop—Brown—Mojave—Los Angeles—Wilmington—Long Beach; US 6.
Nevada Line to Long Beach, 355.2 *m.*

Paved roadbed throughout.
Southern Pacific Line parallels route between Nevada Line and Bishop.
Accommodations limited between Nevada Line and Bishop.

In 1937 US 6 was extended westward to form a single numbered route between Cape Cod and southern California. The sections in Nevada and Utah have been improved.

Section a. NEVADA LINE to BISHOP; 41.5 m. US 6.

South of the Nevada Line US 6 sweeps through Benton Valley, a basin lying between three sections of Inyo National Forest, in the shadow of the 13,000-foot White Mountain Range. Crossing the range, the route emerges into Chalfant Valley, once the stronghold of the Paiute, and still one of the most primitive regions in California. At many points the faces of cliffs and granite mountain walls still bear the marks of aborigines in the form of petroglyphs (stone carvings) and pictographs (picture writings). The area is sparsely settled.

US 6 crosses the Nevada Line, 0 *m.*, 71 miles west of Tonopah, Nev., cuts southeastward across the valley, and aims directly for a towering granite spur which seemingly bars the way straight ahead. The highway rounds this cliff and swings due south through an opening between the mountains.

BENTON STATION, 7.5 *m.* (5,393 alt.), is a small railroad shipping center and an outfitting point for fishing and pack trips into INYO NATIONAL FOREST, just south of the town. The jagged crest of MOUNT DUBOIS (13,545 alt.) rears conspicuously (L) above the mass of the White Mountains, a range nearly as high and fully as precipitous as the main mass of the Sierra Nevada, which bulks straight ahead. The White Mountain district of Inyo National Forest contains rugged areas equaling those in certain sections of the high Sierra and regions white men have not penetrated.

The bordering ranges move in toward highway and railroad in a wide pass at 10 *m.*

At 17 *m.* is the junction with a dirt road.

Right on this road 0.3 *m.* to the INDIAN TRACK CARVINGS, accessible by a short climb up the steep bluff (L). The carvings consist of hundreds of tracks, the origin of which living Indians are unable to explain. They extend for 100 yards along the crest of a broken ridge. Prominent among them are the tracks of an infant's feet and the heavy marks of bears, dogs, coyotes, and cats. About 1888 the Paiute went on a rampage and destroyed many of the markings, which were made, they said, by "evil little men who crept from the rocks at night."

At 25 *m.* is the junction with a dirt road.

Right on this road through Chidalgo Canyon to the INDIAN PICTURE CARVING LABYRINTH, 1 *m.*, a maze of rock carvings made in the remote past. Some of them are geometrical designs and some are of human figures, but most are crude renderings of animals: deer, bear, bighorn sheep, lizards, snakes, and a dragonfly.

As the mountain ranges again draw apart, the highway enters Chalfant Valley. The CHALFANT GROUP OF INDIAN WRITINGS (R), 27 *m.*, visible from the road, extend high on an almost unscalable wall for a half mile. The group consists of both petroglyphs and pictographs, some representing individual figures and others, large connected groups.

At 34.5 *m.* is the junction with a dirt road.

Right on this steep road, which winds to the top of a volcanic tableland, to the PAIUTE INDIAN RESERVATION, 5 *m.*, where dwell the remnants of the Indian tribes that once ranged over the region. At 5.5 *m.* are large INDIAN PETROGLYPHS. The most conspicuous unit is about 5.5 feet in diameter; others, linked with the largest unit by almost indistinguishable lines, constitute a series some 20 feet long. There is also an "inscriptive wall"—a projecting rock carved with designs described variously as a sun dial, a flood gage, and a calendar. The projection, extending 8 feet beyond the rock wall, is accessible only by ladder.

LAWS, 37 *m.* (4,200 alt., 84 pop.), is a railroad water stop; its inhabitants are chiefly railroad workers.

The highway crosses the OWENS RIVER, 37.4 *m.*, whose swift current flows through a deep gorge in the volcanic tableland.

At BISHOP, 41.5 *m.* (4,147 alt., 2,875 pop.) (*see TOUR 6b*), is the junction with US 395 (*see TOUR 6c*). South of Bishop US 6 and US 395 are united for 123.5 miles (*see TOUR 6b*). State 168 runs west from Bishop to Kings Canyon National Park.

Section b. BROWN to LONG BEACH; 190.2 m. US 6.

This section of US 6 sweeps southward across the flat basin of the Mojave Desert toward the distant San Gabriel Mountains in an almost level course. The arid desert is relieved at intervals by the green, irrigated farming districts of lower Antelope Valley. It crosses the San Gabriel, winds through Mint Canyon, and descends into San Fernando Valley. Through straggling metropolitan outskirts, it continues into Los Angeles and southward across the almost flat coastal plain.

South of its junction with US 395 (*see TOUR 6b*), 0 *m.*, on the western outskirts of Brown, US 6 skims along the western edge of broad, arid Indian Wells Valley through typical Mojave Desert country sparsely grown with creosote and wolfbush. Distantly (L) the ARGUS MOUNTAINS rise from the sandy plain—dun gray in the shadows, brilliant lavender where the sun's rays strike. To the west (R) rise the majestic ridges of the High Sierra. OWENS PEAK, (8,475 alt.), 1 *m.*, highest of the east slope Sierra pinnacles in this region, is conspicuous among sister peaks.

Unbroken desert landscape unrolls as US 6 moves across the heat-blistered valley. Paradoxically, a hundred yards from the road (R), winds the water-filled open concrete channel of the LOS ANGELES AQUEDUCT, which redeemed San Fernando Valley from a desert as arid as this.

At 20.6 *m.* is the junction with Hartz Rd.

Left on this road through LAST CHANCE CANYON to ROARING RIDGE PETRIFIED FOREST, 7 *m.* (*adm. 50¢, children free*), where tree sections of a petrified forest are scattered over hills that once teemed with animal life, as fossil discoveries—including tusks of a rhinoceros and an elephant—have proved. A series of dugouts in the canyon wall at 11 *m.* is a reminder of the gold-mining activity that once flourished in this gorge. In the dugouts, now fitted with doors and windows, dwell a strange community of miners who search for nuggets and "dust" in the surrounding hills.

RICARDO, 30.2 *m.* (2,443 alt.), with gas station, garage, and refreshment stand, is hemmed in by the EL PASO MOUNTAINS (L) and the foothills (R) of the PAIUTE MOUNTAINS.

From the desert US 6 now winds through a deep lateral gorge of RED ROCK CANYON, whose towering cliffs are of reddish-brown sandstone, carved by wind and weather into fantastic shapes, and into the desert again.

At 37.6 *m.* is the junction with graveled Jawbone Canyon Rd.

Right on this road through Jawbone Canyon to the HILL OF BLUE-GREEN STONE, 5 *m.*, a rock mass of bright and bizarre colors.

At 5.5 *m.* on the main side road is the junction with a dirt road; L. here 2.6 *m.* to a mountain of stone, white as snow. Known as "vitrox," this formation is used in the manufacture of dishes, vitrified tile, and enamel for bathroom fixtures.

At 52 *m.* proceed left 8 *m.* to CALIFORNIA CITY (14,000 pop. est. 1966), a rapidly growing development, north of Edwards Air Force Base. East 2 *m.* is Castle Butte, 3,145 ft.

The highway at 53 *m.* is bordered by an elfin forest of greasewood, where 10-foot yucca trees lift distorted trunks. The leaves of the greasewood bush give off a strong, acrid, yet rather pleasant odor when crushed. In the spring the bush is covered with yellow blossoms.

At 57 *m.* is the junction with US 466-91 (*see TOUR 11b*).

In MOJAVE, 58.3 *m.* (2,751 alt., 1,845 pop.), J. W. S. Perry built 10 huge wagons in 1883 to haul borax out of Death Valley, and for five following years these conveyances, drawn by 20-mule teams, plied between Death Valley and Mojave. Mojave turned to industry when it obtained the Marine Air Base (closed 1959) for Kern County Airport No. 7, with two mile-long strips and a runway of 9,870 ft.

South of Mojave, the peaks of the High Sierra, which have towered close to the highway throughout, recede to merge with the lower TEHACHAPI MOUNTAINS. The intervening area—partly flat, sandy desert, partly rolling foothills and occasional low peaks—constitutes the Mojave mining district, where greatest activity centers about SOLEDAD MOUNTAIN, 62 *m.* (4,183 alt.), a dark gray rounded mass.

At 62.1 *m.* is the junction with a dirt road.

Right on this road to a lone, one-story frame building housing a filling station and lunchroom and grandiosely styling itself GOLDTOWN, 1 *m.* (2,600 alt., 4 pop.).

GOLDEN QUEEN MINE, 2 *m.*, with its extensive galvanized iron sheds, spidery loading shoots and great ore dumps, disfigures the slope of Soledad Mountain. This mine startlingly revived dwindling activity in the once fabulously rich district when discovered in 1935 by George Holmes, a prospector. He sold his find for $3,500,000.

ROSAMOND, 71.9 *m.* (2,310 alt., 204 pop.), is a supply center for ranching and mining interests. The hotel, built of quartz, has a fireplace studded with gold ore. White, desolate ROSAMOND DRY LAKE glitters (L) in the sun across 5 miles of desert.

US 6 crosses ANTELOPE VALLEY, 75 miles long. About a third of the narrow valley is under cultivation. By 1966 the Valley was making a bid for factories, advertising that 20,000 acres of "choice land" was available for light and heavy industry. Part of the area is used by the Fairmont Butte Gun Club.

In LANCASTER, 84.6 *m.* (2,356 alt., 1,550 pop.), branches of Los Angeles County administrative departments are maintained.

PALMDALE, 92.6 *m.* (2,669 alt., 1,224 pop.), is the business center of southern Antelope Valley, a fruit and alfalfa growing district, where 4,000 tons of Bartlett pears are shipped annually.

PALMDALE RESERVOIR (R), 95.4 *m.,* is one of several storage basins supplied from deep wells.

US 6 ascends the grade leading to MINT CANYON over foothills sparsely covered with chaparral and mountain juniper.

At 104.3 *m.* is the junction with Crown Valley Rd.

Right on this unpaved road to GOVERNOR MINE, 1 *m.,* formerly the Old New York Mine. The mine was opened about 1889 and was worked sporadically until the early 1920's. An increase in the price of gold resulted in its reopening in 1932. The main tunnel penetrates 600 feet.

Mounting steadily between forested slopes, US 6 achieves the summit of MINT CANYON GRADE, (3,429 alt.), 107.5 *m.,* then descends into SIERRA PELONA VALLEY.

At 124.4 *m.* is the junction with Soledad Canyon Rd.

Left on this paved road to SOLEDAD GORGE, 6.3 *m.,* a deep, rocky gash between cliffs 500 to 800 feet high.

At 8.3 *m.* is the junction with Agua Dulce Canyon Rd.; L. here 2.5 *m.* through a geological wonderland of tilted rock formations, to a junction with a private dirt road; R. here to VASQUEZ CAVES, 3.5 *m.* (*adm. 25¢*), reputed hideout of the bandit, Tiburcio Vásquez, who in the 1860's terrorized coastal California. The caves are a series of small caverns gouged by wind and rain in the sandstone cliffs of ESCONDIDO CANYON. The VASQUEZ ROCKS, 5.3 *m.* (*adm. 50¢*), are a grotesque jumble. They, too, perpetuate the memory of the bandit chief, for tradition asserts Vásquez and his highwaymen led pursuing posses many a merry chase through this maze, invariably losing them before they had penetrated to his hideout in the caves.

The SANTA CLARA RIVER, 125.6 *m.,* usually a tiny trickle of water down a graveled bed, is a roaring, destructive torrent during the rainy season.

HILL RANCH (L), 129 *m.,* formerly the property of cowboy actor Hoot Gibson, is the scene of annual rodeos (*time and adm. vary*).

SAUGUS, 130.6 *m.* (1,171 alt., 151 pop.), is a Southern Pacific R.R. division point.

At 132.7 *m.* is the junction with Placeritas (Sp., little gold diggings) Canyon Rd.

Left on this road to the OAK OF GOLDEN LEGEND (L), 3.6 *m.* Under this spreading oak the first gold discovery in California was made in 1842 by Don Francisco López. Noting bright particles clinging to the roots of wild onions he had dug for his lunch, he had a test made, which revealed particles of gold. The discovery caused widespread excitement and brought prospectors from as far away as Mexico. In November 1842, the first California gold was shipped from the Placeritas mines to the U. S. Mint in Philadelphia.

At 133.4 *m.* is the junction with a dirt road.

Right on this road to the junction with a private dirt road, 0.2 *m.;* L. here to the NEWHALL REFINERY, 0.4 *m.,* California's first, built in 1876. The old stills, retorts, and petroleum vats have been carefully restored.

NEWHALL, 133.6 *m.* (1,273 alt., 4,705 pop.), now a trade center for ranchers, farmers, and oil producers, was the point from which oil exploitation in southern California began. Since discovery of oil in the district in 1861, the fields have produced steadily. Overlooking the town from a hill (R) is the estate of the late WILLIAM S. HART, star of the Western motion pictures.

US 6 moves up the NEWHALL GRADE and through NEWHALL PASS, where paths of the eighteenth- and nineteenth-century trail blazers converged. NEWHALL TUNNELL (*use lights; drive slowly*), 136.1 *m.,* is a narrow bore through the ridge.

BOULDER MONUMENT (L), 136.4 *m.,* marks the entrance to a narrow canyon known as OLD FREMONT PASS, because this way in January, 1847, came Capt. John Charles Frémont on his way from Santa Barbara to Los Angeles. Later the pass was operated as a toll road, the owner exacting a stiff fee from hapless travelers. So steep was the grade (29%) that the tollmaster kept horses ready to pull wagons over the crest. At the top the wagons—and even early automobiles —had their wheels chained fast for the downward slide.

At 137.7 *m.* is the junction with US 99 (*see TOUR 3c*), which unites with US 6 between this point and Los Angeles to become Interstate 5.

LOS ANGELES, 164.9 *m.* (*q.v.*)

Los Angeles is at the junction with US 66 (*see TOUR 12c*), US 99 (*see TOUR 3c*), and US 60-70 (*see TOUR 13b*).

US 6 follows Figueroa St. south through city outskirts into nondescript rural sections with many tracts of unused land.

The narrow SHOESTRING STRIP, 176.8 *m.,* is a neck of Los Angeles City property that runs 8 miles through county territory to link the city with its harbor (*see TOUR 2B*); it was incorporated as Los Angeles territory in the city's movement to annex the harbor district and the seaport towns of Wilmington and San Pedro. The maneuver met frantic opposition from subdivision developers and civic independence supporters in the little municipalities whose territory lay in the way. Once the strategic strip was joined to Los Angeles, the city was ready to annex the towns of Wilmington and San Pedro and their harbors, which it did in 1909.

The tall steel and wood derricks of the ATHENS AND ROSECRANS OIL FIELDS, 177.2 *m.,* opened in 1923, rise from low hills on either side of the highway. The northern wells are "stripping" wells—which draw the last remaining oil from the field—and the southern ones are "flowing" wells—which produce oil without pumping.

At 179.2 *m.* is the junction with One Hundred Sixty-First St.

Right on this asphalt-paved road is GARDENA, 1 *m.* (45 alt., 48,000 pop.). In and about its community center are substantial brick and concrete business and civic buildings of the Spanish-California type. The town is gradually filling with small manufacturing plants and businesses and with the stucco homes of those who like rural living near a large metropolitan center. Here is a junction of the major Long Beach and San Diego Freeways, US 6 and US 101-Alt.

At 180.8 *m.,* the highway forks; US 6 turns L. on Main Street past occasional marshy, reed-grown ponds where coots and killdeer come to feed. As roadside stands, small farms, and dwellings increase in number, big aluminum-painted oil-storage tanks, pumping plants and oil derricks appear among fields and pastures.

WILMINGTON, 184.6 *m.* (0-38 alt., 14,907 pop.), (*see TOUR 2B*) is at the junction with US 101 Alt. (*see TOUR 2B*), which unites with US 6 to Long Beach.

The forests of oil derricks thicken as the highway passes great refineries and tank farms. Multiple-armed steel towers carrying high-tension power lines cross the road.

LONG BEACH, 190.2 *m.* (*q.v.*)

Tour 8

Alturas—Burney—Redding—Weaverville—Jct. US 101; 290.3 *m.,* US 299.

Paved roadbed throughout, open all season; during periods of heavy snow, obtain road information at Alturas and Redding.

West of Alturas, in the northeastern corner of California, US 299 cuts diagonally across harsh lava country and the Cascade Range, paralleling the Pit River. West of Redding, in the northern end of the Sacramento Valley, it follows the Trinity River across the Coast Range through hydraulically scarred mountainsides and deserted mining camps.

Section a. ALTURAS to REDDING; 143.5 m. US 299.

ALTURAS, 0 *m.* (4,446 alt., 2,819 pop.), seat of Modoc County, at a junction with US 395 (*see TOUR 6a*), is the center of a region whose history is one of violent and bloody Indian warfare (*see TOUR 8A*). Alturas is principally a commercial center for ranchers who raise stock, potatoes, and alfalfa hay. The short residential streets are really wide, unpaved country lanes, their poplar-shaded lengths branching from the main thoroughfare. The town until 1874 was called Dorris Bridge, for its first permanent white settler, James Dorris, who in 1869 built a crude wooden bridge across the narrow creek at the east end of town and erected a house that became a shelter for travelers.

The creation of Modoc County was in the nature of a secession from the mother county, Siskiyou. The citizens of Dorris Bridge, angered by the refusal of county supervisors to build a road over the mountains between that settlement and Cedarville (*see TOUR 6a*), not only constructed the desired road themselves, but elected a representative to the legislature pledged to the creation of a new and separate county. Enemies of the bill succeeded in changing the name of the proposed county from Canby to Modoc, and rather than lose it entirely, they accepted the hated name of the local Indians, called Moa Docks (near southerners) by the Klamaths to the north.

At MODOC NATIONAL FOREST HEADQUARTERS, on Main St., information on camping, hunting, and fishing in the forest is available. The forest, extending northward is the home of mule-tail deer which weigh up to 350 pounds, dressed. Ducks, quail, geese, and sage hens are abundant. The 1,500,000-acre Modoc Forest is a range for about 94,000 head of sheep and more than 30,000 head of cattle and horses.

The huge barn (L) on Main St., which for many years has housed the ALTURAS TRADING POST, looks today much as it did in the early 1870's when it was the Alturas Livery Stable. The town's most imposing structure, the stately gray stone building at the edge of town (L), constructed at a cost of $60,000 as headquarters of a narrow-gage railway line, has been deserted for years.

The long, low-roofed buildings occupying an extensive tract at 1.4 *m.,* are the $1,000,000 PICKERING LUMBER MILLS, built before the business depression of the 1930's, which stood idle many years for lack of capital.

Crossing Rattlesnake Creek at 5 *m.,* the highway meets the PIT RIVER, so called because the local Indians formerly dug conical pits in which to trap game and hostile tribesmen. The small openings to the pits were hidden by brush, and sharp stakes were placed in the bottoms to impale the victims.

West of here the broad expanse traversed is broken at intervals by small cultivated areas; tall windmills dominate tiny clusters of tree-sheltered farm buildings.

At KELLEY HOT SPRINGS (L), 16.2 *m.,* (*cabins*) the mineral waters bubble from the ground in clouds of white steam and flow into a stream whose bed is colored vividly by the action of the minerals, its bright greens, red, and yellows softened by deposits of dusty gray and blue at the water's edge.

CANBY, 18.2 *m.* (4,351 alt., 75 pop.), named for Gen. Edward S. Canby of Modoc War fame (*see TOUR 8A*), lies among fields of alfalfa and timothy hay.

At 18.7 *m.* is a junction with a paved highway (*see TOUR 8A*).

US 299 follows Pit River, winding through a region of small farms and yellow pastureland. A hilly spur of the Warner Range looms larger now, roughly paralleling the highway. US 299 swings abruptly south across the west-bound river to CANBY BRIDGE (FOREST SERVICE) PUBLIC CAMPGROUND and leaves its south bank to wind up the Adin Mountains through hemlocks and pines. Small lumber camps reveal their positions in the hills by clouds of blue smoke that drift above distant treetops.

From the RONEY FLAT CAMPGROUND (R) at ADIN PASS SUMMIT, 28.2 *m.* (5,196 alt.), US 299 descends past INDIAN SPRINGS (*fuel, camping accommodations*), 33 *m.,* winding along sparkling Rush Creek through thinning woods to the farms of Big Valley. Harvest time finds these smooth reaches dotted by great stacks of yellow hay.

In ADIN, 39.9 *m.* (4,271 alt., 220 pop.), a small lumbering and farming community, is the BIG VALLEY RANGER STATION (R). Tall poplars line the roadway in front of scattered weather-beaten frame buildings and a tiny, white-steepled wooden church.

US 299 swings over a plateau covered with brush to BASSET HOT SPRINGS, 50.1 *m.,* in another of the small cultivated areas that break the monotony of the semiarid expanse, where mineral waters are piped into pools and tubs.

BIEBER, 52.2 *m.* (4,169 alt., 150 pop.), where the erratic Pit River is again crossed by the highway, was founded in 1877 by its first merchant and journalist, Nathan Bieber; it had been called Chalk Ford because of the numerous chalk deposits in the vicinity. ODD FELLOWS' HALL (1879), a two-story frame building on Main St., with its old-fashioned board walk and ancient iron pump, is a sharp contrast to its nearest neighbor, a modern building of gleaming white stucco.

NUBIEBER, 55 *m.* (4,169 alt., 200 pop.), is a small railroad and lumbering community which helped itself to its older neighbor's name.

The highway gradually ascends the foothills of the Bieber Mountains, passing the débris from logging activities: large piles of brush and dead wood which are gathered in spring and summer, when fire hazards are high, to be burned in winter. Rounding a sudden curve at 60.8 *m.,* the road swings high above a scene of rare grandeur. Behind are the treetops of the Bieber Ridge; in the distance ahead, the Pit River winds silver blue toward snow-capped Mount Shasta; and to the left the jagged bulks of Lassen Peak and Mount Burney rise above a vast checkerboard of green and gold—the neat farmlands of the Fall River Canyon.

McARTHUR, 70.2 *m.* (3,342 alt., 100 pop.), is within the SHASTA-TRINITY NATIONAL FOREST, which covers a huge area in the northern part of the State.

FALL RIVER MILLS, 74 *m.* (3,307 alt., 200 pop.), a small lumbering community, is at the junction of the Pit and Fall Rivers, where the daily flow of water exceeds one billion gallons—more than enough to supply all the cities of the San Francisco Bay area for a year. Here from a 500-foot-long diversion dam on Fall River, an intake canal carries the water 1,000 feet to the base of the nearby hills, where a two-mile tunnel conveys it through an intervening hill to a point on the Pit River Canyon above the site of the Pit Power House No. 1 (*see below*).

Below this point the descent of the Pit becomes more precipitous. Subterranean waters, common in volcanic regions, are the chief sources of its tributaries.

US 299 here traverses the McCloud-Pit River mining region. The snow-white substance that frequently replaces the familiar rock and clay on roadside banks is diatomite or diatomaceous earth. The extensive deposits in this vicinity occur both in a solid form, which can be sawed into blocks, and as a white powder; but thus far it has been little exploited, although its commercial uses are many and varied. It is used for filtration, insulation, the making of sound records, dental pastes and powders, fingernail polish, building and refractory bricks, and as an ingredient of polish for metals, glasses, and lacquered surfaces.

At 78.2 *m.* is a junction with a narrow dirt road.

Left on this road 1 *m.* to the PIT POWER HOUSE NO. 1 of the Pacific Gas and Electric Company. A sudden turn in the winding road discloses red-roofed structures, like doll houses, in the canyon below; these are the homes of employees. At the far end of the valley is the gray-walled power house from which double transmission lines, supported by a series of twin towers, carry power more than 200 miles to a substation near Vacaville, from which it is delivered to Bay area consumers. From the Fall River tunnel (*see above*) the water is dropped 454 feet through vertical steel turbines, 14 feet in diameter. During construction of the power project, lack of transportation necessitated laying a 33.5-mile railroad between McCloud and Bartle. Construction of the Fall River tunnel was completed Sept. 30, 1922.

At 83.4 *m.* on US 299 is a junction with a dirt road.

Left on this road, crossing Hat Creek, to CASSEL (*post office, store, museum*), 3.7 *m.* (2,850 alt.), a settlement with a white population of 25 in a vicinity where between 50 and 60 families of the Hat Creek tribe live. The women are skillful basket-makers; their products are on sale at the store, which was built in 1876 by H. E. Williams. Ancient muskets, arrowheads, and arrows are on display here, as are scores of the Indian photographs for which Williams was known nationally.

Southwest of the junction, US 299 cuts through a thick stand of grass pine between banks of rich red clay to a junction with an oiled road, State 89 (*see TOUR 5a*), 89.5 *m.*

Burney, 90 *m.* (3,159 alt., 50 pop.), is a trading center for a lumbering, farming, and stock-raising district. Wheat, potatoes, and hay are Burney Valley's principal products. The town is named for Samuel Burney, an early English settler killed in an Indian raid in 1857.

West of Burney, US 299 begins the slow ascent of the Hatchet Mountains to the HATCHET MOUNTAIN LOOKOUT, 96 *m.*, commanding a splendid view of a vast forested area, near the summit (4,368 alt.), 97.8 *m.*

Beyond the rustic APPLE SHANTY, 106.9 *m.* (*mountain apples and cider for sale*), the forest climbs the foothills of the Snow Mountains (R).

South of MONTGOMERY CREEK, 109.3 *m.* (2,200 alt., 200 pop.), the road widens, following a downward grade through a less thickly wooded region. The tall pine forests give way to scattered oak, and sheep and cattle graze in wide green pastures beside the highway. At ROUND MOUNTAIN, 113 *m.* (2,200 alt., 100 pop.), US 299 enters COW CREEK CANYON, roughly paralleling the stream for which the district is named.

Beside the road in INGOT, 123 *m.* (2,000 alt., 100 pop.), on terraces (L) above Seaman's Gulch, are the deserted mine and smelter of the Afterthought Mining Company, built about 1922. After vain attempts to extract copper from the Copper Hill lode, the million-dollar plant, tramway system, and colony of company houses on the 1650-acre holdings were finally abandoned. In 1925 the California Zinc Company resumed operations, using the 300-ton selective flotation plant and the reverberating smelter, but with no more success than its predecessor, as the zinc and copper sulphides could not be cleaned separately. In a final desperate attempt to make the mine pay, the zinc company transported ore to their Bully Hill smelter over an 8.5-mile tramway system. Gaunt reminders of a more speculative era are the corroded tin roofs sagging above crumbling brick and stone foundations, the huge blast furnaces, and the rusted tram tracks that skirt the mountainside. The vertical workings are more than 800 feet deep. One main working runs east for 600 feet, then continues as a drift for 500 feet on the Afterthought lode and 400 feet northwest on the Copper Hill lode.

West of Ingot, US 299 traverses a level stretch of countryside, passing occasional small farmhouses, whose comfortable front porches are half hidden by honeysuckle vines.

At 141.5 *m.* is a junction with US 99 (*see TOUR 3a*), which unites with US 299 to REDDING, 143.5 *m.* (537 alt., 12,773 pop.), (*see TOUR 3a*).

Section b. REDDING to JUNCTION with US 101; 146.8 m., US 299.

West of REDDING, 0 *m.*, US 299 climbs into the foothills of the TRINITY MOUNTAINS, roughly following the path blazed by Indians which trappers and miners of the 1840's beat into the well-traveled Trinity Trail. To and from the mining camps of the mountainous hinterland, mail and bullion were carried on horseback. In the late 1850's horse and mule pack trains gave way to stagecoaches, as the Buckhorn-Grass Valley Creek toll road was built from Redding to Weaverville. The entrance of the first coach into remote, mountain-hemmed Weaverville was such a gala occasion that, according to a contemporary report: "Trinity County citizens went out in buggies and on horseback, led by the German brass band, to greet and escort it into town."

SHASTA, 6 *m.* (990 alt., 150 pop.), onetime seat of Shasta County, was known until 1850 as Reading Spring, in honor of Maj. Pierson Reading (*see TOUR 3a*). In 1849 Reading Spring was a tent city of several hundred inhabitants—a lively trading post serving a mining region. By November of that year one Milton Magee attained local distinction as the only owner of a log cabin; and by 1852 Shasta had grown so rapidly that two frame hotels, the St. Charles and the Trinity House, were erected. The town was soon called the head of "Whoa Navigation" by the Shasta *Courier;* at one time more than 2,000 pack mules carried supplies to the northern mines, and as many as 100 freight teams were housed at Shasta in a single night. At the height of its prosperity, local merchants were sending out $100,000 in gold dust each week.

Frequent Indian outbreaks added to the general turbulence of the town's early days. On many occasions during the 1850's, bands of painted braves in fighting regalia staged war dances on Main Street. The *Courier* of March 2, 1853, reported it "unsafe to travel over any exposed portion of the country unarmed. The . . . Indians . . . are infesting the Sacramento River trail in such numbers and with such determined fierceness as to render it almost certain death to pass over that road." But later the constant threat of Indian hostilities was overshadowed by a more immediate catastrophe, for on June 14 the entire business district was wiped out by fire in a brief half hour. Flimsy pine buildings, lined with cotton cloth, were quickly reduced to ashes; but before the ashes cooled lumber had been ordered to rebuild the town. In 1855 the *Courier* announced: "There are 28 brick buildings in Shasta at a cash value of $225,000."

With a decline of mining Shasta lost its importance. Then, when the California-Oregon railroad was planned, it was decided that Shasta lay 3 miles too far west, with 400 feet too much altitude, to be included in the route, and today the brick buildings of which the *Courier* boasted stand deserted and crumbling. Only two are in use; one is the MASONIC HALL, the first in the State, in whose vaults repose the first charter of Western Star Lodge, No. 2, brought from Missouri by Peter Lassen in the early 1850's by ox train; the other is the LITSCH STORE (L), built in 1853. This old-fashioned general store, in addition to serving as post office, houses "Litsch's Free Museum of Historic Pioneer Relics." Here are firearms; paper cuffs and collars; gold dust pokes; scales that have weighed millions of dollars worth of "dust" since 1853; crude handmade wooden stirrups, brought across the plains in 1843; and one of the State's first music boxes. There are also relics of an unfortunate pioneer group, the Chinese: a hand-carved opium pipe said to be 150 years old and an iron pitchfork used in the tong war at Weaverville.

North of Redding on US 99 is the huge SHASTA DAM, impounding the waters of the Sacramento, Pit and McCloud Rivers for the vast Central Valley irrigation system. *Daily tours of dam and powerhouse.*

The dam is 602 ft. tall, 3,500 ft. long at the top, and its spillway drops water 480 ft. The backed-up waters form LAKE SHASTA, 35 *m.* long, covering 29,500 acres, with a shoreline of 365 *m.* The power plant has a capacity of 375,000 kilowatts.

A few miles north of Redding KESWICK DAM crosses the Sacramento River. The dam and power plant are smaller than the ones at Shasta. About 15 *m.* north of Redding is the PITT RIVER BRIDGE, a double-deck structure carrying the four-lane highway and the tracks of the Southern Pacific. On Burney Creek, a tributary of Pitt River are the McArthur-Burney Falls, 165 ft. tall.

West of Shasta US 299 ascends by easy grades to the summit of SHASTA DIVIDE, 8 *m.* (1,390 alt.), which separates the watersheds of the Sacramento and Trinity Rivers. The view extends over miles of treetops to massive mountain barriers. Descending on the other side the road enters WHISKEYTOWN, 10.8 *m.* (1,091 alt.), settled by miners on the trail to Oregon near Whiskey Creek, so named when a barrel of whiskey fell off a pack mule into the creek. Postal authorities have called it Blair, Stella, and Schilling, but Whiskeytown has been applied to the dam at this point.

TOWER HOUSE, 17 *m.* (1,247 alt., 5 pop.), dating from 1852, was a stage station on the toll road built from Shasta in the late 1850's and a point of departure for pack and bullock trains bound for the northern mines and for Oregon.

Right from Tower House on an oiled road to FRENCH GULCH, 3 *m.* (1,346 alt., 618 pop.), depot on the Shasta-Yreka Turnpike, where the first mining is said to have been done here by a party of Frenchmen in 1849 or 1850. The discovery of the Washington Quartz Mining Company claim prompted the Shasta *Courier* to report in March 1852 that "such rich diggings have been struck that miners are tearing down their houses to pursue the leads which extend under them."

West of Tower House US 299 follows the route of the old Buckhorn Toll Road, winding sharply into the thickly forested Trinity mountains along the high rims of canyons. From Buckhorn Mountain, 25.5 *m.* (3,212 alt.), it descends to BUCKHORN, 29.6 *m.* (2,500 alt.), among the small farms of Grass Valley.

The trading center of DOUGLAS CITY, 41.9 *m.* (1,700 alt.), lies beside the Trinity River near the mouth of Reading's Creek, where

Pierson B. Reading made the first discovery of gold in Trinity County in 1848. As Reading told the story: "My party consisted of three white men, one Delaware, one Chinook, and about sixty Indians from the Sacramento Valley. . . . I had 120 head of cattle with an abundant supply of other provisions. After about six weeks' work, parties came in from Oregon, who at once protested against my Indian labor." Having already taken out $80,000 in gold, Reading abandoned his diggings.

Reading named Trinity River on a trapping expedition in 1845, under the misapprehension that old Spanish maps indicated it flowed into Trinidad Bay. The river banks for 65 miles are scared from the effects of ruthless hydraulic mining, which has left mounds of gravel. From 1849 until the outbreak of World War I when machinery and labor costs became prohibitive, Trinity River and its tributaries were the scenes of various types of mining operations, from the simplest forms of placering to gigantic hydraulic operations. In 1933, when the price of gold rose, miners' shacks and equipment reappeared.

This Trinity County area is the site of several dams, reservoirs and hydroelectric plants. Trinity River Dam, completed 1962, has a 415 ft. fall. It impounds the big CLAIRE ENGLE LAKE, which has a shoreline of 122 *m.* and dense forests on both banks. The TRINITY RIVER DIVERSION, completed 1964, carries water from the smaller Lewiston Dam and Lake, Whiskeytown Dam and Spring Creek Tunnel to the Sacramento River at Keswick.

Left from Douglas City on an oiled road climbing steadily 19 *m.* to a junction with a narrow dirt road; L. here 6.2 *m.* through green meadows to a picnic ground (*free*), from which a footpath leads between white moss-covered boulders to a gray-white limestone NATURAL BRIDGE, spanning a narrow gulch, SITE OF THE BRIDGE GULCH MASSACRE. In March 1852, following the killing of a man named Anderson by Wentoon Indians, miners from Weaverville (*see below*) overtook and killed about 100 Indians encamped here, pursued the handful of women and children huddled beneath the bridge and knifed them to death. A contemporary reported that a young miner who sickened of the slaughter, saved one Indian girl, took her to Weaverville and entrusted her to a "motherly" woman acquaintance—who promptly sold her to a teamster from Shasta for $45.

WEAVERVILLE, 47.7 *m.* (2,407 alt., 1,736 pop.), seat of Trinity County, has lost nothing of the charm of the old, "easy" West, though the vine-covered frame cottages along the locust-bordered main street now have lawns protected by white picket fences. Over the tree-shaded sidewalks of the business section project broad, second-story galleries, reached by outside circular stairways, which give upper-story tenants private entrances. Only the occasional boisterous activity of the early mining town is absent. John Martin's livery stable became Miller's Garage with little outward change; but today's automobile turns in the wide street without the shouts that attended the maneuvering of the three-wagon freight train, drawn by ten or twelve horses controlled with a deft jerk rein and an astonishing profanity.

Weaverville's social gatherings are no longer punctuated with the roar of "six-shooters," nor is its justice the simple hearty violence once encountered by an unfortunate miner named Seymour. The latter was accused of stealing the money of a visitor from Klamath who had drunk too freely and fallen asleep in a convenient pigsty. According to an early writer, Seymour "was taken up by the authorities, then rescued by the people, then left to his own fate, and then, finally . . . the money was found in the pigsty, where the drunken curse had sought shelter and company." More deserved was the treatment of one John Fehly in 1857; Fehly, drunk in front of the Diana Saloon and threatening darkly that his partner had "better look out," drew his revolver and shot into the crowd, killing Dennis Murray. Led to the gallows, Fehly struggled violently, but was forced to submit by the "Infant," a youth of 16, 6 feet 6 inches tall. In a contemporary account the Infant is described as "well known in Weaverville . . . and supposed to be a descendant of him whose blindness was so fearfully avenged at Gaza upon the Philistines." Overcome, Fehly remarked, "Since I've got to go, give me a coal on my pipe," and went to the scaffold smoking.

Despite several early fires, many old buildings remain, among them the WEEKLY TRINITY JOURNAL OFFICE (the *Journal,* first published in 1856, is one of the four oldest newspapers in the State). MEMORIAL HALL (L), a museum (*open daily; free*), displays relics of '49—ore specimens from well-known mines, a wide variety of old firearms, and an Indian scalp.

Possibly a dozen Chinese remain in Weaverville today to care for the old CHINESE JOSS HOUSE (*open*); but in 1852 half of the town's population of 3,000 was Chinese. In the dim, candle-lit interior of the tiny building, fine, hand-painted tapestries hang beside an altar 3,000 years old. A gully (R), Five-Cent Gulch, is the SITE OF THE CHINA WAR; here, in 1852, members of two rival tongs, encouraged by amusement-seeking white miners, met in a battle. Cheered on by the miners, 300 Ah Yous were badly routed by 500 Young Wos. Soon only two warriors remained on the field; for fifteen minutes they stood, calmly stabbing each other with their crude iron forks, until one fell dead. The other man died after two weeks of suffering. No other serious casualty was reported. (The use of firearms was forbidden to orientals by the whites.)

Right from Weaverville on a dirt road 15 *m.* to TRINITY ALPS CAMP (*cabins, dining room, horses, and guide service*), a resort in the Trinity Mountains. A few miles north of the resort is the SALMON-TRINITY ALPS PRIMITIVE REGION, a part of Trinity National Forest (*see GENERAL INFORMATION*), accessible only by foot and pack trip. Tallest of the jumble of mountains within the area is THOMPSON (SAWTOOTH) PEAK (8,936 alt.), on which are several glaciers.

US 299 ascends OREGON GULCH MOUNTAIN, where the old Baron La Grange Mine, opened in 1851—for years the largest operating hydraulic mine in the world—once blasted away the mountainside.

JUNCTION CITY, 57.1 *m.* (1,471 alt., 300 pop.), on Canyon Creek, is a trading center of ranchers and miners.

At about 61 *m.* (L) on Trinity River is the SITE OF ARKANSAS DAM, one of the earliest "wing" dams, constructed by some 60 miners in 1850 and twice rebuilt after being washed away by floods. "Wing damming" was a system of mining utilized at low water, whereby the top rock and sand of the river bed were removed down to the pay gravel.

BIG BAR, 74 *m.* (1,248 alt., 73 pop.), was settled, according to an early historian, "by one Jones in 1849, who got rich, left the locality, and is no more with the memory of his fellows . . . Big Bar was the first place in Trinity County where the first white woman ever made the first johnny cake or dumpling and there lived and dwelt, and this woman's name was Mrs. Walton, who, with her husband, settled here in 1850 . . ." Still another resident of Big Bar, in 1855, was "Commodore" Ligne, who, "in a spirit of acquisitiveness, used to sell claims to Chinamen and others, and upon payment being duly received, would in the most harmless way 'practicable' drive them away. It seems that some obstinate occupants were foolish enough to remonstrate . . . but the Commodore . . . put an end to this by a dexterous display of shotgun . . . and became a terror to the locality." After an unsuccessful attack on a miner, Ligne "prudently retreated to the woods and did not reappear before the County . . . sent a deputation . . . and invited him to headquarters." Eventually, the activities of the Commodore were confined to San Quentin.

DEL LOMA (Sp., of the mountain), 78.8 *m.* (*tourist cabins*), once an active mining camp known as Taylor's Flat, is in the vicinity of French, Little French, and Canadian Creek, supposedly named for a band of French Canadian trappers who left Oregon at the news of Reading's strike on the Trinity.

US 299 continues past the HAYDEN FOREST SERVICE CAMPGROUND (*improved*), 79.8 *m.,* to the settlement of BURNT RANCH, 91.1 *m.* (2,000 alt., 30 pop.), where farmhouses were destroyed by fire in an Indian raid of 1853. From SALYER, 100.5 *m.* (553 alt., 59 pop.), site of another FOREST SERVICE CAMPGROUND, it descends to less rugged country where trees shade the road.

At 106 *m.* is a junction with State 96 (*see TOUR 3a*).

US 299 again enters a rugged region of forest and canyon, passing BOISE CREEK PUBLIC CAMPGROUND, 107.7 *m.* (R), and EAST FORK PUBLIC CAMPGROUND, 112.2 *m.* (L), both maintained by the Forest Service.

The highway leaves the confines of a narrow mountain gorge to command a wide vista of valley grazing lands edged by rolling yellow hills to the south; beyond it climbs low hills and descends to canyons where ferns grow tall and green by small, deep streams.

At 139.6 *m.* is a junction with an oiled road.

Left here 2.5 *m.* to the OLD ARROW TREE. When this dead redwood was young and strong, peace was made in its immediate vicinity between two warring tribes, who thereafter considered it the boundary between their domains. When a member of the tribes passed the spot, he shot an arrow into the bark as a token of peace. As the original meaning fell into obscurity, it became an altar for worship and a place of prayer. Passers thrust sharpened sticks into the bark and prayed for luck. Squaws, after striking their legs with redwood sprigs, threw them against the tree and cried, "I leave you with all my sickness."

BLUE LAKE, 139.7 *m.* (40 alt., 1,234 pop.), is the trade center of a farming and dairying area, distinguished by neat picket fences, tall windmills and silos.

At 146.8 *m.* is a junction with US 101 (*see TOUR 2a*), at a point 2 miles north of Arcata.

Tour 8A

Canby—Lava Beds National Monument—Medicine Lake—Bartle; 118.8 *m.* St 139 and unnumbered roads.

Roadbed paved for first 25.7 miles. Elsewhere dirt and gravel, deeply rutted, covered with powdered pumice and sharp lava fragments in places; travel advised from May to October only. At least one extra tire should be carried. Accommodations only at Canby and Medicine Lake; Forest Service camps at Lava Beds National Monument and Medicine Lake.

The rough and inhospitable waste of lava beds, underground caverns, and cinder cones through which this route leads is a part of a 250,000-square mile volcanic region. The pioneers knew it as the "Dark and Bloody Battleground of the Pacific," where the last and bloodiest of California's Indian Wars was fought. It is now part of Modoc National Forest.

State 139 branches northwest from its junction with US 299 (*see TOUR 8a*), 0 *m.*, 0.5 miles west of Canby, and runs into a lonely, almost unsettled region. The Forest Service HOWARD'S GULCH CAMP GROUND, 7 *m.*, lies at the edge of pleasant meadows. At 25.7 *m.* is a junction with a dirt road; L. here on the dirt road, now the main route.

The entrance to MAMMOTH CAVE (R), 31.2 *m.*—50 yards from the road—is marked by massed lava rocks. The cave stretches for hundreds of yards underground.

At 32.2 *m.* is the junction with an unpaved road; L. here; then R. at 36.2 *m.*, crossing a boundary of the LAVA BEDS NATIONAL MONUMENT, 38.6 *m.*, a rectangular area about 10 miles long and 8 miles wide, set aside in 1925. The region is geologically young; its last lava flow is estimated to have taken place only 5,000 years ago. Seen from a distance, the lava beds are a dark, comparatively level terrain, broken only by crater-pitted cinder cones. On closer inspection, they are revealed as a rugged labyrinth of caves and chasms, where the molten lava has congealed in innumerable strange shapes—fumaroles (chimneys) of gas-inflated lava, smoothly arched bridges, fantastically sculptured shapes, some of them resembling animals. The caves, of which more than a hundred have been explored, are found in the type of billowing lava known as pahoehoe; they were formed when the surface of the lava hardened and the still molten core drained away. In many of these underground galleries, delicate lace-like tracings and Indian pictographs in red, green, or yellow ochre decorate ceilings and walls; in some of the deeper ones, frost crystals, frozen waterfalls, or rivers of ice appear.

When white immigrants were settling northern California, the lava beds were the stronghold of the warlike Modoc. The Nation's costliest Indian conflict, the Modoc War of 1872-73, was required to subdue them. For 5 months a band of Modoc warriors, never numbering more than 60, held off some 1,200 soldiers. The cost of subduing them was more than half a million dollars and the lives of 83 whites.

The trouble began in 1869, when the Modocs were moved to the reservation in Oregon of their hereditary enemies, the Klamaths. When they got no satisfaction from the Indian Agent after their complaints against the persecution by Klamath, a small band led by young Kientpoos—usually called Captain Jack—and his sister, "Queen Mary," left the reservation and returned to their old hunting ground, the Lost River country, just across the line in California. Over the protests of Brig. Gen. E. R. S. Canby, Department Commander, cavalry from Fort Klamath was sent by demand of the Indian Agent to bring the Modoc back to the reservation—"peaceably if you can, forcibly, if you must."

After a fight in which several were killed and wounded on both sides, the Indians entrenched themselves in an almost impregnable position among caves and crevices here, while the soldiers settled down nearby to wait for reinforcements. On January 17, 1873, when the military force had grown to about 400 men, Colonel Wheaton, the commander, launched an attack against the Indians' main position, Captain Jack's stronghold. At the end of the day, the soldiers had worn themselves out chasing the elusive Modoc without ever having seen an Indian, and 39 whites had been killed.

Backed by more reinforcements, General Canby attempted to arrange a settlement of the dispute. On April 11, 1873, he met Captain Jack and several of his chiefs under a flag of truce. Suddenly, in the midst of the parley, he was shot down and killed by Captain Jack; another peace commissioner, Dr. Eleazer Thomas, was slain by "Boston Charley." The other three members of the commission escaped. The Indians had lost two of their staunchest champions.

The troops rallied for a methodical advance, occupying successive positions until they had surrounded the Modocs and cut off their water supply. On April 17 they assaulted the Indian stronghold and drove the defenders out. Four days later the Indians, now completely surrounded, separated into two groups for an attempt to escape. As soon as they tried to fight in the open, they found the odds against them; nonetheless, they cut Capt. Evan Thomas' detachment to pieces before they were finally rounded up. Captain Jack himself eluded capture until June 1. On October 4, he was hanged at Fort Klamath, with three of his aides, for the murder of the commissioners. The rest of the band was moved to a reservation in Kansas.

INDIAN WELL, 42.1 *m.* (*information available from ranger during summer*), campground and administrative headquarters for the monument, is a center for exploration. (*Gasoline lanterns for use in caves available; in larger caves, guide line of strong twine should be fastened at entrance and unwound as exploration progresses. Stout hiking boots essential; ordinary clothing suffices.*)

Right from Indian Well on a dirt road 0.4 *m.* to the junction with a dirt road; R. here 0.3 *m.* to the junction with another dirt road; L. on this road 0.6 *m.* to POST OFFICE CAVE and DRAGON'S HEAD CAVE.

Opposite the ranger station, at the entrance to the LABYRINTH (L), a 2-mile maze of tunnels, is the DEVIL'S MUSH POT, a huge kettle of rock from which molten lava once flowed. Inside the labyrinth are JUPITER'S THUNDERBOLT, a smooth formation extending from wall to wall, and the MENAGERIE, a collection of animal likenesses formed by lava.

Left from the labyrinth on a narrow graveled road, past Frog, Sunshine, Jupiter, and Sentinel Caves, to CATACOMB CAVERN (R), 0.6 *m.*, whose smooth-floored lava passages wind for nearly 1.5 miles, adorned with wall niches resembling the burial places of early Christians in Rome. In some sections are traceries like delicate lace on walls and ceiling.

Opposite the entrance to the Catacombs is CRYSTAL CAVE. From the red lava walls of its "jewel chambers" frost crystals flash, diamond-like.

At 43.6 *m.* is the junction with an unpaved road; R. here on a narrow, winding route, now the main route, pitted with sharp lava fragments, which loops around the monument (*travel advised only for experienced drivers, with tires in good condition*). At 45 *m.* is the junction with a narrow unpaved road.

Left here 0.6 *m.* to BIG PAINTED CAVE (L) and LITTLE PAINTED CAVE (L), on whose walls are many Indian pictographs.

SKULL CAVE (L), 45.2 *m.*, was so named because numerous skulls of pronghorn antelope and bighorn sheep were found in it; both animals are now extinct in this region. The domed roof of Skull Cave's main level is 75 feet high. In the third, and lowest, cavern, in a river of ice, are the bones of many animals. A short distance west of Skull Cave are (R) SHIP CAVERN and WHITE LACE CAVE. On the walls of the latter are natural, white, lace-like patterns.

The ice in FROZEN RIVER CAVE, 46.3 *m.*, rarely melts; the depth of its frozen river is unknown.

A high LAVA BRIDGE is crossed near the entrance to CAPTAIN JACK'S ICE CAVE (R), 46.7 *m.,* one of the Indian leader's retreats during the Modoc War.

The road continues northward through lava wastes to FERN CAVE (R), 52.7 *m.* (4,114 alt.), which derives its name from the luxuriant masses of ferns that grow just within its entrance, kept green the year round in a natural hot bed formed in the rich soil by many steam vents. On cold winter days clouds of steam obscure the entrance to the cave. The walls here also bear Indian pictographs.

Behind the natural fortification of HOSPITAL ROCK (R), 54.6 *m.* (4,051 alt.), wounded soldiers were sheltered. The cavalry camped nearby in 1873.

The road crosses the northern monument boundary at Hospital Rock, then parallels its irregular line for about 8 miles. Around CAPTAIN JACK'S STRONGHOLD (L), 57.1 *m.,* centered much of the Modoc War activities. The caves, the lava trenches, and the natural rock fortifications look today as they did in 1873.

CANBY CROSS (L), 60.6 *m.,* erected by his soldiers, marks the spot where General Canby was murdered (*see above*).

At 62 *m.* is the junction with a dirt road; L. here, again crossing the northern boundary, 62.3 *m.* Nearby is the site of the Army headquarters during the Modoc War. In GILLEM'S GRAVEYARD (R), 62.6 *m.,* about 100 white soldiers were buried.

The route follows the base (R) of GILLEM BLUFF (4,500-4,700 alt.) for nearly 4 miles. Two large groups of fumaroles (L) are passed at about 67.6 *m.* These "chimneys" of gas-inflated lava are similar to the fire fountains of the Hawaiian volcano, Kilauea. About 0.5 miles east of the chimneys (*no trail*) is the place where the force led by Captain Thomas was badly defeated; two-thirds of the command and four (of five) officers were killed or wounded.

SCHONCHIN BUTTE (L), 70.1 *m.* (5,293 alt.), is the largest of the 11 cones of scoriaceous cinder in the southern part of the monument; it was named for a Modoc chief.

At 71.1 *m.* is the junction with a dirt road.

Right on this road 1.5 *m.* to BEARPAW CAVE (R), whose frozen waterfall and river of ice never melt.

At the junction (L) at 71.2 *m.* the loop road completes its rough circle; straight ahead (south) here, to a junction with a dirt road at 71.6 *m.;* R. here passing HIPPO BUTTE (R), 72.1 *m.* (5,488 alt.), and crossing the southern boundary of the Monument, 74.3 *m.,* at the base of 400-foot deep MAMMOTH CRATER (R). At 74.8 *m.* is the junction with a dirt road; the route turns R. here, and L. at 77.8 *m.*

The deep blue waters of MEDICINE LAKE (*private lodge and public camp*), 85.8 *m.,* fill a crater that was once the center of local volcanic activity. The lake is well stocked with game fish. Its sandy beaches are fringed with tall lodgepole pines.

Five miles northeast is GLASS MOUNTAIN (7,649 alt.), whose

bulk of jet black obsidian gleams when struck by the sun. Steam issuing from a 30-foot vent west of the top-most peak bears evidence of the underground fires still smouldering within. The pumice surface nearby is too hot to be touched. So light in weight are the volcanic rock fragments that a man can toss a piece the size of a horse. Glass Mountain, Clear Lake Reservoir and environs are in MODOC NATIONAL FOREST, 1,688,789 acres.

At Medicine Lake is the junction with an unpaved road; here the main route turns R., then L. at 91.8 *m.*

BARTLE (*see TOUR 5a*), 118.8 *m.*, is at junction with State 89.

State 139 proceeds due north and runs east of Lava Beds National Park into a region of wild life refugees. CLEAR LAKE RESERVOIR is on the east and TULE LAKE on the west. The CLEAR LAKE NATIONAL WILD LIFE REFUGE covers 35,600 acres and is the home of Canada geese, ducks and pelicans. The TULE LAKE WILD LIFE REFUGE of 35,000 acres is another home for waterfowl. The LOWER KLAMATH NATIONAL WILDLIFE REFUGE is a 30,000 acre terrain around LOWER KLAMATH LAKE. In the center of this protected area is TULELAKE (4,143 alt., 1,000 pop.). The area is directly south of the Oregon border. The reservations are administered by the Fish & Wildlife Service of the U. S. Dept. of the Interior, headquarters at Tule Lake Refuge, where permits for hunting must be obtained.

Tour 9

(Reno, Nev.)—Truckee—Auburn—Roseville—Sacramento—Vallejo—San Francisco; US 40, Interstate 80. 219.9 *m.*

US 40, most traveled artery between the East and central California, hurdles the sheer, rocky wall of the Sierra Nevada into the valley that stretches to the Golden Gate. From the Great American Basin it climbs to the granite heights of Donner Pass, traverses the boulder-piled Yuba River bottoms, coasts toboggan-like through forests and along river gorges offering hazy vistas of mountain ranges.

Section a. *NEVADA LINE to SACRAMENTO, 124.9 m. US 40; Interstate 80.*

The mountain heights were a forbidding barrier to the pioneers. With prodigious labor they forced their lumbering ox-drawn schooners over tortuous trails and through brush-choked canyons, over knifelike ridges, between gaping chasms. Often, when a sheer cliff blocked the way, oxen had to be unyoked so the ponderous wagons could be lifted or lowered with ropes. The devious California Trail through

Architecture

LA CASA GRANDE, HEARST MONUMENT, SAN SIMEON

SAN CARLO DE BORROMEO MISSION, CARMEL

SAN JUAN CAPISTRANO MISSION

MISSION DOLORES, SAN FRANCISCO

SANTA BARBARA MISSION AND FOUNTAIN

MISSION SAN LUIS REY, OCEANSIDE

NEO-SPANISH BUILDINGS, BALBOA PARK, SAN DIEGO

STATE EMPLOYMENT BUILDING (1955) SACRAMENTO

LELAND STANFORD HOME (1857) AND STATE RESOURCES BUILDING (1964) SACRAMENTO

THE PAVILION, LOS ANGELES MUSIC CENTER

LOS ANGELES COUNTY MUSEUM OF ART

CITY HALL, PASADENA

COURT HOUSE, SANTA BARBARA

Truckee Pass ran a few miles distant from what is now US 40, crossed it near Donner Lake, and recrossed it at Emigrant Gap. Over this trail in the autumn of 1844, 81-year-old Caleb Greenwood, mountaineer and trapper, led the 12 wagons of the Stevens-Murphy Party, first caravan on wheels to cross the Sierra. Others followed with terrible hardship in the autumn of the next year—the Swazey-Todd Party of horsemen; trappers on foot; the Grigsby-Ide Party of more than 100 men, women, and children led by Greenwood; and John C. Frémont on his third exploring expedition. In October 1846 the Donner Party, acting on vague advice, made the Salt Desert crossing and, arriving too late to scale the terrible pass that now bears their name, were caught by the snows.

A saga of transportation fully as exciting followed in 1864-66 when gangs of Chinese coolies swarmed the mountain, laying the rails of the Central Pacific eastward in a race with the Union Pacific, pushing westward from Omaha, that culminated in the completion of the first transcontinental railroad in Utah (1869). In June 1864 the "Big Four," Stanford, Hopkins, Huntington, and Crocker, had opened the Dutch Flat and Donner Lake Wagon Road. A road had reached Colfax, head of "wagon navigation," as early as 1849 and Dutch Flat a few years later; now the way lay open to the Comstock Lode mines in Nevada. For a brief interval stagecoaches raced over it, bearing passengers and freight; but as the Central Pacific was pushed forward—reaching Clipper Gap in June 1865 and Colfax in September—it killed all competition.

US 40 crosses the State Line, 0 *m.,* 15 miles southwest of Reno, Nev., following the Truckee River, which is bordered by steep, craggy slopes. Frémont, camping at its mouth in January 1844 with his second expedition to the Far West, named it the Salmon Trout River because of the fine fish the Indians brought him. Later in the same year, the Stevens-Murphy Party gave it the name it now bears—that of the Paiute chief who guided them out of the burning alkali desert to the river's banks and pointed out the pass into California. His answer to all questions was "truckee," his equivalent for "okeh."

The boundary of TAHOE NATIONAL FOREST is crossed at 0.5 *m.*; the road runs for more than 50 miles through the forest. A reserve of more than 1,000,000 acres, it embraces an ever changing panorama of Sierra grandeur.

At 6.1 *m.* is a junction with a dirt road.

Left on this road is FLORISTON, 0.5 *m.* (5,200 alt., 384 pop.), where rise the jumbled brick walls and smokestacks of a paper pulp mill—once the greatest producer in California. Red-roofed white frame cottages cling to the precipitous slopes.

From the winding canyon, US 40 climbs to a high, wide tableland, rimmed by mountains.

At 11.9 *m.* is a junction with a dirt road.

Right on this road, crossing the river, to BOCA (mouth), 0.5 *m.* (5,535 alt., 25 pop.), once a lumbering town, now merely a handful of ramshackle

shanties around a bleak railroad station at the mouth of the Little Truckee River.

TRUCKEE, 19.8 *m.* (5,820 alt.), located at the junction (R) with State 89 (*see TOUR 5b*), straggles along the banks of the river. Roundabout, pine forests cover the slopes with deep green, but Truckee itself, once a lumbering camp, now a railroad and stock-raising supply center, lacks even a sprig of green. Its ramshackle frame houses and weather-stained brick buildings sprawl over rocky slopes. On Saturday nights the cheap saloons and gambling halls overflow with lumberjacks, cow-punchers, and shepherds. In winter, when great glistening drifts fill the streets, the nearby snow-clad slopes resound with the shouts of skiers. The first California ski club was organized here in 1913. There are a variety of slides, a 1,000-foot toboggan slide with a power pull-back, and an ice rink illuminated for night skating. During the season are the weekly ski-jumping programs and the annual Sierra dog derbies. Often during the winter film companies work here.

At 20.8 *m.* is the western junction with State 89 (*see TOUR 10a*).

South of Truckee the highway plunges into a corridor of nut pines. A stone monument (L), 21.5 *m.,* faces north toward the low pass through which came the California Trail. About 90 yards south is the SITE OF THE GRAVES CABIN, erected by part of the Donner Party in 1846. At 22 *m.* another stone monument indicates the point where the emigrant trail turned south again, up Cold Creek Canyon, on its way to the summit. Here the Donner Party lost the trail.

DONNER MEMORIAL STATE PARK, 22.3 *m.,* is a 353-acre tract set apart as a memorial to the Donner Party. At the end of October 1846, the vanguard of the train arrived at this point. Storm clouds were already gathering as they struggled up the rocky canyon of the Truckee; they made haste, but men and animals were footsore and exhausted; George Donner, the leader, had injured a leg. On October 28, a month earlier than usual, snow began to fall, burying the faint trail. They tried to go on, but could not, and turning back, pitched camp. Winter broke; November ended in four days and nights of continuous snow; December began with furious wind, sleet, and rain.

Led by well-to-do George and Jacob Donner and James F. Reed, the party had set out from Illinois in April. At Fort Bridger, on the strength of an open letter sent to travelers by the dare-devil mountaineer Lansford W. Hastings, who did not meet them as his letter had promised, they had taken the fatal step of breaking off from their companions of the trail to follow the unknown route later called the Hastings Cut-off. Even when they were compelled to break their way a few feet at a time through the thicket-choked canyons of the Wasatch Mountains in Utah, they had stubbornly refused to admit their mistake and turn back. By the time they had struggled for days and nights without water across the deserts beyond the Great Salt Lake, it was too late to correct the error. Utterly exhausted, their animals dying, they had rested for several days in the green meadows along the Truckee on the site of Reno; this final delay was their undoing.

In the camps at Alder Creek and Donner Lake, the snow became 20 feet deep. Starvation faced them, for their cattle had wandered off, and were buried in the drifts. By December 10, Jacob Donner and three others in the Donner huts on Alder Creek were dead. Soon the immigrants were living on the few mice they could catch, and on bark and twigs. In a desperate effort to escape, 10 men and 5 women, known as the "Forlorn Hope," started west on foot with provisions enough for 6 days, scaled the summit, and followed the ridge north of the North Fork of the Yuba River. Within a week one man had died; by Christmas Day, after 4 days without food, they could go no farther. A terrific storm had burst upon them, and for a whole week they were snowbound, huddling in terrible misery about their campfire. When the storm lifted, four more men had died; the starving survivors devoured their bodies. Another man died; he, too, was eaten—and so were the two Indian guides, killed when they began to falter. Struggling on, these survivors ate their moccasins, the strings of their snowshoes, a pair of old boots. Thirty-two days after they had set out, the 5 women and 2 remaining men reached an Indian village and were dragged on to the Johnson Ranch at Wheatland, leaving a trail of bloody footprints.

Relief parties set out from the Sacramento Valley. The first, seven men on foot, reached the camp at Donner Lake February 19, to find it buried under snowdrifts; the survivors, with nothing left to eat but hides, were in a torpor. With 21, mostly children, three of whom died on the way out, the rescuers started back, meeting on the way the second relief expedition, which reached the camp March 1. Seventeen more started back with it, only to be trapped by another blizzard, which held them in the snow for a week. Their feet and hands froze; three more died, and their bodies were eaten by the survivors. The third relief party found nine still alive at the camp, three of them too near death to travel; they, too, had been driven to eat the bodies of the dead. With five of these survivors, the relief party started back, but Tamsen Donner, renouncing her last chance to escape, said goodbye to her two little daughters and struggled back over the five miles of drifts to the camp at Alder Lake to nurse her dying husband. She had died, and her body had vanished, when the fourth relief party, seeking whatever of value remained at the camp for plunder, arrived to find one man still living in hideous squalor among the bones of his fellow travelers. Of the 81 who had pitched camp here in November 1846, only 45 crossed the mountains.

Left from the entrance to the PIONEER MONUMENT, 0.1 *m.*, a great stone block bearing the heroic bronze figures of a man, woman, and child. Here is the SITE OF THE SCHALLENBERGER CABIN, built by members of the Stevens-Murphy Party in November 1844, where young Moses Schallenberger, ill and unable to scale the pass, spent the winter alone, guarding the goods until he was rescued in March of the next year. When the vanguard of the Donner Party camped here November 1, 1846, the Breen family—only one to escape without losing at least one member—moved in.

Right from the end of the road on a foot trail, through a grove of nut pines, 0.3 *m.* to the huge granite boulder marking the SITE OF THE MURPHY

CABIN, in which lived Grandmother Murphy, her family, and others. The crude lean-to was built against the great smooth-faced rock, which served as the fireplace wall. Under the Murphy cabin in June 1847 Gen. Stephen W. Kearny's band buried the bones found in the vicinity.

US 40 curves around the edge of gleaming blue DONNER LAKE, 22.7 *m.* On August 25, 1846, Edwin Bryant, with eight companions, found the going so boggy here in many places that the mules sank to their bellies in the mire. The arresting outlines of CASTLE PEAK (9,139 alt.), rise (R) at the head of a long draw. Bryant, in the journal of his California trip, marveled at its "cyclopean magnitude, the . . . apparently regular and perfect . . . construction of its walls, turrets, and bastions."

In long curves US 40 begins to scale the all but perpendicular wall of granite that lifts ahead beneath the rock-ribbed crowns of DONNER PEAK (8,315 alt.) and LINCOLN PEAK (8,403 alt.). Blasted through solid rocks, the road twists in and out between overhanging ledges. Up the sheer precipice wriggles the long black caterpillar of the railroad snowsheds, worming at intervals through granite tunnels.

To young adventurers like Edwin Bryant and his companions, these heights offered a challenge; but they dismayed those who faced the passage with cumbersome, ox-drawn wagons loaded with belongings and women and children. Old Caleb Greenwood, however, guiding the Stevens-Murphy Party over the mountains in November 1844, exercised rare judgment and skill. They struggled upward until a rise of 10 or 12 feet blocked the way. Greenwood discovered a narrow crevice; through it the oxen were half pushed, half dragged, with men below and ropes above; the household goods were carried up piece by piece. Then, with the use of levers, log chains, and six or eight ox teams, the wagons were lifted over the face of the cliff. With inconceivable labor, several other barriers scarcely less difficult were conquered in the same way.

The highway climbs to the summit of DONNER PASS (7,135 alt.), 28.9 *m.* Here on an exposed point, where the wind seldom stops blowing, a UNITED STATES WEATHER BUREAU OBSERVATORY makes records of wind direction and velocity for the aid of air navigators.

In Summit Valley is LAKE VAN NORDEN (L), an artificial reservoir dotted with decaying stumps, and the winter resort of NORDEN (*hotel*), 30.2 *m.* (6,880 alt.), an ideal locale for winter sports because of its heavy snowfall, long season, and open slopes. SODA SPRINGS, 31.7 *m.* (6,784 alt.), is a handful of rustic resort cabins around a hotel.

Left from Soda Springs on a dirt road, through a mountain wilderness, to a junction with a dirt road, 8 *m.;* L. here 2 *m.* to SODA SPRINGS (5,975 alt.), long known as Hopkins's Springs, the resort that Mark Hopkins and Leland Stanford opened in the 1870's. A trail leads upstream about 1 *m.* to PAINTED ROCK, on the south side of the river, which shows prehistoric Indian pictographs.

At 18.2 *m.* is a junction with a dirt road; R. from this junction 5 *m.* to LAST CHANCE (4,500 alt.). A group of prospectors pushed their way to this place in 1850 and lingered on, greedy for the rich gold deposits, until their provisions were exhausted. Staking all on a last chance, one of the men went into the forest and, as luck would have it, shot a large buck. The camp survived. By 1852 it had grown into a real town, and by 1859 it had three lodge halls. Remnants of its short-lived glory appear in the old hotel and the scattered cabins that still shelter a handful of miners in this isolated wilderness.

US 40 runs through the canyon of the South Fork of the Yuba River, past a string of vacation camps, inns, and public camp grounds. The stream cascades down broken granite slopes; deep among the bordering conifers are piles of gray-white granite, ground from the mountain flanks by ancient glaciers. Here and there tiny meadows, bright with alpine flowers, soften the austerity.

At 44.8 *m.* is a junction with State 20.

Right on State 20, which drops down gently between long lanes of fir and cedar to BEAR RIVER, 4.5 *m.,* flowing through a narrow meadow at the upper end of Bear Valley (4,500 alt.) among high mountains. In these rich grasslands trains of weary, half-starved immigrants paused to refresh their gaunt livestock. The valley is the summer range for cattle from the dry foothill country and the scene of the annual "stampede" of the Flying J Ranch.

At Bear River is a junction with a dirt road; R. here 0.2 *m.* to a junction with a second dirt road; R. again to LAKE SPAULDING, 1.2 *m.* impounded by a stone and concrete dam built in 1912. In a huge cave blasted from the canyon wall stands the power house. A tunnel bored through rock carries the lake water to the electric turbine and back to Drum Canal, where it flows down Bear Canyon to the foothill orchards about Auburn. North of Lake Spaulding are a string of man-made lakes that drain into Lake Spaulding. Since the gold days these reservoirs have been turned to many uses: sluice mining in the 1850's, large-scale hydraulic mining in the 1860's and 1870's, irrigation of farms and orchards from the 1850's on, and generation of electric power since the turn of the century.

Westward, US 40 passes between the high earth embankments of YUBA PASS CUT, and coasts down the crest of the ridge into the meadows of Wilson Valley. On both sides the slopes fall away, R. into Bear River Valley 600 feet below, L. into the wooded canyon of the North Fork of the American River. Through this notch in the ridge, EMIGRANT GAP, the North Fork once spilled into the Yuba, but the upheaval of the Sierra Range lifted up the giant causeway between them; over this the highway now runs.

At 49.1 *m.* is a junction with an improved road.

Left on this road to EMIGRANT GAP (*cabins, hotel*), 0.5 *m.* (5,250 alt., 164 pop.). Up the steep canyon sides from the railroad track are mill buildings, lumber yards, and shanties. At the western edge of Tahoe National Forest, the village is a popular fishing and hunting center.

Emigrant Gap marks the boundary between the High Sierra and the gold-veined foothills. For 40 miles southward the highway rides the crest of the long, tapering ridge, affording magnificent panoramas of steep wooded canyons and gorges. Files of conifers appear on the receding crests.

At 58.4 *m.* is the junction with a graded road.

Right on this road is ALTA, 1.7 *m.* (3,602 alt., 113 pop.), a railroad station, where old-fashioned white cottages stand among apple orchards near a tiny lake. The road passes between broken picket fences, behind which are old houses surrounded in spring by bowers of blossoming fruit trees, purple lilac, and scarlet quince. On the heights above the village, summer visitors have built houses.

DUTCH FLAT, 3.5 *m.* (3,399 alt., 90 pop.), played a vivid part in the history of the northern mines. Settled in 1851 by German miners, Joseph Doranbach and his companions—"Dutchmen" in miners' parlance—it was soon crowded with thousands who flocked in to the placer deposits. It was a Dutch Flat man, Daniel W. Strong, who pointed out Donner Pass to Theodore Dehone Judah, young Central Pacific engineer, and Dutch Flat subscribed money generously for the railroad. After 1863, when the railroad "Big Four" opened their Dutch Flat and Donner Lake Wagon Road, Concord coaches and teamsters stopped at hostelries here. Dutch Flat had lost its importance as a stage center by the fall of the next year, when the railroad stretched 20 miles beyond to Cisco; but the spectacular hydraulic operations of the next decade, when one company worked as many as 32 claims at once on a gigantic scale, kept millions of dollars in gold passing over its counters.

CHINA STORE (L), a massive, fortresslike structure with tiny porthole windows, half stone and adobe and half wood, is all that remains of a Chinatown that housed 1,000 coolies of the railroad construction days.

Right from Dutch Flat a road winds through the DUTCH FLAT HYDRAULIC DIGGINGS, a waste of rugged, man-made ravines choked with rocks and rubble. In these many-hued stony acres, hexagonal crystals, fossils, and even at times a gold nugget, are found.

Crossing Bear River, the road climbs a steep canyon wall to LITTLE ROCK, 2.8 *m.* (2,900 alt.), once a mining town, YOU BET, 6 *m.* (3,000 alt.), and RED DOG, 7.7 *m.* (2,850 alt.), around which 2,000 miners dug for gold in the 1870's.

Descending from the dense forest between banks of reddish earth and crumbling rock, the highway, at 61.7 *m.,* cuts into a deep gash along the face of bluffs tinged with vivid hues of russet, buff, and rose red. Over the canyon bottoms (L) sprawl mounds of reddish earth and gravel, half overgrown with scrub pines; beyond rise crumbling pinnacles and blasted palisades. From the river channels of this region, where streams of the Tertiary era deposited a bed of auriferous gravel, 250 feet deep, 2 miles long, and a half mile wide, was washed out more than $15,000,000 worth of gold during the titanic hydraulic operations of the 1870's and 1880's.

In the midst of rose-streaked debris is GOLD RUN, 63.1 *m.* (3,224 alt., 114 pop.), active in the 1860's and 1870's. The log cabin on stilts beside the highway (L) at 63.7 *m.,* is the GOLD RUN TRADING POST, now a museum for pioneer relics.

US 40 runs through narrow defiles, where interlacing branches hide the sun. The forest of pine and cedar begins to thin out, making way for spreading black oaks, with tattered, yellow-green foliage, and for bushy gray-green manzanita.

COLFAX, 73.2 *m.* (2,422 alt., 912 pop.), is strung out along railroad sidings where boxcars wait by warehouses. On the hillsides are old-fashioned cottages. Nearby are orchards of Bartlett pears and Hungarian prunes, and vineyards of Tokay grapes. Called Alden

Grove by its settlers of 1849, the place was renamed Illinoistown in the early 1850's, and Colfax about 1869. During those turbulent years it was a head of "wagon navigation" to the gold mines; here goods were transferred to muleback for the journey to remote camps.

At 73.8 *m.* is a junction with a dirt road.

Left on this road, down an easy grade, to MINERAL BAR, 5 *m.* (1,200 alt.), on the North Fork of the American River, where Judas trees grow in moist gullies of the steep canyon walls. The river rushes past abandoned diggings. When summer dries the river to a low-water level, a few miners still wash out a little yellow dust where scores once made their "piles."

Steep and narrow, the way leads up the canyon wall to IOWA HILL, 10 *m.* (3,200 alt., 100 pop.), high on the neck of a ridge. Only one of the old brick buildings on the main street has survived the ravages of fire, and only a handful of white frame cottages remain to look primly out from lilac trees and hollyhocks. Over the grassy slopes above the village spread the remnants of deserted gardens and orchards; at the top of the hill, delicate manzanita bells fall softly on moss-grown headstones in the old cemetery. All about are signs of the treasure-trove which, first discovered in 1853, had yielded an estimated $20,000,000 by 1880. Long ago the rich conglomerate of the Blue Lead Channel, which ran under the town, was drifted out; tremendous hydraulic onslaughts all but washed the townsite away.

US 40 traverses an area of tiny orchards in forest clearings; the pears and apples grown here have unusually fine flavor and texture. The old type of diversified farming, which has all but disappeared from much of the State, lingers here in the small holdings given over to orchards, vineyards, and pastures for dairy cattle.

AUBURN, 90.5 *m.* (1,360 alt., 2,661 pop.) (*see TOUR 4a*) is at a junction with State 49 (*see TOUR 4a*).

At 91.8 *m.* is a junction with an oiled road.

Right on this road is OPHIR, 1.5 *m.* (850 alt., 250 pop.), which began life as Spanish Corral, but exchanged it for the name of King Solomon's treasure trove. In 1852 the most populous town in Placer County, Ophir is still the trade center of its chief quartz-mining district. Among the nearby orchards and vineyards are the scars of old diggings, abandoned mining pits, dumps, and stamp mill foundations.

Diving into a long tunnel US 40 runs under NEWCASTLE, 94.6 *m.* (970 alt., 750 pop.), a hilltop town. Newcastle was the only one of the many camps at the head of Secret Ravine to survive. As orchards replaced the spent placer mines in the 1870's and 1880's, it became a fruit-packing and shipping center.

At 97.3 *m.* is a junction with a paved road.

Right on this road is PENRYN, 0.2 *m.* (635 alt., 300 pop.), which wears (under slightly altered spelling) the name of the Welsh town, Penrhyn, from which came Griffith Griffith, who opened a granite quarry here in 1864. Penryn is today a fruit-shipping center.

LOOMIS, 99.9 *m.* (400 alt., 319 pop.), is the successor of Pine, which took its name from Pine Grove in nearby Secret Ravine, where mining began in 1850. Along the highway are strung squat business buildings, and along the railroad, the great fruit warehouses.

As the last orchard-mantled slopes of the Sierra foothills taper off, US 40 strikes over a level straightaway across the edges of the Sacramento Valley. Ahead lies open country, spotted with clumps of cottonwoods and willows. Into the distance sweep grassy knolls, dappled with groves of immense spreading oaks. This parklike oak-covered country, California's most characteristic landscape, encircles the great Central Valley and spreads over much of the Coast Range.

A large percentage of the population of ROCKLIN, 102.8 *m.* (248 alt., 724 pop.), is foreign-born—predominantly Finnish. The Finns maintain their own library and choral society, and work in the granite quarries from which Rocklin takes its name.

ROSEVILLE, 106.7 *m.* (163 alt., 13,421 pop.), at the junction with US 99E (*see TOUR 3B*), is a freight-clearing center clustering about the extensive Southern Pacific shops and yards. Its shady side streets of frame dwellings are lined by olive, maple, and fig trees. Heart of an area that grows plums, berries, almonds, and grapes, it ships large quantities of table fruits to Eastern markets.

NORTH SACRAMENTO, 124.9 *m.* (34 alt., 12,922 pop.), is evidence of the expanding Sacramento community.

US 40 continues southward, and at 125.7 *m.* spans the tree-lined American River, on whose south fork James W. Marshall discovered gold in 1848.

SACRAMENTO, 124.9 *m.* (34 alt., 249,300 pop.) (*see SACRAMENTO*). Sacramento is at junctions with US 99 (*see TOUR 3a*) and State 24 (*see TOUR 9A*).

Section b. SACRAMENTO to SAN FRANCISCO, 95 m. US 40; Interstate 80.

Southwest of the State capital, the Victory Highway crosses the broad farmlands of the Sacramento Valley, passes through the low, orchard-covered foothills of the Coast Range, and follows the shores of San Pablo and San Francisco Bays into San Francisco. The first Yankee immigrants found much of the land already claimed by Mexican dons, but this did not hinder their attempts to possess it. After gold seekers swarmed in from coast to foothills in the 1850's, roads, railroads, and river boat lines were developed; later the land was divided into farms, pastures, and orchard plots. Along winding side roads in the quiet back country appear the crumbling adobes of the dons and the later frame mansions of the Yankee invaders; but the main highway runs through a countryside embellished with signboards, gas stations, and soda pop stands, through vast corporation-controlled farms, and through bleak company towns at factory gates.

At the western end of SACRAMENTO, 0 *m.*, is the 737-foot-long TOWER BRIDGE, a vertical lift span over the broad Sacramento River. Excavation for the bridge abutment on the city side brought to light, 20 feet underground, the tracks of the Central Pacific, first transcontinental railroad. The Sacramento, California's largest river, served for decades as the main artery of travel between the early San Francisco and this valley. The great flat-bottomed, paddle wheel river boats of rival lines raced each other from port to port, sometimes tying down safety valves until steam pressure rose to the explosion point. The Combination Line's *Pearl,* with 93 aboard, was racing the Citizens' Line's *Enterprise* on January 27, 1855, when her boilers exploded near Sacramento, killing 56 persons. Similar disasters on the *Washoe,* the *Yosemite,* and the *Belle* finally dampened the competitive spirit. In the end, the railroad put most of the river steamers out of business: today only one line operates between San Francisco and Sacramento.

At 2.1 *m.* is a junction with a paved road.

Right on this road is BRODERICK, 0.5 *m.* (550 pop.), which was the seat of Yolo County for two brief periods (1851-57 and 1861-62), but lost the honor under stress of fire, flood, and political storm. Chief point of activity today is the waterfront, with launches and fishing craft tied up near boathouses and cabins on the riverbank; here old river steamboats, the *Red Bluff* and the *Dover,* lie beached. A weatherbeaten GREEK ORTHODOX CHURCH testifies to the many Russians who settled here more than a century ago.

Through the roadside clutter of city outskirts—billboards, auto camps, gas stations, fruit and vegetable stands, truck garden plots, and ranchers' shanties—US 40 strikes out across the level floor of the Sacramento Valley.

Yolo Bypass, 5.5 *m.,* is a 3-mile causeway over desolate, tule-grown marshlands—in springtime carpeted with yellow baeria, in autumn the haunt of wild geese. The deeply rutted road that formerly meandered over these swampy acres was impassable in spring when the swollen Sacramento flooded it with a muddy torrent. So yielding was the terrain that the railroad, paralleling the highway on a high embankment, used to sag until the tracks looked like a sway-backed horse. Unnumbered tons of rock had to be dumped into the swamp before the engineers won their fight. Today, when the river threatens to break its levees, flood gates are thrown open and this whole region is inundated.

The Sacramento Valley, level as the Midwest prairies, stretches ahead, an irregular checkerboard of tomato patches, strips of blue-green alfalfa, immense tracts of sugar beets, fields green in spring with waving grain and tawny with stubble in autumn. Scattered farm buildings huddle among trees.

By 1856, at DAVIS, 14.8 *m.* (54 alt., 8,910 pop.), Jerome C. Davis had 400 acres in wheat and barley, great herds of livestock, and orchards and vineyards.

The drooping feathery branches of gray-green tamarisks hide the fields and orchards of the University of California's DAVIS BRANCH OF THE COLLEGE OF AGRICULTURE. From the entrance (L), 15.4 *m.,* a driveway leads to a cluster of classrooms and dormitories, athletic facilities, a model dairy, and pedigreed-livestock and poultry pens. The

1,076-acre farm includes 150 acres of orchard with more than 1,000 varieties of fruit, 27 acres of experimental vineyards, 250 acres planted with forage crops, 130 acres of pasture, and a 10-acre poultry plant.

At 16.2 *m.* is a junction with US 99 (*see TOUR 3a*).

DIXON, 26.3 *m.* (67 alt., 2,970 pop.), is a farming and dairying center, named for its founder, Thomas Dickson, whose name it soon forgot how to spell. The town grew up soon after the advent of the first transcontinental railroad, superseding the village of Silveyville, on the stage road to the gold mines.

Swinging southwest toward low tumbled hills, US 40 is skirted by neatly trimmed fields, orchards and pastures.

At 36.7 *m.* is a junction with a paved road.

Right on this road is VACAVILLE, 0.7 *m.* (166 alt., 10,898 pop.), on gentle knolls, where orchards adjoin rear gardens or look down on housetops. The warm, frostless slopes of hill-sheltered Vaca, Plesants, and Laguna Valleys, each spring produce early cherries, apricots, peaches, plums and pears for the eastern market. Before 1850, when Vacaville was founded, the land roundabout was the range for the great cattle herds of Rancho de los Putos. As Yankee settlers began to arrive the lands were broken up into small farms and orchards. In this region are laid the opening scenes of Charles G. Norris' novel, *Brass*. On Merchant St. stands (L) the old frame PENA HOUSE; in the yard gnarled and blackened fig trees mingle with oleanders and aged pear, walnut, and poplar trees. Vacaville's Chinatown is a survival of the 1880's and 1890's, when Chinese workers flocked in to work in the fruit industry.

US 40 winds up into the smooth Vaca hills. Along the sheltered valley bottoms are apricot orchards, fragrant with delicate blossoms in spring, heavily laden with ripening fruit in summer. With the coming of the first winter rains, the parched hillsides are mantled with tall grass—lush grazing for sheep and cattle.

In this region, the poet, Edwin Markham, best known for his "The Man with the Hoe," spent his boyhood and early youth (1861-1870) as sheepherder and vaquero on his mother's cattle ranges. Here he plowed "the little valleys between the ridges for wheat and barley," and followed the "threshing machine in the time of harvesting." Somewhere in the vicinity, he shared in the "exhilarating spectacle" of the rodeo held yearly by the cattle king of the region.

The PENA ADOBE (L), 40.1 *m.,* sits in a garden under the shade of the fig and walnut trees, the pomegranate, the mission grape, and the Castilian rose cherished by early Spanish settlers. The old structure is now under a board siding.

US 40 descends and crosses level Suisun Valley. At 45 *m.* is a junction with State 12. Left on State 12 2 *m.* to FAIRFIELD (12 alt., 14,968 pop.), seat of Solano County, trade center of farmers and fruit raisers. Two sea captains, Robert H. Waterman and Archibald A. Ritchie, founded the town in 1859 and obtained the transfer of the county seat from Benicia by donating land for a townsite.

Right from Fairfield on Union Ave. to Suisun City, 0.8 *m.* (12 alt., 1,000 pop.) once connected with Fairfield by a mile-long plank road over a swamp, today divided from it only by the railroad tracks. All the way to this inland point, up winding Suisun Slough through the great salt-water marshes north of Suisun Bay, watercraft sailed in the 1850's to the landing place for Mexican ranchos. The old landing (L), near the Plaza, is still owned by a grandson of Capt. Josiah Wing, who began running watercraft here in 1850, erected a warehouse two years later, and in 1854 laid out the town. The clapboarded WING HOUSE, built in 1858, stands on a narrow street. Today canneries and packing plants, taking advantage of shipment both by water and rail, cluster here. The salt-water marshes, owned by sportsmen's clubs, offer excellent duck shooting and bass fishing.

Where today's wide roadbed runs westward past trim white farmhouses, the earliest travelers across Suisun Valley followed a dim trail through wild oats so tall that riders on horseback were hidden by them. Sometimes wild Spanish cattle attacked the wayfarer, or antelope and elk broke into a wild stampede.

The bronze STATUE OF FRANCISCO SOLANO is on a low knoll (R), 51.1 *m.* Solano was overlord of all the Indian tribes north of Suisun and San Pablo Bays. Born Sem Yoto, chief of the Suisune, he was baptized with a Spanish name by the fathers of Misión San Francisco de Solano at Sonoma, from whom he learned many of the arts of the white men. At the time Lieut. Mariano G. Vallejo conquered the Suscol in 1835, he won the friendship of Chief Solano, who aided him in the peace deliberations. On Rancho Suisun, granted to him by the Spaniards, Solano built a large adobe for himself and smaller ones for his people; the Yankees who passed this way in 1846 and 1847 found it occupied by the Indian, Jesús Molino, and surrounded by fields of peas, wheat, and other crops. His alliance with Chief Solano, who could muster a force of 1,000 plumed and painted warriors, was Vallejo's chief protection for the struggling new outposts of Spanish civilization against the depredations of the natives. When Solano County was organized in 1850, it was named at Vallejo's request for his Indian ally, whom he admired and trusted as a friend. "A splendid figure of a man" (he was 6 feet 7 inches tall), as Vallejo described him, Solano was "a keen, clearheaded thinker, readily grasped new ideas, learned to speak Spanish with ease and precision, and was so ready to debate, that few cared to engage with him in a contest of wits." The bronze likeness represents him as an Indian of magnificent physique, with one hand upraised in a gesture of peace—somewhat riddled by the potshots of latter-day hunters.

At 51.9 *m.* is a junction with a paved road.

Right on this road is ROCKVILLE, 1.6 *m.* with a country store. Nearby, a substantial stone METHODIST EPISCOPAL CHURCH, erected in 1856, stands among flowering shrubs and trees in a graveyard.

Left from Rockville, 0.5 *m.*, to the great stone MARTIN HOUSE (R), with peaked roof and gables, projecting eaves, and deep window sills, in the style of an English manor-house. In 1850 Samuel Martin arrived, having crossed the plains from Pennsylvania with his family in a covered wagon, with men on horseback driving his herds. He gradually acquired land until he owned an estate of 11,000 acres. In the year Martin's party arrived, the last of the Suisune, whose chief village stood here, were retreating into the mountains, carrying sacks of grain on their heads. On a wooded knoll back of the mansion are great, deeply embedded boulders that reveal mortar holes worn smooth during the grinding of grass seeds and acorns. The Suisune were an independent and self-reliant people, as the Spanish soldier, Gabriel Moraga, sent to subdue them in 1810, discovered. Many of them, retreating after a bloody encounter set fire to their huts and perished in the flames. Again, when José Sanchez led a second attempt at subjugation in 1817, they followed their chief, Malaca, in a resistance that was defeated only by the superior weapons of the Spanish. In a field (L) opposite the Martin House was a Suisune Burial Ground; a buckeye tree marks the GRAVE OF FRANCISCO SOLANO, who succeeded Malaca.

At 53.2 *m.* is a junction with State 12 (*see TOUR 2A*).

Climbing out of Suisun Valley, US 40 winds into AMERICAN CANYON, a shallow cleft through low, rounded hills dappled with clumps of oak and bay. Grass-covered in spring, these hill slopes give pasture to herds of sheep and cattle, which leave them close-cropped and bare throughout the summer and fall. As the highway leaves the hills, the tang of salt air greets the nostrils. Beyond is a panorama of Vallejo's city blocks (*see TOUR 2A*), topped by the steel masts of the Naval Radio Station, the silver Napa River, and Mount Tamalpais beyond San Pablo Bay.

At 62.3 *m.* is a junction with an oiled road (*see TOUR 2A*).

At 63.1 *m.* US 40 meets a paved road.

Left on this road to SOUTHAMPTON BAY, 3.7 *m.,* where the James Corbett-Joe Choynski prize fight was staged June 5, 1889, on a barge. The encounter was not quite an engagement with bare fists, for Corbett's hands were encased in 3-ounce mitts, Choynski's in driving gloves. Although Corbett broke both hands early in the fight, he went on pummeling Choynski until he knocked him out in the twenty-eighth round. Both had to be carried from the barge. Corbett's seconds were Thomas Williams and Porter Ashe; Choynski's were Nat Goodwin, Eddie Graney, and Jack Dempsey, the "Nonpareil."

BENICIA, 5.5 *m.* (6,070 pop.), along the northern shore of Carquinez Strait, at the base of low hills that slope down gently to the water's edge, is a monument to Gen. Mariano Vallejo's town-founding passion. It rose on his Rancho Suscol, which he deeded in 1846 to Dr. Robert Semple, editor of the *Californian,* and Thomas O. Larkin, only United States consul to Alta California (1844-46). Named Santa Francisca in honor of Vallejo's wife, the town dreamed for a brief interval of becoming the metropolis of San Francisco Bay; but when the citizens of Yerba Buena appropriated the name of the bay, Vallejo and his associates fell back on Señora Vallejo's second name, Benicia. In 1849, Benicia was a thriving waypoint on a main traveled road to the mines; in 1853-54 it was important enough to be the State capital. An army post, established here in 1849, and later an arsenal, made it a military headquarters of the Far West. When the Pacific Mail Steamship Company built wharves and shops along the deep-water harbor, it became a coaling and repair stop for river boats. As miners drifted back after the gold rush and settled in the valleys, Benicia became a quiet, orderly town, proud of its schools and churches.

Today its citizens preserve the landmarks of its past. Each year, during the Old Timers' Festival, a marker is placed to commemorate some feature of early days. California's OLDEST MASONIC HALL (1851) stands (R) on West J St. near 1st St.; it housed the State Legislature on its ground floor when the town was for a short time the State capital.

ST. PAUL'S EPISCOPAL CHURCH (1885), stands on the site of the first Episcopal cathedral (1860) of the Diocese of Northern California. On I St. near 1st St. is the SITE OF THE BENICIA YOUNG LADIES' SEMINARY, founded in 1852; in the 1860's it was bought by Dr. and Mrs. Cyrus Mills who changed the name to Mills Seminary, and moved it to Oakland.

The STATE CAPITOL, restored, is an example of the Greek influence in formal architecture. Originally erected in three months to be ready for the Legislature on Jan. 1, 1853, it cost $24,800; in 1956-57 it was restored by the State Division of Architects for $230,000. It has the form of a Greek temple with two-column porch set flush with the front elevation. The walls are salmon brick, now reinforced by concrete. Original interior wood columns were made from ships masts. The original sheet iron and tin roof has been replaced. After the Legislature moved it was the Solano County Court House, then school, city hall and police station.

Typical of California's early inns is the SOLANO HOTEL (R), corner of 1st and E Sts., stopping place for famous men and women of the 1850's and scene of gay social events. The VON PFISTER ADOBE STORE (R), near 1st and D Sts., was built in 1847; here, where a certain Mr. Von Pfister set up shop in 1848 with merchandise brought from Honolulu, Charles Bennett announced James Marshall's discovery of gold.

The United States Government located an arsenal at the foot of M St., in 1851, where a large square clock tower of sandstone became a familiar landmark after 1869. In recent decades a large supply depot was developed with railway sidings and warehouses. When the government evacuated the depot in 1961-64 the city of Benicia bought the site and zoned it for an industrial park. About 350 buildings, many modern, are on the property. There is a dock area of 6 acres with deep water wharf facilities and 500 acres will be developed for shipping uses. An airport of 55 acres for executive planes is part of the plan.

The MUNICIPAL PARK, 1st St. between K and L Sts., is on the site of California's first Presbyterian Church, organized April 1849. California's first convent school was ST. CATHERINE'S SEMINARY, L St. near 1st St., established at Monterey in 1851 by Dominican sisters and moved to Benicia in 1854. BISHOP WINGFIELD'S HOUSE, M and 2nd Sts., at the end of a tree-lined drive was one of the buildings of St. Augustine Episcopal College. ST. DOMINIC'S CHURCH, I and 5th Sts., was organized in 1854 by Father Villarosa, Dominican priest.

Left from 5th and M Sts. in Benicia on State 21 to ST. DOMINIC'S CEMETERY, 0.5 *m.*, where Doña María Concepción Argüello, who taught in St. Catherine's Seminary from 1854 until her death in August 1857, lies buried. Doña Concepción, immortalized in poetry by Bret Harte and in the novel *Rezanov* by Gertrude Atherton, was not quite 16 when handsome Russian Count Rezanov came to visit her father, Don Luís, comandante of the San Francisco Presidio; Rezanov's mission was to secure food, potatoes especially, for the Russian settlement at Sitka, Alaska. He fell in love with Doña María Concepción and she with him; but Rezanov said the Tsar's consent to the marriage had to be obtained and he set off on the long journey across the Pacific and over Siberia to St. Petersburg. Years passed without word from him, but the girl waited. Giving up hope, she had joined the Dominican Sisterhood before word came back that Rezanov had died on the way home.

The BENICIA BARRACKS (R), 0.8 *m.* in the old Army reservation that included the United States Arsenal stands on a site chosen by Gen. Persifer Smith in 1849. Established in April of that year as State Army headquarters under command of Commodore Robert Stockton, they served the post until 1908.

At 64.5 *m.* is a junction with State 29 (*see TOUR 2A*).

At 65.2 *m.* is the tollhouse of CARQUINEZ CANTILEVER BRIDGE.

The original span was completed in 1927 at a cost of $8,000,000. The sum was considered excessive at the time and the inexperience of the builders, Avon Hanford and Oscar Klatt, was blamed. Hanford died in 1926 before the bridge was completed. A parallel span was erected in 1958 to take care of the enlarging traffic. The approaches are 4,482 ft. long; the towers rise 325 ft. above the water; the central span is 1,100 ft. long.

Through narrow Carquinez Strait, 8 miles long and 1 mile wide at its narrowest point, pour the waters from the Central Valley's two great rivers, the Sacramento and the San Joaquin. Bayard Taylor, visiting California in the gold days, compared the strait to the Bosporous, but said it had a greater natural beauty with its "bold shores" and its "varying succession of bays and headlands on either side." Through it passed the immense river traffic to the gold mines; today it carries products of the orchards, dairies, and grain farms of the hinterland.

Beyond the southern end of Carquinez Bridge rise the Contra Costa hills, bearing the name which the Spanish gave to the whole region lying across San Francisco Bay.

At 66.1 *m.* is a junction with a paved road.

Left on this road is CROCKETT, 0.5 *m* (0-100 alt., 4,800 pop.), spreading over a ravine and up the hillsides from the shore. A company town, Crockett huddles about the concrete CALIFORNIA-HAWAIIAN SUGAR REFINING CORPORATION PLANT (*visiting hours 10-1*). One of the fleet of great seagoing freighters, which bring 500,000 tons of raw cane sugar annually from the Hawaiian Islands, usually lies unloading at the dock. On the shell mound left behind by prehistoric aborigines, who hunted shellfish in a sheltered cove here, the first American settler, Thomas Edwards, built in 1867 the first house, a rough board and batten structure, on his 1,800-acre ranch. The town was laid out with the advent of the railroad in 1877 and christened with the name of a one-time California Supreme Court Justice. In 1882 a foundry was built, and two years later the Starr flour mill, largest on the Pacific Coast at the time. This, superseded by a beet sugar refinery in 1898, became in 1906 the first unit of the C-H sugar refinery.

At Crockett is a junction with State 4 (*see TOUR 9A*).

A parkway (R), 66.4 *m.*, looks out over the cool gray waters of San Pablo Bay. At the foot of the bluffs, beyond the small wharves that jut out from shore, the tiny craft of bass fishermen bob in the swift tide. Pleasant as the scene is, Capt. Pedro Fages and Father Juan Crespi, the first Europeans to see it, March 30, 1772, gazed northward in dismay, for the stretch of water blocked their search for a land route to Point Reyes (*see TOUR 1a*). Both they and the men led by Capt. Juan Bautista de Anza, who arrived here April 2, 1776, watched with interest the Indian fishermen crossing the bay in rafts from their fishing coves on the southern shore to their villages on the opposite side.

Following the contours of receding highlands, US 40 runs now in a southwesterly direction through the succession of factory towns that string in a continuous chain along the shores of San Pablo Bay, where Mexican vaqueros once herded cattle.

At 67.7 *m.* is a junction with a paved road.

Right on this road is SELBY, 0.2 *m.* (0-100 alt., 250 pop.), owned and operated by the American Smelting and Refining Company. Since 1885 the Selby smelter has produced many millions of dollars worth of gold, silver, lead, and antimony. Although lead is its main product, gold from California's Mother Lode and silver from Nevada's Comstock Lode have also been smelted here for many years.

OLEUM, 68 *m.* (0-100 alt., 217 pop.), a Union Oil Company town, huddles about a great refinery in the midst of gigantic oil tanks, at night lit up by great floodlights. Long trains of tank cars show black against the silver sheen of the bay. On the waterfront, oil cargoes are loaded and discharged at five modern piers. There are 900 tanks and the plant can refine 35,000 barrels of petroleum products daily.

RODEO, 69 *m.* (12 alt., 6,500 pop.), preserves in its name the days of Spanish dominion when ranchers held their yearly cattle rodeos in nearby Rodeo Valley. The town of Rodeo sprang up when two Irishmen, John and Patrick Tormey, started a meat-packing plant here on a 9,000-acre cattle ranch.

At 72 *m.* is a junction with an improved road.

Right on this road is HERCULES, 0.3 *m.* (8 alt., 365 pop.), overlooking San Pablo Bay from parklike hilltops. In the ravines, which run down to the water's edge, stand the powder mills, acid house, mixing and packing house, and magazine of the Hercules Powder Works, each stationed in a separate gully as a result of a safety campaign occasioned by disastrous explosions. Established here in 1869 to manufacture dynamite for the mines, the plant covers 3,000 acres. At peak capacity it manufactures a quarter of a million pounds of explosives each month.

PINOLE, 72.4 *m.* (11 alt., 10,230 pop.), is on the former Rancho el Pinole, christened according to legend with the name that a little group of half-starved Spanish soldiers, who struggled out of the Canadá del Hombre to the bayshore, bestowed on the meal made out of acorns and grass seeds that friendly Indian squaws gave them to eat. Ignacio Martínez, comandante of the San Francisco Presidio 1822-32, had occupied these lands by 1829, long before they were officially granted to him in 1842; and he continued his aggressive cattle-raising activities under the American flag. A young Englishman, Dr. Samuel J. Tennant, physician to the king of the Hawaiian Islands, landed at Don Ignacio's embarcadero on one of his gold-prospecting voyages to the interior in 1849. Tennant was so captivated by the beauty of young Señorita Rafaela that he gave up both Hawaii and gold to marry her here Sept. 8, 1849. After Don Ignacio's death, when the Rancho el Pinole was divided among his children, Tennant, through his wife, came into control of a vast part of it, and later laid out the town of Pinole.

EL SOBRANTE, 2 *m.* south of Pinole on State 40 (Interstate 80) has expanded to 12,400, est., population since World War II. Nearby are the giant oil tanks of a huge Standard Oil Company tank farm. Although it produces no oil, this region is an important oil storage center. From fields, the nearest of them 250 miles away, oil is brought by pipe line, tank cars, and tankers.

At 74.8 *m.* is a junction with a paved road.

Right on this road is GIANT, 1 *m.* (10-30 alt., 90 pop.), the company-owned town of the Giant Powder Company, on a peninsula jutting into San Pablo Bay. Half-hidden by the eucalyptus groves of Giant Park are the mills and warehouses. The Giant Powder Company, incorporated in 1867, was the first company in California to manufacture dynamite; for a number of years it held exclusive American rights to the manufacture under Nobel patents.

When De Anza and his expedition passed San Pablo Creek, 77 *m.*, on April 1, 1776, they found an Indian village on its banks, deep among live oaks and sycamores. The friendly inhabitants turned out to gape at the strange procession. Behind the group's chaplain, in his brown Franciscan habit, chanting the *Alabado* (a Spanish hymn in praise of the sacrament), followed De Anza at the head of a cavalcade of soldiers on horseback, muleteers driving beasts of burden, and Indian servants. In exchange for gifts of glass beads, the Indians offered a feast of roasted cacomites, a species of wild iris.

SAN PABLO, 77.5 *m.* (35 alt., 489 pop., 1930; 19,689, 1960; 39,858, 1965, est.), lies between San Pablo and Wildcat creeks. Here, before 1820, Misión San Francisco de Asís pastured great herds of cattle and sheep and built an adobe ranchhouse for the overseer. The whole region roundabout was part of the Rancho San Pablo of Francisco María Castro. Castro, who died in 1831, three years before the grant was confirmed, left half the estate to his wife and half to his 11 children, precipitating a title controversy.

San Pablo Creek and tributaries east of the San Pablo Ridge furnish water for SAN PABLO RESERVOIR.

The CASTRO ADOBE, built about 1838 by Don Francisco's son, Antonio, stands at the northwest corner of US 40 and Church St. Here, from 1849 until his death in 1882, lived Juan Bautista Alvarado, Mexican governor of Alta California (1836-1842), husband of Martina Castro. The little house, now a storeroom, has 30-inch walls that have been hidden by a wooden superstructure. The rear is a remnant of the orchard that Alvarado planted; it still sends forth a shower of pear and apple blossoms in the spring.

RICHMOND, 79.5 *m.* (12 alt., 71,854 pop., 1960; 79,800, 1963 est.), stretches westward over a headland jutting into San Francisco Bay. An industrial city on a deep-water harbor, Richmond is Contra Costa's outlet for farm and factory produce. It was established in 1899, when the Santa Fe Railway purchased a right-of-way to the bayshore here, and grew up around branch plants of such industrial corporations as the Standard Oil Company of California, which employs 3,768; Beckman Instruments, and numerous ship maintenance facilities.

The PORT OF RICHMOND handles nearly 18,000,000 tons of cargo annually, ranking second in California, after Los Angeles, and 15th in the nation. Interstate 80 passes through Richmond going east and State 17 uses the RICHMOND-SAN RAFAEL BRIDGE (built 1957, cantilever, span 1,070 ft.) from Richmond to Pt. San Quentin.

Richmond's Inner Harbor Channel, at the foot of 10th St., where George Ellis, schooner-operator, opened Ellis Landing in 1859, has been developed for the use of industries by dredging and filling.

In Henry Kaiser's shipyard cargo vessels were built by thousands of workers during World War II. The area is now occupied by the Kaiser Aircraft & Electronics plant.

Separated from the downtown area by a thinly settled, industrial section reaching southward, is POINT RICHMOND. From here RICHMOND POINT juts into the Outer Harbor with a 6-mile deep-water frontage sheltered by a sea wall. Standard Ave. leads north from Point Richmond to the STANDARD OIL REFINERY.

EL CERRITO (the little hill), 80.6 *m.* (5-1,000 alt., 25,437 pop. 1960; 27,900, 1965 est.), adjacent to the Eastshore Freeway, is a residential community between Richmond and Berkeley. Many older bungalows have made way for multiple dwellings, but the hilly area, 250 acres, has one-family houses. At the El Cerrito embarcadero, in early days, travelers landed from Don Victor Castro's whaleboat ferry, first to connect San Francisco with Contra Costa; Don Victor acquired it in exchange for his vegetable crop. Travelers were entertained in the 14-room CASTRO ADOBE, which stands in a eucalyptus grove on San Pablo Ave. Its older part, the low south wing, is said to have been built by Don Francisco himself before 1831. In the 1850's his son Victor built the two-story, balconied central section. Walls four feet thick guard a fountain, a winding staircase, and a tiny chapel, where mission fathers sometimes said mass (now altered beyond recognition).

El Cerrito Creek, 82.3 *m.*, forms the boundary between Contra Costa and Alameda Counties and between the cities of El Cerrito and Albany. On Aug. 16, 1820, Sgt. Luís Peralta, his sons, Domingo and Antonio, his friend, Lieut. Ignacio Martínez, and two or three soldiers, arrived here on horseback. "Unto this point, Señor, I wish possession," said Peralta to Martínez, the Governor's representative. Thus the Arroyo del Cerrito de San Antonio (Sp., gulch of the little hill of St. Anthony) was decreed the boundary between Rancho San Pablo (*see above*) and Peralta's 48,000-acre Rancho San Antonio, within whose boundaries lie today the cities of Albany, Berkeley, Emeryville, Piedmont, Oakland, Alameda, and San Leandro.

ALBANY, 83 *m.* (0-300 alt., 14,804 pop.), incorporated in 1908 as Ocean View, one year later took the name of its popular mayor's New York birthplace.

BERKELEY, 84.6 *m.* (*q.v.*)

South of Berkeley, US 40 follows a high-speed roadbed laid across the brackish mudflats of the bayshore, where the stagnant odor of decay wafts inland. The grayish waters lap half-rotted pilings and sway small dories.

EMERYVILLE, 86.5 *m.* (0-60 alt., 2,686 pop.). Even before the advent of the Spanish, a populous settlement lay near here, as shown by the Indian shell mound. This mound, largest of the 425 in the San Francisco Bay region, yielded many artifacts when it was leveled for factory sites in 1924. Where generations of Indians plundered the great beds of shellfish at the mouth of Temescal Creek, Vincente Peralta before 1836 had built an embarcadero for Rancho San Antonio. In the late 1840's Yankees began to settle here on the Peralta acres; in later years the place became a popular racing center, the home of the California Jockey Club Track.

OAKLAND, 87.3 *m.* (*see OAKLAND*).

A turn here leads (R), through labyrinthine overpasses and underpasses, to a junction with US 50 (*see TOUR 10b*) at the entrance to the San Francisco-Oakland Bay Bridge (*see ARCHITECTURE*). From the plaza (*toll 25¢ an auto*), 88.6 *m.*, US 40 follows the great series of spans across San Francisco Bay, midway crossing Yerba Buena Island, through an unusually high double decker tunnel.

SAN FRANCISCO, 95 *m.* (*see SAN FRANCISCO*).

Tour 9A

Sacramento—Rio Vista—Antioch—Pittsburg—Concord—Oakland; 118.4 *m.* State 160. Junctions with State 24, 21, and US 40.

Roadbed paved.
Southern Pacific R. R. roughly parallels route between Antioch Toll Bridge and Walnut Creek.
Accommodations only in larger towns.

State 160 cuts across the southern end of the Sacramento Valley, following the top of the levee which hems in the Sacramento River's slow-moving waters. On one side, it overlooks the river's busy traffic; on the other, the rich farmlands of the reclaimed delta region—a crazy-quilt of "islands" encompassed by meandering sloughs and sheltered by dikes as in the low country of the Netherlands. It strikes through the industrial towns clustering at the confluence of the Sacramento and the San Joaquin Rivers and follows the windings of quiet, orchard-shaded valleys.

State 160 strikes southward from US 40 (*see TOUR 9*) in SAC-

RAMENTO, 0 *m.*, across level farmlands to FREEPORT, 9.3 *m.* (30 alt., 120 pop.), a village on the eastern bank of the Sacramento; then hugs the river's edge, traveling the crest of the levee.

The first known white men to see the Sacramento River were in the Pedro Fages expedition, whose explorations of the eastern shore of San Francisco Bay in 1772 carried them as far as the confluence of the Sacramento and San Joaquin Rivers, where "from a point of vantage," they saw the river's winding lower reaches. The first recorded navigation of the river was in 1811, when Jose Antonio Sanchez proceeded a little way upstream from the mouth in a small boat. The development of the great waterway into an important artery of travel and trade was left to the Argonauts of 1848, who pressed up from San Francisco on their way to the "diggin's" in vessels of every description—from tiny sail-craft to lumbering stern-wheelers. The ambitious realtors of the day, quick to capitalize on the growing stream of travelers, optimistically laid out a series of river towns at steamboat landings—all but a few of which scarcely outlived the flush years of the gold rush.

HOOD, 17.5 *m.* (25 alt., 150 pop.), is a shipping point on the Sacramento River for fruits and vegetables. COURTLAND, 21.7 *m.* (14 alt., 750 pop.), among pear orchards and truck farms, is a canning, packing and shipping center.

At 29.5 *m.* is the junction with a paved road.

Left on this road, across the river, to WALNUT GROVE, 0.2 *m.* (9 alt., 631 pop.), clustering under the levee. Walnut Grove's ramshackle CHINATOWN is a reminder of the days when the Chinese were reclaiming the Sacramento delta region. Today Japanese comprise the majority of the population.

The rich delta region lying southward is crisscrossed with dikes that hem in winding sloughs, sheltering the lush black-loam farmlands, often far below water level, from floods. It was settled in the early 1850's by disappointed gold seekers who squatted here to raise their own food. In the 1870's, the Chinese coolies who had built the Central Pacific Railroad were put to work reclaiming the delta region at low wages. With wheelbarrows they built up the first levees. Gradually the whole 425,000-acre region of tule marshes was reclaimed by an elaborate system of levees, drainage canals, and pumping plants. Today the incalculably fertile black peat soil is irrigated by flood gates from the levees. Asparagus, pears, hops, beans, onions, celery, potatoes and grains are farmed in enormous holdings.

RYDE, 32.1 *m.* (8 alt., 150 pop.), clusters at the river's edge on Grand Island, largest of the delta "islands," encompassed by the Sacramento on one side and Steamboat Slough on the other. It is a port of call for river boats carrying agricultural products.

The broad, placid Sacramento is thronged with a motley parade of water craft—the tiny, light-draught boats of the so-called "mosquito fleet," slow-moving barges laden with sacks of grain, chugging sternwheelers, motorboats, launches, and houseboats.

The highway bridges the river to ISLETON, 39 *m.* (14 alt., 1,039 pop.), a prosperous farm trading and canning town that calls itself "the Asparagus Center of the World." In the green pastures of the surrounding dairy farms, herds of sleek Guernseys and Holsteins browse on alfalfa.

From the Sacramento delta region comes nearly one-half of the United States asparagus crop and about 90 per cent of the world's canned asparagus. In the level, far stretching fields Filipino field hands, decked in veils to keep the region's swarming gnats out of their eyes, cut the white and tender asparagus stalks as soon as they appear above ground. They thrust their long-handled knives into the earth to slice off the plants underground. The work starts at the break of dawn. The asparagus stalks are left in neat little piles along the furrows by the hands working with feverish haste lest the tender stalks wilt in the sun. In the old days, the piles of asparagus were picked up by small wagons drawn by horses shod with wide plates like snowshoes; now tractors are used, outfitted with tires wide enough to keep them from sinking into the soft earth. The asparagus is rushed to nearby canneries, where it is cleaned, canned, cooked, and sealed—all within a few hours from the time it is cut.

At 43.6 *m.* is the junction with State 12, paved.

Right on State 12, crossing the river to RIO VISTA, 1 *m.* (11 alt., 2,616 pop.), in the center of a large natural gas field. It is a river shipping point for fruits and vegetables. The tule-fringed sloughs and marshlands of the vicinity are a mecca, in season, for hunters and fishermen. During the migration season the marshes abound with ducks—chiefly mallard, green-winged teal, and sprig. The Wilson snipe, called hereabout the Jacksnipe, is abundant; and the spoonbill, red-head widgeon, and canvas-back are also found. Every year in October Rio Vista decks itself out in flags to receive visitors on its Bass Derby Day, when it awards a prize to the angler who lands the largest bass. Here, by the river's edge in the fall of 1857, Col. N. H. Davis laid out the settlement of Brazos del Rio (arms of the river) and in the following year built a wharf. A thriving town grew up—renamed in 1860 Rio Vista—only to be washed down the river, house by house, in the January floods of 1862, as the shivering inhabitants looked on from the hillsides. The settlers retreated to the Montezuma Hills, farther back from the waterfront, to rebuild.

State 160 diverges from the river at about 50.5 *m.* and strikes southward across Brannan Island to the ANTIOCH TOLL BRIDGE, 54.6 *m.* over the San Joaquin River. It parallels the San Joaquin westward for a short distance.

At 56.2 *m.* is the junction with State 4, paved.

Left on State 4 is BRENTWOOD, 7.9 *m.* (77 alt., 6,600 pop.), among large fruit and nut orchards. Its chief industries are packing and shipping agricultural products. Right from Brentwood on a paved road, 5.3 *m.* to the junction with paved Marsh Creek Road; R. here 4 *m.* to the STONE HOUSE, in the midst of wide fields, built by "Doctor" John Marsh in 1856. Marsh, a Massachusetts immigrant, was granted a license to practice medicine by the Los Angeles *ayuntamiento* (council) when he displayed his Harvard diploma. He bought Rancho los Medanos (the sandhills) and settled in the shadow of Mount Diablo in 1837. The Indians of the region, in return for his kindness in healing their sick and teaching them to trap bear and otter, helped him to build a crude, thatch-roofed, floorless adobe of four rooms and attic with a tule-roofed portico; plow a field and sow it with wheat; and plant an orchard of figs, pears, and olives, and a vineyard of grape cuttings brought from Mission San Jose. With his vaqueros, who acted as his bodyguard, Marsh lived in solitary state, prizing his little store of books and reading them—stretched out before the fireplace—until he knew them by heart.

Toward everyone but his Indian neighbors, Marsh was sharp, even niggardly. When the Bidwell-Bartleson Party, first to cross the Sierra into California, came to his place in 1841, he charged them a good fee for his hospitality. For his services as a doctor—he was the first and for several years the only one in the San Joaquin Valley—he was paid in cattle, the number of cows expected as a fee depending on the number of miles he had to travel. A shrewd business man, he bargained on California's future by sending regular letters East, chiefly to Missouri newspapers, to induce immigration.

The "Stone House" was built for Marsh's bride, Abbie Tuck of Massachusetts, whom he married in 1851; but she died before it was finished. It was built of local cream-colored freestone, quarried nearby, in "the old English domestic style of architecture—a pleasing and appropriate union of Manor House and Castle"—according to the San Francisco *Daily Evening Bulletin* of July 1856. "The arched windows, the peaked roofs and gables, the projecting eaves, the central tower sixty-five feet in height . . . must be acknowledged a most felicitous deviation from the prevailing style of rural architecture. . . . The building . . . is three stories in height, with three gabled windows in the attic looking east, west, and south. On three sides of the building is a piazza . . . supported by beautiful octagon pillars; over this is a walk on a level with the second floor, enclosed by an elaborately finished balustrade. . . ."

Marsh himself scarcely outlived the completion of his house. On September 24, 1856, he was murdered on the road to Martinez by four of his Mexican neighbors, reputedly in revenge for his stingy and scornful attitude.

On State 4 at 12.8 *m.* is the junction with paved Byron Rd.; R. (straight ahead) is BYRON, 14.1 *m.* (30 alt., 1,200 pop.), shipping center for large apricot orchards, where an Apricot Festival is celebrated each spring.

At 16.1 *m.* on Byron Rd. is the junction with a dirt road; R. here 0.8 *m.* to BYRON HOT SPRINGS (*hotel, cottages; golf*), known to Yankee settlers since 1849. The springs are of three types—a "hot Salt Spring" (122°), a "Liver and Kidney Spring" (58°), and a "White Sulphur Spring" (70°).

At 29.2 *m.* is the junction of Byron Rd. with US 50 (*see TOUR 10b*).

State 4 reaches ANTIOCH, 60.2 *m.* (42 alt., 23,480 pop., est. 1965 pop.), on the San Joaquin's southern bank in rich silt deposited by the rivers, a shipping point for industrial and agricultural products. Antioch's first settlers were twin brothers, Joseph H. and W. W. Smith, both carpenters and ordained ministers, who arrived here from Boston with their families on the schooner *Rialto* in July 1849 and took up quarter-sections of land. In September 1850 W. W. Smith induced a shipload of New England families arriving from Maine on the *California Packet* to settle here; he gave each family a lot on which to build a home. At a Fourth of July picnic in 1851 the question of naming the town arose. Smith suggested a Biblical name in honor of his brother, who had meanwhile died; the name of Antioch (Syria), where Christ's followers were first called Christians, was adopted.

Antioch provides easy access to the Delta Region for sportsmen. The area has 1,000 miles of navigable waterways between the Sacramento and San Joaquin Rivers, home of the striped bass, October-April. Also sturgeon, black bass. Waters are patroled.

PITTSBURG, 64.8 *m.* (25 alt., 19,063 pop., 1960), was so named in 1911 when a mill of the Columbia Steel Co. was built here. It is now the Columbia-Geneva Works of United States Steel Corp., employing up to 3,000 men. Other industries represented here are Continental Can Co., Dow Chemical Co., Johns Manville Products Corp., H. K. Porter Co., Pacific Gas & Electric Co., Shell Chemical Co. and Stanley Steel Co. Several corporations maintain docks for ocean-going freighters. Although facilities for making thin tin plate and galvanized steel sheets have been increased, they are still inadequate for State needs.

The new Yankee owners of part of Rancho los Medanos persuaded William Tecumseh Sherman of later Civil War fame to lay out a city here in 1849. With a small crew, Sherman sailed up the river from San Francisco, making soundings to chart the best channel through Suisun Bay to the confluence of the rivers. The city he platted was so large in area that one of the owners, Col. Jonathan D. Stevenson, who brought the First Regiment of New York Volunteers to California in March 1847, named it New York of the Pacific. Stevenson hoped it would grow into a prosperous seaport, but to globe trotter Bayard Taylor it was still in 1850 only a "three-house city."

For three decades after discovery of coal in the 1850's on the slopes of Mount Diablo, the town was a busy coal shipping port, renamed in 1863 Black Diamond for the best known of the mines. A railroad was built to the coal towns that grew up around the mines, settled by immigrants from the coal fields of England and Wales. The coal was of such poor grade, however, that the industry went into a slump after 1880, strangled by competition and too-lavish outlays of capital. Until 1910, when industries arrived, the population supported itself by fishing for shad, bass and salmon. The quiet waters at the mouth of the Sacramento and San Joaquin Rivers were thronged with boats with lateen sails of Italian and Greek fishermen. Their nets sometimes stretched halfway across the channel. Jack London described the Italian and Greek fishermen in *Tales of the Fish Patrol.* By 1950 commercial fishing was ended by industrial use of the waterfront.

State 24, which connects with State 4 and runs as far as Oakland, rounds the northern slopes of the Mount Diablo Range and

route blazed in the spring of 1772 by the first white men in the region, the Pedro Fages expedition, accompanied by Fray Juan Crespi, 12 soldiers, a muleteer, and an Indian servant. In search of a land route to Point Reyes, they had followed the eastern shore of San Francisco and the southern shore of San Pablo Bay, always finding their passage to the north blocked by water. On March 30, they entered Concord Valley. Climbing one of the hills of the Mount Diablo Range, they looked out over a vast expanse never before seen by any European—

the Sacramento Valley. Next day they returned to Monterey. State 4 meets Interstate 680 near Concord.

CONCORD, 94.6 *m.* (65 alt., pop. 1950, 6,953; 1960, 36,208; 1965, 64,000, est.). Second largest city of fast-growing Contra Costa County, east of the Bay. Don Salvio Pacheco (born 1793 in Monterey) received the grant of the Monte del Diablo ranch in 1834. He built his adobe, now Adobe Restaurant, in 1846. His son Fernando, who lived until 1884 and repeatedly weighed 450 lbs., in 1845, built the adobe now used by the Contra Costa Horsemen's Assn. Fernando and his brother-in-law, Francisco Galindo, laid out the village of Todos Santos (All Saints) in 1868; name changed to Concord, 1869. It was a center for nuts, fruits and poultry until the industrial and commuting flood obliterated farming. It is the northeastern terminus of the Bay Area Rapid Transit development, for which President Lyndon B. Johnson broke ground at Concord in June, 1964. Contra Costa, Alameda and San Francisco counties authorized $792,000,000 in bonds for this 75-mile network, which will take commuters at high speeds on surface, elevated and tunnel tracks to San Francisco, Berkeley and Oakland. The DIABLO TEST TRACK, a research center, is located at Monument Blvd. and Detroit Ave. DIABLO VALLEY COLLEGE (two-year course) in 1964 had 4,280 day and 4,100 evening students. DIABLO FOUNDATION, INC., supported by business, gives substantial sums for college tuition annually to high school seniors whose grades are not A but B-plus, the latter being considered better prospects for leadership. BUCHANAN FIELD, county airport, has berths for 250 private planes. Greyhound Bus provides a service every 20 minutes to San Francisco, 34 *m.* northwest, and Oakland, 26 *m.* southwest. The U. S. NAVAL AMMUNITION DEPOT is a site for testing missiles. It has a 2,000,000-volt x-ray instrument, a 10,000,000-volt linear accelerator, a quality evaluation laboratory and a large computer system.

PLEASANT HILL, west of Concord and State 21 (pop., 1950, 5,686; 1960, 23,844) was little more than a one-room schoolhouse amid orchards until 1937, when the Broadway Tunnel made it accessible for commuters. Incor. 1961, it plans to keep industries out and encourage living comforts. Is in the Bay Area Rapid Transit system.

Right from Concord, 1.8 *m.* is PACHECO, settled partly on Don Salvio Pacheco's ranch after the Gold Rush. The Wallrath House, built in 1853, stands on a hill. The village, laid out in 1857, suffered from disastrous floods that silted up Walnut Creek; after the earthquake of 1868 the villagers moved to Concord, which offered them free sites.

MARTINEZ, 6.9 *m.* right from Pacheco on State 21, is the seat of Contra Costa County since 1850 (pop. 1960, 9,604; 1964, 13,000 est.). U. S. Post Office here serves also ALHAMBRA VALLEY, PACHECO, MT. VIEW and VINE HALL and estimates community pop. at 28,000. City faces Suisan Bay from the mouth of Alhambra Valley. Contra Costa County has grown from 298,984 in 1950 to 409,030 in 1960 and 493,200, est., in 1964. Don Ignacio Martinez, general and commandant of the Presidio of San Francisco, 1828-31, received a grant on this site that he called Rancho El Pinole. Fages, the first explorer, 1772, received seeds called pinole from the Indians. Col. Wm. L. Smith laid out the town for the Martinez family in 1849. Martinez is northern headquarters for Shell Oil and Shell Chemical, a terminal of the important Benicia-Martinez bridge across Carquinez Strait, crossed by US 680 and State 21. East of Martinez runs the Contra Costa Canal (completed 1946), part of the Central Valley Project, providing water for irrigation and industrial use. The tallest office building in the county houses part of the County Administration.

Left from Martinez on Alhambra Ave., which becomes Franklin Canyon Road (State 4), 2 *m.* to the JOHN MUIR RANCH, where a large old-fashioned ranch house, the home of John Muir, overlooks orchards from a knoll in a wooded valley. Muir, who came to the United States from Scotland in 1849 at the age of 11, spent most of his life studying the mountains, valleys, glaciers, and wild life of California and Alaska, discovering and naming many of the peaks and glaciers along the Pacific Coast. One of the foremost advocates of National parks, he publicized the region's natural beauties in his books. Muir died December 24, 1914, aged 76. He was buried on the banks of Alhambra Creek, a few miles south of his home.

West of Martinez State 21 follows the southern slope of Carquinez Strait (*see TOUR 9b*) to PORT COSTA, 12.9 *m.* (11 alt., 280 pop.), today almost abandoned, but once a port from which grain was shipped direct to Europe.

At 14.3 *m.* on State 21 is CROCKETT (*see TOUR 9b*), at the junction with US 40 (*see TOUR 9b*).

WALNUT CREEK, 101.3 *m.* (147 alt., had 2,420 pop. in 1950, 9,903 in 1960, and claimed 56,714 in 1965). Besides being a shipping center for soft-shell walnuts, it supplied facilities for orchards and chicken ranches.

Left from Walnut Creek on State 21, paved, through the orchard-shaded acres of San Ramon Valley, once part of Rancho San Ramon, granted to Jose Maria Amador in 1833.

ALAMO (poplar), 3.2 *m.* (356 alt., 1,791 pop., 1960; 5,530, 1965 est.), where poplars once grew thickly, is shaded today by giant maples. In 1848 or 1849 the first adobe was built here; and in 1854, two stores that drew their trade from the neighborhood's Spanish population. The sturdy FOSTER HOUSE (R), on the main street, was built by James Foster, who opened a wheelwright shop here in 1857.

DANVILLE, 6.3 *m.* (370 alt., 3,585 pop., 1960; 9,775, 1965 est.), stands on land once owned by Daniel Inman, who first visited the site in 1852 and took up wheat farming in 1858.

Left from Danville on a paved road 3.4 *m.* to the south entrance to MOUNT DIABLO STATE PARK (*2,168 acres; open all year around; camp sites; picnicking facilities*). The road winds upward past the jumbled rocks of the DEVIL'S SLIDE and through the animal-like rock formations of the GARDEN OF THE JUNGLE GODS—including La Rana (the frog), La Ballena (the whale), and El Perro (the dog)—to PARK HEADQUARTERS, 8.5 *m.*

At 10.5 *m.* is the junction with a paved road; L. here 8 *m.* to the junction with a paved road; L. on this road 1.5 *m.* to the junction with another paved road; L. on this road 2.1 *m.* to Walnut Creek.

The main side route goes R. from the junction with the Walnut Creek road to the summit of MOUNT DIABLO (3,849 alt.), 15.1 *m.*, a rugged peak rising alone from a level plain to dominate the countryside for great distances in every direction. From the earliest days, explorers and pioneers set their course by its conical, volcano-like outline. Its summit was chosen in 1851 as the base point for United States surveys in California; the positions of all lands of the State, outside southern California and the Humboldt region, are still determined by reference to this point. Of the view from the crest, William Henry Brewer, who in 1862 climbed the mountain with the Whitney geological survey party, wrote: "Probably but few views in North America are more extensive—certainly nothing in Europe. . . . I made an estimate . . . that the extent of land and sea embraced between the extreme limits of vision amounted to eighty thousand square miles." A powerful revolving beacon light on the summit today guides aviators.

The mountain's ruggedly fantastic grandeur—with its two cones divided by a gap through which, wrote Brewer, "the wind roared with a violence almost terrific at times . . . and at intervals the clouds rushed through like a torrent"—gave rise to legends of its being haunted by a demon. Gen. Mariano G. Vallejo reported to the Legislature in 1850 that a battle between soldiers from San Francisco led by Gabriel Moraga and the Bolgones Indians, in 1806, was about to be decided in favor of the Indians, when "an unknown personage, decorated with the most extraordinary plumage and making divers movements, suddenly appeared. . . . The Indians were victorious. . . . The defeated soldiers, on ascertaining that the spirit went through the same ceremony daily and at all hours, named the mount 'Diablo.' . . . In the aboriginal tongue 'Puy' signifies 'Evil Spirit'; in Spanish it means 'Diablo,' and doubtless it signifies 'Devil' in the Anglo-American language." Whether or not the Puy may have been a medicine man impersonating the spirit of the mountain, he failed to prevail against the Spanish, who subdued the Bolgones in a second campaign the same year.

State 21 continues southward from Danville to SAN RAMON, 9.5 *m.* (470 alt., 300 pop.), where John White put up the first house in 1852, originally named Lynchville and nicknamed Limerick, because most of the settlers roundabout were Irish. SAN RAMON VILLAGE, served by the Dublin Post Office, 2,117 pop.

DUBLIN (*see TOUR 10b*), 15.4 *m.*, is at the junction with US 50 (*see Tour 10b*).

State 24 turns westward from Walnut Creek into low hills.

LAFAYETTE, 106.6 *m.* (280 alt., pop. 27,743, 1965, est.), lies on the western fringe of the Concord Lowland and is one of the fastest growing residential areas in Contra Costa County. In 1940 it had 750 people; in 1960, 7,114. It benefited from improved transportation from Oakland via the Broadway Tunnel through the Contra Costa hills and attracted commuters from there and San Francisco. The place was settled in February, 1848, by Elam Brown, after his arrival in California as captain of a 16-wagon immigrant company. To save carrying his grain by ox team to the flour mill at San Jose, Brown in 1849 set up a horsepower mill near his house.

State 24 winds between residential areas to the BROADWAY TUNNEL, a low, level, double-bore tunnel, four lanes wide, 1.8 miles long, completed in 1937 at a cost of $4,500,000.

In OAKLAND (*see OAKLAND*), 118.4 *m.*, is the junction with US 50 (*see TOUR 10b*).

Tour 10

(Carson City, Nev.)—Lake Tahoe—Placerville—Sacramento—Stockton—Tracy—San Francisco; US 50.
Nevada Line to San Francisco, 239.5 *m.*

Section a. *NEVADA LINE to SACRAMENTO; 110 m. US 50.*

For more than a decade this section of US 50 was part of an important overland route into California. Centuries before the gold rush it was an Indian trail. In the period 1850-1857 it was an immigrant course, rough, rocky, and almost unmarked. In 1858 the first crude wagon track was built over the mountains by public funds. In that same year, George Chorpenning carried the first overland mail over this thoroughfare from Salt Lake City across the Sierra to Placerville. When the rich Comstock Silver Lode was discovered in Washoe Valley, Nev., in 1859, it brought boom times to the Placerville Road. As many as 3,000 freight wagons, with 25,000 animals, used it at one time, according to contemporary accounts. The cumbrous wagons carried loads of from three to eight tons, drawn by teams of six to ten horses or mules. Hay for the animals cost four to six cents a pound, and barley in proportion. Private companies got franchises on sections of the route from the State; grades were made easier, bridges built, and the road widened. In some places the profit from tolls during a single year was twice the original investment. In 1862 one section of the road, assessed at $14,000, collected $75,000 in tolls. Total toll charges for a 6-mule team and wagon over the whole length were from $32 to $36.

Lake Tahoe, by the crest of the Sierra Nevada, is of glacial origin, 21.6 miles long and 12 miles wide, with a surface area of 193 square miles. Meadow land and forested valley sweep from its blue waters to the summits of MONUMENT PEAK (10,085 alt.), MOUNT FREEL (10,900 alt.), and JOB'S SISTER (10,820 alt.). In *Roughing It,* Mark Twain wrote of Lake Tahoe: "We plodded on, and at last the lake burst upon us, a noble sheet of blue water . . . walled in by a rim of snow-clad peaks that towered aloft full 3,000 feet higher still. As it lay there with the shadows of the mountains brilliantly photographed upon its still surface, I thought it must surely be the fairest picture the whole earth affords."

The first white men to report seeing Lake Tahoe were John C. Frémont and his topographer, Charles Preuss, on February 14, 1844, who discovered it from the heights of STEVEN'S PEAK (10,100 alt.). Frémont first called it "Mountain Lake," but on the map that he made of his first passage of the Sierra Nevada, it appears as "Lake Bonplan" (for the French botanist and traveling companion of Alexander von Humboldt). On the first official map of California this body of water was called "Lake Bigler," for the third governor of California. In 1862 William Henry Knight, map maker of the U. S. Department of the Interior, restored the Indian name of Tahoe, and subsequent maps have retained it, in spite of the fact that the legislative act of 1870 legalizing the name of Lake Bigler has never been repealed.

BIJOU (jewel), 1.6 *m.* (6,225 alt.), formerly a large lumber camp, named by appreciative French lumberjacks 50 or 60 years ago, is now a summer resort. The long wharf at which lumber schooners docked while loading pine and fir logs to be freighted to the mills at Glenbrook, Nev., has been replaced by a pier for pleasure craft. Most of the great pine and fir forests about Lake Tahoe were cut over between 1860 and 1900. Their timbers built Virginia City, Nev., and its neighbor camps, and lined the shafts and tunnels of the Comstock Lode and other Nevada mines.

EL DORADO COUNTY PUBLIC PARK (*sanitary conveniences*), 2.1 *m.*, is one of several public lake resorts.

AL TAHOE, 2.3 *m.* (6,225 alt., 150 pop.), a resort, known as Rowland's before 1912, is typical of the many recreation centers strung along the west side of the lake.

At 4.9 *m.* is a junction with State 89, a paved road.

Right on State 89 along the shore of the lake through an almost continuous line of resorts, to CASCADE LAKE (L), 5.2 *m.*, one of the many small glacial lakes in the DESOLATION VALLEY PRIMITIVE AREA, a glacially eroded region of thick forests and granite crags.

BAY VIEW PUBLIC CAMP and INSPIRATION POINT (*camping and trailer accommodations*), 7.7 *m.*, offer a fine view of EMERALD BAY. RUBICON POINT STATE PARK, 12 *m.*, a promontory of the lake, is kept in its natural beauty by the State. MEEKS BAY (*hotel, cabin, camps*), 16.5 *m.*, has an excellent sandy beach, regarded as one of the best swimming spots on the lake. Launches and speedboats leave from here.

The road containues through TAHOE NATIONAL FOREST (L) of 576,227 acres.

TAHOE CITY (*all accommodations*), 27.5 *m.* (150 pop.), centers the resort area at the upper end of the lake. Many wealthy families have summer estates near here.

SQUAW VALLEY STATE RECREATION AREA, 1,029 acres, alt. 6,203 ft., is located near State 89, 8 miles northwest of Tahoe City. It has ski runs and lifts and facilities for winter sports. The 1960 Winter Olympic Games were held here.

TRUCKEE (*see TOUR 9a*), 41.5 *m.*, is at the junction with US 40, Interstate 80 (*see TOUR 9a*), and a section of State 89 (*see TOUR 5b*).

South of Lake Tahoe, US 50 goes over the ridges of the glacial moraine that dams up the lower end of the lake.

Here, too, is the usual PONY EXPRESS MONUMENT with a bronze bas-relief of a rider. The story of how these daredevil riders united East and West was described by Mark Twain in *Roughing It:* "Saint Joe to Sacramento, 1900 miles in eight days! The pony-rider was usually a little man. He rode fifty miles without stopping, by daylight, moonlight, starlight, or through the blackness of darkness—just as it happened. He rode a splendid horse, born from a racer, and fed and lodged like a gentleman; kept him at his utmost speed for ten miles, and then, as he came crashing to the station where stood two men holding fast a fresh, impatient steed, the transfer of rider and mail bag was made in the twinkling of an eye, and away flew the eager pair, and were out of sight before the spectator could get hardly the ghost of a look. Both rider and horse went 'flying light.' The rider's dress was thin and fitted close; he wore a 'roundabout' and a skull-cap, and tucked his pantaloons into his boot tops like a race-rider. He carried no arms—he carried nothing that was not absolutely necessary, for even the postage on his literary freight was worth five dollars a letter. He got but little frivolous correspondence to carry—his bag had business letters mostly. His horse was stripped of all unnecessary weight, too. He wore light shoes, or none at all. The little flat mail-pockets strapped under the rider's thighs would each hold about the bulk of a child's primer. They held many and many an important business chapter and newspaper letter, but they were written on paper as airy and thin as gold-leaf, nearly, and thus bulk and weight were economized. The stagecoach traveled about a hundred to a hundred and twenty miles a day (twenty-four hours), the pony-rider about two hundred and fifty. There were about eighty pony-riders in the saddle all the time, night and day, stretching a long, scattering procession from Missouri to California, forty flying eastward, and forty toward the west, and among them making four hundred gallant horses earn a stirring livelihood and see a deal of scenery every single day in the year."

At ECHO CREEK, 10.6 *m.,* is the junction with a paved road, State 89.

Left on State 89 through **LUTHER PASS, 9** ***m.*** (7,800 alt.), to a junction with State 88, 11.5 *m.*; L. on State 88–89, 6.5 *m.* to WOODFORD'S (5,634 alt.) a small resort at the junction with State 4 (*see TOUR 4*).

The side route returns on St. 88-89, then takes St. 88 at eastern end of Luther Pass. St. 88 now rises to cross the main range of the Sierra Nevada through **CARSON PASS** (8,600 alt.), 19.8 *m.* Capt. John Frémont and "Kit" Carson found this pass in the winter of 1843-44 while looking for a direct route to California from the East; it was used by thousands of gold seekers. State 8 descends through a forested region, past several small mountain resorts and the sites of once active mining camps.

PINE GROVE, 74.6 *m.* (2,100 alt., 126 pop.), is at the junction with a dirt road; R. 3 *m.* on this to VOLCANO (2,150 alt., 150 pop.), in a crater-like hollow, rimmed by fir-covered hills. Once one of the richest and most populous towns of the Mother Lode, the dust of its main street is now rarely disturbed. Occasionally an old-timer shuffles along on his way to buy something at the store. Only the wind in the fir trees or the tinkle of cowbells, as the cows come home at sunset, breaks the stillness. In the spring of 1848 Gen. John A. Sutter mined on the creek that bears his name, and the next year men from Stevenson's regiment of New York Volunteers found rich deposits in SOLDIERS' GULCH, where in later years hydraulic washing left weird outcroppings of limestone. By 1850 Volcano was going full blast. Besides the saloons and fandango halls that every camp boasted, it had a Thespian Society and a Miners' Library Association. Opened in 1862 and still doing business is the dignified ST. GEORGE HOTEL (R); its lobby has been modernized, but the bar is the original, and here are kept a collection of miners' relics, among them a Swiss music box that still tinkles plaintively. At the end of the main street are the iron-shuttered ADAMS EXPRESS BUILDING (R) and the ODD FELLOWS AND MASONIC LODGE BUILDING. Here Angelo Rossi, Mayor of San Francisco (1939), was born. He sometimes visits his birthplace in a big car such as miners never dreamed of, escorted by motorcycle police with their sirens wailing.

JACKSON, 83.2 *m.* (*see TOUR 4*), is at a junction with State 49 (*see TOUR 4*).

US 50 begins its climb up through the pines to ECHO SUMMIT, 12.7 *m.* (7,365 alt.), at the top of JOHNSON'S PASS. From Summit Lodge is a magnificent view.

At 12.9 *m.* is a junction with a paved road.

Right on this road to ECHO LAKE, 1 *m.* There are many camps in this section from which foot trails lead into the DESOLATION VALLEY PRIMITIVE AREA, a region preserved in its natural state. The tract of 41,380 acres at the headwaters of the Rubicon River is a picturesque region of granite peaks and blue alpine lakes.

The highway sweeps down the long western slope of the Sierra Nevada through luxuriant groves of pine, fir, and cedar with many free Forest Service camps and summer home tracts. At other choice places are privately owned homes and resorts.

The entrance to AUDRAIN LAKE CAMP (R) is passed at 13.3 *m.* LAKE AUDRAIN is a main source of the South Fork of the American River, down the canyon of which the highway passes. The old Hawley Grade into Lake Valley, used from 1859 to 1861, began at this point.

Audrain's station once stood on the site of the present DARRINGTON'S STORE (R), 14.3 *m.,* on the edge of HAY PRESS MEADOWS, where an old-fashioned hay press baled the tall wild grasses that were cut here for sale at $90 to $100 a ton at Virginia City, Nev.

Another alpine meadow marks the site of the onetime stage station of PHILLIP'S (R), 15.4 *m.* (7,000 alt.), now a summer resort. Nearby is the ASPEN CREEK TRACT for summer homes, within a forest of fir.

TOLL HOUSE FLAT, 17.3 *m.,* is the SITE OF SWAN'S UPPER TOLL HOUSE, where long lines of cumbersome wagons once waited to pay tolls.

The bold, granite ridge that bulwarks the canyon of the South Fork on the north forms the southern extension of Desolation Valley. A U-shaped, rocky gorge here comes down precipitously from the Desolation Valley Primitive Area, leaps the long, white cascades of HORSE TAIL FALLS (R), and follows the glacier-polished basin of Pyramid

Creek to the South Fork. High in the northwest is the austere, grey summit of PYRAMID PEAK (10,020 alt.).

To the south, LOVER'S LEAP (6,985 alt.) rises (L) a sheer 1,285 feet above the river. There is the usual legend of the Indian girl who plunged from that height because her love was unrequited. At the base of the cliff was the SLIPPERY FORD HOUSE. In 1863, Brewer, in *Up and Down California,* wrote: "Clouds of dust arose, filling the air, as we met long trains of ponderous wagons, loaded with merchandise, hay, grain—in fact everything that man or beast uses. We stopped at the Slippery Ford House. Twenty wagons stopped there, driving over a hundred horses or mules—heavy wagons, enormous loads, scarcely any less than three tons. The harness is heavy, often with a steel boy over the hames, in the form of an arch over each horse, and supporting four or five bells, whose chimes can be heard at all hours of the day. The wagons drew up on some small level place, the animals were chained to the tongue of the wagon, neighing or braying for their grain. They are well fed although hay costs four to five cents a pound, and barley accordingly—no oats are raised in this State, barley is fed instead.

"We are at an altitude of over six thousand feet, the nights are cold, and the dirty dusty teamsters sit about the fire in the barroom and tell tales—of how this man carried so many hundredweight with so many horses, a story which the rest disbelieve—tell stories of marvelous mules, and bad roads, and dull drivers, of fights at this bad place, where someone would not turn out, etc., until nine o'clock, when they crawl under their wagons with their blankets and sleep to be up at early dawn to attend to their teams."

A good ski course is at TWIN BRIDGES (R), 19.5 *m.*

At 21.2 *m.* is a junction with a graveled road.

Left on this road to the STRAWBERRY HOUSE, 0.3 *m.*, a stopping place on the later Placerville Road and now a summer resort. The simple dignity of the old house, the commodious barns, and hay lofts nearby are vivid reminders of stagecoach days. They stand at the edge of a well-watered meadow, with the granite walls of the canyon close at hand.

PYRAMID RANGER STATION (R) and PYRAMID PUBLIC CAMP (L) are passed at 23.7 *m.*

For several miles US 50 follows the north bank of the river in long graceful curves; but the old stage road, bits of which can still be traced up and down the mountainside, was hewn from the canyonsides.

Passing the SITE OF LEON'S STATION (R), 28 *m.*, often known as Mother Weltie's in the old days, US 50 enters KYBURZ, 30.2 *m.* (4,700 alt., 10 pop.). Where DICK YARNOLD'S TOLL HOUSE once reaped a remunerative harvest, a store, hotel, and post office now serve tourists and summer campers. Southward, along the Silver Fork of the Truckee River, trails lead into the beautiful SILVER LAKE region (7,250 alt.).

In the shadow of SUGAR LOAF ROCK (R), 31 *m.*, the SUGAR LOAF HOUSE once stood, and the SITE OF PERRIN'S TOLL HOUSE is passed at 31.7 *m.* Ruins of an Old Bridge (L) across the river at 35.6 *m.* mark the eastern end of Oglesby Grade, one of four or five rival detours over which traffic to Virginia City sought easier and quicker routes in the sixties.

The fast rigs of Baker's stage line changed to horses at RIVERTON, 39.6 *m.* (3,300 alt.), formerly Moore's Station. The old hotel stands at the river's edge surrounded by poplar and locust trees.

Between 1859 and 1864, stages used the Brockliss Grade (R), the beginning of which is passed at 43.7 *m.* There is an old orchard and green meadow at the PACIFIC HOUSE, 43.8 *m.* (3,375 alt., 20 pop.).

At BULLION BEND, 47.6 *m.*, a stone monument (R) marks the spot where two stages were held up and robbed on the night of June 30, 1864. The road was then a narrow grade; six men leaped out from a clump of trees, covered Blair, driver of the first stage, with pistols and shotguns, seized his lead team, and ordered him to halt. When they called, "Throw down your treasure box!" Blair said he had none. Then they demanded the stage's cargo of bullion. "Come and get it," Blair answered. Two of the road agents covered him with their guns, and two others took out the bullion. Blair asked them not to rob the passengers, and the men replied, "We don't intend to. All we want is the treasure."

Just then the second stage came around the bend, driven by a man named Watson, from whom the robbers took three sacks of bullion and a small treasure box. The leader of the band, before they galloped off, wrote the following receipt: "This is to certify that I have received from Wells, Fargo & Co., the sum of $—— cash for the purpose of outfitting recruits enlisted in California for the Confederate States army. R. Henry Ingrim, Captain Commd'g Co. C. S. A. June, 1864."

Two of the highwaymen were arrested next morning in the THIRTEEN MILE HOUSE, 49.2 *m.*, where they had overslept, and were taken to Placerville and jailed. When officers surprised others at Somerset House, 12 miles south, the cornered men unlimbered their pistols, killed the sheriff, wounded a constable, robbed them, and escaped. Thomas Poole, most notorious of the band, was captured later and executed.

SPORTSMAN'S HALL, which stood where the SNOW LINE AUTO CAMP (L), 51.2 *m.*, now is, was a stage and freight station with stable room for 500 horses.

Orchards climb the slopes above CAMINO, 54.7 *m.* (3,200 alt., 516 pop.), a busy town with mills and extensive lumberyards fragrant with stores of yellow pine.

SMITH'S FLAT (R), 59.4 *m.* (2,200 alt., 150 pop.), was a rich mining camp of the 1850's.

PLACERVILLE, 62.3 *m.* (1,848 alt., 4,439 pop.) (*see TOUR 4a*), is at a junction with State 49 (*TOUR 4a*), which unites with US 50 between this point and EL DORADO, 69.2 *m.* (1,610 alt., 417 pop.) (*see TOUR 4a*).

Between El Dorado and Sacramento, US 50 follows the old Carson Emigrant Trail, over which many of the forty-niners came into California. The trail was blazed by members of the Mormon Battalion on their way to Salt Lake City in the summer of 1848.

At SHINGLE SPRINGS, 74.4 *m.* (1,425 alt., 119 pop.), site of a shingle mill in 1849, many travel-worn gold seekers stopped for rest and refreshment. The gulches about the springs were filled with miners' cabins in 1850. The Shingle Spring House, now LOCUST TREE INN (L), 74.7 *m.*, was built in 1850 of lumber brought around the Horn.

The ruins of CLARKSVILLE, 82.5 *m.* (L) (600 alt., 25 pop.), are overgrown with ailanthus trees. Here and there, old iron doors still hang to broken stone walls. Roofs and windows are gone. Beyond the village signs of abandoned placer diggings are seen in the fields. From here, the broad, treeless hills slope gently westward to the great Sacramento Valley.

WHITE ROCK, 85.7 *m.*, once a stage stopping place known as the White Rock House, is now merely a railroad flag station. From here the original Placerville Road continued to Sacramento in a westerly direction instead of turning northwest as the present highway does.

The eastern end of a ten-mile gold-dredging area is at 90 *m.* Gigantic piles of smooth round stones line both sides of the highway, stones torn from the bed of an ancient river channel by the huge dredgers, which float on artificial ponds. Operations have been going on since 1880. Since 1922 dredging in Sacramento County has yielded $1,300,000 in gold.

At FOLSOM, 91.8 *m.* (200 alt., 3,925 pop.), a row of brick stores, erected in the late 1850's and early 1860's, line the highway. Negro miners in 1849 dug for gold at this spot on the American River, and their camp was called Negro Bar. The camp, laid out on the Mexican rancho, Río de los Americanos, on the Coloma Road, was granted to William A. Leidesdorff, U. S. vice-consul, Oct. 8, 1844. A mere trail in 1847, when traced by Captain Sutter from his fort (*see SACRAMENTO*) to his sawmill on the South Fork of the American River, Coloma Road became the first route to the gold fields after Marshall's discovery in 1848 (*see TOUR 4*).

After the death of Leidesdorff in 1848, Capt. Joseph L. Folsom, assistant quartermaster of Stevenson's New York Volunteers, traveled to the West Indies to purchase the 35,000-acre estate on the American River from Leidesdorff's heirs. Its acquisition, together with other properties in San Francisco, made Folsom one of California's wealthiest men. In 1855 a town was surveyed by Theodore D. Judah, tireless advocate of a transcontinental railroad. In 1856 it became the temporary terminus of the Sacramento Valley Railroad, the first in California. In 1857 construction of the California Central Railroad between Folsom and Marysville was begun. From 1856 to 1864 Folsom prospered as a point of departure to the mining area. It was also an important freight depot, connected with the Placerville Road by a branch road to White Rock—the route followed by the present highway.

On April 4, 1860, Harry Roff, a curly haired lad of fifteen, galloped through Folsom on the initial run of the Pony Express. The run between Sacramento and Placerville, a distance of 55 miles, was made in 2 hours, 59 minutes.

Folsom is located at the southern end of Folsom Lake, which stretches northeast 20 miles to Auburn and is made by the waters of the American River impounded at Folsom Dam. The dam, 280 ft. high, was built by the Corps of Engineers, U. S. A., and the hydroelectric plant by the Bureau of Reclamation. The units began operation in 1955.

Right from Folsom on a graveled road to FOLSOM STATE PENITENTIARY, 2 *m.*, a fortress-like, unwalled structure built on the south bank of the American River in 1880. The site was selected because of its proximity to a granite quarry, a cheap water and power supply, and adjacent fertile fields. The first electric power plant in central California was constructed on the American River above the penitentiary, and the first power was transmitted from Folsom to Sacramento, July 13, 1895.

NATOMA, 93.5 *m.* (143 alt., 106 pop.), has been a mining center since 1880, the amount of gold taken out by its dredges totaling more than $40,000,000.

At ROUTIER, 103.1 *m.* Leidesdorff's adobe ranch house stood on the bank of the American River (R). Orchards and vineyards now cover the acres of the former Rancho Río de los Americanos.

Hopfields appear at 106.1 *m.,* their green vines spread out over acres of trellises that look like an immense summerhouse. The barnlike buildings with round cupolas are hopkilns.

PERKINS, 107.9 *m.* (48 alt., 200 pop.), with its fruit-packing sheds, is a center of diversified horticulture and viticulture.

The Brighton gristmill was erected on the south bank of the American River in 1847 by Mormons in the employ of Captain Sutter. The news of gold called the millers away, but in 1849 a group of speculators laid out a town at the site and called it BRIGHTON, 109.2 *m.* The first site of Brighton, by the river, was abandoned in 1852, and the present tract by the railroad was opened up in 1861.

SACRAMENTO, 110 *m.* (*see SACRAMENTO*).

In Sacramento are junctions with US 99 (*see TOUR 3a*), US 99E (*see TOUR 3b*), US 40 (*see TOUR 9a*), and Interstate 80.

Between Sacramento and Stockton US 50 and US 99 are united (*see TOUR 3b*).

Section b. STOCKTON to SAN FRANCISCO; 86.5 m. US 50.

STOCKTON, 0 *m.* (*q.v.*), is at the junction with US 99 (*see TOUR 3b*), which continues southward. US 50 also continues southward, a few miles west of US 99.

At 5.2 *m.* is a junction with a paved road.

Left on this road is FRENCH CAMP, 0.5 *m.* (15 alt., 248 pop.), named for the Hudson's Bay Company trappers, many of them French-Canadians, who camped in this vicinity and hunted beaver and other fur-bearing animals along the river and its sloughs from 1830 to 1845.

For several miles US 50 passes through what was once the Rancho Campo de los Franceses, covering some hundred square miles of land between the Calaveras and San Joaquin Rivers. This estate was granted Jan. 13, 1844, to Wiliam Gulnac, a New Yorker, who had married a Mexican girl and become a Mexican citizen. Gulnac had hoped to colonize his land, but hostile Indians, a smallpox epidemic, and primitive conditions discouraged settlement. In 1845 he sold his entire estate to Capt. Charles M. Weber for $60, the amount of a grocery bill he owed to Weber. Weber induced settlement on his rancho only by practically giving the land away.

Across the ancient ford at the SAN JOAQUIN RIVER, 13.2 *m.* (*fishing and boating resort*), now bridged by railroad and highway, passed the Spanish cavalcade of Gabriel Moraga and Father Viader in 1810, and succeeding expeditions, following the well-worn Indian trail through the Tulares (place of tules) in search of mission sites. On these journeys, the Spanish saw Indians fishing and named the ford El Paso del Pescadero (the passage of the fishing place). The main trail from the Sierra Nevada to the coast valleys of California came to the ford as early as 1844. In seasons of high water, Indians ferried travelers across in *tule balsas* (Sp., rafts). After 1848 John Doak and Jacob Bonsell transported miners across the river in a yawl. When traffic to the mines became heavy, a substantial ferryboat was constructed, and high carrying rates—$8 for a team and wagon, $3 each for horsemen, and $1 each for pedestrians—netted the partners enormous profits from 1849 to 1852.

TRACY, 21 *m.* (49 alt., 11,289 pop.), was a stopping place of travelers from the time it was laid out on the Southern Pacific Railroad in the early 1870's. Its main street, lined with restaurants, cafés, and soft drink stands, is a popular stopping place for the truck drivers who pilot big loads through Altamont Pass. A barge canal runs from Tracy to San Francisco Bay Area and Delta Waterways. H. J. Heinz Co. employs up to 1,200 in canning. State 33 connects with US 50 at Tracy. US 50 continues west to the Bay area.

On the site of the MOUNTAIN HOUSE, 29.7 *m.* (193 alt., 20 pop.), a blue tent was pitched in 1849 to supply refreshments to miners.

At 31 *m.* is a junction with a road.

Take latter (R) 4 *m.* to ALTAMONT (740 alt., 64 pop.), at the summit of ALTAMONT PASS, settled in 1868, when the Southern Pacific Railroad was built. Ranch cattle feed on the smooth hills.

At 4.5 *m.* is a junction with an unimproved road.

Right on this road, through two gates (*please close*), to a farmhouse, 3 *m.* (*arrange here to enter a third gate; 25¢ a car*) at the base of BRUSHY PEAK (1,675 alt.), in the early 19th century called La Loma Alta de las Cuevas (high hill of the caves). The crest is grown with clumps of scrub oak, and on the lower eastern slope is an unusual group of water-worn sandstone caves. Like many another rocky outcropping in

California's remote highlands, these caves are said to have been a hide-out for Joaquin Murrieta and his robber band in the early 1850's. MURRIETA'S POST OFFICE is pointed out as the place where the bandit left messages for his gang. The post office is also a tomb; the body of a pioneer of the region was buried in the rocky crypt at his own request.

At 42 *m.* is a junction with a paved road.

LIVERMORE, 1 *m.* left on this road (487 alt., 17,000 pop.) is shipping center for extensive wineries, including Wente, Concannon, Cresta Blanca. It ships livestock and a rodeo is held in June. LIVERMORE LABORATORY of Lawrence Radiation Laboratory of the University of California was established here in 1952 by Dr. Ernest A. Lawrence and Dr. Edward Teller at the request of the Atomic Energy Commission. It is devoted to applied research of nuclear energy and conducts non-nuclear tests east of Livermore. The Plowshare program for moving earth was initiated here.

PLEASANTON, 6.2 *m.* (352 alt., 4,203 pop.) holds a wine festival, La Fiesta del Vino, in the fall. The Alameda County Fair race track is used by trainers. Here is the VALLECITOS ATOMIC LABORATORY, a $30,000,000 component part of General Electric's Atomic Power Equipment Dept. of San Jose. First privately-financed nuclear electric power was attained here by boiling water reactor in 1957. Its 30,000 thermal-watt test reactor for irradiation testing of nuclear fuels obtained AEC license No. 1 in 1959.

ON US 50, 43.7 *m.* a monument commemorates Robert Livermore, a Briton who deserted from his ship in 1822, married Josefa Higuera, and with Jose Noriega received a land grant of 8,880 acres, Rancho los Positas. The ranch was named for perennial springs. In 1851 Livermore built a frame house with timbers shipped around the Horn, later used in Livermore House, an inn.

At 48.9 *m.* is a junction with a paved road.

Right on this road 7 *m.* to the rolling hills and pastures of TASSAJARA (Sp., a place where jerked beef is hung) VALLEY. Bret Harte came here in the autumn of 1856. An old-time religious camp meeting was in progress, which was later described in Harte's short story, *An Apostle of the Tules,* the scene of which was Tassajara Valley.

DUBLIN, 52.7 *m.* (367 alt., 200 pop.), is in a valley visited in 1811 by José María Amador, a young private in the San Francisco Company. Later, while major-domo at Mission San José, he drove the mission flocks and herds onto it. In 1834, in recompense for his military services, he was granted 16,517 acres. For a decade, Amador's adobe served as an inn on the Hayward-Stockton road to the mines, becoming the nucleus of a village. It is said that the village was christened Dublin by James Witt Dougherty, who bought the Amador house in 1852. Asked by a stranger what the place was called, he replied that the post office had been designated Dougherty's Station, but since there were so many Irish there, the settlement might as well be called Dublin.

Left from Dublin on State 21, along the base of the Sierra de San José at the western edge of Livermore Valley, to ST. RAYMOND'S CHURCH, 0.1 *m.,* dating from 1859. The JEREMIAH FALLON HOUSE (L), 1 *m.,* was erected in 1850 of timbers cut from the San Antonio redwoods and of lumber brought around the Horn. Beneath a giant oak tree (L), 3 *m.*, is the ALVISO ADOBE, built in 1845 by Francisco Alviso, major-domo of Rancho Santa Rita. The well-preserved BERNAL ADOBE (R), 3.5 *m.*, was constructed in 1852 by Augustin Bernal, one of the grantees of Rancho el Valle de San José.

SUNOL, 8.5 *m.* (300 alt., 600 pop.), is among the hills at the edge of Sunol Valley's walnut groves. In the 1840's the adobe ranch house of Antonio Sunol stood amid alders and sycamores near where the San Francisco water system's Grecian-style WATER TEMPLE now stands—south of the main road at the end of an avenue in a luxuriant garden.

The way leads westward into Niles Canyon, a winding gorge cut deep by Alameda Creek, whose "very deep pools . . . many sycamores, cottonwoods, and some live oaks and other trees" were noted by Padre Pedro Font of the Juan Bautista de Anza expedition in 1776.

NILES (*see below*), 15.9 *m.*, is at the junction with Foothill Boulevard.

West of Dublin, US 50 crosses the wooded San Jose Range through HAYWARD PASS. At 60 *m.* is the junction with a paved road. Left on this road is HAYWARD (116 alt., 72,000 pop., 1960) center of farm production led by vegetable crops, cut flowers, livestock and poultry. A major terminal in Bay Area Rapid Transit, its Air Terminal traffic ranks third in Bay Area. In 1838 Guillermo Castro began raising livestock on his hacienda here. In 1852 he invited William Hayward to sell merchandise, beginning in a tent, and sold town lots. The city was incorporated 1876. CALIFORNIA STATE COLLEGE has a 365-acre campus and more than 2,500 students. CHABOT COLLEGE (Junior) has a new campus on Hesperian Blvd. *See San Leandro.*

Left from Hayward on Foothill Boulevard.

NILES, 11.3 *m.* (77 alt., 1,525 pop.), known as Vallejo Mills in the 1850's, grew up around a flour mill on José de Jesús Vallejo's Rancho Arroyo de la Alameda (ravine of the tree-lined road). The stone foundations remain east of the railroad tracks and north of Niles Canyon Road, which is paralleled a mile up the canyon by the mill's stone aqueduct. The VALLEJO ADOBE, one of the Vallejo ranch houses, stands in the CALIFORNIA NURSERY, founded in 1865, where trees and shrubs cover hundreds of acres.

MISSION SAN JOSÉ, 15.8 *m.* (48 alt., 525 pop.), clusters around La Misión del Gloriosísimo Patriarcha Señor San José de Guadalupe (the most glorious patriarch, St. Joseph), the fourteenth mission in order of time and one of the most prosperous, founded June 11, 1797, by Padre Fermín Francisco Lasuen. The mission structures were first built of timber roofed with grass and later rebuilt of adobe. The German explorer Georg Heinrich von Langsdorff, visiting here in May 1806, wrote: "The quantity of corn in the granaries far exceeded my expectations . . . and a proportionate quantity of maize, barley, pease, beans, and other grains. The kitchen garden is extremely well laid out, and kept in very good order . . ."

Strategically situated, Mission San José was a starting point for exploration of the interior and for punitive expeditions against the Indians; travelers often stopped here. One of them, the praying Methodist trapper Jedediah Strong Smith, got a cold reception when he appealed in May 1827 for food and clothing to continue his journey. Father Durán, unimpressed by a letter signed ". . . your strange, but real friend and Christian brother, J. S. Smith," put him in the guardhouse. For two weeks he was a prisoner, until the Governor sent a guard to escort him to Monterey. Returning on November 24, he and his men spent two weeks at the mission repairing their guns, drying meat, baling goods, and rounding up horses before departure. Smith made a note in his journal about the music at the Mission services, which "consisted of 12 or 15 violins, 5 base vials and one flute."

The last mission but one to be secularized, San José was so neglected that by 1846, when Governor Pío Pico ordered it sold, its value had fallen to $12,000. Subsequently it was returned to the Church, passing into the keeping of the Sisters of St. Dominic. Of the original structures, only a section of the living quarters remains, a vine-clad adobe building partly restored in 1916. In the belfry of the modern, steepled parish chapel on the site of the old church hang two of the old bells, one dated 1815 and the other, 1826; inside are old vestments, silver relics, and a font of hammered copper. The ungainly three-story brick CONVENT OF THE DOMINICAN SISTERS stands in the rear. The mission gardens, with their statue of St. Dominic, have orange, lemon, fig, apricot, almond, and olive trees, some of them planted by the mission Indians for the padres.

Right from Mission San José on a paved road 1 *m.* to the OLHONE BURIAL GROUND, where a granite monument on a grassy knoll stands over the spot in which are buried about 4,000 Olhone who helped build Mission San José. Foothill Boulevard continues southward to WARM SPRINGS (*see below*), 19.5 *m.*, at the junction with State 17 (*see below*).

US 50 continues to SAN LEANDRO, 65.3 *m.* (48 alt., 65,962 pop. 1960; 70,000, 1965, est.). It adjoins Oakland on the east shore of the Bay, 20 *m.* SE of San Francisco. The town was settled in the early 1950's by squatters, who tried to drive Jose Joaquin Estudillo and his family off their Rancho San Leandro by attacking their white cattle. The Estudillos forced the squatters to pay for the land and laid out the town. At one time San Leandro was the center of cherry production. More recently it has acquired many diversified industries. Among the most profitable is producing flowers for the cut flower market, especially in the Bay area. In the fall the growers hold the Dahlia Show.

About one-fifth is of Portuguese ancestry and holds religious and other festivals, one commemorating the discovery of the Azores. In 1961 the community started CHABOT COLLEGE (Junior) in wooden buildings bought from the Oakland School District for $225; in 1963 it approved $17,200,000 bonds for a permanent campus and two branches. In 1965 Chabot had 5,034 students and obtained a new 93 acre site on Hesperian Blvd. in nearby Hayward. San Leandro has a new Community Library Center and a marina on its three-mile shore front. In 1965 it reduced its tax rate for the seventeenth consecutive year.

Adjoining San Leandro on State 17 is SAN LORENZO (23,773 pop.), a suburb of San Leandro on the banks of San Leandro Creek, which was first known as Squattersville in 1851, when the squatters who overran Estudillo's land settled here.

At MOUNT EDEN, 7.2 *m.* (25 alt.), is the junction with a paved road; R. here 3 *m.* to the SAN MATEO-HAYWARD BAY BRIDGE across San Francisco Bay, with an over-all length of 12 miles and a length over water of 7.1 miles. All the cement that went into its construction was manufactured from oyster shells dredged from the bottom of the bay. At 15.8 *m.* is the junction with Bayshore Highway (US 101-Alt.) (*see TOUR 2b*).

State 17 continues to ALVARADO, 10.2 *m.* (11 alt., 1,907 pop.), site of an early beet sugar factory and of the CALIFORNIA SALT COMPANY WORKS.

Here in the spring of 1853 were not one but three ambitious cities—J. M. Horner's Union City, founded in 1850; Henry C. Smith's New Haven, founded in 1851; and Alvarado, named for Mexican Gov. Juan Bautista Alvarado, founded in 1852-3—all rivaling one another and attempting to rival San Francisco. The name of the third survived for all three. The H. G. SMITH HOUSE, at the head of Vallejo Street, built in 1852, was the home of the founder of the first.

In the midst of vegetable gardens and orange groves is CENTERVILLE, 13.6 *m.* (1,700 pop.), a farm community on the site of one of the 425 Indian shell mounds discovered around San Francisco Bay.

IRVINGTON, 16.8 *m.* (72 alt., 1,000 pop.), was "Washington Corners" in the 1870's and 1880's, when Washington College, one of the State's pioneers in industrial education, was in operation.

At the junction with Foothill Boulevard (*see above*) is WARM SPRINGS, 20.3 *m.* (48 alt., 75 pop.), where the Spanish-Californian women of the neighborhood used to come to do their washing at hot springs. A resort opened here in 1850 was a gay watering place until the earthquake of 1868 destroyed it.

At 21.2 the junction with a private dirt road; L. here 1 *m.* to the HIGUERA ADOBE HOUSES, erected by José Higuera, owner of Rancho los Tularcitos. The older one, built about 1822, is a roofless fire-blackened ruin with crumbling walls, doorways, partitions, and hearth. The second, and larger, building, erected about 1831, is protected by an earthquake-warped wooden superstructure.

In MILPITAS (Sp., little maize patches), 24.8 *m.* (13 alt., 460 pop.), the residents of Pueblo San José used to hold harvest-time merry-makings in their corn, pepper, and squash patches along Penitencia Creek. The claimant of Rancho Milpitas, Nicolas Beryessa, suffered one misfortune after another: Frémont's battalion plundered his cattle and killed his brother, José de Reyes; squatters seized his land; and finally the Land Commission confirmed the rancho to a rival claimant, José María Alviso. He died insane in 1863.

1. Left from Milpitas on paved Calaveras Road, 1.5 *m.* to the ALVISO ADOBE, only remaining one of the four adobe ranch buildings that Alviso built about a century ago. The house, painted white with gay trimmings, stands in a flower garden amid orchards; its second story is encased in modern weather-boarding.

2. Right from Milpitas 4 *m.* on State 9, paved, is ALVISO (8 alt., 381 pop.), a shipping port at the head of Alviso Slough and a starting point for duck hunts in the sloughs and marshes of the vicinity. El Embarcadero de Santa Clara at this point was a port of call for Yankee hide traders from 1835 to 1850, serving the missions and ranchos of the Bay region. Richard Henry Dana wrote in *Two Years Before the Mast:* "Large boats, or launches, manned by Indians . . . are attached to the missions, and sent to the vessels with hides, to bring away goods in return." Here in 1840 settled Ignacio Alviso, mayordomo at Mission Santa Clara, who gave his name to the place. As trade increased with the gold rush, a steamer line from San Francisco was established—fare one way, $35 (and $10 additional for the stage trip to San Jose). After 1865 the railroads began to divert trade, and Alviso dwindled. Today its cove shelters transportation boats and the yachts of the South Bay Yacht Club.

At 8.3 *m.* is the junction of State 9 with US 101-Alt., the Bayshore Highway (*see TOUR 2b*).

SUNNYVALE (*see TOUR 2b*), 10.3 *m.*, is at the junction of State 9 with US 101 (*see TOUR 2b*).

The main side route (State 17) continues southward to SAN JOSE (*see SAN JOSE*), 32.3 *m.*, at the junction with US 101 (*see TOUR 2b*).

OAKLAND, 67 *m.* (0-1,550 alt., 385,900 pop.) (*see OAKLAND*).

Left from 6th and Harrison Sts. in Oakland, through the POSEY TUBE (*see OAKLAND*), 0.6 *m.* to ALAMEDA (25 alt., 63,855 pop., 1960; 71,000 by special U. S. Census, 1964), the Island City, located in San Francisco Bay contiguous to Oakland and separated from it by a tidal canal or estuary developed to provide an inner harbor. When Luis Peralta received the land as part of his huge Rancho San Antonio, it was a peninsula. It was settled by Yankees; became a town in 1854 and a city in 1872. Its position on the Bay has stimulated shipbuilding, fishing and shipping, but its principal appeal has been residential and its fine, tree-lined streets and provisions for golfing, yachting and water ski-ing have drawn many Oakland and San Francisco commuters.

Extending along Alameda's southwestern shore are several beach resorts and amusement parks; largest—and northernmost—is 125-acre ALAMEDA MEMORIAL STATE BEACH PARK, at the foot of Webster St. The resort has an outdoor salt water plunge 300 feet long and 75 feet wide. Adjacent to the resorts are many fine residences and the small homes of retired seafarers. In LINCOLN PARK, at the foot of Santa Clara Ave., is a stone monument erected 1000 feet east of a place where an Indian shell mound, 400 feet long, 150 feet wide, 14 feet high, was excavated in 1908. The remains of 450 Indians, with stone implements and shell ornaments, were found there.

At the western end of the island is the U. S. NAVAL AIR STATION, built 1939-40, which has meant large income for the city. The BETHLEHEM SHIPBUILDING CORPORATION has one of its largest units on the Inner Harbor. The winter quarters of the Alaska Packers' Association are located at Alaska Basin on the estuary.

The city of Alameda extends southwest to BAY FARM ISLAND, from which it is separated by San Leandro Channel. It is occupied by dwellings, truck farms, the Municipal Golf Course and the Model Airplane Field. Reclamation of tidelands is expected to add 800 acres to the city's terrain and add about 4,000 new homes. Part of the island is within the city limits of Oakland and here is located the METROPOLITAN OAKLAND INTERNATIONAL AIRPORT.

Yacht clubs and marinas make their headquarters in the sheltered Inner Harbor and San Leandro Channel. Encinal and Aeolian Yacht Clubs are the best known. The Alameda Yacht Harbor faces the Government Island of the U. S. Coast Guard that lies between Oakland and Alameda. The Posey Tube, the Webster Ave. Tube, and bridges over the Tidal Canal at Park St., Tilden Ave. and High St. connect with the Nimitz Freeway (US 17), in Oakland.

The eastern approach to the SAN FRANCISCO-OAKLAND BAY BRIDGE is at 78.1 *m.,* at the junction with US 40 (*see TOUR 9b*), which unites with US 50 westward.

SAN FRANCISCO, 86.5 *m.* (18 alt., 755,122 pop.) (*see SAN FRANCISCO*).

Tour 11

(Las Vegas, Nev.)—Baker—Barstow—Mojave—Tehachapi—Atascadero—Morro Bay; US 91-466.
Nevada Line to Morro Bay, 381.4 *m.*

Paved roadbed, open all seasons; extremely hot in summer.
Union Pacific R.R. parallels route between Midway and Yermo; Santa Fe R.R. between Barstow and Wasco.
Accommodations limited; extra supplies advisable in desert.

Section a. NEVADA LINE *to* BARSTOW; *117.2 m.* US 466. *Interstate 15 to Barstow.*

South of the Nevada Line and the sere, cracked mud flats of Ivanpah Lake, this section of US 91-466 stretches over the wind-eroded mountains and the vast, torrid valleys of the Mojave Desert. It roughly parallels the route taken by the Mormon pioneers, who, traveling from Salt Lake City to the founding of San Bernardino, toiled in the heat with their cumbersome, crawling oxcarts. The land is still as they found it: inhospitable, ovenlike, feral, bleak, and deceptively lifeless, yet filled with the beauties of its far vistas and its variegated, ever-changing colors.

Great, quadrangular Mojave Desert, ranging from 2,000 to 5,000 feet in altitude—50 blistering valleys and scarred by worn-down mountains—is a land neither as old nor as cadaverous as it appears. Covered at least twice in ancient geologic times by an encroaching blanket of ocean, the region later raised itself to a precipitous land of sheer, needlelike volcanic crags and inconceivable chasms. Volcanic ash, mud, lava, and the effects of erosion at last filled the land's gorges; then great chains of warm lakes, linked by the now trickling Mojave River, supported a life of tropical verdure and roaming prehistoric animals. Slowly, with the rising of encompassing mountain ranges, the Mojave became a cooped-in desert. At the time when Columbus discovered America the last great vestiges of today's bone dry lakes were probably still in existence. Nor is the land dead, for there are innumerable insects, birds, reptiles, and mammals.

The route cuts across the parched bottom of Soda Lake, follows the dry Mojave River for long stretches, passes numerous small desert towns —many of them no more than filling stations with a scattering of desert shacks—rises several times over high and low mountain chains, all studded with the weirdly gesturing Joshua trees.

US 91-466 crosses the Nevada Line, 0 *m.,* 42 miles west of Las Vegas, Nev. Stretching far out on every side is a land yellow, cracked, glittering—a wide expanse of dry clay deposits washed by sudden, though infrequent, rains from the mountains into the shallow basin of dry and torrid IVANPAH LAKE. In the distance saw-toothed, desolate mountains meet the sky. Across the dry lake bed the road runs straight as a die, and then ascends the long upgrade, covered with tumbleweed, to the mountains.

LAKEVIEW STATION, 10.8 *m.,* is a solitary gasoline station sweltering by the roadside.

WHEATON SPRINGS, 12.1 *m.,* in hills dotted with green clumps of Spanish-bayonet, consists of little more than a service station and a cluster of cabins.

The Spanish-bayonet abounding on the nearby hillsides was extremely useful to the desert Indians. Soap was made from the roots, cordage from fibers in the leaves, and the flowers and fruit were considered a great delicacy.

Still ascending, US 466 skirts mountain washes, from the rocky beds of which mesquite trees—the water-searching marvels of the desert—raise their slender, delicately green, and tortuously gnarled shapes. These trees oftentimes send their roots hundreds of feet into the earth in search of moisture. On their delicate limbs grow the mesquite bean, once another food staple of the desert Indians.

MOUNTAIN PASS (*fuel, cabins*), 16.1 *m.* (4,700 alt.), snuggling between yellow hills and black crags, marks the summit of the pass between CLARK MOUNTAIN and the MESCAL RANGE. Clark Mountain, immense in the distance (R), yielded the bearded Mormon pioneer miners immense quantities of gold before the deposits were exhausted.

At CLARK MOUNTAIN STATION, 19.1 *m.,* the customary frame building housing a gas station and general store, backed by cabins, looks across the highway at a squat, barnlike structure with a corrugated metal roof—the gathering place of Death Valley Post 2884, Veterans of Foreign Wars. Every two weeks the building is noisy with the meeting of a group that has boasted as many as 150 members, many of whom ride from such distant points as Death Valley and San Bernardino.

US 466 slopes gently into the wide and arid wastes of a valley dotted with the pearly whiteness of saltbrush and pearlweed, and the olive green of creosote brush—that imperishable shrub which no hot sun can kill and which lends a deceptive appearance of verdancy to the desert floor.

WINDMILL STATION, 24.9 *m.,* is decorated, as its name implies, with a dusty windmill. Around the barren stand of buildings, the yucca, most fantastic of all the desert's growths, begins to make its appearance.

PASO ALTO (high pass), 31.7 *m.* (4,200 alt.) (*accommodations limited*), walled by jagged, crumbling spurs of volcanic rock, is at the summit of a pass; westward the road undulates through the hills in a gentle descent.

YUCCA GROVE, 32.9 *m.* (4,000 alt.) (*restaurant*), is amid a grove of yucca trees. In the daytime the fantastic posturing of yucca limbs seems to mock the traveler. At night the dusky shadows of the contorted arms, backed by the star-crowded Mojave skies and the looming black bulks of hills, lend an air of deeper mystery to the desert.

The yucca, protected by law from destruction, was called the Joshua tree by Mormon pioneers. Its reproduction is accomplished by a tiny white moth that lives only in its blossoms.

SODA LAKE, the great dry sink of dying Mojave River, lies far below in clear view, ringed with faint blue mountains.

BAKER, 50.7 *m.* (921 alt., 109 pop.), a sweltering desert hamlet, once a busy mining town, is at a junction with the Tonopah and Tidewater Line to Death Valley. Northeast of Baker 21 *m.* in the Mojave Desert the Pacific Telephone Corp. in 1965 built its Turquoise Microwave Sta., a focal point for transcontinental radio and television routes entering southern California. Cost was $2,000,000. Before the advent of the railroad, Baker was on the line of march of the carriers of borax from Death Valley—the "20-mule" borax teams.

Right from Baker on State 127 through brush-studded desert. The road now runs north. The Armagosa River, here a tiny stream, is crossed at 32 *m.*

SHOSHONE (*hotel*), 57 *m.*, consists of a few scattered buildings; willows and low bushes indicate the presence of springs. The highway enters increasingly mountainous country.

At DEATH VALLEY JUNCTION, 83 *m.* (2,000 alt., 200 pop.), is the junction with State 190; L. on this route which ascends an easy grade to a low mountain pass, then descends to the eastern entrance of Death Valley National Monument, 100 *m.* (*see DEATH VALLEY Tour 1*).

West of Baker on US 91, Interstate 15, the typical expanse of desert valley is still the glittering, dry surface of Soda Lake. Here, in 1860, on this hard playa of baked sediment from the mountains, Indians and dragoons of the U. S. Army battled. Under the command of Lieutenant Carr, a small body of men rode into the mesquite ambushes in search of Indians who had been waylaying desert travelers. In the ensuing battle three braves were killed, one seriously wounded, and a squaw was taken prisoner. The Army maintained a camp on the floor of the dry lake until 1866.

BEACON STATION, 63.7 *m.* (*accommodations limited*), is named for the red airplane beacon half a mile away (L).

Visible straight ahead is one of the desert's fantastic formations—CAT MOUNTAIN. This mass, resembling a great sleeping cat, is one of a number of rain-fissured hills in the Mojave which, through the years, have assumed the shapes of animals. Below Cat Mountain are the flat and placid CRONISE LAKES, often mistaken for mirages. Frequently dry in summer, the lakes nevertheless made human habitation possible here long before the days of railroads. Not visible from the highway are small farms, kept alive by judicious husbanding of the lake waters and by a thin trickle from deep-sunk wells; they produce a single crop—watermelons.

CRONISE STATION, 68.5 *m.*, is a straggling collection of frame houses.

The route descends into an immense and narrow valley, where the dry Mojave River parallels the highway. Weather-worn pearl-pink hills are passed; green mounds give way to black, crumbling ones atop residual slopes of yellow.

MIDWAY, 81.3 *m.*, is a scattering of buildings in a widening of the valley.

The CALIFORNIA STATE AGRICULTURAL QUARANTINE STATION and DIVISION OF REGISTRATION OF MOTOR VEHICLES, 102.7 *m.*

In YERMO, 103.6 *m.* (1,935 alt.), are division headquarters of the Union Pacific Ry. Right from Yermo to CALICO, 4.1 *m.*, in the Calico mountains, "the ghost town that came back." Before 1896 Calico mines produced silver ore valued at that time at $86,000,000. In 1950 Walter Knott, whose uncle, as sheriff of San Bernardino had grubstaked three prospectors who found silver, converted Calico into a museum.

Left from Calico 0.5 *m.* through ODESSA CANYON to the SITE OF BISMARCK, a mining camp contemporaneous with Calico. Here lived Dorsey, a dog trained to carry the mail to and from Calico. In this section were found the first deposits of the volcanic clays with a crystalline borate of lime content, named colemanite for their discoverer, W. T. Coleman. They were extensively mined for production of borax until purer deposits were found elsewhere.

The CACTUS GARDENS, 111.2 *m.*, on a barren knoll adjoining the road, maintained by the State, contain a representative collection of all the varieties of the region's cacti, which are protected by law.

Pallid mountains loom in the distance; sycamores and alders spring into view; and clusters of houses become visible.

At 116 *m.* is the junction with US 66 (*see TOUR 12a*) which unites with US 91-466 into BARSTOW, 117.2 *m.* (*see TOUR 12a*).

Section b. BARSTOW to MORRO BAY; 264.2 m. US 466.

This section of US 466 strikes westward across a parched and monotonous plateau to the desert metropolis of Mojave, then winds over the Tehachapi Range, to the great fertile expanse of the San Joaquin Valley. West of Bakersfield US 466 runs directly westward through fertile ranches to a barren land studded with oil derricks. The highway traverses mountain valleys, and crosses the Santa Lucia Range to the Pacific Coast.

West of BARSTOW, 0 *m.*, US 466 winds in and out of low, rugged hills toward flat plains, which extend monotonously for miles ahead, relieved only by stunted sagebrush, creosotebush, and saltweed.

HAWES, 21.5 *m.* (2,495 alt., 30 pop.), is a railroad station surrounded by a few dusty houses.

KRAMER, 34.2 *m.* (2,482 alt., 100 pop.), is the sunburned home of the main plant of a large borax company. The one small hotel and a few gas stations are concessions to the tourist trade.

At 37.7 *m.* is the junction with a dirt road.

Right on this road to the PACIFIC COAST BORAX WORKS, 3.1 *m.* In these dust-filmed buildings, the borax is separated from clay mined at a depth of about 150 feet in the surrounding country. The separation is achieved by an ingenious system of blowers, the steady roar of which shatters the desert silence for miles around. The crystals are transported to Wilmington for refining. Borax, prized as a cleansing agent, has long been of commercial importance; for many years it was imported into the United States from Tibet, where it was obtained from saline lakes, then the only known deposit in the world. About a century ago, the chemical was discovered in the hot springs of Tuscany, Italy, recovered by evaporation, and sold to all parts of the world. It was not until 1856 that borax was discovered in California. The deposits at Kramer were discovered in 1912, when a homesteader drilling for water brought up white crystals. In 1925 it was learned that vast deposits of pure borax and sodium borate were concealed under the layer of desert sand. This compound, rasorite, has not been found elsewhere.

ROGERS DRY LAKE, (L) 48.7 *m.,* is a barren, hard-packed, playa. The perfectly flat and saucerlike depression has been found ideal for automobile speed tests, and is used frequently for experimentation, both by racing car owners and manufacturers of stock models. The mud formed in the lake bed during the rainy seasons has been found peculiarly adaptable to the needs of oil-well drillers, and thousands of tons of the substance, known as rotary mud, are shipped annually to southern California drillers. East of the lake is the Flight Test Center of EDWARDS AIR FORCE BASE.

West of MOJAVE, 67.5 *m.* (2,755 alt., 1,845 pop.) (*see TOUR 7b*), at a junction with US 6 (*see TOUR 7b*), US 466 plunges into more desert. Ahead loom the towering TEHACHAPI MOUNTAINS. Since 1964 drilling has been proceeding for the Carley V. Porter Tunnel, part of the State Water Project, which will carry water from northern California 300 *m.* to Los Angeles. Huge pumps will raise the water over the Tehachapi mountain barrier. The state is expected to cooperate with private power companies to build a nuclear power station near the pumps.

The LOS ANGELES AQUEDUCT, 75 *m.,* lies like an immense snake along the base of the mountains. US 466 traverses a canyon of the Tehachapis. A forest of yucca grows thicker as the canyon walls narrow, only to disappear as the walls reach vertiginous heights.

CAMERON, 79 *m.* (3,789 alt.), ships fruit.

Through TEHACHAPI PASS, 80.5 *m.* (3,793 alt.), thousands of early travelers forged their way. The gorge is beautiful, with black escarpments of mountain rock and the coloring contrast of trees.

As the pass grows wider, the highway nears the blue, tree-girt expanse of PROCTOR LAKE, (L), 84 *m.,* whose waters are highly saline. During the summer salt is frequently shoveled from the dry lake.

MONOLITH, 85 *m.* (3,928 alt., 261 pop.), clusters around the giant MONOLITH CEMENT MILLS (L), which provided all the cement used in the construction of the Los Angeles Aqueduct.

TEHACHAPI, 88 *m.* (3,966 alt., 3,161 pop.), is the business and social center of a surrounding fruit area. Gold in the China Hill placers in 1854 drew the first settlers to Tehachapi—then 3 miles east; in 1876, when the railroad passed Tehachapi by, the town, Mohammed-like, went to the railroad.

At 91.4 *m.* is the junction with a paved road.

Left on this road to the CALIFORNIA INSTITUTE FOR WOMEN, 6 *m.* (*visiting hours of inmates, 9-3 daily*), a pleasing array of Normandy architecture rising in Cummings Valley. Around the penitentiary are no high walls; only a wire fence and one lone guard are required for approximately 175 women prisoners and 1,600 acres of ground. There is little of a penal atmosphere. Cottages resembling châteaus each accommodate about 40 women and one matron; inmates wear no uniforms and are permitted to decorate their rooms as they like. Opened in 1932, the institution is one of the most successful experiments in penology.

US 466 leaves Tehachapi Valley, and winds perilously into mountains clothed with gnarled and ancient oak trees.

The LOOP, 98.9 *m.*, high on a mountainside (R), is a shining thread of railroad tracks swinging in a huge circle, rapidly descending; 78 feet from the top it disappears into the mouth of a tunnel. A complete circle, 3,795 feet in circumference, has been completed.

WOODFORD, 100.6 *m.* (2,710 alt., 164 pop.), is a railroad point for scattered fruit farms.

The Tehachapi grade ends at 111.8 *m.;* the mountains are behind, imposing, verdant, dulled with purple; ahead is the flat and monotonous plain of the San Joaquin Valley. The tops of grimy oilwell derricks loom in the distance; as the road continues westward, their numbers increase, dotting the checkerboard of tilled fields like black-headed pins on a field map.

EDISON, 125.8 *m.* (572 alt., 50 pop.), is in one of the thickest forests of oilwell derricks in the district.

BAKERSFIELD, 133.5 *m.* (420 alt., 63,500 pop.) (*see TOUR 3b*), is at the junction with US 99 (*see TOUR 3b*), with which US 466 unites at FAMOSO, 145.8 *m.* (330 alt., 140 pop.).

West of Famoso US 466 cuts through tilled lands producing grapes, grain, and cotton; a few isolated homes and trees break the flatness.

WASCO, 152.3 *m.* (335 alt., 7,815 pop. 1962), serves as the community center for the nearby ranchers. A U. S. Agriculture Experiment Station is located 4 *m.* south. Vitamins and insecticides are products.

The land gradually changes its aspect, although not its contours. Divergent roads become fewer; tilled fields give way to grazing lands. The derricks of oilwells point upward from the fields in the distance.

LOST HILLS, 173.8 *m.* (300 alt., 200 pop.), is a community of oilfield workers and farmers.

ANTELOPE PLAIN stretches dismally ahead; dust devils, little whirlwinds swirling the cloying dust about briefly, are plentiful. Inflexibly straight, US 466 stretches ahead to the hazy bulk of the TEMBLOR RANGE. Rising gradually, the highway at last abandons its straight path to follow the curves of POLONIO PASS, 206.3 *m.* (1,850 alt.).

CHOLAME, 210.9 *m.* (1,200 alt., 571 pop.), in a high valley, is a supply center for ranchers.

SHANDON, 217.6 *m.* (1,035 alt., 112 pop.), by the San Juan River, consists of a business street several blocks long.

Zigzagging southwestward across the valley, past orchards, cornfields, and cattle ranges, US 466 again rises into hills. Occasionally visible are white, red-topped farm buildings. Winding and dipping, the highway at last emerges from the rich irregularity of the hills into Creston Valley, 232.3 *m.*

US 466 passes the rambling Spanish type buildings of a resort, 236 *m.* (*luxury class*), and then turns abruptly (L) into the SANTA LUCIA RANGE.

ATASCADERO (Sp., deep miry place), 246.2 *m.* (853 alt., 2,042 pop.) (*see TOUR 2b*), is at the junction with US 101 (*see TOUR 2b*).

In a twisting ascent of the Santa Lucia Range US 466 reaches the summit, 251.5 *m.* Below and far distant is the sheen of the Pacific, meeting the sky in a blue-white haze. Nearer are the tumbled, rain-eroded slopes down which the highway twists, paralleling the course of the jagged cut named MORRO CREEK. Gradually, the land softens, and the road approaches the ocean on a high, level plateau.

US 466 reaches its terminus at a junction with State 1, 264.2 *m.* (*see TOUR 1c*), 1.2 miles north of Morro Bay.

Tour 12

(Kingman, Ariz.)—Needles—Barstow—San Bernardino—Los Angeles—Santa Monica; US 66.
Arizona Line to Santa Monica, 314.8 *m.*

Atchison, Topeka & Santa Fe Ry. parallels route between Needles and Victorville.
Accommodations scanty in desert sections; gas, oil, water, and food available at desert hamlets, but extra supplies should be carried; sleeping accommodations limited to tourist camps, except in larger towns.
Paved roadbed; extreme high temperatures between Kingman and San Bernardino in midsummer, occasional heavy windstorms in March and April.

West of the green banks of the Colorado River, US 66 traverses the arid Mojave Desert, a bleak plateau furrowed by scores of untillable valleys, shimmering in the fierce sunlight. The road mounts and dips in and out of these sinks, unrelieved in their desolation except after rare rains, when a thorny mantle of delicate-hued vegetation blazes into flower. Ahead rises the blue bulk of the San Gabriel Mountains. The highway runs steadily toward them, between hills of jumbled beauty, passing through widely spaced "towns"—mere groups of tourist cabins about gas stations and lunch rooms—to the desert city of Barstow.

Section a. ARIZONA LINE to BARSTOW; 167 m. US 66.
This highway is followed by Interstate 40.

US 66 crosses the Arizona Line, 0 *m.,* 54 miles west of Kingman, Ariz., on TOPOCK (Ind., bridge) BRIDGE, which spans the deceptive Colorado (ruddy) River—here a lazy-looking stream that periodically goes on a bridge-smashing rampage—and drops US 66 onto California soil. The route follows the mesa edging the river with its fringe of green willows and sycamores.

At 1 *m.* is the junction with a dirt road.

Left on this road to ROCK MAZE, 0.5 *m.,* above the river on a high bluff. Resembling at a distance a plowed field, this work of prehistoric aborigines is believed to have been a place for funeral rites. Rows of brownish pebbles, paralleling one another several feet apart in intersecting, roughly concentric patterns, extend over several acres. A local legend relates that the spirits of the dead, floating down the Colorado, entered the maze and shook off the evil spirits chasing them by losing them in the tangled pathways.

At 9.6 *m.* is the junction with a dirt road.

Left on this road to NEEDLES MUNICIPAL AIRPORT, 0.8 *m.*

NEEDLES, 15.5 *m.* (481 alt., 4,000 pop.), spreading over a flood plain, is an oasis approached through a lane of tamarind and pepper trees. Founded as a way station after the Santa Fe tracks were laid in 1883, it was named for an isolated group of needle-like spires visible 15 miles southeastward in Arizona.

The railroad yards provide Needle's chief occupation. The mines that honeycomb the mountains roundabout yield gold, nonmetallic ores, and semiprecious stones, such as agates, moonstones and turquoise. In the fertile bottom lands, where Mojave Indians till many of the ranches, date palms lift waving fronds beside green truck gardens and citrus orchards.

A sub-tropical city, Needles seeks shelter from the sweltering heat in the shade of the palms, cottonwoods, tamarisks, and pepper trees that border its streets. A miscellany of business buildings surrounds the park that fronts the grayish stucco railroad station. In this torrid square, where temperatures often are 112° at midnight, regal Washingtonia palms, pepper trees, and alders border the grass. Couples stroll here, children play around the ornamental cannon, and swarthy Mojaves, garbed in gaudy scarlets, blues, and yellows, loiter about.

US 66 climbs the low bleached hills back of Needles and plunges westward into the MOJAVE DESERT (*see TOUR 11a, also NATURAL SETTING*), a region of fantastic formations, once swamped under ocean waters, then upheaved to bold heights, and finally buried under lava, mud, and ashes. The unevenly sloping, valley-furrowed desert floor is ringed with mountain chains whose changing hues—sepia, gray, lavender—fuse in the distance into a dull blue. Their bristling outlines and the glistening salt flats of occasional dry lakes provide the only variation to the parched, monotonous wastes. Cacti rear their rigid, spiny leaves in profusion. Here and there jut the stark branches of the Spanish bayonet and Joshua tree. After the heavy rains which come at infrequent intervals, the desert blazes with colorful flowers.

JAVA, 22 *m.* (936 alt.), a knot of dull reddish frame buildings bordering the tracks, is a desert railroad stop.

The cacti now become more conspicuous. Most widespread is the commonest western variety, the cholla. In times of great drought the cholla is eaten by cattle; a single spark of fire will ignite the whole plant, burn off the spines, and leave juicy green fodder. Interspersed are barrel cacti, stout cylinders sometimes six feet high. These contain a fibrous pulp, which, after the top is cut off, can be pounded to yield a liquid that assuages thirst; the Indians cooked their meat in the liquid with the aid of hot stones placed in the open barrels.

SOUTH PASS, 33 *m.* (2,700 alt.), is a cluster of adobe buildings offering tourist accommodations with a sign, "Water Free With Purchases Only."

From the slope west of the pass, the flat desert floor seems to be a fertile plain because of the deceptive greenery of the creosote bush. At intervals appear the tall towers of the Boulder Dam Power Line, webbed with glinting strands of cables, and at 37.3 *m.*, the trim, red-roofed stucco bungalows (L) of the section workers on the line.

US 66 gradually climbs a craggy pass through the PAIUTE RANGE.

MOUNTAIN SPRINGS, 43 *m.* (2,720 alt.). High up at the summit, is a collection of neat little buildings—gas station, lunch room, and tourist cabins—near an irrelevant cross (L) on the hilltop, erected solely to induce passing tourists to stop and ask questions.

As the highway rounds the hump of the pass, snow-crowned MOUNT ANTONIO (Old Baldy) appears, towering in the far distance above the nearer mountain ranges.

At 55.5 *m.* is a junction with a dirt road.

Right on this road 22.3 *m.* to MITCHELL'S CAVERNS (*guide*). The chilly, cavernous chambers, hollowed out of carboniferous limestone, form an underground labyrinth requiring three hours to traverse. More than 20 entrances have been discovered. Within, the ashes of aboriginal cave dwellers' campfires cover the floors. The caverns are frescoed, pillared, and ornamented with stalagmites and stalactites. Exploration should not be attempted without a guide.

The area around ESSEX, 56 *m.* (1,700 alt., 30 pop.), sparsely vegetated though it appears, pastures herds of cattle with its greasewood and bunch-grass.

US 66 continues over a vast plain, sparsely studded with the desert brush—a dull-toned land of mystic hues and vistas—between bright-colored CLIPPER MOUNTAINS (R), a jumble of volcanic rock turned yellow and brown by oxidation, and the OLD WOMAN MOUNTAINS (L), where the almost extinct Nelson mountain sheep clamber over the crags.

DANBY, 65 *m.* (1,353 alt.), a cluster of tourist cabins about a service station, is at the foot of a gradual ascent over bleak, unchanging terrain. SUMMIT, 73.5 *m.,* and CHAMBLESS, 76.5 *m.* (800 alt.), mere dots on the desert's face, provide gas pumps, tourist camps, cafés.

Far out on the desert (L), at 84.5 *m.,* wisps of smoke are seen rising from the stacks of the CALIFORNIA SALT WORKS, which mine pure salt in 40-foot shafts. Eastward, like a ruined castle, rises an abandoned GYPSUM MILL. Nearby (L) are the yellowish, salt-incrusted flats of BRISTOL DRY LAKE, its dry parts covered by puffy, powder-like "self-rising" soil in which a man would sink knee-deep. The lake's spectacular mirages, often create the illusion of a sheet of shimmering water, sometimes of cathedrals, cities, and mountain peaks floating through the sky.

AMBOY, 87 *m.* (614 alt., 95 pop.), is another typical highway stop, blistered by temperatures that often soar above 120° in mid-summer.

At 87 *m.* is the junction with a dirt road.

Left on this road to the base of AMBOY CRATER, 1.5 *m.,* where a footpath leads up the side of the cone of dark gray pumice and lava, rising 200 feet above craggy lava beds.

On the vast desert, here and there, lies an abandoned auto, sometimes on its back like an upturned turtle, or an occasional little pile of rocks, marking the boundary claims of some hopeful prospector or, topped with a weathered cross, the resting place of some luckless wanderer.

BAGDAD (*accommodations*), 95.5 *m.* (787 alt., 20 pop.), is merely a shell of the rip roaring camp that thrived here when the War Eagle and Orange Blossom gold mines to the north were active. The few old buildings that escaped destruction by fire in 1918 are threatened by fierce desert winds, as a huge oil tank with its sides blown in attests. Except for one other spot, Bagdad has less rain than any other place in or near the Mojave Desert—a mean annual average of but 2.3 inches; in four out of 20 years it has had no rainfall at all.

For 20 miles westward US 66 covers a desolate terrain almost as primitive as it was thousands of years ago. The railroad tracks are dotted with lonely stops without accommodations, which bear such curiously incongruous names as Siberia and Klondike.

In comparison with neighboring "towns," LUDLOW, 115.5 *m.* (1,782 alt., 150 pop.), is a metropolis. Here two narrow-gage railroads of the Tonopah & Tidewater connect with the Santa Fe.

The SLEEPING BEAUTY, a formation resembling a dormant, smiling human face is outlined by the crest of the CADY MOUN-

TAINS, northwestward. Directly north appears the yellowish blotch of LUDLOW DRY LAKE, where experiments have been made in processing the lake bed's fine "flour" gold.

The landscape at 129.5 *m.* suddenly losing its vegetation, darkens from gray to coal black. Above a 6-mile-wide lava field looms MOUNT PISGAH (L), an extinct volcano with a deep crater in the summit of its symmetrical 250-foot cone.

MOJAVE WATER CAMP, 136.6 *m.* and GUYMAN, 137 *m.*, each has its small knot of sun-bleached buildings. Northwest of Guyman lies the glittering, salt-crusted bed of TROY DRY LAKE.

In a dry-farming region is NEWBERRY, 146.5 *m.* (1,631 alt.), a place once named simply Water. At the spring that flows beneath the overhanging black precipices of the NEWBERRY MOUNTAINS, early travelers quenched their thirst. Newberry is a refreshing green oasis of alders, willows, and cottonwoods clustering about that desert rarity—a swimming pool. Water is a prized commodity. Trains of eighteen and twenty 10,000-gallon tank cars hauled it daily as far as Bagdad for use in locomotive boilers. Melons, alfalfa, and apricots are shipped from here.

DAGGETT, 158 *m.* (2,006 alt., 102 pop.), a gay camp when gold and silver mines in this region were working at capacity is virtually rejuvenated by the proximity of Daggett Airport, 1,082 acres, and business coming from the U. S. MARINE CORPS SUPPLY CENTER, near Yermo and 3 *m.* from Barstow. This covers 4,550 acres, has normally 1,700 Marines on duty and employs 2,300 civilians.

Along the tree-lined dry wash of the Mojave River, US 66 travels to BARSTOW, 167 *m.* (2,106 alt., 14,960 pop., 1965), a mining center of the 1880-1890 period, located at the junction of US 466 and US 91, and has US 66—National Old Trails Highway—as its Main St. It is diesel headquarters of the A. T. & S. F. Ry. In 1920 the Santa Fe bought the low-lying business district for its needs and business houses moved uphill and rebuilt.

Section b. BARSTOW to SAN BERNARDINO; 77.1 m. US 66.

This route is followed by Interstate 15.

This section of US 66 traverses barren and burning expanses of desert and crosses the heaped masses of the San Bernardino Mountains.

South of BARSTOW, 0 *m.*, US 66 runs through billowing desert country; only the cottonwoods and willows on the banks of the Mojave River (R) relieve the tedium of rolling sandscape.

LENWOOD, 5.2 *m.* (2,229 alt., 200 pop.), is encircled by slate-colored elevations blanketed with desert growths. In the surrounding country are a few sprawling alfalfa fields and an occasional chicken ranch.

HODGE, 11.7 *m.* (2,150 alt., 102 pop.), is a supply center. Its brick grammar school, perched on a slight rise (L) serves the far-flung desert district between Barstow and Oro Grande (*see below*).

The Mojave River nears the highway in Hodge. Screwbean mesquite and green desert willows grow abundantly along its banks. Distantly, the harsh, jagged mass of IRON MOUNTAIN protrudes (R) islandlike from the sand and gravel wastes.

HELENDALE, 21.8 *m.* (2,424 alt., 150 pop.), is encompassed, oasis-like, by waving alfalfa and corn.

Beyond the cultivated circle of the Helendale district, the tawny desert spreads away. SHADOW MOUNTAIN, holding turquoise-bearing porphyry deposits that were worked by desert-dwelling Indians—predecessors of the Mojave—looms indistinctly on the western horizon (R) of HELENDALE MESA.

In the shade of wide-spreading sycamores and spirelike poplars, ORO GRANDE, 31.7 *m.* (2,648 alt., 600 est. pop.), sprawls along the highway, dreaming of the prosperous gold-boom days of the 1880's. Its population skyrocketed to 2,000 when in 1878 gold was discovered in the OLD SILVER MOUNTAINS and the GRANITE MOUNTAINS (L). After 1885 gold production diminished until in 1928 the last of the mines closed.

Crossing the Mojave River, here a turgid, muddy stream with a year-round flow, US 66 sweeps south through alfalfa fields and cattle ranges rich with foot-high bunch grass. As the highway mounts a long, easy grade, the fantastic yucca trees appear. The countryside soon takes on a verdant, cultivated appearance. The Mojave River becomes a stream of respectable size, serpentining between broad acres of farm land, fruit groves, and chicken and turkey ranches.

VICTORVILLE, 35.7 *m.* (2,716 alt., 9,655 pop.), is a curious blend of the present and the past—a past carefully preserved. This was Mormon Crossing from 1878 to 1885, until the river camp, by then grown to a roaring mining town, was named Victor—later Victorville. Mining here had dwindled to insignificant proportions by 1900, but the characteristic false front frontier buildings remained, attracting, a decade and a half later, the attention of the young motion-picture industry. From 1914 to 1937 the town and its "wild West" back country were used as the locale for more than 200 films. The first picture made here was a William S. Hart "quickie" in 1914. From 1916 to 1924 Hart made an average of two pictures a year in the vicinity, among them *Wild Bill Hickok, O'Malley of the Mounted,* and *Tumbleweed.* When the resemblance of D Street to the popular conception of Main Street in a "roaring" western town attracted other film producers such western stars as Harry Carey, Tom Mix and Hoot Gibson began performing in "horse operas" here. Will Rogers, too, was a frequent and popular visitor. The town's first talking picture, *In Old Arizona,* starred Warner Baxter. A Boulder Dam high-tension power line was used as a "prop" in *Slim,* starring Pat O'Brien and Henry Fonda as linesmen. Victorville admits its attempts to recapture its waning movie trade. The atmosphere of OLD TOWN across the railroad tracks (L) is zealously preserved. Even when new ranch houses, corrals, and stables are built in the cattle range back country, they are constructed in the old style to meet the demands of location scouts. Meanwhile, Victorville pursues a more prosaic destiny as the trade center for irrigated farming and poultry and cattle ranches, and as headquarters for mining interests. West of US 66 is GEORGE AIR FORCE BASE.

In the heart of town is (L) the ARENA OF THE VICTORVILLE NON-PROFESSIONAL RODEO (*two days each Oct.; adm. $1*), sponsored by 50 Hollywood writers and actors. Only cowboys working on the ranges of the Southwest are permitted to compete.

US 66 crosses rolling desert country again toward immense blue ranges.

MILLERS CORNER, 47.7 *m.* (3,050 alt.), has a gas station and a few cabins for motorists.

US 66 now sweeps across BALDY MESA (3,000 alt.), a vast, sun-scorched expanse of mesquite and scattered yucca trees.

At 49.6 *m.* is a junction with US 395 (*see TOUR 6d*), which unites with US 66 for about 30 miles southward.

As the desert surrenders to chaparral-covered foothills, US 66 crosses at 53 *m.* a boundary of the SAN BERNARDINO NATIONAL FOREST, a preserve of 804,045 acres, containing more than a billion board feet of merchantable timber—sugar and Jeffrey pine, big cone spruce, incense cedar and tamarack, among other species. It is maintained principally for watershed protection. A number of streams and lakes in the high country furnish water for hydroelectric power and irrigation, and afford excellent trout fishing.

The mountains draw together to form the mouth of CAJON (Sp., box) PASS, through the Sierra Madre Mountains, for nearly a century the southeastern gateway for overland travel to the coast, since William Wolfskill blazed the Spanish Trail from Santa Fe to Los Angeles through it in 1831. From the summit, 54 *m.* (4,301 alt.), is an inspiring view over mountains, deserts, orchards, and vineyards. US 66 makes its descent in a series of twisting slopes.

At 59 *m.* is a junction with State 2.

1. Left here to LAKE ARROWHEAD, 24.3 *m.* (*see below*).

2. Right on State 2 through a region of grotesque sandstone formations, tooled by centuries of wind and weather into freak shapes, pockmarked with windholes and caves.

At 1.4 *m.* is a junction with a dirt road.

Left here to LONE PINE CANYON, 2 *m.* The main course of the SAN ANDREAS RIFT, one of the two geologic faults in the forest, runs up this canyon. Hundreds of thousands of years ago a rolling, uneven plain spread out where the San Gabriel Mountains now rise. This ancient plain represented the eroded remnants of a range dating from a still more remote period. A mighty underground earth movement began which obliterated the remaining traces of the earlier range and eventually thrust upward a vast, jagged mass. The upward thrust came between two faults whose courses are still clearly defined. The north fault, San Andreas Rift, extends from Cajon Pass through Lone Pine Canyon. Characterized as a "live" or "active" fault, it extends two-thirds the length of California, for an unknown distance northwest under the Pacific Ocean, and across Mexico into the Caribbean Sea. It has caused many California earthquakes, including the San Francisco disturbance in 1906.

Stunted yuccas appear in the sagebrush fields away from the highway toward the base of the range. The road here parallels the route of the Mormons, who in 1851 came through the mountains to settle San Bernardino Valley. Attempting to follow the trail blazed through Cajon Pass in 1847 by an earlier detachment of the Mormon Battalion, the party could not maneuver its heavy wagons through "The Narrows" and was compelled to seek this route farther west.

At 8.3 *m.* the main route turns L. from State 138 on Wildhorse Canyon Rd.

WRIGHTWOOD, 13.5 *m.* (6,000 alt., 50 pop.), in Swartout Valley is a conifer sheltered year-round resort (*all accommodations*). From 500 to 600 persons vacation here during the height of the seasons.

BIG PINES CIVIC CENTER, 16.9 *m.* (6,864 alt.), is the heart of the recreational and administrative activities of BIG PINES RECREATION PARK (*all types of accommodations and recreational facilities*), supervised by the Los Angeles County Department of Playgrounds and Recreation. It consists of two divisions: Big Pines (2,700 acres) and Prairie Fork (1,620 acres) with 17 public campgrounds (*25¢ a day; $2.50 for calendar year*).

A NATURE THEATER, in a natural, pine-rimmed bowl (L) near the Civic Center, is used for group meetings, free lectures, campfire programs, and picnics.

Left from Civic Center 12 *m.* on the left fork of Blue Ridge Rd. to PRAIRIE FORK COUNTY PARK (*no accommodations*).

At 17 *m.* on the main road is a junction with Table Mountain Rd.

Right on this road 1.5 *m.* ascending through conifer forests to the SMITHSONIAN INSTITUTION SOLAR OBSERVATORY, the primary purpose of which is the recording of sunspots (*open 1:30-5 Thurs.*).

A well-equipped public campground, 18.8 *m.*, lies (R) in a parklike grove of giant pines and incense cedars.

JACKSON LAKE, 19.6 *m.*, in a steep-sided gulch (L) rimmed with tall conifers, is the water-sports center of the area.

CAMP MANZANITA, 20 *m.*, is fragrant with the sweetish aroma of the manzanita tree.

West of the WEST GATE RANGER STATION, 20.1 *m.* (6,150 alt.), at the Big Pines Recreation Park, the road goes down SHOEMAKER CANYON. As the 2,000-foot line is passed, hot desert air is felt.

At 28 *m.* is a junction with a paved road.

Left on this road 0.7 *m.* through BIG ROCK CREEK CANYON to a junction with a foot trail; R. on this trail 0.5 *m.* to DEVIL'S PUNCH BOWL, a region of vast, jumbled masses of sandstone. At some places marine fossils are embedded in the boulders of the creek bed.

At 2.2 *m.* on the paved road is one of a string of Forest Service camps.

VALYERMO (*cabins*) 28.2 *m.* (3,920 alt., 65 pop.), is a settlement in a bend of alder-grown Big Rock Creek.

VALYERMO CAMP FOR UNDERPRIVILEGED CHILDREN, (L), 29.2 *m.*, is maintained by the Los Angeles Police Department. Officers of the Crime Prevention Division of the Police Department, detailed to the camp, serve as cooks, waiters, gardeners, and handy men.

PALMDALE, 52.2 *m.* (2,669 alt., 1,224 pop.), is at the junction with US 6 (*see TOUR 7b*).

US 66, roughly following (R) alder-grown CAJON CREEK, rolls smoothly downgrade to DEVORE, 67.1 *m.* (2,025 alt., 153 pop.). MOUNT SAN GORGONIO (11,485 alt.), looms into view.

The ARROWHEAD, a natural phenomenon on the face of Arrowhead Peak, is visible at 75.2 *m.*

At 75.3 *m.* Barstow Freeway (Interstate 15, combining US 66, 91, 395) turns R, dividing the business district of SAN BERNARDINO (1,073 alt., 91,922 pop. 1960; 100,300 est. 1964), seat of San Bernardino County. The name was given by a party from the San Gabriel Mission under Padre Francisco Dumetz, who entered the valley on May 20, 1810, the feast day of San Bernardino of Siena. In 1851, Capt. Jefferson Hunt arrived with 500 Mormons from Salt Lake, who bought Rancho San Bernardino for $77,000 in 1852, and laid out a city along the broad lines of Salt Lake City. The Mormons remained here until 1857, when Brigham Young recalled them.

The National Orange Show is held here annually beginning the third Thursday in February.

SAN BERNARDINO VALLEY COLLEGE, 701 S. Mt. Vernon Ave., had 9,311 students in 1965. The state college system was extended to San Bernardino in 1965-1966.

NORTON AIR FORCE BASE, until recently the largest military establishment in the Southwest in terms of personnel and payroll, is located between Third St. and the Santa Ana River in southeast San Bernardino. Some contraction of personnel was caused by the phase-down program of the Dept. of Defense, but the Base is planning for development of its dispensary and a 56-unit quarters for unmarried officers.

Directly south of San Bernardino Freeway (Interstate 10) and east of Riverside Freeway is the small community of LOMA LINDA, where new buildings have been erected for Loma Linda University, which has medical, dental and nursing schools and hospital and sanitarium facilities. Owned by the Seventh Day Adventist Church, it was formerly College of the Medical Evangelist.

Left from San Bernardino on Third St. to Sierra Way (State 18); L. here in a serpentine ascent of WATERMAN CANYON in SAN BERNARDINO NATIONAL FOREST, 5.6 *m.*

At 7.9 *m.* is a junction with a private road; R. here 1 *m.* to a hot springs resort (rates reasonable), commanding a view of San Bernardino, Pomona, and San Gabriel Valleys from a 1,800-acre park. Dominating the region is the natural phenomenon from which the springs take their name—the huge arrowhead on the slope of ARROWHEAD PEAK, behind and above the spa. The arrowhead consists of outcroppings of quartz and gray granite, grown over with whitish weeds and grass; it covers an area of seven-and-a-half acres.

An Indian legend relates that the Great White Father sent an "arrow of fire" to guide the Cahuilla westward after they had been driven from their homes by aggressive neighbors. The arrowhead finally rested on a mountainside with its point toward a fertile valley (San Bernardino) and boiling hot springs. A legend of Mormon origin is to the effect that Brigham Young had a vision in which he saw a mountain with a strange device upon it. When he learned of the discovery of the arrowhead by the Mormon Battalion in 1851, he knew it to be the mountain of his dream and ordered the establishment of a Mormon settlement in the valley below.

State 18 continues to a junction with the old Mormon Road, 16.7 *m.*, marked (R) by the WAGON WHEEL MONUMENT. The 11-mile roadway was built in 1851 to facilitate transportation of timber for building homes in the valley; its terminus was a sawmill that the Mormons had established near the present Camp Seeley.

The route leaves the chaparral of the lower slopes and enters the big timber country of pine, spruce, and oak.

At 17 *m.* State 18 becomes Rim of the World Drive and curves steadily eastward, 8,000 feet above sea level.

At 17.1 *m.* is a junction with Crestline Rd.

Left on this road 3 *m.* to CAMP SEELEY, a year-round playground maintained by the city of Los Angeles. The lodge (*rates reasonable*) is the social and musical center.

From ARROWHEAD HIGHLAND SUMMIT, 19.9 *m.* (5,174 alt.), slopes billow down to green valleys far below. A few yards from the road (R) is SPHINX ROCK, a corroded formation, 50 feet high, in the shape of a human head.

At 20 *m.* is a junction with an unpaved road.

Left on this winding road to CRESTLINE VILLAGE, 3 *m.* (4,850 alt., 500 pop.), shopping center for nearby resorts. Lying in a forest of evergreens and pines, the village has a view of desert, valley, and mountains. CRESTLINE BOWL is frequently the setting for plays, pageants, and dramas. Nearby are two ski tracks, a quarter-mile toboggan slide, and three ashcan slides.

The route continues upward on the well-banked highway. (*Drinking fountains every 2 miles.*)

BAYLIS PARK PICNIC GROUNDS is at 20.8 *m.* (*free; tables grilles, water, sanitary conveniences*).

RIM OF THE WORLD MONUMENT, 21.6 *m.* (6,150 alt.), is a crude boulder pile, dedicated by John Steven McGroarty, poet laureate of California, in 1932 in delayed commemoration of the highway's completion.

From CREST SUMMIT, 23.7 *m.* (5,756 alt.), is a comprehensive view southward. The drive drops gradually, circling buttes and plunging through parallel walls of dense forests. Here and there, through openings in the pines, the roofs of distant lodges and resorts appear.

At 24.5 *m.* is a junction with Arrowhead Lake Rd.

Left on this well-paved road 2 *m.* through a narrow, heavily timbered canyon to ARROWHEAD VILLAGE (5,109 alt., 510 pop.), on the south shore of LAKE ARROWHEAD, made up of shops, hotels, theaters, cafés, and dance halls, largely in the Norman style. Roads and footpaths radiate to the innumerable resorts on the forested rim of the lake, from auto camps to luxury hotels.

Right from Arrowhead Village 3 *m.* on Lake Rd. to ARROWHEAD DAM. Before the dam was built in 1901 to impound spring and drainage waters, the present lake region was a tree-studded, dry basin.

SKY FOREST POST OFFICE, (L), 25.6 *m.*, on the main road, serves a few scattered mountain homes (*garage and fueling accommodations*).

At 33.6 *m.* is a junction with an unpaved road.

Right on this road 0.7 *m.* to tiny ARROWBEAR LAKE, on the shores of which is the VILLAGE OF ARROWBEAR (7,800 alt., 75 pop.), with a post office and general store. Scattered along the lake front are numerous camps (*cabins; low rates*).

Rim of the World Drive rolls eastward in a succession of curves along the crest of the San Bernardino Range.

At 34.8 *m.* is a junction with Green Valley Rd.

Left on this paved road 4.5 *m.* to GREEN VALLEY LAKE, a tiny body of water in the verdant depths of Green Valley. Hundreds of privately owned lodges cling to the steeply sloping mountainsides beneath the tall cone-bearing pines and spruce.

At 37.6 *m.* is a junction with a hiking trail.

Right 3 *m.* on this trail up forested slopes to a lookout station on KELLER PEAK (7,863 alt.).

Rim of the World Drive drops gradually, for about 6 miles, and at 45.6 *m.* crosses the top of BIG BEAR DAM. There was a small natural lake here in 1845, when Benjamin Davis Wilson, for whom Mount Wilson was named, and a party of men searched the region for Indians who had been stealing cattle from ranchers. They shot 22 bears on the trip; from this incident the valley received its name. Gold was discovered here in 1860, and in the subsequent brief gold rush roads were built and scores of small shacks sprang up. By 1880 the district was almost deserted. The first dam was built in 1884 to provide San Bernardino Valley cities with water. The present dam, 6,750 feet high, was completed in 1911. The reservoir, BIG BEAR LAKE, seven miles long, has become the center of a very popular year-round playground thronged with resorts.

From PINE KNOT VILLAGE, 50.1 *m.* (6,750 alt., 750 pop.), commercial center of the Big Bear resort district, radiate a network of roads and trails.

Rim of the World Drive roughly follows the contour of Big Bear Lake to BIG BEAR CITY, 54.9 *m.* (6,860 alt., 500 pop.), on part of the former 27,000-acre Baldwin estate.

BALDWIN LAKE, 58.6 *m.* (6,674 alt.), a natural catch basin for mountain water, is normally about one-fifth the size of Big Bear Lake, but, unlike the latter, has attracted few visitors.

Section c. SAN BERNARDINO to SANTA MONICA; 71.1 m. US 66.

West of San Bernardino US 66 runs along the base of the Sierra Madre Mountains through the heart of a picture post card landscape—orange groves overlooked by snowcapped peaks.

West of SAN BERNARDINO, 0 *m.*, is RIALTO, 3 *m.* (1,203 alt., 18,567 pop.), with several orange-packing plants.

FONTANA, 5.2 *m.* (1,242 alt., 14,657 pop. 1960; 17,400 est. 1964), 49 *m.* east of Los Angeles, was made a citrus and poultry center by A. Blanchard Miller in 1905, named Fontana in 1913. Orange and lemon groves disappeared when Henry J. Kaiser located his steel plant there in 1942. It has four blast furnaces, nine open hearths, three oxygen process furnaces and employs 8,200. The Kaiser Foundation Hospital (nonprofit) serves the community. In 1921 Miller signed a ten-year contract with the City of Los Angeles to move its garbage to his ranch in the Jurupa Hills, 2½ *m.* southwest of Fontana, to feed hogs. This enterprise lasted until 1950. At its peak Miller produced 67,300 hogs annually.

CUCAMONGA, 14.5 *m.* (1,220 alt., 6,954 pop.), named for CUCAMONGA PEAK (8,911 alt.), the shopping center of a grape- and olive-growing district, has several wineries.

UPLAND, 17 *m.* (1,210 alt., 15,918 pop.), is a citrus-packing community with plants packing oranges and lemons.

CLAREMONT, 19.4 *m.* (1,155 alt., 12,633 pop.), has a citrus-packing plant and a number of small factories, but is essentially a college town. POMONA COLLEGE was founded here in 1887 by the Reverend Charles B. Sumner, a New England Congregational minister. The following January the Santa Fe Railway gave 500 acres of land for a campus here; an unfinished hotel, now Sumner Hall, was the first college building. In 1894 the enrollment was 47; in 1964 it was 1,090. The buildings, of various architectural styles, are scattered over 24 tree-shaded city blocks. In 1927 Pomona became the sponsor of a plan for a group of affiliated colleges, Claremont Colleges, Inc., of which Pomona (co-educational) was the first unit and Scripps College, the second. Another is HARVEY MUDD COLLEGE (private).

The 50-acre campus of SCRIPPS COLLEGE FOR WOMEN is cut by city streets. Scripps became a unit of Claremont Colleges, Inc., through a gift of Miss Ellen Scripps. The enrollment (323) is limited by scholastic requirements. The buildings are of modified Spanish design.

LA VERNE, 21.4 *m.* (1,050 alt., 6,516 pop.), is a citrus-packing center. Retail stores depend heavily on the patronage of LA VERNE COLLEGE students. La Verne originated at Lordsburg during the 1890 land boom. When the bubble burst, the promoters found themselves burdened with a $75,000 three-story hotel. In 1891 the Santa Fe Railway induced new settlers to come to the region, among them a group of Dunkards who later bought the hotel and founded Lordsburg College. In 1916, when the town changed its name to La Verne, the college did likewise. It is a co-educational institution with an enrollment of 565 in 1964.

At 23.3 *m.* is a junction with San Dimas Canyon Rd.

Right on this road to SAN DIMAS CANYON PARK, 0.5 *m.*, 110 acres of naturally wooded land (*picnicking facilities*) at the foot of the San Gabriel Mountains. Hiking trails lead into the mountains. The motor road winds up the canyon to WOLFSKILL FALLS, 11 *m.*

GLENDORA, 25.9 *m.* (776 alt., 20,752 pop.), another citrus-packing community, was founded in 1887 by George Whitcomb, a Chicago manufacturer, who coined the name from the word "glen" and his wife's name, "Ledora." The first commercial orange grove here, planted by John Cook in 1866, is still productive.

AZUSA, 27.6 *m.* (611 alt., 20,752 pop. 1960; 26,096, 1964), at the gateway of San Gabriel Canyon calls itself the Canyon City. Its name is derived from Asuksagna, an Indian village, and "the Azusa" appears in Father Crespi's diary of 1769.

DUARTE, 31.9 *m.* (600 alt., 13,962, pop. 1960), is named for Andreas Duarte, grantee of a 4,000-acre site, who brought water in a ditch from San Gabriel Canyon. South of Duarte are the SANTA FE FLOOD CONTROL BASIN and the SANTA FE DAM.

MONROVIA, 34.3 *m.* (560 alt., 27,079 pop. 1960; 30,200, 1964), was laid out in $100 lots in 1884 by W. N. Monroe. The large gains in population of this area are attributed to Los Angeles commuters, who are displacing the orange groves.

The LYON PONY EXPRESS MUSEUM, (L) 36.4 *m.* (*8-6; adm. 40¢, children 20¢*), where the highway widens triangularly to meet Huntington Drive, houses an exhibit of pioneer relics.

ARCADIA, 39.5 *m.* (560 alt., 41,005 pop. 1960) is a rapidly growing community of Los Angeles commuters. It supports an extensive recreation program for citizens of all ages, who may register for competitive sports including golf, bridge, dancing, even baton-twirling and cheer leading. Arcadia was developed on "Lucky" Baldwin's Santa Anita Ranch. Baldwin's house has been preserved in the grounds of the Los ANGELES STATE AND COUNTY ARBORETUM, where acres are devoted to the cultivation of trees, shrubs and flowers and there are an herb garden, a tropical jungle and a bird sanctuary. The house has a large horticultural library, and historical objects are shown in the Coach Barn. *Open free daily, 9 a.m.-5:30 p.m. Guided tours.*

At 36.4 *m.* is a junction with Huntington Drive.

Left here to the main entrance (R) of SANTA ANITA RACE TRACK, 0.5 *m.* (*racing 1:30 p.m. weekdays, except Mon., during 2-month winter season beginning about New Year's Day*), named for the "Lucky" Baldwin ranch, which it occupies. It is the $1,000,000 home of midwinter racing events that include four $100,000 purses. Blue buildings with white roofs spread along one side of the 1-mile oval track, backed by the lofty peaks of the San Gabriel Range. The grandstand seats 30,000; the stables can hold 1,500 horses. During the racing season the 500-acre scene is one of excitement and holiday bustle, thronged with notables, among them Hollywood celebrities. The season's wagers amount to more than $25,000,000.

SAN MARINO, 4.6 *m.* (557 alt., 13,685 pop. 1960), has attractive bungalows and palatial mansions on the flat basin of San Gabriel Valley and in the surrounding foothills. The city, named by the railroad magnate and patron of arts, Henry E. Huntington, bans all structures except one-family residences and supervises even the cost and architectural design of proposed buildings.

At 4.7 *m.* is a junction with Monterey Rd.; R. here 0.5 *m.* to the HUNTINGTON LIBRARY AND ART GALLERY (*open Tuesday through Saturday 1 p.m.-4:30 p.m., free. Closed October and certain holidays*).

Massive wrought-iron gates on Oxford Road give access to the 207-acre landscaped park that surrounds the two white marble buildings. Founded by Henry E. Huntington, the institution was established with a self-perpetuating board of trustees in 1919 to house a valuable collection of paintings and art objects assembled by Huntington over a period of years. It has three principal divisions: the Library, the Art Gallery, and the Botanical Gardens. The latter contain Japanese, rose, and cactus gardens and a spacious sloping lawn planted with sub-tropical trees and shrubs.

The library contains more than 250,000 rare printed books and more than 2,000,000 letters and documents confined to English and American literature, history and incunabula. Of special interest are the Gutenberg Bible of the year 1455; the Ellesmere Chaucer on vellum; the manuscripts of Milton's *Comus,* Tennyson's *Idylls of the King,* Stevenson's *A Child's Garden of Verses,* Ruskin's *Seven Lamps of Architecture,* Kipling's *Recessional,* Mark Twain's *The Prince and the Pauper,* and Benjamin Franklin's *Autobiography.* Among letters and documents are a large collection of Lincolniana, a collection of Washington's letters, and Christopher Columbus' letter announcing the discovery of America. The Art Gallery, housed in what was once Huntington's mansion, contains a world-famous collection of 18th century British paintings, including a large group of Gainsboroughs, among them *The Blue Boy,* Sir Joshua Reynolds' *The Tragic Muse, Pinkie* by Sir Thomas Lawrence, and canvases by such painters as Raeburn, Romney, Hoppner, Turner, Cotes, Constable; and the Americans, Copley, and West. Flemish and Italian paintings of the 15th and 16th centuries, and French sculpture, porcelains and tapestries are in the west wing of the library.

SIERRA MADRE, 36.9 *m.* (835 alt., 11,300 pop. est. 1965), is a foothill town surrounded by orange groves. A trellised wistaria on private grounds, Carter St. and Hermos Ave., planted in 1894, extends over an acre of ground and has been the occasion for festivals. A water ballet is held annually.

At 40.2 *m.* is a junction with San Gabriel Blvd.

Left on San Gabriel Blvd. to Mission Dr. 4 *m.;* R. on Mission Dr. to SAN GABRIEL, 5.1 *m.* (426 alt., 22,561 pop.), a town populated largely by Los Angeles commuters. San Gabriel's chief business is serving tourists. It was from San Gabriel that the Spanish governor, Felipe de Neve, marched with 44 soldiers, Indians, and Mexican colonists in September 1781 to found the pueblo of Nuestra Señora la Reina de Los Angeles.

MISSION SAN GABRIEL ARCANGEL, 314 Mission Dr. (*open 8-6; adm. 25¢*), now a parish church, was founded in 1771 by order of Fray Junípero Serra. The small band of monks sent from San Diego to establish San Gabriel were met upon arrival by a large band of painted savages who swooped down upon them. Resourcefully, the priests unfurled their banner on which was a painting of the Madonna, whereupon—so the story goes—the warriors threw down their weapons, made submission, and brought gifts. With energy the priests constructed their mission and founded a prosperous community. Their chief troubles were due to the turbulence of the Spanish soldiers stationed with them. When one of them stole an Indian's wife and beheaded the husband for objecting, "it was only," remarks a commentator, "by miraculous power that the missionaries were able to keep the Indians in hand." The original building, swept by river floods, was replaced by the present church in 1800-1806, and partly rebuilt after the 1812 earthquake. The church, rectangular in plan, is without the customary front towers. The exterior is relieved only by slender buttresses that line the long side walls and rise above the roof, forming pointed finials. The façade is unadorned except for religious figures set in arched niches high in the face of the two front buttresses. The gable end has a simple classic pediment. At the rear is a part of the patio, with one wing of the mission house and a vine-covered gable belfry. The main altar of the church is the most decorative feature of the mission. Its ornate retablo and reredos antedate the church and were probably brought from Mexico. The painted and gilded figures of the Saints, the elaborate scrolls and other ornaments were executed by Indians. Other notable features include the dado and curious dado crestings, painted on the walls.

SAN GABRIEL CIVIC AUDITORIUM, 320 S. Mission Dr., is a replica of the San Antonio de Padua Mission and was built originally for productions of the Mission Play. Remodelled 1946, it seats 1,492. Next door is the GRAPEVINE ADOBE, advertised as the birthplace of Ramona. In a walled patio is a grapevine with a spread of 100 feet, bearing the date of 1771; it was actually planted in 1861.

Among San Gabriel's remaining adobes of early days are the PURCELL ADOBE, 308 Mission Dr., which was built in 1768 to house the Spanish comandante and the padres while the mission was being erected; the MAY PLACE, 725 Carmelita St., built in 1851 by J. R. Evertson, Los Angeles' first census taker; the VIGARE ADOBE, 616 Ramona St., built by a soldier of the mission guard; and LA CASA VIEJA DE LÓPEZ (the old house of López), 330 N. Santa Anita St., believed to be one of the mission buildings.

West of its junction with San Gabriel Blvd., US 66 runs along Colorado St., through charming residential outskirts to the business center of Pasadena. *See PASADENA.*

Right from Colorado Ave. in Pasadena on Linda Vista Ave., which becomes La Cañada Rd., to Michigan Ave.; L. on Michigan Ave., which becomes Foothill Blvd., to Haskell Ave.; R. on Haskell Ave. to Angeles Crest Highway; L. on Angeles Crest Highway, which at this point enters the San Gabriel Mountains.

The highway follows a ledge over winding ARROYO SECO (Sp., dry creek). It proceeds upward toward the chaparral-covered slopes through spruce, oak, and bay trees, and rocky stretches partly covered with yucca. As the ascent continues gorges become deeper, canyon sides steeper, and the contorted strata of the early part of the route yield to the vast granite core of the San Gabriel Mountains.

NINO CANYON LOOKOUT STATION, 8.8 *m.*, provides an excellent view (R) of BROWN MOUNTAIN (4,485 alt.), named for two sons of John Brown, the anti-slavery revolutionist, who settled in one of its remote glens after the Civil War.

WOODWARDIA CANYON, 11.4 *m.* (2,700 alt.), is an idyllic, fern-strewn gorge, named for the *Woodwardia radicans.*

At 11.9 *m.* is the junction with Dark Canyon trail.

Left on this trail 1.5 *m.* through GRIZZLY BEAR FLATS to BIG TUJUNGA CANYON, a wild region abounding in summer with rattlesnakes.

GEORGE'S GAP, 15.6 *m.* (3,750 alt.), provides a view westward. The name is a tribute to the culinary achievements of a cook stationed at an early ranger camp here. Across the canyon rises the gray granite peak of MOUNT JOSEPHINE (5,520 alt.). Beyond in the distance is CONDOR PEAK (5,430 alt.), named for the rare California condor, which rivals in size the great vulture-like Andean condor.

LADYBUG CANYON, 18 *m.* (3,650 alt.), is so-called because it is a State-maintained hibernation refuge for the red and black beetle used in combating scale and other citrus pests.

RED BOX DIVIDE, 20.3 *m.* (4,666 alt.), is the crest of the ridge dividing the drainage areas of the San Gabriel and Arroyo Seco Rivers. The divide was named for a large red box, a landmark for many years, in which early forest rangers kept fire-fighting equipment. The steep, rocky headlands of Mount Wilson in the distance are seen from this place; below lies the jumbled maze of torrent-cut canyons wrought by the two forks of the twisting San Gabriel River. Far beyond is MOUNT BALDY (10,080 alt.), snowcapped half the year, and behind it a saw-toothed sky line.

1. Right from Red Box Divide 2 *m.* on a trail to a secondary trail; L. here to the SUMMIT of MOUNT SAN GABRIEL, 5.5 *m.*

2. Right from Red Box Divide on a paved road is OPID'S RESORT, 1.5 *m.* (*rates reasonable*), in picturesque West Fork Canyon of the San Gabriel River.

Right from Angeles Crest Highway at Red Box Divide on Mount Wilson Rd. around sharp, blind curves.

The SADDLE, 22.7 *m.* (5,116 alt.), is a narrow ledge of rock bridging Mounts San Gabriel and Wilson; R. from the Saddle 2.9 *m.* on a trail to MOUNT LOWE.

CREST LOOKOUT, 24.1 *m.* (5,622 alt.), offers a sweeping view on three sides of jagged peaks and deep canyons; R. from Crest Lookout 1.2 *m.* on a hiking trail to MOUNT HARVARD. In 1889 a telescope was mounted on this peak by observers from Harvard University. The site was abandoned after one year because of the number of rattlesnakes.

MOUNT WILSON, 25.7 *m.* (5,710 alt.) (*small fee charged cars and pedestrians is refunded to overnight guests; hotel rates reasonable*), is topped by a 1,050-acre, much eroded plateau in Angeles National Forest that is owned by the Mount Wilson Hotel Company, which has leased the land occupied by the Mount Wilson Observatory (*see below*) to the

Carnegie Institution for 100 years at a nominal sum. Benjamin Davis Wilson, for whom the peak is named, blazed a trail in 1864 to the summit, where he found two abandoned cabins on the plateau. Wilson's trail, early popular with hikers and riders, was used until the Mount Wilson Toll Road Company built the paved toll road. The hotel is a low, rustic building of stucco, with a wide veranda looking out over the vast valley.

MOUNT WILSON OBSERVATORY, whose white buildings, towers, and domes are scattered among the trees, is operated by the Carnegie Institution. The first telescope was set up in 1904. Originally planned primarily for solar research, "the necessity for seeking, among the stars and nebulae, for evidence as to the past and future stages of solar and stellar life" soon became evident, early resulting in a broadening of the scope of the observatory. Today eight telescopes are in use. There is a technical library of 13,000 volumes and 10,000 pamphlets. The institution maintains a large laboratory and optical shop in Pasadena.

The 100-INCH TELESCOPE (*shown daily from 1:30-2:30*) with a huge reflecting mirror, is housed in a large white dome northeast of the hotel. This telescope has helped to push the boundaries of the known universe out to about 100,000,000 light years. It admits 250,000 times more light than the unaided human eye and 2,500 times as much light as did the telescope with which Galileo began the modern era of astronomy in 1610. Images seen by the giant eye are recorded on photographic plates.

The 60-INCH TELESCOPE (*open for cosmic observation Fri. at 7:30-8:30 p.m.; adm. passes obtained by mail from the Carnegie Institution, Pasadena, two weeks in advance*) also has a reflecting mirror. Its dome is near the 100-inch telescope.

The 12-INCH TELESCOPE (*open for cosmic observation with accompanying lecture daily, except Mon., at 7:30 p.m.*) is in the museum. It is principally used to demonstrate the public lectures.

The SUN TELESCOPE (*not open*) is housed in a 150-foot tower that rises above the pine trees near a cluster of low, snub-nosed domes. This instrument, which produces an image of the sun 16 inches in diameter, is elevated to prevent reflected heat from interfering with the accurate operation of the delicate mirrors. To increase the steadiness of the lenses and mirrors at so great a height, an inner tower supports the instruments, an outer tower the dome.

The MUSEUM (*open 1:30-2 p.m.*), adjacent to the 100-inch telescope dome, shows transparencies of the more spectacular photographs made with the giant telescope.

LOS ANGELES, 54.2 *m.* (*see LOS ANGELES*).

Right from Figueroa St. on Sunset Blvd. to a junction with Santa Monica Blvd.; L. on Santa Monica Blvd. to HOLLYWOOD (*q.v.*).

BEVERLY HILLS, 62.2 *m.* (260 to 325 alt., 34,230 pop.). Standing like a small island almost surrounded by Los Angeles, it is an independent city laid out in 1907 by a former resident of Beverly, Mass. Beverly Hills remains popular as the residence of famous stars, although many have built in more outlying places. Pickfair, 1145 Summit, is the home of Mary Pickford. Tony Curtis lives at 1152 San Ysidro; Glen Ford has a $400,000 dwelling at 1012 Cove; Harold Lloyd is at 1225 Benedict Canyon, and James Stewart at 918 N. Roxburg.

Beverly Hills announces that it has no low-cost housing projects; that residences sell from about $45,000 for older homes up to $350,000 and more for luxurious homes. The area for single residences is almost completely developed and residential lots are available from $60,000. In recent years a number of apartment buildings have been erected, with rentals from $100 to $1,000 a mo. and the available supply is limited.

Luxuriously appointed hotels make the city a favorite place for guests with business at television and motion picture studios. The best known are the Beverly Hills, 312 rooms; the Beverly Hilton, 500 rooms; the Beverly-Wilshire, 350 rooms; the Beverly Carlton, 90 rooms; the Beverly Crest, 54 rooms; the Beverly Rodeo, 100 rooms; the Wilshire House, 100 rooms; the Beverly Terrace Motor Hotel.

WESTWOOD (32,300 pop., 1963 est.) adjoins Beverly Hills on the southwest. It is part of Los Angeles with civic autonomy. North is the University of California at Los Angeles (*q.v.*). At 10741 Santa Monica Blvd. is the largest MORMON TEMPLE, with a white tower 257 ft. tall surmounted by a 15-ft. angel in gold leaf. (*Tours, 9 a.m.-5 p.m.*)

PALMS (50,500 pop., 1963) is a residential part of Los Angeles covering 4.9 square miles just south of Beverly Hills and north of Culver City.

CULVER CITY (33,163 pop.) is an independent municipality south of Palms. (*See HOLLYWOOD.*)

SANTA MONICA, 71.1 *m.*, is at the junction with US 101 Alt. (*see TOUR 2B*).

Tour 13

(Quartzsite, Ariz.)—Blythe—Indio—Beaumont—Riverside—Los Angeles; US 60; Interstate 10.
Arizona line to Los Angeles, 231.7 *m.*

Caution: For side trips on desert roads, carry good tires and extra food, gasoline and water. After cloudbursts, when water fills highway dips (indicated by pairs of stakes marked with two black bands), wait for water to run off.

The red, white and blue shields of Interstate 10 are joining the markers of US 60, with which it is identical. The highway crosses a changing waste, now lumped into derby-shaped hills, now swept into sand piles that are often leveled smooth by the wind. At Indio, as the mountain barrier comes into closer focus, the highway strikes boldly northwest toward the San Bernardino and San Jacinto ranges. From San Gorgonio Pass, it coasts down the long slope through pastures, orchards and vineyards.

The route was opened in 1862, when a stage line began carrying prospectors from California to the newly-discovered gold fields in western Arizona. The first stage, a 6-horse Concord of the type familiar to devotees of Western films, traveled from Los Angeles to La Paz in 12 days, carrying passengers for $40 apiece. Its drivers were murdered by a bandit whom they had attempted to hang; he later pleaded self-defense and was acquitted. Another party of travelers on the route was massacred by Indians. The stage line was operated until the railroad put it out of business in 1877. Over this road, in 1908, was run the first Los Angeles-to-Phoenix auto race, which the Los Angeles *Times* of that day hailed as "the most hazardous race ever undertaken." The victor, a steam-driven car, negotiated the 455 miles in 30 hours. The sparse population along the route engages chiefly in catering to travelers, occasionally in farming, mining, or health-seekers.

Section a. ARIZONA LINE *to* EASTERN JUNCTION US 99; *101 m. US 60-70. The latter is partly identical with Interstate 10.*

Between the Colorado River and Indio US 60-70 sweeps across the desert between distant mountain ranges, for long stretches in a virtually straight line.

The Arizona line, 0 *m.,* marked by the COLORADO (*ruddy*) RIVER, is crossed 19 miles west of Quartzsite, Ariz. The Colorado River was called Ahan Yava Kothickwa (all the water there is) by the Indians.

At 2.8 *m.* is a junction with US 95.

Right on US 95 to a series of spectacular INDIAN INTAGLIOS, 18 *m.* On the mesa extending westward from the bluffs of the river lie three sets of colossal figures of men and animals, cut into the white rock substratum by Indians in the remote past. The great size of these figures —the largest man is 167 feet long and has an arm spread of 164 feet—prevented their character from being recognized until they were first photographed from the air in 1932. The human figures are well proportioned, but the animals excite conjecture: they resemble horses, antelopes, or deer, but their identification is complicated by long buffalo tails. These figures are centuries old; beyond that little is known of their history or meaning.

BLYTHE, 4.5 *m.* (6,023 pop.), dates from 1910. Until then all the settlers in the region clustered on the Arizona bank of the river. In 1877 Thomas Blythe, an Englishman, filed claims on 40,000 acres along the river under the Swamp and Overflow Act, believing that this area could be turned into a new Nile Valley; but before he could develop it, he died of a paralytic stroke. His estate, in litigation for 21 years, was finally sold to the syndicate that platted the town in 1910. The farmers who were drawn to Blythe by the lure of rich overflow land soon had to contend with the source of the land's fertility. The powerful Colorado went on a yearly spree, cutting through the levees. Blythe residents fought the river, sank $4,500,000 in the

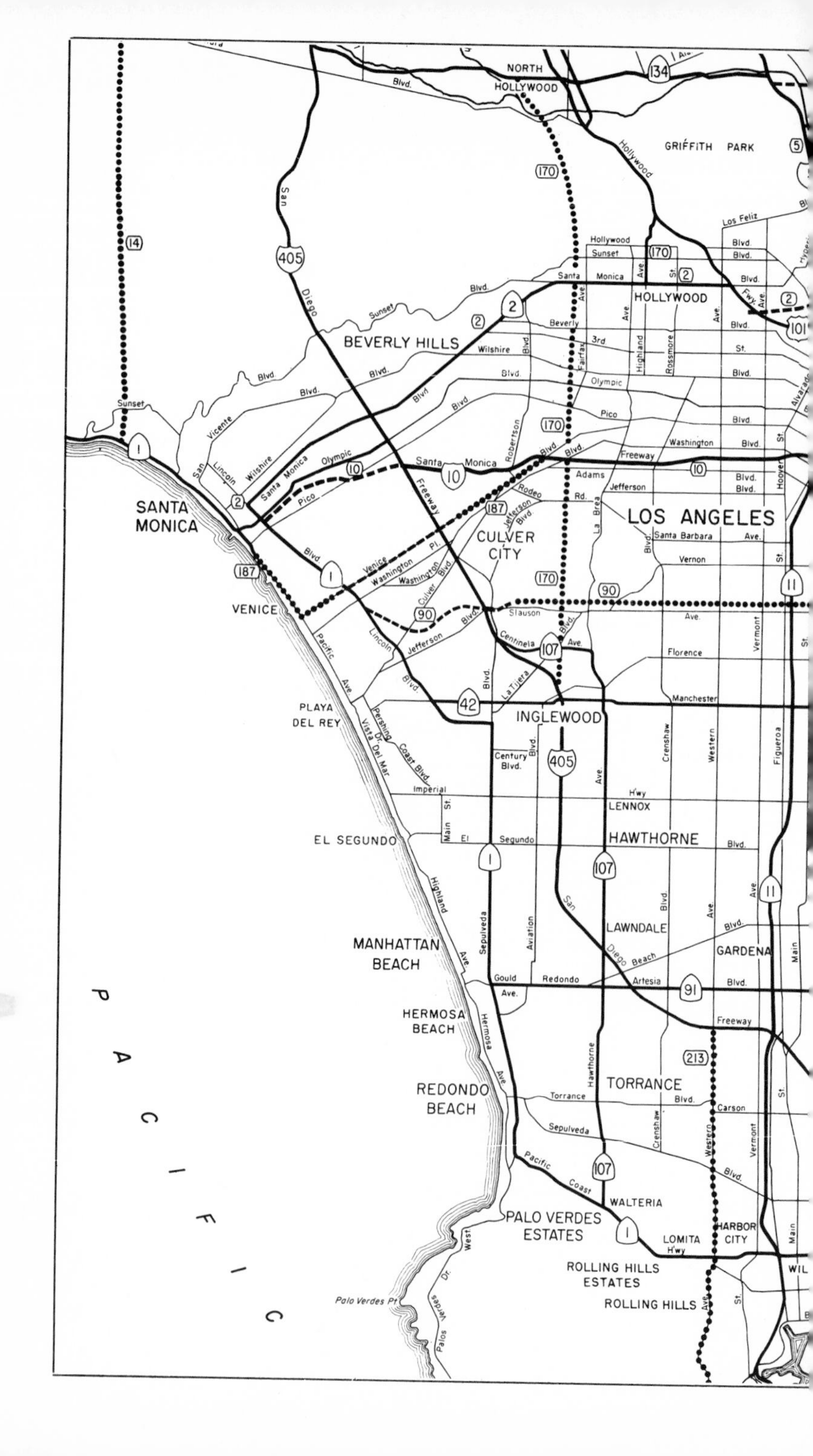

NORTH HOLLYWOOD
GRIFFITH PARK
HOLLYWOOD
BEVERLY HILLS
SANTA MONICA
LOS ANGELES
CULVER CITY
VENICE
PLAYA DEL REY
INGLEWOOD
LENNOX
EL SEGUNDO
HAWTHORNE
LAWNDALE
MANHATTAN BEACH
GARDENA
HERMOSA BEACH
TORRANCE
REDONDO BEACH
WALTERIA
PALO VERDES ESTATES
LOMITA
HARBOR CITY
ROLLING HILLS ESTATES
ROLLING HILLS
Palo Verdes Pt
P A C I F I C

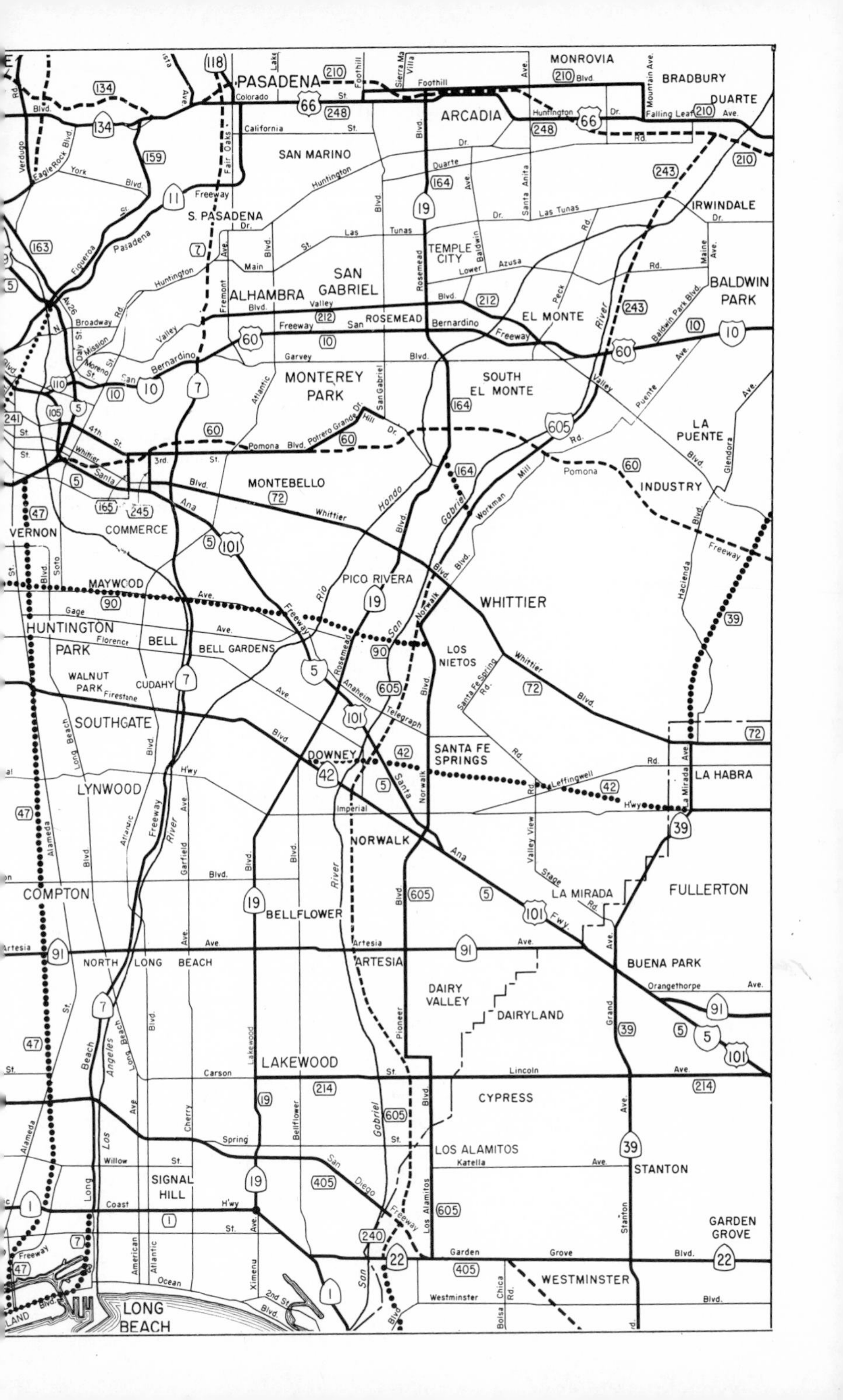

PASADENA
MONROVIA
BRADBURY
DUARTE
ARCADIA
SAN MARINO
S. PASADENA
IRWINDALE
TEMPLE CITY
SAN GABRIEL
ALHAMBRA
BALDWIN PARK
ROSEMEAD
EL MONTE
MONTEREY PARK
SOUTH EL MONTE
LA PUENTE
MONTEBELLO
INDUSTRY
VERNON
COMMERCE
MAYWOOD
PICO RIVERA
WHITTIER
HUNTINGTON PARK
BELL
BELL GARDENS
LOS NIETOS
WALNUT PARK
CUDAHY
SOUTHGATE
DOWNEY
SANTA FE SPRINGS
LA HABRA
LYNWOOD
NORWALK
LA MIRADA
FULLERTON
COMPTON
BELLFLOWER
NORTH LONG BEACH
ARTESIA
BUENA PARK
DAIRY VALLEY
DAIRYLAND
LAKEWOOD
CYPRESS
LOS ALAMITOS
STANTON
SIGNAL HILL
GARDEN GROVE
WESTMINSTER
LONG BEACH

battle, and in time lost 90 percent of their property to the State because of inability to liquidate the immense debt the river had forced upon them. The building of Boulder Dam ended the unequal struggle.

Blythe presents the first California superlative on this route; the fields lining the highway and running southward into the Palo Verde Valley grow long staple cotton, modestly proclaimed to be the best in the world. The highway traverses a plain checkered with snowy white patches and rich green alfalfa fields, and runs between flanking feathery green tamarisks.

Over a low ridge suddenly appears the COLORADO DESERT. This low sink was formed ages ago when the Coast Range—then twice its present height—cracked, the rear slope falling flat at the base of the western half, which remained erect. In time the scouring of wind and rain will wear down the ranges and plane off the region until it has the drab monotony of the older deserts. In the meantime—a two-million-year meantime—travelers may enjoy the cruel beauties of a desert in its youth, the pale pastels of its encircling mountains blending with the delicate grays and dusty tans of its rolling floor.

At 36.5 *m.* is a junction with a dirt road.

Left on this road, which strikes off through the brush toward the CHUCKWALLA MOUNTAINS to CORN SPRINGS, 8 *m.*, an ancient Indian campsite, still strewn with shards and artifacts. Indians lived on the desert at unknown periods, spacing their camps over a long arc that curves to the northwest, apparently adhering to the course of a dried-up river.

DESERT CENTER, 49 *m.* (905 alt., 125 pop.), is a way station (*all accommodations*) where travelers mail, over a 70-day period, an average of 18,000 post cards purchased on the spot.

The COLORADO RIVER AQUEDUCT, 49.4 *m.*, veers toward the highway. Although at that point it runs underground and can be detected only by the long scar it has left on the plain, it lies within view, a great silver snake ascending the backbone of a ridge (Eagle Mountain Pumps Lift) to the northwest. Constructed at a cost of $220,000,000, the aqueduct carries water to 13 southern California cities; five pumping stations lift the water 1,000 feet, the highest point being the Hayfield Lift (R), 62.2 *m.*, and syphons draw it down the western slope to its original level. The pipe keeps pace with the highway, running for 17 miles under the COACHELLA MOUNTAINS until it cuts across the road 2 miles east of Cabazon and dives through the towering San Jacintos.

The highway passes between the OROCOPIA (abundance of gold) MOUNTAINS and the EAGLE RANGE, named for a rumored colony of eagles.

At 101 *m.* is the junction with US 99 (*see TOUR 3e*).

Section b. JUNCTION US 99 *to* LOS ANGELES; *130.7 m.* *US 60-70-99. Interstate 10.*

Northwest of Indio US 60-70 traverses sand plains, rises through San Gorgonio Pass and descends into the fertile valleys and populous areas of the coast.

US 99 (*see TOUR 3e*) unites northward with US 60-70, 0 *m.,* 2 miles south of Indio.

INDIO, 2 *m.* (—22 alt., 13,450 pop. 1964 est.) is the distribution point for the rich fruit products of the Coachella Valley and calls itself the Date Capital of the United States. Production in 1963 had a value of nearly $44,000,000 from 60,629 acres. Grapes, chiefly Thompson seedless, were grown on 12,967 acres; dates ranked second in value, citrus third. Since 1960 agriculture has been greatly aided by water from the Colorado River routed via the All-American Canal. The National Date Festival of Riverside County is held annually in February in Indio.

The marketing possibilities of date production are exemplified by the output of local packing plants. Cultivation of dates is explained for interested visitors at Shield's Date Gardens, 3½ *m.* northwest of Indio. *Open daily, 8 a.m. to 6 p.m., free.*

Left from Indio on State 111—the Palm Springs Highway—which crosses Coachella Valley through date orchards and orange groves. The desert is encircled by mountains that are snow-covered during the winter and spring months. Long rows of date palms flank the roadway west of Indio. Far to the west the San Jacinto (Sp., St. Hyacinth) Mountains fall away into foothills and disappear.

SAN JACINTO PEAK (10,987 alt.) is visible for a hundred miles.

The highway is bordered by tamarisk trees (also known as athel trees), and by subtropical date palms looking like planted pompons, mingling their fronds high overhead. Just under the green density of the natural ceiling, the heavy date clusters hang, encased in paper or burlap cones during the early fall months when the fruit is ripening. The cones themselves, transparent in the filtered sunlight, resemble great bell-like blossoms, suspended above the long, still alleys of the groves. In spring bloom, the trees display whitish, candelabra-like spikes under the dome of the overhanging foliage. Principally of the Deglet Noor variety, these trees were imported and set out in this region in 1904. They require unusual care, their natural thirst necessitating heavy irrigation at least twice a month. Dates can be purchased from the many roadside stands—along this route disguised as pyramids, Bedouin tents, or Persian shrines.

INDIAN WELLS, 8 *m.,* are fresh-water springs.

Tamarisks, tall cottonwoods, hung with bright green clusters of mistletoe in winter, border the highway at intervals. Many small irrigation canals, with water brought from deep wells, weave back and forth under the road. Gaudy signboards invite a visitor to the ranches, to sample the fruit, and to view the exhibits of date culture.

PALM DESERT, 13 *m.* (1,295 pop.), is a new settlement at the junction of State 111 with State 74. At 15 *m.* is RANCHO MIRAGE. At DEEP CANYON is the DESERT RESEARCH CENTER of the University of California, 2,800 acres owned, and 7,200 operated by arrangement with the U. S. Bureau of Land Management.

The ANCIENT BEACH LINE is plainly visible (R) along mountain slopes and canyon walls (*see TOUR 3e*). Along the beach line, fossils and other evidence of former life are found.

CATHEDRAL CITY, 20.3 *m.* at the mouth of Cathedral Canyon is in the Palm Springs resort area and has hotel and motel accommodations.

At 25.6 *m* is the junction with a poor dirt road; L. on it 2.3 *m.* through great groves of palms, to the mouth of ANDREAS CANYON (R), and past MURRAY CANYON (R), 2.7 *m.*, to PALM CANYON NATIONAL MONUMENT, 5 *m.* At the head of Palm Canyon is a grove of 5,000 California fan palms, the only palm native to the western United States, ranging in age from seedlings to trees thousands of years old. Many of the giant palms show traces of fire on their lower trunks. Legend has it that the Cahuillas, who picked the clusters of berries from the palms for food, always burned the trees that belonged to a single family when the head of the family died, to enable the deceased person to carry his berry clusters with him on his journey.

At 12 *m.* is the junction with a canyon road; L. 0.5 *m.* on this road, which winds under the shadow of TAHQUITZ PEAK (8,826 alt.) to a parking space, from which a trail leads into TAHQUITZ CANYON. At the entrance to the canyon is TAHQUITZ BOWL, an outdoor amphitheater in which a play is given each summer. This canyon, like Palm, Murray, and Andreas Canyons, is in the Agua Caliente Reservation and the Indians exact minor tariffs from the tourists.

PALM SPRINGS, 27.2 *m.* (430 alt., 18,250, May, 1964) is one of the most fashionable resort towns on the Pacific Coast. Its population expands greatly in the salubrious climate of spring, fall and winter, when it has 50,000 residents; at the same time thousands of weekend and short-term transients enjoy its 18 golf courses, its 3,350 swimming pools, its tramway rides and its luxury shops.

Playground of rich Americans, Palm Springs has been successively the domain of Cahuilla Indians, a stagecoach stop, and a desert home for the convalescent and tubercular. Today this ultrasmart winter resort for movie stars and for people who like and can afford to live where and as movie stars live, gleams as brightly as a new toy village. Its buildings are uniformly of California pseudo-Spanish architecture: the white, lemon, or buff-colored dwellings, entered by doors painted bright red, blue, or yellow, are surmounted by red tile roofs and enclosed by wooden fences, bordered by rows of pink and white oleanders or the green feathery plumage of tamarisk trees. Here are branches of the most expensive New York and Los Angeles shops; golf courses and hotels that range from the palatial to the modestly magnificent; private and public schools and no lack of masseurs and masseuses; dude cowhands for atmosphere and branch brokerage offices for the bigger businessmen.

First known as Agua Caliente (hot water)—so named by De Anza in 1774 because of its hot springs—the settlement dates from 1876, when the Southern Pacific first laid down its tracks through Coachella Valley. Until 1913 Palm Springs remained a sleepy little hamlet with a single store and a roadside inn on a poor desert road. A nearby gold mine, the Virginia Dale, offered the only attraction to people in search of a living, and the warm dry climate (average noontime temperature, 81°; average evening temperature, 45°) made it an excellent but little known health resort. Then Hollywood, early in the 1930's, discovered the climatic and topographical charm of the little village resting on a shelf of the San Jacinto Mountain base at the edge of the desert. A new highway was cut through, Los Angeles and New York promoters got to work, and the modern town sprang up almost with the speed of a movie set.

The popularity of Palm Springs has brought about a large expansion of building construction, especially of resort hotels—the word "motel" is frowned on by the community fathers, who prefer villa, lodge, inn, manor and guest ranch. In 1964 building permits reached a value of $22,835,590, an increase of more than $9,000,000 over 1963. Two convention hotels, the Riviera and El Mirador, enlarged their facilities, as did the older Oasis. The Canyon Club Inn, of the Canyon Country Club, adjoins the 19th hole. In the heart of Palm Springs the Spa Hotel and Bathhouse has been built on the site of the mineral wells and natural hot springs that have 104° F. A major development is Sky Mountain, 640 acres on Highway 111 between Palm Springs and Palm Desert, extending into the Santa Rosa Mountains, which was planned for 500 home sites costing from $20,000 to 8-acre mountain top locations costing $150,000.

Instead of trailer parks in undesirable areas Palm Springs provides twenty "mobile parks" with recreational facilities, with room for 1,500 units. These "mobile homes" are the aristocracy of living on wheels; sometimes one may represent a $50,000 investment, and there is record of one that had a $10,000 mobile home in the rear for the maid.

An unusual aerial experience is provided by the PALM SPRINGS AERIAL TRAMWAY, which starts from Highway 111 just north of downtown and moves for two and one-half miles over cliffs and chasms to the Mountain Station on Mt. San Jacinto, 8,516 alt., a ride of great scenic beauty. A Swiss-like chalet and restaurant serve the visitor at the top and uncrowded ski slopes are available in winter.

PALM SPRINGS MUNICIPAL AIRPORT is served by Western and Bonanza Air Lines. There are about 65,000 takeoffs and landings a year. In 1965-66 it erected a 60-ft. control tower at a cost of $400,000.

The town's SULPHUR SPRINGS lie just east of the business district on the AGUA CALIENTE RESERVATION; here the Indians who own the property maintain a bathhouse on the spot where their forefathers camped for hundreds of years.

Directly west of Palm Springs is TACHEVAN (dry rock) CANYON; a short walk leads to a mountain stream trickling among brightly painted rocks—an ideal picnic site.

Southeast of Palm Springs on State 111 are located a number of fashionable resort hotels, often with golf links and catering to privacy. They include CATHEDRAL CITY, 7 *m.;* RANCHO MIRAGE, 10 *m.;* PALM DESERT, 13 *m.;* BERMUDA DUNES, 18 *m.* and LA QUINTA, 16 *m.,* considered the most luxurious in the Palm Springs area.

Northwest of Indio US 60 runs over gradually rising sand plains past whitish billows of mesquite-spotted sand dunes to WHITEWATER, 30.1 *m.* (1,130 alt., 100 pop.), a service stop under a grove of cottonwoods.

At 33.6 *m.* is a junction with State 111 (*see above*).

The highway rises westward through low SAN GORGONIO PASS, between the peaks of SAN JACINTO and SAN GORGONIO, both of which are usually snowcapped. Indians had used the pass for their winter trek to lower levels, but hostile tribes at top and bottom did much to discourage white travel. In 1862 the stage route to Ehrenberg started down the grade, but later shifted its terminus to Yuma and selected another route that took the stages through more miles of mountainous territory before dropping them to the hot floor of the desert. Passengers had protested at the frequency with which they had to clamber out and help the stages out of the sand.

Through this pass winds blow ceaselessly from the ocean to the desert. East winds seldom occur, but when they do Los Angeles wilts under blistering "Santa Anas," dry wind storms. These reversals usually occur on the rare occasions when rain strikes the desert. Foggy days on the coast intensify the rush of air toward the east, causing the harsh desert winds of April and May.

Along the highway long freight trains are frequently seen, two engines herding them slowly and powerfully up the grade. The Southern Pacific laid its tracks through here in 1875, after first winning the friendliness of the Cahuilla by promising them free rides.

CABAZON, 40.1 *m.* (1,791 alt., 400 pop.), is a Southern Pacific R.R. town with a roundhouse for the auxiliary engines that help push passenger and freight trains over the pass. The name, a corruption of the Spanish word *cabezón* (big head), was given by Spaniards, according to legend, for a local Indian chief with a very large head. The almost continuous winds from the pass have given the town a sandblasted appearance.

Upward through country increasingly green as the summit (2,559 alt.) is approached, the highway climbs into BANNING, 46.1 *m.* (2,350 alt., 13,000 pop.), laid out in 1883 by Phineas Banning, stagecoach operator. Banning is in a fruit-producing area.

Within recent years a camel frisked about the neighborhood of Banning, finally making such a nuisance of himself that he was hunted down by a posse and shot. This was undoubtedly an aged survivor of the government caravans that crossed the desert prior to the Civil War. For various reasons—the camels terrified horses, their feet were cut by stones, and the officers in charge hated the slow bad-tempered beasts—the experiment was closed; but the participating animals remained a problem for many years afterward. Some wild camels were sighted on the desert as late as 1890, and even now newcomers are solemnly assured they can expect to run into them at any moment.

1. Right from Banning on a dirt road to the MORONGO INDIAN RESERVATION, 5.5 *m.* (*appointments to visit reservation should be mad. with Indian Agent at Banning*). On this reservation, named for "Captain John" Morongo, a great Indian leader, live 294 Indians of the Cahuilla tribe and Serrano clan.

2. Right from Banning on a good dirt road through Banning Canyon. where desert and mountain meet in vivid contrast, to HIGHLAND SPRINGS, 2 *m.*, a resort among mountains, streams, canyons, and pine forests. On the hotel grounds is an INDIAN GRINDING STONE, left by prehistoric tribes.

At INTERNATIONAL PARK, 2.5 *m.*, another resort area, a Japanese cherry blossom festival is held each spring.

BOGART BOWL, a natural amphitheater adjacent to International Park, draws picnickers and receives wide notice as the focal point of an ambitious international peace program and as the setting of an annual Easter sunrise service.

BEAUMONT, 52.1 *m.* (2,559 alt., 4,432 pop.), spreads over a plain at the summit of the pass. The region was opened to Yankees in 1853 by a herd of straying cattle. Fast in pursuit, Dr. I. W. Smith, a Mormon, followed the errant beeves from San Bernardino to their final stopping point, directly north of the townsite. Smith at once began to pioneer in the region, beating off grizzlies and fleeing from Indians, until the area became somewhat sophisticated by the addition of a stage station and later a hotel (1884). Beaumont, called San Gorgonio from 1884 to 1887, did not grow with the customary California speed until a development company shifted the townsite and publicized the regional potentialities for fruit-growing. Apples were planted in vain; but cherries and almonds prospered; each spring thousands drive to see the cherry blossoms that cover the region with a sea of white billows. A cherry picking festival is held in June.

At the western margin of Beaumont the stately eucalyptus tree appears. These trees, imported from Australia in the belief that they could be exploited for their oil and hardwood, were set out in great tracts all over the State. But the trees proved practically valueless and they are now merely graceful additions to the coastal scenery. They grow with extreme rapidity, sometimes reaching a height of 175 feet in 25 years. The highway begins a winding passage through the MORENO BADLANDS, a country that hardly deserves the name. Through gullies and canyons appear foothills green with live-oaks. In spring these hills are gay with blue lupin, yellow wild mustard, and golden California poppies. US 60 next traverses a broad expanse of open land; great tracts of the reddish soil are used for the growing of grain, principally wheat, oats, and barley.

RIVERSIDE, 76.5 *m.* (851 alt., 84,382 pop., 1960; 127,869 est. 1965), with wide tree-lined streets, has street lights shaped like the Indian rain cross—one light is at the top of the tall pole, two others are suspended from the crossarm. The first white man to tramp this soil came from over the Sierra. In 1774 Captain Juan de Anza, on his way from Arizona to Monterey, built a bridge across the flowing Santa Ana River here. The area was part of the Jurupa Rancho of Don Juan Bandini, mentioned by Dana in his *Two Years Before the Mast.* The discovery of gold brought in the Yankees, who soon elbowed the Mexicans out of their holdings. A flood in 1862, followed by two years of drought, decimated the herds and forced intensive development of water resources and agriculture. In 1870 the Southern California Colony Association purchased the ranchos and surveyed the Riverside townsite.

The city's showplace is the block-square, privately owned GLENWOOD MISSION INN, bounded by Sixth, Seventh, Main, and Orange Sts. Its Patio Dining Court, Spanish Cloisters, and Courts of the Birds and the Bells are proudly displayed to visitors.

East of Riverside at Day St. and US 60 is the 600-acre RIVERSIDE INTERNATIONAL RACEWAY for automotive sports, opened 1957, with a full-length course of 3.27 *m.* and a 5,400 ft. Straightaway. Major events are a 500 *m.* stock car road race (purse $65,000) January; a sports car race, April; the 200-mile Los Angeles Times' Grand Prix for sports cars, October; the Golden State 440-mile race for American-made stock cars for Grand National Championship, November.

In the year 1873 a Riverside woman obtained from the Department of Agriculture at Washington two cuttings of a "sport" variety of orange, imported from Brazil. This orange had suddenly appeared on one tree in Bahia, secreting more juice and abandoning its ancient custom of reproducing by seed. It was the first seedless orange. Abetted in its waywardness by orchardists, it has been widely propagated. When the first trees were set out the shoots were so much in demand that a price of $1 a dozen was set upon them. Thieves stole so many shoots that it became necessary to surround the trees with a high barbed wire fence. The surviving PARENT NAVEL ORANGE TREE stands at the corner of Magnolia and Arlington Aves.

UNIVERSITY OF CALIFORNIA, RIVERSIDE, occupies 1,200 acres. Originally the Citrus Research Center and Agricultural Experiment Station, founded 1907, was conducted by the university. In 1954 the College of Letters & Science was added, and since the College of Agriculture has been formed to provide a four-year course in agricultural science. In 1962 the Air Pollution Research Center was added. Enrollment in 1965 was 2,629.

Left from Riverside on Magnolia Ave. (State 18) to the SHERMAN INDIAN INSTITUTE, (L) 5.2 *m.,* a large resident, co-educational school for Indian children more than 14 years old; it occupies a 40-acre tract. Academic instruction is given to the twelfth grade, with special emphasis on vocational training.

CORONA, 13.9 *m.* (700 alt., 23,000 pop. 1965 est.), formerly South Riverside, has a circular boulevard. In 1913, when the automobile was still new, Ralph De Palma, Barney Oldfield, and Earl Cooper participated in a 300-mile, free-for-all race on the boulevard; Cooper, the winner, averaged a breath-taking 75 miles per hour.

ARLINGTON, near Riverside, is the seat of LA SIERRA COLLEGE, a Seventh Day Adventist institution enrolling 1,160.

West of Riverside, US 60 passes under a graystone archway, above which green terraces (L) circle the lower slopes of MOUNT RUBIDOUX (1,337 alt.). According to tradition, this hill was once the altar of Cahuilla and Serrano sun worship. Pagan influence has been obliterated by the erection of an immense cross on the peak, honoring Fra Junipero Serra, founder of the California missions. The cross, visible across the valleys, is the focal point of Easter sunrise services attended by thousands. On the side of the mountain stands the WORLD PEACE TOWER, dedicated by the people of Riverside to Frank A. Millar, proprietor of the Mission Inn, active in the promotion of international peace. At the foot of the mountain is the ST. FRANCIS SHRINE, with a fountain for birds.

MIRA LOMA (lookout hill), 86 *m.* (787 alt., 250 pop.), is a little settlement surrounded by acres of vineyards. Grapes thrive in the sandy, irrigated soil. At 95.7 *m.* is the junction with Central Ave.

Left on Central Ave. to CHINO (713 alt., 10,305 pop.), 3 *m.,* the trade center of a citrus and beet-sugar growing area. It was founded in 1887 when Richard Gird subdivided half of his 47,000-acre Rancho Del Chino into 10-acre farms and laid out a mile square townsite in the center.

The CALIFORNIA INSTITUTION FOR MEN occupies a tract between Edison, Central, Euclid, and N. Robles Aves. Completed in 1939 it has room for

2,100 inmates. It does not stress punishment but gives offenders an opportunity to return to normal lives, and its record for success is extraordinary.

POMONA, 100.2 *m.* (861 alt., 67,157 pop., 1960; 80,802, 1964, est.), was named for the Roman goddess of fruits at its founding in 1875. The sheep of Mission San Gabriel, tended by Indian herdsmen, grazed here at the close of the eighteenth century, when Rancho San José was part of the mission's holdings. In 1830 the rancho's 2 square leagues were granted to Ygnacio Palomares and Ricardo Vejar, but there were few settlers here until the Southern Pacific arrived. Only in 1882, with the formation of the Pomona Land and Water Company, did steady growth begin. Since 1960 Pomona had been building a Civic and Governmental Center on a 12-block site, to include a new City Hall, a Superior Courts building, a civic auditorium seating 2,500 and a government office building. Completed are the Public Safety Building (fire and police headquarters) and the Main Library. In October, 1962, Pomono opened its MALL, a nine-block shopping area devoid of traffic, landscaped, with fountains and textured concrete pavement.

The PALOMARES ADOBE, 1569 N. Park Ave., back from the street among oleanders, jasmines, roses, and peach and orange trees, was built soon after 1837. Its low roof projects over a broad *corredor* and its outside stairway climbs through vines and creepers. Nearby, at 1475 N. Park Ave., hidden among flowers and fruit trees is the ALVARADO ADOBE, once the residence of Palomares' close friend, Ygnacio Alvarado, who had a private chapel in his home. Not far away, at 548 S. Kenoak Dr., is the CHRISTIAN OAK, under which, on Nov. 18, 1873, Father Zalvidea of Mission San Gabriel conducted a dedicatory mass for Rancho San José, and pronounced benediction on the Palomares and Vejar families.

GANESHA PARK, between Walnut and Val Vista Sts., named for a Hindu god, and planted to resemble a natural woodland, holds a large open-air GREEK THEATER and sport facilities. LOOKOUT POINT commands a fine panorama.

Adjoining Ganesha Park on the northwest are the 175-acre LOS ANGELES COUNTY FAIR GROUNDS, since 1922 the scene of what the exposition's literature calls "the biggest county fair in America." Half-a-million visitors flock in annually in September to see agricultural, horticultural, livestock, domestic arts, educational, machinery, and arts and crafts exhibits, as well as poultry, rabbits, and diary produce.

North of the Fair Grounds is PUDDINGSTONE RESERVOIR, and west of the Reservoir is the STATE PARK.

Left from Pomona on Hamilton Ave. to a junction with Valley Blvd. at 2.3 *m.;* R. to SPADRA, 3.5 *m.* (711 alt., 275 pop.), named for his native home in Arkansas, Spadra Bluffs, by William Rubottom, first Yankee settler in Pomona Valley, who conducted a tavern.

On the DIAMOND BAR RANCH (L) 4.5 *m.,* against the rugged lower slopes of the Rocky Hills, is the two-story VEJAR ADOBE, second ranch house of Ricardo Vejar. Built in 1850, just two years before Yankee business methods reduced Vejar to poverty, the ranch house is one of the best examples of the adobe mansion in southern California, its broad corridors and handmade doors unmarred by the passage of time.

At 103.8 *m.* is an asphalt paved road. Left on this road to the grounds of the W. K. KELLOGG INSTITUTE OF ANIMAL HUSBANDRY, 1 *m.* a unit of California State Polytechnic College. The 750-acre farm was presented with an endowment of $600,000.

At 108.5 *m.* is a junction with Citrus Ave. Right on Citrus Ave. is COVINA, 1.2 *m.* (555 alt., 20,124 pop.) originally a fruit shipping center. East of Covina is the Voorhis unit of California STATE POLYTECHNIC COLLEGE.

West of Covina State 39, which runs through the Puenta Hills and proceeds due north from West Covina to Azusa, then to become the Canyon Road, bounds an area to the west that has multiplied its population since World War II. Small communities, formerly centers for orange and citrus fruits, have become cities. They lie in the area with the San Bernardino Freeway (US 60-70) on the south and the Foothill Boulevard on the north, and include West Covina, Baldwin Park, El Monte, Temple City and Rosemead, the last adjoining Alhambra.

WEST COVINA (60,329 pop. 1964), had only 4,499 people in 1950. Its growth is ascribed to its central location in the East San Gabriel Valley, which adds about 6.2% a year, and its business enterprise. It was incorporated 1923. The U. S. Navy's anti-submarine weapon system is made by Honeywell Corp. here.

BALDWIN PARK (381 alt., 42,950 pop., est. 1964), in the East San Gabriel Valley sector of Los Angeles County, was named Vineland in 1880 and changed to Baldwin Park in 1906 in order to dissuade "Lucky" Baldwin from starting a competing town nearby. Chiefly residential, it has 400 acres zoned for light industry.

EL MONTE, 13 *m.* east of Los Angeles Civic Center (13,163 pop., 1960), is southwest of Baldwin Park. It was settled by pioneers who found water here at the end of the Santa Fe Trail. In its early decades vigilantes called the Monte Boys were formed to keep order. The first permanent jail (1880) is preserved by the El Monte Historical Society. The area developed hop fields and walnut groves. A large industrial development is in SOUTH EL MONTE, a separate community.

TEMPLE CITY (31,838 pop., 1960; 37,750 est. 1965), incor. 1960, is 15 *m.* from Los Angeles Civic Center and lies south of Arcadia, east of San Gabriel and north of Rosemead and El Monte. It calls itself the Home of Camellias. The Temple City Camellia Society, org. 1946, promotes camellia culture. The annual Camellia Festival in February features a children's parade, with more than 5,000 participating.

ALHAMBRA (450 alt., 54,807 pop. 1960; 59,940 est. 1964), 8 *m.* northeast of Los Angeles Civic Center is adjacent to SOUTH PASADENA, SAN MARINO, SAN GABRIEL and MONTEREY PARK. Incor. 1903, it is served by Southern Pacific, Metropolitan Transit, Greyhound Bus Lines. Hi Neighbor Days celebration takes place the second week in May. The site was part of a grant given Mission San Gabriel in 1771. Benjamin D. Wilson, "Don Benito," who owned vast tracts of present Los Angeles County, subdivided the land in 1874; his family, after reading Washington Irving, named it Alhambra and gave streets Moorish names; settlers found Boabdil difficult and renamed it Main Street.

SOUTH PASADENA (659 alt., 20,899 pop. 1963), incorporated

1888, is two miles south of Pasadena and 10 from the center of Los Angeles via the Pasadena Freeway. Its strict supervision of landscaping, parks and playgrounds make it a popular residential city. Its principal industries are electronic.

MONTEREY PARK, 121.9 *m.* (475 alt., 37,821 pop.), incorporated in 1916, is in the low, rolling Monterey hills. The community is proud of its many old trees, which include the pepper, the California live-oak, pink-flowering white ironbark, Australian beech, Torrey pine, various types of palm, and several kinds of cork oak.

LOS ANGELES, 130.7 *m.* (*see LOS ANGELES*).

Tour 14

(Yuma, Ariz.)—El Centro—Jacumba—La Mesa—San Diego, US 80. Arizona Line to San Diego, 177.6 *m.* Identical with Interstate 8.

US 80 runs westward across the thinly populated southernmost section of California. In altitude the route varies from below sea level to an elevation of 4,050 feet. Spanish explorers first called this road the Jornada de la Muerte (journey of death) and El Camino del Diablo (devil's highway).

The first Butterfield stage, carrying mail from St. Louis to San Francisco, used this route in 1858. Twice a week after this until 1861 the long trip was made, the only scheduled service from East to West and one of the longest stage routes in history. The Butterfield line used 100 Troy and Concord coaches, 1,000 horses, 500 mules and 800 men. The eastern part of the route was menaced by Comanche, the middle part by Apache and the western section by Yuma and Mojave. Because road agents regarded the stages as legitimate prey, the coaches became traveling forts. Advertisements of the line in the East read, "Fare $200 in gold. Passengers are advised to provide themselves with a Sharp's rifle and one hundred rounds of ammunition. Through service from St. Louis to San Francisco by the quickest route. Twenty-four days travel in our luxurious stages and you will arrive in 'The Land of Gold'!" The outbreak of the Civil War ended the Southern control of Congress, which had forced the use of this hard route.

West of the Colorado River lies the great Colorado Desert, the dark brown Chocolate Mountains flanking its northern and eastern rims; a great series of shifting sand dunes lies diagonally across it, intersecting the highway between Yuma and Holtville. The Imperial Valley area, much of which is below sea level, lies in the heart of the desert—irrigated land in the white waste.

West of the desert, the highway climbs the forested mountains of the Peninsular Range; as the road winds down the western slope to the Pacific Coast, orange, lemon, and grapefruit groves, vineyards, and avocado orchards appear.

Section a. ARIZONA LINE *to* EL CENTRO; *56.6 m. US 80. Interstate 8.*

The COLORADO (Sp., ruddy) RIVER, 0 *m.,* separates Arizona and California. It is crossed by a steel bridge, one end of whose span is in Yuma, Arizona.

The Colorado River was discovered, supposedly, by Hernando de Alarcón in August 1540; he named it Río Grande de Buena Esperanza (great river of good hope). For 15 days Alarcón and his party explored the river in small boats. Two land expeditions, commanded by Francisco Vásquez de Coronado, explored this territory in 1540 and 1541. Melchior Díaz, chief of one of the parties, named the river Río del Tizón (river of the firebrand) for the sticks of burning wood carried by the native Indians. Father Kino, the Jesuit priest, called it Río de los Martires (river of the martyrs), when he crossed the desert in 1701. The present name was given it by Governor Juan de Oñate of New Mexico in 1905 and was suggested by the reddish color imparted by silt suspended in the water.

When bridges were built across the Colorado in the 1870's and 1880's, an adventurous period of river history ended. Early explorers and adventurers crossed the river by fording or swimming their horses. In 1849, when men were crossing the continent to California by every possible route, two ferries were in operation. One was run by Indians using a boat abandoned by Gen. Alex Anderson of Tennessee; the other was operated by a Dr. Lincoln, in partnership with an outlaw, John Glanton. To quell competition, Glanton had his men kill Lincoln and wreck the Indians' boat. The Indians in turn wiped out Glanton and his gang and had the river to themselves until 1850. In that year L. J. (Don Diego) Iaeger led a group across the desert from San Diego. The men nearly died of thirst and heat before reaching the banks of the Colorado on July 10, 1850. Work on a ferryboat was started as soon as they recovered. Beams were hewed from cottonwoods, and fastened with wooden pegs.

West of the bridge the road traverses small Yuma or Bard Valley. The derivation of "Yuma" is uncertain, but was first mentioned by Father Kino. It may come from the word Yahmayo (Ind., sons of the captain, or, sons of the river).

The YUMA INDIAN SCHOOL and INDIAN AGENCY, 0.1 *m.,* on a hill (R), is on the site of the old Fort Yuma. A STATUE OF FATHER GARCÉS stands in front of the chapel. In 1781 the Mission Puríssima Concepción, which had been built on this spot, was destroyed by Indians who objected to the serflike conditions the fathers imposed. In 1880 an Indian school was opened on Don Diego's farm, somewhat northwest, but in 1890 the Catholic Church established this school on a

Government subsidy. The Indians are still reluctant to send their children to the white man's school.

The YUMA INDIAN MISSION (R), 0.2 *m.*, is on the Fort Yuma Indian Reservation.

At 0.3 *m.* is the junction with a graveled road.

Right on this road through the FORT YUMA INDIAN RESERVATION, covering 8,350 acres, where the dwindling members of the Yuman tribe have been living and farming since the 1880's. Scattered on each side of the road are the small adobe huts of the Indians with thatched roofs and ramadas (arbors) where the owners rest after work and hang their vegetables, skins, and utensils. Although the reservation is irrigated by a canal from Laguna Dam, the smaller farms are not well kept and look parched. It is noticeable that white inhabitants of the reservation outnumber the Indians.

ROSS CORNER, 5 *m.*, is a small trading center; L. here 19 *m.* on a sand road to the old PICACHO (summit) MINES. Gold was found here in 1860 by an Indian. Twenty years later Mexican prospectors struck a rich lode, and the town of PICACHO, 21.5 *m.*, sprang up. The gold brought gay times in its wake. Bull fights, fandangos, and fiestas were frequent. Enterprising Yankees built a stamp mill and a small railroad. The "bankroll" period ended when the mines were exhausted early in the twentieth century.

On the main side road is the UNITED STATES EXPERIMENTAL FARM (L), 6.5 *m.*, under the direction of the Division of Plant Introduction in Washington. The station has developed a disease-resistant alfalfa, now widely used.

BARD, 7 *m.* (87 alt., 550 pop.), is a settlement of the Bard Irrigation District and starting point for fishing in the Colorado.

POT HOLES, 10 *m.*, named for the pot holes in the river at this point, is little more than a group of houses occupied by the caretakers of the Laguna Dam.

The LAGUNA DAM, 12 *m.*, formerly called the Yuma Dam, is a diversion dam for irrigation purposes, the first undertaken by the United States Reclamation Service in California. The weir was completed in 1909, and the water turned into the canals the following year.

IMPERIAL DAM, 17.5 *m.*, is the headquarters for the All-American Canal (*see below*), part of the Hoover Dam Project, authorized by Congress in 1928. The dam, of the floating, or Indian weir type, has raised the surface of the Colorado River 22 feet.

WINTERHAVEN, 0.9 *m.* (85 alt., 1,100 pop.), a bustling, businesslike spot, was headquarters for the All-American Canal construction work.

ARAZ STAGE STATION, 5.8 *m.*, once a way-station for freight wagons, now provides gas and food for motorists.

The ALL-AMERICAN CANAL is crossed at 6.3 *m.* This irrigation canal is 80 miles long, 232 feet wide, and 21.6 feet deep at the water level. "Walking dredges," with booms 185 feet long and buckets that raised 14 cubic yards of sand at a scoop, were used to cut through the sand dunes. The canal can deliver 15,000 cubic feet of water per second. A branch to the Coachella Valley is 48.9 miles long.

PILOT KNOB (1,000 alt.) is visible (L) along this stretch of the highway. It is a single, dark butte, composed of granite overlaid by volcanic rock and lava flows, and stands just north of the Mexican border.

At 91.1 *m.* is the junction with a dirt road.

Right on this road is OGILBY, 3.9 *m.* (365 alt., 60 pop.), once a booming mining town and now only a station on the Southern Pacific R.R. Gold was mined from 1879 to 1918 in the CARGO MUCHACHO (errand boy) MOUNTAINS, whose twin cones (2,000 alt.) rise 5 miles northeast of the town.

TUMCO, 10.5 *m.*, named for the initials of the United Mine Company, is another ghost of a mining camp. Early in the 1900's, Tumco had a population of 3,000, four saloons along Stingaree Gulch, and an occasional gunfight. The Tumco gold deposits were discovered by a Swedish track-walker, who apparently deviated from his line of duty.

Desert sand hills are visible on both sides of the highway at 14.2 *m.* Formed by the wind, the shifting dunes, some 200 feet high, sometimes bury the road. It was through such land that De Anza and his party struggled—the first white men known to have crossed the Colorado Desert. Juan Bautísta de Anza was captain of cavalry of the Royal Presidio of Tubac, Sonora, Mexico. Determined to find an overland route to the Pacific, he assembled his first expedition at Mission San Xavier del Bac, Mexico, and began his journey in 1774. He set up his first camp near the present Yuma. Wtih him were Father Garcés and Capt. Juan Valdéz, who has been called the Spanish "Kit Carson." Hostile Apaches stole many of his best horses, but a chief of another tribe helped the party immeasurably with food and guides. In 1775 De Anza organized a second expedition larger and better equipped, which crossed the desert and took colonists to San Francisco.

Vestiges of the Old Plank Road (L), parallel the highway at 14.6 *m.* This was built in 1914 and followed the contours of the sand hills across the desert; the present highway skirts their tops. Constant shifting of the sand made this narrow plank road expensive and difficult to maintain.

LITTLE VALLEY, 17 *m.*, a small valley amid shifting dunes, is one of Hollywood's favorite desert locations. Many a sheik picture and epic of the Foreign Legion has been shot here.

GRAY'S WELL, 20.4 *m.* (*restaurant, gasoline, cabins*), was named for Newt Gray of Holtville, who worked hardest for the extension of the desert road to Yuma.

West of Gray's Well is an arid country, part of the SALTON SINK (*see TOUR 3e*), typical of Imperial Valley before it was reclaimed by irrigation.

MIDWAY WELL, 29.6 *m.* (*service station and cafe*), is on the edge of the Imperial Valley.

At 40.6 *m.* parts of the ANCIENT BEACH LINE of Lake Cahuilla are visible. The beach line is a hard, elliptic ridge of sand around the Imperial Valley.

US 80 crosses the EAST HIGHLAND CANAL, 41.1 *m.*, which extends along the rim of the valley to a point north of Niland, and, with the Main Central Canal and the West Side Main Canal supplies water to the valley. The highway drops gradually here.

DATE CITY, 42.5 *m.* (15 alt.), is headquarters of a date-growing district. Fresh dates are on sale at stands lining the road.

HOLTVILLE, 48.7 *m.* (— 10 alt., 3,080 pop.), is the chief trading point for eastern Imperial Valley. It has a large Swiss colony and its major interest is dairying. The town was named for W. F. Holt, who bought large blocks of stock in the Imperial Water Companies from the California Development Company. The town has a strategic position next to a 40-foot drop in the Alamo River, where it was possible to build a hydroelectric plant and generate power.

The Alamo River was greatly enlarged by the flood of 1905, which cut a channel 50 feet deep in the sandy soil in places. It flows into the Salton Sea. Holtville calls itself the world's carrot capital.

At MELOLAND, 52.7 *m.* (21 alt., 460 pop.), is the EXPERIMENTAL STATION (*open to qualified persons*) of the University of California's College of Agriculture. Here experiments are made in the application of scientific principles to the cultivation of land, especially in the production of field crops. Meloland is also a packing and shipping point.

EL CENTRO, 56.6 *m.* (— 52 alt., 18,390 pop.), is one of the chief trade, storage and shipping centers of Imperial Valley. It is also the seat of Imperial County, and second largest town below sea level inside the United States. The town is in the heart of the most arid area in the country—reclaimed from the desert by irrigation.

Efforts of Imperial County investors to base their prosperity on more than agriculture are led by El Centro, which has been adding good roads and motels and increasing hospital facilities. Valley Plaza is a new shopping center costing $4,000,000; a new city hall to cost $400,000 and an office building for Imperial Valley Irrigation District, to cost $500,000, were designed.

The valley has already become the scene of literature; the first novel celebrating it was Harold Bell Wright's *Winning of Barbara Worth;* Jefferson Worth's prototype was W. F. Holt, the town's founder. Its first sheriff, Mobley Meadows, who had an uncanny ability in tracking down horse-thieves, has been memorialized, indirectly, in THE PLAINSMAN, a statue for which he posed; it stands in the courthouse square.

A government scientist in 1853 saw the agricultural possibilities of the 600 square mile depression, which was called "the hollow of God's hand" by the Indians, but little attempt was made to irrigate it until 1900 when a private corporation began construction of a canal from the Colorado River. The first settlers appeared in the valley in 1901 in spite of the imperfect workings of the canal system.

El Centro had been laid out in 1905 by Holt, but the population did not reach 5,000 until 1920. The Boulder (later Hoover) dam project authorized 1928 contained provision for irrigating the Valley. This started a boom and the big crop returns brought new investors and prosperity.

El Centro and other towns in the Valley are below sea level. The growing season is 12 months, and the growers boast of 365 days of sunshine a year. Average rainfall is 3 inches. Actual irrigated crop land consists of 430,398 acres divided into 5,114 farms, the average farm area being 95¼ acres. Field crops in 1963 yielded 1,182,000 tons of sugar beets, valued at $19,762,000 but alfalfa hay, next with 1,046,000 tons, yielded $32,216,000, while cotton lint, 177,900 bales, brought $29,024,-000. Lettuce was the largest vegetable crop, 300,200 tons worth $23,660,000. Nearest in tonnage were watermelons, 46,000 tons, but their value, $2,369,000, was exceeded by that of 21,000 tons of cantaloupes, $2,818,000. Next in order of income were carrots, dehydrated onions, tomatoes; even garlic brought in more than $500,000. In addition to diversified vegetables the Valley had substantial crops of fruits and nuts, livestock and dairy returns worth $65,000,000; profitable seed crops and apiary products, yielding a total income in 1963 of $213,-370,750, of which field crops provided nearly one-half.

The first water delivery in Imperial Valley came from the Colorado River in 1901 through the Alamo Canal, which had practically its entire length in Mexico. In 1905 the Colorado flooded the Valley and formed the Salton Sea. In 1911 agricultural interests formed the Imperial Irrigation District, which in 1922 acquired the properties of thirteen mutual water companies, which had been distributing water. The situation improved in 1928 when Congress authorized construction of Hoover Dam and the All-American Canal. First water was stored behind the dam in 1935. Imperial Dam on the Colorado, 20 m. north of Yuma, Ariz., in 1938 was another asset for the Valley. It is 181 ft. above sea level. The All-American Canal is 79.7 m. long and the Coachella Branch is 48.9 m. long. The Imperial Irrigation District also produces hydroelectric power and has acquired most of the competing power company properties in Imperial County and Coachella Valley. Revenue from power exceeds that from water.

In the 1930's the Valley suffered from labor troubles, brought about by the great influx of unemployed workers. Many of them were farmers ruined in the Dust Bowl of the Plains states. Organizers complicated the situation by making demands on farm owners who themselves suffered from the depressed market. A strike of 8,000 lettuce pickers in 1934 brought losses to the perishable crops.

The Valley also employed Mexican migrant workers, whose labor was prohibited by Federal law beginning in 1965. This brought new difficulties, for automation was not yet far enough advanced to cope with the huge crops of the Valley.

El Centro is at the junction with US 99 (*see TOUR 3e*).

Section b. EL CENTRO *to* SAN DIEGO; *121 m. US 80, also designated Interstate 8.*

Westward, this route traverses typical desert, then climbs out of Imperial Valley and penetrates the dry, brush-covered mountains that separate the depressed interior basin of southern California from the coastal slope. The verdure of parts of this mountain area is a refreshing contrast to the glaring white and monotonous brown of the desert.

The route is attractive at night, when the lights of the villages are visible for unbelievable distances, and the many weird rock and sand formations so strikingly resemble box cars that the stranger thinks he is entering a railroad yard. Driving in the night is advisable in the summer, for then the intense heat of the day evaporates and the temperature drops twenty degrees or more.

SEELEY, 8.1 *m.* (26 alt., 158 pop.), supplies farmers with necessities. It was founded in 1911, near the site of Silsbee, a town destroyed by the 1905-07 flood.

The channel of NEW RIVER, 8.5 *m.,* was eroded by swollen floodwaters of the Colorado. This was one of the few watering spots on the old desert trail between Fort Yuma and San Diego. Small humps of earth by the road, sculptured by the flood of 1905, look like the homes of cave dwellers.

DIXIELAND, 13.1 *m.,* was established in 1909 in anticipation of a new high-line canal west of the present ones.

South of Dixieland, stretching into Mexico, is the YUHA PLAIN, a bleak and desolate land of modified sand hills and outcroppings of mica. De Anza crossed this bare stretch in 1774 and found a watering place which he called Pozo de Santa Rosa de las Lajas (wells of St. Rose of the flat stones), now called Yuha Wells. The YUHA BADLANDS are a three-mile strip along the Mexican border. In them are fossil beds of oyster shells, fish and reptile casts, all deposited when this region was part of the sea.

The summit of SIGNAL MOUNTAIN (2,262 alt.), a knob (L) of old granite and schist in Mexico, is topped by an aviation beacon.

SUPERSTITION MOUNTAIN (750 alt.), 17.3 *m.,* rises (R) from the desert floor. It is probably an old volcano; the Indians say it is inhabited by evil spirits whose voices used to make weird rumblings.

At 17.8 *m.* is the junction with a desert road.

Right on this road, called the Butterfield Trail, is CARRIZO (reed grass), 19 *m.,* a small water hole fed by Carrizo Creek, which is usually dry. The SITE OF A BUTTERFIELD STAGE STATION is marked here by a pile of weathered timber and crumbling adobe. Eighty yards west is the grave of an alleged cattle rustler. A headstone reads, "Frank Fox. Killed 1882." He was shot while bathing in the creek and thus died with his boots off.

VALLECITO (little valley), 38 *m.* (*water and campsite*). The adobe BUTTERFIELD STAGE STATION here was rebuilt in 1934. Two men killed each other on this spot, according to the San Diego *Star* of Nov. 2, 1870, which wrote: "Two immigrants had an altercation on Monday last at Vallecito, resulting in the death of both . . . they took two shots apiece, both of the first shots missing and both of the second shots taking effect . . . one in the breast and the other in the abdomen . . . they were brothers-in-law and their families were present at the time of the difficulty."

At 57 *m.* is the junction with State 78 (*see TOUR 6e*).

PLASTER CITY, 21.1 *m.* (786 lat., 80 pop.), is the settlement and loading point of a company making plaster of paris from gypsum extracted 25 miles north, in Fish Creek Canyon.

At 25.1 *m.* is the junction with a desert road.

Right on this road to the PAINTED GORGE, 7 *m.*, a cleft in the COYOTE MOUNTAINS, with highly colored walls in reds, pinks, delicate greens and hybrid hues. Coral deposits and beds of fossil oystershells are at the upper end.

COYOTE WELLS, 26.1 *m.* (*service station and cafe*), was for many years a watering place for desert travelers. The well was hidden in a mesquite clump and so named because the coyote can smell out water around the roots of the mesquite.

Dry washes and gullies in this area are signs of the infrequent but heavy summer rains that batter down the vegetation and erode the sandy cliffs. As the road gains altitude, such desert flora as the ocotillo, ironwood, cactus, and creosotebush give way to the juniper, the yucca, manzanita, and oak.

At SHEPARD'S BRIDGE, 31.6 *m.* (2,784 alt.), the highway begins the ascent of the Mountain Springs Grade through INCOPAH GORGE, a rocky gap at the southern rim of Imperial Valley. The grade which takes the jump from the desert floor to the uplands was regarded as an engineering triumph when it was finished in April 1913. It eliminates the hazards of the old Devil's Canyon route.

Near MOUNTAIN SPRINGS GAS STATION, 37.5 *m.* (2,886 alt.), some fine gems have been found in the essonite garnet deposits. They are sold as "California hyacinth." These hills also contain deposits of marble, feldspar, magnesia, and silica.

BOULDER PARK, 39.2 *m.*, the summit of Mountain Springs Grade, offers a far view of Imperial Valley and the desert.

JACUMBA (Ind., hut by the water), 48 *m.* (2,800 alt., 400 pop.), is a resort spa, 200 yards from the Mexican border, with mineral springs and baths. Chief patrons of this resort are residents of Imperial Valley, who have summer cabins on the mountain slopes. Jacumba was first a mail station on the Yuma route, established by James McCoy in 1852. McCoy built a fort and at one time held off an attack by 500 Indians.

About 3 miles north of Jacumba is the mouth of the 11-mile Carrizo Gorge, one of the most spectacular valleys of the region. The light-colored granite walls have been gullied and worn into deep lateral crevices and contain odd formations. High up on the precipitous slopes is a narrow contorted shelf bearing the tracks of a branch of the Southern Pacific Railroad.

ECKENER PASS HAMLET, 55.3 *m.*, was named for Dr. Hugo Eckener, who passed over this point in his Zeppelin flight across America, Aug. 26, 1929.

West of EL RICO, 59.2 *m.* (3,008 alt.), the highway traverses the CLEVELAND NATIONAL FOREST.

At LA POSTA, 65.6 *m.* (3,119 alt.), a service station, is a junction with a dirt road.

Right on this road to LA POSTA INDIAN RESERVATION, 0.8 *m.*, an area of 3,879 acres peopled by only three Indians.

BUCKMAN SPRINGS, 71.1 *m.* (3,225 alt.), was named for Col. Amos Buckman, who hunted for gold in this region in the 1860's.

Left from Buckman Springs on a dirt road to MORENA LAKE and MORENA RESORT, 8.5 *m.* (*boats and tackle for hire*). The lake is part of the San Diego water system. Its dam, 171 feet high, is of the rock-fill type, cheaper than concrete construction and more resistant to earthquake. A professional rainmaker, C. M. Hatfield, used this site for his precipitation plant in 1915, to try to bring rain after several dry seasons. Disastrous floods followed his experiments, washed away a dam, and filled Morena Lake to within 18 inches of capacity. Hatfield threatened to sue the city of San Diego for payment. The city answered with a threat to sue him for the damage done by the floodwaters. Both suits were dropped.

At LAGUNA JUNCTION, 73.9 *m.* (4,050 alt.), is a service station and restaurant.

Right from Laguna Junction on an oiled road to the LAGUNA MOUNTAINS in Cleveland National Forest (*cabins, public campgrounds*), one of the best recreational districts in the region. The forest covers 815,000 acres, and is a game refuge. It takes in two parallel ranges of mountains, many lakes, and peaks. The road winds over mountains topped with chaparral and through valleys thick with pines. There are many varieties of pine and oak, incense cedar, big cone willow, and alder. The watershed brush cover is mainly chemical, redshank, lilac, and scrub oak, with some chinquapin, sumac, California laurel, toyon, wild cherry, and mountain mahogany.

At 8.8 *m.* is a junction with a dirt road; R. here 1 *m.* to the CUYAPAIPE INDIAN RESERVATION. Once many Indians ran away from the white men to this region, but today this reservation, covering 5,000 acres, harbors only five Indians and these intermittently.

On the main side road is LAGUNA POST OFFICE, 10 *m.*, headquarters for the resort area.

At the FOREST RANGER STATION, 10.4 *m.* information is available on the forest. The road skirts rocky slopes on the edge of the desert, with the wilderness spread out like a relief map almost a mile below. At 12.4 *m.* is a junction with a footpath; R. on this path 1.5 *m.* to MONUMENT PEAK (6,321 alt.).

PINE VALLEY, 74.4 *m.* (4,016 alt.), is a resort (*hotel, cabins, trails*). Maj. William H. Emory who settled here in 1869 is said to have bought the land from an old Indian for a saddle horse and a pack horse.

GUATAY (Ind., large house), 77.1 *m.* (*gasoline, restaurant*), was named for GUATAY MOUNTAIN (5,300 alt.); this peak (L) looked like an enormous wigwam.

CUYAMACA (Ind., rain above) JUNCTION, 78.7 *m.*, is at the junction with State 79 (*see TOUR 6C*).

The VIEJAS INDIAN RESERVATION (R), 81.7 *m.*, is in a small valley named El Valle de Las Viejas (the valley of the old women) by the Spanish, because the Indian men ran away to the hills and left the women and children. About 80 Indians live here now on modern farms. Before the reservation was set aside, the valley was used for breeding and pasturing race horses.

ALPINE, 90.4 *m.* (1,860 alt. 3,624, 1965 pop.), is a resort. Above the coastal fog belt, its crisp mountain air attracts people with bronchial trouble.

The road leaves the rough, mountainous district, and descends rapidly into a farming and horticultural region in El Cajon Valley. Wine and table grapes, oranges and lemons grow well in the fertile soil of this valley, where Spaniards pastured their stock.

FLINN SPRINGS, 97 *m.* (1,300 alt.), a resort (*camping and picnic grounds*), is just east of the old Rancho La Cañada de los Coches (colloq. Sp., valley of the hogs ranch), the smallest grant in California. In mission days the padres had a hog ranch here, and, until recently, it was quite common to see hogs running wild.

Left on US 80 up Magnolia Ave. (State 67) to SANTEE, site in 1964-65 of scientific experiments in removing pollution from small lakes, adjudged successful.

EL CAJON, 104.1 *m.* (450 alt., 5,600 pop., 1950; 37,618, 1960), occupies part of the old El Cajon Rancho, opened in 1869. Industries are locating here.

At 106.7 *m.* is the junction with an oiled road.

Left on this road to MOUNT HELIX (Lat., spiral), 1 *m.* (1,380 alt.), topped by a cross 30 feet high. Sunrise services are held on Easter morning in the amphitheater. The name was suggested by the corkscrew road that winds to the summit, where there is a view of San Diego and the Pacific Ocean.

LA MESA, 108.3 *m.* (538 alt., 10,940 pop., 1950; 35,350 1965, est.), is served by US 80 and State 67. Incorporated 1912. In the 1870's this section, called Allison's Springs, was used for grazing sheep. Wild game was so plentiful that a woman who ran a boarding house in the vicinity was able to supply her table with venison, rabbits, duck, and quail killed on her daily ramble near the house. La Mesa has an annual fall flower festival.

At 108.5 *m.* is the junction with a paved road. Right on this road to MURRAY DAM AND LAKE, 1 *m.*

South of La Mesa on the Helix Parkway is LEMON GROVE (19,348 pop., 1960). To the south the Sweetwater River fills LAKE LOVELAND, 27,700 acre-ft., completed 1945.

At 112.3 *m.* US 80 meets College Ave.

Right on College Ave. to SAN DIEGO STATE COLLEGE, 1.1 *m.* The buildings, of modified Moorish design, with creamy white walls and red tile roofs, are built around a square. In one corner of the quadrangle against the vine-covered arches of the buildings is *The Aztec,* a crouching figure cut from native diorite by Donal Hord, of the Federal Art Project. The AZTEC BOWL (L) is a football stadium. East of the campus is BLACK MOUNTAIN (1,590 alt.), and with a giant "S" on its side. In 1964 the College enrolled 14,195 students.

SAN DIEGO, 120.5 *m.,* is at the junction with US 101 (*TOUR 2d*).

Death Valley National Monument

Season: Open all year; regular season, Oct. 15 to May 15.
Automobile Routes: From the South, US 6 or US 395, northeast through Trona; US 60 from Barstow northeast to Baker on US 91, north on State 127 to Shoshone, west into park through Jubilee Pass, or continue north on 127 to Death Valley Junction, then west on State 190 to Furnace Creek Visitor Center. From the north, US 395 to Lone Pine, east on State 190 enter park at Towne's Pass. From the east: US 95 from Las Vegas to Lathrop Wells, then south to Death Valley Junction.
Railroad and Bus Routes: Union Pacific to Las Vegas, Nevada, then bus lines; Union Pacific and Santa Fe to Barstow, then bus; Santa Fe and Southern Pacific to Bakersfield, then bus. Bus service also from San Francisco and Los Angeles by Wanderlust Death Valley Tours. Conducted motor tours by the latter and by Tanner-Gray Line.
Air Service: Bonanza, Pacific, TWA, United and Western airlines to Las Vegas. Airport for private planes at Furnace Creek Ranch.
Accommodations: Furnace Creek Inn (Fred Harvey) American plan, has 67 rooms and provides swimming, golf, tennis, riding and one-day saddle trips, some with fees. Reservations with deposit necessary and cancellations require two to four day notice for refund. Furnace Creek Ranch (Fred Harvey) European plan, has 140 rooms in cabins and cottages, numerous facilities (fees for some), same conditions of reservation. Address Manager, Death Valley, or Fred Harvey Reservations Office, 530 W. 6th St., Los Angeles. Wildrose Station, open all year, operated under government franchise, has 4 cabins and some facilities. Stove Pipe Wells Hotel, 65 units, Scotty's Castle has 12 rooms operated by Gospel Foundation.
General Information: No fees are asked for admission, automobile or camping. There are two major and 9 secondary campgrounds. Texas Spring Campground has 85 sites, one level for house trailers. Furnace Creek campground has 10 sites. Hunting is illegal and firearms must remain unloaded. Pets must be kept on leash or under control. Firewood is available at Furnace Creek Ranch. The government does not have a guide service. Automobiles should have a full supply of oil, gasoline and water. In the winter warm clothing is recommended for chilly nights and at all times practical clothes and shoes.
Climate, Temperature: Fair weather, minimum rainfall, low humidity prevail.

Average number of clear days in one year is 283, although 351 days were once recorded. Annual precipitation average at headquarters in 15 years was 2.03 inches. Summer temperatures maximum 134° F.; usually above 100°, nights 80° to 100° with low humidity. Temperature seldom below freezing in winter. *Warnings and Regulations:* 45-mile speed limit. Register at ranger stations; travel only on roads that are open and patrolled, or make inquiries of rangers or at park headquarters. Carry abundant water for drinking purposes and radiator. Check gas and oil before all trips and carry additional supplies if long itinerary does not include service stations. Do not attempt to walk in valley during summer.

Death Valley National Monument, established in 1933, covers 2,981 square miles, 550 of which are below sea level. The narrow trough of the valley curves for 140 miles between steep mountains of naked rock that are striped and patched with barbaric colors. The Panamint Range, rising 6,000 to 11,000 feet above sea level, gives Death Valley its western wall, and the Grapevine, the Funeral, and the Black Mountains, rising 4,000 to 8,000 feet, form its eastern wall. The heavy rains that sometimes fall on the mountains run swiftly off the steep barren slopes and cascade into the valley, where the water quickly vanishes. The valley floor, once the bed of an ancient lake, is streaked with white salt, gray clay, and yellow sand; in the glimmering salt beds is the lowest spot in North America. From June to September, while the days are a blaze of light and the rocks radiate stored heat at night, Death Valley is one of the hottest places in the world. The Indian name for the valley was Tomesha—ground afire. In the winter, snow lies lightly on the ranges, except on Telescope Peak, which is white until May, and sunny days in the valley are delightful. The great charm of the area lies in its magnificent range of color, which varies from hour to hour.

In contrast with a verdant land Death Valley seems completely barren, but it has a varied vegetation. In a wet year, when from two to five inches of rain fall, canyons, washes, and even the valley floor are tinted with gray desert holly and green creosote-bush, with cigarette plant, paperbag-bush, ephedras, sprucebush, saltbush, brittle-bush, and wetleaf. Two varieties of mesquite, cottonwoods, and willows are found. High on the Panamints and Grapevines are piñon, juniper, mountain mahogany, Rocky Mountain maple, and bristlecone and limber pine.

The animals include the Nelson bighorn sheep in the mountains, the desert coyote and kit fox, and the wild burro. The bushy-tailed antelope, ground squirrel, and the trade and wood rat are often encountered. Among the birds that live in the valley the year round are big black ravens, road runners, prairie falcons, the beautiful Le Conte thrasher and busy rock wrens. There are several varieties of lizards: whip-tailed, gridiron, horned toads, and chuckwallas. Rattlesnakes are rare. An inch-long killifish, found at Salt Creek and Saratoga Springs, is a leftover from the days when Death Valley was occupied by a lake.

Death Valley contains rock of all the great divisions of geologic time; without exaggeration it can be called the geologists' paradise. At one time this was probably a region of low mountains and wide valleys

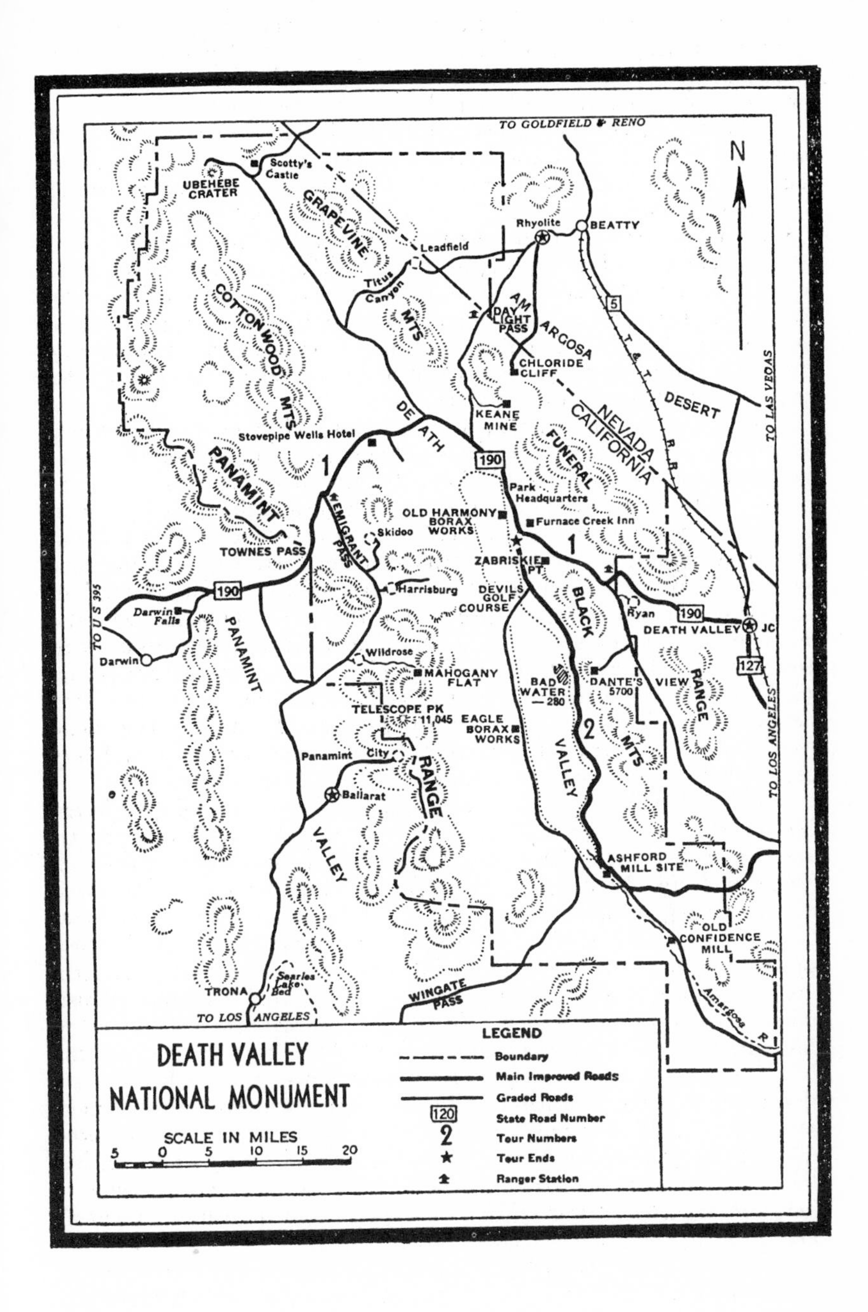
TO GOLDFIELD & RENO
N
Scotty's Castle
UBEHEBE CRATER
GRAPEVINE
Leadfield
Rhyolite
BEATTY
Titus Canyon
MTS
COTTONWOOD MTS
DAY LIGHT PASS
AMARGOSA
CHLORIDE CLIFF
KEANE MINE
DESERT
NEVADA
CALIFORNIA
FUNERAL
DEATH
TO LAS VEGAS
Stovepipe Wells Hotel
PANAMINT
190
Park Headquarters
OLD HARMONY BORAX WORKS
Furnace Creek Inn
EMIGRANT PASS
Skidoo
TOWNES PASS
ZABRISKIE PT
Harrisburg
DEVILS GOLF COURSE
BLACK
Ryan
DEATH VALLEY JC
TO U S 395
Darwin Falls
Darwin
Wildrose
MAHOGANY FLAT
BAD WATER —280
DANTE'S VIEW 5700
RANGE
127
TELESCOPE PK 11,045
EAGLE BORAX WORKS
VALLEY
MTS
TO LOS ANGELES
Panamint City
Ballarat
ASHFORD MILL SITE
OLD CONFIDENCE MILL
Searles Lake Bed
TRONA
TO LOS ANGELES
WINGATE PASS
Amargosa R
DEATH VALLEY NATIONAL MONUMENT
SCALE IN MILES
5 0 5 10 15 20
LEGEND
Boundary
Main Improved Roads
Graded Roads
120 State Road Number
2 Tour Numbers
Tour Ends
Ranger Station

with streams and lakes. Then followed a period of great earth disturbance accompanied by volcanic actions; Death Valley was formed by faulting. At the end of the glacial period the valley was a lake, but the waters gradually evaporated, and the land became a desert.

The barrenness and furnace heat of this gash in the earth's crust took grim toll of life from prospectors and immigrants seeking short-cuts into California in pioneer days. The valley got its name when an immigrant train turned off the Los Angeles-Salt Lake Trail in 1849 into the sandy wastes. As they went on, the train broke up into two companies, each struggling for its life. The first, a party composed chiefly of unmarried men who called themselves the Jayhawkers, went ahead, leaving the men with women and children to struggle along as best they could. As they toiled across the valley, they too broke up into smaller groups. Some Jayhawkers died in the valley, one in the Mojave Desert; the majority survived.

A clergyman named Brier, his wife, and their three small children, who traveled with the Jayhawkers at intervals, struggled safely out of the area after harrowing hardships. As Mrs. Brier afterward told the story: "One valley ended in a canyon with great walls rising up almost as high as we could see. There seemed no way out, for it ended almost in a straight wall . . . In the morning the men returned with the same story: 'No water.' Even the stoutest heart sank then, for nothing but sagebrush and dagger trees greeted the eye. My husband tied little Kirk to his back and staggered ahead. The child would murmur occasionally, 'Oh, father, where's the water?' . . . I staggered and struggled wearily behind with our other two boys and the oxen. The little fellows bore up bravely and hardly complained, though they could barely talk, so dry and swollen were their lips and tongues . . . Every step I expected to sink down and die. I could hardly see."

In late December 1849, the 15 men and women and 7 children left behind by the Jayhawkers, known as the Bennett-Arcane Party, camped at Furnace Creek—too weary, hungry, and thirsty to go on. William Lewis Manly and John Rogers, volunteering to go ahead for aid and provisions, got across the valley and the bare mountains beyond to San Fernando. Returning with some pack animals, they passed a member of the party, lying dead beside his empty canteen. At the camp, they saw no sign of life. Manly fired his gun. A man climbed from under a wagon. In Manly's words, ". . . he threw up his arms high over his head and shouted—'The boys have come, the boys have come!' . . . The great suspense was over and our hearts were first in our mouths, and then the blood all went away and left us almost fainting . . . Bennett and Arcane caught us in their arms and embraced us with all their strength, and Mrs. Bennett . . . fell down on her knees and clung to me like a maniac in the great emotion that came to her, and not a word was spoken." Escaping from the valley, the party arrived finally at San Fernando March 7, 1850.

A few years later much search was made for a silver mine one of these immigrants had reported; it became known as the Lost Gunsight.

The French and George parties, who explored the valley in 1860 in an unsuccessful search for the mine, bestowed many of its present place names. Later prospectors also failed to find the Gunsight, but discovered deposits of gold, silver, copper, and other minerals. A few large deposits were found, and lusty towns flared briefly. A building or two, bottles, adobe walls, and a clutter of worthless debris mark their sites. Borax, however, provided the foundation for an industry that thrived for a number of years and was responsible for the development of the region.

PARK TOUR 1

Western Entrance at Towne's Pass to Eastern Entrance in Furnace Creek Wash; 56 *m.*, State 190.

Oiled roadbed.

State 190 descends the steep eastern face of the Panamints to the pale, brush-dotted floor of Death Valley, runs south for about 25 miles between high bordering ranges fantastically streaked with vivid colors, then continues east through the low pass of Furnace Creek Wash.

East of the monument boundary, at the summit of Towne's Pass (5,500 alt.), 0 *m.* (*see TOUR 6c*), State 190 crosses the rolling summits of the tawny Panamint Range and descends Emigrant Wash. The rocky wash, speckled with desert shrubs, was named for the Jayhawkers, who climbed it in their escape from Death Valley.

A RANGER STATION, 7.7 *m.* (1,542 alt.), which registers incoming cars, is at the junction with Emigrant Canyon Rd.

Right on Emigrant Canyon Rd., traversing Emigrant Canyon, 1.8 *m.*, a sandy wash between rounded hills, to EMIGRANT SPRING (R), 4.8 *m.* (4,045 alt.), which flows from a little cove above the road. For many years prospectors camped below the green thicket and pastured their burros on scanty growth in the wash. Drinking water for Stove Pipe Wells Hotel comes from this spring.

The road winds up the canyon to emerge on HARRISBURG FLATS, 9.5 *m.* (4,500 alt.); the cone-shaped mountain directly ahead is Telescope Peak.

At 10.3 *m.* is the junction with a road; L. here 8 *m.* to the SITE OF SKIDOO (5,500 alt.), a camp that grew up after a discovery of 1906 and produced more than $3,000,000 in gold. At its peak, 500 persons lived here, and it had a telephone line across Death Valley to Rhyolite.

At 11.7 *m.* is the junction with a road; L. here 6 *m.* to AUGUERRE-BERRY POINT (6,000 alt.; *view best in afternoon*). Below is Death Valley, cradled in colored, wrinkled mountains; eastward, beyond tawny and brown desert ranges is Charleston Peak (11,910 alt.), 80 miles away in Nevada.

The main side route continues across the brushy Panamints and descends through a narrow winding defile into WILD ROSE CANYON, to the junction with a dirt road 20.4 *m.*; L. here to SUMMER PARK HEADQUARTERS, 0.3 *m.* The road (*impassable when snow-covered*) continues to the stone CHARCOAL KILNS, 7.3 *m.*, built in the 1870's by George Hearst. MAHOGANY FLATS, 8.5 *m.* (8,133 alt.), on a saddle of the wooded Panamints, is at the end of the automobile road. A foot trail (R) leads 6 *m.* to TELESCOPE PEAK (11,045 alt.), so named because of the magnificent widespreading view it affords.

WILD ROSE SPRINGS, 21.7 *m.* (3,617 alt.), on the main side route was named for the roses the George party found here in 1860.

South of WILD ROSE STATION, 22.3 *m.* (*service station, lunch-room, grocery store*), the road descends a narrow canyon, and passes the park boundary at 25.1 *m.* The road continues to TRONA, 60 *m.*, and INYOKERN, 93 *m.*, on US 395 (*see TOUR 6d*).

Northeast of the ranger station State 190 descends steadily. On the eastern side of the white valley the Grapevine Mountains, banded in red and black, rise above huge gravel fans, deposits of alluvial débris swept from canyon mouths high above the valley floor.

At 16.8 *m.* is the junction with a road.

Right to MOSAIC CANYON, 3 *m.*, whose floor of vari-colored pebbles embedded in gray conglomerate has been worn to a mosaic-like surface.

STOVE PIPE WELLS HOTEL (sea level), 17 *m.*, is a simple resort. The OLD WAGON (L) was abandoned in the rough, sandy country 25 miles north by employees of the borax company in 1889.

South of the hotel, some 25 square miles of yellow SAND DUNES (L) rise in sharply sculptured lines. The mesquite, creosote, and four-winged saltbrush around the dunes are havens for small animal life. The road curves east across the SALT MARSHES and passes the DEVIL'S CORNFIELD, where the arrowweed resembles shocks of tied corn. This is the only definitely known Death Valley camp of the Jayhawkers in 1849.

At 25 *m.* is the junction with the oil-surfaced Ubehebe Crater Rd.

Right on this road 2.8 *m.* to the junction with a dirt road; L. here 0.4 *m.* to STOVEPIPE WELLS (—49 alt.), now a rock well. The water holes in this sandy waste saved many lives. Drifting sand often filled the holes, so the spot was marked by a stovepipe, now at Stove Pipe Wells Hotel.

The Ubehebe Crater Rd. parallels the Grapevines and climbs gradually; at about 24 *m.* the brush-grown alluvial slopes of the Grapevines and Panamints merge.

At 33.5 *m.* is the junction with a dirt road; R. here, through Grapevine Canyon, 3 *m.*, to SCOTTY'S CASTLE (*admin. fee; rooms*), entered by a bridge. The handsome Provincial Spanish style buildings are believed to have cost Scotty and his wealthy partner, Albert M. Johnson, $2,000.000. The two main houses, separated by a patio, are flanked by a clock tower (L) and a guest house (R). Walter Scott, said to have been a cowboy who rode in Buffalo Bill's Wild West Show, lived here more than 30 years. His spending sprees in Los Angeles—where he was lavish with $100 bills, despite the fact that he was without known resources—made him a widely-known eccentric.

At 35.8 *m.* on the main road is the junction with a short one-way loop road; L. here through low mud cliffs and up cinder hills to the lip of UBEHEBE CRATER (2,900 alt.). The east wall of the 800-foot deep crater is spectacularly striped in brilliant red and orange.

The highway crosses a low range of hills. The Funeral Mountains (L) and the Panamint Range (R) hem in the valley. Snowcapped Telescope Peak (11,331 alt.) is the highest mountain in the area. The valley is stony and sparsely covered with brush.

At 31 *m.* is the junction with the Daylight Pass Rd.

Right on the Daylight Pass Rd., over oddly colored hills, 6 *m.* to the junction with a rough dirt road; R. here 5.3 *m.* to ruins of the property of the KEANE WONDER MINE (3,000 alt.), built after gold was found here in 1903. The mine was up the canyon.

HELL'S GATE, 10.5 *m.* (2,263 alt.), on the Daylight Pass Rd., was named by teamsters in 1905; when they left the protecting walls of the canyon, the tender noses of their horses were scorched by the burning summer winds of Death Valley. Toward the south are the white Salt Beds; bright colored mountains rise steeply on either side.

East of Hell's Gate, Boundary Canyon—between the Grapevine and Funeral Mountains—is narrow and steep. CORKSCREW PEAK (5,000 alt.) is circled by broad bands of gray and red.

A RANGER STATION, 17 *m.*, at the summit (4,317 alt.) of Daylight Pass, checks incoming cars. The monument boundary, 17.5 *m.*, is at the California-Nevada Line, 10 miles from the ghost town of Rhyolite, Nev.

At 38.7 *m.* on State 190 is the junction with an oiled road.

Left 0.6 *m.* to the DEATH VALLEY NATIONAL MONUMENT HEADQUARTERS (*information and maps*).

At 39.6 *m.* is the junction with a dirt road.

Left 0.2 *m.* to a parking station, from which a short footpath descends to the GNOME'S WORKSHOP; here, in the midst of acres of odd alkaline formation, is a tiny bitter-tasting stream forming miniature waterfalls.

The RUINS OF THE HARMONY BORAX WORKS are visible on a low bluff (R) at 40.5 *m.* Of this first borax works in Death Valley only some rusty machinery and a few adobe walls remain. Borax was discovered here by Aaron Winters, a prospector, who sold his claims in 1882 to W. T. Coleman and F. M. "Borax" Smith. The refined borax was hauled by the famous mule teams through Wingate Pass to Mojave, 160 miles away. Each outfit consisted of a lead wagon and a trailer carrying 20 tons of borax, which were drawn by teams of twelve to twenty mules, guided by a check line 125 feet long. The works were closed in 1887 when the price of borax fell, but the picturesque mule teams had become identified with the product. In 1890 colemanite, another form of borax, was discovered near Furnace Creek, and the industry again boomed. The largest mine at Ryan was operated until 1927, when a new type of borax that could be produced more economically was discovered on the Mojave Desert.

Here are located the VISITORS' CENTER and FURNACE CREEK RANCH. The green oasis of the Ranch is irrigated by water from springs in Furnace Creek Wash. In early days alfalfa and hay for the borax teams were grown here, and date palms and long rows of tamarisk were planted to break the hot desert winds. Today, dates from these palm trees are sold. Bellerin Teck, the first white man to settle in Death Valley, began farming here in 1870. Opposite the ranch is DEATH VALLEY AIRPORT.

FURNACE CREEK CAMP, 42.2 *m.*, is in the southeast corner of the ranch. Beside the road are a borax wagon, trailer, and steam tractor used here in the eighties.

Right from Furnace Creek Camp to the INDIAN VILLAGE, 0.7 *m.*, where adobes have been built for about 30 Shoshone who live in the park. Before the whites came, the Indians wandered over the country in search of food. Although their culture was primitive, the women made very good baskets from materials found here.

At 42.5 *m.* is the junction with a road.

Left here 0.5 *m.* to TEXAS SPRINGS PUBLIC CAMPGROUND (*water, stone fireplaces, sanitary facilities*), on a bench above the valley.

Near Furnace Creek Jc., 43.3 *m.*, the meeting place with oiled East Highway (*see PARK TOUR 2*), is FURNACE CREEK INN (sea level), 43.4 *m.*, with its garage, service station, and small store. This is a luxury resort.

At 46.5 *m.* is the junction with an oiled road.

Right here 0.3 *m.* to Zabriskie Point, which reveals a far-reaching panorama of stark peaks and ridges. MANLY BEACON, a sharply pointed hill, is surrounded by eroded yellow clay hills.

At 47.8 *m.* is the exit of a one-way road (*see below*). Westward is an impressive view of Death Valley and the rugged mountains. The road continues east between cleanly sculptured brown and yellow hills.

At 50.6 *m.* is the junction with a one-way looping road.

Right here through TWENTY-MULE TEAM CANYON. The road dips and winds between low clay hills, scarred from mining ventures. The road rejoins State 190 at 4.4 *m.*

State 190 continues eastward, past the hot tan and rosy brown Funeral Mountains, banded with somber gray; striped PYRAMID PEAK (6,725 alt.) raises its blunt tip at the eastern end of the range. The Black Mountains (R) receive their name from the lava cap that crowns them.

At 54.5 *m.* is the junction with the Dante's View Rd.

Right on this road up a canyon wash where the flanks of the Black Mountains appear brilliantly streaked with rose, fawn, and milky green, 2.6 *m.*, to the junction with an oiled road; L. 2 *m.* to RYAN (2,500 alt.), a model company town while the Pacific Coast Borax Co. operated mines here from 1914 to 1928. Sightseeing trains (*fare $1*) run to borax mines, 7 miles distant.

The road continues up GREENWATER VALLEY (4,000 alt.), a shallow depression in the mountains; here is a junction with a dirt road 7.9 *m.*; L. here 3.8 *m.* to a junction with a dirt road; R. 1.9 *m.* to the SITE OF GREENWATER, a ghost town marked by widely scattered debris. Greenwater boomed in 1905 when copper was discovered; but deposits were scanty, and its saloons, its banks, and newspaper soon had no community to serve.

The main side route continues R. from the junction to DANTE'S VIEW (*light effects best in morning*), 13.9 *m.* (5,220 alt.), overlooking Death Valley from the summit of the Black Mountains. The two extremes of altitude in the 48 States are visible from this point. More than a mile below is Badwater (—279.6 alt.), and westward over the Panamints in the snowy Sierra Nevada is Mt. Whitney (14,495 alt.). Snow-capped Telescope Peak just opposite in the Panamints towers more than two miles above the bottom of Death Valley and a mile above this spot. White salt areas in the valley are sharply outlined against the gravel slopes. Mesquite thickets make green patches at Mesquite and Bennett Wells (L) and Furnace Creek Ranch (R). The steep, rugged mountain walls that baffled the Death Valley party of 1849 stretch north and south, and the Avawatz and Owlshead Mountains block the valley southward. Eastward, beyond the twisted slopes of the Black Mountains, are the barren desert ranges of southern Nevada.

A RANGER STATION, 55 *m.*, registers incoming cars. The boundary of the monument (2,000 alt.) is crossed at 56 *m.* (*see TOUR 12a*).

PARK TOUR 2

Furnace Creek Jct.—Badwater—Ashford Jct.—Saratoga Springs Jct.; 63 *m.*, East Highway.

Oiled roadbed between Furnace Creek Jct. and Ashford Jct.; remainder graded dirt road.

The East Highway explores the eastern side and extreme southern end of Death Valley, passing the Salt Beds and the lowest point in North America. The West Highway is an alternate of this road.

Branching south from State 190 at FURNACE CREEK JUNCTION, 0 *m.* (*see PARK TOUR 1*) the East Highway curves along the base of the Black Mountains.

MUSHROOM ROCK, 4.6 *m.*, is an oddly shaped formation carved by wind and sand.

At 5 *m.* is the junction with an unpaved one-way road, looping back to the main road.

Left on this road 0.2 *m.* to VOLCANIC DRIVE, which winds upward through hills tinted with rose, pale green, ocher, and sienna. At the summit brilliantly colored strata tilt skyward. Descending through yellow walls, the road turns sharply at 3.3 *m.*, becoming Artist's Drive, and climbs again. At the second summit it turns south, passes brightly tinted hills, and reaches a point overlooking Death Valley, whose floor is seen streaked with white, soft purple, and brown. The route rejoins the East Highway at 9 *m.*

The DEVIL'S GOLF COURSE lies below the highway, opposite the entrance to Volcanic Drive. The Salt Beds here are covered with ridges and pinnacles of salt, crystallized into a hard substance that breaks with a ringing sound.

At 6.3 *m.* is the junction with the graded West Highway.

Right here, across the Salt Beds, 13 *m.* to the GRAVES OF SHORTY HARRIS AND JIM DAYTON. Harris, a well-known Death Valley prospector, asked to be buried beside Jim Dayton, a borax company employee who died here in 1898 and was buried on the spot.

At 13.5 *m.* is the junction with a dirt road; L. here 0.2 *m.* to the EAGLE BORAX WORKS, the earliest borax extraction plant in Death Valley, of which nothing remains but a boiling pan which was hauled 140 miles over the desert.

At 17 *m.* on the West Highway is the junction with a road; L. here 0.3 *m.* to BENNETT'S WELL, a watering station on the early borax route.

The Bennett-Arcane party camped within a few miles of this spot in 1849-50.

At ASHFORD JUNCTION, 39 *m.,* the West Highway rejoins the East Highway (*see below*).

BADWATER, 16.8 *m.* (— 279.6 alt.), is the lowest point in North America; a marker on the mountain (L) indicates sea level. The salty pools (R) are fed by the Amargosa River, a small stream that flows south in Nevada and California, rounds the Black Mountains, and flows north in Death Valley.

ASHFORD JUNCTION, 45 *m.,* received its name from a mill built nearby in 1914 by the Ashford brothers. Here is the junction with the West Highway (*see above*). The route continues southeastward over gently rounded mesas and a dry lake bed, or playa.

SARATOGA SPRINGS JUNCTION is at 63 *m.*

Left on a rough road 2 *m.* to SARATOGA SPRINGS; in the swampy marshes are a few pools, the home of the Tiny Death Valley fish. Migrant waterfowl rest here during the winter.

The East Highway continues beyond the monument 27 miles to a junction with State 127 (*see TOUR 11a*).

Sequoia and Kings Canyon National Parks

Season: Principal camping period, May through October; parks open all year, but some highways are closed during winter.
Administration: Headquarters of Superintendent for both parks at Ash Mountain entrance. Otherwise, Three Rivers, Calif., 93271.
Admission: Two main entrances are on the west side. By car, State 180 from Fresno or State 198 and 65 from Visalia to Big Stump Entrance, General Grant Grove, Kings Canyon National Park (52 mi.) or State 198 to Ash Mountain Entrance, Sequoia National Park (34 mi.). Generals Highway connects the two parks. From Tulare or Visalia, accessible by train or bus, sightseeing buses operate to Giant Forest in summer, taxicabs in winter. By raidroad, Southern Pacific to Tulare; Santa Fe to Hanford (bus to Visalia); by bus, Western Greyhound to Tulare or Visalia, Continental Trailways to Tulare; by air, United Air Lines to Visalia. Sequoia Motor Coach schedule, May-September, leaves Tulare South Pacific depot, 1:50 p.m., Visalia Bus Depot, 2:25 p.m. (stops at all railroad and bus depots). Section of Generals Highway may be closed in winter. Car fees collected at entrance by Rangers: a 15-day permit costs $2, a year's, $4.
Accommodations: Giant Forest Lodge (American and European plans), open from late May to late October; Grant Grove Lodge (European plan), open late May to mid-September; Camp Kaweah, motel-type rooms (European plan); housekeeping cabins open all year; Meadow Camp and General Grant Grove, open late May to mid-September; Bearpaw Meadow Camp (wooden platforms, tents) 11 mi. from Giant Forest on High Sierra Trail, no road, open from late June to early September; Pinewood Camp, (partially equipped cabins) on Generals Highway open early June to early September. Free campgrounds on designated sites at Giant Forest, Lodgepole, Dorst Creek, Grant Grove and Cedar Grove, have running water, fireplaces, toilets, tables. Lodgepole and Cedar Grove camps designed for trailers, but have no electricity or sewer connections.

Coffee shops at Giant Forest (open all year) General Grant Grove, May to mid-September, lunch service in winter; Cedar Grove, summer only. Supplies at most camps; camping and hiking at Pinewood and Giant Forest; snowshoes and ski at Wolveeton; skates at Lodgepole. Mail, telephone and telegraph services at most lodges and camps; gasoline stations easily accessible. Roman Catholic and Protestant church services every Sunday at Giant Forest, General Grant Grove and Cedar Grove, mid-June to September.

Sightseeing Tours: Three-day tour from Tulare or Visalia, including meals and lodging at Giant Forest Lodge; all-day tour including Giant Forest, General Grant Grove, Kings Canyon; two-hour tour, Giant Forest. Inquire of Sequoia & Kings Canyon National Parks Co., Sequoia National Park, California; winter address, 129 East Center St., Visalia. Horseback rides with guide can be arranged at Wolverton and Mineral King. Backpack trips on foot over wilderness trails can be planned at Chief Ranger's headquarters.

Naturalist Service: Park naturalists at Giant Forest Administration Building and the Plaza in charge of guide service. Lectures daily and campfire programs nightly in summer at Sequoia. Lectures and campfire programs two or three times weekly and concerts three times weekly in summer at General Grant.

Warnings and Regulations: Camping, smoking, and building of fires permitted only in designated areas. Do not harm or frighten animals. Do not feed bears or leave food within their reach or where they can break into containers.

Fishing: Rainbow, Loch Leven, eastern brook, German brown, and golden trout. Persons over 18 must procure license; available at Giant Forest. Fee for California residents, $5; for non-residents, 10 days, $5. Tackle available at Giant Forest. Bulletins containing regulations for current year supplied by rangers. Fishing permitted in all waters of parks unless prohibited by sign.

Bathing: Pools at Lodgepole Camp, Bridge Camp, Hospital Rock Camp, and Heather Lake, Sequoia. Swimming in Sequoia Lake near southwestern boundary of General Grant. Bathing permitted only in designated areas.

SEQUOIA and KINGS CANYON NATIONAL PARKS cover the wildest country on the western slopes of the Sierra Nevada. Sequoia stretches from the headwaters of the Kings River on the north to the headwaters of the Tule River on the south. The tallest peaks of the High Sierra—barely dominated by Mount Whitney (14,495 alt.), highest point in the 48 States—bound it on the east and the foothills of the Sierra on the west. Bisecting the park from north to south is a jagged granite ridge, the Great Western Divide. West of the divide are the park's major accommodations and most popular attractions, its motor roads, and shorter trails. Here, in the 4,000- to 8,000-feet elevations, are the groves and forests of California big trees (*Sequoia gigantea*) for which Sequoia is best known. Paralleling the divide for 25 miles, about halfway between it and the crest of the Sierra, are the 3,000-foot walls of the Kern River Canyon (*see TOUR 3b*). In the eastern section of Sequoia are high mountain lakes—of glacial origin, as are the mountainsides of exposed rock and the great, irregular granite ridges, cleared of their earth and vegetation by ice thousands of years ago. The Sierra—as a distinct range—dates from the latter part of the Jurassic period, when it began to rise from the receding Logan Sea; it is approximately 120,000,000 years old.

When Hale D. Tharp, a farmer from Three Rivers, visited Sequoia in 1856 in search of pastures and a ranch site, he was met by peaceful Yokut. "The Indians liked me," Tharp said later, "because I was good to them. I liked the Indians, too, for they were honest and kind to each other. I never knew of a theft or murder among them."

In 1858 the Indians led Tharp to the big trees and thus he became one of the first white men to see the gigantic forest. In 1862 Joseph Thomas came upon the woods now known as the General Grant Grove, which contains the General Grant Tree. In 1879 James Wolverton, a trapper, sized up the huge General Sherman Tree and named it for his Civil War commander. John Muir, the naturalist who became a powerful spokesman for conservation, gave the name Giant Forest to the great trees. The dimensions of the two famous trees in feet are as follows:

	General Sherman	*General Grant*
Height above mean base	272.4	267.4
Circumference at base	101.6	107.6
Maximum diameter at base	36.5	40.3
Mean diameter at base	32.2	33.3
Diameter 60 feet above ground	17.5	18.8
Diameter 120 feet above ground	17.0	15.0
Diameter 180 feet above ground	14.0	12.9
Height to first large branch	130.0	129.0
Diameter of largest branch	6.8	4.5
Weight of trunk (est.)	625 tons	565 tons
Total volume of trunk	50,010 cu. ft.	45,232 cu. ft.

The settlers who followed Tharp to the Three Rivers region were less popular with the Yokut; by 1862 their increasing numbers were forcing the Indians to retreat into the canyons of the lower ranges. At this time they contracted their first white man's diseases—smallpox, measles, and scarlet fever—and perished by the hundreds, crawling, unless restrained by force, into their sweat houses to die. Very simply, one of their leaders, Chief Chappo, had Tharp ask the white men to go away. When told by his friend that the settlers refused to leave, Tharp said that the chief and his braves "sat down and cried." By 1865 the last of the tribe had retreated into the mountains.

An abortive gold strike in the fifties opened the area's first trails, across the southern panhandle to the Kern River. The first official exploration took place in the north, in the Kings River country, in 1861; it was led by William H. Brewer of the U. S. Geological Survey. Three years later Clarence King, who accompanied Brewer, crossed the Kings-Kern Divide and climbed Mount Tyndal. In 1873 Mount Whitney was climbed for the first time by A. H. Johnson, C. D. Begole, and John Lucas. Eager miners again crossed the southern panhandle—this time to found the noisy, short-lived settlement of Beulah.

Then, in 1885, a stunned registrar in Visalia reported to Washington the application in one day of 55 claims of 160 acres each to timber lands above the Marble Fork of the Kaweah. Applicants were members of a socialistic group who planned a Utopia in the wilderness (*see TOUR 3b*). Immediately after filing, they organized the Tulare Valley and Giant Forest Railroad Company and laid plans for the development of the land and the lumbering of sugar and yellow pine. They visited Giant Forest, vowed solemnly never to cut the giant Sequoias, and named the outstanding big trees after the heroes of the Paris Commune

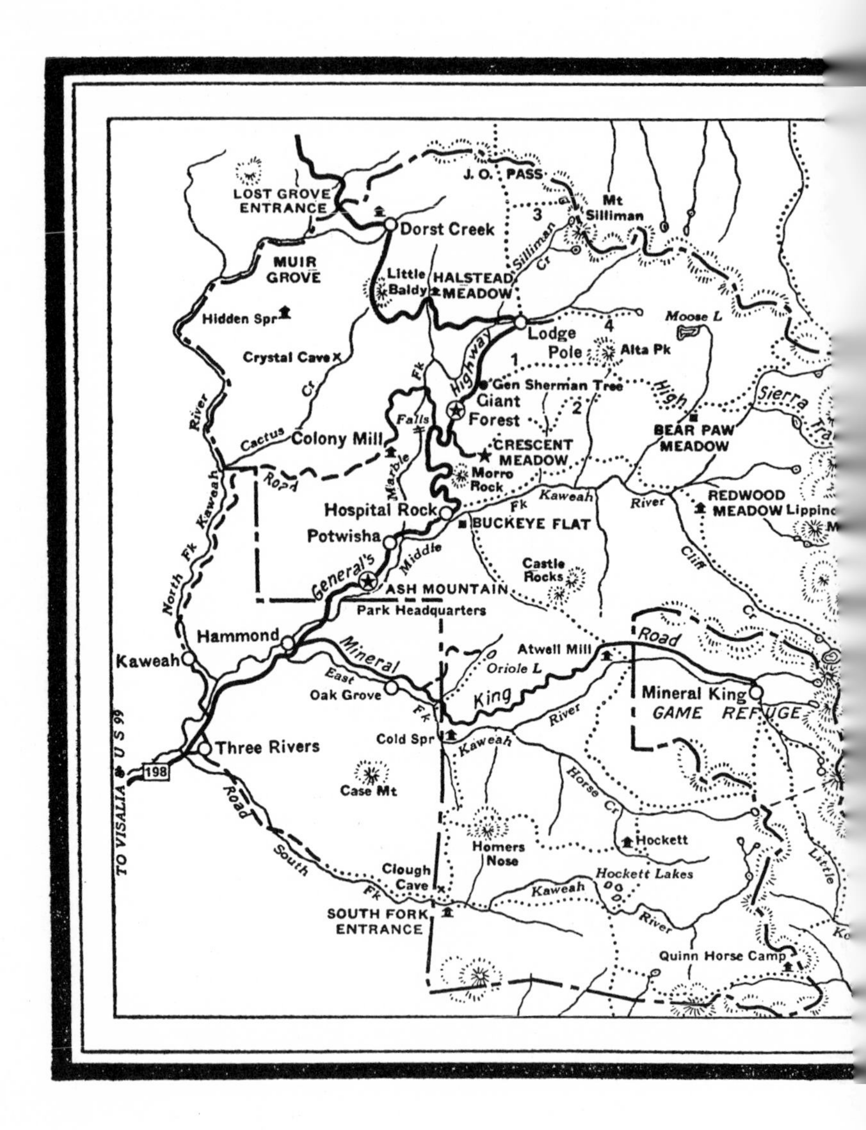

LOST GROVE ENTRANCE
J. O. PASS
Mt Silliman
Dorst Creek
MUIR GROVE
Little Baldy
HALSTEAD MEADOW
Silliman Cr
Hidden Spr
Lodge Pole
Moose L
Alta Pk
Crystal Cave
Gen Sherman Tree
Giant Forest
High Sierra Trail
BEAR PAW MEADOW
Colony Mill
Cactus Cr
Falls
CRESCENT MEADOW
Morro Rock
Marble Fk
Road
Hospital Rock
Kaweah
Fk
River
REDWOOD MEADOW
BUCKEYE FLAT
Potwisha
Middle
General's
ASH MOUNTAIN
Park Headquarters
Castle Rocks
Cliff Cr
North Fk Kaweah River
Hammond
Kaweah
Mineral
East Fk
Oak Grove
Atwell Mill
Oriole L
King
River
Road
Mineral King
GAME REFUGE
Three Rivers
198
TO VISALIA US 99
Cold Spr
Kaweah
Case Mt
Horse Cr
Road
South Fk
Homers Nose
Hockett
Hockett Lakes
Clough Cave
SOUTH FORK ENTRANCE
Kaweah
River
Quinn Horse Camp
Little

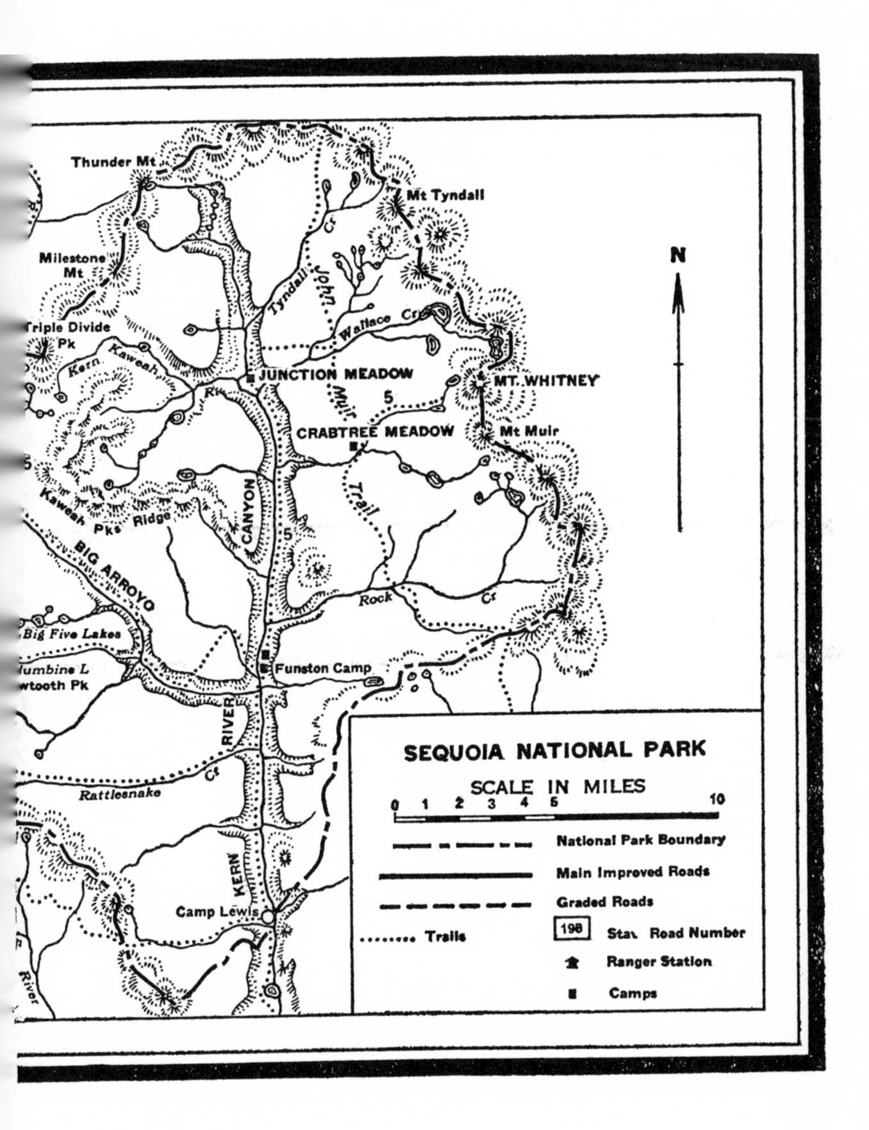
Thunder Mt
Mt Tyndall
Milestone Mt
N
Tyndall Cr
John Muir
Wallace Cr
Triple Divide Pk
Kern Kaweah R
JUNCTION MEADOW
MT. WHITNEY
5
CRABTREE MEADOW
Mt Muir
Trail
Kaweah Pks Ridge
CANYON
BIG ARROYO
Rock Cr
Big Five Lakes
Funston Camp
RIVER
KERN
Rattlesnake Cr
Camp Lewis
River
SEQUOIA NATIONAL PARK
SCALE IN MILES
0 1 2 3 4 5 10
National Park Boundary
Main Improved Roads
Graded Roads
Trails
198
State Road Number
Ranger Station
Camps

and of American socialism. When the government suspended the entire district from entry pending investigation, the colonists launched a series of litigations to secure title to their land. In the following year they reorganized their company and formed the Kaweah Co-operative Commonwealth Colony. A sawmill and a railroad were built to Colony Mill.

There were commotions and feuds in the Kaweah Valley, dissension within the colony, debates in the State Legislature, articles in the newspapers all over the State, and finally over the country. Then, in September 1890, Sequoia National Park was created by Congress and troops were sent to expel the colonists and administer the new park. The colonists retreated with what grace they could muster, but their vocabulary was adopted by their neighbors. The Act of Congress was completely unexpected by the settlers in the vicinity, who for 30 years had hunted and grazed over the area at will. The local papers blossomed with phrases like "Cossack Terrorism" and the soldiers were booed in the streets of the villages and shot at from ambuscades in the forests.

Captain Dorst, the first military administrator of Sequoia and General Grant National Parks, created in 1890, found himself faced with a formidable array of problems. The sheep and cattlemen, who had come to look on the high country as their own, did not take kindly to the invitation of Congress to give up their privileges, yet no penalties had been provided for infractions of the rules. Just as an efficient administration was being developed the Spanish-American War broke out and the troops left. Sheep and cattle roamed at will, deer were slaughtered by hundreds. The First Utah Volunteer Company was sent to restore order, and replaced by Regulars one month later.

The next 15 years were a period of gradual development. The hard-worked administrators built roads and improved services with the limited funds at their disposal. Slowly the park began to take on its present appearance. Outstanding among the military administrators were Lt. Hugh S. Johnson and Capt. Charles Young, commander of a troop of Negro cavalry known as the Black Battalion. In 1914 a civilian administration replaced the military. In 1926 the Mt. Whitney-Kern River district was added and the area increased from 252 to 604 square miles.

The Kings Canyon area proved so much larger and more diverse than the restricted General Grant section that in 1940 the name was changed to Kings Canyon National Park, of which General Grant Grove is a part. The Generals Highway winds through the Giant Sequoia belt and covers 47 miles from Ash Mountain entrance to the Grant Grove. From the Grove the visitor travels 28 miles on State 180 through Sequoia National Forest and along the south fork of Kings River to Cedar Grove. The road continues for six miles through the canyon to Copper Creek, where solid granite walls tower thousands of feet above the canyon. Cedar Grove is a base for many trips on trails into the high country.

The park's outstanding attraction—the big tree—was present thousands of years ago, when the present coal deposits were marshes crowded with dense and luxuriant vegetation. In time the Sequoia came to be one of the dominating trees of the northern hemisphere. Then, because of the Ice Age and changes in topography, its range became more and more restricted. The first known white men to discover it were the Joseph R. Walker party in 1839. After considerable controversy, it was named *Sequoia gigantea,* to distinguish it from the related coast species, the *Sequoia sempervirens.*

The size of the big tree has often been exaggerated. The greatest height measured is about 330 feet. Heights of 350 feet or more are theoretically possible, but almost all the largest trees have been broken off at the top by lightning. The extreme diameter above the root swellings averages about 27 feet. Similarly exaggerated stories of the great age of the trees are common. Most of the largest living specimens are probably from 1,500 to 3,000 years old. The big tree is scale-leafed and somewhat resembles the incense cedar, but its sprays are rounded rather than flattened, a light green in color. In youth the branches are slender, the tree is pyramidal in form, and the leaves are needle-like and of a bluish cast. After about 200 years the lower branches begin to drop off, the leaves flatten to scales and change in color, and the bark begins to split into ridges and turn from a purple grey to a cinnamon red. On mature trees the bark is from one to two feet thick, separated into long parallel ridges that make the trunk seem like an immense fluted Doric column; the few remaining branches, of enormous size, gnarled and twisted, add to the impression of massive power. The 32 big tree groves within the park stand at the edge of the Transition Zone (*see FLORA AND FAUNA*). The display of flowers, particularly in June and early July is profuse and brilliant.

General Grant Grove was set aside principally to preserve the giant Sequoia for which it is named, which stands in one of its two splendid big tree groves. The Grove lies lower on the slopes of the Sierra than does Sequoia, at an average altitude of 6,500 feet. Its eastern boundary is the "hogback," Grant Grove Ridge (7,350-7,750 alt.).

SEQUOIA PARK TOUR

Ash Mountain Entrance—Hospital Rock—Giant Forest Village—Lost Grove Entrance—General Grant South Entrance Station; **45.8** *m.,* Generals Highway.

Oil-surfaced roadbed throughout.

The Generals Highway, a continuation of State 198 (*see TOUR 3b*) crosses the boundary of Sequoia National Park, 0 *m.,* and continues to ASH MOUNTAIN PARK HEADQUARTERS, 1 *m.,* where incoming travelers are registered and automobile fees collected.

Northeast of Ash Mountain the highway climbs steadily, winding above the KAWEAH RIVER. Western sycamores grow along the canyon floor, and in early summer the grass is full of yellow blazing stars, crimson godetia, and wild hyacinth.

Before 1865 CAMP POTWISHA, 2.8 *m.,* a CCC camp in 1938, was the principal village of the Pot-wi-sha tribe of the Yokut. The mortar holes in which they ground their meal of acorns or buckeyes are still visible in the flat rocks below the road, and several well preserved pictographs are on the cliffs above. The Yokut, who had no knowledge of the tribes that left the prehistoric paintings, asked Hale Tharp to explain them.

East of Potwisha the road follows the twisting course of the Middle Fork of the Kaweah. The nearby mountains are characterized by unusual rock formations. Long, jagged granite scars cut the sides of some; others are walled by smooth, almost polished surfaces.

HOSPITAL ROCK CAMP (2,600 alt.), 5.2 *m.* (*all-year cabins; campsites*), is popular in winter with those who prefer to sleep in a warmer climate and drive to snow for the day. Opposite the camp is HOSPITAL ROCK. In the cave beneath this huge boulder (*reached by a short path*) the Indians stored their food, held pow-wows, and ministered to the sick and injured. Castle Rocks (R) and Moro Rock (L) tower 4,000 feet above, Hospital Rock like an altar before them. The place was a religious sanctuary for several tribes; numerous pictographs, their exact significance unknown, attest to the importance of the place in the lives of Indians.

Right from Hospital Rock on a narrow dirt road to BUCKEYE FLAT CAMPGROUND, 0.5 *m.*, and MORO CREEK, 2 *m.*

North of Hospital Rock the Generals Highway begins the long climb up the ridge between the Marble and Middle Forks of the Kaweah in a series of hairpin curves. The Yucca give way to buckeyes, which soon dominate the chaparral. The road cuts again and again through limestone, honeycombed with caves (*not safely accessible*). As the road continues to climb the limestone rocks are replaced by schists, the tightly pressed, metamorphosed stone that lies along the surface of the Sierra.

AMPHITHEATER POINT (4,450 alt.), 10 *m.* (*parking space*), overlooks the canyons below and the distant San Joaquin Valley.

At 13.4 *m.* is the junction with Colony Mill Rd.

Left on this rough dirt road, which the Kaweah Colonists (*see above*) were four years constructing, to the SITE OF THE KAWEAH COLONY SAWMILL, 6 *m.*, on the Marble Fork. A RANGER STATION is here today.

The buildings of GIANT FOREST VILLAGE, 16.6 *m.* (6,412 alt.), residential and recreational center, stretch for a half mile along the highway. Trails radiate from the village into the eastern section of the park. Giant Forest Lodge, Camp Kaweah, Pinewood Camp, and several campgrounds are in the immediate vicinity.

The ADMINISTRATION BUILDING (*open 8-12, 1-5*) at the eastern end of the village (L) houses an administrative office, a TULARE COUNTY LIBRARY (*open 1-4:30 summer season*), and a MUSEUM. In the museum are a few Indian artifacts and a large relief map of the park showing all major roads and trails.

In front of the building is a SECTION OF A SEQUOIA, 9 feet 8 inches in diameter, that is estimated to have been about 1,705 years old at the time of its death. Markers on some of the annual rings indicate events that occurred during the life of the tree, beginning with the invasion of Rome by the Goths in the year 250 and ending with the start of the World War in 1914.

Opposite the Administration Building is the CHURCH OF THE SEQUOIAS on a gently sloping clearing; here, in the shadow of giant Sequoias, are the log benches and stump pulpit used for outdoor religious services.

1. Left from Giant Forest Village on a footpath to BEETLE ROCK, 0.2 *m.*, a huge, smooth-topped granite promontory resembling the back of a beetle, high above Marble Canyon and overlooking the distant San Joaquin Valley.

2. Right from Giant Forest Village on an oil-surfaced road to the junction with a paved road, 0.1 *m.;* L. here 0.4 to BEAR HILL (*bears fed daily at 2:30 p.m.; ranger naturalist talk at 3*), a large enclosed area in which are the park incinerator and bear-feeding platform.

At 0.8 *m.* on the main side road is the AUTO LOG (L), a huge fallen big tree with a flattened top onto which an automobile may be driven.

At TRINITY CORNER, 1.2 *m.*, the main route goes R. MORO ROCK, 1.6 *m.* (6,719 alt.), is climbed by a winding rock and concrete stairway. The view from its summit is magnificent, particularly at sunset. A sheer 4,000 feet below is the Kaweah River, a shining thread. In the east the pale moon rises above the slowly darkening, irregular ridges of the Great Western Divide. Westward miles of hilltops reach toward the San Joaquin Valley, and in the far distance the red sun sinks behind the Coast Range.

At 2.1 *m.* is the junction with a paved road; R. here, past the PARKER GROUP, 2.2 *m.*, a small grove of giant Sequoias.

The main side road ends in a parking area at the southern end of CRESCENT MEADOW, 3.4 *m.*, filled with flowers and rich grass and surrounded by Sequoias. This is one of the finest meadows of the middle altitudes in the entire Sierra.

a. Right from Crescent Meadow at a point a few yards north of the parking area on a trail 0.6 *m.* to THARP'S CABIN, a large, hollow big tree log that was converted into living quarters by Hale Tharp, who pastured his cattle in the nearby meadow. The log house is preserved in its early state, and the carven "Hale Tharp, 1858" is protected by glass. John Muir spent several nights here in 1875 and wrote enthusiastically of the "noble den."

b. Right from the parking area on the High Sierra Trail (here a pedestrian path; stock should leave from corral and join trail 4 miles east). This requires an average of 14 days as a round trip hike, and leads through wild, spectacular country, through deep canyons, past mountain lakes and meadows, to the Sierra's highest peaks. Overnight camps are maintained in summer at approximate 10-mile intervals.

BEARPAW MEADOW (7,900 alt.), 12 *m.*, overlooks the Kaweah River Valley. Beautiful HAMILTON LAKE (9,500 alt.) is at 16 *m.*, and the suspension bridge over HAMILTON GORGE in the Great Western Divide at 18 *m.* KAWEAH GAP (10,800 alt.), 20 *m.*, offers sweeping views of the Kaweah Peaks Ridge and the main crest of the Sierra Nevada. East of MORAINE LAKE (9,450 alt.), 30 *m.*, is the KERN RIVER CANYON at FUNSTON MEADOWS (6,700 alt.), 34.5 *m.* The trail follows the canyon northward from Funston Meadows, past KERN HOT SPRINGS (6,900 alt.), 37 *m.*, to JUNCTION MEADOW (8,200 alt.). From here it swings east for about 3.5 miles, then southward to CRABTREE MEADOW,

54.5 *m.*, Mount Whitney base camp. The summit of MOUNT WHITNEY (14,495 alt.) is at 62 *m.* Right from the summit on a trail 13 *m.* down the eastern slope to the junction with an automobile road (*see TOUR 6c*).

3. Left from Giant Forest on the Alta Trail (for pedestrians and equestrians). The KEYHOLE (L), 1.3 *m.*, is a big tree, fire-hollowed at the base, in whose sides have been burned openings resembling giant "keyholes."

At 1.6 *m.* at the junction with the Rim Rock Trail (L) stands the LINCOLN TREE (R), a rugged Sequoia 259 feet high with a diameter of 31 feet.

At 1.8 *m.* is the junction with a short trail; R. here 0.1 *m.* to CIRCLE MEADOW, on whose northern rim are the FOUNDERS GROVE, the CONGRESS GROUP, and the 250-foot PRESIDENT TREE.

At 2.3 *m.* on the Alta Trail is the junction with the Wolverton Corral Trail; R. here 4 *m.* to a junction with the High Sierra Trail (*see above*).

The Alta Trail continues northeastward, past lovely MEHRTEN MEADOW (*shelter cabin*) (9,000 alt.), on Mehrten Creek. ALTA MEADOWS (*good camping and pasture*), 7 *m.*, lies at the southern base of ALTA PEAK, 9.2 *m.* (11,211 alt.). The peak is the nearest point to Giant Forest from which Mount Whitney is visible. The view is exceptional.

The Generals Highway continues eastward to a junction with a graveled road, 17.1 *m.*

Right 0.1 *m.* to the LOWER CORRAL, one of the park's two stables (*horses available*). The HAZELWOOD PICNIC AREA is at 0.3 *m.*

At 17.5 *m.* on the highway is the junction with a trail.

1. Left on this trail through SUNSET ROCK CAMPGROUND, 0.5 *m.*, to SUNSET ROCK, 0.6 *m.*, a huge granite boulder offering the same view as Beetle Rock (*see above*). It is popular for sunset picnics.

2. Right 0.9 *m.* on Rim Rock Trail, along a granite ridge, to the junction with the Alta Trail (*see above*).

At 18.7 *m.* is the junction with an oil-surfaced road.

Right to a large parking area, 0.2 *m.* Opposite the area, in a park-like clearing (L), is the GENERAL SHERMAN TREE, the largest living thing. It was called the Karl Marx Tree by the Kaweah Colonists. The name was changed after James Wolverton, trapper and hunter, asserted he had discovered it in 1879, several years before the advent of the colonists. The height of the Sherman Tree, above the mean base, is 272.4 feet; its base circumference is 101.6 feet. Its age is estimated at between 3,000 and 4,000 years.

At 1 *m.* on the side road is a junction with an oil-surfaced road; L. 0.9 *m.* to a junction with a graded road; L. here 0.1 *m.* to the GOVERNMENT CORRAL (*horses*), called the upper corral.

At 20.9 *m.* on Generals Highway is the junction with a graded road.

Right here into LODGEPOLE CAMPGROUND (*swimming pool*), 0.2 *m.*, on the Marble Fork of the Kaweah. The special campsites for trailers are at 0.3 *m.*

1. Right 2.5 *m.* from Lodgepole Camp on a trail (for pedestrians or equestrians) to TOKOPAH FALLS. The trail ascends the canyon of the

Marble Fork of the Kaweah River through forests and meadows. The river gorge narrows between precipitous slopes, offering an example of the U-shaped glaciated valley of the Yosemite type.

2. Left 5 *m.* from Lodgepole Camp on Twin Lakes Trail (for pedestrians or equestrians) to CLOVER CREEK. TWIN LAKES (*fine trout fishing*), 7 *m.* (9,750 alt.), are at the base of SILLIMAN SHOULDER (10,500 alt.), which offers a fine mountain panorama.

At 27.4 *m.* on the Generals Highway is an attractive PICNIC AREA at the edge of HALSTEAD MEADOW.

DORST CREEK CAMPGROUND (L), 31.7 *m.,* one of the most pleasant of Sequoia's camping areas, occupies a thick forest of fir along the banks of Dorst Creek.

LOST GROVE CAMPGROUND, 32.3 *m.,* marks the northwestern boundary of Sequoia National Park.

The Generals Highway continues northwestward, sweeping around one of the lateral ridges (6,500 average alt.) of the Sierra Nevada in long, easy curves. (*Road frequently closed by winter snows.*) The views of the San Joaquin Valley are numerous and impressive, and the road is lined with a mixed coniferous forest of great splendor. White fir, incense cedar, sugar pine, and western yellow pine predominate, and there are occasional Sequoias. The western dogwood, with large white blooms, is common beneath the larger trees and the steep slopes that fall away to the San Joaquin Valley are densely clothed with wild cherry, ceanothus, and other flowering shrubs.

At 45.8 *m.* is the southern boundary of General Grant Grove.

GENERAL GRANT GROVE TOUR

South Entrance Station—Northern Junction with State 180; 2.5 *m.,* Generals Highway.

The Generals Highway crosses the boundary of General Grant Grove, 0 *m.,* at the SOUTH ENTRANCE RANGER STATION.

At 0.2 *m.* is the junction with State 180 (*see TOUR 3b*).

At the PLAZA, 1 *m.,* near the center of the Grove are cabins, campgrounds, a store, the CORRAL, and the CHURCH OF THE SEQUOIAS, where outdoor religious services are held.

Right from the Plaza on oiled Rocking Rock Rd., 2.5 *m.* to ROCKING ROCK, a 48-ton granite slab balanced on the edge of PARK RIDGE. About 100 yards east is PANORAMIC POINT (7,500 alt.). Both vantages offer impressive views of KINGS RIVER CANYON, 20 miles north.

At 1.2 *m.* on Generals Highway is the junction with an oiled road.

Left on this road 0.1 *m.* to the junction with an oiled road; R. here 0.9 *m.* to the GRANT GROVE of big trees, largest of which is the GENERAL GRANT TREE, under which services are held at high noon each Christmas Day. The ancient tree is 267 feet high with a maximum base diameter of 40.3 feet and a diameter of 12 feet 200 feet above the ground.

The park's second outstanding group of Sequoias, NORTH GROVE (R), is passed at 1.5 *m.* From here the route swings sharply southward to the eastern shore of SEQUOIA LAKE (*trout and bass fishing*), 4 *m.* (5,300 alt.), 0.2 miles outside the park's western boundary. The 100-acre body of water is owned by the Y.M.C.A., which maintains several boys' camps on its shores.

The Generals Highway continues northward from the Plaza and crosses into Kings Canyon National Park, where it becomes the Kings River Canyon Highway to Cedar Grove and beyond, open only in summer.

The JOHN MUIR TRAIL, the longest and most varied route in the two parks, enters Kings Canyon Park at Pavilion Dome (11,846 ft.), its northwest corner. Passing Glacier Divide, it moves over an extraordinary terrain, ringed by sharp-toothed mountains, glaciers, deep-blue glacial lakes, wooded ridges, moving southeast into Sequoia. Within the horizon are Mt. Goethe (13,227 ft.) Mt. Darwin (13,830 ft.) and Darwin Glacier; Mt. Gilbert, 13,103; numerous peaks in the 12,000-13,000 ft. range. The waters of the numerous creeks eventually reach the forks of Kings River on the border of Sequoia. The Canyon has tall, steep walls through which the River moves for nine miles. Cedar Grove is the base for trips into Zumwalt Meadows, Roaring River Falls, Mist Falls and Paradise Valley.

Yosemite National Park

Season: Summer, May 1 to Oct. 1; winter, Dec. 15 to late April.
Administration: National Park Service, U. S. Dept. of the Interior. Superintendent at Headquarters, Yosemite Village, is in charge of the Park, which is controlled by the uniformed Park Rangers. Uniformed Park Naturalists interpret the wonders of the park. Campgrounds are free; hotels, lodges and cabins are operated by the Yosemite Park & Curry Co., under contract with the Dept. of the Interior, with offices in the Park and at 55 Grant Ave., San Francisco, and 514 So. Grand Ave., Los Angeles.
Admission: Vehicle permit fees, $3 for 15 days; $6 for year, are collected at entrance stations. By automobile: From the West, Arch Rock entrance via State 140, open all year; Crane Flat, Big Oak Flat Road via State 120, is closed in winter. From the South (Fresno), south entrance via State 41, open all year. From the East, Tioga Pass entrance via State 120, closed in winter. Big Oak Flat Road west of Crane Flat is not suitable for buses, trucks or house-trailers. The new Tioga Pass Road from Lee Vining, on Highway 395, open from late May till the first snows, is an easy drive through a panorama of granite peaks.
Transportation: Southern Pacific and Santa Fe trains and Pacific Greyhound and Continental Trailways buses operate to Merced and Fresno from the north and south. The Yosemite Transportation System of the Yosemite Park & Curry Company carries passengers to Yosemite Valley from Fresno and Lake Tahoe in summer and from Merced all year. Write to the Yosemite Transportation System, Yosemite National Park, Calif., for timetables. United Air Lines serves Merced and Fresno on Los Angeles-San Francisco flights; Trans World Airlines serves Fresno.

ACCOMMODATIONS

WAWONA HOTEL, near South entrance—Fresno Road, State 41. American plan, May 18 to Sept. 3. Room, one person, with connecting bath, $19; two persons, each $15, three persons, each $13; private baths, $21-$23, $17-$18, $15-$16. Rooms without bath are $3 to $4 cheaper. THE AHWAHNEE, in Yosemite Valley, "a hotel in the grand manner," with cottages, tennis, swimming pool. American plan, open all year, except late Nov., early Dec. Dancing and movies. One person in room, $26-$29; two in room, each $22-$24; three in room, $21-$23 each. Private sitting room, $18. GLACIER POINT HOTEL, 3,200 ft. above Yosemite Valley, European Plan (meals not included), open May 18-Sept. 9. One person in room, with bath, $11, without $7; two persons or more, each $7 and $4.50. MOUNTAIN HOUSE, next door to hotel, one person in room, $6, two persons, each $3.50.

YOSEMITE LODGE, near Village and Yosemite Falls. European Plan, meals not included. Open all year. Five buildings, 122 rooms, tent cabins, swimming pool, restaurant, cafeterias, cocktail lounge, post office. Hotel type rooms with bath, one person, $11.50-$14; two in room, each, $7.50-$9; three in room each, $5.75-$7; four, each $5-$6. Without bath, single, $7.50, double, $5 each. Redwood cabins with bath, electric heat, $8 for one, $6 for two each. Cabins, without running water, with oil heaters, single, $4.75, double, $3 each. Summer tent cabins, $3.75 for one, $2.75 each for two.

CAMP CURRY, base of Glacier Point. Open May 15-Sept. 7. Hotel-type rooms and tent cabins. Prices and facilities similar to Yosemite Lodge. BIG TREES LODGE, in Mariposa Grove, Wawona Road, open May 10-Oct. 12. European Plan. Room, single with bath, $10, without bath, $7; two or more in room, each $6.50 and $4.50. TUOLUMNE MEADOWS LODGE, Tioga Pass Rd. near east gate in the High Sierra, 8,600 ft., open June 14-Sept. 8. European plan. Canvas cabins without bath, single, $5, two or more each, $3.50; cabin for 4, $3 each. WHITE WOLF LODGE, off Tioga Pass Road, 8,000 ft. Rustic cabins and tents, pack and saddle animals. Meals not included but available. Cabin with bath, one person, $10, two, each, $6.50; three, each, $5; four each, $4. Tents, $5 and $3.50. HOUSEKEEPING CAMP, 1 *m.* west of Camp Curry along Merced River. Open May 25-Sept. 4. Duplex units, one to 4 persons, $8 daily, $44 weekly. Furnished tents are lower. Housekeeping cabins, completely furnished with lights, bath, one room for two, $12 daily, $66 weekly; two rooms, four persons, $20 daily, $110 weekly; without running water, one room for two, $8.50 daily, $46.75 weekly.

CAMPGROUNDS: Free. Located chiefly in the Yosemite Valley, but also at Glacier Point, Bridal Veil Creek, Wawona, Tuolumne Meadows, White Wolf and Crane Flat. Camp 4¼ *m.* west of Yosemite Lodge, accommodates 175 camps and is open all year except during heavy snowfall. Others open from about May 15 to Sept. 15. Bridal Veil Creek, Crane Flat and White Wolf campgrounds are Mission 66 projects with new tables, fireplaces and comfort stations. Register at entrances. House trailers admitted, but there are no electrical or other facilities. For trail trips get campfire permits from Chief Ranger's Office, Yosemite Village. Dogs and cats must be kept on leash. Housekeeping equipment may be rented, likewise saddle horses, with guides.

Winter Sports: Ski-ing, skating. Badger Pass has 4 T-bar lifts, runs for beginners and experts; ski school, for children 6 and up, and adults. Snowmobile tours to Glacier Pt. Mountain House, 10 *m.*, Ostrander Lake Ski Hut, 9 *m.* Accommodations at The Ahwahnee after Dec. 22, rates: 1 person with private bath and meals, $26-$28; two in room, $21-$23; Yosemite Lodge, 1 person, bath, no meals, $11.50-$14; two in room, $7.50-$9; cabins with bath, 1 person $8-$9; two in room, $5.50-$6. There are special rates for week-end and midweek groups, and for children.

Regulations: Camping is limited to 14 days in one year and to 10 days in Yosemite Valley during June, July, August. California licenses are required for fishing; limit is 10 fish. Cats and dogs are not allowed on beaches, trails or in public buildings. No smoking is allowed on hikes or horseback rides and all fires must be extinguished with water. Hunting is not permitted. Wild animals must not be approached or fed but left at a distance.

INFORMATION ABOUT TOURS

Tours and Trips: The Yosemite Park & Curry Co. provides a number of package tours with bus service and overnight accommodations for one or two nights, with lodging, transportation and taxes included, and meals except in case of the

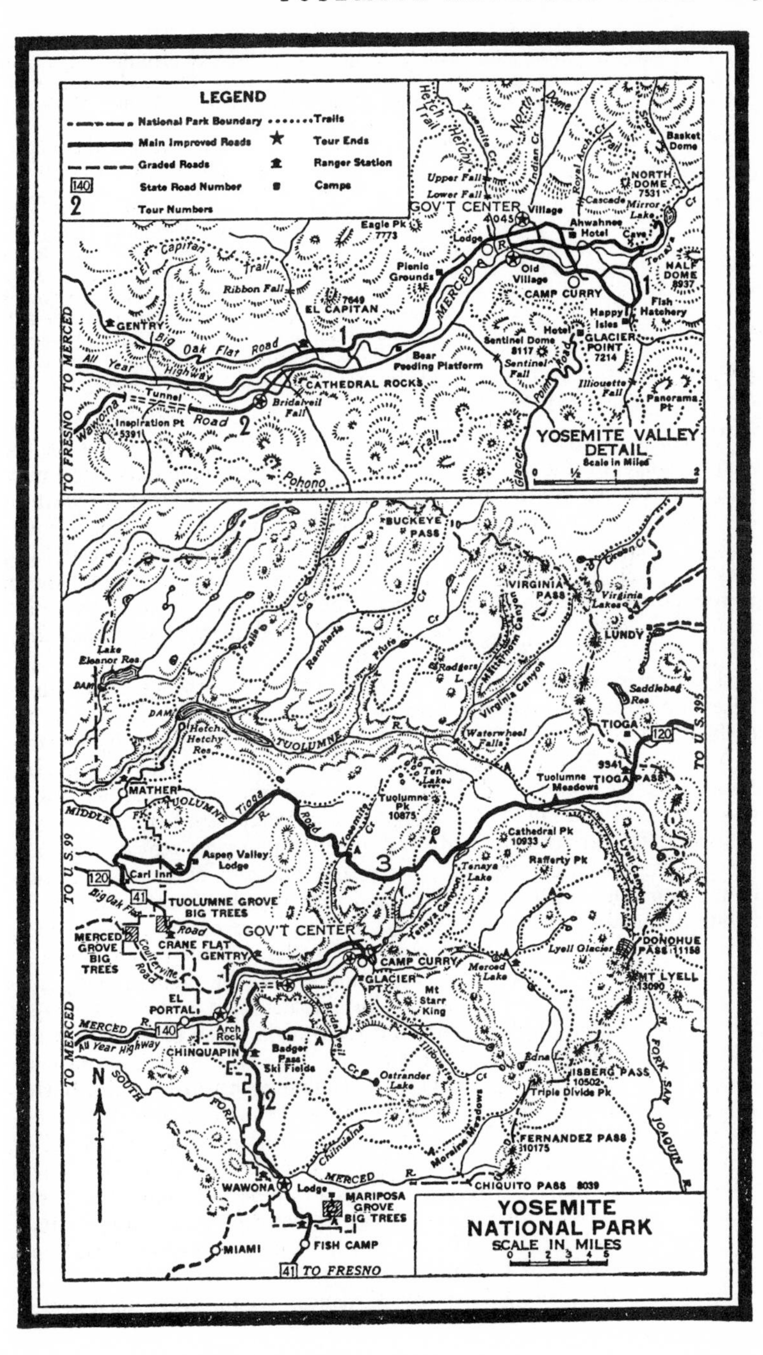
LEGEND
National Park Boundary
Trails
Main Improved Roads
Tour Ends
Graded Roads
Ranger Station
State Road Number
Camps
Tour Numbers
YOSEMITE VALLEY DETAIL
Scale in Miles
GOV'T CENTER
EL CAPITAN
CATHEDRAL ROCKS
CAMP CURRY
GLACIER POINT
HALF DOME
NORTH DOME
TO FRESNO
TO MERCED
YOSEMITE NATIONAL PARK
SCALE IN MILES
TUOLUMNE GROVE BIG TREES
MERCED GROVE BIG TREES
CRANE FLAT
EL PORTAL
CHINQUAPIN
WAWONA
MARIPOSA GROVE BIG TREES
FISH CAMP
MIAMI
TIOGA PASS
Tuolumne Meadows
MT LYELL
DONOHUE PASS
ISBERG PASS
FERNANDEZ PASS
CHIQUITO PASS
VIRGINIA PASS
LUNDY
TIOGA
MATHER
BUCKEYE PASS
TO U.S. 395
TO U.S. 99

socalled Economy Tour. Most tours depart from Merced; one also from Fresno. The trips include a Four-day Saddle Trip, and a Six-day Saddle Trip, with charges for meals, lodging, guide and animals in the fee. There also are available a Seven-Day Hiking Trip and Pack Trips from stables in Valley, Wawona, Tuolumne Meadows, White Wolf and Mather.
Reservations and Information: Reservations should be made months in advance. Include date of arrival, number of persons in party, type of accommodations desired, length of stay, and how arriving (bus or private car). If possible give alternate dates or type of accommodations. Deposit amounting to first day's rate required. Reservations may be made through the Yosemite Park & Curry Co. (authorized by the National Park Service), or through travel agent. Children under 3 are lodged free; and from 3 to 9 are charged half rates. Pets must be kept on leashes at all times and are not allowed in accommodations. Kennels are located in Yosemite Valley. Rates do not include 4% sales tax on meals and 4% county tax on lodgings except as shown for tours. Call or write: Yosemite Park & Curry Co., Yosemite National Park, California 95389 (209) 372-4671; 55 Grant Avenue, San Francisco 94108 (415) 982-9912; 514 S. Grand Avenue, Los Angeles 90017 (213) 626-0515.

Yosemite National Park, a spectacular mountain region, lies on the western slope of the Sierra Nevada. Its eastern boundary, 40 miles from the Nevada State line, is on the crest of the Sierra; its western edge is in the dry foothills where the mountains merge with San Joaquin Valley. The park has 752,744 acres of mountains and forests, 429 lakes, a chain of mountain peaks averaging 10,000 feet and more, granite domes and monoliths, many trout streams, glaciers, and high mountain meadows. Five great waterfalls and many lesser ones drop over perpendicular cliffs as high as 1,612 feet.

Two main canyons bisect the park from east to west: Yosemite Valley to the south and the Grand Canyon of the Tuolumne to the north—both gouged down thousands of feet into solid rock by streams.

Yosemite Valley (3,985 alt.), with waterfalls, Mirror Lake, Half Dome, Glacier Point, and El Capitán, was the first area opened to travel; most of the campgrounds, hotels and lodges are here.

Over 230 varieties of birds and many mammals, reptiles and amphibians have been catalogued. Tree squirrels and chipmunks scamper about camps looking for food; bear and deer wander over the valley unafraid of man. The Big Trees (*Sequoia gigantea*) of the Mariposa Grove, giants left over from the Age of Reptiles, are the park's outstanding trees. The valley floors, the mountain slopes, the ravines, and gorges are grown with luxuriant stands of pine, fir, hemlock, cedar, oak, maple, western yellow pine, Jeffrey pine, and California black oak. The 1,200 kinds of flowering plants and ferns are at their best in spring and early summer, when meadows are alive with Indian paintbrush, lupin, buttercups, wild geranium, leopard lilies, camas, and shooting stars.

For centuries before white men arrived, Indians lived in the valley, where they had at one time 40 villages. They belonged to the Miwok tribe, superior to most California Indians and especially skilled in basket making. Their name for the valley was Ahwahnee (deep grassy valley). "Yosemite" came from the Miwok word, Uzumati (grizzly bear), one of the tribal divisions or totems.

The first Americans to see the Yosemite region were members of

Joseph Walker's California expedition of 1833, who failed to impress its wonders on the world, although they told of "many small streams, which would shoot out from under high snow banks, and after running a short distance in deep chasms, precipitate themselves from one lofty precipice to another." The discovery of Yosemite Valley was accidental. Major James D. Savage and Dr. L. H. Bunnell, leading the "Mariposa Battalion" in an expedition to track down the warlike Miwok Indians and force them to sign a treaty, stumbled out of the forest on March 25, 1851, at Inspiration Point, where Yosemite Valley lay spread out before them. Bunnell was so impressed that he forgot the Indians. At the campfire on the bank of the Merced River that night, he suggested the name Yosemite. Afterward he spread the fame of the valley in his book, *Discovery of the Yosemite.*

In 1864 Congress granted Yosemite Valley and Mariposa Grove to California as a State park; in 1890 it established Yosemite National Park and in 1906 Yosemite State Park was incorporated in it. The park was guarded in summer by Federal troops and in winter by civilian rangers until the creation of the National Park Service in 1916.

YOSEMITE TOUR 1

Arch Rock Entrance Station—Yosemite Lodge—Government Center—Indian Caves—Happy Isles—Old Village; 17.2 *m.,* All-Year Highway, El Capitán Rd.

Oil-treated macadam roadbed throughout.

The All-Year Highway continues northeast from State 140 (*see TOUR 3b*) from the ARCH ROCK ENTRANCE STATION, 0 *m.* (2,855 alt.).

ARCH ROCK, 0.1 *m.,* at the western end of forest-choked Merced Canyon, is formed by two great granite boulders. The cliff-bound route runs through yellow pines, incense cedars, and Douglas firs. Across riotous Merced River (R), innumerable waterfalls tumble down the canyonside in a series of cascades. CASCADE FALLS, visible (L) at 3.1 *m.,* shoot from the rim of the canyon 594 feet above the road. INSPIRATION POINT (5,391 alt.) juts from the rim of the gorge high above granite PULPIT ROCK (4,195 alt.) across the river (R) at 4.8 *m.*

At 5.8 *m.* is the junction with Pohono Rd.

Right on Pohono Rd., crossing the Merced River, 0.2 *m.* to FERN SPRINGS (R), bubbling cold and clear into a small stream where a unique species of salamander is found.

The road winds through the grassy reaches of pine- and oak-dotted BRIDAL VEIL MEADOWS. On the far side (L) at 0.6 *m.* a plaque in a grove of trees marks the GRAVES OF ROSE AND SHURBORN, prospectors from Coarse Gold Gulch ambushed here by Yosemite Indians May 26, 1852. The United States Cavalry expedition that buried their bodies shot five warriors in revenge and chased the rest southwestward into the Mono Lake region.

BRIDAL VEIL FALL (R), 1.5 *m.,* drops 620 feet from the cliff above in a delicate lacy pattern clouded with fine, rain-like spray. The mist that hangs in the air roundabout is tinged by the sun in late afternoon—the best time to see the fall—with changing rainbows of rich reds, blues, purples, and yellows, which the observer can reach out and touch.

The twin shafts of CATHEDRAL ROCKS (R), 2.4 *m.*, towering 2,154 feet above the valley floor, challenged mountain climbers until 1935, when youthful members of the Sierra Club inched up them with ropes and spikes like human flies under the lenses of motion picture cameras.

A concourse of shaggy brown or black Yosemite bears—and sometimes an uninvited coyote or skunk—usually gather at the concrete troughs of the BEAR FEEDING PITS (L), 3.3 *m.* (*feeding time 8 p.m.*) in the glare of floodlights when garbage is brought in from hotels and lodges. The bears, whose sense of smell is keener than their eyesight, come tracking down the scent of sugar or honey, set out to attract them. Mother bears sometimes bring their cubs and chase them up a tree for safety while they investigate the evening menu. The bears will eat almost everything but grapefruit rinds.

The route continues along the southern rim of the valley, past a meadow (R), 4.3 *m.*, starting point for the 4-mile trail to Glacier Point (*see Trail 3*), to OLD VILLAGE, 5.1 *m.* (*see below*).

East of the junction with Pohono Rd. the route continues as El Capitán Rd. through the Gates of the Valley, formed by the sheer precipices of El Capitán (L) and Cathedral Rocks (R). Beyond these towering granite portals lie the peaceful forest-fringed green meadows of the valley floor beneath the colossal upthrust of cliffs. High up on the north wall of the canyon (L), the slim stream of 1,602-foot RIBBON FALLS, highest in Yosemite, appears when snows are melting. Across the meadows, behind the projecting cliff of Glacier Point (R), bald granite Half Dome soars aloft. The stony peak in the distance is CLOUDS' REST, highest point on Yosemite's walls.

At EL CAPITAN CONTROL STATION, 7.2 *m.*, is the junction with Big Oak Flat Rd. (*see YOSEMITE TOUR 3*).

EL CAPITAN (L), 7.8 *m.*, bulking above the valley floor 3,604 feet, is the world's largest monolith of exposed granite. In volume it equals four Gibraltars; in height, three Empire State Buildings.

At ROCKY POINT (L), 9.9 *m.*, the road winds beside rough boulders that fell from the cliff above in February 1923, during one of the biggest rock slides in the valley's recent history. Just above the tree line, the face of the cliff bears evidence of glacial polish and rock striations that give it the appearance of rough concrete. Higher than Rocky Point are the THREE BROTHERS, three peaks looking like steps as they rise one above the other.

In the trim, Government-supplied modern cottages of the NEW INDIAN VILLAGE (L), 10.1 *m.* (*visitors welcome*), live about 60 Indians, descendants of the original Yosemite, Paiute, Mono, and other Indian peoples. The Indians sell baskets, beadwork, and other articles; they are given work preference in the valley.

YOSEMITE LODGE is at 10.6 *m.* (*cottages, cabins, swimming pool, lounge*).

At 10.7 *m.* is the junction with a paved road.

Left on this road 0.2 *m.* for the best view of YOSEMITE FALLS, which plummet downward in two separate falls—Upper and Lower Yosemite—in a ponderous 2,425 foot drop from the rim of the canyon wall to the floor. The Upper Yosemite Fall, highest free leaping waterfall in the world, drops 1,430 feet—nine times as far as Niagara. The 910-foot plunge of Middle Cascade carries it to 310-foot Lower Yosemite Fall.

From the end of the road, Lost Arrow Trail winds through black oaks, goldencup oaks, and California laurels to the foot of the lower fall, 0.7 *m.*

The route continues through a fine grove of glossy-leaved California black oaks to YOSEMITE VILLAGE, 11.2 *m.* (4,045 alt.), administrative headquarters of Yosemite National Park. The YOSEMITE MUSEUM (*open daily 9-12, 1-5*) houses a library and exhibits devoted to all the park's major aspects; geology, flora and fauna, Indians, and history. A special feature is the relief scale models of Yosemite—with all trails, roads, falls, peaks, and glaciers marked—where riding and hiking trips can be planned in detail. Behind the museum are two Indian bark huts, a *temescal* (Ind., sweat house), and two grain storage stacks; the local Indians here demonstrate how their ancestors pounded acorns, harvested wild seeds, made baskets, and built huts. At the eastern edge is LEWIS MEMORIAL HOSPITAL.

At the entrance (L) to the $1,000,000 AHWANEE HOTEL, 11.7 *m.*, Yosemite's luxury hostelry, built in rustic style of concrete blocks and rough timber in a forested nook of the valley, the route turns R. to a junction at 12 *m.* and then L. across two bridges over the Merced River.

The flat face of HALF DOME (8,927 alt.) towers straight ahead nearly a mile above the valley floor. Above the meadows (R) rises GLACIER POINT (7,214 alt.) (*see YOSEMITE TOUR 2*). The ROYAL ARCHES appear (L) at 12.6 *m.* on the perpendicular valley wall—the main arch measuring 1,000 feet in height and 1,800 feet in width—and behind them, the granite mass of NORTH DOME (7,531 alt.). The WASHINGTON COLUMN (5,912 alt.) lifts its top almost half a mile above the road right of the Royal Arches.

INDIAN CAVES (L), 13.1 *m.*, is the site of the old Indian village Lah-koo-hah (come out). Great granite boulders, knocked off by slides and glaciers, lie in a jumble at the foot of the cliff, forming several big caves and many small ones. Indians used these caves for storage, for shelter from storms, and sometimes for hide-outs. In front of the caves on a large flat rock are some bark huts where the Indians dance and chant in summer for the benefit of tourists.

At 13.2 *m.* is the junction with a paved road.

Left on this road through TENAYA CANYON to MIRROR LAKE, 0.5 *m.*, a favorite subject of painters and photographers. The lake is best before 9 a.m., when its glassy surface reflects Half Dome and CLOUDS REST (9,924 alt.) on the south side of Tenaya Canyon, and MOUNT WATKINS (8,600 alt.) and BASKET DOME (7,602 alt.) on the north side. Mirror Lake was formed by a rockslide from the walls of Tenaya Canyon which dammed up Tenaya Creek. Now the creek is slowly filling the lake bottom with silt.

At 13.5 *m.* is a junction; the route continues L. here, to cut through boulder-piled MEDIAL MORAINE, formed by glaciers during the Ice Age.

Around the rocky HAPPY ISLES, 14.1 *m.* (*parking space, picnicking, hiking*), foams the Merced River, shooting down from Vernal and Nevada Falls, with an unceasing roar. The STATE FISH HATCHERY (*open 8-12; 1-5*) has an annual output of nearly a million and a half Loch Levan and eastern brook trout. Long troughs inside the building are filled with growing fish that will stock park streams. The Concrete Footbridge (L) is the starting point for many short trail trips and long hikes.

CAMP 14 (R), 14.8 *m.,* is one of the largest of the park's free campgrounds.

The OLD APPLE ORCHARD (L), 14.9 *m.,* was planted by James Lamon, first homesteader in the valley, in the early 1860's.

CAMP CURRY, 15.1 *m.* (*lodge, cabins, tent houses*), is open during summer only, but in winter operates a skating rink, ashcan slide, and beginners' ski slopes.

JOSEPH LECONTE MEMORIAL LODGE (L), 15.5 *m.,* is a stone building erected by the Sierra Club and friends to the memory of the geologist who loved Yosemite so much. "Joe" Le Conte—as his students called him—was a professor at the University of California. Inside the lodge are a lounge, a library on Yosemite and the Sierra, and a collection of photographs.

In the OLD VILLAGE, 16.2 *m.* (*general stores, café*), formerly park headquarters, is CEDAR COTTAGE (L), built in 1859 of handsawn boards and shakes split from local pine. Back of the building is the BIG TREE ROOM, built around a large incense cedar that grows up through the roof.

YOSEMITE PARK TOUR 2

South Entrance Gate—Mariposa Big Tree Grove—Wawona—Glacier Point—Wawona Tunnel—Jct. with Pohono Bridge Rd.; 25.4 *m.*

At the SOUTH ENTRANCE GATE, 0 *m.* (*see TOUR 3b*), is the junction with a paved road.

Right on this road to the MARIPOSA GROVE OF BIG TREES, 2 *m.* (*lodge, museum, campground*), a 2-mile tract in which grow 617 giant Sequoias—some 200 of them measuring 10 feet or more in diameter at breast height. The grove was discovered in May 1857 by Galen Clark, whose cabin, built soon after the discovery, is reproduced in the BIG TREE GROVE MUSEUM.

The oldest tree in the grove is the gnarled GRIZZLY GIANT, estimated to be 3,800 years of age, which has a girth of 96.5 feet and a height of 209 feet. The fallen MASSACHUSETTS TREE near the museum, 280 feet long and 28 feet in diameter, which broke into sections when it was blown over in 1927, affords an opportunity to study the growth rings of the Sequoia. The FALLEN MONARCH, nearly 300 feet long and 26 feet in diameter, is of such size that a six-horse stage, a row of automobiles—and even a troop of United States cavalry—have been photographed on its trunk at various times. The TELESCOPE, 175 feet high and 18 feet in diameter, is still bearing cones, although its heart was burned out by fire. The road through the grove goes through a tunnel 8 feet high and 11 feet wide, cut through the base of the 231-foot WAWONA TREE in 1881.

WAWONA POINT (6,980 alt.), at the end of the loop road through the grove, overlooks the panoramic expanses of Wawona Basin and South Fork Canyon.

WAWONA (4,096 alt.), 4 *m.* (*hotel, store, campground, garage; tennis courts, golf course, and swimming pools; saddle and pack animals*), stands in a wide mountain meadow, fringed with a forest of pine and fir. The surrounding country, known as the Wawona Basin, was added to the park in 1932.

At CHINQUAPIN JUNCTION (6,050 alt.), 16 *m.* (*gas pump and lunch counter*), is a ranger station (*information*).

Right from Chinquapin Junction on Glacier Point Rd. 5.2 *m.* to BADGER PASS SKI HOUSE (*fireplace, restaurant, ski rental rooms*). The skiing here, comparing favorably with that at St. Moritz, ranges from easy practice slopes for beginners to steep, 11-mile downhill runs for experts. Four T-bar lifts tow skiers to the Ski Top (7,950 alt.), starting point of the downhill runs.

GLACIER POINT (7,214 alt.), 16.1 *m.* (*hotel, restaurant*), commands the climax of all Yosemite views. From the top of the cliff beside OVERHANGING ROCK (L), Yosemite Valley lies, 3,254 feet below. The Merced River winds in a thin silver course through the deep green of the meadows and forests, which startlingly resemble a well-kept park in the midst of wild crags and mountains. Automobiles look like crawling beetles, and human beings are hardly visible. From the rim of the valley the High Sierra stretches out and up with all the domes, snowy peaks, waterfalls, cascades, and gorges distinct in the clear mountain air. Half Dome dominates the landscape. Vernal and Nevada Falls are visible (R).

At 23.2 *m.* the Wawona Road enters WAWONA TUNNEL, drilled and blasted through solid granite to avoid scarring the outside of the cliff. The tunnel was finished in 1933 at a cost of $837,000. It is 4,230 feet long, 28 feet wide, and 19 feet high. It includes the latest types of automatic ventilating and safety devices.

At 24 *m.* is the EAST PORTAL (4,408 alt.) of Wawona Tunnel. A wide parking area commands a sweeping panorama of Yosemite Valley, with massive El Capitán (L) and Half Dome looming in the distance. Right are the THREE GRACES and Bridal Veil Fall. A photograph near the granite curb gives the names and altitudes of the peaks and falls visible from this point.

At 25.4 *m.* is the junction with Pohono Rd. (*see YOSEMITE TOUR 1*).

YOSEMITE PARK TOUR 3

Junction with Big Oak Flat Rd.—Aspen Valley Entrance Station—Tenaya Camp—Tuolumne Meadows—Tioga Pass Entrance Station—Jct. with US 395; 66.4 *m.*, State 120.

Oil-treated macadam roadbed between junction with Big Oak Flat Road and Aspen Valley Entrance Station, elsewhere partly dirt and partly paved; open from about July 1 to October 1.

The Tioga Pass road follows the divide between the Merced and Tuolumne Rivers, passing chasms truly startling in size and reached so abruptly that the traveler is totally unprepared for their appearance; below, the rivers shine and glint in their gorges. Along this divide between the two rivers Captain Joseph Reddeford Walker, the first white man in the Yosemite region, led an exploring party in November 1833, coming through from the east. This route penetrates the heart of the High Sierra near the headwaters of the Tuolumne River in the northern sector of Yosemite National Park. This formerly little-known region includes, as John Muir wrote, "snowy mountains soaring into the sky twelve and thirteen thousand feet . . . separated by tremendous canyons and amphitheaters; gardens on their sunny brows, avalanches thundering down their long white slopes, cataracts roaring gray and foaming in the crooked, rugged gorges, and glaciers in their shadowy recesses . . ."

At 0 *m.* is the junction of State 120 (*see TOUR 3b*) with oil-surfaced Big Oak Flat Rd.

Right on Big Oak Flat Rd. 0.3 *m.* to CARL INN (*fuel and refreshments*).

The TUOLUMNE GROVE ENTRANCE STATION, 5.3 *m.* (*open 6-9*) is in the TUOLUMNE GROVE OF BIG TREES, near the headwaters of Crane Creek. The grove contains many of the best specimens left standing of the *Sequoia gigantea*, or Big Tree. The DEAD GIANT is one of the grove's "tunnel trees," originally 120 feet and still more than 100 feet in circumference, although several times damaged by forest fires. Through it passes a full-sized road. Also in this grove are two trees that have become united about 20 feet from the ground, named the SIAMESE TWINS. A little below the Dead Giant in a small ravine are the remains of what once was a titanic tree, estimated to be perhaps 4,000 years old.

Big Oak Flat Rd. skirts sweeping vistas of Yosemite Valley as it passes through fine stands of fir and sugar pine. East of the GENTRY RANGER STATION, 15.3 *m.* (*telephone, information*), the road is under one-way control (*in-bound on odd hours only, out-bound on even hours only*).

At 23.3 *m.* is the junction with El Capitán Rd. in Yosemite Valley (*see YOSEMITE TOUR 1*).

At 2.1 *m.* on State 120 is the junction with an oiled road.

Left here 6 *m.* to the MATHER ENTRANCE STATION (*open 6-9:30*) to Yosemite.

The highway runs along the rim of Poopenaut Valley to the great concrete expanse of O'SHAUGHNESSY DAM, 12.2 *m.*, 20 years in the building, which blocks the mouth of Hetch Hetchy Valley, impounding a reservoir that supplies water and power to San Francisco. It is about 3 miles long and from a quarter to three-quarters of a mile wide, carved out by a huge glacier. For 2,000 feet the sheer granite walls of the canyon rise almost perpendicularly; the crests of pinnacles and domes in the range surrounding it rise twice that high above the lake. HETCH HETCHY FALLS, or the Wapama, on the north side below North Dome, form a cascade, not quite perpendicular. The varied battlements of rock resemble those of Yosemite Valley, although less impressive because smaller in size. A hunter by the name of Joseph Screech discovered this valley in 1850. The spelling of the name customarily used at the time was Hatchatchie, an Indian word for a kind of grass used for food.

At 8 *m.* is the ASPEN VALLEY ENTRANCE STATION (*open 6-9:30*) to Yosemite National Park.

From Aspen Valley, populous with silvery aspens, the Tioga Pass Road leads eastward across the Sierra highlands through magnificent stands of fir and pine. It traverses Long Gulch to YOSEMITE CREEK CAMPGROUND, on the upper waters of the torrent which plunges 5 miles downstream into Yosemite Valley.

Along the base (L) of MOUNT HOFFMAN (10,836 alt.) it proceeds, crossing Porcupine and Snow Flats, to TENAYA LAKE, 28 *m.* Indians called it Pyweack (Lake of the Shining Rocks) because of the glacier-burnished granite in the depths of the lake and on the shores. Here old Chief Tenaya and his band of Yosemites were chased down and captured in May 1851.

The road continues northeast, running along the base of Polly Dome, huge and massive, devoid of vegetation except for a great juniper tree perched jauntily high on its side. It passes Fairview Dome, another typical cupola of granite, and runs on into TUOLUMNE MEADOWS (8,594 alt.), 45.1 *m.* (*gas station, garage, store, post office, lodge, free campground*), in the basin of an ancient lake, a grassy floor of shining green in summer. Before the basin held a lake it was filled by a glacier from which poured streams of melting ice through Tuolumne, Hetch Hetchy, and Tenaya Canyons. In the northern section of the meadows is SODA SPRINGS, an especially popular spot for camping. The story is that biscuits made with the spring water have no equal for lightness; the carbon dioxide in the water raises them as baking-powder.

Tuolumne Meadows is a starting place for trips by foot or horseback down the Tuolumne River Gorge by way of Glen Aulin High Sierra Camp to Waterwheel Falls, where the river, John Muir wrote, "is one wild, exulting, onrushing mass of snowy purple bloom . . . gliding in magnificent silver plumes, dashing and foaming against huge boulder dams, leaping high in the air in wheel-like whirls. . . ." Farther down the canyon of the Tuolumne are half-mile-deep Muir Gorge and meadowed Pate Valley with its overhanging rocky walls where are Indian pictographs. Other trails from Tuolumne Meadows lead to Mount Conness and to glacier-shrouded, 13,090-foot Mount Lyell, dominating the upper Tuolumne Region from the east.

The road continues northeastward from Tuolumne Meadows. MOUNT CONNESS (12,560 alt.) rises ahead, named for the Senator who in 1864 secured legislation making the Yosemite region a public reserve.

TIOGA PASS (9,941 alt.), 52.1 *m.,* the highest elevation of the route, marks the crest of the Sierra. Here the road leaves Yosemite National Park at the TIOGA PASS ENTRANCE STATION (*open 6-9:30*).

The rugged snow-mottled flanks of red MOUNT DANA (13,055 alt.), named for James Dwight Dana, American mineralogist and geologist, tower high to the southeast. On its slope lies TIOGA LAKE (R). Northeast of the lake is the old Tioga Mine, whose owners hired Chinese laborers to build the Tioga Road in the early '80's. ELLERY LAKE (R) lies at the head of Leevining Canyon, through which the road swoops downward in breath-taking curves. The descent is abrupt, twisting on narrow ledges carved on the edge of mountain slopes—and even from bare granite cliffs.

At 66.4 *m.* is the junction with US 395 (*see TOUR 6b*).

YOSEMITE PARK TRAILS

Information on trails, fishing, and camping, and park maps available at Yosemite Village and ranger stations. Hikers should avoid short cuts from designated trails, should start early on long hikes and return before dark, should register at chief ranger's office before starting on trips to isolated sections. Taxi service available to and from start of trails in upper half of valley; telephones available at base of all trails. Inquire at hotels or stables about daily saddle trips to trail points.

TRAIL 1: Happy Isles—Sierra Point—Vernal Fall—Nevada Fall; 3.5 *m.* (*Horses not allowed off valley floor without guide.*) The easiest of the trails off the valley floor, wide and paved, affording fine views of trees, rocks and waterfalls. From the footbridge in HAPPY ISLES, 0 *m.*, the trail winds along the cascading Merced River.

At 0.3 *m.* is the junction with a trail.

Left here 0.5 *m.* over a series of switchbacks on the flank of Grizzly Peak to SIERRA POINT, where five great waterfalls—Upper and Lower Yosemite, Vernal, Nevada, and Illilouette—can be seen at one time.

As the main trail rounds the base of GRIZZLY PEAK, the mouth of ILLILOUETTE CANYON and 370-foot ILLILOUETTE FALL appear at its head (R).

VERNAL BRIDGE, 1.5 *m.*, offers a good view of Vernal Fall (*see below*).

At 1.7 *m.* is the junction with a horse trail, an alternate route.

Right here 2.4 *m.*, to NEVADA FALL, GLACIER POINT, 7.3 *m.*, and MERCED LAKE, 12.9 *m.*

The main route, known for a short distance as the Mist Trail, where a fine rain descends from Vernal Fall, continues up the right side of the canyon in a series of steps across the steep cliffs and across the face of a perpendicular cliff on a narrow ledge (*iron handrails*).

The waters of 317-foot VERNAL FALL, 2.3 *m.*, one of the great falls of Yosemite, thunder from the top of the granite cliff in a wide sheet.

From Vernal Fall the trail sometimes zigzags up rocky slopes and then plunges into cool pine woods.

At 2.5 *m.* is the junction with a horse trail; the route continues L., crosses the river at DIAMOND CASCADES, climbs to SNOW FLAT, below LIBERTY CAP (7,072 alt.), and continues its ascent (L) another 500 feet, by the NEVADA FALL ZIGZAGS.

NEVADA FALL, 3.5 *m.* (5,910 alt.), drops 594 feet with a curious twist, breaking off at the edges to separate in what are known as spray rockets.

TRAIL 2: Happy Isles—Nevada Fall—Panorama Cliff—Glacier Point (7,214 alt.); 8.3 m. One of the easiest of the longer hikes, requiring about 6 hours.

TRAIL 3: Bridal Veil Rd.—Union Point—Glacier Point; 4.6 *m.* One of the most popular Yosemite trails, rewarding the climber with superb views; it rises rapidly in switchbacks up the valley rim through forests of Douglas fir and sugar pine.

TRAIL 4: Happy Isles—Vernal Fall—Nevada Fall—Half Dome (8,852 alt.); 8.3 *m.* A difficult climb, safe but requiring caution; canteens should be carried. Half Dome Summit, 4,892 feet above the valley floor, affords a view of the San Joaquin Valley and the Coast Range on clear days.

TRAIL 5: Happy Isles—Vernal Fall—Nevada Fall—Little Yosemite Valley—Clouds Rest; 10.5 *m.* A one-day climb to and from Clouds Rest (9,930 alt.), highest of the peaks enclosing Yosemite Valley.

TRAIL 6: Yosemite Valley—Columbia Point—Yosemite Point and Falls; 3.16 *m.* John Muir, a man "almost beside himself with the glory of the Sierras," described the view from Yosemite Point as one of the most impressive phenomena in the area.

TRAIL 7: Mirror Lake to North Dome, 8 *m.* This trail is very steep, and an early start should be made. Splendid views from North Dome, especially west to the Gates of the Valley and to the east where the colossal bulk of Half Dome dominates the scene.

TRAIL 8: Yosemite Valley—Happy Isles—Merced Lake—Vogelsang Pass and Camp—Tuolumne Meadows—Glen Aulin—Waterwheel Falls—May Lake—Mirror Lake; 70 *m.* This hike, led by a ranger naturalist every Monday morning during July and August (*reservations at Yosemite Museum; no charge*), covers the route in easy stages with a stop each night at a High Sierra camp (*cots and bedding $1; meals $1*). Provision is made for fishing, swimming, and walks to scenic point of interest along the route.

PART IV

Appendices

Chronology

1533 Pilot Fortuno Ximenes discovers an "island" (Lower California) west of Mexico. Killed while trying to land.

1535 May 5. Cortés lands where Ximenes was killed. Calls place Santa Cruz (possibly the later La Paz). Names the country *California.*

1539 Francisco de Ulloa surveys both shores of "Sea of Cortés" (later Gulf of California); misses mouth of Colorado River, but discovers that Baja (Lower) California is a peninsula.

1540 Hernando de Alarcón ascends gulf; discovers Colorado River. To contact him, Melchor Diaz traverses Arizona, and crosses Colorado River, near Yuma. He, or de Alarcón, first white man to set foot in Alta (Upper) California.

1542 Sept. 28. Juan Rodríguez Cabrillo sails into San Diego Bay, which he names "San Miguel."

1579 June 15. Francis Drake enters Drake's Bay; holds California's first Christian service, and claims "Nova Albion" in the name of Her Majesty, Queen Elizabeth.

1602 Nov. 10. Sebastián Vizcáino enters San Miguel Bay; renames it San Diego de Alcala. On Dec. 16 he anchors in Monterey Bay.

1697 Jesuits, under Father Juan María Salvatierra, begin mission at Loreta—first permanent colony in Baja California.

1701 Nov. 21. Father Eusebio Francisco Kino, Jesuit missionary, crosses southeastern corner of California, working among Indians of Pimeria Alta.

1767 Carlos III, of Spain, issues decree banishing Jesuits from all Spanish colonies.

1768 Father Junipero Serra, "patron saint of California," arrives at Loredo, with 16 Franciscan monks. Jesuit missions in Lower California surrendered.

1769 April. Two vessels arrive San Diego Bay, with supplies, to equip colony.

May-June. Settlers and soldiers under Gov. Gaspar de Portolá, Capt. Fernando de Rivera y Moncada and Father Junipero Serra come overland from Lower California, with cattle.

July 16. Father Serra blesses site of Misión San Diego de Alcala, first of 21 missions established in California within 54 years.

Aug. 2. Portolá camps at site of Los Angeles; continues northward in search of Monterey Bay, but, unaware, passes it, Oct. 2. José Artego, of advance guard, sights San Francisco Bay (Nov. 2), but expedition turns back, reaching San Diego Jan. 24, 1770.

1770 June 3. Father Serra founds Misión San Carlos de Monterey (renamed San Carlos Borromeo de Carmelo in 1771, when removed). Governor Portolá establishes Monterey presidio (fort) and takes formal possession of country, in name of Carlos III.

1771 Two missions founded: San Antonio de Padua and San Gabriel Arcangel.

1772 March 20. Captain Pedro Fages leads expedition from Monterey to explore San Francisco Bay. First white men to see San Joaquin and Sacramento Valleys.

Sept. 1. Father Serra founds Misión San Luis Obispo de Tolosa.

1773 Aug. 19. Thirty miles S. of present Mexican border, Father Francisco Palóu sets cross to mark Baja and Alta California boundary.

1774 March 22. Juan Bautista de Anza reaches San Gabriel Mission from Sonora by overland route.

1775 Aug. 1. First white man to enter San Francisco Bay: Juan Manuel de Ayala, in ship *San Carlos*.

Aug. 16. By royal decree Monterey becomes capital of California.

Nov. 4. 800 Indians attack San Diego mission, kill Father Luis Jaume and burn buildings.

Dec. 24. Woman of Anza's party of colonists, en route from Sonora to San Francisco, gives birth to son, Salvator Ignacio Linares, first white child born in California.

1776 March 28. Anza, with 247 colonists, reaches site of San Francisco. The presidio of San Francisco founded Sept. 17. Two missions, San Francisco de Asis (Dolores) and San Juan Capistrano, founded.

1777 Jan. 12. Misión Santa Clara founded.

Nov. 29. San José de Guadalupe, first pueblo in California, founded.

1779 June 1. Governor Felipe de Neve drafts regulations for government of California.

1781 Sept. 4. Los Angeles founded.

1782 March 31. Misión San Buenaventura founded.

1784 Aug. 28. Father Serra dies and is buried at San Carlos Borromeo Mission.

1786 Sept. 14. Jean François Galaup de la Pérouse brings French scientific expedition into Monterey. Severely condemns mission system.

1786–91 Father Fermin Francisco de Lasuen, who succeeds Serra, founds four missions: Santa Barbara, La Purisma Concepcion, Santa Cruz, and La Soledad.

1791 Sept. 13. John Groeham (Graham), first American in California, reaches Monterey, ill; dies same day.

1792 Nov. 14. Captain George Vancouver, in British sloop *Discovery*, reaches San Francisco—first of his three visits to California.

1793 Pueblo of Branciforte founded on site of Santa Cruz.

1796 Oct. 29. Yankee skipper, Ebenezer Dorr, brings his ship, the *Otter*, into Monterey Bay—first American ship in California waters.

1797–98 Five missions founded: San Jose, San Juan Bautista, San Miguel Arcangel, San Fernándo Rey de Espana, and San Luis Rey de Francia.

1803 June 26. Father de Lasuen dies at Misión San Carlos.

1804 Baja and Alta California separated. Monterey is made capital of Alta California. Santa Ines (Ynez) Mission founded.

1806 April 5. Nikolai Petrovich Rezanof comes to San Francisco to buy supplies for Russian trading post at Sitka. Presidio commander reluctantly permits sale.

1810 Mexicans revolt against Spain.

1811 Oct. 15. San Joaquin and Sacramento rivers explored for first time by water.

1812 Fort Ross, Russian trading post, is built less than 100 miles north of San Francisco.

1816 Jan. 15. Thomas Doak lands from *Albatross,* near Santa Barbara; becomes California's first American settler.

1817 Dec. 14. Misión San Rafael Arcangel founded.

1818 Nov. 20. Hippolyte de Bouchard brings two ships of war into Monterey Bay. Captures city Nov. 22, sacks it; later attacks other coast towns, is resisted at San Diego; sails away.

1820 Population of Upper California 3,270. Neophyte (Indian slave) population of missions 20,500.

1821 Feb. Augustin Iturbide leads rebel army into Mexico City. Becomes ruler.
Oct. Luis Arguello makes first extensive exploration of Sacramento Valley.

1822 April. Monterey and other California garrisons lower Spanish flag; recognize Iturbide regency.
Sept. 26. California formally proclaimed province of Empire of Mexico.
Nov. 9. First provincial legislature elected; meets in Monterey.
Nov. 22. Luis Antonio Arguello, first native-born California governor, elected.

1823 July 4. San Francisco Solano, 21st and last of Alta California missions, founded.

1824 Jan. 7. News of abdication of Emperor and of establishment of the Republic of Mexico is received in Monterey.

1825 March 26. California formally becomes a territory of the Mexican Republic.

1826 Nov. 6. Captain Frederick William Beechey, of British Navy, maps San Francisco Bay.
Nov. 27. Jedediah S. Smith, with trappers, arrives at Misión San Gabriel; first Americans to make overland trip to California.

1830 Population of California 4,256.

1831 Nov. 29. Pio Pico, Juan Bandini, and José Antonio Carrillo lead revolt against Governor Manuel Victoria, forcing his resignation.

1833 Aug. 17. Mexican Congress decrees secularization of missions. (Completed in 1837.)

1834 Sept. 1. Two hundred Mexican colonists arrive at San Diego from San Blas.

1835 May 23. Los Angeles is raised from pueblo to city status.

June 24. Algerez Mariano G. Vallejo founds presidio and pueblo at Misión San Francisco Solano. Names settlement *Sonoma.*

United States offers to buy California.

1836 Nov. 3. Don Juan Bautista Alvarado and José Castro lead revolt. Castro becomes governor.

Nov. 8. California *diputacion* issues declaration of independence. California remains free State for eight months.

1839 July 1. John Augustus Sutter, Swiss, lands in San Francisco. In 1840 acquires 11 sq. leagues of land, comprising New Helvetia, and builds Sutter's Fort. In 1841 acquires Russian property, Fort Ross, ending Russian encroachment in California.

1840 March 10. First Supreme Court of California, *Tribunal de Justicia,* is formed.

April 7. Arrest of Isaac Graham, American trapper settled in California, is followed by imprisonment of all "foreigners" (47) not married to California women. Graham acquitted of treason and released June 1841.

1841 Aug. 14. Wilkes expedition (first U. S. scientific expedition) arrives in San Francisco. James A. Dana, of party, writes of gold found in American River, confirming earlier reports.

Nov. 4. First overland immigrant train (Bidwell-Bartleson party), from midwestern U. S., arrives in California.

1842 March 9. Francisco Lopez, sheep-herder, discovers placer gold in Santa Feliciana (Placercita) Canyon. Twenty ounces of gold dust sent (Nov. 22) to Philadelphia mint, by Abel Stearns and Alfred Robinson, is first gold sent out of California.

Oct. 19. Believing U. S. and Mexico are at war, Commodore Thomas Catesby Jones, U. S. N., seizes Monterey and raises American flag. Two days later he apologizes and departs.

1843 May 1. Thomas Larkin, first and only U. S. consul to California (1843-46), appointed.

1844 March 8. Capt. John Charles Frémont, U. S. Army, arrives at Sutter's Fort.

Nov. 14. Californians start revolt which forces Governor Micheltorena's abdication.

Dec. 13. First wagon train over Truckee and Donner Lake route reaches Sutter's Fort.

1845 July 10. Further immigration of Americans to California forbidden by Mexican Government.

First wedding of Americans in California: Mary Peterson and James Williams, both from Missouri.

1846 March 6. Frémont raises American flag on Gabilan Peak, near Monterey. Ordered to leave California, he retreats on third night to Sutter's Fort.

May 13. War between U. S. and Mexico declared.

June 3. Col. Stephen W. Kearny is ordered to march to California from Santa Fé and take command of U. S. forces there.

June 14. The Bear Flag of "California Republic" is raised at Sonoma. Frémont takes command of Bear Flag revolt on July 5, declaring California's independence.

July 7. Commander John D. Sloat raises American flag at Monterey; California formally declared a possession of U. S.

July 9. American flag replaces Bear Flag at Sonoma.

Aug. 15. *Californian,* first California newspaper, is published in Monterey.

Sept. 23. American garrison at Los Angeles attacked by rebellious Californians. General uprising, led by Captain José María Flores, follows. In first battle, at Chino Rancho, Americans are routed.

Oct. 31. Donner party of immigrants halted by heavy snows at Donner Lake. Before rescue party arrives 39 of 87 members die.

1847 Jan. 9. At La Mesa last battle of rebellion won by U. S. forces; articles of capitulation signed at Rancho Cahuenga 4 days later.

Jan. 19. John C. Frémont becomes first American Governor of California, appointed by Com. Robert F. Stockton. Forty days later he is removed by Kearny; in August, is arrested at Fort Leavenworth, Kans., court-martialed and sentenced to dismissal from Army. Sentence later remitted but Frémont declines clemency and resigns.

1848 Jan. 24. James Wilson Marshall, building sawmill for Sutter on the American River, discovers the gold which starts California "gold-rush."

Feb. 2. Treaty of Guadalupe Hidalgo, ending war with Mexico, is signed. United States acquires California, New Mexico, Nevada, Utah, most of Arizona and part of Colorado.

Oct. 14. Town of Sacramento is founded by Sutter's son.

Nov. 9. First U. S. post office in California opened in San Francisco. White population of California about 15,000.

1849 Feb. 28. *California,* first steamer to bring "Gold Rush" passengers, comes into San Francisco with 365 passengers.

Aug. 17. Regular service between San Francisco and Sacramento begun, with *George Washington.*

Sept. 1. Forty-eight delegates convene in Colton Hall, Monterey, to draft State Constitution (adopted Oct. 10, signed Oct. 13, ratified Nov. 13).

Oct. 25. Democratic Party organized. Peter H. Burnett nominated for Governor; elected Nov. 13.

Dec. 24. San Francisco's first great fire destroys 50 houses.

1850 Feb. 4. Jayhawkers, an immigrant party, reach San Francisquito Ranch, after great suffering in Death Valley.

Feb. 18. Legislature creates original 27 counties.

April 22. Legislature passes law to protect rights of Indians.

Aug. 14. Armed squatters riot in Sacramento in dispute over validity of Sutter's grant.

Sept. 9. President Fillmore signs act of Congress admitting California as a State into the Union.

Population 92,597 (U. S. Census).

1851 March 3. Land commission appointed.
March 25. Chasing band of raiding Indians, Major James D. Savage discovers Yosemite Valley.
June 9. Vigilantes organize in San Francisco under Sam Brannan.
Aug. 31. *Flying Cloud,* famous Yankee clipper, arrives in San Francisco 81 days, 21 hours, after leaving New York.

1853 March 3. Congress authorizes survey of railroad route from Mississippi River to Pacific.

1854 Feb. 25. Sacramento made capital of State.
April 3. United States opens branch mint at San Francisco ending private coinage of gold.

1855 Feb. 22. Run on bank of Page, Bacon and Co. begins panic. Stringency becomes State-wide and lasts for two years.
Aug. 7. "Know Nothing" or "American" party hold State convention in Sacramento.

1856 Feb. 22. California's first railroad, Sacramento to Folsom, opened.
May 15. San Francisco Vigilantes reorganize on day after James King, crusading newspaper editor, is murdered by James Casey, politician; they hang Casey and drive corrupt city officials from office.
Aug. 23. First wagon road across Sierra Nevada is opened.

1857 Aug. 31. First overland stage reaches San Diego from San Antonio.

1859 Sept. 13. Senator David C. Broderick and Judge David S. Terry fight duel. Broderick is killed.

1860 April 14. San Franciscans get their first Pony Express mail. Postage $5 a half ounce.

1861 April 1. Vines of 1,400 different varieties, shipped from Europe, become nucleus of State's vineyards.
May 17. News of Fort Sumter's surrender received; California pledges its loyalty to the Union.
June 28. Central Pacific R.R. Co. of California is organized.
Oct. 24. First transcontinental telegraph line completed, ending need for Pony Express.

1867 May 13. San Francisco workers demand eight-hour day. Established for skilled labor in 1868.

1868 March 23. University of California is founded.

1869 May 10. First transcontinental railroad system, the Central Pacific and Union Pacific, completed; final spike, connecting the two railroads, is driven at Promontory, Utah.

1871 Oct. 24. "Chinese Massacre" in Los Angeles. Results in Chinese exclusion act of 1882.

1873 Jan. to June. Modoc War, California's last Indian trouble.

1877 Sept. 12. Workingmen's Party of California organized under Dennis Kearney.

1880 University of Southern California founded.

1885 Oct. 24. Orange Growers Protective Union of Southern California organized.

1886 Feb. 14. First trainload of oranges leave Los Angeles for the East.

1891 Oct. 1. Leland Stanford Jr. University opened.
1900 April 3. Work on reclamation of Imperial Valley by irrigation begins.
1903 Feb. 16. The golden poppy (*Eschscholtzia*) voted state flower.
1906 April 18. San Francisco partly destroyed by earthquake and fire.
1909 March 24. Direct Primary Law.
1910 Oct. 1. Los Angeles *Times* building dynamited.
1911 Jan. 3. Hiram W. Johnson becomes Governor; in 1917, Senator.
April 3. Initiative, Referendum and Recall Act is approved.
April 8. Workmen's Compensation Act passed.
Nov. 14. First woman voter in California votes at Stockton.
1913 May 19. Land Act prohibits Japanese ownership of farm land.
1915 Panama-California Exposition at San Diego and Panama-Pacific International Exposition at San Francisco.
1916 July 22. Preparedness Day parade bombed in San Francisco.
1926 Oct. 25. University of California at Los Angeles dedicated.
1929 Herbert Clark Hoover, of Palo Alto, becomes President.
1932 July. Tenth Olympic games at Los Angeles.
1933 Feb. 11. President Herbert Hoover sets aside 1,601,800 acres in Inyo County, as Death Valley National Monument.
June. California adopts Amendment repealing Prohibition.
1934 Dec. 16. Work on All-American Canal begun.
1936 Nov. San Francisco-Oakland Bay Bridge opened.
1937 May. Golden Gate Bridge opened.
1939 Feb. 18. Golden Gate International Exposition, San Francisco.
1945 April 25-June 26. United Nations Org. formed at San Francisco.
1951 Sept. 8. Japan signs treaty of peace with 49 nations at San Francisco.
1955 Floods damage Marysville and Yuba City.
1956 Aug. 22. Republican convention at San Francisco nominates Eisenhower and Nixon.
1957 New York Giants move to San Francisco; Brooklyn Dodgers to Los Angeles.
1960 Voters authorize bond issue of $1,750,000,000 for water project.
Olympic Winter Games held at Squaw Valley.
July 14. Democrats at Los Angeles nominate John F. Kennedy and Lyndon B. Johnson.
1962 Gov. Edmund G. Brown authorizes $120,000,000 for Oroville Dam.
1964 January. State Dept. of Finance announces California first state in population with 17,973,000. U. S. Bureau of the Census July 1 reports California pop. 18,084,000, including 121,000 U. S. Armed Forces, 17,749,000 without. New York 17,915,000 without Forces.
July 13-16. Republicans at San Francisco nominate Barry M. Goldwater, who won lead by California primaries, June 2.
1965 Aug. 11-16. Rioting by 5,000 to 8,000 Negroes in Watts section of Los Angeles, with looting and burning of stores; 34 persons killed.
1966 November 8. Ronald Reagan, Republican, motion picture actor, elected governor over Edmund G. Brown, Democrat, incumbent.

A Select Reading List of California Books

Asbury, Herbert. *The Barbary Coast.* New York, 1933.

Atherton, Gertrude. *Adventures of a Novelist.* New York, 1932.

——— *California, an Intimate History.* New York, 1914 (rev. ed. 1927).

Audubon, John Woodhouse. *Audubon's Western Journal: 1849–1850.* Cleveland, 1906.

Austin, Mary. *The Land of Little Rain.* Boston, 1903. Describes the Mojave desert and adjoining regions.

Bancroft, Hubert Howe. *History of California.* 7 vols. San Francisco, 1884–90.

——— *Popular Tribunals.* San Francisco, 1890. Describes the vigilance committees of early California.

Banning, Capt. William, and George H. *Six Horses.* New York, 1930. Describes a coaching trip.

Bell, Major Horace. *On the Old West Coast.* New York, 1930.

——— *Reminiscences of a Ranger.* Los Angeles, 1881 (later ed. 1935, indexed).

Bolton, Herbert Eugene. *Anza's California Expeditions.* 5 vols. Berkeley, 1930.

——— *Spanish Exploration in the Southwest, 1542–1706.* New York, 1925.

Burbank, Luther, with Wilbur Hall. *The Harvest of the Years.* Boston, 1927. Burbank's own account of his experiments in plant breeding.

California. Works Progress Administration. Federal Writers' Project. *Death Valley.* New York, 1939.

——— *Los Angeles.* New York (1939).

——— *San Diego, a California City.* San Diego, 1937.

——— *San Francisco and the Bay Area.* New York (1939).

Carr, Harry. *Los Angeles, City of Dreams.* New York, 1935.

Chalfant, Willie A. *The Story of Inyo.* Chicago, 1922 (rev. ed. 1933). A record of Death Valley days.

Chapman, Charles E. *A History of California: the Spanish Period.* New York, 1921 (rev. ed. 1928).

Chase, J. Smeaton. *California Desert Trails.* Boston, 1919.

Chittenden, Hiram Martin. *The American Fur Trade of the Far West.* 3 vols. New York, 1902 (later ed. 1935).

Cleland, Robert Glass. *A History of California: the American Period.* New York, 1922.

Cleland, Robert Glass, and Osgood Hardy. *March of Industry.* Los Angeles, 1929.

Clemens, Samuel L. (Mark Twain). *Autobiography.* 2 vols. New York, 1924.
——— *Roughing It.* New York, 1872 (later ed. 1934). Describes a journey from St. Louis across the plains through Nevada, Utah, and California in the early 1860's.
Coblentz, Stanton A. *Villains and Vigilantes.* New York, 1936.
Colton, Walter. *Three Years in California.* New York, 1850.
Coolidge, Mary Roberts. *Chinese Immigration.* New York, 1909.
Cowan, Robert Ernest. *Forgotten Characters of Old San Francisco.* Los Angeles, 1938.
Cowan, Robert Ernest, and Robert Granniss Cowan. *A Bibliography of the History of California.* 3 vols. San Francisco, 1933.
Coyner, David H. *The Lost Trappers.* New York, 1892. Purports to be a description of adventures in the western country in the early 19th century.
Cross, Ira. B. *A History of the Labor Movement in California.* Berkeley, 1935.
Dana, Julian. *Sutter.* New York, 1934.
Dana, Richard Henry. *Two Years Before the Mast.* Boston, 1840 (later ed. 1936).
Davis, William Heath. *Seventy-five Years in California.* San Francisco, 1929.
Dawson, William Leon. *The Birds of California.* Los Angeles, 1921 (later ed. in 4 vols. 1923).
Denis, Alberta Johnston. *Spanish Alta California.* New York, 1927.
Derby, George H. *Phoenixiana.* San Francisco, 1856 (later ed. 1937).
Dobie, Charles Caldwell. *San Francisco, a Pageant.* New York, 1933.
Drake, Sir Francis, and others. *The World Encompassed.* A history of Sir Francis Drake's voyage by his nephew of the same name. London, 1628 (later ed. 1926).
Drury, Aubrey. *California, an Intimate Guide.* New York, 1935.
Eakle, Arthur Starr. *Minerals of California.* Sacramento, 1923.
Eldredge, Zoeth Skinner, ed. *History of California.* 5 vols. New York, 1915.
Englehardt, Fr. Zephyrin. *The Missions and Missionaries of California.* 4 vols. and index. San Francisco, 1908–16 (2d ed. in 2 vols. 1929–30).
Ferrier, William Warren. *Ninety Years of Education in California, 1846–1936.* Berkeley, 1937.
Forbes, Alexander. *A History of Upper and Lower California.* London, 1839 (later ed. San Francisco, 1937).
Forbes, Mrs. Harrie R. Piper. *California Missions and Landmarks.* Los Angeles, 1903 (8th ed. 1925).
Frémont, Mrs. Jessie Benton. *The Story of the Guard: a Chronicle of the War.* Boston, 1863.
Frémont, John Charles. *Memoirs of My Life.* Chicago, 1887.
Garrison, Myrtle. *Romance and History of California.* San Francisco, 1933.

Genthe, Arnold. *As I Remember.* New York, 1936.

Graves, Jackson A. *My Seventy Years in California, 1857–1927.* Los Angeles, 1927.

Gray, A. A. *History of California from 1542.* Boston, 1934.

Greenhow, Robert. *History of Oregon and California.* Boston, 1844 (4th ed. 1847).

Grinnell, Joseph, Joseph S. Dixon, and Jean M. Linsdale. *Fur-bearing Mammals of California.* 2 vols. Berkeley, 1937.

Gudde, Erwin G. *Sutter's Own Story.* New York, 1936.

Hampton, Benjamin Bowles. *A History of the Movies.* New York, 1928.

Hanna, Phil Townsend. *California through Four Centuries.* New York, 1935.

——— *Libros Californianos, or, Five Feet of California Books.* Los Angeles, 1931.

Hannaford, Donald R. *Spanish Colonial or Adobe Architecture of California, 1800–1850.* New York, 1931.

Harlan, Jacob Wright. *California '46 to '48.* San Francisco, 1888 (later ed. 1896). The writer crossed the plains in 1846, part of the way with the Donner party.

Harte, Bret. *The Letters of Bret Harte.* Boston, 1926.

Hittell, Theodore H. *History of California.* 4 vols. San Francisco, 1885-97.

Holder, Charles Frederick. *Life in the Open; Sport with Rod, Gun, Horse, and Hound in Southern California.* New York, 1906.

Hoover, Mildred Brooke. *Historic Spots in California.* Vol. III: *Counties of the Coast Range.* Stanford University, 1937.

Hopkins, Ernest Jerome. *What Happened in the Mooney Case.* New York, 1932.

Hunt, Rockwell D., and Nellie van de Grift Sanchez. *A Short History of California.* New York, 1929.

Irwin, Will. *Herbert Hoover.* New York, 1928.

Jackson, Mrs. Helen Hunt. *A Century of Dishonor; a Sketch of the United States Government's Dealings with Some of the Indian Tribes.* Boston, 1881 (later ed. 1905).

Jaeger, Edmund Carroll. *The California Deserts.* Stanford University, 1933.

James, George Wharton. *In and out of the Old Missions of California.* Boston, 1905 (rev. ed. 1927).

Jepson, Willis Linn. *A Manual of the Flowering Plants of California.* Berkeley, 1925.

——— *The Trees of California.* San Francisco, 1909 (2d ed. 1923).

Jordan, David Starr, and Vernon L. Kellogg. *The Scientific Aspects of Luther Burbank's Work.* San Francisco, 1909.

Kemble, Edward C. *A History of California Newspapers.* New York, 1927.

Kroeber, Alfred L. *Handbook of the Indians of California.* Washington, D. C., 1925. (Bureau of American Ethnology. Bulletin No. 78.)

Lewis, Oscar. *The Big Four*. New York, 1938. Describes Huntington, Stanford, Hopkins, and Crocker, and the building of the Central Pacific Railroad.

London, Mrs. Charmian. *The Book of Jack London*. 2 vols. New York, 1921.

Lyman, George D. *Ralston's Ring; California Plunders the Comstock Lode*. New York, 1937.

Manly, William Lewis. *Death Valley in '49*. San Jose, 1894 (rev. ed. 1929).

Mayo, Morrow. *Los Angeles*. New York, 1933.

Mitchell, Sydney B. *Gardening in California*. Garden City, N. Y., 1923.

Morley, Prof. S. Griswold. *The Covered Bridges of California*. Berkeley, 1928.

Muir, John. *John of the Mountains; the Unpublished Journals of John Muir*. Boston, 1938.

——— *The Mountains of California*. New York, 1894 (later ed. 1911).

——— *The Yosemite*. New York, 1912.

Neville, Amelia. *The Fantastic City San Francisco*. Boston, 1932.

Nevins, Allan. *Frémont, the World's Greatest Adventurer*. New York, 1928.

Newmark, Harris. *Sixty Years in Southern California, 1853–1913*. New York, 1916 (3d ed. 1930).

Older, Fremont. *My Own Story*. New York, 1919 (3d ed. 1926). The author was editor of the San Francisco *Call-Bulletin*.

Paine, Albert Bigelow. *Mark Twain, a Biography*. 3 vols. New York, 1912.

Palou, Francisco. *Historical Memoirs of New California*. 4 vols. Berkeley, 1926. First printed in 1857 in the *Diario oficial* of Mexico.

——— *Life and Apostolic Labors of the Venerable Father Junipero Serra, Founder of the Franciscan Missions of California*. Pasadena, 1913. First printed in 1787 in Mexico.

Parsons, Mary Elizabeth. *The Wild Flowers of California*. San Francisco, 1902 (later ed. 1914).

Pattie, James O. *Personal Narrative*. Cincinnati, 1831 (later ed. 1930). Purports to be an account of six years' hazardous travel from St. Louis to the Pacific Ocean and back through Mexico.

Powell, L. C. *An Introduction to Robinson Jeffers*. Dijon, France, 1932.

Reed, Ralph Daniel. *Geology of California*. Tulsa, Okla., 1933.

Rensch, H. E., E. G. Rensch, and Mildred Brooke Hoover. *Historic Spots in California*. Vol. I: *The Southern Counties*. Vol. II: *Valley and Sierra Counties*. Stanford University, 1932 and 1933.

Richman, Irving Berdine. *California under Spain and Mexico, 1535–1847*. Boston, 1911.

Rider, Fremont, ed. *Rider's California; a Guidebook for Travelers*. New York, 1925.

Ritchie, Robert Welles. *The Hell-roarin' Forty-niners*. New York, 1928.

Robinson, Alfred. *Life in California before the Conquest*. New York, 1846 (later ed. 1925).

Rourke, Constance M. *Troupers of the Gold Coast; or, The Rise of Lotta Crabtree.* New York, 1928.
Royce, Josiah. *California from the Conquest in 1846 to the Second Vigilance Committee in San Francisco (1856).* Boston, 1886 (later ed. 1892).
Sanchez, Nellie van de Grift. *Spanish and Indian Place Names of California, Their Meaning and Their Romance.* San Francisco, 1914.
Saunders, Charles Francis. *Finding the Worth While in California.* New York, 1916 (5th ed. 1937).
Scherer, James A. B. *The Japanese Crisis.* New York, 1916.
Shinn, C. H. *Mining-camps. A Study in American Frontier Government.* New York, 1885.
Sinclair, Upton. *American Outpost; a Book of Reminiscences.* New York, 1932.
——— *The EPIC Plan for California.* New York, 1934. EPIC stands for "End Poverty in California."
Soulé, Frank, John H. Gihon, and James Nisbet. *The Annals of San Francisco.* New York, 1855.
Starrett, Vincent. *Ambrose Bierce.* Chicago, 1920.
Stewart, George Rippey. *Ordeal by Hunger; the Story of the Donner Party.* New York, 1936.
——— *Bret Harte, Argonaut and Exile.* Boston, 1931.
Stewart, J. A. *Robert Louis Stevenson, His Work and His Personality.* 2 vols. New York, 1924.
Stoddard, Charles Warren. *In the Footprints of the Padres.* San Francisco, 1902 (rev. ed. 1912).
Stone, Irving. *Sailor on Horseback: the Biography of Jack London.* Boston, 1938.
Talbot, Clare. *Historic California in Book-plates.* Los Angeles, 1936.
Taylor, Bayard. *Eldorado; or, Adventures in the Path of Empire.* New York, 1850 (rev. ed. 1882).
Thomas, William H. *On Land and Sea; or, California in the Years 1843, '44, and '45.* Chicago, 1892.
Vancouver, George. *A Voyage of Discovery to the North Pacific Ocean, and Round the World . . . in the Years 1790-95.* 3 vols. London, 1798.
Walker, Franklin. *Frank Norris.* Garden City, N. Y., 1932.
White, Stewart Edward. *The Forty-niners; a Chronicle of the California Trail and El Dorado.* New Haven, 1918.
——— *The Story of California.* Garden City, N. Y., 1932 (later ed. 1937).
Woon, Basil. *Incredible Land: a Jaunty Baedeker to Hollywood and the Great Southwest.* New York, 1933.
——— *San Francisco and the Golden Empire.* New York, 1935.
Young, John P. *History of San Francisco.* 2 vols. Chicago, 1913.
——— *Journalism in California.* San Francisco, 1915.

Recent Books About California

Ainsworth, Edward Maddin. *Beckoning Desert.* Prentice-Hall, 1962.

Altrocchi, Julia Cooley. *The Spectacular San Franciscans.* Dutton, 1949.

Ault, Phillip H. *How to Live in California.* Dodd, Mead, 1961.

Book Club of California. *California on Canvas* (8 parts) Grabhorn, 1941; *The California Poetry Folios* (12 parts). 1947.

Bunzel, John H., and Lee, Eugene C. *The California Democratic Delegates of 1960.* University of Alabama Press, 1962.

Busch, Niven. *California Street* (novel). Simon & Schuster, 1959.

Caen, Herb. *The San Francisco Book.* Photos by Max Yavno. Houghton, Mifflin, 1948. *San Francisco.* Doubleday, 1965.

Crow, John Armstrong. *California as a Place to Live.* Scribner's, 1953.

Crowther, Bosley. *Hollywood Rajah; the Life and Times of Louis B. Mayer.* Holt, Rinehart & Winston, 1960.

Dane, George Ezra. *Ghost Town.* (Beatrice J. Dane, Collab.) Knopf, 1941.

Dasmann, Raymond F. *The Destruction of California.* Macmillan, 1965.

Davis, Bernice Freeman (with Al Hirshberg). *The Desperate and the Damned.* Foreword by Clinton T. Duffy. Thos. Y. Crowell, 1961.

Delkin, James L., ed. *Monterey Peninsula.* American Guide Series. Stanford University Press, 1941.

Doyle, Kathleen. *Californians: Who, Whence, Whither.* Haynes Foundation, 1956.

Drury, Aubrey. *California, an Intimate Guide.* Harper. (Revised, 1947.)

Duffy, Clinton T. (with Al Hirshberg). *Eightyeight Men and Two Women.* Doubleday, 1962. *The San Quentin Story* (with Dean Jennings). Doubleday, 1950.

Duffy, Gladys (with Blaise W. Lane). *Warden's Wife.* Appleton-Century-Crofts, 1959.

Ehrlich, J. W., ed. *Howl of the Censor* (The People vs. Lawrence Ferlinghetti). Nourse Publishing Co., San Carlos, 1961.

Gleason, Joe Duncan. *The Islands and Ports of California.* Devin-Adair, 1958.

Harris, John Pratt. *California Politics.* (3rd edition). Stanford University Press, 1961.

Jackson, Joseph Henry. *My San Francisco; An Appreciation.* Thos. Y. Crowell, 1953.

Jacobson, Dan. *No Further West; California Revisited.* Macmillan, 1959.

Johnson, Paul C., and editors of Sunset Books and Sunset Magazine. *National Parks of the West.* Lane Magazine & Book Co., 1965.

Kahn, Gordon. *Hollywood on Trial.* Foreword by Thomas Mann. Boni & Gaer, 1948.

Lee, William Storss. *The Great California Deserts.* Illus. by Edward Sanborn. Putnam, 1963. *The Sierra,* same, 1962.

MacCann, Richard Dyer. *Hollywood in Transition.* Houghton, Mifflin, 1962.

Lewis, Oscar. *High Sierra Country.* (American Folkways). Duell, Sloan & Pearce-Little Brown, 1955.
——— *Sutter's Fort; Gateway to the Gold Fields.* Prentice-Hall, 1966.
Lipset, Seymour M. and Sheldon S. Wolin, eds. *The Berkeley Student Revolt; Facts and Interpretations.* Anchor Books, 1965.
Lowenthal, Marjorie Fiske. *Lives in Distress; the Paths of the Elderly to the Psychiatric Ward.* Langley Porter Neuropsychiatric Institute, San Francisco. Basic Books, 1964.
McClure, James D. *California Landmarks.* A Photographic Guide. Stanford University Press, 1948.
McWilliams, Carey. *Factories in the Field.* Little, Brown, 1939.
——— *Southern California Country.* Duell, Sloan & Pearce, 1946.
Miller, Max. *It Must Be the Climate.* McBride, 1941.
——— *And Bring All Your Folks.* Doubleday, 1959.
Monroe, Keith. *California; How to Live, Work and Have Fun in the Golden State.* Dutton, 1963.
Moore, Truman E. *The Slaves We Rent.* Random House, 1965.
Muench, Josef. *Along Yosemite Trails.* Hastings House, 1948.
Nadeau, Remi A. *California: the New Society.* David McKay, 1963.
O'Brien, Robert. *This Is San Francisco.* McGraw-Hill, 1948.
Phillips, John. *Inside California.* Foreword by Ruth Comfort Mitchell. Murray & Gee, 1959.
Potter, Elizabeth Gray. *The San Francisco Skyline.* Dodd, Mead, 1939.
Powers, Alfred. *Redwood Country.* Duell, Sloan & Pearce, 1949.
Raitt, Helen. *Exploring the Deep Pacific.* (Capricorn Expedition, Scripps Institute of Oceanography). Norton, 1956.
Rubin, Robert, ed. *Olympic Winter Games, Squaw Valley, 1960.* California Olympic Commission, Sacramento.
Scudder, Kenyon J. *Prisoners Are People.* (Chino). Doubleday, 1952.
Shippey, Lee. *The Los Angeles Book.* Photos by Max Yavno. Houghton, Mifflin, 1950.
Shutes, Milton H. *Lincoln and California.* Stanford University Press, 1943.
Steinbeck, John (with Edward F. Ricketts). *The Sea of Cortez.* Viking Press, 1941. *The Grapes of Wrath.* (Novel). Viking Press.
Stone, Adolf, Ph.D., ed. *et al. California Information Almanac.* Lakewood, Calif. Doubleday, distributor. (Annual)
Taylor, Katherine Ames. *Yosemite Tales & Trails.* Stanford University Press, 1948.
Thompson, Warren S. *Growth and Changes in California's Population.* Haynes Foundation, 1965.
Turner, Henry A. and John A. Vieg. *The Government and Policies of California.* McGraw-Hill, 1964.
Weston, Charis Wilson (and Edward Weston). *California and the West.* A U. S. Camera Book. Duell, Sloan & Pearce, 1940.
Wheeler, Horace J., ed. *Mariner Mission to Venus.* Foreword by W. H. Pickering. Jet Propulsion Laboratory, California Institute of Technology. Pasadena. McGraw-Hill, 1963.

INDEX TO MAP SECTIONS

KEY

U. S. HIGHWAYS 70

STATE HIGHWAYS 25

POINTS OF INTEREST 12

SECTIONAL DIVISION OF STATE MAP

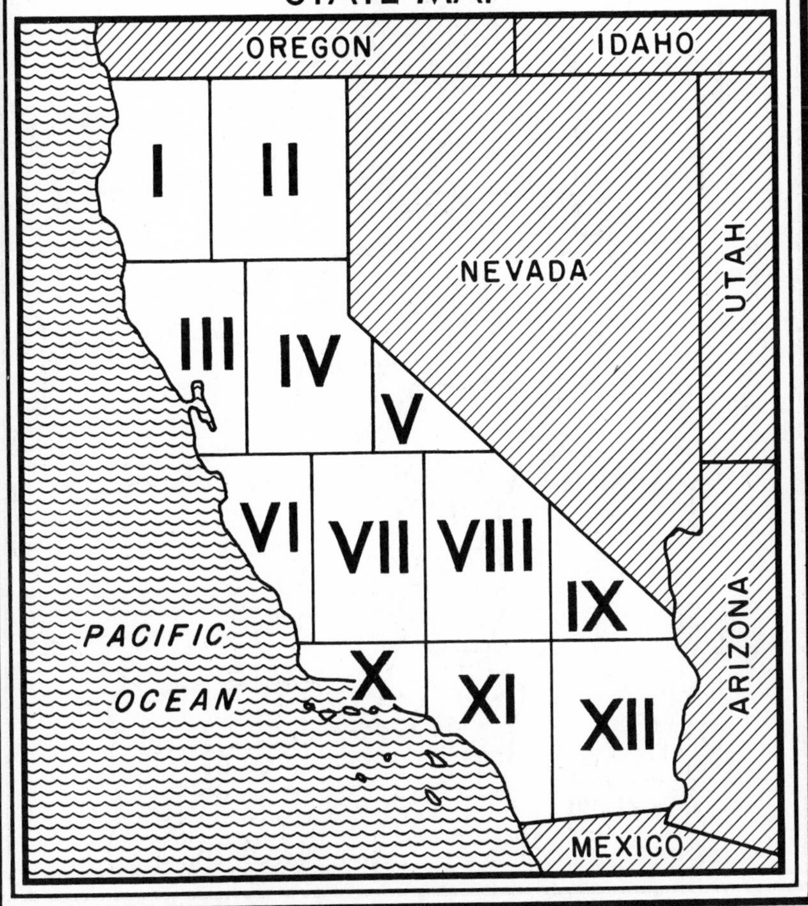

SECTION I

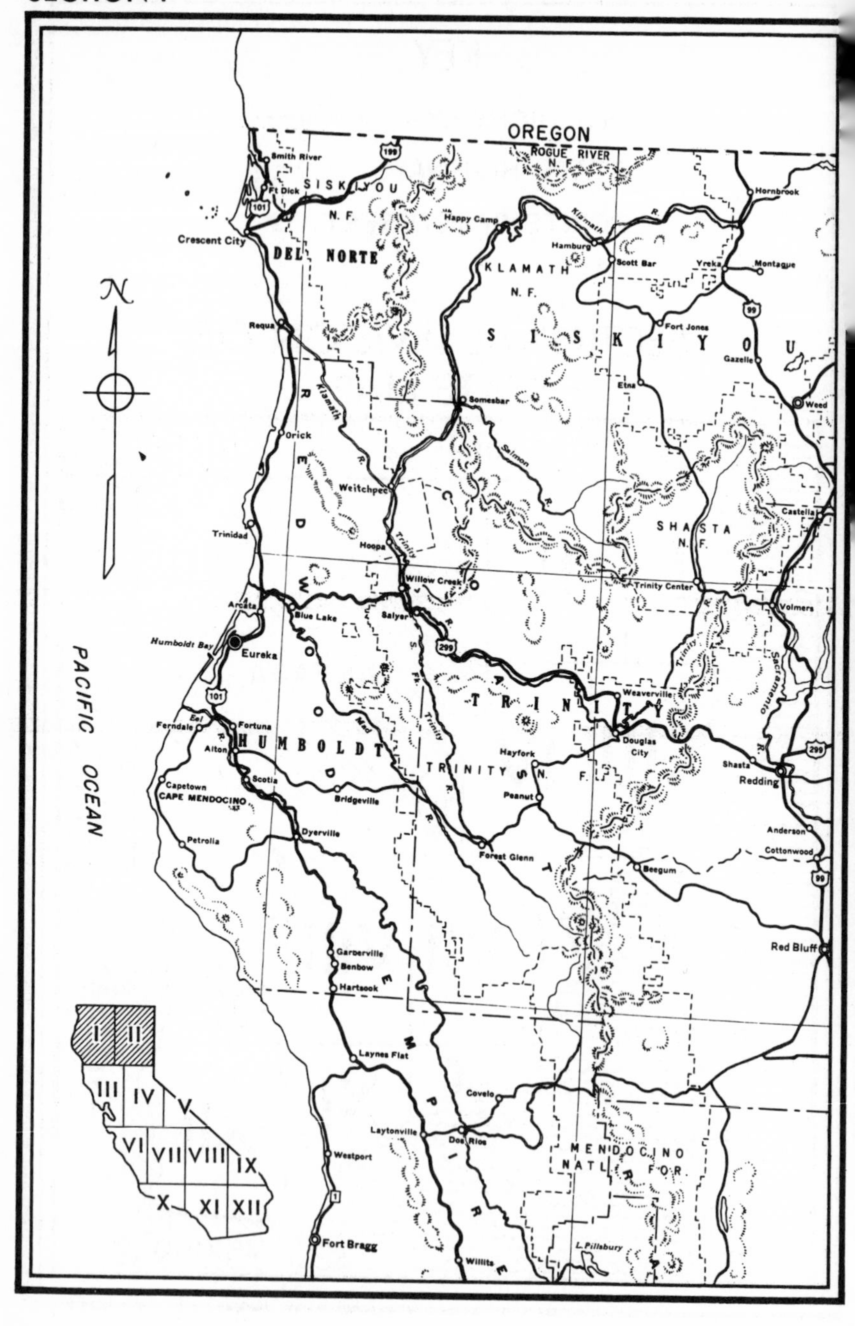

OREGON
ROGUE RIVER N.F.
Smith River
Ft Dick
SISKIYOU N.F.
Crescent City
DEL NORTE
Happy Camp
Klamath R.
Hamburg
Scott Bar
Hornbrook
Yreka
Montague
KLAMATH N.F.
SISKIYOU
Fort Jones
Gazelle
Requa
Etna
Weed
Somesbar
Salmon R.
Orick
Klamath R.
Weitchpec
Trinidad
Hoopa
SHASTA N.F.
Castella
Willow Creek
Trinity Center
Arcata
Blue Lake
Salyer
Volmers
Humboldt Bay
Eureka
PACIFIC OCEAN
Weaverville
TRINITY
Eel R.
Ferndale
Fortuna
Alton
HUMBOLDT
Douglas City
Hayfork
Shasta
Redding
Sacramento R.
TRINITY N.F.
Capetown
CAPE MENDOCINO
Scotia
Bridgeville
Peanut
Anderson
Petrolia
Dyerville
Forest Glenn
Cottonwood
Beegum
Red Bluff
Garberville
Benbow
Hartsook
Laynes Flat
Covelo
Laytonville
Dos Rios
MENDOCINO NATL. FOR.
Westport
Fort Bragg
Willits
L. Pillsbury
I
II
III
IV
V
VI
VII
VIII
IX
X
XI
XII

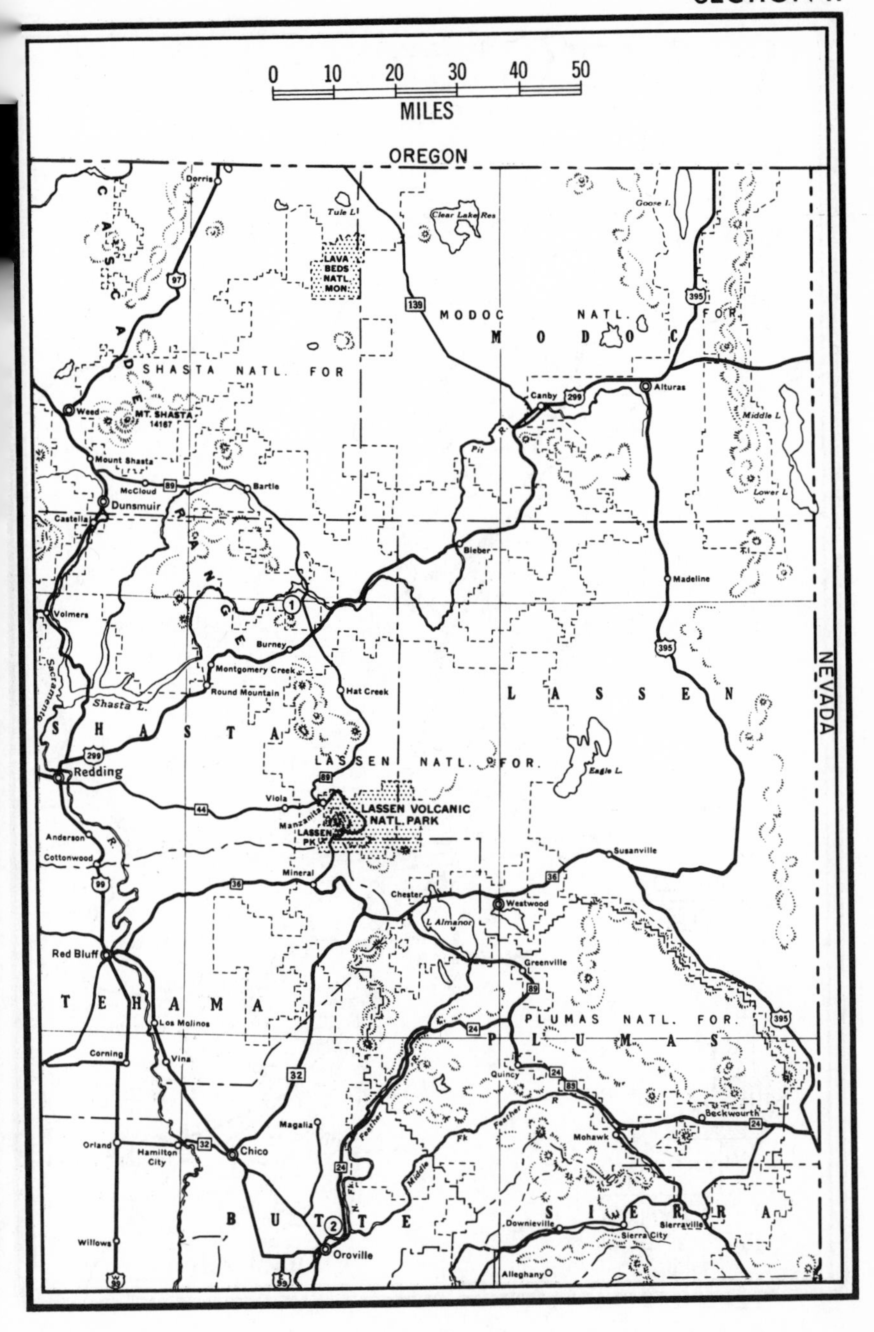
0 10 20 30 40 50
MILES
OREGON
NEVADA
Dorris
Tule L.
Clear Lake Res
Goose L.
LAVA BEDS NATL. MON.
MODOC NATL. FOR.
MODOC
SHASTA NATL. FOR
MT. SHASTA 14167
Weed
Mount Shasta
McCloud
Bartle
Dunsmuir
Castella
Canby
Alturas
Middle L.
Lower L.
Pit R.
Bieber
Madeline
Volmers
Burney
Montgomery Creek
Round Mountain
Hat Creek
Shasta L.
Sacramento
SHASTA
LASSEN
LASSEN NATL. FOR.
Eagle L.
Redding
Viola
Manzanita
LASSEN VOLCANIC NATL. PARK
LASSEN PK.
Anderson
Cottonwood
Mineral
Susanville
Chester
Westwood
L. Almanor
Red Bluff
Greenville
TEHAMA
PLUMAS NATL. FOR.
PLUMAS
Los Molinos
Corning
Vina
Quincy
Beckwourth
Magalia
Mohawk
Orland
Hamilton City
Chico
Feather
Middle Fk
N. Fk
BUTTE
SIERRA
Downieville
Sierra City
Sierraville
Willows
Oroville
Alleghany
97
139
395
299
89
44
36
99
32
24
W 99
E 99

SECTION III

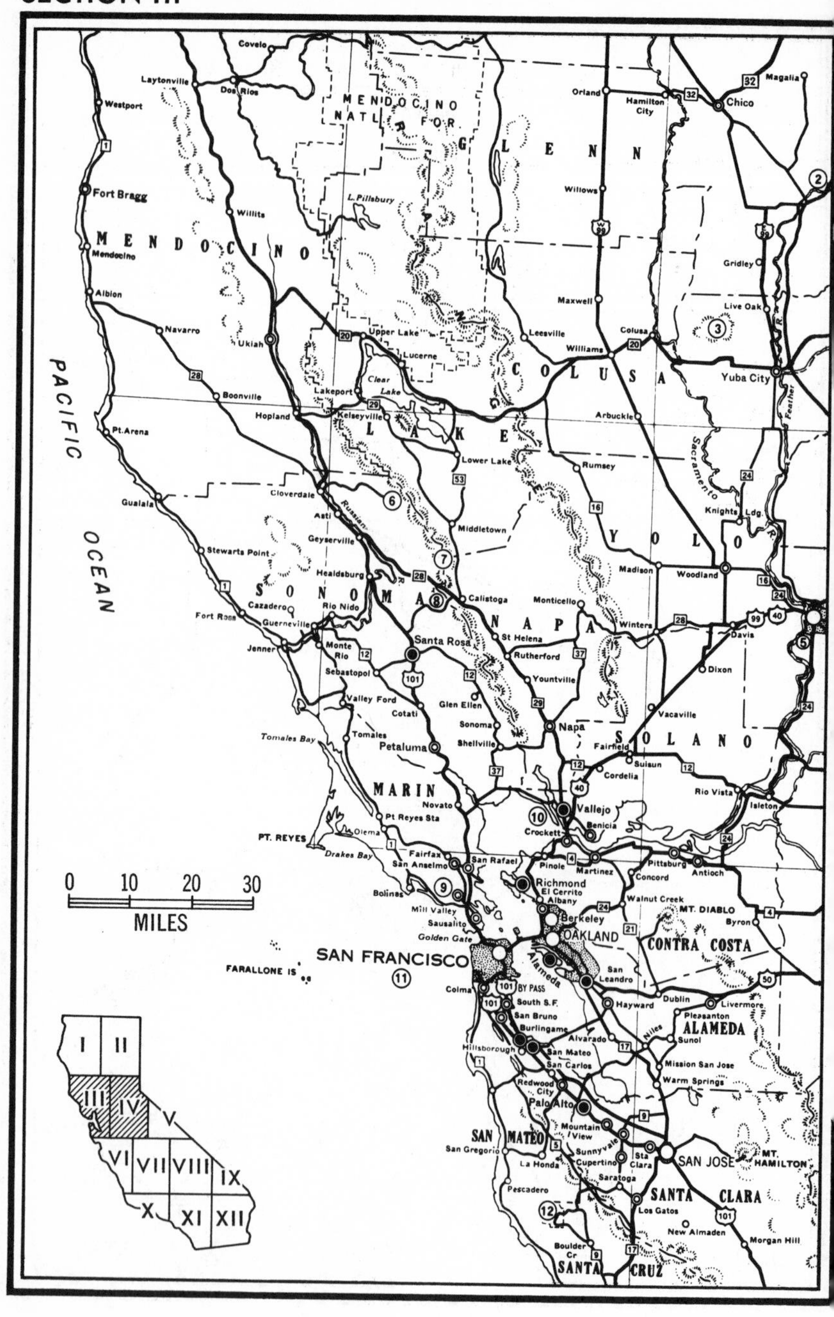
PACIFIC
OCEAN
MENDOCINO NATL. FOR.
M E N D O C I N O
G L E N N
C O L U S A
L A K E
Y O L O
S O N O M A
N A P A
S O L A N O
MARIN
CONTRA COSTA
ALAMEDA
SAN MATEO
SANTA CLARA
SANTA CRUZ
SAN FRANCISCO
FARALLONE IS
PT. REYES
Drakes Bay
Tomales Bay
Golden Gate
Clear Lake
L. Pillsbury
Sacramento
Feather
Russian
MT. DIABLO
MT. HAMILTON
0 10 20 30
MILES
Fort Bragg
Ukiah
Santa Rosa
Petaluma
Napa
Vallejo
Richmond
Berkeley
OAKLAND
SAN JOSE
Chico
Yuba City
Palo Alto
I II III IV V VI VII VIII IX X XI XII

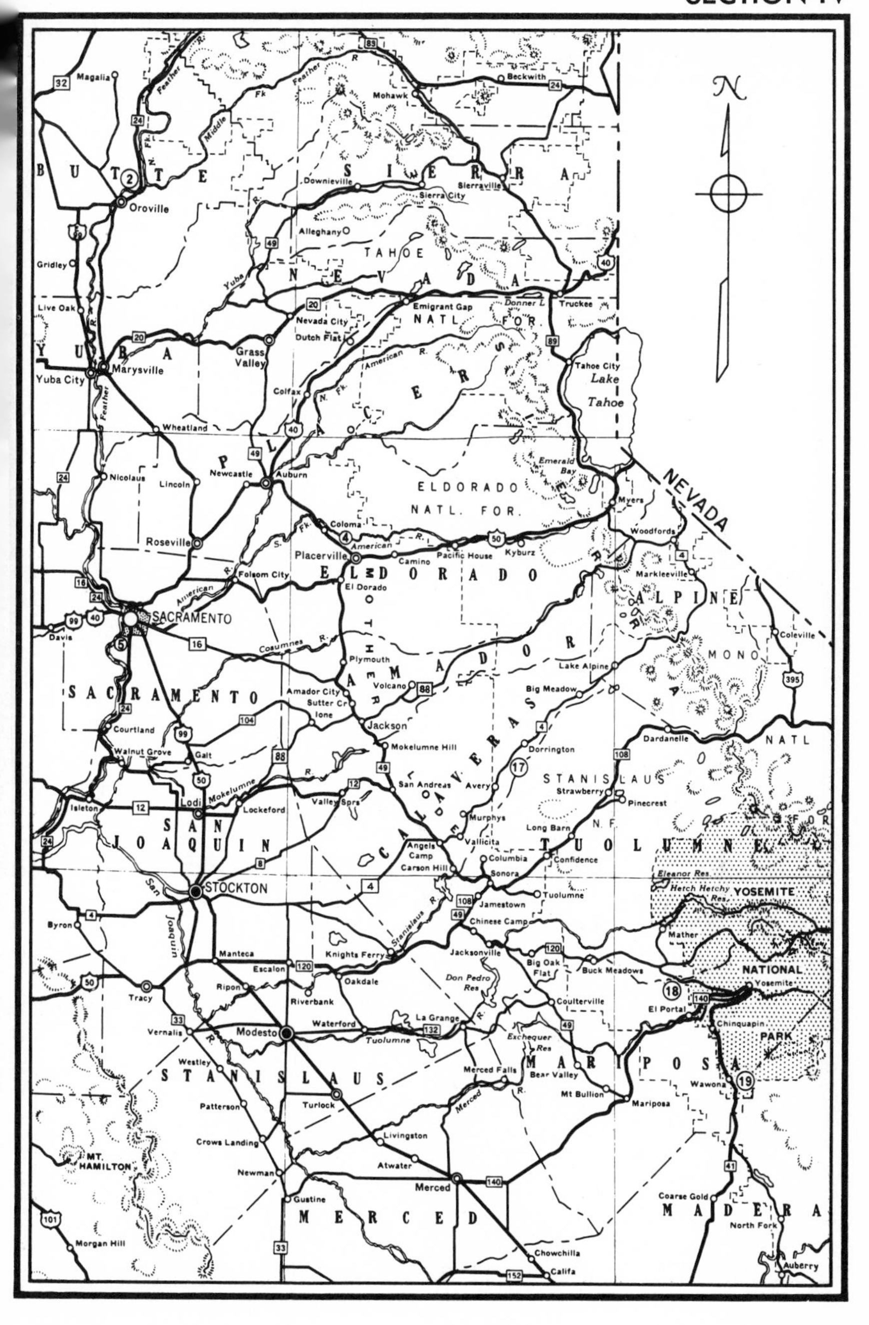

N
BUTTE
SIERRA
NEVADA
TAHOE
NATL FOR.
PLACER
ELDORADO
ELDORADO NATL. FOR.
YUBA
SACRAMENTO
AMADOR
ALPINE
MONO
CALAVERAS
STANISLAUS N.F.
TUOLUMNE
SAN JOAQUIN
STANISLAUS
MARIPOSA
MERCED
MADERA
YOSEMITE NATIONAL PARK
NEVADA
Magalia
Oroville
Gridley
Live Oak
Yuba City
Marysville
Wheatland
Nicolaus
Lincoln
Newcastle
Auburn
Roseville
Folsom City
Sacramento
Davis
Courtland
Walnut Grove
Isleton
Lodi
Lockeford
Galt
Stockton
Byron
Tracy
Manteca
Escalon
Ripon
Vernalis
Modesto
Westley
Patterson
Crows Landing
Newman
Gustine
Mt. Hamilton
Morgan Hill
Beckwith
Mohawk
Downieville
Sierra City
Sierraville
Alleghany
Truckee
Emigrant Gap
Donner L.
Nevada City
Dutch Flat
Grass Valley
Colfax
Tahoe City
Lake Tahoe
Emerald Bay
Myers
Woodfords
Markleeville
Coleville
Coloma
Placerville
Camino
Pacific House
Kyburz
El Dorado
Plymouth
Amador City
Sutter Cr
Ione
Volcano
Jackson
Mokelumne Hill
San Andreas
Valley Sprs
Lake Alpine
Big Meadow
Dorrington
Avery
Murphys
Vallicita
Angels Camp
Carson Hill
Columbia
Sonora
Jamestown
Chinese Camp
Tuolumne
Confidence
Long Barn
Strawberry
Pinecrest
Dardanelle
Eleanor Res.
Hetch Hetchy Res.
Mather
Knights Ferry
Oakdale
Riverbank
Jacksonville
Don Pedro Res.
Big Oak Flat
Buck Meadows
Coulterville
El Portal
Yosemite
Chinquapin
Wawona
Waterford
La Grange
Exchequer Res
Merced Falls
Bear Valley
Mt Bullion
Mariposa
Turlock
Livingston
Atwater
Merced
Chowchilla
Califa
Coarse Gold
North Fork
Auberry
Feather R.
Middle Fk
N. Fk.
Yuba R.
American R.
S. Fk.
Cosumnes R.
Mokelumne R.
San Joaquin R.
Stanislaus R.
Tuolumne R.
Merced R.

SECTION V

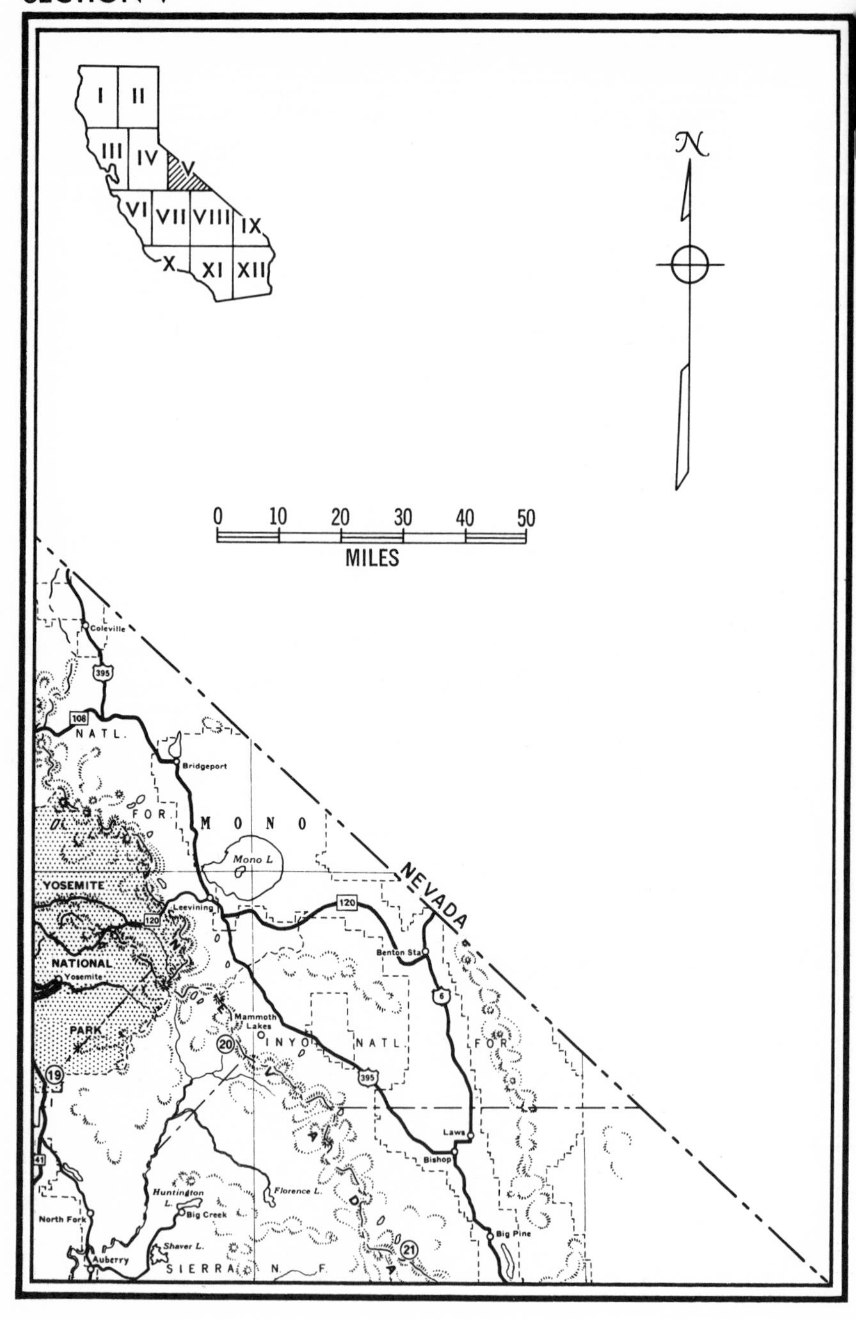

I
II
III
IV
V
VI
VII
VIII
IX
X
XI
XII
N
0 10 20 30 40 50
MILES
NEVADA
Coleville
395
108
NATL.
Bridgeport
FOR.
MONO
Mono L
YOSEMITE
Leevining
120
NATIONAL
Yosemite
PARK
Benton Sta
6
Mammoth Lakes
INYO
NATL.
FOR.
20
19
395
Laws
Bishop
41
Florence L.
Huntington L.
Big Creek
North Fork
Shaver L.
Auberry
Big Pine
21
SIERRA
N.
F.

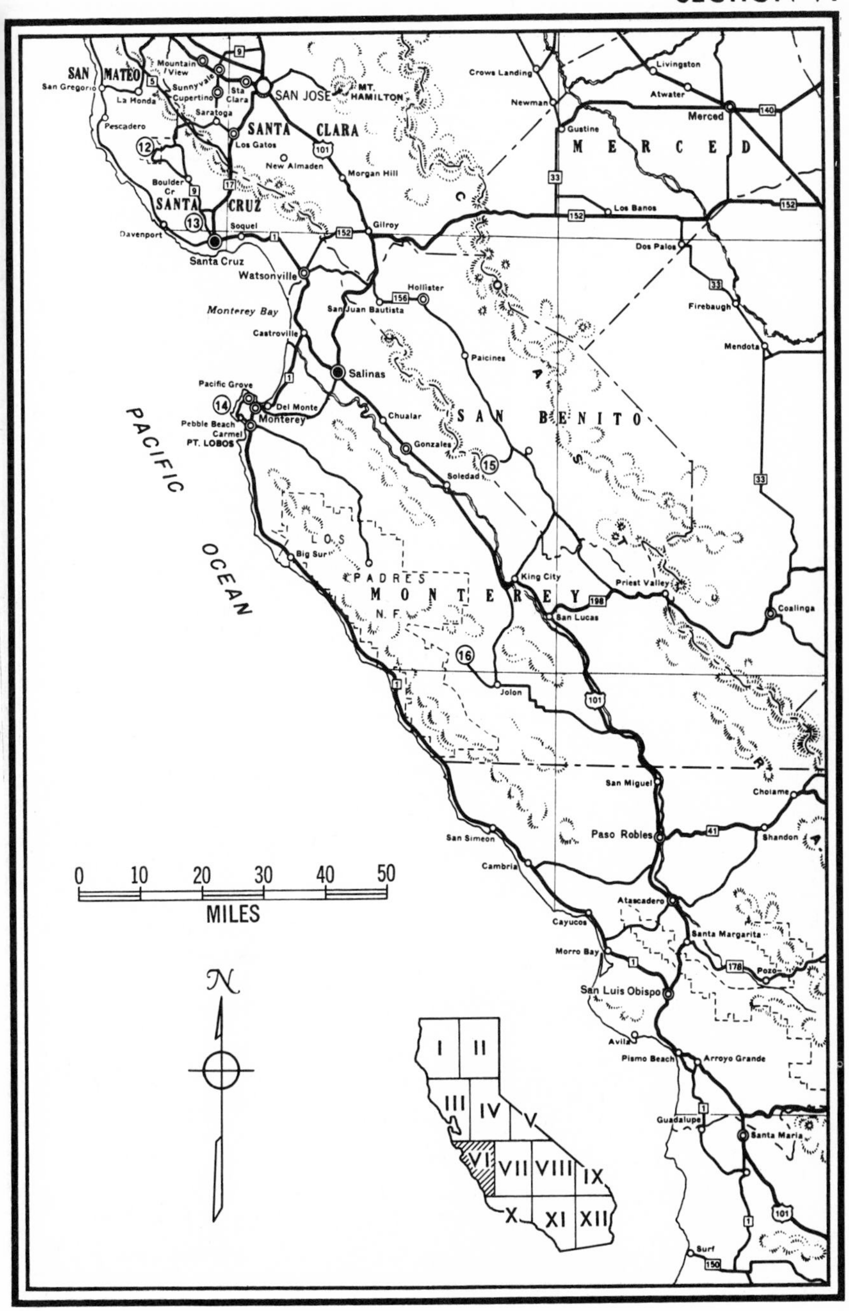
SAN MATEO
San Gregorio
La Honda
Pescadero
Mountain View
Sunnyvale
Cupertino
Sta Clara
Saratoga
SAN JOSE
MT. HAMILTON
SANTA CLARA
Los Gatos
New Almaden
Morgan Hill
Boulder Cr
SANTA CRUZ
Davenport
Soquel
Santa Cruz
Gilroy
Watsonville
Monterey Bay
Hollister
San Juan Bautista
Castroville
Paicines
Salinas
Pacific Grove
Del Monte
Monterey
Pebble Beach
Carmel
PT. LOBOS
Chualar
Gonzales
Soledad
SAN BENITO
PACIFIC OCEAN
LOS PADRES N.F.
Big Sur
MONTEREY
King City
San Lucas
Priest Valley
Coalinga
Jolon
Crows Landing
Newman
Gustine
Livingston
Atwater
Merced
MERCED
Los Banos
Dos Palos
Firebaugh
Mendota
San Miguel
Cholame
Paso Robles
Shandon
San Simeon
Cambria
Atascadero
Cayucos
Santa Margarita
Morro Bay
Pozo
San Luis Obispo
Avila
Pismo Beach
Arroyo Grande
Guadalupe
Santa Maria
Surf
0 10 20 30 40 50
MILES
N
I II III IV V VI VII VIII IX X XI XII

SECTION VII

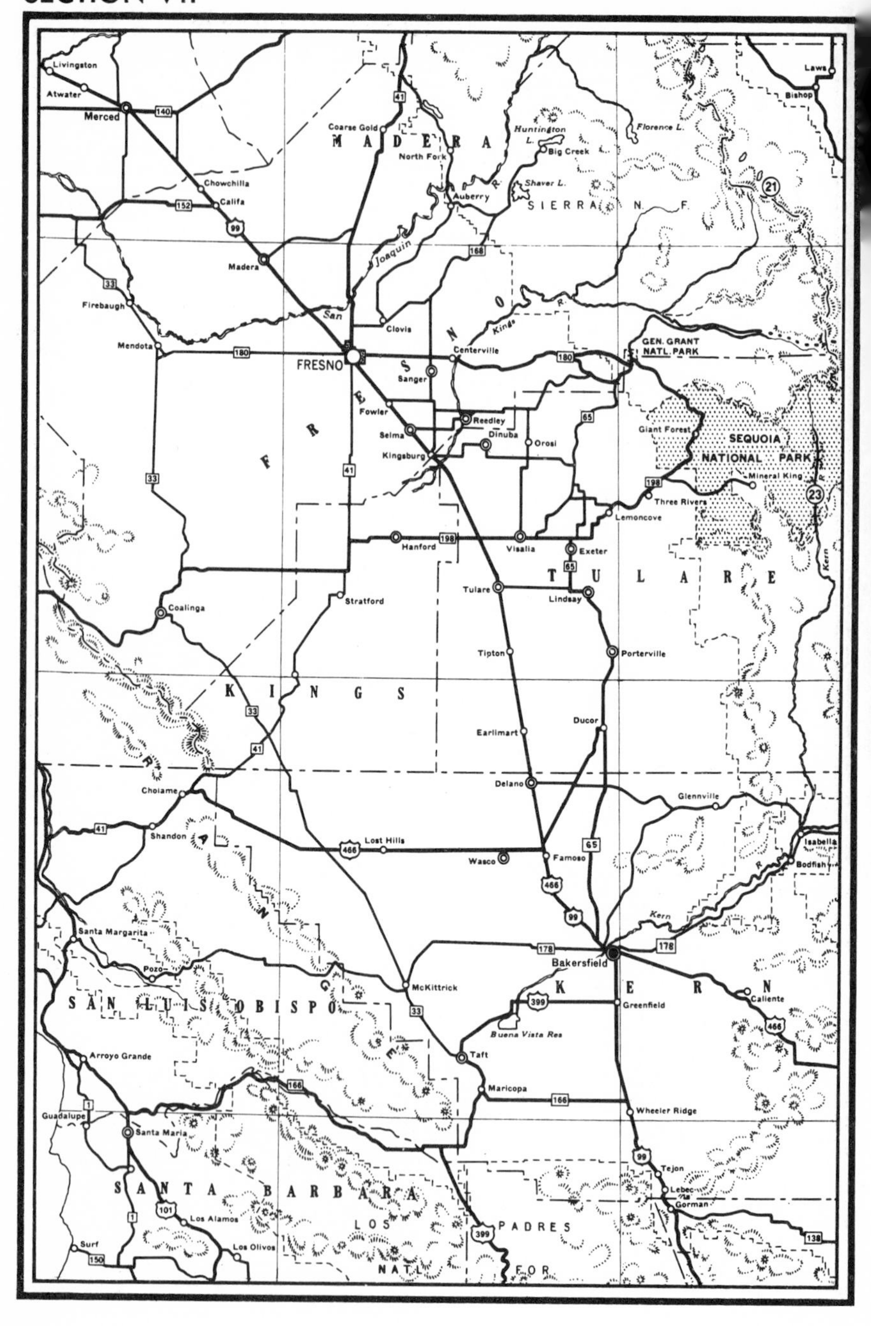

Livingston
Atwater
Merced
140
Chowchilla
Califa
152
99
Madera
33
Firebaugh
Mendota
180
FRESNO
Coarse Gold
41
M A D E R A
North Fork
Auberry
Huntington L.
Big Creek
Shaver L.
Florence L.
SIERRA N. F.
21
San Joaquin
168
Clovis
Kings R.
Centerville
Sanger
180
GEN. GRANT NATL. PARK
F R E S N O
Fowler
Reedley
Selma
Dinuba
Orosi
Kingsburg
65
Giant Forest
SEQUOIA NATIONAL PARK
Mineral King
198
Three Rivers
23
Lemoncove
41
Hanford
198
Visalia
Exeter
65
T U L A R E
Kern
Tulare
Lindsay
Stratford
Coalinga
Tipton
Porterville
K I N G S
33
41
Earlimart
Ducor
Choiame
Delano
Glennville
41
Shandon
466
Lost Hills
Wasco
Famoso
65
Isabella
Bodfish
466
99
Kern R.
Santa Margarita
178
178
Bakersfield
Pozo
McKittrick
K E R N
Caliente
399
Greenfield
S A N L U I S O B I S P O
33
Buena Vista Res
466
Arroyo Grande
Taft
Maricopa
166
166
1
Guadalupe
Wheeler Ridge
Santa Maria
99
Tejon
Lebec
Gorman
S A N T A B A R B A R A
101
1
Los Alamos
L O S
399
P A D R E S
138
Surf
150
Los Olivos
N A T L
F O R.

SECTION VIII

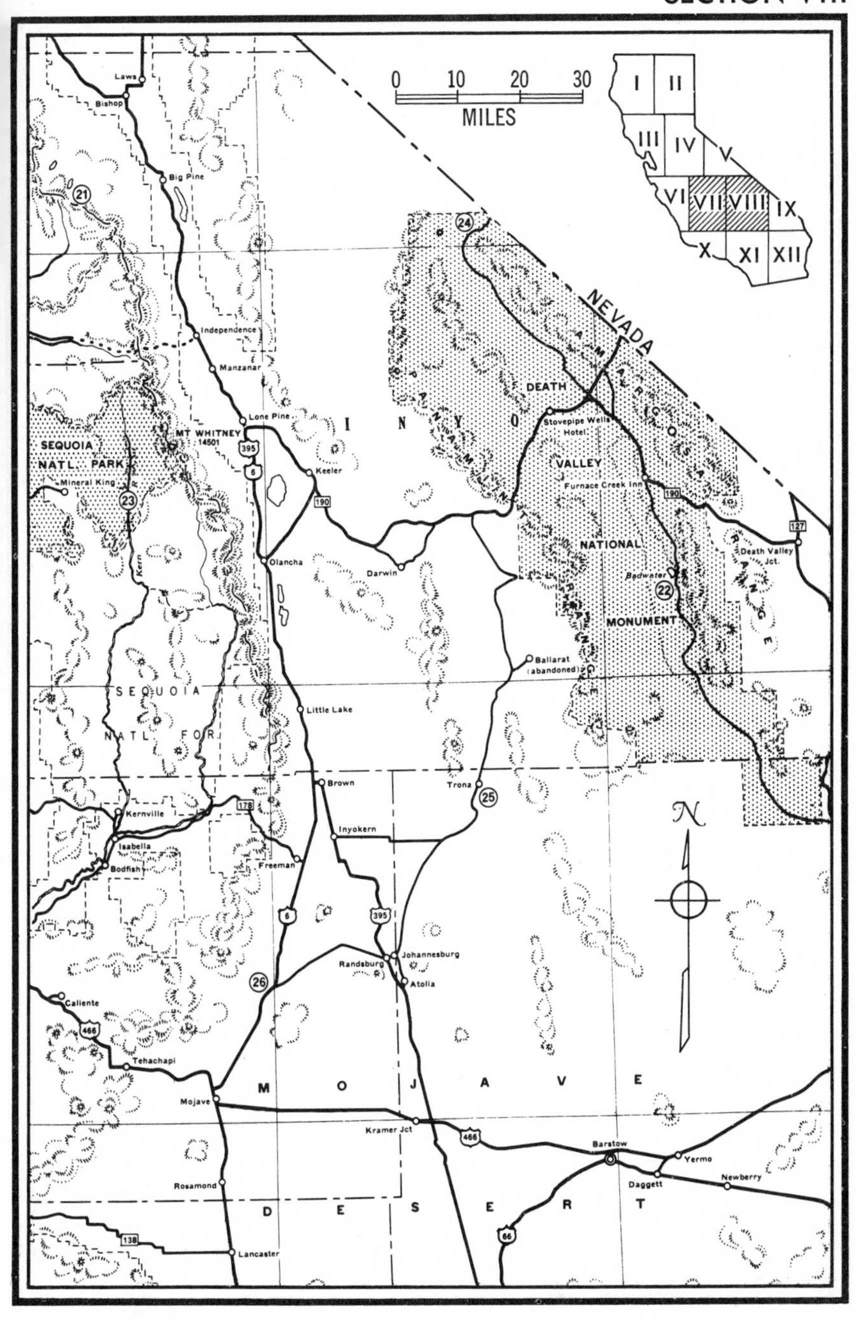

SECTION IX

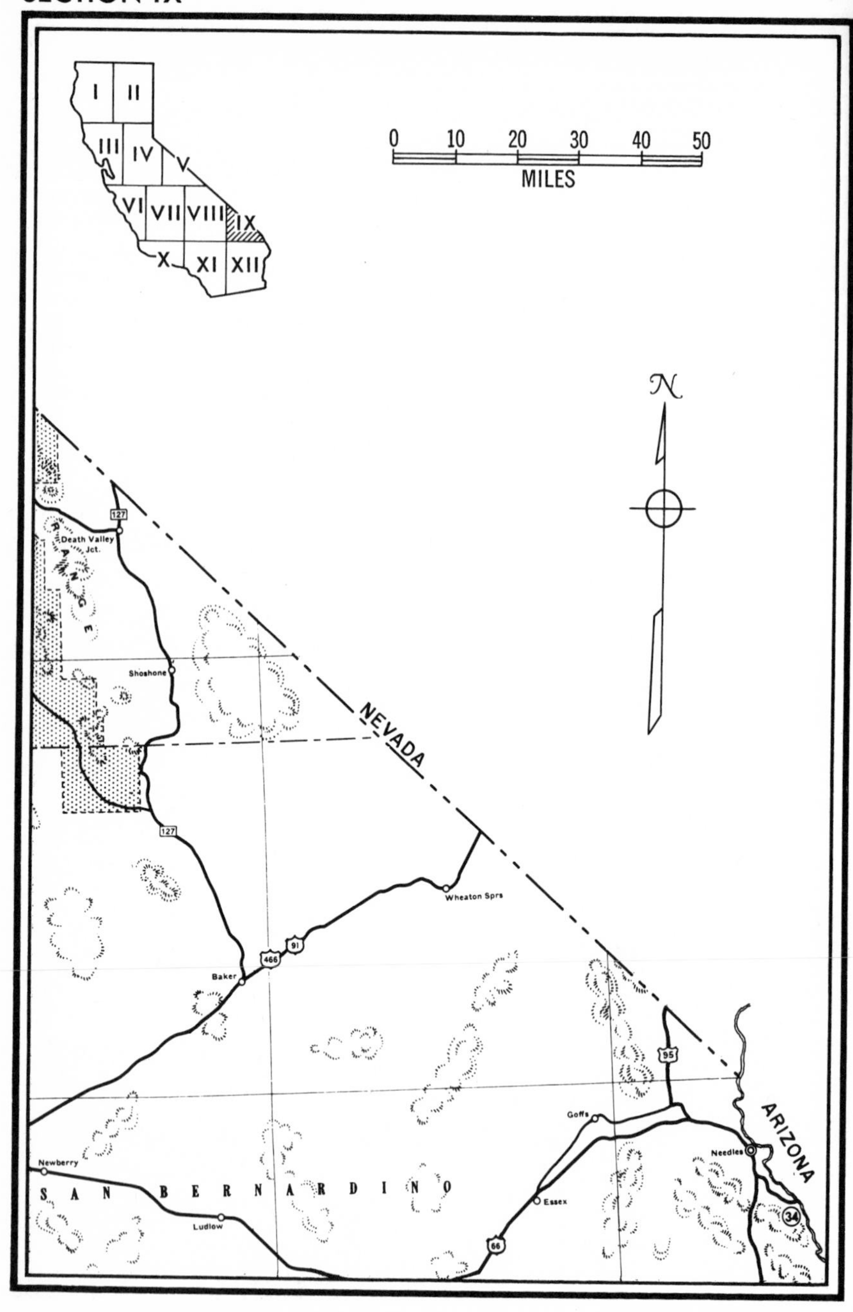

I
II
III
IV
V
VI
VII
VIII
IX
X
XI
XII
0
10
20
30
40
50
MILES
N
127
Death Valley Jct.
Shoshone
NEVADA
Wheaton Sprs
466
91
Baker
95
Goffs
Needles
ARIZONA
34
Newberry
SAN BERNARDINO
Ludlow
Essex
66

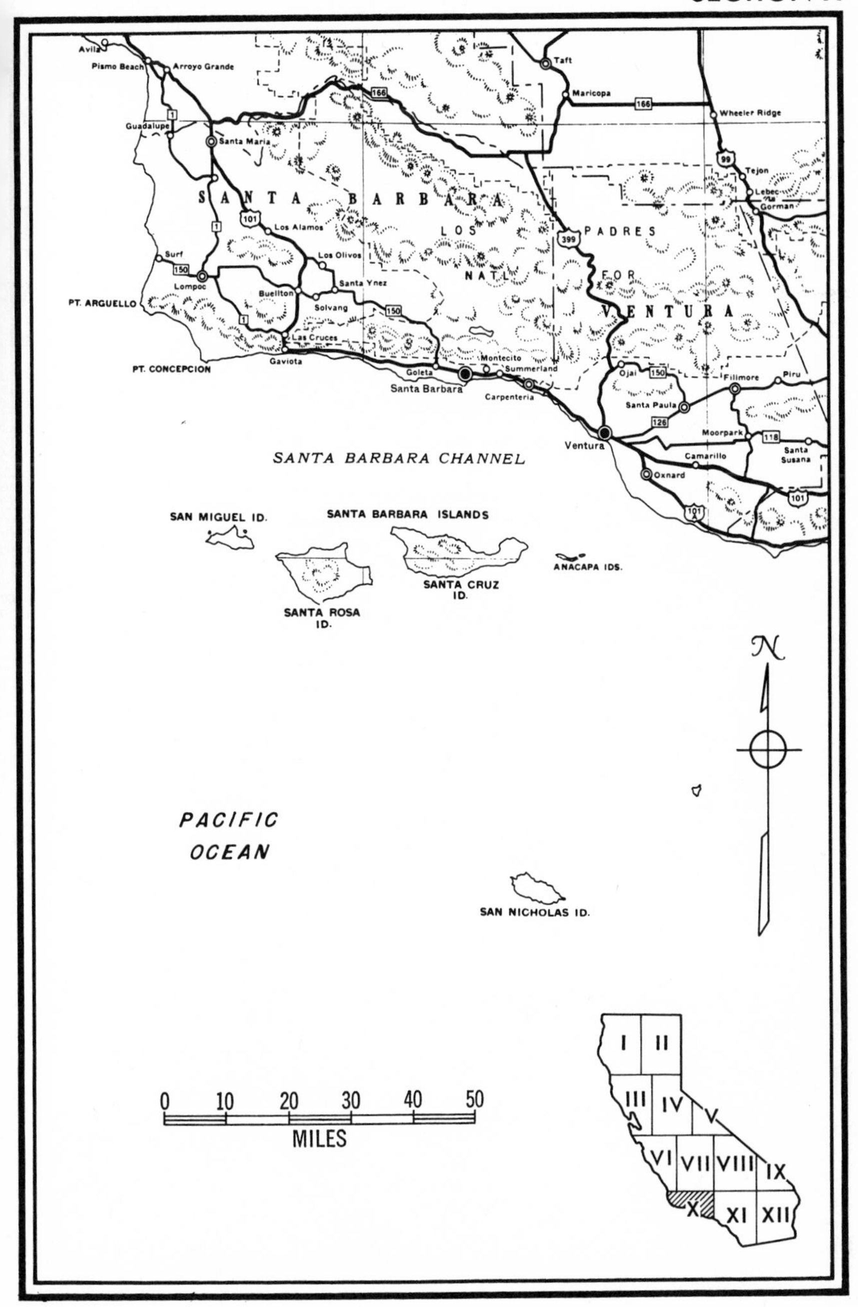
Avila
Pismo Beach
Arroyo Grande
Guadalupe
Santa Maria
Taft
Maricopa
Wheeler Ridge
Tejon
Lebec
Gorman
S A N T A B A R B A R A
LOS PADRES NATL FOR
VENTURA
Los Alamos
Los Olivos
Santa Ynez
Solvang
Buellton
Surf
Lompoc
PT. ARGUELLO
Las Cruces
Gaviota
PT. CONCEPCION
Goleta
Santa Barbara
Montecito
Summerland
Carpenteria
Ojai
Santa Paula
Fillmore
Piru
Moorpark
Ventura
Camarillo
Santa Susana
Oxnard
SANTA BARBARA CHANNEL
SAN MIGUEL ID.
SANTA BARBARA ISLANDS
SANTA ROSA ID.
SANTA CRUZ ID.
ANACAPA IDS.
PACIFIC OCEAN
SAN NICHOLAS ID.
N
0 10 20 30 40 50
MILES
I II III IV V VI VII VIII IX X XI XII

SECTION XI

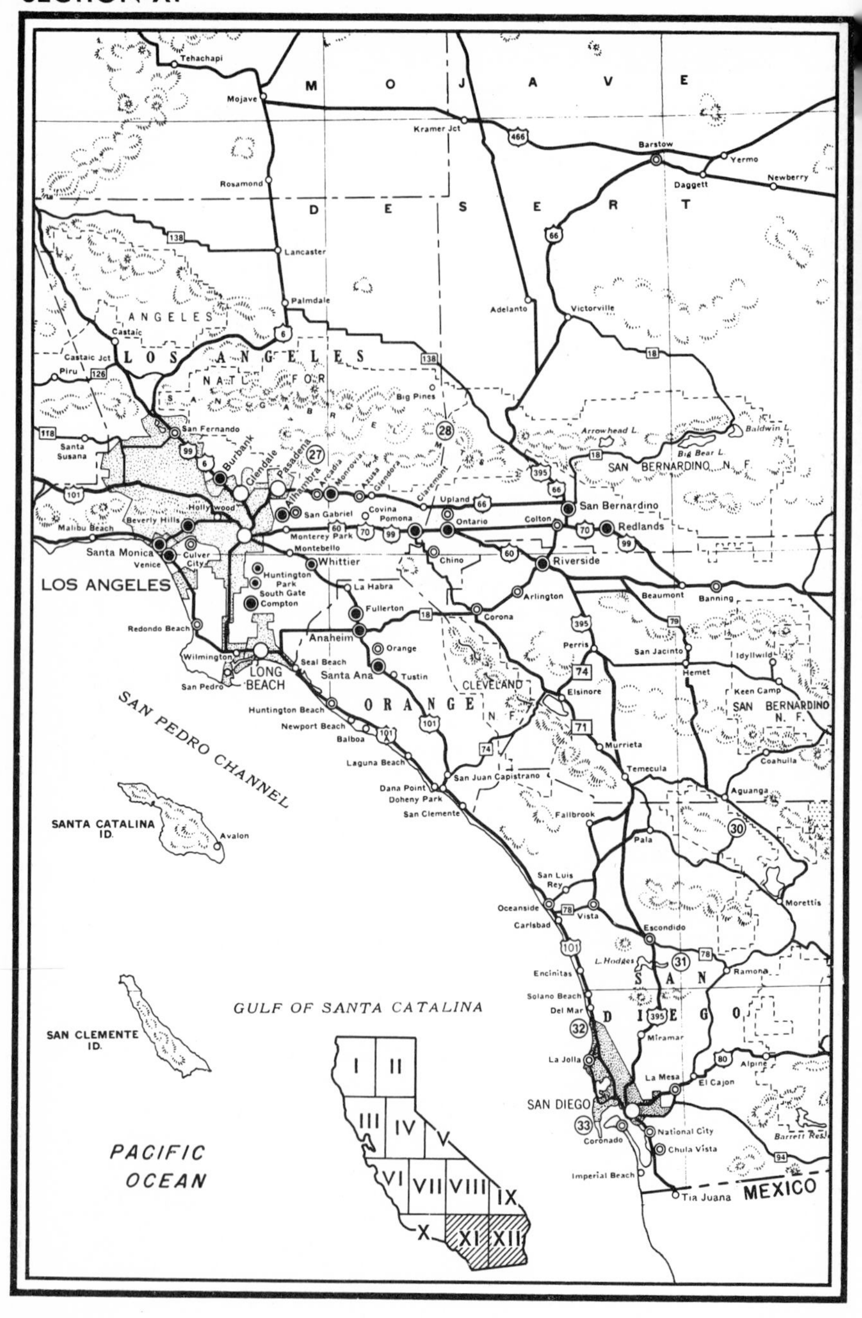

MOJAVE DESERT
Tehachapi
Mojave
Kramer Jct
Barstow
Yermo
Daggett
Newberry
Rosamond
Lancaster
Palmdale
Adelanto
Victorville
ANGELES
Castaic
Castaic Jct
Piru
LOS ANGELES NAT'L FOR
SAN GABRIEL
Big Pines
San Fernando
Santa Susana
Burbank
Glendale
Pasadena
Alhambra
Arcadia
Monrovia
Azusa
Glendora
Claremont
Upland
San Bernardino
Redlands
Hollywood
Beverly Hills
Malibu Beach
San Gabriel
Covina
Pomona
Ontario
Colton
Monterey Park
Montebello
Santa Monica
Venice
Culver City
Whittier
Chino
Riverside
Huntington Park
South Gate
Compton
LOS ANGELES
La Habra
Fullerton
Arlington
Corona
Beaumont
Banning
Redondo Beach
Anaheim
Orange
Wilmington
Seal Beach
Santa Ana
Tustin
LONG BEACH
San Pedro
Perris
San Jacinto
Idyllwild
Hemet
Arrowhead L.
Big Bear L.
Baldwin L.
SAN BERNARDINO N. F.
CLEVELAND N. F.
Elsinore
Keen Camp
SAN BERNARDINO N. F.
Huntington Beach
ORANGE
Newport Beach
Balboa
Murrieta
Laguna Beach
San Juan Capistrano
Temecula
Coahuila
Dana Point
Doheny Park
San Clemente
Aguanga
Fallbrook
SAN PEDRO CHANNEL
SANTA CATALINA ID.
Avalon
Pala
San Luis Rey
Oceanside
Carlsbad
Vista
Escondido
Morettis
L. Hodges
Encinitas
Solano Beach
Del Mar
SAN DIEGO
Ramona
Miramar
La Jolla
La Mesa
El Cajon
Alpine
SAN DIEGO
Coronado
National City
Chula Vista
Imperial Beach
Tia Juana
MEXICO
Barrett Res.
SAN CLEMENTE ID.
GULF OF SANTA CATALINA
PACIFIC OCEAN
I
II
III
IV
V
VI
VII
VIII
IX
X
XI
XII

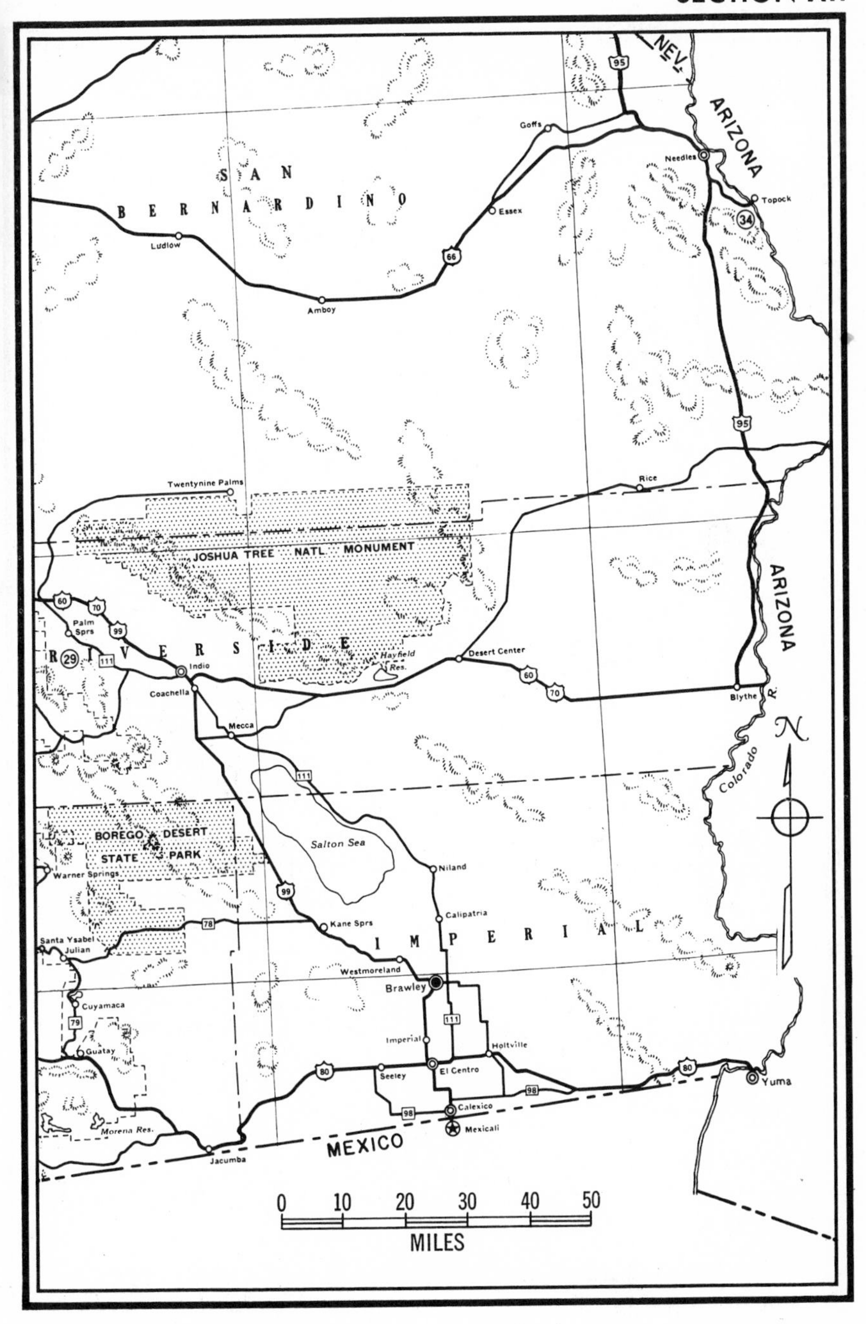
NEV.
ARIZONA
SAN
BERNARDINO
Goffs
Needles
Topock
Essex
Ludlow
Amboy
95
34
66
Twentynine Palms
Rice
JOSHUA TREE NATL MONUMENT
ARIZONA
60
70
99
Palm Sprs
RIVERSIDE
29
111
Indio
Coachella
Hayfield Res.
Desert Center
Blythe
Mecca
Colorado R.
N
Salton Sea
BOREGO DESERT STATE PARK
Warner Springs
Niland
Calipatria
Kane Sprs
78
Santa Ysabel
Julian
IMPERIAL
Westmoreland
Brawley
Cuyamaca
79
Guatay
Imperial
Holtville
80
Seeley
El Centro
98
Calexico
Mexicali
Yuma
Morena Res.
Jacumba
MEXICO
0 10 20 30 40 50
MILES

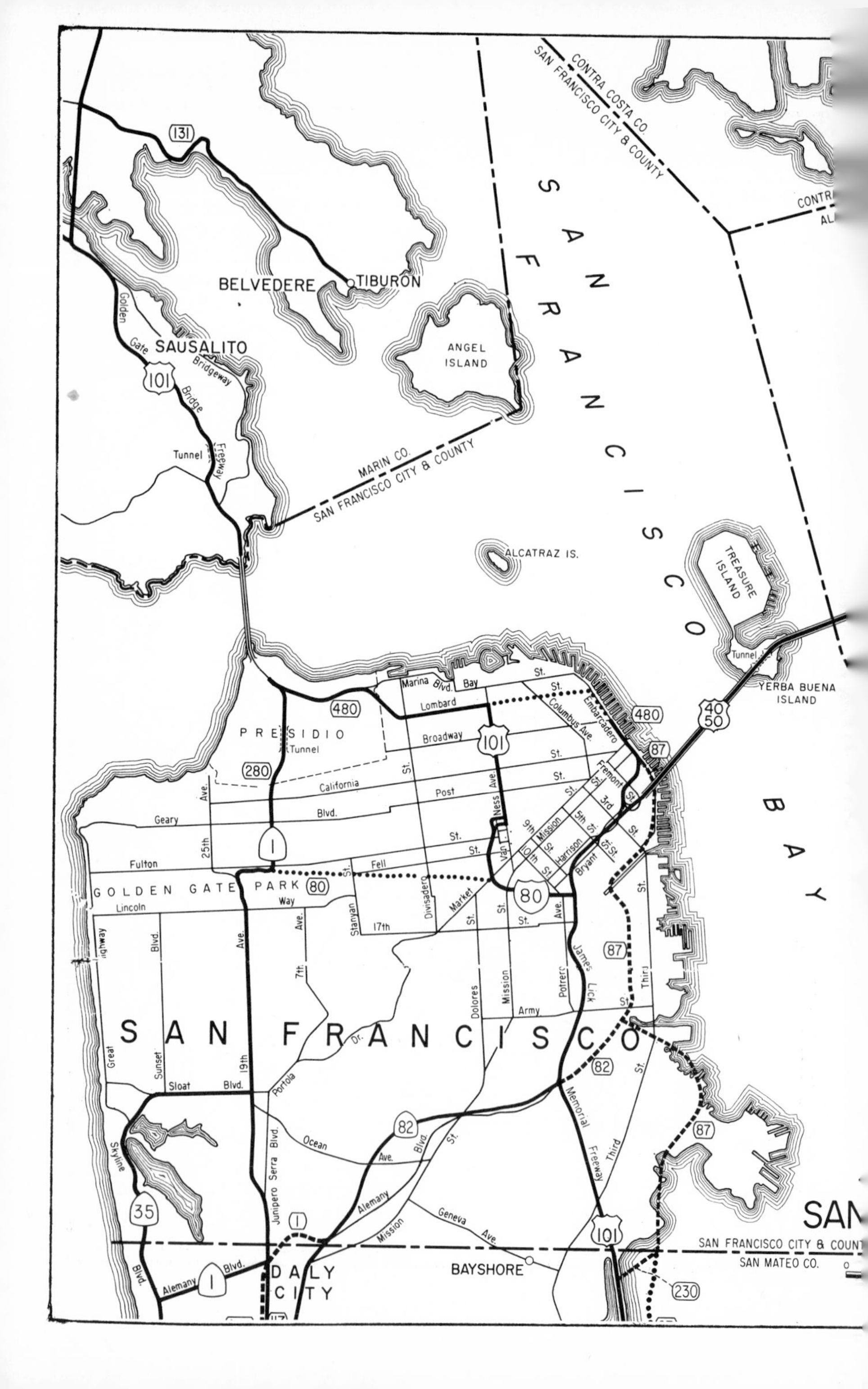

CONTRA COSTA CO.
SAN FRANCISCO CITY & COUNTY
131
SAN FRANCISCO
BELVEDERE
TIBURON
SAUSALITO
ANGEL ISLAND
Golden Gate Bridge
Bridgeway
101
Tunnel
Freeway
MARIN CO.
SAN FRANCISCO CITY & COUNTY
ALCATRAZ IS.
TREASURE ISLAND
Tunnel
YERBA BUENA ISLAND
40
50
Marina Blvd.
Bay St.
Lombard
480
PRESIDIO
Tunnel
Broadway
Columbus Ave.
Embarcadero
87
280
California
Post
Van Ness Ave.
Fremont
3rd
Geary Blvd.
Mission
5th St.
9th
10th St.
Harrison
Bryant
6th St.
Fulton
Fell St.
25th Ave.
1
GOLDEN GATE PARK
80
Way
Lincoln
Stanyan
17th
Divisadero
Market
BAY
Highway
Sunset Blvd.
7th Ave.
Dolores St.
Mission St.
Potrero Ave.
James Lick
Third St.
Army St.
SAN FRANCISCO
Great
Dr.
Sloat Blvd.
19th Ave.
Portola
82
Memorial Freeway
Ocean Ave.
Junipero Serra Blvd.
Skyline Blvd.
35
Alemany Blvd.
Mission
Geneva Ave.
Third
SAN
SAN FRANCISCO CITY & COUNTY
SAN MATEO CO.
DALY CITY
BAYSHORE
230

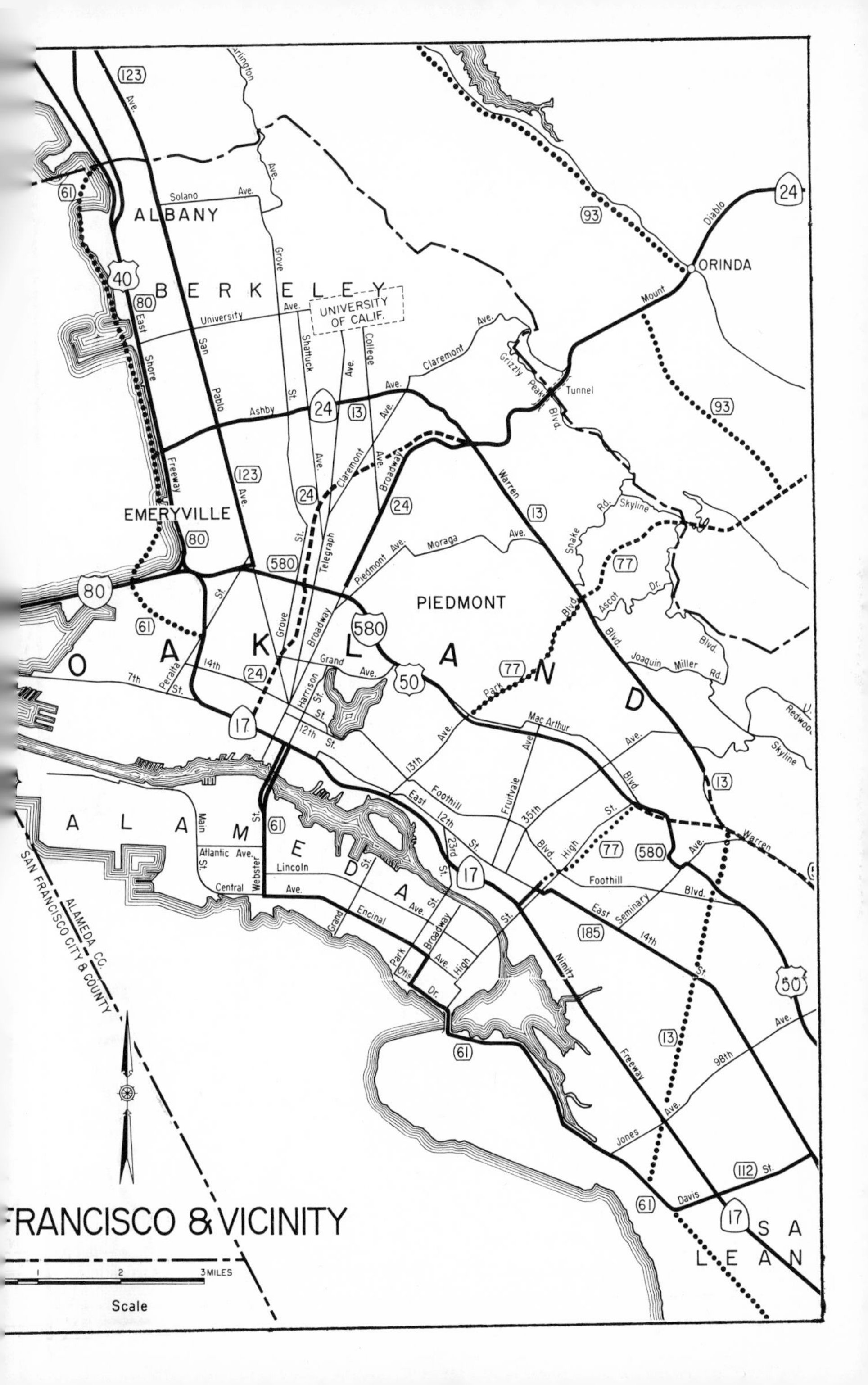

FRANCISCO & VICINITY
OAKLAND
BERKELEY
ALBANY
EMERYVILLE
PIEDMONT
ALAMEDA
ORINDA
UNIVERSITY OF CALIF.
SAN FRANCISCO CITY & COUNTY
ALAMEDA CO.
SAN LEAN
1
2
3 MILES
Scale

SAN FRANCISCO BAY AREA
LEGEND
U. S. Highways
State Highways
SCALE OF MILES
SONOMA
NAPA
YOLO
SOLANO
MARIN
CONTRA COSTA
ALAMEDA
SAN MATEO
SANTA CLARA
SANTA CRUZ
PACIFIC OCEAN
SAN PABLO BAY
SAN FRANCISCO BAY
Ft.Ross
Rio Nido
Hilton
Guerneville
Monte Rio
Fulton
CALISTOGA
Monticello
ST.HELENA
WINTERS
SEBASTOPOL
SANTA ROSA
Kenwood
Rutherford
Oakville
Yountville
Valley Ford
Bay
Cotati
El Verano
SONOMA
VACAVILLE
Elmira
NAPA
FAIRFIELD
PETALUMA
Tomales
Shellville
SUISUN
Cordelia
Inverness
Novato
VALLEJO
Pt.Reyes Sta.
Ignacio
BENICIA
Lagunitas
Pt. Reyes
SAN RAFAEL
MARTINEZ
PITTSBURG
CONCORD
RICHMOND
WALNUT CREEK
Clayton
SAUSALITO
BERKELEY
OAKLAND
ALAMEDA
SAN FRANCISCO
FARALLON ISLANDS
Dublin
BY PASS
HAYWARD
Rockaway Beach
SAN MATEO
Moss Beach
Half Moon Bay
REDWOOD CITY
PALO ALTO
ALVISO
San Gregorio
SAN JOSE
Pescadero
Saratoga
LOS GATOS
Big Basin
Boulder Cr.
Pt. Ano Nuevo
Ben Lomond
Felton
Davenport
Soquel
SANTA CRUZ
Capitola

Index

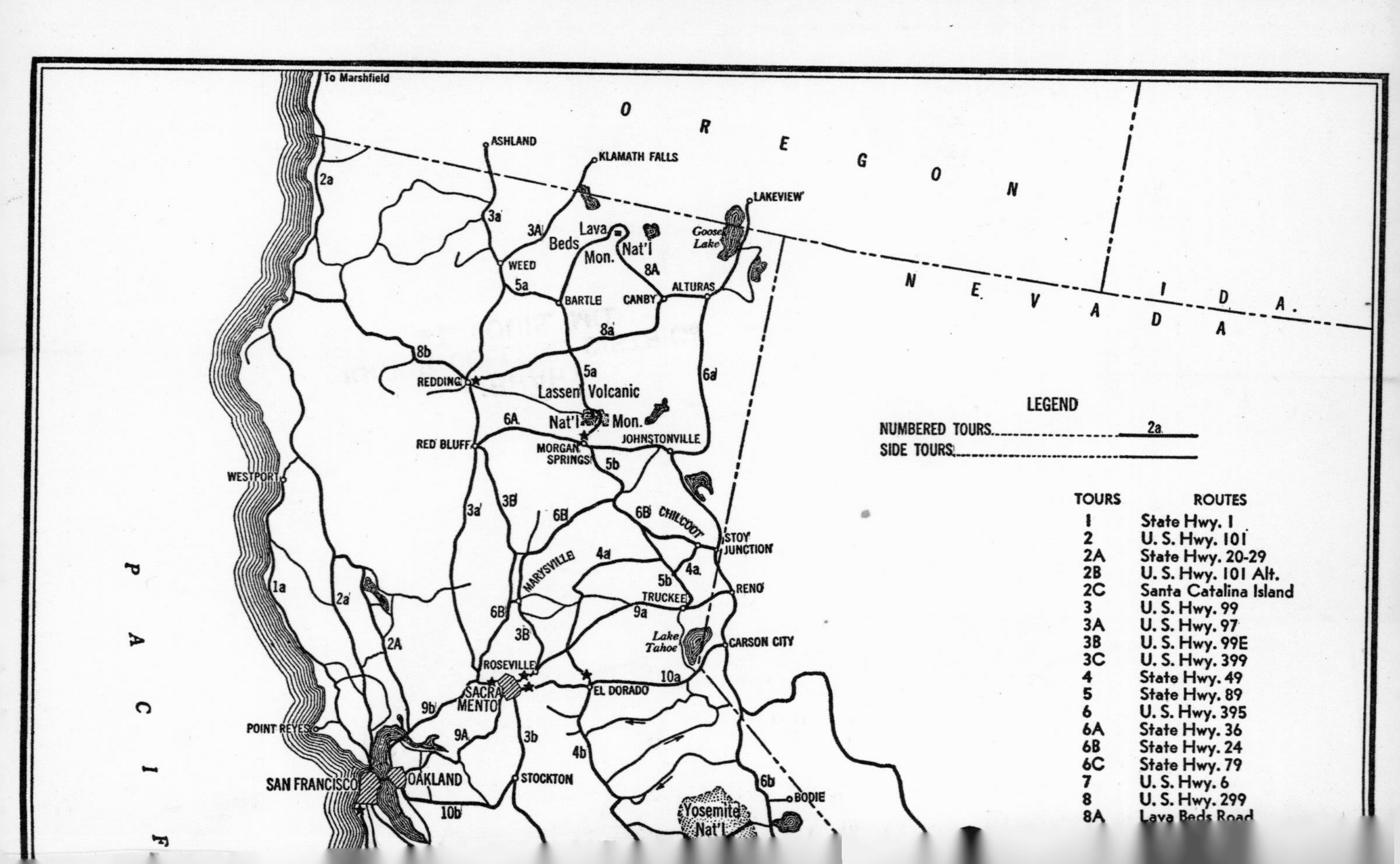

OREGON
NEVADA
IDA.
DA
PACIF
To Marshfield
ASHLAND
KLAMATH FALLS
LAKEVIEW
Goose Lake
Lava Beds Nat'l Mon.
WEED
BARTLE
CANBY
ALTURAS
REDDING
Lassen Volcanic Nat'l Mon.
RED BLUFF
MORGAN SPRINGS
JOHNSTONVILLE
WESTPORT
CHILCOOT
STOY JUNCTION
MARYSVILLE
TRUCKEE
RENO
Lake Tahoe
CARSON CITY
ROSEVILLE
SACRAMENTO
EL DORADO
POINT REYES
SAN FRANCISCO
OAKLAND
STOCKTON
BODIE
Yosemite Nat'l
2a
3a
3A
5a
8A
8a
8b
5a
6a
6A
5b
3a
3B
6B
6B
4a
4a.
5b
1a
2a
6B
9a
2A
3B
10a
9b
9A
3b
4b
6b
10b
LEGEND
NUMBERED TOURS 2a
SIDE TOURS
TOURS ROUTES
1 State Hwy. 1
2 U. S. Hwy. 101
2A State Hwy. 20-29
2B U. S. Hwy. 101 Alt.
2C Santa Catalina Island
3 U. S. Hwy. 99
3A U. S. Hwy. 97
3B U. S. Hwy. 99E
3C U. S. Hwy. 399
4 State Hwy. 49
5 State Hwy. 89
6 U. S. Hwy. 395
6A State Hwy. 36
6B State Hwy. 24
6C State Hwy. 79
7 U. S. Hwy. 6
8 U. S. Hwy. 299
8A Lava Beds Road